Coral Reefs of the World

Volume 8

More information about this series at http://www.springer.com/series/7539

Peter W. Glynn · Derek P. Manzello
Ian C. Enochs
Editors

Coral Reefs of the Eastern Tropical Pacific

Persistence and Loss in a Dynamic Environment

Springer

Editors
Peter W. Glynn
RSMAS MBE
University of Miami
Miami, FL
USA

Ian C. Enochs
RSMAS CIMAS - NOAA/AOML
University of Miami
Miami, FL
USA

Derek P. Manzello
NOAA/AOML
Miami, FL
USA

ISSN 2213-719X ISSN 2213-7203 (electronic)
Coral Reefs of the World
ISBN 978-94-017-7498-7 ISBN 978-94-017-7499-4 (eBook)
DOI 10.1007/978-94-017-7499-4

Library of Congress Control Number: 2016933478

Cover illustration: Healthy *Pocillopora* reef, Uva Island, Gulf of Chiriquí, Panama. Photograph taken 16 March 2005 by Tyler Smith.

Printed on acid-free paper

This Springer imprint is published by Springer Nature
The registered company is Springer Science+Business Media B.V. Dordrecht

For Jerry (Gerard M.) Wellington,
a pioneer and devoted student of eastern
Pacific coral reefs

Tribute

Peter W. Glynn, Patriarch of Coral Reef Research in the Eastern Tropical Pacific

By his students and friends,

Peter Glynn grew up on Coronado Island, California, and began scuba diving in the late 1940s, making regular trips with his high school friends into México to explore marine life in the Gulf of California and other nearby sites. Heading inland for college, he earned his B.A. at the University of South Dakota in 1955 returning to the Pacific to earn a M.A. and Ph.D. from Stanford University in 1960 and 1963, respectively, working at the Hopkins Marine Station on the trophic relationships of an intertidal community, with Donald P. Abbott serving as his advisor. Peter started working on coral reefs in the Caribbean in 1959 when he offered his services to a Mexican fishing crew in exchange for passage to the Belizean barrier reef. In 1960, he became a professor at the University of Puerto Rico, and 7 years later moved to Panama to join researchers at the Smithsonian Tropical Research Institute (STRI), where he initiated a detailed and systematic study of eastern tropical Pacific (ETP) coral reefs that continues today. His unprecedented and seminal research legacy of contributions began with the basic description of the Pacific coral reefs of Panama (Glynn et al. 1972; Glynn and Stewart 1973; Glynn 1976), and continued with a prolific series of contributions, substantively changing our knowledge and perspectives of these systems.

These publications address a wide variety of topics, including Pacific coral reef Holocene growth history (Glynn and Macintyre 1977), interactions among reef organisms, such as the corallivorous seastar *Acanthaster planci* (Glynn 1973, 1974, 1977) and the symbiosis between pocilloporid corals and crustacean symbionts (Glynn 1976, 1980, 1987), the first papers on El Niño warming impact and subsequent biological responses (Glynn 1983, 1990; Glynn et al. 1988, 2001a), coral bioerosion (Glynn 1988, 1997), coral reproduction (Glynn et al. 1991, 1994, 1996a, 2000, 2008, 2011, 2012), description of new coral species (Glynn 1999, Glynn et al. 2001b) and a possible coral extinction (Weerdt and Glynn 1991; Glynn and Feingold 1992; Glynn et al. 2001a), plus other investigations too numerous to mention here (see references throughout this book). He also did pioneering coral reef research in the Galápagos Islands (Glynn et al. 1979; Glynn and Wellington 1983; Glynn 1994, 2003), Colombia (Glynn et al. 1982), Costa Rica (Glynn et al. 1983), Clipperton Atoll (Glynn et al. 1996b), southern Mexico (Glynn and Leyte-Morales 1997), and Easter Island (Rapa Nui) (Glynn et al. 2003, 2007).

Peter Glynn's careful documentation of the causes, short-term impacts, and long-term ecological consequences of one of the first mass coral bleaching events to severely impact reefs regionally (the 1982–1983 El Niño-Southern Oscillation in the ETP) has proven invaluable in helping reef scientists understand the growing global threat of climate change on reef ecosystems.

Peter Glynn has been the mentor of, and an inspiration for, many coral reef scientists who have worked in the ETP: Andrew Baker, Iliana B. Baums, Charles Birkeland, Mitch Colgan, Susan Colley, Jorge Cortés, C. Mark Eakin, Ian Enochs, Joshua Feingold, Peggy Fong, Gerardo Leyte-Morales, Derek Manzello, Priscilla Martínez, Juan Maté, Héctor

Reyes-Bonilla, Bob Richmond, Fernando Rivera, Tyler Smith, Bernardo Vargas-Ángel, and Jerry Wellington. In recognition of his unparalleled contributions to coral reef science, Peter Glynn was awarded the Darwin Medal in 1992, which is the most prestigious award granted by the International Society for Reef Studies (ISRS). He has also received two "Best Paper Awards" for articles published in the journal Coral Reefs (Glynn and D'Croz 1990 and Glynn 1993). As a distinguished professor at the University of Miami's Rosenstiel School of Marine and Atmospheric Science, he has shaped the scientific education of countless undergraduates who have taken his invertebrate zoology and coral reef classes. He launched and inspired decades of research on reefs of the ETP, revealing an entirely new coral reef environment and exposing many of the secrets they held.

His seminal contributions continue to this day, as he just published a manuscript documenting the > 500 year growth history of a coral reef in the far Northern Galápagos Island of Darwin and its recovery from recurrent El Niño bleaching (Glynn et al. 2015). He aptly named this reef "Wellington Reef" in honor of his life-long friend and colleague, Jerry Wellington, with whom he discovered the coral reefs of the Galápagos Islands in the 1970s (Glynn and Wellington 1983).

References

Glynn PW (1973) *Acanthaster*: effect on coral reef growth in Panama. Science 180:504–506

Glynn PW (1974) The impact of *Acanthaster* on corals and coral reefs in the eastern Pacific. Environ Conserv 1:295–304

Glynn PW (1976) Some physical and biological determinants of coral community structure in the eastern Pacific. Ecol Monogr 46:431–456

Glynn PW (1977) Interactions between *Acanthaster* and *Hymenocera* in the field and laboratory. Proceedings of 3rd International Coral Reef Symposium, vol 1, Miami, pp 209–215

Glynn PW (1980) Defense by symbiotic Crustacea of host corals elicited by chemical cues from predator. Oecologia 47:287–290

Glynn PW (1983) Extensive bleaching and death of reef corals on the Pacific coast of Panamá. Environ Conserv 10:149–154

Glynn PW (1987) Some ecological consequences of coral-crustacean guard mutualisms in the Indian and Pacific Oceans. Symbiosis 4:301–324

Glynn PW (1988) El Niño warming, coral mortality and framework destruction by echinoid bioerosion in the eastern Pacific. Galaxea 7:129–160

Glynn PW (1990) Coral mortality and disturbances to coral reefs in the tropical eastern Pacific. In: Glynn PW (ed) Global ecological consequences of the 1982–1983 El Niño-Southern Oscillation. Amsterdam, Elsevier, p 55–126

Glynn PW (1993) Coral reef bleaching: ecological perspectives. Coral Reefs 12:1–18

Glynn PW (1994) State of coral reefs in the Galápagos Islands: natural vs anthropogenic impacts. Mar Poll Bull 29:131–140

Glynn PW (1997) Bioerosion and coral-reef growth: a dynamic balance. In: Birkeland C (Ed) Life and death of coral reefs, Chapman & Hall, New York, 68–95

Glynn PW (1999) *Pocillopora inflata*, a new species of scleractinian coral (Cnidaria: Anthozoa) from the tropical eastern Pacific. Pac Sci 53:168–180

Glynn PW (2003) Coral communities and coral reefs of Ecuador. In: Cortés J (ed), Latin American coral reefs. Elsevier, Amsterdam, p 449–472

Glynn PW, Wellington GM (1983) Corals and coral reefs of the Galápagos Islands. University California Press, Berkeley, 330 p

Glynn PW, D'Croz L (1990) Experimental evidence for high temperature stress as the cause of El Niño coincident mortality. Coral Reefs 8:181–191

Glynn PW, Feingold JS (1992) Hydrocoral species not extinct. Science 257:1845

Glynn PW, Leyte-Morales, GE (1997) Coral reefs of Huatulco, West Mexico: reef development in upwelling gulf of Tehuantepec. Rev Biol Trop 45:1033–1047

Glynn PW, Macintyre IG (1977) Growth rate and age of coral reefs of the Pacific coast of Panamá. Proceedings of 3rd International Coral Reef Symposium, vol 2, Miami, pp 251–259

Glynn PW, Stewart RH (1973) Distribution of coral reefs in the Pearl Islands (Gulf of Panama) in relation to thermal conditions. Limnol Oceanogr 18:367–379

Glynn PW, Stewart RH, McCosker JE (1972) Pacific coral reefs of Panamá: structure, distribution and predators. Geol Rundsch 61:483–519

Glynn PW, Wellington GM, Birkeland C (1979) Coral reef growth in the Galápagos: limitation by sea urchins. Science 203:47–49

Glynn PW, von Prahl H, Guhl F (1982) Coral reefs of Gorgona Island, Colombia, with special reference to corallivores and their influence on community structure and reef development. An Inst Invest Mar Punta Betín 12:185–214

Glynn PW, Druffel EM, Dunbar RB (1983) A dead Central American coral reef tract: possible link to the Little Ice Age. J Mar Res 41:605–637

Glynn PW, Cortés J, Guzmán HM, Richmond RH (1988) El Niño (1982–83) associated coral mortality and relationship to sea surface temperature deviations in the tropical eastern Pacific. Proceedings of 6th Int Coral Reef Symp, Australia 3:237–243

Glynn PW, Gassman NJ, Eakin CM, Cortés J, Smith DB, Guzmán HM (1991) Reef coral reproduction in the eastern Pacific: Costa Rica, Panamá, and Galápagos Islands (Ecuador), Part I—Pocilloporidae. Mar Biol 109:355–368

Glynn PW, Colley SB, Eakin CM, Smith DB, Cortés J, Gassman NJ, Guzmán HM, Rosario JB, Feingold J (1994) Reef coral reproduction in the eastern Pacific: Costa Rica, Panamá, and Galápagos Islands (Ecuador)-II. Poritidae. Mar Biol 118:191–208

Glynn PW, Colley SB, Gassman NJ, Black K, Cortés J, Maté JL (1996a) Reef coral reproduction in the eastern Pacific: Costa Rica, Panamá, and Galápagos Islands (Ecuador). III. Agariciidae (*Pavona gigantea* and *Gardineroseris planulata*). Mar Biol 125:579–601

Glynn PW, Veron JEN, Wellington GM (1996b) Clipperton Atoll (eastern Pacific): oceanography, geomorphology, reef-building coral ecology and biogeography. Coral Reefs 15:71–99

Glynn PW, Colley SB, Ting JH, Maté JL, Guzmán HM (2000) Reef coral reproduction in the eastern Pacific: Costa Rica, Panamá, and Galápagos Islands (Ecuador)—IV. Agariciidae, recruitment and recovery of *Pavona varians* and *Pavona* sp.a. Mar Biol 136:785–805

Glynn PW, Maté JL, Baker AC, Calderón MO (2001a) Coral bleaching and mortality in Panamá and Ecuador during the 1997–1998 El Niño-Southern oscillation event: Spatial/temporal patterns and comparisons with the 1982–1983 event. Bull Mar Sci 69:79–101

Glynn PW, Maté JL, Stemann TA (2001b) *Pavona chiriquiensis*. A new species of zooxanthellate coral (Cnidaria: Anthozoa: Agariciidae) from the eastern tropical Pacific. Bull Biol Soc Washington 10:210–225

Glynn PW, Wellington GM, Wieters EA, Navarrete SA (2003) Reef-building coral communities of Easter Island (Rapa Nui), Chile. In: Cortés J (ed) Latin American coral reefs. Elsevier, Amsterdam, p 473–494

Glynn PW, Wellington GM, Riegl B, Olson DB, Borneman E, Wieters EA (2007) Diversity and biogeography of the scleractinian coral fauna of Easter Island (Rapa Nui). Pac Sci 61:67–90

Glynn PW, Colley SB, Maté JL, Cortés J, Guzmán HM, Bailey RL, Feingold JS, Enochs IC (2008) Reproductive ecology of the azooxanthellate coral *Tubastrea coccinea* in the Equatorial Eastern Pacific: Part V. Dendrophylliidae. Mar Biol 153:529–544

Glynn PW, Colley SB, Guzman HM, Enochs IC, Cortés J, Maté JL, Feingold JS (2011) Reef coral reproduction in the eastern Pacific: Costa Rica, Panamá, and the Galápagos Islands (Ecuador). VI. Agariciidae, *Pavona clavus*. Mar Biol 158:1601–1617

Glynn PW, Colley SB, Maté JL, Baums IB, Feingold JS, Cortés J, Guzman HM, Afflerbach JC, Brandtneris VW, Ault JS (2012) Reef coral reproduction in the equatorial eastern Pacific: Costa Rica, Panamá, and the Galápagos Islands (Ecuador). VII. Siderastreidae, *Psammocora stellata* and *Psammocora profundacella*. Mar Biol 159:1917–1932

Glynn PW, Riegl B, Purkis S, Kerr JM, Smith TB (2015) Coral reef recovery in the Galápagos Islands: The northernmost islands (Darwin and Wenman). Coral Reefs 34:421–436

Weerdt WH de, Glynn PW (1991) A new and presumably now extinct species of *Millepora* (Hydrozoa) in the eastern Pacific. Zool Meded 65:267–276

Preface

Just a few decades ago it was uncertain whether structural coral reefs were present in the eastern tropical Pacific (ETP) region. Several renowned scientists who characterized the global development of coral reefs, for example Charles Darwin, James Dana, A.E. Verrill, T. Wayland Vaughan, and David R. Stoddart embraced the erroneous idea that coral reefs were absent from western American shores. Ironically, these authorities spent little or no time researching corals or coral reefs on site in the ETP, but obtained their information from marine charts and secondhand often from the cursory observations of others.

Beginning in the 1960s and 1970s, studies were initiated (and continue) to document the widespread occurrence of structural coral reefs along western tropical American shores and offshore islands. Compared to coral reef provinces in the Indo-Pacific and broader Caribbean Sea, however, these ETP reefs are relatively small and discontinuous. Also, the numbers of reef-building species are few, exhibiting a limited variety of colony morphologies, and reef framework structures. Additionally, extant eastern Pacific reefs are generally restricted to shallow depths, sheltered shores, and exhibit little zonation. One of the reasons they have gone virtually unnoticed is their lack of emergent reef structures such as algal ridges, drying reef flats, rubble berms, and lagoons. Despite the absence of such geomorphologic features, these marginally developed reefs share essentially the same kinds of biotic interactions and ecological functions as coral reefs in the more diverse and celebrated Coral Seas' provinces. In this book we offer a comprehensive synthesis of recent progress in ETP coral reef research, providing a framework for understanding of the structure and dynamics of these ecosystems as they approach an uncertain future in the face of natural and human-related disturbances.

This volume is organized into three principal subject areas (physical environmental conditions, biology and ecology of coral reef biota, and impacts and management), and begins with a review by Peter W. Glynn on the history of coral and coral reef studies in the ETP region (Chap. 1). Surprisingly, despite an apparent depauperate reef-building biota and modest reef development, the eastern Pacific region boasts a relatively rich history of coral research. Following the historical framework, five chapters of the book's first section examine the physical setting of the eastern Pacific basin with particular attention to those areas supporting reef-building coral populations. The geology and paleoecology of early eastern Pacific coral reefs are traced by Andrés López-Pérez from the Eocene epoch and closure of the Isthmian Seaway to the development, distribution, biogeographic character, and evolution of Holocene reefs (Chap. 2). Basin-wide oceanographic conditions that control coral distributions and reef development, including ocean currents, upwelling centers, storm tracks, and location of the Intertropical Convergence Zone, are examined by Paul C. Fiedler and Miguel Lavín in Chap. 3. Chunzai Wang and co-authors trace the development, evolution, and spatial occurrence of El Niño warming events in Chap. 4. With the current recognition of two classes of El Niño, eastern and central Pacific events, our understanding of the spatial and temporal extent of disturbances to ETP coral reefs should now come into better focus. Peter Glynn and co-authors present an overview in Chap. 5 of the biogeographic occurrence of coral species, reef types, morphology, zonation, and their biotic composition in different eastern Pacific regions. This synoptic treatment is the product of several workers who have contributed their knowledge of particular reef areas. Information on the coral fauna and coral reefs of Mexico is provided by Héctor Reyes-Bonilla. Recently discovered coral assemblages in El Salvador and

Nicaragua are characterized by H. Reyes-Bonilla and Juan J. Alvarado, respectively. Brief descriptions of equatorial eastern Pacific coral habitats are offered by Jorge Cortés and Carlos Jiménez (Costa Rica), Juan L. Maté (Panama), Fernando A. Zapata and Bernardo Vargas-Ángel (Colombia), and Stuart Banks, Joshua S. Feingold, Peter W. Glynn, Priscilla Martínez, and Fernando Rivera (Ecuador). Evie A. Wieters and Sergio A. Navarrete have provided information on the coral communities of remote Rapa Nui (Easter Island), now recognized as a distant outpost of the ETP region. This first section concludes with an overview of the distribution, development, and centennial- to millennial-scale history of structural reefs in the eastern Pacific by Lauren T. Toth and co-authors in Chap. 6.

The second thematic section of this volume, comprising 13 chapters, treats a variety of topics focused on the biology and ecology of eastern Pacific coral reef ecosystems. These chapters are framed in the context of functional ecology, i.e., processes and interactions that influence or control the structure and dynamics of coral communities and reef building. The species richness and diversity of reef-associated organisms, their habitats, niches, and varied ecological interactions in structuring reef communities are addressed by Jorge Cortés and co-authors in Chap. 7. In Chap. 8, P. Glynn and co-authors examine ENSO impacts on reefs with special attention to the severe 1982–1983 and 1997–1998 disturbances. In Chap. 9, I. Enochs and P. Glynn describe the major coral reef trophic guilds, their connectivity and contributions to reef function. Food webs demonstrating energy and nutrient inputs are examined and evaluated in terms of flow (gains and losses), regeneration and recycling processes. Coral community structure, dynamics and reef accretion are greatly influenced by corallivores in the eastern Pacific. The high diversity of invertebrate and fish corallivores, relative to their coral prey, is examined by I. Enochs and P. Glynn in Chap. 10. A process-focused account of algal dynamics and herbivory is presented in Chap. 11 by P. Fong and colleagues. This chapter considers algal diversity, patterns in distribution and abundance, the ecological controlling forces that may produce these patterns, and presents a simulation model based on empirical data that demonstrate that ETP reefs may currently exist as alternate stable states. In Chap. 12, Juan Alvarado and co-authors document the high rates of bioerosion that are exacerbated following ENSO-related coral mortality, and underscore the significance of this potentially important process that contributes to reef degradation in the ETP. The rates and effects of diverse excavating metazoan taxa are reviewed, from Mexico to several sites in the equatorial eastern Pacific of Central and South America.

The roles of symbiotic zooxanthellae in eastern Pacific reef-building corals are examined in two chapters. The first (Chap. 13) by Andrew C. Baker and colleagues focuses on the occurrence of endosymbionts in different coral host species and environments, e.g., upwelling and non-upwelling areas, and the temporal stability of these distributions. The presence of certain symbiont types impart a degree of thermal tolerance to host corals, which is considered in light of the adaptive bleaching hypothesis. In Chap. 14, Jorge H. Pinzón addresses the biogeographic patterns of corals and their symbiont types, coral host morphometrics, and species boundaries vis-à-vis symbiont associations. Molecular genetics and morphometric evidence are shown to be highly discordant in defining species boundaries in *Pocillopora*, with several fewer, but highly supported, species recognized by application of the former methodology.

Like the previous topics on coral/algal holobionts, several of the remaining chapters have relevance to coral resilience, i.e., the capacity of reef coral populations to recover from disturbances. This capacity for recovery is manifest in the reproductive potential of reef corals, their genetic connectivity within the eastern Pacific region, as well as the existence and possible effectiveness of deep reef refugia that could play a role in the repopulation of impacted shallow reef habitats. Both sexual and asexual modes of reproduction are critical life history processes in the maintenance and recovery of coral communities. Sexual reproductive patterns, spawning modes, seasonal timing of spawning, fecundity and recruitment are examined in the major reef-building coral species by P. Glynn and co-authors in Chap. 15. The prevalence of asexual propagation, commonly caused by foraging fishes, is also reviewed for branching and massive corals in Mexico, Panama, and the Galápagos Islands. Gene flow and population connectivity between major reef regions within the eastern tropical Pacific and across the eastern Pacific Barrier are examined by H.A. Lessios and I.B. Baums in Chap. 16. They provide an analysis of all available DNA data, expressed as the genetic divergence between populations (F_{ST}), to better understand the connectivity of conspecific populations of reef-building corals, and reef-associated molluscs, echinoderms, and fishes. Tyler B. Smith and co-authors in Chap. 17 examine refuges and refugia for ETP corals and coral reefs in a highly dynamic ocean climate. The survival of deep coral communities and their ability to reproduce and promote recruitment to devastated shallow reef habitats, embraced in the "deep reef refugia hypothesis", is evaluated in this chapter in terms of eastern Pacific recovery processes.

In light of potentially stressful conditions affecting ETP coral reefs—high and low thermal swings, high CO_2 concentration, and demonstrated rapid bioerosion of carbonate structures—Derek P. Manzello and co-authors examine coral/algal calcification and the integrity of reef frameworks in Chap. 18. They conclude this treatment with consideration of present-day species-specific responses to such challenges, and the risks to eastern Pacific coral reefs in the coming decades. In Chap. 19, Julia E. Cole and Alexander W. Tudhope observe climate variability over multidecadal and century timescales from coral-based paleoclimate reconstructions during the Holocene. Coral-based reconstructions of ocean temperatures are examined in detail from geochemical records obtained in the Galápagos Islands, Panama, and Clipperton Atoll.

The third and last section of this book addresses issues stemming from the uses and abuses of coral reefs by humankind, from local and regional impacts to global-scale degradation. Direct and indirect sources of damage to coral reefs, e.g., fishing-related effects, coastal alteration, and deforestation, are examined by J. Cortés and H. Reyes-Bonilla in Chap. 20. Significant advances have been made during the past few decades in the establishment of marine protected areas (MPAs). Eastern Pacific MPAs harboring coral communities and coral reefs are characterized by J. Alvarado and co-authors in Chap. 21. Attention is directed to the location and size of MPAs, the types of coral communities and reefs present, and their degree of protection in relation to stressors and threats. A case is made in this final chapter for the role of MPA networks in the protection of coral refugia, source populations, and the potential for repopulation of degraded reef areas.

To assist in the recognition and systematic status of abundant and ecologically important species mentioned throughout the book, a pictorial field guide, assembled by Juan L. Maté and colleagues, is included in Chap. 22. This section is organized into brief species' accounts from each of the major metazoan phyla and algal divisions. Photographs and accompanying information sources, with the assistance of systematists familiar with the taxa considered, are offered.

Several research initiatives in the ETP have been shown to have universal relevance, thus serving to advance an understanding of coral reef ecology worldwide. For example, the cause and effect of elevated sea temperatures and widespread coral bleaching, first recognized in the ETP in the early 1980s, has subsequently been observed globally, signaling a clarion warning of reef degradation in step with climate change. Related coral bleaching research, initiated in the ETP, has demonstrated differential susceptibility of algal symbiont types to changes in sea temperature, rendering certain coral hosts relatively resistant to temperature extremes. Such environmentally induced responses have now been documented in other coral reef regions. The recognition that eastern Pacific reefs occur under naturally low pH conditions has provided important insights into how the structure and function of coral reefs might be affected by ocean acidification. These pioneering contributions to coral reef science, and others initiated in the eastern tropical Pacific, are considered in detail in the relevant chapters in this volume.

The principal ETP coral reef study sites examined in this book extend from the Gulf of California in northern Mexico through Central America to coastal Ecuador in South America (Fig. 1). Oceanic island sites are also identified, from the Revillagigedo Islands off Mexico to remote Rapa Nui, ~4,000 km west of Chile. Throughout this book, Latin American countries and major cities are anglicized following National Geographic's Collegiate Atlas of the World (2006). Most other geographic names retain their Spanish spelling, which should be helpful in

Fig. 1 Eastern Pacific ocean basin from the Gulf of California, Mexico to Rapa Nui (Easter Island), Chile showing major oceanic islands and continental sites with coral reef development. Modified from the World Physical Map. *Source* US National Park Service (http://www.arcgis.com/home/item.html?id=c4ec722a1cd34cf0a23904aadf8923a0)

the location of in-country coral reef study sites. Recently some Latin American authors have dropped accents from their names. Current literature citations often list both, accented and unaccented names. Since it is difficult to determine the year the omissions occurred, we have attempted to retain accented names throughout for the sake of consistency and as a guide for correct pronunciation. Acronyms representing institutions and agencies are used according to their country of origin. With respect to cultural preferences, paternal and maternal surnames (in that order) identify most Latin American workers. In a few instances, accents have been omitted from proper names according to individual preference.

Nearly all of the information contained in this book has originated from research conducted in Latin America and Clipperton Atoll, an overseas territory of France. Without the insights, dedication, and camaraderie of local researchers, students, and technicians, the body of research making up this volume would not have been possible. The editors wish to thank many individuals who have made significant contributions to eastern Pacific coral reef studies, and who have benefitted this volume at several levels, including those who offered help beyond the chapters they co-authored: **Mexico**—Luis E. Calderon-Aguilera, David A. Paz-García, and H. Reyes-Bonilla; **Clipperton Atoll**—Lt. Col. Graeff (Chief, Bureau d'Etudes, Haut-Commissariat de la République en Polynésie Française), and J.D. Jackson; **Costa Rica**—J.J. Alvarado, J. Cortés, Ana C. Fonseca-Escalante, Héctor M. Guzmán, Carlos E. Jiménez, Ian G. Macintyre, and Manuel M. Murillo, Michael J. Risk, and Paula J.B. Scott; **Panama**—Lawrence G. Abele, Irwin Bethancourt Acosta, Rebecca Albright, Charles Birkeland, Peter Castro, Thomas F. Dana, Luis D'Croz, Juan B. Del Rosario, Ellen R.M. Druffel, Robert B. Dunbar, J. Wyatt Durham, Caitlin R. Fong, Sandra L. Gilchrist, Héctor M. Guzmán, Leslie Harris, Gordon Hendler, Raymond C. Highsmith, Gena Hockensmith, Jeremy B.C. Jackson, Nancy Knowlton, Martin Moynihan, William A. Newman, James W. Porter, Robert H. Richmond, D. Ross Robertson, Adrienne Romanski (Adrienne M.S. Correa), Ira Rubinoff, Aníbal Velarde, Matthew J. Wartian, and Ingo S. Wehrtmann; **Colombia**—Henry von Prahl, Bernardo Vargas-Ángel, and Fernando A. Zapata; **Ecuador**—Stuart A. Banks, Margarita Brandt, Victor W. Brandtneris, Andrew Bruckner, Rodrigo H. Bustamante, Ángel I. Chiriboga, Lenín Cruz, Graham J. Edgar, René Espinosa, Priscilla C. Martínez, Thomas Okey, Marjorie L. Reaka, Günter Reck, Philip Renaud, Fernando Rivera, Gerard M. Wellington, John W. Wells, and Jon Witman; **Easter Island**—Eric H. Borneman, Henri and Michel Garcia, Dennis K. Hubbard, and G.M. Wellington.

Technical assistance in producing illustrations and tables, and help with sundry editorial tasks was kindly offered by Rafael Araujo, Maria Luisa M. Estevanez, Michael Fuller, Carmen S. Glynn, Benjamin Grassian, Joshua Levy, Luigi Lombardi, and Megan K. Williams.

Research and travel funding was provided by numerous national and international agencies and institutions that are acknowledged in Chap. 1 (Table 1.2), and in the respective chapters of the book that benefitted from this support. We are grateful to Bernhard M. Riegl for his invitation to join this series of reviews on the Coral Reefs of the World, and the editorial staff of Springer SBM Publishing, in particular Martine van Bezooijen, Alexandrine Cheronet, and Judith Terpos. Finally, the recommendations of the following reviewers greatly improved the presentations: Andrew Baird, Brian Beck, Ray Berkelmans, Giacomo Bernardi, Charles Birkeland, Deron Burkepile, José D. Carriquiry, Iliana Chollet, Julia Cole, Mitchell W. Colgan, Jorge Cortés, Paul Fiedler, Pamela Fletcher, Walter M. Goldeberg, James Guest, James C. Hendee, Patricia A. Hutchings, Carlos Jiménez, Joshua Idjadi, Michael Kramer, David Lea, Sang-Ki Lee, Aaron O'Dea, Thomas Okey, María Del Mar Palacios, Héctor Reyes Bonilla, Cynthia Riginos, Michael J. Risk, Scott Santos, Christine Schoenberg, Michael Stat, Daniel Thornhill, Lauren Toth, Francisco Benitez Villalobos, Jon Witman, and Kimberly Yates.

Reference

National Geographic Society (2006) Collegiate Atlas of the World. Willar, Ohio, RR Donnelley & Sons Co, 384 p

In Memoriam

We gratefully acknowledge the various research efforts of Ana Cecilia Fonseca Escalante and Priscilla Zamora (Costa Rica), Miguel F. Lavín (Mexico), Henry von Prahl (Colombia), and Gerard M. Wellington (USA) whose lives ended much too soon but nonetheless significantly advanced our understanding of eastern Pacific coral reefs.

Contents

1 **History of Eastern Pacific Coral Reef Research** 1
Peter W. Glynn

2 **Revisiting the Cenozoic History and the Origin of the Eastern Pacific Coral Fauna** 39
Andrés López-Pérez

3 **Oceanographic Conditions of the Eastern Tropical Pacific** 59
Paul C. Fiedler and Miguel F. Lavín

4 **El Niño and Southern Oscillation (ENSO): A Review** 85
Chunzai Wang, Clara Deser, Jin-Yi Yu, Pedro DiNezio and Amy Clement

5 **Eastern Pacific Coral Reef Provinces, Coral Community Structure and Composition: An Overview** 107
Peter W. Glynn, Juan J. Alvarado, Stuart Banks, Jorge Cortés, Joshua S. Feingold, Carlos Jiménez, James E. Maragos, Priscilla Martínez, Juan L. Maté, Diana A. Moanga, Sergio Navarrete, Héctor Reyes-Bonilla, Bernhard Riegl, Fernando Rivera, Bernardo Vargas-Ángel, Evie A. Wieters and Fernando A. Zapata

6 **Holocene Reef Development in the Eastern Tropical Pacific** 177
Lauren T. Toth, Ian G. Macintyre and Richard B. Aronson

7 **Marine Biodiversity of Eastern Tropical Pacific Coral Reefs** 203
Jorge Cortés, Ian C. Enochs, Jeffrey Sibaja-Cordero, Luis Hernández, Juan José Alvarado, Odalisca Breedy, José Antonio Cruz-Barraza, Octavio Esquivel-Garrote, Cindy Fernández-García, Alicia Hermosillo, Kirstie L. Kaiser, Pedro Medina-Rosas, Álvaro Morales-Ramírez, Cristian Pacheco, Alejandro Pérez-Matus, Héctor Reyes-Bonilla, Rafael Riosmena-Rodríguez, Celeste Sánchez-Noguera, Evie A. Wieters and Fernando A. Zapata

8 **El Niño-Southern Oscillation: Effects on Eastern Pacific Coral Reefs and Associated Biota** 251
Peter W. Glynn, Alissa B. Mones, Guillermo P. Podestá, Angela Colbert and Mitchell W. Colgan

9 **Trophodynamics of Eastern Pacific Coral Reefs** 291
Ian C. Enochs and Peter W. Glynn

10 **Corallivory in the Eastern Pacific** 315
Ian C. Enochs and Peter W. Glynn

11 **Algal Dynamics: Alternate Stable States of Reefs in the Eastern Tropical Pacific** 339
Peggy Fong, Tyler B. Smith and Ranjan Muthukrishnan

12 **Coral Reef Bioerosion in the Eastern Tropical Pacific** 369
Juan José Alvarado, Benjamin Grassian, Jaime Ricardo Cantera-Kintz, José Luis Carballo and Edgardo Londoño-Cruz

13 **Diversity, Distribution and Stability of *Symbiodinium* in Reef Corals of the Eastern Tropical Pacific** 405
Andrew C. Baker, Adrienne M.S. Correa and Ross Cunning

14 **Insights from the Application of Genetics on *Pocillopora-Symbiodinium* Associations in the Eastern Tropical Pacific** 421
Jorge H. Pinzón C.

15 **Coral Reproduction in the Eastern Pacific** 435
Peter W. Glynn, Susan B. Colley, Eugenio Carpizo-Ituarte and Robert H. Richmond

16 **Gene Flow in Coral Reef Organisms of the Tropical Eastern Pacific** 477
H.A. Lessios and Iliana B. Baums

17 **Thermal Refuges and Refugia for Stony Corals in the Eastern Tropical Pacific** 501
Tyler B. Smith, Juan L. Maté and Joanna Gyory

18 **Effects of Global Warming and Ocean Acidification on Carbonate Budgets of Eastern Pacific Coral Reefs** 517
Derek P. Manzello, C. Mark Eakin and Peter W. Glynn

19 **Reef-Based Reconstructions of Eastern Pacific Climate Variability** 535
Julia Cole and Alexander W. Tudhope

20 **Human Influences On Eastern Tropical Pacific Coral Communities and Coral Reefs** 549
Jorge Cortés and Héctor Reyes-Bonilla

21 **Coral Reef Conservation in the Eastern Tropical Pacific** 565
Juan José Alvarado, Octavio Aburto-Oropeza, Rubén Abad, Enrique Barraza, Margarita Brandt, Jaime Cantera, Priscila Estrada, Carlos F. Gaymer, Ana Gloria Guzmán-Mora, James J. Herlan and Juan L. Maté

22 **Field Guide to Select Eastern Pacific Corals and Associated Coral Reef Biota** 593
Juan L. Maté, Margarita Brandt, Benjamin Grassian and Ángel Chiriboga

General Index 639

Taxonomic Index 651

Contributors

Rubén Abad Museo de Zoología QCAZ, Sección Invertebrados, Pontificia Universidad Católica del Ecuador, Quito, Ecuador

Octavio Aburto-Oropeza Marine Biology Research Division, Scripps Institution of Oceanography, La Jolla, CA, USA

Juan José Alvarado Centro de Investigación en Ciencias del Mar y Limnología (CIMAR), Universidad de Costa Rica, San José, Costa Rica; Escuela de Biología, Universidad de Costa Rica, San José, Costa Rica

Richard B. Aronson Department of Biological Sciences, Florida Institute of Technology, Melbourne, FL, USA

Andrew C. Baker Department of Marine Biology and Ecology, Rosenstiel School of Marine and Atmospheric Science, University of Miami, Miami, FL, USA; Marine Program, Wildlife Conservation Society, Bronx, NY, USA

Stuart Banks Marine Ecosystem Research Programme, Charles Darwin Research Station, Quito, Ecuador

Enrique Barraza Ministerio de Medio Ambiente y Recursos Naturales, Col. y Calle Las Mercedes, San Salvador, El Salvador

Iliana B. Baums The Pennsylvania State University, University Park, PA, USA

Margarita Brandt Colegio de Ciencias Biológicas y Ambientales, Universidad San Francisco de Quito, Diego de Robles s/n y Pampite, Cumbayá, Quito, Ecuador

Odalisca Breedy Centro de Investigación en Ciencias del Mar y Limnología (CIMAR), Universidad de Costa Rica, San José, Costa Rica; Escuela de Biología, Universidad de Costa Rica, San José, Costa Rica

Jaime Ricardo Cantera-Kintz Departamento de Biología, Universidad del Valle, Cali, Colombia

Jaime Cantera Departamento de Biología, Universidad del Valle, Cali, Colombia

José Luis Carballo Instituto de Ciencias del Mar y Limnologia (estación Mazatlán), Universidad Nacional Autónoma de Mexico, Mazatlán, Mexico

Eugenio Carpizo-Ituarte Laboratorio de Ecología y Biología Del Desarrollo, Instituto de Investigaciones Oceanológicas, Universidad Autónoma de Baja California, Baja California, Mexico

Ángel Chiriboga PIXEL, De los Eucaliptos y De los Pinos, Cumbayá, Quito, Ecuador

Amy Clement Rosenstiel School of Marine and Atmospheric Science, University of Miami, Miami, FL, USA

Angela Colbert Department of Ocean Sciences, Rosenstiel School of Marine and Atmospheric Science, University of Miami, Miami, FL, USA; Patricia and Phillip Frost Museum of Science, Miami, FL, USA

Julia Cole Department of Geosciences and Atmospheric Sciences, University of Arizona, Tucson, AZ, USA

Mitchell W. Colgan Department of Geology and Environmental Geosciences, College of Charleston, Charleston, SC, USA

Susan B. Colley Stanley S. Scott Cancer Center, LSU Health Science Center, New Orleans, LA, USA

Adrienne M.S. Correa Department of Marine Biology and Ecology, Rosenstiel School of Marine and Atmospheric Science, University of Miami, Miami, FL, USA; Biosciences at Rice, Rice University, Houston, TX, USA

Jorge Cortés Centro de Investigación en Ciencias del Mar y Limnología (CIMAR), Universidad de Costa Rica, San Pedro, San José, Costa Rica; Escuela de Biología, Universidad de Costa Rica, San Pedro, San José, Costa Rica

José Antonio Cruz-Barraza Instituto de Ciencias del Mar y Limnología, Universidad Nacional Autónoma de México (Unidad Académica Mazatlán), Mazatlán, Sinaloa, Mexico

Ross Cunning Department of Marine Biology and Ecology, Rosenstiel School of Marine and Atmospheric Science, University of Miami, Miami, FL, USA; Hawaii Institute of Marine Biology, University of Hawaii, Kaneohe, HI, USA

Clara Deser National Center for Atmospheric Research, Boulder, CO, USA

Pedro DiNezio International Pacific Research Center, University of Hawaii, Honolulu, HI, USA

Ian C. Enochs Cooperative Institute of Marine and Atmospheric Studies, Rosenstiel School of Marine and Atmospheric Science, University of Miami, Miami, FL, USA; Atlantic Oceanographic and Meteorological Laboratory (AOML), Ocean Chemistry and Ecosystem Division (OCED), National Oceanic and Atmospheric Administration (NOAA), Miami, FL, USA

Octavio Esquivel-Garrote Centro de Investigación en Ciencias del Mar y Limnología (CIMAR), Universidad de Costa Rica, San José, Costa Rica; Escuela de Biología, Universidad de Costa Rica, San José, Costa Rica

Priscila Estrada Marine Biology Research Division, Scripps Institution of Oceanography, La Jolla, CA, USA

Joshua S. Feingold Department of Marine and Environmental Sciences, Halmos College of Natural Sciences and Oceanography, Nova Southeastern University, Fort Lauderdale, FL, USA

Cindy Fernández-García Centro de Investigación en Ciencias del Mar y Limnología (CIMAR), Universidad de Costa Rica, San José, Costa Rica; Escuela de Biología, Universidad de Costa Rica, San José, Costa Rica

Paul C. Fiedler Marine Mammal and Turtle Division, NOAA/NMFS/Southwest Fisheries Science Center, La Jolla, CA, USA

Peggy Fong Department of Ecology and Evolutionary Biology, University of California Los Angeles, Los Angeles, CA, USA

Carlos F. Gaymer Departamento de Biología Marina, Millennium Nucleus for Ecology and Sustainable Management of Oceanic Islands (ESMOI) and Centro de Estudios Avanzados en Zonas Áridas (CEAZA), Universidad Católica del Norte, Larrondo 1281, Coquimbo, Chile

Peter W. Glynn Department of Marine Biology and Ecology, Rosenstiel School of Marine and Atmospheric Science, University of Miami, Miami, FL, USA

Benjamin Grassian Rosenstiel School of Marine and Atmospheric Science, Marine Biology and Ecology, University of Miami, Miami, FL, USA; Graduate School of Oceanography, University of Rhode Island, Narragansett, RI, USA

Ana Gloria Guzmán-Mora Marine Program, Conservation International-Costa Rica, San Pedro, Costa Rica

Joanna Gyory Center for Marine and Environmental Studies, University of the Virgin Islands, US Virgin Islands, USA; Department of Ecology and Evolutionary Biology, Tulane University, New Orleans, LA, USA

James J. Herlan Millennium Nucleus for Ecology and Sustainable Management of Oceanic Islands (ESMOI), Departamento de Biología Marina, Universidad Católica del Norte, Larrondo 1281, Coquimbo, Chile

Alicia Hermosillo Natural History Museum of Los Angeles County, Los Angeles, CA, USA

Luis Hernández Laboratorio de Sistemas Arrecifales, Universidad Autónoma de Baja California Sur, La Paz, Baja California Sur, Mexico

Carlos Jiménez Energy, Environment and Water Research Center (EEWRC), The Cyprus Institute (CyI), Nicosia, Cyprus

Kirstie L. Kaiser Santa Barbara Museum of Natural History, Santa Barbara, CA, USA

Miguel F. Lavín Departamento de Oceanografía Física, Centro de Investigación Científica y de Educación Superior de Ensenada, Ensenada, BC, Mexico

H.A. Lessios Smithsonian Tropical Research Institute, Balboa, Panama

Edgardo Londoño-Cruz Departamento de Biología, Universidad del Valle, Cali, Colombia

Andrés López-Pérez Laboratorio de Ecosistemas Costeros, Departamento de Hidrobiología, UAM-Iztapalapa, Iztapalapa, Ciudad de México, México

Ian G. Macintyre Department of Paleobiology, National Museum of Natural History, Smithsonian Institution, Washington DC, USA

Derek P. Manzello Atlantic Oceanographic and Meteorological Laboratories (AOML), National Oceanic and Atmospheric Administration (NOAA), Miami, FL, USA

James E. Maragos Department of Geography, University of Hawaii at Manoa, Honolulu, HI, USA

C. Mark Eakin Coral Reef Watch, National Environmental Satellite Data and Information Services, National Oceanic and Atmospheric Administration, Silver Spring, MD, USA

Priscilla Martínez Instituto Nazca de Investigaciones Marinas, Salinas, Ecuador

Juan L. Maté Smithsonian Tropical Research Institute, Balboa, Ancón, Panama

Pedro Medina-Rosas Centro Universitario de La Costa, Universidad de Guadalajara, Puerto Vallarta, Jalisco, Mexico

Diana A. Moanga Department of Environmental Science, Policy, and Management, University of California, Berkeley, Berkeley, CA, USA

Alissa B. Mones Marine and Atmospheric Science Program, University of Miami, Coral Gables, FL, USA; College of Veterinary Medicine, University of Illinois at Urbana-Champaign, Urbana, IL, USA

Álvaro Morales-Ramírez Centro de Investigación en Ciencias del Mar y Limnología (CIMAR), Universidad de Costa Rica, San José, Costa Rica; Escuela de Biología, Universidad de Costa Rica, San José, Costa Rica

Ranjan Muthukrishnan Department of Ecology and Evolutionary Biology, University of California Los Angeles, Los Angeles, CA, USA

Sergio Navarrete Estación Costera de Investigaciones Marinas, Advanced Studies in Ecology and Biodiversity, Departamento de Ecología, Facultad de Ciencias Biológicas, Pontificia, Universidad Católica de Chile, Santiago, Chile

Cristian Pacheco Instituto de Ciencias del Mar y Limnología, Universidad Nacional Autónoma de México (Unidad Académica Mazatlán), Mazatlán, Sinaloa, Mexico

Jorge H. Pinzón C. Department of Biology, University of Texas at Arlington, Arlington, TX, USA

Guillermo P. Podestá Department of Ocean Sciences, Rosenstiel School of Marine and Atmospheric Science, University of Miami, Miami, FL, USA

Alejandro Pérez-Matus Subtidal Ecology Laboratory and Marine Conservation Center, Estación Costera de Investigaciones Marinas, Pontificia Universidad Católica de Chile, Santiago, Chile

Héctor Reyes-Bonilla Laboratorio de Sistemas Arrecifales, Departamento Académico de Biología Marina, Universidad Autónoma de Baja California Sur, La Paz, BCS, Mexico

Robert H. Richmond Kewalo Marine Laboratory, University of Hawaii at Manoa, Honolulu, HI, USA

Bernhard Riegl Department of Marine and Environmental Sciences, Halmos College of Natural Sciences and Oceanography, Nova Southeastern University, Dania Beach, FL, USA

Rafael Riosmena-Rodríguez Programa de Investigación en Botánica Marina, Departamento de Biología Marina, Universidad Autónoma de Baja California Sur, La Paz, Baja California Sur, Mexico

Fernando Rivera Instituto Nazca de Investigaciones Marinas, Salinas, Ecuador

Jeffrey Sibaja-Cordero Centro de Investigación en Ciencias del Mar y Limnología (CIMAR), Universidad de Costa Rica, San José, Costa Rica; Escuela de Biología, Universidad de Costa Rica, San José, Costa Rica

Tyler B. Smith Center for Marine and Environmental Studies, University of the Virgin Islands, US Virgin Islands, USA

Celeste Sánchez-Noguera Centro de Investigación en Ciencias del Mar y Limnología (CIMAR), Universidad de Costa Rica, San José, Costa Rica; Leibniz Center for Tropical Marine Ecology (ZMT), Bremen, Germany

Lauren T. Toth Department of Biological Sciences, Florida Institute of Technology, Melbourne, FL, USA; USGS Coastal and Marine Science Center, Saint Petersburg, FL, USA

Alexander W. Tudhope School of GeoSciences, Grant Institute, University of Edinburgh, Edinburgh, UK

Bernardo Vargas-Ángel Coral Reef Ecosystem Division, Pacific Islands Fisheries Science Center, NOAA Inouye Regional Center, Honolulu, HI, USA

Chunzai Wang NOAA/Atlantic Oceanographic and Meteorological Laboratory, Miami, FL, USA

Evie A. Wieters Estación Costera de Investigaciones Marinas, and Departamento de Ecología, Facultad de Ciencias Biológicas, Pontificia Universidad Católica de Chile, Santiago, Chile

Jin-Yi Yu University of California at Irvine, Irvine, CA, USA

Fernando A. Zapata Departamento de Biología, Universidad del Valle, Cali, Colombia

History of Eastern Pacific Coral Reef Research

1

Peter W. Glynn

Abstract

During the Nineteenth and early Twentieth Centuries several renowned researchers—e.g., C. Darwin, J. Dana, A. Agassiz, A.E. Verrill, and T.W. Vaughan—remarked on the absence of coral reefs in the eastern tropical Pacific (ETP), concluding this was due mainly to the influence of cool currents and upwelling centers. From later surveys in southern Mexico in the 1920s (R.H. Palmer), and the Allan Hancock Expeditions across the eastern tropical Pacific in the 1930s (C. Wyatt Durham), workers began to recognize the presence of modest structural coral reef development at several sites in the ETP. A new era of coral reef studies began in the 1970s, and has increased in pace and intellectual depth to the present. This latest surge in coral science was a result of locally active researchers, especially in Mexico, Costa Rica, Panama, and Colombia, and more recently in Ecuador and Chile. Progress in coral research in El Salvador and Nicaragua was thwarted from the 1970s to the early 1990s because of political turmoil and violence in this sector of the eastern Pacific. The establishment of national research funding agencies in Latin American countries, such as CONACYT (Mexico), CONICIT (Costa Rica), COLCIENCIAS (Colombia), and FONDECYT (Chile), has greatly benefitted coral reef research as well as scholarship opportunities for resident students to pursue advanced studies in marine science. The availability of vessels to support research in remote areas and at offshore sites has also facilitated ETP coral studies. Contrary to early views, structural coral reefs were discovered in major upwelling centers—southern Mexico (Huatulco), Costa Rica (Guanacaste), and Panama (Gulf of Panama). The number of known coral species approximately doubled from the late 1940s to 2000, from 60 (total scleractinians) and <20 (reef-building or zooxanthellate corals) to >120 and >40 species, respectively. Additionally, reports of the appearance or disappearance of extralimital coral species (e.g., in the genera *Montipora*, *Siderastrea*, *Acropora*, and *Millepora*) have added to the excitement and adventure of eastern Pacific studies. On site observations during two very strong El Niño-Southern Oscillation (ENSO) disturbances, 1982–83 and 1997–98, demonstrated that acute temperature stress can cause widespread coral bleaching and mortality and interference with reef growth. More recently, ETP waters have been found to have low pH and aragonite saturation state, which can also weaken carbonate structures and interfere with reef development. Recognizing the economic, scientific, and esthetic value of ETP coral reefs, several national and international agencies are now supporting coral research, which will have a salutary effect in increasing our knowledge and understanding of these long-neglected ecosystems.

P.W. Glynn (✉)
Department of Marine Biology and Ecology, Rosenstiel School of Marine and Atmospheric Science, 4600 Rickenbacker Causeway, Miami, FL 33149, USA
e-mail: pglynn@rsmas.miami.edu

P.W. Glynn et al. (eds.), *Coral Reefs of the Eastern Tropical Pacific*,
Coral Reefs of the World 8, DOI 10.1007/978-94-017-7499-4_1

Keywords
Early coral studies • Latin America • Hancock expeditions • Gulf of California • Galápagos Islands

1.1 Introduction

> The western shores of America appear to be entirely without coral-reefs: south of the equator the survey of the *Beagle*, and north of it the published charts show that this is the case. Even in the Bay of *Panama*, where corals flourish, there are no true coral-reefs, as I have been informed by Mr. Lloyd. There are no coral-reefs in the *Galapagos* archipelago, as I know from personal inspection: and I believe there are none on the *Cocos*, *Revilla-gigedo*, and other neighboring islands. (Darwin 1842, p. 156, Appendix).

Coral reef research off western America can be divided into three periods, (1) a time of discovery and general characterization of the region, primarily during the Nineteenth Century; (2) mostly expeditionary activities during the early to mid-Twentieth Century; and (3) on-site field and laboratory studies by resident national and international workers during the final quarter of the Twentieth Century to the present.

This historical review will trace the earliest discoveries of corals and coral reefs in the eastern Pacific and note current rapidly developing research initiatives in Meso-America and South America. All areas within the eastern Pacific region will be considered, from the subtropical Gulf of California and Easter Island (Rapa Nui) to the oceanic islands and continental shores of tropical western America. Information is primarily from the published scientific literature, but theses and popular accounts with relevant information are also included and referenced. The growing inventory of scleractinian species is traced from the first published records near the mid nineteenth century to the present. Attention is also directed to some perplexing coral records, with the intent of clarification regarding their authenticity. Opportunities, resources, and advances in coral research in each Latin American country are briefly summarized. Finally, some projections are offered on future research directions, with attention to regional environmental variability and initiatives vis-à-vis coral reef conservation and management.

1.2 Coral Formations: Reefs Versus Coral Communities

Beginning with the Nineteenth Century perceptions of Darwin (1842), Couthouy (1842), and Dana (1843), the emerging mantra, "the western shores of America appear to be entirely without coral reefs", has been perpetuated for nearly 150 years (Vaughan 1919; Yonge 1940; Stoddart 1969). Since uncertainty on whether coral reefs are present in the eastern Pacific has pervaded the literature over the years, brief attention to the nature of coral formations is worth early consideration. Studies beginning in the 1970s have shown that coral reefs do occur in the eastern Pacific, and can generally be characterized as: (a) small in size, (b) highly dispersed spatially, (c) lacking emergent reef flats or algal ridges, (d) shallow development, <10 m in depth, (e) paucispecific, and (f) with relatively thin structural frameworks (Cortés 1997; Glynn 2001a). The latter criterion corresponds with the critical distinction between coral communities and structural reefs proposed by Wainwright (1965). That is, eastern Pacific coral reefs are constructional, accreting vertically on carbonate substrates produced during Holocene sea-level rise. Unlike massive Indo-Pacific and Caribbean reefs, they seldom form emergent features such as reef flats and algal ridges. This probably contributed to their not being discovered by early researchers.

Suboptimal temperatures due to upwelling and cool currents (California Current, Peru Coastal Current) moving into tropical latitudes were considered by early workers to be responsible for the depauperate coral fauna and absence of reefs from western America. Later supporting evidence for the negative effects of low temperatures was provided by studies of coral feeding and reproduction. For example, Mayor (1916) and Edmondson (1928) demonstrated that low temperatures depress heterotrophic feeding, and Yonge (1940) suggested that coral reproductive activity could similarly be compromised at low temperatures. More recent research, however, indicates that thermal conditions in the eastern tropical Pacific are highly variable, resulting in markedly different biological responses depending on location and season. Coral skeletal growth, for example, is generally high with reduced but positive rates during seasonal upwelling periods (e.g., Glynn and Stewart 1973; Glynn 1977a). At Bahía Culebra, in the upwelling center of Papagayo (Costa Rica), coral growth rates were actually highest for some species during the upwelling season, possibly related to elevated nutrient levels at such times (Jiménez and Cortés 2003). Also, corals are reproductively active at least part of the year and often year-round in all areas investigated, e.g. Mexico (Chávez-Romo and Reyes-Bonilla 2007; Carpizo-Ituarte et al. 2011), Costa Rica, Panama, and the Galápagos Islands (e.g., Glynn et al. 2011). Coral reproduction has even been observed to continue unabated in some areas affected by the 1997–98 ENSO warming event (Colley et al. 2006). Finally, core-drilling and radiocarbon dating have shown significant reef accretion in different areas of the equatorial eastern Pacific. For example, in Costa Rica and Panama reef frameworks range between 4 and 12 m in thickness and are 4000–5000 years in age (Glynn and Macintyre 1977; Macintyre et al. 1993; Cortés et al. 1994).

1.3 Earliest Observations

Possibly the first description of corals in Mexico, and perhaps in the entire eastern Pacific, was by the Jesuit missionary Miguel del Barco. This intriguing account was first brought to light by Reyes-Bonilla (2003). [Much of the information on the historical development of coral research in the eastern Pacific was obtained from the sources listed by country in Table 1.1. Many of these references provide additional information at specific localities.] del Barco (1757) provided a thorough description of colony structure and the disposition of shallow coral formations present at Mulegé, inside the Gulf of California near the entrance to Concepción Bay (see Chap. 5, Glynn et al. for locator maps with place names). He referred to the corals as '*rizo*', meaning curl, frizzle or ringlet (Cuyás 1972), a colloquial term still in use today in parts of Mexico. It is likely this appellation is due to the curly or twisted appearance of the coral's branch tips. Additional comments on the coral's morphology (corallite sizes and distribution, branching pattern, possession of sharp projections [most likely verrucae], and colony fragility) strongly suggest that del Barco was describing a species of coral in the genus *Pocillopora*. Del Barco's description of colony growth and how this relates to adjacent colonies is suggestive of incipient framework building by *Pocillopora* spp., which has been described subsequently in the gulf by several workers (e.g., Squires 1959; Reyes-Bonilla 2003). The coral buildups described by del Barco ranged from 10 to 40 cm in thickness. Somewhat puzzling was his reference to the white color of colonies although the detail provided strongly suggests they were alive. Living colonies of *Pocillopora* spp. are typically brown or golden brown, occasionally with traces of green pigmentation. Could these corals at that time have suffered a recent mortality or bleaching event when colonies turn white?

During the 19th Century, scientists who began to characterize the extent of coral development in the eastern Pacific, and the composition of its fauna, actually spent little or no time in the region. Charles Darwin remained in the Galápagos Islands for slightly more than two months in 1835, during the voyage of the *HMS Beagle* (1831–1836). He barely mentioned corals except to remark on their paucity. Darwin's perception of the absence of reef development in the eastern Pacific is encapsulated in the quotation above (Sect. 1.1) from, "*The Structure and Distribution of Coral*

Table 1.1 Publications with historical information on reef-building corals and reef studies in the eastern tropical Pacific

Area	Authorities
Mexico	Squires (1959)
	Reyes-Bonilla (2003)
	Brusca et al. (2005)
	Reyes-Bonilla et al. (2005)
	Arizpe (2008)
Clipperton Island (France)	Sachet (1960, 1962a, b)
	Skaggs (1989)
	González Avelar (1992)
	Carricart-Ganivet and Reyes-Bonilla (1999)
	Jost (2005)
	Salvat et al. (2008)
	Charpy (2009)
El Salvador	Orellana-Amador (1985)
	Reyes-Bonilla and Barraza (2003)
Costa Rica	Hertlein (1963)
	Rodríguez-S (2001)
	Cortés and Jiménez (2003)
	Cortés (2008, 2009)
Panama	Maté (2003)
	Robertson et al. (2009)
Colombia	Durham and Barnard (1952)
	Birkeland et al. (1975)
	von Prahl et al. (1979)
	von Prahl and Erhardt (1985)
	Escobar (1990)
	Ramos and Lemaitre (1991)
	Brando et al. (1992)
	Barrios and López-Victoria (2001)
	Zapata and Vargas-Ángel (2003)
	Ricardo and Kintz (2008)
Galápagos Islands	Durham and Barnard (1952)
	Slevin (1959)
	Durham (1966)
	Glynn and Wellington (1983)
	Wells (1983)
	Glynn (2003)
Easter Island/Rapa Nui (Chile)	Glynn et al. (2003)
General (eastern tropical Pacific)	Durham (1947)
	Guzmán and Cortés (1993)
	Cortés (1997)
	Glynn (2001b)
	Cortés (2011)

Reefs". Wells (1983) was puzzled over this neglect, after having personally observed an abundance of living corals in the shallow waters there and the large numbers of coral boulders and deposits of carbonate debris on Galápagos beaches. One coral collected by Darwin, an azooxanthellate species, was described 41 years later by Duncan (1876) as *Placopsammia darwini.* Unfortunately, this specimen was lost but appears to have been synonymous with *Tubastraea coccinea*, described by Lesson in 1829 (see Wells 1983).

Close on the heels of the *Beagle*, the US Exploring Expedition, a squadron of six ships with the sloop-of-war flagship *Vincennes* in the lead, departed from New England in 1838. This expedition of exploration and science spanned five years (1838–1842), but did not include in its itinerary any part of the eastern tropical Pacific (Stanton 1975). Among its scientific corps were James D. Dana and Joseph P. Couthouy (Fig. 1.1), the first workers to articulate the occurrence of coral reefs, but only under equable tropical thermal conditions (see below). These scientists also entertained keen systematic interests in corals, molluscs, and other invertebrates, which greatly advanced our knowledge of diverse coral community taxa. Dana produced outstanding monographic works on the scleractinian corals and crustaceans collected during the Exploring Expedition (Dana 1846, 1849, 1852–53, 1853, 1855; see Gilman 1899 for complete citations to these studies). Incredibly, 229 of the 483 corals described by Dana were new species, and several of these contribute importantly to reef building in the eastern Pacific. Dana's continuing work on the expedition's collections also resulted in the naming of 650 species of crustaceans. Couthouy, a recognized conchologist, worked on the mollusc collections sent to Washington DC, but was unable to complete these studies. Due to the mismanagement of the collections by an inexperienced curator (loss of labels and specimens), and a severe reduction in salary, Couthouy was forced to resign from the expedition's scientific corps and seek employment elsewhere—as captain of a merchant vessel. A.A. Gould, a member of the Boston Society of Natural History, completed the expedition's report on the Mollusca (Johnson 1946).

For reasons of ill health and "dissension" (according to Captain Charles Wilkes, the expedition's leader), Couthouy was expelled prematurely from the expedition at Hawaii. Upon his early return to the USA, Couthouy read a paper before the Boston Society of Natural History on the control of coral distributions by low sea temperatures. This perception, published in 1842, was proposed to explain the supposed absence of coral reefs in the eastern tropical Pacific. This was an important deduction, but one that has been subject to substantial revision in recent years. Couthouy's publication resulted in turmoil with Dana, who accused the former of plagiarism (Dana 1843; Stoddart 1994). Upon further inquiry, this claim was later retracted by Dana (Dana and Couthouy 1844), who eventually acknowledged that both discovered this relationship independently. The professional lives of Dana and Couthouy diverged notably in the years following the US Exploring Expedition. In 1856, Dana was offered a Silliman Professorship at Yale University and enjoyed many productive years of teaching and research until his death in 1895 at age 82. Couthouy was not so fortunate, having suffered financial hardships and then caught up in the maelstrom of the US Civil War. While commanding the ironclad steamer USS *Chillicothe* on the Red River, he was mortally wounded by sniper fire and died 9 h later in 1864 at age 56 (Joiner and Sandefur 2007).

The next publications of note were by Milne Edwards and Haime (1848a, b), who described a fossil coral from the Galápagos Islands, and Grewingk (1848) who recorded a specimen of *Porites* from the Gulf of California. In a rapid succession of papers by Addison E. Verrill (Fig. 1.2) from 1866 to 1870, six genera and 12 species of reef-building corals were described from the Gulf of California and Panama (e.g., Verrill 1866, 1868–70, 1870). Four species of *Porites*, including *Porites californica* (= *panamensis*), were described from La Paz (Gulf of California) as well as *Montipora fragosa*, the latter causing considerable confusion for several years (see Sect. 1.6). *Porites panamensis* and *Psammocora stellata* (Verrill's original generic designation of *Psammocora* was *Stephanocora* = *Stephanaria*) were described from specimens collected in the Pearl Islands, Panama. Verrill (1870) also noted the occurrences of *Pavona gigantea* and *Pavona clivosa* (=*clavus*), two presumably Indo-Pacific species, in the Pearl Islands. [*Pavona gigantea* is considered endemic to the ETP by Veron (2000) and in this book in Chap. 5, Glynn et al.] All of these corals were collected along Central American shores by F.H. Bradley for the Museum of Yale College, and at Cape San Lucas by John Xantus for the Smithsonian Institution. While Verrill did not engage in field work in the ETP, he nonetheless contributed enormously to the taxonomic knowledge of marine invertebrates in this region. He is credited with describing more than 1000 species in such disparate taxa as the Porifera, Cnidaria, Polychaeta, Mollusca, and Echinodermata.

Pourtalès (1875) listed five species described by Verrill from the Galápagos Islands, from beach collections made in 1873 during the visit of the US Coast Survey steamer R.V. *Hassler*. These included species in the Indo-Pacific genera *Pavona* and *Pocillopora*. By the end of the Nineteenth Century, a total of 35 scleractinian corals were known from the eastern tropical Pacific. Fifteen of these belonged to zooxanthellate or reef-building species (Fig. 1.3). In spite of these advances, several celebrated researchers continued to perpetuate the myth of the absence of coral reefs in this region. For example, Verrill (1866) concluded, "Consequently at Panama, and on the entire western coast of Central America within the tropics, no true coral reefs occur." As

Fig. 1.1 **a** James D. Dana (1813–1895) "The abundant opportunities that the Expedition afforded Dana to study the three great types of coral islands—atolls, barrier reefs, and fringing reefs—enabled him independently to confirm Darwin's theory [of coral reef subsidence]." (Stanton 1975). "I have not for some years been so much pleased as I have just been by reading your most able discussion on coral reefs…" (Darwin to Dana correspondence, Gilman 1899). **b** Joseph P. Couthouy (1808–1864), conchologist of the scientific corps, US Exploring Expedition. A personal plea to President Andrew Jackson to serve as a member of the expedition—"Well General, I'll be hanged if I don't go, if I have to go before the mast." (Johnson 1946)

noted above, this perception was conceived secondhand mainly from the observations of collectors. Thanks to the introduction of scuba diving technology in the 1950s, the pace and breadth of coral research accelerated in the ETP as well as globally (Lang et al. 2013).

1.4 Early Twentieth Century Expeditions

At the turn of the century, from 1888 to 1911, the US Fish Commission Steamer *Albatross* conducted marine biological surveys, mostly deep water dredging, around the Galápagos Islands and at several other localities in the ETP. Alexander Agassiz accompanied two of these expeditions (1891, 1904) as chief scientist. In his report on the 1891 *Albatross* expedition, he noted that nearly all Galápagos sandy beaches contained reef rock consisting of coral fragments and the remains of other invertebrates (Agassiz 1892, p 70). *Pocillopora* fragments dominated the cemented limestone, and extensive patches of these corals (presumably living) were observed off the sand beaches. Agassiz further noted that these coral patches, as was well known, never form true coral reefs in the Panamic district. He surmised that high siltation resulting from terrigenous deposits limited reef development more than low temperatures. Other expeditions to the Galápagos Islands, e.g. the Hopkins-Stanford Expedition (1898–1899), the California Academy of Sciences Expedition (1905–1906), the Crane Pacific Expedition (1929), and the Templeton Crocker Expedition (1932), directed their scientific efforts to the terrestrial fauna and flora or marine

Fig. 1.2 Addison Emery Verrill (1839–1926), a pioneer coral taxonomist who described several eastern Pacific scleractinian coral species, "… our excellent Yale Zoölogist, Professor A.E. Verrill, who now stands first in the country in the department of Zoöphytes." (J.D. Dana 1872). Courtesy of the Peabody Museum of Natural History, Yale University

vertebrates (Slevin 1959) with no significant advances in coral research. Vaughan (1906) identified one species of *Pocillopora* from Taboguilla Island (Gulf of Panama) and one species each of *Pocillopora* and *Porites* from Easter Island, collected during the *Albatross* Expedition.

In the 1920s and 1930s, William Beebe and colleagues embarked on four marine biological expeditions that covered the length and breadth of the eastern tropical Pacific. These expeditions were conducted by the Department of Tropical Research, New York Zoological Society (now the Wildlife Conservation Society). The results and adventures were popularized in several articles and books by Beebe, with the technical accounts published in the Society's scientific journal *Zoologica*. The first of these, a relatively brief venture aboard the steam yacht *Noma*, traveled to the Galápagos Islands in 1923 (Beebe 1924). In 1925 the *Arcturus* Expedition carried out studies at Cocos Island and the Galápagos Islands utilizing hardhat diving technology (Beebe 1926). The cruise tracks of the first (1936) and second (1938) *Zaca* Expeditions included sites in northwestern Mexico (Beebe 1938), and numerous collecting stations spanning over nearly 30° of latitude from Baja California, Mexico to Gorgona Island, Colombia respectively (Beebe 1942). Tantalizing comments on the presence of corals and even coral reefs occur in these books. For example, mention was made of large coral boulders ("Gargantuan mushrooms") and coral reefs in the Galápagos Islands and at Cocos Island. It was noted repeatedly that coral chunks were dislodged and broken apart in search of fishes and cryptic invertebrates. A brief description was given of a "real coral reef" at Huatulco, Mexico. Unfortunately, these various cursory observations were never documented because coral reef geologists or biologists did not participate in the expeditions.

The first two eastern Pacific studies dedicated to coral research were carried out in the mid to late 1920s, the SY *St. George* Expedition in the equatorial eastern Pacific with Cyril Crossland (1927) as chief scientist, and Robert H. Palmer's reconnaissance of the southwestern Mexican coast off Oaxaca (Palmer 1928). Even though Crossland's team visited several sites in Panama (Gulf of Chiriquí), Colombia (Gorgona Island), and the Galápagos Islands where coral reefs are now known to be present, he dismissed the occurrence of coral reefs in the region, following Couthouy and Dana's pronouncements 85 years earlier. Palmer (1928), however, should probably be credited with first recognizing and documenting the presence of structural coral reefs in the eastern Pacific. Although Palmer's descriptions of the pocilloporid reefs were incomplete, he provided enough information to substantiate their constructional nature.

Indubitably, the most comprehensive and sustained Twentieth Century effort in marine biological investigations in the eastern tropical Pacific was initiated in 1931 by the *Velero III* expeditions of the Hancock Foundation for Scientific Research, University of Southern California (Fig. 1.4). The *Velero III* completed ten expeditions over a 10-year period (1931–1941), ranging from Baja California, Mexico to the Galápagos Islands, Ecuador and the Peruvian coast (Fraser 1943a, b, c). A typical major research cruise, and the research sites occupied during nine months in 1938, is exemplified by the Seventh Expedition off Mexico, Central America, the Galápagos Islands and Peru (Fig. 1.5). Captain Allan Hancock (Fig. 1.6), a consummate explorer and benefactor of marine science, financed the establishment of the Allan Hancock Foundation and construction of the motor cruiser *Velero III*, which was donated to the University of Southern California in 1939. *Velero III*, a steel, Diesel-propelled cruiser of 59 m overall length and with a cruising radius of 15,300 km, was designed specifically for marine research. She had ample laboratory space, small boat support, trawling and dredging capabilities, and comfortable living quarters, including a library, and elegant dining and recreational rooms (Fraser 1943a). *Velero III* evoked considerable pride from her crew and scientific personnel, as

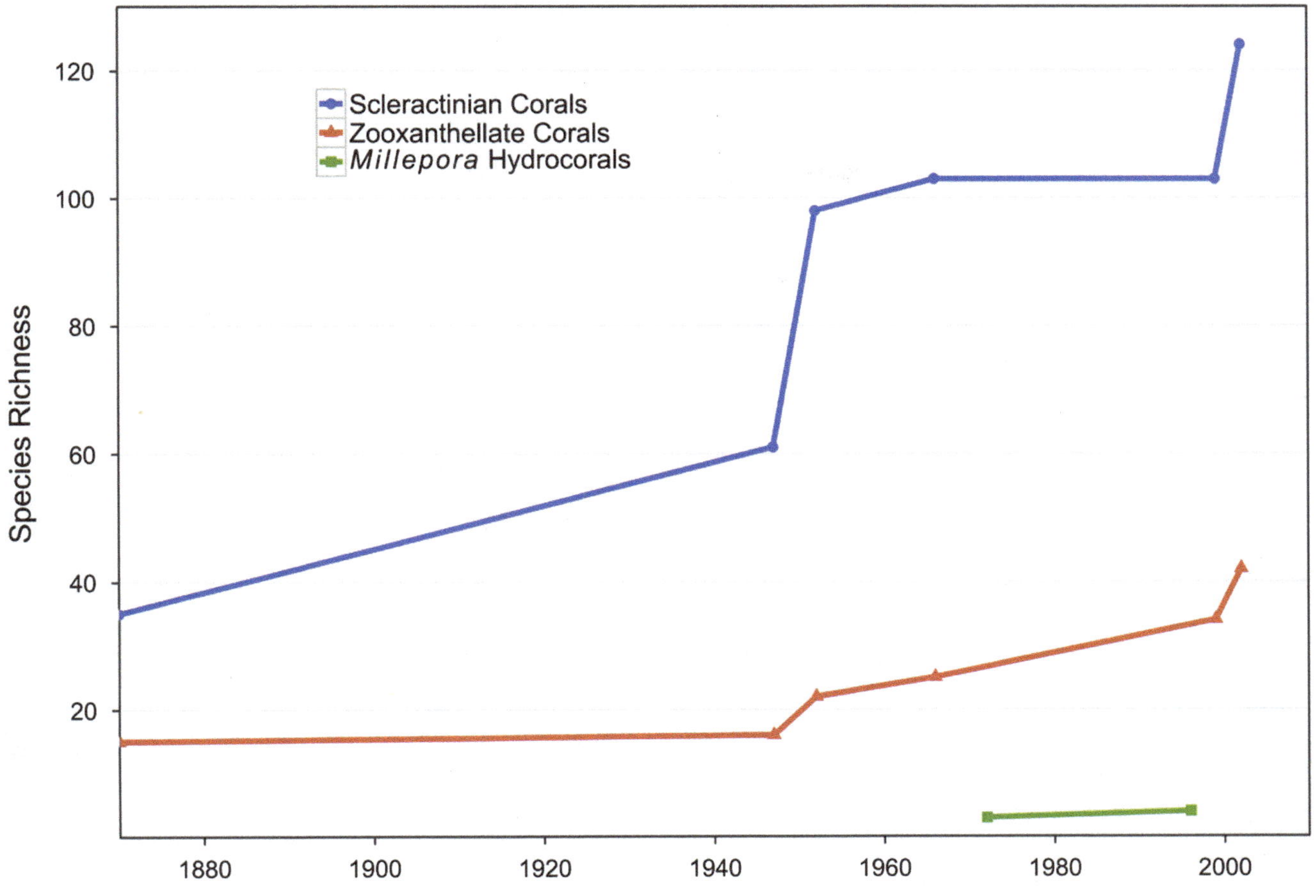

Fig. 1.3 Increasing trends in species richness of eastern Pacific scleractinian and milleporid corals based on comprehensive taxonomic reviews. Scleractinian corals include both azooxanthellate and zooxanthellate species. Zooxanthellate species are phototrophic and confined to relatively shallow depths. Milleporid hydrocorals are also zooxanthellate. Key: (1) Verrill (1870); (2) Durham (1947); (3) Durham and Barnard (1952); (4) Durham (1966); (5) Cairns et al. (1999); (6) Reyes-Bonilla (2002); (7) Glynn (1972), Porter (1972a); (8) Glynn et al. (1996)

Fig. 1.4 M/V *Velero III* of the Allan Hancock Foundation, University of Southern California, lying at anchor off Puerto Utría Colombia, Third Expedition, 15 February 1934. Courtesy of the Allan Hancock Foundation

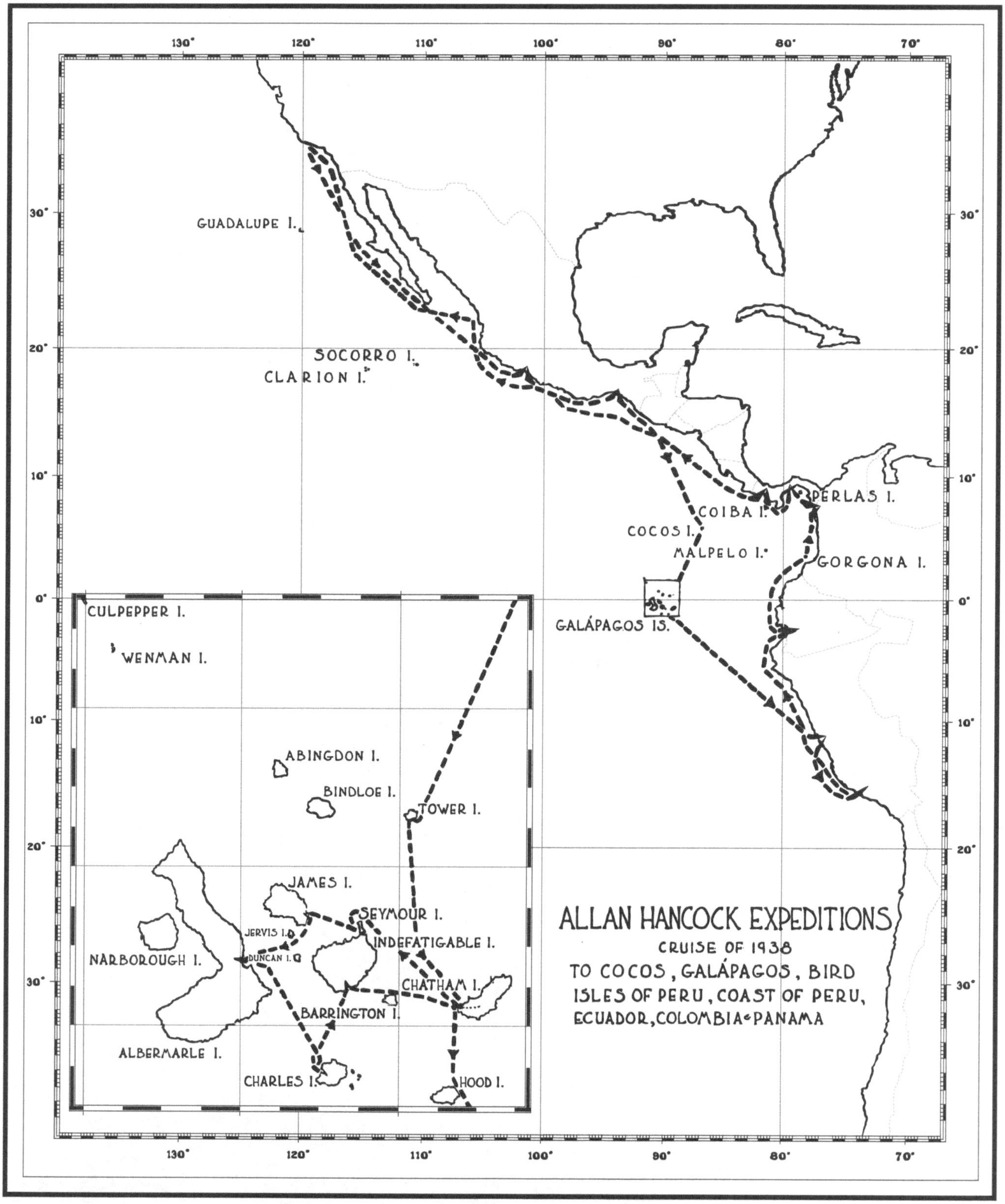

Fig. 1.5 Cruise tracks of the *Velero III* Seventh Expedition (1938) to Central and South America (modified reproduction from Meredith 1939)

Fig. 1.6 Captain Allan Hancock standing beside the barrel post office at Post Office Bay, Floreana Island, Galápagos (undated, 1930s)

reflected in the first of 51 stanzas of the *Rhyming Log of Velero III*

> Velero III, that master craft
> Lacking in nothing, fore and aft,
> Was due to cruise to famed La Paz—
> None quainter south of Alcatraz.
> By Captain Allan Hancock bid,
> His guests, of shore engagements rid
> Assembled promptly, eager-eyed,
> To have their longings satisfied.
> Stanza I (Clover 1932)

Shore parties from the *Velero* discovered numerous vibrant coral communities at several sites in the eastern Pacific. An extensive field of small massive corals, possibly *Porites lobata*, was observed at Espíritu Santo Island in the Gulf of California (Fig. 1.7). Numerous *Pocillopora* spp. colonies were collected in Panama, the Galápagos Islands and elsewhere, but primarily for the molluscs, crustaceans, fishes and other associates seeking refuge within the sheltering branches of these coral hosts (Figs. 1.8 and 1.9). One photographic image actually shows a skiff traversing the reef flat of a true structural coral reef in the Gulf of Chiriquí, Panama (Fig. 1.10). Some forty years later a study of the growth history of this km-long fringing coral reef indicated a maximum reef-frame thickness of 12.0–13.4 m, and an age likely in excess of 5000 years BP (Glynn and Macintyre 1977). Surprisingly, coral geologists or biologists were never among the many distinguished members of the *Velero III* scientific teams. However, coral collections were obtained and these plus other miscellaneous collections were worked up later by various specialists, particularly J. Wyatt Durham (Fig. 1.11). The notable increase in scleractinian coral richness in the 1940s and 1950s (Fig. 1.3), exceeding 100 species, was due in large measure to Durham's systematic efforts (Durham 1947, 1962, 1966; Durham and Barnard 1952). Of the total corals, 25 of these belonged to zooxanthellate species, mostly in the genera *Porites*, *Pocillopora* and *Pavona*. In Durham's (1966) last systematic study, he lamented that corals were collected only incidentally during the Allan Hancock Pacific Expeditions. He offered the prescient conclusion that "… much is yet to be learned about the eastern Pacific coral fauna". The Tenth Expedition of *Velero III*, to the outer coast of Baja California (Cedros Island) in 1941, was the last engaged in Latin American studies. *Velero III* was replaced by *Velero IV* after the Second World War with a new research focus in the temperate and subarctic waters from California to Alaska (1948–1985).

By the mid-20th Century, enough had been observed and published by competent workers for Durham (1966) to conclude that structural coral reefs, albeit small compared with their Indo-Pacific counterparts, were present in the eastern Pacific region. The pocilloporid reefs briefly described by Palmer (1928) along the Huatulco (Oaxaca) coast of southern Mexico, may represent the first scientific documentation of coral reefs in the eastern Pacific. A little more than a decade later, Steinbeck and Rickets (1941) offered a

Fig. 1.7 Abundant corals, probably *Porites* sp., and worm tubes during an extreme low tidal exposure at Espíritu Santo Island, Gulf of California (undated, possibly March 1936)

popular account of a prominent pocilloporid reef at Cabo Pulmo in the Gulf of California. Squires (1959), a coral paleontologist-member of the Puritan-American Museum Expedition in 1957, mapped and described the coral formations in El Pulmo Bay. Unlike Steinbeck and Rickets (1941), he concluded that this was not a structural coral reef because the living corals formed only a thin veneer on nearshore intrusive dikes. At other sites, however, for example at San Gabriel Bay (Espíritu Santo Island), Puerto Escondido, and Salinas Bay (Carmen Island), Squires considered these coral formations to represent incipient constructional reefs. Squires found that the tidally-exposed *Porites* colonies in San Gabriel Bay (Fig. 1.7), observed earlier during a *Velero III* survey, extended further seaward forming a constructional reef.

Villalobos (1960) and colleagues were the first Mexican researchers to describe a Pacific coral community, in the offshore Revillagigedo Islands. Their attention was mainly directed toward non-coral metazoan associates, but they did make reference to a coral reef ("seguramente muy antiguo", definitely very old) present on the south side of Socorro Island. More recent studies in the 1990s have substantiated the occurrence of modest reef building, a relatively species rich coral fauna, and the biogeographic importance of these oceanic islands (Reyes-Bonilla 1993a; Ketchum and Reyes-Bonilla 1997, 2001).

Clipperton Island is the eastern-most coral atoll in the Pacific Ocean, situated about 1300 km southwest of mainland Mexico (Acapulco). Its date of discovery has been inconsistently reported, noted in one historical account as early as 1526 by a Spanish vessel under the command of Alvaro Saavedra (Carricart-Ganivet and Reyes-Bonilla 1999), and in another historical review in 1705 by the English buccaneer John Clipperton (Salvat et al. 2008). Étienne (2009) noted that the first accurate description of Clipperton's location was made in 1711 by personnel aboard the French frigates *La Découverte* and *La Princesse*. On this occasion the island was named "île de la Passion". Clipperton has been variously claimed and exploited in the Nineteenth Century by Mexico, France, and the USA. The atoll was annexed by Mexico from 1897 to 1917 and occupied by a garrison, including some soldier's families. Tragically, chaos following the 1914 Mexican revolution led to the malicious neglect and death of many innocent souls marooned on Clipperton (Skaggs 1989). Mexico lost sovereignty over Clipperton in 1931 when King Victor Manuel of Italy, with Benito Mussolini possibly exercising some influence in the decision, awarded the island to France (González Avelar 1992). Clipperton is now directly administered by the High Commissioner of French Polynesia.

Clipperton's isolation in the eastern Pacific, located some 4260 km from Pukapuka Atoll (in the eastern Tuamotus), the nearest South Pacific atoll, has attracted considerable scientific interest (Sachet 1960, 1962a; Salvat et al. 2008; Charpy 2009). The first scientific expedition to visit Clipperton, with terrestrial and marine biological research objectives, was the Hopkins Stanford Galápagos Expedition of 1898–1899. Samples collected from Clipperton Rock in 1935, by a landing party from the French training vessel *Jeanne d'Arc*, proved this structure to be an extrusive volcanic dome (Lacroix 1939). The presence of this residual

Fig. 1.8 Collecting party working from M/V *Velero III* at Bahía Honda, Panama, Third Expedition, 21 February 1934. Fred C. Ziesenhenne, an echinoderm systematist, is at the oars and Waldo L. Schmitt, a crustacean systematist, is examining *Pocillopora* colonies brought up by local divers *au naturel*. Waldo L. Schmitt was the first to alert Ziesenhenne and John S. Garth, a noted student of coral crabs, to the possibilities of crustacean discoveries in *Pocillopora* coral colonies. Quoting from Garth (pers. comm.), "Dr. Schmitt's instructions to us were to make one last grab at the nearest coral head before leaving the beach and bring it aboard, where it could be worked over at leisure. On one occasion, when all his buckets were full, he took off his shirt and carefully wrapped it around a head of coral. The type locality of one or more of our new species might well have been 'Schmitt's shirtful of coral'." Courtesy of the Allan Hancock Foundation

trachytic prominence allows Clipperton to be properly classified as an "almost atoll". In 1938 the F.D. Roosevelt Presidential Cruise, supported by the USS *Houston* of World War II fame, visited Clipperton briefly with Waldo Schmitt aboard as the marine biologist. Dr. Schmitt was able to make modest collections of a variety of marine species.

Various physical measurements and biological reconnaissances were carried out in the 1950s by US expeditions before and during the International Geophysical Year, 1957–1958 (Sachet 1962b). The most notable of these was the Doldrums Expedition (1958) of the Scripps Institution of Oceanography. The R/V *Spencer F Baird* landed and provided logistical support for a research party for nearly seven weeks. While the marine biologists were chiefly concerned with collecting molluscs and other non-coral invertebrates (Hertlein and Emerson 1953, 1957; Allison 1959), sufficient information was published to indicate that geomorphologically this island was an actively accreting coral atoll. Scuba diving by Conrad Limbaugh and others, to 40 m depth, revealed distinct reef front coral zonation similar to that described for atolls in the southwestern Pacific (Allison 1959).

Charpy (2009) summarized the history of French expeditions to Clipperton, including the several Bougainville expeditions in the late 1960s, the Cousteau expedition in 1978, and others to the more recent 2004–2005 expedition organized by J-L Étienne. Jost's (2005) bibliography, covering the period from 1711 to 2005, lists 411 references on Clipperton, including 324 scientific publications.

Fig. 1.9 Waldo L. Schmitt aboard *Velero III* extracting animals sheltering in *Pocillopora* corals, Bahía Honda, Panama, 1934

Fig. 1.10 Rowing across a tidally exposed reef flat in the Secas Islands, Gulf of Chiriquí, Panama, Fourth *Velero III* Expedition (February 1935). Bleached summits of *Pocillopora* colonies are visible in foreground

1.5 Recent Advances (1970s to Present)

A plethora of coral and reef-related studies were published in the 1970s, ushering in the modern era of eastern Pacific coral reef research. Nearly 80 % of 28 coral-related publications were completed by resident scientists and students during this decade. This was an unprecedented move toward local in-country research. Among the research themes were coral systematics (Wells 1972), reef community species' composition and ecology (Greenfield et al. 1970; Glynn et al. 1972; Glynn 1973; Porter 1972a, 1974; Bakus 1975; Birkeland et al. 1975; Wellington

Fig. 1.11 J. Wyatt Durham (1907–1996), a pioneering coral systematist in the eastern Pacific. "With respect to the absence of coral reefs on the Pacific shores of the Americas and the adjacent islands, it may possibly be more apparent than real." (Durham and Barnard 1952)

1975; Birkeland 1977), reef structure, distribution and growth (Glynn 1972, 1977a; Glynn and Stewart 1973; Dana 1975; Glynn and Macintyre 1977; von Prahl et al. 1979), crustacean symbiont communities of corals (Abele 1972, 1974, 1976a, b, 1979; Abele and Patton 1976; von Prahl et al. 1978), and *Acanthaster* corallivory (Dana and Wolfson 1970; Barham et al. 1973; Glynn 1974, 1977b; de Alba 1978) with investigations ranging from Mexico to Costa Rica (Cocos Island), Panama, Colombia (Gorgona and Malpelo Islands), Ecuador (Galápagos Islands), and Chile (Easter Island). This 1970s publication surge has continued unabated in most Latin American countries, with a growing cadre of national scientists and students. Now there may be several hundred resident scientists and students pursing coral reef studies in the eastern Pacific region. A sampling of these researchers, their opportunities and accomplishments are now briefly highlighted. A latitudinal progression from Mexico to Chile follows, with some brief comments on four Central American countries where coastal conditions are generally unfavorable to support coral growth. The primary literature sources consulted are listed by country in Table 1.1.

1.5.1 Mexico

Beginning in the 1970s marine research on the Pacific coast of Mexico, particularly in the Gulf of California and at some mainland sites along the coast of Nayarit, advanced rapidly. Investigators from three Mexican academic and research institutions in Baja California, and researchers from Arizona and California were the principal scientists involved in these initiatives. In the mid-1980s and early 1990s, Oscar Arizpe at the Universidad Autónoma de Baja California Sur (UABCS) in La Paz, and José Carriquiry at the Universidad Autónoma de Baja California (UABC), Ensenada began their studies of corals and coral community ecology. Arizpe, primarily concerned with coral conservation, proposed guidelines for eco-tourism at the Pulmo reef rather than the prevalent destructive practices of extraction (Arizpe and Álvarez 1987; Arizpe 2004, 2008). Employing isotopic analyses, Carriquiry and colleagues quantified coral responses (e.g., skeletal composition, tissue bleaching, and mortality) to oceanographic changes during El Niño events (Carriquiry et al. 1988, 1994, 2001; Carriquiry and Villaescusa 2010). Also active at UABCS since the early 1990s, Héctor Reyes-Bonilla, colleagues and students began adding significantly to the inventory and geographic distributions of Mexican coral species (e.g., Reyes-Bonilla 1992, 1993a, b; Carriquiry and Reyes-Bonilla 1997; Ketchum and Reyes-Bonilla 1997; Reyes-Bonilla and López-Pérez 1998; Reyes-Bonilla et al. 1999; Pérez-Vivar et al. 2006), which is evident in the marked increase in species richness in 2000 (Fig. 1.3). Additionally, these researchers published numerous studies on coral community composition, and disturbances (e.g., El Niño warming and upwelling cold water bleaching events) along the entire Mexican mainland coast and offshore islands (Revillagigedo Islands and Clipperton Island). CICIMAR (Centro Interdisciplinario de Ciencias Marinas), founded in 1976 and located in La Paz, is the marine research arm of the Instituto Politécnico Nacional. It publishes the journal *Oceánides*, a current version of *Investigaciones Marinas*, which was established in 1985. Dinorah Herrero, an active researcher at CICIMAR, has investigated the population biology of an endangered holothurian inhabiting coral communities in the southern gulf region (Herrero-Pérezrul et al. 1998, 1999).

Studies focused on coral community structure and dynamics in the Gulf of California were also published in the 1990s by Mexican researchers (Reyes-Bonilla 1993b; Reyes-Bonilla et al. 1997a, b; Reyes-Bonilla and Calderón-Aguilera 1999) and investigators from Arizona (University of Arizona) and California (Scripps Institution of Oceanography) (Robinson and Thompson 1992; Baynes

1999). An earlier study by Brusca and Thompson (1975) concluded that the Pulmo Reef was a true constructional coral reef. This view was challenged recently on the basis of carbonate production rates and losses. Riegl et al. (2007) demonstrated that while the Pulmo calcifying community was productive, it was not contributing to vertical framework construction due to off reef carbonate transport. The *Acanthaster* survey by Dana and Wolfson (1970) offers an early baseline on the effects of this corallivore in the Gulf of California.

In the mid-1990s, coral studies were initiated by Gerardo Leyte at the Universidad del Mar (UMar) in the Mexican state of Oaxaca. Cenozoic coral formations and recent coral population dynamics are subject areas being pursued by Andrés López-Pérez at UMar. During the latter half of the 1990s, Pedro Medina and Amílcar Cupul-Magaña began coral investigations off the state of Jalisco, at the Puerto Vallarta campus of the Universidad de Guadalajara. The numbers of investigators and students pursuing coral studies in these institutions have grown exponentially during the last three decades.

In order to promote research and cooperative initiatives on the Atlantic and Pacific coasts of Mexico, several scientists in 2004 participated in founding a national organization, La Sociedad Mexicana de Arrecifes Coralinos (SOMAC), dedicated to coral reef studies. There are presently about 90 members in this society. Meetings (Congresos Mexicanos de Arrecifes de Coral) are held approximately biennially, alternating between the Pacific and Atlantic coasts of Mexico. The first national meeting was convened in 2000, under the auspices of Universidad Veracruzana and Acuario de Veracruz, A.C., at Veracruz, in the state of Veracruz. The second meeting was hosted in 2003 by the Universidad del Mar in Huatulco, Oaxaca. A third meeting was held in 2006 at Cancún, Quintana Roo, organized by UNAM (Universidad Nacional Autónoma de México) and ECOSUR (El Colegio de la Frontera Sur), then in 2007 by UABCS in La Paz, Baja California Sur. A fifth meeting was convened in 2009 in Tuxpan de Rodríguez Cano, Veracruz, hosted by Universidad Veracruzana, and a sixth meeting in Ensenada, Baja California in 2011, hosted by UABC. The seventh congress was successfully convened in 2013 at Mérida, Yucatán, organized by UNAM and CINVESTAV (Centro de Investigación y de Estudios Avanzados). And a 10-year celebration was successfully conveved by SOMAC in 2015 at Puerto Vallarta, with an opening review of coral reef research in Mexico by Héctor Reyes Bonilla.

Coastal and offshore marine research in Mexico has benefitted from national ship support and governmental and non-governmental funding organizations. *El Puma*, a national oceanographic vessel operated by UNAM, began supporting research off the Pacific coast in 1981. This ocean-going platform, with the port of Mazatlán serving as her home base, has supported coral research in the 1990s and 2000s at the remote Revillagigedo Islands and Clipperton Atoll. Competitive research support was established by the national government in 1971 via CONACYT (Consejo Nacional de Ciencia y Tecnología) and since the mid-1990s from CONABIO (Comisión Nacional para el Conocimiento y Uso de la Biodiversidad) (Table 1.2). The latter foundation has funded research in coral taxonomy, systematics, and biogeography. Several non-governmental organizations have also supported studies in Mexico related to coral reef conservation and education, e.g. the Walton Family Foundation, the David and Lucile Packard Foundation, the World Wildlife Fund, and The Nature Conservancy.

The last decade has seen a new research theme encompassing the ecology, biogeography and evolution of holobionts (*Symbiodinium* communities, their coral hosts and associated microbes) beginning in the Gulf of California. This work has been the result of a collaboration between U.S. and Mexican investigators (e.g., Iglesias-Prieto et al. 2004; LaJeunesse et al. 2007, 2008; Paz-García et al. 2008; Pettay and LaJeunesse 2009; LaJeunesse et al. 2010; Pinzón and LaJeunesse 2010). These studies are providing insights into the responses of corals to environmental stressors, such as temperature and solar radiation extremes, the taxonomic identities of coral hosts (particularly *Pocillopora* spp.), and the regional distribution patterns of eastern Pacific holobionts.

A current awareness of the value of coral communities and coral reefs in Mexico has prompted the establishment of protective measures to safeguard this ecosystem. Six marine protected areas off western Mexico support coral formations, namely the islands of the Gulf of California, the Revillagigedo Islands, Cabo Pulmo reef, Los Cabos region, Los Arcos (Puerto Vallarta), and Huatulco (Reyes-Bonilla 2003). The largest of these MPAs, extending over nine degrees of latitude (23°00′–32°00′N) and including 124 major islands, is the Gulf of California Biosphere Reserve, created in 1995. While full protection is afforded the island habitats and species, enforcement in remote areas is virtually nonexistent. Two coral rich areas that receive moderate protection, principally through enforcement by local stakeholders, are the Cabo Pulmo reef (Reyes-Bonilla 1997) and the Huatulco reef tract (Bezaury-Creel 2005). The Biosphere Reserve at Cabo Pulmo, established in 1995 and covering over 7000 ha (Reyes-Bonilla 1997), supports one of the richest biotas in the Gulf of California (Brusca et al. 2005).

1.5.2 Clipperton Island (France)

Several studies centering on the marine environment and biota of Clipperton began in the 1990s and have continued to the present day (Jost 2005; Salvat et al. 2008; Charpy 2009). For the first time, these have involved investigations

Table 1.2 Funding organizations supporting coral reef research in the eastern Pacific

Country	Agency
Mexico	CONACYT (Consejo Nacional de Ciencia y Tecnología)
	CONABIO (Comisión Nacional para el Conocimiento y Uso de la Biodiversidad)
	Fondo Mexicano para la Conservación de la Naturaleza
Costa Rica	UCR (Universidad de Costa Rica)
	CIMAR (Centro de Investigación en Ciencias del Mar y Limnología)
	CONARE (Consortium of Public Universities of Costa Rica)
	CONICIT (Consejo Nacional para Investigaciones Cientificas y Tecnológicas)
	CI (Conservation International)
	FFEM (French Funds for the World Environment)
	ACMIC (Área de Conservación Isla del Coco)
	SINAC (Sistema Nacional de Áreas de Conservación)
Panama	Smithsonian Tropical Research Institute (STRI), Marine Fellowship Program
	Smithsonian Institution (Research Foundation, Research Awards Program, Scholarly Studies Program, Office of Environmental Studies)
	University of Panama, Centro de Ciencias del Mar y Limnologia
	SENACYT (National Secretariat of Science, Technology and Innovation)
	INDICASAT (Instituto de Investigaciones Científicos y Servicios de Alta Tecnología)
	Public Interest Association—Research Center for Biomedicine
	ICBG (International Cooperative Biodiversity Group)
	AVINA (Avina Foundation, Panama)
	German Academic Exchange Service (DAAD)
	Leibniz Centre for Tropical Marine Ecology (ZMT)
	Project AWARE Foundation (PADI)
	Liquid Jungle Lab (LJL) Panama
Colombia	COLCIENCIAS (Departamento Administrativo de Ciencia, Tecnología e Innovación)
	Fondo Colombiano de Investigaciones Científicas y Proyectos Especiales "Francisco José de Caldas"
	Special Administrative Unit of the System of National Natural Parks (UASPNN) Ministry of the Environment
	Universidad del Valle
	Universidad de los Andes
	Sociedad Alemana de Cooperación Técnica (GTZ)
	Banco de Occidente
	Fundación Malpelo
Ecuador	Charles Darwin Foundation for the Galápagos Islands
	Charles Darwin Research Station
	Galápagos National Park Service
	Instituto Nazca de Investigaciones Marinas
	Wildlife Conservation Society
	United Kingdom DEFRA (Department for Environment, Food and Rural Affairs) Darwin Initiative Project
Chile	FONDECYT (Fondo Nacional de Desarrollo Científico y Tecnológico, Comisión Nacional de Investigaciones Científica y Tecnológica
Regional	Conservation International (CI)
	US National Geographic Society
	US National Science Foundation (International Program)
	Nature Conservancy (NC)
	J William Fulbright Foreign Scholarship Program
	US Peace Corps
	La Sociedad Mexicana de Arrecifes Coralinos
	Exxon Mobil Undergraduate Scholarship Programme
	Fundación MarViva
	US National Institutes of Health (NIH), bioprospecting

specifically aimed at corals, e.g. species inventories, zonation, community structure, framework development, and the systematics and ecology of coral-associated organisms (Glynn et al. 1996; Robertson and Allen 1996, 1997; Carricart-Ganivet and Reyes-Bonilla 1999; Flot and Adjeroud 2009). Oxygen isotope analyses of Clipperton corals have also provided information on climate variability and El Niño activity in this sector of the eastern Pacific (Linsley et al. 1999, 2000).

Several new coral records and the discovery of a new species resulted from the 1997 Mexican SUR-PACLIP-1 R/V *El Puma* cruise organized by Vivianne Solis-Weiss of UNAM. *Porites arnaudi*, named after the last commander of the Mexican Navy garrison who drowned in 1916 attempting a rescue, contributes importantly to the coral framework construction at Clipperton (Reyes-Bonilla and Carricart-Ganivet 2000). The most comprehensive recent assessment of Clipperton's biota was performed during the Jean-Louis Étienne 2004–2005 Expedition. Among the numerous research themes were studies that focused on the physical environment (e.g., the origin of the island, the oceanographic setting, and the biogeochemistry of the unique meromictic lagoon), and the biology and ecology of the island's terrestrial and marine biota (namely plankton communities, benthic algae, corals, crustaceans, molluscs, echinoderms, annelid worms, bryozoans, and fishes) (Charpy 2009). Although Clipperton's reef-building coral diversity is very low, in terms of its geomorphologic development and active framework construction it compares favorably with most western Pacific atolls. The most recent comprehensive survey lists 17 zooxanthellate cnidarians, including one rare or locally extinct scleractinian (*Pavona gigantea*) and the hydrocoral species *Millepora exaesa* (Flot and Adjeroud 2009).

1.5.3 Guatemala, Honduras

No zooxanthellate coral communities or coral reefs have been reported from the Pacific coasts of Guatemala or Honduras (Glynn 2001a; Spalding et al. 2001; Cortés 2003). This approximately 1000 km coastal stretch, from Tangola Tangola Bay, Huatulco (Oaxaca) to the Gulf of Fonseca, is dominated by lagoons, sandy beaches, estuaries and mangrove ecosystems. First recognized by Springer (1958) from the absence of rocky shore fish records, it became known as the "Pacific Central American faunal gap". Long considered a major biogeographic break, this coastal expanse was assumed to separate coral faunas and associated species from southern Mexico to Costa Rica (Briggs 1974). Unidentified stony corals, possibly zooxanthellate species, were noted by Gotuzzo (1996) to be present at Maculis, El Salvador, at the entrance of the Gulf of Fonseca (see Sect. 1.5.4). If these records are substantiated, then it is possible that some coral species may also occur on Honduran islands in the Gulf of Fonseca.

1.5.4 El Salvador, Nicaragua

Until recently, only fragmentary information was available on the occurrence of corals along the Pacific shores of El Salvador and Nicaragua. As early as the mid-1970s, Gierloff-Emden (1976) noted the presence of corals and reef rock off Punta Remedios, about 8 km south of Acajutla, El Salvador. A catalog of reef-inhabiting fishes collected off Los Cóbanos (El Salvador), published by Orellana-Amador (1985), supported Gierloff-Embden's brief account. Accompanying the fish study was a map denoting four, approximately 1-km long coral patches just off the coast. In the next record of note, unidentified photographs of octocorals and scleractinian corals were illustrated by Gotuzzo (1996) in a popular two volume series, *Historia Natural y Ecología de El Salvador*. The scleractinian corals included in this account appeared to be *Pocillopora* sp., *Porites* sp, and *Tubastraea* sp. A photograph of a possible hydrocoral species (*Millepora*) was also illustrated.

Adding to unpublished government and student reports up to the year 2000, an ecological survey of the coral formations off Los Cóbanos in 2001 by Reyes-Bonilla and Barraza was published in *Latin American Coral Reefs* (2003). These workers reported well developed pocilloporid communities, consisting of eight zooxanthellate scleractinian species. No *Millepora* hydrocorals were observed. Outcrops of basalt rock served as substrates for the pocilloporid and other scleractinian species, but no structural frameworks were reported.

Until mid-2000, the occurrence of small patches of solitary corals near the Costa Rican border was all that was known of coral development on the Pacific coast of Nicaragua (Ryan and Zapata 2003). This changed dramatically in 2009, following the surveys conducted by Juan Alvarado and co-workers. They reported the occurrence of constructional reefs along the south coast of Nicaragua, with reef frameworks consisting of massive (*Pavona gigantea, Gardineroseris planulata*) and branching (*Pocillopora elegans*) species (Alvarado et al. 2010, 2011a). A total of 13 scleractinian corals were found and nine of these were zooxanthellate species. Like Huatulco in southern Mexico, this region in the Gulf of Papagayo is subject to intense seasonal upwelling, but nonetheless supportive of reef development.

In retrospect, two factors that delayed coral research along much of the Central American coast from the 1960s to the 1990s were the social-political turmoil that evolved into brutal civil wars in Guatemala (1960–1996), El Salvador (1980–1992) and Nicaragua (1979–1990), and an absence of local scientists dedicated to coral studies. Surely, the atmosphere of violence must have discouraged field studies and the pursuit of marine science careers.

1.5.5 Costa Rica

Marine studies on the Pacific coast of Costa Rica were greatly facilitated with the establishment of the Centro de Investigación en Ciencias del Mar y Limnología (CIMAR), at the University of Costa Rica, San José (Rodríguez-S. 2001, www.cimar.ucr.ac.cr). This multidisciplinary research center opened in 1979 with seven scientists, one administrative assistant, and a modest skiff for marine coastal and freshwater studies. CIMAR was the brainchild of Professor Manuel M. Murillo and faculty co-founders José A. Vargas, Carlos Villalobos, William Bussing and Myrna López. After three decades, CIMAR scientists numbered 22. Some notable accomplishments include over 700 journal publications, several books, and 102 student theses, one-third of the latter by foreign students.

A peer-reviewed venue for the publication of tropical biological studies, including coral research, was established in 1953 at the University of Costa Rica, San José. Julián Monge-Nájera has served for over two decades as the chief editor of the bi-lingual journal, *Revista de Biología Tropical* (*International Journal of Tropical Biology and Conservation*). Several reef-related, eastern Pacific studies in Mexico, Costa Rica, Panama and Colombia have appeared in *Revista de Biología Tropical*, particularly beginning in the 1980s (Cortés and Nielsen 2002). One special issue in 2008 (Cortés 2008) and two in 2012 (Cortés 2012a, b) were dedicated to comprehensive reviews of the history of marine research at Culebra Bay and Isla del Coco (Cocos Island). These journal issues contained sundry coral studies, species inventories and information on environmental conditions relevant to corals and coral reefs (e.g. Cortés 1996–1997; Cortés and Blum 2008). For a chronological summary of marine exploration and scientific studies at Cocos Island, from the first documented stopover of William Dampier in 1697 to the CIMAR-COCO expedition in 2008, see Cortés (2008).

Bakus (1975) conducted the first quantitative subtidal survey at Cocos Island in 1972 during the R/V *Searcher* research cruise off Pacific Central America. The first coral reef surveys on the Costa Rican mainland coast were initiated in the late 1970s by Glynn et al. (1983) and Jorge Cortés (Fig. 1.12), and colleagues from 1982 to 1984 (Cortés et al. 1984; Cortés and Murillo 1985). These early surveys were particularly timely because they were performed immediately before and during the 1982–1983 El Niño warming disturbance, which resulted in severe coral mortality and the regional extinction of an Indo-Pacific coral species (Guzmán et al. 1987a; Guzmán and Cortés 1993; Cortés and Guzmán 1998). A comprehensive study of the growth history of the Punta Islotes coral reef was completed in 1989 (Cortés 1990). Financial support for the coastal surveys was provided by the University of Costa Rica (UCR), CONICIT (Consejo Nacional para Investigaciones Científicas y Tecnológicas), and non-governmental organizations: Conservation International and The Nature Conservancy, and from private funds; the offshore work at Cocos Island by UCR, FFEM (Fonds Français pour l'Environnement Mondial), CONARE (Consejo Nacional de Rectores), ACMIC (Área de Conservación Marina Isla del Coco), and Conservation International (Table 1.2).

Fig. 1.12 Jorge Cortés, pioneer Costa Rican coral reef researcher surveying the coastline at Parque Nacional Isla del Coco, Costa Rica, October 2007. Photograph courtesy of Jaime Nivia

Three of Cortés' students who have contributed importantly to coral studies in Costa Rica are Héctor M. Guzmán, beginning in the mid-1980s, Carlos Jiménez in the mid-1990s, and Juan José Alvarado in the early 2000s. Guzmán investigated a variety of topics, including coral community structure, El Niño disturbance effects and coral reef recovery, corallivore interactions, coral growth rates, and the meiofauna composition of reef sediments (Guzmán et al. 1987b; Guzmán 1988a, b, 1991; Guzmán and Cortés 1989a, b, 1992, 2001, 2007). Jiménez discovered some unique coral communities at Culebra Bay, an upwelling environment in the Gulf of Papagayo, and quantified El Niño-related mortality in this region (Jiménez 1997; Jiménez and Cortés 2001; Jiménez et al. 2001, 2010). More recently, Alvarado investigated the impact of anthropogenic effects on coral reefs of the central Pacific coast of Costa Rica (Alvarado et al. 2005, 2009). Guzmán and Jiménez teamed up with Cortés to document the time course and patterns of coral community recovery following ENSO disturbances at coastal and offshore sites (Guzmán and Cortés 1992, 2001, 2007; Jiménez and Cortés 2001; Jiménez et al. 2001). These various research activities received support from the Consejo Nacional para Investigaciones Científicas y Tecnológicas (CONICIT), CIMAR, UCR and other private and public organizations (Table 1.2).

1.5.6 Panama

Thanks to the vision and efforts of Martin Moynihan and Ira Rubinoff, marine research at the Smithsonian Tropical Research Institute (STRI) in Panama began in the early 1960s with the establishment of laboratories on the Pacific and Caribbean coasts of Panama. These laboratories supported studies on the evolutionary divergence of species with the emergence of the Panamanian isthmus and inter-oceanic separation approximately 3 Ma. Also in the 1960s, serious consideration was being given to a controlled nuclear detonation/excavation project for the construction of a sea-level canal across the Isthmus of Panama. This raised great concerns among biologists and environmentalists regarding the likelihood of radionuclide contamination and potentially devastating interoceanic invasions (e.g., Rubinoff 1968; Briggs 1969). Even with the well-established fresh-water barrier of Gatun Lake, some marine species had already migrated through the Panama Canal (Rubinoff and Rubinoff 1969), and fouling organisms on ship's bottoms were shown to have the potential to transit the canal alive (Menzies 1968). A special symposium was convened in Panama in 1971—The Panamic Biota: Some Observations Prior to a Sea-level Canal (Jones 1972)—to consider the ecological risks of such a project. A major conclusion of this symposium was how little was known of the Central American marine biota. This was a strong justification for renewed marine biological studies and the provision of funding and facilities support for research near the existing canal and in other Panamanian coastal areas. The major programs of the Smithsonian Institution supporting this work are identified in Table 1.2. STRI's first research vessel, the RV *Tethys* (Fig. 1.13), provided access to all coastal and offshore island localities on the Pacific and Caribbean sides of Panama.

With the establishment of the Departamento de Biología Marina (DBM) at the University of Panama (UP) in 1966, early research initiatives were forged between STRI and UP that accelerated interest in studies of coral biology and ecology. Luis Howell-Rivero, a renowned ichthyologist and Cuban exile, was a founding member of DBM and its first director until his resignation in 1974 when his duties shifted to an UNESCO fisheries science advising position for the government of Panama. Carlos Arellano, another early member of DBM, encouraged student involvement in marine biological studies and helped to solidify cooperative research projects between DBM and STRI. Through its fellowship and internship programs, STRI has provided coral research support to students and scientists from UP, other Latin American universities, and worldwide (Table 1.2). This support has allowed students to earn advanced degrees, and has contributed to the development of a cadre of resident coral scientists.

For more than three decades, Luis D'Croz has served a central role in fostering marine science at UP. D'Croz was first appointed to a position in DBM in 1977. He then became director of the newly founded Centro de Ciencias Marinas y Limnología (CCML) in 1982 and remained in this position until 1991. D'Croz continued with his research and teaching at UP, always with a sharp eye on identifying exceptional students. Collaboration between UP and STRI was further enhanced with D'Croz's appointment to STRI's

Fig. 1.13 RV *Tethys*, former yacht of Panama Canal Zone governor, became first research vessel of the Smithsonian Tropical Research Institute. Years of service, 1970–1972

research staff in 1994. D'Croz, his steadfast technical assistant Juan B. del Rosario and students have focused on characterizing the oceanographic conditions and plankton ecology off Pacific Panama (D'Croz et al. 1991; D'Croz and Robertson 1997; D'Croz and O'Dea 2007). This line of research has contributed to an understanding of coral responses associated with trophic resources in upwelling and non-upwelling environments, particularly during ENSO disturbance events (Glynn and D'Croz 1990; Maté 1997; D'Croz et al. 2001; Hueerkamp et al. 2001; D'Croz and Maté 2004; Schlöder and D'Croz 2004; Anlauf et al. 2011). D'Croz retired from STRI in 2009, and continued teaching at UP until 2015.

Crossland's (1927) perception of the meager development of coral formations off the Pacific coast of Panama was seriously challenged by the findings of several workers associated with STRI. Numerous structural coral reefs were found in both upwelling and nonupwelling environments in the early 1970s and 1980s (Glynn 1972; Glynn et al. 1972; Porter 1972a; Glynn and Stewart 1973; Dana 1975; Wellington 1982). These discoveries and other early reef studies were greatly aided by Aníbal Velarde, a stalwart research technician at STRI. Relatively large numbers of *Acanthaster planci* also were found feeding on corals in the nonupwelling Gulf of Chiriquí (Glynn 1972, 1973, 1974; Porter 1972b) shortly after Chesher (1969) described massive devastation of corals at Guam in the Mariana Islands. This raised concerns due to the proximity of this sea star corallivore to the Panama Canal and possible access to corals in the Caribbean Sea should the freshwater canal barrier be breached or removed by construction of a sea-level canal. Mesocosm feeding experiments in Panama demonstrated that *Acanthaster* has a strong preference for certain Caribbean coral prey (Glynn 1982).

As coral research progressed, some provocative controversies arose that had a salutary effect in advancing understanding of coral reef biogeographic relationships. One of these began with T. Dana's (1975) hypothesis explaining the origin of contemporary corals and reef building in the eastern Pacific. He suggested that present-day Panamanian reefs owe their existence to west-to-east long distance dispersal of coral propagules across the eastern Pacific Barrier (EPB). This hypothesis was countered by McCoy and Heck (1976), and Heck and McCoy (1978) who proposed that eastern Pacific corals were derived from a pan-Tethyan western Atlantic fauna that has been modified by speciation and extinctions since its isolation. Recent molecular genetic studies have demonstrated a strong biogeographic connection across the EPB among several species of echinoids (Lessios et al. 1996), and fishes (Rosenblatt et al. 1972; Lessios and Robertson 2006). West to east gene flow may be enhanced during ENSO events with an elevated influx of Central Pacific surface waters at such times (Johnson et al. 2002). Genetic evidence also indicates significant east-to-west connections for some species (Lessios and Robertson 2006; Duda and Lessios 2009; see Chap. 16, Lessios and Baums). In terms of morphospecies similarities, the eastern Pacific and Indo-west Pacific coral faunas show a very strong relationship (Cortés 1986; Glynn and Ault 2000; Reyes-Bonilla 2002). To what extent this reflects genetic similarity remains the subject of future studies (see Chap. 14, Pinzón). Darwin's (1872) notion of an impassable eastern Pacific barrier is presently under serious revision.

The unprecedented and widespread coral bleaching/mortality event associated with the very strong 1982–83 El Niño disturbance (Glynn 1983) spawned numerous studies in Panama and at other equatorial eastern Pacific localities—Costa Rica (Cortés et al. 1984; Carriquiry et al. 1988), Colombia (von Prahl 1983), and Ecuador (Robinson 1985). An equally strong El Niño event in 1997–98 (Enfield 2001) precipitated a second round of coral bleaching and mortality. These disturbances have stimulated several research questions in Panama that have focused on, for example, immediate and long-term community effects (Glynn 1990), local and regional extinctions (Glynn and de Weerdt 1991; Glynn 1997), inter- and intra-specific differences in coral bleaching resistance (Glynn et al. 2001; Maté 2003a), algal symbiont diversity (*Symbiodinium* spp.) and its effects on bleaching susceptibility (Glynn et al. 2001; Baker 2003; Baker et al. 2004; LaJeunesse et al. 2007; van Oppen et al. 2009), bioerosion (Glynn 1988; Eakin 1992, 1996, 2001), coral sexual and asexual reproduction (e.g., Colley et al. 2006; Glynn and Colley 2008; Glynn et al. 2011), reef fish dispersal and recruitment (Wellington and Viktor 1985; Lessios et al. 1998; Viktor et al. 2001), coral-based paleoclimate reconstructions (Linsley et al. 1994; Wellington and Dunbar 1995), and coral community resistance, resilience, and recovery (Glynn et al. 2001; Baker et al. 2004, 2008; Wellington and Glynn 2007). These first documented coral bleaching and mortality events, caused by stressful high sea temperatures (Glynn and D'Croz 1990), foreshadowed more recent bleaching in the eastern Pacific (Carriquiry et al. 2001; Glynn et al. 2001), and at numerous localities in all global coral seas (e.g., Hoegh-Guldberg et al. 2007; Baker et al. 2008; Wilkinson 2008; Wilkinson and Souter 2008).

In June 1996 STRI and the University of Panama hosted the Eighth International Coral Reef Symposium (ICRS) in Panama City, Panama. This venue, on the shores of the equatorial eastern Pacific, provided an unprecedented opportunity for Latin American scientists and students to participate in the six day conference. Attendance exceeded 1200 persons from 62 countries, the largest ICRS gathering up to that time. Several eastern Pacific coral studies were presented among the 15 plenary lectures, 790 oral presentations and 140 poster exhibits. The plenary lectures were published in 1997, the Year of the Reef, in *Coral Reefs*, the flagship journal of the International Society for Reef Studies. Over 350 of the oral presentations, all peer-reviewed, were published in two symposium volumes edited by Harilaos A. Lessios and Ian G. Macintyre. Three of 14 post-symposium field trips were conducted in coral reef areas on the Pacific side of Panama and in the Galápagos Islands.

Since the 1970s, several studies have demonstrated that coral communities in the Gulf of Chiriquí support a high diversity of metazoan taxa. Many of the species in this remote and oceanographically unique region (D'Croz et al. 1991; Macias Mayorga 2002) are undescribed, represent new distributional records or are Indo-Pacific migrants. Representative examples include cnidarian hydrocoral, octocoral and scleractinian species (de Weerdt and Glynn 1991; Glynn and Ault 2000; Breedy and Guzmán 2003; Maté 2003b), polychaete worms (López García et al. 1997), opisthobranch molluscs (Hermosillo 2004), crustaceans (Kim and Abele 1988), echinoderms, and fishes (Rosenblatt et al. 1972; Rosenblatt and Waples 1986; Robertson and Allen 2008). These baseline inventories have served an important role in the marine zoning plan of the Coiba National Park (ANAM 2009). The mapping of diversity foci ('hot spots'), including the presence of endemic, rare, and endangered species, is an essential element in the design of priority conservation units in the Coiba National Park (Guzmán et al. 2004, 2008).

Fig. 1.14 Henry von Prahl (1948–1989), pioneer Colombian naturalist and coral reef researcher. Photograph undated. *Han pasado casi 20 años y Henry aún hace falta* (Cantera-Kintz 2008)

1.5.7 Colombia

Terrestrial and marine investigations of oceanic Malpelo Island, located about 500 km west of Buenaventura, Colombia, were carried out in 1972 by a joint team of scientists from STRI (Panama), the US Navy, and the Smithsonian Institution, Washington DC (Graham 1975). The subtidal surveys by Birkeland et al. (1975) represented the first quantitative description of coral community structure in Pacific Colombia. This work also included contributions to the coral crustacean symbionts of corals (Abele 1975), and coral-associated asteroids (Downey 1975), and fishes (McCosker and Rosenblatt 1975).

In the late 1970s, Henry von Prahl and colleagues from the Universidad de los Andes began investigating the coral reefs of Gorgona Island, located about 30 km off the southern Pacific coast of Colombia (von Prahl et al. 1979; Glynn et al. 1982). These investigators organized and led the *Sula* III Expedition to Gorgona Island in 1979, which resulted in the publication of several coral-related articles in the Anales del Instituto de Investigaciones Marinas de Punta de Betín (known today as Boletín de Investigaciones Marinas y Costeras), Colombia's most celebrated marine science journal (von Prahl et al. 1979; Castro 1982; von Prahl 1982; Glynn et al. 1982; Werding and Haig 1982). These baseline surveys have allowed for accurate assessments of coral mortality coincident with the 1982–83 El Niño heat stress event (von Prahl 1983, 1985, 1986) and later disturbances (Vargas-Ángel et al. 2001).

von Prahl was a gifted artist and talented naturalist with interests in mangroves, ornithology, crustaceans and corals, the latter his true passion (Fig. 1.14). In a relatively short period (1979–1989), he published some 80 journal articles and eight books, six of them on the corals and coral reefs of Colombia (Fig. 1.15). One of these books, *Malpelo, Isla Oceánica de Colombia* (Brando et al. 1992), which featured stunning photographs by Aldo Brando and texts by Jaime R. Cantera K., listed von Prahl as a co-author to honor his memory. von Prahl met an untimely death in November 1989, an unwitting victim of a cowardly terrorist's attack. Nonetheless, he managed to leave an indelible legacy. After joining the faculty of the Marine Biology Section in the Biology Department at Universidad del Valle (Cali) in 1982, von Prahl and his new colleagues, most notably Jaime Cantera (a malacologist and ecologist) and Efraín Rubio (an ichthyologist) played a pivotal role in eastern Pacific coral reef research. Cantera's contributions in coral reef ecology (Cantera 1983; Cantera and Contreras 1988) and numerical taxonomy (Cantera et al. 1989) complemented von Prahl's research. By calling attention to the well-developed coral reefs of Gorgona Island, von Prahl and colleagues were instrumental in promoting the establishment of the Gorgona Island National Natural Park in 1985. Their writings about Malpelo Island, a World Heritage Site since 2006, would also help diving enthusiast and conservationist Sandra

Fig. 1.15 Two of Henry von Prahl's six books on the corals and coral reefs of Colombia, *Isla de Gorgona* (1986), edited by von Prahl and M. Alberico, and *Malpelo, La Roca Viviente* (1989) by von Prahl. Courtesy of the Universidad del Valle, Cali, Colombia

Bessudo to capture the Colombian government's attention and designate Malpelo a Sanctuary of Flora and Fauna in 1995. A measure of the continuing scientific interest in this oceanic MPA is evident in a recent special issue of the Bulletin of Marine and Coastal Research (López-Victoria 2011a, b) dedicated to diverse studies at Malpelo Island. von Prahl mentored several students, including Bernardo Vargas-Ángel, who would later describe coral formations in the northern sector of the Colombian Pacific coast (Vargas-Ángel 1988, 1996), and elucidate the historical development of coral reefs under recurrent natural disturbances (Vargas-Ángel 2001, 2003). For additional information on von Prahl's legacy, see Escobar (1990) and Ramos and Lemaitre (1991).

von Prahl's position at the Universidad del Valle was filled in 1991 by Fernando Zapata, a reef-fish ecologist with broad interests in coral reefs. While his studies in the Colombian Pacific naturally tended to focus on fishes (e.g., Zapata and Morales 1997; Zapata 2001a; Zapata and Herrón 2002; Lozano and Zapata 2003; Zapata and Robertson 2007), he and his colleagues (particularly Jaime Cantera with whom he created a group for the study of coastal ecosystems), and students also worked on several aspects of coral biology and ecology, including the quantitative description of coral formations (Zapata 2001b; Zapata and Vargas-Ángel 2003; Zapata et al. 2008), coral bleaching and the impact of natural perturbations such as El Niño, and subaerial exposures during low tides (Vargas-Ángel et al. 2001; Zapata et al. 2001, 2011), associated fauna (Cantera and Arnaud 1995), coral reef bioerosion (Cantera et al. 2001, 2003; Londoño-Cruz et al. 2003; Toro-Farmer et al. 2004), and corallivory (Narváez and Zapata 2010). Currently the group is carrying out studies on coral reproduction (both sexual and asexual) and recruitment.

Two additional developments that advanced reef studies were the initiation of biological and physical monitoring programs. The first was the establishment in 1998 of the National Monitoring System of Coral Reefs in Colombia (SIMAC), an initiative developed under the vision and leadership of Jaime Garzón-Ferreira of the Instituto de Investigaciones Marinas y Costeras (INVEMAR, Garzón-Ferreira and Rodríguez-Ramírez 2010). Under this initiative, INVEMAR and Universidad del Valle joined efforts and established four stations for annual monitoring at Gorgona Island's largest

Fig. 1.16 ARC *Malpelo*, one of four Colombian oceanographic vessels, commissioned in April 1981. Bahía Solano, Pacific coast of Colombia, February 2007

coral reef (see Navas-Camacho et al. 2010; Zapata et al. 2010). Two more stations were established in 2002 on another coral reef and six more were added in 2010 for a total of 12 stations on three reefs. Other monitoring stations were set up by INVEMAR at Malpelo Island and at Ensenada de Utría on the coast in 2002. The results of this monitoring effort have been published by INVEMAR since 2002 in an annual report, the latest of which describes the state of marine and coastal environments and resources in Colombia until 2013 (INVEMAR 2014).

The second development was an increased capacity for oceanographic studies in the Colombian Pacific basin. While the Colombian Navy began undertaking oceanographic cruises in 1965 for the ACENTO and EASTROPAC programs led by the InterAmerican Tuna Commission, beginning in 1970 the navy has been carrying out their own cruises with scientific personnel from the Centro de Investigaciones Oceanográficas e Hidrográficas (CIOH), and from the Centro Control Contaminación del Pacífico since 1994. These cruises have been normally performed biannually using the ARC *Malpelo* oceanographic vessel (Fig. 1.16) as the research platform and have contributed to the understanding of the spatial and temporal variability in oceanographic conditions in the Colombian Pacific (Tejada et al. 2002). At a more localized scale, scientists in the Marine Biology Section at Universidad del Valle began investigating the oceanographic conditions in the vicinity of Gorgona Island in 2005. These studies, led by Alan Giraldo, have substantially improved our knowledge of the physical and biological oceanographic conditions and their relevance to coral reefs around the island (Escarria et al. 2007; Giraldo 2008; Giraldo et al. 2008).

1.5.8 Ecuador

Coral reef research with an ecological bias began in Ecuador in the mid-1970s with Gerard M. Wellington, when he served as a Peace Corps volunteer at the Charles Darwin Research Station (CDRS), Santa Cruz Island, Galápagos Islands (Fig. 1.17). During his two-year internship (1973–1975) he was tasked with identifying and describing Galápagos coastal marine environments. The product of this work (Wellington 1975) served as a blueprint for nearshore shelf areas in need of protection by the Galápagos National Park Service. All coral assemblages making up the iconic patch reef inside the Onslow Island drowned crater were meticulously mapped by Wellington (Fig. 1.18). Wellington teamed up with John Wells, Chuck Birkeland and Peter Glynn in 1975, who systematically surveyed all major island sites, focusing on coral community species' composition, zonation, biotic interactions, and reef structure. The existence of several incipient reef formations also was an important finding during this survey. *Beagle III*, the Darwin station's research vessel, served as the working platform for this archipelago-wide study (Fig. 1.19). This work was continued in 1976 with the major results published in *Corals and Coral Reefs of the Galápagos Islands* (Glynn and Wellington 1983). Feeding observations and quantification of the effects of the echinoid corallivore *Eucidaris galapagensis* in limiting coral growth were also ecologically significant findings in these early initiatives (Glynn et al. 1979).

The very strong 1982–83 El Niño had a devastating effect on Galápagos coral communities and island-wide marine life (Robinson and del Pino 1985; Glynn et al. 1988; Glynn 1990). Coral cover suffered 95–99 % mortality at most

Fig. 1.17 Gerard M. Wellington ('Jerry'), at the helm of an inflatable boat near Darwin's Arch, Darwin (Culpepper) Island, Galápagos, circa 1974

shallow sites, and then experienced intense bioerosion by *Eucidaris*, lithophagine bivalve molluscs and other bioeroders (Glynn 1988; Scott et al. 1988). By 1993, all monitored coral reef frameworks in the southern and central islands had been reduced to sand and rubble or were otherwise severely degraded (Glynn 1994). Mobile coral communities in deeper water (>14 m), subject to cooler temperatures and reduced light, mostly survived the 1982–83 warming event (Feingold 1996).

From the early 1990s, the sedulous lead of the BioMar (CDRS) research team with Rodrigo Bustamante, Graham Edgar and Stuart Banks in the forefront, has championed the need to protect and conserve unique Galápagos marine communities. Due to a notable increase in tourist diving activities in the Galápagos Marine Reserve (GMR), beginning as early as the mid-1970s, quantitative baseline surveys were mandated by the Directorate of the Galápagos National Park and the Charles Darwin Foundation in order to assess and safeguard the shelf biota. The biotic surveys performed by BioMar in all biogeographic zones were carried out in 2000 and 2001, to ~15 m depth island-wide at more than 200 sites, including all known coral and reef areas (Danulat and Edgar 2002; Heylings et al. 2002). Information from these assessments is also included in a companion document on the physical and biotic condition of 15 marine tourist diving sites in the GMR (Danulat et al. 2003). Directed more specifically to coral conservation concerns, the 2009 issue of the journal *Galapagos Research* gives a view of recent research initiatives in the northern-most islands of Wolf and Darwin (Dawson et al. 2009). The GMR, encompassing ~138,000 km^2, was established by the Ecuadorian government in 1998. It is presently a World Heritage Site and the largest marine protected area in the equatorial eastern Pacific.

Coral studies along the Ecuadorian mainland coast have lagged behind those in the Galápagos Islands. The first mention of a constructional coral reef on the Ecuadorian coast was by Glynn and Wellington (1983), who briefly described a fringing coral reef in 1975 at Machalilla, about 15 km north of Puerto López. Seven species of zooxanthellate corals were observed on this reef during a brief reconnaissance with the predominant frame-building species in the genus *Pocillopora*. This reef was estimated to cover about 1 ha and to have built vertically 3–4 m. Later coastal surveys in 1991 and 1998 revealed small constructional reefs at Cabuya, Los Frailes and La Plata Island (Glynn 2003). All mainland sites observed in 1991 exhibited mostly dead and eroded, branching and massive corals, with high abundances of the sea urchin *Diadema mexicanum*. The larger *Diadema* aggregations ranged in size from 20 to 50 individuals m^{-2}, and these sea urchins were mostly large with test diameters of 10–12 cm. The extent of coral mortality and reef framework degradation along the coast was very similar to that observed in the Galápagos Islands in the aftermath of the 1982–83 El Niño disturbance, and was likely caused by the same sea-warming impact. Repeated surveys at several coastal sites in 1998, toward the end of the 1997–98 El Niño event, revealed widespread coral bleaching, but also significant recovery since 1991.

This central coastal area of Ecuador, including Machalilla and all of the coral sites previously noted, lies within the bounds of the Machalilla National Park (MNP), established in 1979 and encompassing 20,000 ha of coastal and marine shelf habitats. Credit is due the Ecuadorian NGO Nazca, with support from Conservation International, for supporting baseline surveys on coral, mobile invertebrates and fishes in this area. A pictorial guide assembled by Rivera and Martínez (2011) illustrates the variety of scleractinian corals and octocorals present in the MNP reef communities. An effort is now underway by the MNP team to map, evaluate, and monitor the coral communities within the park. A developing awareness of the value of coral reefs on the Ecuadorian coast will hopefully serve to bring them under more effective protection, as is currently the case in the Galápagos Islands.

1.5.9 Easter Island (Rapa Nui), Chile

The only Chilean areas reported to support zooxanthellate coral populations are the oceanic Easter (Rapa Nui) and

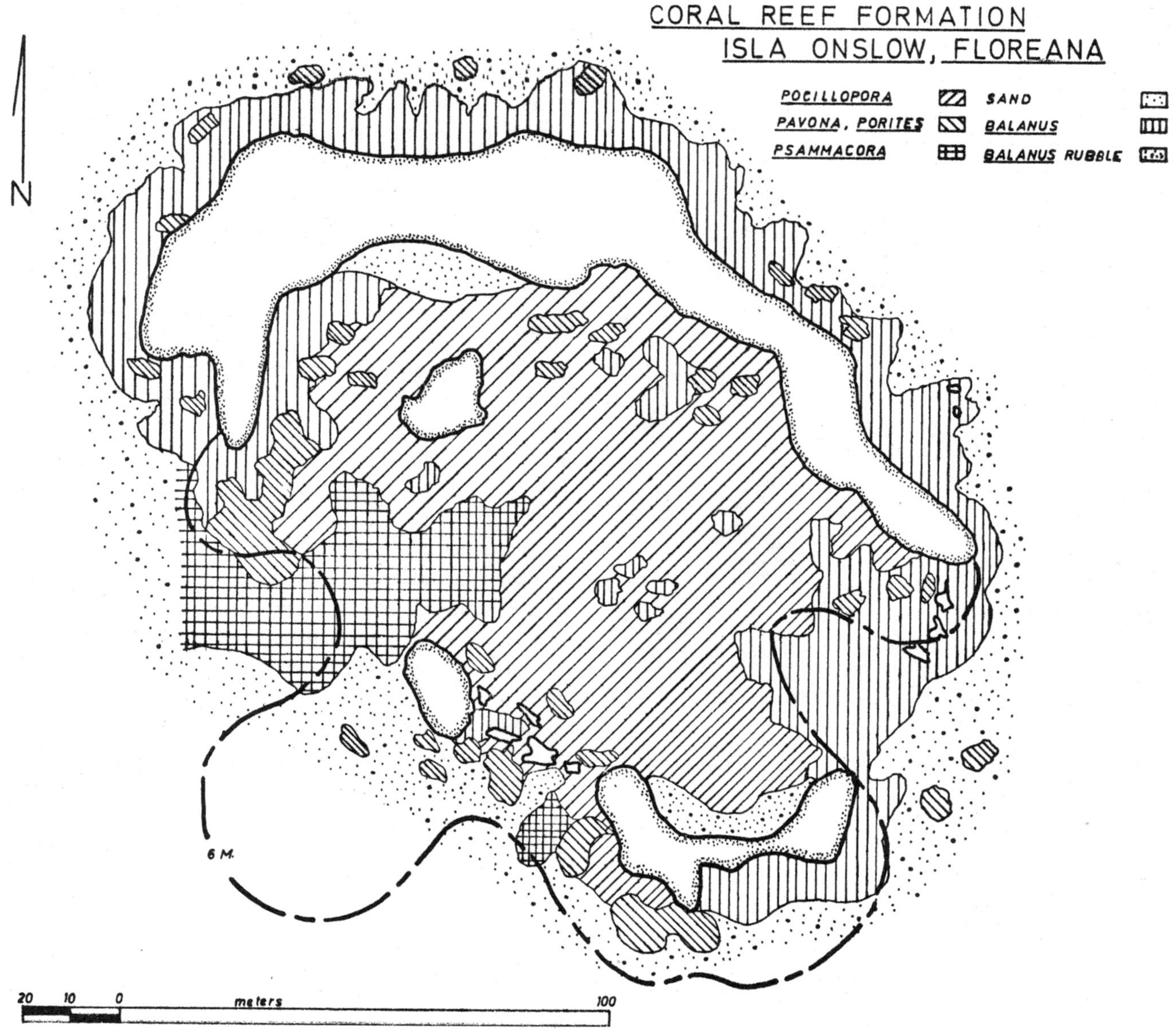

Fig. 1.18 Wellington's map of the coral and barnacle communities at Isla Onslow (La Corona del Diablo), a highly eroded volcanic cone, Floreana Island, Galápagos Islands. Planar view, basalt rock is depicted as an unfilled *white* rim (slightly modified from Wellington 1975)

Salas y Gómez Islands (Castilla and Rozbaczylo 1987), both located in the relatively warm and stable thermal conditions that characterize the South Pacific subtropical gyre. Long assumed to belong to the southeast Pacific biogeographic region, along with the Pitcairn Islands and the Hawaiian Islands (e.g., Veron 1995), new coral species records from Easter Island demonstrate a strong affinity with eastern Pacific oceanic coral faunas (Glynn et al. 2007; Veron et al. 2015; see Chap. 5, Glynn et al.). Further, recent surface current studies have demonstrated a strong east to west flow along the Easter/Salas y Gómez corridor suggesting a barrier to recruitment from the west.

Vaughan (1906) described the first two coral species collected at Easter Island during the cruise of the *Albatross* in 1904–1905. This collection and others made in the 1950s and more recently were examined by Wells (1972) who increased the zooxanthellate coral richness to seven species (Glynn and Wellington 1983, see p. 176; Wells 1983). An unprecedented coral bleaching event at Easter Island in 1980, presumably due to heavy rains and sediment-laden runoff, was reported by Cea Egaña and DiSalvo (1982). These investigators, along with John E. Randall, performed the first quantitative surveys, to 60 m depth, in 1985 and 1986 (DiSalvo et al. 1988). This work, supported by the US National Geographic Society and the Charles Engelhard Foundation, resulted in thorough collections of several taxa. Coral zonation and relative abundances were described from quantitative transect sampling with eight zooxanthellate

Fig. 1.19 M/Y *Beagle III*, research vessel of the Charles Darwin Research Station, Galápagos Islands, January 1975. Photograph courtesy of C. Birkeland

coral species reported. By the early 2000s, 11 species were recognized (DiSalvo et al. 1988; Glynn et al. 2003), and more recently 13 species from collections acquired to the year 2005 (Glynn et al. 2007).

Wellington and co-workers documented a second island-wide coral bleaching event in 2000 that was attributed to anomalous sea warming (Wellington et al. 2001). In spite of two major bleaching events over a 30-year period, recent surveys have reported vibrant coral communities island wide (Glynn et al. 2003; Hubbard and Garcia 2003). Hubbard and Garcia (2003) identified a site in La Pérouse Bay, a leeward shelf area, with notable massive coral frameworks.

The first National Marine Parks in Chile, three in number and located at Easter Island in coral habitats nearshore, were not created until April 2000. These MPAs have a combined area of only $\sim$48 km^2, but do contain unique coral assemblages. As recently as 1999, pocilloporid corals were collected avidly for the curio market at Easter Island (Glynn et al. 2003). *Pocillopora eydouxi*, a strongly targeted species, became rare in the early 2000s. Hopefully such over-harvesting will be recognized by the current conservation initiatives of the Chilean government, and the uncontrolled extraction of corals and other marine life be prohibited. In February 2011, the National Geographic Society and Oceana partnered with the Chilean Navy to conduct baseline monitoring studies at Salas y Gómez (Motu Motiro Hiva) and Easter Islands (Rapa Nui) with the prime objective being the assessment of the current state of conservation and the need for new protection measures. Motu Motiro Hiva Marine Park, encompassing 150,000 km^2, was created in October 2010 at uninhabited Salas y Gómez Island and surrounding waters.

1.6 Enigmatic Species

In the course of discovery of new coral species and distributional records in the eastern Pacific, the status of some species has remained elusive. One of the earliest of these, the acroporid *Montipora fragosa* described by Verrill (1870) from La Paz, Baja California, has not been observed anywhere else in the eastern Pacific since the 1860s. The holotype specimen is all that is known of this species (Fig. 1.20). Although La Paz was listed as the collection locality, Squires (1959) suggested that it may have been collected elsewhere and simply shipped to the USA from La

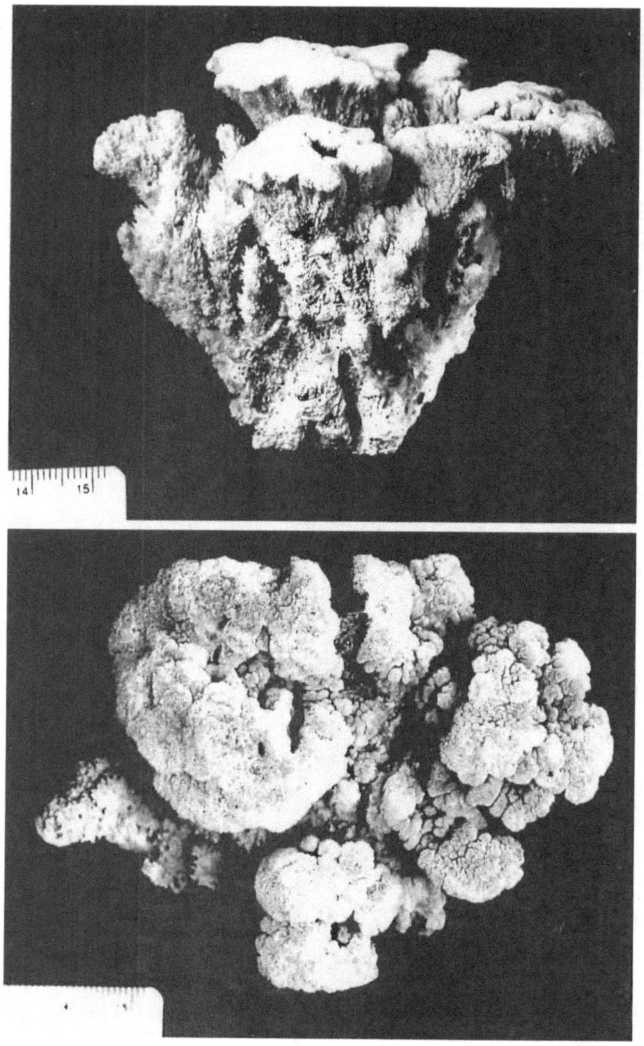

Fig. 1.20 Side (upper) and top (lower) views of *Montipora fragosa* Verrill, holotype YPM No. 3057, purportedly collected at La Paz, Gulf of California by Maj. William Rich. Species described in 1870 and probably collected in late 1860s although locality and date unknown

Paz. Diligent searching for this species in the southern sector of the Gulf of California during the Puritan-American Museum Expedition failed to produce additional material. Over a century after its description, Veron and Wallace (1984) concluded that *M. fragosa* is probably synonymous with *Montipora aequituberculata* Bernard 1897, which is distributed widely from the Red Sea to the Tuamotus in French Polynesia. Therefore, Squires may have correctly concluded that it never occurred in the eastern Pacific.

Oddly, the reef-building alcyonarian blue coral *Heliopora coerulea*, ostensibly collected by E.C. Allison, was listed in Sachet's (1962b) catalog of Clipperton's marine fauna. The nearest documented occurrence of this species is in American Samoa (Zann and Bolton 1985), and Babcock (1990) suggested that its mode of development does not favor long-distance dispersal. Inquiries made to several collection managers at institutional repositories in California, likely to hold material collected by Allison, failed to reveal specimens of this species from Clipperton.

Acropora valida, another acroporid species reported from the eastern Pacific, has raised several questions. Unexpectedly, three colonies of *A. valida* were found at Gorgona Island, Colombia in 1983 by von Prahl and Mejía (1985). The Gorgona collection site lies more than 6000 km east of known populations in the Pitcairns Islands and French Polynesia (Wallace 1999). Guzmán (1988c) criticized the - von Prahl and Mejía report because of the lack of key information that would have better substantiated the biogeographic implications of this intriguing find. For example, maximum colony sizes were not reported, which would have allowed age estimates, and thus an idea of the time when the coral larvae settled. In addition, the *Acropora* colonies were observed near the end of the 1982–83 El Niño bleaching event when coral communities in general suffered extreme mortality at Gorgona Island (von Prahl 1983). If long distance dispersal from central Pacific populations occurred, then Guzmán suggested that Cocos Island off Costa Rica, which is in the path of the North Equatorial Counter Current, would have been a more likely recruitment site. Coral populations at Cocos Island were also decimated during the 1982–83 El Niño disturbance (Guzmán and Cortés 1992), suggesting that survival of extralimital recruits would have been unlikely. Lamentably, Henry von Prahl lost his life in a terrorist's attack in 1989. An attempt to obtain additional information from co-author Ángela Mejía after von Prahl's death was unsuccessful because of severe mental impairment she suffered after a motorbike accident. A single branch-tip of *A. valida* collected by these workers, ~6 cm in length and identified by C.C. Wallace (Reg No G27106) is in the coral collections of the Museum of Tropical Queensland (Northern Queensland Branch), Australia.

Several other corals representing new records or newly described species in the 1970s and early 1980s also were impacted by disturbances caused during the 1982–83 El Niño event. For example, three species of the hydrocoral *Millepora* bleached and suffered high mortality in Panama, and two of these species disappeared and have not been observed in the eastern Pacific since (Glynn and de Weerdt 1991; Glynn 2011). The discovery of *Porites* (*Synarea*) *rus* in Costa Rica in late 1983 was the first record of this species in the eastern Pacific (Cortés and Murillo 1985) and it, too, disappeared, presumably a casualty of El Niño heat stress. Several other reef-building coral species experienced extreme reductions in population size and even local extinctions (Glynn 1997). *Leptoseris papyracea* at Culebra Bay (Costa Rica), that formed a unique patch framework before the 1997–98 ENSO event, suffered 95 % live tissue mortality due to this disturbance (Jiménez et al. 2001). Finally, azooxanthellate scleractinians and numerous reef-associated metazoan taxa suffered severe declines, disappearances, and possibly extinctions in the Galápagos Islands after 1983 (Edgar et al. 2010).

Five colonies of a *Siderastrea* species, a previously unknown genus in the eastern Pacific, were found by Héctor Guzmán in 1992 in 7–8.5 m depth at one island site (Urabá) in the Taboga Islands, Gulf of Panama. Based on skeletal morphology, Budd and Guzmán (1994) concluded that this small population represented a new species, which they named *Siderastrea glynni*. The colonies were spherical and unattached (coralliths), 7–10 cm in diameter. Exhaustive searches in the Taboga Islands area and elsewhere in Panama since Guzmán's discovery have failed to reveal additional specimens. Only one other species in this genus is known from the Indo-Pacific, where it is of uncommon occurrence in the western Pacific and Indian Ocean (Veron 2000). Three species of *Siderastrea* are known from the western Atlantic, and two of these occur commonly on the Caribbean side of Panama. This distribution pattern provokes the question, is the most recent common ancestor of *S. glynni* in the Indo-Pacific or Caribbean regions? Examination of molecular phylogenetic relationships among ITS region sequences of all five *Siderastrea* species demonstrated a close affinity between *S. glynni* and *S. siderea*, a Caribbean species (Forsman et al. 2005). This suggests that *S. glynni* originated by a relatively recent breach of the Isthmus ($\sim$2 mya) or is a contemporary transisthmian introduction. Due to their deteriorating condition during the 1997–98 ENSO, the only known four remaining colonies of *S. glynni* were collected and transferred to aquaria at the Smithsonian laboratory on Naos Island. As of March 2011, only a single small live fragment remained (H. Guzmán, pers. comm.).

1.7 Future Directions

From the late 1940s to the present, progress in coral reef research and environmental protection in the eastern Pacific were hampered by severe forms of civil unrest. Most notable is the period of *La Violencia* in Colombia (1948–1958), the Nicaraguan Revolution (1960s–1970s), the Salvadoran Civil War (1980s), leftist guerrilla and rightist paramilitary violence in Colombia (1960s to present), and more recently drug cartel violence in Colombia (1980s to present) and Mexico (present). These unfortunate disruptions added to the risk of conducting field studies in remote areas, and also diverted national and local resources from research, and the protection and management of critical marine habitats. Nonetheless, in spite of these significant distractions there have been noteworthy advances in coral reef science and conservation efforts. To conclude this historical survey, comments follow on some of the (a) advantages offered in the study of eastern Pacific coral assemblages, and (b) important gains that have been made recently in the protection and conservation of coral habitats (see also Chap. 21, Alvarado et al.).

1.7.1 Eastern Pacific Environmental Conditions

Investigations of long-recognized environmental conditions that can limit coral growth in the eastern tropical Pacific have offered insights into the critical factors and processes that promote reef building, accretion and resilience. Extreme physical conditions characterize most eastern Pacific coral reef environments across a range of spatial and temporal scales. The effects of short-term fluctuations in sea temperature, for example during seasonal upwelling events, and less frequent longer term temperature excursions accompanying El Niño/La Niña episodes, have been quantitatively documented at several sites in the eastern Pacific, serving to better understand such stressors globally during the past several decades of widescale coral bleaching (see Chap. 8, Glynn et al.). Studies in the Panamanian eastern Pacific (Baker and Rowan 1997; Baker 1999; Glynn et al. 2001) and Caribbean (Rowan and Knowlton 1995; Rowan et al. 1997) revealing that coral species hosted diverse algal symbionts (*Symbiodinium* spp.) that helped explain differences in bleaching susceptibility, were among the first to offer support for Buddemeier and Fautin's (1993) adaptive bleaching hypothesis. Coral reef development in the eastern Pacific is also noteworthy in occurring under low and variable pH (Manzello et al. 2008; Manzello 2009, 2010a, b), which may offer insights to responses under stressful conditions likely to prevail in all tropical seas before the end of the century.

The demonstrated impacts of other stressors in the eastern Pacific, such as nutrient pulses, increased sedimentation, noxious plankton blooms, corallivory, bioerosion, and overfishing, have also greatly advanced understanding of the effects of these factors (e.g. Edgar et al. 2010, 2011). Additional research areas directly or indirectly associated with acute or chronic environmental disturbances were initiated in the 1970s, gaining momentum particularly after the ENSO impacts of 1982–83 and 1997–98. Some of the more actively pursued initiatives are paleoclimatology, coral population modeling and community recovery, reproduction and recruitment, refuge populations, biogeography and connectivity, and trophic structure and dynamics. These research themes are being investigated over a range of eastern Pacific environments spanning 34° of latitude, from nearshore habitats subject to upwelling, seasonal wet and dry season periods, cyclone activity, and extreme river discharge and flooding, and to relatively stable oceanic island settings. On one hand, these diverse conditions offer opportunities for gaining insights into biotic responses, but on the other hand lead to complications via additive and multiplicative interactions, many of which will be explored in the relevant chapters that follow in this book.

1.7.2 Conservation and Management Initiatives

A salutary change from the late 1970s to the present has been the establishment of marine protected areas (MPAs) across the eastern Pacific region. This is a result of the recent recognition of the value of diverse and productive marine ecosystems and the threat of current and projected global-scale disturbances (see Chap. 20, Cortés and Reyes Bonilla and Chap. 21, Alvarado et al.). By 2000, 22 existing and six proposed MPAs with coral communities or coral reefs were known in the eastern Pacific, present from Mexico to Ecuador (Glynn 2001a). Just three years later, 32 MPAs with coral habitats were recognized by various authors in the edited volume *Latin American Coral Reefs* (Cortés 2003). Costa Rica alone has 17 MPAs located in its coastal and offshore Pacific waters (Alvarado et al. 2012). Six of the eastern Pacific MPAs are UNESCO World Heritage Sites, all listed from the late 1990s to the mid 2000s, encompassing 100s–1000s km^2 of critical coral reef habitat: Gulf of California islands, 244 islands, islets and coastal areas (Mexico); Coiba National Park (Panama); Cocos Island National Park (Costa Rica); Malpelo Fauna and Flora Sanctuary (Colombia); Galápagos Islands (Ecuador); Easter Island (Chile) (Fig. 1.21). Four of these MPAs support robust coral communities and were selected to serve as critical nodes in marine dispersal pathways (Costa Rica, Panama, Colombia, Ecuador). This international network was conceived and aided by the UNESCO World Heritage Centre and Conservation International to promote long-term management and conservation of marine protected areas within the ETP region. The San José Declaration, signed in April 2004, formally established the Marine Conservation Corridor of the Eastern Tropical Pacific (MCCETP) as an instrument for the conservation and sustainable use of the biological diversity of this equatorial region.

Besides the many charismatic species present in the equatorial eastern Pacific, such as sea turtles, sharks, tuna, and cetaceans, a multitude of less well known biota also contribute importantly to the biodiversity and productivity of

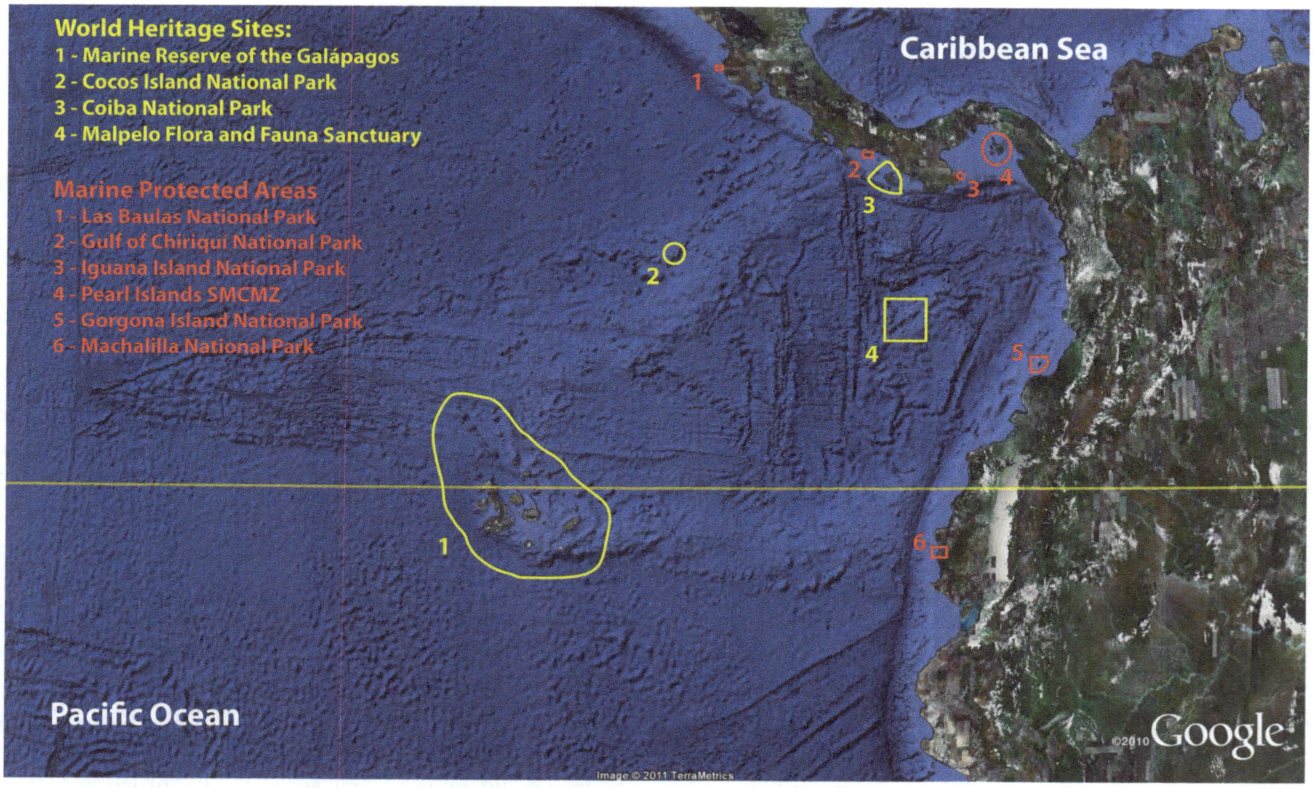

Fig. 1.21 Locations of some of the better known marine protected areas (MPAs) in the equatorial eastern Pacific marine conservation corridor. Each MPA is bounded by its respective protection zone. Boundaries of UNESCO World Heritage Sites (WHS) are denoted in *yellow* and other MPAs in *red*. National affiliation and year of establishment: WHSs—*1* Galápagos Islands, Ecuador, 1978; *2* Cocos Island, Costa Rica, 1997; *3* Coiba Island, Panama, 2005; Malpelo Island, Colombia, 2006. Other MPAs—*1* Las Baulas, Costa Rica, 1991; *2* Gulf of Chiriquí, Panama, 1994; *3* Iguana Island, Panama, 1980; *4* Pearl Islands, Panama, Special Zone for Coastal and Marine Management, 2007; *5* Gorgona Island, Colombia, 1985; *6* Machalilla, Ecuador, 1979

this region. As Guzmán and colleagues have urged, it is vital to include species-rich coastal sites in the conservation network (Guzmán et al. 2008). Scleractinian corals, octocorals and many associated taxa are concentrated on islands near mainland shores, so called 'diversity hotspots'. Recently discovered species-rich coral assemblages in Nicaragua (Alvarado et al. 2010, 2011b) and Ecuador (Rivera and Martínez 2011), known 'hotspots' at Caño Island in Costa Rica and the Pearl Islands, Panama (Guzmán et al. 2008), are sites that ideally should be included in the MCCETP network. Biological surveys and inventories should be encouraged and supported to continue to locate and characterize biotically diverse areas for protective management action (e.g. Cortés 1996-1997). The recent gap identification analysis completed in Costa Rica is another procedure designed to promote ecosystem connectivity and resilience in the long-term conservation of marine biodiversity (Alvarado et al. 2011b). Optimistically such undertakings will be a continuing trend, proceeding effectively and timely in the unfolding history of eastern Pacific coral studies.

Acknowledgments For information relating to particular countries, I thank Héctor Reyes-Bonilla (Mexico, El Salvador), Jean-François Flot (Clipperton Atoll), Juan José Alvarado (Nicaragua, Costa Rica), Jorge Cortés (Costa Rica), Héctor Guzmán and Juan Maté (Panama), Jaime R. Cantera Kintz, Edgardo Londoño, Bernardo Vargas Ángel, and Fernando Zapata (Colombia), Stuart Banks, Graham J. Edgar, Joshua S. Feingold, Priscilla Martínez and Fernando Rivera (Ecuador), and Dennis K. Hubbard, Sergio Navarrete and Evie Wieters (Easter Island). Thanks are due Eric A. Lazo-Wasem, Collections Manager, Peabody Museum of Natural History, Yale University, for arranging the loan of *Montipora fragosa*. Patricia Wellington kindly supplied the photograph of Gerard (Jerry) Wellington. Angela Clark, Ann Campbell and Elizabeth Fish, Rosenstiel Marine School Library, provided much of the referenced literature. I also wish to thank Rachelle B. Smith, Special Collections, Doheny Memorial Library, University of Southern California for making available pictures from the Hancock Expeditions, and Oris Sanjur, Smithsonian Tropical Research Institute, for information on funding agencies. Finally, I thank Victor W. Brandtneris, Benjamin Grassian, and Michael Fuller who facilitated several tasks needed to complete this chapter.

References

Abele LG (1972) Comparative habitat diversity and faunal relationships between the Pacific and Caribbean Panamanian decapod Crustacea: a preliminary report, with some remarks on the crustacean fauna of Panamá. Bull Biol Soc Wash 2:125–138 (Jones ML (ed))

Abele LG (1974) Species diversity of decapod crustaceans in marine habitats. Ecology 55:156–161

Abele LG (1975) The macruran decapod crustaceans of Malpelo Island. Smithson Contrib Zool 176:69–85

Abele LG (1976a) Comparative species richness in fluctuating and constant environments: coral-associated decapod crustaceans. Science 192:461–463

Abele LG (1976b) Comparative species composition and relative abundance of decapod crustaceans in marine habitats of Panamá. Mar Biol 38:263–278

Abele LG (1979) The community structure of coral-associated decapod crustaceans in variable environments. In: Livingston RJ (ed) Ecological processes in coastal and marine systems. Plenum Press, New York, pp 265–287

Abele LG, Patton WK (1976) The size of coral heads and the community biology of associated decapod crustaceans. J Biogeogr 3:35–47

Agassiz A (1892) General sketch of the expedition of the "Albatross", from February to May, 1891. The Galapagos Islands. Bull Mus Comp Zoöl 23:56–74 Harvard Coll

Allison EC (1959) Distribution of *Conus* on Clipperton Island. Veliger 1:32–34

Alvarado JJ, Cortés J, Fernández C, Nivia J (2005) Comunidades y arrecifes coralinos del Parque Nacional Marino Ballena, costa del Pacífico de Costa Rica/Coral communities and reefs of Ballena Marine National Park, Pacific coast of Costa Rica. Cienc Mar 31:641–651

Alvarado JJ, Fernández C, Cortés J (2009) Water quality conditions on coral reefs at the Marino Ballena National Park, Pacific Costa Rica. Bull Mar Sci 84:137–152

Alvarado JJ, Reyes-Bonilla H, Buitrago F, Aguirre-Rubí J, Álvarez del Castillo Cárdenas PA (2010) Coral reefs of the Pacific coast of Nicaragua. Coral Reefs 29:201

Alvarado JJ, Aguirre-Rubí J, Ayala A, Buitrago F, Fernández C, Reyes-Bonilla H, Sánchez A (2011a) Characterization of the coral communities off San Juan del Sur, Pacific Nicaragua. Bull Mar Sci 87:129–146

Alvarado JJ, Herrera B, Corrales L, Asch J, Paaby P (2011b) Identificación de las prioridades de conservación de la biodiversidad marina y costera en Costa Rica. Rev Biol Trop (Int J Trop Biol) 59:829–842

Alvarado JJ, Cortés J, Fernanda Esquivel M, Sala E (2012) Costa Rica's marine protected áreas: status and perspectives. Rev Biol Trop (Int J Trop Biol) 60:129–142

ANAM (2009) Plan de manejo del Parque Nacional Coiba. Compiladores Maté JL, Tovar D, Arcia E, Hidalgo Y, Ciudad de Panamá, República de Panamá, 168 pp

Anlauf H, D'Croz L, O'Dea A (2011) A corrosive concoction: the combined effects of ocean warming and acidification on the early growth of a stony coral are multiplicative. J Exp Mar Biol Ecol 397:13–20

Arizpe OC (2004) El turismo como alternative a la pesca en el manejo de un arrecife coralino. Caso Cabo Pulmo, Golfo de California. In Rivera-Arriaga E, Villalobos GJ, Azuz-Adeath I, Rosado-May F (eds) El manejo costero en México. EPOMEX/SEMARNAT/CETYS/Universidad de Quintana Roo, Campeche, pp 573–588

Arizpe OC (2008) Caracterización del arrecife coralino de Cabo Pulmo. In Gámez AE (ed) Turismo y sustentabilidad en Cabo Pulmo, Baja California Sur, México. San Diego State University and Universidad Autónoma de Baja California Sur, San Diego, pp 53–74

Arizpe OC, Álvarez F (1987) Distribución y abundancia de los corales del arrecife coralino de Cabo Pulmo, B.C.S. Resumen VII Congreso Nacional de Oceanografía. Ensenada BC, Instituto Nacional de Pesca, 437 p

Babcock R (1990) Reproduction and development of the blue coral *Heliopora coerulea* (Alcyonaria: Coenothecalia). Mar Biol 104:475–481

Baker AC (1999) The symbiosis ecology of reef-building corals. PhD diss, Univ Miami, Coral Gables, 120 p

Baker AC (2003) Flexibility and specificity in coral-algal symbiosis: diversity, ecology and biogeography of *Symbiodinium*. Annu Rev Ecol Syst 34:661–689

Baker AC, Rowan R (1997) Diversity of symbiotic dinoflagellates (zooxanthellae) in scleractinian corals of the Caribbean and eastern Pacific. In: Proceedings of 8th International Coral Reef Symposium, vol 2, Panamá, pp 1301–1306

Baker AC, Starger CJ, McClanahan TR, Glynn PW (2004) Corals' adaptive response to climate change. Nature 430:741
Baker AC, Glynn PW, Riegl B (2008) Climate change and coral reef bleaching: an ecological assessment of long-term impacts, recovery trends and future outlook. Estuar Coast Shelf Sci 80:435–471
Bakus GJ (1975) Marine zonation and ecology of Cocos Island off Central America. Atoll Res Bull 179:1–9
Barham EG, Gowdy RW, Wolfson FH (1973) *Acanthaster* (Echinodermata, Asteroidea) in the Gulf of California. Fish Bull (Seattle) 71:927–942
Barrios LM, López-Victoria M (eds) (2001) Gorgona marina, contribución al conocimiento de una isla única. INVEMAR, Serie Publicaciones Especiales No. 7, Santa Marta, Colombia, 160 p
Baynes TW (1999) Factors structuring a subtidal encrusting community in the southern Gulf of California. Bull Mar Sci 64:419–450
Beebe W (1924) Galápagos: world's end. GP Putnam's Sons, New York, xxi + 443 p
Beebe W (1926) The Arcturus adventure: an account of the New York Zoological Society's First Oceanographic Expedition. GP Putnam's Sons, New York, 439 p
Beebe W (1938) Zaca venture. Harcourt, Brace and Co, New York, 308 p
Beebe W (1942) Book of bays. Harcourt, Brace and Co, New York, xviii + 302 p
Bezaury-Creel JE (2005) Protected areas and coastal and ocean management in México. Ocean Coastal Manage 48:1016–1046
Birkeland C (1977) The importance of rate of biomass accumulation in early successional stages of benthic communities to the survival of coral recruits. In: Proceedings of 3rd International Coral Reef Symposium, vol 1, Miami, pp 15–21
Birkeland C, Meyer DL, Stames JP, Buford CL (1975) Subtidal communities of Malpelo Island. Smithson Contrib Zool 176:55–68
Brando A, von Prahl H, Cantera JR (1992) Malpelo, isla oceánica de Colombia. Banco de Occidente, Cali, Colombia 195 p
Breddy O, Guzmán HM (2003) A new species of *Pacifigorgia* (Coelenterata: Octocorallia: Gorgoniidae) from Panama. Zootaxa 128:1–10
Briggs JC (1969) The sea-level Panama Canal: potential biological catastrophe. Bioscience, Wash DC 19:44–47
Briggs JC (1974) Marine zoogeography. New York: McGraw-Hill, 475 pp
Brusca RC, Thomson DA (1975) Pulmo Reef: the only "coral reef" in the Gulf of California. Cienc Mar 2:37–53
Brusca RC, Findley LT, Hastings PA, Hendrickx ME, Torre Cosio J, van der Heiden AM (2005) Macrofaunal diversity in the Gulf of California. In: Cartron J-LE, Ceballos G, Felger RS (eds) Biodiversity, ecosystems, and conservation in northern Mexico. Oxford University Press, Oxford, pp 179–202
Budd AF, Guzman HM (1994) *Siderastrea glynni*, a new species of scleractinian coral (Cnidaria: Anthozoa) from the eastern Pacific. Proc Biol Soc Wash 107:591–599
Buddemeier RW, Fautin DG (1993) Coral bleaching as an adaptive mechanism—a testable hypothesis. Bioscience 43:320–326
Cairns SD, Hoeksema BW, van der Land J (1999) Appendix: list of extant stony corals. In: Cairns SD (ed) Species richness of recent Scleractinia, pp 13–46 (Atoll Res Bull 459-1-12)
Cantera JR (1983) Distribution des peuplement de scléractiniaires sur un récif de l'ile de Gorgona (cote Pacífique de Colombie). Tethys 11:25–31
Cantera JR, Arnaud PM (1995) Coral associated molluscan assemblages in Gorgona Island, Colombian Pacific Ocean. Haliotis 24:87–100
Cantera JR, Contreras YR (1988) Bivalvos perforadores de corales escleractiniarios en la Isla de Gorgona, Colombia. Rev Biol Trop 36:151–158
Cantera JR, von Prahl H, Escobar JC, Peña YE (1989) Sistemática de corales del género *Pocillopora* del Pacífico colombiano utilizando taxonomía numérica. Rev Biol Trop 37:23–28
Cantera JR, Zapata FA, Forero P, Francisco V, Jiménez JM, Londoño E, Narváez K, Neira R, Orozco CA, Toro-Farmer G (2001) Organismos bioerosionadores en el arrecife de Playa Blanca, Isla Gorgona. In: Barrios LM, López-Victoria M (eds) Gorgona marina: contribución al conocimiento de una isla única. INVEMAR, Serie Publicaciones Especiales 7, Santa Marta, pp 51–64
Cantera JR, Orozco C, Londoño-Cruz E, Toro-Farmer G (2003) Abundance and distribution patterns of infaunal associates and macroborers of the branched coral (*Pocillopora damicornis*) in Gorgona Island (Eastern Tropical Pacific). Bull Mar Sci 72:207–219
Cantera-Kintz JR (2008) Henry von Prahl Bauer: científico, artista y persona. Día Nacional de los Arrecifes Coralinos. Cartagena, Diciembre 2008 (an oral tribute)
Carpizo-Ituarte E, Vizcaíno-Ochoa V, Chi-Barragán G, Tapia-Vázquez O, Cupul-Magaña AL, Medina-Rosas P (2011) Evidence of sexual reproduction in the hermatypic corals *Pocillopora damicornis*, *Porites panamensis*, and *Pavona gigantea* in Banderas Bay, Mexican Pacific. Cienc Mar 37:97–112
Carricart-Ganivet JP, Reyes-Bonilla H (1999) Clipperton, Île de la Passion. Reef Encounter 26:16–19
Carriquiry JD, Reyes-Bonilla H (1997) Estructura de la comunidad y distribución geográfica de los arrecifes coralinos de Nayarit. Cienc Mar 23:227–248
Carriquiry JD, Villaescusa JA (2010) Coral Cd/Ca and Mn/Ca records of El Niño variability in the Gulf of California. Clim Past Discuss 6:63–85
Carriquiry JD, Risk MJ, Schwarcz HP (1988) Timing and temperature record from stable isotopes of the 1982–1983 El Niño warming event in eastern Pacific corals. Palaios 3:359–364
Carriquiry JD, Risk MJ, Schwarez HP (1994) Stable isotope geochemistry of corals from Costa Rica as proxy indicator of the El Niño/Southern Oscillation (ENSO). Geochim Cosmochim Acta 58:335–351
Carriquiry JD, Cupul-Magaña AL, Rodríguez-Zaragoza F, Medina-Rosas P (2001) Coral bleaching and mortality in the Mexican Pacific during the 1997–98 El Niño, and predictions from a remote sensing approach. Bull Mar Sci 69:237–249
Castilla JC, Rozbaczylo N (1987) Invertebrados marinos de Isla de Pascua y Sala y Gómez. In: Castilla JC (ed) Islas Oceánicas Chilenas: conocimiento científicos y necesidades de investigaciones. Ediciones Universidad Católica de Chile, Santiago, pp 191–215
Castro P (1982) Notes on symbiotic decapod crustaceans from Gorgona Island, Colombia, with a revision of the eastern Pacific species of *Trapezia* (Brachyura, Xanthidae), symbionts of scleractinian corals. An Inst Invest Mar Punta Betín 12:9–17
Cea Egaña A, DiSalvo LH (1982) Mass expulsion of zooxanthellae by Easter Island corals. Pac Sci 36:61–63
Charpy L (2009) Clipperton: environnement et biodiversité d'un microcosme océanique. Muséum national d'Histoire naturelle, Paris, IRD, Marseille, p 420 (Patrimoines naturels; 68)
Chávez-Romo HE, Reyes-Bonilla H (2007) Reproducción sexual del coral *Pocillopora damicornis* al sur del Golfo de California, México. Cienc Mar 33:495–501
Chesher RH (1969) Destruction of Pacific corals by the sea star *Acanthaster planci*. Science 165:280–283
Clover ST (1932) A pioneer heritage. Saturday Night Pub Co., Los Angeles, p 291
Colley SB, Glynn PW, May AS, Maté JL (2006) Species-dependent reproductive responses of eastern Pacific corals to the 1997–1998 ENSO event. In: Proceedings of 10th International Coral Reef Symposium, vol 1, Okinawa, pp 61–70

Cortés J (1986) Biogeografía de corales hermatípicos: El Istmo Centro Americano. An Inst Cienc Mar Limnol, UNAM 13:297–304

Cortés J (1990) The coral reefs of Golfo Dulce, Costa Rica: distribution and community structure. Atoll Res Bull 344:1–37

Cortés J (1996–1997) Biodiversidad marina de Costa Rica: filo Cnidaria. Rev Biol Trop 44/45:323–334

Cortés J (1997) Biology and geology of eastern Pacific coral reefs. Coral Reefs 16(suppl):S39–S46

Cortés J (ed) (2003) Latin American coral reefs. Elsevier, Amsterdam, p 497

Cortés J (sci ed) (2008) Investigaciónes marinas en la Isla del Coco, Costa Rica/Marine research at Isla del Coco (Cocos Island), Costa Rica/Recherches marine a l'Ile du Coco, Costa Rica. Rev Biol Trop 56(suppl 2):1–217

Cortés J (2009) A history of marine biodiversity, scientific research in Costa Rica. In: Wehrtmann IS, Cortés J (eds) Marine biodiversity of Costa Rica, Central America. Springer Science + Business Media BV, Berlin, pp 47–80

Cortés J (2011) Eastern tropical Pacific coral reefs. In: Hopley D (ed) Encyclopedia of modern coral reefs: structure, form and process. Encyclopedia Earth Science Series, Springer, Berlin

Cortés J (sci ed) (2012a) Bahía Culebra: scientific research of the marine and coastal zone, Guanacaste, North Pacific of Costa Rica. Rev Biol Trop 60(suppl 2):1–244

Cortés J (sci ed) (2012b) Marine research at Isla del Coco National Park, Costa Rica. Rev Biol Trop 60(suppl 3):1–392

Cortés J, Blum S (2008) Life to 450 m depth at Isla del Coco, Costa Rica. Rev Biol Trop (Int J Trop Biol ISSN-0034-7744) 56(suppl 2): 189–206

Cortés J, Guzmán HM (1998) Organismos de los arrecifes coralinos de Costa Rica: descripción, distribución geográfica e historia natural de los corales zooxantelados (Anthozoa: Scleractinia) del Pacífico. Rev Biol Trop 46:55–92

Cortés J, Jiménez C (2003) Corals and coral reefs of the Pacific of Costa Rica: history, research and status. In: Cortés J (ed) Latin American coral reefs. Elsevier, Amsterdam, pp 361–385

Cortés J, Murillo MM (1985) Comunidades coralinas y arrecifes del Pacífico de Costa Rica. Rev Biol Trop 33:197–202

Cortés J, Nielsen V (2002) Las ciencias del mar en la Revista de Biología Tropical en su 50 aniversario. Rev Biol Trop 50:903–907

Cortés J, Murillo MM, Guzmán HM, Acuña J (1984) Pérdida de zooxantelas y muerte de corales y otros organismos arrecifales en el Caribe y Pacífico de Costa Rica. Rev Biol Trop 32:227–232

Cortés J, Macintyre IG, Glynn PW (1994) Holocene growth history of an eastern Pacific fringing reef, Punta Islotes, Costa Rica. Coral Reefs 13:65–73

Couthouy JP (1842) Remarks upon coral formations in the Pacific; with suggestions as to the causes of their absence in the same parallels of latitude on the coast of South America. Boston J Nat Hist 4(66–105):137–162

Crossland C (1927) The expedition to the South Pacific of the S.Y. "St. George". Marine ecology and coral formations in the Panama region, the Galapagos and Marquesas Islands, and the Atoll of Napuka. Trans Royal Soc Edinb 55:531–554

Cuyás A (1972) Appleton's New Cuyás English-Spanish and Spanish-English dictionary. 5th edition revised, Appleton-Century-Crofts, New York p 589

D'Croz L, Maté JL (2004) Experimental responses to elevated water temperature in genotypes of the reef coral *Pocillopora damicornis* from upwelling and non-upwelling environments in Panama. Coral Reefs 23:473–483

D'Croz L, O'Dea A (2007) Variability in upwelling along the Pacific shelf of Panama and implications for the distribution of nutrients and chlorophyll. Est Coast Shelf Sci 73:325–340

D'Croz L, Robertson DR (1997) Coastal oceanographic conditions affecting coral reefs on both sides of the Isthmus of Panama. In: Proceedings of 8th International Coral Reef Symposium, vol 2, Panama, pp 2053–2058

D'Croz L, Del Rosario JB, Gómez JA (1991) Upwelling and phytoplankton in the Bay of Panamá. Rev Biol Trop 39:237–245

D'Croz L, Maté JL, Oke JE (2001) Responses to elevated seawater temperature and UV radiation in the coral *Porites lobata* from upwelling and non-upwelling environments on the Pacific coast of Panama. Bull Mar Sci 69:203–214

Dana JD (1843) On the temperature limiting the distribution of corals. Amer J Sci 45:130–131

Dana TF (1975) Development of contemporary eastern Pacific coral reefs. Mar Biol 33:355–374

Dana JD, Couthouy JP (1844) Acknowledgments relative to a charge of plagiarism. Am J Sci 47:122–123

Dana TF, Wolfson A (1970) Eastern Pacific crown-of-thorns starfish populations in the lower Gulf of California. Trans San Diego Soc Nat Hist 16:83–90

Danulat E, Edgar GJ, (eds) (2002) Reserva Marina Galápagos. Línea base de la biodiversidad. Fundación Charles Darwin/Servicio Parque Nacional Galápagos, Santa Cruz, Galápagos, Ecuador, 484 pp

Danulat E, Brandt M, Zárate P, Montesinos, M, Banks S (2003) El turismo en la Reserva Marina de Galápagos. Análisis global y condiciones en sitios de visita marinos seleccionados. Diciembre de 2003. Fundación Charles Darwin y Dirección Parque Nacional Galápagos. Puerto Ayora, Santa Cruz, Galápagos, p 140

Darwin CR (1842) The structure and distribution of coral reefs. Being the first part of the geology of the voyage of the *Beagle*, under the command of Capt. Fitzroy, R.N. during the years 1832 to 1836. Smith Elder and Co, London

Darwin C (1872) The origin of species by means of natural selection, 6th edn. Doubleday & Co, Garden City, New York

Dawson TP, Henderson SJ, Banks S (2009) Galapagos coral conservation: impact mitigation, mapping and monitoring. Galapagos Res 66:3–5

de Alba C (1978) Predación de coral por la estrella de mar corona de espinas *Acanthaster planci* (Gray) en el área de Bahía de La Paz. Mem I Simp Biol Mar, UABCS 1:45–51

del Barco M (1757) Historia natural y crónica de la Antigua California. 1973 edit, Instituto de Investigaciones Históricas, Universidad Nacional Autónoma de México, p 464

DiSalvo LH, Randall JE, Cea A (1988) Ecological reconnaissance of the Easter Island sublittoral marine environment. Nat Geogr Res 4:451–473

Downey ME (1975) Asteroidea from Malpelo Island with a description of a new species of the genus *Tamaria*. Smithson Contrib Zool 176:86–90

Duda TF Jr, Lessios HA (2009) Connectivity of populations within and between major biogeographic regions of the tropical Pacific in *Conus ebraeus*, a widespread marine gastropod. Coral Reefs 28:651–659

Duncan PM (1876) Notices of some deep-sea and littoral corals from the Atlantic Ocean, Caribbean, Indian, New-Zealand, Persian Gulf and Japanese, etc., Seas. Proc Zool Soc London 1876:428–442

Durham JW (1947) Corals from the Gulf of California and the North Pacific coast of America. Geol Soc Am Mem 20:1–46

Durham JW (1962) Corals from the Galápagos and Cocos Islands. Proc Calif Acad Sci 4(32):41–56

Durham JW (1966) Coelenterates, especially stony corals from the Galápagos and Cocos Islands. In: Bowman RI (ed) The Galápagos. University of California Press, Berkeley, pp 123–135

Durham JW, Barnard JL (1952) Stony corals of the eastern Pacific collected by the *Velero III* and *Velero IV*. Allan Hancock Pac Exped 16:1–110

Eakin CM (1992) Post-El Niño Panamanian reefs: less accretion, more erosion and damselfish protection. In: Proceedings 7th International Coral Reef Symposium, vol 1, Guam, pp 387–396

Eakin CM (1996) Where have all the carbonates gone? A model comparison of calcium carbonate budgets before and after the 1982-83 El Niño at Uva Island in the eastern Pacific. Coral Reefs 15:109–119

Eakin CM (2001) A tale of two ENSOs: carbonate budgets and the influence of two warming events and intervening variability, Uva Island, Panamá. Bull Mar Sci 69:171–186

Edgar GJ, Banks SA, Brandt M, Bustamante RH, Chiriboga A, Earle SA, Garske LE, Glynn PW, Grove JS, Henderson S, Hickman CP, Miller KA, Rivera F, Wellington GM (2010) El Niño, grazers, and fisheries interact to greatly elevate extinction risk for Galápagos marine species. Glob Change Biol 16:2876–2890

Edgar GJ, Banks SA, Bessudo S, Cortés J, Guzmán HM, Henderson S, Martinez C, Rivera F, Soler G, Ruiz D, Zapata FA (2011) Variation in reef fish and invertebrate communities with level of protection from fishing across the eastern tropical Pacific seascape. Glob Ecol Biogeogr. doi:10.1111/j.1466-8238.2010.00642.x

Edmondson CH (1928) The ecology of an Hawaiian coral reef. Bull Bernice P Bishop Mus 45:1–64

Escarria E, Beltrán-León BS, Giraldo A, Zapata FA (2007) Ichthyoplankton in the National Natural Park Isla Gorgona (Pacific Ocean of Colombia) during September 2005. Invest Mar 35:127–133

Escobar JC (1990) Henry von Prahl: el científico y el hombre. Revista de Ciencias, Facultad de Ciencias, Univ del Valle 2(junio):5–22

Étienne J-L (2009) De la découverte de I'île de Clipperton à nos jours. In: Charpy L (coord) Clipperton, environnement et biodiversité d'un microcosme océanique. Muséum national d'Histoire naturelle, Paris; IRD, Marseille, pp 11–20

Feingold JS (1996) Coral survivors of the 1982-83 El Niño-Southern Oscillation, Galápagos Islands, Ecuador. Coral Reefs 15:108

Flot JF, Adjeroud M (2009) Les coraux de Clipperton. In: Charpy L (coord) Clipperton: environnement et biodiversité d'un microcosme océanique. Muséum national d'Histoire naturelle, Paris; IRD, Marseille (Patrimoines naturels; 68), pp 155–162

Forsman ZH, Guzman HM, Chen CA, Fox GE, Wellington GM (2005) An ITS region phylogeny of *Siderastrea* (Cnidaria: Anthozoa): is *S. glynni* endangered or introduced? Coral Reefs 24:343–347

Fraser CM (1943a) General account of the scientific work of the *Velero III*, in the eastern Pacific, 1931–41, Part I, historical introduction, *Velero III*, personnel. Allan Hancock Pac Exped 1:1–48

Fraser CM (1943b) General account of the scientific work of the *Velero III* in the eastern Pacific, 1931–1941, Part II: geographic and biological associations. Allan Hancock Pac Exped 1:49–258

Fraser CM (1943c) General account of the scientific work of the *Velero III* in the eastern Pacific, 1931–1941, Part III: a ten-year list of the *Velero III* collecting stations (charts 1–115). With an appendix of collecting stations of the Allan Hancock Foundation for the year 1942. Allan Hancock Pac Exped 1:259–431

Garzón-Ferreira J, Rodríguez-Ramírez (2010) SIMAC: development and implementation of a coral reef monitoring network in Colombia. Rev Biol Trop 58(suppl 1):67–80

Gierloff-Emden HG (1976) La Costa de El Salvador: monografía morfológica-oceanográfica. Dirección de Publicaciones, San Salvador, El Salvador, Ministerio de Educación 285 p

Gilman DC (1899) The life of James Dwight Dana: scientific explorer, mineralogist, geologist, zoologist, Professor in Yale University. Harper & Brothers Pub, New York and London, p 409

Giraldo A (2008) Variabilidad espacial de temperatura, salinidad y transparencia en el ambiente pelágico del PNN Gorgona durante septiembre 2007 y marzo 2008. Bol Cient CIOH 26:157–163

Giraldo A, Rodríguez-Rubio E, Zapata FA (2008) Condiciones oceanográficas en Isla Gorgona, Pacífico oriental tropical de Colombia. Lat Am J Aquat Res 36:121–128

Glynn PW (1972) Observations on the ecology of the Caribbean and Pacific coasts of Panamá. In: Jones ML (ed) The Panamic biota: some observations prior to a sea level canal. Bull Biol Soc Wash 2:13–20

Glynn PW (1973) *Acanthaster*: effect on coral reef growth in Panama. Science 180:504–506

Glynn PW (1974) The impact of *Acanthaster* on corals and coral reefs in the eastern Pacific. Environ Conserv 1:295–304

Glynn PW (1977a) Coral growth in upwelling and nonupwelling areas off the Pacific coast of Panamá. J Mar Res 35:567–585

Glynn PW (1977b) Interactions between *Acanthaster* and *Hymenocera* in the field and laboratory. In: Proceedings of 3rd International Coral Reef Symposium, vol 1, Miami, 209–215

Glynn PW (1982) Coral communities and their modifications relative to past and prospective Central American seaways. Adv Mar Biol 19:91–132

Glynn PW (1983) Extensive bleaching and death of reef corals on the Pacific coast of Panamá. Environ Conserv 10:149–154

Glynn PW (1988) El Niño warming, coral mortality and reef framework destruction by echinoid bioerosion in the eastern Pacific. Galaxea 7:129–160

Glynn PW (1990) Coral mortality and disturbances to coral reefs in the tropical eastern Pacific. In: Glynn PW (ed) Global ecological consequences of the 1982–1983 El Niño-Southern Oscillation. Elsevier, Amsterdam, pp 55–126

Glynn PW (1994) State of coral reefs in the Galápagos Islands: natural vs anthropogenic impacts. Mar Poll Bull 29:131–140

Glynn PW (1997) Eastern Pacific reef coral biogeography and faunal flux: Durham's dilemma revisited. In: Proceedings of 8th International Coral Reef Symposium, vol 1, Panama, pp 371–378

Glynn PW (2001a) Eastern Pacific coral reef ecosystems. In: Seeliger U, Kjerfve B (eds) Coastal marine ecosystems of Latin America. Springer, Berlin, pp 281–305

Glynn PW (2001b) Eastern Pacific coral reefs: new revelations in the twentieth century. Atoll Res Bull, Golden Issue 1951–2001 494:118–158

Glynn PW (2003) Coral communities and coral reefs of Ecuador. In: Cortés J (ed) Latin American coral reefs. Elsevier, Amsterdam, pp 449–472

Glynn PW (2011) In tandem reef coral and cryptic metazoan declines and extinctions. Bull Mar Sci 87:767–794

Glynn PW, Ault JS (2000) A biogeographic analysis and review of the far eastern Pacific coral reef region. Coral Reefs 19:1–23

Glynn PW, Colley SB (2008) Survival of brooding and broadcasting reef corals following large scale disturbances: is there any hope for broadcasting species during global warming? In: Proceedings of 11th International Coral Reef Symposium, Ft Lauderdale, sess 11, pp 361–365

Glynn PW, D'Croz L (1990) Experimental evidence for high temperature stress as the cause of El Niño-coincident coral mortality. Coral Reefs 8:181–191

Glynn PW, de Weerdt WH (1991) Elimination of two reef-building hydrocorals following the 1982-83 El Niño warming event. Science 253:69–71

Glynn PW, Macintyre IG (1977) Growth rate and age of coral reefs of the Pacific coast of Panamá. In: Proceedings of 3rd International Coral Reef Symposium, vol 2, Miami, pp 251–259

Glynn PW, Stewart RH (1973) Distribution of coral reefs in the Pearl Islands (Gulf of Panamá) in relation to thermal conditions. Limnol Oceanogr 18:367–379

Glynn PW, Wellington GM (1983) Corals and coral reefs of the Galápagos Islands. University of California Press, Berkeley, p 330
Glynn PW, Stewart RH, McCosker JE (1972) Pacific coral reefs of Panamá: structure, distribution and predators. Geol Rundsch 61:483–519
Glynn PW, Wellington GM, Birkeland C (1979) Coral reef growth in the Galápagos: limitation by sea urchins. Science 203:47–49
Glynn PW, von Prahl H, Guhl F (1982) Coral reefs of Gorgona Island, with special reference to corallivores and their influence on community structure and reef development. An Inst Invest Mar Punta Betín 12:185–214
Glynn PW, Druffel EM, Dunbar RB (1983) A dead Central American coral reef tract: possible link with the Little Ice Age. J Mar Res 41:605–637
Glynn PW, Cortés J, Guzmán HM, Richmond RH (1988) El Niño (1982–83) associated coral mortality and relationship to sea surface temperature deviations in the tropical eastern Pacific. In: Proceedings of 6th International Coral Reef Symposium, vol 3, Townsville, pp 237–243
Glynn PW, Veron JEN, Wellington GM (1996) Clipperton Atoll (eastern Pacific): oceanography, geomorphology, reef-building coral ecology and biogeography. Coral Reefs 15:71–99
Glynn PW, Maté JL, Baker AC, Calderón MO (2001) Coral bleaching and mortality in Panamá and Ecuador during the 1997–1998 El Niño-Southern Oscillation event: spatial/temporal patterns and comparisons with the 1982–1983 event. Bull Mar Sci 69:79–101
Glynn PW, Wellington GM, Wieters EA, Navarrete SA (2003) Reef-building coral communities of Easter Island (Rapa Nui), Chile. In: Cortés J (ed) Latin American coral reefs. Elsevier, Amsterdam, pp 473–494
Glynn PW, Wellington GM, Riegl B, Olson DB, Borneman E, Wieters EA (2007) Diversity and biogeography of the scleractinian coral fauna of Easter Island (Rapa Nui). Pac Sci 61:67–90
Glynn PW, Colley SB, Guzmán HM, Enochs IC, Cortés J, Maté JL, Feingold JS (2011) Reef coral reproduction in the eastern Pacific: Costa Rica, Panamá, and the Galápagos Islands (Ecuador). VI. Agariciidae *Pavona clavus*. Mar Biol 158:1601–1617
González Avelar M (1992) Clipperton, Isla Mexicana. Fondo de Cultura Económica, México DF, p 249
Gotuzzo R (1996) Celenterados In: Serrano F (ed) Historia natural y ecología de El Salvador. Ministerio de Educación, Editorial Offset S.A., Xochimilco, D.F., vol 2, pp 51–64
Graham JB (ed) (1975) The biological investigation of Malpelo Island, Colombia. Smithson Contrib Zool 176:1–8
Greenfield DW, Hensley D, Wiley JW, Ross ST (1970) The Isla Jaltemba coral formation and its zoogeographical significance. Copeia 1:180–181
Grewingk C (1848) Beitrag zur Kenntniss der geognostischen Beschaffenheit Californiens. Verhandl Russich K MinGesell St Petersburg, for 1847, pp 142–162
Guzmán HM (1988a) Distribución y abundancia de organismos coralívoros en los arrecifes coralinos de la Isla del Caño, Costa Rica. Rev Biol Trop 36:191–207
Guzmán HM (1988b) Feeding behavior of the gastropod corallivore *Quoyula monodonta* (Blainville). Rev Biol Trop 36:209–212
Guzmán HM (1988c) Primer informe de un coral acropórido para el Pacífico Oriental: crítica a Prahl y Mejía. Rev Biol Trop 36:163–166
Guzmán HM (1991) Restoration of coral reefs in Pacific Costa Rica. Conserv Biol 5:189–195
Guzmán HM, Cortés J (1989a) Coral reef community structure at Caño Island, Pacific Costa Rica. PSZNI Mar Ecol 10:23–41
Guzmán HM, Cortés J (1989b) Growth rates of eight species of scleractinian corals in the eastern Pacific (Costa Rica). Bull Mar Sci 44:1186–1194
Guzmán HM, Cortés J (1992) Cocos Island (Pacific of Costa Rica) coral reefs after the 1982–83 El Niño disturbance. Rev Biol Trop 40:309–324
Guzmán HM, Cortés J (1993) Arrecifes coralinas del Pacífico Oriental Tropical: revisión y perspectivas. Rev Biol Trop 41:535–557
Guzmán HM, Cortés J (2001) Changes in reef community structure after fifteen years of natural disturbances in the eastern Pacific (Costa Rica). Bull Mar Sci 69:133–149
Guzmán HM, Cortés J (2007) Reef recovery 20-yr after the 1982–83 El Niño massive mortality. Mar Biol 151:401–411
Guzmán HM, Obando VL, Cortés J (1987a) Meiofauna associated with a Pacific coral reef in Costa Rica. Coral Reefs 6:107–112
Guzmán HM, Cortés J, Richmond RH, Glynn PW (1987b) Efectos del fenómeno del Niño Oscilación Sureña 1982/83 en los arrecifes coralinos de la Isla del Caño, Costa Rica. Rev Biol Trop 35:325–332
Guzmán HM, Guevara CA, Breedy O (2004) Distribution, diversity, and conservation of coral reefs and coral communities in the largest marine protected area of Pacific Panama (Coiba Island). Environ Conserv 31:111–121
Guzmán HM, Benfield S, Breedy O, Mair JM (2008) Broadening reef protection across the marine conservation corridor of the eastern tropical Pacific: distribution and diversity of reefs in Las Perlas Archipelago, Panama. Environ Conserv 35:46–54
Heck KL, McCoy ED (1978) Long-distance dispersal and the reef-building corals of the eastern Pacific. Mar Biol 48:348–356
Hermosillo A (2004) Opisthobranch mollusks of Parque Nacional de Coiba, Panamá (tropical eastern Pacific). Festivus 36:105–117
Herrero-Pérezrul MD, Reyes Bonilla H, García Domínguez F (1998) Casual hermaphroditism in gonochoric *Isostichopus fuscus* (Ludwig, 1875) (Echinodermata: Holothuroidea) in the southern Gulf of California, Mexico. Bull Mar Sci 63:611–615
Herrero-Pérezrul MD, Reyes Bonilla H, García Domínguez F, Cintra-Buenrostro CE (1999) Reproduction and growth of *Isostichopus fuscus* (Echinodermata: Holothuroidea) in the southern Gulf of California, México. Mar Biol 135:521–532
Hertlein LG (1963) Contribution to the biogeography of Cocos Island, including a bibliography. In: Proceedings of the California Academy of Science 4th series, vol 32, pp 219–289
Hertlein LG, Emerson WK (1953) Mollusks from Clipperton Island (eastern Pacific) with the description of a new species of gastropod. Trans San Diego Soc Nat Hist 11:345–364
Hertlein LG, Emerson WK (1957) Additional notes on the invertebrate fauna of Clipperton Island. Am Mus Novit 1859:1–9
Heylings P, Bensted-Smith R, Altamirano M (2002) Zonificación e historia de la Reserva Marina de Galápagos. In: Danulat E, Edgar GJ (eds) Reserva Marina de Galápagos. Línea base de la biodiversidad. Santa Cruz, Galápagos, Fundación Charles Darwin/Servicio Parque Nacional Galápagos, pp 10–21
Hoegh-Guldberg O, Mumby PJ, Hooten AJ, Steneck RS, Greenfield P, Gomez E, Harvell CD, Sale PF, Edwards AJ, Caldeira K, Knowlton N, Eakin CM, Iglesias-Prieto R, Muthiga N, Bradbury RH, Dubi A, Hatziolos ME (2007) Coral reefs under rapid climate change and ocean acidification. Science 318:1737–1742
Hubbard DK, Garcia M (2003) The corals and coral reefs of Easter Island—a preliminary look. In: Loret J, Tanacredi JT (eds) Easter Island: scientific exploration into the world's environmental problems in microcosm. Kluwer Academic/Plenum Publishers, New York, pp 53–77
Hueerkamp C, Glynn PW, D'Croz L, Maté JL, Colley SB (2001) Bleaching and recovery of five eastern Pacific corals in an El Niño-related temperature experiment. Bull Mar Sci 69:215–236
Iglesias-Prieto R, Beltrán VH, LaJeunesse TC, Reyes-Bonilla H, Thome PE (2004) Different algal symbionts explain the vertical distribution of dominant reef corals in the eastern Pacific. Proc R Soc Lond B 271:1757–1763

INVEMAR (2014) Informe del estado de los ambientes y recursos marinos y costeros en Colombia: Año 2013. Serie de Publicaciones Periódicas No 3, Santa Marta p 192

Jiménez CE (1997) Corals and coral reefs of Culebra Bay, Pacific coast of Costa Rica: anarchy in the reef. In: Proceedings of 8th International Coral Reef Symposium, vol 1, Panama pp 329–334

Jiménez CE, Cortés J (2001) Effects of the 1991–1992 El Niño on scleractinian corals of the Costa Rican central Pacific coast. Rev Biol Trop 49(supl 2):239–250

Jiménez CE, Cortés J (2003) Growth of seven species of scleractinian corals in an upwelling environment of the eastern Pacific (Golfo de Papagayo, Costa Rica). Bull Mar Sci 72:187–198

Jiménez CE, Cortés J, León A, Ruiz E (2001) Coral bleaching and mortality associated with the 1997–98 El Niño in an upwelling environment in the eastern Pacific (Gulf of Papagayo, Costa Rica). Bull Mar Sci 69:151–169

Jiménez CE, Bassey G, Segura A, Cortés J (2010) Characterization of the coral communities and reefs of two previously undescribed locations in the upwelling region of Golfo de Papagayo (Costa Rica). Rev Cien Mar Cost 2:95–108

Johnson RI (1946) Joseph Pitty Couthouy—a bibliography and catalogue of his species. In: Occasional Papers on Mollusks, vol 1, pp 33–40

Johnson GC, Sloyan BM, Kessler WS, Taggart KE (2002) Direct measurements of upper ocean currents and water properties across the tropical Pacific during the 1990s. Prog Oceanogr 52:31–61

Joiner GD, Sandefur JH (2007) Joseph Pitty Couthouy: the death of a sailor-scientist. In: Joiner GD (ed) Little to eat and thin mud to drink: letters, diaries, and memoirs from the Red River Campaigns, 1863–1864. University of Tennessee Press, Knoxville, pp 219–229

Jones ML (ed) (1972) The Panamic biota: some observations prior to a sea-level canal. Bull Biol Soc Wash 2:viii + 270

Jost C (2005) Bibliographie de I'île de Clipperton – Île de La Passion (1711–2005). J Soc Océanides 120–121:181–197. http://jso.revues.org/481

Ketchum JT, Reyes-Bonilla H (1997) Biogeography of the hermatypic corals of the Revillagigedo Archipiélago, México. In: Proceedings of 8th International Coral Reef Symposium, vol 1, Panama, pp 471–476

Ketchum JT, Reyes-Bonilla H (2001) Taxonomía y distribución de los corales hermatípicos (Scleractinia) del Archipiélago de Revillagigedo, Pacífico de México. Rev Biol Trop 49:803–848

Kim W, Abele LG (1988) The snapping shrimp genus *Alpheus* from the eastern Pacific (Decapoda: Caridea: Alpheidae). Smith Contrib Zool 454:1–119

Lacroix A (1939) Clipperton, îles de Páques et Pitcairn, esquisse lithologique. Ann Inst Océanogr 18:290–330

LaJeunesse TC, Reyes-Bonilla H, Warner ME (2007) Spring "bleaching" among *Pocillopora* in the Sea of Cortez, eastern Pacific. Coral Reefs 26:265–270

LaJeunesse TC, Reyes Bonilla H, Warner ME, Wills M, Schmidt GW, Fitt WK (2008) Specificity and stability in high latitude eastern Pacific coral-algal symbioses. Limnol Oceanogr 53:719–727

LaJeunesse TC, Smith R, Walther M, Pinzón J, Pettay DT, McGinley M, Aschaffenburg M, Medina-Rosas P, Cupul-Magaña AL, López-Pérez A, Reyes-Bonilla H, Warner ME (2010) Host-symbiont recombination versus natural selection in the response of coral-dinoflagellate symbioses to environmental disturbance. Proc R Soc B 277:2925–2934

Lang MA, Marinelli RL, Roberts SJ, Taylor PR (2013) Research and discoveries: the revolution of science through scuba. Smith Contrib Mar Sci 39:258

Lessios HA, Robertson DR (2006) Crossing the impassable: genetic connections in 20 reef fishes across the eastern Pacific barrier. Proc R Soc Lond B Biol Sci 273:2201–2208

Lessios HA, Kessing BD, Wellington GM, Graybeal A (1996) Indo-Pacific echinoids in the tropical eastern Pacific. Coral Reefs 15:133–142

Lessios HA, Kessing BD, Robertson DR (1998) Massive gene flow across the world's most potent marine biogeographic barrier. Proc R Soc Lond B 265:583–588

Lesson RP (1829) Voyage autour de Monde sur … *La Coquille*, pendant les Années 1822, 1823, 1824, et 1825. Zoologie, 2, Part 2, Zoophytes, p 151

Linsley BK, Dunbar RB, Wellington GM, Mucciarone DA (1994) A coral-based reconstruction of intertropical convergence zone variability over Central America since 1707. J Geophys Res 99 (C5):9977–9994

Linsley BK, Messier RG, Dunbar RB (1999) Assessing between-colony oxygen isotope variability in the coral *Porites lobata* at Clipperton Atoll. Coral Reefs 18:13–27

Linsley BK, Ren L, Dunbar RB, Howe SS (2000) El Niño Southern Oscillation (ENSO) and decadal-scale climate variability at 10°N in the eastern Pacific from 1893 to 1994: a coral-based reconstruction from Clipperton Atoll. Paleoceanography 15:322–335

Londoño-Cruz E, Cantera JR, Toro-Farmer G, Orozco C (2003) Internal bioerosion by macroborers in *Pocillopora* spp. in the tropical eastern Pacific. Mar Ecol Prog Ser 265:289–295

López García E, San Martín Peral G, Cladera de Codina P, Capa Corrales M (1997) La fauna de anélidos poliquetos del Parque Nacional de Coiba (Panamá). In: Castroviejo S, Velayos M (eds) Flora y fauna del Parque Nacional de Coiba (Panamá): inventario preliminar. Madrid, AECI, Agencias Española de Cooperación Internacional, pp 57–73

López-Victoria M (ed) (2011a) Suplemento especial Malpelo. Bol Invest Mar Cost (Bull Mar Coast Res) 40(supl esp):1–199

López-Victoria M (ed) (2011b) Más de un siglo de investigaciones cientificos en la isla Malpelo. Bol Invest Mar Cost (Bull Mar Coast Res) 40(supl esp):13–17

Lozano S, Zapata FA (2003) Short-term temporal patterns of early recruitment of coral reef fishes in the tropical eastern Pacific. Mar Biol 142:399–409

Macias Mayorga DM (2002) Estructura trófica del plankton en el Pacífico de Panamá. Thesis, Escuela de Biología, Facultad de Ciencias Naturales, Exactas y Tecnología, Universidad de Panamá, Panamá

Macintyre IG, Glynn PW, Cortés J (1993) Holocene reef history in the eastern Pacific: mainland Costa Rica, Caño Island, Cocos Island, and Galápagos Islands. In: Proceedings of 7th International Coral Reef Symposium, vol 2, Guam, pp 1174–1184

Manzello DP (2009) Reef development and resilience to acute (El Niño warming) and chronic (high-CO_2) disturbances in the eastern tropical Pacific: a real-world climate change model. In: Proceedings of 11th International Coral Reef Symposium, vol 1, Ft Lauderdale, pp 1299–1304

Manzello DP (2010a) Ocean acidification hotspots: spatiotemporal dynamics of the seawater CO_2 system of eastern Pacific coral reefs. Limnol Oceanogr 55:239–248

Manzello DP (2010b) Coral growth with thermal stress and ocean acidification: lessons from the eastern tropical Pacific. Coral Reefs 29:749–758

Manzello DP, Kleypas JA, Budd DA, Eakin CM, Glynn PW, Langdon C (2008) Poorly cemented coral reefs of the eastern tropical Pacific: possible insights into reef development in a high-CO_2 world. Proc Nat Acad Sci 105:10450–10455

Maté JL (1997) Experimental responses of Panamanian reef corals to high temperature and nutrients. In: Proceedings of 8th International Coral Reef Symposium, vol 1, Panama, pp 515–520

Maté JL (2003a) Ecological, genetic, and morphological differences among three *Pavona* (Cnidaria: Anthozoa) species from the Pacific coast of Panama. Mar Biol 142:427–440

Maté JL (2003b) Corals and coral reefs of the Pacific coast of Panamá. In: Cortés J (ed) Latin American coral reefs. Elsevier, Amsterdam, pp 387–417

Mayor AG (1916) The lower temperatures at which reef-corals lose their ability to capture food. Carnegie Inst Wash Yearb 14:212

McCosker JE, Rosenblatt RH (1975) Fishes collected at Malpelo Island. Smithson Contrib Zool 176:91–93

McCoy ED, Heck KL (1976) Biogeography of corals, seagrasses and mangroves: an alternative to the center of origin concept. Syst Zool 25:201–210

Menzies RJ (1968) Transport of marine life between oceans through the Panama Canal. Nature 220:802–803

Meredith DW (1939) Voyages of the *Velero III*, a pictorial version with historical background of scientific expeditions through tropical seas to equatorial lands aboard M/V *Velero III* Los Angeles, Bookhaven Press, p 294

Milne Edwards H, Haime J (1848a) Recherches sur les Polypiers. Premier Mémoire. Structure et développement des polypiers en general. Annales des Sciences Naturelles, 9:37–89

Milne Edwards H, Haime J (1848b) Recherches sur les Polypiers. Deuxième Mémoire. Monographie des Turbinolides. Ann Sci Nat 9:211–344

Narváez K, Zapata FA (2010) First record and impact of the crown-of-thorns starfish, *Acanthaster planci* (Spinulosida: Acanthasteridae) on corals of Malpelo Island Colombian Pacific. Rev Biol Trop 58(suppl 1):139–143

Navas-Camacho R, Rodríguez-Ramírez A, Reyes-Nivia C (2010) Agents of coral mortality on reef formations of the Colombian Pacific. Rev Biol Trop 58(suppl 1):133–138

Orellana-Amador JJ (1985) Fishes of El Salvador: marine fishes of Los Cóbanos. Sigma Foundation, New York, p 126

Palmer RH (1928) Fossil and recent corals and coral reefs of western Mexico. Proc Amer Philos Soc Philadelphia 67:21–31

Paz-García DA, LaJeunesse TC, Chávez-Romo HE, Correa-Sandoval F, Reyes-Bonilla H (2008) Differences in the distribution of *Symbiodinium* spp. among morphotypes and genotypes of *Porites panamensis* from the Gulf of California, Mexico. In: Proceedings of 11th International Coral Reef Symposium, vol 1, Ft Lauderdale, pp 167–170

Pérez-Vivar TL, Reyes-Bonilla H, Padilla C (2006) Corales pétreos (Scleractinia) de las Islas Marías, Pacífico de México. Cienc Mar 32:259–270

Pettay DT, Lajeunesse TC (2009) Microsatellite loci for assessing genetic diversity, dispersal and clonality of coral symbionts in 'stress-tolerant' clade D *Symbiodinium*. Mol Ecol Res 9:1022–1025

Pinzón JH, LaJeunesse TC (2010) Species delimitation of common reef corals in the genus *Pocillopora* using nucleotide sequence phylogenies, population genetics and symbiosis ecology. Mol Ecol. doi:10.1111/j.1365-294X.2010.04939.x

Porter JW (1972a) Ecology and species diversity of coral reefs on opposite sides of the Isthmus of Panamá. In: Jones ML (ed) The Panamic biota: some observations prior to a sea level canal, vol 2, pp 89–116 (Bull Biol Soc Wash)

Porter JW (1972b) Predation by *Acanthaster* and its effects on coral species diversity. Am Nat 106:487–492

Porter JW (1974) Community structure of coral reefs on opposite sides of the Isthmus of Panamá. Science 186:543–545

Pourtalès LF (1875) Corals of the Galapagos Islands. Am J Sci Arts 10:282–283

Ramos GE, Lemaitre R (1991) Henry von Prahl (1948–1989) J Crust Biol 11:473–476

Reyes-Bonilla H (1992) New records for hermatypic corals (Anthozoa: Scleractinia) in the Gulf of California, México, with an historical and biogeographical discussion. J Nat Hist 26:1163–1175

Reyes-Bonilla H (1993a) Biogeografía y ecología de los corales hermatípicos (Anthozoa: Scleractinia) del Pacífico de México. In: Salazar-Vallejo S, González NE (eds) Biodiversidad marina y costera de México. CONABIO/CIQRO, Chetumal, pp 207–222

Reyes-Bonilla H (1993b) Corales hermatípicos (Anthozoa: Scleractinia) de la region de Los Cabos, Baja California Sur. Rev Inv Cient UABCS 4:1–9

Reyes-Bonilla H (1997) Cabo Pulmo reef: a new marine reserve in the Gulf of California. Conserv Biol 11:838

Reyes-Bonilla H (2002) Checklist of valid names and synonyms of stony corals (Anthozoa: Scleractinia) from the eastern Pacific. J Nat Hist 36:1–13

Reyes-Bonilla H (2003) Coral reefs of the Pacific coast of México. In: Cortés J (ed) Latin American coral reefs. Elsevier, Amsterdam, pp 331–349

Reyes-Bonilla H, Barraza JE (2003) Corals and associated marine communities from El Salvador. In: Cortés J (ed) Latin American coral reefs. Elsevier, Amsterdam, pp 351–360

Reyes-Bonilla H, Calderón-Aguilera LE (1999) Population density, distribution and consumption rates of three corallivores at Cabo Pulmo reef, Gulf of California, Mexico. Mar Ecol 20:347–357

Reyes-Bonilla H, Carricart-Ganivet JP (2000) *Porites arnaudi*, a new species of stony coral (Anthozoa: Scleractinia: Poritidae) from oceanic islands of the eastern Pacific Ocean. Proc Biol Soc Wash 113:561–571

Reyes-Bonilla H, López-Pérez A (1998) Biogeography of scleractinian corals of the Pacific coast of México. Cienc Mar 24:211–224

Reyes-Bonilla H, RiosmenaRodríguez R, Foster MS (1997a) Hermatypic corals associated with rhodolith beds in the Gulf of California. Pac Sci 51:328–337

Reyes-Bonilla H, Sinsel-Duarte F, Arizpe-Covarrubias (1997b) Gorgonias y corales pétreos (Anthozoa: Gorgonacea y Scleractinia) de Cabo Pulmo, México. Rev Biol Trop 45:1439–1443

Reyes-Bonilla H, Pérez-Vivar TL, Ketchum-Mejía JT (1999) Distribución geográfica y depredación de *Porites lobata* (Anthozoa: Scleractinia) en la costa occidental de México. Rev Biol Trop 47:66–80

Reyes-Bonilla H, Calderón-Aguilera LE, Cruz-Piñón G, Medina-Rosas P, López-Pérez RA, Herrero-Pérezrul MD, Leyte-Morales GE, Cupul-Magaña AL, Carriquiry Beltrán JD (2005) Atlas de corales pétreos (Anthozoa: Scleractinia) del Pacífico mexicano. Triciclo, Centro Univ de la Costa, Univ de Guadalajara, p 124

Ricardo J, Kintz C (2008) Henry von Prahl Bauer: científico, artista y persona. Día Nacional de los Arrecifes Coralinos. Cartagena, Diciembre, p 24

Riegl BM, Halfar J, Purkis SJ, Godinez-Orta L (2007) Sedimentary facies of the eastern Pacific's northernmost reef-like setting (Cabo Pulmo, Mexico). Mar Geol 236:61–77

Rivera F, Martínez PC (2011) Guía fotográfica de corales y octocorales: Parque Nacional Machalilla y Reserva de Producción Faunística Marino Costera Puntilla de Santa Elena. NAZCA and Conservación Internacional, Ecuador, p 78

Robertson DR, Allen GR (1996) Zoogeography of the shorefish fauna of Clipperton Atoll. Coral Reefs 15:121–131

Robertson DR, Allen GR (1997) An annotated checklist of the fishes of Clipperton Atoll, tropical eastern Pacific. Rev Biol Trop 45:813–843

Robertson DR, Allen GR (2008) Shorefishes of the tropical eastern Pacific online information system. Version 1.0, Smithsonian Tropical Research Institute, Balboa, Panamá. www.neotropicalfishes.org/sftep

Robertson DR, Christy JH, Collin R, Cooke RG, D'Croz L, Kaufmann KW, Heckadon Moreno S, Maté JL, O'Dea A, Torchin ME (2009) The Smithsonian Tropical Research Institute: marine research, education, and conservation in Panama. Smith Contrib Mar Sci 38:73–93

Robinson G (1985) The influence of the 1982–83 El Niño on Galápagos marine life. In: Robinson G, del Pino EM (eds) El Niño in the Galápagos Islands: the 1982–1983 event. Fund Charles Darwin para Las Islas Galápagos, Quito, Ecuador, pp 153–190

Robinson JA, Thomson DA (1992) Status of the Pulmo coral reefs in the lower Gulf of California. Environ Conserv 19:261–264

Rodríguez-S K (2001) CIMAR: pasado, presente y futuro de un centro de investigación. Ambientes Acuáticos 1:2–6

Rosenblatt RH, Waples RS (1986) A genetic comparison of allopatric populations of shore fish species from the eastern and central Pacific Ocean: dispersal or vicariance? Copeia 2:275–284

Rosenblatt RH, McCosker JE, Rubinoff I (1972) Indo-west Pacific fishes from the Gulf of Chiriquí, Panamá. Contrib Sci Nat Hist Mus LA Co 234:1–18

Rowan R, Knowlton N (1995) Intraspecific diversity and ecological zonation in coral-algal symbiosis. Proc Nat Acad Sci 92:2850–2853

Rowan R, Knowlton N, Baker A, Jara J (1997) Landscape ecology of algal symbionts creates variation in episodes of coral bleaching. Nature 388:265–269

Rubinoff I (1968) Central American sea level canal: possible biological effects. Science 161:857–861

Rubinoff RW, Rubinoff I (1969) Observations on migration of a marine goby through the Panama Canal. Copeia 1969:395–397

Ryan J, Zapata Y (2003) Nicaragua's coral reefs: status, health and management strategies. In: Cortés J (ed) Latin American coral reefs. Elsevier, Amsterdam, pp 203–222

Sachet M-H (1960) Histoire de l'île Clipperton. Cahiers du Pacifique 2:3–32

Sachet M-H (1962a) Geography and land ecology of Clipperton Island. Atoll Res Bull 86:1–115

Sachet M-H (1962b) Monographie physique et biologique de l'île de Clipperton. Annales de l'Institut Océanographique 40(1):107

Salvat B, Adjeroud M, Charpy L (2008) Les récifs coralliens de Clipperton. Rev Ecol (Terre Vie) 63:179–187

Schlöder C, D'Croz L (2004) Responses of massive and branching coral species to the combined effects of water temperature and nitrate enrichment. J Exp Mar Biol Ecol 313:255–268

Scott PJB, Risk MJ, Carriquiry JD (1988) El Niño, bioerosion and the survival of east Pacific reefs. In: Proceedings of 6th International Coral Reef Symposium, vol 2, Australia, pp 517–520

Skaggs JM (1989) Clipperton: a history of the island the world forgot. Walker and Company, New York 318 p

Slevin JR (1959) The Galápagos Islands, a history of their exploration. In: Occas Papers California Academy of Sciences, vol 25, pp 1–150

Spalding MD, Ravilious C, Green EP (2001) World atlas of coral reefs. Univ California Press, Berkeley, p 424

Springer VG (1958) Systematics and zoogeography of the clinid fishes of the subtribe Labrisomini Hubbs. Pubs Inst Mar Sci Univ Texas 5:417–492

Squires DR (1959) Corals and coral reefs in the Gulf of California. Bull Am Mus Nat Hist 118:370–431

Stanton W (1975) The great United States exploring expedition of 1838–1842. University of California Press, Berkeley, p 433

Steinbeck J, Rickets EF (1941) Sea of Cortez, a leisurely journal of travel and research. The Viking Press, New York x + p 598

Stoddart DR (1969) Ecology and morphology of recent coral reefs. Biol Rev 44:433–498

Stoddart DR (1994) "This coral episode": Darwin, Dana, and the coral reefs of the Pacific. In: MacLeod R, Rehbock PF (eds) Darwin's laboratory: evolutionary theory and natural history in the Pacific. University of Hawai'i Press, Honolulu, pp 21–48

Tejada C, Devis A, García L, Málikov I, Villegas NL (2002) Compilación Oceanográfica de la Cuenca Pacífica Colombiana. Centro Control Contaminación del Pacífico Colombiano. Ed DIMAR, Serie Publicaciones Especiales, San Andrés de Tumaco, vol 1, p 124

Toro-Farmer G, Cantera JR, Londoño-Cruz E, Orozco C, Neira R (2004) Patrones de distribución y tasas de bioerosión del erizo *Centrostephanus coronatus* (Diadematoida: Diadematidae) en el arrecife de Playa Blanca, Pacífico colombiano. Rev Biol Trop 52:67–76

van Oppen MJH, Baker AC, Coffroth MA, Willis BL (2009) Bleaching resistance and the role of algal endosymbionts. In: van Oppen MJH, Lough JM (eds) Coral bleaching: patterns, processes, causes and consequences. Ecol Stud 205, Springer, Berlin, pp 83–102

Vargas-Ángel B (1988) Contribución al conocimiento de las formaciones coralinas del litoral Pacífico colombiano. Mem VI Sem Nal Ciencias del Mar. pp 562–570. Comisión Colombiana de Oceanografía, Bogotá, Colombia

Vargas-Ángel B (1996) Distribution and community structure of the reef corals of Ensenada de Utría, Pacific coast of Colombia. Rev Biol Trop 44:643–651

Vargas-Ángel B (2001) Corals and coral reefs of the Pacific coast of Colombia with special reference to spatial and temporal patterns of environmental disturbances. Ph.D. dissert, Univ Miami, Coral Gables, Florida

Vargas-Ángel B (2003) Coral community structure off the Pacific coast of Colombia: onshore vs offshore coral reefs. Atoll Res Bull 499:1–21

Vargas-Ángel B, Zapata FA, Hernández H, Jiménez JM (2001) Coral and coral reef responses to the 1997–98 El Niño event on the Pacific coast of Colombia. Bull Mar Sci 69:111–132

Vaughan TW (1906) Reports on the scientific results of the expedition to the eastern tropical Pacific in charge of Alexander Agassiz, by the U.S.Fish Commission steamer "Albatross," from October, 1904, to March, 1905, Lieut. Commander L.M. Garrett, U.S.N., Commanding. VI. Madreporaria. Bull Mus Comp Zoöl, Harvard Coll 50:61–72

Vaughan TW (1919) Corals and the formation of coral reefs. Annu Rep Board Regents Smith Inst 17:189–238

Veron JEN (1995) Corals in space and time: the biogeography and evolution of the Scleractinia. Cornell Univ Press, Ithaca, London, p 321

Veron JEN (2000) Corals of the world. Townsville, Aust Inst Mar Sci 2:429

Veron JEN, Wallace CC (1984) Scleractinia of eastern Australia, Part V family Acroporidae. Aust Inst Mar Sci Monogr 6:1–485

Veron J, Stafford-Smith M, DeVantier L, Turak E (2015) Overview of distribution patterns of zooxanthellate Scleractinia. Front Mar Sci 1:81. doi:10.3389/fmars.2014.00081

Verrill AE (1866) On the polyps and corals of Panama with descriptions of new species. Proc Boston Soc Nat Hist 10:323–357

Verrill AE (1868–70) Review of the corals and polyps of the west coast of America. Trans Conn Acad Arts Sci 1(6):377–567

Verrill AE (1870) Notes on Radiata. Part 7. On the geographical distribution of the polyps of the west coast of America. Trans Conn Acad Arts Sci 1:558–570

Victor BC, Wellington GM, Robertson DR, Ruttenberg BI (2001) The effect of the El Niño-Southern oscillation event on the distribution of reef-associated labrid fishes in the eastern Pacific Ocean. Bull Mar Sci 69:279–288

Villalobos AF (1960) Notas acerca del aspecto hidrobiológico de la isla. In: Adem J, Cobo E, Blázquez L, Villalobos A, Miranda E,

Herrera T, Villa B, Vásquez L (eds) La Isla Socorro, Archipiélago de las Revillagigedo. Monogr Inst Geofis 2:155–180

von Prahl H (1982) Las formaciones de pústulas coralinas por el cangrejo *Hapalocarcinus marsupialis* Stimpson (Crustacea: Decapoda) en Gorgona, Colombia. An Inst Invest Mar Punta Betín 12:97–103

von Prahl H (1983) Blanqueo masivo y muerte de corales en la Isla Gorgona, Pacífico Colombiano. Cespedesia 45–46:125–129

von Prahl H (1985) Blanqueo y muerte de corales hermatípicos en el Pacífico colombiano atribuidos al fenómeno de El Niño 1982–83. Boletín ERFEN 12:22–24

von Prahl H (1986) Mortandad de corales hermatípicos en América a consequencia del fenómeno El Niño 1982–1983. Boletín Erfen 17:14–15

von Prahl H, Erhardt H (1985) Colombia: Corales y arrecifes coralinos. Editorial Presencia Ltda, Bogotá, p 295

von Prahl H, Mejía A (1985) Primer informe de un coral acropórido, *Acropora valida* (Dana, 1846) (Scleractinia: Astrocoeniida: Acroporidae) para el Pacífico americano. Rev Biol Trop 33:39–43

von Prahl H, Guhl F, Grogl M (1978) Crustáceos decapodos comensales del coral *Pocillopora damicornis* L. en la Isla de Gorgona, Colombia. An Inst Inv Mar Punta Betin 10:81–93

von Prahl H, Guhl F, Grogl M (1979) Gorgona. Futura Grupo Editorial Ltd, Bogotá, Colombia, p 279

Wainwright SA (1965) Reef communities visited by the Israel South Red Sea Expedition, 1962. Israel South Red Sea Expedition, 1962, Rep No 9, Bull Sea Fish Res Sta, Haifa, vol 38, pp 40–53

Wallace C (1999) Staghorn corals of the world. CSIRO Pub, Collingwood, Australia, p 422

Weerdt WH de, Glynn PW (1991) A new and presumably now extinct species of *Millepora* (Hydrozoa) in the Eastern Pacific. Zool Med Leiden 65:267–276

Wellington GM (1975) The Galápagos coastal marine environments: a resource report to the Department of National Parks and Wildlife, Quito, Ecuador, p 357

Wellington GM (1982) Depth zonation of corals in the Gulf of Panamá: control and facilitation by resident reef fishes. Ecol Monogr 52:223–241

Wellington GM, Dunbar RB (1995) Stable isotopic signature of El Niño-Southern Oscillation events in Eastern tropical Pacific reef corals. Coral Reefs 14:5–25

Wellington GM, Glynn PW (2007) Responses of coral reefs to El Niño-Southern Oscillation sea warming events. In: Aronson RB (ed) Geological approaches to coral reef ecology: placing the current crisis in historical context. Ecol Stud 192, Springer, Berlin, pp 342–385

Wellington GM, Victor BC (1985) El Niño mass coral mortality: a test of resource limitation in a coral reef damselfish population. Oecologia (Berl) 68:15–19

Wellington GM, Glynn PW, Strong AE, Navarrete SA, Wieters E, Hubbard D (2001) Crisis on coral reefs linked to climate change. Eos Trans Am Geophys Union 82(1):5

Wells JW (1972) Notes on Indo-Pacific scleractinian corals. Part 8: Scleractinian corals from Easter Island. Pac Sci 26:183–190

Wells JW (1983) Annotated list of the scleractinian corals of the Galápagos. In: Glynn PW, Wellington GM (eds) Corals and coral reefs of the Galápagos Islands. University of California Press, Berkeley, pp 212–295

Werding B, Haig J (1982) The porcellanid crabs of the Isla Gorgona, Pacific coast of Colombia, with a description of *Clastotoechus gorgonensis* sp. nov. (Crustacea: Anomura). An Inst Inv Mar Punta Betín 12:57–70

Wilkinson CR (2008) Status of coral reefs of the world: 2008. Global Coral Reef Monitoring Network and Reef and Rainforest Research Centre, Townsville, p 304

Wilkinson CR, Souter D (2008) Status of Caribbean coral reefs after bleaching and hurricanes in 2005. Global Coral Reef Monitoring Network and Reef and Rainforest Research Centre, Townsville, p 152

Yonge CM (1940) The biology of reef-building corals. Great Barrier Reef Expedition, 1928–29. British Mus Nat Hist Sci Repts 1:353–389

Zann LP, Bolton L (1985) The distribution, abundance and ecology of the blue coral *Heliopora coerulea* (Pallas) in the Pacific. Coral Reefs 4:125–134

Zapata FA (2001a) Ecología de peces arrecifales en la Isla de Gorgona: composición, abundancia, diversidad e historia de vida temprana. In: Barrios LM, López-Victoria M (eds) Gorgona marina: Contribución al conocimiento de una isla única. INVEMAR, Serie Publicaciones Especiales No 7, Santa Marta, pp 111–122

Zapata FA (2001b) Formaciones coralinas de Isla Gorgona. In: Barrios LM, López-Victoria M (eds) Gorgona marina: Contribución al conocimiento de una isla única. INVEMAR, Serie Publicaciones Especiales No 7, Santa Marta, pp 27–40

Zapata FA, Herrón PA (2002) Pelagic larval duration and geographic distribution of tropical eastern Pacific snappers (Pisces: Lutjanidae). Mar Ecol Prog Ser 230:295–300

Zapata FA, Morales YA (1997) Spatial and temporal patterns of fish diversity in a coral reef at Gorgona Island, Colombia. In: Proceedings of 8th International Coral Reef Symposium, vol 1, Panama, pp 1029–1034

Zapata FA, Robertson DR (2007) How many shore fishes are there in the tropical eastern Pacific? J Biogeogr 34:38–51

Zapata FA, Vargas-Ángel B (2003) Corals and coral reefs of the Pacific coast of Colombia. In: Cortés J (ed) Latin American coral reefs. Elsevier, Amsterdam, pp 419–447

Zapata FA, Vargas-Ángel B, Garcón-Ferreira J (2001) Salud y conservación de las comunidades coralinas. In: Barrios LM, López-Victoria M (eds) Gorgona marina: Contribución al conocimiento de una isla única. INVEMAR, Serie Publicaciones Especiales No 7, Santa Marta, pp 41–50

Zapata FA, Garcia JL, Tobón A (2008) Formaciones coralinas en las localidades de Piñas (Punta Cruces) y El Acuario (Cabo Marzo). In: Giraldo A, Valencia B (eds) Chocó: paraíso por naturaleza – Punta Cruces y Cabo Marzo. Departamento de Biología, Universidad del Valle, Cali, Colombia, pp 23–32

Zapata FA, Rodríguez-Ramírez A, Caro-Zambrano C, Garzón-Ferreia J (2010) Mid-term coral-algal dynamics and conservation status of a Gorgona Island (Tropical Eastern Pacific) coral reef. Rev Biol Trop 58(suppl 1):81–94

Zapata FA, Jaramillo-González J, Navas-Camacho R (2011) Extensive bleaching of the coral *Porites lobata* at Malpelo Island, Colombia, during a cold water episode in 2009. Bol Invest Mar Cost 40(Spec Suppl):185–193

Revisiting the Cenozoic History and the Origin of the Eastern Pacific Coral Fauna

2

Andrés López-Pérez

Abstract

Distribution, composition and functioning of eastern Pacific (EP) coral communities and reefs have resulted from modern and ancient processes. While much has been learned from living systems, still many questions remain regarding how pre-Holocene events shaped modern EP coral communities and reefs. From the late Paleocene to late Miocene, fossil outcrops were spatially restricted to the Washington-Seattle and California regions, whereas from the late Miocene to late Pleistocene corals have been mainly recovered from the Gulf of California, but scarcely (just two outcrops) from western Mexico and Central America. From the Paleocene to Recent, 191 reef-building species, including living taxa without fossil representatives, have inhabited the eastern Pacific region. Of the 53 identified genera, 11 are living and 42 are regional (i.e., currently restricted to the Indo-Pacific or Caribbean and western Atlantic) or globally extinct; 49 species are living and 142 are extinct. Fourteen of the 48 living EP species have fossil records that extend back to the early (*Porites panamensis*) and middle (*Pocillopora capitata*) Pliocene, while the remainder appeared during the middle to late Pleistocene in the Gulf of California. Considering the age and number of taxa per assemblage, genera and species ranged from 1 to 21 (mean = 2.98 ± 3.75 SD), and 1–39 (mean = 3.87 ± 6.12 SD) respectively. The highest numbers of genera (>11) and species (>12) correspond with the middle Eocene to early Miocene of Central Chiapas, and the late Eocene of Panama. As such, EP coral communities and reefs probably never attained high species richness; in addition, depositional evidence and paleoenvironmental reconstructions of coral-bearing deposits from Washington-Seattle, Central California and the Gulf of California suggest that from late Paleocene to late Pleistocene, reefs were small and paucispecific as are today's formations. Based on coral occurrences, species richness increased during the Late Eocene/Early Oligocene (55 species), decreased during the early and middle Pleistocene (5 species), rose again during the late Pleistocene (13 species) and finally peaked again during Recent times (46 species). Bootstrap tests of the temporal changes in species richness indicate that except for the numbers recorded in Late Eocene-Early Oligocene, Late Pleistocene and Recent, changes elsewhere were non-significant. From the spatio-temporal distribution of the fossil outcrops, our knowledge regarding events directly related to the origin and evolution of the current coral fauna is biased since it represents the evolutionary history of subtropical coral communities and reefs from the Gulf of California. Quantitative analysis of presence/absence coral data suggests

A. López-Pérez (✉)
Laboratorio de Ecosistemas Costeros, Departamento de Hidrobiología, UAM-Iztapalapa, San Rafael Atlixco 186, Col. Vicentina, Iztapalapa, Ciudad de México 09340, México
e-mail: alopez@xanum.uam.mx

P.W. Glynn et al. (eds.), *Coral Reefs of the Eastern Tropical Pacific*, Coral Reefs of the World 8, DOI 10.1007/978-94-017-7499-4_2

that during the last six million years, Gulf coral communities and reefs experienced dramatic turnover, in particular (a) the extinction of Caribbean-related regional endemics, and (b) since the middle Pleistocene, the steady arrival of Indo-Pacific taxa likely via the North Equatorial Counter Current. Lastly, morphologic, electrophoretic and molecular analyses clearly support a strong affinity of the eastern Pacific endemics, *Porites panamensis* and *Porites sverdrupi*, with the Caribbean and western Atlantic *Porites* clades, suggesting that the eastern Pacific poritid fauna resulted from local speciation of Caribbean/Atlantic populations and long-distance dispersal of Indo-Pacific taxa (e.g., *Porites australiensis*, *Porites lutea*). Hence, the vicariance and dispersal hypotheses regarding the origin of the eastern Pacific coral fauna, at least for some taxa, are complementary.

Keywords

Pacific American corals • Paleoecology • Faunal turnover • Dispersal and vicariance • Evolution

2.1 Introduction

Modern coral reefs in the eastern Pacific extend from the Gulf of California, Mexico to Ecuador, and the oceanic islands off western Mexico, Costa Rica, Colombia, Ecuador, and Chile (Cortés 2003). Yet, their distribution, composition and functioning have been intricately crafted by modern and ancient processes. Dana (1975) claimed that the modern eastern Pacific coral fauna may have resulted from the trans-Pacific dispersal of Indo-Pacific taxa. Today it is well known that ocean currents are associated with ongoing gene flow among populations of trans-Pacific species across a variety of taxonomic groups (e.g., echinoids, Lessios et al. 2003; fishes, Lessios and Robertson 2006; corals, Combosch et al. 2008; see Chap. 16, Lessios and Baums), but also with the potential of bringing new taxa into the eastern Pacific via the North Equatorial Counter Current during very strong ENSO events, as likely occurred during El Niño 1982–83 (molluscs, Emerson and Chaney 1995; echinoids, Lessios et al. 1996; fishes, Robertson and Allen 1996; and references therein). It is fair to say that the genetic similarity and introductions of many species are well in agreement with Dana's hypothesis.

By contrast, while much has been learned from living systems, many questions remain about how pre-Holocene events shaped modern coral communities and reefs in the eastern Pacific. Two facts are undeniable: (1) the pre-Pliocene reef coral faunistic similarity between the Pacific and Caribbean-Atlantic side of America (Vaughan 1917), and (2) the current faunistic similarity between the Indo-west and eastern Pacific regions (Squires 1959; Dana 1975; Glynn and Ault 2000; Glynn et al. 2007). However, when, where and how this change happened needs to be properly addressed in light of recent published data and current research. In order to reach this goal, this chapter will address several related issues: (1) an analysis of the large-scale evolutionary changes experienced by the coral fauna in the eastern Pacific over the last 60 million years up until the late Pleistocene, (2) a description and characterization of the composition and distribution of early (Late Miocene to Pleistocene) eastern Pacific reefs, and (3) a review of the role of dispersal and vicariance in shaping reef faunas following eastern Pacific isolation.

2.2 Large-Scale Temporal Patterns of the Eastern Pacific Coral Fauna

Seminal papers regarding large-scale spatio-temporal patterns of the eastern Pacific coral fossil fauna were published by Durham and Allison (1960), Durham (1966), Dana (1975), Heck and McCoy (1978), Colgan (1990), and Cortés (1997); few details apart, the studies addressed geographic affinities at a large spatial scale (i.e., eastern Pacific vs. Caribbean vs. Indo-Pacific), used epochs (i.e., Paleocene, Eocene) as the finest temporal resolution, and genera as the lowest taxonomic resolution. In an attempt to increase spatial, temporal and taxonomic resolution of the data, detailed databases from material collected at nearly 250 sites on the Pacific side of America (see López-Pérez 2005, 2008, 2012, for detailed information) were assembled (Tables 2.1 and 2.2) with the aid of recent published (López-Pérez 2005, 2008, 2012; Baron-Szabo 2006, 2008; Carreño and Smith 2007; López-Pérez and Budd 2009; Reyes-Bonilla et al. 2012; Paleobiology database http://paleodb.org/; and references therein) and unpublished information.

Table 2.1 summarizes the relative and/or absolute age of fossil assemblages, the geographic areas where corals have been recorded in the eastern Pacific during the Cenozoic, and major sources of information. It is apparent that from the late Paleocene to late Miocene fossil outcrops were spatially restricted to the

Table 2.1 Stratigraphic unit and relative/absolute ages of the 71 fossil assemblages of Tertiary and Quaternary eastern Pacific reef corals

Loc. ID	Stratigraphic unit	Formation	Geologic age	Age (Ma)	Source
1	Central California, CA	Martinez	L. Paleocene	58.7–55.8[o]	Gabb (1864, 1869), Dickerson (1914), Faustino (1931)
2	Mount Diablo, CA	Meganos	E. Eocene	55.8–48.6[m]	Dickerson (1914), Vaughan (1919)
3	Salem, Oregon	Capay	E. Eocene	55.8–48.6[m]	Baldwin (1964)
4	Kern County, CA	Domengine	E/M. Eocene	50–48[s]	Durham (1942b, 1943)
5	San Ignacio-San Juanico, BCS	Bateque	M. Eocene	48.6–40.4[n]	Squires and Demetrion (1992)
6	Central California, CA	Tejon	M. Eocene	48.6–40.4[p]	Nomland (1916)
7	Little Tar Spring, CA	Avenal	M. Eocene	48.6–40.4[q]	Wells (1940)
8	Western Washington	Crescent	M. Eocene	48.6–37.2[m]	Durham (1942a)
9	Northwest Peru	Chira	M. Eocene	46–43[r]	Vaughan (1922)
10	Central Chiapas	San Juan	M. Eocene	46–40[a]	Frost and Langenheim (1974)
11	Washington	Lincoln	L. Eocene/E. Oligocene	37.2–28.4[m]	Durham (1944)
12	Lake Alahuela, Panama	Gatuncillo	L. Eocene	37.2–33.9	Woodring (1957), Budd et al. (1992)
13	Brito, Nicaragua	Brito	L. Eocene	37.2–33.9	Vaughan (1919)
14	Central Chiapas	Ixtaclum	L. Eocene	36–33[a]	Frost and Langenheim (1974)
15	Contra Costa County, CA	–	E. Oligocene	33.9–28.4[m]	Nomland (1916)
16	Central Chiapas	Rancho Berlin	E. Oligocene	32–28[a]	Frost and Langenheim (1974)
17	Western Washington	Sooke	L. Oligocene	28.4–23[m]	Durham (1942a)
18	Seattle, Washington	Blakeley	L. Oligocene	28.4–23[m]	Durham (1942a, 1944)
19	Gatun Lake, Panama	Caimito	L. Oligocene	28.4–23	Woodring (1957)
20	Central Chiapas	La Quinta	L. Oligocene	28–24[a]	Frost and Langenheim (1974)
21	Canal Zone, Panama	La Boca	E. Miocene	23–16	Woodring (1957)
22	Central Chiapas	Santa Ana	E. Miocene	22–15[a]	Frost and Langenheim (1974)
23	Washington/Oregon	Temblor	E/M. Miocene	20.4–13.8[u]	Nomland (1917)
24	Nazca Ridge Guyot, Southern Pacific	–	M. Miocene	16–11.6	Allison et al. (1967), Durham (1980)
25	Valiente Peninsula, Panama	Valiente	M. Miocene	16–11	Coates et al. (2005)
26	Isla Montserrat, BCS	Carmen	L. Miocene/E. Pliocene	8–6[v]	López-Pérez (2008), López-Pérez and Budd (2009)
27	Imperial Valley, CA	Imperial	L. Miocene/E. Pliocene	6.5–6[b, 1]	Vaughan (1917)
28	Los Algodones, BCS	El Refugio	E. Pliocene	5.3–3.6[v]	López-Pérez (2008), López-Pérez and Budd (2009)
29	Punta Chivato, BCS	San Marcos	L. Pliocene	3.6–2.6[v]	Simian and Johnson (1997), Johnson and Ledezma-Vazquez (2001)
30	Isla María Madre, Islas Marías, Nayarit	–	L. Pliocene	3.6–2.6[v]	Jordan and Hertlein (1926), Hertlein and Emerson (1959)
31	Central California	San Diego	L. Pliocene	3.6–2.6[m]	Hertlein and Grant (1960)
32	Isla San José, BCS	–	E-M Pliocene	>3[c]	López-Pérez (2008), López-Pérez and Budd (2009)
33	San Nicolas, BCS	San Nicolas	M-L Pliocene	<3.3–2[d, v]	López-Pérez (2008), López-Pérez and Budd (2009)
34	Las Barracas/B. Marquer, BCS	Marquer	L. Pliocene	3–2[e, v]	Durham (1950a), López-Pérez (2008), López-Pérez and Budd (2009)
35	Baltra Island, Galápagos Islands	–	L. Pliocene/Pleistocene	–	Hertlein (1972), Durham (1980)
36	Isla Coronados-1, BCS	–	Pleistocene	–	López-Pérez (2008), López-Pérez and Budd (2009)

(continued)

Table 2.1 (continued)

Loc. ID	Stratigraphic unit	Formation	Geologic age	Age (Ma)	Source
37	La Ventana-1, BCS	–	Pleistocene	–	López-Pérez (2008), López-Pérez and Budd (2009)
38	Isla Coronados-2, BCS	–	Pleistocene	–	López-Pérez (2008), López-Pérez and Budd (2009)
39	La Ventana-2, BCS	–	Pleistocene	–	López-Pérez (2008), López-Pérez and Budd (2009)
40	Isla Coronados-3, BCS	–	Pleistocene	–	López-Pérez (2008), López-Pérez and Budd (2009)
41	La Ventana-3, BCS	–	Pleistocene	–	López-Pérez (2008), López-Pérez and Budd (2009)
42	Isla Coronados-4, BCS	–	Pleistocene	–	López-Pérez (2008), López-Pérez and Budd (2009)
43	Isla Coronados-5, BCS	–	Pleistocene	–	López-Pérez (2008), López-Pérez and Budd (2009)
44	La Ventana-4, BCS	–	L. Pleistocene	>0.3[f]	López-Pérez (2008), López-Pérez and Budd (2009)
45	Isla Coronados-6, BCS	–	L. Pleistocene	–	López-Pérez (2008), López-Pérez and Budd (2009)
46	Arroyo Blanco, I. Carmen	–	L. Pleistocene	–	López-Pérez (2008), López-Pérez and Budd (2009)
47	Cabo Pulmo, BCS	–	L. Pleistocene	0.14–0.12[g]	Squires (1959), López-Pérez (2008), López-Pérez and Budd (2009)
48	Cañada Coronados, BCS	–	L. Pleistocene	0.127 ± .006[h]	Durham (1950b), López-Pérez (2008), López-Pérez and Budd (2009)
49	Puerto Balandra, I. Carmen	–	L. Pleistocene	0.117 ± .008[i]	Durham (1950b), López-Pérez (2008), López-Pérez and Budd (2009)
50	Timbabichi, BCS	–	L. Pleistocene	–	Squires (1959), López-Pérez (2008), López-Pérez and Budd (2009)
51	La Ventana-5, BCS	–	L. Pleistocene	–	López-Pérez and Budd (2009)
52	Punta Baja, I. Carmen	–	L. Pleistocene	0.117 ± .008[i]	Durham (1950b), López-Pérez (2008), López-Pérez and Budd (2009)
53	Bahía Oto, I. Carmen	–	L. Pleistocene	0.117 ± .008[i]	Durham (1950b), López-Pérez (2008), López-Pérez and Budd (2009)
54	Las Animas, BCS	–	L. Pleistocene	0.13–0.128[j]	De Diego-Forbis et al. 2003, López-Pérez (2008), López-Pérez and Budd (2009)
55	San Antonio, BCS	–	L. Pleistocene	0.13–0.12[k]	Johnson and Ledesma-Vázquez (1999)
56	South Punta Chivato, BCS	–	L. Pleistocene	0.117 ± .008[i]	Johnson and Ledesma-Vázquez (2001)
57	El Bajo, Loreto	–	L. Pleistocene	–	Johnson et al. (2007)
58	Isla San José, BCS	–	L. Pleistocene	–	Squires 1959
59	Isla San Diego, BCS	–	L. Pleistocene	–	Squires (1959)
60	Isla María Madre, Islas Marías, Nayarit	–	L. Pleistocene	–	Squires (1959)
61	Isla Cerralvo, BCS	–	L. Pleistocene	–	Squires (1959)
62	Puerto Escondido, Oaxaca	–	L. Pleistocene	–	Palmer (1928), Durham (1947)
63	Bahía Santa Inez, BCS	–	L. Pleistocene	0.117 ± .008[i]	Durham (1947), Hertlein (1957)
64	Isla María Cleofas, Islas Marías, Nayarit	–	L. Pleistocene	–	Hertlein and Emerson (1959), Squires (1959)
65	Isla Guadalupe, BC	–	L. Pleistocene	–	Durham (1980)
66	Isla Tiburón, Sonora	–	L. Pleistocene	–	Durham (1947, 1950b)
67	Isla San Marcos, BCS	–	L. Pleistocene	0.117 ± .008[i]	Durham (1947, 1950b)
68	Isla Monserrat, BCS	–	L. Pleistocene	–	Durham (1947, 1950b)
69	Bahía Magdalena, BCS	–	L. Pleistocene	–	Durham (1947)
70	Isla Coronados-8, BCS	–	L. Pleistocene	–	López-Pérez (2008), López-Pérez and Budd (2009)
71	Isla Coronados-9, BCS	–	L. Pleistocene	–	López-Pérez (2008), López-Pérez and Budd (2009)

Error terms are standard deviations
[a]Budd (2000), [b]Eberly et al. (1978), [c]Schwennicke (2005), [d]Ledesma-Vázquez (2002), [e]Dorsey et al. (2001), [f]Sirkin et al. (1990), [g]Muhs et al. (1994), [h]Johnson (2005), [i]Libbey et al. (1997), [j]DeDiego-Forbis et al. (2004), [k]Johnson et al. (1999), [l]McDougall et al. (1999), [m]Paleobiology database, [n]Schweitzer et al. (2006), [o]Brabb et al. (2008), [p]Lindberg and Squires (1990), [q]Dibblee (1973), [r]Kulm et al. (1990), [s]Squires (1997), [t]Prothero et al. (2009), [u]Bridges and Castle (2003), [v]Carreño and Smith (2007)

Washington-Seattle and California regions, whereas from the early Pliocene to late Pleistocene corals have been mainly recovered from the Gulf of California (see Fig. 1 in López-Pérez 2005). Nevertheless, aside from isolated late Pleistocene fossil coral occurrences at Guadalupe Island (Durham 1980), and Puerto Escondido (Palmer 1928) in western Mexico, and Baltra Island (Hertlein 1972; Durham 1980) in the Galápagos Islands, there is a lack of deposits in western Mexico and Central America. Also, another set of coral-bearing deposits has been recovered from central Chiapas (Eocene-Miocene: La Quinta Fm., Ixtaclum Shale Fm., San Juan Fm., Lecheria Fm., Rancho Berlin Fm., Río Lajas Fm., Santa Ana Fm., Balumtun Fm., Frost and Langenheim 1974), Nicaragua [Eocene: Brito Fm., Vaughan 1919), Costa Rica (Miocene: Limon Fm., Vaughan 1919), Panama (Eocene: Gatuncillo Fm. (Budd et al. 1992), Oligocene: Caimito Fm., Miocene: La Boca Fm., (Woodring 1957); Valiente Fm., (Coates et al. 2005)] and Colombia (Miocene: Usiacuri, Vaughan 1919). These units were deposited at the thermal equator during Eocene to Miocene times and, according to paleocoordinates (http://paleodb.org/), not in the eastern Pacific per se but rather midway between the Pacific and Caribbean Seas.

Whether these Central American assemblages formed part of the Cenozoic eastern Pacific fossil fauna deserves further scrutiny. For example, faunistic affinity between northwestern and Central America during Eocene to Miocene times can be considered relatively high at the genus level (32 %, n = 53) when compared with the low percentage of shared taxa (5.7 %, n = 11) at the species level. Regional species-level faunistic differences between northwestern and Central America are not unexpected; however, the extent to which this regional pattern is true or the result of poor taxonomy (Baron-Szabo 2006, 2008) needs to be carefully addressed, especially because Central American isthmus models suggest that, with an open passage, as much as 10^6 m^3 s^{-1} of surface water would pass from the Atlantic to the Pacific (Hay et al. 2002). This passage would connect Atlantic and Central American communities with northwestern (Seattle, Washington, California, and the Gulf of California, Mexico) communities and produce a continuous marine province and therefore a much more homogeneous coral fauna. Contrary to this view, recent geologic reconstructions (Kirby et al. 2008; Montes et al. 2012), based on new lithostratigraphic, biostratigraphic and strontium chemostratigraphic analyses, suggest that ocean circulation and the biogeographic connection between the Pacific and Caribbean was much more constricted in the early Neogene (Eocene to Miocene) than previously understood, and therefore relatively large faunistic differences between coral faunas may have occurred.

Table 2.2 depicts the systematic list of the 191 reef-building species, including living taxa without fossil representatives that inhabited the eastern Pacific (including

Table 2.2 Species list of Cenozoic hermatypic corals, eastern Pacific

Order Scleractinia
Family Acroporidae
Genus *Acropora*
†*Acropora panamensis* Vaughan 1919 (20)
†*Acropora saludensis* Vaughan 1919 (22)
Acropora valida (Dana 1846) (L)
Genus *Alveopora* (L)
†*Alveopora chiapanacae* Frost and Langenheim 1974 (20)
Genus *Astreopora* (L)
†*Astreopora antiguensis* Vaughan 1919 (19)
†*Astreopora duwamishensis* Durham 1942a (8)
†*Astreopora esperanzae* Frost and Langenheim 1974 (10)
†*Astreopora gatuncilloensis* Budd et al. 1992 (12)
†*Astreopora occidentalis* Nomland 1917 (23)
†*Astreopora sanjuanensis* Durham 1942b (4, 8)
†*Astreopora spinulosum* Frost and Langenheim 1974 (14)
Genus *Dendracis* (†)
†*Dendracis cantabrigiensis* Vaughan 1899 (12)
Genus *Montipora* (L)
†*Montipora schencki* Durham 1942a (8)
Family Actinacididae
Genus *Actinacis* (†)
†*Actinacis alabamensis* (Vaughan 1900) (12)
†*Actinacis barretti* Wells 1934 (10, 20)
Family Agariciidae
Genus *Cyathoseris* (†)
†*Cyathoseris allisoni* Frost and Langenheim 1974 (10)
†*Cyathoseris formosa* d'Achiardi 1875 (12)
Genus *Gardineroseris* (L)
Gardineroseris planulata (Dana 1846) (44, L)
Genus *Leptoseris* (L)
Leptoseris papyracea (Dana 1846) (L)
†*Leptoseris portoricensis* Vaughan 1919 (20)
Leptoseris scabra Vaughan 1907 (L)
†*Leptoseris sinuata* (Nomland 1916) (3)
Leptoseris solida (Quelch 1886) (L)
Genus *Pavona* (L)
Pavona chiriquiensis Glynn et al. 2001 (L)
Pavona clavus (Dana 1846) (37, 39, 41, 44, 48, 51, L)
Pavona duerdeni Vaughan 1907 (L)
Pavona frondifera (Lamarck 1816) (L)
Pavona gigantea Verrill 1869 (35, 42, 48, L)
Pavona maldivensis (Gardiner 1905) (L)
Pavona minuta Wells 1954 (L)
†*Pavona modeloensis* Frost and Langenheim 1974 (22)

(continued)

Table 2.2 (continued)

Order Scleractinia
†*Pavona trinitatis* Vaughan and Hoffmeister 1926 (22)
Pavona varians Verrill 1864 (L)
Pavona xarifae Scheer and Pillai 1974 (L)
Genus *Trochoseris* (†)
†*Trochoseris aperta* Duncan 1864 (14)
†*Trochoseris gerthi* Wells 1941 (9)
Family Agathiphylliidae
Genus *Agathiphyllia* (†)
†*Agathiphyllia antiguensis* (Duncan 1863) (12, 16)
†*Agathiphyllia anguillensis* (Vaughan 1919) (20)
†*Agathiphyllia browni* (Vaughan 1919) (12)
†*Agathiphyllia hilli* (Vaughan 1919) (16, 20)
†*Agathiphyllia robusta* Budd et al. 1992 (12)
†*Agathiphyllia roxboroughi* (Vaughan 1919) (20, 22)
†*Agathiphyllia splendens* (Vaughan 1919) (12)
Genus *Pattalophyllia* (†)
†*Pattalophyllia grumi* Catullo 1852 (10)
Family Astrocoeniidae
Genus *Astrocoenia* (†)
†*Astrocoenia decaturensis* Vaughan 1919 (12)
†*Astrocoenia dilloni* Durham 1942b (2)
†*Astrocoenia guantanamensis* Vaughan 1919 (10)
†*Astrocoenia jukesbrownei* Wells 1945 (12)
†*Astrocoenia portoricensis* Vaughan 1919 (16)
Genus *Madracis* (L)
†*Madracis crescentensis* Durham, 1942 (8)
†*Madracis duncani* Wells 1945 (12)
†*Madracis stewarti* Durham, 1942 (8)
†*Madracis vaughani* Wells, 1941 (9, 12)
†*Madracis wellsi* Bentson 1943 (4)
Genus Stephanocoenia (L)
†*Stephanocoenia storrsi* Wells 1941 (9)
Family Calamophylliidae
Genus *Antilloseris* (†)
†*Antilloseris cantabrigiensis* (Vaughan 1899) (10,14)
†*Antilloseris vaughani* Wells 1940 (7)
Family Columnastreidae
Genus *Haimesastrea* (†)
†*Haimesastrea conferta* Vaughan 1900 (9,10)
†*Haimesastrea petrosa* (Gabb 1864) (1)
Family Faviidae
Genus *Antiguastrea* (†)
†*Antiguastrea cellulosa* (Duncan 1863) (12, 16, 20)
†*Antiguastrea prava* Budd et al. 1992 (12)
Genus *Colpophyllia* (L)

(continued)

Table 2.2 (continued)

Order Scleractinia
†*Colpophyllia duncani* Wells 1935 (12)
†*Colpophyllia elegans* Budd et al. 1992 (12)
†*Colpophyllia mexicanum* Frost and Langenheim 1974 (20, 22)
†*Colpophyllia nicholasi* Squires and Demetrion 1992 (5)
†*Colpophyllia reagani* Durham 1942a (8)
†*Colpophyllia willoughbiensis* (Vaughan 1919) (12,16, 20)
Genus *Diploastrea* (L)
†*Diploastrea aequalis* Budd et al. 1992 (12)
†*Diploastrea bacata* Budd et al. 1992 (12)
†*Diploastrea crassolamellata* (Duncan 1863) (12,16, 20)
Genus *Diploria* (L)
†*Diploria bowersi* (Vaughan 1917) (27, 32)
†*Diploria sarasotana* Weisbord 1974 (29)
†*Diploria wellsi* (Durham 1942b) (4)
Genus *Favia* (L)
†*Favia dominicensis* Vaughan and Hoffmeister 1925 (22)
†*Favia favioides* (Wells 1945) (12)
†*Favia hannai* Durham 1942b (4)
†*Favia maitreyiae* López-Pérez 2012 (26)
†*Favia tulsidasi* López-Pérez 2012 (29)
†*Favia weisbordi* Wells 1934 (12)
Genus *Favites* (L)
†*Favites crassiseptata* Stemann 1992 (12)
†*Favites magnificata* Stemann 1992 (12)
†*Favites polygonalis* (Duncan 1863) (20)
Genus *Goniastrea* (L)
†*Goniastrea canalis* Vaughan 1919 (12, 20)
†*Goniastrea longula* Budd et al. 1992 (12)
†*Goniastrea muralis* Budd et al. 1992 (12)
Genus *Leptastrea* (L)
†*Leptastrea hertleini* Durham 1942b (4)
Leptastrea transversa Klunzinger, 1879 (L)
Leptastrea purpurea (Dana 1846) (L)
Genus *Leptoria* (L)
†*Leptoria laxa* Budd et al. 1992 (12)
†*Leptoria pauca* Budd et al. 1992 (12)
Genus *Montastraea* (L)
†*Montastraea antilliana* Wells 1945 (10)
†*Montastraea bainbridgensis* (Vaughan 1919) (16, 20)
†*Montastraea canalis* (Vaughan 1919) (25)
†*Montastrea costata* (Duncan 1863) (21)
†*Montastrea eocenica* Budd et al. 1992 (12)
†*Montastraea gabbi* (Vaughan 1919) (20)
†*Montastrea imperatoris* (Vaughan 1919) (25)
†*Montastraea laurae* Squires and Demetrion 1992 (5)

(continued)

Table 2.2 (continued)

Order Scleractinia
†*Montastrea nodosa* Budd et al. 1992 (12)
†*Montastrea prima* Budd et al. 1992 (12)
†*Montastrea rotunda* Budd et al. 1992 (12)
†*Montastraea tampaensis* (Vaughan 1919) (20,22)
Genus *Oulophyllia* (L)
†*Oulophyllia californica* Durham 1942b (4)
Genus *Plesiastrea* (L)
Plesiastrea versipora (Lamarck 1816) (24, L)
Genus *Solenastrea* (L)
†*Solenastrea fairbanksi* (Vaughan 1900) (27–30, 32)
Family Fungiidae
Genus *Cycloseris* (L)
Cycloseris curvata (Hoeksema 1989) (L)
Cycloseris distorta (Michelin 1842) (L)
Cycloseris vaughani (Boschma 1923) (L)
Family Meandrinidae
Genus *Dichocoenia* (L)
†*Dichocoenia eminens* Weisbord 1974 (27, 29)
†*Dichocoenia merriami* (Vaughan 1900) (27, 29, 32, 34)
Genus *Eusmilia* (L)
†*Eusmilia bainbridgensis* Durham 1942a (18)
†*Eusmilia carrizensis* Vaughan 1917 (27)
Genus *Meandrina* (L)
†*Meandrina alahuelensis* Budd et al. 1992 (12)
†*Meandrina antiguensis* (Vaughan 1919) (16)
Family Montlivaltiidae
Genus *Placosmilia* (†)
†*Placosmilia copoyensis* Frost and Langenheim 1974 (10)
†*Placosmilia*? *aliciae* López-Pérez 2012 (29, 33)
Family Mussidae
Genus *Cyclomussa* (†)
†*Cyclomussa concinna* Wells 1941 (9)
Genus *Leptomussa* (†)
†*Leptomussa mexicanae* Frost and Langenheim 1974 (14)
Genus *Mycetophyllia* (L)
†*Mycetophyllia bullbrooki* Vaughan and Hoffmeister 1926 (16)
Genus *Syzygophyllia* (†)
†*Syzygophyllia hayesi* Vaughan 1919
Genus *Trachyphyllia* (L)
†*Trachyphyllia batequensis* (Squires and Demetrion 1992) (5)
†*Trachyphyllia bullbrooki* (Hoffmeister 1926) (20, 22)
†*Trachyphyllia clarki* (Durham 1946) (14)
† *Trachyphyllia hadleyi* (Wells 1934) (12)
†*Trachyphyllia heueri* (Frost and Langenheim 1974) (20)
†*Trachyphyllia sawkinsi* (Vaughan 1926) (20, 22)

(continued)

Table 2.2 (continued)

Order Scleractinia
Family Pocilloporidae
Genus *Pocillopora* (L)
Pocillopora capitata Verrill 1864 (26, 40, 42–45, 47, 50–52, 54, 68, 71, L)
Pocillopora damicornis (Linnaeus 1758) (47, 50, 51, 54, L)
Pocillopora danae Verrill 1864 (L)
Pocillopora effusus Veron 2000 (L)
Pocillopora elegans Dana 1846 (42, 50, 61, L)
Pocillopora grandis Dana 1846 (L)
†*Pocillopora guadalupensis* Durham 1980 (65)
Pocillopora inflata Glynn 1999 (L)
Pocillopora ligulata Dana 1846 (L)
†*Pocillopora maddenensis* Budd et al. 1992 (12)
Pocillopora meandrina Dana 1846 (47, L)
Pocillopora verrucosa (Ellis and Solander 1786) (50, L)
Pocillopora woodjonesi Vaughan 1918 (L)
Genus *Stylophora* (L)
†*Stylophora cambridgensis* Wells 1934 (10)
†*Stylophora chaneyi* Durham 1942b (4, 5)
†*Stylophora chiraensis* Wells 1941 (9)
†*Stylophora goethalsi* Vaughan 1919 (19)
†*Stylopora granulata* Duncan 1864 (25)
†*Stylopora imperatoris* Vaughan 1919 (19)
†*Stylophora lindavistae* Frost and Langenheim 1974 (20)
Family Poritidae
Genus *Goniopora* (L)
†*Goniopora cascadensis* Vaughan 1919 (21)
†*Goniopora elegans* (Leymerie, 1846) (9)
†*Goniopora hilli* Vaughan 1919 (16, 20)
†*Goniopora imperatoris* Vaughan 1919 (5, 6, 20, 22)
†*Goniopora reussiana* (Duncan, 1865) (10, 12)
Genus *Porites* (L)
Porites arnaudi Reyes-Bonilla and Carricart-Ganivet 2000 (L)
Porites australiensis Vaughan 1918 (L)
†*Porites baracoaensis* Vaughan 1919 (20, 22)
Porites baueri Squires 1959 (L)
†*Porites carrizensis* Vaughan 1917 (26, 27, 29, 34)
†*Porites douvillei* Vaughan 1919 (16, 20)
Porites lichen Dana 1846 (L)
Porites lobata Dana 1846 (47, L)
Porites lutea Milne Edwards and Haime 1860 (L)
†*Porites macdonaldi* Vaughan 1919 (20)
Porites panamensis Verrill 1866 (26, 29, 34, 36–60, 62–64, 66, 67, 69–71, L)
†*Porites portoricencis* (Vaughan 1919) (20)
Porites rus (Forskal, 1775) (L)

(continued)

Table 2.2 (continued)

Order Scleractinia
Porites sverdrupi Durham 1947 (52, 53, 58, 60, 63, 64, 66, L)
†*Porites trinitatis* Vaughan and Hoffmeister 1926 (20, 22)
†*Porites waylandi* Foster 1986 (20, 25)
Family Rhipidogyridae
Genus Barysmilia (†)
†Barysmilia aenigma (Budd et al. 1992) (12)
Family Siderastreidae
Genus *Coscinaraea* (L)
†*Coscinaraea colei* Frost and Langenheim 1974 (22)
†*Coscinaraea panamensis* Stemann 1992 (12)
Genus *Pironastrea* (†)
†*Pironastrea antiguensis* Vaughan 1919 (20)
Genus *Psammocora* (L)
Psammocora brighami (Vaughan 1907) (L)
Psammocora contigua (Esper, 1797) (L)
Psammocora haimeana Milne Edwards and Haime 1851 (L)
Psammocora obtusangula Lamarck 1816 (L)
Psammocora profundacella Gardiner 1898 (L)
Psammocora stellata (Verrill 1866) (39, 47, 49, 65, L)
Psammocora superficialis Gardiner 1898 (L)
Genus *Siderastrea* (L)
†*Siderastrea annae* López-Pérez 2012 (29)
†*Siderastrea clarki* Nomland 1916 (15)
†*Siderastrea conferta* (Duncan, 1863) (16, 20, 21, 31)
Siderastrea glynni Budd and Guzmán 1994 (L)
†*Siderastrea mendenhalli* Vaughan 1917 (27)
†*Siderastrea radcliffi* Faustino 1931 (1)
†*Siderastrea scotia* Wells 1945 (10)
Siderastrea siderea (Ellis and Solander 1786) (20, 22, L)
†*Siderastrea vancouverensis* Vaughan 1923 (11, 17)
†*Siderastrea washingtonensis* Durham 1942a (18)
Genus *Sideroseris* (†)
†*Sideroseris durhami* Wells 1945 (10)
Family Synastreidae
Genus *Leptophyllastrea* (†)
†*Leptophyllastrea vaughani* Durham 1942a (8)
Family Thamnasteriidae
Genus *Thamnasteria* (†)
†*Thamnasteria sinuata* Nomland 1916 (6)

Valid names according to Foster (1986), Budd (1987, 1991), Budd and Johnson (1999), Reyes-Bonilla (2002), Baron-Szabo (2006, 2008), López-Pérez (2008, 2012), López-Pérez and Budd (2009). Numbers in parenthesis refer to stratigraphic unit/time where corals were recovered (see Table 2.1). † = Extinct, L = Living. Families, genera and species follow alphabetic order

Eocene to Miocene Central American communities) since the late Paleocene. Of the 53 identified genera, 11 are living and 42 are regionally extinct (currently restricted to Indo-Pacific or Caribbean and western Atlantic waters), or globally extinct. Among the living genera, except for the circumtropically distributed *Acropora*, *Porites*, *Siderastrea* and *Leptoseris*, the remaining taxa are currently restricted to the Pacific Ocean (Veron 2000). Among the regionally extinct genera, *Stephanocoenia*, *Dichocoenia*, *Meandrina*, *Eusmilia*, *Mycetophyllia*, *Colpophyllia*, *Solenastrea*, and *Diploria* are currently restricted to the Caribbean and western Atlantic (Budd 2000); whereas *Astreopora*, *Montipora*, *Diploastrea*, *Favites*, *Goniastrea*, *Leptastrea*, *Leptoria*, *Oulophyllia*, *Plesiastrea*, *Fungia*, *Trachyphyllia*, *Stylophora*, *Alveopora*, *Goniopora*, *Coscinaraea* and *Psammocora* are currently restricted to the Indo-Pacific (Veron and Kelley 1988). Finally, *Madracis* and *Siderastrea*, along with members of the paraphyletic Faviidae (*Favia* and *Montastraea*) (Fukami et al. 2004), are widely distributed across the Caribbean, western Atlantic, and Indo-Pacific. The remainder of the genera are globally extinct (Table 2.2).

Of the 191 species, 49 are living and 142 are extinct (Table 2.2). Of the living species, 48 are restricted to the Pacific whereas *Siderastrea siderea*, which was recorded in the late Oligocene La Quinta Formation and early Miocene Santa Ana Formation in Central Chiapas (Frost and Langenheim 1974), currently inhabits the Caribbean and western Atlantic area (Sekar et al. 2008). Fourteen of the 48 living eastern Pacific species ($\sim$30 %) have fossil records that extend back to the early (*Porites panamensis*) and middle (*Pocillopora capitata*) Pliocene, while the remainder appeared during the middle to late Pleistocene in the Gulf of California (López-Pérez 2008; López-Pérez and Budd 2009). Among the taxa with a fossil record, several of the major reef building species in the eastern Pacific are included, such as *Pocillopora damicornis* (Fig. 2.1a), *Pocillopora meandrina* (Fig. 2.1c), *Pocillopora elegans*, *Pocillopora verrucosa* (Fig. 2.1b), *Porites lobata* (Fig. 2.1l), *Pavona gigantea*, *Gardineroseris planulata* (Fig. 2.1j) (Glynn and Ault 2000; Reyes-Bonilla 2003), the more restricted but common *Psammocora stellata*, *Pavona clavus* (Fig. 2.1h), and the Gulf of California endemic *Porites sverdrupi* (López-Pérez et al. 2003).

Considering the age and number of genera and species per assemblage, the number of genera ranged from 1 to 21 (mean = 2.98 ± 3.75 SD), whereas the species ranged from 1 to 39 (mean = 3.87 ± 6.12 SD) (Fig. 2.2 a, b). The highest numbers of genera (>11) and species (>12) correspond with the middle Eocene to early Miocene of Central Chiapas (Frost and Langenheim 1974) and the late Eocene of Panama (Budd et al. 1992) whereas the remainder of the assemblages had less than seven genera (Kern County, California; Northwestern Peru; Punta Chivato, Baja California Sur, Mexico; Table 2.1) and up to nine species (Punta Chivato; Table 2.1).

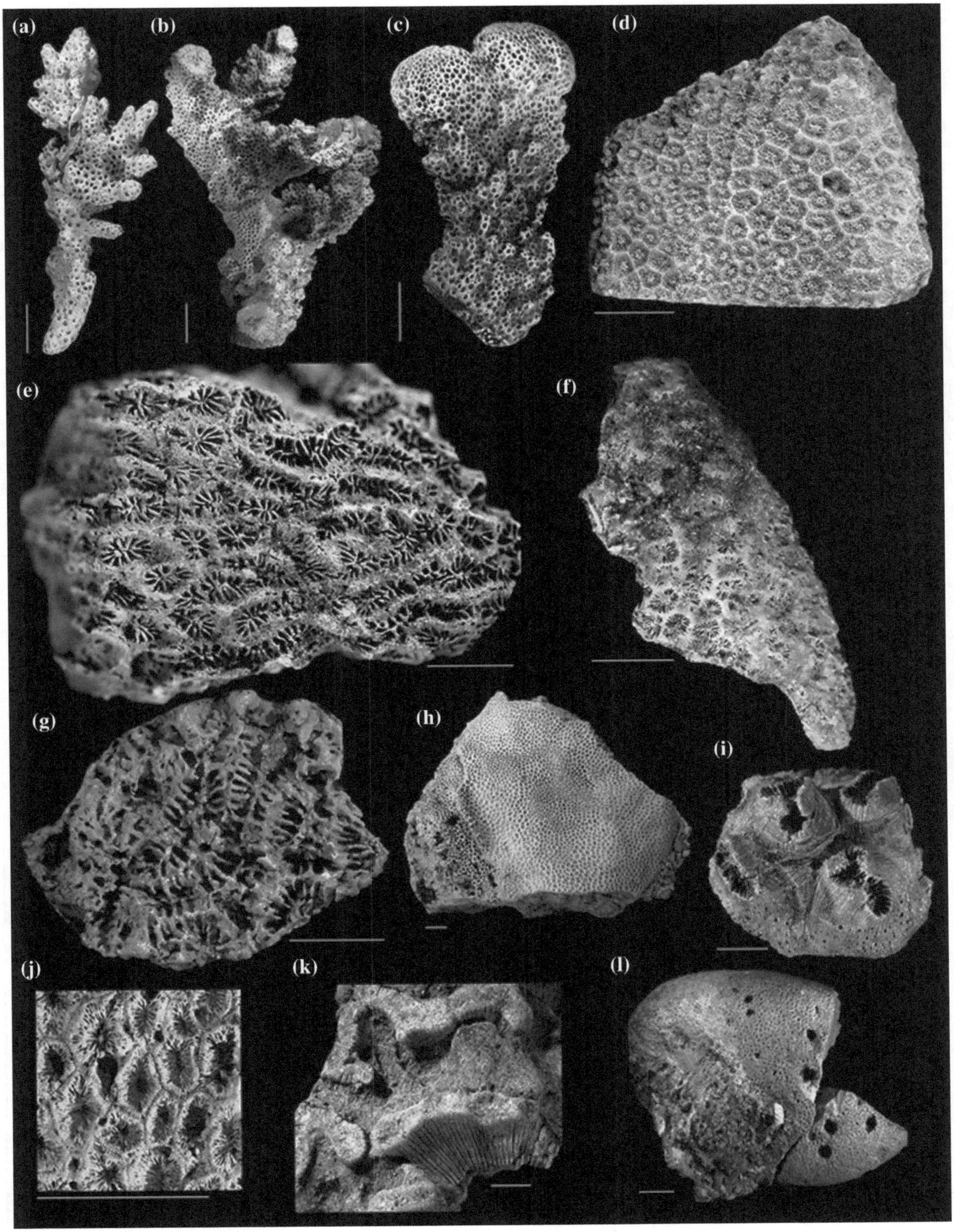
(a)
(b)
(c)
(d)
(e)
(f)
(g)
(h)
(i)
(j)
(k)
(l)

◀ **Fig. 2.1** New records of fossil corals in the Gulf of California. **a** *Pocillopora damicornis* (Linnaeus 1758). SUI 100625. Late Pleistocene, locality BC 5, Timbabichi, Baja California Sur, Mexico. **b** *Pocillopora verrucosa* (Ellis and Solander 1786). SUI 100659. Late Pleistocene, locality BC 5, Timbabichi, Baja California Sur, Mexico. **c** *Pocillopora meandrina* Dana (1846). NHMLAC 11740. Late Pleistocene, locality BC 8, Cabo Pulmo, Baja California Sur, Mexico. **d** *Siderastrea annae* López-Pérez 2012. Holotype, SUI 100674. Early Pliocene San Marcos Formation, locality BC 15 Puerto de la Lancha, Isla Carmen, Gulf of California, Mexico. **e** *Favia maitreyiae* López-Pérez 2012. Holotype SUI 100686. Middle Pliocene Carmen Formation, locality BC 27, Isla Montserrat, Gulf of California, Mexico. **f** *Favia tulsidasi* López-Pérez 2012. Paratype SUI 100687. Early Pliocene San Marcos Formation, locality BC 15, Puerto de la Lancha, Isla Carmen, Gulf of California, Mexico. **g** *Diploria sarasotana* Weisbord, 1974. SUI 100690, Early Pliocene San Marcos Formation, locality BC 35, Ensenada El Muerto, Baja California Sur, Mexico. **h** *Pavona clavus* (Dana, 1846). SUI 100866. Late Pleistocene, locality BC 10, La Ventana, Baja California Sur, Mexico. **i** *Dichocoenia eminens* Weisbord, 1974. NHMLAC 38953. Late Miocene, northeastern Coyote Mountains, locality UCLA 631, Carrizo Creek, Imperial Formation, California. **j** *Gardineroseris planulata* (Dana 1846). SUI 100660, Middle Pleistocene, locality BC 11, La Ventana, Baja California Sur, Mexico. **k** *Placosmilia*? *aliciae* López-Pérez 2012. Holotype SUI 100680. Middle Pliocene, San Nicolas Formation, locality BC 4, San Nicolas, Baja California Sur, Mexico. **l** *Porites lobata* Dana, 1846. NHMLAC 11739. Late Pleistocene, locality BC 8, Cabo Pulmo, Baja California Sur, Mexico. Scale bars = 1 cm

This suggests that assemblages other than those that developed at the thermal equator during Eocene to Miocene times, were relatively less diverse through the entire Cenozoic. In light of recent Central America geologic reconstructions (Kirby et al. 2008; Montes et al. 2012) and the low Eocene to Miocene faunistic affinity observed between northwestern and Central American assemblages (see above), it can be argued that the eastern Pacific has never experienced extensive reef building or relatively high species diversity. Depositional accounts and paleoenvironmental reconstructions of coral-bearing deposits from Seattle (Washington), Central California (Durham 1942a, b, 1950a), and the Gulf of California support instead [except for relatively large Pleistocene reef patches at Carmen and Coronado islands (Fig. 2.3a) in the Gulf (Durham 1950b; Johnson et al. 2007; López-Pérez 2008)], small low-diversity (Fig. 2.3c, d) biostromal reefs

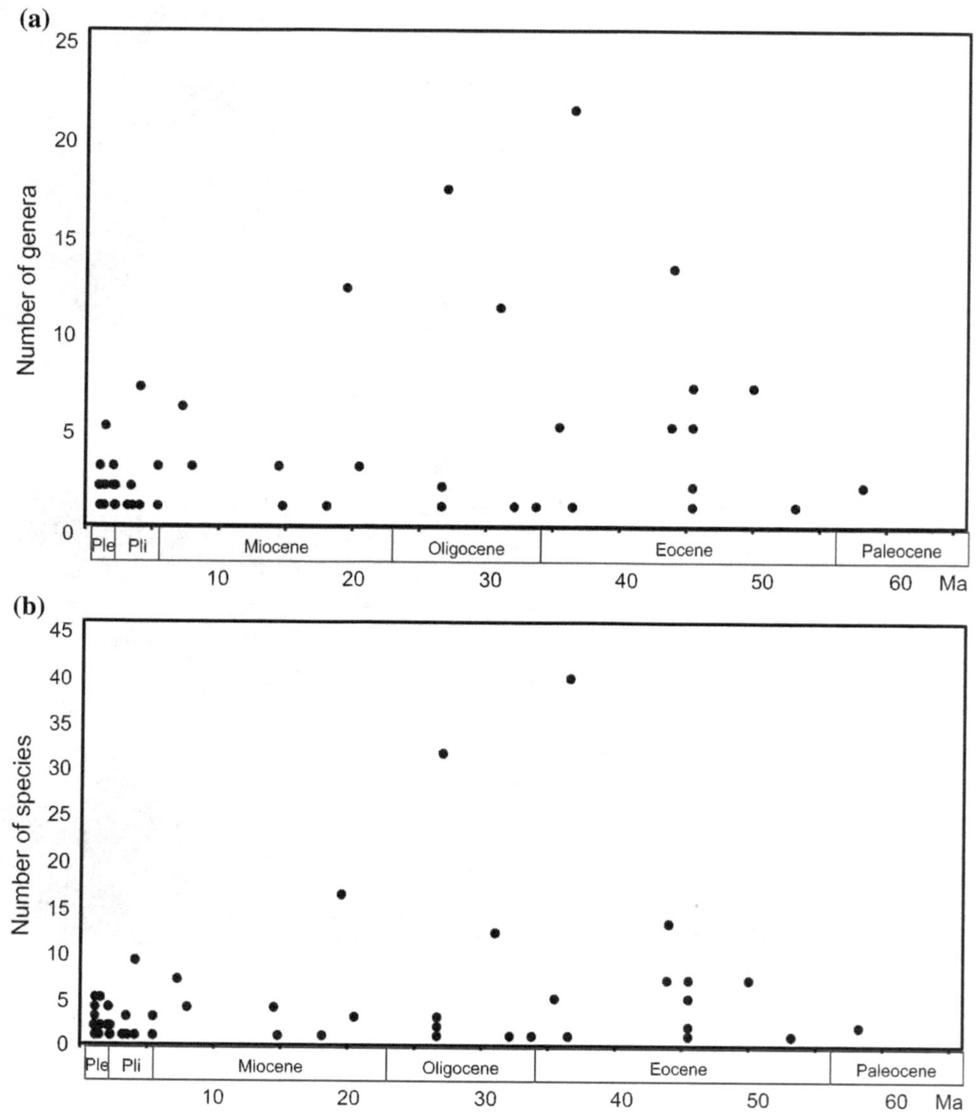

Fig. 2.2 Number of taxa recorded in 71 Cenozoic fossil assemblages, eastern Pacific region. Each point represents an assemblage and denotes locality richness. The geologic age plotted for each assemblage is the midpoint between oldest and youngest age estimates noted in Table 2.1. **a** Number of genera per assemblage, **b** number of species per assemblage

Fig. 2.3 Fossil outcrops and in situ corals in the Gulf of California, Mexico. **a** Cañada Coronados reef. Late Pleistocene *Porites panamensis* reef deposited in a large protected embayment at Isla Coronados (26°N, 111°W). **b** Punta Chivato fringing patch reef. Early Pliocene *Solenastrea fairbanksi* reef deposited in the open, exposed, high-energy environment on the north side of the Punta Chivato promontory (27°N, 111°W). **c** In situ Late Pleistocene *Porites panamensis* reef. West side of Cañada Coronados, Isla Coronados. Colonies may reach a maximum height of 1.1 m. **d** In situ Early Pliocene *Solenastrea fairbanksi* colonies. Early Pliocene reef, north side of Punta Chivato promontory (27°N, 111°W)

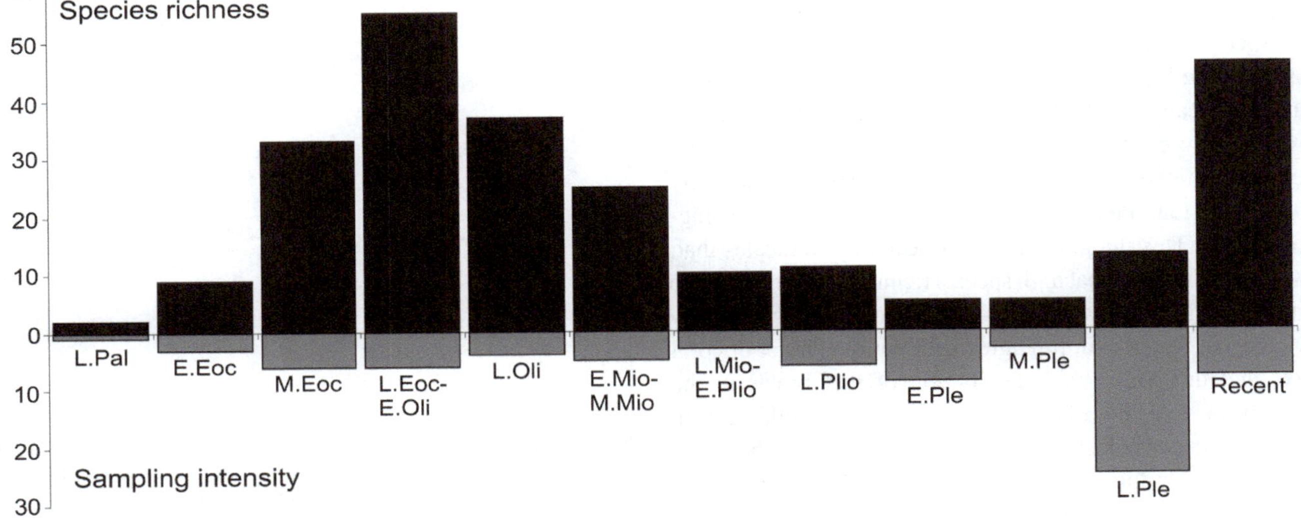

Fig. 2.4 Species richness and sampling intensity (i.e., number of sampled formations per time) of eastern Pacific coral reef species during the Cenozoic. *L.Pal* Late Paleocene, *E.Eoc* Early Eocene, *M.Eoc* Middle Eocene, *L.Eoc-E.Oli* Late Eocene-Early Oligocene, *L.Oli* Late Oligocene, *E.Mio-M.Mio* Early Miocene-Middle Miocene, *L.Mio-E.Plio* Late Miocene-Early Pliocene, *L.Plio* Late Pliocene, *E.Ple* Early Pleistocene, *M.Ple* Middle Pleistocene, *L.Ple* Late Pleistocene, *Recent* Recent time

formed on low-angle ramps or flat-lying terraces usually deposited in open, exposed, high-energy environments (Fig. 2.3b) or well preserved units deposited in small protected embayments. Although more data are needed, this information suggests that from the late Paleocene to the late Pleistocene, reefs were small and low in diversity as are today's coral communities and reefs in the eastern Pacific.

Large-scale temporal analysis suggests that, notwithstanding the well-known worldwide collapse in reef building at the end of the Cretaceous (Rosen 2000), numerous reef coral genera (73.3 %) and species (52 %) crossed the K/T boundary (Baron-Szabo 2006, 2008). Members of K/T surviving genera, *Haimesastrea petrosa* and *Siderastrea radclifi* were collected from Central California, and can be considered to be the oldest (late Paleocene) Cenozoic species known in the eastern Pacific (Table 2.2). Based on these coral occurrences, species richness increased during the Late Eocene/Early Oligocene (55 species), decreased during the early and middle Pleistocene (5 species), rose again during the late Pleistocene (13 species) and finally peaked during Recent times (46 species) (Fig. 2.4). Along with this trend, reef corals expanded and contracted their distributions poleward and vice versa, such that they were able to extend northward to Vancouver Island (~49.8°N) (http://paleodb.org/) during the late Oligocene (Durham 1942a, b, 1950a). Data suggest that these expansions and compressions of climatic belts, associated with global climate warming and cooling respectively (Hay et al. 2002; Thomas et al. 2006), were accompanied by increases-decreases in species richness in the eastern Pacific (Spearman $r = 0.739, n = 11, p < 0.05$), as has been reported for reef building corals elsewhere during the Cenozoic (Indo-Pacific, Veron 1995; Caribbean, Budd 2000).

Thus, temporal species richness trends in the eastern Pacific apparently responded to the climatic changes experienced during the Cenozoic. However, bootstrap tests (using 999 bootstrap samples) of species richness mean values for every sampled time interval (same as Fig. 2.4) indicated that the species richness observed in every time interval (except Late Eocene-Early Oligocene, Late Pleistocene and Recent) is not significantly different than the mean species richness expected by chance, therefore suggesting that the richness recorded in the eastern Pacific during Late Eocene-Early Oligocene, Late Pleistocene and Recent represents meaningful signals. This finding is relevant because it indicates that: (a) it is not unlikely that high species richness developed at the eastern Pacific thermal equator during the Late Eocene-Early Oligocene, which is in full agreement with the burst of coral taxa and the widespread distribution of reef biotas in the Caribbean and Indo-Pacific (Veron 1995; Budd 2000), (b) coral taxa were lost in the eastern Pacific and reached their lowest number during the Late Pleistocene (López-Pérez and Budd 2009), and (c) following the Late Pleistocene species richness rose toward the Recent with the arrival of Indo-Pacific taxa (Glynn and Ault 2000; Glynn et al. 2007). Nonetheless, aside from these three signals (Late Eocene-Early Oligocene, Late Pleistocene and Recent), clear evidence is lacking regarding true species richness in the eastern Pacific during most of the Cenozoic. Therefore, the large-scale historical trend suggested for most of the Cenozoic (Durham 1966; Frost 1977; Colgan 1990; Cortés 1997; Glynn and Ault 2000) is still debatable, and more data are needed.

2.3 Late Miocene to Modern Times: Shaping the Recent Eastern Pacific Coral Communities and Reefs

Large-scale spatio-temporal data have delineated major evolutionary trends in Cenozoic reef corals in the eastern Pacific (Durham 1966; Frost 1977; Colgan 1990; Cortés 1997; Glynn and Ault 2000, see Sect. 2.2). Nevertheless, new records of coral species (López-Pérez 2008, 2012; López-Pérez and Budd 2009); the lithology, age and paleoenvironment of buildups (Johnson and Ledesma-Vázquez 1999; Mayer and Vincent 1999; Halfar et al. 2001; DeDiego-Forbis et al. 2004; Johnson et al. 2007), and the relative absence of information from western Mexico and Central America (Palmer 1928; Hertlein 1972; Durham 1980), clearly imply that the only clues to the origin and evolution of the modern eastern Pacific coral fauna, come from the Late Miocene to Late Pleistocene buildups recovered from the Gulf of California. It is fair to say, however, that information from the gulf may be biased because it potentially represents the evolutionary history of subtropical coral communities and reefs, while information provided by eastern tropical Pacific (ETP) reefs (western Mexico and Central America) is lacking. Cortés (1993) observed that the

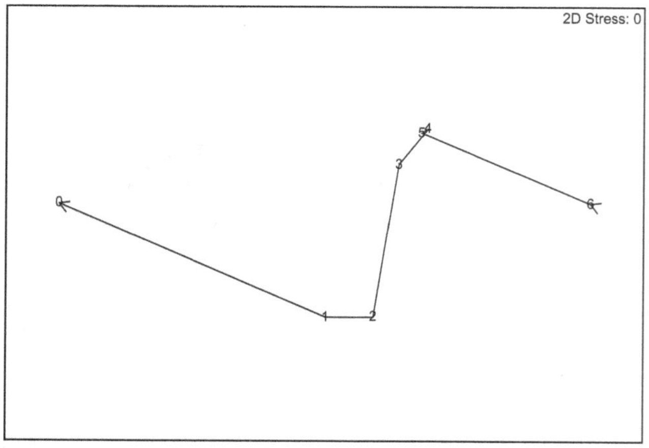

Fig. 2.5 Qualitative changes of coral reef communities and reefs in the Gulf of California, Mexico during the last six million years. Dissimilarities among time intervals (one million years duration; *6* oldest, *0* Recent) were generated with a Bray-Curtis index from an incidence matrix. Distance map produced by multidimensional scaling

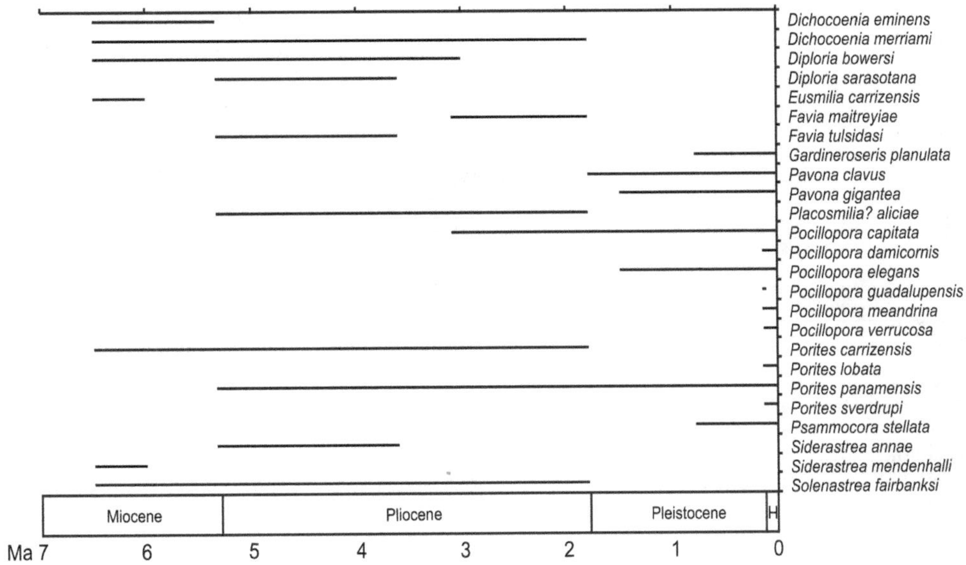

Fig. 2.6 Ranges of all known reef coral species (total = 25) in the Gulf of California during Late Miocene and recent times. Ranges are based on occurrences in 47 Gulf of California localities listed in López-Pérez (2008, Table 1). Ends of range lines are maximum age estimates for localities in which first and last occurrences take place. Only species with fossil occurrences are plotted. *H* Holocene

absence of ETP fossil reefs may be related to the poor structural integrity of reef frameworks and crustose coralline algal pavements, the relative absence of submarine cementation, and the high rates of bioerosion in this region. Along with the aforementioned negative effects, Greenstein and Pandolfi (2003) studied alteration in reef-coral death assemblages in the Florida Keys, and demonstrated that physico-chemical degradation (abrasion and dissolution) was greatest in reef-crest and patch-reef environments. It is possible that the relative absence of eastern Pacific fossil outcrops may be explained by the combined effect of high degradation of patch-reef environments coupled with one or several of the conditions advanced by Cortés (1993). Nevertheless, the data suggest that the apparent lack of fossil outcrops is deeply rooted in the relative absence of extensive and detailed studies in the eastern Pacific, as was recently demonstrated for the Gulf of California (Johnson and Ledesma-Vazquez 1999, 2001; DeDiego-Forbis et al. 2004; Johnson et al. 2007; López-Pérez 2008).

Quantitative analysis of published (López-Pérez 2008, 2012; López-Pérez and Budd 2009) and unpublished presence/absence coral data, collected during multidisciplinary and inter-institutional field expeditions to the Baja California Peninsula since 2002, suggest that during the last six million years, gulf coral communities and reefs experienced dramatic changes (Fig. 2.5). Several elements are particularly relevant. Six million years ago (late Miocene) the stony coral fauna was entirely distinct from the fauna currently inhabiting the Gulf of California and eastern Pacific (Glynn and Ault 2000; Glynn et al. 2007; see Chap. 5, Glynn et al.); indeed, the two temporally separated faunas are markedly different. During late Miocene time, the fauna was mainly composed of regional endemics (*Dichocoenia merriami*, *Diploria bowersi*, *Eusmilia carrizensis*, *Porites carrizensis*, *Siderastrea mendenhalli* and *Solenastrea fairbanksi*), and Caribbean species (*Dichocoenia eminens*) (Fig. 2.1i) (López-Pérez and Budd 2009) (Fig. 2.6). In spite of the large numbers of regional endemics, all of these species belonged to genera currently inhabiting the Caribbean. Figure 2.6 also shows the existence of three abrupt faunal shifts coupled with two periods of relative stasis in species composition. The shifts occurred at 6 to 5 Ma, 3 to 2 Ma, and 1 Ma to the Recent, with the last shift, by far, the most dramatic. While the first two shifts were related to the emergence of Caribbean-related endemic species in the gulf (*Siderastrea annae*, *Favia maitreyiae*, *Favia tulsidasi*, *Placosmilia? aliciae*, *Diploria sarasotana*; Fig. 2.1d–f, k, i), along with the appearance of *Porites panamensis* (Early Pliocene) and *Pocillopora capitata* (Middle Pliocene) (López-Pérez 2008) (Fig. 2.6), the last shift involved the extinction of regional endemics and the arrival of Indo-Pacific species currently inhabiting coral communities and reefs in the ETP region. Regarding faunal stasis, this occurred between 5 to 3 Ma and 2 to 1 Ma as a consequence of expansion and dominance of Caribbean-derived species in the Gulf of California (López-Pérez and Budd 2009). In summary, the change experienced in the Gulf during the last six million years is related to the extinction of Caribbean-related regional endemic species, and the steady arrival of Indo-Pacific taxa likely via the North Equatorial Counter Current (Fig. 2.6).

From the foregoing, two elements converged in the abrupt shift that shaped the current composition of coral communities and reefs in the Gulf and eastern Pacific: the extinction pulses that drove the local removal of Caribbean-related fauna between 3 and 1 Ma (López-Pérez and Budd 2009), and the immigration of species into the Gulf from the Indo-Pacific (Dana 1975; Glynn and Ault 2000; see also Chap. 5, Glynn et al.). As demonstrated by López-Pérez and Budd (2009), both pre-turnover (Caribbean-related) and post-turnover (Indo-Pacific) taxa do not coincide in space or time, thereby

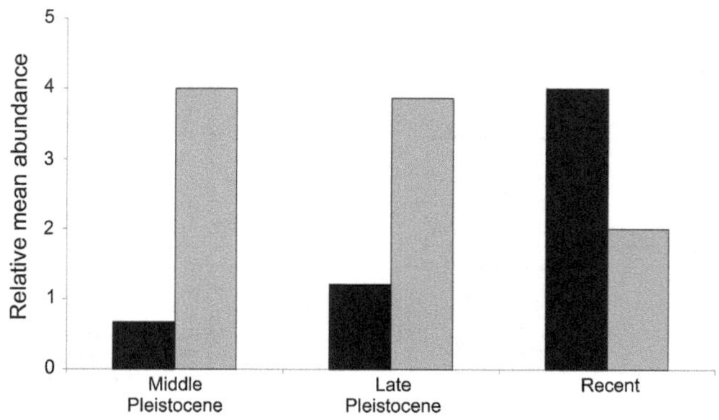

Fig. 2.7 Relative abundance variations of stony corals from middle Pleistocene to Recent times in coral communities and reefs, Gulf of California, Mexico. Gray = *Porites panamensis*, black = *Pocillopora capitata*. Relative values: *1* rare, *2* common, *3* abundant, *4* very abundant

ruling out the role of biological interactions or ecological replacement as playing the leading role in species extinctions. On the contrary, Indo-Pacific species probably invaded the Gulf at a steady but sporadic rate after the arrival of *Pavona clavus* during the lower to middle Pleistocene near the mouth of the Gulf (López-Pérez 2008). It is worth mentioning that, except for living *Porites panamensis* and *Pocillopora capitata* that spatio-temporally coexisted with pre-turnover taxa, no single regional endemic Caribbean-related coral species, or its descendant, has been recorded in the Gulf of California (Reyes-Bonilla and López-Pérez 2009) or elsewhere in the eastern Pacific (Glynn and Ault 2000). Nevertheless, related species of *Diploria, Dichocoenia, Solenastrea* and *Eusmilia* still persist on living Caribbean reefs (Budd et al. 1994), and *Favia* and *Siderastrea* also inhabit Indo-Pacific and Atlantic-Caribbean reefs. Recently, *Siderastrea glynni* was described from the Pacific side of Panama by Budd and Guzmán (1994), but later found to share identical sequence types with *Siderastrea siderea* currently living in the Atlantic. Therefore, it is unlikely that *S. glynni* represents the remnants of a population that was divided by the closure of the Central American Seaway (Forsman et al. 2005).

Relative abundance analysis of Gulf of California coral communities and reefs illustrates the emergence of modern reefs after the demise of Caribbean-related coral species (Fig. 2.7). The oldest record of *Porites panamensis* is in the Early Pliocene of Punta Chivato and in Puerto de La Lancha at Isla Carmen (López-Pérez 2008), where relatively small but abundant massive and branching colonies built small patch reefs along with Caribbean-related coral taxa. After the demise of the Caribbean-related coral fauna during the late Pliocene, *P. panamensis* flourished and constructed relatively large monospecific patch reefs from Punta Chivato to Isla Cerralvo during Pleistocene times (Johnson et al. 2007). In the Recent, this species became the most widespread taxon in the Gulf of California and eastern Pacific coral communities and reefs (Glynn and Ault 2000; Reyes-Bonilla and López-Pérez 2009). Figure 2.7 further illustrates the shift from *P. panamensis* to *Pocillopora* spp. in the Gulf. In particular, it is remarkable how *Pocillopora capitata*, first recorded during the middle Pliocene (López-Pérez 2008), has increased in abundance since the middle Pleistocene to Recent, while *P. panamensis* (thought still to be a common species) has become a relatively minor contributor to living Gulf of California south and central coral communities and reefs (Reyes-Bonilla and López-Pérez 2009). From this it can be hypothesized that the dominance of *Pocillopora* spp. in today's reefs may have a causal connection with the slightly historic decrement of *P. panamensis*, and also with the burst of speciation and ecological success of pocilloporid holobionts in the ETP. In other words, the relative high abundance of *Pocillopora* spp. over *Porites* in today's living coral communities and reefs in the gulf may be explained mainly by a dilution effect and not by the competitive displacement of *P. panamensis* by *Pocillopora* species.

2.4 The Modern Eastern Pacific Coral Fauna: A Tale of Dispersal and Vicariance

There are two hypotheses regarding the origin of the eastern Pacific coral fauna, commonly termed dispersal and vicariance. The dispersal hypothesis proposes that during the Plio-Pleistocene glaciations, and after the rise of the Isthmus of Panama, all of the eastern Pacific corals that were mainly Atlantic in origin became extinct and were replaced by an Indo-Pacific fauna (Dana 1975). In contradistinction, Heck and McCoy (1978) suggested that the modern eastern Pacific fauna was derived from previous Atlantic-Caribbean coral taxa through vicariance following the rise of the Isthmus of Panama. Evidence for the dispersal hypothesis was based on the relative absence of Plio-Pleistocene coral reef fossil outcrops on the Pacific side of America, the morphologic similarity between Indo-Pacific and eastern Pacific species, and the absence of transisthmian sibling taxa. Although Heck and McCoy (1978) criticized Dana's approach in favor of a more parsimonious explanation, they were unable to provide any alternative analysis, or unambiguously address the Atlantic-Caribbean origin of the eastern Pacific taxa.

Today it is widely accepted that the eastern Pacific coral reef fauna is almost entirely Indo-Pacific in origin (Glynn and Ault 2000; Glynn et al. 2007; Veron et al. 2015; see Chap. 5, Glynn et al.). In fact, based on the comparison of traditional morphologic characters (Squires 1959; Wells 1983; Ketchum-Mejía and Reyes-Bonilla 2001) as well as morphometric (Weil 1992; López-Pérez et al. 2003) and genetic (Weil 1992; López-Forment 2004; Forsman 2005, 2010; Combosch et al. 2008) analyses, it is estimated that approximately 66 % of the eastern Pacific coral fauna consists of Indo-Pacific immigrants. There are, however, studies suggesting that the ancestry of some eastern Pacific taxa may be closely related to Caribbean and western Atlantic species. For example, based on a morphometric comparison of selected taxa and a generic-level data analysis of Caribbean, eastern Pacific and Indo-Pacific coral species, Budd (1989) found a close relationship between the late Miocene Imperial Valley species and Plio-Pleistocene Caribbean faunas. In a similar analysis, based on the uninterrupted fossil record of *Porites* species in the Gulf of California during the Plio-Pleistocene, Reyes-Bonilla (1992) suggested that some Caribbean related species may have survived the Plio-Pleistocene extinctions that occurred in the eastern Pacific (López-Pérez and Budd 2009).

A few years ago, the genera *Siderastrea* and *Porites* were the best candidates for testing the hypothesis regarding the Caribbean origin for some eastern Pacific taxa because of their modern circumtropical distributions (Veron 2000), common occurrences in the fossil record (Veron and Kelley 1988; Budd 2000; López-Pérez 2008) and, at least for *Porites*, survival to the Plio-Pleistocene regional extinctions on both sides of the Central American Isthmus (Budd 2000; Lopez-Perez and Budd 2009). With a Central American Seaway connecting the Atlantic and Pacific, *Siderastrea mendenhalli* and *Siderastrea annae* were recorded in the Late Miocene/Pliocene of the Gulf of California, but they disappeared during the Pliocene/Pleistocene regional extinction (López-Pérez and Budd 2009). Recently, *Siderastrea glynni* appeared in the Pacific of Panama (Budd and Guzmán 1994), initiating a debate about the Atlantic origin of this species. Recent genetic analysis, however, demonstrated that this species shares identical sequence types with the Caribbean *S. siderea,* and therefore it is unlikely that *S. glynni* represents the remnants of a population that was divided by the closure of the Central American Seaway ($\sim$3.4 Ma; Coates and Obando 1996). Instead, *S. glynni* may have originated in response to a breach of the isthmus ($\sim$2 Ma), or a contemporary introduction by ship (Forsman 2005).

Contrary to *Siderastrea* species, several recent studies (Weil 1992; López-Forment 2004; Forsman 2005, 2010) clearly support the closest relation of *Porites panamensis* and *Porites sverdrupi* with the Caribbean and western Atlantic clades. For example, the electrophoretic analysis carried out by Weil (1992) suggests that *P. panamensis* is more closely related to *Porites colonensis* and *Porites astreoides* than to the Pacific species *Porites lobata*. In the same way, the maximum likelihood and parsimony analysis of the cytochrome oxidase 1 gene performed by López-Forment (2004) supported a close relationship between Gulf of California *P. sverdrupi* and Caribbean *Porites furcata*, *Porites porites* and *Porites divaricata*. Similarly, Forsman's (2006, 2010) analysis of the nuclear and ribosomal ITS region found that *P. panamensis* and *P. sverdrupi* are more closely related to the *P. furcata*-*P. divaricata* clade than to any Indo-Pacific poritid taxa. Finally, preliminary (although poorly supported; López-Pérez, unpublished data) phylogenetic analysis of morphometric data suggests the inclusion of *P. panamensis* and *P. sverdrupi* in Caribbean and western Atlantic clades, but unrelated to Indo-Pacific poritids. Therefore, the agreement among independent studies of genetic and morphologic data (Weil 1992; López-Forment 2004; Forsman's 2006, 2010), clearly supports the presence of a mix of Indo-Pacific (*P. lobata, P. lutea, P. lichen, P. australiensis, P. arnaudi, P. baueri*) and Caribbean/Atlantic derived *Porites* (*P. panamensis, P. sverdrupi*) in extant eastern Pacific coral communities and reefs. From the foregoing, the eastern Pacific poritid fauna resulted from long-distance dispersal of Indo-Pacific taxa, and from local speciation of Caribbean/Atlantic populations. Therefore, the hypotheses regarding the origin of the eastern Pacific coral fauna proposed by Dana (1975), and Heck and McCoy (1978), at least for some taxa, are complementary.

2.5 Summary and Future Research Agendas of Fossil Coral Fauna

Today both the pre-Pleistocene stony coral faunistic affinity between the Pacific and Caribbean-Atlantic side of America (Vaughan 1917) and the Pleistocene turnover with a bias towards an Indo-Pacific and eastern Pacific similarity (Dana 1975) is undeniable. New data, however, have helped to further disentangle the large-scale spatio-temporal patterns reported in a series of seminal papers published since 1960 (Durham and Allison 1960; Durham 1966; Dana 1975; Heck and McCoy 1978; Colgan 1990; Cortés 1997). So far, based on paleocoordinates of the coral deposits (http://paleodb.org/), and a much more restricted connection between the eastern Pacific and Atlantic/Caribbean during Eocene to Miocene times than previously envisaged (Kirby et al. 2008; Montes et al. 2012), the coral-bearing deposits recovered in Central America (Central Chiapas to Colombia) (Vaughan 1919; Woodring 1957; Frost and Langenheim 1974; Budd et al. 1992; Coates et al. 2005), once considered present at the thermal equator of the eastern Pacific region (Colgan 1990), should best be considered part of the Eocene-Miocene Caribbean Province. If future geologic studies support the hypothesis that Eocene to Miocene deposits from Central America are part of the Caribbean Province, then there is no reason to suggest that the eastern Pacific coral fauna was richer and supported more extensive reef building in pre-Holocene times (Durham 1966; Frost 1977; Colgan 1990; Cortés 1997; Glynn and Ault 2000) compared with eastern Pacific living coral communities and reefs. In support of this alternative view, data suggest that most of the eastern Pacific coral formations during the entire Cenozoic were small, low-diversity biostromal reefs deposited in low-angle ramps or flat-lying terraces in open, exposed, high-energy environments, or in small protected embayments (Durham 1942a, b, 1950b; Johnson et al. 2007; López-Pérez 2008). In summary, they are much like today's eastern Pacific coral communities and reefs.

Further bootstrap analysis of detailed spatio-temporal data demonstrated that the temporal trend of species richness in the eastern Pacific during the Cenozoic, currently suggested by large scale studies (Durham 1966; Frost 1977; Colgan 1990; Cortés 1997; Glynn and Ault 2000; López-Pérez 2005), may be spurious and, except for the richness recorded during the Late Eocene-Early Oligocene, Late Pleistocene and Recent times, they are not significantly different from a trend generated by chance. In other words, we lack clear evidence regarding the true species richness in the eastern Pacific (except for the above mentioned time slices), and therefore the large-scale trend through most of the Cenozoic era is still debatable.

Recently published information suggests that during the past six million years, Gulf of California coral communities and reefs experienced dramatic changes. The late Miocene coral fauna in the gulf was composed of regionally endemic species closely related to the Atlantic-Caribbean fauna (López-Pérez 2008; López-Pérez and Budd 2009) as a consequence of the existence of the Central American Seaway between the Atlantic and Pacific Oceans. During the late Miocene to Pliocene, an Atlantic/Caribbean-related fauna flourished throughout the gulf and constructed relatively large coral patches and reefs (López-Pérez 2008; López-Pérez and Budd 2009), until long after the connection between the Atlantic and Pacific was closed (~3.4 Ma; Coates and Obando 1996). Finally, the Atlantic/Caribbean-related coral fauna disappeared during the late Pliocene (~1.8 Ma) extinction pulses in the gulf and was replaced by the steady arrival of Indo-Pacific coral taxa (López-Pérez and Budd 2009) probably via the North Equatorial Counter Current.

Nevertheless, the late Pliocene extinction (López-Pérez and Budd 2009) pulses were apparently unable to eliminate some taxa (*Porites panamensis* and *Porites sverdrupi*), which strongly suggests that the hypotheses regarding the origin of the eastern Pacific coral fauna proposed by Dana (1975), and Heck and McCoy (1978) are to some extent complementary. There is a strong agreement between independent genetic (Weil 1992; López-Forment 2004; Forsman 2006, 2010) and morphologic analyses (López-Pérez, unpublished), which supports the phylogenetic relationship between the above species and Caribbean/Atlantic *Porites*. Thus, current information supports the presence of a mix of Indo-Pacific (*P. lobata*, *P. lutea*, *P. lichen*, *P. australiensis*, *P. arnaudi*, *P. baueri*) and Caribbean/Atlantic derived *Porites* (*P. panamensis*, *P. sverdrupi*) in modern eastern Pacific coral communities and reefs, indicating that contemporary eastern Pacific *Porites* resulted from long-distance dispersal of Indo-Pacific taxa and from local speciation of Caribbean/Atlantic ancestral populations of *Porites*.

Finally, the evolutionary history of the eastern Pacific coral fauna over the past 6 Ma (Late Miocene to Late Pleistocene) has been based on subtropical coral communities from the Gulf of California and, therefore, may, to some extent, be biased. It is therefore urgent to fully research coral-bearing fossil outcrops unambiguously deposited at the thermal equator (western Mexico and Central America) in the eastern Pacific during the geological past. In addition, several other research programs are required. For example, except for the taxonomic studies conducted by Baron-Szabo (2006, 2008) for the Paleogene, and by López-Pérez (2012) for the Late Miocene to Recent Gulf of California coral fauna, the taxonomy of a large number of eastern Pacific coral taxa needs to be carefully examined in order to unambiguously assess species richness and affinities among faunas (Budd et al. 1994). At the same time, a more intensive and widespread sampling design based on ecological quantitative techniques, similar to those used for sampling modern coral reef communities (i.e. transect and quadrat methods), must be implemented in new coral-bearing outcrops and previously studied exposures. This is especially important because recent published research conducted in the Gulf of California (López-Pérez 2008) has resulted in new localities and species records, even for previously studied exposures.

References

Allison EC, Durham JW, Mintz LW (1967) New southeast Pacific echinoids. Occ Pap Cal Acad Sci 62:1–23

Baldwin EM (1964) Geology of the Dallas and Valsetz quadrangles, Oregon. Bull State Dep Geol Miner Ind 35:1–61

Baron-Szabo RC (2006) Corals of the K/T-boundary: scleractinian corals of the suborders Astrocoeniina, Faviina, Rhipidogyrina and Amphiastraeina. J Syst Palaeontol 4(1):1–108

Baron-Szabo RC (2008) Corals of the K/T-boundary: scleractinian corals of the suborders Dendrophylliina, Caryophylliina, Fungiina, Microsolenina, and Stylinina. Zootaxa 1952:1–244

Brabb EE, Ristau D, Bukry D, McDougall K, Almgren AA, Saul L, Sanfilippo A (2008) Newly discovered Paleocene and Eocene rocks near Fairfield, California, and correlation with rocks in Vaca Valley and the so-called Martinez formation or stage. US Geol Survey Open-File Report 2008-1228

Bridges RA, Castle JW (2003) Local and regional tectonic control on sedimentology and stratigraphy in a strike-slip basin: Miocene Temblor Formation of the Coalinga area, California, USA. Sed Geol 158:271–297

Budd AF (1987) Neogene paleontology in the northern Dominican Republic 4. The genus *Stephanocoenia* (Anthozoa: Scleractinia: Astrocoeniidae). Bull Am Paleontol 93(328):1–22

Budd AF (1989) Biogeography of Neogene Caribbean reef corals and its implications for the ancestry of Eastern Pacific reef corals. Mem Ass Aust Palaeontol 8:219–230

Budd AF (1991) Neogene paleontology in the northern Dominican Republic 11. The family Faviidae (Anthozoa: Scleractinia). Part I. The genera *Montastraea* and *Solenastrea*. Bull Am Paleontol 101:5–83

Budd AF (2000) Diversity and extinction in the Cenozoic history of Caribbean reefs. Coral Reefs 19:25–35

Budd AF, Guzmán HM (1994) *Siderastrea glynni*, a new species of scleractinian coral (Cnidaria: Anthozoa) from the eastern Pacific. Proc Biol Soc Wash 107:591–599

Budd AF, Johnson KG (1999) Neogene paleontology in the northern Dominican Republic 19. The family Faviidae (Anthozoa: Scleractinia). Part II. The genera *Caulastrea*, *Favia*, *Diploria*, *Thysanus*, *Hadrophyllia*, *Manicina*, and *Colpophyllia*. Bull Am Paleontol 356:1–83

Budd AF, Stemann TA, Johnson KG (1994) Stratigraphic distributions of genera and species of Neogene to Recent Caribbean reef corals. J Paleontol 68:951–977

Budd AF, Stemann TA, Steward RH (1992) Eocene Caribbean reef corals: a unique fauna from the Gatuncillo Formation of Panamá. J Paleontol 66:570–594

Carreño AL, Smith JT (2007) Stratigraphy and correlation for the ancient Gulf of California and Baja California Peninsula, Mexico. Bull Am Paleontol 371:1–146

Coates AG, Obando JA (1996) The geologic evolution of the Central American Isthmus. In: Jackson JBC, Budd AF, Coates AG (eds) Evolution and environment in tropical America. University of Chicago Press, London, pp 21–56

Coates AG, McNeill DF, Aubry MP, Berggren WA, Collins LS (2005) An Introduction to the geology of the Bocas del Toro Archipelago, Panama. Caribb J Sci 41(3):374–391

Colgan MW (1990) El Niño and the history of Eastern Pacific reef building. In: Glynn PW (ed) Global ecological consequences of the 1982–83 El Niño-Southern Oscillation. Elsevier, Amsterdam, pp 183–232

Combosch DJ, Guzman HM, Schuhmacher H, Vollmer SV (2008) Interspecific hybridization and restricted trans-Pacific gene flow in the tropical eastern Pacific *Pocillopora*. Mol Ecol 17:1304–1312

Cortés JC (1993) Comparison between Caribbean and eastern Pacific coral reefs. Rev Biol Trop 41(supl 1):19–21

Cortés JC (1997) Biology and geology of eastern Pacific coral reefs. In: Proceedings of the 8th International Coral Reef Symposium, vol 1, Panama, pp 57–64

Cortés JC (2003) Coral reefs of the Americas: an introduction to Latin American coral reefs. In: Cortés J (ed) Latin American coral reefs. Elsevier, Amsterdam, pp 1–7

Dana TF (1975) Development of contemporary eastern Pacific corals reefs. Mar Biol 33:355–374

DeDiego-Forbis T, Douglas R, Gorsline D, Nava-Sanchez E, Mack L, Banner J (2004) Late Pleistocene (Last Interglacial) terrace deposits, Bahía Coyote, Baja California Sur, México. Quatern Int 120:29–40

Dibblee TW (1973) Stratigraphy of the southern coast ranges near the San Andreas Fault from Cholame to Maricopa, California. Geol Sur Prof Paper 764:1–45

Dickerson RE (1914) Fauna of the Martinez Eocene of California. California Univ Publ Geol Sci 8:61–180

Durham JW (1942a) Eocene and Oligocene coral faunas of Washington. J Paleontol 16:84–104

Durham JW (1942b) Reef corals from the California Middle Eocene. Proc Cal Acad Sci 23:503–510

Durham JW (1943) Pacific coast Cretaceous and Tertiary corals. J Paleontol 17:196–202

Durham JW (1944) Megafaunal zones of the Oligocene of Northwestern Washington, California. Univ Publ Geol Sci 27:101–211

Durham JW (1947) Corals from the Gulf of California and the north Pacific coast of America. Geol Soc Am Mem 20:1–68

Durham JW (1950a) Cenozoic marine climates of the Pacific coast. Geol Soc Am Bull 61:1243–1264

Durham JW (1950b) 1940 E.W. Scripps cruise to the Gulf of California. Part II. Megascopic paleontology and marine stratigraphy. Geol Soc Am Mem 43:1–216

Durham JW (1966) Coelenterates, especially stony corals, from the Galápagos and Cocos Islands. In: Bowman RL (ed) the Galápagos. University of California Press, Berkeley, pp 123–135

Durham JW (1980) A new fossil *Pocillopora* (coral) from Guadalupe Island, México. In: Powers DM (ed) The California Islands, vol 1, pp 63–70. Proceedings symposium, Santa Barbara

Durham JW, Allison EC (1960) The geologic history of Baja California and its marine faunas. Syst Zool 9:47–91

Emerson WK, Chaney HW (1995) A zoogeographic review of the Cypraeidae (Mollusca: Gastropoda) occurring in the eastern Pacific Ocean. Veliger 38:8–21

Faustino LA (1931) Two new madreporarian corals from California. Philipp J Sci 44:285–289

Forsman ZH, Guzmán HM, Chen CA, Fox GE, Wellington GM (2005) An ITS region phylogeny of *Siderastrea* (Cnidaria: Anthozoa): is *S. glynni* endangered or introduced? Coral Reefs 24:343–347

Forsman ZH, Hunter CL, Fox GE, Wellington GM (2006) Is the ITS region the solution to the 'species problem' in corals? Intragenomic variation and alignment permutation in *Porites*, *Siderastrea* and outgroup taxa. In: Proceedings of 10th International Coral Reef Symposium, vol 1, Okinawa, pp 14–23

Forsman ZH, Barshis DJ, Hunter CL, Toonen RJ (2010) Shape-shifting corals: molecular markers show morphology is evolutionarily plastic in *Porites*. BMC Evol Biol 9:45. doi:10.1186/1471-2148-9-45

Frost SH, Langenheim RL (1974) Cenozoic reef biofacies: Tertiary larger Foraminifera and scleractinian corals from Chiapas, México. North Illinois University, De Kalb p 388

Frost SH (1977) Miocene to Holocene evolution of Caribbean province reef-building corals. In: Proceedings of 3rd International Coral Reef Symposium, vol 1, Miami, pp 353–359

Foster AB (1986) Neogene paleontology in the northern Dominican Republic 3. The family Poritidae (Anthozoa: Scleractinia). Bull Am Paleontol 90:45–123

Fukami H, Budd AF, Paulay G, Solé-Cava A, Chen CA, Iwao K, Knowlton N (2004) Conventional taxonomy obscures deep divergence between Pacific and Atlantic corals. Nature 427:832–835

Gabb WM (1864) Description of the Cretaceous fossils, paleontology. Geol Surv Cal 1:57–217

Gabb WM (1869) Cretaceous and Tertiary fossils, paleontology. Geol Surv Cal 2:1–276

Glynn PW, Ault JS (2000) A biogeographic analysis and review of the far eastern Pacific coral reef region. Coral Reefs 19:1–23

Glynn PW, Wellington GM, Riegl B, Olson DB, Borneman E, Wieters EA (2007) Diversity and biogeography of the scleractinian coral fauna of Easter Island (Rapa Nui). Pacific Sci 61:67–90

Greenstein BJ (2003) Taphonomic alteration of reef corals: effects of reef environment and coral growth form II: the Florida Keys. Palaios 18:495–509

Halfar J, Godinez-Orta L, Goodfriend GA, Mucciarone DA, Ingle JC, Holdern P (2001) Holocene-Late Pleistocene non-tropical carbonate sediments and tectonic history of the western rift basin margin of the southern Gulf of California. Sed Geol 144:149–178

Hay WW, Soeding E, DeConto RM, Wold CN (2002) The Late Cenozoic uplift–climate change paradox. Int J Earth Sci (Geol Rundsch) 91:746–774

Heck KL, McCoy ED (1978) Long-distance dispersal and the reef-building corals of the Eastern Pacific. Mar Biol 48:349–356

Hertlein LG (1957) Pliocene and Pleistocene fossils from the southern portion of the Gulf of California. Bull Soc Cal Acad Sci 56:57–75

Hertlein LG (1972) Pliocene fossils from Baltra (South Seymour) Island, Galapagos Islands. Proc Cal Acad Sci 39:25–46

Hertlein LG, Emerson WK (1959) Results of the Puritan-American Museum of Natural History Expedition to western México. 5. Pliocene and Pleistocene megafossils from Tres Marías Islands. Am Mus Novit 1940:1–15

Hertlein LG, Grant US (1960) The geology and paleontology of the marine Pliocene of San Diego, California. Part 2a, Paleontology (Coelenterata, Bryozoa, Brachiopoda, Echinodermata). Mem San Diego Soc Nat Hist 1:73–133

Johnson ME, Ledesma-Vázquez J (1999) Biological zonation on a rocky-shore boulder deposit: upper Pleistocene Bahía San Antonio (Baja California Sur, México). Palaios 14:569–584

Johnson ME, Ledesma-Vázquez J (2001) Pliocene-Pleistocene rocky shorelines trace coastal development of Bahía Concepción, gulf coast of Baja California Sur (México). Palaeogeogr Palaeoclim Palaeoecol 166:65–88

Johnson ME, López-Pérez RA, Ransom CR, Ledesma-Vázquez J (2007) Late Pleistocene coral-reef development on Isla Coronados, Gulf of California. Cien Mar 33(2):1–16

Jordan EK, Hertlein LG (1926) Expedition to the Revillagigedo Islands México, in 1925, IV. Proc Cal Acad Sci 15:209–217

Ketchum JT, Reyes-Bonilla H (2001) Taxonomía y distribución de los corales hermatípicos (Scleractinia) del Archipiélago de Revillagigedo, Pacífico de México. Rev Biol Trop 49:803–848

Kirby MX, Jones DS, MacFadden BJ (2008) Lower Miocene stratigraphy along the Panama Canal and its bearing on the Central American Peninsula. PLoS ONE 3(7):e2791. doi:10.1371/journal.pone.0002791

Kulm LD, Thornburg TM, Suess E, Resig J, Fryer P (1990) Clastic, diagenetic, and metamorphic lithologies of a subsiding continental block: Central Peru Forearc. Proc Ocean Drill Program Part B Sci Results 112:1–738

Lessios HA, Kessing BD, Wellington GM, Graybeal A (1996) Indo-Pacific echinoids in the tropical eastern Pacific. Coral Reefs 15:133–142

Lessios HA, Robertson DR (2006) Crossing the impassable: genetic connections in 20 reef fishes across the eastern Pacific barrier. Proc R Soc B 273:2201–2208

Lessios HA, Kane J, Robertson DR (2003) Phylogeography of the pantropical sea urchin *Tripneustes*: contrasting patterns of population structure between oceans. Evolution 57(9):2026–2036

Lindberg DR, Squires RL (1990) Patellogastropods (Mollusca) from the Eocene Tejon Formation of Southern California. J Paleontol 64 (4):578–587

López-Forment MB (2004) DNA phylogeny and biogeography of the corals *Porites* and *Pocillopora* in the Gulf of California. University of California Berkeley, Berkeley p 750

López-Pérez RA (2005) The Cenozoic hermatypic corals in the eastern Pacific: history of research. Earth-Sci Rev 72:67–87

López-Pérez RA (2008) Fossil corals from the Gulf of California, Mexico: still a depauperate fauna but it bears more species than previously thought. Proc Cal Acad Sci 59(13):515–531

López-Pérez RA (2012) Late Miocene to Pleistocene reef corals in the Gulf of California. Bull Am Paleontol 383:1–78

López-Pérez RA, Budd AF (2009) Coral diversification in the Gulf of California during Late Miocene to Pleistocene. In: Johnson ME, Ledesma-Vasquez J (eds) Atlas of coastal ecosystems in the Gulf of California: past and present. University of Arizona Press, Tucson, pp 58–71

López-Pérez RA, Reyes-Bonilla H, Budd AF, Correa-Sandoval F (2003) The taxonomic status of *Porites sverdrupi*, an endemic coral of the Gulf of California. Cien Mar 29:677–691

Mayer L, Vincent KR (1999) Active tectonics of the Loreto area, Baja California Sur, Mexico. Geomorphology 27:243–255

Montes C, Cardona A, McFadden R, Morón SE, Silva CA, Restrepo-Moreno S, Ramírez DA, Hoyos N, Wilson J, Farris D, Bayona GA, Jaramillo CA, Valencia V, Bryan J, Flores JA (2012) Evidence for middle Eocene and younger land emergence in central Panama: implications for isthmus closure. Geol Soc Am Bull 124:780–799

Nomland JO (1916) Corals from the Cretaceous and Tertiary of California and Oregon. Calif Univ Publ Geol Sci 9:59–76

Nomland JO (1917) New fossil corals from the Pacific coast. California University Publ Geol Sci 10:185–190

Palmer RH (1928) Geology of southern Oaxaca, México. Geol J 36:718–734

Prothero DR, Draus E, Nesbitt EA, Smiley TM, Burns C (2009) Paleomagnetism and counterclockwise tectonic rotation of the Eocene-Oligocene Lyre, Quimper, and Marrowstone Formations, NE Olympic Peninsula, Washington State, USA. Newsl Stratigr 43 (2):127–138

Reyes-Bonilla H (1992) New records for hermatypic corals (Anthozoa: Scleractinia) in the Gulf of California, México, with an historical and biogeographical discussion. J Nat Hist 26:1163–1175

Reyes-Bonilla H (2002) Checklist of valid names and synonyms of stony corals (Anthozoa: Scleractinia) from the eastern Pacific. J Nat Hist 36:1–13

Reyes-Bonilla H (2003) Coral reefs of the Pacific coast of México. In: Cortés J (ed) Coral reefs of Latin America. Elsevier, Amsterdam, pp 310–330

Reyes-Bonilla H, López-Pérez RA (2009) Corals and coral-reef communities in the Gulf of California. In: Johnson ME, Ledesma-Vasquez J (eds) Atlas of coastal ecosystems in the Gulf of California: past and present. University of Arizona Press, Tucson, pp 45–57

Reyes-Bonilla H, López-Pérez RA, Mohedano A (2012) Discrepancia taxonómica y riqueza de corales zooxantelados del Pacífico oriental durante el Cenozoico. Rev Mex Biol 83:939–945

Robertson DR, Allen GR (1996) Zoogeography of the shorefish fauna of Clipperton Atoll. Coral Reefs 15:121–131

Rosen BR (2000) Algal symbiosis, and the collapse and recovery of reef communities: Lazarous corals across the K-T boundary. In: Culver SJ, Rawson PF (eds) Biotic response to global change. Cambridge University Press, London, pp 164–180

Sekar R, Kaczmarsky LT, Richardson L (2008) Microbial community composition of black band disease on the coral host *Siderastrea siderea* from three regions of the wider Caribbean. Mar Ecol Prog Ser 362:85–98

Schweitzer CE, Feldmann RM, González-Barba G, Cosovic V (2006) New Decapoda (Anomura, Brachyura) from the Eocene Bateque and Tepetate Formations, Baja California Sur, México. Bull Mizunami Fos Mus 33:21–45

Simian ME, Johnson ME (1997) Development and foundering of the Pliocene Santa Ines Archipelago in the Gulf of California: Baja California Sur, México. In: Johnson ME, Ledesma-Vázquez J (eds) Pliocene carbonates and related facies flanking the Gulf of California, Baja California, México. Geol Soc Amer Spec Pap 318:25–38

Squires DF (1959) Results of the Puritan-American Museum of Natural History Expedition to western México. 7 Corals and coral reefs in the Gulf of California. Bull Am Mus Nat Hist 118:367–432

Squires RL (1997) Taxonomy and distribution of the buccinid gastropod *Brachysphingus* from Uppermost Cretaceous and Lower Cenozoic marine strata of the Pacific Slope of North America. J Paleontol 71(5):847–861

Squires RL, Demetrion RA (1992) Paleontology of Eocene Bateque Formation, Baja California Sur, México. Contrib Sci Nat Hist Mus Los Angeles Co 434:1–55

Thomas E, Brinkhuis H, Huber M, Röhl U (2006) An ocean view of the early Cenozoic greenhouse world. Oceanography 19(4):94–103

Vaughan TW (1917) The coral reef fauna of Carrizo Creek, Imperial County, California, and its significance. Geol Sur Prof Pap 98:355–386

Vaughan TW (1919) Fossil corals from Central America, Cuba, and Porto Rico with an account of the America Tertiary, Pleistocene, and Recent coral reefs. Bull US Nat Mus 103:189–524

Vaughan TW (1922) Description of the species of coral from the Negritos Formation. In: Bosworth TO (ed) Geology of the Tertiary and Quaternary periods in the north-west part of Peru. Macmillan, London, pp 126–135

Veron JEN (1995) Corals in space and time. Comstock/Cornell, Ithaca p 321

Veron JEN (2000) Corals of the world, Vol. 1–3. Aus Inst Mar Sci, Townsville

Veron JEN, Kelley R (1988) Species stability in reef corals of Papua New Guinea and the Indo-Pacific. Mem Ass Aus Palaeont 6:1–69

Veron JEN, Stafford-Smith M, DeVantier L, Turak E (2015) Overview of distribution patterns of zooxanthellate Scleractinia. Front Mar Sci. doi:10.3389/fmars.2014.00081

Weil E (1992) Genetic and morphological variation in Caribbean and eastern Pacific *Porites* (Anthozoa: Scleractinia). Preliminary results. In: Proceedings of 7th International Coral Reef Symposium, vol 2, Guam, pp 643–656

Wells JW (1940) Two new corals from the Avenal Formation (Eocene) of California. J Wash Acad Sci 30:374–376

Wells JW (1983) Annotated list of the scleractinian corals of the Galápagos Islands. In: Glynn PW, Wellington GM (eds) Corals and coral reefs of the Galápagos Islands. University of California Press, Berkeley, pp 212–295

Woodring WP (1957) Geology and paleontology of Canal Zone and adjoining parts of Panama. Geol Surv Prof Pap 306(A):360

Oceanographic Conditions of the Eastern Tropical Pacific

3

Paul C. Fiedler and Miguel F. Lavín

Abstract

The eastern Pacific warm pool supports reef-building corals, as well as distinct communities of plankton, fishes, marine mammals and birds. This habitat is characterized by warm, low-salinity surface water lying on top of a strong, shallow thermocline. It is bounded by the South Equatorial Current and equatorial cold tongue to the south, cooler and more saline subtropical water to the northwest, and cold eastern boundary currents to the north and south (California and Peru Currents). The continental boundary influences atmospheric forcing by gap winds during winter and by causing the rainy Intertropical Convergence Zone to be located north of the equator and over the warm pool. Patterns of waves, tides and tropical cyclones impinging on coral reefs are described. The structure and variability of water masses and circulation are determined by solar and atmospheric processes, both within and outside of the region. To the west of the Galápagos, surface circulation is predominantly the east-west equatorial currents. Near the coast, surface circulation is modified by the coastal boundary, local winds, eddies, and interaction with eastern boundary currents. Primary productivity depends on oceanic upwelling along the equator and local centers of upwelling and wind mixing in coastal waters. Eastern tropical Pacific surface waters are moderately productive. Phytoplankton productivity is limited by a lack of the micronutrient dissolved iron, except where local coastal processes provide a source, so that macronutrients such as nitrate are never depleted. Seasonal changes in solar forcing, winds, rainfall, surface temperature and salinity, and other environmental characteristics are described, although seasonality in this region is not as pronounced as at higher latitudes. In contrast, interannual variations caused by the El Niño-Southern Oscillation across the entire tropical Indo-Pacific are very important in this region (Chap. 4). Oxygen depletion is extreme below the sharp thermocline, with consequences for mesopelagic and subthermocline benthic organisms. Surface waters are relatively low pH and marginally carbonate-saturated. Climate change is predicted to lead to future

Miguel F. Lavín—deceased (27 January 2014)

P.C. Fiedler (✉)
Marine Mammal and Turtle Division, NOAA/NMFS/Southwest Fisheries Science Center, 8901 La Jolla Shores Drive, La Jolla, CA 92037-1509, USA
e-mail: paul.fiedler@noaa.gov

M.F. Lavín
Departamento de Oceanografía Física, Centro de Investigación Científica y de Educación Superior de Ensenada, Ensenada, BC, Mexico

P.W. Glynn et al. (eds.), *Coral Reefs of the Eastern Tropical Pacific*,
Coral Reefs of the World 8, DOI 10.1007/978-94-017-7499-4_3

oceanographic changes in this region: warming and acidification of surface waters, increased stratification and reduced productivity, and upwelling/mixing of hypoxic waters into the surface layer. These changes are likely to affect organisms and populations living in the eastern tropical Pacific.

Keywords
Hydrography • Circulation • Productivity • Oxygen • Climate change

3.1 Introduction

The eastern tropical Pacific Ocean (ETP, Fig. 3.1) lies between the subtropical gyres of the North and South Pacific and encompasses both the eastern terminus of the equatorial current system of the Pacific and the eastern Pacific warm pool that forms half of the western hemisphere warm pool straddling Central America (Wang and Enfield 2001). The warm pool supports reef-building corals, as well as distinct communities of plankton, fishes, marine mammals and birds. Eastern boundary currents flow into the region from the north (California Current) and south (Peru or Humboldt Current). The structure and variability of water masses and circulation are determined by solar and atmospheric processes, both within and outside of the region. In a physical-biological or ecological sense, it is part of the low-latitude Trades Biome influenced by trade winds and by zonal equatorial current systems (Longhurst 2007). The coast and orography of Mexico and Central America greatly modify the processes forced by these wind and current systems, especially in the coastal zone where most coral reefs are located.

3.2 Atmospheric Forcing

Solar radiation supplies the energy that drives winds and ocean currents. In the tropics, compared to higher latitudes, the annual total of solar radiation is larger, its annual range is smaller and its diurnal cycle by far exceeds the annual range (Hastenrath 1991). In general, the tropics show a net heat gain, while the higher latitudes experience a net loss; this

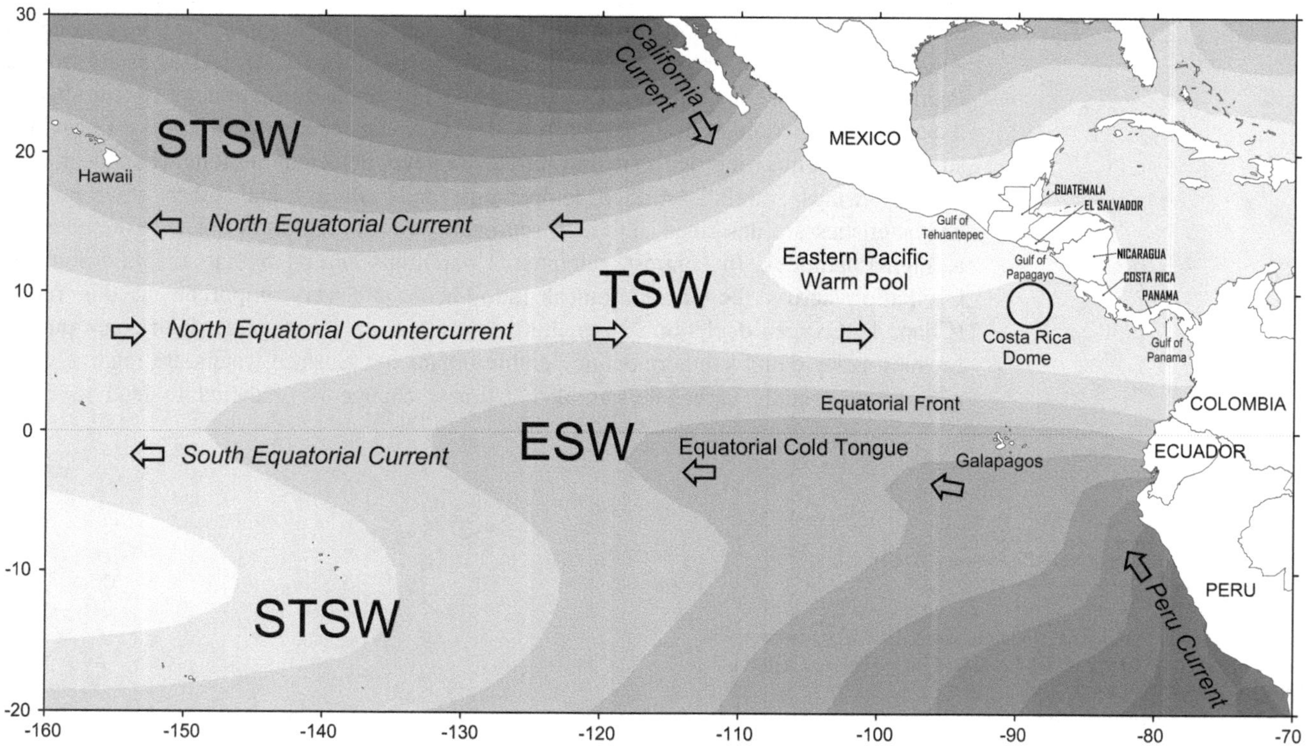

Fig. 3.1 Major geographic and oceanographic features of the ETP Ocean. Surface water masses: *STSW* subtropical surface water, *TSW* tropical surface water, *ESW* equatorial surface water. Background shading is mean surface temperature (Shea et al. 1992) ranging from 18 to 29 °C (dark to light)

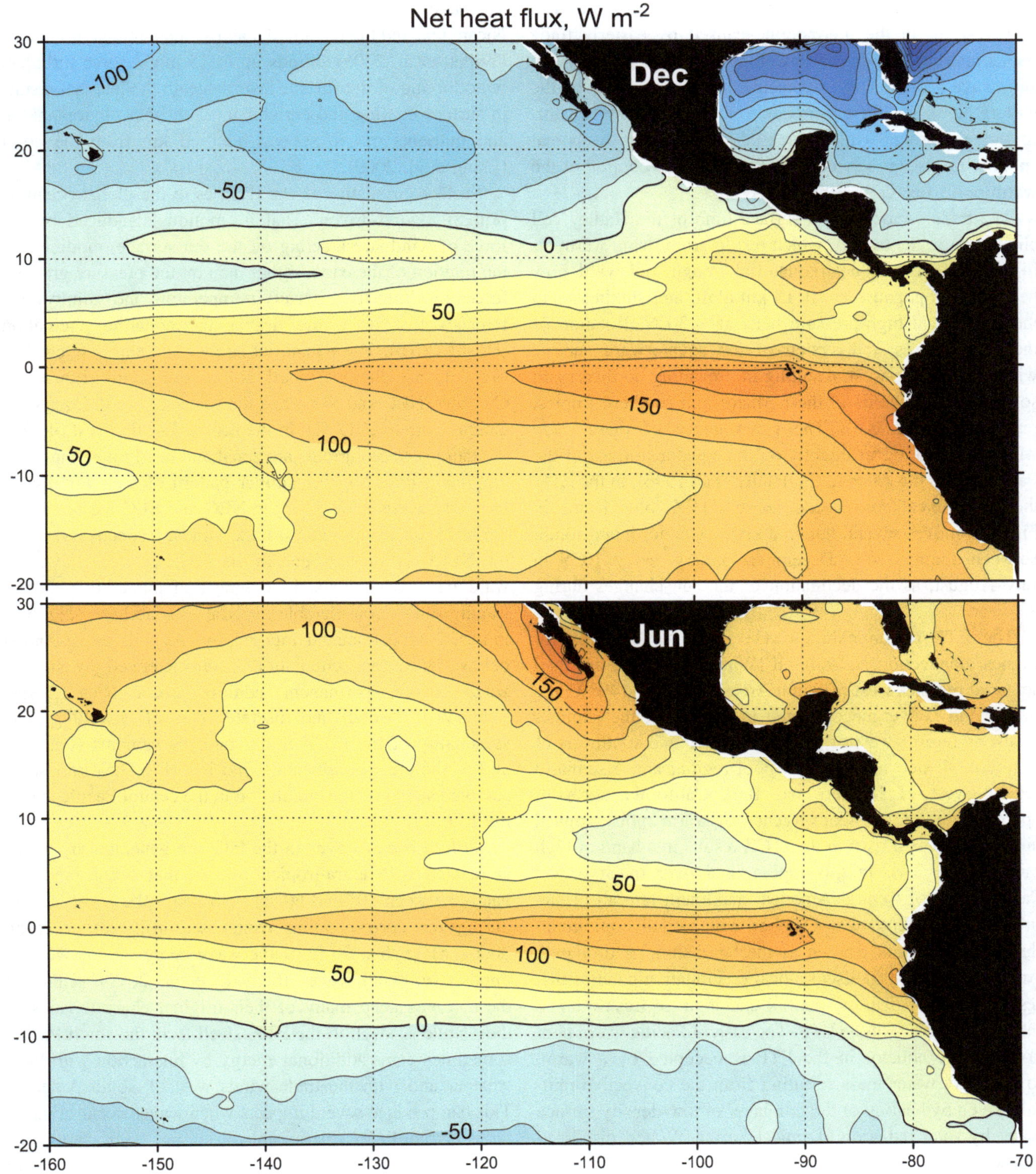

Fig. 3.2 Net heat flux into the ocean, W m^{-2}, December and June climatologies, 1984–2007 (WHOI OAFlux Project, http://oaflux.whoi.edu)

difference is responsible for much of the observed spatial distribution of temperature and surface pressure, and is also the driving force for many oceanic and atmospheric circulations (Amador et al. 2006). At any latitude, the surface of the earth gains heat during summer and loses heat during winter. However, net heat flux (incoming short-wave radiation minus outgoing long-wave radiation and latent and sensible heat flux) is into the ocean year-round near the equator (Fig. 3.2), due to relatively low water temperature and cloudiness. As will be discussed below, tropical waters are colder in the eastern than in the western Pacific, due to an east-west tilt in the thermocline caused by easterly trade winds and to wind-driven upwelling and mixing. Clear-sky shortwave radiation is responsible for the warming of the eastern North Pacific warm pool, to the north of the equator, in boreal spring, while increased cloud cover in boreal summer results in reduced solar radiation and cooling of the warm pool (Lee et al. 2007).

The trade winds are the surface limb of the Hadley cell circulation in the tropics, which results from rising warm air in the low-pressure near-equatorial trough, poleward flow below the tropopause at 10–15 km aloft, and sinking air in the subtropical high-pressure belts at ~30°N. Rotation of the earth causes a diversion of the equatorward surface winds toward the west, resulting in prevailing northeast and southeast trade winds in the northern and southern tropics, respectively (Fig. 3.3). The convergence of these trade winds does not occur directly on the equator in most of the Pacific nor in the Atlantic (Philander et al. 1996). In the ETP, the intertropical convergence zone (ITCZ, blue bands in Fig. 3.3) moves several hundred kilometers north and south, between about 4°N in December–January and 9–10°N in July–August, as the declination of the sun changes during the year.

The ETP region exhibits several distinct regimes for tropical convection and rainfall (Wang 1994): the arid and semiarid regimes west of South America to near 140°W; the trade wind convergence regime in the ITCZ; the summer monsoon regime west of Central America and southwest of southern Mexico in the warm pool region; and a semiarid regime west of Mexico and Baja California north of approximately 15°N. Convergence of warm moist tropical air that cools as it rises in the ITCZ results in a band of high rainfall (Fig. 3.4). In this region, and over much of the Pacific slope of Central America, maximum rainfall occurs in June and in September–October, with a relative minimum during July–August known as the "mid-summer drought". Amador et al. (2006) discuss local precipitation patterns that result in locally extreme rainfall and input of freshwater to the surface waters of the ETP. The eastern Pacific warm pool north of the equator in the ETP is occupied by a warm, low-salinity water mass resulting from the convective rainfall driven by heating of the thin layer of low-density surface water by solar radiation (Tropical Surface Water, discussed below).

The coastal boundary of the ETP has profound effects on wind forcing (Fig. 3.5). Mean winds are relatively weak due to the continental barrier and convergence in the ITCZ. However, from November through March, high-pressure polar air masses moving into the Gulf of Mexico result in intense wind bursts, collectively called "Nortes". These wind jets are channeled through three gaps in the Central American cordillera: the Tehuantepec Jet across the Isthmus of Tehuantepec, the Papagayo Jet over the central lowlands of Nicaragua, and the Panama Jet across the Isthmus of Panama (Willett et al. 2006). Winds up to 15 m s^{-1} force turbulent wind mixing of the surface layer beneath the jets that results in surface cooling and enrichment with both phytoplankton and nutrients out to several hundred km from the coast (Liang et al. 2009).

Surface circulation in the ETP, as in the global ocean, is primarily wind-driven. That is, motion is induced by the force of wind stress acting on the sea surface, modified by the rotation of the earth, rather than by the pressure gradient forces resulting from density (temperature and salinity) differences that drive the deeper thermohaline circulation (Huang 2010). The equatorial surface currents, sketched in Fig. 3.1, are the westward North and South Equatorial Currents (NEC and SEC) and the eastward North Equatorial Countercurrent (NECC). In the central Pacific, west of 120° W, these surface currents are zonal and nearly invariant with longitude, reflecting the similar pattern of their forcing by the trade winds (Fig. 3.3): the NEC and SEC are forced by the northeast and southeast trade winds, respectively, while the NECC is located beneath the ITCZ between the westward trade wind belts. In contrast, in the ETP the wind forcing has considerable spatial variability (Fig. 3.5) imposed by continental topography. As will be discussed below, its surface circulation is characterized by smaller zonal scales, permanent eddies (Costa Rica Dome, Tehuantepec Bowl) and discrete areas of upwelling. The seasonality of the ITCZ north-south movements (Fig. 3.3) and of the cross-continental wind jets (Fig. 3.5) imposes a seasonal signal on the circulation in the central Pacific and in the ETP.

Surface ocean waves in the ETP are generated by winds associated with extra-tropical storms that occur predominately between 30° and 60° in both hemispheres, especially during the corresponding winter season (Amador et al. 2006). More than 90 % of the wave energy is in swell, as opposed to wind waves (Semedo et al. 2011). Although these waves keep many of their original characteristics as they move as swell to the ETP, swell from the south Pacific sometimes gains additional energy by interaction with local weather and with the trade winds west of South America. This can produce severe impacts when waves occur in phase with extraordinary astronomical spring tides along the Central American and Mexican coasts. During northern winter (Fig. 3.6), high waves come as swell from the northwest, generated in the North Pacific Ocean. Waves with mean climatological heights of over 2.5 m are observed in the northwestern portion of the ETP, although wave height is diminished by the time they reach the Central American coast. The seasonal wind jets discussed above are associated with local ocean wave formation on the Pacific side of Mexico and Central America. During boreal summer, the

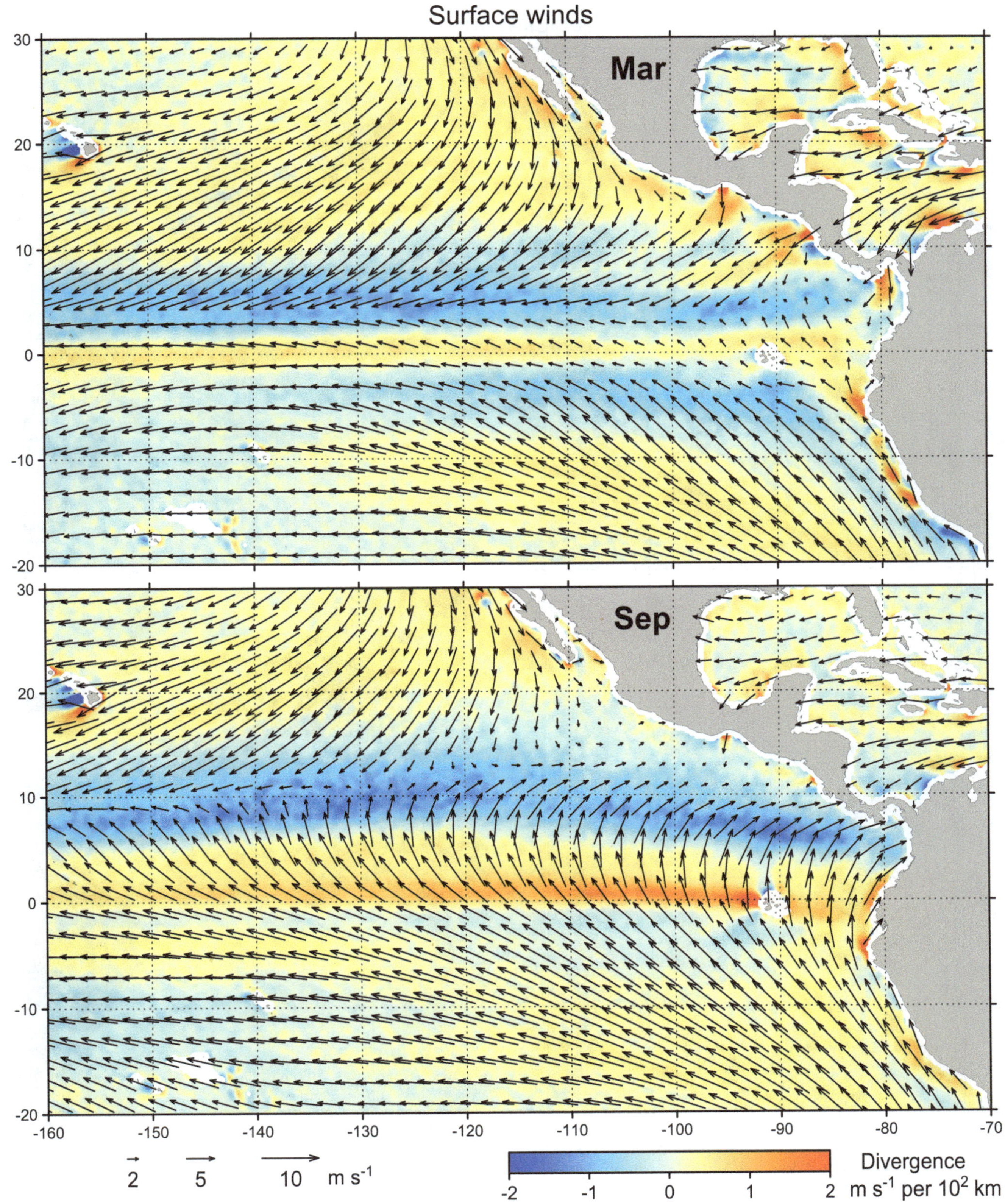

Fig. 3.3 Surface wind vectors and divergence (negative is convergence), March and September climatologies 1999–2009 (Risien and Chelton 2008, http://cioss.coas.oregonstate.edu/scow)

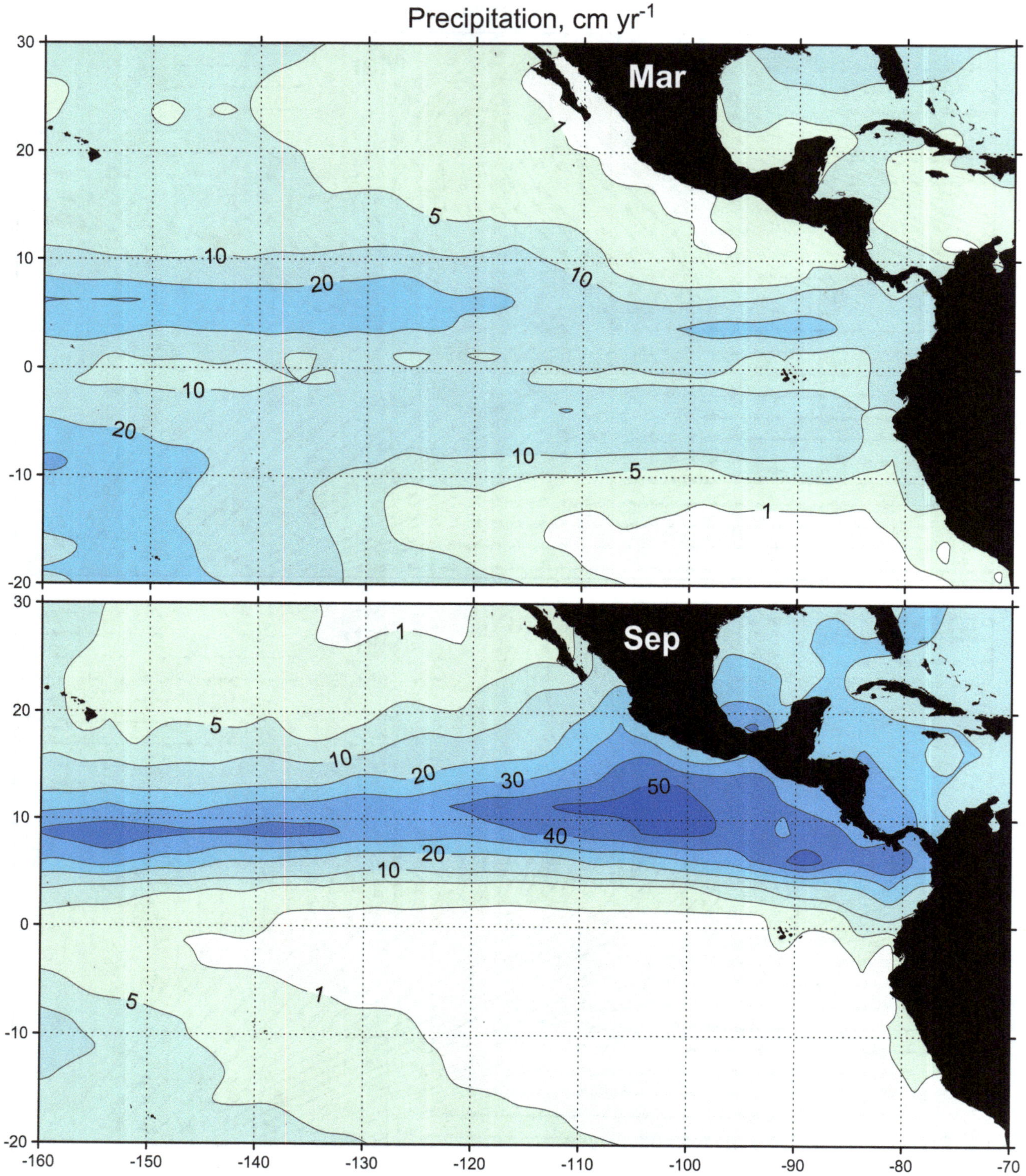

Fig. 3.4 Precipitation rate, March and September climatologies (CPC merged analysis of precipitation enhanced, http://www.esrl.noaa.gov/psd/data/gridded/data.cmap.html)

energy of the waves is smaller than in the winter, and they approach the ETP, on the average, as swell from the southern Pacific Ocean and Southern Ocean. Tropical cyclones, both in the north Pacific and in the Caribbean are also important events that generate significant wave energy in the ETP region (Amador et al. 2006). Effects of surface waves on coral reefs and the coastline must be considered in the context of the tidal range (Fig. 3.7), which is quite large

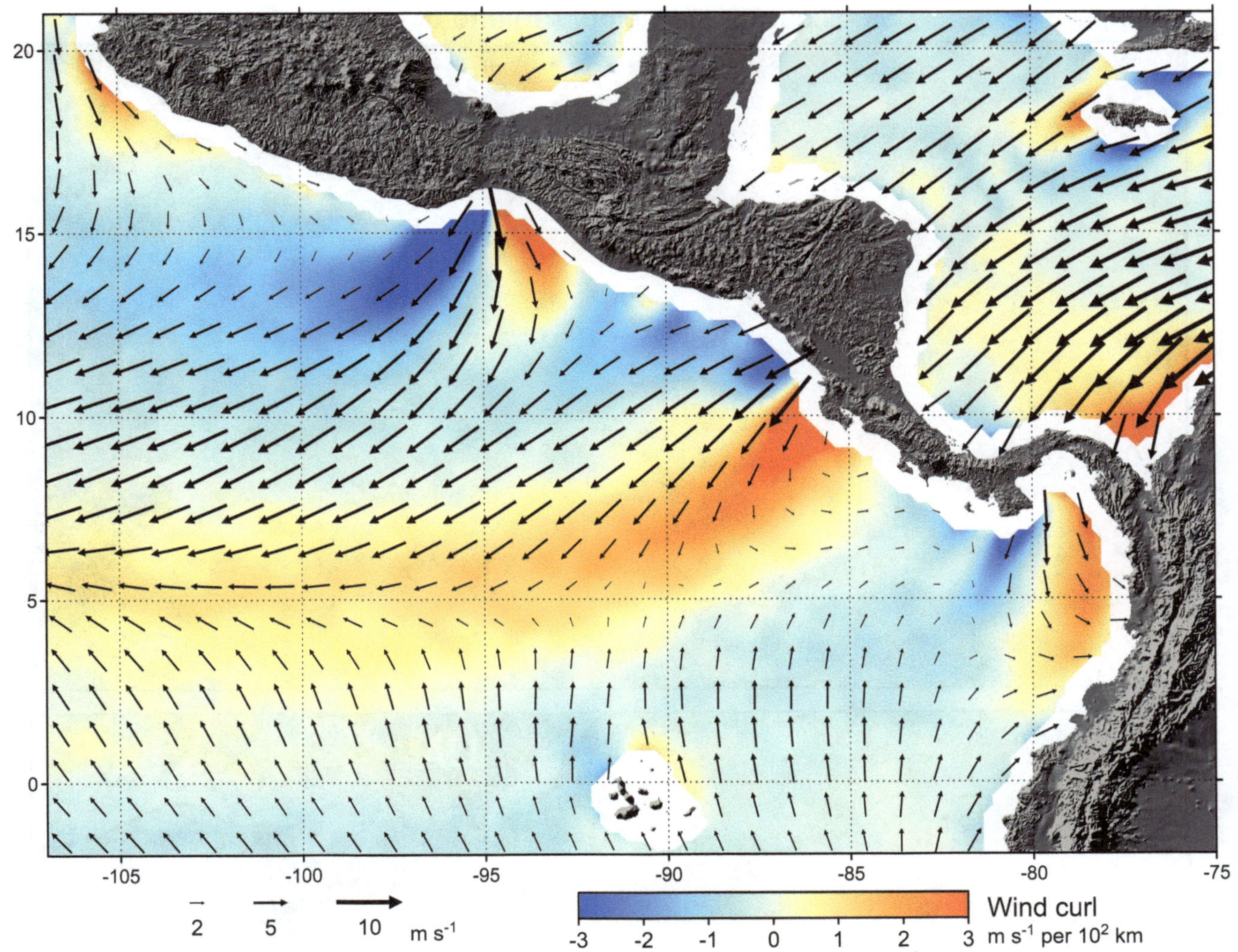

Fig. 3.5 Wind vectors and curl (related to rotation, colors), January climatology 1999–2009 (Risien and Chelton 2008, http://cioss.coas.oregonstate.edu/scow). Three prominent wind jets with associated anticyclonic (negative) and cyclonic (positive) wind curl are, from northwest to southeast, Tehuantepec, Papagayo, and Panama (see Fig. 3.1)

in the Gulf of Panama. In addition, local geography can alter wave paths and energy at island and coastal coral reefs.

The ETP area of cyclone formation is relatively small compared with the tropical Atlantic and Indo-Pacific, but it accounts for nearly 17 % of the global total of tropical storm development (Amador et al. 2006). Tropical cyclones arise over the eastern Pacific warm pool (Fig. 3.1, discussed below) in June–October. For the recent forty years of reliable records (1971–2010), 15.2 tropical storms (range 7–25), 8.5 hurricanes (range 3–16), and 3.9 (range 0–9) major hurricanes per year have formed in the eastern Pacific region (NWS/Climate Prediction Center, http://www.cpc.ncep.noaa.gov/products/Epac_hurr/background_information.html). Conditions for genesis of tropical cyclones begin with vertical convection of warm, moist air in an unstable atmosphere with low vertical wind shear, with intensification of vorticity by a low-level disturbance such as an easterly wave. Cyclones do not form within about 5° of the equator because the Coriolis effect is not sufficient to induce vorticity from the atmospheric motions. Once formed, the cyclonic storms or hurricanes move with the easterly trade winds and tend to curve northward over colder water, which leads to weakening (Fig. 3.8). Coral reefs on the Mexican coast and islands and on Clipperton Island and in the Revillagigedo Archipelago are on the path of some of these cyclones; storm waves generated by the high winds can reach the coast of Costa Rica and Panama. Genesis of hurricanes and tropical storms in the eastern Pacific is modulated between years by the El Niño (Romero-Vadillo et al. 2007; see Chap. 8, Glynn et al.) and within the warm season by the intraseasonal mode of tropical atmospheric variability known as the Madden-Julian Oscillation (Madden and Julian 1994; Maloney and Hartmann 2000), but these effects are less significant

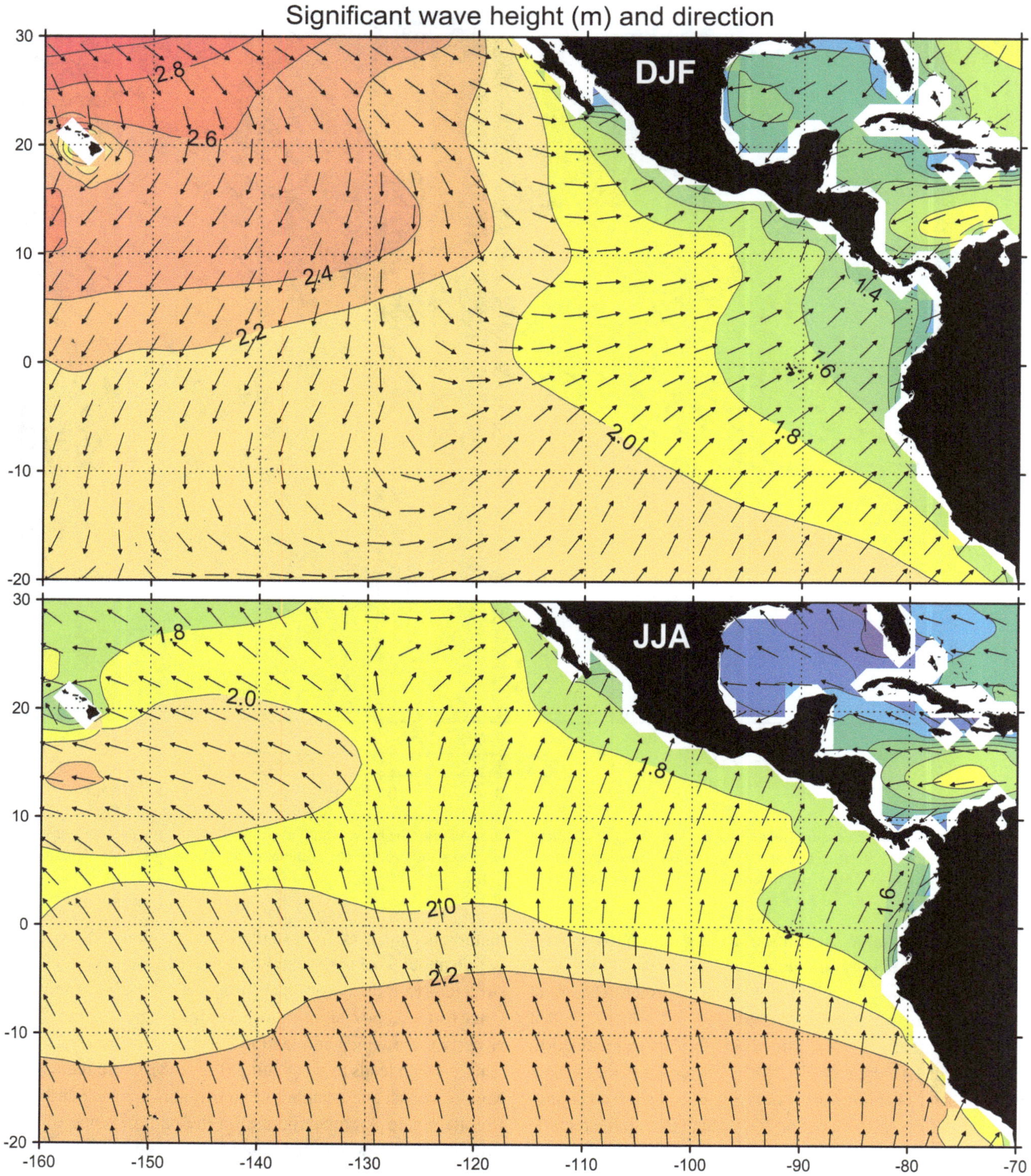

Fig. 3.6 Climatological wave height (m) and direction for December–February (*above*), and June–August (*below*), 1957–2002. Based on Semedo et al. (2011), data kindly provided by Alvaro Semedo

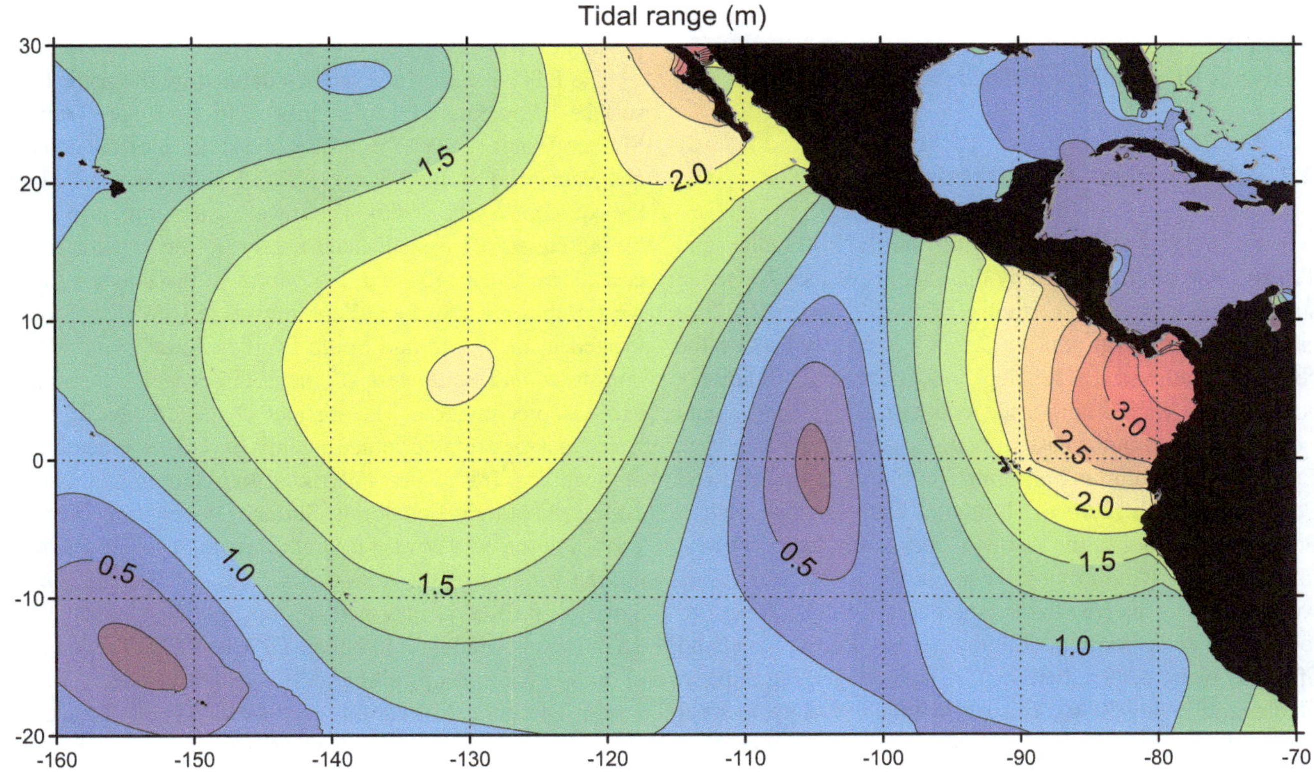

Fig. 3.7 Tidal range (m), from FES2004 data produced by Legos and CLS Space Oceanography Division and distributed by Aviso, with support from Cnes (http://www.aviso.oceanobs.com/)

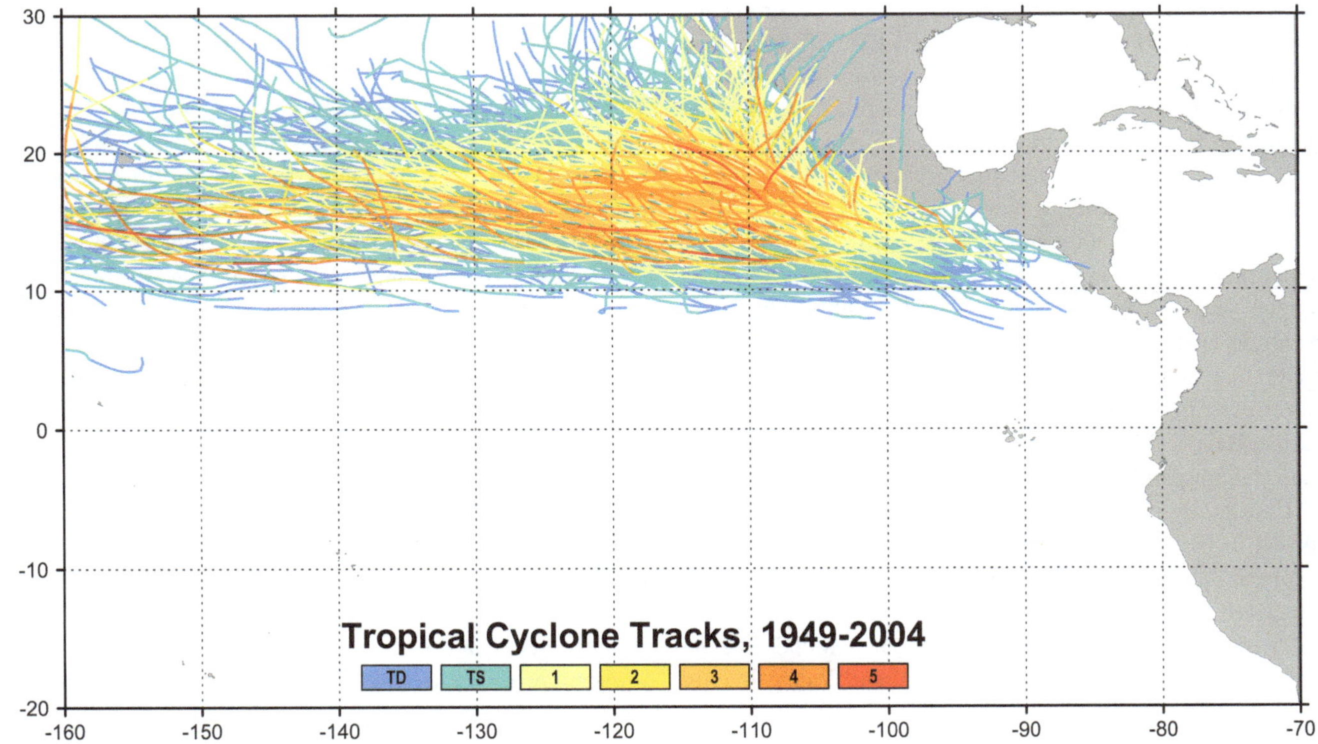

Fig. 3.8 Eastern Pacific tropical cyclone tracks, 1949–2004. Data from NOAA National Hurricane Center (http://nationalatlas.gov/atlasftp.html). Saffir-Simpson Hurricane Intensity Scale: *TD* tropical depression, *TS* tropical storm, *1–5* = hurricane categories

for the cyclones that impact the coastline (Camargo et al. 2008). Destructive effects of tropical storms on eastern Pacific coral reefs have been documented (Lirman et al. 2001).

3.3 Water Mass Characteristics

General hydrographic features of the ETP (Fiedler and Talley 2006) are illustrated by the patterns of monthly mean surface temperature and salinity (Figs. 3.9 and 3.10). The eastern Pacific warm pool (>27.5 °C) is centered along the coast of southwestern Mexico and Guatemala. Relatively cold surface waters define the equatorial cold tongue and extend poleward in the eastern boundary currents along Baja California to the north (California Current), and Ecuador and Peru to the south (Peru or Humboldt Current). The eastern Pacific warm pool is the result of seasonally large net heat flux and weak wind mixing (Wang and Enfield 2001). This warm pool is connected to the larger and slightly warmer (up to 30 °C) Pacific warm pool in the western Pacific by a band of 27 °C water centered along 7°N latitude (see Fig. 3.6a in Fiedler and Talley 2006). This warm band marks the thermal equator in the Pacific: the amplitude of the seasonal cycle is low there, except near the coast of Central America.

The equatorial cold tongue is centered directly on the equator west of 120°W, and slightly south of the equator to the east. Surface water properties here "are determined by seasonal advection of cooler water from the Peru Current and by equatorial upwelling" (Wyrtki 1966, 1981). Temperature of the cold tongue decreases towards the east as progressively cooler waters are upwelled from the Equatorial Undercurrent that shoals as it flows from west to east (Kessler 2006). The cold tongue never looks like the mean of Fig. 3.9 (September); tropical instability waves markedly distort the fronts along the northern and southern edges so that the "tongue" looks more like an "auger" (Willett et al. 2006). The cold tongue seasonal amplitude is ±1–3 °C, with coldest temperatures in September–October. Eastern boundary current waters are marked by thermal fronts where they reach the eastern Pacific warm pool off Ecuador and the southern end of the Baja California peninsula. These waters have high seasonal amplitudes of surface temperature (± >3 °C).

Eastern tropical Pacific surface salinity (Fig. 3.10) is lowest in the Gulf of Panama (29–31 psu), where extreme local rainfall systems enhance the summer monsoon rainfall along the Pacific slope of Central America (Amador et al. 2006). Another mass of low-salinity water is found off Baja California, a signature of California Current Water. Salinity of surface waters increases towards the subtropical gyres to the south and northwest of the ETP. Low-salinity surface waters (<34.5) are found in a zonal band centered along 10–11°N west of 110°W and 5–7°N east of 100°W. This low-salinity water does not coincide with the warm-water band described above. Rather, it lies beneath the ITCZ, where precipitation is high (Fig. 3.4) and exceeds evaporation.

The ETP surface water masses defined by the patterns of surface temperature and salinity (Fig. 3.1) are Tropical Surface Water (TSW: T > 25°, S < 34) north of the equator and Equatorial Surface Water (ESW: T < 25°, S > 34) along the equator (Wyrtki 1966). These water masses are separated by the equatorial front located nearly on the equator to the east of the Galápagos and 2–3° north of the equator to the west. Subtropical Surface Water masses (STSW: S > 35) are located in the North and South Pacific central gyres, to the northwest and southwest of the ETP, where evaporation exceeds precipitation. The distinct characteristics of these water masses are determined mainly by local surface fluxes of heat and freshwater (Wyrtki 1966; Large and Nurser 2001). Although Equatorial Surface Water lies between TSW and the STSW of the South Pacific, it is not a simple mixing product of those two water masses (Wyrtki 1967). Salinity of ESW is intermediate, but for most of the year, ESW is considerably colder than TSW and STSW, the result of both equatorial upwelling and advection of Peru Current water. The cold, low-salinity surface water off Baja California is California Current Water, originating in the subarctic North Pacific. Gulf of California Water (S > 34.5) is present only inside and very close to the Gulf entrance, and is formed by high evaporation.

3.4 Upper Ocean Hydrographic Structure

The pycnocline is a layer of water in which density changes more rapidly with depth than it does in the layers above (surface mixed layer) or below (deep water). It is a physical gradient that affects buoyancy, heat budgets, currents and transport. The pycnocline is also an ecological boundary both because it may include a physiological temperature limit and because it often corresponds to gradients in nutrients, oxygen, or other limiting factors. Tropical pycnoclines are relatively strong (high vertical density gradient) and shallow; the pycnocline in the ETP is an extreme case. It is primarily a thermocline, in that most of the density stratification is due to a vertical temperature gradient, but is reinforced by a strong halocline beneath the warm, low-salinity TSW (Fig. 3.11). The region is characterized by a permanent shallow thermocline (<80 m deep; Wyrtki 1964; Fig. 3.7a of Fiedler and Talley 2006) with a temperature gradient that usually exceeds 1 °C per 10 m and a large temperature difference between the top and bottom of the thermocline.

The thermocline deepens from east to west along the equator, in response to the westward wind-driven transport of surface waters (Fig. 3.11, upper left); the pressure gradient associated with the equatorial thermocline tilt is the

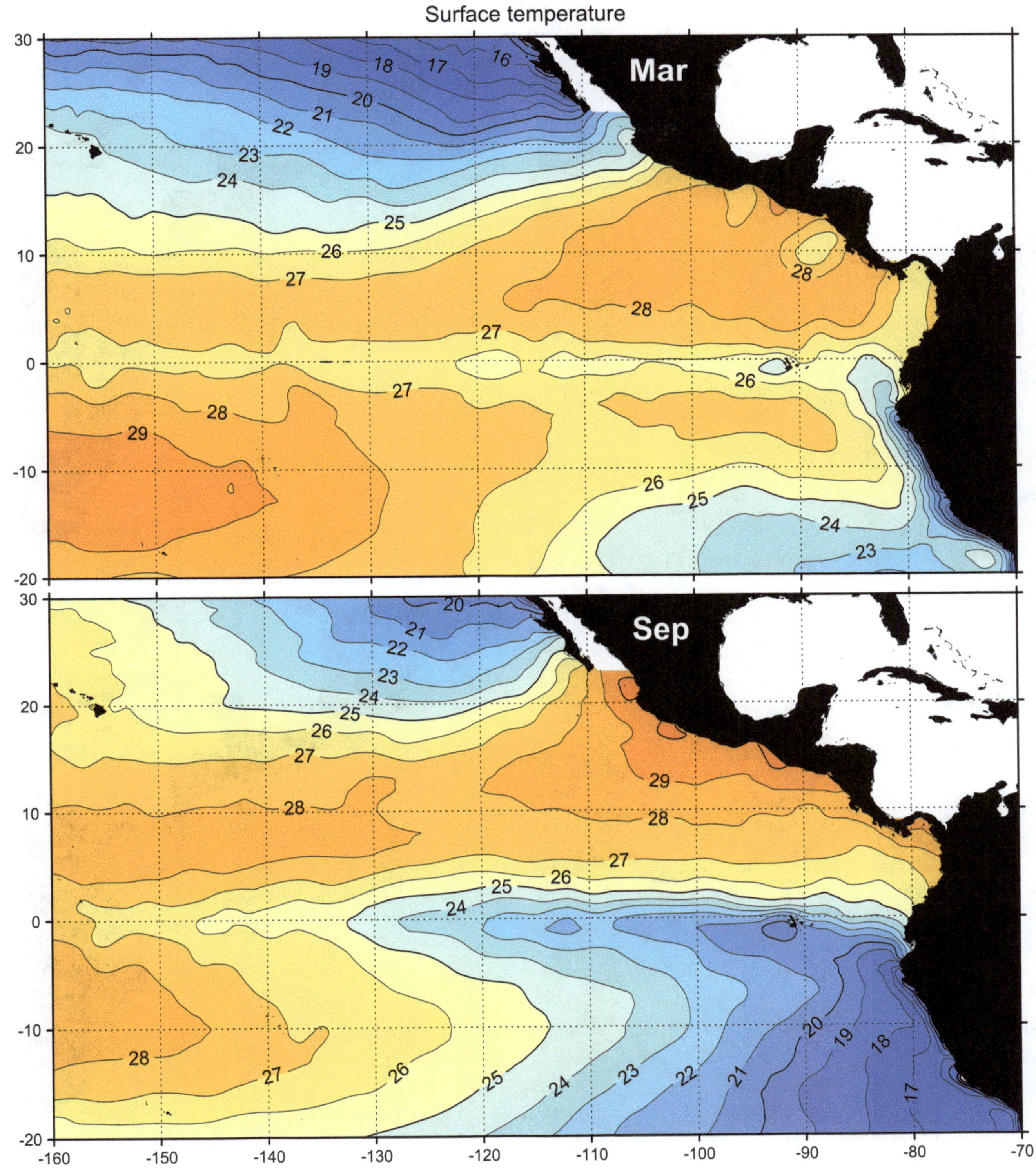

Fig. 3.9 Surface temperature (°C), March and September climatologies. Based on Fiedler and Talley (2006)

driving force of the Equatorial Undercurrent (below). Two zonal thermocline ridges (Fig. 3.11, lower left) are consistent with the surface circulation that is predominantly zonal to the west of 110°W. The equatorial thermocline ridge lies along the equator, coincident with the equatorial cold tongue. The countercurrent thermocline ridge along 10°N is named for the North Equatorial Countercurrent that flows at the surface along its southern flank. The countercurrent

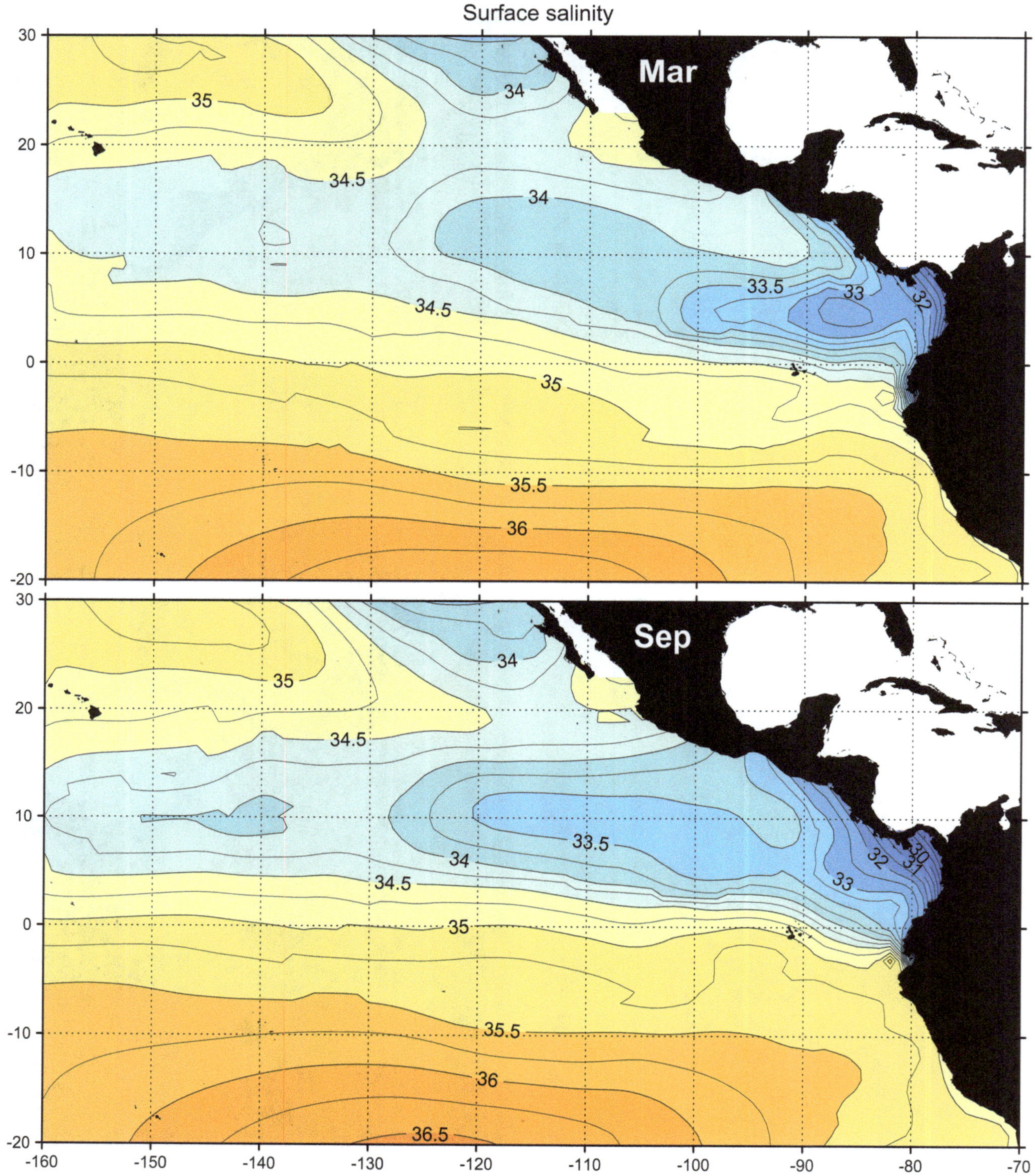

Fig. 3.10 Surface salinity (psu, practical salinity units), March and September climatologies. Based on Fiedler and Talley (2006)

thermocline ridge has a local peak at 89°W known as the Costa Rica Dome, an important oceanic center of production in the region (Fiedler 2002). The thermocline is strongest between 5°S and 15°N, except directly on the equator; the vertical spreading of isotherms in the equatorial thermocline is associated both with the geostrophy of the Equatorial Undercurrent and with equatorial divergence and upwelling. Deepening of the thermocline towards the subtropical gyres

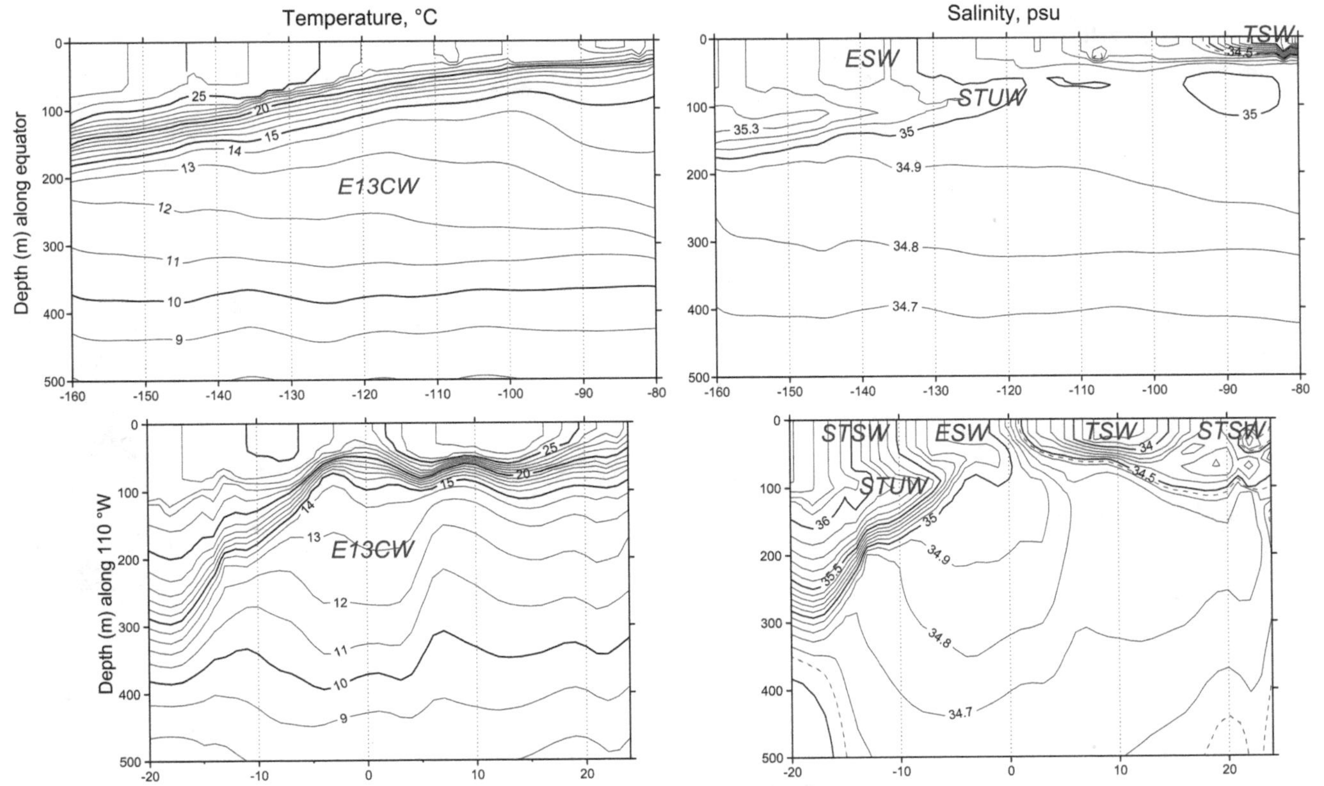

Fig. 3.11 Mean sections of potential temperature (*left*) and salinity (*right*) along the equator (*top*) and 110°W (*bottom*), adapted from Fiedler and Talley (2006)

reflects the geostrophic balance of the large-scale basin circulation (below).

Thermocline and sub-thermocline (intermediate and deep) water masses were reviewed by Fiedler and Talley (2006). The most important are labeled in Fig. 3.11 (right panels). Subtropical Underwater (STUW) is high-salinity South Pacific Subtropical Surface Water subducted into the thermocline beneath Equatorial Surface Water. The thermostad beneath the equatorial thermocline is identified as a water mass (Equatorial 13 °C Water or E13CW), but is related to circulation at these depths rather than to a distinct, remote origin that characterizes a water mass. Low-salinity Antarctic Intermediate Water (AIW) and North Pacific Intermediate Water (NPIW) are subducted from higher latitudes and mostly lie deeper than 500 m off the equator in the ETP.

3.5 Upper Ocean Circulation

Although wind-driven equatorial surface currents are predominantly zonal in the central Pacific (Fig. 3.1), wind forcing in the ETP has considerable spatial variability (Fig. 3.5) imposed by continental topography, and seasonality due to changes in local winds (Fig. 3.12). The properties of the local winds that determine wind-driven circulation patterns are velocity (speed and direction) and differences in velocity over the sea surface: divergence and curl (spin or vorticity). The motion of surface waters induced by the frictional force of wind (wind stress) is altered by the rotation of the earth. Thus, winds blowing parallel to a coastline force transport of surface waters to the right or left of the wind direction in the northern and southern hemisphere, respectively. This offshore or onshore transport results in coastal upwelling or downwelling. Wind stress curl, which is either cyclonic or anticyclonic if the vorticity is in the same or opposite direction as the rotation of the earth, forces upwelling or downwelling, respectively, without a coastal barrier. These processes are described in more detail in any basic physical oceanography textbook (e.g., Talley et al. 2011).

In the large bight between southern Mexico and Peru, several hydrographic/circulation features are formed by the influence of the strong meridional component of the wind stress (arrows in Fig. 3.5) and its short spatial variability, especially as it affects the distribution of the wind-stress curl (curl(τ); colors in Fig. 3.5). This is because the vertical Ekman pumping velocity (w_E = curl(τ)/*f*ρ, where τ is wind stress, *f* is the local Coriolis parameter and ρ is water density) distorts the thermocline, forming the pressure gradients that balance the local rotation of the Earth in geostrophic

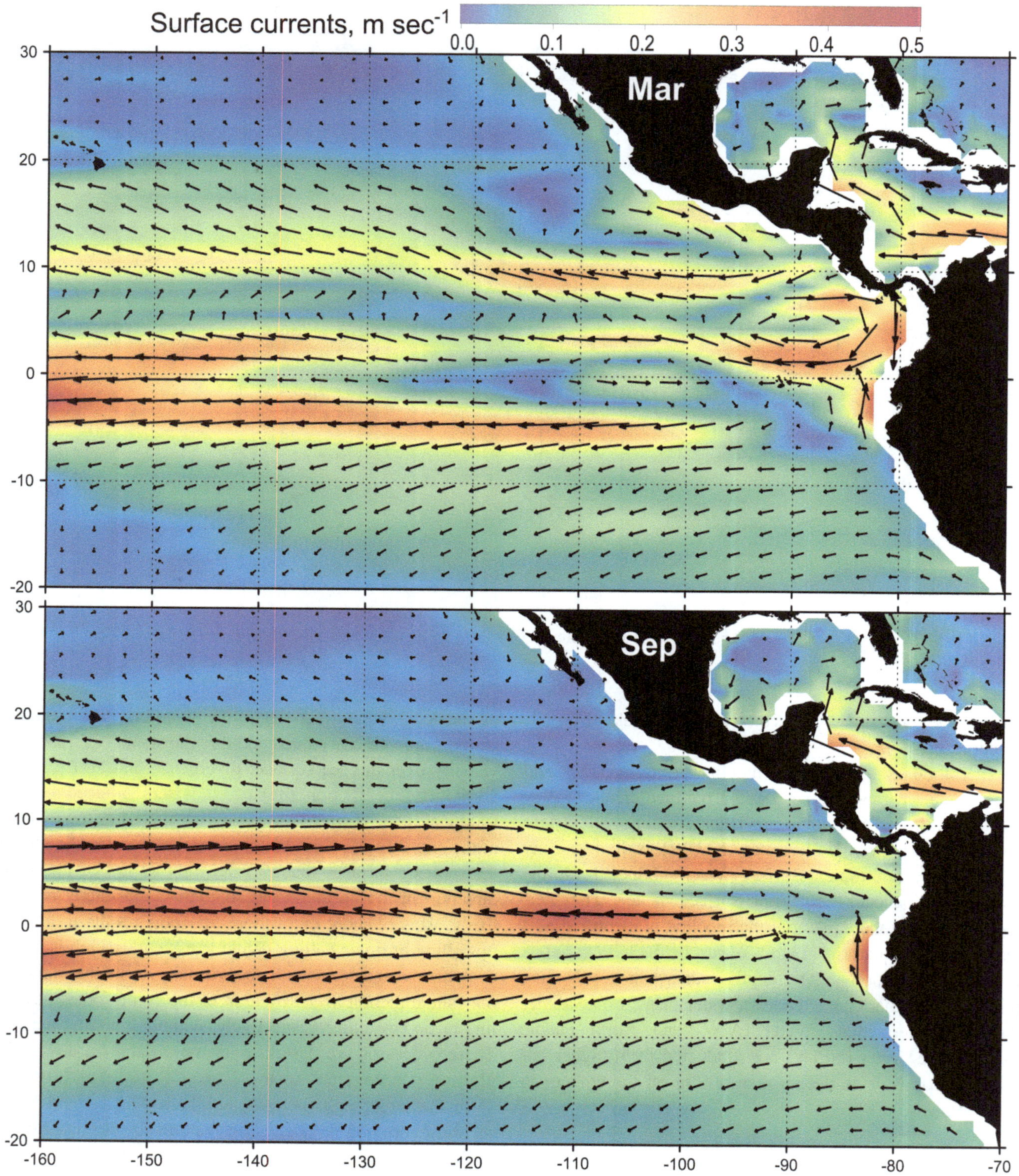

Fig. 3.12 Surface currents 1993–2010, March and September climatologies (NASA's ocean surface current analysis-real time project, OSCAR, http://www.oscar.noaa.gov/)

circulation. Positive curl (red color in Fig. 3.5) produces positive uplift of the thermocline ($w_E > 0$) and negative curl depresses the thermocline ($w_E < 0$). If the thermocline distortion is limited in space, geostrophic eddies can form (Willett et al. 2006): anticyclonic eddies under areas of negative curl/thermocline depression, and cyclonic eddies under positive curl/thermocline uplift. This is how anticyclonic eddies are formed on the poleward side of the Tehuantepec, Papagayo and Panama Jets. At a larger spatial scale, when Ekman pumping stretches or squeezes a water column, it will be forced to move northward or southward, respectively, to conserve its total vorticity (see e.g. Kessler 2006); this is expressed by the Sverdrup balance βM_y = curl (τ), where M_y is meridional mass transport and β is the change of *f* with latitude. This is how the zonal troughs and ridges that mark the equatorial currents are formed by the ITCZ, as well as the Tehuantepec Bowl and the Costa Rica Dome (Kessler 2002) by the wind jets.

3.5.1 Equatorial Currents

Seasonal extremes of surface currents, estimated from satellite surface wind and sea surface height measurements, are illustrated in Fig. 3.12. In March, the NEC is centered at 8–10°N east of 130°W and has speeds of ~20 cm s^{-1}. It is fed by a southeastward current (10–20 cm s^{-1}) off the southern coast of Mexico; this is the northern limb of the Tehuantepec Bowl anticyclonic circulation (Kessler 2006), strengthened by the California Current, which has its southernmost extension at this time of year. The NECC is absent from March to May, its ridge at 10°N flattened by the absence of the ITCZ, which at this time is found at its southernmost position. The SEC is centered at 5°S to 5°N with speeds 20–40 cm s^{-1} north and south of the equator. There is a minimum-speed band near the equator west of 110°W, which narrows and accelerates from near 0 cm s^{-1} at 114°W to 35 cm s^{-1} at 160°W. From 112 to 98°W at the equator the surface current is to the east, due to surfacing of the Equatorial Undercurrent (EUC).

In September (Fig. 3.12, bottom), the NEC is weaker than in March, with speeds of 10–20 cm s^{-1}. The NECC, well-developed by ITCZ-induced thermocline uplifting (Fig. 3.11, bottom left panel) at ~8–12°N, covers approximately from 5 to 10°N and has maximum speeds ~40 cm s^{-1}. At ~120°W it weakens to ~20 cm s^{-1}, but to the east it is strengthened again in the southern flank of the anticyclonic circulation around the Costa Rica Dome. At its terminus at ~85°W, it contributes to the coastal circulation off Central America and to the SEC. The SEC covers from 7°S to ~3°N; it is strongest (~40 cm s^{-1}) north of the equator and the minimum-speed band (20 cm s^{-1}) is at ~0–3°S. The southern lobe of the SEC has a speed of 25–35 cm s^{-1}, and south of 3°S it has a southward component. The SEC is split into northern and southern lobes due to poleward divergence of the eastward flow and to the EUC just below the surface along the equator.

The mean vertical distribution of the equatorial currents and their relationship with the mean subsurface thermal structure, are shown in Fig. 3.13 (adapted from Fig. 3.6 in Kessler 2006). The geostrophic character of these currents (beyond 3° of latitude from the equator) is well exemplified by the NECC located on the thermocline slope between the 5°N trough and the 10°N ridge. It is ~120 m deep, and has its maximum speed (10–20 cm s^{-1}) at the surface. The NEC is ~150 m deep; its speed at the surface is ~5 cm s^{-1} (surface speeds tend to be lower in this vertical section than in the surface distributions in Fig. 3.12 because of the different data sources and processing). The SEC straddles the equator, with surface speeds up to 15 cm s^{-1}.

The main subsurface equatorial currents are the EUC and the northern and southern Tsuchiya Jets (Fig. 3.13), also called Equatorial Subsurface Countercurrents. All these currents flow due east, basically driven by a pressure gradient

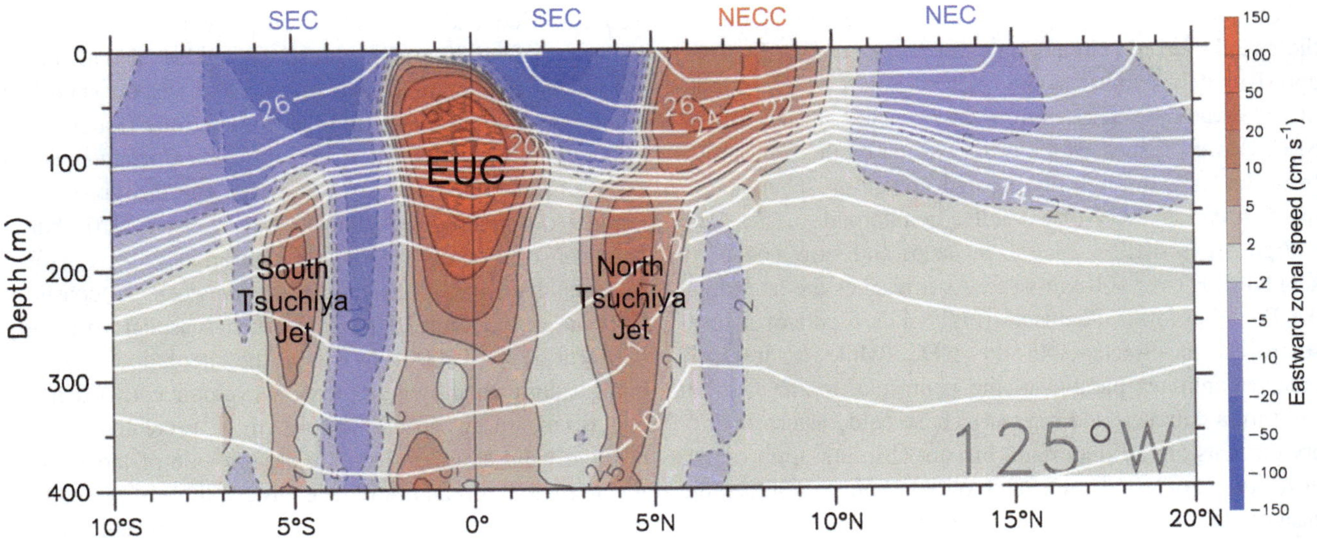

Fig. 3.13 Equatorial currents at 125°W, between 10°S and 20°N. Mean speed in cm s^{-1} shown in color (scale, *right side bar*): eastward in *red*, westward in *blue*. Mean temperature (°C) is shown as white lines. Within ±8° of latitude currents were measured directly with acoustic profilers, and elsewhere were calculated by geostrophic balance. Adapted from Kessler (2006)

force due to the tilting of the equatorial thermocline (Cromwell et al. 1954). The EUC is located within the thermocline from 2°S to 2°N; it therefore shoals from west to east (see Fig. 3.11, top left). At 125°W (Fig. 3.13), it is found from near the surface to ~300 m depth and its core speed (at ~80 m) is 100 cm s^{-1}. The Tsuchiya Jets, which originate in the far western Pacific, are located below the thermocline between 4 and 5° of latitude. At 125°W, they are at about 125–400 m depth with core speeds of ~10 cm s^{-1} in the north and 20 cm s^{-1} in the south. Like the EUC, these currents shoal with the thermocline from west to east.

As the eastward currents (NECC, EUC and Tsuchiya Jets) approach the American continent, their flow is transferred to adjoining, nascent westward and coastal currents, although the details of these transitions are mostly unknown (see Fig. 3.5 in Kessler 2006). Most of the flow of the NECC feeds into the NEC and to the northern lobe of the SEC at ~118°W, while the remaining flow continues near the surface until it reaches the southern side of the Costa Rica Dome and continues toward the Panama Bight. The EUC contributes to the two lobes of the SEC along its equatorial path, especially west of the Galápagos. East of 90°W, it discharges mostly to the southern hemisphere (Lukas 1986) where it contributes to the Peru Undercurrent and once more to the SEC through coastal upwelling (Fiedler and Talley 2006). The southern Tsuchiya Jet appears to contribute mostly (through upwelling) to the SEC. The northern Tsuchiya Jet turns north at ~95°W, and then flows under and around the Costa Rica Dome; thence, it contributes to the Costa Rica Coastal Current and to the NEC, and probably to the Mexican Coastal Current (described below).

The other contributors to the formation of the NEC and SEC are two eastern boundary currents: the California Current in the north and the Peru (Humboldt) Current in the south (Fig. 3.12). The Peru Current is equatorward flow that is both part of the southern subtropical gyre circulation and the wind forced circulation along the coast. It flows approximately parallel to the coast of South America south of 15°S where it separates from the coast to contribute to the SEC. However, most of the water carried by the SEC comes from the EUC through mixing and upwelling. The California Current is the North Pacific counterpart of the Peru Current. It is shallow (~150 m deep) and slow (<4–5 cm s^{-1}) but it is over 500 km wide. As it reaches the latitude of the Baja California Peninsula (Fig. 3.12), a part of it turns southwest to directly join the NEC, while a "tropical" branch continues parallel to the peninsula, to the Gulf of California entrance and beyond. These cold eastern boundary currents are an important, but not absolute, limit on the distribution of coral reefs and other tropical biota in the eastern Pacific (Kleypas et al. 1999a).

3.5.2 Circulation Near the American Coast

In winter (Fig. 3.12, top), the tropical branch of the California Current extends southeast off the coast to join, successively, the southeastward flow in the northeastern limb of the anticyclonic flow around the Tehuantepec Bowl, the westward flow in the northern rim of the cyclonic circulation around the Costa Rica Dome, and finally the beginning of the NEC (Fig. 3.7 in Kessler 2006). A similarly convoluted circulation is present in the neighborhood of the Costa Rica Dome and in the Panama Bight. The Costa Rica Dome is created by upward Ekman pumping under the southern side of the Papagayo Jet (Fig. 3.5; Fiedler 2002) and the northeasterly trade winds. Its formation starts in March when it has its shortest diameter (500 km) and is attached to the coast, and then develops along the year in accordance to the movement of the ITCZ, reaching its maximum westward elongation (1000 km) in November, when the ITCZ is furthest south.

The Costa Rica Coastal Current (CRCC) is the poleward flow between the Dome and the American coast. The surface speed of the cyclonic current around the Dome is ~25–40 cm s^{-1} (Brenes et al. 2008). Over the continental shelf of Central America, a poleward coastal current with speeds of 20–25 cm s^{-1} and 100–150 m deep was observed in winter and summer, and influenced by low salinity due to runoff during the rainy season. This current appears to be separate from the flow around the Dome. The formation of the CRCC can be quite tortuous because of the presence of mesoscale eddies (e.g., Papagayo and Tehuantepec eddies). Between the Gulf of Tehuantepec and Cabo Corrientes (at the Gulf of California entrance), the poleward Mexican Coastal Current (MCC) is generated by local winds through Sverdrup balance (Godínez et al. 2010). Observations just south of Cabo Corrientes (Lavín et al. 2006) show that the MCC is 90–180 km wide and 250–400 m deep, with surface speeds between 15 and 30 cm s^{-1}. Mesoscale eddies also affect the continuity of the MCC and appear to transport modified California Current water into the coastal zone (Lavín et al. 2006). The fate of the MCC beyond Cabo Corrientes is not known, since the Gulf of California entrance is usually populated by mesoscale eddies (Godínez et al. 2010), but at least part of it enters the Gulf of California.

Circulation in the Panama Bight (between the coasts of Panama and Colombia) has a marked seasonal variability (Chaigneau et al. 2006). In winter, the circulation inside the bight is dominated by a geostrophic cyclonic eddy (diameter ~400 km, mean speeds 30–60 cm s^{-1}) created by the positive wind stress curl on the eastern side of the Panama Jet (Figs. 3.3, top and 3.5). The eastern limb of the eddy is the northward-flowing coastal Colombia Current (speeds

$\sim$50 cm s^{-1}, >100 km wide); as it leaves the coast after turning anticlockwise off Panama it is joined by the branch of the NECC that did not turn north around the Costa Rica Dome. The winter Panama wind jet forces a fast southward surface current ($\sim$50 cm s^{-1}, 200–300 km wide), with negative curl creating an anticyclonic eddy to the west, thus forming a dipole with the Panama eddy. The southward current between the two eddies occupies the top 100 m of the water column and extends from 4–5°N to 2–3°N, where it feeds the northern lobe of the SEC. In spring, the cyclonic eddy is weaker than in winter because both the Panama Jet and NECC are weaker. In summer and fall, the Panama Jet is not present and the southeast trade winds veer to southerly or even westerly as they cross the equator when the ITCZ moves north. The surface circulation is weaker ($\sim$30–40 cm s^{-1}) than in winter, but still exhibits mesoscale variability. During autumn, surface currents in the NECC and in the cyclonic eddy increase to $\sim$30 cm s^{-1}.

The upper circulation of the ETP Ocean is strongly influenced by seasonal variations as described here, and also by interannual El Niño/Southern Oscillation events (Bonjean and Lagerloef 2002; Kessler 2006; see Chap. 4, Wang et al.).

3.6 Primary Productivity

Primary productivity in the surface waters of the ocean is the conversion of inorganic carbon (dissolved CO_2), macronutrients (nitrate, phosphate and silicate), and micronutrients such as iron, into phytoplankton biomass by photosynthesis. Rates of phytoplankton production in the ETP are average for the world's oceans—greater than in the oligotrophic subtropical gyres and less than in the eutrophic waters of eastern boundary current coastal upwelling systems or seasonally productive polar waters (Pennington et al. 2006). However, significant variations occur within the region. Productivity at low latitudes is generally limited by the availability of macronutrients, primarily nitrogen, in the well-lighted surface layer, whereas light tends to be limiting at higher latitudes. Macronutrients are supplied from deeper waters below the pycnocline. The close relationship between the distribution of nitrate and temperature in the water column can be seen by comparing the sections in Figs. 3.11 and 3.14. As a result, the supply of nutrients to surface waters varies with the depth of the pycnocline, which is generally highly stratified in the ETP as described above.

Lack of the micronutrient iron limits phytoplankton growth over large portions of the open-ocean ETP, depressing rates of primary production and resulting in the so-called high nitrate-low chlorophyll condition (Pennington et al. 2006). This means that nitrate and other macronutrients are never depleted in surface waters. In coastal waters, dissolved iron is input by mixing from bottom sediments on the narrow continental shelf and by river runoff during the rainy season. Thus, very high rates of phytoplankton production can be observed in these waters when upwelling or mixing brings nutrient-rich water from depth. Surface chlorophyll concentration, an index of phytoplankton biomass in near-surface waters, is high beneath the three Central American wind jets and in the Peru Current coastal upwelling system during northern winter (Fig. 3.15, January). In contrast, during northern summer, chlorophyll is high along the equator due to upwelling driven by the southeast trade winds and at the Costa Rica Dome where the thermocline is shallow.

Limiting nutrients in this region are supplied by vertical mixing and upwelling. Regional wind-driven upwelling and downwelling (Fig. 3.16) show a pattern that corresponds very closely to the distribution of phytoplankton production (Fig. 3.15). Upwelling is high (up to 1 m d^{-1}) along the equator, where trade winds and rotation of the earth force a divergence of surface waters. To the north and south of the equator are zonal bands of downwelling. Oceanic divergence and upwelling also occur between 10 and 15°N. Coastal upwelling occurs in the eastern boundary current systems, Peru Current and California Current, where strong winds blow equatorward along the coast during spring/summer and force surface waters offshore. Along the coast of Central America, local centers of upwelling and downwelling are associated with the winter wind jets discussed above and with the Costa Rica Dome. Other centers of seasonal upwelling may be important for coral reefs, notably Cabo Corrientes on the Mexican Pacific coast ($\sim$20°N). The algal symbionts of reef corals are ultimately limited by the same light and nutrient factors as phytoplankton.

3.7 Oxygen Minimum Layer

While the surface waters of the world ocean, in the mixed layer above the pycnocline, are generally nearly saturated with dissolved oxygen, ETP subpycnocline waters are very low in oxygen. Lowest oxygen concentrations are found beneath the Peru Current off northern Peru, south of the equator, and beneath the eastern Pacific warm pool off the coast of SW Mexico (Fig. 3.17). In the world ocean, deep waters of the eastern subtropical North Pacific and northern Indian Ocean (Arabian Sea and Bay of Bengal) are also low in oxygen (Fiedler and Talley 2006; Paulmier and Ruiz 2009). However, the ETP oxygen minimum layer, lying between the pycnocline and intermediate waters, is remarkable for its size and degree of hypoxia. The large and thick oxygen minimum layer in the North Pacific is deeper and less depleted and is clearly the result of lack of ventilation since no deep water is formed in the North Pacific.

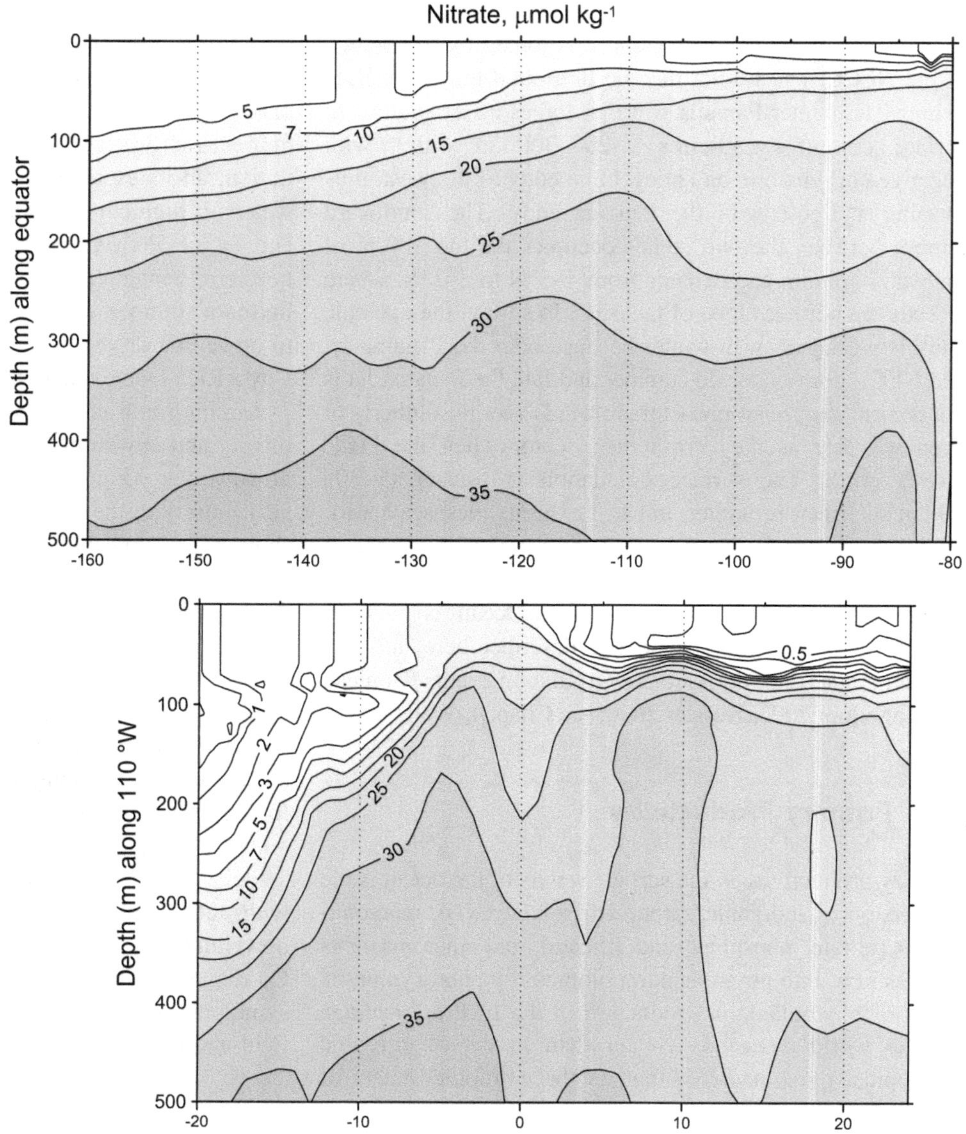

Fig. 3.14 Mean sections of nitrate (NO_3) along the equator (*top*) and 110°W, adapted from Fiedler and Talley (2006)

The extreme oxygen deficiency in the ETP oxygen minimum zone is attributable to several factors: (1) high phytoplankton production at the surface; (2) a sharp permanent pycnocline that prevents local ventilation of subsurface waters; (3) a sluggish and convoluted deep circulation and therefore old "age" of subpycnocline waters. The subpycnocline waters of the ETP, lying between the eastern equatorward borders of two subtropical gyres, is a "shadow zone" that is not directly ventilated by surface water subducted to thermocline depths within the gyres (Karstensen et al. 2008).

3.8 Impacts of Climate Change

Oceanographic characteristics of the ETP Ocean vary seasonally, as shown above. Interannually, with a warm/cold cycle of 3–7 years, variables including surface temperature, winds, and thermocline structure undergo changes associated with the El Niño-Southern Oscillation (see Chap. 4, Wang et al.). Longer-term climate change, caused by natural changes and anthropogenic global warming, is also occurring in the region. Sea surface temperature has increased by 0.4–1.0 °C since 1900 (Deser et al. 2010). Warming is expected to continue in the tropical Pacific, by ~2 °C for a doubling of atmospheric CO_2 (DiNezio et al. 2009), although not to the same degree as at higher latitudes (IPCC 2007). Walker circulation and the associated zonal surface winds along the equator are expected to weaken (Vecchi and Soden 2007). Model projections of future climate change show that the equatorial thermocline is expected to shoal, but the temperature gradients across the thermocline become steeper (Collins et al. 2010). Fiedler et al. (2013) show that, along with warming surface temperatures, the thermocline has shoaled in recent years in this region, and the

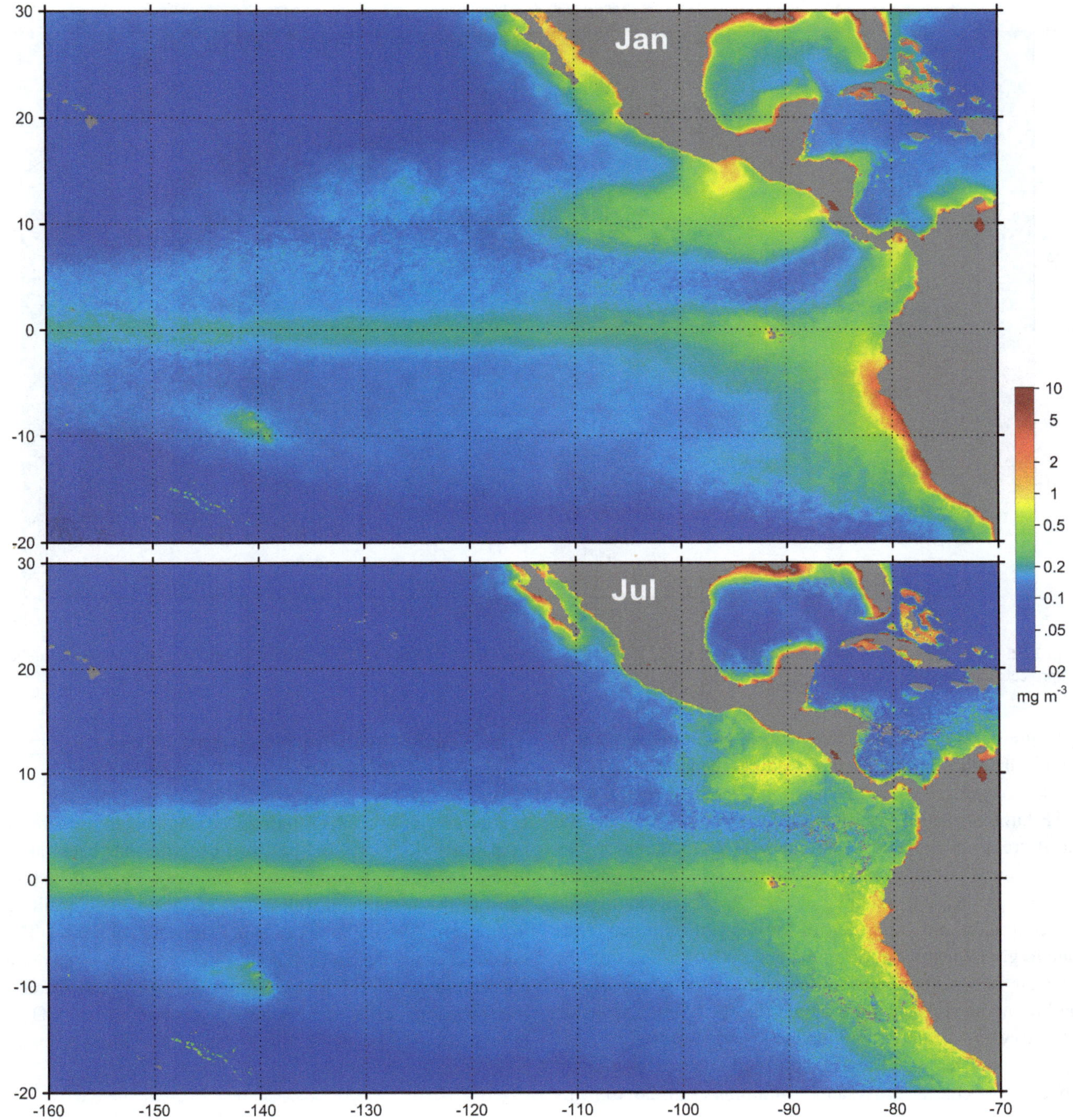

Fig. 3.15 SeaWiFS chlorophyll (mg m^{-3}) January and July climatologies. Data from monthly Level-3 standard mapped archive monthly composites, September 1997 to December 2010, NASA Goddard Space Flight Center Ocean Data Processing System, (http://oceandata.sci.gsfc.nasa.gov/)

stratification of near-surface waters has tended to decrease (Fig. 3.18). The current consensus of model projections is that phytoplankton production in tropical oceans will decrease with global warming as stratification increases and nutrient input to surface waters is reduced (Steinacher et al. 2010).

Sea level is rising globally, and will continue to rise, due to ocean warming and melting of glaciers and ice caps (IPCC 2007). This has the potential to "drown" a coral reef, due to reduced light levels at depth, if the sea level rise exceeds the reef accretion rate (Smith and Buddemeier 1992). Secular trends of <1 cm y^{-1} increasing wave height

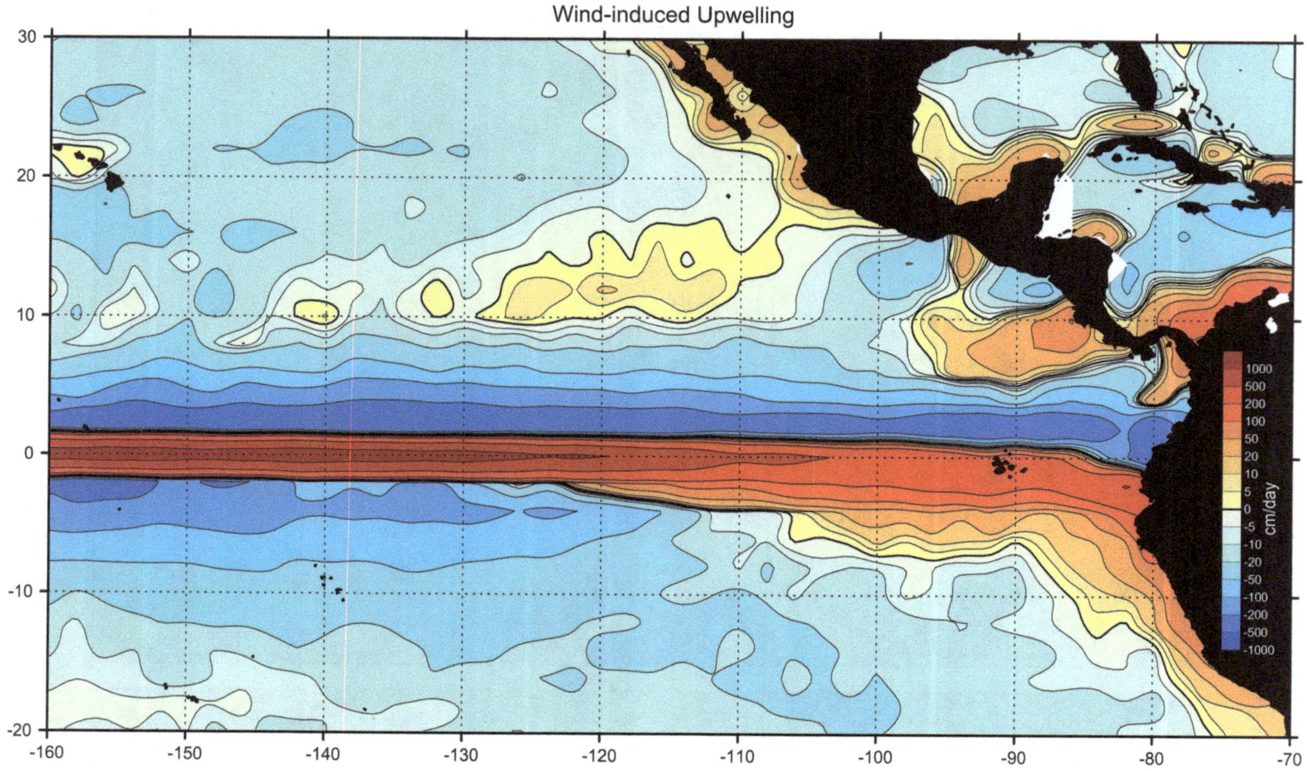

Fig. 3.16 Annual mean wind-driven upwelling, cm d^{-1}, 1950–1997 (Xie and Hsieh 1995). Data provided by Liusen Xie, http://www.ocgy.ubc.ca/projects/clim.pred/Upwell/

are significant (Sterl and Caires 2005), but not as large as at higher latitudes in the North Pacific (Semedo et al. 2011).

The oxygen minimum zones at intermediate depths in the ETP have expanded vertically during the past 50 years, but no decrease in minimum levels can be detected, since they cannot go lower than anoxia (Stramma et al. 2008). Climate models predict a continuing decline in dissolved oxygen concentration and expansion of the oxygen minimum zone due to global warming (Stramma et al. 2010). Effects of low oxygen on organisms and communities in the water column and in the benthos are well-known (Levin 2003). Expansion of oxygen minimum layers and upwelling of oxygen-depleted water onto continental shelves have raised concern for coastal ecosystems (Stramma et al. 2010). This is a potential pathway for low-oxygen waters to impact coral reef communities in the ETP.

Globally, reef-building corals are threatened by climate change through warming SST and an increase in coral bleaching, increasingly marginal carbonate saturation states due to ocean acidification, increases in temperature-related coral diseases, and possibly increased disturbance from greater storm intensity, and a myriad of local effects (Baskett et al. 2010; see Chaps. 8 and 18, Glynn et al. 2016 and Manzello et al. 2016 respectively).

3.9 Carbonate Chemistry and Ocean Acidification

Up to one-half of the CO_2 introduced into the atmosphere by human activities over the past two centuries has been taken up by the ocean (Raven et al. 2005; LeQuéré et al. 2009). Most of this remains in well-mixed surface waters, resulting in an overall decline in pH of ~ 0.1 (Raven et al. 2005). The net effect of the dissolution of CO_2 in seawater is to increase concentrations of H^+, H_2CO_3 (carbonic acid), and HCO_3^- (bicarbonate), while decreasing concentrations of CO_3^{2-} (carbonate), according to the chemical equilibria summarized as follows:

$$CO_2 + H_2O \leftrightarrow H_2CO_3 \leftrightarrow H^+ + HCO_3^- \leftrightarrow 2H^+ + CO_3{}^{2-}$$

Carbonate ions tend to buffer changes in ocean pH by reacting with H^+ to form bicarbonate ions. Although atmospheric CO_2 levels have been higher than today in the geological past, the carbonate buffer could effectively minimize changes in ocean pH because changes in dissolved CO_2 and H^+ were slow and exchange with deep waters could dissolve carbonate from deep-ocean sediments (Raven et al. 2005). At the present time, rapid acidification and reduction

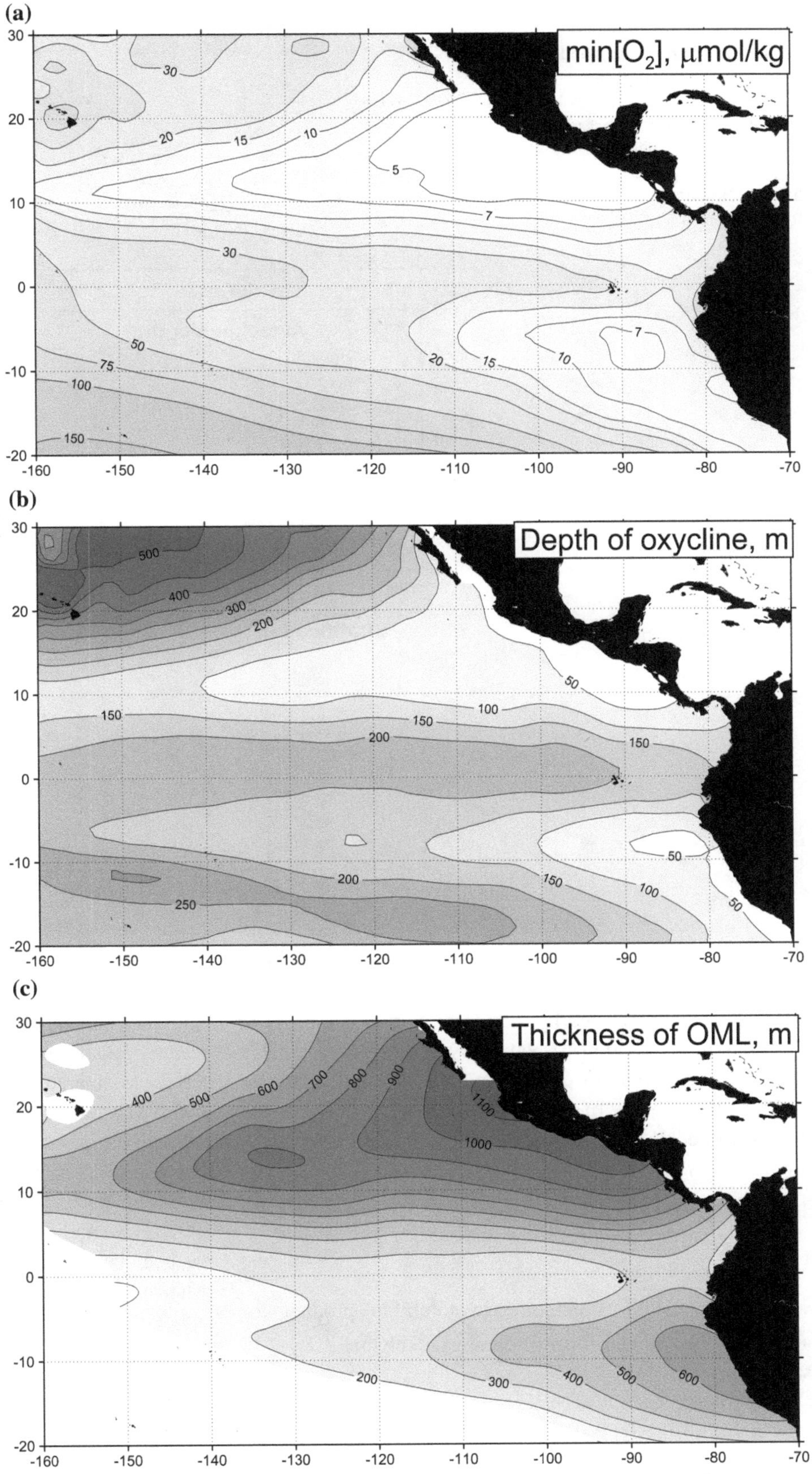

Fig. 3.17 Eastern tropical Pacific oxygen minimum layer: **a** minimum dissolved oxygen concentration (μmol kg^{-1}); **b** depth of oxycline (45 μmol kg^{-1}) at the top of the OML; **c** thickness of OML (<45 μmol kg^{-1}). From Fiedler and Talley (2006)

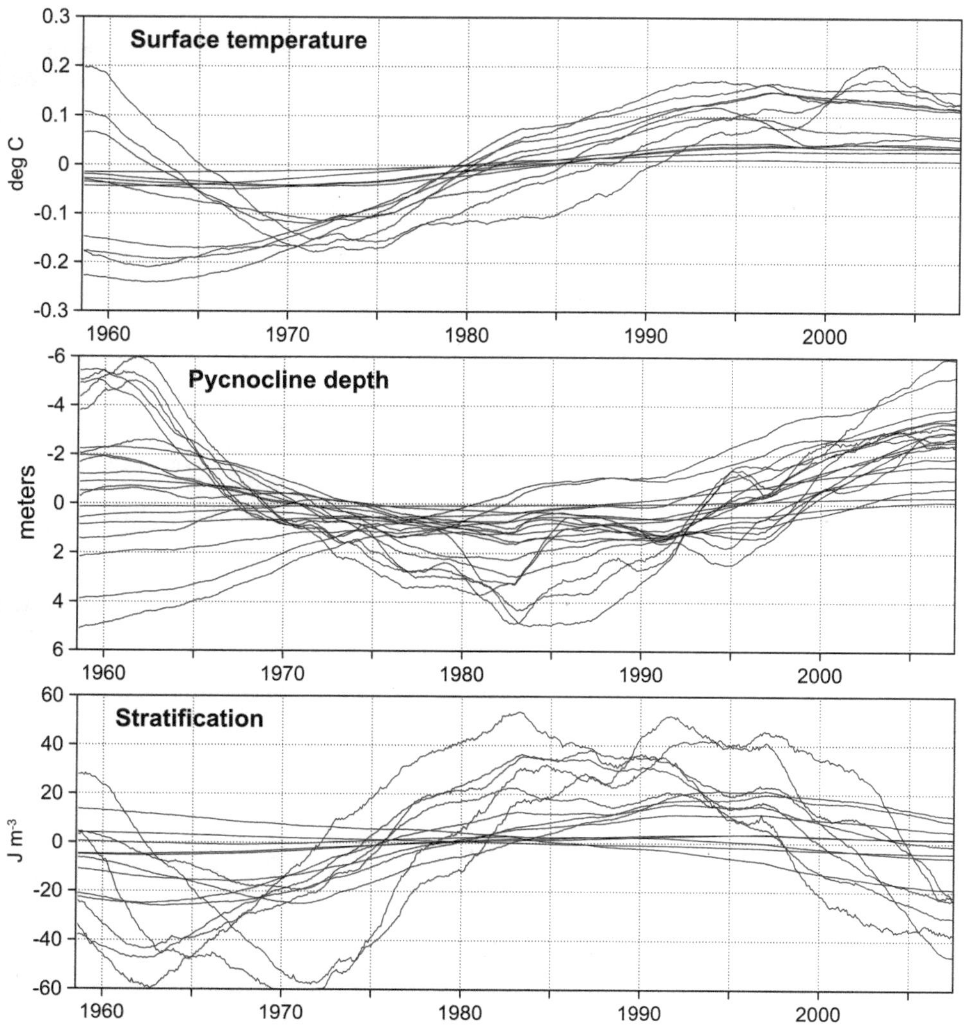

Fig. 3.18 Trends in surface temperature, depth of the pycnocline (thermocline), and density stratification of the 0–300 m surface layer (potential energy anomaly) in the eastern equatorial Pacific and ETP warm pool. Plots for 5-degree squares, 5°S–20°N and east of 110°W, from Fiedler et al. (2013)

of the concentration of carbonate in surface waters can have important consequences for the chemistry of carbonate minerals commonly used by marine biota to form shells or skeletons (e.g., calcium carbonate by reef-building corals). The ability of corals to precipitate carbonate efficiently, as well as the stability of biogenic carbonate that forms the structure of coral reefs, depends on the saturation state (Ω) of ambient seawater:

$$\Omega_{arag} = \frac{[Ca^{2+}][CO_3^{2-}]}{K'_{arag}}$$

where K'_{arag} is the stoichiometric solubility product for aragonite, the mineral form of coral exoskeletons (Ω = 1 at saturation).

The lowest surface pH values in the world's oceans are now found in the ETP (Raven et al. 2005), because deeper waters with high pCO_2, and thus low pH, are upwelled from below the shallow thermocline. For the same reason, the ETP is a major source for atmospheric CO_2, except during El Niño events, rather than a sink like most of the world's ocean (Takahashi et al. 2009). Due to the effect of low surface temperature on calcium carbonate solubility, the lowest Ω values are currently found in the Southern Ocean. Projections based on the "business-as-usual" scenario are that ETP waters will remain saturated (Fig. 3.19), but the saturation state will change from Ω ~3 to 2 (Kleypas et al. 1999b; Cao and Caldeira 2008). Low ambient Ω in the ETP has been correlated with limited reef structural development (Manzello et al. 2008; see Chap. 18, Manzello et al. 2016) and declines in coral growth have already been documented over the past 20–40 years (Manzello 2010). It has been suggested that ETP reefs provide a present-day, real-world analog of future reef structure and function with warming and acidification resulting from recurrent El Niño events and naturally corrosive waters, respectively (Manzello 2009).

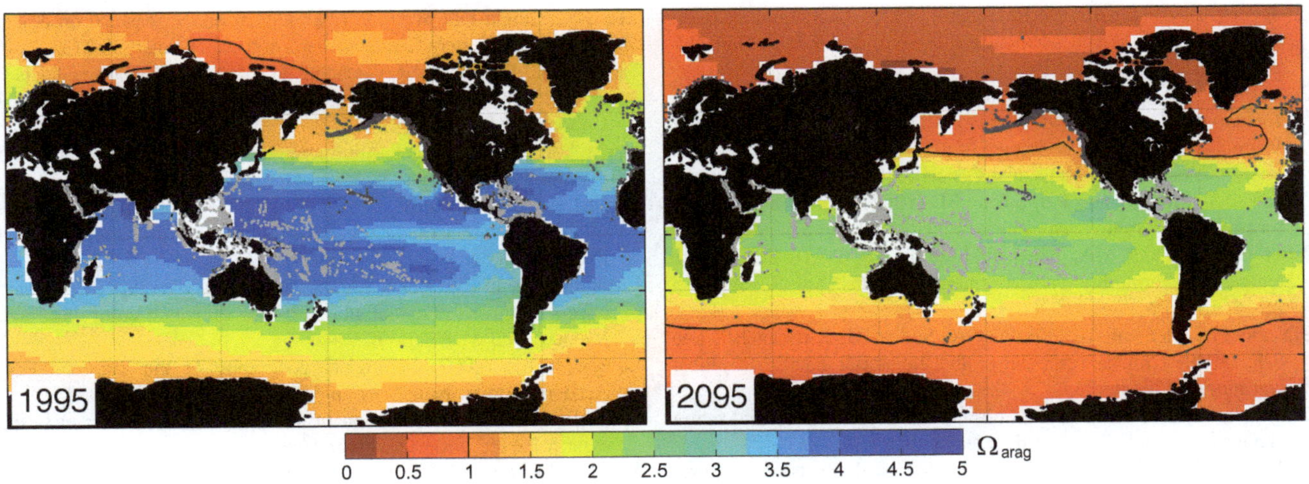

Fig. 3.19 Modeled decadal mean Ω_{arag} at the sea surface, centered on the years 1995 and 2095 (National Center for Atmospheric Research Community Climate System Model 3.1). Deep coral reefs are indicated by *darker gray dots*; shallow-water coral reefs are indicated with *lighter gray dots*, adapted from Feely et al. (2009)

Further climate change and ocean acidification are expected to negatively impact these reefs, which are already at the environmental limits for reef development (Kleypas et al. 1999b; Hoegh-Guldberg et al. 2007; Feely et al. 2009; see Chap. 18, Manzello et al. 2016).

3.10 Summary

Coral reefs of the eastern Pacific largely reside in the eastern Pacific warm pool. This environment is characterized by warm, low-salinity surface water lying on top of a strong, shallow thermocline. It is bounded by the equatorial cold tongue to the south, cooler and more saline subtropical water to the northwest, cold eastern boundary currents to the north and south, and the coast of the Americas. The continental boundary influences atmospheric forcing by trade winds through gap winds during winter and the location of the Intertropical Convergence Zone north of the equator and over the warm pool. To the west of the Galápagos, surface circulation is predominantly the east-west equatorial currents. Near the coast, surface circulation is modified by the coastal boundary, local winds, eddies, and interaction with eastern boundary currents. Primary productivity depends on oceanic upwelling along the equator and local centers of upwelling and wind mixing in coastal waters. Eastern tropical Pacific surface waters are moderately productive. Phytoplankton productivity is limited by a lack of the micronutrient dissolved iron, except where local coastal processes provide a source, so that macronutrients such as nitrate are never depleted.

The annual cycle of the declination of the sun results in seasonal changes in winds, rainfall and other environmental characteristics in this region, although seasonality is not as pronounced as at higher latitudes. In contrast, interannual variations caused by the El Niño-Southern Oscillation across the entire tropical Indo-Pacific are very important in this region. Oxygen depletion is extreme below the sharp thermocline, with consequences for mesopelagic and subthermocline benthic organisms. Surface waters are relatively low pH and marginally carbonate-saturated. Climate change is predicted to lead to future oceanographic changes in this region, such as warming and acidification of surface waters, increased stratification and reduced productivity, and upwelling/mixing of hypoxic waters into the surface layer. These changes are likely to affect organisms and populations inhabiting the ETP region.

References

Amador JA, Alfaro EJ, Lizano OG, Magaña VO (2006) Atmospheric forcing of the eastern tropical Pacific: a review. Prog Oceanogr 69 (2–4):101–142

Baskett ML, Nisbet RM, Kappel CV, Mumbey PJ, Gaines SD (2010) Conservation management approaches to protecting the capacity for corals to respond to climate change: a theoretical comparison. Global Change Biol 16:1229–1246

Bonjean F, Lagerloef GSE (2002) Diagnostic model and analysis of the surface currents in the tropical Pacific Ocean. J Phys Oceanogr 32:2938–2954

Brenes C, Lavín MF, Mascarenhas A Jr (2008) Geostrophic circulation between the Costa Rica Dome and Central America. Deep-Sea Res I 55:608–629

Camargo SJ, Robertson AW, Barnston AG, Ghil M (2008) Clustering of eastern North Pacific tropical cyclone tracks: ENSO and MJO effects. Geochem Geophys Geosyst 9(6). doi:10.1029/2007GC001861

Cao L, Caldeira K (2008) Atmospheric CO_2 stabilization and ocean acidification. Geophys Res Lett 35. doi:10.1029/2008GL035072

Chaigneau, A, Abarca del Río R, Colas F (2006). Lagrangian study of the Panama Bight and surrounding regions. J Geophys Res 111. doi:10.1029/2006JC003530

Collins M, An S-I, Cai W, Ganachaud A, Guilyardi E, Jin F-F, Jochum M, Lengaigne M, Power S, Timmermann A, Vecchi G, Wittenberg A (2010) The impact of global warming on the tropical Pacific Ocean and El Niño. Nat Geosci 3:391–397

Cromwell T, Montgomery RB, Stroup ED (1954) Equatorial undercurrent in Pacific Ocean revealed by new methods. Science 119:648–649

Deser C, Alexander MA, Xie S-P, Phillips AS (2010) Sea surface temperature variability: patterns and mechanisms. Ann Rev Mar Sci 2:115–143

DiNezio PN, Clement AC, Vecchi GA, Soden BJ, Kirtman BP, Lee S-K (2009) Climate response of the equatorial Pacific to global warming. J Climate 22:4873–4892

Feely RA, Doney SC, Cooley SR (2009) Present conditions and future changes in a high-CO_2 world. Oceanography 22(4):37–47

Fiedler PC (2002) The annual cycle and biological effects of the Costa Rica Dome. Deep-Sea Res I 49:321–338

Fiedler PC, Talley LD (2006) Hydrography of the eastern tropical Pacific: a review. Prog Oceanogr 69(2–4):143–180

Fiedler PC, Mendelssohn R, Palacios DM, Bograd SJ (2013) Pycnocline variations in the eastern tropical and North Pacific, 1958–2008. J Climate 26:583–599

Glynn PW, Mones AB, Podestá GP, Colbert A, Colgan MW (2016) El Niño-Southern Oscillation: Effects on Eastern Pacific Coral Reefs and Associated Biota. In: Glynn PW et al. (eds) Coral reefs of the eastern tropical Pacific. doi:10.1007/978-94-017-7499-4_8

Godínez VM, Beier E, Lavín MF, Kurczyn JA (2010) Circulation at the entrance of the Gulf of California from satellite altimeter and hydrographic observations. J Geophys Res 115. doi:10.1029/2009JC005705

Hastenrath S (1991) Climate dynamics of the tropics. Kluwer, Dordrecht, p 488

Hoegh-Guldberg O, Mumby PJ, Hooten AJ, Steneck RS, Greenfield P, Gomez E, Harvell CD, Sale PF, Edwards AJ, Caldeira K, Knowlton N, Eakin CM, Iglesias-Prieto R, Muthiga N, Bradbury RH, Dubi A, Hatziolos ME (2007) Coral reefs under rapid climate change and ocean acidification. Science 318:1737–1742

Huang RX (2010) Ocean circulation: wind-driven and thermohaline processes. Cambridge University Press, Cambridge. Online version available at: http://www.knovel.com/web/portal/browse/display?_EXT_KNOVEL_DISPLAY_bookid=3237

Intergovernmental Panel on Climate Change (2007) Fourth assessment report. http://www.ipcc.ch/publications_and_data/ar4/wg1/

Karstensen J, Stramma L, Visbeck M (2008) Oxygen minimum zones in the eastern tropical Atlantic and Pacific oceans. Prog Oceanogr 77:331–350

Kessler WS (2002) Mean three-dimensional circulation in the northeast tropical Pacific. J Phys Oceanogr 32:2457–2471

Kessler WS (2006) The circulation of the eastern tropical Pacific: a review. Prog Oceanogr 69(2–4):181–217

Kleypas JA, McManus JW, Meñez LAB (1999a) Environmental limits to coral reef development: where do we draw the line? Am Zool 39:146–159

Kleypas JA, Buddemeier RW, Archer D, Gattuso J-P, Langdon C, Opdyke BN (1999b) Geochemical consequences of increased atmospheric carbon dioxide on coral reefs. Science 284:118–120

Large WG, Nurser AJG (2001) Ocean surface water mass transformation. In: Siedler G, Church J, Gould J (eds) Ocean circulation and climate: observing and modeling the global ocean. Academic Press, New York, pp 317–336

Lavín MF, Beier E, Gómez J, Godínez VM, García J (2006) On the summer poleward coastal current off SW México. Geophys Res Lett 33. doi:10.1029/2005GL024686

Lee S-K, Enfield DB, Wang C (2007) What drives seasonal onset and decay of the Western Hemisphere warm pool? J Climate 20:2133–2146

Le Quéré C, Raupach MR, Canadell JG, Marland G, Bopp L, Ciais P et al (2009) Trends in the sources and sinks of carbon dioxide. Nat Geosci 2:831–836

Levin LA (2003) Oxygen minimum zone benthos: adaptation and community response to hypoxia. Oceanogr Mar Biol 41:1–45

Liang J-H, McWilliams JC, Gruber N (2009) High-frequency response of the ocean to mountain gap winds in the northeastern tropical Pacific. J Geophys Res 114. doi:10.1029/2009JC005370

Lirman D, Glynn PW, Baker AC, Leyte Morales GE (2001) Combined effects of three sequential storms on the Huatulco coral reef tract, Mexico. Bull Mar Sci 69:267–278

Longhurst AR (2007) Ecological geography of the sea, 2nd edn. Academic Press, New York, p 542

Lukas R (1986) The termination of the equatorial undercurrent in the eastern Pacific. Prog Oceanogr 16:63–90

Madden RA, Julian PR (1994) Observations of the 40–50-day tropical oscillation—a review. Mon Weather Rev 122:814–837

Maloney ED, Hartmann DL (2000) Modulation of eastern North Pacific hurricanes by the Madden-Julian Oscillation. J Climate 13:1451–1460

Manzello DP (2009) Reef development and resilience to acute (El Niño warming) and chronic (high-CO_2) disturbances in the eastern tropical Pacific: a real-world climate change model, vol 2, pp 1299–1304. In: Proceedings of 11th International Coral Reef Symposium, Ft Lauderdale

Manzello DP (2010) Coral growth with thermal stress and ocean acidification: lessons from the eastern tropical Pacific. Coral Reefs 29:749–758

Manzello DP, Kleypas JA, Budd DA, Eakin CM, Glynn PW, Langdon C (2008) Poorly cemented coral reefs of the eastern tropical Pacific: possible insights into reef development in a high-CO_2 world. Proc Natl Acad Sci USA 105:10,450–10,455

Manzello DP, Mark Eakin C, Glynn PW (2016) Effects of Global Warming and Ocean Acidification on Carbonate Budgets of Eastern Pacific Coral Reefs. In: Glynn PW et al. (eds) Coral reefs of the eastern tropical Pacific. doi:10.1007/978-94-017-7499-4_18

Paulmier A, Ruiz-Pino D (2009) Oxygen minimum zones (OMZs) in the modern ocean. Prog Oceanogr 80:113–128

Pennington JT, Mahoney KL, Kuwahara VS, Kolber DD, Calienes R, Chavez FP (2006) Primary production in the eastern tropical Pacific: a review. Prog Oceanogr 69(2–4):285–317

Philander SGH, Gu D, Halpern D, Lambert G, Lau N-C, Li T, Pacanowski RC (1996) Why the ITCZ is mostly north of the equator. J Climate 9:2958–2972

Raven J, Caldeira K, Elderfield H, Hoegh-Guldberg O, Liss P, Riebesell U, Shepherd J, Turley C, Watson A (2005) Ocean acidification due to increasing atmospheric carbon dioxide. Policy document 12/05, The Royal Society, London

Risien CM, Chelton DB (2008) A global climatology of surface wind and wind stress fields from eight years of QuikSCAT scatterometer data. J Phys Oceanogr 38:2379–2413

Romero-Vadillo E, Zaytsev O, Morales-Pérez R (2007) Tropical cyclone statistics in the northeastern Pacific. Atmosfera 20:197–213

Semedo A, Sušelj K, Rutgersson A, Sterl A (2011) A global view on the wind, sea and swell climate and variability from ERA-40. J Climate 24:1461–1479

Shea TJ, Trenberth KE, Reynolds RW (1992) A global monthly sea surface temperature climatology. J Climate 5:987–1001

Smith SV, Buddemeier RW (1992) Global change and coral reef ecosystems. Ann Rev Ecol Syst 23:89–118

Steinacher MF, Joos F, Frölicher TL, Bopp L, Cadule P, Cocco V, Doney SC, Gehlen M, Lindsay K, Moore JK, Schneider B, Segschneider J (2010) Projected 21st century decrease in marine productivity: a multi-model analysis. Biogeosciences 7:979–1005

Sterl A, Caires S (2005) Climatology, variability and extrema of ocean waves: the web-based KNMI/ERA-40 wave atlas. Int J Climatol 25:963–977

Stramma L, Johnson GC, Sprintall J, Mohrholz V (2008) Expanding oxygen-minimum zones in the tropical oceans. Science 320:655–658

Stramma L, Schmidtko S, Levin LA, Johnson GC (2010) Ocean oxygen minima expansions and their biological impacts. Deep-Sea Res I 57:587–595

Takahashi T, Sutherland SC, Wanninkhof R, Sweeney C, Feely RA, Chipman DW et al (2009) Climatological mean and decadal change in surface ocean pCO_2, and net sea–air CO_2 flux over the global oceans. Deep-Sea Res II 56:554–577

Talley LD, Pickard GL, Emery WJ, Swift JH (2011) Descriptive physical oceanography: an introduction, 6th edn. Academic Press, London, p 560

Vecchi GA, Soden BJ (2007) Global warming and the weakening of the tropical circulation. J Climate 20:4316–4340

Wang B (1994) Climatic regimes of tropical convection and rainfall. J Climate 7:1109–1118

Wang C, Enfield D (2001) The tropical western hemisphere warm pool. Geophys Res Lett 28:1635–1638

Willett CS, Leben R, Lavín MF (2006) Eddies and mesoscale processes in the eastern tropical Pacific: a review. Prog Oceanogr 69(2–4):218–238

Wyrtki K (1964) The thermal structure of the eastern Pacific Ocean. Deut Hydrograph Zeit, Ergänzungsheft A 6:84

Wyrtki K (1966) Oceanography of the eastern equatorial Pacific Ocean. Oceanogr Mar Biol Ann Rev 4:33–68

Wyrtki K (1967) Circulation and water masses in the eastern equatorial Pacific Ocean. Int J Oceanol Limn 1:117–147

Wyrtki K (1981) An estimate of equatorial upwelling in the Pacific. J Phys Oceanogr 11:1205–1214

Xie L, Hsieh WW (1995) The global distribution of wind-induced upwelling. Fish Oceanogr 4:52–67

El Niño and Southern Oscillation (ENSO): A Review

4

Chunzai Wang, Clara Deser, Jin-Yi Yu, Pedro DiNezio, and Amy Clement

Abstract

The ENSO observing system in the tropical Pacific plays an important role in monitoring ENSO and helping improve the understanding and prediction of ENSO. Occurrence of ENSO has been explained as either a self-sustained and naturally oscillatory mode of the coupled ocean-atmosphere system or a stable mode triggered by stochastic forcing. In either case, ENSO involves the positive ocean-atmosphere feedback hypothesized by Bjerknes. After an El Niño reaches its mature phase, negative feedbacks are required to terminate growth of the mature El Niño anomalies in the central and eastern Pacific. Four negative feedbacks have been proposed: reflected Kelvin waves at the ocean western boundary, a discharge process due to Sverdrup transport, western Pacific wind-forced Kelvin waves, and anomalous zonal advections. These negative feedbacks may work together for terminating El Niño, with their relative importance varying with time. Because of different locations of maximum SST anomalies and associated atmospheric heating, El Niño events are classified as eastern and central Pacific warming events. The identification of two distinct types of El Niño offers a new way to examine global impacts of El Niño and to consider how El Niño may respond and feedback to a changing climate. In addition to interannual variations associated with ENSO, the tropical Pacific SSTs also fluctuate on longer timescales. The patterns of Pacific Decadal Variability (PDV) are very similar to those of ENSO. When SST anomalies are positive in the tropical eastern Pacific, they are negative to the west and over the central North and South Pacific, and positive over the tropical Indian Ocean and northeastern portions of the high-latitude Pacific Ocean. Many mechanisms have been proposed for explaining PDV. Changes in ENSO under global warming are uncertain. Increasing greenhouse gases change the mean states in the tropical Pacific, which in turn induce ENSO changes. Due to the fact that the change in mean tropical condition under global warming is quite uncertain, even during the past few decades, it is difficult to say whether ENSO will intensify or weaken, but it is very likely that ENSO will not disappear in the future.

C. Wang (✉)
NOAA/Atlantic Oceanographic and Meteorological Laboratory, 4301 Rickenbacker Causeway, 33149 Miami, FL, USA
e-mail: Chunzai.Wang@noaa.gov

C. Deser
National Center for Atmospheric Research, Boulder, CO, USA

J.-Y. Yu
University of California at Irvine, Irvine, CA, USA

P. DiNezio
International Pacific Research Center, University of Hawaii, Honolulu, HI, USA

A. Clement
Rosenstiel School of Marine and Atmospheric Science, University of Miami, Miami, FL, USA

P.W. Glynn et al. (eds.), *Coral Reefs of the Eastern Tropical Pacific*,
Coral Reefs of the World 8, DOI 10.1007/978-94-017-7499-4_4

Keywords
Ocean-atmospheric interaction • Climate variability • Pacific decadal oscillation • Climate impacts

4.1 Introduction

El Niño is a large-scale oceanic warming event in the tropical Pacific Ocean that occurs every few years. The Southern Oscillation is characterized by an interannual seesaw in tropical sea level pressure (SLP) between the western and eastern Pacific, consisting of a weakening and strengthening of the easterly trade winds over the tropical Pacific. Bjerknes (1969) recognized that there is a close connection between El Niño and the Southern Oscillation (ENSO) and they are two different aspects of the same phenomenon. Bjerknes hypothesized that a positive ocean-atmosphere feedback involving the Walker circulation is a cause of ENSO. (The Walker circulation consists of the surface trade winds blowing from the east to the west across the tropical Pacific Ocean, the rising air in the tropical western Pacific, the upper-level winds blowing from the west to the east, and the sinking air returned back to the surface in the tropical eastern Pacific). An initial positive sea surface temperature (SST) anomaly in the equatorial eastern Pacific reduces the east-west SST gradient and hence the strength of the Walker circulation (Gill 1980; Lindzen and Nigam 1987), resulting in weaker trade winds around the equator. The weaker trade winds in turn drive the ocean circulation changes that further reinforce the SST anomaly. This positive ocean-atmosphere feedback leads the equatorial Pacific to a warm state, i.e., the warm phase of ENSO—El Niño. During Bjerknes' pioneering studies, the causes of a turnabout from a warm phase to a cold phase (La Niña) was not known (Philander 1990).

After Bjerknes published his hypothesis, ENSO was not intensively studied until the 1980s. The intense warm episode of the 1982–83 El Niño, which was not recognized until it was well developed, galvanized the tropical climate research community to understand ENSO and ultimately predict ENSO. This motivated the ten-year international TOGA (Tropical Ocean-Global Atmosphere) program (1985–94) to study and predict ENSO. TOGA successfully built an ENSO observing system (McPhaden et al. 1998) and greatly advanced our understanding of ENSO by focusing on the interaction between the tropical Pacific Ocean and atmosphere [e.g., see ENSO overviews by Philander (1990); Neelin et al. (1998); and Wang and Picaut (2004)].

After TOGA, the ENSO research community focused on different types of ENSO events, ENSO low-frequency variability and ENSO variability under global warming. All of these can change ocean temperatures in the tropical eastern Pacific, which in turn affect coral reefs. For example, coral reefs in the eastern tropical Pacific experienced catastrophic coral mortality during the 1982/83 El Niño event (e.g., Glynn 1985; Chap. 8, Glynn et al.). Therefore, an understanding of ENSO-related ocean temperature variability in the eastern tropical Pacific is very important. This chapter provides a brief overview of ENSO studies. Section 4.2 describes observations of ENSO. Section 4.3 reviews the understanding of ENSO mechanisms. Section 4.4 summarizes the recent development of the central versus eastern Pacific ENSO events. Section 4.5 reviews ENSO low-frequency variations known as Pacific decadal variability. ENSO variability under global warming is discussed in Sect. 4.6.

4.2 ENSO Observations

Modern observational data associated with ENSO can go back to the late 19th Century. Because actual observations are sparse in the tropical Pacific during early periods, data sets are normally produced using sparse observations, models, and statistical methods. Figure 4.1a–c shows ENSO indices from the beginning of the 20th Century: the SST anomalies in the Nino3 (150°W–90°W, 5°S–5°N) and Nino4 regions (160°E–150°W, 5°S–5°N), and the zonal wind anomalies in the Nino4 region. With reference to Fig. 4.1, several points can be made. First, the SST and zonal wind anomalies are highly correlated, indicating that ENSO is a coupled ocean-atmosphere phenomenon. Second, these ENSO indices show an oscillatory behavior with a 3–5 year preferred timescale in spite of considerable irregularity in the oscillation. Third, ENSO events show asymmetry between El Niño warm events and La Niña cold events, with anomalies of El Niño larger than those of La Niña. Fourth, Fig. 4.1b shows that the central Pacific warm events (as represented by the Nino4 index) occur more frequently during the last few decades (see Sect. 4.4 for details). Another feature that cannot be clearly seen in Fig. 4.1 is that ENSO is phase-locked to the seasonal cycle. That is, ENSO events tend to mature in the boreal winter (e.g., Rasmusson and Carpenter 1982). In the equatorial eastern Pacific (e.g., Costa Rica, Panama, Colombia, and Ecuador), this is during the dry season with low cloud cover and high solar insolation.

Variations of the thermocline are very important in ENSO events, but measurements of subsurface ocean temperature

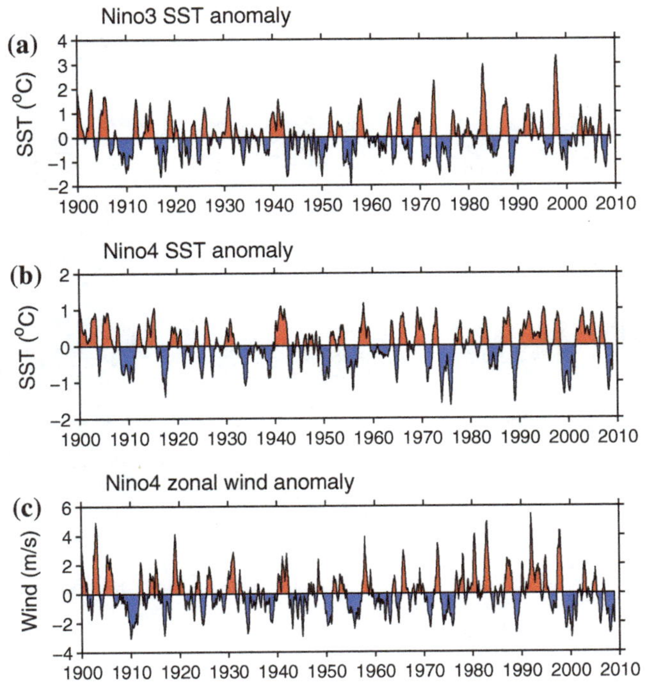

Fig. 4.1 ENSO indices. Shown are **a** SST anomalies in the Nino3 region (5°S–5°N, 150°W–90°W), **b** SST anomalies in the Nino4 region (5°N–5°S, 160°E–150°W), and **c** zonal wind anomalies in the Nino4 region. All time series are three-month running means. SST is from the NOAA extended reconstructed SST version 3 (Smith et al. 2008) and wind is from the newly developed NOAA Earth System Research Laboratory (ESRL) 20th Century Reanalysis (20CR) (Compo et al. 2011)

have been sparse in the past. One of the planned accomplishments of the TOGA program was to build an ENSO observing system in the tropical Pacific Ocean. The backbone of the ENSO observing system is now called the TAO/TRITON array, consisting of about 70 moored buoys (Hayes et al. 1991; McPhaden 1995), which is now supported and maintained by the USA and Japan. Most of the buoys are equipped with a 500-m thermistor chain and meteorological sensors. At the equator, five to seven moorings are equipped with ADCP (Acoustic Doppler Current Profiler) and current meters (McPhaden et al. 1998). The success of moored arrays in the tropical Pacific Ocean led to subsequent deployment of similar buoys in the tropical Atlantic and Indian Oceans (Fig. 4.2). Observed data are transmitted via satellites to data centers where they are compiled and made available to researchers and forecast centers in near-real time.

The evolution of El Niño and La Niña can be seen in the SST, zonal wind, and 20 °C isotherm depth (a proxy for thermocline depth) anomalies of the TAO/TRITON array data along the equator. Here we use the 1997/98 El Niño event to demonstrate the evolution of the eastern Pacific warm event (see Sect. 4.4 for the detail of differences between the eastern and central Pacific warm events). The TAO/TRITON moored data in Fig. 4.3 show that there is a close relationship among zonal wind anomalies, SST anomalies and thermocline depth anomalies. The importance of subsurface variation is clearly seen. Even one year in advance of the maximum surface warming, the precursor of El Niño is visible subsurface in the western Pacific (with positive thermocline anomalies) associated with westerly wind anomalies. The depression of the thermocline extends slowly from the west to the east along the equator. When warm subsurface temperature anomalies caused by the thermocline deepening reach the east, they are carried by equatorial upwelling to the surface. Once SST becomes anomalously warm, the Bjerknes feedback begins. The

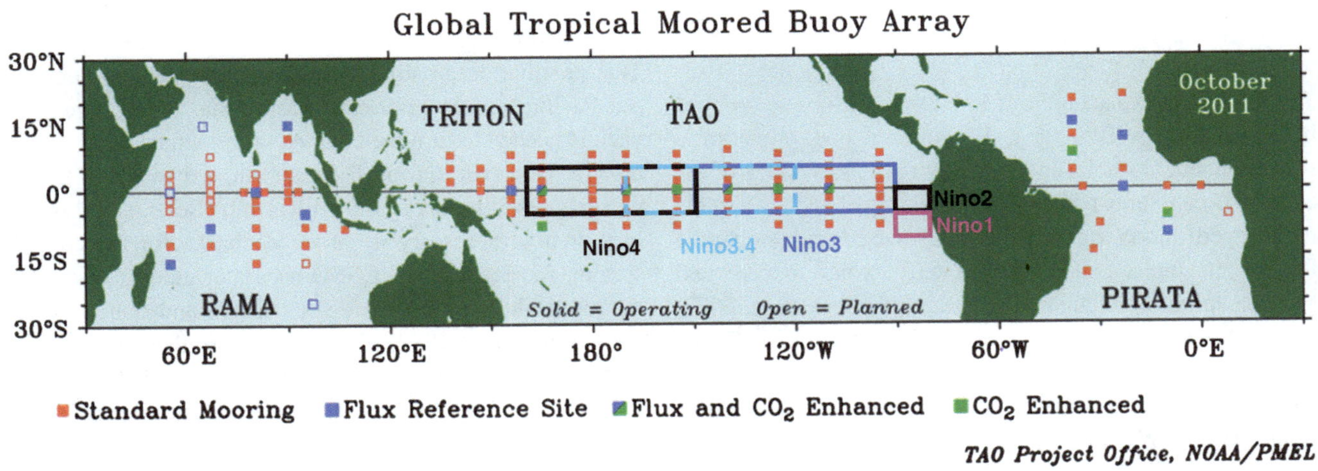

Fig. 4.2 The global tropical moored buoy array. The ENSO observed array in the tropical Pacific is called the TAO/TRITON array. The instruments deliver, in near-real time, data on surface and subsurface temperature and salinity, wind speed and direction, current velocity, and air-sea fluxes. Boxes outline the Nino3 region of 150°W–90°W, 5°S–5°N, the Nino4 region of 160°E–150°W, 5°S–5°N, the Nino3.4 region of 170°W–120°W, 5°S–5°N, the Nino1 region of 90°W–80°W, 5°S–10°S, and the Nino2 region of 90°W–80°W, 0°–5°S. The figure is modified from one provided by Dr. Michael McPhaden at NOAA/PMEL

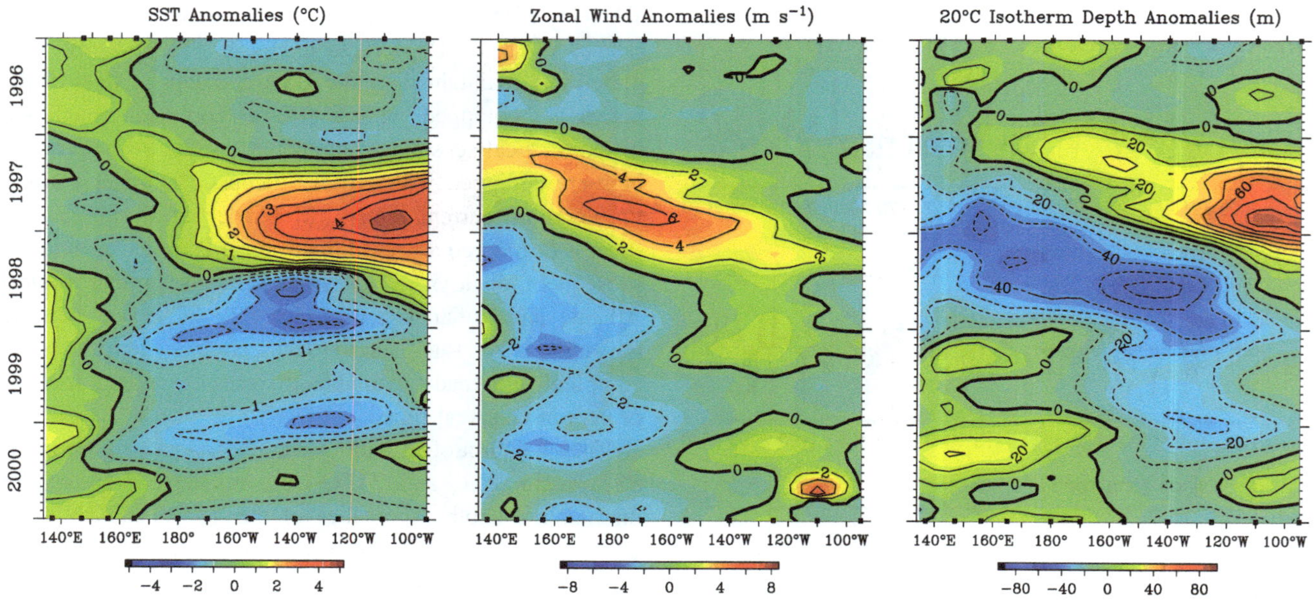

Fig. 4.3 Time-longitude sections of monthly SST, zonal wind, and 20 °C isotherm depth anomalies between 2°S to 2°N from January 1996 to December 2000. Data provided by the TAO/TRITON array

westerly wind anomalies in the central Pacific cause the eastern Pacific thermocline to deepen further, inducing additional warming. However, when the surface water is warm in the eastern Pacific, the shallower thermocline is seen as subsurface cold anomalies in the western Pacific (Fig. 4.3). Even during the development of an El Niño event, the seeds of its destruction are being sown in the western Pacific. The eastward extension of the subsurface cold anomalies causes the gradual erosion of the surface warm anomalies. This initiates a reversal of the chain of events of the Bjerknes feedback, which acts to drive the coupled system into a La Niña event.

4.3 ENSO Mechanisms

The theoretical explanations of ENSO can be loosely grouped into two frameworks. First, El Niño is one phase of a self-sustained, unstable, and naturally oscillatory mode of the coupled ocean-atmosphere system. Second, El Niño is a stable (or damped) mode triggered by or interacted with stochastic forcing or noise such as westerly wind bursts and Madden-Julian Oscillation events (e.g., Gebbie et al. 2007) and the tropical instability waves in the eastern Pacific Ocean (e.g., An 2008). In either framework, ENSO involves the positive ocean-atmosphere feedback of Bjerknes (1969). The early idea of Wyrtki's (1975) sea level "buildup" in the western Pacific warm pool treats El Niño as an isolated event. Wyrtki suggested that prior to El Niño, the easterly trade winds strengthened, and there was a "buildup" in sea level in the western Pacific warm pool. A "trigger" is a rapid collapse of the easterly trade winds. When this happens, the accumulated warm water in the western Pacific would surge eastward in the form of equatorial downwelling Kelvin waves to initiate an El Niño event.

4.3.1 Self-Sustained Oscillators of ENSO

Bjerknes (1969) first hypothesized that interaction between the atmosphere and the equatorial eastern Pacific Ocean causes El Niño. In Bjerknes' view, an initial positive SST anomaly in the equatorial eastern Pacific reduces the east-west SST gradient and hence the strength of the Walker circulation, resulting in weaker trade winds around the equator. The weaker trade winds in turn drive the ocean circulation changes that further reinforce the SST anomaly. This positive ocean-atmosphere feedback leads the equatorial Pacific to a never-ending warm state. A negative feedback is needed to turn the coupled ocean-atmosphere system around. However, during Bjerknes' time, it was not known what causes a turnabout from a warm phase to a cold phase. In search of necessary negative feedbacks for the coupled system, four conceptual ENSO oscillator models have been proposed: the delayed oscillator (Suarez and Schopf 1988; Battisti and Hirst 1989), the recharge oscillator (Jin 1997a, b), the western Pacific oscillator (Weisberg and Wang 1997; Wang et al. 1999), and the advective-reflective oscillator (Picaut et al. 1997). These oscillator models emphasized respectively the negative feedbacks of reflected Kelvin waves at the ocean western boundary, a discharge process due to Sverdrup transport, western Pacific wind-forced Kelvin waves, and anomalous zonal advection. These negative feedbacks may work together for terminating El Niño warming, as suggested by the unified oscillator (Wang 2001).

4.3.1.1 The Delayed Oscillator

Suarez and Schopf (1988) introduced the conceptual delayed oscillator as a candidate mechanism for ENSO, by considering the effects of equatorially trapped oceanic wave propagation. Based on the coupled ocean-atmosphere model of Zebiak and Cane (1987), Battisti and Hirst (1989) formulated and derived a version of the Suarez and Schopf (1988) conceptual delayed oscillator model and argued that this delayed oscillator model could account for important aspects of the numerical model of Zebiak and Cane (1987). The positive ocean-atmosphere feedback occurs in the equatorial eastern Pacific, leading the Nino3 SST anomaly to a warm state. The delayed negative feedback is by free Rossby waves generated in the eastern Pacific coupling region that propagate to and reflect from the ocean western boundary, returning as Kelvin waves to reverse the Nino3 SST anomalies in the eastern Pacific coupling region. The delayed oscillator assumes that the western Pacific is an inactive region for air-sea interaction and that ocean eastern boundary wave reflection is unimportant, emphasizing the significance of wave reflection at the ocean western boundary.

4.3.1.2 The Recharge Oscillator

Wyrtki (1975, 1985) first suggested a buildup in the western Pacific of warm water as a necessary precondition to the development of El Niño. Prior to El Niño upper ocean heat content or warm water volume over the entire tropical Pacific tends to build up (or charge) gradually, and during El Niño warm water is flushed toward (or discharged to) higher latitudes. After the discharge, the tropical Pacific becomes cold (La Niña) and then warm water slowly builds up again (recharge) before the occurrence of the next El Niño. The concept of the recharge and discharge processes is further emphasized by Jin (1997a, b). Based on the coupled model of Zebiak and Cane (1987), Jin (1997a, b) formulated and derived the recharge oscillator model. During the warm phase of ENSO, the divergence of Sverdrup transport associated with equatorial central Pacific westerly wind anomalies and equatorial eastern Pacific warm SST anomalies results in the discharge of equatorial heat content. The discharge of equatorial heat content leads to a transition phase in which the entire equatorial Pacific thermocline depth is anomalously shallow due to the discharge of equatorial heat content. This anomalous shallow thermocline at the transition phase allows anomalous cold waters to be pumped into the surface layer by climatological upwelling, leading to the cold phase. The converse occurs during the cold phase of ENSO. It is the recharge-discharge process that makes the coupled ocean-atmosphere system oscillate on interannual time scales.

4.3.1.3 The Western Pacific Oscillator

Observations show that ENSO displays both eastern and western Pacific interannual anomaly patterns (e.g., Rasmusson and Carpenter 1982; Wang et al. 1999; Wang and Weisberg 2000). During the warm phase of ENSO, warm SST anomalies in the equatorial eastern Pacific are accompanied by cold SST and shallow thermocline depth anomalies in the off-equatorial western Pacific. Also, while the zonal wind anomalies over the equatorial central Pacific are westerly, those over the equatorial western Pacific are easterly. Consistent with these observations, Weisberg and Wang (1997) and Wang et al. (1999) developed a conceptual western Pacific oscillator model for ENSO. This model emphasizes the role of the western Pacific in ENSO that had been overlooked in the delayed oscillator. In particular, off-equatorial SST anomalies (and off-equatorial anomalous anticyclones) in the western Pacific induce equatorial western Pacific wind anomalies that affect the evolution of ENSO. Condensation heating due to convection in the equatorial central Pacific (e.g., Deser and Wallace 1990) induces a pair of off-equatorial cyclones with westerly wind anomalies on the equator. These equatorial westerly wind anomalies act to deepen the thermocline and increase SST in the equatorial eastern Pacific, thereby providing a positive feedback for anomaly growth. In contrast, the off-equatorial cyclones raise the thermocline there via Ekman pumping. Thus, a shallow off-equatorial thermocline anomaly expands over the western Pacific leading to a decrease in SST and an increase in SLP (sea level pressure) in the off-equatorial western Pacific (Wang et al. 1999; Wang 2000). During the mature phase of El Niño, the off-equatorial anomalous anticyclones initiate equatorial easterly wind anomalies in the western Pacific. These equatorial easterly wind anomalies cause upwelling and cooling that proceed eastward as a forced ocean response providing a negative feedback, allowing the coupled ocean-atmosphere system to oscillate.

4.3.1.4 The Advective-Reflective Oscillator

Picaut et al. (1997) proposed a conceptual advective-reflective oscillator model for ENSO. In this conceptual model, they emphasize a positive feedback of zonal currents that advect the western Pacific warm pool toward the east during El Niño. Three negative feedbacks tend to push the warm pool back to its original position in the western Pacific, namely: anomalous zonal currents associated with wave reflection at the western boundary, anomalous zonal currents associated with wave reflection at the eastern boundary, and mean zonal currents converging at the eastern edge of the warm pool. During the warm phase of ENSO, equatorial westerly wind anomalies in the central Pacific produce upwelling Rossby and downwelling Kelvin waves that propagate westward and eastward, respectively. The westward propagating upwelling Rossby waves reflect as upwelling Kelvin waves after they reach the

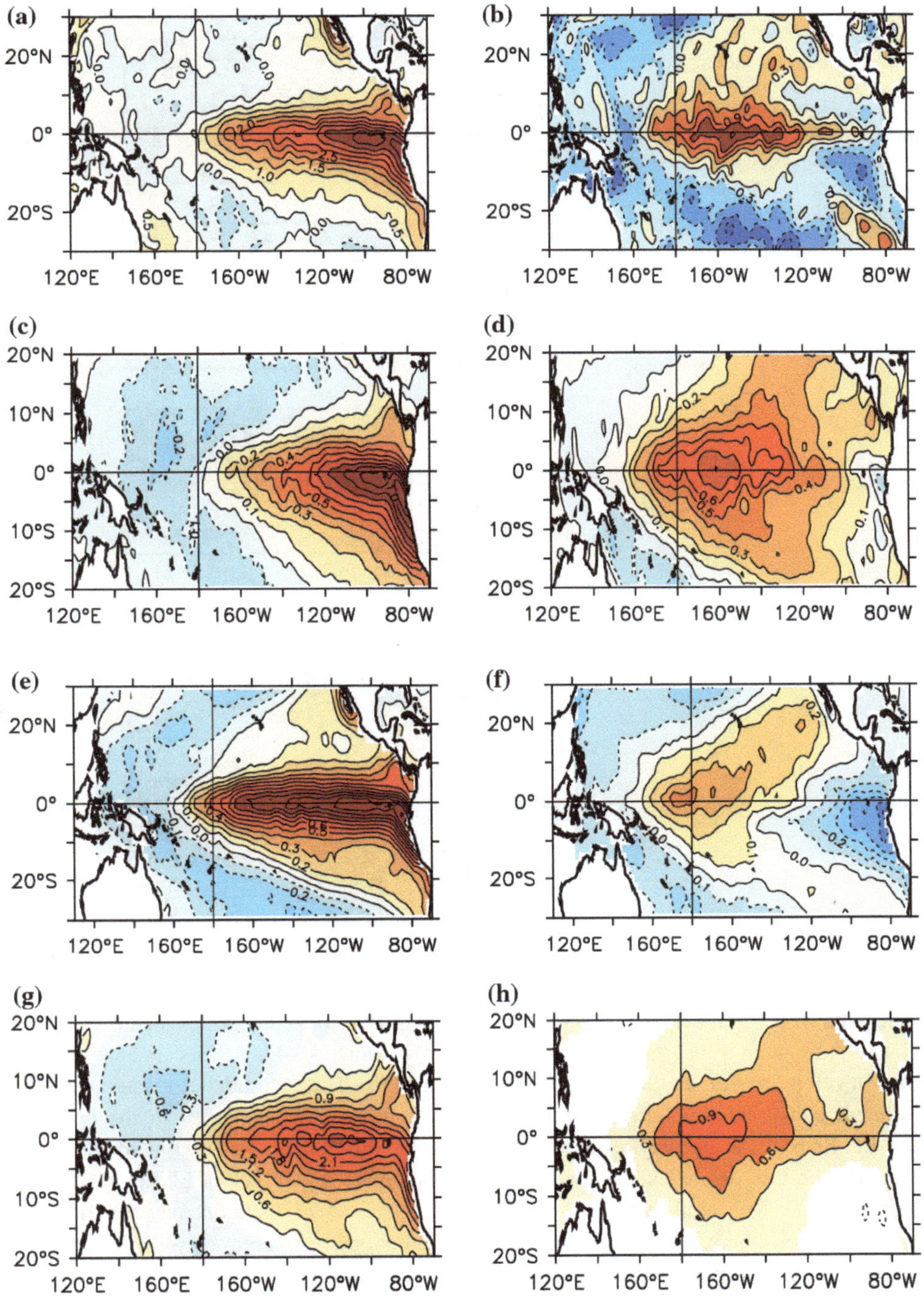

Fig. 4.4 SST anomaly patterns for **a** the 1997–98 El Niño (anomalies averaged from November 1997 to January 1998); **b** the 1977–78 El Niño (anomalies averaged from November 1977 to January 1978); **c–d** the EP-ENSO and CP-ENSO obtained from the EOF-regression method of Kao and Yu (2009); **e–f** the 1st and 2nd EOF that representing the conventional El Niño and El Niño Modoki obtained from the regular EOF analysis of Ashok et al. (2007); and **g–h** the Cold Tongue and Wam Pool El Niño composed by Kug et al. (2009). SST data from HadISST (1970–2005) were used in the calculation

statistical methods have been used to identify the typical SST anomaly patterns of the EP and CP types of ENSO, including the standard Empirical Orthogonal Function (EOF) analysis of Ashok et al. (2007), the regression-EOF method of Kao and Yu (2009), and the composite analysis of Kug et al. (2009). The spatial patterns obtained by all three methods (Fig. 4.4c–h) for the two types of ENSO are similar. In particular, the CP ENSO patterns obtained by all three methods exhibit a poleward extension of SST anomalies from the central Pacific into both the northern and southern subtropics. The connection to the northern subtropics seems to be stronger, and the SST anomalies in the subtropical northeastern Pacific precede those in the equatorial central Pacific (Fig. 4.5). In addition, the propagating feature of the SST anomalies is weaker and less clear in the CP type of ENSO than in the EP type of ENSO. Below the ocean surface, while the EP ENSO is known to be characterized by subsurface temperature anomalies propagating across the Pacific basin (Fig. 4.3) similar to those described by the delayed oscillator theory of ENSO (Battisti and Hirst

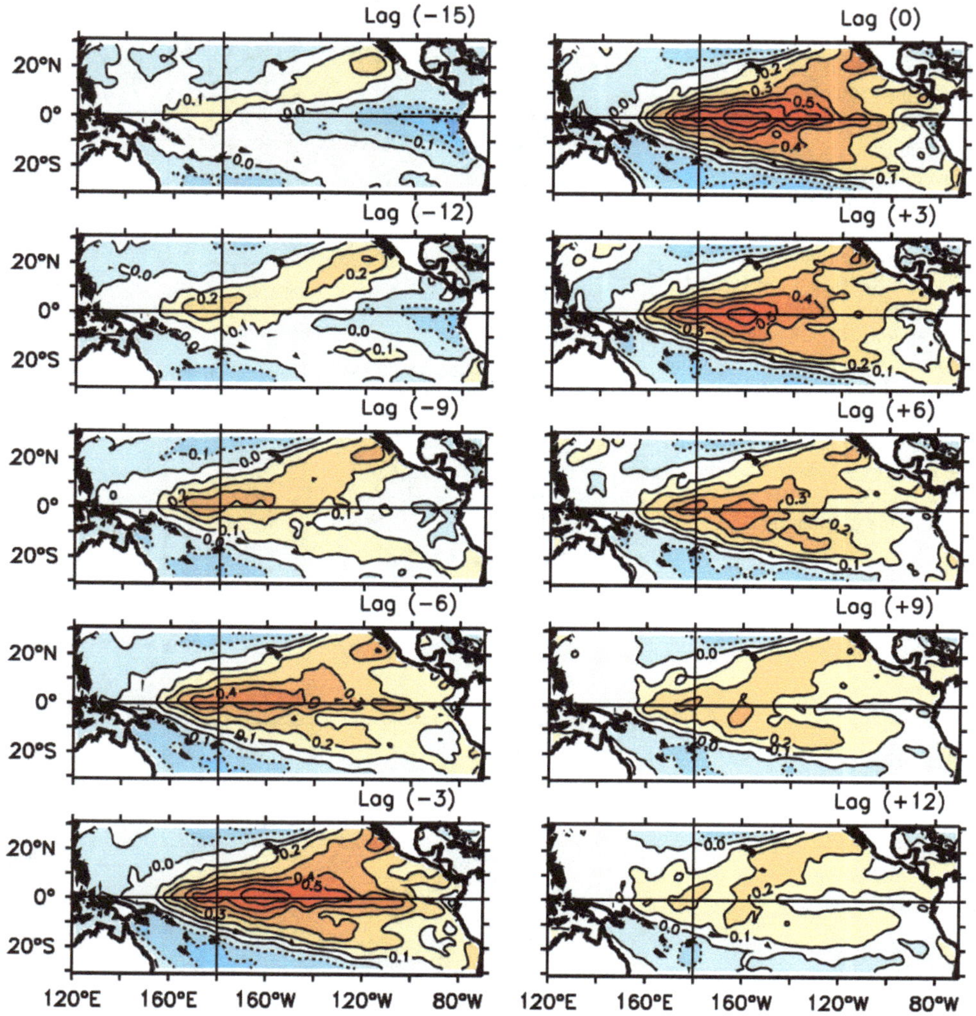

Fig. 4.5 Evolution of SST anomalies of the CP ENSO. Values shown are the lagged (in months) correlation coefficients between the principal components of the CP-ENSO obtained from the EOF-regression method of Kao and Yu (2009), and tropical Pacific SST anomalies. Contour intervals are 0.1

1989 and Schopf and Suarez 1988), the CP ENSO is associated more with subsurface ocean temperature anomalies that develop in situ in the central Pacific (Fig. 4.6). The different subsurface evolution indicates that, in contrast to the EP ENSO, the underlying dynamics of the CP ENSO seems to be less dependent on thermocline variations (Kao and Yu 2009; Kug et al. 2009).

In the atmosphere, wind stress and precipitation anomaly patterns associated with these two types of ENSO are also different. While the EP El Niño is associated with significant westerly anomalies covering a large part of the tropical Pacific, the westerly anomalies associated with the CP El Niño have a smaller spatial scale and are centered in the equatorial central-to-western Pacific (Kao and Yu 2009; Kug et al. 2009). This more westward location of the westerly anomalies in the CP El Niño is consistent with the location of its SST anomalies. Significant easterly anomalies also appear over the tropical eastern Pacific during the CP El Niño. In terms of precipitation, positive anomalies associated with the EP El Niño typically extend from the equatorial eastern to central Pacific where the largest SST anomalies occur. For the CP El Niño, the precipitation anomalies are characterized by a dipole pattern, with positive anomalies located mainly in the western Pacific and negative anomalies in the eastern Pacific (Kao and Yu 2009; Kug et al. 2009). The different precipitation patterns of these two types of ENSO imply that the associated convective heating locations and the mid-latitude teleconnections could be different as well.

Several methods have been proposed to identify the two types of ENSO, which include using SST or subsurface ocean temperature information. Although there are some inconsistencies among the events identified by these various methods, there are several events commonly identified as the CP type by these methods, which include the following major El Niño events that have occurred since the 1960s: 1968–69, 1977–78, 1986–87, 1991–92, 1994–95, 2002–03, 2004–05, and 2009–10. The 2009–10 El Niño is known to be one of the strongest CP El Niño events in recent decades (Lee and McPhaden 2010). This list reveals a tendency for the CP El Niño to occur more often in recent decades (Ashok et al. 2007; Kao and Yu 2009; Kug et al. 2009; Lee and McPhaden

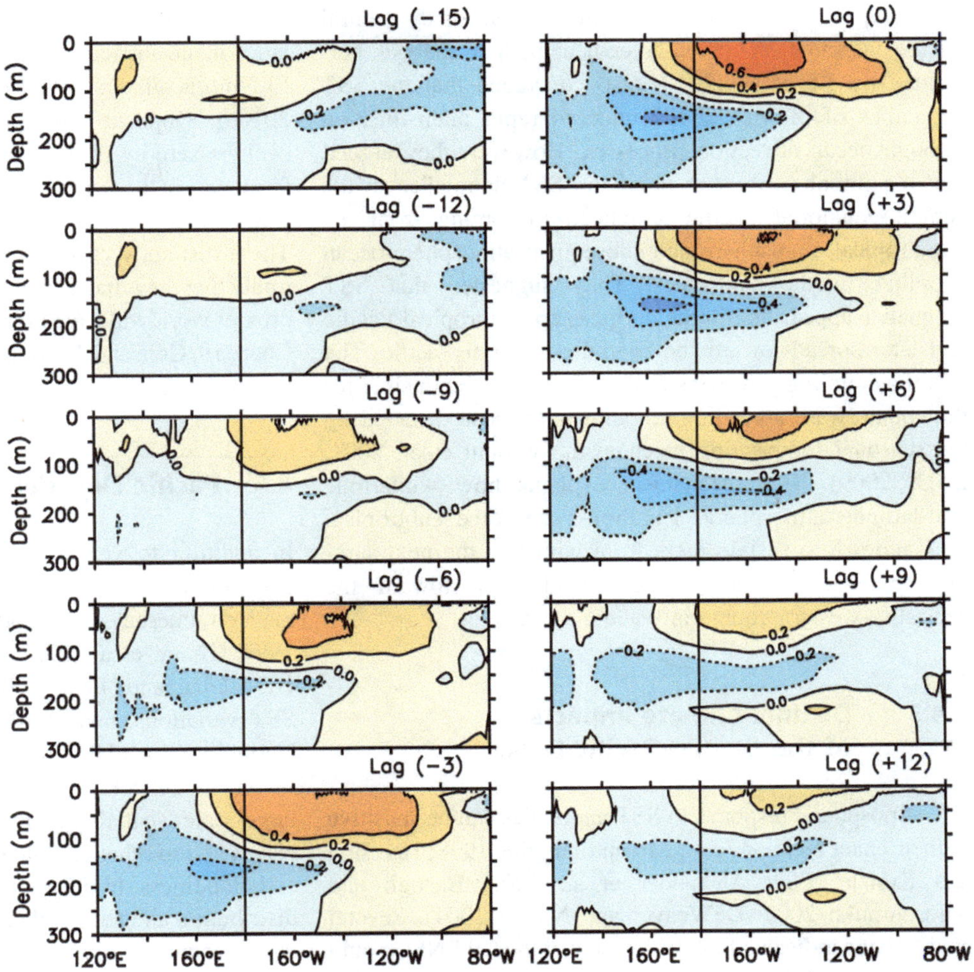

Fig. 4.6 Evolution of subsurface ocean temperature anomalies of the CP ENSO along the equator (5°S–5°N). Values shown are the lagged correlation coefficients between the principal components of the CP ENSO obtained from the EOF-regression method of Kao and Yu (2009) and temperature anomalies. Contour intervals are 0.2. Ocean temperature assimilation from GECCO (1970–2001) was used in the calculation

2010). It is interesting to note that at least three of the four El Niño events in the 21st century (i.e., the 2002/03, 2004/05, and 2009/10 events) have been of the CP type. Yeh et al. (2009) compared the ratio of the CP to EP type of El Niño events in Coupled Model Intercomparison Project phase 3 (CMIP3) model simulations and noticed that the ratio is projected to increase under a global warming scenario. They argued that the recent increase in the occurrence of the CP El Niño is related to a weakening of the mean Walker circulation and a flattening of the mean thermocline in the equatorial Pacific, which might be a result of global warming (Vecchi et al. 2007). However, it was also argued that the increasing occurrence of the CP El Niño in recent decades could be an expression of natural multidecadal variability and not necessarily a consequence of anthropogenic forcing (Newmann et al. 2011; McPhaden et al. 2011).

4.4.2 Dynamics of the Central-Pacific El Niño

A specific generation mechanism for the CP ENSO has yet to be fully agreed upon, and there are ongoing debates as to whether the CP ENSO should be considered as a completely different entity from the EP ENSO or simply a different expression of the same EP ENSO dynamics. As mentioned, for the CP El Niño the equatorial westerly anomalies appear to the west of the positive SST anomalies in the central Pacific and the equatorial easterly anomalies to the east. Ashok et al. (2007) argued that the thermocline variations induced by this wind anomaly pattern are responsible for the generation of the CP ENSO. The equatorial westerly anomalies induce downwelling Kelvin waves propagating eastward and the equatorial easterly anomalies induce downwelling Rossby waves propagating westward and, together, they deepen the thermocline in the central Pacific to produce the CP El Niño. Kug et al. (2009) emphasized the fact that the equatorial easterly anomalies can suppress warming in the eastern Pacific during a CP El Niño event by enhancing upwelling and surface evaporation. However, they also argued that the mean depth of the thermocline in the central Pacific is relatively deep and the wind-induced thermocline variations may not be efficient in producing the CP SST anomalies. Instead, they suggested that ocean advection is responsible for the development of the central Pacific warming. A mixed-layer heat budget analysis performed by Yu et al. (2010) also concluded that the SST anomalies of the CP ENSO undergo rapid intensification through ocean advection processes. However, they argued that the initial establishment of the SST anomalies in the central equatorial Pacific is related to forcing from the extratropical atmosphere and subsequent atmosphere-ocean coupling in the subtropics. They suggested that SST anomalies appear first in the northeastern subtropical Pacific and later spread toward the central equatorial Pacific. The specific coupling processes in the subtropics responsible for the equatorward spreading are similar to those depicted by the seasonal footprinting mechanism (Vimont et al. 2001, 2003, 2009). This mechanism explains how wintertime mid-latitude atmospheric variations can force subtropical SST anomalies, sustain them from winter to the next summer, and at the same time spread them toward the central-to-western equatorial Pacific.

4.4.3 Distinct Climate Impacts of the Central-Pacific ENSO

The atmospheric response to SST anomalies can be sensitive to their exact location (e.g., Mo and Higgins 1998; Hoerling and Kumar 2002; Alexander et al. 2002; Basugli and Sardeshmukh 2002; DeWeaver and Nigam 2004). Several studies have indicated that the impacts of the CP ENSO could be distinctly different from those of the EP ENSO. For example, the impact of the CP ENSO on the winter temperatures of the USA is characterized by an east-west dipole pattern rather than by the well known north-south dipole pattern conventionally associated with the EP ENSO (Mo 2010). The western North Pacific summer monsoon has a stronger relationship with the CP ENSO than the EP ENSO (Weng et al. 2011), and rainfall variations in southern China are different for the two types of ENSO (e.g., Feng and Li 2011; Zhang et al. 2011). Australian rainfall also has been suggested to be more sensitive to the CP than to the EP type of ENSO (Wang and Hendon 2007; Taschetto and England 2009). The results shown in Kumar et al. (2006) also imply that the CP El Niño can reduce the Indian monsoon rainfall more effectively than the EP El Niño. In the Southern Hemisphere, the CP ENSO has been shown to have a stronger impact on storm track activity than the EP ENSO (Ashok et al. 2007). The 2009 CP El Niño event was argued to have an influence far south as to contribute to the melting of Antarctica ice by inducing a stationary anticyclone outside the polar continent and enhancing the eddy heat flux into the region (Lee T et al. 2010b). The influence of the CP El Niño on Atlantic hurricanes may also be different from the conventional EP El Niño (Kim et al. 2009), but it has been shown that the anomalous atmospheric circulation in the hurricane main development region during the CP El Niño is similar to that during the EP El Niño (Lee et al. 2010a). Opposite impacts were noticed for the tropical cyclone activity in the western Pacific: the tropical cyclone frequency in the South China Sea increases during CP El Niño years but decreases during EP El Niño years (Chen 2011). These distinctly climate impacts of the EP and CP ENSOs imply that they may leave different signatures in paleoclimate proxies worldwide, including corals, (e.g., coral skeletons, see Chap. 19, Cole and Tudhope), but these signatures have yet to be explored.

4.5 Pacific Decadal Variability

In addition to year-to-year variations associated with the ENSO phenomenon, SSTs in the tropical Pacific also fluctuate on timescales of decades and longer (e.g., Mantua et al. 1997; Zhang et al. 1997; Power et al. 1999; Deser et al. 2004; Guan and Nigam 2008). These tropical Pacific decadal SST variations, henceforth referred to as "Pacific Decadal Variability" or PDV (also called the Pacific Decadal Oscillation, or PDO in the literature), are organized into large-scale spatial patterns with linkages to other ocean basins and to other climate parameters such as rainfall, wind, and cloudiness. In this section, we describe the geographical distribution and temporal characteristics of PDV, and discuss the mechanisms that contribute to PDV.

4.5.1 Geographical and Temporal Characteristics of PDV

Figure 4.7 shows maps of the standard deviation (σSST) of SST fluctuations on timescales less than 8 years (referred to as interannual) and greater than 8 years (referred to as decadal) based on the Hadley Centre Sea Ice and SST version 1 (HadISST1) dataset (Rayner et al. 2003) for the period of 1900–2010. Similar results are also obtained with other data sets (not shown). Prior to computing the σSST, the mean seasonal cycle was removed by subtracting the long-term monthly means from each month and the trend was removed using a quadratic fit to the time series. The largest values of interannual σSST (exceeding 1 °C) occur along the equatorial Pacific and coastal Peruvian upwelling zones, with additional prominent maxima along the Gulf Stream, Kuroshio Current and coastal Argentina. The same regions exhibit high values of decadal σSST but with weaker amplitude (0.4°–0.6°C). Notably, the contrast between minimum and maximum values of σSST is much less on decadal compared to interannual timescales, especially

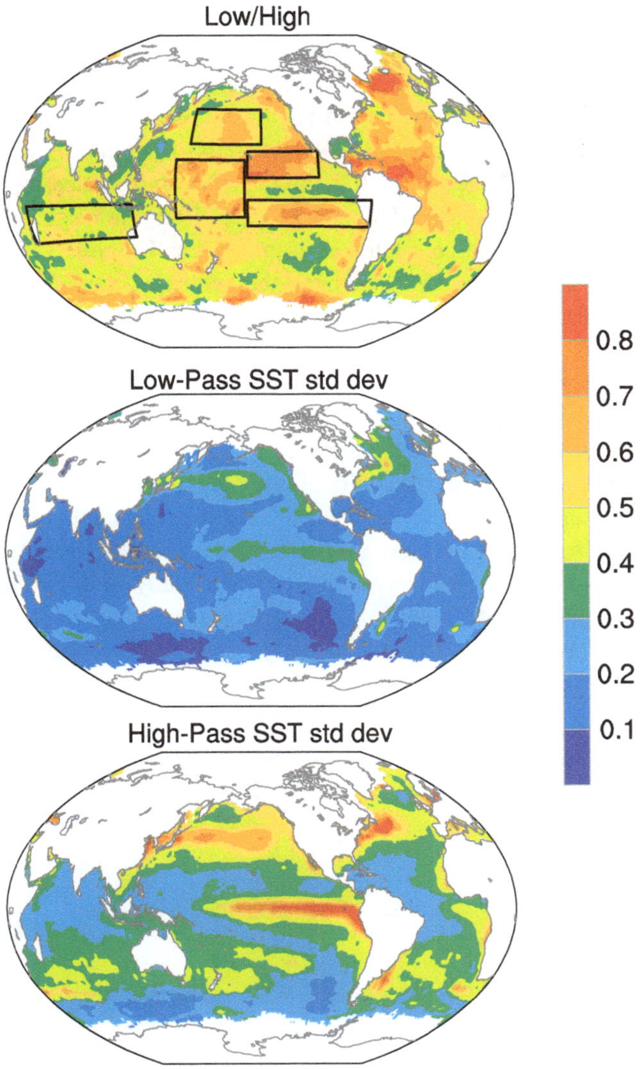

Fig. 4.7 The standard deviation (°C) of SST fluctuations on time scales less than 8 years (bottom panel) and greater than 8 years (middle panel) based on the HadISST1 dataset (Rayner et al. 2003) for the period 1900–2010 using the filter in Zhang et al. (1997). Prior to computing the standard deviations, the mean seasonal cycle was removed by subtracting the long-term monthly means from each month and the trend was removed using a quadratic fit to the time series. The top panel shows the ratio between the middle panel and the bottom panel. Boxes outline regions used in Fig. 4.8

within the tropics. The ratio of decadal σSST to interannual σSST (the top panel of Fig. 4.7) reveals that decadal variability is comparable in magnitude to interannual variability throughout much of the tropical Pacific, except along the eastern equatorial upwelling zone where it is considerably weaker. Thus, decadal SST fluctuations are more readily apparent outside of the canonical ENSO region (often termed the Nino3.4 region).

To further illustrate this point, we show a sequence of SST time series for selected regions in the tropical Indo-Pacific arranged from the east to west (Fig. 4.8). The top panel shows the canonical ENSO region, and the remaining panels

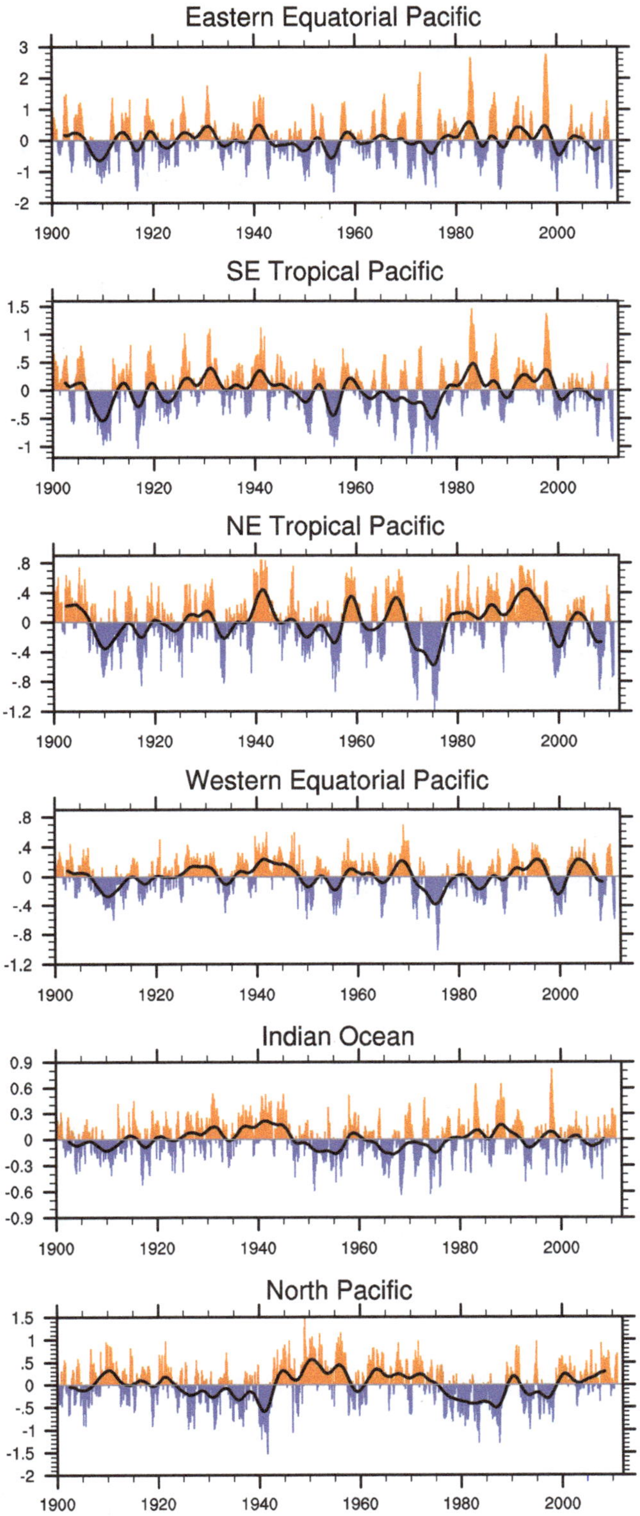

Fig. 4.8 Regional SST time series in the tropical Indo-Pacific, arranged from the east to west (regions are outlined in the top panel of Fig. 4.7). Unfiltered monthly SST anomalies are displayed as colored bars and low-pass filtered anomalies are shown as black curves; all time series have been quadratically detrended. Regions are defined as follows: Eastern Equatorial Pacific (5°N–5°S, 160°W–90°W); SE Tropical Pacific (5°S–20°S, 160°W–70°W); NE Tropical Pacific (5–20°N, 160°E–110°W); Western Equatorial Pacific (15°S–15°N, 150°E–160 W°); Indian Ocean (5°S–25°S, 45°E–120°E); North Pacific (25° N–45°N, 160°E–150 W°)

show regions where the ratio of decadal-to-interannual variability is a maximum (regions are outlined in Fig. 4.7 and latitude/longitude boundaries are given in the figure caption). The bottom panel shows the canonical PDV region in the central North Pacific (see Mantua et al. 1997). Unfiltered monthly SST anomalies are displayed as colored bars and low-pass filtered anomalies are shown as black curves; all time series have been quadratically detrended.

The regional SST records exhibit similarities and differences. Interannual fluctuations associated with ENSO are most readily evident in the eastern equatorial Pacific index, but also appear in the other records, albeit less prominently due to the enhanced low-frequency variability. Decadal and longer timescale fluctuations are most apparent in the central North Pacific and tropical Indian Ocean indices (note that these have opposite signs), but similar behavior is evident in the other records. These low-frequency variations are characterized by relatively cold conditions during approximately 1910–1925 and 1947–1976, and by relatively warm conditions during approximately 1926–1945 and 1977–1998; however, the exact timing of the warm and cold intervals varies with region (similar epochs are evident in the central North Pacific record but with opposite sign). In addition to these inter-decadal variations, shorter decadal-scale fluctuations are evident, especially in the northeastern Pacific and western equatorial Pacific indices but also in the other records to varying degrees. For example, the 1947–1976 cold epoch is punctuated by a warm interval in the 1960s, and the 1925–1946 warm epoch is interrupted by a cold interval in the 1930s. The time series shown in Fig. 4.8 illustrate the complex nature of low-frequency SST variations at a particular location, and highlight that some aspects of decadal variability are common to all regions while others are unique to a particular area.

EOF analysis (e.g., von Storch and Zwiers 2002) is a standard statistical technique used to identify preferred patterns of temporal variability. Figure 4.9 shows the global SST anomaly pattern associated with the leading EOF of SST anomalies over the tropical Indo-Pacific (20°E–80°W, 30°N–30°S) based on unfiltered data (left) and 8-year low-pass filtered data (right). As before, results are based on quadratically detrended monthly anomalies from the HadISST1 dataset during 1900–2010. Although the EOF

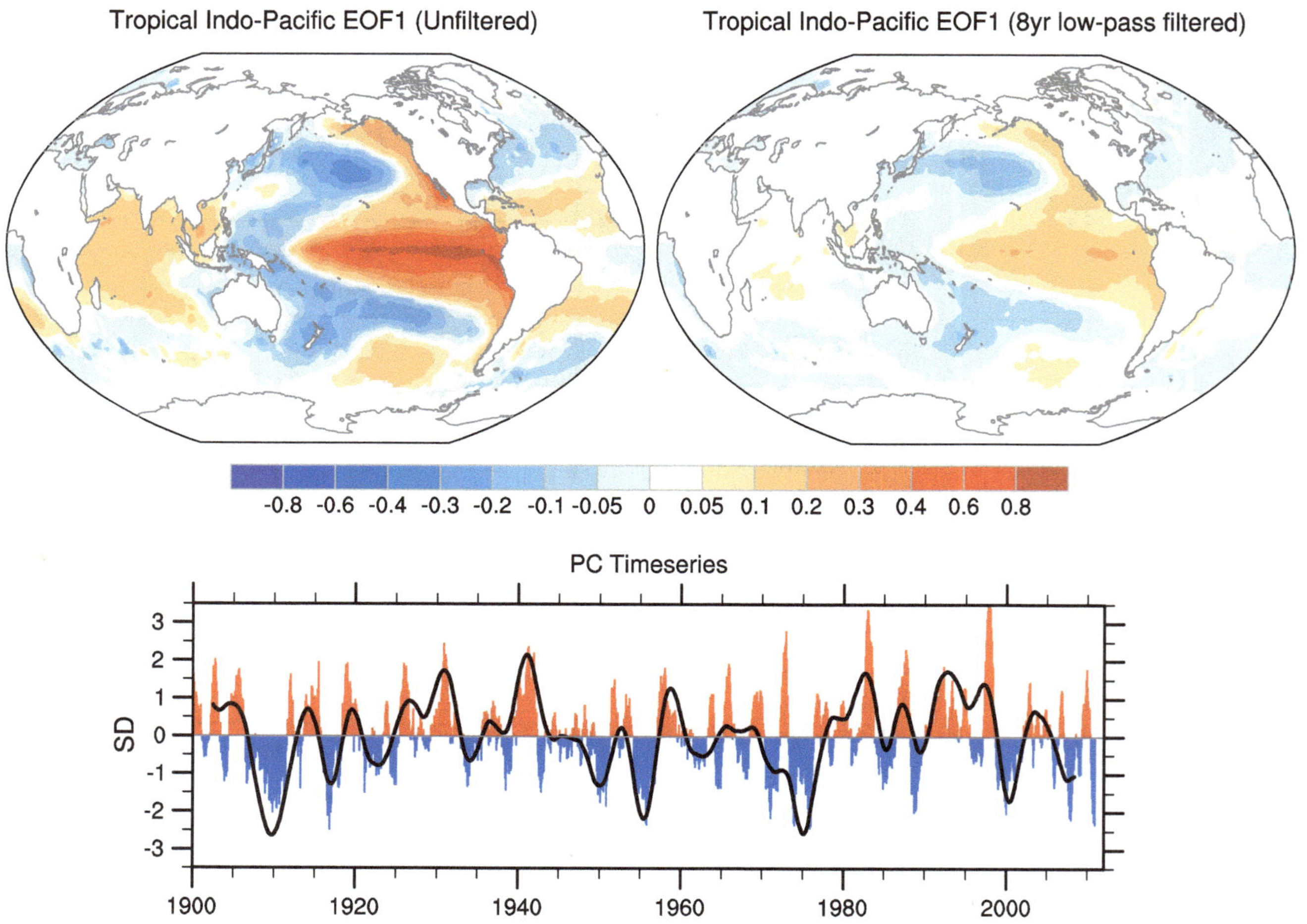

Fig. 4.9 (*Top panels*) Global SST anomaly pattern associated with the leading EOF of SST anomalies over the tropical Indo-Pacific (30°N–30°S, 20°E–80°W) based on unfiltered data (*left*) and 8-year low-pass filtered data (*right*) based on quadratically detrended monthly anomalies from the HadISST1 dataset during 1900–2010. Units are °C per standard deviation of the PC time series. (Bottom panel) Standardized PC time series (colored bars for unfiltered data and black curve for low-pass filtered data)

calculation was restricted to the tropical Indo-Pacific domain, the patterns are displayed globally by regressing the detrended monthly SST anomalies at each location upon the associated PC time series. These regression maps display the amplitude of the SST anomalies (°C) associated with a one standard deviation departure of the PC time series.

The EOF based on unfiltered data depicts ENSO and its global SST teleconnections, while that based on low-pass filtered data represents PDV (e.g., Mantua et al., 1997; Zhang et al. 1997; Power et al. 1999). These EOFs account for 49 % and 52 % of the unfiltered and low-pass filtered variance, respectively; compared to EOF2, which accounts for 10 % and 13 %, respectively. The ENSO and PDV patterns are very similar, not only over the tropical Indo-Pacific but also globally. In particular, when SST anomalies are positive in the tropical eastern Pacific, they are negative to the west and over the central North and South Pacific, and positive over the tropical Indian Ocean and northeastern portions of the high-latitude Pacific Ocean. The primary difference between the PDV and ENSO patterns is that PDV lacks the narrow equatorial Pacific maximum that is the hallmark of ENSO. For this reason, PDV is often referred to as a "broadened ENSO pattern" (e.g., Zhang et al. 1997; Vimont 2005). It may also be noted that PDV resembles the areas of maximum decadal σSST (recall Fig. 4.7). The unfiltered PC time series (colored bars in the bottom panel of Fig. 4.9) is dominated by the interannual sequence of ENSO events, while the low-pass filtered PC (black curve) highlights the decadal variability in the unfiltered PC. This decadal variability is similar to that described in connection with the regional SST time series shown in Fig. 4.8. However, the decadal PC record should not be mistaken for the actual SST time series at a given location, which often shows more complex behavior (Fig. 4.8). Nor should the EOF pattern be assumed to represent the actual spatial distribution of SST anomalies during a given warm epoch (such as 1926–1945 and 1977–1998) or cold epoch (such as 1910–1925 and 1947–1976) due to additional sources of variability as documented in Deser et al. (2004). By design, EOF analysis oversimplifies the actual spatial and temporal characteristics of the data.

Rainfall patterns are affected by ENSO and PDV, as shown by the regression maps of monthly precipitation anomalies upon the SST PC time series based on unfiltered and low-pass filtered data in Fig. 4.10 (left and right panels, respectively). The top panels show results based upon the globally-complete but short satellite record since 1979 from the Global Precipitation Climatology Project (GPCP; Adler et al. 2003), and the bottom panels show results based upon the spatially-incomplete but longer (1900–1998) land station data records (Hulme et al. 1998). There is remarkable consistency between the two types of rainfall data where they overlap in space, lending confidence to the results. As for SST, the patterns of rainfall anomalies associated with ENSO and PDV are generally similar and consist of positive anomalies in the equatorial Pacific and negative anomalies over the maritime continent and the southwestern tropical Pacific. Smaller amplitudes of positive and negative rainfall anomalies occur in the eastern Indian Ocean and western tropical Atlantic, respectively. Regional precipitation time series since 1900 are shown in Deser et al. (2004).

4.5.2 Mechanisms of PDV

There is much ambiguity regarding the physical origin of PDV and whether it is separable from ENSO. Some studies suggest that PDV, unlike ENSO, is not a single physical phenomenon or "mode", but a superposition of several phenomena including ENSO, random atmospheric variability, and oceanic processes (e.g., Newman et al. 2003; Vimont 2005; Schneider and Cornuelle 2005; Newman 2007; Alexander 2010). Other workers indicate that PDV is the result of deterministic ocean-atmosphere interactions between the tropical Indo-Pacific and higher latitudes of the Pacific Ocean that produce a preferred timescale, although the mechanisms put forth differ in terms of which latitudinal region is key [see the recent review of Liu (2012)]. Some research suggests that there is no preferred timescale for PDV, proposing instead that it reflects a first-order autoregressive (or "red noise") process that is stochastically driven from either the extratropical Pacific, the tropical Pacific, or both (Pierce 2001; Dommenget and Latif 2008; Dommenget 2010; Clement et al. 2011). In this view, the limited temporal record from the instrumental data ($\sim$100–150 years) is not sufficient to provide robust evidence of a spectral peak in the power spectrum, although nominal timescales of about 20 and 50 years have been proposed (Minobe 1997, 1999; Deser et al. 2004). Indeed, attempts at reconstructing PDV from a variety of paleoclimate proxy records fail to show a robust timescale before the instrumental era (e.g., Biondi et al. 2001; D'Arrigo et al. 2005). Perhaps the most important message emerging from studies of PDV is that analysis techniques that pre-suppose a particular timescale for PDV through band-pass or low-pass filtering may not be isolating any underlying physical phenomenon but simply sampling a continuum of low-frequency variability, albeit with a well-defined geographical pattern. This cautionary message does not preclude the importance of reconstructing PDV back in time from paleoclimate proxy records, but it may alter the interpretation of such reconstructions.

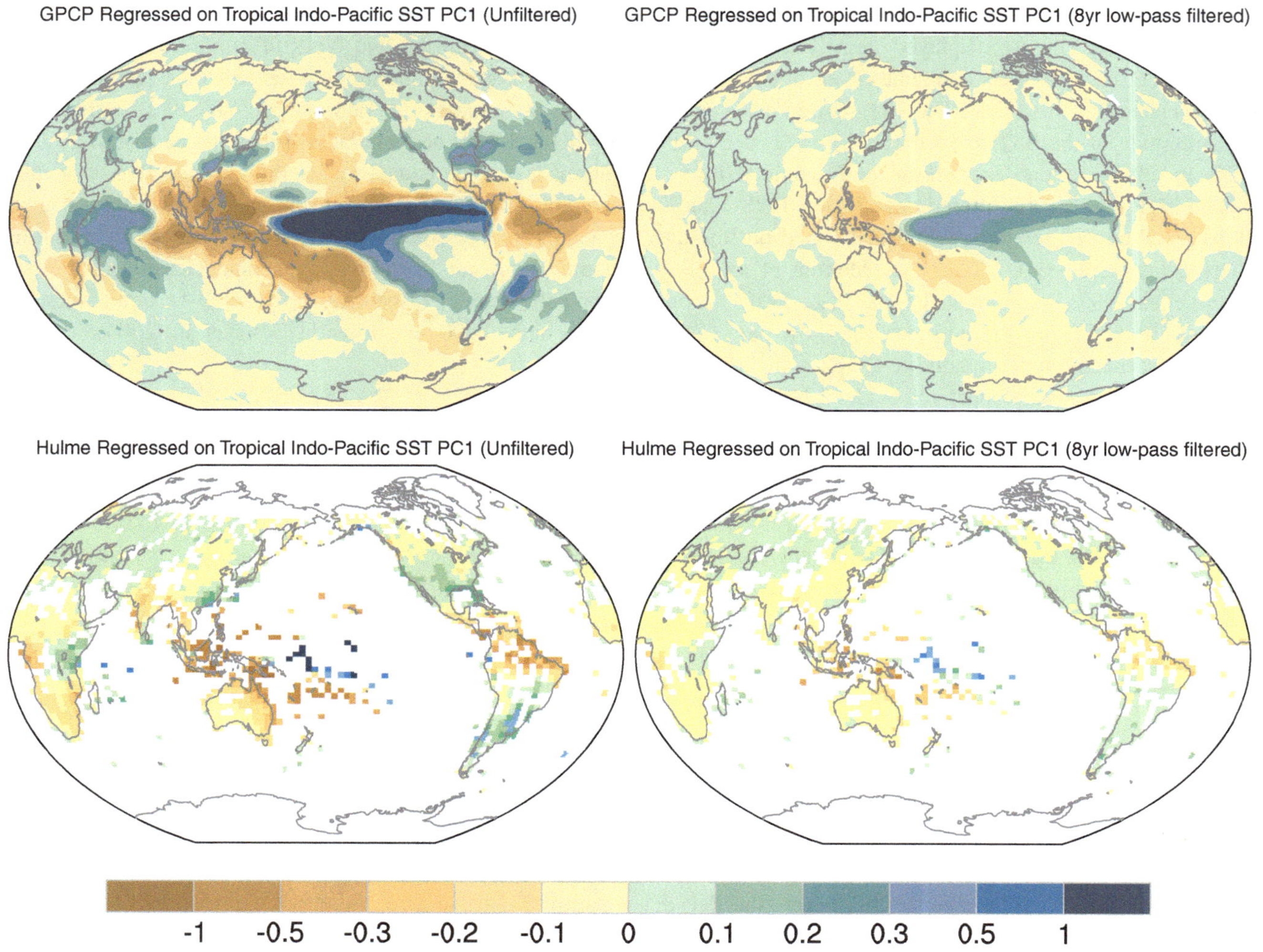

Fig. 4.10 Regression maps of monthly precipitation anomalies upon the SST PC time series based on unfiltered and low-pass filtered data in Fig. 4.9 (*left* and *right* panels, respectively). The *top panels* show results based upon the satellite-based Global Precipitation Climatology Project for the period 1979–2011; *bottom panels* show results based upon land station data records during 1900–1998. Units are mm/day per standard deviation of the PC time series

PDV as defined in this chapter is only one measure of low-frequency SST variability in the Pacific sector. Other recurring patterns include the "North Pacific mode" (Deser and Blackmon 1995; Nakamura et al. 1997; Barlow et al. 2001; Guan and Nigam 2008), which is closely related to PDV albeit with less amplitude in the tropical Indo-Pacific, and a "North Pacific Gyre Oscillation mode" with connections to the central equatorial Pacific (Di Lorenzo et al. 2008). Recently, Messie and Chavez (2011) have suggested that PDV is a combination of the ENSO and North Pacific modes. Like PDV, whether these patterns are true physical modes is not well understood. PDV is generally poorly represented in coupled ocean-atmosphere climate models, particularly in terms of linkages between the tropics and extratropics (Lienert et al. 2011; Deser et al. 2012).

PDV and ENSO share many common features, such as the spatial patterns, although it is premature to state that PDV is oscillatory in nature. Not surprisingly, their mechanisms may also have some similarities. As summarized in Sect. 4.3, many oscillator concepts have been proposed for the oscillatory and self-sustained nature of ENSO. Several authors have extended three oscillator concepts of ENSO to explain the decadal variability in the tropical Pacific. Off-equatorial Rossby waves are at the root of the modified delayed action oscillator of White et al. (2003) and the recharge oscillator of Jin (2001a). Yu and Boer (2004) note the resemblance of the ENSO western Pacific oscillator of Weisberg and Wang (1997) with their findings on decadal variability and heat content anomalies in the western North and South Pacific.

4.6 ENSO Under Global Warming

The summary and discussion in the previous sections are all focused on studies based on modern data and models. In this section, we briefly describe ENSO from the perspectives of global warming and paleoclimatic records.

4.6.1 Climate Response of the Equatorial Pacific to Global Warming

Paleoclimatic records suggest that the strong east-west SST contrast of the annual-mean conditions in the equatorial Pacific may not be a stable and permanent feature. Average SST contrast across the equatorial Pacific was about 2 °C, similar to the difference during a modern El Niño event (Wara et al. 2005) and during the warm early Pliocene (~4.5–3.0 million years ago). This mean state may have occurred during the most recent interval with a climate warmer than today, suggesting that the equatorial Pacific could undergo similar changes as the Earth warms up in response to increasing greenhouse gases.

Competing theories anticipate either a stronger or weaker east-west SST contrast in response to warming. The eastern Pacific could warm up more due to cloud feedbacks (Meehl and Washington 1996), evaporation feedbacks (Knutson and Manabe 1995), or a weakening of the Walker circulation (Vecchi and Soden 2007). But the ocean could also oppose warming in the east because increased stratification enhances the cooling effect of upwelling (Clement et al. 1996; Seager and Murtugudde 1997). The balance between these processes is not known, therefore it is unclear whether the SST gradient will strengthen or weaken in the future. For instance, the SST signature of these mechanisms has been difficult to detect in the simulations, modern observations, or proxies. Modern observations do not show a robust pattern of El Niño-like warming (Vecchi et al. 2008; Deser et al. 2010) (Fig. 4.11), despite evidence for a weakening of tropical atmospheric circulation (Vecchi et al. 2006; Zhang and Song 2006). However, there is robust evidence for warming of the eastern equatorial Pacific during the 20th Century (Bunge and Clarke 2009). The tropical eastern Pacific SST trend may also be caused by the Atlantic warming (Kucharski et al. 2011) through the mechanisms of the Walker circulation across equatorial South America or the inter-basin SST gradient and ocean dynamics (Wang 2006; Wang et al. 2009; Rodriguez-Fonseca et al 2009).

Climate models project a weak reduction of the SST gradient into the 21st Century (Knutson and Manabe 1995; Collins et al. 2005; Meehl et al. 2007). The lack of robust evidence for El Niño-like warming in models and observations could be due to cancellation among the mechanisms listed above, especially among the enhanced warming due to slower currents driven by a weaker Walker circulation and the enhanced cooling due to a more stratified ocean (DiNezio et al. 2009). Moreover, due to basic equatorial dynamics the adjustment of the thermocline to changes in the trade winds renders the Bjerknes feedback ineffective to amplify an initial El Niño-like warming (DiNezio et al. 2010; Clarke 2010). For these reasons, a "permanent El Niño" in response to global warming is very unlikely, even if the Walker circulation weakens. Instead, climate models indicate than the equatorial Pacific may just warm up slightly more than the tropics (Fig. 4.12) due to the effect of the weakening of the Walker circulation on equatorial currents and a differential in evaporative damping with the off-equatorial tropics (Liu et al. 2006; DiNezio et al. 2009).

4.6.2 Sensitivity of ENSO to Global Warming

Paleoclimate records and climate models overwhelmingly indicate that the Pacific will continue to be characterized by large seasonal and interannual variability as the Earth warms up. Seasonally-resolved tropical Pacific paleoclimate records from periods in the Earth's history that were both warmer and colder than today show that interannual variability was present. Available Pliocene records, for example, show that ENSO frequency and amplitude were not significantly different from today (Watanabe et al. 2011; Scroxton et al. 2011). Moreover, glacial climate also exhibited large seasonal and interannual variability as suggested by isotopic measurements on individual foraminifers at the Last Glacial Maximum (Koutavas and Joanidis 2009), and coral records from prior glacial stages (Tudhope et al. 2001; see Chap. 19, Cole and Tudhope). Climate models have not thus far been able to render ENSO inactive in either warmer (Huber and Caballero 2003; Galeotti et al. 2010; von der Heydt et al. 2011) or cooler climates (Zheng et al. 2008).

Neither climate models and observations nor proxies provide a conclusive answer on whether ENSO is going to become stronger or weaker as the tropics warm up in response to increasing greenhouse gases (GHGs). Climate change simulations coordinated by the CMIP3 project a wide range of responses from weaker to stronger. Whether ENSO has changed due to recent observed warming is also controversial according to the observational record (e.g., Trenberth and Hoar 1997; Harrison and Larkin 1997; Rajagopalan et al. 1997). For these reasons, the Intergovernmental Panel on Climate Change (IPCC) Fourth Assessment Report (AR4) concluded that there is no consistent indication of discernible changes in ENSO amplitude in response to increasing GHGs (Meehl et al. 2007; Guilyardi et al. 2009).

The direct cause of ENSO changes in response to climate changes is generally not straightforward. For instance, the CMIP3 models largely agree in the response of the

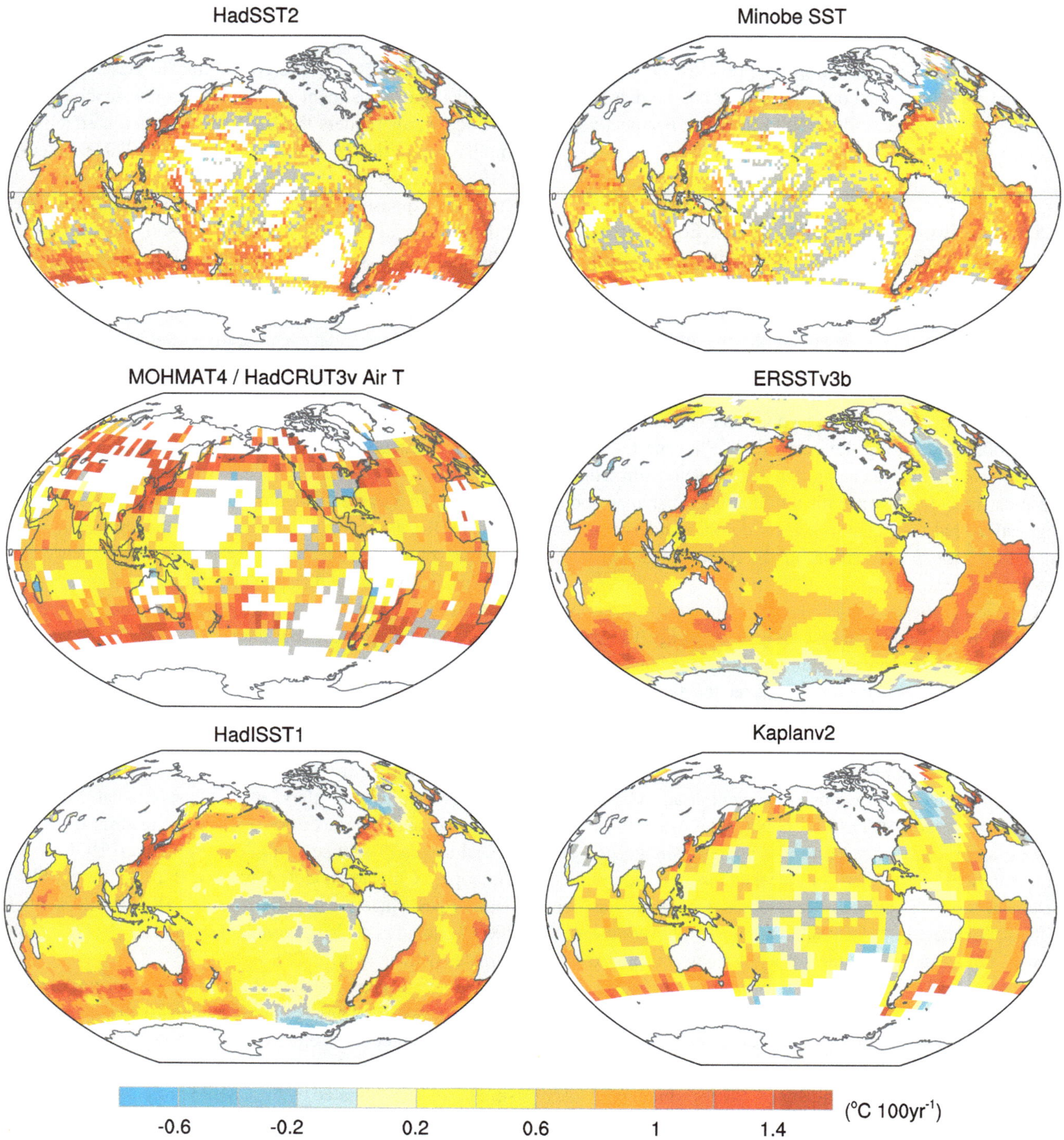

Fig. 4.11 Observed Twentieth Century SST trends (°C per century) computed from monthly anomalies since 1900 using various SST datasets. *White grid boxes* denote insufficient data, and *gray boxes* indicate trends that are not statistically significant at the 95 % confidence level (from Deser et al. 2010)

background ocean conditions over which ENSO variability occurs, but they do not agree on whether or not ENSO will strengthen (Fig. 4.13). The projected changes in the mean climate include a shoaled, less tilted, and sharper thermocline; weaker zonal currents; and weaker upwelling (Vecchi and Soden 2007; DiNezio et al. 2009). ENSO theory indicates that any of these changes in the mean climate can lead to changes in ENSO amplitude.

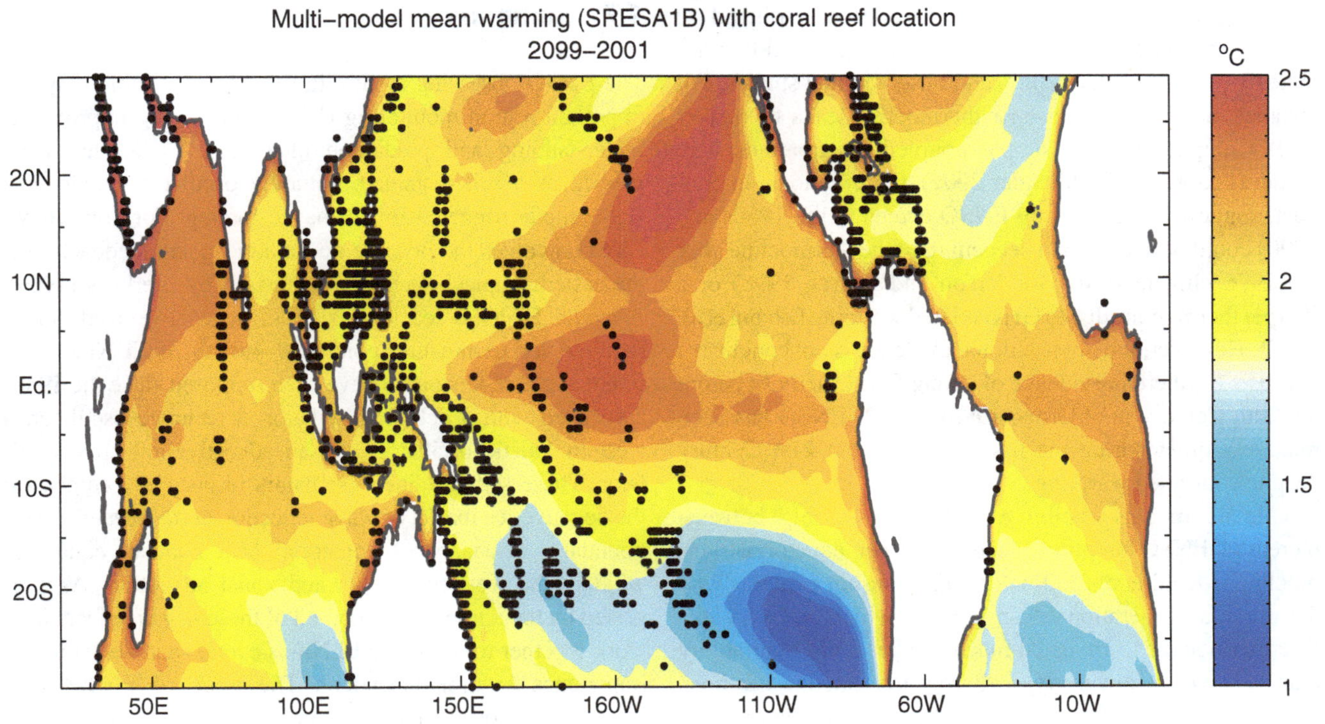

Fig. 4.12 SST change over the twenty-first Century, averaged over 22 climate models for a mid-range emissions scenario (SRES A1B). Reef locations are indicated by *black circles* (data from www.reefbase.org) (from Clement et al. 2010)

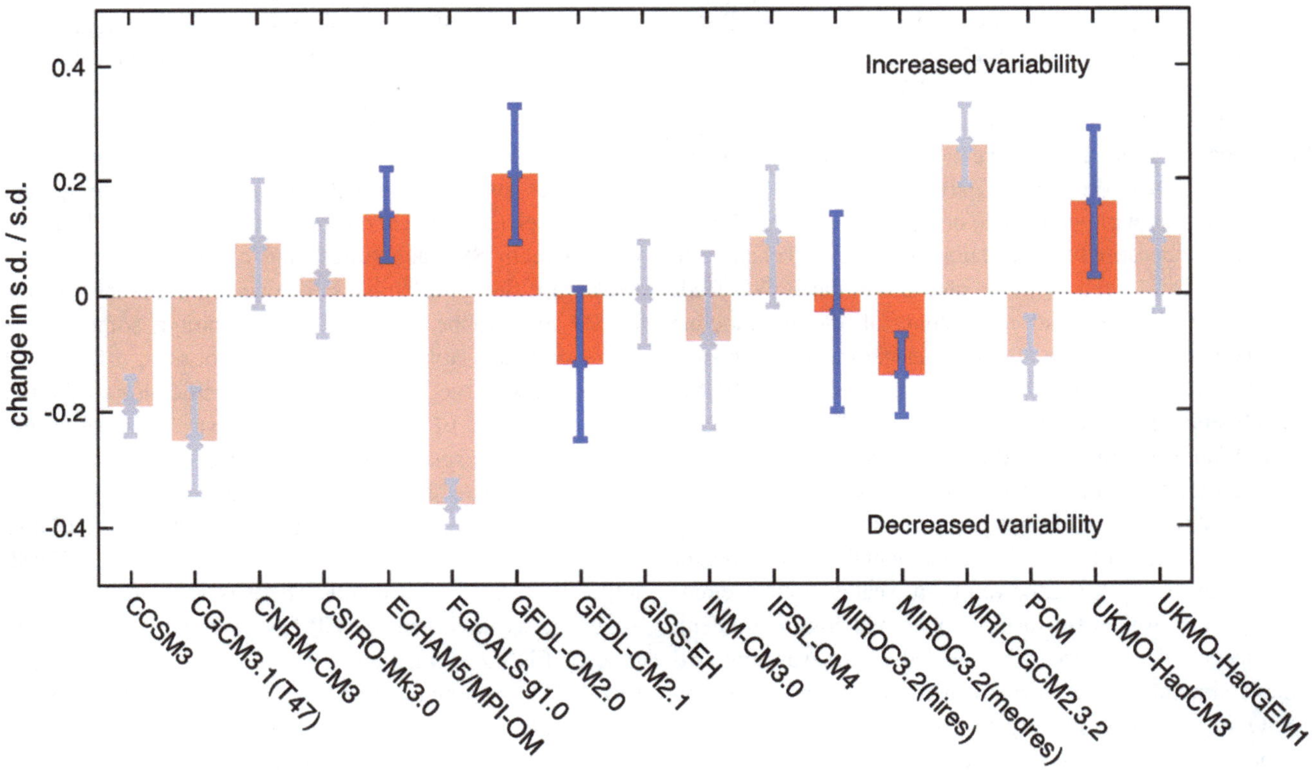

Fig. 4.13 Projected changes in the amplitude of ENSO variability, as a response to global warming, from the CMIP3 models. The measure is derived from the interannual standard deviation (s.d.) of a mean sea-level-pressure index, which is related to the strength of Southern Oscillation variations. Positive changes indicate a strengthening of ENSO, and negative changes indicate a weakening. Statistical significance is assessed by the size of the blue bars, and the bars indicated in bold color are from those CMIP3 CGCMs that are judged to have the best simulation of present-day ENSO characteristics and feedbacks (from Collins et al. 2010)

Changes in ENSO amplitude have been attributed to changes in the depth and sharpness of the equatorial thermocline by theoretical, modeling, and observational studies. For instance, a sharper and deeper thermocline leads to weaker ENSO amplitude in a simple coupled ocean-atmosphere model (Fedorov and Philander 2001). Observations, in contrast, suggest that the strong ENSO events in the 1980s and 1990s could be a result of a deepening of the thermocline after the 1976 climate shift (Guilderson and Schrag 1998) or a sharper thermocline due to GHG related warming (Zhang et al. 2008). However, the observational evidence is not conclusive because: (1) there is evidence of strong ENSO activity before the 20th Century (e.g. Grove 1998) and (2) ENSO has been relatively quiet during the first decade of the 21st Century despite continued warming.

Climate models exhibit a robust relationship between increased ENSO amplitude and a sharper equatorial thermocline (Meehl et al. 2001). This relationship explains why the previous generation ocean models, which had very diffuse thermoclines, simulated much weaker ENSO variability than observed. Conversely, a sharper thermocline has been invoked to explain the increased ENSO amplitude in some increasing GHG experiments (e.g., Timmermann et al. 1999; Park et al. 2009). All models participating in CMIP3 simulate a sharper thermocline in response to increasing GHGs, yet not all of them simulate a stronger ENSO. Other physical processes, such as the shoaling of the thermocline, weaker upwelling, or warmer mean SST could also have an amplifying or damping effect on ENSO. Thus, it is reasonable to hypothesize that depending on the balance of these changes, ENSO could strengthen or weaken (e.g., Guilyardi et al. 2009; Vecchi and Wittenberg 2010; Collins et al. 2010).

Climate model projections do not agree on whether ENSO will increase because the interaction of ENSO and changes in the mean climate lead to subtle changes in the ENSO feedbacks. The shoaling and sharpening of the thermocline enhance ENSO variability, but the warmer mean SST results in stronger atmospheric damping, thus weakening ENSO (van Oldenborgh et al. 2005; Kim and Jin 2011). Weakening of the Walker circulation and the increased thermal stratification associated with surface intensified ocean warming are both robust features of climate projections that play opposing roles. This interaction results fundamentally from weaker climatological upwelling, driven by a weaker Walker circulation, and a stronger subsurface zonal temperature gradient, associated with the surface intensified ocean warming. All models simulate these mechanisms, yet their net effect on ENSO is not equal leading to the wide range of ENSO responses (e.g., Figure 4.13). Overall, these studies show that there is a substantial amount of cancellation among the effect of the changes in the different ENSO feedbacks. As a result, the sensitivity of ENSO simulations to increasing greenhouse gases is much reduced.

4.7 Summary

The ENSO observing system in the tropical Pacific plays an important role in monitoring ENSO and helping improve the understanding and prediction of ENSO. ENSO has been viewed as a self-sustained, naturally oscillatory mode or a stable mode triggered by stochastic forcing. For both views, ENSO involves the positive ocean-atmosphere feedback over the eastern tropical Pacific hypothesized by Bjerknes (1969). After an El Niño reaches its peak, a negative feedback is required for terminating continued growth of El Niño. Different negative feedbacks have been proposed since the 1980s associated with a delayed oscillator, a recharge oscillator, a western Pacific oscillator, and an advective-reflective oscillator. These self-sustained oscillators respectively emphasize the negative feedbacks of wave reflection at the ocean western boundary, discharge of equatorial heat content, equatorial wind in the western Pacific, and zonal advection. As suggested by the unified oscillator, all of these negative feedbacks work together to terminate El Niños events, and their relative importance is time-dependent. The issue of ENSO as a self-sustained oscillation mode or a stable mode triggered by random forcing is not settled. It is possible that ENSO is a self-sustained mode during some periods, a stable mode during others, or a mode that is intermediate or mixed between the former and the latter. The predictability of ENSO is more limited if ENSO is a stable mode triggered by stochastic forcing than if ENSO is a self-sustained mode, because the former depends on random disturbances.

A recent development or focus for ENSO study is to separate ENSO into the EP and CP ENSO events. These two types of ENSO events have different climate and weather-related impacts on the globe because the locations of maximum SST anomalies (and associated atmospheric heating) differ. Mechanisms for causing different types of ENSO need to be further studied although some studies have proposed different physical processes of the CP compared with the EP ENSO. The identification of the two distinct types of El Niño offers a new way to consider how El Niño may respond and the nature of its feedback to a changing climate. In addition to considering how climate changes may affect the amplitude and frequency of ENSO, we should also consider that the dominant type of ENSO might be altered as a result. There is still much to learn about the dynamics of the CP ENSO and what causes the type of ENSO to vary and alternate over time. Nevertheless, ENSO may be changing and there is a need to prepare for the possible emergence of a new dominant type of ENSO and to revise existing modeling and prediction strategies that were developed primarily for the conventional EP type of ENSO. It is unfortunate that there are only a few CP El Niño events available in the observational records (less than 12 since the 1950s, depending on the way a CP event is defined). While much can still be learned from this limited number of events, we should look for assistance from long-term coupled climate model simulations, as well as paleoclimate records, to obtain a better understanding of the emerging CP type of El Niño.

Low-frequency SST variability in the Pacific sector is termed PDV. The ENSO and PDV patterns are very similar, not only over the tropical Indo-Pacific but also globally. When SST anomalies are positive in the tropical eastern Pacific, they are negative to the west and over the central North and South Pacific, and positive over the tropical Indian Ocean and high-latitude northeastern portions of the Pacific Ocean. The primary difference between the PDV and ENSO patterns is that PDV lacks the narrow equatorial Pacific maximum that is the hallmark of ENSO. The temporal variations of PDV are characterized by relatively cold conditions in the tropics as was demonstrated from approximately 1910–1925 and 1947–1976, and by relatively warm conditions from approximately 1926–1945 and 1977–1998. There is much ambiguity regarding the physical origin of PDV and whether it is separable from ENSO. Some studies suggest that PDV, unlike ENSO, is not a single physical phenomenon or "mode", but a superposition of several phenomena including ENSO, random atmospheric variability, and oceanic processes. Other studies indicate that PDV is the result of deterministic ocean-atmosphere interactions between the tropical Indo-Pacific and higher latitudes of the Pacific Ocean that produce a preferred timescale, although the mechanisms put forth differ in terms of which latitudinal region is key.

ENSO changes under global warming are uncertain. The tropical Pacific response to global warming has been suggested to be neither El Niño-like nor La Niña-like since the mechanisms for these changes are different from that of ENSO events—the Bjerknes feedback. Increasing greenhouse gases change the background mean states in the tropical Pacific Ocean and atmosphere, which in turn induce ENSO changes. However, the response of the mean states to increasing greenhouse gases is uncertain. For example, the tropical Pacific zonal SST contrast under global warming is reported to be either strengthened or weakened. The uncertainty in the eastern Pacific warming may also be caused by the Atlantic warming. Due to the fact that the change in tropical mean condition under global warming is quite uncertain even during the past few decades, it is difficult to say whether ENSO is going to intensify or weaken, but it is very likely that ENSO will not disappear in the future.

Tropical warm waters are normally favorable for coral reef development and growth. However, extremely warm waters can result in coral bleaching and mortality. A combination of interannual, decadal and anthropogenic ENSO variations can induce large ocean temperature changes in the tropical eastern Pacific, which then affects coral reefs directly and through a cascade of disturbances. As detailed in this chapter, significant advances have been made in our understanding of ENSO since the past decades. However, there are many issues and uncertainties that are still under debate and need to be further investigated. In particular, the improvement of our understanding and prediction of natural and anthropogenic ENSO variations is an important task for the ENSO community of investigators.

Acknowledgments CW thanks Ms. L. Zhang for plotting Fig. 4.1 and helping modify Fig. 4.2 provided by Dr. M. McPhaden. CD would like to thank Dr. Toby Ault for useful discussions and Mr. Adam Phillips for technical assistance with the figures. We thank Dr. Paul Fiedler and an anonymous reviewer for their comments and suggestions. CW is supported by grants from NOAA's Climate Program Office, and the base funding of NOAA AOML. NCAR is sponsored by the National Science Foundation (NSF). JYY acknowledges the support from NSF Grant ATM-0925396 and NOAA-MAPP Grant NA11OAR4310102. The findings and conclusions in this report are those of the author(s) and do not necessarily represent the views of the funding agencies.

References

Adler RF, Huffman GJ, Chang A, Ferraro R, Xie P-P, Janowiak J, Rudolf B, Schneider U, Curtis S, Bolvin D (2003) The version-2 global precipitation climatology project (GPCP) monthly precipitation analysis (1979-present). J Hydrometeorol 4:1147–1167

Alexander MA (2010) Extratropical air-sea interaction, sea surface temperature variability, and the Pacific Decadal Oscillation. (PDO). In: Sun D, Bryan F (eds) Climate dynamics: why does climate vary, AGU Monogr 189, Washington D.C., pp 123–148

Alexander MA, Bladé I, Newman M, Lanzante JR, Lau N-C, Scott JD (2002) The atmospheric bridge: the influence of ENSO teleconnections on air-sea interaction over the global oceans. J Climate 15:2205–2231

An S-I (2008) Interannual variations of the tropical ocean instability wave and ENSO. J Climate 21:3680–3686

Ashok K, Behera SK, Rao SA, Weng H, Yamagata T (2007) El Niño Modoki and its possible teleconnection. J Geophys Res 112: C11007. doi: 10.1029/2006JC003798

Barlow M, Nigam S, Berbery EH (2001) ENSO, Pacific decadal variability, and U.S. summertime precipitation, drought, and stream flow. J Climate 14:2105–2128

Barsugli JJ, Sardeshmukh PD (2002) Global atmospheric sensitivity to tropical SST anomalies throughout the Indo-Pacific basin. J Climate 15:3427–3442

Battisti DS, Hirst AC (1989) Interannual variability in a tropical atmosphere-ocean model: influence of the basic state, ocean geometry and nonlinearity. J Atmos Sci 45:1687–1712

Biondi F, Gershunov A, Cayan DR (2001) North Pacific decadal climate variability since AD 1661. J Climate 14:5–10

Bjerknes J (1969) Atmospheric teleconnections from the equatorial Pacific. Mon Weather Rev 97:163–172

Bunge L, Clarke AJ (2009) A verified estimation of the El Niño index NINO-3.4 since 1877. J Climate 22(14):3979–3992

Chen G (2011) How does shifting Pacific Ocean warming modulate on tropical cyclone frequency over the South China Sea? J Climate 24:4695–4700

Clarke AJ (2010) Analytical theory for the quasi-steady and low-frequency equatorial ocean response to wind forcing: the "tilt" and "warm water volume" modes. J Phys Oceanogr 40(1):121–137

Clement AC, Seager R, Cane MA, Zebiak SE (1996) An ocean dynamical thermostat. J Climate 9:2190–2196

Clement AC, Baker AC, Leloup J (2010) Climate change: patterns of tropical warming. Nature Geosci 3:8–9. doi:10.1038/ngeo728

Clement A, DiNezio P, Deser C (2011) Rethinking the ocean's role in the Southern Oscillation. J Climate 24:4056–4072

Collins M CMIP Modelling group (BMRC Australia, CCC Canada, CCSR/NIES Japan), and 12 others (2005) El Niño-or La Ñina-like climate change? Climate Dynam 24:89–104

Collins M, An S-I, Cai W, Ganachaud A, Guilyardi E, Jin F-F, Jochum M, Lengaigne M, Power S, Timmermann A, Vecchi G, Wittenberg A (2010) The impact of global warming on the tropical Pacific Ocean and El Niño. Nature Geosci 3:391–397

Compo GP, Whitaker JS, Sardeshmukh PD, Matsui N, Allan RJ, Yin X, Gleason BE, Vose R, Rutledge G, Bessemoulin P (2011) The twentieth century reanalysis project. Q J Roy Meteor Soc 137:1–28

D'Arrigo R, Wilson R, Deser C, Wiles G, Cook E, Villalba R, Tudhope A, Cole J, Linsley B (2005) Tropical North Pacific climate linkages over the past four centuries. J Climate 18:5253–5265

Deser C, Blackmon ML (1995) On the relationship between tropical and North Pacific sea surface temperature variations. J Climate 8:1677–1680

Deser C, Blackmon ML (1993) Surface climate variations over the North Atlantic Ocean during winter: 1900-1989. J Climate 6:1743–1753

Deser C, Wallace JM (1990) Large-scale atmospheric circulation features of warm and cold episodes in the tropical Pacific. J Climate 3:1254–1281

Deser C, Phillips AS, Hurrell JW (2004) Pacific interdecadal climate variability: linkages between the tropics and the North Pacific during boreal winter since 1900. J Climate 17:3109–3124

Deser C, Phillips AS, Alexander MA (2010) Twentieth century tropical sea surface temperature trends revisited. Geophys Res Lett 37: L10701. doi:10.1029/2010GL043321

Deser C, Phillips AS, Tomas RA, Okumura YM, Alexander MA, Capotondi A, Scott JD, Kwon Y-O, Ohba M (2012) ENSO and Pacific decadal variability in the Community Climate System Model version 4. J Climate 25:2622–2651

DeWeaver E, Nigam S (2004) On the forcing of ENSO teleconnections by anomalous heating and cooling. J Climate 17:3225–3235

Di Lorenzo E, Schneider N, Cobb K, Franks P, Chhak K, Miller A, McWilliams J, Bograd S, Arango H, Curchitser E (2008) North Pacific Gyre Oscillation links ocean climate and ecosystem change. Geophys Res Lett 35:L08607. doi: 10.1029/2007GL032838

DiNezio PN, Clement AC, Vecchi G, Soden B, Kirtman B, Lee S-K (2009) Climate response of the equatorial Pacific to global warming. J Climate 22:4873–4892

DiNezio PN, Clement AC, Vecchi GA (2010) Reconciling differing views of tropical Pacific climate change. Eos Trans AGU 91 (16):141–142

Dommenget D, Latif M (2008) Generation of hyper climate modes. Geophys Res Lett 35:L02706. doi:10.1029/2007GL031087

Dommenget D (2010) The slab ocean El Niño. Geophys Res Lett 37: L20701. doi:10.1029/2010GL044888.1

Fedorov AV, Philander SG (2001) A stability analysis of tropical ocean–atmosphere interactions: bridging measurements and theory for El Niño. J Climate 14:3086–3101

Feng J, Li J (2011) Influence of El Niño Modoki on spring rainfall over south China. J Geophys Res-Atmos 116:D13102. doi:10.1029/2010JD015160

Galeotti S, Von der Heydt A, Huber M, Bice D, Dijkstra H, Jilbert T, Lanci L, Reichart G-J (2010) Evidence for active El Nino Southern Oscillation variability in the late Miocene greenhouse climate. Geology 38:419–422

Gebbie G, Eisenman I, Wittenberg AT, Tziperman E (2007) Modulation of westerly wind bursts by sea surface temperature: a semi-stochastic feedback for ENSO. J Atmos Sci 64:3281–3295

Gill AE (1980) Some simple solutions for heat-induced tropical circulation. Q J Roy Meteorol Soc 106:447–462

Glynn PW (1985) El Niño-associated disturbance to coral reefs and post disturbance mortality by *Acanthaster planci*. Mar Ecol Prog Ser 26:295–300

Grove RH (1998) Global impact of the 1789-93 El Niño. Nature 393:318–319

Guan B, Nigam S (2008) Pacific sea surface temperatures in the twentieth century: an evolution-centric analysis of variability and trend. J Climate 21:2790–2809

Guilderson TP, Schrag DP (1998) Abrupt shift in subsurface temperatures in the tropical Pacific associated with changes in El Niño. Science 281:240–243

Guilyardi E, Wittenberg A, Federov A, Collins M, Wang C, Capotondi A, Van Oldenborgh GJ, Stockdale T (2009) Understanding El Niño in ocean-atmosphere general circulation models: progress and challenges. Bull Am Meteorol Soc 90:325–340

Harrison DE, Larkin NK (1997) Darwin sea level pressure, 1876–1996: evidence for climate change? Geophys Res Lett 24:1775–1782

Hayes SP, Mangum L, Picaut J, Sumi JA, Takeuchi K (1991) TOGA-TAO: a moored array for real-time measurements in the tropical Pacific Ocean. Bull Am Meteorol Soc 72:339–347

Hoerling MP, Kumar A (2002) Atmospheric response patterns associated with tropical forcing. J Climate 15:2184–2203

Holland CL, Mitchum GT (2003) Interannual volume variability in the tropical Pacific. J Geophys Res-Oceans 108:3369. doi:10.1029/2003JC001835

Huber M, Caballero R (2003) Eocene El Nino: evidence for robust tropical dynamics in the "hothouse". Science 299:877–881

Hulme M, Osborn TJ, Johns TC (1998) Precipitation sensitivity to global warming: comparison of observations with HadCM2 simulations. Geophys Res Lett 25:3379–3382. doi:10.1029/98GL02562

Jin F-F (1997a) An equatorial ocean recharge paradigm for ENSO. Part I: conceptual model. J Atmos Sci 54:811–829

Jin F-F (1997b) An equatorial ocean recharge paradigm for ENSO. Part II: a stripped-down coupled model. J Atmos Sci 54:830–847

Jin F-F, Kimoto M, Wang X (2001) A model of decadal ocean-atmosphere interaction in the North Pacific Basin. Geophys Res Lett 28:1531–1534

Kao H-Y, Yu J-Y (2009) Contrasting Eastern-Pacific and Central-Pacific types of ENSO. J Climate 22:615–632

Kessler WS (2002) Is ENSO a cycle or a series of events? Geophy Res Lett 29(23):2125. doi:10.1029/2002GL015924

Kim H-M, Webster PJ, Curry JA (2009) Impact of shifting patterns of Pacific Ocean warming on North Atlantic tropical cyclones. Science 325:77–80

Kim S-T, Jin F-F (2011) An ENSO stability analysis. Part II: results from the twentieth and twenty-first century simulations of the CMIP3 models. Climate Dynam 36:1609–1627

Koutavas A, Joanidis S (2009) El Nino during the last glacial maximum. Geochim Cosmochim Acta 73:A690–A690

Knutson TR, Manabe S (1995) Time-mean response over the tropical Pacific to increased CO_2 in a coupled ocean-atmosphere model. J Climate 8:2181–2199

Kucharski F, Kang I-S, Farneti R, Feudale L (2011) Tropical Pacific response to 20th century Atlantic warming. Geophys Res Lett 38: L03702. doi:10.1029/2010GL046248

Kug J-S, Jin F-F, An S-I (2009) Two-types of El Niño events: cold tongue El Niño and warm pool El Niño. J Climate 22:1499–1515
Kumar KK, Rajagopalan B, Hoerling M, Bates G, Cane M (2006) Unraveling the mystery of Indian monsoon failure during El Niño. Science 314:115–119
Landsea CW, Knaff JA (2000) How much skill was there in forecasting the very strong 1997-98 El Niño? Bull Am Meteorol Soc 81:2107–2119
Larkin NK, Harrison DE (2005) Global seasonal temperature and precipitation anomalies during El Niño autumn and winter. Geophys Res Lett 32:L16705. doi:10.1029/2005GL022860
Lau K-M (1985) Elements of a stochastic-dynamical theory of the long-term variability of the El Niño-Southern Oscillation. J Atmos Sci 42:1552–1558
Lee S-K, Wang C, Enfield DB (2010a) On the impact of central Pacific warming events on Atlantic tropical storm activity. Geophys Res Lett 37:L17702. doi:10.1029/2010GL044459
Lee T, McPhaden MJ (2010) Increasing intensity of El Niño in the central-equatorial Pacific. Geophys Res Lett 37:L14603. doi:10.1029/2010GL044007
Lee T, Hobbs WR, Willis JK, Halkides D, Fukumori I, Armstrong EM, Hayashi AK, Liu WT, Patzert W, Wang O (2010b) Record warming in the South Pacific and western Antarctica associated with the strong central-Pacific El Niño in 2009–10. Geophys Res Lett 37:L19704. doi: 10.1029/2010GL044865
Lienert F, Fyfe JC, Merryfield WJ (2011) Do climate models capture the tropical influences on North Pacific sea surface temperature variability? J Climate 24:6203–6209
Lindzen RS, Nigam S (1987) On the role of sea surface temperature gradients in forcing low-level winds and convergence in the tropics. J Atmos Sci 44:2418–2436
Liu Z, Vavrus SJ, He F, Wen N, Zhong Y (2006) Rethinking tropical ocean response to global warming: the enhanced equatorial warming. J Climate 18:4684–4700
Liu Z (2012) Dynamics of interdecadal climate variability: a historical perspective. J Climate 25:1963–1995
Mantua NJ, Battisti DS (1994) Evidence for the delayed oscillator mechanism for ENSO: the "observed" oceanic Kelvin mode in the far western Pacific. J Phys Oceanogr 24:691–699
Mantua NJ, Hare SR, Zhang Y, Wallace JM, Francis R (1997) A Pacific interdecadal climate oscillation with impacts on salmon production. Bull Am Meteorol Soc 78:1069–1079
McPhaden MJ (1995) The tropical atmosphere ocean array is completed. Bull Am Meteor Soc 76:739–741
McPhaden MJ, Busalacchi AJ, Cheney R, Donguy JR, Gage KS, Halpern D, Ji M, Julian P, Meyers G, Mitchum GT (1998) The Tropical Ocean-Global Atmosphere observing system: a decade of progress. J Geophy Res-Oceans 103:14,169–14,240
McPhaden MJ, Lee T, McClurg D (2011) El Niño and its relationship to changing background conditions in the tropical Pacific. Geophys Res Lett 38:L15709. doi:10.1029/2011GL048275
Meehl GA, Gent PR, Arblaster JM, Otto-Bliesner BL, Brady EC, Craig A (2001) Factors that affect the amplitude of El Niño in global coupled climate models. Climate Dynam 17:515–526
Meehl GA, Washington WM (1996) El Niño-like climate change in a model with increased atmospheric CO_2 concentrations. Nature 382:56–60
Meehl GA, Stocker TF, Collins WD, Friedlingstein P, Gaye AT, Gregory JM, Kitoh A, Knutti R, Murphy JM, Noda A (2007) Global climate projections. Climate change 2007: the physical science basis. Contribution of Working Group I to the fourth assessment report of the intergovernmental panel on climate change. Cambridge University Press, pp 747–845
Méssie M, Chavez F (2011) Global modes of sea surface temperature variability in relation to regional climate indices. J Climate 24:4314–4331
Minobe S (1997) A 50–70 year climatic oscillation over the North Pacific and North America. Geophys Res Lett 24:683–686
Minobe S (1999) Resonance in bidecadal and pentadecadal climate oscillations over the North Pacific: role in climatic regime shifts. Geophys Res Lett 26:855–858
Mo KC, Higgins RW (1998) Tropical convection and precipitation regimes in the western United States. J Climate 11:2404–2423
Mo KC (2010) Interdecadal modulation of the impact of ENSO on precipitation and temperature over the United States. J Climate 23:3639–3656
Moore AM, Kleeman R (1999) Stochastic forcing of ENSO by the intraseasonal oscillation. J Climate 12:1199–1220
Nakamura H, Lin HG, Yamagata T (1997) Decadal climate variability in the North Pacific in recent decades. Bull Am Meteorol Soc 78:2215–2226
Neelin JD, Battisti DS, Hirst AC, Jin F-F, Wakata Y, Yamagata T, Zebiak SE (1998) ENSO theory. J Geophys Res-Oceans 103:14,262–14,290
Newman M, Compo G, Alexander MA (2003) ENSO-forced variability of the Pacific Decadal Oscillation. J Climate 16:3853–3857
Newman M (2007) Interannual to decadal predictability of tropical and North Pacific sea surface temperatures. J Climate 20:2333–2356
Newman M, Shin S-I, Alexander MA (2011) Natural variation in ENSO flavors. Geophys Res Lett L14705. doi: 10.1029/2011GL047658
Park W, Keenlyside N, Latif M, Ströh A, Redler R, Roeckner E, Madec G (2009) Tropical Pacific climate and its response to global warming in the Kiel climate model. J Climate 22:71–92
Penland C, Sardeshmuhk P (1995) The optimal growth of tropical sea surface temperature anomalies. J Climate 8:1999–2024
Philander SG (1985) El Niño and La Niña. J Atmos Sci 42:2652–2662
Philander SG (1990) El Niño, La Niña, and the Southern Oscillation. Academic Press, London, p 289
Philander SG, Fedorov A (2003) Is El Niño sporadic or cyclic? Annu Rev Earth Planet Sci 31:579–594
Picaut J, Masia F, du Penhoat Y (1997) An advective-reflective conceptual model for the oscillatory nature of the ENSO. Science 277:663–666
Picaut J, Hackert E, Busalacchi AJ, Murtugudde R, Lagerloef GSE (2002) Mechanisms of the 1997–1998 El Niño–La Niña, as inferred from space-based observations. J Geophys Res-Oceans 107. doi: 10.1029/2001JC000850
Pierce DW (2001) Distinguishing coupled ocean-atmosphere interactions from background noise in the North Pacific. Prog Oceanogr 49:331–352
Power SB, Casey T, Folland C, Colman A, Mehta V (1999) Interdecadal modulation of the impact of ENSO on Australia. Climate Dynam 15:319–324
Rajagopalan B, Lall U, Cane MA (1997) Anomalous ENSO occurrences: an alternative view. J Climate 10:2351–2357
Rasmusson EM, Carpenter TH (1982) Variations in tropical sea surface temperature and surface wind fields associated with the Southern Oscillation/El Niño. Mon Weather Rev 110:354–384
Rayner NA, Parker DE, Horton EB, Folland CK, Alexander LV, Rowell DP, Kent EC, Kaplan A (2003) Global analyses of sea surface temperature, sea ice, and night marine air temperature since the late nineteenth century. J Geophys Res-Atmos 108(D14):4407. doi:10.1029/2002JD002670
Rodriguez-Fonseca B, Polo I, García-Serrano J, Losada T, Mohino E, Mechoso CR, Kucharski F (2009) Are Atlantic Niños enhancing

Pacific ENSO events in recent decades? Geophys Res Lett 36: L20705. doi: 10.1029/2009GL040048

Schneider N, Cornuelle BD (2005) The forcing of the Pacific Decadal Oscillation. J Climate 18:4355–4373

Schopf PS, Suarez MJ (1988) Vacillations in a coupled ocean-atmosphere model. J Atmos Sci 45:549–566

Scroxton N, Bonham SG, Rickaby REM, Lawrence SHF, Hermoso M, Haywood AM (2011) Persistent El Niño–Southern Oscillation variation during the Pliocene Epoch. Paleoceanography 26:PA2215. doi:10.1029/2010PA002097

Seager R, Murtugudde R (1997) Ocean dynamics, thermocline adjustment, and regulation of tropical SST. J Climate 10:521–534

Suarez MJ, Schopf PS (1988) A delayed action oscillator for ENSO. J Atmos Sci 45:3283–3287

Taschetto AS, England MH (2009) El Niño Modoki impacts on Australian rainfall. J Climate 22:3167–3174

Timmermann A, Oberhuber J, Bacher A, Esch M, Latif M, Roeckner E (1999) Increased El Niño frequency in a climate model forced by future greenhouse warming. Nature 398:694–697

Trenberth KE, Hoar TJ (1997) El Niño and climate change. Geophys Res Lett 24:3057–3060

Trenberth KE, Stepaniak DP (2001) Indices of El Niño evolution. J Climate 14:1697–1701

Tudhope AW, Chilcott CP, McCulloch MT, Cook ER, Chappell J, Ellam RM, Lea DW, Lough JM, Shimmield GB (2001) Variability in the El Nino-Southern Oscillation through a glacial-interglacial cycle. Science 291:1511–1517

van Oldenborgh GJ, Philip SY, Collins M (2005) El Niño in a changing climate: a multi-model study. Ocean Sci 1:81–95

Vecchi GA, Soden BJ, Wittenberg AT, Held IM, Leetmaa A, Harrison MJ (2006) Weakening of tropical Pacific atmospheric circulation due to anthropogenic forcing. Nature 441:73–76

Vecchi GA, Soden BJ (2007) Global warming and the weakening of the tropical circulation. J Climate 20:4316–4340

Vecchi GA, Clement A, Soden BJ (2008) Examining the tropical Pacific's response to global warming. EOS, Trans Am Geophys Union 89:81–83

Vecchi GA, Wittenberg AT (2010) El Niño and our future climate: where do we stand? Wiley Interdisc Rev: Climate Change 1:260–270. doi: 10.1002/wcc.33

Vimont DJ, Battisti DS, Hirst AC (2001) Footprinting: a seasonal connection between the tropics and mid-latitudes. Geophys Res Lett 28:3923–3926

Vimont DJ, Wallace JM, Battisti DS (2003) The seasonal footprinting mechanism in the Pacific: implications for ENSO. J Climate 16:2668–2675

Vimont DJ (2005) The contribution of the interannual ENSO cycle to the spatial pattern of ENSO-like decadal variability. J Climate 18:2080–2092

Vimont DJ, Alexander M, Fontaine A (2009) Midlatitude excitation of tropical variability in the Pacific: the role of thermodynamic coupling and seasonality. J Climate 22:518–534

Von der Heydt AS, Nnafie A, Dijkstra HA (2011) Cold tongue/warm pool and ENSO dynamics in the Pliocene. Clim Past Discuss 7:997–1027. doi:10.5194/cpd-7-997-2011

von Storch H, Zwiers FW (2002) Statistical analysis in climate research. Cambridge University Press, Cambridge, UK

Wang C (2000) On the atmospheric responses to tropical Pacific heating during the mature phase of El Niño. J Atmos Sci 57:3767–3781

Wang C (2001) A unified oscillator model for the El Niño-Southern Oscillation. J Climate 14:98–115

Wang C (2006) An overlooked feature of tropical climate: Inter-Pacific-Atlantic variability. Geophys Res Lett 33:L12702. doi:10.1029/2006GL026324

Wang G, Hendon HH (2007) Sensitivity of Australian rainfall to inter-El Nino variations. J Climate 20:4211–4226

Wang C, Weisberg RH (2000) The 1997-98 El Niño evolution relative to previous El Niño events. J Climate 13:488–501

Wang C, Kucharski F, Barimalala R, Bracco A (2009) Teleconnections of the tropical Atlantic to the tropical Indian and Pacific Oceans: a review of recent findings. Meteorol Z 18:445–454

Wang C, Weisberg RH, Virmani JI (1999) Western Pacific interannual variability associated with the El Niño-Southern Oscillation. J Geophys Res 104:5131–5149

Wang C, Picaut J (2004) Understanding ENSO physics—a review. In: Wang C, Xie S.-P, Carton J (eds) Earth's climate: The Ocean-Atmosphere Interaction (pp 21–48). American Geophysical Union

Wara MW, Ravelo AC, Delaney ML (2005) Permanent El Niño-like conditions during the Pliocene warm period. Science 309:758–761

Watanabe T, Suzuki A, Minobe S, Kawashima T, Kameo K, Minoshima K, Aguilar YM, Wani R, Kawahata H, Sowa K, Nagai T, Kase T (2011) Permanent El Niño during the Pliocene warm period not supported by coral evidence. Nature 471:209–211. doi:10.1038/nature09777

Weisberg RH, Wang C (1997) A western Pacific oscillator paradigm for the El Niño-Southern Oscillation. Geophys Res Lett 24:779–782

Weng H, Wu G, Liu Y, Behera SK, Yamagata T (2011) Anomalous summer climate in China influenced by the tropical Indo-Pacific Oceans. Climate Dynam 36:769–782. doi:10.1007/s00382-009-0658-9

White WB, Tourre YM, Barlow M, Dettinger M (2003) A delayed action oscillator shared by biennal, interannual, and decadal signals in the Pacific Basin. J Geophys Res-Oceans 108, 3070. doi:10.1029/2002JC001490

Wyrtki K (1975) El Niño—the dynamic response of the equatorial Pacific Ocean to atmospheric forcing. J Phys Oceanogr 5:572–584

Wyrtki K (1985) Water displacements in the Pacific and the genesis of El Niño cycles. J Geophys Res 90:7129–7132

Yeh S-W, Kug J-S, Dewitte B, Kirtman B, Jin F-F (2009) Recent changes in El Niño and its projection under global warming. Nature 461:511–515

Yu B, Boer GJ (2004) The role of the western Pacific in decadal variability. Geophys Res Lett 31:L02204. doi:10.1029/2003GL018471

Yu J-Y, Kao H-Y (2007) Decadal changes of ENSO persistence barrier in SST and ocean heat content indices: 1958-2001. J Geophys Res 112:D13106. doi:10.1029/2006JD007654

Yu J-Y, Kao H-Y, Lee T (2010) Subtropics-related interannual sea surface temperature variability in the equatorial central Pacific. J Climate 23:2869–2884

Zebiak SE, Cane MA (1987) A model El Niño-Southern Oscillation. Mon Weather Rev 115:2262–2278

Zhang M, Song H (2006) Evidence of deceleration of atmospheric vertical overturning circulation over the tropical Pacific. Geophys Res Lett 33:L12701. doi:10.1029/2006GL025942

Zhang Q, Guan Y, Yang H (2008) ENSO amplitude change in observation and coupled models. Adv Atmos Sci 25:361–366

Zhang W, Jin F-F, Li J, Ren H-L (2011) Contrasting impacts of two-type El Niño over the western north Pacific during boreal autumn. J Meteor Soc Jap 89:563–569

Zhang Y, Wallace JM, Battisti DS (1997) ENSO-like interdecadal variability: 1900–93. J Climate 10:1004–1020

Zheng W, Braconnot P, Guilyardi E, Merkel U, Yu Y (2008) ENSO at 6 ka and 21 ka from ocean–atmosphere coupled model simulations. Climate Dynam 30:745–762

Eastern Pacific Coral Reef Provinces, Coral Community Structure and Composition: An Overview

5

Peter W. Glynn, Juan J. Alvarado, Stuart Banks, Jorge Cortés, Joshua S. Feingold, Carlos Jiménez, James E. Maragos, Priscilla Martínez, Juan L. Maté, Diana A. Moanga, Sergio Navarrete, Héctor Reyes-Bonilla, Bernhard Riegl, Fernando Rivera, Bernardo Vargas-Ángel, Evie A. Wieters, and Fernando A. Zapata

By consensus, authors are arranged alphabetically after lead author P. W. Glynn

P.W. Glynn (✉)
Department of Marine Biology and Ecology, Rosenstiel School of Marine and Atmospheric Science, 4600 Rickenbacker Causeway, Miami, FL 33149, USA
e-mail: pglynn@rsmas.miami.edu

J.J. Alvarado
Escuela de Biología/Museo de Biología, Centro de Investigación en Ciencias del Mar y Limnología (CIMAR), Universidad de Costa Rica, 11501-2060 San Pedro, San José, Costa Rica
e-mail: juan.alvarado@ucr.ac.cr

S. Banks
Marine Ecosystem Research Programme, Charles Darwin Research Station, Galápagos, Ecuador; Conservation International, c/o Christian Lavoie, Av. Catalina Aldáz N34-181 y Portugal, Edificio Titanium II, Piso 4, of. 402, Quito, Ecuador
e-mail: sbanksocean@gmail.com

J. Cortés
Centro de Investigación en Ciencias del Mar y Limnología (CIMAR), Escuela de Biología, Ciudad de la Investigación, Universidad de Costa Rica, 2060 San Pedro, San José, Costa Rica
e-mail: jorge.cortes@ucr.ac.cr

J.S. Feingold
Department of Marine and Environmental Sciences, Halmos College of Natural Sciences and Oceanography, Nova Southeastern University, 3301 College Avenue, Fort Lauderdale, FL 33314-7796, USA
e-mail: Joshua@nova.edu

C. Jiménez
Energy, Environment and Water Research Center (EEWRC), The Cyprus Institute (CyI), P.O. Box 274561645 Nicosia, Cyprus
e-mail: c.jimenez@cyi.ac.cy

J.E. Maragos
Department of Geography, University of Hawaii at Manoa, Kilauea, P.O. Box 1338Honolulu, HI 96754, USA
e-mail: jimmaragos@yahoo.com

P. Martínez
Instituto Nazca de Investigaciones Marinas, Avenida 12 de Octubre, # 1615, entre calle 20 y 21, Salinas, Ecuador
e-mail: pmartinez@institutonazca.org

J.L. Maté
Smithsonian Tropical Research Institute, 0843-03092 Panama, Republic of Panama
e-mail: matej@si.edu

D.A. Moanga
Department of Environmental Science, Policy, & Management, University of California, Berkeley, # 326 Mulford Hall, Berkeley, 94720, CA, USA
e-mail: dianamng@berkeley.edu

S. Navarrete
Estación Costera de Investigaciones Marinas, Advanced Studies in Ecology and Biodiversity, Departamento de Ecología, Facultad de Ciencias Biológicas, Pontificia, Universidad Católica de Chile, Casilla 114-d, Alameda 340, Santiago, Chile
e-mail: snavarrete@bio.puc

H. Reyes-Bonilla
Departamento Académico de Biología Marina, Universidad Autónoma de Baja California Sur, Carretera al sur km 5.5. col. El Mezquitito, 23080 La Paz, B.C.S, Mexico
e-mail: hreyes@uabcs.mx

B. Riegl
Department of Marine and Environmental Sciences, Halmos College of Natural Sciences and Oceanography, Nova Southeastern University, 8000 North Ocean Drive, Dania, FL 33004, USA
e-mail: rieglb@nova.edu

F. Rivera
Instituto Nazca de Investigaciones Marinas, Avenida 12 de Octubre # 1615, entre calle 20 y 21, Salinas, Ecuador
e-mail: frivera@institutonazca.org

B. Vargas-Ángel
Pacific Islands Fisheries Science Center, Coral Reef Ecosystem Division, NOAA Inouye Regional Center, 1845 Wasp Blvd, Bldg #176, Honolulu, HI 96818, USA
e-mail: bernardo.vargasangel@noaa.gov

E.A. Wieters
Estación Costera de Investigaciones Marinas, and Departamento de Ecología, Facultad de Ciencias Biológicas, Pontificia Universidad Católica de Chile, La Alameda 340, Santiago, Chile
e-mail: ewieters@bio.puc

F.A. Zapata
Departamento de Biología, Universidad del Valle, Apartado Aéreo 25360, Cali, Colombia
e-mail: fernando.zapata@correounivalle.edu.co

P.W. Glynn et al. (eds.), *Coral Reefs of the Eastern Tropical Pacific*,
Coral Reefs of the World 8, DOI 10.1007/978-94-017-7499-4_5

Abstract
Advances in our knowledge of eastern tropical Pacific (ETP) coral reef biogeography and ecology during the past two decades are briefly reviewed. Fifteen ETP subregions are recognized, including mainland and island localities from the Gulf of California (Mexico) to Rapa Nui (Easter Island, Chile). Updated species lists reveal a mean increase of 4.2 new species records per locality or an overall increase of 19.2 % in species richness during the past decade. The largest increases occurred in tropical mainland Mexico, and in equatorial Costa Rica and Colombia, due mainly to continuing surveys of these under-studied areas. Newly discovered coral communities are also now known from the southern Nicaraguan coastline. To date 47 zooxanthellate scleractinian species have been recorded in the ETP, of which 33 also occur in the central/south Pacific, and 8 are presumed to be ETP endemics. Usually no more than 20–25 zooxanthellate coral species are present at any given locality, with the principal reef-building genera being *Pocillopora*, *Porites*, *Pavona*, and *Gardineroseris*. This compares with 62–163 species at four of the nearest central/south Pacific localities. Hydrocorals in the genus *Millepora* also occur in the ETP and are reviewed in the context of their global distributions. Coral community associates engaged in corallivory, bioerosion, and competition for space are noted for several localities. Reef framework construction in the ETP typically occurs at shallow depths (2–8 m) in sheltered habitats or at greater depths (10–30 m) in more exposed areas such as oceanic island settings with high water column light penetration. Generally, eastern Pacific reefs do not reach sea level with the development of drying reef flats, and instead experience brief periods of exposure during extreme low tides or drops in sea level during La Niña events. High rates of mortality during El Niño disturbances have occurred in many ETP equatorial areas, especially in Panama and the Galápagos Islands during the 1980s and 1990s. Remarkably, however, no loss of resident, zooxanthellate scleractinian species has occurred at these sites, and many ETP coral reefs have demonstrated significant recovery from these disturbances during the past two decades.

Keywords
Species distributions • Biogeography • Eastern Pacific • Coral occurrences • Species richness

5.1 Introduction

Geographic isolation, variable oceanographic conditions, and a peninsula-like shallow shelf spanning a relatively narrow latitudinal corridor, have all figured prominently in the formation and development of the eastern tropical Pacific coral reef region. The vast majority of eastern Pacific corals belong to species occurring in the central and south Pacific (e.g., Wells 1983; Cortés 1986; Grigg and Hey 1992; Veron 1995; Glynn and Ault 2000). Dana (1975) recognized this faunal affinity, and suggested the existence of genetic connectivity between the central Pacific Line Islands and the equatorial eastern Pacific by larval dispersal via the North Equatorial Counter Current (NECC). A few species are eastern Pacific endemics, and there is a near absence of Caribbean and western Atlantic species along western American shores. The narrowing and eventual closing of the Central American isthmus, 3.5-3.0 Ma (Coates and Obando 1996), brought about extinctions of once-shared transoceanic species, and the invasion of coral species from the west by long-distance dispersal (see Chap. 2, López-Pérez, and Chap. 16, Lessios and Baums; López-Pérez and Budd 2009).

The occurrence of reef-building corals along coastal western America is highly skewed toward the northern hemisphere, with a strong presence of corals from the northern Gulf of California, Mexico ($\sim$30°N) to just a few degrees south of the equator along coastal Ecuador (2°S). Modern zooxanthellate coral communities have not been reported on continental shores south of the Gulf of Guayaquil, Ecuador, which are under the influence of the cool, north-flowing Peru Coastal Current (Fig. 5.1). Five relatively isolated oceanic island groups, from the northern-most Revillagigedo Islands off Mexico to the Galápagos Islands lying astride the equator, also belong to the eastern Pacific coral reef region. Vibrant coral communities and incipient

Fig. 5.1 Eastern Pacific coral and coral reef localities, Gulf of California, Mexico to Ecuador, and remote Rapa Nui (Easter) and Salas y Gómez Islands, Chile. *Red diamonds* identify oceanic localities supporting coral communities and coral reefs

reef development also appear far to the south and west, about 4000 km from the Galápagos Islands and coastal Ecuador. Rapa Nui (Easter) and Salas y Gómez Islands, at 27°S and over 1000 km west of Chile, represent the most isolated ETP coral outpost (Glynn et al. 2003, 2007; Hubbard and Garcia 2003; see Chap. 6, Toth et al.).

The most recent overview of the eastern tropical Pacific (ETP) coral reef region was published over 10 years ago, in *Latin American Coral Reefs* edited by Jorge Cortés (2003). This work comprised seven chapters dedicated to the ETP, describing coral communities and coral reefs from mainland Mexico to Ecuador, including offshore islands from the Revillagigedo Islands, Mexico to Rapa Nui, Chile. These summaries included historical sketches of coral reef research, site descriptions, species lists, natural and anthropogenic disturbances, and issues pertaining to protection and management. The present chapter updates Cortés (2003), and several earlier reviews of eastern Pacific coral reef biology (e.g., Durham and Barnard 1952; Squires 1959; Durham 1966; Glynn and Wellington 1983; Cortés 1986, 1997, 2011; Glynn 2001).

An updated biogeographic analysis of zooxanthellate Scleractinia is examined first to review species relationships within the eastern Pacific region, and secondly extra-regional faunal affinities. This analysis will serve as a framework for brief introductions to the main coral reef localities from Mexico to Ecuador, including the Galápagos Islands. An overview of the more distant eastern tropical Pacific coral outposts of Rapa Nui and Salas y Gómez Islands will conclude the chapter. Eastern Pacific *Millepora* records are also reviewed, and compared to hydrocoral occurrences in the central and southeastern Pacific. Each eastern Pacific coral reef locality is examined in terms of its community structure, species richness, physical environmental conditions, and chief ecological interactions. The treatments are admittedly unbalanced, with more information provided for newly discovered or less well-known areas. It is hoped that this overview will serve to stimulate further research, and provide information for comparisons with other coral reef regions. A comprehensive review of ETP Holocene reef development—including coral framework structures, time of reef growth initiation, thicknesses, and accumulation rates—can be found in Toth et al., Chap. 6.

5.2 Eastern Pacific Coral Reef Region

The eastern tropical Pacific region consists of a distinct biotic assemblage of marine species isolated from the Caribbean Sea by the Central American land bridge. The west American marine biota is also separated from the central/western Pacific region by a wide expanse (5000–8000 km) of open ocean waters, namely the eastern Pacific Barrier or Ekman's Barrier (e.g., Darwin 1880; Ekman 1953; Briggs 1974; Grigg and Hey 1992; Veron 1995). Abundant recent evidence indicates that this barrier, which Darwin considered impassable, is best viewed as a filter bridge since numerous shallow-living species from several major taxa—scleractinian corals (Glynn and Ault 2000), bryozoans (Banta 1991), crustaceans (Garth 1991), molluscs (Scheltema 1988; Emerson 1991), echinoderms (Lessios et al. 1998), and fishes (Robertson and Allen 1996; Robertson et al. 2004; Lessios and Robertson 2006; see Chap. 16, Lessios and Baums)—have crossed this deep water stretch by larval dispersal or rafting. In particular, many marine species shared between the eastern Pacific and central/western Pacific are common members of coral communities and coral reefs. Indeed, the majority of ETP reef-building coral taxa—species in the genera *Pocillopora, Porites, Pavona, Gardineroseris*—are conspecific with those found west of the ETP filter bridge.

Within the eastern Pacific region, four subtropical/tropical biogeographic provinces were early recognized: (1) Cortez Province (middle to upper Gulf of California), (2) Mexican Province (lower Gulf of California to the Gulf of Tehuantepec), (3) Panamanian Province (Tangola-Tangola Bay, Gulf of Tehuantepec to the Gulf of Guayaquil), and (4) the Galápagos Islands (Briggs 1974). More recent analyses of coral distributions have recognized an eastern Pacific oceanic islands fauna, with coral affinities shared among Clipperton Atoll, Malpelo Island, Easter Island, and Tabuaeran (=Fanning Atoll) in the Line Islands group (Glynn and Ault 2000; Glynn et al. 2007). Another recent biogeographic analysis, based on the distributions of 1135 resident shore fishes, also recognized the oceanic islands as a distinct faunal subdivision, but concluded that the continental coast is best represented by two provinces, namely (1) the Cortez, comprising the Gulf of California and lower Pacific Baja California areas, and (2) the Panamic, southward to the Gulf of Guayaquil, Ecuador (Robertson and Cramer 2009).

With an increasing knowledge of zooxanthellate scleractinian distributions, particularly during the last two decades, several workers have examined patterns of coral faunal distributions along the eastern tropical Pacific (Glynn 1997; Glynn and Ault 2000; Glynn et al. 2007; Alvarado et al. 2011), and within specific eastern Pacific countries and locations (Reyes-Bonilla 1993a; chapters in Cortés 2003; Reyes-Bonilla and López-Pérez 1998, 2009). The existence of a dispersal barrier within the eastern Pacific, the long-recognized Pacific Central American faunal gap (PCAFG), has been supported mainly by the absence of corals (Glynn and Ault 2000) and reef-associated fishes (Springer 1958; Briggs 1974) between Tangola Tangola (southern Mexico) south to the Gulf of Fonseca (El Salvador/Honduras/Nicaragua). However, the faunal similarity of newly discovered coral communities and reef-associated fishes in Nicaragua and El Salvador with central Mexican coastal areas (Michoacán, Colima) is very high, suggesting some degree of genetic connectivity across the PCAFG (Robertson and Cramer 2009; Alvarado et al. 2011).

5.2.1 Zooxanthellate Scleractinian Fauna

Here we present an updated biogeographic analysis of zooxanthellate scleractinian coral faunas from 15 ETP localities and four central/south Pacific localities. In the present data-base, the listed ETP fauna consists of 47 species in 11 genera, and the central/south Pacific localities 225 species in 40 genera (Appendix). The central/south Pacific localities include the Hawaiian Islands, Johnston Atoll, Pitcairn Islands, and Palmyra Atoll, and will be employed for analysis of and comparison with the ETP fauna. The species in this analysis were delineated chiefly from gross morphometric characteristics; the sources of this information are listed in the caption to Fig. 5.7. In light of current molecular genetic studies (Flot et al. 2010; Pinzón and LaJeunesse 2011; Pinzón et al. 2013; Boulay et al. 2014; see Chap. 14, Pinzón), and evidence from a combination of genetic and morphological analyses—including both macro- and micromorphometric characteristics (Schmidt-Roach et al. 2014)—it is evident that species delineations across the Indo-Pacific region are still unresolved. On the one hand, genetic studies offer evidence of fewer species of *Pocillopora* in the ETP than recognized morphometrically, three instead of seven species (Pinzón et al. 2013; see Chap. 14, Pinzón). On the other hand, new previously unrecognized cryptic species have come to light in the *Pocillopora damicornis* complex on the Great Barrier Reef (Schmidt-Roach et al. 2014).

A major three-way grouping of coral faunas is evident from an agglomerative hierarchical cluster analysis: (a) dominantly equatorial eastern Pacific (EEP) localities (PAN-GLN), (b) ETP oceanic islands (MAL-REV), and (c) central/southern Pacific localities (Fig. 5.2). [Locality acronyms, e.g. PAN = Panama, follow those employed in Fig. 5.2.] A notable east to west increase in species richness is apparent with eastern Pacific localities demonstrating impoverished faunas compared with potential source areas to the west of Ekman's filter bridge, a long recognized pattern (e.g., Stehli and Wells 1971; Veron 1995). All ETP reef communities consist of fewer than 30 species, whereas central/south Pacific localities exhibit from 44 to 163 species. At Palmyra Atoll, which is in the path of the North Equatorial Counter Current (Maragos et al. 2008a, b), 163 coral species have been recorded. The North Equatorial Counter Current has been considered a likely means of transport across the filter bridge (e.g., Dana 1975; Glynn and Ault 2000), and species-rich Palmyra is a potential source locality. In the EEP cluster, the low species occurrences in the southern Galápagos Islands (GLS) and Gulf of California (GOC) are likely due, in large part, to their subtropical locations. Two species-poor mainland sites (NIC, ELS) cluster with the ETP oceanic islands. The recently-discovered coral communities in Nicaragua may reveal additional species upon further study; El Salvador, however, is a marginal coral area with highly turbid waters and is not expected to show a significant increase in species richness. Additional comments on ETP faunal relationships are noted below, in site-specific sections.

Eight species in this updated inventory are likely ETP endemics, suggesting a moderate level of speciation in this region (Table 5.1). This number represents 19.5 % of the total ETP fauna reported to date. Latitudinally, all eight coral endemics are concentrated in Mexico, in the northern-most eastern Pacific region. Three *Porites* endemics, *P. arnaudi*, *P. sverdrupi*, and *P. baueri*, demonstrate restricted distributions, known from only one or two localities. All eastern Pacific endemic species are absent from Rapa Nui (Easter Island). Three endemics, *Pavona gigantea*, *Pocillopora effusus*, and *Porites panamensis*, are widely distributed throughout the eastern Pacific, presently reported from eight to 14 localities.

5.2.2 *Millepora*

Zooxanthellate hydrocorals in the genus *Millepora* can be important contributors to reef frameworks and coral reef growth. *Millepora* has a very restricted distribution in the ETP. Three species of *Millepora* were present on reefs in the Gulf of Chiriquí (Panama) before the 1982–83 El Niño event: *Millepora intricata* Milne-Edwards 1857, *Millepora boschmai* Weerdt and Glynn 1991, and *Millepora platyphylla* Hemprich and Ehrenberg 1834. All three species bleached and experienced high mortality, during and immediately following the period of elevated sea temperatures (Glynn and de Weerdt 1991). Living colonies of *M. platyphylla* could no longer be found in Panama after 1983 (as of 2015), and are now apparently absent from the ETP (Glynn 2011). *Millepora boschmai* also disappeared from survey sites at Uva Island after 1983, but a small population was found nearby in 1992 (Glynn and Feingold 1992). All colonies from this remnant population disappeared in early 2000, as well as a few known colonies at Coiba Island (Brenes et al. 1993), and none have been observed in the eastern Pacific since that time (Maté 2003). This species was recognized in collections made from two localities in Indonesia by Razak and Hoeksema (2003), including South Sulawesi in 1994 and Sumba in 1984. In Panama, *Millepora intricata* re-appeared a few years after 1983, ostensibly by sexual recruitment from deep refuge populations (Smith et al. 2014; see Chap. 17, Smith et al.); it is again an important member of coral communities in the Gulf of Chiriquí, Panama (Fig. 5.3). Unlike *M. platyphylla* and *M. boschmai*, with populations restricted to shallow habitats (usually $\leq$6–8 m depth), *M. intricata* occurs in deep

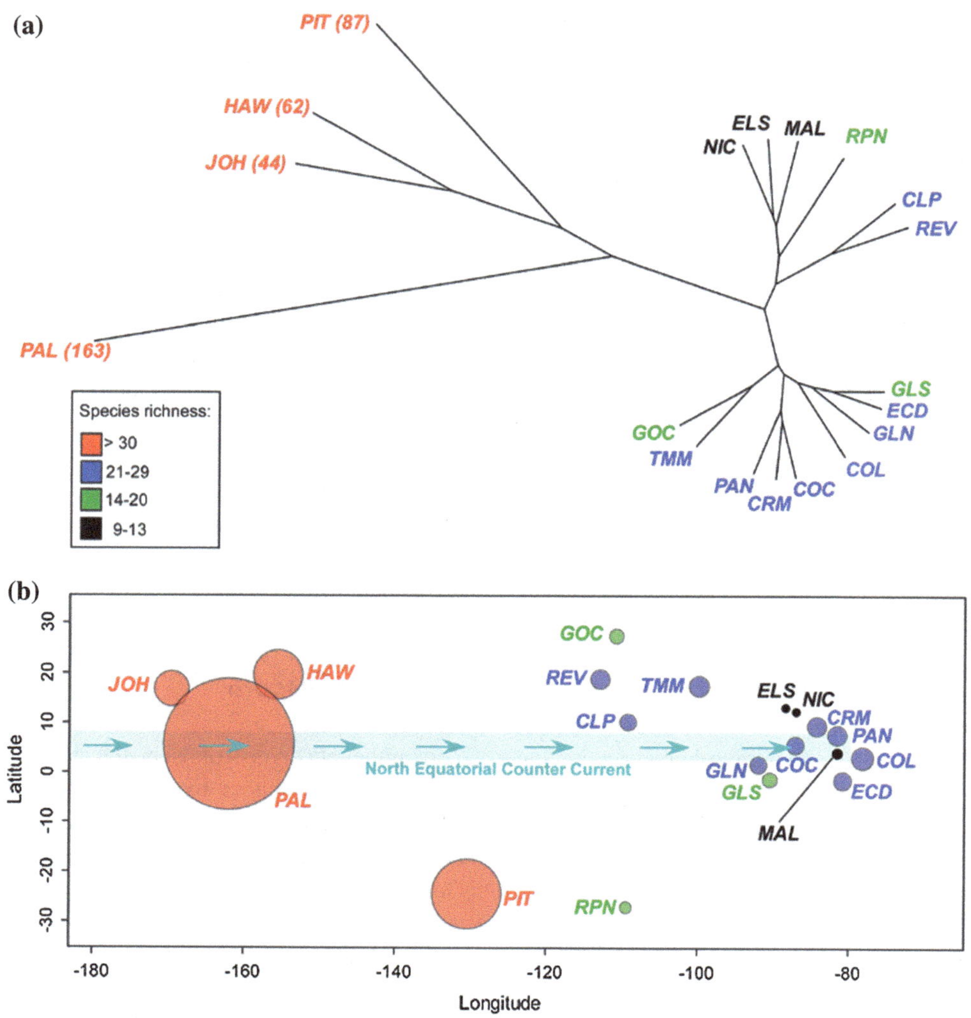

Fig. 5.2 **a** Relationships of zooxanthellate scleractinian coral faunas at 15 eastern Pacific and four central and south Pacific localities. "Spatial patterns" of an agglomerative, hierarchical cluster analysis were portrayed using a Euclidean distance measure and group-average method for cluster formation. Constructed from presence/absence data in Appendix. **b** Geographic pattern of coral species richness across EP filter bridge in relation to NECC eastward flow. *Circle areas* are proportional to species richness. GOC, Gulf of California; TMM, tropical Mexican mainland; REV, Revillagigedo Islands; CLP, Clipperton Atoll; ELS, El Salvador; NIC, Nicaragua; CRM, Costa Rican mainland; COC, Cocos Island; PAN, Panama; COL, Colombia; MAL, Malpelo Island; ECD, Ecuador; GLS, southern Galápagos Islands; GLN, northern Galápagos Islands; RPN, Rapa Nui (Easter) and Salas y Gómez Islands; HAW, Hawaiian Islands; JOH, Johnston Island; PIT, Pitcairn (Henderson, Oeno, Ducie Islands); PAL Palmyra Atoll

(>11 m depth) as well as shallow reef zones. This allowed *M. intricata* in deep-water habitats to survive two of the strongest El Niño events on record (1982–83, 1997–98), and then to re-populate shallow reef sites (Smith et al. 2014).

Millepora exaesa is known only from Clipperton Atoll in the eastern Pacific, and colonies are uncommon there (Glynn et al. 1996a, b; Carricart-Ganivet and Reyes-Bonilla 1999; Flot and Adjeroud 2009; Glynn, unpublished observation). The nearest reported central Pacific record of *M. exaesa* is at Wake Island (167°E), over 5500 km west of Clipperton (Table 5.2).

Early surveys showed that *Millepora* spp. are widespread throughout the Indo-Pacific region, and this has been confirmed by recent surveys. However, their relatively infrequent reproduction and short-lived dispersive phase (gonochoric medusae) appear inconsistent with their widespread global occurrence (Lewis 1989; see also Chap. 15 , Glynn et al.). Veron (2000) shows *Millepora* present in all parts of the Indian Ocean and the western and southern Pacific (Fig. 5.4). Some early and more recent surveys show that various *Millepora* species also occur widely across the central Pacific region, e.g. Johnston Atoll (Maragos and Jokiel 1986), the Phoenix Islands (Maragos et al. 2008a, b), Tabuaeran (formerly Fanning, Maragos 1974) and other Line Islands (Maragos et al. 2008a, b); Marquesas (Chevalier 1978), Tuamotus (Ricard 1985; Harmelin-Vivien 1985; Montaggioni 1985), Society Islands (Galzin and Pointier 1985), and the Pitcairn Islands (Irving and Dawson 2012;

Table 5.1 Regional occurrence of eastern tropical Pacific endemic zooxanthellate Scleractinia at 14 localities

Species	Localities	
	Present (no. localities)	Absent
Pavona gigantea	MEX EEP (14)	RPN
Pocillopora effusus	MEX EEP (8)	RPN
Pocillopora inflata	MEX EEP (6)	RPN, CLP
Porites panamensis	MEX EEP (9)	RPN, CLP
Porites arnaudi	MEX CLP (2)	RPN, EEP
Porites sverdrupi	MEX (2)	EEP, RPN, CLP
Porites baueri	MEX (1)	EEP, RPN, CLP
Psammocora brighami	MEX (2) EEP (5)	RPN

MEX (Mexico)—Gulf of California, Revillagigedo Is., tropical Mexican mainland; *EEP* (equatorial eastern Pacific)—El Salvador, Nicaragua, Costa Rica, Cocos Is, Panama, Colombia, Malpelo, Ecuador, Galápagos Iss; *RPN*—Easter Is; *CLP*—Clipperton Atoll

USNM collection records) (Fig. 5.4; Table 5.2). Subfossil fragments of *Millepora intricata* recovered from reef cores revealed radiocarbon dates of ~5000 cal year BP, indicating the presence of this hydrocoral species in Panama for much of the Holocene Epoch (Smith et al. 2014). The highly restricted occurrence of *Millepora* in the ETP is mirrored in the eastern Atlantic where it is known from only two island groups, the Canary and Cape Verde Islands (Clemente et al. 2011). It is hypothesized to be a recent introduction to the eastern Atlantic region (Clemente et al. 2011).

5.3 Intra-regional Précis

Coral faunas and coral communities within the eastern tropical Pacific region are reviewed below, from northern Mexico to the southern-most mainland assemblages of Ecuador. The eastern Pacific oceanic islands—Revillagigedo Islands, Clipperton Atoll, Isla del Coco, Galápagos Islands, and Rapa Nui—are also considered along a N–S latitudinal

Fig. 5.3 *Millepora intricata* prominence in a coral community at Brincanco Island, Gulf of Chiriquí, Panama (5.4 m depth, 4 April 2012, courtesy Irving Bethencourt). *Pocillopora* sp. in foreground ~30 cm in diameter

Table 5.2 Documented collection records of *Millepora* species in the central and south eastern Pacific Ocean

Location	Longitude	Taxa	Authority
Wake Is.	167°E	*Millepora* 2 spp.	Lobel and Lobel (2008)
		Millepora tuberosa (exaesa)[a]	Maragos, unpublished record, 1979–2005
Phoenix Iss.	172°E	*Millepora* sp.	Maragos (1974)
Johnson Atoll	169°E	*Millepora dichotoma*	Maragos et al. (2008a, b)
		Millepora tenera	Maragos and Jokiel (1986)
Cook Iss.	160–162° W	*Millepora*	Maragos (1974)
Line Iss.	157–169° W	*Millepora*	Maragos et al. (2008a, b)
Kingman reef	162°W	*Millepora platyphylla*	Maragos, personal communication
Tabuaeran	159°W	*Millepora platyphylla*	Maragos (1973)
Moorea Is.	150°W	*Millepora platyphylla*	Galzin and Pointier (1985)
Tuamotus Iss.	135–150° W		
Makatea Is.	148°W	*Millepora* sp.	Montaggioni (1985)
Mataiva Atoll	148°W	*Millepora platyphylla*	Delasalle (1985)
Tikahau Atoll	148°W	*Millepora platyphylla*	Harmelin-Viven (1985)
Rangiroa Atoll	147–148° W	*Millepora platyphylla*	Riccard (1985)
Takapoto Atoll	145°W	*Millepora platyphylla*	Salvat and Richard (1985)
Marquesas Iss.	138–141° W	*Millepora platyphylla*	Chevalier (1978)
Pitcairn Iss.	124–131° W	*Millepora* sp.	Irving and Dawson (2012)
Pitcairn Is.	130°W	*Millepora platyphylla*	D. Devaney USNM No. 79568
Henderson Is.	128°W	*Millepora platyphylla*	G. Paulay USNM No. 91571, 91572

[a]*Millepora tuberosa* is a junior synonym of *M. exaesa* (see Razak and Hoeksema 2003)

gradient. This overview will hopefully provide basic coral community-level information to help evaluate intra-regional similarities and differences. We recognize that these brief intra- and inter-regional treatments would reveal only approximate comparisons at best due to the often great differences in available habitat area and habitat diversity. Regional Holocene reef development, including the coral taxa that construct frameworks, times of reef initiation, framework thickness, and accumulation rates are reviewed by Toth et al. in Chap. 6.

5.3.1 Gulf of California (Mexico)

The northern-most limit of zooxanthellate corals is in the upper reaches of the 1000 km-long Gulf of California (Fig. 5.5). Corals occur commonly over most of the length of the gulf, with increasing species richness in the southerly warmer waters. The higher thermal conditions in the southern gulf are exemplified by satellite imagery of sea surface temperatures (SSTs) in January 2006, with upper gulf SSTs of 14.5–18.3 °C and lower gulf SSTs of 20.1–22.8 °C (Ledesma-Vázquez et al. 2009). Corals are virtually absent from the western shores of the Baja California peninsula due to the southerly-flowing, cool California Current (see Chap. 3, Fiedler and Lavín). Distributional analyses based on presence/absence data have revealed two coral faunas within the Gulf of California, one in the northern and one in the southern gulf (Reyes-Bonilla and López-Pérez 2009). The northern fauna ranges from Punta Peñasco (31.3°N) to Punta Prieta (27°N), and consists predominantly of encrusting, monospecific *Porites panamensis* populations. Occasional colonies of *Porites sverdrupi* are also present on rocky substrates, and *Cycloseris curvata*, the latter in deeper rhodolith communities. The southern fauna ranges from Loreto to Cabo San Lucas. This fauna is more diverse, consisting of 20 species with a close affinity with tropical mainland Mexico (Fig. 5.2). According to recent assessments, the gulf coral fauna has demonstrated an increase from 16 to 20 species (e.g., Reyes-Bonilla et al. 2005, 2010; Reyes-Bonilla and López-Pérez 2009) (Fig. 5.6). Sixteen of these species belong to the Indo-Pacific fauna, three to the eastern Pacific fauna, and one (*P. sverdrupi*) is endemic to the gulf and mainland Mexico.

The coral assemblages and associated biota in the gulf were classified into four biotopes by Reyes-Bonilla and López-Pérez (2009): (1) isolated colonies or patches, (2) corals in rhodolith communities and other soft bottom habitats, (3) coral communities, and (4) coral reefs. Biotopes 1 and 2 occur in the northern gulf, and coral communities and coral reefs in the southern gulf. Corals, predominantly *Porites panamensis*, are highly dispersed in type 1 assemblages, usually amounting to <1 % cover where present. Even at low abundance, however, Halfar et al. (2005) demonstrated a potentially high carbonate sediment production rate for *P. panamensis* at Bahía de Los Ángeles (Fig. 5.5). Only a single rhodolith community, with associated fungiid corals, was reported in the northern gulf, while five were reported in the southern gulf. Three of these communities occurred at 25–30 m depth between La Paz and Monserrat Island near Loreto. *Diaseris distorta* is predominant with the occasional presence of *Cycloseris curvata*

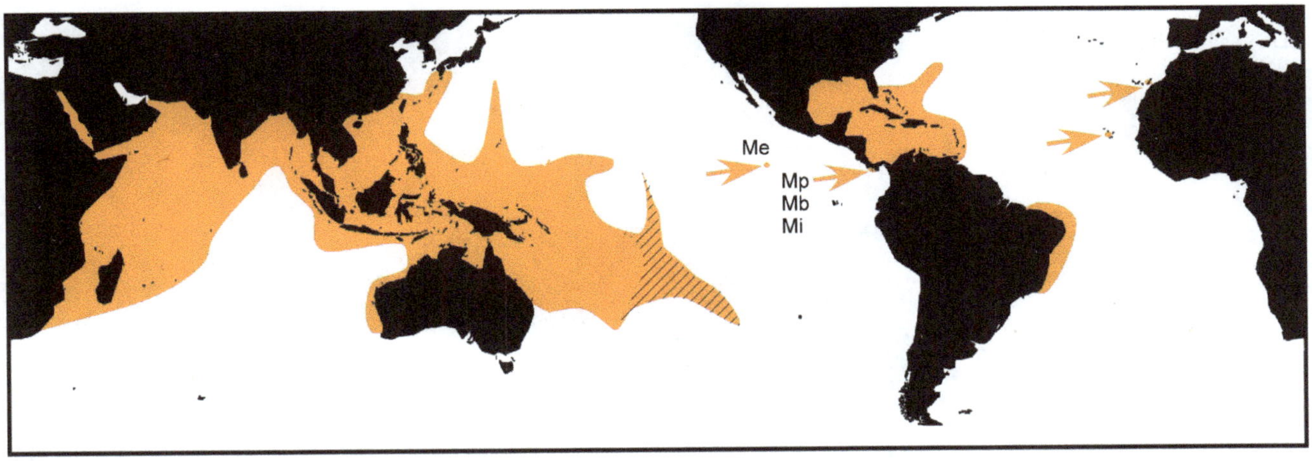

Fig. 5.4 Known global distribution of *Millepora* spp. as of 2014. In the eastern Pacific: Me, *M. exaesa*; Mp, *M. platyphylla*; Mb, *M. boschmai*; Mi, *M. intricata*. Mp and Mb now regionally extinct. Eastern Atlantic, Canary and Cape Verde Islands, *Millepora* sp., possibly *M. alcicornis*. Cross hatching in south-central Pacific denotes records of *Millepora* occurrences in Line Islands and Pitcairn Islands

and mobile species of *Psammocora*, *Porites*, and rarely *Pavona* (Table 5.3). Present taxonomic evidence indicates that the coralline flora of rhodolith communities consists of four species in four genera (Steller et al. 2009). This biotope provides shelter and trophic resources for a diverse assemblage of non-coral invertebrates (bryozoans, molluscs, crustaceans, and echinoderms), fishes, and cryptic biota (Steller et al. 2003, 2009).

Coral community biotopes occur commonly at shallow depths (≤ 10 m) along the southwestern peninsular gulf coast. The predominant coral genera in these communities are *Pocillopora* and *Pavona*. While buildups can attain several meters in thickness (height), framework structures per se are not evident. Particularly robust coral communities, e.g. at Cabo Pulmo and La Paz, demonstrated mean coral cover values of 15–20 % with 4–6 species present. Both *Pocillopora damicornis* and *Pocillopora verrucosa* contribute substantially to these communities. The structural complexity of these coral communities offers numerous shelter sites that greatly increase the diversity of macroinvertebrates, fishes, and cryptic biota. In addition, metazoan coral symbionts and diverse coral bioeroding organisms inhabit these communities. Gastropod, echinoderm, and fish corallivores are ubiquitous in the gulf and actively prey on all coral species (e.g., Reyes-Bonilla 2003; Hendrickx et al. 2005; Álvarez-Filip et al. 2006; Reyes-Bonilla and López-Pérez 2009). *Acanthaster planci* is widespread, especially in the lower half of the gulf where corals are in high abundance (Dana and Wolfson 1970; Barham et al. 1973). Reyes-Bonilla (1993a, b, 2003) reported relatively high population densities of *A. planci* at Cabo Pulmo, 16 to >40 ind ha^{-1} over a 13-year period, but with no indication of local coral depletion.

True coral reefs, i.e. those with wave-resistant framework structures, were reported by Reyes-Bonilla and López-Pérez (2009) to be present at five sites in the extreme southern sector of the Baja California peninsula. These reefs are built predominantly by *Pocillopora* spp., with coral frameworks reaching 2 m in height. The largest and most thoroughly studied reef is located at Cabo Pulmo (Brusca and Thomson 1975; Reyes-Bonilla et al. 1997a, b). It is often cited as the northern-most eastern Pacific coral reef. It covers 150 ha and contributes to its surroundings an estimated 13,500–31,000 tons of $CaCO_3$ annually (Reyes-Bonilla and López-Pérez 2009). This reef has formed on intrusive dikes and rocky hardgrounds. Even though the carbonate production rate of this reef is high, the coarser, reef-derived sediments are transported off-reef, especially during the passage of tropical cyclones, with no appreciable sediment retention. This loss greatly limits vertical reef accretion. Riegl et al. (2007) concluded that the Cabo Pulmo formation, sedimentologically, functions only in a limited way like a coral reef.

Compared to natural biotic effects, such as predation and bioerosion, extreme physical conditions have demonstrated the greatest negative influence on coral communities in the Gulf of California. Hurricanes and tropical storms regularly frequent the southern-most sector of the gulf (see Chap. 3, Fiedler and Lavín), and can limit reef development as noted above at Cabo Pulmo. Storm-generated turbulence and sediment transport are responsible for local coral breakage and the burial of fungiid populations (Reyes-Bonilla and López-Pérez 2009). During the past few decades, ENSO-related high and low seawater temperature extremes have caused significant coral bleaching and mortality in the lower gulf (Reyes-Bonilla 2001; Iglesias-Prieto et al. 2004; Paz-García et al. 2012). These thermal-bleaching events can lead to widespread and spatially uniform mortality of *Pocillopora* spp., approaching 20 %, over 10s–100s of km of coastline. While noteworthy, this level of mortality is significantly lower than that reported in other ETP localities. Reyes-Bonilla (2001) hypothesized

Fig. 5.5 Gulf of California (GOC) and Revillagigedo Islands (REV), Mexico. *Red diamonds* identify surveyed sites

that the relatively low level of thermal bleaching/mortality is possibly a result of local upwelling during the summer season, and proximity of lower gulf waters to the California Current system that contributes to the formation of a cool front at the entrance to the gulf.

5.3.2 Tropical Mainland Mexico

Beyond the Gulf of California, the tropical mainland coast of Mexico extends toward the southeast from southern Sinaloa to Chiapas states, a distance of nearly 2000 km (Fig. 5.7). Only two poritid species have been reported from Sinaloa (*Porites panamensis*, *Porites sverdrupi*) and, excluding extensive Miocene-Pliocene fossil reefs, no zooxanthellate corals are known from Chiapas. Both of these regions are dominated by coastal lagoons, sandy beaches, estuaries and mangrove ecosystems (Flores-Verdugo et al. 2001). The greatest reef development occurs in Nayarit and Oaxaca, each with 20 and 16 coral species, respectively (Reyes-Bonilla 2003; Reyes-Bonilla et al. 2005). Both of these areas are near upwelling centers; the northern center is located at Banderas Bay, Nayarit (López-Sandoval et al. 2009), and the southern upwelling area occurs along the Huatulco coast, Oaxaca (Steenburgh et al. 1998), which borders the Gulf of Tehuantepec (see Chap. 3, Fiedler and Lavín). Pocilloporid fringing

Table 5.3 Occurrence of zooxanthellate scleractinian corals in three Pacific Mexican regions

Species	Location		
	Gulf of California	Tropical Mainland Mexico	Revillagigedo Iss
Cycloseris curvata (Hoeksema, 1989)	R	R	R
Fungia vaughani Boschma, 1923	R		
Diaseris distorta (Michelin, 1843)	R	R	R
Gardineroseris planulata (Dana, 1846)		R	
Leptoseris papyracea (Dana, 1846)	R	R	
Pavona clavus (Dana, 1846)	C	C	C
Pavona duerdeni Vaughan, 1907	R		R
Pavona gigantea Verrill, 1869	A	C	C
Pavona maldivensis (Gardiner, 1905)			R
Pavona minuta Wells, 1954			R
Pavona varians Verrill, 1864	R	R	R
Pocillopora damicornis (Linnaeus, 1758)	C	A	R
Pocillopora effusus Veron, 2002		R	
Pocillopora elegans Dana, 1846	C	A	C
Pocillopora eydouxi Milne Edwards and Haime, 1860	R	R	C
Pocillopora inflata Glynn, 1999	R	R	
Pocillopora meandrina Dana, 1846	A	C	C
Pocillopora capitata Verrill, 1864	A	C	R
Pocillopora verrucosa (Ellis and Solander, 1786)	A	A	A
Pocillopora woodjonesi Vaughan, 1918			R
Porites arnaudi Reyes-Bonilla and Carricart-Ganivet, 2000			R
Porites australiensis Vaughan, 1918			R
Porites baueri Squires, 1959		R	
Porites evermanni Vaughan, 1907		R	
Porites lichen Dana, 1846			C
Porites lobata Dana, 1846		R	A
Porites lutea Milne Edwards and Haime, 1851			R
Porites panamensis Verrill, 1866	A	R	R
Porites sverdrupi Durham, 1947	R		
Psammocora brighami (Vaughan, 1907)	R		
Psammocora haimeana Milne Edwards and Haime, 1851	R		
Psammocora profundacella Gardiner, 1898		R	R
Psammocora stellata Verrill, 1864	C	C	C
Psammocora superficialis Gardiner, 1898		R	R
Total number species	20	22	24

A abundant (>50 % presence at surveyed sites); *C* common (20–50 % presence); *R* rare (<20 % presence). Records are from Reyes-Bonilla et al. (2010). Species names in this and following tables from Wells (1983), Reyes-Bonilla et al. (2005) and Veron (2000)

reefs occur in both of these areas, usually at sheltered sites; reef frameworks attain vertical heights of up to 6 m, live coral cover commonly ranges between 20 and 50 %, and extends over several ha at several sites (Carriquiry and Reyes-Bonilla 1997; Reyes-Bonilla 2003). Coral communities and small patch reefs are also present in other west Mexican areas, particularly along coastal Guerrero and Michoacan states. Recent exploration has dramatically increased the species richness of zooxanthellate corals over a relatively brief period, from 19 (Reyes-Bonilla 2003) to 26 (Reyes-Bonilla et al. 2005, 2010; Glynn et al. 2007; Boulay et al. 2014), due mainly to continued exploration (Fig. 5.6; Appendix).

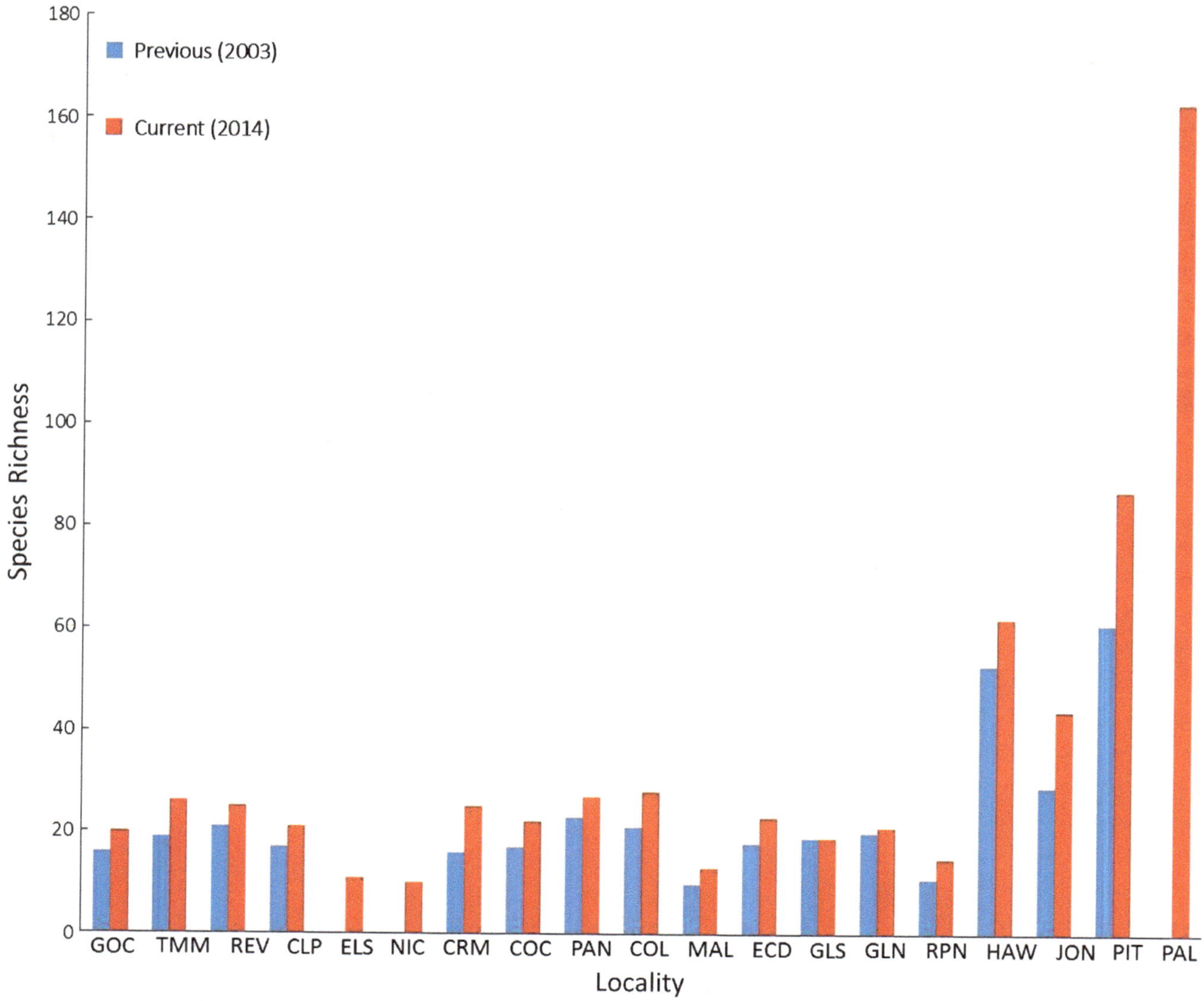

Fig. 5.6 Differences in zooxanthellate scleractinian coral species richness at eastern and central Pacific localities before (*blue*) and after (*red*) major surveys. Most 'previous' records were published in 2003 (Cortés 2003), but some are from major works as early as the latter half of the 1980s. Single bars (ELS, NIC, PAL) denote a lack of data from before the time periods examined. *Sources Previous/Current* GOC 1/2; TMM 1/2,3; REV 1/2; CLP 4/3,4; ELS -/5; NIC -/6; CRM 7,8,9/3,10; COC 8/3,10; PAN 11, 12/3,12; COL 13/14; MAL 10,13/12,14; ECD 15/16; GLS 15/17; GLN 15/3,4,18; RPN 19/4,20; HAW 21/22,23; JOH 24/25,26; PIT 27/28; PAL -/29. (1) Reyes-Bonilla (2003); (2) Reyes-Bonilla et al. (2010); (3) Boulay et al. (2013); (4) Glynn et al. (2007); (5) Reyes-Bonilla and Barraza (2003); (6) Alvarado et al. (2010), (2011); (7) Cortés and Guzmán (1998); (8) Cortés and Jiménez (2003); (9) Jiménez (2007a, b); (10) Cortés (updated checklist); (11) Holst and Guzmán (1993), Maté (2003); (12) Maté (updated checklist); (13) Zapata and Vargas-Ángel (2003); (14) Zapata (updated checklist); (15) Glynn (2003); (16) Rivera and Martínez (2011); (17) Glynn (updated checklist); (18) Hickman (2008); (19) Glynn et al. (2003); (20) Wieters and Navarrete (updated checklist); (21) Maragos (1995); (22) Maragos (updated checklist); (23) Maragos et al. (2004); (24) Maragos and Jokiel (1986); (25) Lobel and Lobel (2008); (26) Maragos (updated checklist); (27) Paulay (1989); (28) Irving and Dawson (2012); (29) Williams et al. (2008), Maragos (updated checklist)

The coral faunas of tropical mainland Mexico (TMM) and the Gulf of California (GOC) are very similar (Fig. 5.2a), however, mainland Mexico is more speciose (26 vs. 20 species), with several reef-building species (in the genera *Pocillopora*, *Porites*, *Gardineroseris*) that have not been reported from the Gulf.

Information on coral community development at the Marías Islands, Nayarit, was lacking until relatively recently, with the delayed publication of field studies conducted in 1997 (Pérez-Vivar et al. 2006). All 10 species of zooxanthellate corals collected there are known from other Mexican mainland areas. *Porites baueri*, long considered an endemic species restricted to the Marías Islands and elsewhere on the Nayarit coast (Reyes-Bonilla 1993b), was not found. This species may be synonymous with *Porites lobata*, which was abundant in the islands in 1997. Including museum collections, the presently known coral fauna of the Marías Islands consists of 16 species. Coral zonation was characterized by

Fig. 5.7 Tropical mainland Mexico (TMM), Sonora to Chiapas. *Red diamonds* identify surveyed sites

shallow populations of *Pocillopora*, and deeper occurrences, to 10 m depth and deeper, of species of *Porites*, *Pavona*, and *Psammocora*. This is a frequent pattern observed throughout Mexico (Reyes-Bonilla and López Pérez 1998) and elsewhere in the eastern Pacific (Glynn 1976; Wellington 1982). The community of algal endosymbionts (*Symbiodinium*) inhabiting particular coral host species was found recently to be a critical factor in defining this depth zonation. From a combination of genetic and photo-physiological analyses of *Pocillopora verrucosa* and *Pavona gigantea*, Iglesias-Prieto et al. (2004) demonstrated that the physiological performance of host-specific zooxanthellae were adapted to different depth-related light regimes. Thus, pocilloporid holobionts exhibited optimal growth at shallow depths, whereas the growth rates of massive colonies were highest in deeper waters.

It is now recognized that the notable coral development of the Huatulco reef tract is subject to a variety of relatively frequent and severe natural disturbances with high inter-annual variability (Glynn and Leyte-Morales 1997; López-Pérez et al. 2002; López-Pérez and Hernández-Ballesteros 2004). These disturbances range from seasonal upwelling events to plankton blooms and high turbidity conditions, hurricanes, ENSO high and low temperature extremes, and sea urchin outbreaks leading to intense grazing and bioerosion (e.g., Leyte-Morales 2000; Lirman et al. 2001; Herrera-Escalante et al. 2005; López-Pérez et al. 2007; Benítez-Villalobos et al. 2008). López-Pérez and Hernández-Ballesteros (2004) have proposed that coral mortality caused by such random spatio-temporal disturbances are likely to produce a mosaic of patches at different stages of recovery, thus contributing to dynamic and diverse coral community structures. It would be of interest to study the coral-rich Nayarit area, also subject to upwelling and severe ENSO activity, to determine if disturbance regimes there similarly influence coral community structure and dynamics.

Several recent molecular genetic studies have helped to elucidate the connectivity between coral populations and

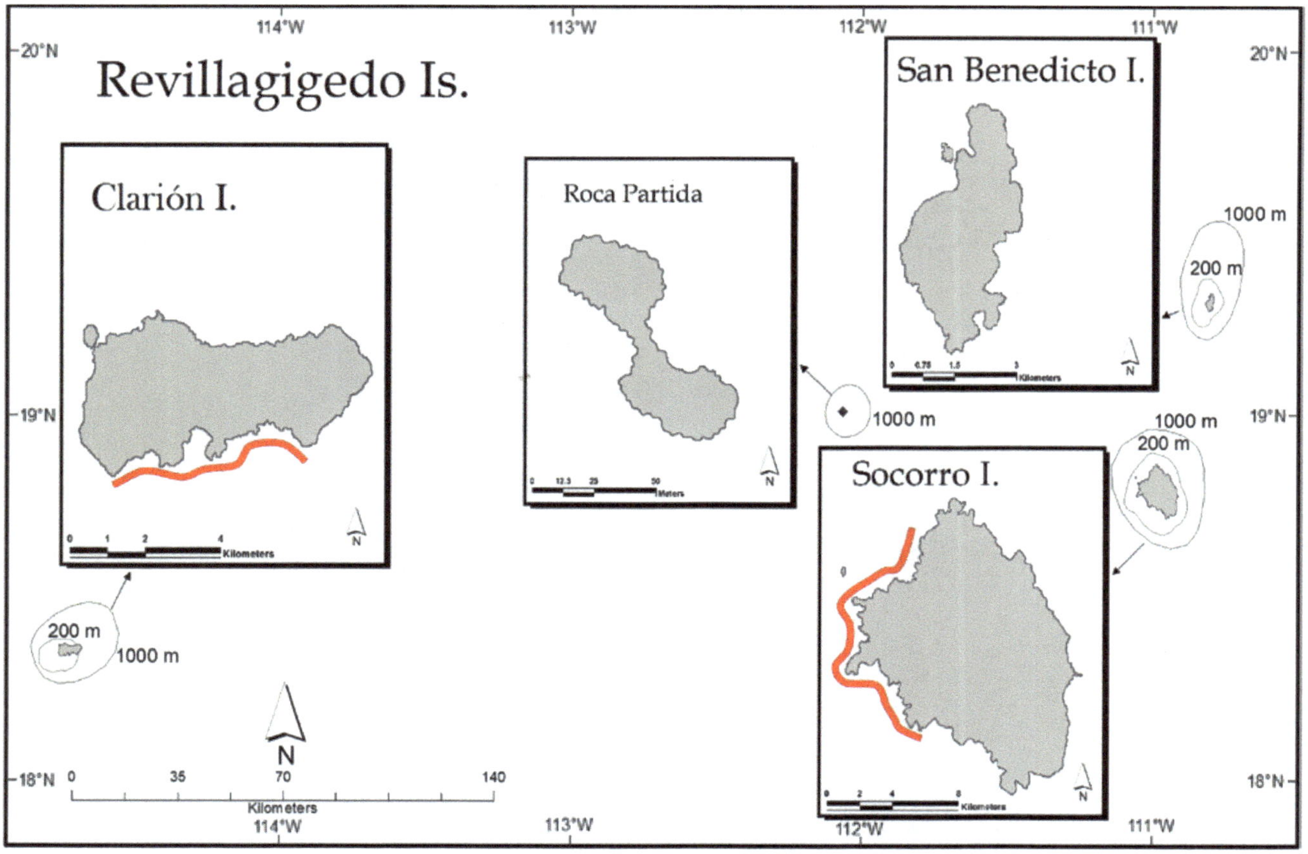

Fig. 5.8 Revillagigedo Islands (REV), Mexico. *Red lines* skirting the Clarión and Socorro coastlines denote abundant coral growth

their biogeographic relationships in Mexican waters and beyond. Most of these studies indicate a high degree of genetic structure in local populations and, presumably, limited gene flow. For example, significant geographic differentiation was reported for *Pocillopora damicornis*, a broadcast spawning species (Chavez-Romo et al. 2008), and *Porites panamensis*, a brooder with a brief larval life (Paz-García et al. 2008). Limited gene flow among most sampled localities was also reported for *Pavona gigantea*, a wide-ranging broadcast spawning species (Saavedra-Sotelo et al. (2011). However, the largest estimated migration rate indicated moderate and unidirectional gene flow from Huatulco to the Marietas Islands in Banderas Bay (Nayarit) and to La Paz. Blancas-López (2009) also observed weak genetic differentiation (i.e. connectivity) among populations of *Pocillopora verrucosa*, suggesting some level of gene flow among populations of this broadcast spawning species. Among the various physical and biotic factors invoked to explain these patterns, the surface circulation of west Mexico is critical. There is a general northerly set of surface currents, involving the Costa Rica Coastal Current (CRCC) moving along the Huatulco coast and the West Mexican Current that flows from Oaxaca to the entrance of the Gulf of California (see Chap. 3, Fiedler and Lavín). This would potentially allow certain source populations in the south to supply larvae that could recruit to downstream Mexican mainland areas and eventually as far north as the Gulf of California. Since the CRCC originates in the equatorial eastern Pacific, it could conceivably serve as a vehicle of transport for coral larvae to reach Mexico from Costa Rica and Panama.

5.3.3 Revillagigedo Islands (Mexico)

The volcanic islands making up the Revillagigedo Islands group are dispersed between 19° to 20°N latitude, and range from 750 km (San Benedicto and Socorro Islands) to 1200 km (Clarión Island) to the west of the Mexican mainland coast (Fig. 5.8). Roca Partida, consisting of two small rock peaks with a very limited shallow shelf, is located to the NE of Clarión Island. The number of coral records in the Revillagigedo Islands increased dramatically in the early 1990s, thanks to scuba surveys conducted by Mexican workers. This increase in species richness, from 12 to 21 then to 25, was due mostly to the discovery of Indo-Pacific corals not present on the west Mexican coast or elsewhere in the eastern tropical Pacific (Ketchum and Reyes-Bonilla 1997, 2001). Several of these species are shared with

Clipperton Atoll, located nearly 1000 km to the south (at 10° N), which is occasionally in the path of the North Equatorial Counter Current (Glynn et al. 1996a, b). The current species list for the Revillagigedos totals 25 zooxanthellate scleractinians (Appendix). The strong coral faunal similarities between the Revillagigedos and Clipperton, also involving brachyuran crabs (Garth 1992) and reef fishes (Robertson and Allen 1996), prompted Ketchum and Reyes-Bonilla (1997, 2001) to suggest that this oceanic region should be recognized as a unique biogeographic subregion of the eastern tropical Pacific. The latter workers noted that surface currents would allow dispersal from Clipperton to the Revillagigedos, especially during the summer season, and that both island areas could serve as stepping stones for the migration of inshore marine species from the central to the more northern eastern tropical Pacific region. Traditionally, the W to E biotic migration route was assumed to be closer to the equator, from the Line Islands to Isla del Coco and Central America (Costa Rica, Panama) via the NECC (Dana 1975; Cortés 1997; Glynn and Ault 2000). Based on distributional evidence, it appears that there may be two dispersal routes, one just north of the equator and one leading to northern Mexican waters via oceanic island stepping stones.

The principal fringing reefs of Socorro and Clarión Islands are generally present in bays on western and southern shores in the lee of westerly-moving cyclones, which frequent this region (Fig. 5.8; Reyes-Bonilla 2003; see Chap. 3, Fiedler and Lavín). *Pocillopora* spp. predominate at shallow depths and massive corals (*Porites, Pavona*) increase in abundance from 5 to 30 m depth; coral rubble also increases with depth. Relatively few coral species were observed at Roca Partida and San Benedicto Island, likely due to the small size of these islands and restricted shelf space. In addition, the Bárcena volcano on San Benedicto Island was active as recently as 1952, and the resulting lava flows, debris, and fine sediments probably devastated nearshore communities and interfered with recovery. The six coral species observed there in 1994 were found principally on large basalt blocks and not on unstable volcanic sediments (Glynn et al. 1996a, b).

Moderate numbers of *Acanthaster planci* were observed at Clarión and San Benedicto Islands in 1994 (Glynn, unpublished observation). On 29 and 30 April, 36 individuals were counted on each of two occasions by a single diver between 6 and 17 m near Roco del Cuervo (18° 21′ 13.7″; 114° 41′ 06.2″), off the southeast shore of Clarión Island. The sea stars were feeding mostly on crustose coralline algae, but also on massive *Porites lobata* colonies, and large and small *Pocillopora eydouxi* colonies as well as broken branches of *Pocillopora* spp. Several *A. planci* also were observed feeding on *P. eydouxi* in coral communities on the northwestern end of San Benedicto Island on 3 May, 1994.

5.3.4 Clipperton Atoll (France)

Clipperton Atoll, the only atoll present in the eastern Pacific region, is located about 1300 km SW of the Mexican mainland and 950 km SSE of the Revillagigedo Islands (Fig. 5.1). Clipperton is volcanic in origin, part of a chain of guyots on the Clipperton fault, bordering the Cocos (south) and Pacific (north) plates. It is an 'almost atoll', with a 29 m high trachyte rock on the SE shore (Rocher Clipperton, Fig. 5.9), and an enclosed meromictic lagoon (Charpy et al. 2009; Trichet 2009). Present-day shallow, stratified brackish waters and H_2S-enriched bottom waters prevent coral growth inside the lagoon. However, before the mid-1800s at least two channels communicated with the open ocean, thus allowing for coral framework construction; fossilized coral deposits are still evident today inside the lagoon (Carricart-Ganivet and Reyes-Bonilla 1999). It is likely that a tectonic uplift event (Trichet 2009) closed off the lagoon between 1839 and 1849 (Charpy et al. 2010). Bottom profiling to 200 m depth around the perimeter of the atoll revealed a prominent 20 m terrace that is generally present at all seaward exposures (Glynn et al. 1996a, b). The 200 m isobath is located relatively close to shore, from 200 to 500 m seaward of the reef flat.

With recent coral studies supported by scuba, the known coral fauna has increased from 16 or 17 species (Carricart-Ganivet and Reyes-Bonilla 1999; Flot and Adjeroud 2009) to 21 species (Fig. 5.7; Appendix). Contributing to this increase is the recognition of additional species of *Pocillopora* (e.g., *P. elegans* and *P. verrucosa*), the naming of a new poritid species (*Porites arnaudi*), and a molecular genetic study revealing the presence of *Porites evermanni* (Boulay et al. 2014). These new species' records are not considered recent introductions, but resident species present for some undetermined time. *Pavona gigantea* was listed from Clipperton by Durham and Barnard (1952), but has not been observed at the atoll for several decades since. The zooxanthellate sceractinian fauna of Clipperton, consisting of mainly (~90 %) Indo-Pacific species is most closely allied with the Revillagigedo Islands in the EP (Fig. 5.2).

In 1994, live coral cover was exceptionally high (10–100 %) at all insular shelf depths from 8 to 60 m, and crustose coralline algae were abundant (5–40 % cover) between 0.5 and 7 m depth (Glynn et al. 1996a, b). *Porites lobata* and *Pavona varians* contributed prominently to live cover, the former up to 50 % of cover between 10 and 30 m, and the latter commonly as high as 100 % between 25 and 60 m and deeper. Monospecific patches of *Pavona varians* were present at 70 m. *Porites arnaudi*, noted as *Porites* sp., was commonly observed, but *P. lobata* predominated in most areas. *Porites arnaudi* was often more common on the near-vertical faces of the 20 m terrace (Fig. 5.10). *Pocillopora* sp. patches, 10s to 100s m^2 in area, were highly

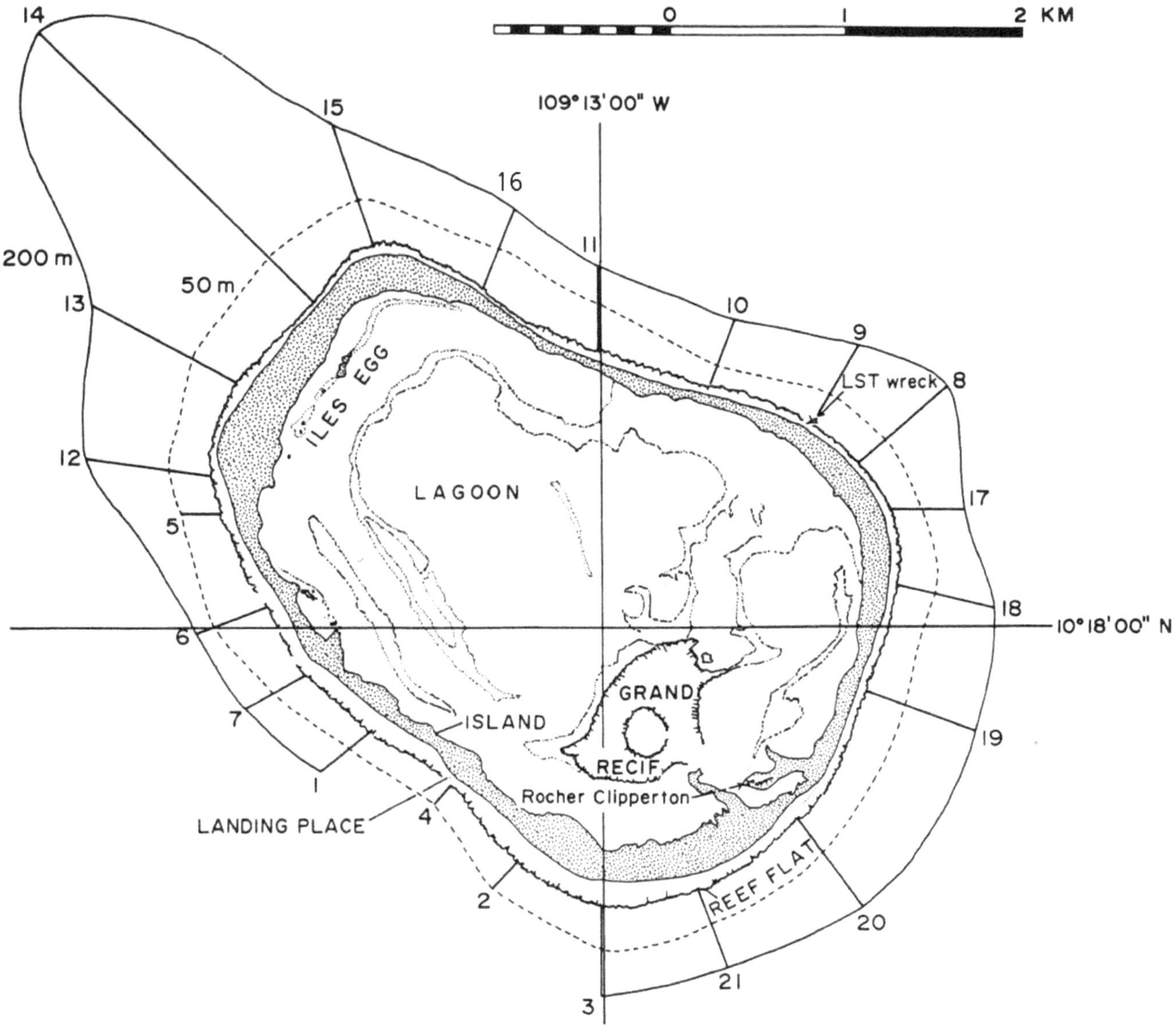

Fig. 5.9 Clipperton Atoll (CLP), modified after Glynn et al. 1996a. 50 and 200 m isobaths are based on the 21 bottom-profiling transects shown

dispersed and usually present between 5 and 20 m depth. The corals in most of these patches were dead and encrusted by the coralline alga *Porolithon onkodes*. The pocilloporid species making up these patches, later identified by Carricart-Ganivet and Reyes-Bonilla (1999), was *Pocillopora meandrina*. More recent workers have not commented on whether the *Pocillopora* patches were dead or recovering. *Pavona minuta*, *Pavona maldivensis*, and *Leptoseris scabra* were generally minor components of the coral communities. Surveys conducted 10 years later indicated no significant change in the overall robust state of Clipperton's coral communities (Flot and Adjeroud 2009). Two permanently-marked monitoring sites were established in 2005 at 10–12 m depth on the north and south reef slopes, thus allowing an opportunity to track coral community condition over time (Salvat et al. 2008).

Like most ETP localities, the coral reef biota of Clipperton exhibits a strong affinity with the central/south Pacific region. In addition to the zooxanthellate coral fauna, high proportions of reef algae (Payri et al. 2009), crustaceans (Poupin et al. 2009), molluscs (Kaiser 2009), echinoderms (Solís-Marín and Laguarda-Figueras 2009), and fishes at Clipperton (Robertson and Allen 1996; Robertson et al. 2004; Béarez and Séret 2009) exhibit transpacific distributions. For coral reef fishes, Robertson et al. (2004) noted that pelagic spawners with strong dispersal ability were highly represented among transpacific species. Also, many reef-associated Indo-Pacific invertebrate taxa inhabiting eastern Pacific reefs possess teleplanic larvae (Scheltema 1988; Glynn and Ault 2000). Molecular studies demonstrate gene flow and recent invasions between the Central Pacific and Clipperton Atoll of coral, echinoderm, and fish species

Fig. 5.10 *Porites arnaudi*, cascading skirt-like growth form, 18 m, NE exposure, Clipperton Atoll, 23 April 1994 (photo credit M. Kazmers, Shark Song)

(Lessios et al. 1998; Lessios and Robertson 2006; Baums et al. 2012). Some of these studies have shown east to west gene flow, and connectivity of reef species among eastern Pacific oceanic islands (see Chap. 16, Lessios and Baums). Thus, it is necessary to know the flow regimes and transit times of the North Equatorial Counter Current (NECC), the primary eastward-flowing current in the central and eastern tropical Pacific.

Satellite-tracked drifting buoy data analyzed over a 15 year period (1979–1993) revealed a mostly eastward flow of the NECC across the E Pacific 'barrier' between 6° and 10°N, occasionally reaching 12–14°N (Glynn et al. 1996a, b). Travel times from the Line Islands in the Central Pacific to Clipperton Atoll were commonly 200 days, although one record during an ENSO event showed the NECC travelling across 46° of longitude in about 60 days. A comparison of west to east distances travelled during non-El Niño and El Niño years indicated a significant increase in NECC velocity during the latter periods. Accordingly, the NECC could offer an effective means of larval transport to Clipperton, which in turn could serve as a stepping stone to other ETP localities, e.g. the Revillagigedo Islands as suggested by Ketchum and Reyes-Bonilla (1997).

5.3.5 El Salvador

The recently described coral assemblages of El Salvador and Nicaragua are populated by a combined total of 14

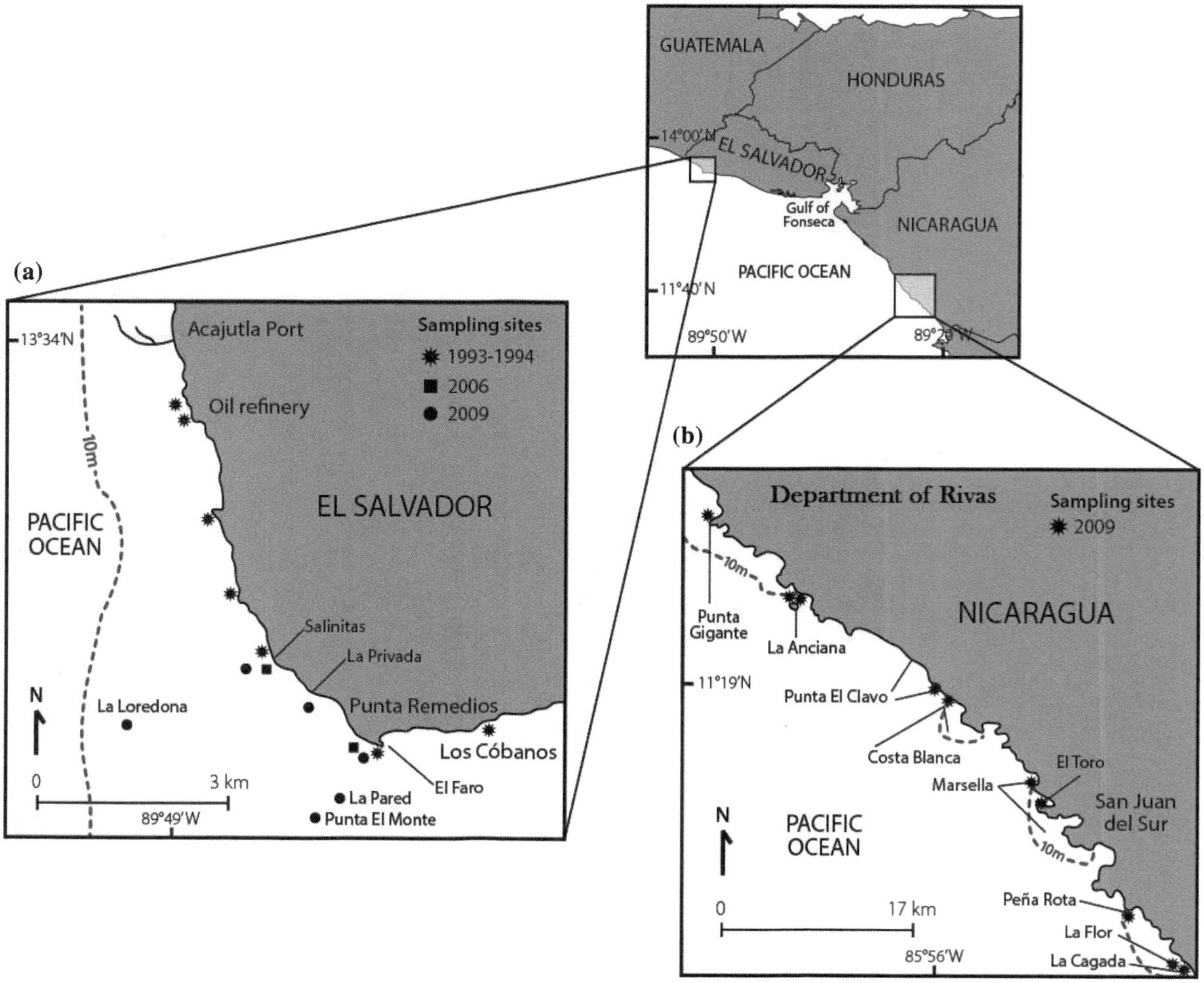

Fig. 5.11 **a** Northwestern El Salvador (ELS); **b** southern Nicaragua (NIC), Central America. Coral development on Nicaraguan coast is within the Department of Rivas. Ten meter isobaths from US Navy Hydrographic chart no. 21026

zooxanthellate species. Notable differences, however, are the occurrences of *Porites lobata* at El Salvador only (Reyes-Bonilla and Barraza 2003), and *Pavona clavus*, *Pavona varians*, and *Gardineroseris planulata* in Nicaragua but not at El Salvador (Alvarado et al. 2011). No structural coral reef development has been reported from these areas (see Chap. 6, Toth et al.), although coral community patches are present.

The coastal stretch between Guatemala and northwestern Nicaragua (Gulf of Fonseca) is known as the "Pacific Central American Faunal Gap (PCAFG)", originally named due to the absence of rocky shore fishes (Springer 1958). The Guatemalan coast and the northern sector of El Salvador consist of sandy beaches and mangrove forests. Until recently the PCAFG was also understood to separate the coral faunas of southern Mexico and Costa Rica (Glynn and Ault 2000). At Los Cóbanos (El Salvador), there is an abrupt change in the orientation of the coastline with the appearance of low rocky outcrops and platforms (Fig. 5.11a). The remainder of the Salvadoran coast consists of high energy sandy beaches or muddy estuaries. The littoral fringe of the Gulf of Fonseca consists of rocky outcrops and one of the most extensive mangrove ecosystems of Central America. The southern Nicaraguan coast consists of a combination of rocky headlands and sandy beaches (Cortés 2007). The occurrence of coral populations within the PCAFG could serve as a dispersal corridor, and explain the close association of southern and northern ETP coral faunas (Fig. 5.2).

The rocky outcrops of Los Cóbanos support the best developed and most diverse coral communities in El Salvador (Reyes-Bonilla and Barraza 2003). Small pocilloporid coral communities occur on basalt outcrops, most no larger than 30 m^2, that occupy a total of 159 km^2 of a large terrace (8000 ha) that borders the coast. Thirteen coral species (eight zooxanthellate and five azooxanthellate) were reported at this site. Lemus et al. (1994) performed the first quantitative coral surveys at Los Cóbanos in 1993 and 1994; these workers sampled *Porites lobata* and reported highly divergent live cover values between 0.41 and 62 % for this species. The mean diameter of the sampled *Porites* colonies was ~50 cm. At Punta Remedios (Fig. 5.11a), they reported 0.42 % live coral cover, 34.6 % algae, and 44.8 % sand and rock. Only dead *Pocillopora* fragments were noted at this site.

New sites at Los Cóbanos (Playa El Faro, Playa Salinitas or Decameron Beach) were sampled in 2006 by Segovia-Prado and Navarrete-Calero (2007). They reported high live cover of large *Porites* colonies: El Faro—33.7 % cover, 161 colonies, mean diameter = 121 cm; Salinitas—65.4 % cover, 75 colonies, mean diameter = 146 cm. Algal cover also was high: El Faro—41.7 %; Salinitas—29.5 %. These mixed coral/algal communities were classified into five distinct assemblages, based on generic prevalence: (1) *Padina-Halimeda-Codium*, (2) *Porites-Padina-Halimeda*, (3) *Porites-Acanthophora-Padina*, (4) *Porites-Padina-Hypnea*, and (5) *Porites-Dictyota-Galaxaura*. Concerns were raised over possible threats to corals by the invasive alga *Acanthophora spicifera*, and a high incidence of bioeroders (*Lithophaga*, *Cliona*) and potential space competitors (vermetid gastropods). Reyes-Bonilla and Barraza (2003) noted other macroalgae, invertebrates (e.g., octocorals, molluscs) and reef-associated fishes observed at Los Cóbanos. A potentially significant physical threat noted by Reyes-Bonilla and Barraza (2003) is the seasonally high rainfall period from May to October with associated severe coastal erosion and transport of suspended sediments.

At Playa el Zope (Los Cóbanos), López and Jiménez (2008) found healthy colonies of *Psammocora obtusangula* and *Psammocora stellata* interspersed with *Porites lobata*. They reported 81 colonies of *P. obtusangula* (mean length × width = 10.8 × 7.2 cm) and two small colonies of *P. stellata* (<6 cm^2). Several *P. obtusangula* colonies bore small wounds resembling fish bite marks.

The most recent near-shore surveys at Los Cóbanos, in 2009 and at the same sites sampled by Lemus et al. (1994), and Segovia-Prado and Navarrete-Calero (2007), failed to reveal living or dead *Pocillopora* and reported only low *Porites lobata* cover of 0.1 and 5.5 % (Alvarado, unpublished data). A live *Pocillopora* community was reported in the early literature at Los Cóbanos, the Punta Remedios site (Orellana-Amador 1985; Gotuzzo 1996), but has not been re-located in recent surveys (Reyes-Bonilla and Barraza 2003). All *Pocillopora* spp. collected in this study were identified from dead beach fragments (Appendix). At relatively deep (10–15 m) offshore sites, only azooxanthellate corals were observed, with *Tubastraea coccinea* predominant. Considering the high variability of live coral cover reported over the years by different workers, and the uncertain location of sampling sites, this coastal sector is in need of a systematic and technologically advanced survey protocol.

5.3.6 Nicaragua

Before 2010, coral surveys along the Pacific coast of Nicaragua indicated the possibility of isolated colonies of *Pocillopora* close to the Costa Rican border in the area of San Juan del Sur (Durham and Barnard 1952; Glynn and Ault 2000; Spalding et al. 2001; Ryan and Zapata 2003; Cortés 2007). The presumed meagre occurrence of corals in this area may be related to severe and prolonged upwelling (Glynn et al. 1983).

In July 2009, Alvarado et al. (2010, 2011) quantitatively assessed the composition of ten coral communities along the southern coast of Nicaragua, including sites in the Department of Rivas, Municipalities of San Juan del Sur, and Tola (11° 23′ N; 86° 02′W and 11° 07′N; 85° 47′W; Fig. 5.11b). They reported the presence of 39 species of macroalgae, 13 scleractinian corals (nine zooxanthellate and four azooxanthellate species), two crustaceans, five molluscs, 11 echinoderms, and 52 reef fishes. Mean coral cover across all sites was 9.0 ± 1.9 (SD)%, with the highest cover at Guacalito-La Anciana (18.5 ± 8.8 %) and Punta Pie de Gigante (16.7 ± 5.3 %). The most common species were *Gardineroseris planulata* and *Pavona gigantea* with notably high cover at Punta Gigante and La Anciana (Fig. 5.11b). A relatively large and vibrant *Pocillopora* incipient reef was present in the Punta El Toro embayment with mean live cover of 14.5 ± 7.9 %.

The Rivas coral biotopes, such as those present at Punta Gigante, La Anciana, and Punta El Toro, were small, patchy, and confined to a series of small coves and islets protected from wave assault. These communities were dominated by *Pavona gigantea*. The coral reef fish assemblages were less diverse than those reported for more developed eastern Pacific reef areas elsewhere, but were similar to those in marginal environments, such as upwelling centers (Alvarado et al. 2011). The coral-rich area between Punta Gigante and La Anciana, a 13-km coastal stretch, was touted by Alvarado and co-workers as a hotspot—despite the lack of reef development—and in urgent need of protection and management.

5.3.7 Mainland Costa Rica

A dramatic increase in coral species richness, from 10 to 26 species, is evident in faunal inventories when passing from Nicaragua to Costa Rica (Table 5.4; Appendix). This

Table 5.4 Reef building corals from Pacific Costa Rica

Location	1	2	3	4	5	6	7	8	9	10	11
Suborder ASTROCOENIINA, Family Pocilloporidae											
1. *Pocillopora capitata* Verrill, 1864				CA	CA	CA				CA	CA
2. *Pocillopora damicornis* (Linnaeus, 1758)		RB	RB	CA	RB	RB	RB	RB	RB	RB	RB
3. *Pocillopora elegans* Dana, 1846		RB	RB	RB	RB	RB	RB	RB	RB	RB	RB
4. *Pocillopora eydouxi* Milne Edwards and Haime, 1860		RB	CA	CA	CA			CA		CA	RB
5. *Pocillopora inflata* Glynn, 1999		RB	RB	CA	CA					CA	CA
6. *Pocillopora meandrina* Dana, 1846		CA	CA	CA	CA						CA
7. *Pocillopora verrucosa* (Ellis and Solander, 1786)[a]				RB	CA					RB	
Suborder Fungiina, Family Poritidae											
8. *Porites evermanni* Vaughan, 1907					RB		RB	RB		RB	CA
9. *Porites lobata* Dana, 1846		CA	CA	CA	RB	RB	RB	RB	RB	RB	RB
10. *Porites panamensis* Verrill, 1866	CA	CA	CA	CA	CA	CA	CA	CA		CA	
11. *Porites rus* (Forskål, 1775)					CA						
Family SIDERASTREIDAE											
12. *Psammocora obtusangula* (Lamarck, 1816)					CA	CA	CA	CA	RB	CA	
13. *Psammocora stellata* Verrill, 1864		CA	RB	RB	CA	CA	CA	CA	CA	RB	CA
14. *Psammocora superficialis* Gardiner, 1898		CA	CA	CA	CA	CA	CA	CA	CA	CA	CA
Family Agariciidae											
15. *Gardineroseris planulata* (Dana, 1846)		RB	RB	CA			CA	CA		RB	RB
16. *Leptoseris papyracea* (Dana, 1846)			CA	RB	CA						CA
17. *Leptoseris scabra* (Vaughan, 1907)			CA								CA
18. *Pavona chiriquiensis* Glynn et al. 2001			CA	CA		CA				CA	CA
19. *Pavona clavus* (Dana, 1846)		CA	RB	RB	CA	CA		CA	CA	RB	RB
20. *Pavona frondifera* (Lamarck, 1816)				CA			CA		RB	CA	CA
21. *Pavona gigantea* Verrill, 1869	RB	RB	CA	RB	CA	CA		CA	CA	CA	CA
22. *Pavona maldivensis* (Gardiner, 1905)		CA	CA	CA		CA	CA			CA	CA
23. *Pavona varians* Verrill, 1864	CA	CA	CA	CA		CA		CA	CA	CA	CA
24. *Pavona minuta* Wells, 1954											CA
Family Fungiidae											
25. *Cycloseris curvata* (Hoeksema, 1989)				CA					CA		CA
26. *Diaseris distorta* (Michelin, 1843)				CA							CA
Total number species	3	14	17	21	17	13	11	13	11	19	22

RB = Main reef builder, *CA* = coral community associate. *Areas 1* = Bahía Salinas, *2* = Península de Santa Elena, *3* = Islas Murciélago, *4* = Bahía Culebra, *5* = Península de Nicoya, *6* = Pacífico Central, *7* = Parque Nacional Marino Ballena, *8* = Península de Osa, *9* = Golfo Dulce, *10* = Reserva Biológica Isla del Caño and *11* = Parque Nacional Isla del Coco. *RB* reef building reported; *CA* coral community associate
[a]This species was not recognized in the field until recently; it is probably more widespread than indicated

disparity is likely due to a combination of factors: (a) the recent discovery of coral communities in Nicaragua, (b) a concentrated research effort on coral reefs in Costa Rica since the 1980s, (c) the relatively large area of varied coastal habitats along Costa Rica's coast, and (d) the stressful conditions associated with upwelling along the southeastern Nicaraguan coast. The mainland coastline of Costa Rica consists of diverse marine ecosystems as well as islets and islands of various sizes, and differences in proximity to the coast (Denyer and Cárdenas 2000; Cortés and Wehrtmann 2009). The presently recognized tally of 26 zooxanthellate coral species from mainland Costa Rica has been relatively stable during the past decade, thanks to the sustained studies of local workers (e.g., Guzmán and Cortés 1989a; Cortés 1990; Jiménez 1997, 2001a, b; Cortés and Guzmán 1998). This inventory takes into account a single new record for Costa Rica, *Porites evermanni*, revealed in a molecular genetic study (Boulay et al. 2014), and *Porites rus*, not

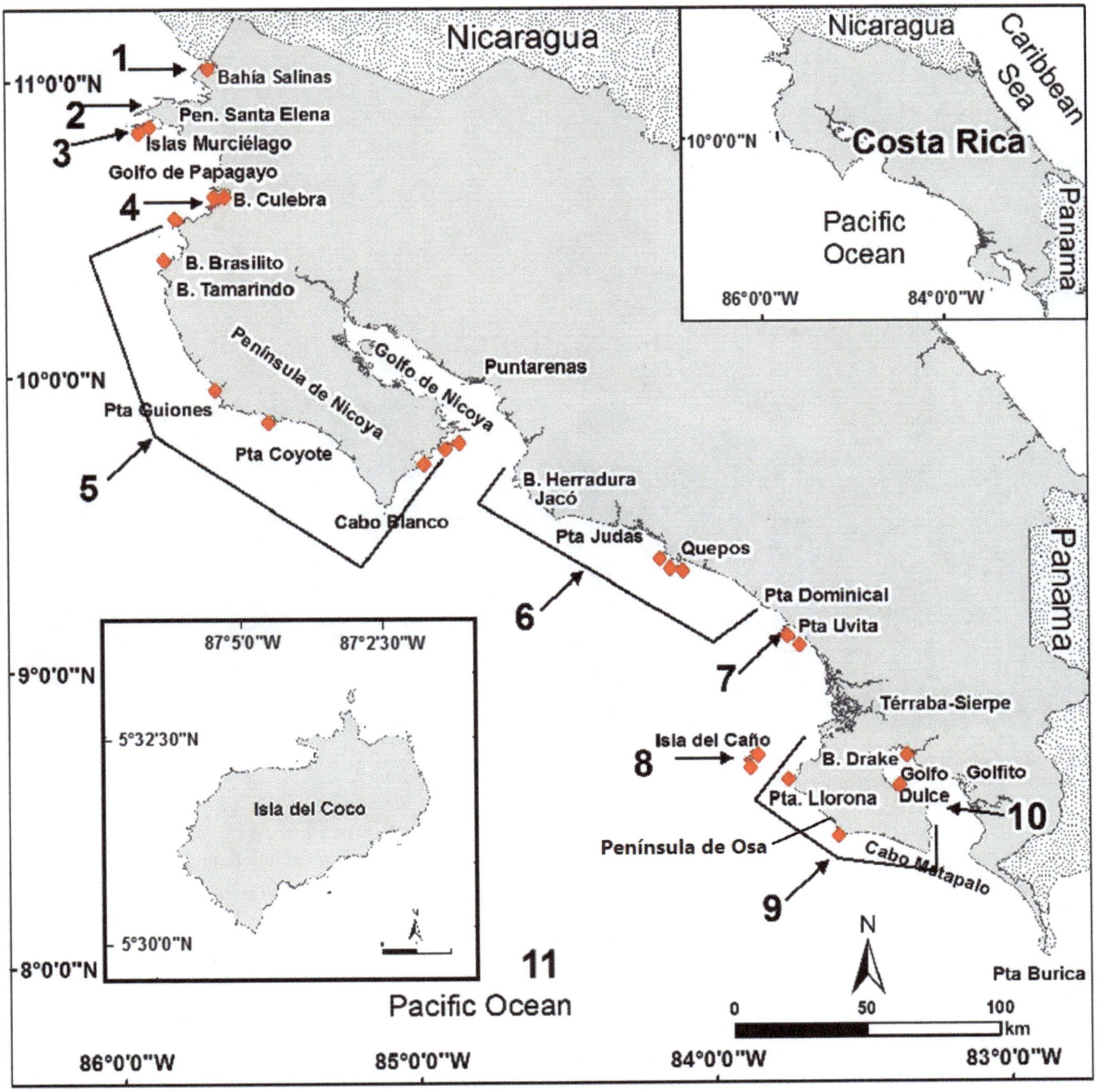

Fig. 5.12 Costa Rica mainland (CRM) partitioned into 10 coastal and one offshore (Isla del Coco) coral-bearing sectors. *Red diamonds* identify sites with abundant corals

observed again since its discovery at Bahía Sámara (sector 5, Fig. 5.12) in the early 1980s (Cortés and Murillo 1985). *Pavona chiriquiensis*, initially described from populations in Panama, is now known to occur at several sites along the Costa Rican coast. New distributional records for *P. chiriquiensis* in the central and south Pacific include the Tokelau Islands (~172°W), American Samoa (~170°W), the northern and southern Phoenix Islands (~170°W), and the Line Islands (155°–160°W) (Maragos, unpublished records). A cluster analysis of presence/absence data demonstrates a close affinity with the coral faunas of Panama and Colombia (Fig. 5.2).

The northwestern Pacific sector of Costa Rica is characterized by a marked dry season (January–April), with strong easterly trade winds that promote seasonal upwelling (Cortés 1996/1997; Jiménez 2001b; Alfaro et al. 2012). Upwelling is most pronounced further north, toward the Nicaraguan border. The dry climate extends south to Golfo de Nicoya,

Fig. 5.13 Matapalo fringing reef, Punta Gorda, Costa Rica (10 February 2007, ~1000 m elevation, courtesy María M. Chavarría). Patches of an invasive green alga, *Caulerpa sertularioides*, seaward of live coral cover, ~10 m depth

the largest estuarine environment in Costa Rica. The central coastal section is a transitional zone with an increasing humid climate to the south. The southeastern coast is subject to year-round high rainfall, which supports a tropical rain forest ecosystem (Jiménez and Soto 1985; Herrera 1986; Kappelle et al. 2002).

For ease of discussion, mainland Costa Rica is divided into 10 sectors where coral communities and reefs occur, from the Nicaraguan border to Golfo Dulce near Panama (Fig. 5.12). Oceanic Isla del Coco (Cocos Island) is assigned to sector 11, and is considered below in Sect. 5.3.8.

The northern-most area, Bahía Salinas (sector 1), is heavily influenced by seasonal upwelling in the Gulf of Papagayo (Alfaro et al. 2012). Water temperature in this area can be as low as 12 °C with low oxygen levels and low pH (recorded by Rixen et al. 2012 in Bahía Culebra; see also Jiménez 2001a; Alfaro and Cortés 2012; Alfaro et al. 2012). Because of these extreme conditions, reef and coral community development are limited; *Pavona gigantea* is predominant with isolated colonies of *Porites panamensis*. Upwelling effects generally decrease to the south and southeast towards the Nicoya Peninsula. At Península de Santa Elena (sector 2), relatively large pocilloporid reefs are present; one of these structural reefs is built mainly by *Pocillopora eydouxi* (Cortés 1996/1997; Bassey-Fallas 2010). Immediately south of Santa Elena, lie the Islas Murciélago (sector 3). Several coral reefs occur in this area; one is constructed of massive *Gardineroseris planulata* colonies. An extensive pocilloporid reef, consisting of large areas of *Pocillopora inflata*, was killed during a phytoplankton bloom in 2007 (Jiménez et al., in preparation). Coral recovery, including sexual and asexual recruitment, is very slow in this cool upwelling zone (Cortés, unpublished data). Bahía Culebra (sector 4) is one of the most intensely studied areas in Pacific Costa Rica (Cortés 2012a). Coral reefs and coral communities have been surveyed and monitored in this large bay and surrounding areas since the early to mid-1980s (Cortés and Murillo 1985; Jiménez 1997, 2001b; Jiménez et al. 2001, 2010; Bezy et al. 2006; Cortés et al. 2010; Cortés 2012a).

The Matapalo fringing reef in the southern part of sector 4, 1.4 km in length, is possibly the largest coral reef bordering mainland Costa Rica (Jiménez 2007a; Fig. 5.13). This reef, first surveyed quantitatively in 1978, was largely degraded at that time; it consisted of large tracts of dead pocilloporid corals, and at most 0.1 % live cover of *Psammocora stellata* (Glynn et al. 1983). [An aerial view of this reef in 1973, with the name Punta Gorda, is shown in Fig. 5.4a in Glynn et al. 1983.] The death of the reef at that time was attributed to cold-water stress due to intense upwelling near the end of the Little Ice Age (150–300 years BP). The overall live cover of pocilloporid corals on this reef amounted to >40 % when surveyed in 2007. Species richness was high, consisting of 16 living species, and 19 species total, including freshly dead coralla of three species (*Cycloseris curvata*, *Diaseris distorta*, and *Leptoseris papyracea*). Carbonate buildups or bioherms of *Leptoseris papyracea* are unique to this area. The Matapalo reef has again experienced recent high coral mortality first observed in 2007, and coincidental with extensive and prolonged harmful algal blooms (Jiménez 2007b). It is likely that recent declines in coral cover are due, at least in part, to coastal development activities and poor waste management promoting algal blooms, which are now nearly monthly occurrences.

Overall, much of the reef building in the extreme northwestern area of Costa Rica (sectors 1–4) has been achieved by massive instead of branching pocilloporid species. For example, the major reef builders north of Islas Murciélago are *Pavona gigantea* and *Gardineroseris planulata.* At Bahía Culebra, *Pavona clavus* is one of the major reef builders (Cortés and Jiménez 2003).

Extensive growths of the invasive green alga *Caulerpa sertularioides* have been reported in northwestern Costa Rica at Bahía Culebra (Fernández and Cortés 2005; Fernández-García 2007; Fernández-García et al. 2012), and at the Matapalo reef (Jiménez 2007a). Since 2001, *C. sertularioides* cover has been expanding with negative impacts on coral reefs and communities. At La Penca, Bahía Culebra, a *Psammocora* reef lost 95 % live cover at the expense of this alga during a 6 year period (Bezy et al. 2006). It is a strong competitor for space, overgrowing and smothering low-profile coral colonies. *Caulerpa racemosa* blooms are also known to cause coral mortality in Panama (see Fig. 8.31 in Chap. 8, and Fong et al., Chap. 11).

Little studied and isolated reefs occur in sector 5 along the outer coastline of Península de Nicoya (Cortés and Jiménez 2003; Bezy et al. 2006; Jiménez et al. 2010). A few small colonies of *Porites rus*, first observed in the eastern Pacific at Bahía Sámara, about 10 km SE of Punta Guiones, disappeared a few years after their discovery (after 1983) and have not been observed anywhere in the EP since. *Porites evermanni* was recently reported in coral communities on the Península de Nicoya, in other areas of Costa Rica, and elsewhere in the equatorial eastern Pacific (Boulay et al. 2014). The identity of this species was determined from a molecular genetic study; judging from its widespread occurrence and large population sizes it is probably a long-term resident of the equatorial eastern Pacific. Reef-building no longer occurs in the Gulf of Nicoya due to high suspended sediment loading and freshwater dilution. Currently only dead pocilloporid frameworks are present in the gulf, usually in sheltered environments.

The central Pacific coastline (sector 6) extends from the east margin of the outer Gulf of Nicoya to Punta Dominical. Isolated reefs and coral communities also have been reported in this area (Cortés and Jiménez 2003). Parque Nacional Marino Ballena (sector 7) is exceptional in the central Pacific coastal area because of its well-developed reefs (Alvarado et al. 2005). These reefs have been impacted by El Niño warming events (Jiménez and Cortés 2001) and excessive terrigenous sedimentation resulting from coastal development (Alvarado et al. 2005, 2009).

The Reserva Biológica Isla del Caño (Caño Island Biological Reserve, sector 8) is located about 15 km west of Península de Osa. Well-developed coral reefs are present at Isla del Caño, from drying reef flats (unusual in the ETP) with abundant *Pocillopora* spp., crustose coralline algae and *Porites* microatolls to deep patch reefs constructed of *Porites lobata* and *Pavona clavus* (Guzmán and Cortés 1989a; Macintyre et al. 1993; Cortés and Jiménez 2003). These coral communities have been intensely studied, including corallivore activity (Guzmán 1988a, b), coral growth rates (Guzmán and Cortés 1989b), zooplankton ecology (Guzmán and Obando 1988), and meiofaunal populations of reef sediments (Guzmán et al. 1987a). Also, coral mortality associated with algal blooms ("red tides") was documented at Isla del Caño following the 1982–83 El Niño warming event (Guzmán et al. 1987b, 1990). Finally, Glynn et al. (1991, 1994, 1996a, b, 2000, 2008, 2011, 2012; see Chap. 15, Glynn et al.) have completed long-term studies on the reproductive activities of eleven reef-building coral species and one abundant azooxanthellate coral at Isla del Caño. This work has demonstrated a potentially important role of sexual reproduction in recovery processes in Costa Rica (Guzmán and Cortés 2001), and at several other areas in the equatorial ETP (Glynn and Colley 2008; Glynn et al. 2012).

Coral communities and reefs occur on the seaward side of the Península de Osa (sector 9), but are generally only present in sheltered locations due to high wave assault (Cortés and Jiménez 1996, 2003). Golfo Dulce (sector 10), a fjord-like embayment (Hebbeln and Cortés 2001), is protected from high seas; numerous coral communities and a large fringing reef are present in this gulf (Cortés 1990; Macintyre et al. 1993; Cortés and Jiménez 2003). One of the upper gulf reefs (Punta Islotes), constructed primarily of *Porites lobata*, initiated growth over 5000 years ago (Cortés et al. 1994). Highest accretion occurred between 2500 and 500 years BP, and then slowed due to changes in the local

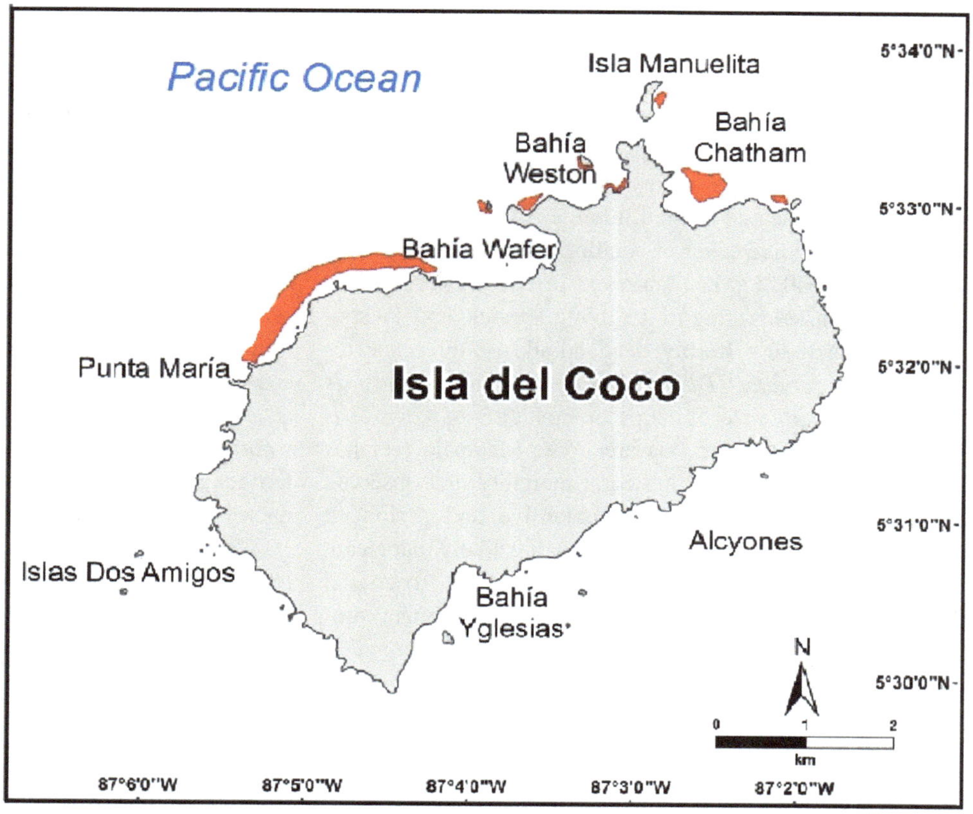

Fig. 5.14 Isla del Coco (COC), Costa Rica. *Red patches* denote location of coral reefs (modified after Cortés and Jiménez 2003)

geological setting that increased freshwater input. Reef growth ceased during the last 60 years due to high terrigenous sediment loads.

5.3.8 Isla del Coco (Costa Rica)

Isla del Coco (sector 11), is located over 500 km offshore of Costa Rica, about half way between the Pacific coast of Central America and the Galápagos Islands (Cortés 2008). It is a volcanic island situated on the northeastern-trending Coco Volcanic Cordillera (Rojas and Alvarado 2012). The island and surrounding waters are a marine protected area—Parque Nacional Isla del Coco—the largest MPA in Costa Rica, and is now imbedded in the Área Marina de Manejo de Montes Submarinos (Sea Mounts Marine Management Area).

Twenty-two zooxanthellate scleractinians make up the Isla del Coco coral fauna, compared to 25 species total for all 10 Costa Rican mainland sectors (Table 5.4; Appendix). Of all eastern tropical Pacific coral localities, the Isla del Coco coral fauna demonstrates its closest affinity with the Costa Rican mainland (Fig. 5.2; Cortés 2012b). *Pavona xarifae*, once considered a unique Isla del Coco record in the eastern tropical Pacific, has been synonymized with *Pavona minuta* by Veron (2000). Veron's synonymy is recognized here; therefore *P. minuta* is now more broadly distributed, considered present at Clipperton Atoll, in the Revillagigedo Islands, and at Palmyra Atoll (Appendix). Verification is necessary, however, ideally from specimens collected at their type localities and subjected to the same genomic and morphological analyses.

Coral reefs are best developed on the northern island shelf (from Punta María to Wafer Bay, and at Weston and Chatham bays), which is protected from southern swells (Guzmán and Cortés 1992; Cortés 2014). The largest Costa Rican coral reefs occur at Isla del Coco, with one reef exceeding 2 km in length (Fig. 5.14). *Porites lobata* is the chief framework species, with colonies of *Pavona clavus*, *Pavona gigantea*, and *Gardineroseris planulata* also contributing to reef building; some reef areas have extensive cover of *Pocillopora* spp. as well (i.e., *P. elegans*, *P. damicornis*, and *P. eydouxi*). Severe coral bleaching and high levels of coral mortality occurred at Isla del Coco during the 1982–83 El Niño warming event (Guzmán and Cortés 1992; Macintyre et al. 1993). Pre-El Niño coral cover of 80-90 % on some reefs declined to 3-26 % after the event. However, by 2002 recovery had been impressive and continues to the present (as of 2015). Since the mid- to late-1990s, *Leptoseris scabra*, a wide-ranging Indo-Pacific species, appeared for the first time on Isla del Coco reefs and now plays a role there in reef consolidation. The appearance of other species not reported from Isla del Coco before the mid-1990s include *Pavona maldivensis*, *Pavona frondifera*, *Pavona chiriquiensis*, and

Pocillopora inflata. All three *Pavona* species are known to occur at Palmyra Atoll, Line Islands (Appendix), and *Pocillopora inflata* may be present in the Phoenix Islands (Obura 2002). The sudden appearance of these species at Isla del Coco, in the downstream path of the North Equatorial Counter Current (NECC), prompted Guzmán and Cortés (1992, 2007) to suggest that long-distance dispersal may be an important factor in the recovery of these eastern Pacific reefs following ENSO disturbances.

Summarizing for all Costa Rican reef sites subject to disturbance and long-term monitoring, a gradient of coral recovery is apparent that is related to increasing anthropogenic impacts and levels of protection (Cortés et al. 2010). The emerging pattern, from coral reefs exhibiting rapid to slow recovery, is as follows: Isla del Coco, Isla del Caño, Parque Nacional Marino Ballena, and Bahía Culebra. Coral cover at oceanic Isla del Coco, the most remote of the monitored sites, exhibited higher levels of recovery than the reefs at Isla del Caño, a moderate distance off-shore and relatively un-disturbed by humans. The lowest levels of coral recovery occurred at Parque Nacional (PN) Marino Ballena and Bahía Culebra, both on-shore areas subject to diverse coastal stresses. The PN Marino Ballena is protected whereas the coral reefs at Bahía Culebra are not. The degree of protection of these four areas is greatest at Isla del Coco and least at Bahía Culebra. A similar pattern was reported by Edgar et al. (2010) who found that eastern Pacific MPAs with higher controls and regulation had greater fish biomass and more coral cover than areas with limited or no protection.

5.3.9 Panama

Panama has been the focus of coral reef ecological studies since the early 1970s. The initial research in Panama during the 1970s was directed toward documenting the presence of structural coral reefs in the eastern Pacific region (see Chap. 1 for a historical perspective). With the discovery of numerous reefs, interests then turned to investigation of the physical and biotic controls of their distributions, particularly in relation to the different oceanographic conditions in the Gulfs of Panama and Chiriquí (Fig. 5.15). As in southern Mexico (Gulf of Tehuantepec), and the border region between Nicaragua and Costa Rica (Gulf of Papagayo), the low elevation of the Central American Cordillera in Panama,

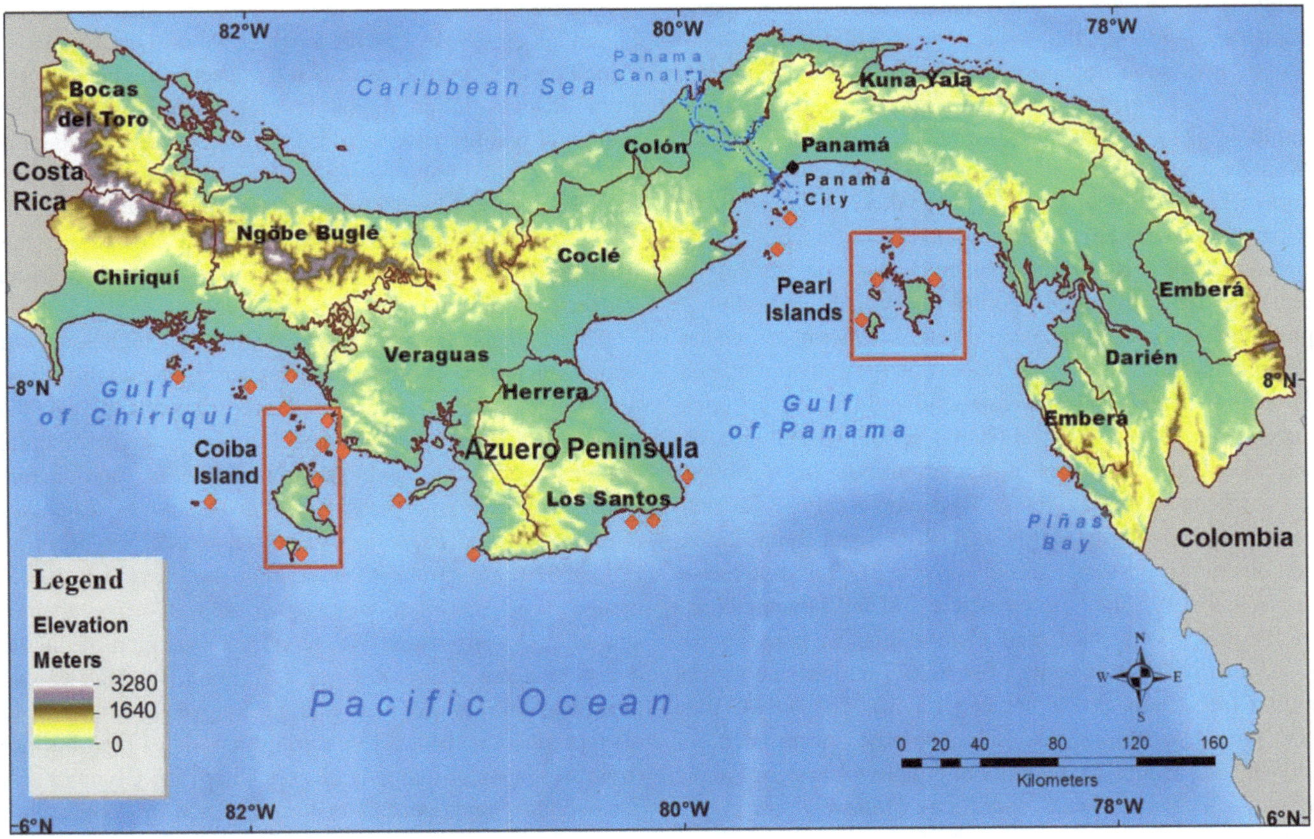

Fig. 5.15 Isthmus of Panama (PAN), framed are Las Perlas MPA, Gulf of Panama, and the Coiba National Park, Gulf of Chiriquí. Red diamonds identify sites with high coral cover. See Figs. 5.16 and 5.17 for actual MPA boundaries

at 80°W, allows seasonal (January–April) NE Trade Wind flow across the isthmus, thus causing upwelling in the Gulf of Panama and further south across the Panama Bight and along coastal Colombia. Upwelling, with accompanying low water temperatures and high nutrient conditions, promotes phytoplankton blooms that can interfere with coral growth. The high Tabasará mountain range in western Panama, reaching 3000 m elevation and more in some areas, reduces by more than two-thirds the Trade wind flow, thus preventing upwelling in the Gulf of Chiriquí, but allowing brief SST shoaling and cooling of the upper surface layer more commonly than previously suspected (D'Croz and O'Dea 2007). The extent of reef building is notable in this warm and more thermally stable enclave. Structural coral reefs are also present in the seasonally variable Gulf of Panama, but are not so well developed as in the Gulf of Chiriquí (see Chap. 6, Toth et al.). Coral community development, and the presence of coral reefs, is greatest on the numerous islands present on the isthmian shelf.

With the exception of the hydrocoral *Millepora intricata*, which often contributes importantly to coral communities and reef frameworks in the Gulf of Chiriquí (Fig. 5.3), the chief frame-building scleractinian species in both gulfs are *Pocillopora damicornis*, *Pocillopora elegans*, *Porites lobata*, *Pavona clavus*, *Pavona gigantea*, and *Gardineroseris planulata*. Zooxanthellate scleractinian species richness is virtually the same in both gulfs (Maté 2003). Three uncommon transitory species, *Leptoseris papyracea*, *Diaseris distorta* and *Cycloseris curvata*, are sometimes found in deeper (12–25 m), offreef, sedimentary habitats. A single, small, living population of *C. curvata* is presently known in the Gulf of Chiriquí; only dead skeletons of the other two species have been reported from Panama.

Like Costa Rica, the Panamanian coral fauna has remained relatively stable during the past decade, with 25 presently recognized zooxanthellate scleractinian species. Recent additions are *Porites evermanni*, *Pavona minuta*, *Pavona duerdeni*, and *Siderastrea glynni* (Fig. 5.7; Appendix). The validity of *S. glynni* is in question; possibly it is not a resident endemic, but a human-introduced Caribbean species (Forsman et al. 2005). Panama's coral species richness is equal to that of Costa Rica, with both faunas demonstrating a strong affinity (Fig. 5.2). It is not inconceivable that the high species richness of Indo-Pacific corals in Panama, Costa Rica, and Colombia is at least in part associated with the inflow of the NECC into this equatorial region (see Chap. 3, Fiedler and Lavín). The NECC may also be responsible for the sporadic occurrence of Indo-Pacific cnidarian vagrants in equatorial eastern Pacific waters, e.g., *Porites rus* (Costa Rica), *Millepora platyphylla* (Panama), and *Acropora valida* (Colombia) as well as ephemeral occurrences of non-cnidarian invertebrate and fish species (Glynn and Ault 2000; see Chap. 16, Lessios and Baums).

Perhaps the most influential environmental driver affecting Panama's coral populations is ENSO. The ENSO warm phase (El Niño) in particular, but also the cool phase (La Niña) can cause coral bleaching and mortality. Mean overall coral mortality in Panama following the 1982–83 El Niño disturbance ranged from 76 to 90 % (Wellington and Glynn 2007). Coral mortality during the 1997–98 El Niño was much lower, amounting to only ~13 % in the Gulf of Chiriquí, and no detectable mortality in the Gulf of Panama. The low coral mortality in the Gulf of Chiriquí was attributed to endosymbiont resistance (see Chap. 13, Baker et al.), and the absence of coral mortality in the Gulf of Panama to upwelling, which prevented high stressful temperatures (Glynn et al. 2001).

Biotic disturbances may follow thermally-induced coral mortality. For example, bioerosion of dead corals can accelerate with increasing sea urchin population densities, which could lead to coral framework loss and the cessation of reef growth (Glynn 1988; Eakin 1992, 1996, 2001). *Acanthaster planci*, where abundant in the Gulf of Chiriquí in the 1970s and early 1980s, posed a threat to some uncommon corals that survived El Niño bleaching. However, this corallivore's numbers decreased dramatically on reefs after the mid-1990s and is now mostly present in coral communities where it is no longer a threat (see Chap. 10, Enochs and Glynn). Past ENSO impacts can be detected from changes in coral colony morphology, increases in non-coral benthic cover, eroded reef structures, and increases in coral rubble accumulations. ENSO-related coral mortality has not been observed in Panama during the last 18 years, and significant recovery of live coral cover has occurred on reefs in the Gulfs of Panama and Chiriquí (Baker et al. 2008).

The demonstrated capacity for sustained high levels of sexual reproduction as well as asexual regeneration (Glynn and Fong 2006) has likely figured importantly in the recovery of coral communities. Continued study of coral sexual reproduction has demonstrated that the reproductive modes and spawning patterns of *Pavona clavus*, *Psammocora stellata* and *Psammocora profundacella* (Glynn et al. 2011, 2012) are similar to those of seven previously investigated species (*Pocillopora*, two spp; *Porites lobata*; agariciids, 4 spp), namely (a) broadcast spawning of minute ova at or a few days following full moon, (b) high annual fecundity due to frequent and prolonged spawning, and (c) reproduction most active during periods of high but not stressful sea water temperature (see Chap. 15, Glynn et al.). Year-round gametogenesis occurred in non-upwelling Gulf of Chiriquí, but ceased during the 3-month upwelling season in the Gulf of Panama. Gametogenesis continued during mild El Niño warming events (Colley et al. 2006). Glynn and Colley (2008) hypothesized that broadcast spawning coral species can better tolerate and survive unstable environmental conditions in the eastern Pacific than brooding species. *Tubastraea coccinea*, a ubiquitous azooxanthellate brooder in Panama (and elsewhere in the eastern Pacific), releases planulae during new and full lunar phases, mostly during high thermal periods (Glynn et al. 2008). This species also is highly fecund, and is reproductively mature when colonies possess as few as 2 to <10 polyps.

During the past decade (early to mid 2000s) detailed mapping of the presence and abundances of coral and octocoral communities has been undertaken in two of Panama's largest MPAs, the Pearl Islands (Gulf of Panama), and Coiba Island (Gulf of Chiriquí) and surroundings (Guzmán et al. 2004, 2008). [As a cautionary note, the live coral cover on these maps (1) includes corals and octocorals, (2) none of the islands supports a continuous fringing reef, and (3) coral cover on the relatively small structural reefs (not shown) often exceeds 40 %. The reader is referred to Guzmán et al. (2004, 2008) for details on the sampling methodology.] Moderate to high coral cover, 20 to >40 %, was reported in both of these archipelagos (Figs. 5.16 and 5.17) with the highest coral cover recorded on the smaller islands (Table 5.5). The high coral cover category in the Coiba Islands (18.4 %) was slightly over twice as high as

Fig. 5.16 Live coral cover in Pearl Islands MPA (168,800 ha), Gulf of Panama, Panama. The ~7 km wide Trollope Bank is shown to the SE of El Rey Island. MPA boundaries in *blue*. Modified after Guzmán et al. (2008)

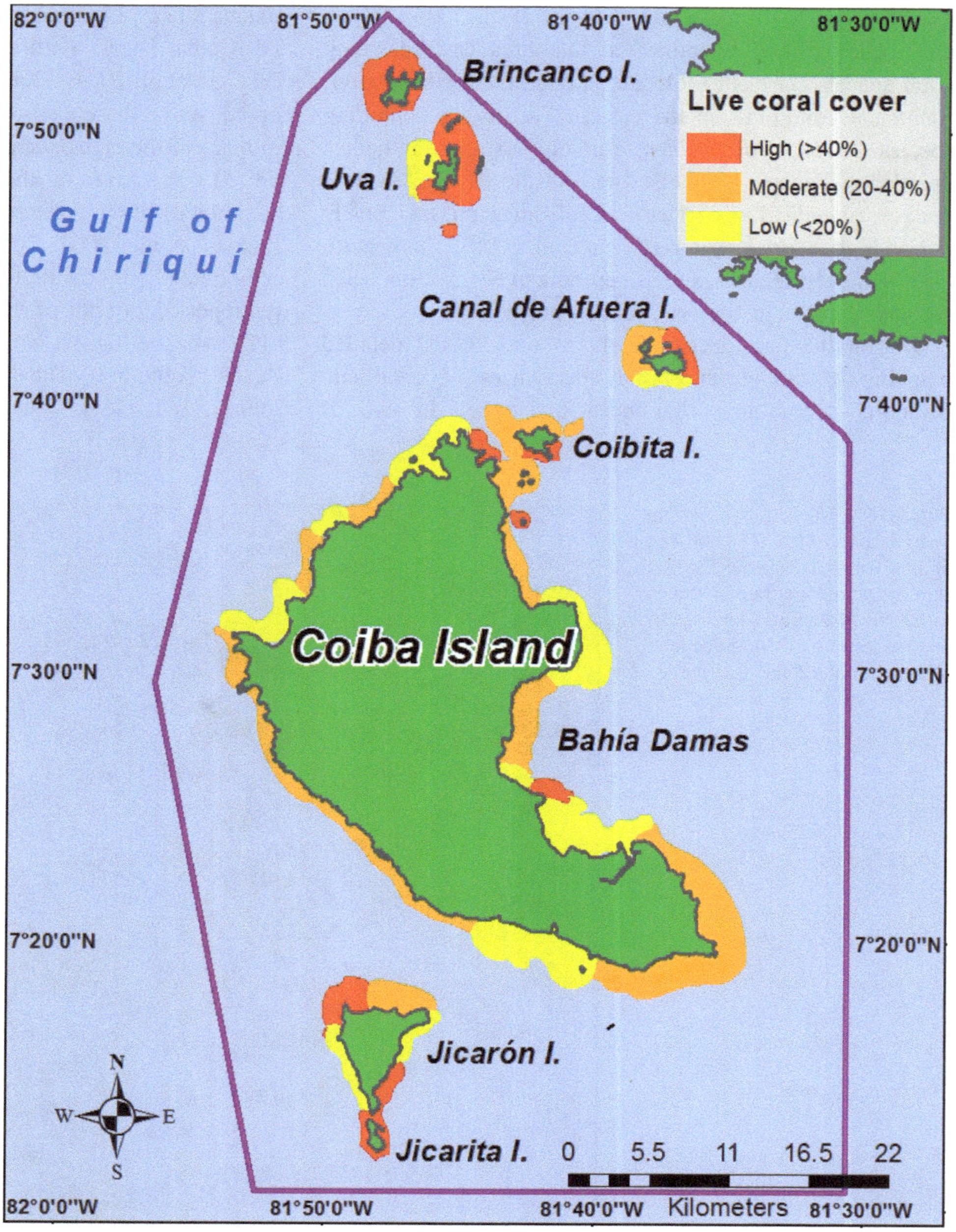

Fig. 5.17 Live coral cover in Coiba MPA (270,125 ha), Gulf of Chiriquí, Panama. MPA boundary in *blue*. Modified after Guzmán et al. (2004)

that recorded in the Pearl Islands (8.1 %). Nearly 80% of the surveyed areas in the Pearl Islands supported low coral cover (<20 %) compared with ~35 % in the Coiba Islands. It is necessary to consider at least three factors contributing to this difference: (a) deforestation, land use and sedimentation, all of which have been greatest in the Pearl Islands, (b) upwelling versus non-upwelling environments, the latter generally promoting more robust coral growth, and (c) phytoplankton blooms causing copious mucus production that can smother corals (Guzmán et al. 1990).

In these island-wide surveys a distinction was made between coral communities (CC) and coral reefs (CR), which revealed some intriguing differences: (a) scleractinian corals and octocorals demonstrated higher species richness in CC compared to CR, (b) coral cover was higher in CC than CR in the Coiba Islands preserve, but higher in CR than CC in the Pearl Islands, (c) uncommon, rare or endangered cnidarians were encountered more frequently in CC than CR, and (d) a negative relationship between live coral cover and species richness emerged in the Coiba Islands, with CC consisting of a greater variety of cnidarian species than CR. A similar trend to that in 'a' above was reported by Benfield et al. (2008) for reef fish assemblages in the Pearl Islands, which were significantly more diverse and species rich in CC than CR. These results prompted Guzmán et al. (2004, 2008) to hypothesize that CC may serve as refugia and

Table 5.5 Live coral cover at various islands surveyed at Coiba and Pearl Islands MPAs

Location	Area of live coral cover (km^2)		
	High (>40 %)	Moderate (20-40 %)	Low (<20 %)
Coiba Islands			
Brincanco	12.237	–	–
Uva	14.587	–	4.747
Canal de Afuera	6.377	7.822	4.951
Coibita	2.384	7.829	–
Coiba	9.531	135.693	93.053
Jicarón	13.532	7.156	13.707
Jicarita	3.227	–	–
Total = 336.84	61.88 (18.4 %)	158.50 (47.1 %)	116.46 (34.6 %)
Pearl Islands			
Pacheca	–	3.261	1.260
Saboga	–	5.037	1.359
Contadora	1.387	2.133	–
Chapera	2.418	1.432	–
Mogo Mogo	4.500	–	–
Gibraleon	–	–	5.898
Bayoneta	–	–	6.947
Casaya	–	–	12.248
La Mina	–	–	7.120
Viveros	–	–	37.279
Señorita	–	2.168	–
Pedro González	3.477	–	9.725
San José	2.920	5.110	13.100
San Telmo	–	–	3.131
Puerco	–	–	3.336
Cane	–	1.560	3.600
El Rey	–	1.800	38.320
Total = 180.52	14.70 (8.1 %)	22.50 (12.5 %)	143.32 (79.4 %)

Estimated areas digitized using Arc Map 10.1 software from Guzmán et al. (2004, 2008)

source areas for larvae that could repopulate CR that become degraded during large-scale disturbances such as El Niño bleaching/mortality events.

Building on Abele's (1976) demonstration of a species-rich community of decapod crustaceans associated with living pocilloporid corals (55 species), several recent studies have revealed an impressive number of motile, cryptic invertebrates and fishes that seek refuge in living and dead coral frameworks and rubble. For example, the biomass and abundances of various invertebrate taxa (Polychaeta, Gastropoda, Crustacea, Holothuroidea, Ophiuroidea) associated with live corals are generally higher than those sheltering among dead corals and coral rubble (Enochs and Hockensmith 2009; Enochs 2012). Species richness, however, is significantly higher in dead and degraded coral rubble (261–370 species) compared with live coral habitats (112–219). Among some of the several key reasons for this are (a) the inimical defenses of live corals (nematocysts, mucus secretions), (b) the agonistic and competitive nature of obligate coral metazoan symbionts, and (c) the greater habitat heterogeneity and niche diversity of rubble substrates. These results show that an abundant and speciose cryptic fauna is still present on highly degraded reefs that can continue to function in various reef processes, e.g. providing prey for reef fishes, scavenging, biodegradation, remineralization, bioerosion, interalia (Enochs et al. 2011; Enochs 2012; Enochs and Manzello 2012a, b).

Recent studies of the diversity, community structure and trophic relationships of coral-associated fishes have increased our understanding of their diverse and pivotal ecological roles in the ETP (Dominici-Arosemena and Wolff 2006; Benfield et al. 2008; Glynn 2008; Robertson and Allen 2008; Glynn et al. 2014). A large number of facultative corallivores, and cryptic fishes residing in coral rubble, many of the latter invertivore and piscivore predators, have been identified (Glynn 2006, 2008). Some of these taxa may play critical roles in limiting the settlement and survival of recruiting species.

5.3.10 Mainland Colombia

In spite of the adverse climatic and oceanographic conditions (Glynn and Ault 2000; Zapata and Vargas-Ángel 2003), isolated colonies of scleractinian corals occur along most of the 1450 km of the Colombian Pacific coast. Corals are most common in the northern half of the coast from the Panama-Colombia border south to Cabo Corrientes, a stretch dominated by igneous rocky shores. South of Cabo Corrientes to the Ecuador-Colombia border the coast is composed of alluvial sediments and dominated by soft bottoms and mangroves intermixed with some stretches of soft sedimentary rock from the Tertiary. The Colombian Pacific coastal region is one of the rainiest regions of the world, with annual precipitation exceeding 10,000 mm at some localities. High freshwater discharge from the western slopes of the Andes and the adjacent lowlands into the Pacific Ocean significantly reduces salinity and increases turbidity and sedimentation, particularly in the southern half of the region (Restrepo and Kjerfve 2000; Alory et al. 2012). Yet, the best developed coral reefs and coral communities of the Colombian Pacific are located in this southern portion at Gorgona Island, which is adjacent to this stretch of coastline, but located sufficiently far offshore to reduce the limiting influence of the coastal environment. Accordingly, Kleypas et al. (1999) listed Gorgona Island and Ensenada de Utría, also in Colombia (see below), as the two lowest salinity localities with coral reefs in the eastern Pacific. Therefore, coral communities and true reefs (carbonate frameworks) are few, small, patchily distributed and poorly developed. Consequently, coral communities and reefs in this area of the eastern Pacific are species-poor (≤ 12 scleractinian coral species per site) and the entire coral fauna is composed of about 28 species, although taxonomic difficulties, particularly within the genus *Pocillopora*, make this number somewhat uncertain (Table 5.6).

Four localities along Colombian Pacific shores have been reported to have significant coral communities or coral reefs: Cabo Marzo and Punta Cruces, Punta Tebada, Ensenada de Utría, and Gorgona Island (Fig. 5.18). The northernmost area is a prominent rocky point located at Cabo Marzo (Fig. 5.19) where a significant coral formation occurs. Known locally as El Acuario, the shallow (≤ 3 m) substrate here is composed of a solid rock bed with boulders dominated by filamentous algae (47 % mean cover), scleractinian corals (33 % mean cover) and octocorals. Branching corals of the genus *Pocillopora* are the most common (27 % mean cover) followed by massive colonies of *Pavona gigantea*. There is, however, little indication of significant reef framework development at this site. Six species of zooxanthellate scleractinian corals occur at El Acuario (Zapata et al. 2008; Table 5.6). A small coral patch consisting of sparse *Pocillopora damicornis* colonies and associated rubble is also located in neighboring Bahía Aguacate. Further south at Punta Cruces, in a cove in front of the hamlet known as Piñas (Fig. 5.19), there also is a coral community dominated by massive and encrusting corals covering nearly 38 % of the rocky substrate.

Coral assemblages characterized by high colony densities, but lacking a framework buildup are common in the Utría and Tebada region. Largely, these communities develop on rocky banks, shoals, and vertical walls, particularly in small embayments and coves, ranging in depth from about 1 to 10 m; moderately strong swells and vigorous water circulation are prevalent in these habitats, however. Despite the high energy of the environment, the dominant taxa are branching *Pocillopora* species, particularly *P. elegans*, and massive colonies of *Porites lobata*, *Pavona clavus*, and *Pavona gigantea*; less abundant to rare are *Gardineroseris planulata*, *Pocillopora damicornis*, *Psammocora stellata*, and *Pavona* sp. aff. *frondifera* (Vargas-Ángel 2001, unpublished data). Together with the scleractinian corals, coralline and turf algae, and other sessile invertebrates including non-zooxanthellate corals, barnacles, sea fans, and clams, inhabit these shallow basalt substrates. Generally high levels of grazing from sea urchins and herbivorous fishes are in part responsible for the limited macroalgal development in these communities. Coral recruitment from sexual propagules or fragmentation is also noticeably low on both the grazed basalt surfaces or the rubble and sand-covered bottom surrounding corals.

Of the coastal coral reefs, La Chola (also known as La Aguada) fringing reef, located at Ensenada de Utría (Fig. 5.18), is the largest, best developed, and probably the most thoroughly studied in this part of Colombia. The reef is located on the central, eastern shoreline of a narrow embayment, characterized by calm and relatively clear waters. Elevated turbidity and runoff events following heavy rainfall, which is common in this region, can reduce visibility to about 1.0–1.5 m, demonstrating that these corals are able to tolerate periods of elevated turbidity and reduced salinity. Thinly-branched ecomorphs of *Pocillopora* are prevalent on the reef, which contrast with the more stout and stubby congeneric colonies developing in the non-framework building communities of outer wave-exposed habitats. When the reef was surveyed in 1988–89 by Vargas-Ángel (1996), the predominant coral species were *Pocillopora damicornis* (80 %) and *Psammocora stellata* (16 %), which formed a terrace-like framework, covering approximately 10.5 ha; uncommon to rare species included *Pocillopora capitata*, *Pavona varians*, and *Pavona gigantea*. The reef at Utría lacks the typical geomorphological zones (i.e., backreef, reef flat, forereef); it is a relatively broad reef flat followed by a seaward slope, where coral development gradually decreases to a depth of approximately 10 m. Reef coring and probing revealed that the uncemented *Pocillopora* framework was at least 2.0–4.5 m in height, with uninterrupted growth throughout the last 2000–3000 years, and mean vertical accretion rates of 2.47 m kiloyear^{-1} ±0.15 (SE) (Vargas-Ángel 2001; see Chap. 6, Toth et al.). Community structure surveys conducted later in 1996–98 (Vargas-Ángel 2001) indicated that mean percent live coral cover was 41 %. A temporal comparison of permanent transects placed since 2002 indicates that coral cover on this reef has declined on average by 65 % from 57 % in 2002 to 37 % in 2013 (M. López-Victoria and F. A. Zapata, unpublished data). In all of these surveys, coral composition was relatively homogeneous throughout the shallow reef flat, with intermingled turf algal patches and pockets of rubble and bioclastic sand. Absence of ecological zonation has been regarded as an indicator of "instability", whereby chronic stress and environmental disturbances may offset the competitive interactions and biological processes leading to species habitat partitioning (Karlson 1999). In addition to chronic siltation and turbidity stress, this reef is subject to subaerial exposure during extreme low tides, recurrent ENSO events, and periodic intense upwelling episodes (Vargas-Ángel 2001; Zapata and Vargas-Ángel 2003).

Located approximately 70 km north of Utría, the Tebada reef is a small structure, covering roughly 4.5 ha, and situated in the southern end of the Gulf of Cupica. When

Table 5.6 Known zooxanthellate scleractinian coral species from the Colombian Pacific Ocean

Species	Malpelo	Gorgona	Ensenada de Utría	Tebada	Bahía Cupica	Cabo Marzo	Comments
F. ACROPORIDAE							
Acropora valida (Dana, 1846)		4, 5, 8					
F. FUNGIIDAE							
Cycloseris curvata (Hoeksema, 1989)		2, 5, 8					
Diaseris distorta (Michelin, 1843)			12				Only dead fragments found outside reefs
F. AGARICIIDAE							
Gardineroseris planulata (Dana, 1846)	1, 5, 8	2, 5, 9	10	12	12		
Leptoseris papyracea (Dana, 1846)		5					
Pavona chiriquiensis Glynn et al., 2001	11, 13	13					
Pavona clavus (Dana, 1846)	1, 5, 8	2, 3, 5, 8	5, 6, 10		12	14	
Pavona frondifera (Lamarck, 1816)		14			17		
Pavona gigantea Verrill, 1869	5, 8	2, 3, 5, 8	5, 6, 10	12, 17	12	14	
Pavona maldivensis (Gardiner, 1905)	18, 14						
Pavona varians Verrill, 1864	1, 5, 8	2, 3, 5, 8	5, 6, 10	12, 17	12		
F. POCILLOPORIDAE							
Pocillopora capitata Verrill, 1864	1, 8	2, 3, 5, 8	5, 6, 10	12		14	
Pocillopora damicornis (Linnaeus, 1758)		2, 3, 5, 8	5, 6, 10	12, 17	12	14	
Pocillopora elegans Dana, 1846	5	2, 5	5, 6, 10		12		
Pocillopora eydouxi Milne Edwards & Haime, 1860	8	2, 3, 5, 8	10			14	
Pocillopora verrucosa (Ellis and Solander, 1786)	20	20					
Pocillopora meandrina Dana, 1846		20					
Pocillopora effusus Veron, 2002	20	20					
Pocillopora ligulata Dana, 1846		20					
Pocillopora woodjonesi Vaughan, 1918		20					
Pocillopora sp.	15						
F. PORITIDAE							
Porites evermanni Vaughan, 1907		16					Inferred from results in Boulay et al. (2014), which indicate that 50 % of massive *Porites* in coastal Ecuador and Panamá are *P. evermanni*
Porites lobata Dana, 1846	1, 5, 8	2, 3, 5, 8	10	12,17	12		
Porites panamensis Verrill, 1866		2, 3, 5, 8	5, 6			14	
F. SIDERASTREIDAE							
Psammocora contigua (Esper, 1797)		9, 19					Based on *Psammocora obtusangula* (Lamarck, 1816) as listed by Zapata and Vargas-Ángel (2003). *P. obtusangula* was synonymized with *P. contigua* (Esper, 1794) by Stefani et al. (2008)

(continued)

Table 5.6 (continued)

Species	Malpelo	Gorgona	Ensenada de Utría	Tebada	Bahía Cupica	Cabo Marzo	Comments
Psammocora profundacella Gardiner, 1898		9, 19		12			Based on *Psammocora superficialis*, Gardiner, 1898, as listed by Zapata and Vargas-Ángel (2003). *P. superficialis* was synonymized with *P. profundacella* Gardiner, 1898 by Benzoni et al. (2010)
Psammocora stellata Verrill, 1864		2, 3, 5 ,8	5, 6, 10	12, 17	12		
Total number species = 27	13	24	12	8	9	6	

Numbers indicate information sources as indicated in the references and sources section. List prepared by Fernando A. Zapata with contributions by Bernardo Vargas-Ángel for Ensenada de Utría, Tebada and Bahía Cupica. 9 February 2014

1. Birkeland et al. (1975)
2. Glynn et al. (1982)
3. Cantera (1983)
4. von Prahl and Mejía (1985)
5. von Prahl and Erhardt (1985)
6. von Prahl (1986a)
7. von Prahl (1986b)
8. von Prahl (1990)
9. Guzmán and Cortés (1993)
10. Vargas-Ángel (1996)
11. Garzón-Ferreira and Pinzón (1999)
12. Vargas-Ángel (2001)
13. Zapata et al. (2007)
14. Zapata et al. (2008)
15. Rodríguez-Ramírez and Zapata (2012)
16. Boulay et al. (2014)
17. B. Vargas- Ángel, personal observation, unpublished data
18. J. Garzón-Ferreira, photographic record
19. H.M. Guzmán, personal communication
20. F.A. Zapata, personal observation

Other references
Stefani et al. (2008)
Benzoni et al. (2010)

surveys were conducted in 1996–1998 mean live coral cover was 39 %. At that time the living coral community was dominated by a mixture of *Psammocora stellata* and *Psammocora profundacella*, which together amounted to nearly 90 % of the live reef cover (Vargas-Ángel 2001), in contrast with the reefs farther south at Ensenada de Utría and Gorgona Island, which were dominated by pocilloporid species (Zapata and Vargas-Ángel 2003). Four more coral species were present at Tebada than further south, including a few small colonies of *Pocillopora damicornis*, *Pavona varians*, *Pavona gigantea*, and *Gardineroseris planulata*; the latter is relatively uncommon on coastal reefs and in coral communities. The evident structural pattern of the Tebada reef was an increase of live coral cover with increasing distance from shore, related to a decrease in scleractinian species richness (Vargas-Ángel 2001, 2003). Reef probing on the reef flat in 1996 revealed that the Tebada reef framework was at least 4 m thick, and coring to a depth of approximately 1 m revealed an uncemented framework composed mainly of *Psammocora* rubble (Vargas-Ángel, unpublished data). Based on its geographic location, it is reasonable to assume that the Tebada reef is subject to similar environmental stressors as those at Utría, particularly ENSO and upwelling stress events. The reef flat at Tebada, however, is approximately 1 m deeper than that at Utría; therefore, it is not subject to subaerial exposure during extreme low tides. Recent inspection of this site in April 2014 revealed that both *Pocillopora* and *Psammocora* were important members of this coral formation and that framework construction was modest (C.G. Muñoz and M. Rodríguez-Moreno, personal communication). While Vargas-Ángel et al. (2001) reported minimal coral bleaching at this reef during June 1998, local fishermen recalled that noticeable coral bleaching occurred here at some time during the 1997–1998 El Niño event.

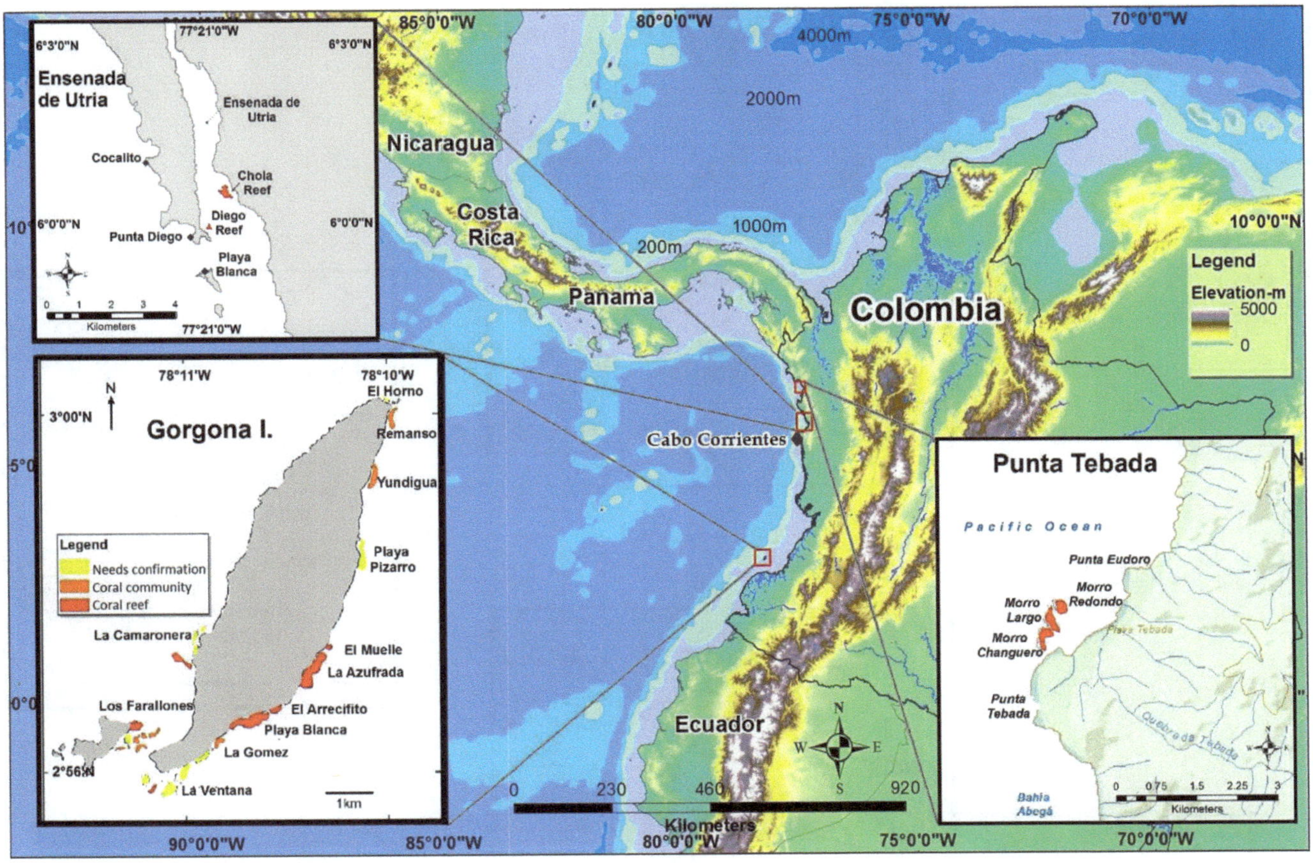

Fig. 5.18 Coral communities and coral reefs at Punta Tebada, Ensenada de Utría and Gorgona Island, Pacific Colombian (COL) coast. *Red patches* at Punta Tebada and Ensenada de Utría denote high coral cover. Gorgona Is legend: *red* = structural coral reef (arrecife); *orange* = coral community (comunidad coralina); *yellow* = coral reef/community in need of confirmation (por confirmar)

The coral reefs and coral communities of Gorgona Island are the largest and best developed in the Colombian Pacific region (Glynn et al. 1982; Zapata and Vargas-Ángel 2003). The largest structural coral reefs of Gorgona Island, La Azufrada and Playa Blanca reefs, are located on the eastern, leeward side of the island (Fig. 5.18). These reefs exhibit high coral cover, are more or less similar in size (~10 ha), structure and zonation, and are composed mainly of pocilloporid species with a significant presence of massive corals of the genera *Pavona* and *Gardineroseris* at the seaward reef base. A total of 24 species has been recorded for Gorgona Island (Table 5.6), but some of these records may have to be confirmed with more detailed work. The report of *Acropora valida* at Gorgona Island, a species known previously from the western Pacific but known today to be widespread across the Indo-Pacific region (Veron 2000), is intriguing: three colonies were found at 10 m depth on a rocky outcrop at the northernmost tip of Gorgona Island in 1983 (von Prahl and Mejía 1985). While the biogeographical implications of this report were potentially major given competing hypotheses regarding the origin of the eastern Pacific coral fauna (Dana 1975; Heck and McCoy 1978), despite repeated efforts to find new specimens, this species has never again been observed at Gorgona Island or any other locality in the eastern Pacific.

Live coral and algal cover on these reefs appear to be highly dynamic, particularly on shallow areas due to recurrent subaerial exposures during extreme low tides (Zapata et al. 2010). While recurrent disturbances on the reef flat can at times reduce live coral cover to very low levels and increase algal turfs, high densities of herbivorous sea urchins and fishes readily remove filamentous turf algae leaving an exposed reef matrix composed of coral rubble cemented by calcareous algae that sexually-produced coral recruits eventually colonize. Coral recovery is further enhanced by asexual recruitment via colony fragmentation, which is very common and facilitated by the feeding activities of some fishes, particularly triggerfishes (Balistidae) and puffers (Tetraodontidae) (Palacios et al. 2014).

Coral formations of Gorgona Island have been studied since the 1970s and have been described in several accounts (von Prahl et al. 1979; Glynn et al. 1982; von Prahl and Erhardt 1985; von Prahl 1986a, b; Zapata 2001; Zapata et al. 2001; Zapata and Vargas-Ángel 2003). Nonetheless, significant coral formations continue to be found around the island. While Glynn et al. (1982) had previously described

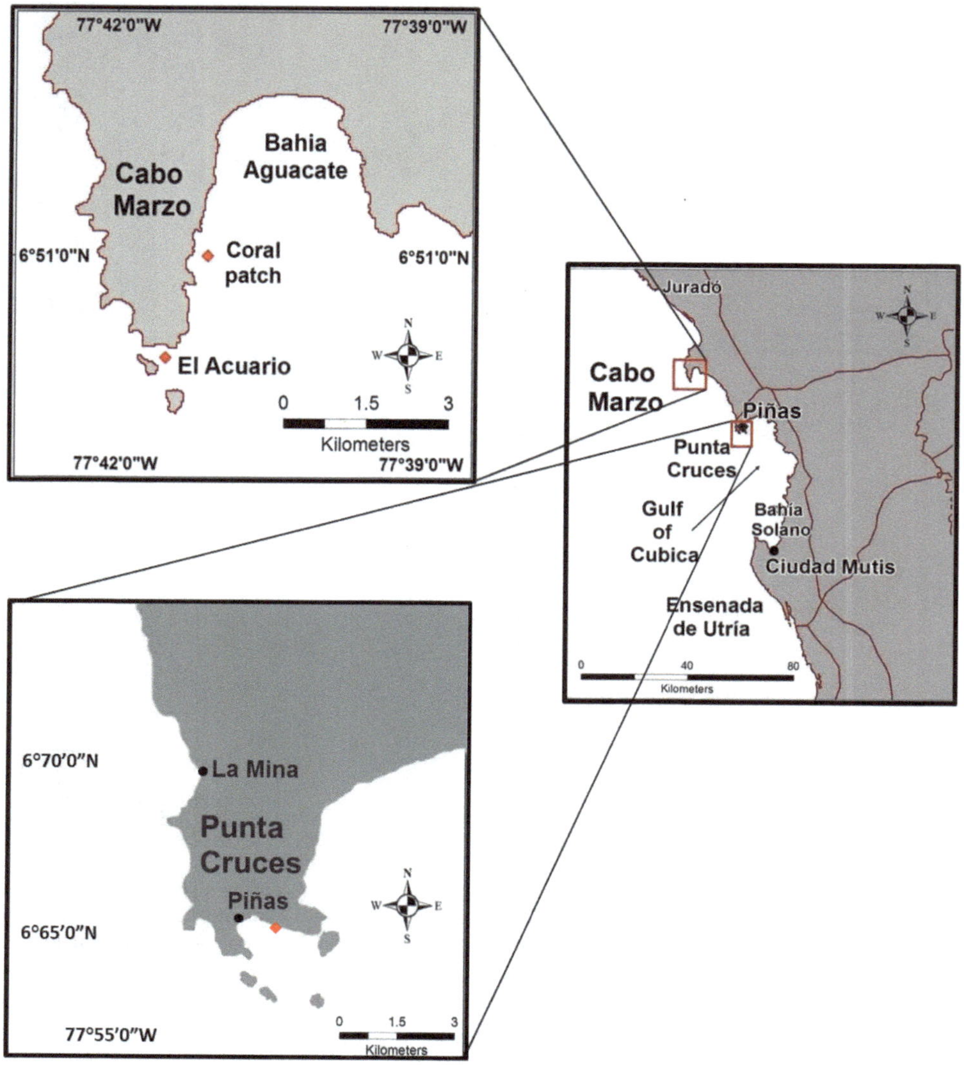

Fig. 5.19 Cabo Marzo and Punta Cruces, northern-most known coral communities on coastal Colombia

an incipient reef formation at La Camaronera, located on the western, wave-exposed coast, as "linear pocilloporid build-ups or spurs and hillocks of various sizes", a different formation was found further offshore. This is a coral community that extends 8 ha and is made up of 11 species of corals, which cover 49 % of the substrate. Massive species of the genera *Pavona*, *Gardineroseris* and *Porites* are dominant, forming high topographic relief that provides habitat for a distinct fish community (Palacios and Zapata 2014). This coral formation contrasts with the incipient coral reef of La Ventana, also mentioned briefly by Glynn et al. (1982). This is a pocilloporid reef located on the south-eastern side of the island that remained undescribed until recently. While this is a small reef covering 1 ha, it exhibits very high live coral cover (73 %) because its low-lying reef framework is never subaerially exposed during extreme low tides. However, this dense pocilloporid stand provides low relief and constitutes a relatively monotonous habitat for a distinct but less equitable fish community than that observed at La Camaronera (Palacios and Zapata 2014).

5.3.11 Malpelo Island (Colombia)

Malpelo Island, the easternmost and least remote of the five oceanic islands or archipelagos in the ETP, is a very small (1.6 km long) volcanic island located nearly 380 km off the Colombian continental coast, 630 km from Isla del Coco and 996 km from San Cristóbal, the nearest of the Galápagos Islands (Fig. 5.20). Malpelo's marine environment is most similar to that of its closest oceanic neighbor, Isla del Coco, but in contrast to the latter, only a total of 13 zooxanthellate scleractinian species has been recorded at Malpelo (Table 5.6). A particularly interesting and possibly new undescribed species of *Pocillopora*, which exhibits an unusual morphology with a tightly submassive, hemispherical corallum (Fig. 5.21), was reported by Rodríguez-Ramírez

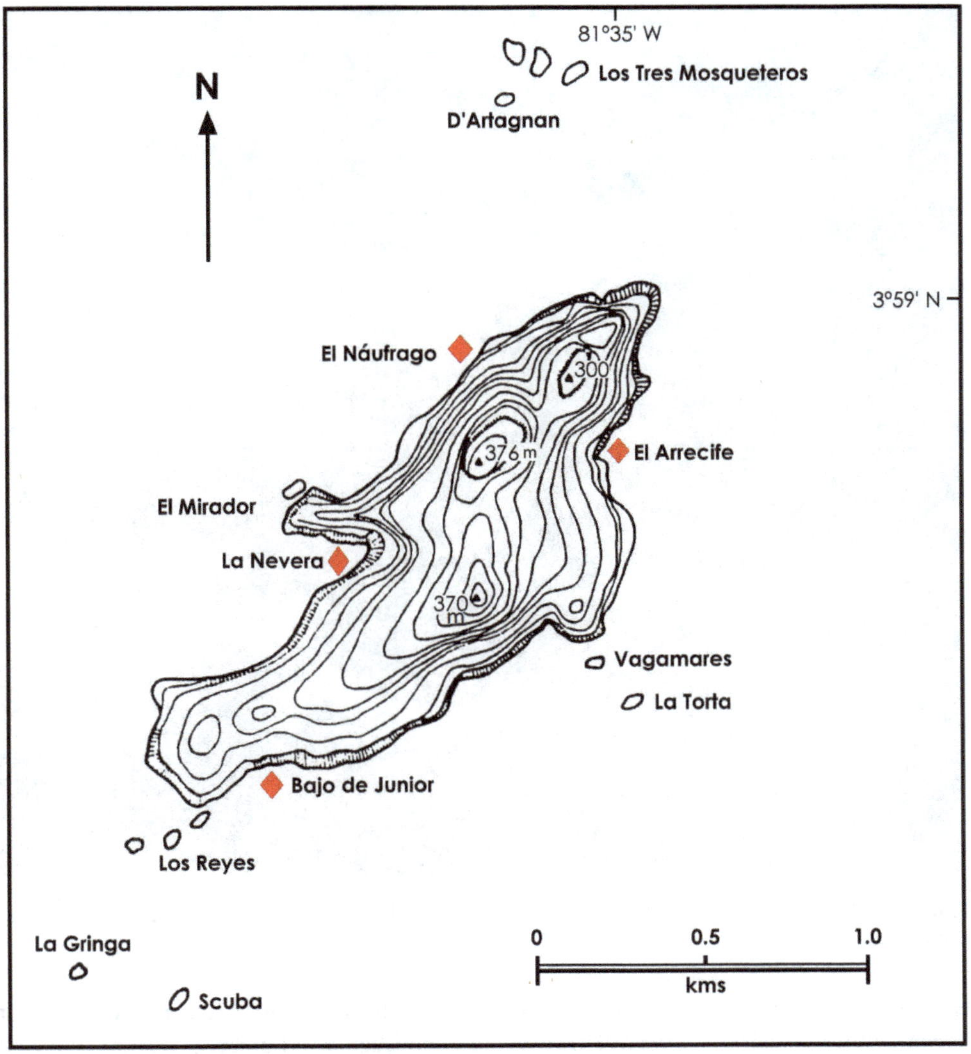

Fig. 5.20 Malpelo Island (MAL), Colombia. *Red diamonds* identify best-developed coral communities

and Zapata (2012). This rare coral, if a valid new species, would represent a possible endemic species for the island. Nonetheless, in terms of coral species composition, Malpelo is more similar to continental localities in El Salvador and Nicragua than to the neighboring oceanic islands of Isla del Coco or the Galápagos (Fig. 5.2). However, this may be a result of poor species richness at Malpelo, El Salvador and Nicaragua rather than a reflection of similar environments with similar faunas. This interpretation is supported by results of similar analyses with the fish fauna in which oceanic islands in general, and Isla del Coco, Isla Malpelo and the Galápagos Islands in particular, were always most similar in species composition, and distinct from continental localities (Robertson and Cramer 2009).

While oceanographic conditions appear to be favorable for vigorous coral growth (but see below), steep rocky walls prevent the buildup of significant reef structures, and corals often exhibit unusual growth forms. For instance, some pocilloporids display horizontally-oriented colony growth with flattened branches, likely the result of strong wave action (Fig. 5.22), while *Porites lobata* very commonly forms shingle-like structures on steeply inclined substrates (Fig. 5.23). Besides small, isolated coral patches and scattered coral colonies of *Pocillopora* spp. and *Porites lobata* along the rocky sublittoral, several sites around the island harbor coral communities, the largest of which is known as El Arrecife (The Reef). This community demonstrates a marked zonation pattern with pocilloporid colonies in shallow areas, *Porites lobata* and *Pavona clavus* at mid-depths and *Gardineroseris planulata* in the deepest areas. Strong oceanic swells and occasional rockfalls appear to upturn and break coral colonies affecting the development of coral formations around the island (Birkeland et al. 1975; Garzón-Ferreira and Pinzón 1999). Clear evidence of the effects of an unknown major disturbance was observed in July 2011 at El Arrecife, suggesting that the lack of reef building at this site may be due to recurrent sediment export during strong turbulence. Bastidas-Salamanca et al. (2014) reported a reduction in coral cover from ~45 % in 2011 to ~35 % in 2012 at El Arrecife and speculated that this may have been

Fig. 5.21 *Pocillopora* sp., possibly an undescribed species, El Arrecife, Malpelo Island, 11 m depth, 24 June 2007. Colony in foreground, approximately 0.5 m in diameter

caused by the mechanical action of a tsunami-like disturbance during a seismic event. However, the noticeable increase in variance around the mean coral cover values reported in 2012 does not lend support to this hypothesis because a major wave would probably have caused more uniform damage and suggests instead the action of a much more localized factor that affected some sites more than others.

Like other localities in the eastern Pacific in general and within Panama Bight in particular, Malpelo Island is exposed to high thermal variability. First, there is marked interannual thermal variability associated with ENSO events that occur at irregular intervals every few years. However, little is known about the effects of the 1982–83 El Niño warming event on Malpelo's coral formations, although they may have been affected as much as other reefs in the region (Glynn et al. 1988). During the 1997–98 El Niño, coral bleaching and mortality were low at Malpelo Island, but such effects were also relatively low everywhere else in the Colombian Pacific compared to levels of bleaching observed in 1982–83 (Vargas-Ángel et al. 2001). Second, there is a markedly seasonal annual cycle with lower and more variable water temperature between January and March than during the rest of the year, as in much of Panama Bight (Forsberg 1969; Stevenson et al. 1970; D'Croz and Robertson 1997; Zapata et al. 2011). During this period, the Intertropical Convergence Zone migrates from ~8–10°N to ~2°N and the north-east trade winds blow more strongly over the bight, producing a cyclonic pattern of water circulation and cold water upwelling in the middle of the bight, with a consequent shallowing of the thermocline (Rodríguez-Rubio et al. 2003; Devis-Morales et al. 2008). While cooling episodes may be common at Malpelo Island, in April 2009 an unusually cold event caused extensive bleaching of *Porites lobata* and several other coral species (Zapata et al. 2011). Thus, coral formations at Malpelo Island may also be limited by extreme temperature variations over several temporal scales.

5.3.12 Mainland Ecuador

Coral reef formations of relatively high diversity occur along the mainland coast of Ecuador from Esmeraldas to Santa Elena. At Esmeraldas, (El Faro point) near Cabo San Francisco, small patches of *Porites lobata* are present in the low

Fig. 5.22 *Pocillopora effusus*, horizontal and flattened growth form, 15 m depth, 24 June 2007, Malpelo Island (photo by F.A. Zapata)

intertidal zone. A *P. lobata* patch reef, about 1250 m^2 between 15 to 20 m depth, is present on the Surrones Bank, located approximately 6 km north of Pedernales (Fig. 5.24). Another small *P. lobata* patch reef occurs at 7 m depth at Ligüique in the Pacoche Wild Life Refuge. Coral diversity and reef development are greatest in the Machalilla National Park (Fig. 5.24). The Machalilla MPA lies within the coastal stretch from 2° 15′S to 1°00′N (Glynn 2003). A fringing reef located near the fishing village of Machalilla, at 1°28′S, is the southern-most known coral reef on the American Pacific coast. A mixed and diverse octocoral/coral community is present at Islote los Ahorcados, just south of the Machalilla reef. *Pavona clavus*, *Pavona gigantea*, *Pocillopora verrucosa*, and *Pocillopora damicornis* are the chief corals at this site. Small pocilloporid patches are also present at Pelado Islet, a new MPA located about 40 km south of the Machalilla MPA. And most recently (personal communication, 2015) Jesenia Zambrano has reported the presence of *Pocillopora* in the Puntilla Santa Elena MPA (see Fig. 21.7 in Chap. 21, Alvarado et al.).

Over the last decade, coral species richness has increased from 18 to 23 species (Fig. 5.6; Appendix). This increase is due to the recognition of four additional species of *Pocillopora* and *Porites evermanni*. These new records probably represent long-term resident species and not recent arrivals. The relatively high species count of 23 for coastal Ecuador and no authenticated coral records for Peru underlines the limiting influence of the cool Peru Coastal Current, which reaches Talara just south of the Gulf of Guayaquil (Fig. 5.1; see Chap. 3, Fiedler and Levín). Pocilloporid corals have been reported (René Espinoza, personal communication), but unconfirmed, at Santa Clara Island, near 3° S, at the entrance to the Gulf of Guayaquil. The species composition of the coral fauna of mainland Ecuador is most similar to that of the southern Galápagos Islands (Fig. 5.2).

As generally observed across the equatorial eastern Pacific, coastal Ecuadorian coral communities were severely impacted by El Niño-induced bleaching in 1982–83 and 1997–98 (Glynn 2003). Due to the remote locations of the coastal coral sites it has not been possible to monitor community responses or recovery after these disturbances. Thanks to the comprehensive cnidarian surveys and inventories of two MPAs compiled by Rivera and Martínez (2011), with the collaboration of Odalisca Breedy, it is possible to infer recovery since 1997–98, which appears to have been significant and widespread. Their photographic

Fig. 5.23 *Porites lobata*, shingle-like growth form on inclined rocky substrate, 17 m depth, 25 June 2007, Malpelo Island (photo by F.A. Zapata)

guide illustrates numerous, robust, zooxanthellate species at several localities and depths.

Coral formations are best developed and most numerous at Isla de La Plata, about 30 km west of the mainland. Several patch and fringing coral reefs are present between 1 and 15 m depth along La Plata's east coast (Fig. 5.25a). Most of these formations are isolated and relatively small, ≤ 200 m in length; however, the largest fringing reef at Acuario is about 500 m long. *Pavona clavus*, some colonies 3–5 m in diameter, is the chief framework-building species at this locality. Farther south, at Palo Santo, *Pocillopora meandrina* is the principal reef-building species. A bathymetric assessment of zooxanthellate corals performed by Martínez et al. (2011) at five sites (Palo Santo, Acuario, Bahía Drake, Tres Piedra, El Faro) showed that *Pavona* and *Psammocora* exhibited a broad depth distribution with marked differences between species. *Pavona clavus* predominated from 1 to 6 m depth, whereas *Pavona gigantea* was present primarily from 6 to 15 m, with a few isolated colonies as shallow as 4 m. *Psammocora superficialis* colonies were found from 2 to 15 m, while *Psammocora stellata* occurred from 6 to 10 m only. *Pocillopora* species occupied mostly shallow depths, especially *Pocillopora ligulata* and *Pocillopora meandrina*, which were not observed deeper than 5 m. *Pocillopora damicornis*, *Pocillopora verrucosa*, and *Pocillopora eydouxi*, however, reached depths of 10–11 m. The only colonies of *Porites lobata* observed at La Plata Island were present at the Tres Piedras site at 5 m depth.

The presence of six species of azooxanthellate scleractinians was also reported by Rivera and Martínez (2011). One of these (*Cladopsammia gracilis*), was especially abundant on the sides and underneath *Pavona clavus* colonies at the Acuario site. As in Panama, numerous gorgonian species were observed, but these generally were not common in coral reef habitats. Especially diverse and densely populated gorgonian communities were present at sites with strong currents, such as the shoals off Puntilla de Santa Elena reserve and south of Machalilla (Fig. 5.24).

Eastern Pacific coral reef development is generally confined to sheltered areas; the satellite image in Fig. 5.25b shows the prevailing swell direction from the SW and high wave assault on the western and southern island shores. Encrusting coralline algal communities are well developed along the exposed rocky coastal areas. The extent of coral reef development on mainland Ecuador now rivals and may exceed that in the Galápagos Islands. This was not the case

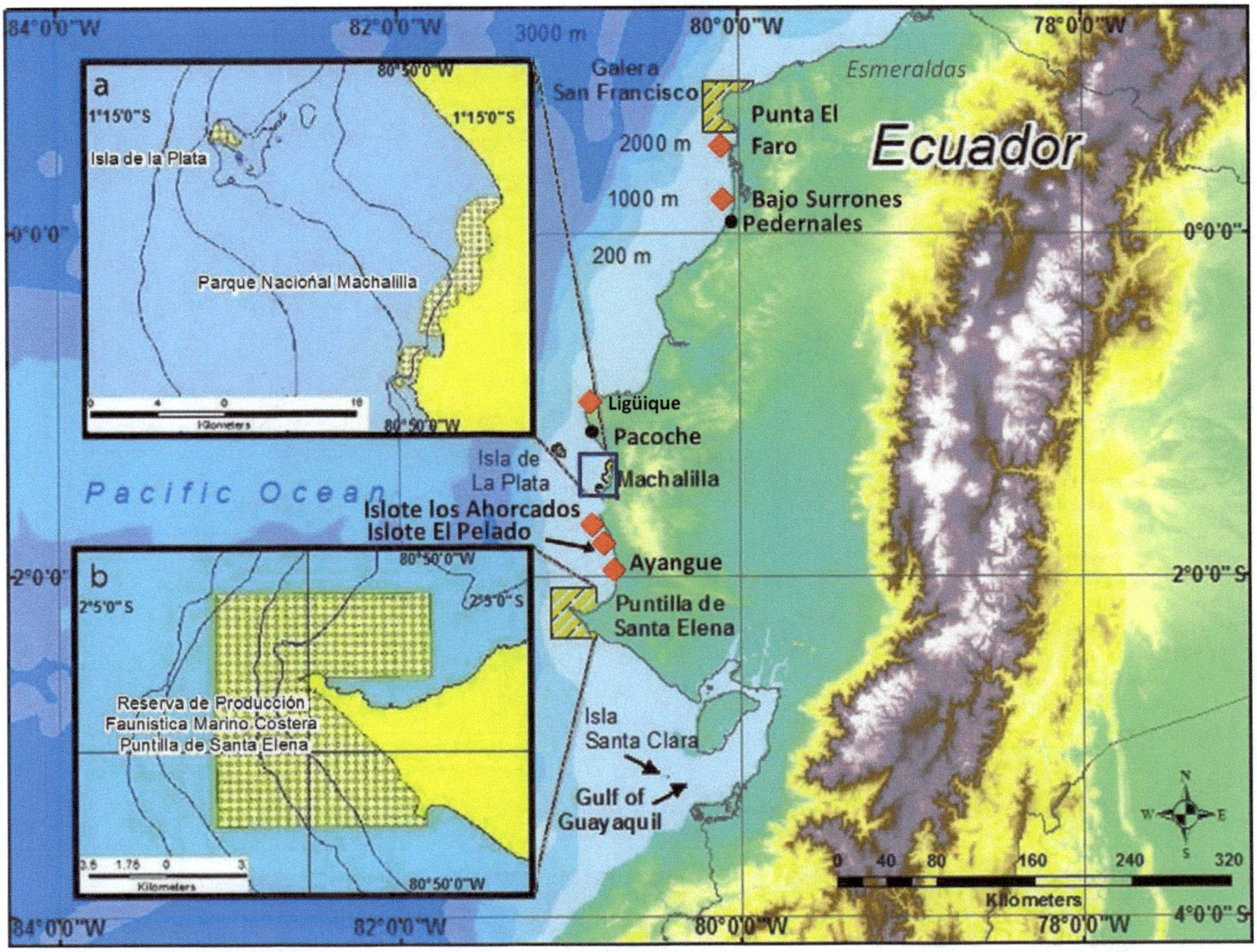

Fig. 5.24 Coastal Ecuador showing locations of three MPAs (Galera San Francisco, Machalilla, Puntilla de Santa Elena) and seven coral-rich sites (*red diamonds*) in Esmeraldas, Manabí, and Guayas provinces

before the ENSO activity of the past few decades, which has resulted in a marked reduction of coral communities and coral reefs in the southern Galápagos Islands. Since 2010, managers of the Machalilla National Park have installed nine permanent moorings in the MPA to prevent anchor damage from increasing tourism.

5.3.13 Galápagos Islands (Ecuador)

The Galápagos Islands are bathed in a confluence of westerly flowing surface currents that vary seasonally in magnitude and in latitude and are collectively referred to as the South Equatorial Current (SEC) in their passage towards the central Pacific. During years considered ENSO "neutral" the northerly island pinnacles of Wenman and Darwin (Fig. 5.26) are typically bathed in constitutively warmer waters from the southwesterly deviation of the NECC extending from the Panama Bight. It strengthens its southerly extension as trade winds from the SE weaken during the equatorial warm season months (Dec-May). The southerly and central islands, however, are more exposed to the northerly extent of the South Pacific gyre, which transports surface water from the SE Peruvian oceanic and coastal systems, often referred to as the Humboldt Current. Separated by 1–2° of latitude from the aforementioned northwestern pinnacles, the central islands are as a result on average 1–2 °C cooler throughout the year (Banks 2002) harboring a greater relative proportion of coral genera such as *Pavona* spp. that persist under a wider range of fluctuating and cooler temperatures compared to *Porites* spp. that dominate the coral communities of warmer northerly islands (Banks et al. 2009).

The northeast to southwest SST gradient is exaggerated during the cold season months (Jun–Nov) and is reduced to less than 1 °C during the warm season (Dec–May) (Figs. 5.27 and 5.28). These regional differences notably disappear during

(a)

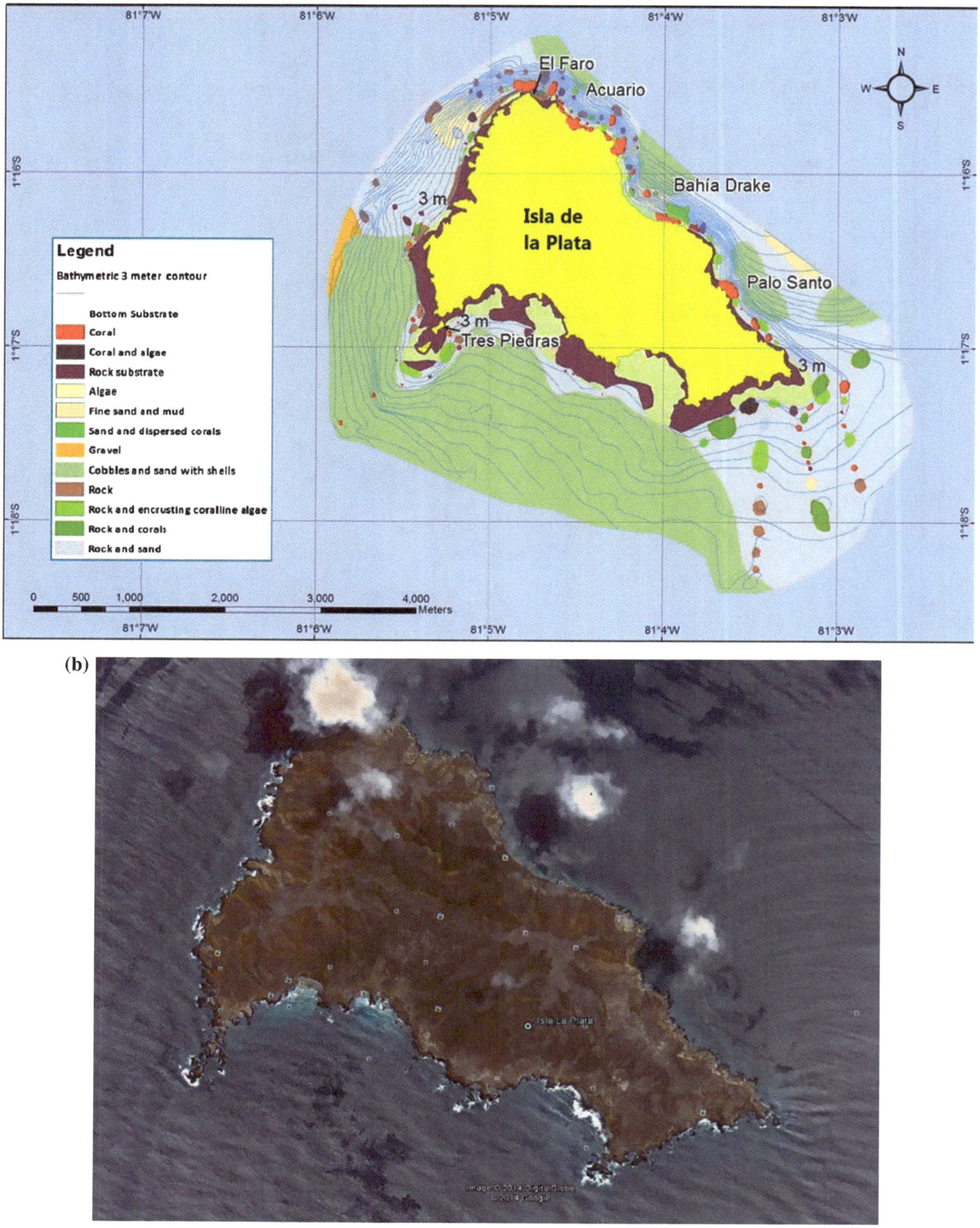

Fig. 5.25 **a** La Plata Island, Machalilla National Park, Ecuador. Substrate types are *color coded* (see legend), isobaths are shown at 3 m intervals. Modified after Rivera and Martínez (2011). **b** Satellite image of La Plata Island (2014 Digital Globe, Google Maps, April 2007)

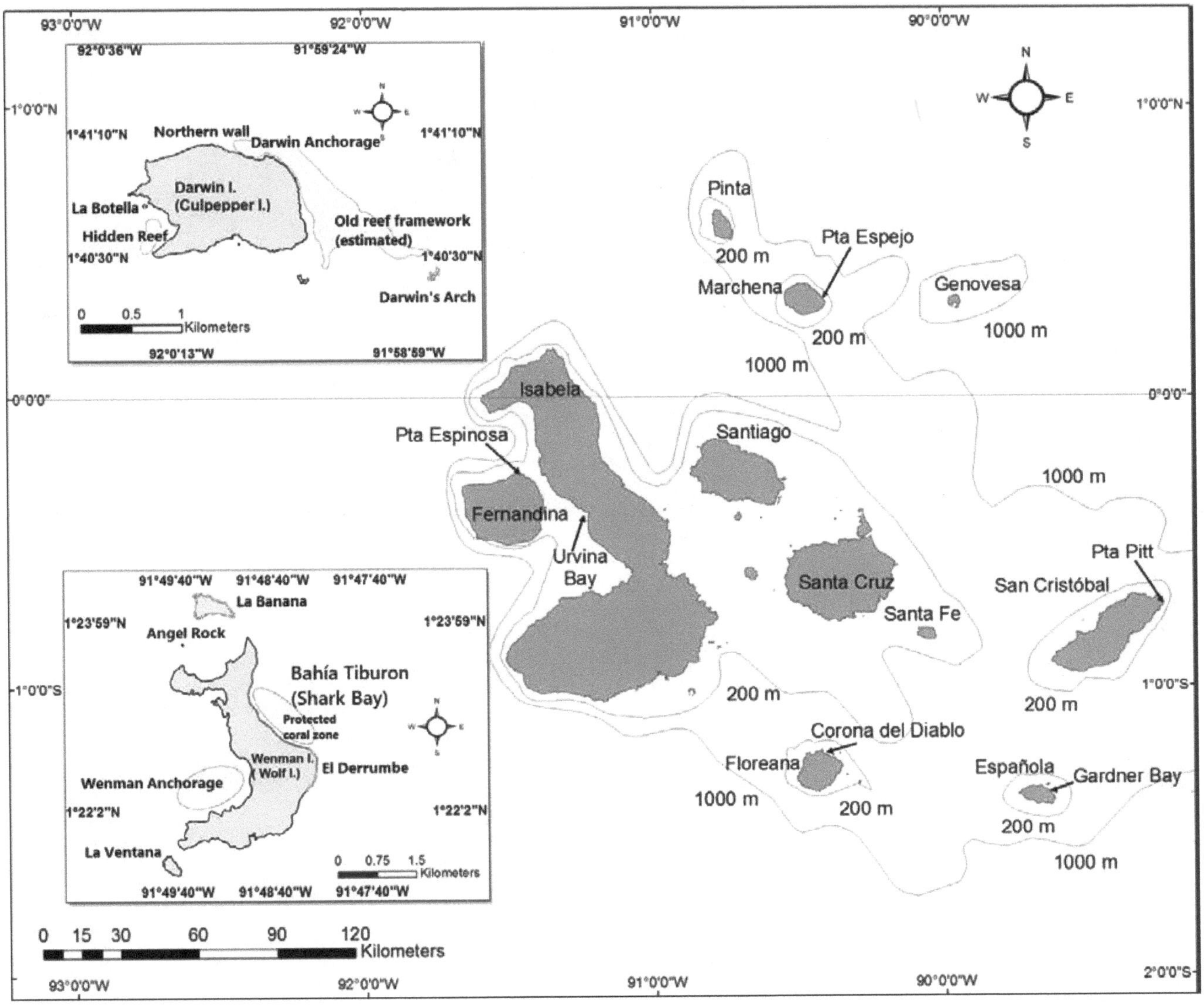

Fig. 5.26 Galápagos Islands (GAL), Ecuador. *Insets* of two northern-most islands Darwin (Culpepper) and Wenman (Wolf), located between 1° 21′N and 1° 41′N

strong El Niño years when surface temperatures are elevated, sustained and homogenized across space and seasons, and conversely emphasized during La Niña episodes when trade winds and Humboldt transport are strengthened.

Coral species richness of the southern and northern Galápagos Islands is similar in spite of the large difference in habitable substrate areas (Dawson et al. 2009). This is likely a result of high connectivity between islands. Nineteen and 21 zooxanthellate scleractinian species are recognized as members of the coral faunas of the southern and northern islands respectively (Appendix). The coral fauna of the southern islands demonstrates its strongest affinity with the Ecuadorian mainland, and that of the northern islands with mainland Colombia (Fig. 5.2). Three species seemingly restricted to the southern islands, *Pavona chiriquiensis*, *Cycloseris curvata*, and *Diaseris distorta*, are not important in reef building. Four species confined to the northern islands, *Leptoseris scabra, Pavona maldivensis, Pocillopora woodjonesi*, and *Porites evermanni*, are widespread Indo-Pacific species. All of these species except *P. woodjonesi*, which is rare, do contribute minimally to reef building in different areas of the eastern Pacific.

Development of coral communities and structural reefs is best expressed at Darwin (Fig. 5.29) and Wenman (Fig. 5.30), the two northern-most islands (Glynn et al. 2015). The largest known coral reef in the Galápagos Islands, located on the insular shelf between Darwin Arch and the eastern shore of Darwin Island, is nearly one km in length, and consists of vertically consolidated poritid stacks towering from 2 to nearly 4 m in height. While the chief frameworks

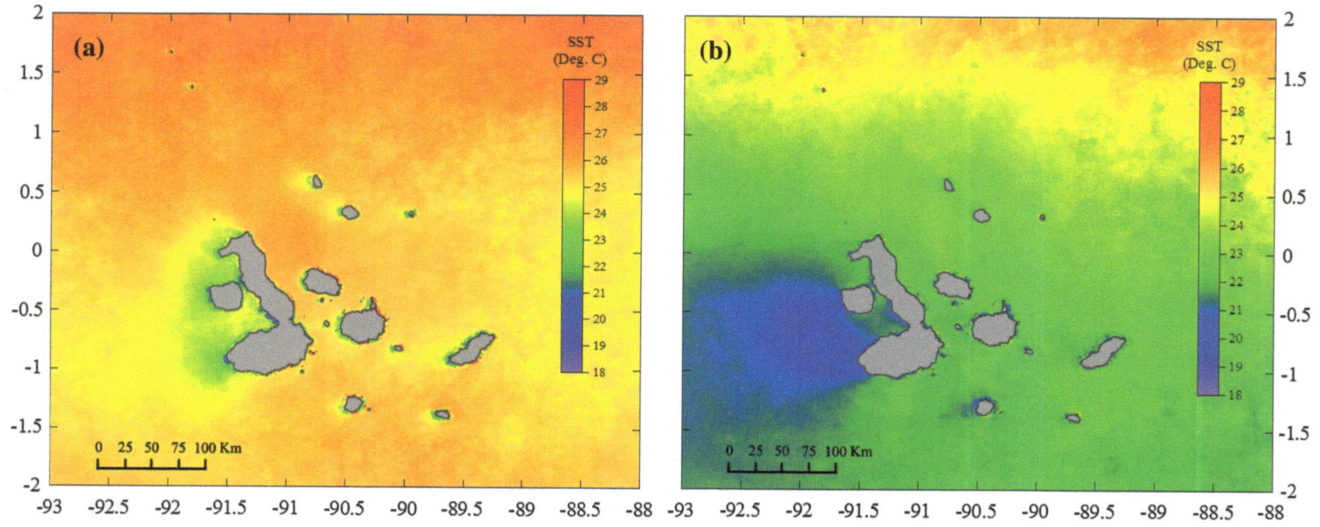

Fig. 5.27 Monthly averaged sea-surface temperatures (SSTs) across the archipelago during **a** the warm season (Dec–May) and **b** the cold season (June–Nov) derived from NASA MODIS-Aqua 1 km^2 remotely sensed data (2003–2008)

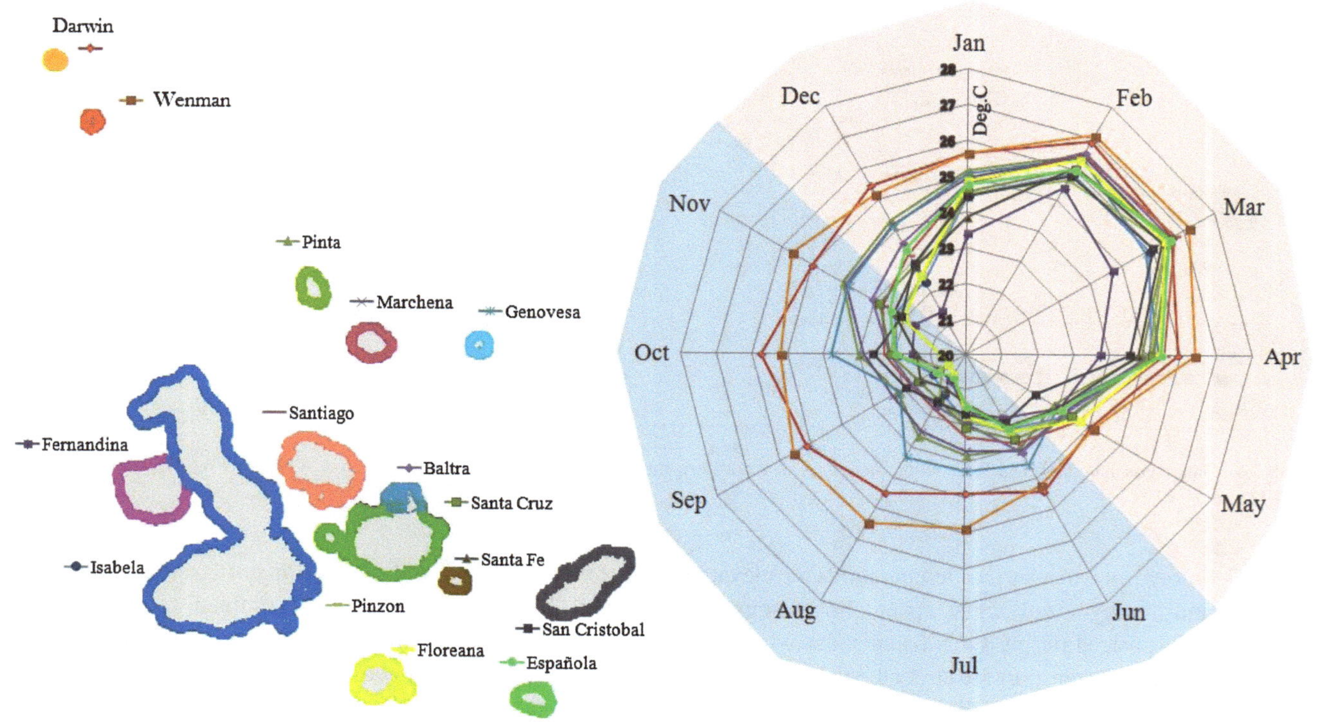

Fig. 5.28 Monthly patterns in mean SST for near-shore environments across each major island (calculated areas up to 5 km from island coastline using NASA MODIS-Aqua 1 km^2 remotely sensed data 2003–2008). Islands and respective monthly SST plots are color coded. Lowest and highest monthly SSTs occurred at Floreana in September (20–21 °C) and at Darwin in March (27–28 °C)

are at 12–16 m depth, live coral cover extends to 28 m depth along the NW/SE trending isobath. This reef has demonstrated notable recovery since the 1982–83 El Niño with live cover now exceeding 25 % over most of its length. The high cover of *Pocillopora* spp. before 1982–83 has virtually disappeared (Glynn, unpublished observation). We propose naming this impressive structure Wellington Reef, in honor of Gerard M. Wellington who performed the first island-wide marine resource surveys in the Galápagos Islands.

Unusually high coral abundances and incipient reef development characterize the narrow insular shelf along a 1 km stretch at Bahía Tiburón, Wenman Island (Fig. 5.30).

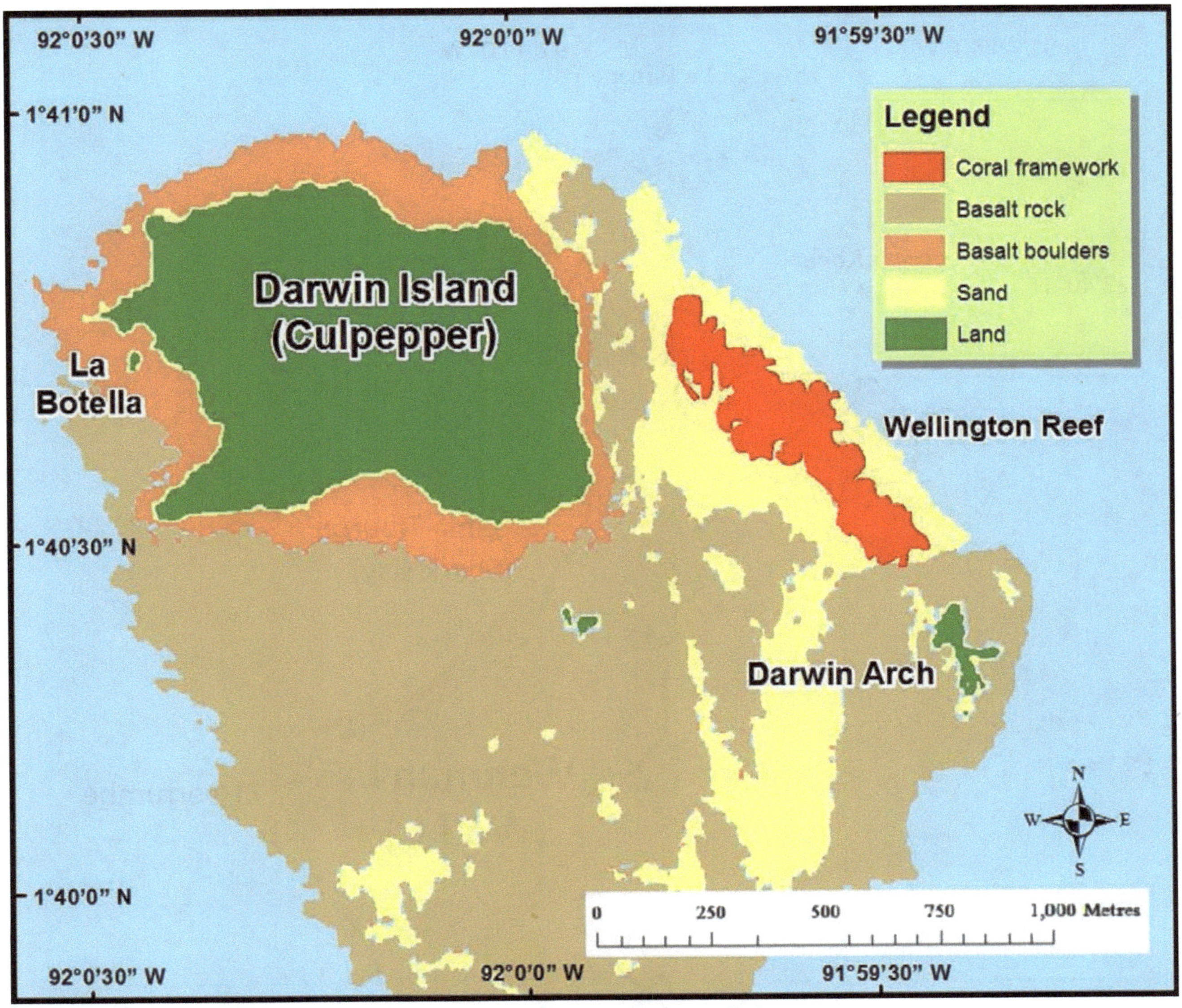

Fig. 5.29 Darwin Island, showing location of recently mapped coral framework of Wellington Reef

As at Darwin Island, massive *Porites* colonies predominate; however, large branching *Pocillopora* spp. colonies are numerous on rock boulders at shallow depths (<8 m), and *Pavona* spp. increase in abundance toward the deep (20–30 m) reef base.

Biogeographic patterns across the archipelago are greatly influenced by meso- and small-scale oceanography. Surface waters mix significantly with subsurface inputs from bathymetrically- and wind-driven upwelling processes across the archipelago. The eastward flowing Equatorial Undercurrent (EUC or Cromwell Current) shoals to 40–80 m depth over and around the Galápagos platform. The EUC brings 13–18°C water into the surface and is considered the most important oceanographic feature in maintaining extraordinary plankton productivity and unique cold water habitats, particularly in the western exposed islands of Fernandina and Isabela (Fig. 5.26).

In contrast to the enabling Panamic influence in the north, the EUC favors solitary azooxanthellate corals and rather restricts the development of reef-building corals in the western regions, generating small scale patterns of cold water mixing around western shores of the central islands. These areas are relatively spatially consistent yet quite ephemeral where EUC filaments shoal around the bathymetry and mix into the upper euphotic zone (Shaeffer et al. 2008). Since widespread mortalities during the El Niño event of 1982–83, the patterning of suitable coral habitat in the southern islands is more heterogeneous compared to that of Wolf and Darwin. Although small patches of coral communities are common across the central isles (Pinta,

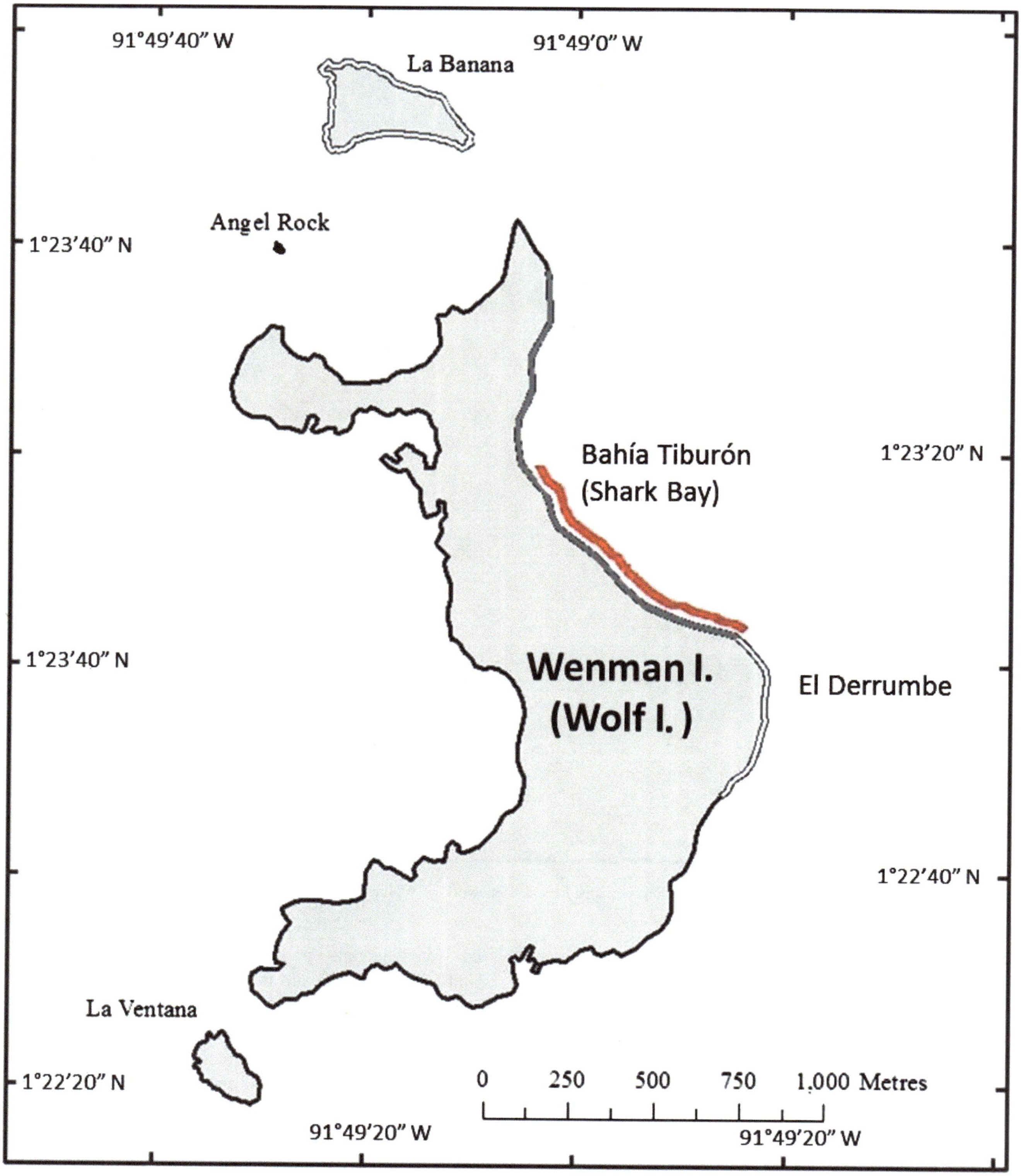

Fig. 5.30 Wenman Island, coral community development (*red line*) greatest along northeastern shelf at Shark Bay

Marchena, Genovesa), the largest extensions of zooxanthellate corals are found across the relatively sheltered northern leeward shores of Floreana Island, protected bays in eastern Santiago, Punta Pitt at the eastern end of San Cristóbal and the southerly extent of Gardner Bay at Española Island (Fig. 5.26).

An interesting feature of the western islands of Isabela and Fernandina is that they show historical evidence of large reef-forming corals in sheltered bays, despite the prevailing EUC upwelling and near absence of reef-building corals today. For example, there are large *Pavona clavus* colonies in Urvina Bay, Isabela, which was uplifted in the 1950s (Glynn and Wellington 1983; Colgan 1990, 1991; Dunbar et al. 1994). There are also reports of *Pocillopora* spp. persisting in shallow coastal pools such as those at Punta Espinosa, Fernandina, and Concha y Perla lagoon near Puerto Villamil, Isabela (Baums et al. 2014; Feingold and Glynn 2014). It seems possible that the strong stratification through solar heating at the surface in sheltered pools with low current could create a sub-habitat suitable for sporadic hermatypic coral growth. The downstream oceanic connectivity between coralline communities in the east and possible incursions of warmer water during strong El Niño years may also present opportunities for new recruitment over decadal scales.

Other unique coralline environments in the islands include extensive beds of free-living *Psammocora stellata* and *Porites lobata* composed of 3–14 cm diameter fragments across sand substrates at Punta Espejo, southeast Marchena Island (Feingold 2001). The fungiid free-living coral *Diaseris distorta* also persisted with a large population (>100,000 individuals) localized in patches across the sand sediment platform at 15–20 m depth east of Corona del Diablo (a small volcanic offshore crater at the northern tip of Floreana Island). Another fungiid, *Cycloseris curvata*, was present there in small numbers (30–40 individuals) (Feingold 1996).

Internal waves also influence local coral ecology. Kelvin waves that propagate to the east and pass in sets from the western Pacific shoal and deepen the thermocline over week-long periods in their passage across the islands. Tropical Instability Waves also form large scale front features with eddies that extend out from the upwelled equatorial cold tongue passing over the archipelago towards the west (Sweet et al. 2009). Both produce short-term changes in the water-column structure within larger seasonal signals. Kelvin waves being part of the El Niño trigger, and wider ENSO dynamic, are usually responsible for anomalous periods of alternating cool and heated surface waters that can last several weeks regardless of the season. Given strong surface heating of the upper mixed layer under the equatorial sun this can cause up to 10–12 °C thermal anomalies at 10 m depth over very short periods (4–6 days) as cold water extends into the shallow coastal zone.

Whereas warm-water stress upon corals in the region was largely evident in the widespread mortalities of the 1982–83 and 1997–98 El Niño events, cold water anomalies are pronounced and also impact coral health when combined with La Niña episodes (see Chap. 8, Glynn et al.). Significant mortalities of *Pocillopora* spp. were observed during the onset of the 2007–08 La Niña event. The upwelling phase of a set of Kelvin waves coincided with the transition into the cold season, further exaggerated by La Niña conditions both exacerbating and prolonging cold-water shock to shallow water communities across the southerly islands. The gradual recovery observed since 1995 of *Pocillopora* spp. colonies monitored in Corona del Diablo (Floreana Island) was greatly impacted between 2007 and 2009 with numbers falling from 167 colonies to just 20 (Feingold and Glynn 2014) and similar patterns were observed across the south. Visual surveys along the northeast coast of Pinta in 2010, however, showed encouraging survivorship of large *Pocillopora* colonies, likely as a result of their positioning in transitory warmer water to the north.

These effects were also observed as far north as Wolf (Wenman) and Darwin islands, which harbor the greatest remaining massive coral frameworks (Glynn et al. 1983, 2003, 2008, 2015; Banks et al. 2009). Those species that persisted after the 1982–83 and the compounding 1997–98 El Niño event exhibited an interesting potential for redundancy in functional roles between alternating periods of warm water and cold-water stress. *Pavona clavus* colonies adjacent to *Porites lobata*, for example, remained healthy during heavy bleaching of *Porites* during cold water incursons in February 2008. This may have allowed for the maintenance of key ecological services (e.g., structure, refuge, and productivity) until the recovery of the more abundant *Porites* as warmer conditions returned. Although not yet observed directly, the reverse is likely to be true under strong heating stress for *P. clavus*. These observations also lend themselves to the possibility of selection for temperature-resistant zooxanthellae clades or "disaster" endosymbionts (Correa and Baker 2011; see Chap. 13, Baker et al.) persisting in those corals that survived temperature shock during recent strong ENSO events in northerly communities.

A reduced and clustered population following El Niño mortalities also presents other concerns. During surveys from 2006–2008 the average prevalence of coral tissue abnormalities across eight sites at Wenman and Darwin was over 23 % (Vera and Banks 2009), which is at least twice as high as that reported in Caribbean and Indo-Pacific regions. The species *Porites lobata* was most affected with 35 % showing symptoms of parasitism or illness. The condition identified as *Porites* trematodiasis was most evident. These findings support observations from other areas correlating the frequency of strong bleaching events and high host densities with disease outbreaks (Bruno et al. 2007; Cróquer and Weil 2009). Background climate-change scenarios of increased ocean acidification in a region already subjected to low pH-upwelled waters also presents unqualified risks for reduced coral calcification, increased infiltration of bioeroders, and diminished reef accretion (Manzello 2010; see Chap. 18, Manzello et al.).

Nonetheless a gradual seven-year recovery trend across the northerly coral frameworks over the 12–25 m platform between Darwin Arch and the east coast of Darwin Island (Glynn et al. 2009, 2015) is also corroborated by local observations of tissue regeneration and coral recruits close to existing colonies at other sites (Banks, Personal Communications). Wenman and Darwin, despite their small geographic size, make a disproportionately large contribution to the overall Panamic and Indo-Pacific subtidal biodiversity in the islands. Galápagos park authorities, in concert with local NGOs and the Ecuadorian Navy, have begun implementing low-impact tourism anchorages to complement zoned-protected areas physically signposted in 2006 (Merlen et al. 2009). Best-practice guidelines for tourism dive operations would help reduce physical damage by inexperienced divers to reefs, while fixed bouy anchorages should help minimize the significant damage caused by anchors and chains to slow growing colonies and encourage their future recovery.

The effect of the increased human footprint in the Galápagos Marine Reserve since the 1980s and its impact on habitat-forming corals is a topic of continuing speculation. Observations since the strong El Niño events suggest that climate disturbance when coupled with certain human disturbances (such as unsustainable extraction of resource species or chronic anchor damage) can cause broad shifts in ecosystem composition and biodiversity state that further inhibit natural recovery processes (Sonnenholzner et al. 2009; Edgar et al. 2010). Significant interactions between overfishing and ENSO may be particularly important in determining extinction risk given that since 1982–83 several azooxanthellate coral species have greatly reduced populations (e.g., *Tubastrea floreana*) or are even possibly already extinct (e.g., *Rhizopsammia wellingtoni*). Significant increases in the proportion of degraded corals and the bioerosion of weakened coral frameworks by the slate pencil sea urchin *Eucidaris galapagensis* have been linked to coral degradation and observed both *in situ* and upon tectonically uplifted colonies (Colgan 1990, 1991). The recovery of coral populations is further compounded by overgrazing scenarios inhibiting recruits through the notable proliferation of the sea urchins *Tripnuestes depressus* and *Lytechinus semituberculatus*; these sea urchins may create barrens if there is a significant reduction in their natural reef predators (Sonnenholzner et al. 2009; Edgar et al. 2010). Any such biotic and indirect cascade effects will influence coral recovery intervals, population viability, and recruitment between strong El Niño events.

The 1997–98 El Niño event coincided with rapid extraction of lobster species *Panulirus gracilis* and *Panulirus penicillatus,* both candidate predators for coral burrowing molluscs such as *Lithophaga* spp. and the corallivore *Coralliophila violacea* at Wenman and Darwin Islands. Targeted and incidental fishing of intermediate-sized predatory fish with regulatory roles such as the leather bass Dermatolepis *dermatolepis*, demersal labrids and other serranids (including the exported insular endemic bacalao *Mycteroperca olfax*) also accompany both lobster and sea cucumber fishing seasons. Multivariate reconstructions of benthic coverage (Edgar et al. 2010) from early (pre-1982–83 El Niño) photographic records and subsequent photo-quadrat and visual census monitoring indicate little change in benthic community composition between 1984 immediately after the strong El Niño event and some 20 years later in 2004. The footprint when removing natural biological controls of grazers through fisheries appears to be widespread in the archipelago and has resulted in high abundance of encrusting calcareous algae such as *Lithothamnium* sp. indicative of sea urchin barrens. Observations of increased *Eucidaris* sea urchin densities near to fishing ports (Sonnenholzner et al. 2009; Edgar et al. 2010) appears to support this hypothesis with analogies reported in other regions such as Hong Kong (Qui et al. 2014).

With new coastal-zoning reforms being considered, GMR coral communities present a curious dichotomy. They are particularly vulnerable before acidification and cascade effects after removal of regulatory species in their vicinity, yet are also a naturally resilient habitat subject to intense changes conditioned by a unique oceanography and millennia of ENSO fluctuations in the region. Moderation and mitigation of these effects through monitoring and feedback into management certainly deserve careful consideration given that novel climate stress is likely to be deterministic in the future survivorship of remaining Galápagos corals and the diverse communities they support.

5.3.14 Rapa Nui (Easter) and Salas y Gómez Islands (Chile)

Rapa Nui (Easter) and Salas y Gómez islands (both included under RPN, Appendix), the latter slightly more than 400 km east of Rapa Nui, share a common and relatively species poor shallow marine biota (Fig. 5.31). Two comprehensive reports by Castilla and Rozbaczylo (1987) and DiSalvo et al. (1988) offer excellent introductions to the marine habitats and biota of Rapa Nui. Studies focusing on the coral faunas and communities became available in the early to mid-2000s (Wellington et al. 2001; Hubbard and Garcia 2003; Glynn et al. 2003, 2007).

Only 15 species of zooxanthellate corals are known from the larger and more speciose Rapa Nui with nearly half of these in the genus *Pocillopora*. *Pocillopora verrucosa* exhibits high cover over mid-depths at several sites on the west coast and at Motu-Iti and Motu-Nui at the southern tip of Rapa Nui (Fig. 5.32). *Porites lobata* is the single poritid species present, and makes up the bulk of the coral cover

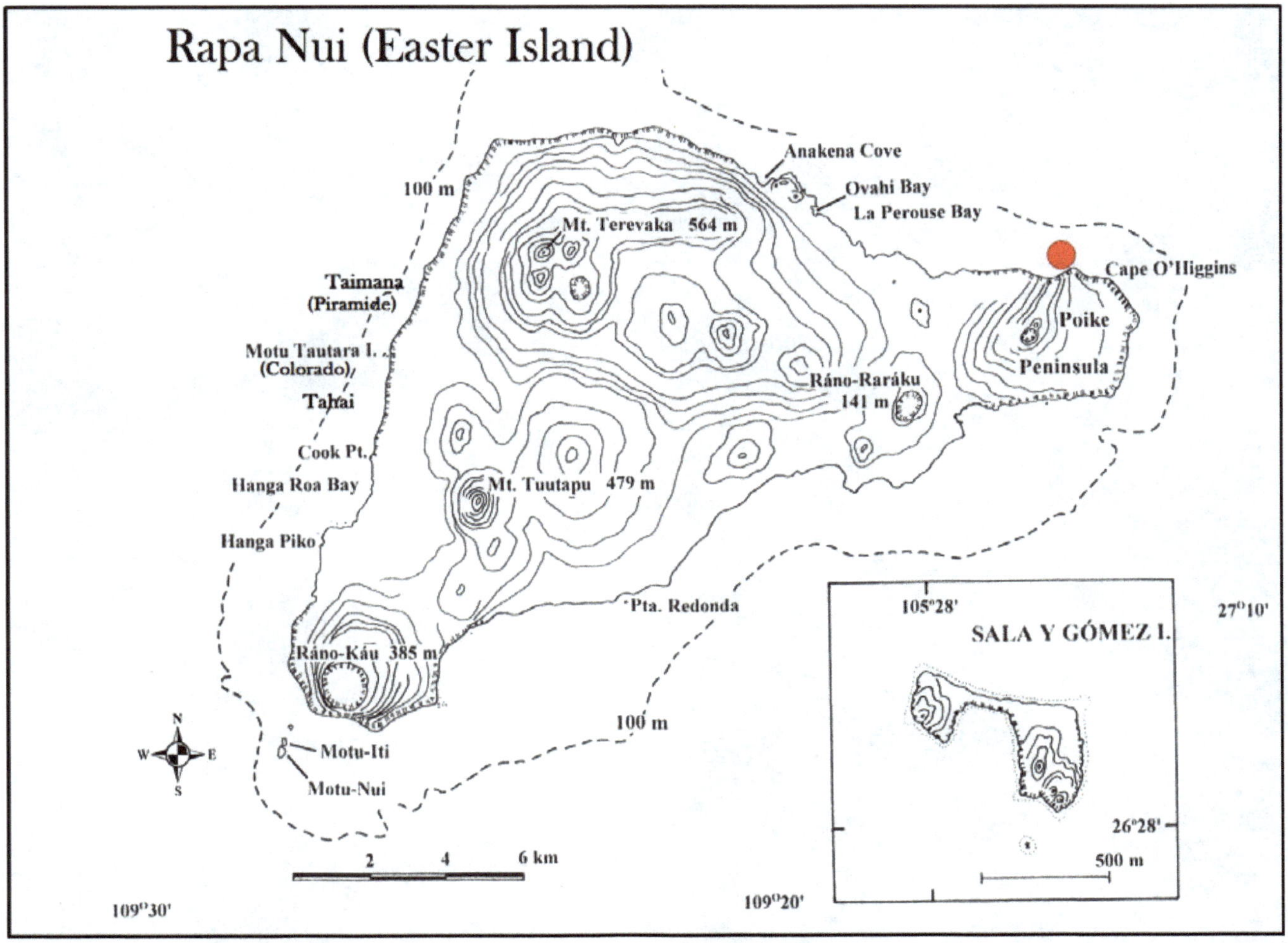

Fig. 5.31 Rapa Nui (Easter Island), Chile. Inset, Salas y Gómez Island, ~500 km east of Easter Island. Red dot west of Cape O'Higgins denotes location of coral reef framework (Hubbard and García 2003)

across all depths. Unlike its colony growth form elsewhere in the eastern Pacific, which is smooth to nodular and dome-shaped, at Rapa Nui the corallum takes on a variety of forms—columns, blades, spikes, cones, and marginal tabulae (Glynn et al. 2003). *Pavona* species, so prominent at other eastern Pacific sites, are absent from the Rapa Nui islands group. However, two species of faviid corals, *Leptastrea purpurea* and *Leptastrea transversa*, are present but they are minor components of RPN coral communities. These are the only members of the Faviidae known from the eastern Pacific region. While endemism is high in some groups (e.g., molluscs, 42 %; fishes, 77 %; Friedlander et al. 2013), all zooxanthellate scleractinian corals, except for the two faviid species, are widely distributed throughout the eastern Pacific and central/south Pacific regions (Appendix). Ongoing molecular genetic studies, however, suggest re-classification of *Pocillopora verrucosa* as a separate endemic species (D. A. Paz-García, E. Wieters, F.J. García-de-León, E.F. Balart, unpublished data).

Traditionally, Rapa Nui has been considered as the eastern-most outpost of the Polynesian or southeast Pacific region (e.g., Veron 1995). However, an analysis of coral species similarities (presence/absence records) between the Pitcairn Island group and Rapa Nui indicates that this locality is more closely aligned with the eastern Pacific faunal group (Glynn et al. 2007). In support of this result, a dispersal barrier between Rapa Nui and the Pitcairn Islands was revealed through satellite-tracked surface drifter buoys. Surface currents were found to flow from east to west, forming an effective barrier to the eastward dispersal of planktonic larvae that occupy the upper surface of the ocean. Most recently, Veron et al. (2015) also concluded that Rapa Nui has its strongest affinity with the far eastern scleractinian zooxanthellate fauna. Reef-building coral species richness is

Fig. 5.32 *Pocillopora verrucosa* prominence in a mid-depth (12–15 m) coral community at Taimana, Rapa Nui west coast, 12 December 2012 (photo courtesy of Alba Medrano)

notably higher in the Pitcairn's Islands than at Rapa Nui. Veron (1995) reported 61 species in the Pitcarins; recent surveys have increased this number to 87 (Appendix).

Vibrant coral communities with high coral cover (40–80 %) characterize most sites along the now well-sampled western and northeast coasts of Rapa Nui, whereas corals appear to be scarcer on the less-studied, wave-battered southeast coast that is directly exposed to major oceanic swells (DiSalvo et al. 1988). Considering the high live coral cover at favorable sites, it is not surprising that incipient reef development has been observed. Massive *Porites* colonies in particular have relatively high skeletal extension rates, about 1 cm year^{-1} (Glynn et al. 2003), are longevous, and attain enormous dimensions (Figs. 5.33, 5.34 and 5.35). *Porites* frameworks were observed by Hubbard and Garcia (2003) on the western and northern sides of the island. The largest of these is found northeast of Poike Peninsula, off Cape O'Higgins. Some coral buildups on elevated basalt substrates appear deceptively like true structural reefs (Fig. 5.34). These shelf locations are protected from severe wave assault from the southeast, originating in high latitude southern seas. Even though pocilloporid corals often occur as dense aggregations at sheltered sites, they do not form interlocking frameworks as at other ETP localities.

In general, two species, *Porites lobata* and *Pocillopora verrucosa*, account for the vast majority (95 %) of total coral cover across the island. While typically scarce, encrusting colonies of *Leptastrea* spp. can reach up to 10 % cover at some locations near Ovahe and Tahai. There they occur intermixed with *Porites* and *Pocillopora* in large patches

Fig. 5.33 Single colony of *Porites lobata*, ~10 m in diameter, off Anakena, 17 m depth. Photographed in August 1998 by Henry Tonnemacher, courtesy Dennis K Hubbard

(10s of square meters) on flat, mid-depth (10–12 m) platforms. Although still nominal in terms of their contribution to total coral cover, a notable increase in the prevalence of *Pocillopora eydouxi* has been observed over the past five years, particularly along the west coast (E. Wieters and H. Garcia, personal observation).

Generally, weak depth zonation patterns characterize changes in the relative abundances of the dominant coral community functional groups (Friedlander et al. 2013; Wieters et al. 2014). Total coral cover is lowest at shallow depths (<5 m), however, where branching *Pocillopora* spp. are nearly absent, massive *Porites* are predominant (average 38 % cover). Below this, *Pocillopora* increase in abundance to peak at mid depths (15–20 m), where their relative contribution to total coral cover is similar to that of *Porites* at most sites (average 38 and 36 %, respectively). Massive *Porites* again dominate in deeper habitats (>20 m), and *Pocillopora* become less abundant, but still reach 10–20 % cover. Abundant live coral extends well below the 30 m depth limit with *Pocillopora* and *Porites* known to occur frequently to 60 m depth.

Across the northeast and west coasts, shifts in the relative abundance of species cause spatial differentiation in the functional structure of coral-dominated benthic communities and fish assemblages (Wieters et al. 2014). Though changes are best described by multivariate characterization of communities based on functional groups, some of the most apparent differences in terms of species are due to relatively higher abundances of *Pocillopora* and some macroalgal groups along the northwestern coast and tip of the island, as well as variable presence and abundances of the reef fish *Centropyge hotumatua*, *Acanthurus leucopareius* and *Chrysiptera rapanui*. Further studies have also identified contrasting communities at similar depths, with a preponderance of either corals or macroalgae (predominantly *Padina*). These alternative, coral/algal community configurations can exist

Fig. 5.34 *Porites lobata* community, off Anakena, 30 m depth. Photographed in August 1998 by Henry Tonnemacher, courtesy Dennis K. Hubbard

side-by-side (within hundreds to thousands of meters), and commonly present sharp boundaries between them, which remain unaltered for relatively long periods of time.

Published studies concerning ecological processes maintaining coral-dominated communities at Rapa Nui are virtually nonexistent. Ongoing experimental and comparative studies at a few locations suggest that, like elsewhere, grazing is a critical process involved in preventing macroalgae from overgrowing live corals. Evidence thus far also suggests grazing is especially effective when corals have been previously damaged and weakened by thermally-induced bleaching (Wieters, unpublished data). At Rapa Nui, where herbivorous fishes are generally small-bodied and in low abundance, sea urchins (*Diadema*) appear to be particularly important. While no previous studies have documented ecological effects or even the diet of benthic herbivores at Rapa Nui, comparisons with nearby Salas y Gomez reveal between-island differences in abundances of sea urchins and macroalgae (Friedlander et al. 2013), further suggesting the potential for a critical top-down role by these benthic herbivores. Whether the effects and relative importance of sea urchins are currently magnified at Rapa Nui, due to the very apparent overfishing that seems responsible for reduced abundances of carnivorous fishes and other large predators, is not known. Obvious sea urchin-induced erosion of dead corals and limestone deposits is limited to a few specific locations of particularly high sea urchin densities (mean 19.0 ind. $m^{-2} \pm 2.6$ SE), but there is no evidence that sea urchins cause mortality of established live colonies (Wieters, personal observation).

Severe, island-wide coral bleaching at Rapa Nui is known to have occurred twice in recent years: (1) following torrential rains and associated increased sedimentation in 1980 (Egaña and DiSalvo 1982), and (2) following prolonged sea warming associated with La Niña in 2000 (Wellington et al. 2001). This exceptionally strong La Niña event, centered at and above the equator in the northern hemisphere, resulted in two month-long elevated SSTs in the southern ocean, that extended over 73° of latitude from Rapa Nui to Fiji. During this event the bleaching threshold at Rapa Nui (26.1 °C) was

Fig. 5.35 Veneer of *Porites lobata* colonies overtopping a lava tube off Motu Tautara, ~25 m depth. Photographed in August 1998 by Henry Tonnemacher, courtesy of Dennis K. Hubbard

exceeded for 9–10 weeks in February and March. This thermally-driven disturbance led to high mortality of *Pocillopora* spp. to 20 m depth throughout the island. Within one year after the disturbance, greater than 80 % of *Pocillopora* colonies inhabiting shallow habitats (<9 m) were dead and a clear pattern of decreased mortality with increasing depth was documented (Fig. 5.36). While *Porites* experienced generally similar levels of bleaching compared with *Pocillopora* in 2000, the former appeared much more resistant with far fewer colonies having succumbed a year later. Initial recovery of *Pocillopora* was uneven spatially, with increased prevalence of coral tissue abnormalities and continued mortality observed up to 5 years after the massive bleaching at some sites, particularly in mid- to deep habitats (>13 m; compare mortality levels in 2001 and 2005, Fig. 5.36). The factors inhibiting or delaying recovery and causing the continued declines of *Pocillopora* at some sites are unknown, but preliminary evidence suggests that depth-dependent resistance and other disturbances can be important. Despite the extensive bleaching, large scale mortality of one of the dominant species, and among-site variability in recovery, overall recovery of coral communities to date has been impressive, with total coral cover now similar to (deep habitats) or above (shallow habitats) the pre-bleaching levels (Wieters et al. 2014).

While no massive bleaching event similar to that documented in 2000 has been observed recently (as of 2014), regular surveys over the past 5 years show that bleaching of *Pocillopora* and *Porites* occurs every year. While the intensity/prevalence varies greatly among sites and years, the regular decline in zooxanthellae densities and/or degraded pigmentation suggests that these corals are frequently subjected to stressful conditions. This relatively mild bleaching rarely appears to cause coral mortality directly, but does appear to have consequences for other species in the community. Bleached colonies of both

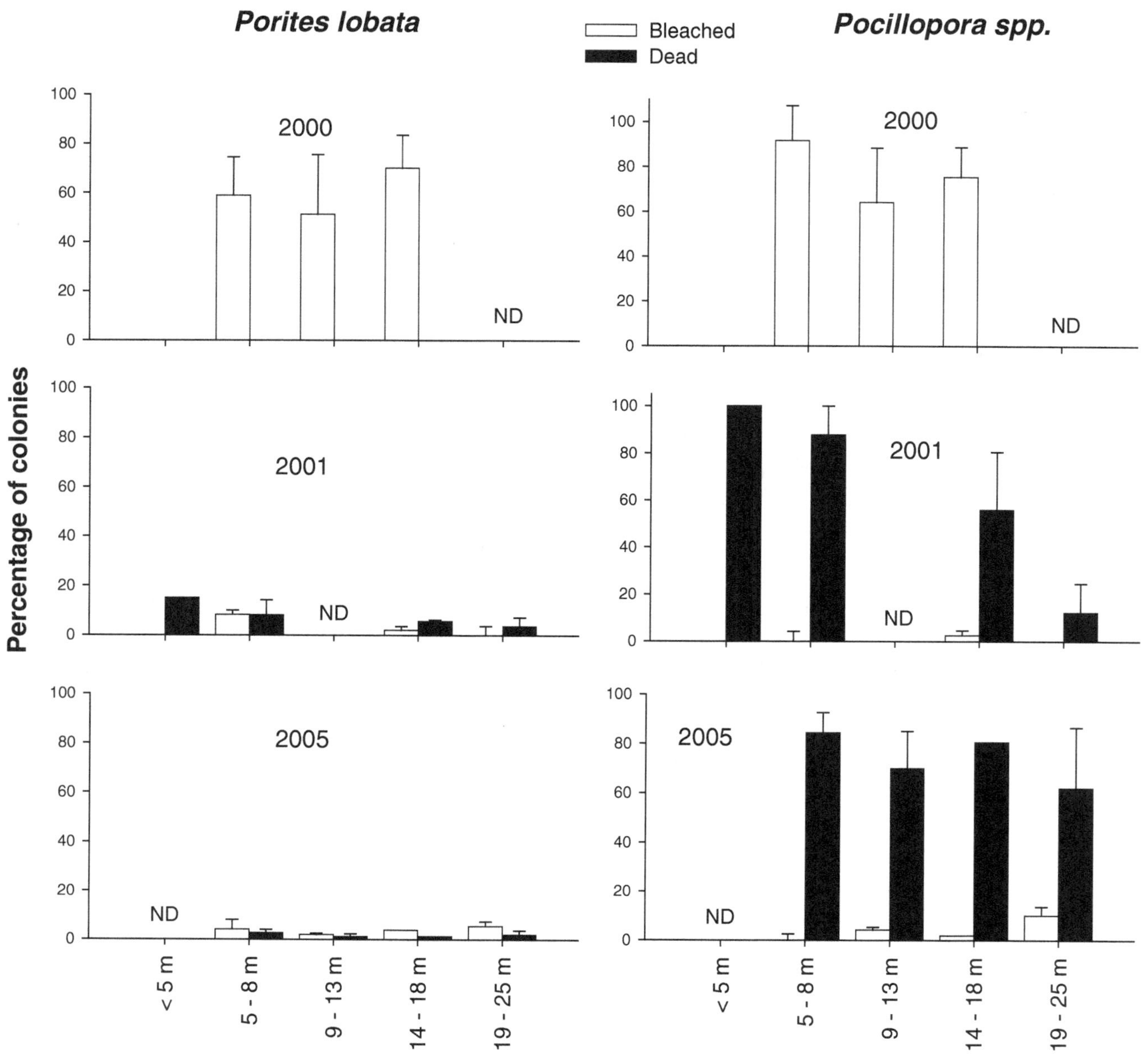

Fig. 5.36 Mean percentage (±SE) of *Porites* and *Pocillopora* colonies bleached or recently dead across varying subtidal depth zones. *Error bars* represent variation among sites 1–10s of km apart. *ND* no data available

Pocillopora and *Porites* are more likely to harbor associated fauna than their neighboring "healthy" (fully pigmented) counterparts. Moreover, the abundance and species richness of fauna are higher on bleached (but live) corals (Fig. 5.37). Increase in macroalgal abundance also occurs during years of more prevalent bleaching, though susceptibility of corals to overgrowth varied among sites and appears to be associated with local coral energy reserves (lipid concentration, Fig. 5.38). These results suggest that disruptive stress that causes coral bleaching changes the habitat-modifying role of branching and massive corals as well as the intensity of competition with algae under natural conditions in the field. These hypotheses are currently being tested.

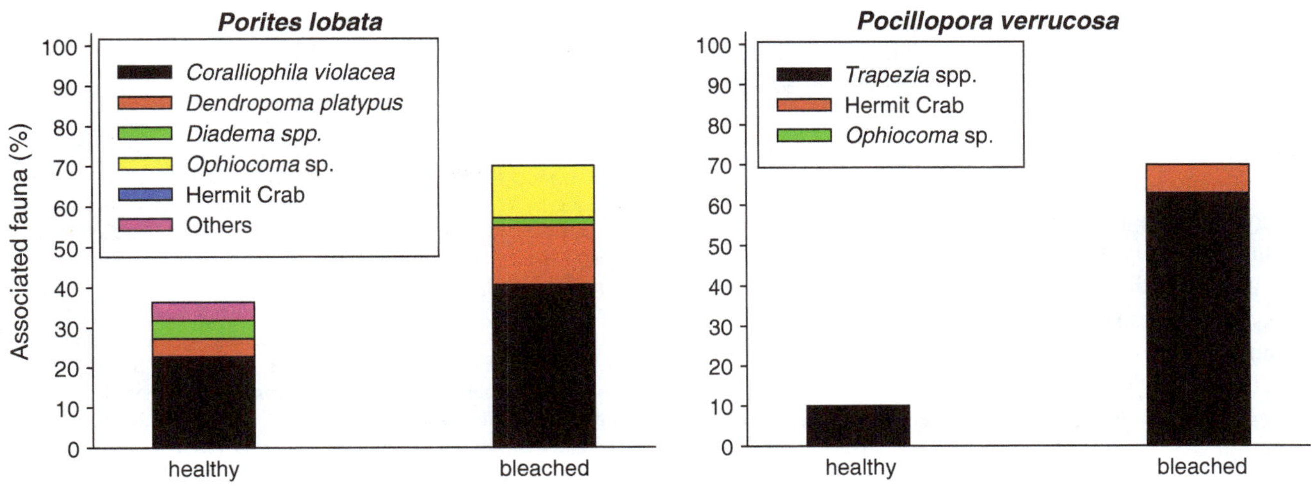

Fig. 5.37 Percentage of healthy and bleached *Porites* and *Pocillopora* colonies with diverse associated species of invertebrate functional groups. 'Others' include infrequently observed taxa such as sponges, polychaete worms, turrid snails, caridean shrimp, and non-trapeziid brachyuran crabs. A minimum of 50 colonies of each species and condition was assessed at mid-depth (10–15 m) across four sites on the west coast of Rapa Nui

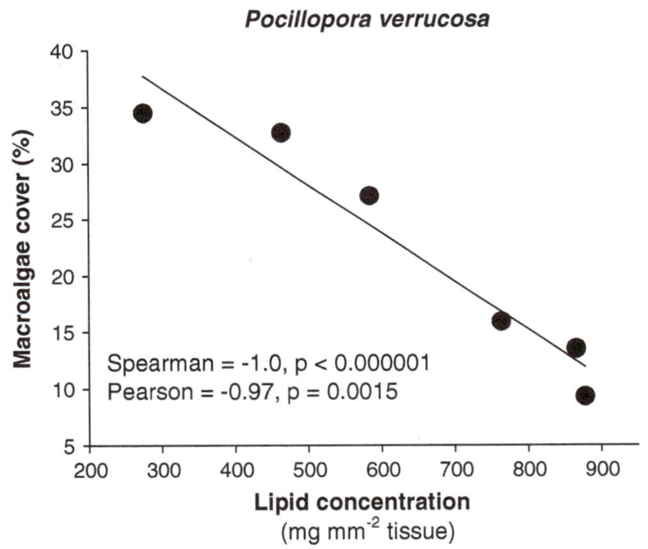

Fig. 5.38 Relationship between mean percentage cover of macroalgae and mean lipid concentration of *Pocillopora verrucosa* colonies at mid-depth (10–15 m) habitats during years of mild bleaching (>30 % cover bleached corals) across western and northeastern shores of Rapa Nui. *Symbols* are sites 1 to 10s of km apart

5.4 Conclusions

Coral reef-building potential exists across the entire eastern tropical Pacific region, from the Gulf of California to Rapa Nui, regardless of low species richness at the northern and southern latitudinal limits of coral distribution. With the exception of newly discovered ETP coral formations, e.g. El Salvador and Nicaragua, and small habitable sites, e.g. Malpelo Island, coral species richness is relatively uniform at 20–25 species per locality across the eastern Pacific region. The presently recognized total species count of zooxanthellate corals in the ETP region is 47.

Concentrated reef development occurs chiefly on islands of the continental shelf and on distant oceanic islands, but also at some environmentally suitable coastal areas such as Huatulco (Mexico), along the shores bordering the southern Gulf of Papagayo and Golfo Dulce (Costa Rica), and at Machalilla (Ecuador). Massive colonies of *Porites*, *Pavona*, and *Gardineroseris* are usually the dominant reef-building corals in ETP oceanic island environments such as Clipperton Atoll, Isla del Coco, the northern Galápagos Islands, and Rapa Nui.

Coral species richness is relatively high in the northern latitudes of mainland Mexico and offshore in the Revillagigedo Islands, and farther south in the equatorial eastern Pacific, from Costa Rica to Ecuador. Reef building occurs in upwelling centers (Huatulco, Gulf of Papagayo, Gulf of Panama), and in several areas of the equatorial eastern Pacific, which experience aperiodic severe ENSO disturbances.

Thomas Dana's hypothesis of the dispersal of central/west Pacific corals into the ETP via the NECC is indirectly supported by analyses of faunal affinities and molecular genetic evidence. Species-rich Palmyra Atoll in the northern Line Islands is in the path of the easterly flowing NECC and could likely serve as a source of coral propagules to be transported to the equatorial eastern Pacific. Surface current transport from the Line Islands via the NECC to Clipperton Atoll (a stepping stone) is also possible.

Potentially important corallivore and bioeroder species have caused greatest damage to corals after El Niño warming/bleaching events. The crown-of-thorns seastar *Acanthaster planci* is moderately abundant in only three ETP areas—Gulf of California and Revillagigedo Islands (Mexico), and Gulf of Chiriquí (Panama), and has not been reported to overwhelm coral cover anywhere in the ETP. Population outbreaks of the black long-spined sea urchin, *Diadema mexicanum*, have been reported at several localities, usually following disturbances and high coral mortality, resulting in extensive bioerosion of reef frameworks and massive coral colonies. The slate pencil sea urchin *Eucidaris galapagensis* occurs at high population densities in the southern Galápagos Islands, where it has eroded reef frameworks and prevented the recovery of coral communities devastated by the severe 1982–83 El Niño event.

Competition for space among corals and macroalgae has been documented locally in El Salvador (*Acanthophora spicifera* vs. *Porites lobata*), Costa Rica (*Caulerpa sertularioides* vs. *Pocillopora* spp., *Psammocora stellata*) and Panama (*Caulerpa racemosa*, *C. sertularioides* vs. *Psammocora stellata*). Coral patches are sometimes overgrown and killed by algae, but such competitive dominance is usually ephemeral, with plant cover eventually subsiding and corals re-populating previously occupied substrates.

Despite diverse natural physical and biotic disturbances at both local (e.g., upwelling centers, bioeroder outbreaks) and regional (severe ENSO episodes) scales, ETP coral communities have generally demonstrated a high degree of resilience, with robust recovery at several localities in recent years.

Acknowledgments The following organizations facilitated research in the ETP: Mexico—Consejo Nacional de Ciencia y Tecnología (CONACYT) project numbers 108302 and 183534; Costa Rica—Centro de Investigación en Ciencias del Mar y Limnología (CIMAR), University of Costa Rica; Panama—University of Panama, Smithsonian Tropical Research Institute (STRI); Colombia—Colciencias, Conservation International, Fundación para la Promoción de la Investigación y la Tecnología del Banco de la República, Instituto de Investigaciones Marinas y Costeras (INVEMAR), Parques Nacionales Naturales, Universidad del Valle, World Wildlife Fund, and members of the Coral Reef Research Group at Universidad del Valle; Ecuador—Charles Darwin Foundation, Galápagos National Park Service, Charles Darwin Research Station; Conservation International; Khaled bin Sultan Living Oceans Foundation, Saudi Arabia; Chile, Rapa Nui—Henri and Michel Garcia (ORCA Diving Center), H. Buck-Weise, I. Burgues, A. Medrano, T. Navarrete Fernandez, A. Perés-Matus, FONDECYT grant no. 1130167, and Center for Marine Conservation Nucleo Milenio Initiative P10-033F; research at several ETP sites was supported by the U.S. National Science Foundation (Biological Oceanography Program) and the U.S. National Geographic Society. Assistance with coral distributional records and species identifications was kindly offered by I.B. Baums, E.H. Borneman, S.D. Cairns, D. Fenner, D.K. Hubbard, and J.E.N. Veron. Joshua Levy helped with the graphics, tables, and organization of the text, and Rafael Araujo with editorial advice. The organization and clarity of this overview benefitted greatly from suggestions offered by A.C. Baker, W.M. Goldberg, and L.T. Toth. Finally, we are all indebted to the late John W. Wells, directly or indirectly, for showing us the way in the study of eastern Pacific corals.

Appendix: Coral Species Presence (1)/Absence (0) Data Matrix

Presence/absence of zooxanthellate scleractinian corals at 15 eastern Pacific and four central/south Pacific localities. Recent sources for each locality are listed in the caption to Fig. 5.7. No species re-assignments were attempted for the central/south Pacific taxa listed. *Pocillopora verrucosa* and *Pocillopora elegans* are treated as separate species. GOC, Gulf of California; TMM, tropical mainland Mexico; REV, Revillagigedo Islands; CLP, Clipperton Atoll; ELS, El Salvador; NIC, Nicaragua; CRM, Costa Rican mainland; COC, Cocos Islands; PAN, Panama; COL, Colombia; MAL, Malpelo Island; ECD, Ecuador; GLS, southern Galápagos Islands; GLN, northern Galápagos Islands; RPN, Rapa Nui (Easter Island); HAW, Hawaii Is., including the NW Is.; JOH, Johnston Atoll; PIT, Pitcairn Islands, including Pitcairn, Henderson, Oeno, and Ducie Islands; PAL, Palmyra Atoll.

Species	Locality																		
	GOC	TMM	REV	CLP	ELS	NIC	CRM	COC	PAN	COL	MAL	ECD	GLS	GLN	RPN	HAW	JOH	PIT	PAL
Acropora abrotanoides	0	0	0	0	0	0	0	0	0	0	0	0	0	0	0	0	0	1	1
Acropora aculeus	0	0	0	0	0	0	0	0	0	0	0	0	0	0	0	0	0	0	1
Acropora acuminata	0	0	0	0	0	0	0	0	0	0	0	0	0	0	0	0	0	1	1
Acropora aspera	0	0	0	0	0	0	0	0	0	0	0	0	0	0	0	0	0	0	1
Acropora austere	0	0	0	0	0	0	0	0	0	0	0	0	0	0	0	0	0	1	0
Acropora cerealis	0	0	0	0	0	0	0	0	0	0	0	0	0	0	0	0	1	1	1
Acropora cf. *solitaryensis*	0	0	0	0	0	0	0	0	0	0	0	0	0	0	0	0	0	1	0
Acropora clathrata	0	0	0	0	0	0	0	0	0	0	0	0	0	0	0	0	0	0	1
Acropora cymbicyathus	0	0	0	0	0	0	0	0	0	0	0	0	0	0	0	0	0	0	0
Acropora cytherea	0	0	0	0	0	0	0	0	0	0	0	0	0	0	0	1	1	1	1
Acropora delicatula	0	0	0	0	0	0	0	0	0	0	0	0	0	0	0	0	0	0	0
Acropora digitifera	0	0	0	0	0	0	0	0	0	0	0	0	0	0	0	0	0	1	1
Acropora divaricata	0	0	0	0	0	0	0	0	0	0	0	0	0	0	0	0	0	1	1
Acropora elseyi	0	0	0	0	0	0	0	0	0	0	0	0	0	0	0	0	1	0	1
Acropora florida	0	0	0	0	0	0	0	0	0	0	0	0	0	0	0	0	0	0	1
Acropora formosa[a]	0	0	0	0	0	0	0	0	0	0	0	0	0	0	0	0	0	0	1
Acropora gemmifera[b]	0	0	0	0	0	0	0	0	0	0	0	0	0	0	0	1	1	1	1
Acropora glauca	0	0	0	0	0	0	0	0	0	0	0	0	0	0	0	0	0	0	0
Acropora globiceps	0	0	0	0	0	0	0	0	0	0	0	0	0	0	0	0	0	1	1
Acropora granulosa	0	0	0	0	0	0	0	0	0	0	0	0	0	0	0	0	0	0	1
Acropora humilis	0	0	0	0	0	0	0	0	0	0	0	0	0	0	0	1	1	1	1
Acropora hyacinthus	0	0	0	0	0	0	0	0	0	0	0	0	0	0	0	0	0	1	1
Acropora latistella	0	0	0	0	0	0	0	0	0	0	0	0	0	0	0	0	0	1	1
Acropora listeri	0	0	0	0	0	0	0	0	0	0	0	0	0	0	0	0	0	1	0
Acropora longicyathus	0	0	0	0	0	0	0	0	0	0	0	0	0	0	0	0	0	0	1
Acropora loripes	0	0	0	0	0	0	0	0	0	0	0	0	0	0	0	1	1	1	1
Acropora lutkeni	0	0	0	0	0	0	0	0	0	0	0	0	0	0	0	0	0	1	0
Acropora microclados	0	0	0	0	0	0	0	0	0	0	0	0	0	0	0	0	0	1	1
Acropora microphthalma	0	0	0	0	0	0	0	0	0	0	0	0	0	0	0	0	0	1	0
Acropora monticulosa	0	0	0	0	0	0	0	0	0	0	0	0	0	0	0	0	0	0	1
Acropora multiacuta	0	0	0	0	0	0	0	0	0	0	0	0	0	0	0	0	0	0	1
Acropora nana	0	0	0	0	0	0	0	0	0	0	0	0	0	0	0	0	0	0	1
Acropora nasuta	0	0	0	0	0	0	0	0	0	0	0	0	0	0	0	1	1	1	1
Acropora palmerae	0	0	0	0	0	0	0	0	0	0	0	0	0	0	0	1	0	0	0

(continued)

Species	Locality																		
	GOC	TMM	REV	CLP	ELS	NIC	CRM	COC	PAN	COL	MAL	ECD	GLS	GLN	RPN	HAW	JOH	PIT	PAL
Acropora paniculata	0	0	0	0	0	0	0	0	0	0	0	0	0	0	0	1	1	0	1
Acropora pocilloporina	0	0	0	0	0	0	0	0	0	0	0	0	0	0	0	0	0	1	0
Acropora polymorpha	0	0	0	0	0	0	0	0	0	0	0	0	0	0	0	0	0	0	0
Acropora polystoma	0	0	0	0	0	0	0	0	0	0	0	0	0	0	0	0	0	0	1
Acropora reticulata	0	0	0	0	0	0	0	0	0	0	0	0	0	0	0	0	0	0	0
Acropora retusa	0	0	0	0	0	0	0	0	0	0	0	0	0	0	0	0	1	0	0
Acropora robusta	0	0	0	0	0	0	0	0	0	0	0	0	0	0	0	0	0	1	1
Acropora samoensis	0	0	0	0	0	0	0	0	0	0	0	0	0	0	0	0	0	1	1
Acropora secale	0	0	0	0	0	0	0	0	0	0	0	0	0	0	0	0	0	1	0
Acropora selago	0	0	0	0	0	0	0	0	0	0	0	0	0	0	0	0	1	0	1
Acropora sp. (Line and Phoenix Is.)	0	0	0	0	0	0	0	0	0	0	0	0	0	0	0	0	0	0	1
Acropora sp. (slanted cones)	0	0	0	0	0	0	0	0	0	0	0	0	0	0	0	0	0	0	1
Acropora sp. B	0	0	0	0	0	0	0	0	0	0	0	0	0	0	0	0	1	0	0
Acropora sp. C	0	0	0	0	0	0	0	0	0	0	0	0	0	0	0	0	1	0	0
Acropora spicifera	0	0	0	0	0	0	0	0	0	0	0	0	0	0	0	0	0	0	1
Acropora squarrosa	0	0	0	0	0	0	0	0	0	0	0	0	0	0	0	0	0	0	1
Acropora striata	0	0	0	0	0	0	0	0	0	0	0	0	0	0	0	0	0	1	0
Acropora subulata	0	0	0	0	0	0	0	0	0	0	0	0	0	0	0	0	0	1	1
Acropora syringodes[c]	0	0	0	0	0	0	0	0	0	0	0	0	0	0	0	0	0	0	0
Acropora tenuis	0	0	0	0	0	0	0	0	0	0	0	0	0	0	0	0	0	0	1
Acropora tortuosa	0	0	0	0	0	0	0	0	0	0	0	0	0	0	0	0	0	1	0
Acropora valida	0	0	0	0	0	0	0	0	0	1	0	0	0	0	0	1	1	1	1
Acropora vaughni	0	0	0	0	0	0	0	0	0	0	0	0	0	0	0	0	0	1	1
Acropora verweyi	0	0	0	0	0	0	0	0	0	0	0	0	0	0	0	0	0	0	1
Acropora yongei	0	0	0	0	0	0	0	0	0	0	0	0	0	0	0	0	1	0	0
Alveopora tizardi	0	0	0	0	0	0	0	0	0	0	0	0	0	0	0	0	0	1	0
Alveopora verrilliana	0	0	0	0	0	0	0	0	0	0	0	0	0	0	0	0	0	0	1
Astreopora cf. *moretonensis*	0	0	0	0	0	0	0	0	0	0	0	0	0	0	0	0	0	1	0
Astreopora expansa	0	0	0	0	0	0	0	0	0	0	0	0	0	0	0	0	0	0	1
Astreopora gracilis	0	0	0	0	0	0	0	0	0	0	0	0	0	0	0	0	0	0	1
Astreopora listeri	0	0	0	0	0	0	0	0	0	0	0	0	0	0	0	0	0	0	1
Astreopora myriophthalma	0	0	0	0	0	0	0	0	0	0	0	0	0	0	0	0	0	1	1
Astreopora ocellata	0	0	0	0	0	0	0	0	0	0	0	0	0	0	0	0	0	0	0

(continued)

Species	Locality																		
	GOC	TMM	REV	CLP	ELS	NIC	CRM	COC	PAN	COL	MAL	ECD	GLS	GLN	RPN	HAW	JOH	PIT	PAL
Astreopora suggesta	0	0	0	0	0	0	0	0	0	0	0	0	0	0	0	0	0	0	1
Caulastrea cf. *furcata*	0	0	0	0	0	0	0	0	0	0	0	0	0	0	0	0	0	1	0
Coscinaraea columna	0	0	0	0	0	0	0	0	0	0	0	0	0	0	0	0	0	1	0
Coscinaraea wellsi	0	0	0	0	0	0	0	0	0	0	0	0	0	0	0	1	0	0	0
Ctenactis echinata	0	0	0	0	0	0	0	0	0	0	0	0	0	0	0	0	0	0	1
Cycloseris curvata	1	1	1	0	0	0	1	1	1	1	0	1	1	1[i]	0	0	0	0	0
Cycloseris cyclolites	0	0	0	0	0	0	0	0	0	0	0	0	0	0	0	0	0	0	1
Cycloseris fragilis	0	0	0	0	0	0	0	0	0	0	0	0	0	0	0	0	0	0	1
Cycloseris sp.	0	0	0	0	0	0	0	0	0	0	0	0	0	0	0	0	0	0	1
Cycloseris tenuis	0	0	0	0	0	0	0	0	0	0	0	0	0	0	0	1	0	0	1
Cycloseris vaughani	1	0	1	0	0	0	0	0	0	0	0	0	0	0	1	1	1	1	0
Cyphastrea ocellina	0	0	0	0	0	0	0	0	0	0	0	0	0	0	0	1	1	0	0
Cyphastrea serailia	0	0	0	0	0	0	0	0	0	0	0	0	0	0	0	0	0	1	1
Diaseris distorta	1	1	1	0	0	0	1	1	1	1[i]	0	1	1	0	0	1	0	0	0
Echinophyllia aspera	0	0	0	0	0	0	0	0	0	0	0	0	0	0	0	0	0	0	1
Echinophyllia echinata	0	0	0	0	0	0	0	0	0	0	0	0	0	0	0	0	0	0	1
Echinophyllia sp.	0	0	0	0	0	0	0	0	0	0	0	0	0	0	0	0	0	0	1
Favia amicorum	0	0	0	0	0	0	0	0	0	0	0	0	0	0	0	0	0	0	0
Favia chinensis	0	0	0	0	0	0	0	0	0	0	0	0	0	0	0	0	0	0	0
Favia favus	0	0	0	0	0	0	0	0	0	0	0	0	0	0	0	0	0	0	1
Favia matthaii	0	0	0	0	0	0	0	0	0	0	0	0	0	0	0	0	0	1	1
Favia pallida	0	0	0	0	0	0	0	0	0	0	0	0	0	0	0	0	0	0	1
Favia rotumana	0	0	0	0	0	0	0	0	0	0	0	0	0	0	0	0	0	1	1
Favia rotundata	0	0	0	0	0	0	0	0	0	0	0	0	0	0	0	0	0	0	1
Favia speciosa	0	0	0	0	0	0	0	0	0	0	0	0	0	0	0	0	0	0	1
Favia stelligera	0	0	0	0	0	0	0	0	0	0	0	0	0	0	0	0	0	1	1
Favites abdita	0	0	0	0	0	0	0	0	0	0	0	0	0	0	0	0	0	0	1
Favites flexuosa	0	0	0	0	0	0	0	0	0	0	0	0	0	0	0	0	0	0	1
Favites halicora	0	0	0	0	0	0	0	0	0	0	0	0	0	0	0	0	0	0	1
Favites pentagona	0	0	0	0	0	0	0	0	0	0	0	0	0	0	0	0	0	0	1
Favites russelli	0	0	0	0	0	0	0	0	0	0	0	0	0	0	0	0	0	0	1
Fungia concinna	0	0	0	0	0	0	0	0	0	0	0	0	0	0	0	0	0	0	1
Fungia fungites	0	0	0	0	0	0	0	0	0	0	0	0	0	0	0	0	0	0	1
Fungia granulosa	0	0	0	0	0	0	0	0	0	0	0	0	0	0	0	1	0	0	1
Fungia horrida	0	0	0	0	0	0	0	0	0	0	0	0	0	0	0	0	0	0	1
Fungia moluccensis	0	0	0	0	0	0	0	0	0	0	0	0	0	0	0	0	0	0	1
Fungia paumotensis	0	0	0	0	0	0	0	0	0	0	0	0	0	0	0	0	0	1	1

(continued)

Species	Locality																		
	GOC	TMM	REV	CLP	ELS	NIC	CRM	COC	PAN	COL	MAL	ECD	GLS	GLN	RPN	HAW	JOH	PIT	PAL
Fungia repanda	0	0	0	0	0	0	0	0	0	0	0	0	0	0	0	0	0	0	1
Fungia scutaria	0	0	0	0	0	0	0	0	0	0	0	0	0	0	0	1	1	1	1
Gardineroseris planulata	0	1	0	0	0	1	1	1	1	1	1	1	1	1	0	1	0	0	1
Goniastrea australensis	0	0	0	0	0	0	0	0	0	0	0	0	0	0	0	0	0	1	0
Goniastrea edwardsi	0	0	0	0	0	0	0	0	0	0	0	0	0	0	0	0	0	0	1
Goniastrea pectinata	0	0	0	0	0	0	0	0	0	0	0	0	0	0	0	0	0	0	1
Goniastrea retiformis	0	0	0	0	0	0	0	0	0	0	0	0	0	0	0	0	0	0	1
Halomitra pileus	0	0	0	0	0	0	0	0	0	0	0	0	0	0	0	0	0	0	1
Herpolitha limax	0	0	0	0	0	0	0	0	0	0	0	0	0	0	0	0	0	0	1
Hydnophora exesa	0	0	0	0	0	0	0	0	0	0	0	0	0	0	0	0	0	0	1
Hydnophora microconos	0	0	0	0	0	0	0	0	0	0	0	0	0	0	0	0	0	0	1
Hydnophora pilosa	0	0	0	0	0	0	0	0	0	0	0	0	0	0	0	0	0	0	1
Hydnophora rigida	0	0	0	0	0	0	0	0	0	0	0	0	0	0	0	0	0	0	1
Isopora brueggemanni	0	0	0	0	0	0	0	0	0	0	0	0	0	0	0	0	1	0	1
Isopora cuneata	0	0	0	0	0	0	0	0	0	0	0	0	0	0	0	0	0	0	1
Isopora palifera	0	0	0	0	0	0	0	0	0	0	0	0	0	0	0	0	0	0	1
Leptastrea agassizi	0	0	0	0	0	0	0	0	0	0	0	0	0	0	0	1	0	0	1
Leptastrea bewickensis	0	0	0	0	0	0	0	0	0	0	0	0	0	0	0	1	0	0	1
Leptastrea hawaiiensis	0	0	0	0	0	0	0	0	0	0	0	0	0	0	0	1	0	0	0
Leptastrea pruinosa	0	0	0	0	0	0	0	0	0	0	0	0	0	0	0	1	0	1	1
Leptastrea purpurea	0	0	0	0	0	0	0	0	0	0	0	0	0	0	1	1	1	1	1
Leptastrea sp. A	0	0	0	0	0	0	0	0	0	0	0	0	0	0	0	0	0	0	1
Leptastrea sp. B	0	0	0	0	0	0	0	0	0	0	0	0	0	0	0	0	0	0	1
Leptastrea sp. C	0	0	0	0	0	0	0	0	0	0	0	0	0	0	0	0	0	0	1
Leptastrea transversa	0	0	0	0	0	0	0	0	0	0	0	0	0	0	1	0	0	1	1
Leptoria phrygia	0	0	0	0	0	0	0	0	0	0	0	0	0	0	0	0	0	0	0
Leptoseris hawaiiensis	0	0	0	0	0	0	0	0	0	0	0	0	0	0	0	1	1	1	0
Leptoseris incrustans	0	0	0	0	0	0	0	0	0	0	0	0	0	0	0	1	1	1	0
Leptoseris mycetoseroides	0	0	0	0	0	0	0	0	0	0	0	0	0	0	0	1	0	0	1
Leptoseris papyracea	1	1	0	0	0	0	1	1	1	1	0	1	0	0	0	1	0	0	0
Leptoseris scabra	0	0	0	1	0	0	1	1	0	0	0	0	0	1	1	1	1	0	0
Leptoseris solida	0	0	0	0	0	0	0	0	0	0	0	0	0	0	1	0	0	1	0
Leptoseris tubulifera	0	0	0	0	0	0	0	0	0	0	0	0	0	0	0	1	0	0	0
Lobophyllia corymbosa	0	0	0	0	0	0	0	0	0	0	0	0	0	0	0	0	0	0	1

(continued)

Species	Locality																		
	GOC	TMM	REV	CLP	ELS	NIC	CRM	COC	PAN	COL	MAL	ECD	GLS	GLN	RPN	HAW	JOH	PIT	PAL
Lobophyllia hemprichii	0	0	0	0	0	0	0	0	0	0	0	0	0	0	0	0	0	1	1
Merulina ampliata	0	0	0	0	0	0	0	0	0	0	0	0	0	0	0	0	0	0	1
Montastrea annuligera	0	0	0	0	0	0	0	0	0	0	0	0	0	0	0	0	0	0	1
Montastrea curta	0	0	0	0	0	0	0	0	0	0	0	0	0	0	0	0	0	1	1
Montipora aequituberculata	0	0	0	0	0	0	0	0	0	0	0	0	0	0	0	0	0	1	1
Montipora australiensis	0	0	0	0	0	0	0	0	0	0	0	0	0	0	0	0	0	1	0
Montipora bilaminata	0	0	0	0	0	0	0	0	0	0	0	0	0	0	0	0	0	1	0
Montipora caliculata	0	0	0	0	0	0	0	0	0	0	0	0	0	0	0	0	0	1	1
Montipora capitata	0	0	0	0	0	0	0	0	0	0	0	0	0	0	0	1	1	1	1
Montipora complanata	0	0	0	0	0	0	0	0	0	0	0	0	0	0	0	0	0	1	0
Montipora crassituberculata	0	0	0	0	0	0	0	0	0	0	0	0	0	0	0	0	0	1	0
Montipora danae	0	0	0	0	0	0	0	0	0	0	0	0	0	0	0	0	0	0	1
Montipora dilatata	0	0	0	0	0	0	0	0	0	0	0	0	0	0	0	1	0	0	1
Montipora efflorescens	0	0	0	0	0	0	0	0	0	0	0	0	0	0	0	0	0	0	1
Montipora elschneri	0	0	0	0	0	0	0	0	0	0	0	0	0	0	0	0	0	0	0
Montipora flabellata	0	0	0	0	0	0	0	0	0	0	0	0	0	0	0	1	1	0	1
Montipora foliosa	0	0	0	0	0	0	0	0	0	0	0	0	0	0	0	0	0	0	1
Montipora foveolata	0	0	0	0	0	0	0	0	0	0	0	0	0	0	0	0	0	1	1
Montipora grisea	0	0	0	0	0	0	0	0	0	0	0	0	0	0	0	0	0	1	0
Montipora hoffmeisteri	0	0	0	0	0	0	0	0	0	0	0	0	0	0	0	0	1	0	1
Montipora informis	0	0	0	0	0	0	0	0	0	0	0	0	0	0	0	0	0	0	1
Montipora lobulata	0	0	0	0	0	0	0	0	0	0	0	0	0	0	0	0	0	1	0
Montipora millepora	0	0	0	0	0	0	0	0	0	0	0	0	0	0	0	0	0	0	1
Montipora monasteriata	0	0	0	0	0	0	0	0	0	0	0	0	0	0	0	0	0	0	1
Montipora nodosa	0	0	0	0	0	0	0	0	0	0	0	0	0	0	0	0	0	1	0
Montipora patula	0	0	0	0	0	0	0	0	0	0	0	0	0	0	0	1	1	0	1
Montipora peltiformis	0	0	0	0	0	0	0	0	0	0	0	0	0	0	0	0	0	0	1
Montipora socialis	0	0	0	0	0	0	0	0	0	0	0	0	0	0	0	0	0	0	0
Montipora sp. 4 cf. *incrassata*	0	0	0	0	0	0	0	0	0	0	0	0	0	0	0	1	1	1	0
Montipora tuberculosa	0	0	0	0	0	0	0	0	0	0	0	0	0	0	0	1	1	1	1

(continued)

Species	Locality																		
	GOC	TMM	REV	CLP	ELS	NIC	CRM	COC	PAN	COL	MAL	ECD	GLS	GLN	RPN	HAW	JOH	PIT	PAL
Montipora venosa	0	0	0	0	0	0	0	0	0	0	0	0	0	0	0	0	0	1	0
Montipora verrilli	0	0	0	0	0	0	0	0	0	0	0	0	0	0	0	1	1	0	1
Montipora verrucosa	0	0	0	0	0	0	0	0	0	0	0	0	0	0	0	1	0	1	0
Pachyseris sp.	0	0	0	0	0	0	0	0	0	0	0	0	0	0	0	0	0	0	1
Pachyseris speciosa	0	0	0	0	0	0	0	0	0	0	0	0	0	0	0	0	0	0	0
Parahalomitra robusta	0	0	0	0	0	0	0	0	0	0	0	0	0	0	0	0	0	0	0
Pavona cactus	0	0	0	0	0	0	0	0	0	0	0	0	0	0	0	0	0	0	1
Pavona chiriquiensis	0	0	0	0	0	0	1	1	1	1	1	1	1	0	0	0	0	0	1
Pavona clavus	1	1	1	1	0	1	1	1	1	1	1	1	1	1	0	1	1	0	0
Pavona divaricata	0	0	0	0	0	0	0	0	0	0	0	0	0	0	0	0	0	0	0
Pavona duerdeni	1	1	1	1	0	0	0	0	1	0	0	0	0	0	0	1	1	0	1
Pavona explanulata	0	0	0	0	0	0	0	0	0	0	0	0	0	0	0	0	0	0	1
Pavona frondifera	0	0	0	0	0	0	1	1	1	1	0	0	0	0	0	0	0	0	1
Pavona gigantea	1	1	1	1	1	1	1	1	1	1	1	1	1	1	0	0	0	0	0
Pavona maldivensis	0	0	1	1	0	0	1	1	1	1	1	0	0	1	0	1	1	1	1
Pavona minuta	0	0	1	1	0	0	0	1	1	0	0	0	0	0	0	0	0	0	1
Pavona sp. cf. *pollicata*	0	0	0	0	0	0	0	0	0	0	0	0	0	0	0	0	0	1	0
Pavona sp. 1	0	0	0	0	0	0	0	0	0	0	0	0	0	0	0	0	0	1	0
Pavona varians	0	1	1	1	0	1	1	1	1	1	1	1	1	1	0	1	1	1	1
Platygyra daedalea	0	0	0	0	0	0	0	0	0	0	0	0	0	0	0	0	0	1	1
Platygyra lamellina	0	0	0	0	0	0	0	0	0	0	0	0	0	0	0	0	0	0	1
Platygyra pini	0	0	0	0	0	0	0	0	0	0	0	0	0	0	0	0	0	0	1
Platygyra ryukyuensis	0	0	0	0	0	0	0	0	0	0	0	0	0	0	0	0	0	0	1
Platygyra sinensis	0	0	0	0	0	0	0	0	0	0	0	0	0	0	0	0	0	0	1
Plesiastrea versipora	0	0	0	0	0	0	0	0	0	0	0	0	0	0	0	0	0	1	1
Pocillopora damicornis	1	1	1	0	1	1	1	1	1	1	0	1	1	1	1	1	1	1	1
Pocillopora danae	0	0	0	0	0	0	0	0	0	0	0	0	0	0	1	0	0	0	0
Pocillopora effusus	0	1	0	1	1	0	0	0	0	1	1	1	1	1	0	0	0	0	0
Pocillopora elegans	1	1	1	1	1	1	1	1	1	1	1	1	1	1	1	0	0	1	0
Pocillopora eydouxi	1	1	1	1	0	0	1	1	1	1	1	1	1	1	1	1	1	1	1
Pocillopora inflata	0	1	0	0	0	0	1	1	1	0	0	1	1	1	0	0	0	0	0
Pocillopora laysanensis	0	0	0	0	0	0	0	0	0	0	0	0	0	0	0	1	0	0	0
Pocillopora ligulata	0	0	0	0	0	0	0	0	0	1	0	1	1[j]	1[j]	1	1	1	0	0

(continued)

Species	Locality																		
	GOC	TMM	REV	CLP	ELS	NIC	CRM	COC	PAN	COL	MAL	ECD	GLS	GLN	RPN	HAW	JOH	PIT	PAL
Pocillopora meandrina	1	1	1	1	1	0	1	1	1	1	0	1	1	1	1	1	1	1	1
Pocillopora molokensis	0	0	0	0	0	0	0	0	0	0	0	0	0	0	0	1	1	0	0
Pocillopora sp.	0	0	0	0	0	0	0	0	0	0	0	0	0	0	0	0	0	0	1
Pocillopora sp. 1	0	0	0	0	0	0	0	0	0	0	0	0	0	0	0	0	0	1	0
Pocillopora sp. 2	0	0	0	0	0	0	0	0	0	0	1	0	0	0	0	0	0	0	0
Pocillopora sp. 11 cf. *capitata*	1	1	1	0	1	0	1	1	1	1	1	1	1	1	0	1	1	0	1
Pocillopora verrucosa	1	1	1	1	1	1	1	0	1	1	1	1	1	1	1	1	0	1	1
Pocillopora woodjonesi	0	0	1	1	0	0	0	0	0	1	0	0	0	1	0	0	0	1	0
Pocillopora zelli	0	0	0	0	0	0	0	0	0	0	0	0	0	0	0	1	0	0	1
Porites aff. *Annae*	0	0	0	0	0	0	0	0	0	0	0	0	0	0	0	1	0	1	0
Porites annae	0	0	0	0	0	0	0	0	0	0	0	0	0	0	0	0	0	0	1
Porites arnaudi	0	0	1	1	0	0	0	0	0	0	0	0	0	0	0	0	0	0	0
Porites australiensis	0	0	1	1	0	0	0	0	0	0	0	0	0	0	0	0	0	1	1
Porites baueri	0	1	0	0	0	0	0	0	0	0	0	0	0	0	0	0	0	0	0
Porites brighami	0	0	0	0	0	0	0	0	0	0	0	0	0	0	0	1	0	0	0
Porites compressa	0	0	0	0	0	0	0	0	0	0	0	0	0	0	0	1	0	0	0
Porites duerdeni	0	0	0	0	0	0	0	0	0	0	0	0	0	0	0	1	0	0	0
Porites evermanni	0	1	0	1	0	0	1	1	1	1	0	1	0	1	0	1	1	0	1
Porites hawaiiensis	0	0	0	0	0	0	0	0	0	0	0	0	0	0	0	1	0	0	0
Porites lichen	0	0	1	0	0	0	0	0	0	0	0	0	0	0	0	0	0	0	0
Porites lobata	0	1	1	1	1	0	1	1	1	1	1	1	1	1	1	1	1	1	1
Porites lutea[d]	0	1	1	1	0	0	0	0	0	0	0	0	0	0	0	0	1	1	1
Porites murrayensis	0	0	0	0	0	0	0	0	0	0	0	0	0	0	0	0	0	0	1
Porites panamensis	1	1	1	0	1	1	1	0	1	1	0	1	0	0	0	0	0	0	0
Porites pukoensis	0	0	0	0	0	0	0	0	0	0	0	0	0	0	0	1	0	0	0
Porites rus	0	0	0	0	0	0	1	0	0	0	0	0	0	0	0	1	0	0	1
Porites solida[e]	0	0	0	0	0	0	0	0	0	0	0	0	0	0	0	0	0	0	1
Porites sp.	0	0	0	0	0	0	0	0	0	0	0	0	0	0	0	0	0	0	1
Porites studeri[f]	0	0	0	0	0	0	0	0	0	0	0	0	0	0	0	1	0	0	0
Porites superfusa	0	0	0	0	0	0	0	0	0	0	0	0	0	0	0	0	0	0	1
Porites sverdrupi	1	1	0	0	0	0	0	0	0	0	0	0	0	0	0	0	0	0	0
Porites vaughani	0	0	0	0	0	0	0	0	0	0	0	0	0	0	0	0	0	0	1
Psammocora albopicta	0	0	0	0	0	0	0	0	0	0	0	0	0	0	0	0	0	1	0
Psammocora brighami	1	1	0	0	0	0	0	0	1	1	0	1	1	1	0	0	0	0	0
Psammocora contigua	0	0	0	0	0	0	0	0	0	0	0	0	0	0	0	0	0	0	1
Psammocora explanulata	0	0	0	0	0	0	0	0	0	0	0	0	0	0	0	1	0	0	0
Psammocora haimeana[g]	1	0	0	0	0	0	0	0	0	0	0	0	0	0	0	0	0	1	1

(continued)

Species	Locality																		
	GOC	TMM	REV	CLP	ELS	NIC	CRM	COC	PAN	COL	MAL	ECD	GLS	GLN	RPN	HAW	JOH	PIT	PAL
Psammocora nierstraszi	0	0	0	0	0	0	0	0	0	0	0	0	0	0	0	1	1	0	1
Psammocora obtusangula	0	0	0	0	1	1	1	0	1	1	0	0	0	0	0	0	0	1	0
Psammocora profundacella	0	1	1	1	0	0	0	0	0	1	0	0	0	0	0	0	0	1	1
Psammocora stellata	1	1	1	1	1	1	1	1	1	1	0	1	1	1	1	1	1	0	1
Psammocora superficialis[h]	1	1	1	1	0	0	1	1	1	1	0	1	1	1	1	0	0	0	1
Psammocora verrilli	0	0	0	0	0	0	0	0	0	0	0	0	0	0	0	1	0	0	0
Sandalolitha robusta	0	0	0	0	0	0	0	0	0	0	0	0	0	0	0	0	0	0	1
Scolymia cf. *vitiensis*	0	0	0	0	0	0	0	0	0	0	0	0	0	0	0	0	0	1	0
Siderastrea glynni	0	0	0	0	0	0	0	0	1	0	0	0	0	0	0	0	0	0	0
Stylocoeniella guentheri	0	0	0	0	0	0	0	0	0	0	0	0	0	0	0	0	0	1	0
Stylophora pistillata	0	0	0	0	0	0	0	0	0	0	0	0	0	0	0	0	0	0	1
Symphyllia recta	0	0	0	0	0	0	0	0	0	0	0	0	0	0	0	0	0	0	1
Turbinaria frondens	0	0	0	0	0	0	0	0	0	0	0	0	0	0	0	0	0	0	1
Turbinaria reniformis	0	0	0	0	0	0	0	0	0	0	0	0	0	0	0	0	0	0	1
Turbinaria sp.	0	0	0	0	0	0	0	0	0	0	0	0	0	0	0	0	0	0	1
	19	26	25	21	11	10	25	22	27	28	13	23	20	22	15	62	44	87	163

[a]Wallace (1999) recognizes this species as *Acropora muricata*
[b]Walsh et al. (2013)
[c]A junior synonym of *A. longicyathus*
[d]This species is probably *Porites evermanni*
[e]A frequently misnamed species
[f]Type locality in Hawaii
[g]Frequently confused with *Psammocora nierstraszi*
[h]Frequently confused with *Psammocora explanulata*
[i]Only dead skeletons present
[j]Requires confirmation (see Hickman 2008)

References

Abele LG (1976) Comparative species richness in fluctuating and constant environments: coral-associated decapod crustaceans. Science 192:461–463

Alfaro EJ, Cortés J (2012) Atmospheric forcing of cold subsurface water events in Bahía Culebra, Costa Rica. Rev Biol Trop 60(suppl 2):173–186

Alfaro EJ, Cortés J, Alvarado JJ, Jiménez C, León A, Sánchez-Noguera C, Nivia-Ruiz J, Ruiz-Campos E (2012) Clima y variabilidad climática de la temperatura subsuperficial del mar en Bahía Culebra, Guanacaste, Costa Rica. Rev Biol Trop 60(Suppl 2):159–171

Alory G, Maes C, Delcroix T, Reul N, Illig S (2012) Seasonal dynamics of sea surface salinity off Panama: the far Eastern Pacific Fresh Pool. J Geophys Res 117:C04028. doi:10.1029/2011JC007802

Alvarado JJ, Cortés J, Fernández C, Nivia J (2005) Coral communities and reefs of Ballena Marine National Park, Pacific coast of Costa Rica. Cienc Mar 31:641–651

Alvarado JJ, Fernández C, Cortés J (2009) Water quality conditions on coral reefs at the Marino Ballena National Park, Pacific Costa Rica. Bull Mar Sci 84:137–152

Alvarado JJ, Reyes-Bonilla H, del Castillo Á, Cárdenas PA, Buitrago F, Aguirre-Rubí J (2010) Coral reefs of the Pacific coast of Nicaragua. Coral Reefs 29:201

Alvarado JJ, Ayala A, Álvarez del Castillo-Cárdenas PA, Fernández C, Aguirre-Rubí J, Buitrago F, Reyes-Bonilla H (2011) Coral communities of San Juan del Sur, Pacific Nicaragua. Bull Mar Sci 87:129–146

Álvarez-Filip L, Reyes-Bonilla H, Calderón-Aguilera LE (2006) Community structure of reef fishes in Cabo Pulmo reef, Gulf of California. Mar Ecol 27:253–262

Asher J, Maragos J, Kenyon J, Vargas-Ángel B, Coccagna E (2012) Range extensions for several species of *Acropora* in the Hawaiian Archipelago and the Papahānaumokuākea Marine National Monument. Bull Mar Sci 88:337–338

Baker AC, Glynn PW, Riegl B (2008) Climate change and coral reef bleaching: an ecological assessment of long-term impacts, recovery trends and future outlook. Estuar Coast Shelf Sci 80:435–471

Banks SA (2002) Ambiente físico. In: Danulat E, Edgar GJ (eds) Reserva Marina de Galápagos. Línea Base de la Biodiversidad. Fundación Charles Darwin/Servicio Parque Nacional Galápagos, Santa Cruz, Galápagos, Ecuador, pp 22–37

Banks SA, Vera M, Chiriboga Á (2009) Characterizing the last remaining reefs: establishing reference points to assess long term change in Galápagos zooxanthellate coral communities. Galapagos Res 66:43–64

Banta WC (1991) The Bryozoa of the Galápagos. In: James MJ (ed) Galápagos marine invertebrates: taxonomy, biogeography, and evolution in Darwin's islands. Plenum Press, New York, pp 371–389

Barham EG, Gowdy RW, Wolfson FH (1973) *Acanthaster* (Echinodermata, Asteroidea) in the Gulf of California. Fish Bull (Seattle) 71:927–942

Bassey-Fallas G (2010) Evaluación ecológica de los arrecifes y comunidades coralinas de las Islas Murciélago y sección norte de la Península de Santa Elena en el Pacífico de Costa Rica. MSc thesis, Universidad Nacional Heredia, Costa Rica

Bastidas-Salamanca M, Rangel-Buitrago N, Morales-Giraldo D, Ricaurte C, Gomez-López DI, Navas-Camacho R, Navarrete SM, Alonso DA, Vivas-Aguas LJ, Obando-Madera PS, Rodríguez-Rodríguez JA, Licero-Villanueva LV, Perdomo LV (2014) Estado del ambiente y los ecosistemas marinos y costeros. In: INVEMAR. Informe del estado de los ambientes y recursos marinos y costeros en Colombia: Año 2013. Serie de Publicaciones Periódicas No. 3, Santa Marta, pp 23–71

Baums IB, Boulay JN, Polato NR, Hellberg ME (2012) No gene flow across the Eastern Pacific Barrier in the reef-building coral *Porites lobata*. Mol Ecol 21:5418–5433

Baums IB, Durante MD, Laing AA, Feingold J, Smith T, Bruckner A, Monteiro J (2014) Marginal coral populations: the densest known aggregation of *Pocillopora* in the Galápagos Archipelago is of asexual origin. Front Mar Sci 1:59. doi:10.3389/fmars.2014.00059

Béarez P, Séret B (2009) Les poisons. In: Charpy L (ed) Clipperton, environnement et biodiversité d'un microcosme océanique. Mus National d'Hist Naturelle, Paris IRD, Marseille, pp 143–154

Benfield S, Baxter L, Guzmán HM, Mair JM (2008) A comparison of coral reef and coral community fish assemblages in Pacific Panama and environmental factors governing their structure. J Mar Biol Assoc UK 88:1331–1341

Benítez-Villalobos F, Domínguez y Gómez MT, López-Pérez RA (2008) Temporal variation of the sea urchin *Diadema mexicanum* population density at Bahías de Huatulco, western Mexico. Rev Biol Trop (Int J Trop Biol) 56(Suppl 3):255–263

Benzoni F, Stefani F, Pichon M, Galli P (2010) The name game: morpho-molecular species boundaries in the genus *Psammocora* (Cnidaria, Scleractinia). Zool J Linnean Soc 160:421–456

Bezy MB, Jiménez C, Cortés J, Segura A, León A, Alvarado JJ, Gillén C, Mejía E (2006) Contrasting *Psammocora*-dominated coral communities in Costa Rica, tropical eastern Pacific. In: Proceedings of 10th International Coral Reef Symposium, vol 1, Okinawa, pp 376–381

Birkeland C, Meyer DL, Stames JP, Buford CL (1975) Subtidal communities of Malpelo Island. Smithson Contrib Zool 176:55–68

Blancas-López AV (2009) Conectividad demográfica de *Pocillopora verrucosa* (Anthozoa: Scleractinia) en el Pacífico mexicano. MSc thesis CICESE, p 83

Boulay JN, Hellberg ME, Cortés J, Baums IB (2014) Unrecognized coral species diversity masks differences in functional ecology. Proc R Soc B 281:20131580

Brenes R, Cuadras J, Durbán M, Fernández-López A, González LM, Miranda A (1993) Plan de manejo del Parque Nacional Coiba (1ª fase). Informe inédito, AECI, ICONA, INRENARE, Panamá, 213 p +19 maps

Briggs JC (1974) Marine zoogeography. McGraw-Hill, New York p 475

Bruno J, Selig E, Casey K, Page C, Willis B, Harvell C, Sweatman HY, Melendy A (2007) Thermal stress and coral cover as drivers of coral disease outbreaks. PLoS Biol. doi:10.1371/journal.pbio.0050124

Brusca RC, Thomson DA (1975) Pulmo Reef: the only "coral reef" in the Gulf of California. Cienc Mar 2:37–52

Cantera JR (1983) Distribution des peuplements des scléractiniaires sur un récif de l'Ille de Gorgona (Côte Pacifique de Colombie). Tethys 11:25–31

Carpenter KE, Abrar M, Aeby G, Aronson RB, Banks SA, Bruckner A, Chiriboga A, Cortés J, Delbeek JC, DeVantier L, Edgar G, Edwards AJ, Fenner D, Guzmán HM, Hoeksema BW, Hodgson G, Johan O, Licuanan WY, Livingstone SR, Lovell ER, Moore JA, Obura DO, Ochavillo D, Polidoro BA, Precht WF, Quibilan MC, Reboton C, Richards ZT, Rogers AD, Sanciangco J, Sheppard A, Sheppard C, Smith J, Stuart S, Turak E, Veron J, Wallace C, Weil E, Wood E (2008) One third of reef-building corals face elevated extinction risk from climate change and local impact. Science 321:560–563

Carricart-Ganivet JP, Reyes-Bonilla H (1999) New and previous records of scleractinian corals from Clipperton Atoll, eastern Pacific. Pac Sci 53:370–375

Carriquiry JD, Reyes-Bonilla H (1997) Estructura de la comunidad y distribución geográfica de los arrecifes coralinos de Nayarit. Cienc Mar 23:227–248

Castilla JC, Rozbaczylo N (1987) Invertebrados marinos de Isla de Pascua y Sala y Gómez. In: Castilla JC (ed) Islas Oceánicas Chilenas: Conocimiento científicos y necesidades de investigaciones. Ediciones Univ Católica de Chile, Santiago, pp 191–215

Cea Egaña A, DiSalvo LH (1982) Mass expulsion of zooxanthellae by Easter Island corals. Pac Sci 36:61–63

Charpy L, Rodier M, Sarazin G (2009) Biogéochimie du lagon. In: Charpy L (ed) Clipperton, environnement et biodiversité d'un microcosme océanique. Muséum national d'Histoire naturelle, Paris, IRD, Marseille. Patrimoines naturels 68:67–79

Charpy L, Rodier M, Couté A, Perrette-Gallet C, Bley-Loëz C (2010) Clipperton, a possible future for atoll lagoons. Coral Reefs 29:771–783

Chavez-Romo HE, Correa-Sandoval F, Paz-Garcia DA, Reyes-Bonilla H, López-Pérez RA, Medina-Rosas P, Hernández-Cortés MP (2008) Genetic structure of a scleractinian coral, *Pocillopora damicornis*, in the Mexican Pacific. Proceedings of 11th International Coral Reef Symposium, vol 1, Ft Lauderdale, pp 429–433

Chevalier J-P (1978) Les coraux des Iles Marquises. In: Les Iles Marquises, geómorphologie, climatologie, faune et flore. Cah Pac No 21 Paris, pp 243–283

Clemente S, Rodríguez A, Brito A, Ramos A, Monterroso Ó, Hernández JC (2011) On the occurrence of the hydrocoral *Millepora* (Hydrozoa: Milleporidae) in the subtropical eastern Atlantic (Canary Islands): is the colonization related to climatic events? Coral Reefs 30:237–240

Coates AG, Obando JA (1996) The geologic evolution of the Central American isthmus. In: Jackson JBC, Budd AF, Coates AG (eds) Evolution and environment in tropical America. University of Chicago Press, Chicago, pp 21–56

Colgan MW (1990) El Niño and the history of eastern Pacific reef building. In: Glynn PW (ed) Global ecological consequences of the 1982–83 El Niño-Southern Oscillation. Elsevier oceanography series 52. Elsevier, Amsterdam, pp 183–232

Colgan MW (1991) El Niño and coral reef development in the Galápagos Islands: a study of the Urvina Bay uplift. In: James MJ (ed) Galápagos marine invertebrates: taxonomy, biogeography, and evolution in Darwin's Islands. Plenum Press, New York, pp 99–120

Colley SB, Glynn PW, May AS, Maté JL (2006) Species-dependent reproductive responses of eastern Pacific corals to the 1997–1998 ENSO event. In: Proceedings of 10th International Coral Reef Symposium, vol 1, pp 61–70

Correa AMS, Baker AC (2011) Disaster taxa in microbially mediated metazoans: how endosymbionts and environmental catastrophes influence the adaptive capacity of reef corals. Glob Change Biol 17:68–75

Cortés J (1986) Biogeografía de corales hermatípicos: El Istmo Centro Americano. An Inst Cienc del Mar y Limnol, Univ Nal Autón Mexico 13:297–304

Cortés J (1990) The coral reefs of Golfo Dulce, Costa Rica: distribution and community structure. Atoll Res Bull 344:1–37

Cortés J (1996/1997) Comunidades coralinas y arrecifes del Área de Conservación Guanacaste, Costa Rica. Rev Biol Trop 44/45:623–625

Cortés J (1997) Biology and geology of coral reefs of the eastern Pacific. Coral Reefs 16(suppl):S39–S46

Cortés J (2003) Latin American coral reefs. Elsevier, Amsterdam, p 497

Cortés J (2007) Coastal morphology and coral reefs. In: Bundschuh J, Alvarado GE (eds) Central America: geology, resources, and hazards, vol 1. Taylor & Francis, Milton Park, pp 185–200

Cortés J (2008) Historia de la investigación marina de la Isla del Coco, Costa Rica. Rev Biol Trop 56(suppl 2):1–18

Cortés J (2011) Eastern Tropical Pacific coral reefs. In: Hopley D (ed) The Encyclopedia of Modern Coral Reefs: Structure, Form and Process. Springer, Berlin pp 351–358

Cortés J (2012a) Historia de la investigación marina en Bahía Culebra, Guanacaste, Costa Rica. Rev Biol Trop 60 (Suppl 2):19–37

Cortés J (2012b) Marine biodiversity of an Eastern Tropical Pacific oceanic island, Parque Nacional Isla del Coco, Costa Rica. Rev Biol Trop 60 (Suppl 3):131–185

Cortés J (2014) Cocos Island marine ecosystems. In: Kappelle M (ed) Costa Rican ecosystems. University of Chicago Press, Chicago (in press)

Cortés J, Guzmán HM (1998) Organismos de los arrecifes coralinos de Costa Rica: descripción, distribución geográfica e historia natural de los corales zooxantelados (Anthozoa: Scleractinia) del Pacífico. Rev Biol Trop 46:55–92

Cortés J, Jiménez CE (1996) Coastal-marine environments of Parque Nacional Corcovado, Puntarenas, Costa Rica. Rev Biol Trop 44 (suppl 3):35–40

Cortés J, Jiménez C (2003) Corals and coral reefs of the Pacific of Costa Rica: history, research and status. In: Cortés J (ed) Latin American coral reefs. Elsevier Science BV, Amsterdam, pp 361–385

Cortés J, Murillo MM (1985) Comunidades coralinas y arrecifes del Pacífico de Costa Rica. Rev Biol Trop 33:197–202

Cortés J, Wehrtmann IS (2009) Diversity of marine habitats of the Caribbean and Pacific of Costa Rica. In: Wehrtmann IS, Cortés J (eds) Marine biodiversity of Costa Rica, Central America. Springer, Berlin, pp 1–45

Cortés J, Macintyre IG, Glynn PW (1994) Holocene growth history of an eastern Pacific fringing reef, Punta Islotes, Costa Rica. Coral Reefs 13:65–73

Cortés J, Jiménez CE, Fonseca AC, Alvarado JJ (2010) Status and conservation of coral reefs in Costa Rica. Rev Biol Trop 58(Suppl 1):33–50

Cróquer A, Weil E (2009) Changes in Caribbean coral disease prevalence after the 2005 bleaching event. Dis Aquat Org 87:33–43

D'Croz L, O'Dea A (2007) Variability in upwelling along the Pacific shelf of Panama and implications for the distribution of nutrients and chlorophyll. Estuar Coastal Shelf Sci 73:325–340

D'Croz L, Robertson DR (1997) Coastal oceanographic conditions affecting coral reefs on both sides of the Isthmus of Panama. In: Proceedings of 8th International Coral Reef Symposium, vol 2, Panama, pp 2053–2058

Dana TF (1975) Development of contemporary eastern Pacific coral reefs. Mar Biol 33:355–374

Dana TF, Wolfson A (1970) Eastern Pacific crown-of-thorns starfish populations in the lower Gulf of California. San Diego Soc Nat Hist Trans 16:83–90

Darwin C (1880) The origin of species by means of natural selection or the preservation of favored races in the struggle for life. John Murray, London, p 458

Dawson TP, Jarvie F, Reitsma F (2009) A habitat suitability model for predicting coral community and reef distributions in the Galapagos Islands. Galapagos Res 66:20–26

Delesalle B (1985) Atoll de Mataiva: archipel des Tuamotu. In: Delesalle B, Galzin R, Salvat B (eds) French Polynesian coral reefs. Proceedings of 5th International Coral Reef Congress, vol 1, Tahiti, pp 269–307

Denyer P, Cárdenas G (2000) Costas marinas. In: Denyer P, Kussmaul S (eds) Geología de Costa Rica. Editorial. Tecnología de Costa Rica, Cartago, pp 185–218

Devis-Morales A, Schneider W, Montoya-Sánchez RA, Rodríguez-Rubio E (2008) Monsoonlike winds reverse oceanic circulation in the Panama Bight. Geophys Res Lett 35:L20607. doi:10.1029/2008GL035172

DiSalvo LH, Randall JE, Cea A (1988) Ecological reconnaissance of the Easter Island sublittoral marine environment. Nat Geogr Res 4:451–473

Dominici-Arosemena A, Wolff M (2006) Reef fish community structure in the tropical eastern Pacific (Panamá): living on a relatively stable rocky reef environment. Helgol Mar Res 60:287–305

Dunbar RB, Wellington GM, Colgan MW, Glynn PW (1994) Eastern Pacific sea surface temperature since 1600 AD: the $\delta^{18}O$ record of climate variability in Galápagos corals. Paleoceanography 9:291–315

Durham JW (1966) Coelenterates, especially stony corals, from the Galápagos and Cocos Islands. In: Bowman RI (ed) The Galápagos, Proceedings of the symposia of the Galápagos international scientific project. University of California Press, Berkeley, pp 123–135

Durham JW, Barnard JL (1952) Stony corals of the eastern Pacific collected by the *Velero III* and *Velero IV*. Allan Hancock Pac Expeditions 16:1–110

Eakin CM (1992) Post-El Niño Panamanian reefs: less accretion, more erosion and damselfish protection. In: Proceedings of 7th International Coral Reef Symposium, vol 1, Guam, pp 387–396

Eakin CM (1996) Where have all the carbonates gone? A model comparison of calcium carbonate budgets before and after the 1982–1983 El Niño. Coral Reefs 15:109–119

Eakin CM (2001) A tale of two ENSO events: carbonate budgets and the influence of two warming disturbances and intervening variability, Uva Island, Panama. Bull Mar Sci 69:171–186

Edgar GJ, Banks SA, Brandt M, Bustamante RH, Chiriboga Á, Earle SA, Garske LE, Glynn PW, Grove JS, Henderson S, Hickman CP, Miller KA, Rivera F, Wellington GM (2010) El Niño, grazers and fisheries interact to greatly elevate extinction risk for Galápagos marine species. Global Change Biol 16:2876–2890

Ekman S (1953) Zoogeography of the sea. Sidgwick and Jackson, London p 417

Emerson WK (1991) First records for *Cymatium mundum* (Gould) in the eastern Pacific Ocean, with comments on the zoogeography of the tropical trans-Pacific tonnacean and non-tannacean prosobranch gastropods with Indo-Pacific faunal affinities in west American waters. The Nautilus 105:62–80

Enochs IC (2012) Motile cryptofauna associated with live and dead coral substrates: implications for coral mortality and framework erosion. Mar Biol 159:709–722

Enochs IC, Hockensmith G (2009) Effects of coral mortality on the community composition of cryptic metazoans associated with *Pocillopora damicornis*. In: Proceedings of 11th International Coral Reef Symposium, vol 2, Ft Lauderdale, pp 1368–1372
Enochs IC, Manzello DP (2012a) Responses of cryptofaunal species richness and trophic potential to coral reef habitat degradation. Diversity 4:94–104
Enochs IC, Manzello DP (2012b) Species richness of motile cryptofauna across a gradient of reef framework erosion. Coral Reefs 31:653–661
Enochs IC, Toth LT, Brandtneris VK, Afflerbach JC, Manzello DP (2011) Environmental determinants of motile cryptofauna on an eastern Pacific coral reef. Mar Ecol Prog Ser 438:105–118
Feingold JS (1996) Coral survivors of the 1982–83 El Niño-Southern Oscillation, Galápagos Islands, Ecuador. Coral Reefs 15:108
Feingold JS (2001) Responses of three coral communities to the 1997–98 El Niño-Southern Oscillation: Galápagos Islands, Ecuador. Bull Mar Sci 69:61–77
Feingold JS, Glynn PW (2014) Coral research in the Galápagos Islands, Ecuador. In: Walsh SJ, Mena CF (eds) The Galápagos marine reserve: a dynamic social-ecological system. Springer Science and Business Media, New York, pp 3–22
Fernández-García C (2007) Propagación del alga *Caulerpa sertularioides* (Chlorophyta) en Bahía Culebra, Golfo de Papagayo, Pacífico norte de Costa Rica. M.Sc. thesis, Universidad de Costa Rica, SanJosé, Costa Rica p 92
Fernández C, Cortés J (2005) Reef site: *Caulerpa sertularioides*, a green alga spreading aggressively over coral reef communities in Culebra Bay, North Pacific of Costa Rica. Coral Reefs 24:10
Fernández-García C, Cortés J, Alvarado JJ, Nivia-Ruiz J (2012) Physical factors contributing to the benthic dominance of the alga *Caulerpa sertularioides* (Caulerpaceae, Chlorophyta) in the upwelling Bahía Culebra, north Pacific of Costa Rica. Rev Biol Trop 60(Suppl 2):93–107
Flores-Verdugo FJ, de la Lanza-Espino G, Contreras-Espinosa F, Agraz-Hernández CM (2001) The tropical Pacific coast of Mexico. In: Seeliger U, Kjerfve B (eds) Coastal marine ecosystems of Latin America. Ecol Stud 144:307–314. Springer, Berlin
Flot JF, Adjeroud M (2009) Les coraux de Clipperton. In: Charpy L (ed) Clipperton: environnement et biodiversité d'un microcosme océanique. Muséum national d'Histoire naturelle, Paris, IRD, Marseille (Patrimoines naturels; 68), pp 155–162
Flot J-F, Couloux A, Tillier S (2010) Haplowebs as a graphical tool for delimiting species: a revival of Doyle's "field for recombination" approach and its application to the coral genus *Pocillopora* in Clipperton. BMC Evol Biol 10:372
Forsbergh ED (1969) On the climatology, oceanography and fisheries of the Panama Bight. Inter-Am Trop Tuna Comm Bull 14:49–385
Forsman ZH, Guzmán HM, Chen CA, Fox GE, Wellington GM (2005) An ITS region phylogeny of *Siderastrea* (Cnidaria: Anthozoa): is *S. glynni* endangered or introduced? Coral Reefs 24:343–347
Friedlander AM, Ballesteros E, Beets J, Berkenpas E, Gaymer CF, Gorny M, Sala E (2013) Effects of isolation and fishing on the marine ecosystems of Easter Island and Salas y Gómez, Chile. Aquatic Conserv: Mar Freshw Ecosyst 23:515–531
Galzin R, Pointier J-P (1985) Ile de Moorea: Archipel de la Société. In: Delesalle B, Galzin R, Salvat B (eds) French Polynesian coral reefs. In: Proceedings of 5th International Coral Reef Congress, vol 1, Tahiti, pp 75–101
Garth JS (1991) Taxonomy, distribution, and ecology of Galápagos Brachyura. In: James MJ (ed) Galápagos marine invertebrates: taxonomy, biogeography, and evolution in Darwin's islands. Plenum Press, New York, pp 123–145
Garth JS (1992) The brachyuran crabs of the Revillagigedo Islands, Colima, Mexico with remarks on insular endemism in the eastern tropical Pacific. Proc San Diego Soc Nat Hist 24:1–6
Garzón-Ferreira J, Pinzón J (1999) Evaluación rápida de estructura y salud de las formaciones coralinas de la Isla de Malpelo (Pacífico colombiano). Bol Invest Mar Cost 28:137–154
Glynn PW (1976) Some physical and biological determinants of coral community structure in the eastern Pacific. Ecol Monogr 46:431–456
Glynn PW (1988) El Niño warming, coral mortality and reef framework destruction by echinoid bioerosion in the eastern Pacific. Galaxea 7:129–160
Glynn PW (1997) Eastern Pacific reef coral biogeography and faunal flux: Durham's dilemma revisited. In: Proceedings of 8th International Coral Reef Symposium, vol 1, Panama, pp 371–378
Glynn PW (2001) Eastern Pacific coral reef ecosystems. In: Seeliger U, Kjerfve B (eds) Coastal marine ecosystems of Latin America. Springer, Berlin, pp 281–305
Glynn PW (2003) Coral communities and coral reefs of Ecuador. In: Cortés J (ed) Latin American coral reefs. Elsevier Science BV, Amsterdam, pp 449–472
Glynn PW (2006) Fish utilization of simulated coral reef frameworks versus eroded rubble off Panama, eastern Pacific. In: Proceedings of 10th International Coral Reef Symposium, vol 10, Okinawa, pp 250–256
Glynn PW (2008) Food-web structure and dynamics of eastern tropical Pacific coral reefs: Panama and Galápagos Islands. In: McClanahan TR, Branch GM (eds) Food webs and the dynamics of marine reefs. Oxford University Press, Oxford, pp 185–208
Glynn PW (2011) In tandem reef coral and cryptic metazoan declines and extinctions. Bull Mar Sci 87:767–794
Glynn PW, Ault JS (2000) A biogeographic analysis and review of the far eastern Pacific coral reef region. Coral Reefs 19:1–23
Glynn PW, Colley SB (2008) Survival of brooding and broadcasting reef corals following large scale disturbances: is there any hope for broadcasting species during global warming? In: Proceedings of 11th International Coral Reef Symposium, Ft. Lauderdale (session 11), pp 1–5
Glynn PW, de Weerdt WH (1991) Elimination of two reef-building hydrocorals following the 1982–83 El Niño warming event. Science 253:69–71
Glynn PW, Feingold JS (1992) Hydrocoral species not extinct. Science 257:1845
Glynn PW, Fong P (2006) Patterns of reef coral recovery by the regrowth of surviving tissues following the 1997–98 El Niño warming and 2000, 2001 upwelling cool events in Panama, eastern Pacific. In: Proceedings of 10th International Coral Reef Symposium, Okinawa, pp 624–630
Glynn PW, Leyte-Morales GE (1997) Coral reefs of Huatulco, west Mexico. Reef development in upwelling Gulf of Tehuantepec. Rev Biol Trop 45:1033–1047
Glynn PW, Wellington GM (1983) Corals and coral reefs of the Galápagos Islands. University of California Press, Berkeley, p 297
Glynn PW, von Prahl H, Guhl F (1982) Coral reefs of Gorgona Island, with special reference to corallivores and their influence on community structure and reef development. An Inst Invest Mar Punta Betín 12:185–214
Glynn PW, Druffel EM, Dunbar RB (1983) A dead Central American coral reef tract: possible link with the Little Ice Age. J Mar Res 41:605–637
Glynn PW, Cortés J, Guzmán HM, Richmond RH (1988) El Niño (1982–83) associated coral mortality and relationship to sea surface temperature deviations in the tropical eastern Pacific. In: Proceedings of 6th International Coral Reef Symposium, vol 3, Brisbane, pp 237–243
Glynn PW, Gassman NJ, Eakin CM, Cortés J, Smith DB, Guzmán HM (1991) Reef coral reproduction in the eastern Pacific: Costa Rica, Panama, and Galápagos Islands (Ecuador), Part I—Pocilloporidae. Mar Biol 109:355–368

Glynn PW, Colley SB, Eakin CM, Smith DB, Cortés J, Gassman NJ, Guzmán HM, Rosario JB, Feingold J (1994) Reef coral reproduction in the eastern Pacific: Costa Rica, Panama, and Galápagos Islands (Ecuador) II. Poritidae. Mar Biol 118:191–208

Glynn PW, Veron JEN, Wellington GM (1996a) Clipperton Atoll (eastern Pacific): oceanography, geomorphology, reef-building coral ecology and biogeography. Coral Reefs 15:71–99

Glynn PW, Colley SB, Gassman NJ, Black K, Cortés J, Maté JL (1996b) Reef coral reproduction in the eastern Pacific: Costa Rica, Panamá, and Galápagos Islands (Ecuador). III. Agariciidae (*Pavona gigantea* and *Gardineroseris planulata*). Mar Biol 125:579–601

Glynn PW, Colley SB, Ting JH, Maté JL, Guzmán HM (2000) Reef coral reproduction in the eastern Pacific: Costa Rica, Panama, and Galápagos Islands (Ecuador)—IV. Agariciidae, recruitment and recovery of *Pavona varians* and *Pavona* sp.a. Mar Biol 136:785–805

Glynn PW, Maté JL, Baker AC, Calderón MO (2001) Coral bleaching and mortality in Panama and Ecuador during the 1997–1998 El Niño-Southern Oscillation event: spatial/temporal patterns and comparisons with the 1982–1983 event. Bull Mar Sci 69:79–109

Glynn PW, Wellington GM, Wieters EA, Navarrete SA (2003) Reef-building coral communities of Easter Island (Rapa Nui), Chile. In: Cortés J (ed) Latin American coral reefs. Elsevier Science BV, Amsterdam, pp 473–494

Glynn PW, Wellington GM, Riegl B, Olson DB, Borneman E, Wieters EA (2007) Diversity and biogeography of the scleractinian coral fauna of Easter Island (Rapa Nui). Pac Sci 61:67–90

Glynn PW, Colley SB, Maté JL, Cortés J, Guzmán HM, Bailey RL, Feingold JS, Enochs IC (2008) Reproductive ecology of the azooxanthellate coral *Tubastrea coccinea* in the Equatorial Eastern Pacific: Part V. Dendrophylliidae. Mar Biol 153:529–544

Glynn PW, Riegl B, Romanski AMS, Baums IB (2009) Rapid recovery of a coral reef at Darwin Island, Galapagos Islands. Galapagos Res 66:6–13

Glynn PW, Colley SB, Guzmán HM, Enochs IC, Cortés J, Maté JL, Feingold JS (2011) Reef coral reproduction in the eastern Pacific: Costa Rica, Panama and the Galápagos Islands (Ecuador). VI. Agariciidae, *Pavona clavus*. Mar Biol 158:1601–1617

Glynn PW, Colley SB, Maté JL, Baums IB, Feingold JS, Cortés J, Guzmán HM, Afflerbach JC, Brandtneris VW, Ault JS (2012) Reef coral reproduction in the equatorial eastern Pacific: Costa Rica, Panama, and the Galápagos Islands (Ecuador). VII. Siderastreidae, *Psammocora stellata* and *Psammocora profundacella*. Mar Biol 159:1917–1932

Glynn PW, Enochs IC, Afflerbach JA, Brandtneris VW, Serafy JE (2014) Eastern Pacific reef fish responses to coral reef recovery following El Niño disturbances. Mar Ecol Prog Ser 495:233–247

Glynn PW, Riegl BM, Purkis S, Kerr J, Smith T (2015) Coral reef recovery in the Galápagos Islands: the northern-most islands (Darwin and Wenman). Coral Reefs 34:421–436. doi:10.1007/s00338-015-1280-4

Gotuzzo R (1996) Celenterados. In: Serrano F (ed) Historia natural y ecología de El Salvador, vol II. Ministerio de Educación, San Salvador, El Salvador, pp 51–64

Grigg RW, Hey R (1992) Paleoceanography of the tropical eastern Pacific Ocean. Science 255:172–178

Guzmán HM (1988a) Distribución y abundancia de organismos coralívoros en los arrecifes coralinos de la Isla del Caño, Costa Rica. Rev Biol Trop 36:191–207

Guzmán HM (1988b) Feeding behavior of the gastropod corallivore *Quoyula monodonta* (Blainville). Rev Biol Trop 36:209–212

Guzmán HM, Cortés J (1989a) Coral reef community structure at Caño Island, Pacific Costa Rica. PSZNI: Mar Ecol 10:23–41

Guzmán HM, Cortés J (1989b) Growth rates of eight species of scleractinian corals in the eastern Pacific (Costa Rica). Bull Mar Sci 44:1186–1194

Guzmán HM, Cortés J (1992) Cocos Island (Pacific of Costa Rica) coral reefs after the 1982–83 El Niño disturbance. Rev Biol Trop 40:309–324

Guzmán HM, Cortés J (1993) Arrecifes coralinos del Pacífico oriental tropical: Revisión y perspectivas. Rev Biol Trop 41:535–557

Guzmán HM, Cortés J (2001) Changes in reef community structure after fifteen years of natural disturbances in the eastern Pacific (Costa Rica). Bull Mar Sci 69:133–149

Guzmán HM, Cortés J (2007) Reef recovery 20-yr after the 1982–83 El Niño massive mortality. Mar Biol 151:401–411

Guzmán HM, Obando VL (1988) Diversidad y abundancia diaria y estacional del zooplankton marino de la Isla del Caño, Costa Rica. Rev Biol Trop 36:139–150

Guzmán HM, Obando VL, Cortés J (1987a) Meiofauna associated with a Pacific coral reef in Costa Rica. Coral Reefs 6:107–112

Guzmán HM, Cortés J, Richmond RH, Glynn PW (1987b) Efectos del fenómeno de "El Niño-Oscilación Sureña" 1982/83 en los arrecifes de la Isla del Caño, Costa Rica. Rev Biol Trop 35:325–332

Guzmán HM, Cortés J, Glynn PW, Richmond RH (1990) Coral mortality associated with dinoflagellate blooms in the eastern Pacific (Costa Rica and Panama). Mar Ecol Prog Ser 60:299–303

Guzmán HM, Guevara CA, Breedy O (2004) Distribution, diversity, and conservation of coral reefs and coral communities in the largest marine protected area of Pacific Panama (Coiba Island). Environ Conserv 31:111–121

Guzmán HM, Benfield S, Breedy O, Mair JM (2008) Broadening reef protection across the marine conservation corridor of the eastern tropical Pacific: distribution and diversity of reefs in Las Perlas Archipelago, Panama. Environ Conserv 35:46–54

Halfar J, Godinez-Orta L, Riegl B, Valdez-Holguin JE, Borges JM (2005) Living on the edge: high latitude *Porites* carbonate production under temperate eutrophic conditions. Coral Reefs 24:582–592

Harmelin-Vivien M (1985) Tikehau Atoll, Tuamotu Archipelago. In: Delesalle B, Galzin R, Salvat B (eds) Proceedings of 5th International Coral Reef Congress, vol 1, Tahiti, pp 211–268

Hebbeln D, Cortés J (2001) Sedimentation in a tropical fyord: Golfo Dulce, Costa Rica. Geo-Mar Lett 20:142–148

Heck KL Jr, McCoy ED (1978) Long-distance dispersal and the reef-building corals of the Eastern Pacific. Mar Biol 48:349–356

Hendrickx ME, Brusca RC, Findley LT (2005) A distributional checklist of the macrofauna of the Gulf of California, Mexico. Part 1. Invertebrates. Tucson, Arizona-Sonora Desert Museum, p 429

Herrera W (1986) Clima de Costa Rica. Edit. UNED, San José, Costa Rica, p 118

Herrera-Escalante T, López-Pérez RA, Leyte-Morales GE (2005) Bioerosion caused by the sea urchin *Diadema mexicanum* (Echinodermata: Echinoidea) at Bahías de Huatulco, western Mexico. Rev Biol Trop 53(Suppl 3):263–273

Hickman PC Jr (2008) A field guide to corals and other radiates of Galápagos. Galápagos marine life series. Sugar Spring Press, Lexington, Virgina p 161

Holst I, Guzmán HM (1993) Lista de corales hermatípicos (Anthozoa: Scleractinia; Hydrozoa: Milleporina) a ambos lados del Istmo de Panamá. Rev Biol Trop 41:871–875

Hubbard DK, Garcia M (2003) The corals and coral reefs of Easter Island—A preliminary look. In: Loret J, Tanacredi JT (eds) Easter Island: scientific exploration into the world's environmental problems in microcosm. Kluwer Academic, New York, pp 53–77

Iglesias-Prieto R, Reyes-Bonilla H, Riosmena-Rodriguez R (2003) Effects of 1997–1998 ENSO on coral reef communities in the Gulf of California, Mexico. Geofísica Int 42:1–5

Iglesias-Prieto R, Beltrán VH, LaJeunesse TC, Reyes-Bonilla H, Thomé PE (2004) Different algal symbionts explain the vertical distribution of dominant reef corals in the eastern Pacific. Proc R Soc Lond B 271:1757–1763

Irving R, Dawson T (2012) The marine environment of the Pitcairn Islands. The Pew Environment Group, Global Ocean Legacy, Dundee University Press, p 61
Jiménez C (1997) Corals and coral reefs of Culebra Bay, Pacific coast of Costa Rica: anarchy in the reef. In: Proceedings of 8th International Coral Reef Symposium, vol 1, Panama, pp 329–334
Jiménez C (2001a) Arrecifes y ambientes coralinos de Bahía Culebra, Pacífico de Costa Rica: aspectos biológicos, económico-recreactivos y de manejo. Rev Biol Trop 49(Suppl 2):215–231
Jiménez C (2001b) Seawater temperature measured at the surface and at two depths (7 and 14 m) in one coral reef at Culebra Bay, Gulf of Papagayo, Costa Rica. Rev Biol Trop 49(2):153–161
Jiménez C (2007a) Evaluación ecológica rápida del arrecife coralino de Playa Matapalo (Golfo de Papagayo), uno de los arrecifes más extensos en la costa del Pacífico de Costa Rica. Informe para Conservación Internacional, San José, Costa Rica, p 42
Jiménez C (2007b) Evaluación rápida del blanqueamiento y mortalidad de corales ocurrida en el Golfo de Papagayo (Septiembre-Diciembre 2007). Unpub report to Vicerrectoría de Investigación de la Universidad de Costa Rica, p 36
Jiménez CE, Cortés J (2001) Effects of the 1991-1992 El Niño on scleractinian corals of the Costa Rican central Pacific coast. Rev Biol Trop 49 (Supl 2):239-250
Jiménez J, Soto R (1985) Patrones regionales en la estructura y composición florística de los manglares de la costa Pacífica de Costa Rica. Rev Biol Trop 33:25–37
Jiménez C, Cortés J, León A, Ruíz E (2001) Coral bleaching and mortality associated with El Niño 1997/98 event in an upwelling environment in the eastern Pacific (Gulf of Papagayo, Costa Rica). Bull Mar Sci 69:151–169
Jiménez C, Sánchez C, Ruíz E, Buxton T (2009) Informe de la re-evaluación de algunos ambientes coralinos de las Islas Murciélagos (ACG), Mayo 2009. Unpubl report to Research Program Área de Conservación Guanacaste (ACG), p 29
Jiménez C, Bassey G, Segura Á, Cortés J (2010) Characterization of the coral communities and reefs of two previously undescribed locations in the upwelling region of Gulf of Papagayo (Costa Rica). Rev Mar Cost 2:95–108
Johnson ME, Backus DH, Riosmena-Rodríguez R (2009) Contribution of rhodoliths to the generation of Pliocene-Pleistocene limestone in the Gulf of California. In: Johnson ME, Ledesma-Vázquez J (eds) Atlas of coastal ecosystems in the western Gulf of California: tracking limestone deposits on the margin of a young sea. University of Arizona Press, Tucson, pp 83–94
Kaiser KL (2009) Les mollusques. In: Charpy L (ed) Clipperton, environnement et biodiversité d'un microcosme océanique. Mus National d'Hist Naturelle, Paris IRD, Marseille, pp 217–234
Kappelle M, Castro M, Acevedo H, González L, Monge H (2002) Ecosistemas del área de Conservación Osa (ACOSA). Edit. INBio, Sto. Domingo de Heredia, Costa Rica, p 500
Karlson RH (1999) Dynamics of coral communities. Kluwer Academic Publisher, Drodrecht, p 290
Ketchum JT, Reyes-Bonilla H (1997) Biogeography of hermatypic corals of the Archipiélago Revillagigedo, Mexico. In: Proceedings of 8th International Coral Reef Symposium, vol 1, Panama, pp 471–476
Ketchum JT, Reyes-Bonilla H (2001) Taxonomía y distribución de los corales hermatípicos (Scleractinia) del Archipiélago de Revillagigedo, México. Rev Biol Trop 49:803–848
Kleypas JA, McManus JW, Meñez LAB (1999) Environmental limits to coral reef development: where do we draw the line? Am Zool 39:146–159
Ledesma-Vázquez J, Johnson ME, Gonzalez-Yajimovich O, Santamaría-del-Ángel E (2009) Gulf of California geography, geological origins, oceanography, and sedimentation patterns. In: Johnson ME, Ledesma-Vázquez J (eds) Atlas of coastal ecosystems in the western Gulf of California: tracking limestone deposits on the margin of a young sea. University of Arizona Press, Tucson, pp 1–10
Lemus LS, Pocasangre JA, Zelaya TD (1994) Evaluación del estado actual de la distribución y cobertura de los arrecifes coralinos en la zona de Los Cóbanos, Departamento de Sonsonate. Licenciatura thesis, Escuela de Biología, Facultad de Ciencias Naturales y Matemática, Univ de El Salvador, San Salvador, p 67
Lessios HA, Robertson DR (2006) Crossing the impassable: genetic connections in 20 reef fishes across the eastern Pacific barrier. Proc R Soc London B 273:2201–2208
Lessios HA, Kessing BD, Robertson DR (1998) Massive gene flow across the world's most potent marine biogeographic barrier. Proc R Soc London B 265:583–588
Lewis JB (1989) The ecology of *Millepora*: a review. Coral Reefs 8:99–107
Leyte-Morales GE (2000) Perturbaciones naturales y antropogénicas en las comunidades coralinas de Oaxaca (1977–1998). Resumen. XII Congreso Nacional de Oceanografía. Huatulco, Oaxaca, p 5
Lirman D, Glynn PW, Baker AC, Leyte-Morales GE (2001) Combined effects of three sequential storms on the Huatulco coral reef tract, Mexico. Bull Mar Sci 69:267–278
Lobel PS, Lobel LK (2008) Aspects of the biology and geomorphology of Johnston and Wake Atolls, Pacific Ocean. In: Riegl BM, Dodge RE (eds) Coral reefs of the USA, Coral reefs of the world 1. Springer, Berlin, pp 655–689
López WA, Jiménez C (2008) Primer registro de *Psammocora obtusangula* y *P. stellata* (Anthozoa: Scleractinia: Siderastreidae) para la costa salvadoreña. Resúmenes XII Congreso de la Sociedad Mesoamericana para la Biología y la Conservación (SMBC), San Salvador, El Salvador. Mesoamericana 12(3):99
López-Pérez RA, Budd AF (2009) Coral diversification in the Gulf of California during the Late Miocene to Pleistocene. In: Johnson ME, Ledesma-Vázquez J (eds) Atlas of coastal ecosystems in the western Gulf of California: tracking limestone deposits on the margin of a young sea. University of Arizona Press, Tucson, pp 58–71
López-Pérez RA, Hernández-Ballesteros LM (2004) Coral community structure and dynamics in the Huatulco area, western Mexico. Bull Mar Sci 75:453–472
López-Pérez RA, Hernández-Ballesteros LM, Herrera-Escalante T (2002) Cambio en la dominancia de la comunidad arrecifal en Chachacual, Bahías de Huatulco, Oaxaca. Cienc Mar 16:33–38
López-Pérez RA, Mora-Pérez MG, Leyte-Morales GE (2007) Coral (Anthozoa: Scleractinia) recruitment at Bahías de Huatulco, western Mexico: implications for coral community structure and dynamics. Pac Sci 61:355–369
López-Sandoval DC, Lara-Lara JR, Álvarez-Borrego S (2009) Phytoplankton production by remote sensing in the region off Cabo Corrientes, Mexico. Hidrobiológica 19:185–192
Macintyre IG, Glynn PW, Cortés J (1993) Holocene reef history in the eastern Pacific: mainland Costa Rica, Caño Island, Cocos Island, and Galápagos Islands. In: Proceedings of 7th International Coral Reef Symposium, vol 2, Guam, pp 1174–1184
Manzello DP (2010) Coral growth with thermal stress and ocean acidification: lessons from the eastern tropical Pacific. Coral Reefs 29:749–758
Maragos JE (1974) Reef corals of Fanning Island. Pac Sci 28:247–255
Maragos JE (1995) Revised checklist of extant shallow-water stony coral species from Hawaii (Cnidaria: Anthozoa: Scleractinia). Bishop Mus Occas Pap 42:54–55
Maragos JE, Jokiel PL (1986) Reef corals of Johnston Atoll: one of the world's most isolated reefs. Coral Reefs 4:141–150
Maragos JE, Potts DC, Aeby G, Gulko D, Kenyon J, Siciliano D, Van Ravenswaay D (2004) 2000–2002 rapid ecological assessment of corals (Anthozoa) on shallow reefs of the northwestern Hawaiian Islands. Part 1: Species and distribution. Pac Sci 58:211–230

Maragos JE, Miller J, Gove J, DeMartini E, Friedlander AM, Godwin S, Musburger C, Timmers M et al (2008a) US coral reefs in the Line and Phoenix Islands, Central Pacific Ocean: history, geology, oceanography, and biology. In: Riegl BM, Dodge RE (eds) Coral reefs of the USA. Springer Science + Business Media BV, Berlin, pp 595–654

Maragos JE, Miller J, Gove J, DeMartini E, Friedlander AM, Godwin S, Musburger C, Timmers M, Tsuda R, Vroom P, Flint E, Lundblad E, Weiss J, Avotte P, Sala E, Sandin S, McTee S, Wass T, Siciliano D, Brainard R, Obura D, Ferguson S, Mundy B (2008b) U.S. coral reefs in the Line and Phoenix Islands, Central Pacific Ocean: history, geology, oceanography, and biology. In: Riegl B, Dodge RE (eds) Coral reefs of the USA. Coral Reefs of the World, vol 1. Springer, Berlin, pp 595–641

Maragos JE, Kenyon J, Aeby G, Vroom P, Vargas-Ángel B, Brainard R, Wedding L, Friedlander A, Asher J, Zgliczynski B, Siciliano D (2009) Benthic communities of the Northwestern Hawaiian Islands. In: Friedlander A, Keller, K, Wedding L, Clarke A, Monaco M (eds) A marine biogeographic assessment of the northwestern Hawaiian Islands. NOAA Tech Memorandum NOS NCCOS 84. Prepared by the NCCOS Biogeography Branch in cooperation with the Office of National Marine Sanctuaries, Silver Spring, Maryland, pp 105–154

Martínez P, Rivera F, Proaño F (2011) Ambientes coralinos del Parque Nacional Machalilla y Reserva de Producción Faunística Marino Costera Puntilla de Santa Elena: Un estudio caso para el manejo de los corales en Ecuador. Reporte Técnico del Instituto Nazca de Investigaciones Marinas y Conservación Internacional, pp 27–50

Maté JL (2003) Corals and coral reefs of the Pacific coast of Panamá. In: Cortés J (ed) Latin American coral reefs. Elsevier, Amsterdam, pp 387–417

Merlen G, Quevedo FO, Pépolas R (2009) Protection of shallow marine ecosystems in Galápagos by permanent moorings. Galapagos Res 66:75–76

Montaggioni LF (1985) Île de Makatea: Archipel des Tuamotu. In: Delesalle B, Galzin R, Salvat B (eds) Vol 1: French Polynesian coral reefs. Proceedings of 5th International Coral Reef Congress, vol 1, Tahiti, pp 103–157

Obura D (2002) Coral species diversity. In: Obura D, Stine GS (eds) Phoenix Island: summary of marine and terrestrial assessment. New England Aquarium, Massachusetts, pp 10–12

Orellana-Amador JJ (1985) Especies marinas de Los Cóbanos. Peces de El Salvador. Mandarin Offset International Ltd, Hong Kong, p 126

Palacios MM, Zapata FA (2014) Fish community structure on coral habitats with contrasting architecture in the Tropical Eastern Pacific. Rev Biol Trop 62(Suppl 1):343–357

Palacios MM, Muñoz CG, Zapata FA (2014) Fish corallivory on a pocilloporid reef and experimental coral responses to predation. Coral Reefs. doi:10.1007/s00338-014-1173-y

Paulay G (1989) Marine invertebrates of the Pitcairn Islands: species composition and biogeography of corals, molluscs and echinoderms. Atoll Res Bull 326:1–28

Payri C, Menon J-L, N'Yeurt A (2009) La flore marine du complexe récifal et quelques aspects de la biodiversité et de la géomorphologie de l'île. In: Charpy L (ed) Clipperton, environnement et biodiversité d'un microcosme océanique. Mus National d'Hist Naturelle, Paris IRD, Marseille, pp 129–141

Paz-García DA, Correa-Sandoval F, Chávez-Romo HE, Reyes-Bonilla H, López-Pérez RA, Medina-Rosas P, Hernández-Cortés P (2008) Genetic structure of the massive coral *Porites panamensis* (Anthozoa: Scleractinia) from the Mexican Pacific. In: Proceedings of 11th International Coral Reef Symposium, vol 1, Ft Lauderdale, pp 456–460

Paz-García DA, Balart EF, García-de-Léon FJ (2012) Cold water bleaching of *Pocillopora* in the Gulf of California. In: Proceedings of 12th International Coral Reef Symposium, vol 9A–10, Cairns, pp 1–6

Pérez-Vivar TL, Reyes-Bonilla H, Padilla C (2006) Corales pétreos (Scleractinia) de las Islas Marías, Pacífico de México. Cienc Mar 32:259–270

Pinzón JH, LaJeunesse TC (2011) Species delimitation of common reef corals in the genus *Pocillopora* using nucleotide sequence phylogenies, population genetics and symbioses ecology. Mol Ecol 20:311–325

Pinzón JH, Sampayo EM, Cox EF, Chauka LJ, Chen CA, Voolstra CR, LaJeunesse TC (2013) Blind to morphology: genetics identifies several widespread ecologically common species and few endemics among Indo-Pacific cauliflower corals (*Pocillopora*, Scleractinia). J Biogeogr 40:1595–1608

Poupin J, Bouchard J-M, Albenga L, Cleva R, Hermoso-Salazar M, Solis-Weiss V (2009) Les crustacés décapodes et stomatopodes, inventaire, écologie et zoogéographie. In: Charpy L (ed) Clipperton, environnement et biodiversité d'un microcosme océanique. Mus National d'Hist Naturelle, Paris IRD, Marseille, pp 163–216

Qiu J-W, Lau DCC, Cheang C-c, Chow W-k (2014) Community-level destruction of hard corals by the sea urchin *Diadema setosum*. Mar Pollut Bull. doi:10.1016/j.marpolbul. 12 Dec 2013

Razak TB, Hoeksema BW (2003) The hydrocoral genus *Millepora* (Hydrozoa: Capitata: Milleporidae) in Indonesia. Zool Verh Leiden 345:313–336

Restrepo J, Kjerfve B (2000) Water discharge and sediment load from the western slopes of the Colombian Andes with focus on Río San Juan. J Geol 108:17–33

Reyes-Bonilla H (1993a) Estructura de la comunidad, influencia de la depredación y biología poblacional de corales hermatípicos en el arrecife de Cabo Pulmo, Baja California Sur. MSc thesis, Centro de Investigación Científica y Educación Superior de Ensenada, Ensenada, p 169

Reyes-Bonilla H (1993b) Biogeografía y ecología de los corales hermatípicos (Anthozoa: Scleractinia) del Pacífico de México. In: Salazar-Vallejo SI, González NE (eds) Biodiversidad marina y costera de México. CONABIO/CIQRO, Chetumal, pp 207–222

Reyes-Bonilla H (2001) Effects of the 1997–1998 El Niño-Southern Oscillation on coral communities of the Gulf of California, Mexico. Bull Mar Sci 69:251–266

Reyes-Bonilla H (2003) Coral reefs of the Pacific coast of México. In: Cortés J (ed) Latin American coral reefs. Elsevier Science BV, Amsterdam, pp 331–349

Reyes-Bonilla H, Barraza JE (2003) Corals and associated marine communities from El Salvador. In: Cortés J (ed) Latin American coral reefs. Elsevier, Amsterdam, pp 351–360

Reyes-Bonilla H, Carricart-Ganivet JP (2000) *Porites arnaudi*, a new species of stony coral (Anthozoa: Scleractinia: Poritidae) from oceanic islands of the eastern Pacific Ocean. Proc Biol Soc Wash 113:561–571

Reyes-Bonilla H, López-Pérez RA (1998) Biogeography of the stony corals of the Mexican Pacific. Ciencias Marinas 24:211–224

Reyes-Bonilla H, López-Pérez RA (2009) Corals and coral-reef communities in the Gulf of California. In: Johnson ME, Ledesma-Vázquez J (eds) Atlas of coastal ecosystems in the western Gulf of California: tracking limestone deposits on the margin of a young sea. University of Arizona Press, Tucson, pp 43–55

Reyes-Bonilla H, Riosmena-Rodriguez R, Foster MS (1997a) Hermatypic corals associated with rhodolith beds in the Gulf of California, México. Pac Sci 51:328–337

Reyes-Bonilla H, Sinsel-Duarte H, Arizpe-Covarrubias F (1997b) Gorgonias y corales pétreos (Anthozoa: Gorgonacea y Scleractinia) de Cabo Pulmo, Mexico. Rev Biol Trop 45:1439–1443

Reyes-Bonilla H, Calderón Aguilera LE, Cruz Piñon G, Medina Rosas P, López Pérez RA, Herrero Pérezrul MD, Leyte Morales GE, Cupul Magaña AL, Carriquiry Beltrán JD (2005) Atlas de corales pétreos (Anthozoa: Scleractinia) del Pacífico mexicano. Universidad de Guadalajara, Guadalajara, TRICICLO triciclocincocinco@yahoo.com.mx, p 124

Reyes-Bonilla H, Calderón-Aguilera LE, Cruz-Piñón G, López-Pérez RA, Medina-Rosas P (2010) Evaluation of gamma diversity of reef corals (Scleractinia) in the Mexican Pacific. Rev Mexicana Biodiversidad 81:113–121

Ricard M (1985) Atoll Rangiroa, Archipel des Tuamotu. In: Delesalle B, Galzin R, Salvat B (eds) Proceedings of 5th International Coral Reef Congress, vol 1, Tahiti, pp 159–210

Riegl BM, Halfar J, Purkis SJ, Godinez-Orta L (2007) Sedimentary facies of the eastern Pacific's northernmost reef-like setting (Cabo Pulmo, Mexico). Mar Geol 236:61–77

Rivera F, Martínez P (2011) Guía fotográfica de corales y octocorales: Parque Nacional Machalilla y Reserva de Producción Faunística Marino Costera Puntilla de Santa Elena. With the contribution of Odalisca Breedy, NAZCA and Conservación Internacional, Ecuador, p 78

Rixen T, Jiménez C, Cortés J (2012) Impact of upwelling events on the seawater carbonate chemistry and dissolved oxygen concentration in the Gulf of Papagayo (Culebra Bay), Costa Rica: implications for coral reefs. Rev Biol Trop (Int J Trop Biol) 60 (Suppl 2):187–195. ISSN-0034-7744

Robertson DR, Allen GR (1996) Zoogeography of the shorefish fauna of Clipperton Atoll. Coral Reefs 15:121–131

Robertson DR, Allen GR (2008) Shorefishes of the tropical eastern Pacific online information system, version 1.0. www.stri.org/sftep. Accessed 23 June 2014

Robertson DR, Cramer KL (2009) Shore fishes and biogeographic subdivisions of the tropical eastern Pacific. Mar Ecol Prog Ser 380:1–17

Robertson DR, Grove JS, McCosker JE (2004) Tropical transpacific shore fishes. Pac Sci 58:507–565

Rodríguez-Ramírez A, Zapata FA (2012) An atypical corallum morphology in a species of *Pocillopora*. Coral Reefs 31:253

Rodríguez-Rubio E, Schneider W, Abarca del Río R (2003) On the seasonal circulation within the Panama Bight derived from satellite observations of wind, altimetry and sea surface temperature. Geophys Res Lett 30:1410. doi:10.1029/2002GL016794

Rojas W, Alvarado GE (2012) Marco geológico y tectónico de la Isla del Coco y la region marítima circunvecina, Costa Rica. Rev Biol Trop 60(Suppl 3):15–32

Ryan J, Zapata Y (2003) Nicaragua's coral reefs: status, health and management strategies. In: Cortés J (ed) Latin American coral reefs. Elsevier, Amsterdam, pp 203–222

Saavedra-Sotelo NC, Calderon-Aguilera LE, Reyes-Bonilla H, López--Pérez RA, Medina-Rosas P, Rocha-Olivares A (2011) Limited genetic connectivity of *Pavona gigantea* in the Mexican Pacific. Coral Reefs 30:677–686

Salvat B, Richard G (1985) Takapoto Atoll: Tuamotu Archipelago. In: French Polynesian coral reefs. Proceedings of 5th International Coral Reef Congress, vol 1, Tahiti, pp 323–362

Salvat B, Adjeroud M, Charpy L (2008) Les récifs coralliens de Clipperton. Rev Écol (Terre Vie) 63:179–187

Schaeffer BA, Morrison JM, Kamykowski D, Feldman GC, Xie L, Liu Y, Sweet W, McCulloch A, Banks SA (2008) Phytoplankton biomass distribution and identification of productive habitats within the Galapagos Marine Reserve by MODIS, a surface acquisition system, and in-situ measurements. Remote Sens Environ 112:3044–3054

Scheltema RS (1988) Initial edvidence for the transport of teleplanic larvae of benthic invertebrates across the east Pacific barrier. Biol Bull 174:145–152

Schmidt-Roach S, Miller KJ, Lundgren P, Andreakis N (2014) With eyes wide open: a revision of species within and closely related to the *Pocillopora damicornis* species complex (Scleractinia; Pocilloporidae) using morphology and genetics. Zool J Linn Soc 170:1–33

Segovia-Prado JV, Navarrete-Calero MT (2007) Biodiversidad a nivel de ecosistema en parches de corales hermatípicos (*Porites lobata*, *Pocillopora* sp.) en la zona intermareal de la Playa Los Cóbanos, Departamento de Sonsonante, El Salvador. Licenciatura thesis, Escuela de Biología, Facultad de Ciencias Naturales y Matemática, Univ de El Salvador, San Salvador, p 68

Smith TB, Glynn PW, Maté JL, Toth LT, Gyory J (2014) A depth refugium from catastrophic coral bleaching prevents regional extinctions. Ecology 95:1663–1673

Solís-Marín F-A, Laguarda-Figueras A (2009) Les échinoderms. In: Charpy L (ed) Clipperton, environnement et biodiversité d'un microcosme océanique. Mus National d'Hist Naturelle, Paris IRD, Marseille, pp 235–247

Sonnenholzner JI, Ladah LB, Lafferty KD (2009) Cascading effects of fishing on Galapagos rocky reef communities: reanalysis using corrected data. Mar Ecol Prog Ser 375:209–218

Spalding MD, Ravilious C, Green EP (2001) World atlas of coral reefs. UNEP/WCMC, Univ California Press, Berkeley, p 424

Springer VG (1958) Systematics and zoogeography of the clinid fishes of the subtribe Labrisomini Hubbs. Pubs Inst Mar Sci Univ Texas 5:417–492

Squires DF (1959) Results of the Puritan-American Museum of Natural History Expedition to western Mexico, 7. Corals and coral reefs in the Gulf of California. Bull Amer Mus Nat Hist 118:367–432

Steenburgh WJ, Schultz DM, Colle BA (1998) The structure and evolution of gap outflow over the Gulf of Tehuantepec, Mexico. Mon Weather Rev 126:2673–2691

Stefani F, Benzoni F, Pichon M, Cancelliere C, Galli P (2008) A multidisciplinary approach to the definition of species boundaries in branching species of the coral genus *Psammocora* (Cnidaria, Scleractinia). Zoolog Scr 37:71–91

Stehli FG, Wells JW (1971) Diversity and age patterns in hermatypic corals. Syst Zool 20:115–126

Steller DL, Riosmena-Rodríguez R, Foster MS, Roberts CA (2003) Rhodolith bed diversity in the Gulf of California: the importance of rhodolith structure and consequences of disturbance. Aquatic Conserv: Mar Freshw Ecosyst 13:S5–S20

Steller DL, Riosmena-Rodríguez R, Foster MS (2009) Living rhodolith bed ecosystems in the Gulf of California. In: Johnson ME, Ledesma-Vázquez J (eds) Atlas of coastal ecosystems in the western Gulf of California: tracking limestone deposits on the margin of a young sea. University of Arizona Press, Tucson, pp 72–82

Stevenson MR, Guillén O, Santoro de Ycaza J (1970) Marine atlas of the Pacific coastal waters of South America. University of California Press, Berkeley, p 23

Sweet WV, Morrison JM, Liu Y, Kamykowski D, Schaeffer BA, Xie L, Banks S (2009) Tropical instability wave interactions within the Galápagos Archipelago. Deep-Sea Res Part I: Oceanogr Res Paps 56:1217–1229

Trichet J (2009) Origine, situation et traits morphologiques généraux de l'île. In: Charpy L (ed) Clipperton, environnement et biodiversité d'un microcosme océanique. Muséum national d'Histoire naturelle, Paris; IRD, Marseille. Patrimoines naturels 68:23–28

Vargas-Ángel B (1996) Distribution and community structure of the reef corals of Ensenada de Utría, Pacific coast of Colombia. Rev Biol Trop 44:643–651

Vargas-Ángel B (2001) Corals and coral reefs of the Pacific coast of Colombia with special reference to spatial and temporal patterns of environmental disturbances. PhD dissert, University of Miami, Coral Gables, Florida, p 209

Vargas-Ángel B (2003) Coral community structure off the Pacific coast of Colombia: onshore vs offshore reefs. Atoll Res Bull 499:1–21

Vargas-Ángel B, Zapata FA, Hernández H, Jiménez JM (2001) Coral and coral reef responses to the 1997–98 El Niño event on the Pacific coast of Colombia. Bull Mar Sci 69:111–132

Vera M, Banks SA (2009) Health status of the coralline communities of the northern Galapagos Islands Darwin, Wolf and Marchena. Galapagos Res 66:65–74

Veron JEN (1995) Corals in space and time: the biogeography and evolution of the Scleractinia. Comstock/Cornell, Ithaca, London, p 321

Veron JEN (2000) Corals of the world, vols 1–3. Austrlian Institute of Marine Science, Townsville

Veron J, Stafford-Smith M, DeVantier L, Turak E (2015) Overview of distribution patterns of zooxanthellate Scleractinia. Front Mar Sci 1:81. doi:10.3389/fmars.2014.00081

von Prahl H (1986a) Corales y arrecifes coralinos. In: von Prahl H, Alberico M (eds) Isla de Gorgona, Biblioteca Banco Popular. Textos Universitarios, Bogotá, Colombia, pp 59–87

von Prahl H (1986b) Crustáceos decápodos asociados a diferentes hábitas de la Ensenada de Utría. Actualidades Biológicas 15:95–99

von Prahl H (1990) Malpelo la roca viviente. Fondo FEN COLOMBIA, Editorial Presencia, Bogotá, Colombia

von Prahl H, Erhardt H (1985) Colombia: corales y arrecifes coralinos. Fondo FEN COLOMBIA, Editorial Presencia, Bogotá, Colombia, p 295

von Prahl H, Mejía A (1985) Primer reporte de un coral acropórido, *Acropora valida* (Dana, 1846), (Scleractinia: Astrocoeniida: Acroporidae) para el Pacífico Americano. Rev Biol Trop 33:39–43

von Prahl H, Guhl F, Grogl M (1979) Gorgona. Futura Grupo Editorial Ltd, Bogotá, Colombia, p 279

Wallace CC (1999) Staghorn corals of the world. CSIRO Publ, Melbourne, p 422

Walsh WJ, Cotton S, Jackson L, Lamson M, Martin R, Osada-D'Avella K, Preskitt L (2013) First record of *Acropora gemmifera* in the main Hawaiian Islands. Coral Reefs. doi:10.1007/s00338-013-1092-3

Wellington GM (1982) Depth zonation of corals in the Gulf of Panama: control and facilitation by resident reef fishes. Ecol Monogr 52:223–241

Wellington GM, Glynn PW (2007) Responses of coral reefs to El Niño-Southern Oscillation sea-warming events. In: Aronson RB (ed) Geological approaches to coral reef ecology. Ecol Stud 192:342–385. Springer, New York

Wellington GM, Glynn PW, Strong AE, Navarrete SA, Wieters E, Hubbard DK (2001) Crisis on coral reefs linked to climate change. EOS, Trans Amer Geophys Union 82(1):5

Wells JW (1983) Annotated list of the scleractinian corals of the Galápagos. In: Glynn PW, Wellington GM (eds) Corals and coral reefs of the Galápagos Islands. University of California Press, Berkeley, pp 212–295

Wieters EA, Medrano A, Pérez-Matus A (2014) Functional community structure of shallow hard bottom communities at Rapa Nui, Easter Island. Lat Am J Aquat Res 42:827–844

Williams GJ, Maragos JE, Davy SK (2008) Characterization of the coral communities at Palmyra Atoll in the remote Central Pacific Ocean. Atoll Res Bull 557:1–30

Zapata FA (2001) Formaciones coralinas de Isla Gorgona. In: Barrios LM, López-Victoria M (eds) Gorgona marina: contribución al conocimiento de una isla única. INVEMAR, Serie Publicaciones Especiales No. 7, Santa Marta, pp 27–40

Zapata FA, Vargas-Ángel B (2003) Corals and coral reefs of the Pacific coast of Colombia. In: Cortés J (ed) Latin American coral reefs. Elsevier Science, Amsterdam, pp 419–447

Zapata FA, Vargas-Ángel B, Garzón-Ferreira J (2001) Salud y conservación de las comunidades coralinas In: Barrios LM, López-Victoria M (eds) Gorgona marina: Contribución al conocimiento de una isla única. INVEMAR, Serie Publicaciones Especiales No. 7, Santa Marta, pp 41–50

Zapata FA, Rodríguez-Ramírez A, Rodríguez-Moreno M, Muñoz C, López-Victoria M (2007) Confirmation of the occurrence of the coral *Pavona chiriquiensis* Cnidaria: Anthozoa: (Agariciidae) in the Colombian Pacific. Boletín de Investigaciones Marinas y Costeras 36:307–312

Zapata FA, García JL, Tobón A (2008) Formaciones coralinas en las localidades de Piñas (Punta Cruces) y El Acuario (Cabo Marzo). In: Giraldo A, Valencia B (eds) Chocó: paraíso por naturaleza - Punta Cruces y Cabo Marzo. Departamento de Biología, Universidad del Valle, Cali, Colombia, pp 23–32

Zapata FA, Rodríguez-Ramírez A, Caro-Zambrano C, Garzón-Ferreira J (2010) Mid-term coral-algal dynamics and conservation status of a Gorgona Island (Tropical Eastern Pacific) coral reef. Rev Biol Trop 58(Suppl. 1):81–94

Zapata FA, Jaramillo González J, Navas Camacho R (2011) Extensive bleaching of the coral *Porites lobata* at Malpelo, Colombia, during a cold water episode in 2009. Bol Invest Mar Cost 40(spec suppl):185–193

Holocene Reef Development in the Eastern Tropical Pacific

6

Lauren T. Toth, Ian G. Macintyre, and Richard B. Aronson

Abstract

Contrary to early assessments, the eastern tropical Pacific (ETP) is not devoid of well-developed reefs. Significant accumulations of Holocene reef framework are present throughout the region, although they tend to be poorly consolidated, lack the submarine cementation common on most reefs elsewhere in the world, and are subject to considerable bioerosion. These reef frameworks began accreting as early as 7000 years ago. The thickest accumulations of *Pocillopora* frameworks occur in coastal areas of Mexico, Costa Rica, Panama, and Colombia, but reefs composed of massive corals—species of *Porites*, *Pavona*, or *Gardineroseris*—are present throughout the region. Reef development in the ETP is limited by a variety of characteristics of the physical environment. Because of high turbidity in most areas, reef development is generally restricted to less than ~10 m depth. The spatial extent of reefs in the ETP is also limited from the combined influences of wave action and upwelling. Most reefs in the ETP are only a few hectares in size and the best-developed reefs generally occur in areas sheltered from strong oceanic influence. Upwelling also influences long-term trends in reef development in the region. There does not appear to be a significant impact of upwelling on the millennial-scale growth rates of Panamanian reefs; however, reefs in upwelling environments typically have thinner frameworks than nearby reefs in non-upwelling environments. Furthermore, upwelling may have contributed to a historic shutdown of reef development in Costa Rica and Panama. Although both ecological and oceanographic disturbances have had some impact on the long-term development of reefs in the ETP, the most important control on reef development in this region throughout the Holocene has most likely been the El Niño–Southern Oscillation (ENSO). ENSO activity—especially that of the 1982–83 and 1997–98 El Niño events—has shaped the landscape of coral reefs across the ETP both in recent decades and

L.T. Toth · R.B. Aronson
Department of Biological Sciences, Florida Institute of Technology, 150 West University Boulevard, Melbourne, FL 32901, USA
e-mail: raronson@fit.edu

I.G. Macintyre
Department of Paleobiology, National Museum of Natural History, Smithsonian Institution, MRC 121, P.O. Box 37012, Washington DC, 20013-7012, USA
e-mail: macintyr@si.edu

L.T. Toth (✉)
USGS Coastal and Marine Science Center, 600 Fourth Street South, Saint Petersburg, FL 33701-4846, USA
e-mail: ltoth@usgs.gov

P.W. Glynn et al. (eds.), *Coral Reefs of the Eastern Tropical Pacific*, Coral Reefs of the World 8, DOI 10.1007/978-94-017-7499-4_6

in the past. Reefs in Pacific Panama and Costa Rica experienced a 2500-year hiatus in vertical growth beginning ~4100 years ago as a result of enhanced ENSO activity. Although the degree of framework accumulation and rate of reef accretion in some parts of the ETP are more similar to that of the western Atlantic than previously thought, the region still remains a marginal environment for reef development. Given the dominant role that climatic variability has played in controlling reef development in the past, the future of reefs in the ETP under accelerating climate change remains uncertain.

Keywords
Reef accretion • Paleoecology • Upwelling • El Niño–Southern Oscillation • Latin American reefs

6.1 Introduction

This chapter reviews the known records of Holocene coral-reef development across the eastern tropical Pacific (ETP), including the reefs of Mexico, Nicaragua, Costa Rica, Panama, Colombia, Ecuador, Clipperton Atoll, and Easter Island (Rapa Nui). Although initial reports suggested a lack of coral-reef development in the ETP (e.g., Darwin 1874), later studies identified some reef formations with rapid accumulation rates that initiated up to 7000 years ago (e.g., Toth 2013; Hendrickson 2014). These reefs exhibit considerable variation in areal extent, degree of framework construction, and depth of formation (Glynn 2000; Cortés 2011). Although we identify a variety of geological and ecological characteristics common to most reefs in the region, considerable variety exists in the Holocene histories of coral reefs of the ETP. We discuss these variations as they relate to geography, environmental changes, and physical conditions, including temperature, turbidity, shelf area, and antecedent topography. By synthesizing the known records of reef development, we evaluate the proposition that reef development is generally poor in the ETP. Finally, we discuss the prospects for reef development over the next decades to centuries under conditions of accelerating climate change.

6.2 Regional Characteristics of Eastern Pacific Coral Reefs

6.2.1 Ecology and Structure

Coral reefs in the ETP are predominantly formed by interlocking colonies of the branching coral *Pocillopora damicornis* and/or near-monogeneric stands of massive corals of the genera *Porites*, *Pavona*, or *Gardineroseris* (Glynn 1976; Guzmán and Cortés 1989a; Macintyre et al. 1992; Cortés 1997; Glynn and Maté 1997; Glynn and Ault 2000). Fringing, *Pocillopora*-dominated reefs are common in the sheltered embayments of offshore islands in the ETP, but they can also be found along the mainland coast. Most of the well-developed reefs throughout the eastern Pacific share a typical zonation pattern. The shallow zone is generally dominated by *Pocillopora* spp., and massive species increase in dominance with depth. On the reef flat, which is the shallowest part of the reef, coral cover is patchy. Living coral cover is generally highest at the seaward edge along the upper reef slope, where *Pocillopora* growth rates are high. This upper reef slope typically has the thickest accumulation of Holocene framework. For example, a cross-reef transect of drill-cores collected from a fringing reef in the Secas Islands, Gulf of Chiriquí, Panama penetrated 2.8 m on the reef flat, 8.2 and 12.0 m on the reef slope, and only 1.2 m at the base of the fore reef (Glynn and Macintyre 1977). Although coral diversity increases in the deeper reef zones, coral cover declines with depth and carbonate accumulation is minimal in the deepest zones (Glynn 1976).

6.2.2 Deposition, Bioerosion, and the Fossil Record

Coral reefs of the ETP typically grow in relatively low-energy environments, and because the saturation state of aragonite is low in the region the reefs show little evidence of the type of submarine lithification that is common in coral reefs of the western Atlantic (Macintyre and Glynn 1976; Glynn and Macintyre 1977; Cortés 1997; Manzello et al. 2008). Likewise, coralline algae, although present in the ETP, do not generally consolidate reef structures to the extent they do elsewhere (Glynn and Maté 1997). Reef frameworks of the ETP are stabilized primarily by interdigitation of the *Pocillopora* branches and by the fine sediments the branches entrain, rather than by cementation. As a result, slumping of small patches of over-steepened framework is common on reef slopes in the ETP (Glynn 1976). The lack of cementation also makes these reefs highly susceptible to the destructive forces of storms and wave action; therefore, reefs that have a record of Holocene accumulation are typically located in sheltered embayments (Dana 1975; Glynn and Ault 2000).

Framework accumulation is also limited by the negative influence of bioerosion (Glynn 1997). Bioerosion is common on all reefs of the ETP, although the relative importance of internal bioeroders (primarily clionid sponges and lithophagine bivalves) and external grazers (primarily echinoids) differs from region to region (Cortés 1997; see Chap. 12, Alvarado et al.). In the Galápagos, bioerosion by the sea urchin *Eucidaris galapagensis* can reduce net accumulation by up to 20 %, even where coral cover is high (Glynn et al. 1979), but the impacts of bioerosion on reef accretion are even more severe following major disturbance events (Glynn 1988, 1997; Eakin 1992, 1996; Cortés 1997; Fonseca et al. 2006).

The 1982–83 El Niño–Southern Oscillation (ENSO) event killed 52–97 % of living coral in the ETP (Glynn et al. 1988, 2001). The associated increase in echinoid activity in the Galápagos, Costa Rica, and Pacific Panama, led to rates of bioerosion that exceeded rates of reef-framework growth (Glynn 1988; Eakin 1992, 1996; Reaka-Kudla et al. 1996; Glynn and Wellington 2007; see Chap. 12, Alvarado et al.). A similar increase in bioerosion was observed on Mexican reefs after the 1987 El Niño (Glynn and Leyte-Morales 1997). Most reefs have since recovered their coral populations, but much of the framework that existed in the Galápagos prior to 1982–83 has been eliminated by bioerosion (Macintyre et al. 1992; Glynn et al. 2001, 2015; Glynn 2003). It is unlikely that bioerosion can disrupt the preservation of coral skeletons once they are buried and stabilized by fine sediment, but bioerosion is clearly important in determining the completeness of the fossil record in the ETP.

6.3 Initiation of Holocene Reef Development

The sea-level history in the eastern Pacific over the last 10,000 years (Curray et al. 1969; Hopley 1982) is generally thought to be similar to that reported for the western Atlantic (e.g., Lighty et al. 1982; Toscano and Macintyre 2003), although a comprehensive sea-level curve has not yet been generated for the region. If the Caribbean sea-level model is applicable, then coral reefs on the continental and insular shelves in the ETP could have become established at least 7000 year BP (Toscano and Macintyre 2003; Toth 2013). In general, it appears that reefs first began to grow in regions of the ETP that are relatively sheltered from oceanic influences (Macintyre et al. 1992). Reef development in more exposed areas did not begin until recent millennia (Cortés 1993).

One of the earliest dates for reef initiation in the region comes from a push-core we extracted from a reef in the Gulf of Panama. Our core EP09-28, from the reef at Playa Larga, off Contadora Island in the Pearl Islands Archipelago (8°37′ N, 79°01′W; Fig. 6.1), penetrated 442 cm to bedrock. A sample of *Pocillopora damicornis* from the base of the core yielded a calibrated radiocarbon date of 6956 cal year BP (calibrated calendar years before 1950; 2-sigma range [95 % confidence interval] 7094–6645 cal year BP), giving a date of reef initiation consistent with the inferred time of shelf-flooding. [Dates from the Gulf of Panama were calibrated using local reservoir corrections that we established (Toth et al. 2012, 2015; Toth 2013). Dates from other locations were calibrated using the global marine reservoir correction because no historical reservoir corrections are available for other parts of the ETP. Dates from Macintyre et al. (1992); Cortés et al. (1994) were also corrected for isotopic fractionation using the assumed value of 0 ± 4.]

Most published reports suggest that significant reef development in the ETP began between 5600 and 4500 cal year BP, when sea level presumably neared its present elevation (Dana 1975; Glynn and Macintyre 1977; Cortés et al. 1994). Another core we collected, EP09-32 from the nearby Saboga Island reef, is more consistent with this conclusion. EP09-32 penetrated 363 cm to bedrock. The recovery was 228 cm, of which the bottom 43 cm consisted of sand and fragments of coralline algae with no coral. The earliest date of active reef growth from this core, determined from the radiocarbon date of a sample of *Pocillopora* directly above the layer of sand and coralline algae, was 4348 cal year BP (2-sigma range 4558–4131 cal year BP). Similar lag times between shelf-flooding and reef initiation have been reported for other geographic regions (Hopley et al. 1978; Davies et al. 1985; Macintyre et al. 1985; Montaggioni 1988). The lags between flooding and reef-framework construction were likely caused by extensive erosion of the soil and sediment cover on the adjacent shorelines, which created conditions hostile to coral settlement and growth.

Finally, many coral reefs in the eastern Pacific exist as incipient structures, which, despite exhibiting significant vertical accumulation, were only established within the last 1000 years. They include reefs composed of massive corals, such as those off Caño and Cocos Islands in Costa Rica (Macintyre et al. 1992), as well as seemingly complex fringing *Pocillopora* reefs off many of the islands of Panama and the Galápagos. For example, the framework of the *Pocillopora*-dominated reef of Mogo Mogo Island in the Pearl Islands is only 50 cm thick and established within the past 50 years.

Our cores suggest that the well-developed reefs in Pacific Panama were dominated by coralline algae in their earliest stages of development. Soon after, the free-living coral *Psammocora* was dominant for a short period before the

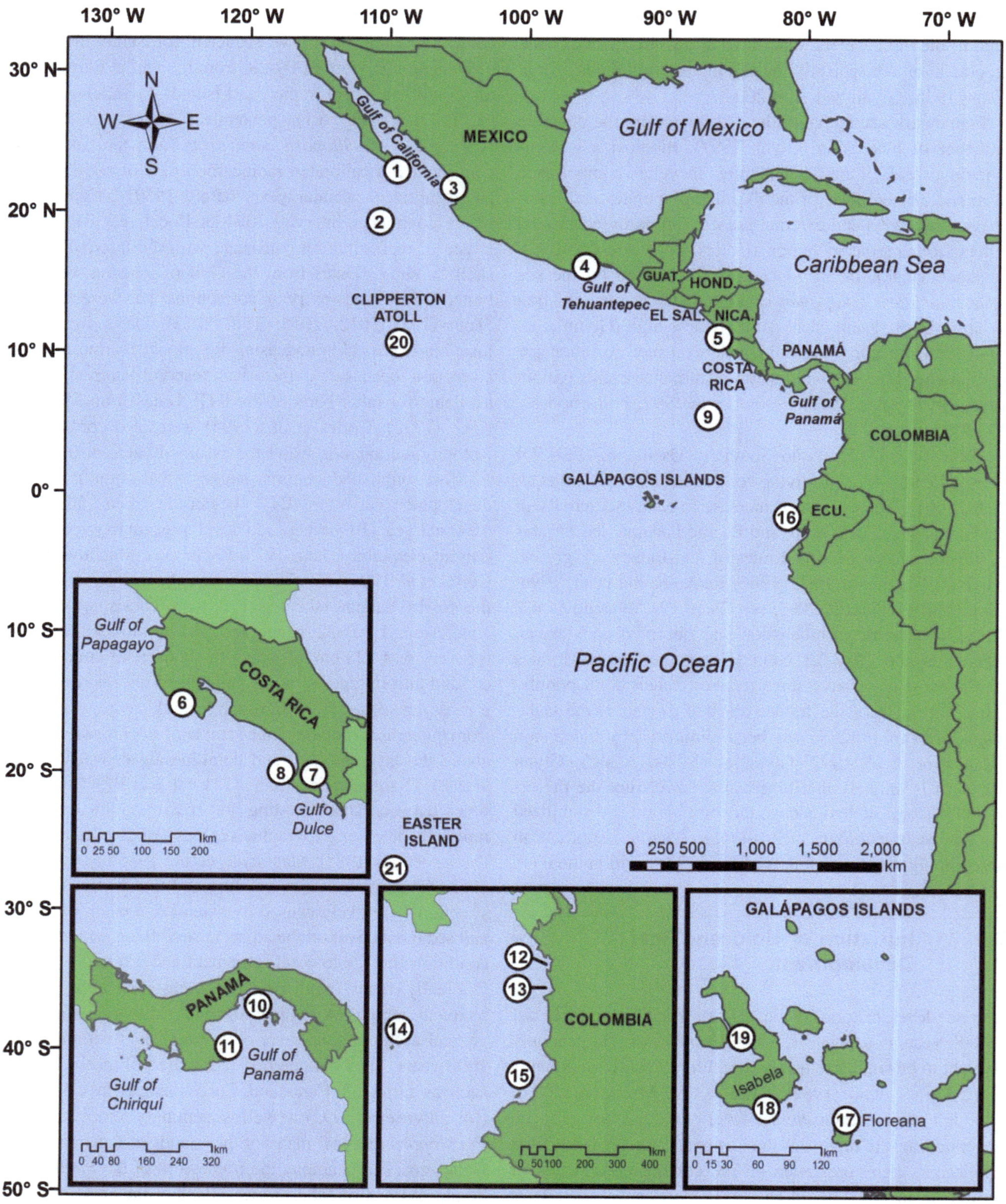

Fig. 6.1 Index map showing location of selected eastern Pacific reefs from Mexico to Ecuador. Numbered sites indicate the locations of reefs discussed in text: (1) Cabo Pulmo, (2) Revillagigedo Archipelago, (3) Nayarit, (4) Huatulco, (5) San Juan del Sur, (6) Sámara, (7) Punta Islotes, (8) Caño Island, (9) Cocos Island, (10) Pearl Islands Archipelago, (11) Iguana Island, (12) Gulf of Cupica, (13) Ensenada de Utría, (14) Malpelo Island, (15) Gorgona Island, (16) Sucre and La Plata Islands, (17) Champion, Onslow, and Floreana Island coral reefs, (18) Villamil coral reef, (19) Urvina Bay, (20) Clipperton Atoll, and (21) Easter Island (Rapa Nui)

Pocillopora-dominated assemblages typical of modern reefs were established. This pattern of reef development is similar to that observed in the uplifted reef section at Urvina Bay, Galápagos (Colgan and Malmquist 1987), and in extant reefs in Costa Rica (Bezy et al. 2006). It is possible, therefore, that a primary successional sequence from coralline algae, to free-living corals like *Psammocora,* to dominance by *Pocillopora* is common to many reefs in the ETP.

6.4 Documented Records of Reef-Framework Construction in the Eastern Pacific

6.4.1 Mexico

Early reports from Pacific Mexico suggested that coral-reef development in this region was minimal (Reyes-Bonilla 2003). Although many Mexican reefs exist only as veneers on the antecedent substrate, further exploration of the region in recent decades has revealed a number of substantial reef formations (Reyes-Bonilla 1992, 1993a, b, 2003; Carriquiry and Reyes-Bonilla 1997; Glynn and Leyte-Morales 1997; Pérez-Vivar et al. 2006), some of which may rival the best-developed coral reefs in the ETP (Glynn and Leyte-Morales 1997). In addition to Holocene reef deposition, some areas of the Mexican Pacific appear to contain continuous fossil records extending back to the Pliocene epoch (Reyes-Bonilla 1992; see Chap. 2, López-Pérez). The coral reefs of Pacific Mexico can be divided into three biogeographic subunits, following Reyes-Bonilla (2003): the Gulf of California, the Revillagigedo Archipelago, and the tropical (coastal) Mexican Pacific.

6.4.1.1 Gulf of California

The Gulf of California contains some of the best-studied reefs in the Mexican Pacific (Reyes-Bonilla 2003). True reef development is limited in this northernmost region of the ETP, but significant reef structures exist in some parts of the gulf (Reyes-Bonilla 1993b, 2003). The largest of these is the reef at Cabo Pulmo (23°20′N, 109°25′W; Fig. 6.1), near the southern tip of the Baja California Peninsula (Reyes-Bonilla 1993a, b, 2003). Although the Cabo Pulmo reef is considered to be the northernmost coral reef in the ETP, there is little retention of carbonate framework; therefore, in many ways it is atypical of coral reefs in other geographic regions (Riegl et al. 2007). Like most coral reefs in the ETP, the 15-ha reef at Cabo Pulmo is dominated by *Pocillopora* spp. (Reyes-Bonilla 1993b), but massive *Porites* and *Pavona* spp. increase in abundance below 6 m depth (Reyes-Bonilla 2003; Carriquiry and Villaescusa 2010). Unlike most areas of the ETP, however, the reef at Cabo Pulmo has been reported as being 'well-cemented' (Carriquiry and Villaescusa 2010), which is surprising considering the low aragonite saturation state at this northerly latitude and the putatively low cementation of the ETP as a whole (Kleypas et al. 1999b; Manzello et al. 2008). The maximum Holocene framework accumulation at Cabo Pulmo was approximately 3 m (Table 6.1; Reyes-Bonilla 2003; Carriquiry and Villaescusa 2010).

6.4.1.2 Revillagigedo Archipelago

The oceanic Revillagigedo Archipelago is located approximately 700 km southwest of the Gulf of California (Fig. 6.1) and includes the islands of Socorro, Clarión, and San Benedicto, and the rock pinnacle Roca Partida (see also Fig. 5.8 in Chap. 5, Glynn et al.). The best-developed reefs in the archipelago are the *Pocillopora*-dominanted fringing reefs that occur along the coasts of Socorro and Clarión Islands (18°20′N, 114°43′W and 18°48′N, 110°57′W, respectively; Reyes-Bonilla 2003), although neither is more than a hectare in area (Ketchum and Reyes-Bonilla 1997). Despite the small size of the reefs, the archipelago contains the greatest number of coral species in the Mexican Pacific (Reyes-Bonilla 2003). Holocene carbonate accumulation in the Revillagigedo Archipelago is similar to that in the Gulf of California, with reef frameworks reaching 3 m in thickness at Clarión Island and 2 m at Socorro Island (Table 6.1; Reyes-Bonilla 2003).

6.4.1.3 Tropical Mexican Pacific

The southernmost region of Pacific Mexico, the tropical Mexican Pacific, extends from 21° to 15°N along the Mexican coastline (Reyes-Bonilla 2003). Much of the coastline is dominated by sandy substrates that do not support reef development, so although isolated patches of coral can be found throughout the region, true coral reefs are confined to the coasts of the states of Nayarit and Oaxaca (Reyes-Bonilla 1993b, 2003).

Small, fringing *Pocillopora* reefs are common around the islands near the southern coasts of Nayarit (20–21°N; Carriquiry and Reyes-Bonilla 1997; Reyes-Bonilla 2003). The reefs at Careyeros, Corral del Risco, and Punta Mita Islands exhibit only 1–3 m of Holocene framework accumulation, likely due to the narrowness of the coastal shelves (Table 6.1; Carriquiry and Reyes-Bonilla 1997). The best-developed reefs are found at nearby Jaltemba Island, where thicker framework accumulation has been reported (Reyes-Bonilla 2003). Unlike the nearshore islands of Nayarit, the oceanic Marías Islands to the west of mainland Nayarit reveal only superficial reef development (Pérez-Vivar et al. 2006).

The southernmost reefs in the Mexican Pacific are located along the southern coast of Oaxaca (15°N), between Puerto Ángel and Huatulco (Reyes-Bonilla 1993b). This region of Mexico includes the Gulf of Tehuantepec, the northernmost of three major upwelling centers along the Central American

Table 6.1 Maximum estimates of Holocene reef accumulation and the dominant coral genera on the ETP reefs described in this chapter

Location	Dominant Genus	Maximum thickness	Reference
Mexico			
Cabo Pulmo, Gulf of California	*Pocillopora*	3 m[u]	Reyes-Bonilla (2003)
Revillagigedo Archipelago	*Pocillopora*	3 m[u]	Reyes-Bonilla (2003)
Nayarit, Tropical Mexican Pacific	*Pocillopora*	3 m[u]	Carriquiry and Reyes-Bonilla (1997)
Huatulco, Tropical Mexican Pacific	*Pocillopora*	6 m[v]	Glynn and Leyte-Morales (1997)
Costa Rica			
Culebra Bay, Gulf of Papagayo	*Pocillopora*	3 m[p]	Glynn et al. (1983)
Punta Islotes, Golfo Dulce	*Pocillopora*	9 m[c]	Cortés et al. (1994)
Los Mogos, Golfo Dulce	*Porites*	11 m[v]	Glynn et al. (1983)
Caño Island	*Pocillopora*	2.4 m[p]	Guzmán (1986)
Cocos Island	*Porites*	3.1 m[c]	Macintyre et al. (1992)
Panama			
Gulf of Chiriquí			
Secas Island	*Pocillopora*	13.4 m[p]	Glynn and Macintyre (1977)
Uva Island	*Pocillopora*	11 m[c]	Glynn and Macintyre (1977)
Gulf of Panama			
Saboga Island	*Pocillopora*	5.6 m[c]	Glynn and Macintyre (1977)
Contadora Island	*Pocillopora*	4.6 m[p]	Glynn and Macintyre (1977)
Iguana Island	*Pocillopora*	6.1 m[p]	Glynn and Macintyre (1977)
Colombia			
Tebada Island, Gulf of Cupica	*Pocillopora*	4 m[u]	Zapata and Vargas-Ángel (2003)
La Chola, Ensenada de Utría	*Pocillopora*	4 m[p,c]	Vargas-Ángel (2001)
La Azufrada, Gorgona Island	*Pocillopora*	8 m[p]	Glynn et al. (1982)
Ecuador			
La Plata Island	*Pavona*	6 m[v]	Glynn and Wellington (1983)
Isabela Island, Galápagos	*Pocillopora*	4 m[v]	Glynn and Wellington (1983)
Bartolomé Island, Galápagos	*Porites*	5 m[v]	Glynn and Wellington (1983)
Champion Island, Galápagos	*Pavona*	5 m[v]	Glynn (2003)
Onslow Island, Galápagos	*Psammocora*	1.5 m[c]	Feingold (1995)
Clipperton Atoll	*Pocillopora*	6 m[v]	Carricart-Ganivet and Reyes-Bonilla (1999)
Easter Island (Rapa Nui)	*Porites*	10 m[v]	Hubbard and Garcia (2003), Glynn et al. (2007)

Superscripts indicate basis of estimates:
[c]Coring
[p]Probing
[v]Visual estimation based on reef topography
[u]Unknown

coast (the others being the Gulf of Papagayo in Nicaragua/Costa Rica, and the Gulf of Panama; Fig. 6.1; Glynn and Leyte-Morales 1997). These reefs are dominated by *Pocillopora* spp. and, like the reefs in the Gulf of Panama, are generally found in small, sheltered embayments. Despite the putative limitation of reef development due to upwelling, coral reefs in this region have attained Holocene framework accumulations comparable to those in other regions of the ETP. The framework thickness of most reefs is between 3 and 4 m, but Holocene framework accumulations of up to 6 m have been recorded (Table 6.1; Glynn and Leyte-Morales 1997), comparable to Holocene framework accumulation in the Gulf of Panama (Glynn and Macintyre 1977).

6.4.2 Guatemala, El Salvador, and Honduras

Few, if any, true coral reefs occur between Mexico and Nicaragua, although there is likely some connectivity between the regions (Glynn and Ault 2000; see Chap. 5,

Glynn et al., and Chap. 16, Lessios and Baums). This 'Central American faunal gap' may be related to the limited amount of hard substrate available for larval settlement along the coasts of Guatemala, El Salvador, and Honduras. Isolated coral assemblages have been found in some areas such as Los Cóbanos, El Salvador (13°32′N, 89°48′W), but there does not appear to be any significant framework accumulation in this region (Reyes-Bonilla and Barraza 2003).

6.4.3 Nicaragua

Recent surveys at San Juan del Sur, Nicaragua (11°15′N, 85° 52′W; Fig. 6.1) suggest that coral assemblages may be more abundant in this area than previously assumed. The dominant coral in this region is *Pavona gigantea*, but pocilloporids are also found in isolated patches (Alvarado et al. 2010, 2011). Coral cover is less than 10 % on average but approaches 20 % at Guacalito (11°22′N, 86°1′W) and Punta Gigante (11°23′N, 86°2′W). These northernmost sites at San Juan del Sur are least affected by seasonal upwelling in the nearby Gulf of Papagayo and appear to have the best development of coral assemblages in the region (Alvarado et al. 2011). It is unclear from Alvarado (2010, 2011) and colleagues' recent reports whether the coral assemblages of San Juan del Sur represent true coral reefs or incipient reef communities; there is, as yet, no information on the degree of carbonate accumulation in this region. Compared with other parts of the ETP, the coral assemblages of Nicaragua and other regions within the Central American faunal gap have been understudied. Perhaps with increased focus on these areas in the future, the faunal gap will be narrowed as more coral assemblages, and possibly even developed coral reefs, are discovered.

6.4.4 Costa Rica

Coral reefs are common along the Pacific coast of Costa Rica, from Santa Elena in the northwest to Golfo Dulce in the east near the Panamanian border. They also are found around the offshore islands of Cocos and Caño (Cortés and Jiménez 2003). As in southern Mexico, reefs in Costa Rica are most common in sheltered embayments where they are protected from wave stress (Glynn et al. 1983). Although the reefs of this region are generally less developed than those of Pacific Panama, considerable framework accumulation has been observed in some areas, including the Gulf of Papagayo, Golfo Dulce, and at Cocos and Caño Islands.

6.4.4.1 Gulf of Papagayo

In the Gulf of Papagayo, which is located in the northwestern region of Costa Rica (10°43′N, 85°48′W; Fig. 6.1), coral reefs are subject to strong seasonal upwelling. The upwelling season in Costa Rica is more protracted and produces lower water temperatures than in the Gulfs of Panama or Tehuantepec (Glynn et al. 1983; Alexander et al. 2012). As a result, the depth-range of the coral reefs is restricted: reef-building corals are abundant to only 5 m depth in the Gulf of Papagayo, compared with 15 m in nearby Golfo Dulce where there is no upwelling (Glynn et al. 1983). Additionally, massive corals in the Gulf of Papagayo have more boreholes and skeletal disconformities than massive corals in non-upwelling areas of Costa Rica and Panama. The numbers of boreholes and skeletal disconformities in the Gulf of Papagayo are also elevated relative to the Gulf of Panama, where there is upwelling (Glynn et al. 1983), supporting the contention that upwelling is more severe in Papagayo (cf. Highsmith 1980). Although upwelling generally slows coral growth in the ETP (Glynn 1977; Glynn et al. 1983), Jiménez and Cortés (2003) found that growth rates of *Pocillopora* spp., *Pavona clavus*, and *Psammocora stellata* were greater in the Gulf of Papagayo than in any other part of the ETP. They suggested that high nutrient levels during upwelling could be responsible for the elevated growth rates in this region (Jiménez and Cortés 2003), and that the effects of increased nutrient availability may override the negative impacts of low sea temperatures. The overall impacts of upwelling on coral reef development may, therefore, be more complex than originally recognized (Chollett et al. 2010).

In a geographic survey, Glynn et al. (1983) found that whereas most reefs in the southern part of Costa Rica were actively accreting, nearly all the reefs in the Papagayo region were dead. The dead reefs appeared to have been formed primarily by interlocking colonies of *Pocillopora damicornis*, which, at the time of the survey were covered by crustose coralline algae and algal turfs. The timing of reef degradation was determined by dating coral samples from dead reefs in the Gulf of Papagayo and Sámara Beach (9°51′ N, 85°52′W; Fig. 6.1), located south of the gulf. Nine radiocarbon dates placed the timing of reef demise between 292 and 154 cal year BP, or 1660–1795 C.E. The apparent agreement between these dates and the timing of the Little Ice Age (LIA), a period of climatic cooling between about 1600 and 1900 C.E., led Glynn et al. (1983) to postulate that the LIA was the cause of coral mortality and reef degradation.

The LIA was a climatic phenomenon associated with centennial-scale southerly displacement of the Inter-Tropical Convergence Zone (ITCZ; Haug et al. 2001; Sachs et al.

2009). The ITCZ is the major driver of interannual variability in the ETP, and its annual southward migration during the boreal winter leads to intensification of the trade winds and the upwelling season in the Gulf of Papagayo and the other upwelling centers in the region (Poveda et al. 2006). Glynn et al. (1983) suggested that a more southerly position of the ITCZ during the LIA could have led to either more intense or more protracted upwelling seasons in the Gulf of Papagayo and Sámara, which may have overwhelmed the capacity of corals to survive. Analysis of $\delta^{18}O$ from the tips of dead coral colonies suggested that the corals died during an extreme low-temperature event, supporting their hypothesis (Glynn et al. 1983).

Reef probing by Glynn et al. (1983) suggested that the maximum Holocene framework accumulation in the Gulf of Papagayo is about 3 m (Table 6.1), which is significantly less than in the non-upwelling Golfo Dulce (discussed below). They concluded that coral reefs in the Gulf of Papagayo are significantly younger than those in the non-upwelling regions of Pacific Costa Rica. Using average coral growth rates from Pacific Panama, they estimated that the l-m-thick reef at Punta Gorda formed at about 800 year BP and the 2.2-m-thick reef at Santa Elena formed about 1700 year BP (Glynn et al. 1983). These could be conservative estimates (i.e., the reefs are probably older) because the actual coral growth rates in the Gulf of Papagayo are limited by seasonal upwelling.

6.4.4.2 Golfo Dulce

The Punta Islotes fringing reef is located in the innermost part of Golfo Dulce, about 50 km from the open ocean (8°28′N, 83°12′ W; Fig. 6.1). Living coral cover is less than 2 %, and most of the reef surface is covered by a filamentous algal mat (Fig. 6.2). During a paleoecological study in the late 1980s, nine core holes were drilled along four short transects on the reef (Fig. 6.3) and four vibracores were collected from the fore-reef talus slope (Cortés 1990a; Macintyre et al. 1992; Cortés et al. 1994). The cores ranged in penetration depth from 2.5–9 m (Table 6.1; Cortés et al. 1994), providing a minimum estimate of Holocene accumulation for this reef. Visual inspection of nearby Los Mogos reef suggested that *Porites*-framework accumulation at that site was ~11 m above the seafloor, which is the thickest massive-coral framework recorded in the ETP (Table 6.1; Glynn et al. 1983).

Fig. 6.2 Punta Islotes reef, Golfo Dulce, Costa Rica, 4 m depth, February 1989. Most of the large overlapping *Porites lobata* colonies are dead and covered with filamentous algae and sediments

Twenty-five radiocarbon dates from Punta Islotes reef, plotted in relation to the sea-level curve (Curray et al. 1969; Hopley 1982), led to the conclusion that the Cretaceous basaltic bench underlying the reef was flooded at least 6000 years ago. Coral debris from the fore-reef talus slope yielded a date of 6162 cal year BP (2-sigma range 6374–5928; Table 6.2), indicating a fairly short lag-time before the establishment of this reef. There is no record of significant reef development during the intervening 3000 years, and accretion rates during this period were only 0.9–1.5 m/1000 year. A secondary reef framework of *Porites lobata*, *P. damicornis*, and *Psammocora stellata* was established on the basaltic base about 2500 year BP and flourished for the next 2000 years. Average accretion rates over this period were 2.4 m/1000 year (Toth et al. 2012); however, during its most rapid stage of development from 1500–500 year BP, the reef at Punta Islotes accreted at a rate of 8.3 m/1000 year, which is comparable to centennial-scale accretion rates in the non-upwelling Gulf of Chiriquí, Panama, discussed below (Glynn and Macintyre 1977; Cortés et al. 1994). An increase in the outflow of nearby rivers caused a deterioration of water quality about 500 years ago. In recent times, deforestation on the adjacent shore resulted in increased sedimentation on the reef, which finally led to its modern, algal-covered state (Cortés et al. 1994).

6.4.4.3 Caño Island

This 3-km^2 island is located 15 km off the coast of Costa Rica (8°42′N, 83°52′W; Fig. 6.1; Guzmán and Cortés 1989a). Most of the reef-flat area currently consists of dead *Pocillopora damicornis* covered by crustose coralline algae, although a few living colonies of *P. damicornis, Pocillopora elegans*, and *Psammocora stellata* can also be found (Guzmán and Cortés 1989a). Probing of the dead *Pocillopora* framework showed that the maximum framework accumulation at Caño Island is about 2.4 m (Table 6.1; Guzmán 1986). The age of the *Pocillopora* reef at Caño Island is unknown.

Microatolls formed by colonies of *Porites lobata* (Fig. 6.4a) cover about 12 % of a 4.2-ha reef flat off the northern end of the island. Seaward of this reef flat are living and dead colonies of *Porites lobata*, on and between buttresses of metasedimentary rock (Guzmán and Cortés 1989a;

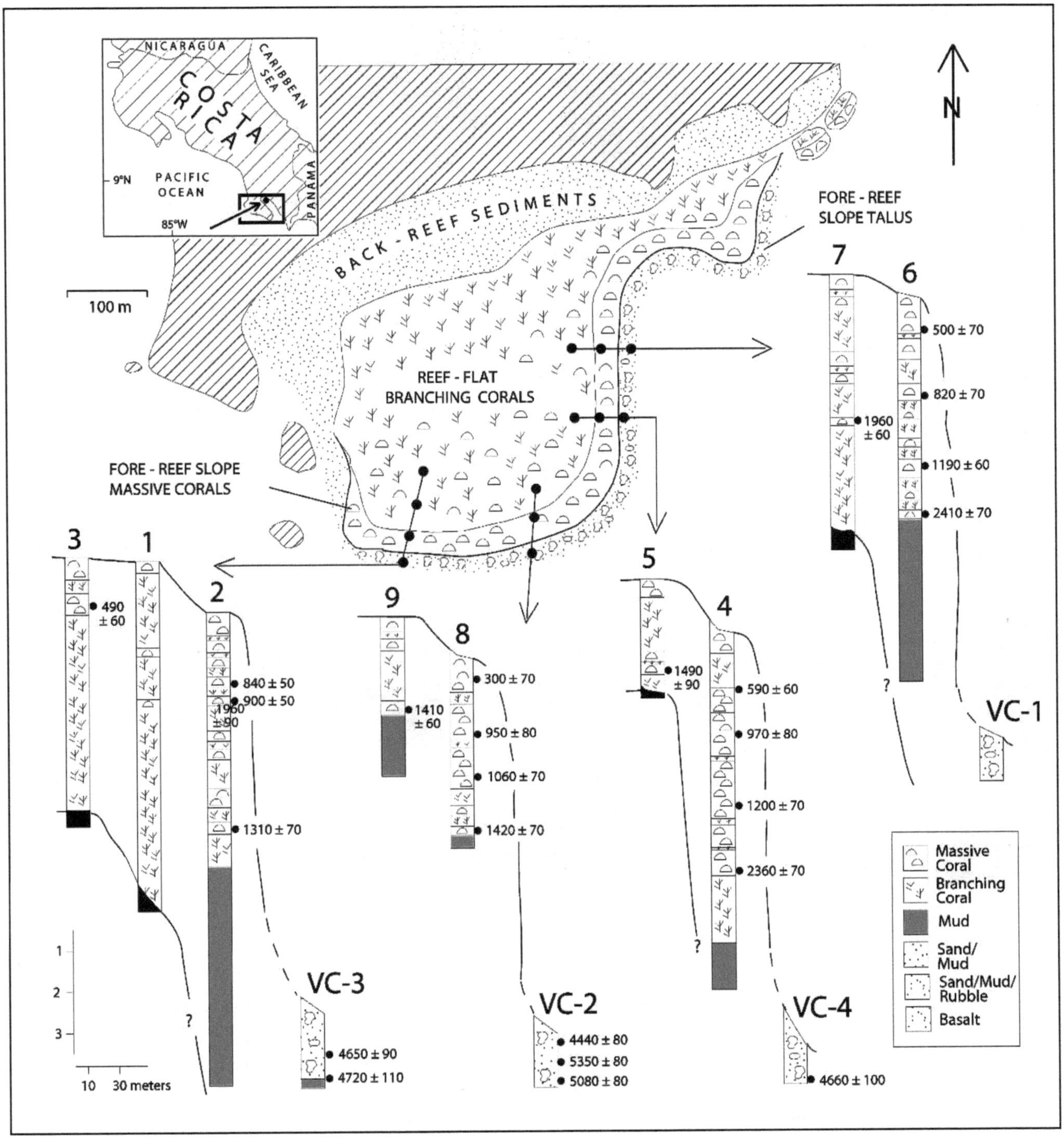

Fig. 6.3 Surface zonation and locations of nine reef cores and four vibracores on the talus slope of Punta Islotes reef. Core logs show the locations of 25 radiocarbon dates (from Cortés et al. 1994)

Macintyre et al. 1992). The buttresses mostly have a thin cover of crustose coralline algae and filamentous algal turf (Fig. 6.4b). Varieties of other corals are found on the reef slope; however, 95 % of the coral in this area was killed during the 1982–83 El Niño warming event (Cortés et al. 1984; Guzmán et al. 1987).

Five microatolls were overturned during a study by Macintyre et al. (1992) and the centers of initial growth at their bases were sampled and dated, yielding an average date of 279 cal year BP (2-sigma range 460–1 cal year BP). The base of a large overturned *Porites lobata* colony on the slope gave a date of 471 cal year BP (2-sigma range 610–

Table 6.2 Earliest records of Holocene reef initiation in the ETP

Location	Reef initiation	Reference
Clipperton Atoll	~5000 year BP	Carricart-Ganivet and Reyes-Bonilla (1999)
Costa Rica		
Punta Islotes, Golfo Dulce	6162 cal year BP	Cortés et al. (1994) (calibrated)
Panama		
Uva Island, Gulf of Chiriquí	5960 cal year BP	Glynn and Macintyre (1977) (calibrated)
Saboga Island, Gulf of Panama	4551 cal year BP	Toth (2013)
Contadora Island, Gulf of Panama	6907 cal year BP	Toth (2013)
Iguana Island, Gulf of Panama	6883 cal year BP	Toth (2013)
Ensenada de Utría, Colombia	3000 cal year	Vargas-Ángel (2001)
Galápagos Islands, Ecuador		
Onslow Island (Devil's Crown)[a]	7330 cal year BP	Toth (2013)

All records were obtained from radiocarbon dating of basal samples of cores except for the record from Clipperton Atoll. The Clipperton date was approximated based on estimates by Carricart-Ganivet and Reyes-Bonilla (1999) of sea-level changes and atoll subsidence. To our knowledge, no bottom dates have been obtained from the bases of coral outcrops on Clipperton Atoll to test these assumptions

[a]Coral formation ~100 m east of Onslow island; this is an unconsolidated biostrome rather than a true reef

Fig. 6.4 Caño Island reef, Costa Rica, February 1989. **a** *Porites lobata* microatolls on the reef flat. **b** Buttresses of algal-covered metasediment and large *Porites lobata* colonies on the reef slope, 4 m depth (*photo* by I.G. Macintyre)

310 cal year BP). Five framework samples attached to the metasedimentary substrate yielded an older average date of 1838 cal year BP (2-sigma range 2284–1522 cal year BP; Macintyre et al. 1992). These data, along with growth data on the microatolls (Guzmán and Cortés 1989b), show that: (1) the reef, which is now mostly dead, began accreting at least 1500 years ago; and (2) the large *Porites lobata* colonies are not more than 200–300 years old (Macintyre et al. 1992).

6.4.4.4 Cocos Island

Cocos Island (Isla del Coco) is a large, basaltic island associated with several smaller islands, located about 500 km southwest of Costa Rica (5°31′N, 87°3′W; Fig. 6.1). Up to 29 species of scleractinian corals have been reported from Cocos Island (Guzmán and Cortés 1992, 2007), but coral cover is patchy (Bakus 1975). The surfaces of dead coral and basalt generally have a thin coating of crustose coralline algae and filamentous algal turfs. Interlocking, massive colonies, which are primarily *Porites lobata*, form an incipient framework in Chatham Bay near Ulloa Island and extensive coral reefs in an embayment south of Pájara Island, and along other sites on the north side of Cocos Island (see Fig. 5.14 in Chap. 5, Glynn et al.).

The reefs of Cocos Island experienced catastrophic coral mortality during the 1982–83 El Niño warming event, and by 1987 living coral cover was only 3.5 % (Guzmán and Cortés 1992). The dead-coral surfaces were extensively bioeroded by the sea urchin *Diadema mexicanum*. The flanks of large *Porites lobata* mounds have survived (Fig. 6.5), and some of the living surfaces have since regrown over the dead patches (Macintyre et al. 1992).

Fig. 6.5 Cocos Island, Costa Rica, 6 m depth, September, 1987. A large colony of *Porites lobata* with its dead surface extensively bioeroded by *Diadema mexicanum* (*photo* by I.G. Macintyre)

Cores were collected from large *Porites lobata* colonies growing around Cocos Island, and bottom samples from these cores were radiocarbon dated. The bases of the longest cores from Chatham Bay (3.1 m penetration) dated to 304 cal year BP (2-sigma range: 473–100 cal year BP) and from Pájara Island (3.0 m penetration) to 455 cal year BP (2-sigma range 626–279 cal year BP; Table 6.1). The dates represent minimum estimates of the timing of reef initiation at these sites because bioerosion has hollowed out the bases of the large colonies (Macintyre et al. 1992). These limited data represent the best records of reef-framework accumulation in the area.

6.4.5 Panama

Some of the best-developed and best-studied reefs in the ETP are found off the Pacific coast of Panama, in the Gulfs of Panama and Chiriquí. Most Panamanian coral reefs are fringing, *Pocillopora*-dominated reefs typical of the ETP region; however, reefs dominated by massive corals are not uncommon. Like many areas of the ETP, the most extensive reefs are found in the sheltered embayments of offshore islands (Glynn and Maté 1997).

Coral reefs in the Gulf of Panama and Gulf of Chiriquí are subjected to distinct oceanographic and ecological conditions. The Gulf of Panama experiences strong, wind-driven, seasonal upwelling, whereas upwelling is weak to absent in the Gulf of Chiriquí (D'Croz and O'Dea 2007; Alexander et al. 2012; Toth et al. 2012; Fig. 6.6). Physical oceanographic conditions appear to play a dominant role in the rate of reef-framework accumulation in the two gulfs, although other limiting factors come into play as well, including the aerial exposure of reef flats during extreme-low tides (Fig. 6.7) and, in the Gulf of Chiriquí, coral mortality from predation by *Acanthaster planci* (Glynn 1976).

6.4.5.1 Gulf of Chiriquí

The reef at Uva Island (7°48′N, 81°45′W) is one of the best-studied reefs in the ETP. This 2.5-ha reef, located in the northern embayment of the island, is dominated by pocilloporid corals, but massive *Gardineroseris planulata* colonies are also abundant (Fong and Glynn 1998; Maté 2003).

Fig. 6.6 Satellite image of sea-surface temperature anomalies during seasonal upwelling, Pacific Panama (modified from Toth et al. 2012)

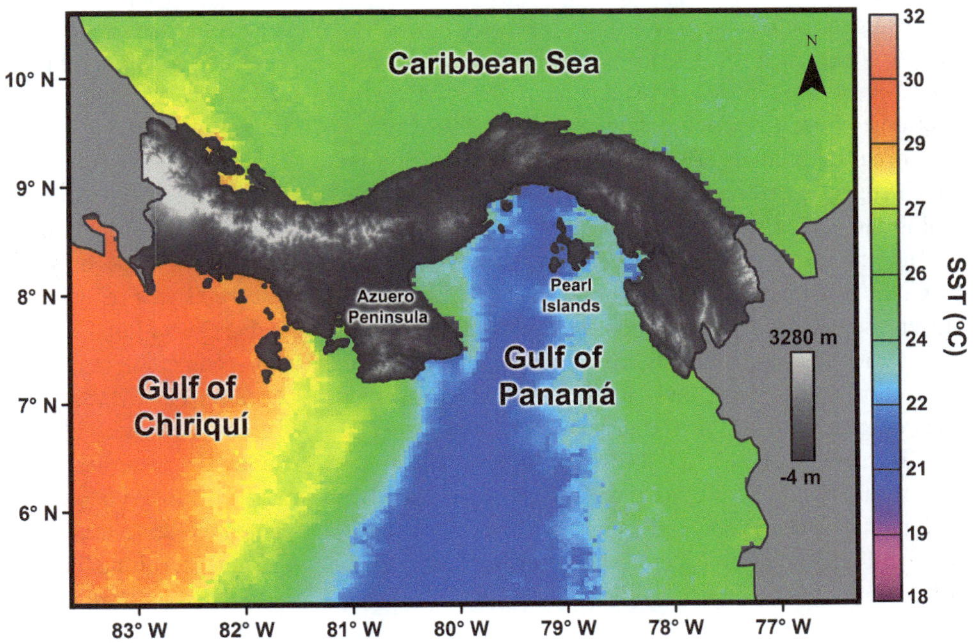

Fig. 6.7 Low tidal exposure of the Uva Island reef flat during the 2010 La Niña, September 2010 (*photo* by L.T. Toth)

Fig. 6.8 *Pocillopora*-dominated reef front at Secas Study Reef, 3 m depth, July 2011. The steep sides of the reef framework are the result of slumping and/or calving of large pocilloporid blocks (*photo* by J.W. Hobbs)

An 11-m drill-core collected from the reef flat suggests that reef initiation in the Gulf of Chiriquí occurred between 6180 and 5810 cal year BP [2-sigma range, calibrated from Glynn and Macintyre (1977); Tables 6.1, 6.2]. Reef-probing at Coiba, Canales de Tierra, and Cavada Islands showed similar degrees of Holocene framework accumulation (up to 9.4, 10.4, and 9.2 m, respectively), but the coastal coral reef at Ensenada de Muertos revealed only 7.9 m of accumulation (Glynn and Macintyre 1977).

The best-developed reef in the Gulf of Chiriquí, Secas Study Reef, occurs along the northwest coast of an unnamed island (7°57′N, 82°0′W) in the Secas Islands group (Glynn and Maté 1997; Maté 2003). This 7.6-ha fringing reef is one of the largest reefs in the ETP (Maté 2003). The Secas Study Reef is dominated by blocks of interlocking colonies of *Pocillopora* spp., many of which have slumped or been overturned by wave action (Glynn and Maté 1997; Fig. 6.8). Glynn and Macintyre (1977) probed 13.4 m into the reef, which suggests that reef accretion has been especially rapid at this site (Table 6.1).

The maximum rates of Holocene reef accretion reported by Glynn and Macintyre (1977) in the Gulf of Chiriquí were all representative of the last thousand years of reef deposition. Centennial-scale records from probing and framework excavation at the Secas Study Reef suggested that accretion rates were as rapid as 2 m/100 year over short time intervals, and averaged 0.74–0.75 m/100 year. At Uva Island reef, centennial-scale rates of reef accretion averaged 0.42 m/100 year. When millennial-scale reef sequences from drill-cores are considered, however, reef accretion is considerably lower. Glynn and Macintyre (1977) calculated an average accretion rate of only 1.6 m/1000 year for the reef at Uva Island over the last six millennia.

We collected a series of push-cores from nearby Canales de Tierra Island (Fig. 6.10). At this site, average accretion rates were ~1 m/1000 year (Toth et al. 2012; Toth 2013); however, our detailed reconstructions of reef development at Canales de Tierra, as well as at Contadora and Iguana Islands in the Gulf of Panama, suggest that previous measurements may have underestimated reef accretion in this region. The reefs at Canales de Tierra, Iguana, and Contadora Islands experienced a 2500-year shutdown in reef development beginning ~4100 cal year BP, most likely as a result of enhanced ENSO activity (Fig. 6.9; Toth et al. 2012, 2015). Accretion in Pacific Panama was not always slow; it just appears to have been slow on average because of a prolonged hiatus in reef growth. Indeed, millennial-scale rates of vertical accretion during times of active reef growth at Canales de Tierra were ~2.4 m/1000 year (Toth 2013). Interestingly, the hiatus in reef development at these sites in Pacific Panama coincided with the period of limited reef accretion in Golfo Dulce, discussed above.

Accretion in the Gulf of Chiriquí during times of active reef growth was only slightly lower than the rate of 3–4 m/1000 year reported for an *Acropora palmata*-dominated reef at Galeta Point, on the Caribbean coast of Panama, but it was comparable to average accretion rates on most Caribbean reefs (Macintyre and Glynn 1976, and references therein; Toth et al. 2012). The highest accretion rate for the Caribbean, measured for uncemented Alacrán Reef, Mexico, was 12 m/1000 year, which, although similar to centennial-scale accretion rates in the Gulf of Chiriquí and Costa Rica (discussed above; Cortés et al. 1994), was much more rapid than any millennial-scale accretion rate measured in the ETP (Macintyre et al. 1977). Accretion in the Gulf of Chiriquí was also generally lower than in most parts of the western Pacific (Montaggioni 2005, and references therein; Toth et al. 2012).

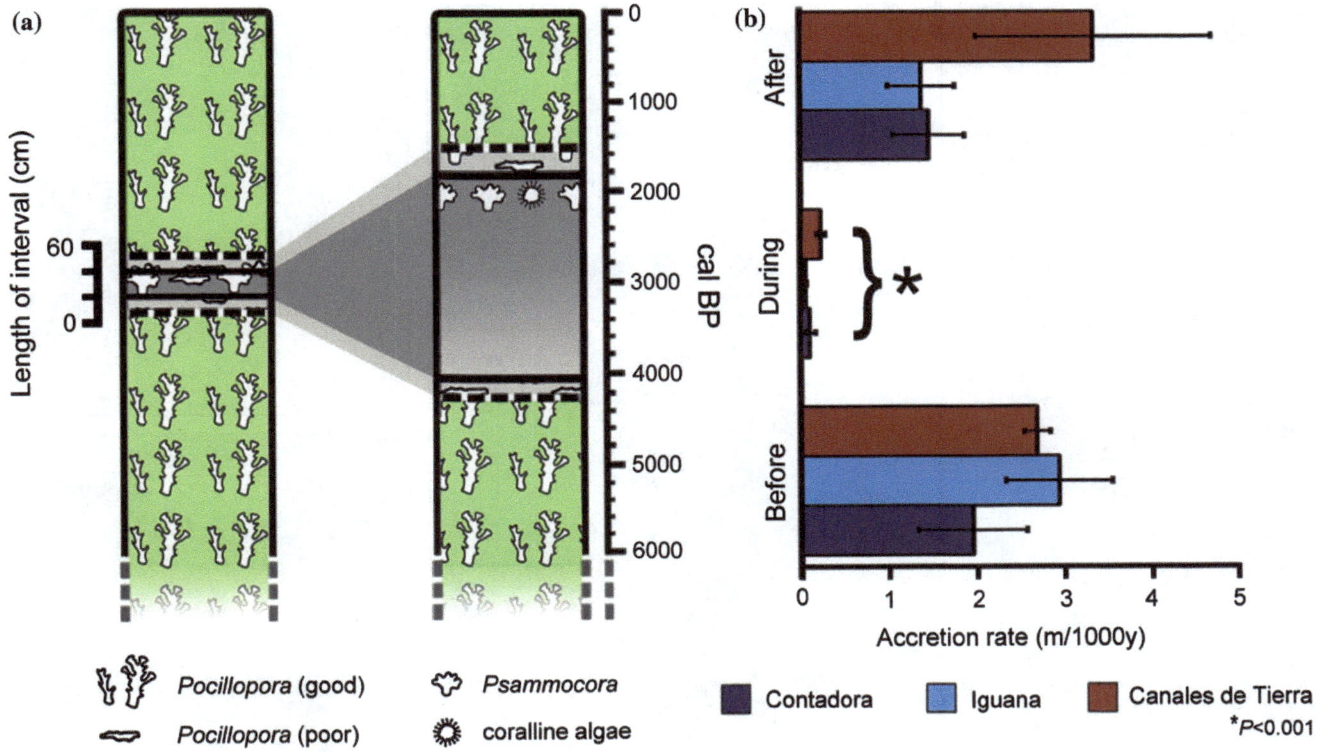

Fig. 6.9 Graphical representation of hiatus in reef development in three locations in Pacific Panama. **a** Composite core log. Green shading with *Pocillopora* in good taphonomic condition indicates periods of active reef growth. Gray shading with *Pocillopora* in poor condition, and presence of *Psammocora* and coralline algae indicate interrupted reef accretion. The length of the hiatus (*gray shading*) is shown in depth in the reef framework (*left*) and expanded to calibrated calendar years before 1950 (*right*). Dark-gray shading represents the most conservative time span for the hiatus; light-gray shading shows the maximum range for all three sites. Note absence of coral dating within all but the beginning and end of hiatus. **b** Mean accretion rates before, during, and after the hiatus. Colors represent upwelling strength: no upwelling (*maroon*), intermediate upwelling (*light blue*), and strong upwelling (*dark blue*). Error bars are standard errors; asterisk indicates significant difference from the other groups (ANOVA, $p < 0.001$). Modified from Toth et al. (2012); reprinted with permission from AAAS

6.4.5.2 Gulf of Panama

Coral reefs in the Gulf of Panama are similar in composition to those in the Gulf of Chiriquí; however, their distribution is strongly influenced by seasonal upwelling (Glynn and Stewart 1973). Most of the coral reefs in the Gulf of Panama are found in the Pearl Islands, located 60 km southeast of Panama City (Fig. 6.1; Glynn and Maté 1997). These islands are exposed to the full force of upwelling currents flowing across the Panama shelf (Toth et al. 2012; Toth 2013), and, therefore, coral reefs in this region are primarily restricted to the sheltered embayments on the northern and eastern sides of the islands (Glynn and Stewart 1973).

Coral assemblages on the exposed sides of the islands are generally incipient structures with little or no geologic history. For example, we collected a push-core from the 1.9-ha pocilloporid reef on the south side of Mogo Mogo Island (8° 33′N, 79°1′W) in the summer of 2009. Despite the high coral cover and large spatial extent of the reef, we were only able to core a maximum of 0.5 m into the coral framework before reaching bedrock. The modern radiocarbon date from a coral sample at the bottom of the core indicated that the reef had been growing for 50 years or less.

In contrast, significant framework accumulations have been observed at some of the more sheltered sites in the Pearl Islands. Based on reef-probing by Glynn and Macintyre (1977), the 5.6-m-thick reef at Saboga Island has the greatest framework accumulation in the Pearl Islands (8°37′ N, 79°3′W; Table 6.1). This 14.3-ha, *Pocillopora*-dominated reef is the largest, best-studied reef in the Pearl Islands (Glynn and Maté 1997). Although living coral once covered over 90 % of the reef surface (Glynn 1976), El Niño-related mortality in 1982–83 resulted in a significant reduction in coral cover at the site (Glynn et al. 2001). Today, living coral cover at the Saboga Island reef is patchy, and most living coral is restricted to the outer edge of the reef.

We collected 13 push-cores from the reef at Saboga Island between 2005 and 2009 (Fig. 6.10). Analysis of subfossil corals in the cores suggests that reef development was spatially patchy and temporally sporadic throughout the Holocene history of the reef. In 2009, we collected two cores

Fig. 6.10 Montage of push-core sampling, Iguana Island reef, September 2007. **a** positioning core tube in reef. **b** core tube in upright position in the reef. **c** adjustable core handles used to rotate core tube into reef framework. **d** hammer-weight for forcing core tube into reef framework. **e** core tube fully penetrated into reef and being prepared for extraction. *Photos* by R.B. Aronson and S.V. Vollmer

that penetrated approximately 3.6 m into the *Pocillopora* framework and reached the basaltic bedrock at the base of the reef. The oldest sections of the cores were dominated by fragments of branching coralline algae, which imply that little or no coral was present during the earliest stages of reef development. Bottom dates from these cores suggest that the pocilloporid reef at this site was established approximately 4348 cal year BP (2-sigma range 4558–4131 cal year BP; Table 6.2). This date is comparable with the bottom date obtained by Glynn and Macintyre (1977) from a drill-core extracted from the reef [2-sigma range 4549–4168 cal year BP, calibrated from Glynn and Macintyre (1977) using the local reservoir correction for the Gulf of Panama]. The timing of reef establishment at Saboga is well after the putative time of shelf-flooding in the region, 7000 years ago (Lighty et al. 1982; Toscano and Macintyre 2003), suggesting that conditions inimical to reef growth persisted for some time after flooding. Based on radiocarbon dating of core samples, Glynn and Macintyre (1977) found the average accretion rate at Saboga Island reef over the last 4500 year to be approximately 1.3 m/1000 year. This millennial-scale accretion rate was less than the 1.6 m/1000 year they reported for Uva Island, which may be due to the negative influence of upwelling on reef development in the Gulf of Panama (Glynn and Macintyre 1977).

Playa Larga reef (8°37′N, 79°1′W), located along the northeast coast of Contadora Island, is less than 3 km from the reef at Saboga Island; however, the Holocene history of reef development at Contadora is very different from that at Saboga. Playa Larga reef is 11.7 ha in area. The reef is dominated by pocilloporid corals, but *Porites lobata* is also common, especially at the southern end of the reef (Glynn and Maté 1997). Probing of the seaward edge of the reef indicated that the maximum framework accumulation is approximately 4.6 m (Table 6.1; Glynn and Macintyre 1977). The deepest push-core we collected at this site, which encountered bedrock, had a very similar penetration depth, namely 4.4 m. Although there is less framework accumulation at the Playa Larga reef than at the Saboga Island reef, a radiocarbon date from the bottom of the deepest push-core we collected at Playa Larga showed that the reef initiated around 6956 cal year BP (2-sigma range 7094–6645 cal year BP), which is more than 2000 years before reef initiation at Saboga Island (Table 6.2). The cores from Playa Larga and Saboga were collected in approximately the same water depth and within the same reef zone, but the reef at Playa Larga apparently began growing shortly after shelf-flooding.

The reef at Contadora experienced the same 2500-year hiatus in development as the reef at Canales de Tierra in the Gulf of Chiriquí (Fig. 6.9). Trends in elemental and isotopic composition of *Pocillopora* branches from a core collected at Contadora suggest that a regime of increased upwelling and precipitation at the onset of the hiatus 3800–3600 cal year BP provided the initial trigger for reef shutdown (negligible to zero accretion). Those conditions indicate an enhanced La Niña regime. Because the hiatus also occurred in the Gulf of Chiriquí, where there is no upwelling, climatic cooling and enhanced turbidity (light limitation) from high precipitation most likely drove regional reef collapse at this time (Toth et al. 2012, 2015). Whereas ENSO variability was the ultimate cause of reef shutdown, upwelling exerted a secondary influence in the Gulf of Panama, leading to a more protracted hiatus at Contadora compared with Canales de Tierra in the Gulf of Chiriquí (Toth et al. 2012).

The overall rate of reef accretion during the Holocene Epoch was lower at Contadora than in the Gulf of Chiriquí: 0.7 versus 1 m/1000 year. Overall accretion rates at Contadora were also lower than at Saboga: 0.7 versus 1.6 m/1000 year; however, when only periods of active reef development are considered, the accretion rate at Contadora was 1.6 m/1000 year, equivalent to the rate at Saboga (Toth 2013).

The other coral reefs probed in the Pearl Islands Archipelago during Glynn and Macintyre's (1977) regional survey, Pedro González and Cañas Islands, had similar levels of Holocene framework accumulation: 4.3 and 4.9 m, respectively. Overall, framework accumulation is significantly lower in the Gulf of Panama than in the Gulf of Chiriquí (independent t-test: $t = 5.664$, df = 9, $P < 0.001$), likely a result of the lower accretion rates in the former area.

The only large coral reef in the Gulf of Panama beyond the Pearl Islands is on the west side of Iguana Island (7°37′N, 80° 0′W), 4 km east of the Azuero Peninsula. The 16-ha reef at Iguana Island is the largest reef in the Gulf of Panama and was declared a Wildlife Reserve in 1981. The shallow habitats of the reef are dominated by pocilloporid corals, but massive corals, including *Porites lobata* colonies up to 400 years old, are abundant at greater depths (Guzmán et al. 1991). Iguana Island experiences some impacts of the annual upwelling in the Gulf of Panama, but upwelling is less intense in this part of the gulf. Based on reef probing, the maximum framework thickness at Iguana Island is 6.1 m (Table 6.1; Glynn and Macintyre 1977). This suggests that the reef at Iguana Island is somewhat better developed than the reefs in the Pearl Islands; however, reef thickness is still greater in the Gulf of Chiriquí (one-sample t-test: $t = 4.324$, df = 5, $P = 0.012$).

We collected 8 push-cores from the reef at Iguana Island in 2006 and 2007. The maximum radiocarbon age from a sample of *Psammocora stellata* at the base of one of our cores indicates that the reef at Iguana Island began accreting around 6762 cal year BP (2-sigma range 6920–6621 cal year BP), at about the same time as the reef at Playa Larga (Table 6.2). The reef at Iguana stopped growing ~4100 cal year BP and vertical accretion was minimal over the next 2500 year, as was the case at Contadora and Canales de Tierra (Fig. 6.9; Toth et al. 2012). Millennial-scale accretion rates at Iguana outside this hiatus were ~1.8 m/1000 year, which are slightly higher than at Contadora but lower than at Canales de Tierra in the Gulf of Chiriquí. The growth rate of *Pocillopora damicornis* is approximately 20 % lower in the Gulf of Panama than in the Gulf of Chiriquí, and this difference is most pronounced during the upwelling season (Glynn 1977). The suppression of pocilloporid growth by upwelling may explain the millennial-scale differences in reef accretion between the two gulfs. In summary, reefs off the Pacific coast of Panama are best developed in the Gulf of Chiriquí, intermediate at Iguana Island, and least developed in the Pearl Islands. These differences reflect the influence of the gradient in upwelling across the three locations.

6.4.6 Colombia

Significant reef formations have been reported in several areas of Pacific Colombia. In the northern part of the country, coral reefs can be found in the Gulf of Cupica (6° 32′N, 77°21′W; Fig. 6.1) and Ensenada de Utría (6°4′N, 77° 23′W; Fig. 6.1). Tebada, the largest reef in the Gulf of Cupica (4.5 ha), is currently dominated by slow-growing *Psammocora* sp.; nevertheless, probing of the reef suggests that Tebada has a Holocene carbonate accumulation of at least 4 m (Table 6.1; Zapata and Vargas-Ángel 2003). The best-developed reef in Ensenada de Utría is the 10.5-ha, *Pocillopora*-dominated reef at La Chola (Vargas-Ángel 1995, 2001; Zapata and Vargas-Ángel 2003). According to the results of probing and coring by Vargas-Ángel (2001), the unconsolidated *Pocillopora* framework at La Chola is 2–4.5 m thick and has been accumulating continuously for the last 2000–3000 years. Average accretion rates at La Chola were 2.5 m/1000 year (Vargas-Ángel 2001), which is similar to rates observed on reefs in other non-upwelling regions in the ETP (i.e., Golfo Dulce and the Gulf of Chiriquí).

Records of large coral aggregations have also been reported around the oceanic island of Malpelo (4°0′N, 81° 36′W; Fig. 6.1), located almost 500 km from the Colombian coast (Birkeland et al. 1975; Garzón-Ferreira and Pinzón

1999; Zapata and Vargas-Ángel 2003). The clear water surrounding Malpelo Island provides good conditions for reef development (Garzón-Ferreira and Pinzón 1999), and corals can be found along the island's slope to a depth of 30 m (Birkeland et al. 1975). Despite the high coral cover at Malpelo (15–89 %; Birkeland et al. 1975; Garzón-Ferreira and Pinzón 1999), there is no evidence of accumulation of fossil and subfossil corals, possibly due to the steep slope of the subtidal substrate (Birkeland et al. 1975). It is generally agreed that the coral aggregations at Malpelo cannot be considered true coral reefs (e.g., Birkeland et al. 1975; Zapata and Vargas-Ángel 2003).

The majority of known coral reefs in Pacific Colombia are found around Gorgona Island (2°59′N, 78°12′W; Fig. 6.1), approximately 35 km off the South American coast (Glynn et al. 1982; Prahl 1986; Zapata and Vargas-Ángel 2003). As at Malpelo Island, the water surrounding Gorgona Island is clear, and corals can occur to depths of 30 m (Zapata and Vargas-Ángel 2003). Fringing coral reefs are best developed on the east side of the island; however, cementation of framework by coralline algae only occurs on the island's western face (Glynn et al. 1982). Most of the reefs at Gorgona display patterns of coral zonation typical of the ETP, with pocilloporids dominating shallow substrates and massive *Porites*, *Pavona*, and *Gardineroseris* species increasing in abundance with depth (Glynn et al. 1982).

The *Pocillopora* framework of La Azufrada reef, the largest coral reef adjacent to Gorgona Island (11.2 ha; Zapata and Vargas-Ángel 2003), was probed by Glynn et al. (1982) and found to be 8 m thick (Table 6.1). Prahl (1986) suggested that the reef at La Azufrada likely formed after stabilization of sea level in the region, around 2500–3000 years ago. Visual inspection of nearby Pier Reef suggested that Holocene framework accumulation at this reef was 6 m above the seafloor. Framework accumulation at La Gómez and La Ventana formations, also nearby, was only 1 m, suggesting that these structures are incipient reefs (Glynn et al. 1982). The maximum framework thickness of the reefs between Gorgona and nearby Gorgonilla Islet was approximately 2 m, based on visual inspection (Prahl 1986; Zapata and Vargas-Ángel 2003).

6.4.7 Ecuador

6.4.7.1 Mainland Ecuador

Ecuador represents the southern limit of reef development along the ETP mainland, and there are few structural coral reefs in this region. During the 1970s, however, several reefs were described along coastal Ecuador (Glynn 2003). Surface inspections of the 1-ha, *Pocillopora*-dominated fringing reef along the leeward shore of Sucre Island (1°28′N, 80°46′W; Fig. 6.1), off the coast of Machalilla, suggested a Holocene framework accumulation of 3–4 m. Several kilometers south of Machalilla, a similar degree of framework accumulation was estimated for the pocilliporid reef adjacent to Cabuya Beach. When both reefs were revisited in the early 1990s, however, there was very little living coral and much of the existing framework had eroded. This was likely the result of the regional devastation of coral reefs by the 1982–83 El Niño event. The only living reefs currently found near the Ecuadorian mainland occur on the northeast coast of La Plata Island (1°16′N, 81°5′W; Fig. 6.1; see also Chap. 5, Fig. 5.25, Glynn et al.), located 31 km west. These reefs are built primarily by massive coral species, including *Pavona clavus*, *Pavona gigantea*, and *Gardineroseris planulata*, and have vertical accumulations of 2–6 m (Table 6.1; Glynn 2003).

6.4.7.2 Galápagos Islands

The Galápagos Islands are located 1050 km west of the Ecuadorian coast (Fig. 6.1). Due to their equatorial position, the Galápagos are subject to the most extreme consequences of ENSO. El Niño in particular has been a major cause of coral mortality in the Galápagos, most notably during the 1982–83 El Niño event (Glynn and Wellington 1983; Glynn et al. 2001; Glynn 2003; Wellington and Glynn 2007). Prior to the 1982–83 El Niño, small coral reefs could be found at many of the central and southern Galápagos Islands. Due to the high water clarity in most areas, corals occurred to depths of 30 m. Reef development on the western sides of Isabela and Fernandina Islands was limited because of localized upwelling, but small patch reefs and assemblages of non-reef-building corals were found in more sheltered areas of these islands. El Niño-related mortality in the early 1980s resulted in the loss of over 90 % of Galápagos reefs, and there has been very little recovery since [Glynn 2003; but see Glynn et al. (2009, 2015) for examples of rapid coral recovery at the northernmost Darwin and Wenman Islands].

Several studies have been conducted on the coral reefs off the north coast of Floreana Island in the southern Galápagos (1°17′S, 90°26′W; Fig. 6.1). The reef community of Champion Island off the northern coast of Floreana, for example, was reported to be flourishing in 1975 and 1976 (Fig. 6.11a). This reef was dominated by *Pavona* spp. and revealed Holocene framework accumulation of up to 5 m (Table 6.1; Glynn 2003). In 1990, the reefs were almost completely bioeroded by the sea urchin *Eucidaris galapagensis* (Fig. 6.11b) following catastrophic coral mortality caused by the 1982–83 El Niño event. The seascape was reduced to extensively bioeroded *Pavona* skeletons on a knobby coral pavement, covered by filamentous algal turf (Fig. 6.11b; Macintyre et al. 1992).

Prior to 1982, there were reports of a *Pocillopora* reef in the embayment of Onslow Island, north of Floreana—an area known as Devil's Crown (1°13′S, 90°25′ W; Fig. 6.1)—although reef-probing suggested the thickness of

Fig. 6.11 Patch reef off Champion Island, Galápagos Islands before and after the 1982–83 El Niño warming event, 6 m depth. **a** Framework dominated by living *Pavona clavus*, January 1975. **b** Same general area with extensive bioerosion of the reef surface by *Eucidaris galapagensis*, November 1990. *Photos* by P.W. Glynn **a**, and I.G. Macintyre **b**

this reef was less than 1 m (Glynn and Wellington 1983; Glynn 2003). The existing framework was completely bioeroded in the wake of the 1982–83 El Niño disturbance (Glynn 2003). Consequently, in 1990 Macintyre et al. (1992) found no significant reef framework at Devil's Crown. The base of one large, toppled, dead colony of *Pavona gigantea* revealed a radiocarbon date of 488 cal year BP (2 sigma range 600–287 cal year BP), suggesting that a thin reef had been growing at this site for a few hundred years. The reefs at Onslow Island and the north coast of Floreana Island, which were flourishing in 1975 and 1976 (Glynn and Wellington 1983), consisted of scattered, small, living colonies of *Porites lobata* and remnants of reef framework that were bioeroded and covered by crustose coralline algae.

Although the inshore reef of Devil's Crown was destroyed, corals persisted in deeper water offshore after the ENSO events of 1982–83 and 1997–98. Surveys in 1999 revealed that an extensive, unconsolidated coral biostrome, dominated by free-living *Psammocora stellata* and *Diaseris distorta*, had covered the sandy bottom at 13–18 m depth, with living coral cover approaching 80 % in some areas. These deep-water assemblages are apparently more resistant to coral bleaching associated with El Niño conditions than the shallow-water assemblages, considering they survived both the 1982–83 and 1997–98 severe events (Feingold 1995, 2001). Three cores collected from Devil's Crown in 1990–91 (Feingold 1995) were analyzed by Hendrickson (2014). The cores demonstrated pronounced shifts in the relative abundance of *Diaseris distorta* and *Psammocora stellata* through time, which may have been related to past climate change. Core penetrations were approximately 1.45, 1.55, and 1.55 m, and the basal dates of the cores were 7330, 5183, and 5116 cal year BP, respectively (Tables 6.1, 6.2), suggesting that the coral biostromes likely began forming soon after shelf-flooding and accreted at an average rate of 0.002–0.003 m/1000 year to the present.

A few coral reefs have been found adjacent to Isabela Island (Fig. 6.1), the largest island in the Galápagos Archipelago. The most significant record of *Pocillopora* framework accumulation, 3–4 m, was observed in the 1970s off Villamil (1°3′S, 91°4′W; Table 6.1; Fig. 6.1), on the southern coast of Isabela. A 1-m accumulation of *Pavona clavus* was observed on the east coast of Isabela (Glynn and Wellington 1983). Like most coral reefs in the Galápagos, however, these reefs succumbed during and following the 1982–83 El Niño event (Glynn 2003).

The living reef in Urvina Bay (0°24′S, 91°15′W; Fig. 6.1), off the west coast of Isabela Island, is covered with knobby mounds of *Pavona clavus* and *Porites lobata* up to 2 m in vertical relief, which protrude from an algae-covered substrate (Fig. 6.12; Macintyre et al. 1992). *Pocillopora-*

Fig. 6.12 Living colonies of *Pavona clavus* protruding above dense algal cover, Urvina Bay, Galápagos Islands, 2 m depth, November 1990. *Photo* by I.G. Macintyre

framework accumulation at Urvina Bay was estimated at 0.5 m (Glynn and Wellington 1983), although this may have been an underestimate of historical accumulations as the reef was extensively bioeroded (Macintyre et al. 1992). A large, subfossil reef, located in the eastern section of Urvina Bay, was uplifted by 6 m in 1954 as a result of volcanism and magmatic intrusion (Glynn and Wellington 1983; Colgan and Malmquist 1987; Colgan 1990; Dunbar et al. 1994). Observations of the uplifted reef showed that calcareous red algae were the first to colonize Urvina Bay, followed by the free-living coral *Psammocora stellata*. In the final stages of reef development, *Pocillopora* spp. and massive coral species also began to recruit and contribute to the framework (Colgan and Malmquist 1987; Colgan 1990).

The only material available for radiocarbon dating in the uplifted section at Urvina Bay came from coral rubble found at the base of a 12-m diameter, uplifted colony of *Pavona clavus*. This sample indicated that reef growth initiated at least 1000 years ago (Colgan 1991). The massive corals in Urvina Bay revealed numerous skeletal discontinuities, suggesting that these coral assemblages were subjected to severe disturbance events in the past, most likely due to ENSO activity (Glynn and Wellington 1983; Dunbar et al. 1994).

Small accumulations of *Pocillopora* framework, 1 m or less in thickness, were also observed at San Cristóbal (0°48′ S, 89°24′W), Española (1°22′S, 89°40′W), Santa Fe (0°49′S, 90°3′W), Fernandina (0°25′S, 91°27′W), Darwin (1°40′N, 92°0′W) and Genovesa (0°19′N, 89°56′W) Islands prior to the 1982–83 El Niño event (Glynn and Wellington 1983). These reefs experienced high levels of bleaching-related mortality in 1982–83, and no pocilloporid reefs have re-established since that time (Glynn 2003). However, at a shallow lagoon near Villamil, Isabela Island, an incipient pocilloporid framework observed recently is forming by asexual propagation (Baums et al. 2014; Feingold and Glynn 2014).

In general, reefs dominated by massive corals were found to have considerably more carbonate accumulation than *Pocillopora*-dominated reefs in the Galápagos Islands. *Porites lobata* reefs off Santiago and Bartolomé Islands had framework accumulations of 2–3 and 4–5 m, respectively (Table 6.1). Bartolomé also had a *Pavona*-dominated reef that was up to 6 m thick, and a 2–3 m thick *Pavona* reef was found at Genovesa (Glynn and Wellington 1983). The vertical accumulation of massive coral skeletons in the Galápagos is comparable to that off the mainland (Glynn and Wellington 1983). Most of the corals on these reefs are now dead and have suffered high levels of bioerosion (Glynn 2003); however, the Darwin Island reef, which was once colonized by both pocilloporid species and massive *Porites lobata*, has shown surprising recovery since the 1982–83 El Niño. Although coral mortality was high, the surviving *Porites lobata* colonies have recolonized the reef since the 1982–83 event; nevertheless, there has been no recovery of *Pocillopora* (Glynn et al. 2009, 2015). This suggests that reefs dominated by massive corals may be more resilient to the impacts of ENSO events, possibly because massive-coral frameworks may be inherently more resistant to the effects of bioerosion. Furthermore, the growth of massive corals is less affected than the branching pocilloporids by upwelling conditions (Manzello 2010), which may be another reason for the high abundance of reefs dominated by massive corals in some parts of the Galápagos Islands.

The rate and extent of reef development in the Galápagos are limited by bioerosion, which is caused primarily by the grazing sea urchin *Eucidaris galapagensis* (Glynn et al. 1979). Although modern growth rates of *Pocillopora* are fairly high in the Galápagos (Glynn and Wellington 1983), the majority of gross carbonate production by pocilloporids is degraded by bioerosion (Glynn et al. 1979). In most areas of the Galápagos, therefore, framework accumulations are relatively thin layers growing atop the basalt bedrock (Glynn and Wellington 1983), and long-term rates of vertical accretion are low.

In summary, nearly all coral reefs in the Galápagos experienced significant loss of framework due to high levels of grazing by *Eucidaris galapagensis* in the wake of the 1982–83 El Niño event (Glynn 2003). It is likely, therefore, that a large portion of the geologic record of coral-reef framework has been degraded and lost. The effects of bioerosion by *Eucidaris galapagensis* are most severe for fragile *Pocillopora* spp. (Glynn et al. 1979), which may explain why most of the pocilloporid reefs in the Galápagos have been reduced to rubble whereas structures built of massive corals are more often partially intact.

6.4.8 Clipperton Atoll

Clipperton Atoll (10°18′N, 109°13′W; Fig. 6.1), located approximately 1100 km west of the Mexican coast, lies at the westernmost edge of the ETP. Because of its isolation, coral diversity is low at Clipperton (Glynn et al. 1996; Glynn and Ault 2000; see Chap. 5, Glynn et al.). Nevertheless, the continuous coral reef that encircles the island, measuring 370 ha in area, is the largest structural reef in the ETP (Dana 1975; Glynn et al. 1996). Water clarity is high compared with other areas supporting reef growth in the ETP (Dana 1975; Glynn et al. 1996), and, as a result, reef development extends to at least 60 m depth (Glynn et al. 1996). In contrast, high turbidity and low light penetration restrict most coral-reef development to depths of 10–15 m or less along the ETP mainland (Dana 1975; Glynn 1976).

Reef-habitats at Clipperton Atoll are dominated by large *Porites* colonies, but monogeneric patches of *Pocillopora* spp. are also present (Glynn et al. 1996). Glynn et al. (1996)

measured living and dead colonies of *Pocillopora* at Clipperton reef and found them to be at least 1–2 m in height, although they suggested that the thickness of Holocene reef accumulation is likely greater than that. Linsley et al. (2000) obtained drill-cores measuring 120, 126, and 245 cm in length from three massive *Porites* colonies on the reef. The basal dates of the cores, 1937, 1947, and 1893 C.E., demonstrated that the growth of *Porites* was rapid at their sampling site. Carricart-Ganivet and Reyes-Bonilla (1999) observed a number of coral terraces approximately 6 m above sea level on Clipperton. Based on the putative sea-level curve for the region, and accounting for subsidence of the atoll, they estimated that these terraces initiated growth approximately 5000 years ago. Their date provides a reasonable minimum estimate for the time of reef initiation at Clipperton Atoll. Another reef terrace, found at approximately 60 m depth, may be the remnant of a shallow fringing reef that formed during a sea-level low-stand in the late Pleistocene (Glynn et al. 1996).

6.4.9 Easter Island (Rapa Nui)

Easter Island, or Rapa Nui as it was first named by colonizing Polynesians, is located in the south-central Pacific, approximately 3750 km west of mainland Chile, to which it belongs (27°07′S, 109°22′W; Fig. 6.1). The coral fauna of Rapa Nui is both geographically and oceanographically isolated from most reefs of the wider Pacific (Glynn et al. 2003; Hubbard and Garcia 2003; Glynn et al. 2007), but the island is generally considered to be the easternmost outpost of the Indo-Pacific bioregion (Glynn et al. 2007; see also Veron et al. 2015). The island is ~2250 km from the Pitcairn Islands group in the central Pacific (24°23′S, 128°19′W), which is more than 1500 km closer than its nearest neighbor in the eastern Pacific: the Galápagos Islands (Glynn et al. 2003, 2007); however, the coral fauna of Rapa Nui is more closely aligned with that of the eastern Pacific than any other bioregion, suggesting that Easter Island should, in fact, be included in the eastern Pacific faunal province (Glynn et al. 2007).

Although coral growth in much of the southeastern Pacific is precluded by low temperatures, the waters surrounding Rapa Nui are warmed by the South Pacific Subtropical Gyre. This warm, subtropical water has allowed extensive coral assemblages to develop around the island (Glynn et al. 2003, 2007; Hubbard and Garcia 2003). Furthermore, because of the high water clarity, corals occur to depths of 60 m (Glynn et al. 2003, 2007; Hubbard and Garcia 2003). There are 13 species of corals living at Rapa Nui (Glynn et al. 2007), but most communities are dominated by *Pocillopora verrucosa* down to 10 m depth and *Porites lobata* in deeper reef zones (DiSalvo et al. 1988; Glynn et al. 2003, 2007; Hubbard and Garcia 2003). The extent of reef development around Rapa Nui is highly variable. Coral cover is low along the southeastern shore due to strong wave action for most of the year, which interferes with coral settlement and growth. In contrast, surveys along the western and northern coasts have revealed that coral cover in some areas can exceed 100 % (Glynn et al. 2003, 2007; Hubbard and Garcia 2003).

True reef development at Rapa Nui is restricted to two leeward-island locations: the Poike Peninsula at the northeastern corner of the island, and Cook Point along the west coast (Glynn et al. 2003; Hubbard and Garcia 2003). At both locations, Hubbard and Garcia (2003) observed *Porites* frameworks representing multiple generations of coral growth. The largest living *Porites* colonies observed were 10 m high, and *Porites* reef frameworks up to 7 m thick were common at these sites (Table 6.1; Glynn et al. 2003; Hubbard and Garcia 2003). During field work at Rapa Nui in 2000, Glynn et al. (2003) collected drill-cores from three relatively small (1–2 m diameter) *Porites lobata* colonies. By measuring the growth rates of these living corals, they estimated that the largest *Porites* colonies were 860–1120 years old. Fossil corals have also been found in some deep-water caves (>50 m depth) along the offshore slopes of the island (DiSalvo et al. 1988), but these corals have not been dated.

6.5 Environmental Controls of Reef Development

6.5.1 Geographic Patterns of Coral-Reef Development in the ETP

The most significant accumulations of *Pocillopora* frameworks are found from about 7–15°N latitude, along the central coastal areas of the ETP (Fig. 6.1; Table 6.1), with the thickest Holocene framework accumulation recorded in the Gulf of Chiriquí, Panama (7°N; Table 6.1). The offshore islands of Clipperton Atoll, Rapa Nui, and to a lesser extent Gorgona Island, also have significant framework accumulation, and because of the high water clarity at these sites corals occur at greater depths than in most regions of the ETP (Table 6.1; Glynn et al. 1982; Carricart-Ganivet and Reyes-Bonilla 1999).

At the extreme northern and southern limits of the region—Mexico and Ecuador—most corals occur as constituents of incipient reef assemblages, and few well-developed reefs can be found (Glynn 2003; Reyes-Bonilla 2003). Framework accumulation is also minimal in the Galápagos Islands, where environmental disturbances are most extreme. Few records from Galápagos reefs have reported *Pocillopora* frameworks thicker than 1 m, and the only significant frameworks consist of isolated patches of massive corals (Glynn and Wellington 1983; Glynn et al. 2009).

6.5.2 Wave Exposure

Throughout the ETP, coral reefs are primarily restricted to sheltered embayments (e.g., Glynn and Stewart 1973; Dana 1975; Glynn and Macintyre 1977; Glynn et al. 1983; Cortés 1990b; Guzmán and Cortés 1992; Glynn and Leyte-Morales 1997; Glynn and Maté 1997). This distributional pattern is due in large part to the protection from wave exposure these areas provide (Dana 1975; Glynn and Maté 1997). At Rapa Nui, for example, few corals are found along the southeastern shore, where wave energy is highest (Glynn et al. 2003; Hubbard and Garcia 2003). Calm, sheltered environments are conducive to the initiation of reef growth and its continuation for several reasons. First, the majority of reefs in the ETP are dominated by pocilloporid corals, and because there is little cementation of reefs in the region (Macintyre and Glynn 1976; Glynn and Macintyre 1977; Manzello et al. 2008), strong wave action can dislodge existing frameworks, precluding significant carbonate accumulation. Calm water also facilitates sedimentation, allowing a buildup of entrained sediments that can stabilize reefs. Reef accretion by carbonate production and sedimentation may in turn extend the habitable shelf area, allowing for further reef development.

6.5.3 Tides and Turbidity

Tidal amplitudes in the ETP range from ~0.5 m off the coast of Mexico and offshore oceanic islands (Sachet 1962; Glynn and Maté 1997; see Chap. 3, Fiedler and Levín) to as much as 7 m in Pacific Panama (Glynn 1976). The large tidal ranges in many parts of the ETP are thought to be an important control on reef development in the region (Glynn 1976; Glynn and Macintyre 1977; Glynn and Maté 1997). Two important impacts of tidal fluctuations on reef development are subaerial exposure of corals during extreme low tides and high turbidity associated with tidal flushing.

Subaerial exposure is most common during the spring tides of La Niña years, when mean sea level in the ETP is reduced (Fig. 6.7; Eakin and Glynn 1996). Subaerial exposure can have a major impact on reefs (Glynn 1976; Eakin and Glynn 1996); Glynn (1976) reported coral mortality of up to 70 % on some reefs after a spring tidal cycle of reef exposures.

Although it is clear that subaerial exposure can cause significant damage in the short term, the high turbidity associated with large tidal amplitudes (Glynn 1976), as well as upwelling (discussed below; Glynn and Stewart 1973) and high seasonal precipitation (Dana 1975), may have an even greater impact on long-term reef development in the region. For example, the depauperate coral assemblages and, in some cases, the complete inhibition of coral development in some areas of the Great Barrier Reef, have been attributed to the elevated turbidity associated with large tidal ranges (Kleypas 1996; van Woesik and Done 1997). High turbidity controls reef development by limiting the amount of light that reaches the phototrophic reef-building corals (Kleypas 1996, 1997; Kleypas et al. 1999a). Light limitation restricts reef development to depths of 10–15 m in many parts of the ETP (Dana 1975; Glynn 1976), whereas regions with low turbidity, such as Clipperton Atoll, Rapa Nui, and Malpelo Island, can support coral growth to 30 m depth and deeper (Glynn et al. 1996; Zapata and Vargas-Ángel 2003).

6.5.4 Upwelling

Seasonal upwelling in the Gulfs of Tehuantepec, Papagayo, and Panama, and equatorial upwelling in some parts of the Galápagos Islands, are associated with the intrusion of cold, nutrient-rich waters into shallow, coastal areas (e.g., Dana 1975; Glynn et al. 1983; D'Croz and O'Dea 2007; Alexander et al. 2012; see Chap. 3, Fiedler and Lavín). Growth rates of pocilloporid corals are reduced during the upwelling season in these areas (Glynn and Stewart 1973); therefore, overall growth rates of *Pocillopora* spp. are generally higher on reefs not exposed to upwelling (Glynn 1977). Interestingly, upwelling does not appear to be as detrimental to the growth of massive corals (Jiménez and Cortés 2003; Manzello 2010).

The negative impact of upwelling on pocilloporid corals is primarily due to the reduction of growth and calcification as a result of low sea temperatures (Glynn and Stewart 1973; Glynn 1977), but enhanced nutrient inputs during upwelling can also be problematic. Elevated levels of some nutrients can be detrimental to coral growth and calcification (Kinsey and Davies 1979; Hallock and Schlager 1986), although inorganic nutrients have also been shown to be beneficial to coral growth (Adkinson et al. 1995; Jiménez and Cortés 2003). Increases in nutrients lead to increases in regional primary productivity (Glynn and Stewart 1973; Martínez et al. 2006; D'Croz and O'Dea 2007), and the resultant plankton blooms increase water turbidity (Glynn and Stewart 1973). High nutrients can also increase the competitive ability of macroalgae and suspension feeders on reefs (Glynn and Stewart 1973; Birkeland 1977; Highsmith 1980). Furthermore, upwelled waters have low pH relative to surface waters, which has resulted in a reduction of the carbonate saturation state in the ETP generally (Manzello et al. 2008). This condition may render reefs of the ETP vulnerable to the impacts of ocean acidification, expected to increase with future global climate change (Kleypas et al. 1999b; Guinotte et al. 2003; Manzello et al. 2008). Given the negative impacts of upwelling on modern coral reefs in the ETP, it is not altogether surprising that upwelling was also important in structuring coral reefs of the past.

Records from coral reefs throughout the ETP indicate that upwelling exerted an important influence on development during the Holocene. Generally, where modern corals persist despite the influence of seasonal upwelling—in the Gulf of Panama, for example—reef development is limited to embayments where the corals are sheltered from the full force of oceanic influences (Glynn and Stewart 1973). Similarly, coral-reef development at Isabela and Fernandina Islands in the Galápagos is primarily restricted to the eastern sides of the islands, where upwelling is less influential (Glynn 2003). In both Panama and Costa Rica, reefs are better-developed, with thicker accumulations of *Pocillopora* frameworks, in non-upwelling areas compared with areas that experience upwelling (Table 6.1; Glynn and Macintyre 1977; Glynn et al. 1983; Macintyre et al. 1992; Cortés et al. 1994). An interesting exception is the Huatulco reef tract in the Gulf of Tehuantepec, one of Mexico's major upwelling centers, which supports greater framework accumulation than the non-upwelling regions (Table 6.1; Glynn and Leyte-Morales 1997).

Although upwelling has limited the magnitude of framework accumulation in the ETP, it may not have significantly reduced long-term accretion rates in the region. We found that upwelling did not influence long-term rates of reef accretion within certain favorable intervals—before and after the observed hiatus in reef development—at Contadora, Iguana, and Canales de Tierra Islands in Pacific Panama. Millennial-scale rates of reef accretion within these periods were not significantly different among sites (Toth et al. 2012; Toth 2013).

Upwelling regimes also did not impact the timing of reef initiation. Despite having significantly thicker frameworks than reefs in the Gulf of Panama, reefs in the non-upwelling Gulf of Chiriquí initiated at least 1000 years later. It is possible therefore, that upwelling was less important in controlling reef development in the early Holocene than at present. Conversely, more intense upwelling during certain periods may have precluded reef development in some areas. In northwestern Costa Rica, which experiences the most protracted and intense seasonal upwelling in the ETP, all reefs have experienced nearly complete mortality in recent centuries, possibly due to enhanced upwelling activity (Glynn et al. 1983). Upwelling also resulted in a longer hiatus in reef development in the Gulf of Panama relative to the Gulf of Chiriquí (Toth et al. 2012, 2015).

Overall, the effects of upwelling on coral-reef development in the ETP are mixed. Upwelling is an important control on the distribution of reefs in the region. Moreover, in many areas, reefs exposed to upwelling have significantly thinner accumulations of Holocene reef-framework. Nevertheless, over millennial scales, upwelling may have been less important in determining long-term rates of reef accretion than previously thought.

6.5.5 ENSO

In recent years, ENSO has devastated coral reefs throughout the ETP. El Niño-related coral mortality, especially during the 1982–83 and 1997–98 events, drastically reduced coral cover (Glynn et al. 2001; Jiménez et al. 2001). In many regions, but most prominently in the Galápagos Islands, the increase in bioerosion after these events has led to significant losses of reef frameworks (Glynn 1988; Eakin 1992; Reaka-Kudla et al. 1996; Fonseca 1999; Glynn et al. 2015). La Niña can be equally problematic for reefs of the ETP. Lowered sea level during La Niña events leads to more frequent subaerial exposure of reefs (Glynn 1976; Eakin and Glynn 1996). In some parts of the ETP, La Niña can also enhance seasonal upwelling events, causing cold shock, coral bleaching and mortality (Poveda et al. 2006; Banks et al. 2009; Alexander et al. 2012).

The dynamics of ENSO have not been constant during the Holocene, and it is likely that its impacts have at times been more severe than at present (e.g., Moy et al. 2002; Riedinger et al. 2002; Toth et al. 2012, 2015). We showed that past increases in ENSO variability ultimately led to a 2500-year collapse of coral-reef ecosystems in Pacific Panama (Toth et al. 2012, 2015; Toth 2013). Our reconstructions demonstrate that ENSO variability was likely the most important control of Holocene reef development in the ETP.

6.6 Summary

Coral reefs of the ETP have long been considered depauperate and poorly-developed compared with other regions of the world (Cortés 1993, 1997; Veron 1995). Despite the relatively low coral diversity and geographic isolation of the ETP (Veron 1995; Glynn and Ault 2000), well-developed coral reefs have been discovered throughout the region in recent decades (e.g., Glynn and Macintyre 1977; Glynn et al. 1982, 1983; Glynn and Wellington 1983; Glynn and Leyte-Morales 1997; Carricart-Ganivet and Reyes-Bonilla 1999; Cortés and Jiménez 2003; Reyes-Bonilla 2003; Zapata and Vargas-Ángel 2003). Although coral reefs in the ETP are generally small in size (Dana 1975; Glynn and Macintyre 1977; Cortés 1997; Glynn and Maté 1997) and restricted to sheltered habitats (Dana 1975; Glynn and Stewart 1973; Glynn et al. 1983), reef accretion in the region has been rapid and has persisted for millennia. Indeed, the maximum levels of Holocene framework accumulation in the ETP—on the order of 10–15 m—rival the thickness of reefs in the western Atlantic [Macintyre and Glynn (1976) for Panama; Macintyre et al. (1977) for Mexico; Westphall (1986) for Belize; Hubbard et al. (2005) for St. Croix; Greer et al. (2009) for the Dominican Republic]. In Panama, both short- and long-term rates of reef accretion were comparable across the Pacific and Caribbean sides of the Isthmus (Glynn and Macintyre 1977; Macintyre and Glynn 1976). Although many reefs in the ETP appear to have initiated only in recent millennia (Cortés 1997), reef development at some locations began as early as 6000 to nearly 7000 years ago, corresponding to the time of reef initiation in the Caribbean, but later than elsewhere in the central and western Pacific (Macintyre 2007; Montaggioni and Braithwaite 2009).

In many ways the trends of Holocene reef development in the ETP are more similar to other regions of the world than originally envisaged; however, differences in the histories of reefs within and across oceanic regions are rapidly becoming moot. The globalization of environmental perturbations in recent decades, most critically rising temperatures, is killing framework-building corals—or at the very least slowing their growth and depositional rates—and homogenizing the composition of living reef assemblages (Hoegh-Guldberg et al. 2007; Pandolfi et al. 2011). The loss of framework-building corals threatens to decelerate vertical accretion at a time when sea-level rise is accelerating, potentially compromising the viability of coral reefs and the ecosystem services they provide (Alvarez-Filip et al. 2013; Perry et al. 2013). Although well-developed coral reefs have persisted in a few locations in the ETP for millennia, the future of coral-reef growth in the region remains uncertain.

Acknowledgments Our work in the eastern tropical Pacific has been supported by grants from the Smithsonian Institution's Marine Science Network, a Graduate Student Research Grant from the Geological Society of America, the Lerner–Gray Fund for Marine Research of the American Museum of Natural History, and most recently by the U.S. National Science Foundation through grant OCE-1535007 to R.B.A. and L.T.T. Field work was carried out under permits from the Republic of Panama. We thank the editors—P.W. Glynn, D.P. Manzello, and I.C. Enochs—for their invitation to contribute to this volume and for their thoughtful suggestions throughout the editorial process. R. van Woesik, M.B. Bush, S.M. Lazarus, and C.L. Pruett also provided valuable comments on earlier drafts of the manuscript. This is contribution number 132 from the Institute for Research on Global Climate Change of the Florida Institute of Technology.

References

Alexander MA, Seo H, Xie SP, Stott JD (2012) ENSO's impact on the gap wind regions of the eastern tropical Pacific ocean. J Climate 25:3549–3565

Alvarado JJ, Reyes-Bonilla H, Buitrago F, Aguirre-Rubí J, Álvarez del Castillo-Cárdenas PA (2010) Coral reefs of the Pacific coast of Nicaragua. Coral Reefs 29:201

Alvarado JJ, Álvarez del Castillo-Cárdenas PA, Fernández C, Aguirre-Rubí J, Buitrago F, Reyes-Bonilla H (2011) Coral communities of San Juan del Sur, Pacific Nicaragua. Bull Mar Sci 87:129–146

Alvarez-Filip L, Carricart-Ganivet JP, Horta-Puga G, Iglesias-Prieto R (2013) Shifts in coral-assemblage composition do not ensure persistence of reef functionality. Sci Rep 3:srep03486

Atkinson MJ, Carlson B, Crow GL (1995) Coral growth in high-nutrient, low-pH water: a case study of corals cultured at the Waikiki Aquarium, Honolulu, Hawaii. Coral Reefs 14:215–233

Bakus GJ (1975) Marine zonation and ecology of Cocos Island, off Central America. Atoll Res Bull 79:1–10

Banks S, Vera M, Chiriboga Á (2009) Establishing reference points to assess long-term change in zooxanthellate coral communities of the northern Galapagos coral reefs. Galapagos Res 66:43–64

Baums IB, Durante MD, Laing AA, Feingold J, Smith T, Bruckner A, Monteiro J (2014) Marginal coral populations: the densest known aggregation of *Pocillopora* in the Galápagos Archipelago is of asexual origin. Front Mar Sci 1:59. doi:10.3389/fmars.2014.00059

Bezy MB, Jiménez C, Cortés J, Segura A, León A, Alvarado JJ, Gillén C, Mejía E (2006) Contrasting *Psammocora*-dominated coral communities in Costa Rica, tropical eastern Pacific. In: Proceedings of 10th International Coral Reef Symposium, vol 1, Okinawa, pp 376–381

Birkeland C (1977) The importance of rate of biomass accumulation in early successional stages of benthic communities to the survival of coral recruits. In: Proceedings of 3rd International Coral Reef Symposium, vol 1, Miami, Florida, pp 15–21

Birkeland C, Meyer DL, Stames JP, Buford CL (1975) Subtidal communities of Malpelo Island. In: Graham JB (ed) The biological investigation of Malpelo Island. Smithson Contrib Zool 176:55–68

Carricart-Ganivet JP, Reyes-Bonilla H (1999) New and previous records of scleractinian corals from Clipperton Atoll, eastern Pacific. Pac Sci 53:370–375

Carriquiry JD, Reyes-Bonilla H (1997) Estructura de la comunidad y distribución geográfica de los arrecifes coralinos de Nayarit, Pacífico de México. Cienc Mar 23:227–248

Carriquiry JD, Villaescusa JA (2010) Coral Cd/Ca and Mn/Ca records of ENSO variability in the Gulf of California. Clim Past 6:401–410

Chollett I, Mumby PJ, Cortés J (2010) Upwelling areas do not guarantee refuge for coral reefs in a warming ocean. Mar Ecol Prog Ser 416:47–56

Colgan MW (1990) El Niño and the history of eastern Pacific reef building. In: PW Glynn (ed) Global ecological consequences of the 1982–83 El Niño-Southern Oscillation. Elsevier Oceanography Series 52. Elsevier, Amsterdam, pp 183–232

Colgan MW (1991) Urvina Bay uplifted coral community history: biological and paleontological implications. In: James MJ (ed) Galápagos marine invertebrates, taxonomy, biogeography and evolution in Darwin's Islands. Plenum, New York, pp 93–120

Colgan MW, Malmquist DL (1987) The Urvina Bay uplift: a dry trek through a Galápagos coral reef. Oceanus 30:61–68

Cortés J (1990a) Coral reef decline in Golfo Dulce, Costa Rica, eastern Pacific: anthropogenic and natural disturbances. PhD Dissertation, University of Miami, Coral Gables, Florida, p 147

Cortés J (1990b) The coral reefs of Golfo Dulce, Costa Rica: distribution and community structure. Atoll Res Bull 344:1–37

Cortés J (1993) Comparison between Caribbean and eastern Pacific coral reefs. Rev Biol Trop 41(Suppl 1):19–21

Cortés J (1997) Biology and geology of coral reefs of the eastern Pacific. Coral Reefs 16(Suppl):S39–S46

Cortés J (2011) Eastern tropical Pacific coral reefs. In: Hopley D (ed) The encyclopedia of modern coral reefs: structure, form and process. Springer, Berlin, pp 351–358

Cortés J, Jiménez C (2003) Corals and coral reefs of the Pacific of Costa Rica: history, research and status. In: Cortés J (ed) Latin American coral reefs. Elsevier, Amsterdam, pp 361–385

Cortés J, Murillo MM, Guzmán HM, Acuña J (1984) Pérdida de zooxantelas y muerte de corales y otros organismos arrecifales en el Atlántico y Pacífico de Costa Rica. Rev Biol Trop 32:227–231

Cortés J, Macintyre IG, Glynn PW (1994) Holocene growth history of an eastern Pacific fringing reef, Punta Islotes, Costa Rica. Coral Reefs 13:1–9

Curray JR, Emmel FJ, Crampton PJS (1969) Holocene history of a strand plain, lagoonal coast, Nayarit, Mexico. In: Castanares A, Phleger FB (eds) Coastal lagoons: a symposium, D.F.: UNAM-UNESCO, Mexico, pp 63–100

Dana TF (1975) Development of contemporary eastern Pacific coral reefs. Mar Biol 33:355–374

Darwin C (1874) The structure and distribution of coral reefs, 2nd edn. Smith, Elder, & Co., London, p 214

Davies PJ, Marshall JF, Hopley D (1985) Relationship between reef growth and sea level in the Great Barrier Reef. In: Proceedings of 5th International Coral Reef Congress, vol 3, Tahiti, pp 95–104

D'Croz L, O'Dea A (2007) Variability in upwelling along the Pacific shelf of Panama and implications for the distribution of nutrients and chlorophyll. Estuar Coast Shelf Sci 73:325–340

DiSalvo LH, Randall JE, Cea A (1988) Ecological reconnaissance of the Easter Island sublittoral marine environment. Natl Geogr Res 4:451–473

Dunbar RB, Wellington GM, Colgan MW, Glynn PW (1994) Eastern Pacific sea surface temperature since 1600 A.D.: the $\delta^{18}O$ record of climate variability in Galápagos corals. Paleoceanography 9:291–315

Eakin CM (1992) Post-El Niño Panamanian reefs: less accretion, more erosion and damselfish protection. In: Proceedings of 7th International Coral Reef Symposium, vol 1, Guam, pp 387–396

Eakin CM (1996) Where have all the carbonates gone? A model comparison of calcium carbonate budgets before and after the 1982–83 El Niño at Uva Island in the eastern Pacific. Coral Reefs 15:109–119

Eakin CM, Glynn PW (1996) Low tidal exposures and reef mortalities in the eastern Pacific. Coral Reefs 15:120

Feingold JS (1995) Effects of elevated water temperature on coral bleaching and survival during El Niño disturbance events. PhD Dissertation, University of Miami, Coral Gables, Florida, p 236

Feingold JS (2001) Responses of three coral communities to the 1997–98 El Niño–Southern Oscillation: Galápagos Islands, Ecuador. Bull Mar Sci 69:61–77

Feingold JS, Glynn PW (2014) Coral research in the Galápagos Islands, Ecuador. In: Denkinger J, Vinueza L (eds) The Galápagos marine reserve: a dynamic social-ecological system. Springer, Cham, pp 3–22

Fong P, Glynn PW (1998) A dynamic size-structured population model: does disturbance control size structure of a population of the massive coral *Gardineroseris planulata* in the eastern Pacific? Mar Biol 130:663–674

Fonseca AC (1999) Bioerosión y bioacreción en arrecifes coralinos del Pacífico sur de Costa Rica. MSc Thesis, University of Costa Rica, San Pedro, Costa Rica, p 206

Fonseca AC, Dean HK, Cortés J (2006) Non-colonial coral macroborers as indicators of coral reef stress in the south Pacific of Costa Rica. Rev Biol Trop 54:101–115

Garzón-Ferreria J, Pinzón JH (1999) Evaluación rápida de estructura y salud de las formaciones coralinas de la Isla de Malpelo (Pacífico colombiano). Bol Invest Mar Cost 28:137–154

Glynn PW (1976) Some physical and biological determinants of coral community structure in the eastern Pacific. Ecol Monogr 46:431–456

Glynn PW (1977) Coral growth in upwelling and nonupwelling areas off the Pacific coast of Panama. J Mar Res 35:567–585

Glynn PW (1988) El Niño warming, coral mortality and framework destruction by echinoid bioerosion in the eastern Pacific. Galaxea 7:129–160

Glynn PW (1997) Bioerosion and coral-reef growth: a dynamic balance. In: Birkeland C (ed) Life and death of coral reefs. Chapman & Hall, New York, pp 68–95

Glynn PW (2000) Eastern Pacific coral reef ecosystems. In: Seeliger U, Kjerfve B (eds) Coastal marine ecosystems of Latin America. Springer, Berlin, pp 281–305

Glynn PW (2003) Coral communities and coral reefs of Ecuador. In: Cortés J (ed) Latin American coral reefs. Elsevier, Amsterdam, pp 449–472

Glynn PW, Stewart RH (1973) Distribution of coral reefs in the Pearl Islands (Gulf of Panamá) in relation to thermal conditions. Limnol Oceanogr 18:367–379

Glynn PW, Macintyre IG (1977) Growth rate and age of coral reefs on the Pacific coast of Panama, In: Proceedings of 3rd International Coral Reef Symposium, vol 2, Miami, pp 251–259

Glynn PW, Wellington GM (1983) Corals and coral reefs of the Galápagos Islands. University of California Press, Berkeley, p 330

Glynn PW, Leyte-Morales GE (1997) Coral reefs of Huatulco, west Mexico: reef development in upwelling Gulf of Tehuantepec. Rev Biol Trop 45:1033–1047

Glynn PW, Maté JL (1997) Field guide to the Pacific coral reefs of Panamá. In: Proceedings of 8th International Coral Reef Symposium, vol 1, Panama, pp 145–166

Glynn PW, Ault JS (2000) A biogeographic analysis and review of the far eastern Pacific coral reef region. Coral Reefs 19:1–23

Glynn PW, Wellington GM, Birkeland C (1979) Coral reef growth in the Galápagos: limitation by sea urchins. Science 203:47–49

Glynn PW, von Prahl H, Guhl F (1982) Coral reefs of Gorgona Island, Colombia, with special reference to corallivores and their influence on community structure and reef development. An Inst Invest Mar Punta Betín 12:185–214

Glynn PW, Druffel EM, Dunbar RB (1983) A dead Central American coral reef tract: possible link to the Little Ice Age. J Mar Res 41:605–637

Glynn PW, Veron JEN, Wellington GM (1996) Clipperton Atoll (eastern Pacific): oceanography, geomorphology, reef-building coral ecology and biogeography. Coral Reefs 15:71–99

Glynn PW, Cortés J, Guzmán HM, Richmond RH (1988) El Niño (1982–83) associated coral mortality and relationship to sea surface temperature deviations in the tropical eastern Pacific. In: Proceedings of 6th International Coral Reef Symposium, vol 3, Townsville, pp 237–243

Glynn PW, Maté JL, Baker AC, Calderón MO (2001) Coral bleaching and mortality in Panama and Ecuador during the 1997–1998 El Niño-Southern Oscillation event: spatial/temporal patterns and comparisons with the 1982–1983 event. Bull Mar Sci 69:79–109

Glynn PW, Wellington GM, Wieters EA, Navarrete SA (2003) Reef-building coral communities of Easter Island (Rapa Nui), Chile. In: Cortés J (ed) Latin American coral reefs. Elsevier, Amsterdam, pp 473–494

Glynn PW, Riegl B, Correa AMS, Baums IB (2009) Rapid recovery of a coral reef at Darwin Island, Galapagos Islands. Galapagos Res 66:6–13

Glynn PW, Riegl B, Purkis S, Kerr JM, Smith TB (2015) Coral reef recovery in the Galápagos Islands: the northernmost islands (Darwin and Wenman). Coral Reefs. doi:10.1007/s00338-015-1280-4

Glynn PW, Wellington GM, Riegl B, Olson DB, Borneman E, Wieters EA (2007) Diversity and biogeography of the scleractinian coral fauna of Easter Island (Rapa Nui). Pac Sci 61:67–90

Greer L, Jackson JE, Curran HA, Guilderson T, Teneva L (2009) How vulnerable is *Acropora cervicornis* to environmental change? Lessons from the Early to Middle Holocene. Geology 37:263–266

Guinotte JM, Buddemeier RW, Kleypas JA (2003) Future coral reef habitat marginality: temporal and spatial effects of climate change in the Pacific basin. Coral Reefs 22:551–558

Guzmán HM (1986) Estructura de la comunidad arrecifal de la Isla del Caño, Costa Rica, y el efecto perturbaciones naturales severas. MSc Thesis, University of Costa Rica, San Pedro, Costa Rica, p 179

Guzmán HM, Cortés J (1989a) Coral reef community structure at Caño Island, Pacific Costa Rica. PSZNI: Mar Ecol 10:23–41

Guzmán HM, Cortés J (1989b) Growth rates of eight species of scleractinian corals in the eastern Pacific (Costa Rica). Bull Mar Sci 44:1186–1194

Guzmán HM, Cortés J (1992) Cocos Island (Pacific, Costa Rica) coral reefs after the 1982–83 El Niño disturbance. Rev Biol Trop 40:309–324

Guzmán HM, Cortés J (2007) Reef recovery 20-years after the 1982–83 El Niño massive mortality. Mar Biol 151:401–411

Guzmán HM, Robertson DR, Díaz ML (1991) Distribución y abundancia de corales en el arrecife del Refugio de Isla Iguana, Pacífico de Panamá. Rev Biol Trop 39:225–231

Guzmán HM, Cortés J, Richmond RH, Glynn PW (1987) Efectos del fenómeno de "El Niño-Oscilación Sureña" 1982/83 en los arrecifes de la Isla del Caño, Costa Rica. Rev Biol Trop 35:325–332

Hallock P, Schlager W (1986) Nutrient excess and the demise of coral reefs and carbonate platforms. Palaios 1:389–398

Haug GH, Hughen KA, Sigman DM, Peterson LC, Röhl U (2001) Southward migration of the Intertropical Convergence Zone through the Holocene. Science 293:1304–1308

Hendrickson KJ (2014) Climatic control of coral communities in the Galápagos Islands, Ecuador. MSc Thesis, Nova Southeastern University, Dania Beach, Florida, p 103

Highsmith RC (1980) Geographic pattern of coral bioerosion: a productivity hypothesis. J Exp Mar Biol Ecol 46:177–196

Hoegh-Guldberg O, Mumby PJ, Hooten AJ, Steneck RS, Greenfield P, Gomez E, Harvell CD, Sale PF, Edwards AJ, Caldiera K, Knowlton N, Eakin CM, Iglesias-Prieto R, Muthiga N, Bradbury RH, Dubi A, Hatziolos ME (2007) Coral reefs under rapid climate change and ocean acidification. Science 318:1737–1742

Hopley D (1982) The geomorphology of the Great Barrier Reef: Quaternary development of coral reefs. John Wiley and Sons, New York, p 453

Hopley D, Mclean RF, Marshall J, Smith AS (1978) Holocene-Pleistocene boundary in a fringing reef: Hayman Island, North Queensland. Search 9:323–325

Hubbard DK, Garcia M (2003) The corals and coral reefs of Easter Island—a preliminary look. In: Loret J, Tanacredi JT (eds) Scientific exploration into the world's environmental problems in microcosm. Springer, New York, pp 55–77

Hubbard DK, Zankl H, Van Heerden I, Gill IP (2005) Holocene reef development along the northeastern St. Croix Shelf, Buck Island, U. S. Virgin Islands. J Sediment Res 75:97–113

Jiménez C, Cortés J (2003) Growth of seven species of scleractinian corals in an upwelling environment of the eastern Pacific (Golfo de Papagayo, Costa Rica). Bull Mar Sci 72:187–198

Jiménez C, Cortés J, León A, Ruiz E (2001) Coral bleaching and mortality associated with El Niño 1997/98 event in an upwelling environment in the eastern Pacific (Gulf of Papagayo, Costa Rica). Bull Mar Sci 69:151–169

Ketchum JT, Reyes Bonilla H (1997) Biogeography of hermatypic corals of the Archipiélago Revillagigedo, Mexico. In: Proceedings of 8th International Coral Reef Symposium, vol 1, Pamana, pp 471–476

Kinsey DW, Davies PJ (1979) Effects of elevated nitrogen and phosphorus on coral reef growth. Limnol Oceanogr 24:935–940

Kleypas JA (1996) Coral reef development under naturally turbid conditions: fringing reefs near Broad Sound, Australia. Coral Reefs 15:153–167

Kleypas JA (1997) Modeled estimates of global reef habitat and carbonate production since the last glacial maximum. Paleoceanography 12:533–545

Kleypas JA, McManus JW, Meñez LAB (1999a) Environmental limits to coral reef development: where do we draw the line? Am Zool 39:146–159

Kleypas JA, Buddemeier RW, Archer D, Gattuso J-P, Langdon C, Opdyke BN (1999b) Geochemical consequences of increased atmospheric carbon dioxide on coral reefs. Science 284:118–120

Lighty RG, Macintyre IG, Stuckenrath R (1982) *Acropora palmata* reef framework: a reliable indicator of sea level in the western Atlantic for the past 10,000 years. Coral Reefs 1:125–130

Linsley BK, Ren L, Dunbar RB, Howe SS (2000) El Niño Southern Oscillation (ENSO) and decadal-scale climate variability at 10°N in the eastern Pacific from 1983 to 1994: a coral-based reconstruction from Clipperton Atoll. Paleoceanography 15:322–335

Macintyre IG (2007) Demise, regeneration, and survival of some Western Atlantic reefs during the Holocene transgression. In: Aronson RB (ed) Geological approaches to coral reef ecology. Springer, New York, pp 181–201

Macintyre IG, Glynn PW (1976) Evolution of modern Caribbean fringing reef, Galeta Point, Panamá. AAPG Bull 60:1054–1072

Macintyre IG, Burke RB, Stuckenrath R (1977) Thickest recorded Holocene reef section, Isla Pérez core hole, Alacran Reef, Mexico. Geology 5:749–754

Macintyre IG, Multer HG, Zankl HI, Hubbard DK, Weiss WP, Stuckenrath R (1985) Growth and depositional facies of a windward reef complex (Nonsuch Bay, Antigua, W.I.). In: Proceedings of 5th International Coral Reef Congress, vol 6, Tahiti, pp 605–610

Macintyre IG, Glynn PW, Cortés J (1992) Holocene reef history in the eastern Pacific: mainland Costa Rica, Caño Island, Cocos Island and Galápagos Islands. In: Proceedings of 7th International Coral Reef Symposium, vol 2, Guam, pp 1174–1184

Manzello DP (2010) Coral growth with thermal stress and ocean acidification: lessons from the eastern tropical Pacific. Coral Reefs 29:749–758

Manzello DP, Kleypas JA, Budd DA, Eakin CM, Glynn PW, Langdon C (2008) Poorly cemented coral reefs of the eastern tropical Pacific: possible insights into reef development in a high-CO_2 world. Proc Natl Acad Sci USA 105:10,450–10,455

Martínez I, Rincon D, Yokoyama Y, Barrows T (2006) Foraminifera and coccolithophorid assemblage changes in the Panama Basin during the last deglaciation: response to sea-surface productivity induced by a transient climate change. Palaeogeogr Palaeoclimat Palaeoecol 234:114–126

Maté JL (2003) Corals and coral reefs of the Pacific coast of Panamá. In: Cortés J (ed) Latin American coral reefs. Elsevier, Amsterdam, pp 387–417

Montaggioni LF (1988) Holocene reef growth history in mid-plate high volcanic islands. In: Proceedings of 6th International Coral Reef Symposium, vol 3, Townsville, pp 455–460

Montaggioni LF (2005) History of Indo-Pacific coral reef systems since the last glaciation: development patterns and controlling factors. Earth-Sci Rev 71:1–75

Montaggioni LF, Braithwaite CJR (2009) Quaternary coral reef systems: history, development processes and controlling factors. Elsevier, Amsterdam, p 532

Moy CM, Seltzer GO, Rodbell DT, Anderson DM (2002) Variability of El Niño/Southern Oscillation activity at millennial timescales during the Holocene epoch. Nature 420:162–165

Pandolfi JM, Connolly SR, Marshall DJ, Cohen AL (2011) Projecting coral reef futures under global warming and ocean acidification 333:418–422

Pérez-Vivar TL, Reyes-Bonilla H, Padilla C (2006) Corales pétreos (Scleractinia) de las Islas Marías, Pacífico de México. Cienc Mar 32:259–270

Perry CT, Murphy GN, Kench PS, Smithers SG, Edinger EN, Steneck RS, Mumby PJ (2013) Caribbean-wide decline in carbonate production threatens coral reef growth. Nature Communications 4:1–7

Poveda G, Waylen PR, Pulwarty RS (2006) Annual and inter-annual variability of the present climate in northern South America and southern Mesoamerica. Palaeogeogr Palaeoclimat Palaeoecol 234:3–27

Reaka-Kudla ML, Feingold JS, Glynn PW (1996) Experimental studies of rapid bioerosion of coral reefs in the Galápagos Islands. Coral Reefs 15:101–107

Reyes-Bonilla H (1992) New records for hermatypic corals (Anthozoa: Scleractinia) in the Gulf of California, Mexico, with an historical and biogeographical discussion. J Nat Hist 26:1163–1175

Reyes-Bonilla H (1993a) 1987 coral reef bleaching at Cabo Pulmo reef, Gulf of California, Mexico. Bull Mar Sci 52:832–837

Reyes-Bonilla H (1993b) Biogeografía y ecología de los corales hermatípicos (Anthozoa: Scleractinia) del Pacífico de México. In: Salazar-Vallejo SI, González NE (eds) Biodiversidad marina y costera de México. Centro de Investigaciones de Quintana Roo, Chetumal, México, pp 205–222

Reyes-Bonilla H (2003) Coral reefs of the Pacific coast of México. In: Cortés J (ed) Latin American coral reefs. Elsevier, Amsterdam, pp 331–349

Reyes-Bonilla H, Barraza JE (2003) Corals and associated marine communities from El Salvador. In: Cortés J (ed) Latin American coral reefs. Elsevier, Amsterdam, pp 351–360

Riedinger MA, Steinitz-Kannan M, Last WM, Brenner M (2002) A 6100 ^{14}C yr record of El Niño activity from the Galápagos Islands. J Paleolimnol 27:1–7

Riegl BM, Halfar J, Purkis SJ, Godinez-Orta L (2007) Sedimentary facies of the eastern Pacific's northernmost reef-like setting (Cabo Pulmo, Mexico). Mar Geol 236:61–77

Robertson DR, Christy JH, Collin R, Cooke RG, D'Croz L, Kaufmann KW, Moreno SH, Maté JL, O'Dea A, Torchin ME (2009) The Smithsonian Tropical Research Institute: marine research, education, and conservation. Smithson Contrib Mar Sci 38:73–93

Sachet M-H (1962) Geography and land ecology of Clipperton Island. Atoll Res Bull 86:1–115

Sachs JP, Sachse D, Smittenberg RH, Zhang Z, Battisti DS, Golubic S (2009) Southward movement of the Pacific intertropical convergence zone AD 1400–1850. Nat Geosci 2:519–525

Toscano MA, Macintyre IG (2003) Corrected western Atlantic sea-level curve for the last 11,000 years based on calibrated ^{14}C dates from *Acropora palmata* framework and intertidal mangrove peat. Coral Reefs 22:257–270

Toth LT (2013) Holocene reef development in the tropical eastern Pacific. PhD Dissertation, Florida Institute of Technology, Melbourne, Florida, p 374

Toth LT, Aronson RB, Cobb KM, Cheng H, Edwards RL, Grothe PR, Sayani HR (2015) Climatic and biotic thresholds of coral-reef shutdown. Nature Climate Change 5:369–374

Toth LT, Aronson RB, Vollmer SV, Hobbs JW, Urrego DH, Cheng H, Enochs IC, Combosch DJ, van Woesik R, Macintyre IG (2012) ENSO drove 2500-year collapse of eastern Pacific coral reefs. Science 337:81–84

van Woesik R, Done TJ (1997) Coral communities and reef growth in the southern Great Barrier Reef. Coral Reefs 16:103–115

Vargas-Ángel B (1995) Distribution and community structure of the reef corals of Ensenada de Utría, Pacific coast of Colombia. Rev Biol Trop 44:643–651

Vargas-Ángel B (2001) Corals and coral reefs of the Pacific coast of Colombia with special reference to spatial and temporal patterns of environmental disturbances. PhD Dissertation, University of Miami, Coral Gables, Florida, p 209

Veron JEN (1995) Corals in space and time. Ithaca, Comstock/Cornell, p 321

Veron J, Stafford-Smith M, DeVantier L, Turak E (2015) Overview of distribution patterns of zooxanthellate Scleractinia. Front Mar Sci 1:81. doi:10.3389/fmars.2014.00081

von Prahl H (1986) Corales y arrecifes coralinos. In: von Prahl H, Alberico M (eds) Isla de Gorgona. Biblioteca Banco Popular. Textos Universitarios, Bogotá, Colombia, pp 59–87

Wellington GM, Glynn PW (2007) Responses of coral reefs to El Niño–Southern Oscillation sea-warming events. In: Aronson RB (ed) Geological approaches to coral reef ecology. Springer, New York, pp 342–385

Westphall MJ (1986) Anatomy and history of a ringed-reef complex, Belize, Central America. MSc Thesis. University of Miami, Coral Gables, Florida, p 135

Zapata FA, Vargas-Ángel B (2003) Corals and coral reefs of the Pacific coast of Colombia. In: Cortés J (ed) Latin American coral reefs. Elsevier, Amsterdam, pp 419–447

Marine Biodiversity of Eastern Tropical Pacific Coral Reefs

7

Jorge Cortés, Ian C. Enochs, Jeffrey Sibaja-Cordero, Luis Hernández, Juan José Alvarado, Odalisca Breedy, José Antonio Cruz-Barraza, Octavio Esquivel-Garrote, Cindy Fernández-García, Alicia Hermosillo, Kirstie L. Kaiser, Pedro Medina-Rosas, Álvaro Morales-Ramírez, Cristian Pacheco, Alejandro Pérez-Matus, Héctor Reyes-Bonilla, Rafael Riosmena-Rodríguez, Celeste Sánchez-Noguera, Evie A. Wieters, and Fernando A. Zapata

J. Cortés (✉) · J. Sibaja-Cordero · J.J. Alvarado · O. Breedy · O. Esquivel-Garrote · C. Fernández-García · Á. Morales-Ramírez · C. Sánchez-Noguera
Centro de Investigación en Ciencias del Mar y Limnología (CIMAR), Universidad de Costa Rica, San Pedro, 11501-2060 San José, Costa Rica
e-mail: jorge.cortes@ucr.ac.cr

J. Sibaja-Cordero
e-mail: jeffrey.sibaja@ucr.ac.cr

J.J. Alvarado
e-mail: juanalva76@yahoo.com

O. Breedy
e-mail: odaliscab@gmail.com

C. Fernández-García
e-mail: cindyfdez28@gmail.com

Á. Morales-Ramírez
e-mail: alvaro.morales@ucr.ac.cr

C. Sánchez-Noguera
e-mail: celeste08@gmail.com

J. Cortés · J. Sibaja-Cordero · J.J. Alvarado · O. Breedy · O. Esquivel-Garrote · C. Fernández-García · Á. Morales-Ramírez
Escuela de Biología, Universidad de Costa Rica, San Pedro, 11501-2060 San José, Costa Rica

I.C. Enochs
Atlantic Oceanographic and Meteorological Laboratories (AOML), NOAA, 4301 Rickenbacker Cswy., Miami, FL 33149, USA
e-mail: ian.enochs@NOAA.gov

L. Hernández · H. Reyes-Bonilla
Laboratorio de Sistemas Arrecifales, Universidad Autónoma de Baja California Sur, Carretera al sur km. 5.5 colonia Mezquitito, C.P, 23080 La Paz, Baja California Sur, Mexico
e-mail: lghm@uabcs.mx

J.A. Cruz-Barraza · C. Pacheco
Instituto de Ciencias del Mar y Limnología, Universidad Nacional Autónoma de México (Unidad Académica Mazatlán), Av. Joel Montes Camarena, s/n, 82000 Mazatlán, Sinaloa, Mexico
e-mail: cristian.pachecosolano@gmail.com

A. Hermosillo
Natural History Museum of Los Angeles County, 900 Exposition Blvd, Los Angeles, CA 90007, USA
e-mail: alicia_hg@prodigy.net.mx

K.L. Kaiser
Santa Barbara Museum of Natural History, 2559 Puesta Del Sol Road, Santa Barbara, CA 93105, USA
e-mail: kirstie.kaiser@gmail.com

P. Medina-Rosas
Centro Universitario de La Costa, Universidad de Guadalajara, Puerto Vallarta, Jalisco, Mexico
e-mail: pedromedinarosas@gmail.com

A. Pérez-Matus
Subtidal Ecology Laboratory & Marine Conservation Center, Estación Costera de Investigaciones Marinas, Pontificia Universidad Católica de Chile, Casilla 114-D, Santiago, Chile
e-mail: aperez@bio.puc.cl

R. Riosmena-Rodríguez
Programa de Investigación en Botánica Marina, Departamento de Biología Marina, Universidad Autónoma de Baja California Sur, Apartado postal 19-B, 23080 La Paz, Baja California Sur, Mexico
e-mail: riosmena@uabcs.mx

C. Sánchez-Noguera
Leibniz Center for Tropical Marine Ecology (ZMT), Fahrenheitstr. 6, 28359 Bremen, Germany

E.A. Wieters
Estación Costera de Investigaciones Marinas, and Departamento de Ecología, Facultad de Ciencias Biológicas, Pontificia Universidad Católica de Chile, Santiago, Chile
e-mail: ewieters@bio.puc.cl

F.A. Zapata
Departamento de Biología, Universidad Del Valle, Apartado Aéreo, 25360 Cali, Colombia
e-mail: fernando.zapata@correounivalle.edu.co

P.W. Glynn et al. (eds.), *Coral Reefs of the Eastern Tropical Pacific*, Coral Reefs of the World 8, DOI 10.1007/978-94-017-7499-4_7

Abstract

The eastern tropical Pacific (ETP) is an isolated oceanic region exposed to extreme oceanographic conditions, including low salinity, low pH, high temperatures during El Niño, and low temperatures during La Niña and seasonal upwelling. The coral reefs in this region have a relatively limited suite of species compared to other coral reef areas of the world, but much like more diverse reefs the species present interact in complex ways. Here we synthezise the knowledge of taxonomic groups of reef organisms from prokaryotes to vertebrates, including algae, sponges, cnidarians, annelids and other worms, molluscs, crustaceans, echinoderms and fishes. We also present summaries on the biodiversity of associated functional groups and habitats, including (a) reef zooplankton and cryptic fauna, and (b) soft benthic environments, rhodolith beds and mesophotic environments. Several factors that structure the biodiversity of ETP coral reefs are explored, including biological, physical and chemical controls. ETP coral reefs are relatively simple systems that can be used as models for studying biodiversity and interactions among species. We conclude this review by highlighting pressing research needs, from very basic inventories to more sophisticated studies of cryptic assemblages, and to investigations on the impacts of natural and anthropogenic effects on ETP coral reef biodiversity.

Keywords

Reef invertebrates • Fishes • Algae • Plankton • Latin America

7.1 Introduction

The eastern tropical Pacific (ETP) is relatively isolated from other regions, yet has an equally diverse range of ecosystems and environments (Fiedler and Lavín 2006; Cortés and Wehrtmann 2009). Climate varies widely across the ETP, ranging from very dry in Baja California, coastal Mexico, northern Central America, southern Ecuador, and most oceanic islands, to very humid in central Costa Rica to northern Ecuador, including Isla del Coco (Amador et al. 2006). Ocean currents that flow within and between the ETP and adjacent regions vary seasonaly, yearly, and on longer time scales. These currents play an important role in the transport and dispersal of marine organisms (see Chap. 16, Lessios and Baums), and changes in nutrient concentrations, salinity and temperature in the region (Fiedler and Talley 2006; Kessler 2006; Lizano 2008).

The ETP region has been divided into separate biogeographic provinces based on the distributions of different taxa. Briggs (1974) divided it into three provinces that include coral reefs: Mexican Province (Gulf of California to Tehuantepec), Panamanian Province (El Salvador to Ecuador), and the Galápagos Islands. Boschi (2000), utilizing decapod crustacean distributions proposed two zoogeographic provinces, namely the Panamic (including the oceanic islands) and the Galápagos Archipelago. Spalding et al. (2007) divided the ETP into 11 ecoregions, and more recently Robertson and Cramer (2009), employing fish distributions, divided the ETP into three provinces, two including coastal areas (Cortez and Panamic) and one the oceanic islands.

7.2 Synopsis of the Biodiversity of ETP Coral Reef Regions

There are few compilations of marine biodiversity from the ETP, and fewer still emcompassing coral reefs. The following examples, listed according to country or region, refer to all coastal and marine ecosystems, including coral reefs where a significant number of the reported species are found (Table 7.1).

Mexico—Reyes-Bonilla et al. (2012) reported 5740 marine species in the Gulf of California, with many of these associated with coral reefs in the southern gulf area. From the current known records from Oaxaca, Bastida-Zavala et al. (2013) reported 2157 species spanning 15 taxonomic groups (ranging from plants to vertebrates). However, it should be recognized that sampling was biased toward some groups. For example, algae and three invertebrate groups (annelids, crustaceans and molluscs) represented over 55 % of the species reported. Since additional taxa are known from adjacent coastal areas, it is highly likely that the number of taxa reported by Bastida-Zavala et al. (2013) is underestimated.

Clipperton—Clipperton Atoll is one of the principal oceanic islands of the ETP (Glynn et al. 1996), with a relatively low biodiversity. This may be attributed to its

Table 7.1 Marine biodiversity of selected eastern tropical Pacific reef regions. Endem, number of endemic species

Regions	Algae	Invertebrates	Fishes	Others[a]	Total	Coral reef	Endem
ETP[1, 2]		85 (corals)				41	26
ETP[3]		2400 (molluscs)				>800	
ETP[3, 4]		850 (crustaceans)					
ETP[5, 6]		722 (echinoderms)					265
ETP[7, 8]			>1300				
Gulf of California, Mexico[9, 10]		4849	891		5740	74 (fishes)	852
Oaxaca, Mexico[11]	242	1153	594	168	2157		several
Mexico-Pacific[12]		76 (sponges)				76	
Clipperton, France[13, 14, 15]	83	576	197		857	290	38
El Salvador[16, 17]		868	100		968		0
Costa Rica-Pacific[18]	156	3379	838	372	4745		85
Golfo Dulce, Costa Rica[19]	11	741	276	199	1028		0
Bahía Culebra, Costa Rica[20, 21]	44	381	82	70	577		0
Isla del Coco, Costa Rica[22]	31	1168	389	100	1688		45
Panama[23]				1157[b]			
Isla Uva, Panama[24, 25]			102			102	
Bahía Honda, Panama[26]			126			88	
Coiba Island, Panama[27]	4	95	69	84	252		
Gorgona Island, Colombia[28]			71				
Ecuador (continental and insular)[29]		862	270	727	1859		
Galápagos Islands, Ecuador[30, 31]		3089	447		3536		565
Costa Rica, Panama, Colombia and Ecuador[32]		5502	1212		6714		122

[a]Foraminifera, phytoplankton, cyanobacteria, fungi, plants, vertebrates other than fishes
[b]Includes both Caribbean and Pacific coasts
Sources [1]Cairns (1991a), [2]Glynn et al. Chap. 6 this book, [3]This chapter, [4]Boschi (2000), [5]Pérez-Ruzafa et al. (2013), [6]Solís-Marín et al. (2013), [7]Robertson and Allen (2008), [8]Eschmeyer (2014), [9]Aburto-Oropeza and Balart (2001), [10]Reyes-Bonilla et al. (2012), [11]Bastida-Zavala et al. (2013), [12]Carballo et al. (2014), [13]Charpy (2009), [14]Payri et al. (2009), [15]Fourriére et al. (2014), [16]Barraza (2000), [17]Barraza (2014a, b), [18]Wehrtmann and Cortés (2009), [19]Morales-Ramírez (2011), [20]Dominici-Arosemena et al. (2005), [21]Cortés J et al. (2012), [22]Cortés (2012), [23]CDB-Panamá (2010), [24]Glynn (2006), [25]Glynn et al. (2014), [26]Dominici-Arosemena and Wolff (2006), [27]Castroviejo (1997), [28]Zapata and Morales (1997), [29]Cruz et al. (2003), [30]Bustamante et al. (2002), [31]Hickman (2009), [32]Miloslavich et al. (2011)

isolation from other land masses and homogeneity of the substrate, which is dominated by steep basaltic rocks and cemented massive corals (Glynn et al. 1996; Charpy 2009; Fourriére et al. 2014). A total of 857 marine species have been reported from Clipperton, and these are predominantly molluscs, fishes and decapods (Table 7.2).

El Salvador—El Salvador, with a relatively short coastal stretch (321 km), consists mainly of sandy beaches and very small coral patches and rocky areas. Only 868 species of invertebrates, mainly molluscs and crustaceans (Barraza 2000, 2014a), and 100 fish species have been reported (Barraza 2014b) from this region.

Costa Rica—The diversity of marine organisms in Costa Rica has been assessed recently by Wehrtmann and Cortés (2009). They reported 4745 species from the Pacific coast of Costa Rica, including 1142 species from Isla del Coco. New records (Cortés 2012) document the occurrence of 1688 species at Isla del Coco, an increase of 32 % in just a few years. Golfo Dulce, an anoxic basin in southern Costa Rica (Hebbeln and Cortés 2001), is also species rich with 1028 described taxa (Morales-Ramírez 2011). Bahía Culebra is a relatively well-studied but small area in northern Costa Rica with 577 documented species (Cortés J et al. 2012). The most speciose groups in Costa Rica are molluscs, fishes and decapods (Wehrtmann et al. 2009).

Panama—Coral reefs occur on the Caribbean and Pacific coasts of Panama, the latter with two contrasting environments, a seasonal upwelling area in the Gulf of Panama and a non-upwelling region in the Gulf of Chiriquí (Glynn 1972a, b; Glynn and Maté 1997; Maté 2003; Guzman et al. 2004, 2008). A total of 1157 species of marine fishes have been reported from both coasts (CDB-Panamá 2010). Glynn (2006) reported 50 species of mostly small cryptic reef fishes and 52 additional species (≥ 15 cm total body length)

Table 7.2 Marine biodiversity of Clipperton Atoll and the biogeographic affinity of nine taxa

Taxon	Coral reef	Biogeographic affinity	Endemics
83 (Algae)[1]	n.r.	81 % West-Central Pacific, 4 % Eastern Pacific, 15 % unidentified species	3
20 (Sponges)[2]	20	90 % Indo-Pacific, 10 % Eastern Pacific	9
22 (Corals)[3]	22	68.4 % Indo-Pacific, 31.6 % Eastern Pacific	0
277 (Molluscs)[4]	n.r.	50 % Indo-West Pacific, 50 % Panamic	0
26 (Polychaetes)[5]	26	42 % Cosmopolitan, 23 % Eastern Pacific, 35 % Panamic	0
14 (Bryozoa)[6]	14	Most with North America	0
190 (Decapods)[7]	95	44 % Indo-West Pacific	19
28 (Echinoderms)[8]	n.r.	46.4 % Indo-Pacific, rest Eastern Pacific	0
197 (Fishes)[9]	106	Most Indo-Pacific	7

n.r. Not reported

Sources [1]Payri et al. (2009), [2]Soest et al. (2011), [3]Flot and Adjeroud (2009), [4]Kaiser (2009), [5]Solís-Weiss and Hernández-Alcántara (2009), [6]Hondt (2009), [7]Poupin et al. (2009), [8]Solís-Marín and Laguarda-Figueras (2009), [9]Fourriére et al. (2014)

observed in swimming transects at the Uva Island coral reef (Glynn et al. 2014). Other than the early taxa-specific reviews in Jones (1972) of marine plants (Earle 1972), crustaceans (Abele 1972), echinoderms (Chesher 1972), and fishes (Robins 1972), no comprehensive reports of the marine biodiversity of Panamanian coral reefs exist. However, several notable advances documenting the biotic richness of reef taxa have been forthcoming in recent decades. One of these, a preliminary inventory of the marine biota of the Coiba Island National Park, reported 252 species ranging from algae to cetaceans (Castroviejo 1997).

Colombia—Colombia, with coasts bordering the Caribbean and ETP, and a marine area of nearly 1,000,000 km^2 encompassing numerous marine habitats, is likely one of the most biologically diverse countries in the Americas (Díaz and Acero 2003). Several Colombian studies (e.g., Alonso et al. 2008) indicate a very high marine biodiversity, but an estimate of the total number of species could not be found.

Ecuador—James (1991) was the first to compile the available information on invertebrate taxonomy and biogeography of the Galápagos Islands. Cruz et al. (2003) reported 1859 species of marine organisms present in continental and insular Ecuador. This contrasts greatly with the 3536 species reported by Hickman (2009) for the Galápagos Islands alone, which was based on an updated list of marine species complied by Bustamante et al. (2002). Of those species, 3089 are invertebrates, predominantly molluscs, crustaceans and echinoderms, and 18.3 % of this fauna is endemic to the Galápagos Islands.

Chile—Two remote volcanic islands on the Easter Fracture Zone are now recognized as belonging to the ETP coral reef region: (1) Rapa Nui or Easter Island (163.6 km^2 in area), located about 4050 km from Tahiti and 3700 km from continental Chile, and (2) Salas y Gómez Island (0.15 km^2 in area), lying about 3210 km west of Chile. Several workers have reported high levels of marine endemism at Rapa Nui (Castilla and Rozbaczylo 1987; DiSalvo et al. 1988; Friedlander et al. 2013), and have also recognized an impoverished coral fauna (Glynn et al. 2007; Veron et al. 2015). Plausible explanations for this are the isolation, small size, and lack of lagoonal ecosystems or protected bays at these oceanic islands.

Multiple countries—Miloslavich et al. (2011) reported 6714 species from Costa Rica, Panama, Colombia and Ecuador. Polychaetes were the most diverse group (1894 species) followed by the usually diverse groups of fishes (1212 spp.), crustaceans (863 spp.) and molluscs (875 spp.). These groups combined made up 47.3 % of the species compiled in this study.

Endemism—Miloslavich et al. (2011) indicated that the percentage of endemics in Costa Rica, Panama, Colombia and Ecuador ranks among the highest globally. This is especially evident among the fishes, with 70 % of the species endemic to the ETP (Robertson and Allen 2008). Cairns (1991a) reported that 41 % of the ahermatypic corals in the Galápagos Islands are endemic, and Hickman (2009) noted that 18.3 % of all marine invertebrates are endemic, ranging from no endemics among the zooxanthellate corals to 71 % of the gorgonians. Carballo et al. (2014) reported a high degree of endemism (29 %) for the sponge fauna of Mexico, but attributed this to a lack of sponge studies elsewhere in the ETP. Finally, Cortés (2012) reported that 2.7 % of the marine organisms at Isla del Coco were endemic to the region, representing 41.2 % of the total number of endemic species known in Costa Rica.

Biogeographically the ETP coral reef fauna has a strong connection with the Central-Western Pacific fauna, although the degree to which this holds true varies considerably among taxa. For example, almost all reef-building corals in the ETP are closely related to those in the Indo-Pacific (Glynn and Ault 2000; see Chap. 5, Glynn et al.), while ETP polychaetes are more closely related to taxa from the American mainland than to species in the Western Pacific (Solís-Weiss and Hernández-Alcántara 2009; Dean et al. 2012). It must be noted, however, that even though ETP reef-building corals have a strong Indo-Pacific affinity, one important ETP reef-builder, *Porites lobata*, is presently isolated from Indo-Pacific populations with no evident gene flow (Baums et al. 2012). Endemism among taxonomic groups also varies within the ETP region. For example, in the Galápagos Islands 71 % of the gorgonians are endemic, while all reef-building corals occur elsewhere (Breedy et al. 2009a; Hickman 2009). The biogeographic affinity of species varies substantially with location. Clipperton Atoll, on the western margin of the ETP, has 277 species of molluscs of which 50 % have Indo-Pacific affinities, and the other 50 % with ties to the Panamic fauna (Kaiser 2009). In contrast, Isla Malpelo, Colombia, at the eastern margin of the ETP, has 341 mollusc species but only 2.6 % of this fauna is related to Indo-Pacific species whereas 97.4 % of the remaining species belong to the Panamic region (Kaiser and Bryce 2001).

In this chapter we consider only the core ETP, traditionally the region between western Mexico and northern Peru. The Rapa Nui bioregion is not included in this review because of its general omission from past ETP biogeographic analyses. The objectives of this chapter are to provide a synthesis of the state of knowledge of ETP coral reef biodiversity, illuminating gaps in our knowledge, and to suggest future lines of research.

7.3 State of Knowledge of ETP Coral Reef Biodiversity

7.3.1 Prokaryotes and Viruses

Little is known concerning ETP reef-associated prokaryotes and viruses; only two published studies are available. In one, water samples were collected for a metagenomic study of the marine planktonic microbiota during the *Sorcerer II* Global Ocean Sampling Expedition (Rusch et al. 2007). Some of these samples were collected near coral reefs in Panama, Costa Rica (Isla del Coco) and the Galápagos Islands, and several species of Archaea and Eubacteria were identified (Rusch et al. 2007). From the same samples, Williamson et al. (2008) recognized viral talC sequences, and P-SSM4-like phages, as well as the cyanobacterium, *Prochlorococcus* sp. Additional studies are needed on these groups and their effects on the health of coral reef organisms.

7.3.2 Macroalgae

In 1945, William R. Taylor published the first and only existing complete key that includes the brown, red and green algae of the ETP, namely the Pacific marine algae of the Allan Hancock Expeditions to the Galápagos Islands (Taylor 1945). Between 1950 and 1965, Elmer Y. Dawson published more than thirty papers describing the marine algal flora of the ETP (Abbott 1966). The introduction of scuba during the 1950s promoted profound changes in the understanding of algal richness throughout the region. Later, from the 1980s to the present, many local and international experts have contributed to our understanding of marine macroalgal biodiversity and ecology, particularly in Mexico, El Salvador, Costa Rica and Panama. More recently, with the advent of molecular biology, there has been even greater resolution of species and their overall richness.

During the past 70 years of ETP macroalgal taxonomy, more than 300 marine species have been documented (Table 7.3). The most diverse group in the region are the red algae, followed by green and brown algae. Algal community diversity in the ETP is dependent on both the geomorphological configuration of the coastline and habitat type (Fernández-García et al. 2011). It is estimated that 60 to 70 % of the ETP macroalgae are associated with coral and rocky reef habitats (Alvarado et al. 2011; Murillo-Muñoz and Peña--Salamanca 2014; Alvarado et al. 2015). Described community richness is also highly dependent on sampling effort and local scientific infrastructure (Fernández-García et al. 2011). For example, in countries such as Mexico, El Salvador, Costa Rica, Panama and Colombia, where local and international phycologists have been studying the marine flora

Table 7.3 Number of ETP marine macroalgal species

	Continental mainland								Oceanic islands			
	MEX[1]	GUA[2]	SAL[2]	NIC[2]	CRC[2]	PAN[2]	COL[3]	ECU[4]	CLI[5]	REV[6]	COC[7]	GAL[8]
Ochrophyta-Phaeophyceae	59	0	14	6	26	25	23	5	11	29	7	46
Chlorophyta	95	5	36	6	44	52	31	8	23	38	10	42
Rhodophyta	78[a]	11	96	12	146	97	81	12	49	123	13	228
Total	232	16	146	24	216	174	135	25	83	190	30	316

Modified from Fernández-García et al. (2011). *MEX* mainland Mexico (Nayarit to Chiapas); *GUA* Guatemala; *SAL* El Salvador; *NIC* Nicaragua; *CRC* Costa Rica; *PAN* Panama; *COL* Colombia; *ECU* Ecuador; *CLI* Clipperton Island (France); *REV* Revillagigedo Islands (Mexico); *COC* Isla del Coco (Costa Rica); *GAL* Galápagos Islands (Ecuador)

[a]There are no updated catalogues of this group; true species richness is expected to be much greater

Sources [1]González-González et al. (1996), Pedroche et al. (2005, 2008); [2]Fernández-García et al. (2011); [3]Schnetter and Bula-Meyer (1982), Bula-Meyer (1995), Peña-Salamanca (2008), Murillo-Muñoz and Peña-Salamanca (2014), [4]Müller-Gelinek and Salazar-Orquera (1996), [5]Sachet (1962), Payri et al. (2009), [6]Serviere-Zaragoza et al. (2007), [7]Fernández (2008), [8]Garske (2002) for decades, there are over a hundred reports of marine macroalgae (Table 7.3). By contrast, in countries such as Guatemala, Nicaragua and Ecuador, there are relatively few local experts, and our knowledge of these regions is less extensive. To further explore macroalgal species diversity and distributions, greater collection effort is needed in these remote and underexplored areas.

Coral reefs in the ETP (<10 m deep) are dominated by algal turf, defined as a typically low lying (several mm to cm high) stratum of algae (Connell et al. 2014) as well as encrusting calcareous and fleshy algae (Fig. 7.1a–d). Algal turfs are composed of a combination of several filamentous and corticated green, red and brown algae. The main turf-forming species are *Asparagopsis taxiformis* (filamentous tetrasporophyte stage), *Polysiphonia, Ceramium, Cladophora, Chlorodesmis*, species within the Order Gelidiales, as well as the red geniculate coralline algae *Jania* and *Amphiroa* (Fig. 7.1a, b). Several fleshy algae such as *Hypnea* sp., *Gracilaria* spp., *Galaxaura rugosa, Halimeda discoidea, Codium isabelae, Padina mexicana* and *Dictyota humifusa* (present from Mexico to Colombia, Fig. 7.1d), are also present in small patches. From about 12 to 17 m depth, the dominant species are encrusting red coralline algae (*Lithohyllum* spp. and *Lithothamnion* spp.), species in the family Peyssoneliaceae, and *Lobophora* spp. (Fig. 7.1c).

Algal diversity of the ETP is lower than in other tropical regions such as the Caribbean Sea, Indian Ocean and West Pacific Ocean. This depauperate flora could be the result of several factors, including the emergence of the Central American Isthmus (Lessios et al. 2001; Wysor 2004), a geologic event resulting in cooler waters associated with seasonal upwelling and overall changes in current patterns (Glynn 1972a; Cortés 2011) causing local species' extinctions that were previously part of a homogeneous regional flora (Wysor 2004). Other factors structuring species' distributions in the region are the lower prevalence of habitats that favor the presence of certain macroalgal species. For instance, the Caribbean and Indo-Pacific regions possess more extensive carbonate platforms, seagrass ecosystems and well developed coral reefs than are present in the ETP (Cortés 2011).

Another factor in need of further study is the effect of herbivorous fishes on the presence or absence of marine macroalgae across the ETP. Edgar et al. (2011) found that overfishing of predatory fishes in the ETP has led to increased population densities of herbivores, which have contributed to regional declines in algal cover. Two common herbivorous guilds that graze on algae are parrotfishes (Scaridae) and sea urchins such as *Diadema mexicanum* and *Eucidaris galapagensis* (Fig. 7.1e, f). An excess of herbivores, coupled with eutrophication or other types of natural (e.g., ENSO events) and/or anthropogenic disturbances, may lead to the presence of fewer macroalgal species (Glynn and Maté 1997; Smith et al. 2010; Fernández-García et al. 2012).

Finally, it is noted that two marine macroalgae may negatively and directly influence coral reefs in the ETP. *Caulerpa sertularioides* has been observed to competitively exclude other algal species and overgrow corals, thereby reducing biodiversity and calcification (Fernández and Cortés 2005; Smith et al. 2010; Fernández-García et al. 2012). *Acanthophora spicifera* is a well-known invasive species in many Pacific environments such as in Hawaii (Russell 1992) and Baja California Sur, Mexico (Ávila et al. 2012). Recent sampling along the Central American Pacific coast has documented the presence of this alga in El Salvador, Panama and Costa Rica (Fong et al. 2006; Segovia-Prado and Naverrete-Calero 2007; C. Fernández-García, pers. obs.). In Panama, Fong et al. (2006) found that the positive interaction between *A. spicifera* and cyanobacterial epiphytes uncoupled herbivore consumers' control of algal community structure, and hypothesized that the epiphytes maintained the macroalgal dominated patches on coral reefs (see Fong et al., Chap. 11). Presently, however, there are no reported direct impacts of *A. spicifera* at other ETP sites.

7.3.3 Metazoans

7.3.3.1 Porifera

Comprehensive assessment of ETP sponge biodiversity has been scarce until recent years. The earliest surveys were conducted during the Albatross expeditions in the late 19th and early 20th centuries (Wilson 1904; Lendenfeld 1910). Research progressed slowly during the last two decades until renewed interest led to the description and documentation of numerous species in the ETP. Currently, Mexico has received greatest attention where a total of 76 species of coral reef sponges has been reported from the mainland coast (Fig. 7.2) and the Revillagigedo Archipelago (Carballo et al. 2004; Carballo and Cruz-Barraza 2006, 2010), followed by the Galápagos Islands with 38 species reported (Chiriboga et al. 2012). Nine species in the Class Homoscleromorpha have been reported for the ETP (Cruz-Barraza et al. 2014) (Table 7.4).

Clipperton Island has 17 documented species in the Class Demospongiae and two species in the Class Calcarea (Table 7.4). Seven of these sponge species exhibit widespread distributions in the Indo-Pacific, suggesting affinities with the Central and West Pacific regions (Soest et al. 2011). Other ETP areas with reports and descriptions of new species are Costa Rica (mainland and Isla del Coco) (Guzmán 1988; Cortés et al. 2009; Pacheco 2012) and Panama (Laubenfels 1936; Diaz et al. 2005; Caballero-George et al. 2010). In Panama, one study revealed that the morphology of sponge species is more cryptic than in the Caribbean, suggesting an adaptation to potentially higher levels of predation (Wulff 1997).

Fig. 7.1 Macroalgae, turf and coralline algae of ETP coral reefs and rocky substrates. **a** A multi-species assemblage of green and red algae, reef base, Coiba Island, Panama; **b** species in the Order Gelidiales, San Juan del Sur, Nicaragua; **c** encrusting Peyssoneliaceae and *Lobophora*, Isla del Coco, Costa Rica; **d** *Dictyota humifusa*, Isla Bolaños, Bahía Salinas, Costa Rica; **e** parrot fish bite marks on turf algae, Isla del Caño, Costa Rica; **f** *Diadema mexicanum* grazing on algae, Acapulco, Mexico. Photographs by Cindy Fernández-García

Fig. 7.2 Sponges of the ETP. **a** *Aplysina clathrata,* Isla Isabel; **b** *Callyspongia californiana,* Isla Isabel; **c** *Haliclona caerulea,* Bahía de Huatulco; **d** *Mycale cecilia,* Bahía de Huatulco; **e** *Chondrilla montanusa,* Bahía de Huatulco; **f** *Cliona californiana,* La Paz, Baja California. Photographs by José Antonio Cruz Barraza, all from Mexico

Table 7.4 Number of ETP sponge species by class from coral reef areas

	Continental mainland			Oceanic islands			
	MEX	PAN	CRC	CLI	REV	COC	GAL
Demospongiae	69	17	16	17	4	2	34
Homoscleromorpha	2	1	0	0	4	0	4
Calcarea	0	0	0	2	0	0	0
Total	71	18	16	19	8	2	38

MEX Mexico; *PAN* Panama; *CRC* Costa Rica; *CLI* Clipperton Island; *REV* Revillagigedo Archipelago; *COC* Isla del Coco; *GAL* Galápagos Islands

Other biogeographic regions have received more taxonomic attention than the ETP. In some cases, revisions are necessary to update inaccurately classified ETP species. However, as others have recognized (e.g., Hooper et al. 2002), there has been increased research effort in the field of sponge systematics in recent years.

7.3.3.2 Cnidaria

Hydrozoa

The hydrozoans (Phylum Cnidaria, Class Hydrozoa) are an abundant group of reef organisms but their generally small size and cryptic nature has contributed to the limited knowledge of this group in the ETP. The chief ETP studies are from the late 1930s and 1940s (Fraser 1938a, b, c, 1948). Calder et al. (2003) reported 96 species of hydroids from the Galápagos Islands, representing the most well studied region in the ETP.

The calcareous Hydrozoa (also known as hydrocorals) are better known than their fleshy counterparts and include the Stylasteridae and Milleporidae. Carins' (1986) report on the stylasterids of the Galápagos Islands included 14 species, and nine of these were newly described. Later he reported five additional species, including two new ones from Isla del Coco (Cairns 1991b). Unlike other regions, all ETP species occupy deep waters (200–500 m) and are related to the western Pacific fauna. There is one known exception described from Panama, *Distichopora robusta,* which has been collected between 5 and 25 m depth (Lindner et al. 2004).

The other group of calcareous hydroids belongs to the milleporid family, which potentially contribute to reef-building in Panama (Maté 2003; see Chap. 5, Glynn et al.). Four species have been identified in the ETP, three from the Gulf of Chiriquí, Panama (Glynn and de Weerdt 1991; de Weerdt and Glynn 1991), and one from Clipperton Atoll (Glynn et al. 1996). The populations of the Panamanian species were greatly reduced by El Niño events. Subsequently, one species (*Millepora boschmai*) is thought to have become regionally extinct (Glynn 2011).

Anthozoa

Octocorallia

A total of 111 species of octocorals have been documented from the ETP (Table 7.5). The majority of these species are gorgonians (Order Alcyonacea) within the families Gorgoniidae (three genera; Fig. 7.3a, b, c) Plexauridae (five genera; Fig. 7.3d) and Clavulariidae (one species, *Carijoa riisei*; Fig. 7.3e). Three families of sea pens (Order Pennatulacea) have been reported from the ETP, but very few studies on this group exist (Table 7.6).

Octocoral studies in the ETP began with the expedition *Voyage autour du monde sur La Frégate La Vénus* 1836–1839 (published in 1846, see Breedy and Guzman 2002), where the first specimens of octocorals from Mazatlán, Mexico were collected and described. Following this expedition, a few sporadic records appeared in the literature. Verrill (1868–70) made the most significant contribution in his paper "Notes on Radiata", synthesizing the existing literature at the time and describing 50 new species. Further contributions have been made in the 20th century by W. Kükenthal, S. J. Hickson and E. Deichmann (see Breedy and Guzman 2002).

From the year 2000 to the present, cooperative research initiatives between the Centro de Investigación en Ciencias del Mar y Limnología (CIMAR), Universidad de Costa Rica, and the Smithsonian Tropical Research Institute (STRI, Panama) have significantly enchanced the knowledge of octocoral biodiversity in the ETP. Related contributions include taxonomic reviews (Breedy and Guzman 2002, 2007, 2011; Williams and Breedy 2004; Breedy 2009; Breedy et al. 2009b; Breedy and Cortés 2014), descriptions of new species (Breedy 2001; Breedy and Guzman 2003, 2004, 2008, 2012 2013, 2014; Guzman and Breedy 2008, 2012; Breedy and Cortés 2011) and biogeographic assessments (Vargas et al. 2008). The majority of this work pertains to shallow water habitats (0–40 m) along the coasts of Costa Rica and Panama. As a consequence, during the last 10 years, 15 new species have been described from Panama, and 17 from Costa Rica. Two species from mesophotic reefs (to 50 m depth) have been described, including *Eugorgia siedenburgae* from Hannibal Bank, Panama, and Santa Elena Bay, Costa Rica (Breedy and Guzman 2013), as well as *Eugorgia beebei* from Baja California, Mexico and Paita, Peru (Breedy et al. 2013).

The most speciose family of octocorals is the Gorgoniidae (76 species), with 14 species of *Eugorgia*, 27 of *Leptogorgia* and 35 in the genus *Pacifigorgia*. Plexauridae is the next most abundant family (31 species) with one species of *Adelogorgia*, four each in the genera *Eumuricea* and *Heterogorgia*, six species of *Psammogorgia* and 16 in the genus *Muricea*. Panama has the greatest number of described species of octocorals (64 spp.), followed by Mexico (46 spp.), and mainland Costa Rica (38 spp.) (Table 7.7). Recent records from Nicaragua (24 spp.) and Ecuador (25 spp.) are found in technical reports (Rivera and Martínez 2011; Cortés-Núñez et al. 2012), and from El Salvador (17 spp.) in an unpublished thesis (Segovia-Prado 2012). Preliminary data from Colombia have been published for Malpelo Island (4 spp.; Sánchez et al. 2011) and for Isla Gorgona (13 spp.; Sánchez et al. 2014). Presently, nothing is known of the octocoral species of Guatemala. A survey of the Revillagigedo Archipelago, Mexico (Hull et al. unpub.) indicates the presence of octocorals, although no species list was included. An analysis of collections from Peru contained 16 species (O. Breedy, pers. obs.).

Scleractinia

The scleractinian corals, both hermatypic and ahermatypic, are probably the most extensively studied taxa in the ETP. Forty-seven species each of reef-building corals (see Chap. 5, Glynn et al.; Table 7.8) and azooxanthellate corals, mostly from deep waters (Cairns 1991a; Cortés 2009), are known from the ETP. Veron et al. (2015) concluded that the reef-building corals throughout the ETP are closely related with the exception of the Clipperton fauna, which is taxonomically closer to the Marquesas Islands.

The main hermatypic scleractinian species of the ETP belong to the genera *Pocillopora*, *Porites* and *Pavona* (Cortés 2011; see Chap. 5, Glynn et al.). The pocilloporids are usually abundant in shallow waters (intertidal to 5 m) while massive species of *Porites* and *Pavona* are generally more abundant in deeper waters (>5 m) (Guzmán and Cortés 1989). While regional species composition is often similar, more local community differences are known to exist. For example, coral reefs in the non-upwelling southern areas of Costa Rica have diverse coral assemblages, including all the prominent reef-building scleractinians. However, in the upwelling northern areas, species community composition and structure change, with the most extreme upwelling sites containing only *Pavona gigantea* and *Porites panamensis* (Cortés and Jiménez 2003).

Until recently *Porites lobata* had been considered ubiquitous along the Costa Rican coast and the only reef-building

Table 7.5 Species list of octocorals from the ETP (see text for references)

Family	Species	Country/continental mainland								Oceanic islands		
		MEX	SAL	NIC	CRC	PAN	COL	ECU	PER	COC	MAL	GAL
Aquaumbridae	*Aquaumbra klapferi*									x		
Clavulariidae	*Carijoa riisei*	x	x	x	x	x	x	x	x	x	x	x
	Carijoa multiflora		x									
Gorgoniidae	*Eugorgia ampla*	x				x						
	Eugorgia aurantiaca	x					x					
	Eugorgia beebei	x							x			
	Eugorgia bradleyi				x							
	Eugorgia excelsa	x										
	Eugorgia daniana	x		x	x	x		x				x
	Eugorgia mutabilis	x			x	x						
	Eugorgia multifida	x										
	Eugorgia nobilis	x		x	x	x		x				
	Eugorgia panamensis					x						
	Eugorgia purpurascens			x		x			x			
	Eugorgia querciformis			x								
	Eugorgia rubens	x			x	x			x			
	Eugorgia siedenburgae	x			x	x						
	Leptogorgia aequatorialis							x				
	Leptogorgia alba	x	x	x	x	x	x	x	x	x	x	x
	Leptogorgia californica	x										
	Leptogorgia clavata	x										
	Leptogorgia cofrini			x	x	x						
	Leptogorgia cortesi				x							
	Leptogorgia christiae					x						
	Leptogorgia cuspidata	x	x	x	x	x		x				
	Leptogorgia diffusa		x	x	x	x	x					
	Leptogorgia ena	x										
	Leptogorgia exigua	x		x	x	x			x			
	Leptogorgia filicrispata	x										
	Leptogorgia flexilis	x	x			x						
	Leptogorgia florae					x						
	Leptogorgia fruticosa					x						
	Leptogorgia ignita				x							
	Leptogorgia laxa	x			x	x		x				
	Leptogorgia obscura							x				
	Leptogorgia parva					x						
	Leptogorgia peruana								x			
	Leptogorgia pumila	x		x	x	x		x	x			
	Leptogorgia ramulus	x	x	x		x	x		x			
	Leptogorgia regis	x	x		x	x						
	Leptogorgia rigida	x	x		x	x		x				
	Leptogorgia taboguillae				x	x		x				
	Leptogorgia tricorata									x		

(continued)

Table 7.5 (continued)

Family	Species	Country/continental mainland								Oceanic islands		
		MEX	SAL	NIC	CRC	PAN	COL	ECU	PER	COC	MAL	GAL
	Leptogorgia labiata	x				x						
	Pacifigorgia adamsii	x		x	x	x	x	x	x			
	Pacifigorgia agassizii	x				x						
	Pacifigorgia bayeri				x	x						
	Pacifigorgia cairnsi		x		x	x	x				x	
	Pacifigorgia catedralensis					x						
	Pacifigorgia cribrum	x										
	Pacifigorgia curta				x					x	x	
	Pacifigorgia dampieri											x
	Pacifigorgia darwinii											x
	Pacifigorgia englemanni	x										
	Pacifigorgia eximia				x	x	x					
	Pacifigorgia exilis	x										
	Pacifigorgia gracilis	x				x						
	Pacifigorgia ferruginea					x						
	Pacifigorgia firma			x	x	x	x	x				
	Pacifigorgia flavimaculata				x							
	Pacifigorgia irene		x		x	x	x	x				
	Pacifigorgia labiata	x										
	Pacifigorgia lacerata				x							
	Pacifigorgia pulchra	x										
	Pacifigorgia marviva					x						
	Pacifigorgia media	x	x	x		x						
	Pacifigorgia rubicunda				x	x	x	x				
	Pacifigorgia rubripunctata											x
	Pacifigorgia rutila	x										
	Pacifigorgia samarensis				x							
	Pacifigorgia sculpta			x		x	x					
	Pacifigorgia senta	x			x	x						
	Pacifigorgia symbiotica											x
	Pacifigorgia smithsoniana					x						
	Pacifigorgia stenobrochis	x	x	x	x	x		x	x			
	Pacifigorgia tabogae					x						
	Pacifigorgia tupperi				x							
	Pacifigorgia rubinoffi					x						
	Pacifigorgia douglasii											x
Ellisellidae	*Ellisella limbaughi*	x										
Plexauridae	*Eumuricea acervata*					x						
	Eumuricea tubigera					x						
	Eumuricea horrida								x			
	Eumuricea squarrosa			x		x		x				
	Heterogorgia hickmani							x				x
	Heterogorgia verrucosa			x	x	x	x	x				x

(continued)

Table 7.5 (continued)

Family	Species	Country/continental mainland								Oceanic islands		
		MEX	SAL	NIC	CRC	PAN	COL	ECU	PER	COC	MAL	GAL
	Heterogorgia tortuosa				x	x						
	Heterogorgia papillosa	x										
	Muricea californica	x										
	Muricea crassa					x	x	x				
	Muricea equinata					x						
	Muricea fruticosa				x	x						x
	Muricea fruticosa var. *miser*	x	x	x	x	x		x				
	Muricea austera	x	x		x	x		x				
	Muricea retusa					x						
	Muricea formosa								x			
	Muricea robusta	x										
	Muricea albida					x						
	Muricea hebes			x		x						
	Muricea purpurea	x	x	x		x		x				
	Muricea appressa	x		x		x		x	x			
	Muricea appressa cf. *flavescens*							x				
	Muricea tenella			x		x			x			
	Muricea aspera					x						
	Psammogorgia arbuscula					x						
	Psammogorgia variabilis									x		
	Psammogorgia fucosa	x										
	Psammogorgia teres					x						
	Psammogorgia hookeri								x			
	Psammogorgia sp.		x									
	Adelogorgia telones											x
Total		46	17	24	38	64	14	25	16	6	4	12

MEX Mexico; *SAL* El Salvador; *NIC* Nicaragua; *CRC* Costa Rica; *PAN* Panama; *COL* Colombia; *ECU* Ecuador; *PER* Peru; *COC* Isla del Coco; *MAL* Malpelo; *GAL* Galápagos Islands

species of that genus (*Porites panamensis* forms only small isolated colonies). Genetic analysis, however, has demonstrated that there are two species in the genus, *P. lobata* and *Porites evermanni*. While these species have similar external morphologies, they display different ecological characteristics, with *P. lobata* more susceptible to bleaching, and *P. evermanni* experiencing higher bioerosion and predation rates (Boulay et al. 2014).

Other Anthozoa (Actinaria, Ceriantharia, Zoanthidea, Antipatharia)

While numerous other anthozoan taxa are known from the ETP, the majority of them are poorly studied (Hickman 2008). Most research on sea anemones (Order Actinaria) was conducted in the 19th and early 20th centuries (Verrill 1868–70; Carlgren 1940). Recent work includes Daly and Fautin (2004) and Fautin et al. (2007) in the Galápagos Islands, in which eight species were reported. Excoffon et al. (2009) and Acuña et al. (2012, 2013) reported six species in Costa Rica, and Garese et al. (2009) 14 species from Panama. Tube anemones (Order Ceriantharia) are even less well known; Fautin et al. (2007) reported two species from the Galápagos Islands. The zoanthids (Order Zoanthidea) have been studied recently in the Galápagos Islands (seven species) (Reimer et al. 2008), but not in other areas of the ETP. There are no recent publications on shallow-water black corals (Order Antipatharia), even though they are common at some sites in the ETP (Hickman 2008; J Cortés, pers. obs.).

7.3.3.3 Polychaeta

The reef-inhabiting marine worms (including the Phyla Annelida, Sipuncula, Nemertea, Platyhelminthes, and

Fig. 7.3 Octocorals of the ETP. **a** *Pacifigorgia stenobrochis*, Isla del Caño, Costa Rica; **b** *Pacifigorgia irene*, Isla Coiba, Panama; **c** *Leptogorgia alba*, Isla del Coco, Costa Rica; **d** *Muricea purpurea*, Bahía Salinas, Costa Rica; **e** *Carijoa riisei*, Isla del Coco, Costa Rica. All photographs by Jorge Cortés except **b** by Graham Edgar

Phoronida) are poorly studied, but play an important role in ecosystem function. Among these worms, the annelids are the dominant group. Dean (2009) cited 19 families and 82 species of coral reef polychaetes (Phylum Annelida, Class Polychaeta), but many were also known to live on other substrates and in different habitats such as soft sediments and rocky bottoms. Bastida-Zavala (1995) recorded 82 species in 21 families inhabiting dead coral, while Blake (1991) noted only ten species in dead and live coral in his checklist for the Galápagos Islands. Enochs and Manzello (2012a) recorded

Table 7.6 Species list of ETP pennatulaceans

Family	Species	Coastal Costa Rica[1]	Oceanic islands	
			Isla del Coco[2]	Galápagos Islands[3]
Pennatulidae	*Ptilosarcus undulatus*		x	x
Renillidae	*Renilla ametystina*	x		
Veretillidae	*Cavernulina* cf. *darwini*			x
Virgulariidae	*Scytalium* sp.			x
Virgulariidae	*Virgularia galapagensis*			x
Virgulariidae	*Stylatula* cf. *elongata*		x	

Sources [1]Breedy (2009), [2]Breedy and Cortés (2008), [3]Hickman (2008)

Table 7.7 Number of species in the most abundant octocoral genera (references in text)

Genus	No. species	Most speciose country
Pacifigorgia	35	Panama (20)
Leptogorgia	27	Panama (17)
Muricea	16	Panama (12)
Eugorgia	14	Mexico (11)
Psammogorgia	6	Panama (2)
Eumuricea	4	Panama (3)
Heterogorgia	4	Costa Rica, Panama Ecuador (2)

Most speciose region and richness indicated in parentheses

Table 7.8 Species richness of ETP hermatypic corals

Country	No. species
Mexico mainland[a]	29
Colombia	28
Panama	27
Costa Rica mainland	25
Revillagigedo	25
Ecuador mainland	23
Galápagos Islands[b]	23
Isla del Coco	22
Clipperton	21
Malpelo	13
El Salvador	11
Nicaragua	10

[a]Includes the Gulf of California
[b]Includes both the northern and southern Galápagos Islands
Source Glynn et al. Chap. 5

21 genera associated with coral exhibiting different degrees of degradation at Playa Larga, Panama, and Dean et al. (2012) reported 62 species from dead coral at Isla del Coco, Costa Rica. A detailed list of polychaetes of the ETP, including species from reef habitats, can be found in Salazar-Vallejo and Londoño-Mesa (2004), and in Dean (2009).

Among the marine worms, the fireworms (Polychaeta: Amphinomidae) are the most conspicuous by their size and coloration. Two species reported from the ETP (Salazar-Vallejo and Londoño-Mesa 2004; Dean 2009) are *Eurythoe complanata* (Fig. 7.4a) and *Notopygos ornata*. The former is considered a cryptic species in the Atlantic Ocean (Barroso et al. 2010), sheltering in crevices and under rocks during the day (Fauchald and Jumars 1979). Other species of this family reported on coral reefs (Dean 2009) are *Chloeia viridis* and *Pherecardia striata*. Species in the family Amphinomidae are traditionally categorized as predators, but *P. striata* also plays an important role as a carrion feeder, and even feeds on the internal organs of injured *Acanthaster planci* (Fauchald and Jumars 1979; Glynn 1984; Jumars et al. 2015). The fireworm *Hermodice carunculata* is common on Caribbean reefs and an active predator of zooxanthellate corals, especially branching species; a single record of this species for the Pacific waters of Costa Rica exists (Treadwell 1941), but the veracity of this record is debated (Dean 2004).

Other families of errant worms are found in coral cavities, in coral rubble, and algae (Dean 2009). Some of these species are known to consume small crustaceans, molluscs or other invertebrates, but others are deposit feeders and omnivores (Jumars et al. 2015). The most active polychaete predators belong to the Eunicidae, and many exhibit endolithic tendencies, boring into coral skeletons with their jaws. *Eunice aphroditois*, *Eunice biannulata* (Fig. 7.4b) and *Palola siciliensis* were found in coral rubble at Playas del Coco, and in dead coral at Isla del Coco, Costa Rica (Dean et al. 2012). The afore-noted species and five more in the genera *Eunice* (*E. antennata*; *E. filamentosa*; *E. reducta*; *E. schemacephala*; and *E. mutilata*) were observed by Bastida-Zavala (1995) at Cabo Pulmo, Mexico. The burrows of these boring taxa are often occupied by additional species, many of them predatory polychaetes in the families Phyllodocidae (*Eulalia* and *Phyllodoce*), as well as scale worms in the family Polynoidae (*Halosydna*, *Harmothoe*, and *Lepidastenia*), including the abundant *Iphione ovata*, a small, ovoid-shaped orange worm that can be found under coral rubble (Fig. 7.4c) (Bastida-Zavala 1995; Dean et al. 2012). Also present are members of the Syllidae with *Haplosyllis*, *Sphaerosyllis, Typosyllis*, and *Exogone* representing genera of high species diversity (Westheide 1974; Bastida-Zavala 1995; Dean 2009). At Isla del Caño and Golfo Dulce, Costa Rica six species of polychaetes in the

Fig. 7.4 Worms of ETP coral reefs. Polychaetes: **a** *Eurythoe complanata* (Amphinomidae); **b** *Eunice biannulata* (Eunicidae); **c** *Iphione ovata* (Polynoidae); **d** *Spirobranchus giganteus* complex (Serpulidae); **e** *Bispira* sp. (Sabellidae); **f** *Lanice* sp. (Terebellidae); **g** spaghetti worm on dead coral (Terebellidae); **h** *Cirriformia* sp. (Cirratulidae) in a sand crevice; **i** *Thalassema steinbecki* (Echiura: Echiuridae) in dead coral; **j** *Aspidosiphon misakiensis* (Sipuncula: Aspidosiphonidae); **a** to **j** from Isla del Coco, Costa Rica; **k** *Baseodiscus mexicanus* (Nemertea: Valenciniidae) from coral colonies at Cabo Blanco, Costa Rica; **l** *Praestheceraeus bellostriatus* (Platyhelminthes: Euryleptidae) from Sándalo reef, Golfo Dulce, Costa Rica. Photographs **e**, **h** and **l** by Kimberly García-Méndez; **d** and **g** by Jorge Cortés, remainder by Jeffrey Sibaja-Cordero

families Eunicidae, Cirratulidae and Dorvilleidae were found on reefs penetrating both live and dead corals. Species in the families Sabellidae, Spionidae and Flabelligeridae were found sheltering in burrows made by species of eunicids (Fonseca et al. 2006).

Another common coral reef polychaete family is the Serpulidae, often called Christmas tree worms because of their multi-colored radial crowns. These sedentary annelids construct their calcareous tubes in dead or live corals. Notable examples include *Hydroides* spp., and the *Spirobranchus giganteus* species complex, often referred to as *Spirobranchus incrassatus* (Fig. 7.4d) (Fiege and Ten Hove 1999; Bastida-Zavala 2008).

The Sabellidae or fan worms are sessile and live in mucus tubes in coral crevices or cavities. Sabellids are solitary or gregarious worms and belong to the suspension-feeding guild (Jumars et al. 2015). Two common genera in the ETP (Dean 2004) are *Bispira* (Fig. 7.4e) and *Hypsicomus*. Capa and López (2004) recorded 11 species in the genera *Amphicorina*, *Amphiglena*, *Branchiomma*, *Notaulax*, *Megalomma* and *Pseudopotamilla* living on dead coral (*Pocillopora*) at Coiba Island, Panama. At Cabo Pulmo, Mexico this family was represented by the genera *Chome*, *Euchone*, *Megalomma* and *Pseudopotamilla* (Bastida-Zavala 1995).

Two additional families consist of conspicuous, sedentary species on ETP coral reefs. The Terebellidae (spaghetti worms), including the genera *Lanice*, *Lanicides* or *Eupolymia* (Fig. 7.4f, g), can be recognized by elongate, usually white tentacles arising from a coral hole or depression. The Cirratulidae, including species in the genus *Cirriformia*, can be recognized by white and reddish tentacles and branchial filaments emerging from crevices or sand patches on coral reefs (Fig. 7.4h). A total of 15 species in this genus is listed for the ETP in Salazar-Vallejo and Londoño-Mesa (2004). These species are surface deposit or suspension feeders (Jumars et al. 2015), and were reported as occurring on ETP coral reefs by Westheide (1977), Bastida-Zavala (2008) and Dean (2009).

Dean et al. (2010a) reported the occurrence of the spoon worm *Thalassema steinbecki* (Fig. 7.4i), now in the subclass Echiura of Annelida sensu Ruggiero et al. (2015), from calcareous rock at Isla del Coco. This echiuran is known to burrow into coral rubble and can be found from Baja California to Ecuador. Fisher (1946) reported specimens of this species from a mud and coral rock habitat with gorgonians and coralline algae in Baja California. Additional records of this species in the ETP include specimens from the intertidal zone (Dean et al. 2010a), and congeneric species are known from rocky and soft bottom habitats (Fisher 1946).

7.3.3.4 Other Marine Worms (Sipuncula, Nemertea, Platyhelminthes, Phoronida)

Many peanut worm species (Phylum Sipuncula) are deposit feeders and are known to bore actively into coral rubble and to occupy preexisting cavities (Fonseca et al. 2006; Dean et al. 2010a). Dean et al. (2010a) found the species *Antillesoma antillarum*, *Aspidosiphon* (*Aspidosiphon*) *misakiensis* (Fig. 7.4j), *Aspidosiphon* (*Aspidosiphon*) *elegans*, *Aspidosiphon* (*Paraspidosiphon*) *fischeri*, *Phascolosoma* (*Phascolosoma*) *agassizii*, *Phascolosoma* (*Phascolosoma*) *nigrescens*, and *Phascolosoma scolops* associated with coral rubble at Isla del Coco. Dean (2001) also reported the species *Siphonosoma vastum* and *P.* (*P.*) *nigrescens* inhabiting coral fragments at Isla del Caño, Costa Rica, and *A.* (*A.*) *elegans* from coral rubble at Sándalo, Golfo Dulce. Pepe (1985) reported the sipunculans *A. antillarum*, *A.* (*P.*) *fischeri* and *Phascolosoma* sp. from coral rubble at Playas de Coco. Fourteen species of sipunculans were reported from the Costa Rican coast, and *A. antillarum, Aspidosiphon* (*Paraspidosiphon*) *parvulus*, *P.* (*P.*) *nigrescens*, and *Siphonosoma vastum* were collected from corals or coral rubble (Cutler et al. 1992; Dean 2001). Other substrates with sipunculans were sands, sandstone, limestone and the undersides of rocks (Dean 2001). Cutler (1994) additionally mentioned the presence of *Aspidosiphon* (*Paraspidosiphon*) *coyi*, *A.* (*P.*) *fischeri*, *A. antillarum*, *Phascolosoma perlucens*, *Thysanocardia catharinae* and *Themiste hennahi* from coastal localities in the Gulf of California, Panama, Ecuador, Galápagos, and Peru; information on habitats was not given. Fonseca et al. (2006) and Dean (2009) reported the above-noted species inhabiting hard bottoms.

Presently, our knowledge of the Nemertea (ribbon worms) occupying reefs is limited (Kirsteuer 1969; Thiel and Kruse 2001). From others studies, it is known that the heteronemerteans prey mainly on polychetes, while the hoplonemerteans consume crustaceans and carrion (Thiel and Kruse 2001). Coe (1940) reported on13 species of ribbon worms from southern California to Ecuador, but only provided a direct reference to the reef-affinity of *Baseodiscus delineatus*, *Baseodiscus mexicanus*, *Baseodiscus punnetti* and *Notospermus geniculatus* (reported as *Lineus geniculatus*) in the Gulf of California. Among these species, *B. mexicanus* is named the zebra worm for its maroon or red and white rings encircling the body (Fig. 7.4k). It attains lengths of 20 cm to 4 m, and occurs on the Pacific coast of the Americas and in the Galápagos Islands (Kajihara et al. 2012).

Flatworms (Phylum Platyhelminthes) associated with live coral, coral rubble, and rocky reefs are primarily polyclads (for other flatworms in the ETP, see Westheide 1991). These worms are usually colorful and often mistaken for nudibranchs. Polyclads are poorly studied in the ETP, although many can be found on coral and rocky reefs. For example, Hyman (1953) mentioned four species of polyclads from rocky substrates in the Galápagos Islands, and Enochs and Manzello (2012a) collected 11 flatworm species from a coral reef in Panama. A specimen of *Praestheceraeus bellostriatus*, or a closely related species (Fig. 7.4l), was found in a *Psammocora stellata* coral patch at Golfo Dulce, Costa Rica (K García-Méndez, pers. comm.). *P. bellostriatus* occurs in California and in the Sea of Cortez (as *Prostheceraeus bellostriatus* in Kerstitch and Bertsch 2007), and is recognized by alternating black and white longitudinal lines (Newman and Cannon 2003; Kerstitch and Bertsch 2007). Presently, no other flatworms have been reported on ETP coral reefs.

Finally, worms in the phylum Phoronida, referred to as horseshoe worms due to the shape of their lophophore, are known from coral reefs but remain poorly studied. *Phoronis hippocrepia* was once the only species in this phylum reported from the ETP. It is known to burrow into coral rubble (Bailey-Brock and Emig 2000; Emig 2009). This species was found in the Miraflores Locks, Panama Canal by Emig (1982). Dean et al. (2010b) reported a second species, *Phoronopsis albomaculata*, that occurs on sandy substrates.

7.3.3.5 Mollusca

The Mollusca is one of the most species-rich phyla of marine invertebrates and representatives may be found across many diverse marine ecosystems, including coral reefs. Many of the mollusc species associated with coral reefs are considered charismatic and have important values, including economic, biomedical, biological and ecological.

Keen (1971) reported more than 2400 species of gastropods and almost 800 species of bivalves from different environments on the west coast of the Americas. The number of known molluscs has increased in the last four decades as many new species have been described, mostly from the micromolluscs (0.45–11 mm, Geiger 2012) and in the subclass Opisthobranchia. Nearly a fourth (>800) of the gastropods, bivalves, cephalopods and opisthobranchs reported from the ETP are known to inhabit coral reefs. Gastropoda is the most species rich class among these, with more than 600 species, followed by Bivalvia with more than 200 species. Currently there are ten known species of Polyplacophora and eight species of Cephalopoda. The number of species reported is as much a function of the physical diversity of the area as the sampling effort. Considering that there are numerous areas of the coastline and islands in the ETP where the molluscan fauna has never been sampled, it is likely that numbers of described species will increase as regions and habitats are further explored.

Molluscs are found associated with numerous substrates (see some examples in Fig. 7.5) including both live and dead coral, coral debris, rocky and sandy substrates, intertidal zones, algal fields and seagrass communities (Landa-Jaime et al. 2013). Thus, higher habitat diversity within a given coral reef may encourage higher molluscan diversity. Table 7.9 shows the numbers of molluscs recorded for coral reef areas of the ETP. Gastropods are the most abundant and diverse taxa, representing 68 % of the total number of species, though it must be noted that studies may be biased towards this class. The next most species rich groups are the bivalves (17 % of spp.) and the opisthobranchs (13 % of spp.); the low percentage of Opisthobranchia is likely due in part to their cryptic nature and the difficulty of studying them in situ. Detailed information on these taxa is available from a limited number of areas (Gosliner 1991; Hermosillo and Camacho-García 2006; Hermosillo et al. 2006; Glynn et al. 2008; Camacho-García 2009).

The Galápagos Islands have the highest numbers of described species of molluscs (800 taxa) associated with coral reefs in the ETP (Kaiser 1997), followed by Isla Coiba (Vega and González 2001, 2002; Hermosillo and Valdés 2004; Hermosillo and Camacho-García 2006) and Isla del Coco (Cortés 2012; Kaiser unpub. data), with more than 500 species each. The Islas Revillagigedo, Clipperton Atoll, Isla de Malpelo, and Islas Marías have approximately 300 species each (for specific values and references see Table 7.9).

There are some mainland Mexican sites that are especially well known for mollusc diversity, as they have been studied for more than ten years. Tenacatita reef is one such site and researchers have catalogued more than one hundred species of associated molluscs (Landa-Jaime et al. 2013). Another exhaustive study completed at Bahía de Banderas, listed one hundred and forty-eight species of opisthobranchs, as a result of intensive field work over a three year period (Hermosillo and Valdés 2004; Behrens and Hermosillo 2005; Hermosillo et al. 2006). Extensive studies of the molluscan fauna of Isla

Fig. 7.5 Selected gastropod molluscs associated with ETP coral reefs, with dead or live corals. **a** Ventral view of *Coralliophila neritoidea*; **b** *Quoyula madreporarum*; **c** *Reliquiaecava robillardi* ventral view and **d** inside a coral skeleton; **e** *Rhizochilus antipathum* ventral view and **f** attached to a black coral; **g** ovulid on a gorgonian colony; **h** *Jenneria pustulata* associated with *Pocillopora* corals. Photographs by Patricia Sadeghian, Île Clipperton, except **g** and **h** by Pedro Medina-Rosas, Islas Marías, Mexico

Table 7.9 Number of species of reef-associated molluscs by ETP area

	MEX[1]	SAL[3]	ALI[4]	MAR	REV	CLI[14]	COI	COC[19]	MAL[21]	GAL[14]
Bivalvia	24	1	9	34[6]	77[9]	42	218[16]	78	46	212
Gastropoda	87	6	32	179[7]	204[10]	212	236[17]	446[20]	256	605
Opisthobranchia	148[2]	–	7[5]	52[8]	42[11]	37[15]	85[18]	22	–	34[22]
Polyplacophora	11	–	–	–	8[12]	1	1[17]	8	6	13
Cephalopoda	1	–	–	–	1[13]	1	–	6	3	8
Total	271	7	48	265	332	293	540	560	311	872

MEX Mainland Mexico (Opisthobranchia from Bahía de Banderas, all others from Tenacatita); *SAL* El Salvador; *ALI* Rocas Alijos, Mexico; *MAR* Islas Marías, Mexico; *REV* Islas Revillagigedo, Mexico; *CLI* Île Clipperton, France; *COI* Isla Coiba, Panama; *COC* Isla del Coco, Costa Rica; *MAL* Isla de Malpelo, Colombia; *GAL* Islas Galápagos, Ecuador

Sources [1]Landa-Jaime et al. (2013), [2]Hermosillo et al. (2006), Behrens and Hermosillo (2005), [3]Hernández (1992), [4]McLean and Coan (1996), Kaiser (2006), [5]Hermosillo (2009a), [6]Bernard et al. (1991), Kaiser unpub. data, [7]Small (1998), Kaiser unpub. data, [8]Hermosillo (2009b), [9]Strong and Hanna (1930), Villalobos (1960), González-Nakawaga and Sánchez-Nava (1986), Bernard et al. (1991), Bautista-Romero et al. (1994), Mille-Pagaza et al. (1994), Emerson (1995), [10]Strong and Hanna (1930), Villalobos (1960), González-Nakawaga and Sánchez-Nava (1986), Chávez-Hernández and Bretado-Aguirre (1990), Holguín-Quiñones et al. (1992), Bautista-Romero et al. (1994), Holguín-Quiñones (1994), Mille-Pagaza et al. (1994), [11]Hermosillo and Gosliner (2008), Behrens et al. (2009), [12]Ferreira (1983), [13]Holguín-Quiñones et al. (1992), [14]Kaiser (2007), [15]Millen and Hermosillo (2012), [16]Vega (2001), [17]Vega (2002), [18]Hermosillo and Valdés (2004), Hermosillo and Camacho-García (2006), [19]Cortés (2012), [20]Cortés (2012), Kaiser unpub. data, [21]Kaiser and Bryce (2001), [22]Gosliner (1991)

del Coco and Isla de Malpelo have reported incidental finds of opisthobranchs (Camacho-García 2009; Rodríguez-Sevilla et al. 2009; Cortés 2012). However, no species lists have presently been published on the opisthobranch fauna of El Salvador, or many other locations in the ETP.

7.3.3.6 Crustacea (Decapoda)

More than 850 species of decapod crustaceans have been recorded from the ETP (Boschi 2000), the majority of which are crabs belonging to the Infraorder Brachyura. Most of the information comes from the coasts of Mexico (Fig. 7.6), Costa Rica, Panama, and Colombia, likely due to greater investment in marine research in these countries. For example, of all of the decapod species known from the ETP, 634 occur in the tropical Mexican Pacific (Hendrickx 1993), representing 77 % of the ETP decapod crustaceans. In Costa Rican waters, there are 438 species (53 % of spp.) according to Vargas and Wehrtmann (2009), and on the Colombian Pacific coast, Lemaitre and Álvarez-León (1992) recorded 378 species (45 % of spp.). Little is known of the decapod species of western Guatemala, El Salvador and Nicaragua, except that some commercially important species are exploited for human consumption, such as shrimps (Penaeoidea), lobsters (Palinuridae), and swimming crabs (Portunidae). While the decapod fauna of the ETP is dominated by brachyuran crabs (Boschi 2000), those exclusively associated with corals (mostly live corals) comprise roughly 120 species and are dominated by caridean shrimps (Table 7.10).

Inventories of decapods in the ETP, including coral associated faunas, began with expeditions from the late 19th and mid 20th centuries (*Albatross, Zaca, Velero*, among others) (Garth 1992; Wicksten and Hendrickx 1992). Most of these early studies included descriptions of new species and annotations to species' ranges. It was not until the late 1970s that ecological investigations were carried out with coral-associated crustaceans. On Panamanian reefs (8°N), Abele (1974) noted that the richness of crustacean species is higher on coralline substrates than in sandy or mangrove habitats. Gotelli and Abele (1983) showed that the number of coral-associated species was correlated with coral colony size. On the Pacific coast of Colombia (2°58′N), research on coral-associated decapods has been less extensive. However, studies made by von Prahl et al. (1978) and Castro (1982) noted the importance of symbiotic decapods as promoters of coral health. In Mexico, Ramírez-Luna et al. (2002) demonstrated that abundance and diversity of coral-associated crabs at Huatulco (16°N) was low in the wet season (from May to October). In two studies conducted by Hernández et al. (2009, 2013), 60 coral-associated decapod species were collected along the tropical Mexican Pacific (from 15°N to 26°N), and 11 of these were considered to be coral symbionts. Alvarado and Vargas-Castillo (2012) recorded 35 species of invertebrates from *Pocillopora* corals at Bahía Culebra, Costa Rica (10°36′N), and 18 (nearly 50 %) were decapod crustaceans. Pocilloporid corals have an Indo-Pacific origin (Cortés 1997) and many of their coral-associated decapod symbionts mirror this biogeographic connection (Garth 1974).

There are differences between crustaceans associated with live and dead corals. Enochs and Manzello (2012a) noted higher species richness associated with dead versus live coral substrates, and the dead coral species' assemblage seems to be more closely related to crustaceans associated with rocky than coral substrates. This is an interesting topic worthy of further research.

Fig. 7.6 Selected ETP coral-associated decapods. **a** *Trapezia digitalis*, Revillagigedo Islands; **b** *Calcinus californiensis*, Ixtapa; **c** *Harpiliopsis spinigera*, Huatulco; **d** *Trizopagurus magnificus*, Ixtapa; **e** *Alpheus lottini*, Ixtapa; **f** *Stenorhynchus debilis*, Acapulco. Photographs by Luis Hernández, all from Mexico

Table 7.10 Total ETP species richness of decapod crustaceans and numbers of species associated with corals

	Total no. species[1]	No. with corals[2]
Anomura	165	31
Brachyura	411	38
Caridea	161	50
Others	88	3
Total	825	122

Sources [1]Boschi (2000), [2]Abele (1974), Prahl et al. (1978), Gotelli and Abele (1983), Lemaitre and Álvarez (1992), Hernández et al. (2009, 2013); Enochs et al. (2011), Alvarado and Vargas-Castillo (2012)

7.3.3.7 Echinodermata

A total of 722 species of echinoderms has been reported from the ETP, encompassing 214 ophiuroids, 195 asteroids, 174 holothuroids, 118 echinoids, and 21 crinoids (Solís--Marín et al. 2013a). Of these, only 93 (13 %) inhabit coral reef environments (sensu Solís-Marín et al. 2013a) (Fig. 7.7). These numbers will likely increase with further collections and assessments. The groups with the highest numbers of reef species are the ophiuroids and holothurians, with 31 and 30 species, respectively (Solís-Marín et al. 2013a). Additionally, 16 species of echinoids and asteroids are known from reefs in the region (Solís-Marín et al. 2013a). There are no reports of crinoids inhabiting coral reefs in the ETP.

The most species rich echinoderm fauna in the ETP is known from Mexico, but only 29 % of Mexican echinoderms occur on coral reefs (Table 7.11). This relatively low percentage of coral reef species is similar in Panama (24 %), the Galápagos Islands (27 %), Ecuador (30 %), and at Isla del Coco (35 %). In other areas, the proportion of reef-dwelling echinoderms is notably higher, for example, El Salvador (60 %), Nicaragua (58 %), Revillagigedo Archipelago (58 %), Costa Rican mainland (49 %), Malpelo Island (42 %), and Colombia (41 %) (Table 7.11). This is likely indicative of a greater research effort in more accessible continental environments relative to remote regions and oceanic islands. Considering El Salvador and Nicaragua, those species that are presently reported are the most conspicuous and easy to identify, suggesting a high likelihood that additional species are present but as yet unaccounted for (Alvarado et al. 2010, 2013).

Of the 93 species of reef-associated echinoderms in the ETP, only six have been reported from 11 sampled regions (Table 7.11). Those species are the asteroid *Phataria unifascialis* (Fig. 7.7a), the holothuroid *Isostichopus fuscus* (Fig. 7.7d), the echinoids *Echinometra vanbrunti* and *Diadema mexicanum* (Fig. 7.7f), and the ophiuroids *Ophiocoma alexandri* and *O. aethiops*. The seastar corallivore *Acanthaster planci* (Fig. 7.7b) is present at eight of the eleven regions, only absent in El Salvador, Nicaragua and the Ecuadorian mainland (Solís-Marín et al. 2013a). Another six species are reported from ten of the 11 regions, including the asteroid *Nidorellia armata*, the ophiuroid *Ophiactis savignyi*, the holothuroid *Holothuria (Halodeima) kefersteini* (Fig. 7.7c) and the echinoids *Eucidaris thouarsii*, *Tripneustes depressus* (Fig. 7.7e) and *Toxopneustes roseus*. Although the total number of studies on reef echinoderms in the ETP is limited, five species, *D. mexicanum*, *Eucidaris galapagensis*, *E. thouarsii*, *A. planci* and *I. fuscus*, are especially important ecologically as bioeroders, corallivores and in bioturbation, and have been widely studied (Alvarado et al. 2013; Benavides-Serrato et al. 2013; Coppard and Alvarado 2013; Solís-Marín et al. 2013b; Sonnenholzner et al. 2013).

7.3.3.8 Fishes

Over 1300 species of marine fishes have been described from shallow-waters (<100 m depth) in the ETP (Robertson and Allen 2008; Eschmeyer 2014). Based on previous analysis of new species descriptions during the 50 years before 2004, the total number of described fishes in the region increased at a rate of 4.7 new species per year during this time (Zapata and Robertson 2007). Nearly 49 % of the species described in this period inhabit soft bottom habitats while 38 % live in reef (either rocky or coral) habitats. New species likely to be found in soft bottom or reef habitats will tend to be small and restricted in both their geographic and depth distribution, relatively localized towards the center of the ETP region, and from relatively deep waters (Zapata and Robertson 2007).

Currently, the entire shallow water marine fish fauna in the region is composed of 111 species of sharks or rays and 1193 species of bony fishes. Nearly 47 % of the species can be found associated with rocky or coral reefs and adjacent rubble, sandy or water-column habitats. Of these, 25 elasmobranchs and 482 bony fishes (39 % of the fauna) can occur on coral or rocky reefs, but only 353 of them (27 % of the total fish fauna) are found exclusively on either type of reef. However, only 157 species (12 % of all species) can be found associated with corals (not necessarily on reefs since isolated coral colonies are common), and a subset of these (115 species) occurs exclusively on coral reefs (Robertson and Allen 2008). Given the scarcity and small size of coral reefs in the ETP region, it is not surprising that the majority of the exclusive reef fish fauna is found on rocky reefs only (238 species) while only a handful (6 species) are found associated only with corals. This, however, seems to be an artifact of the extremely restricted geographic distribution of these few species, five of which are endemic to the only coral reef atoll in the ETP (Clipperton Atoll: Robertson and Allen 2008).

Therefore, nearly all of the fishes found on coral reefs in the ETP are also found on rocky reefs but not vice versa. Thus, throughout the remainder of this section we will use the term "reef fishes" to refer in general to fishes found on either rocky or coral reefs, while the term "coral reef fishes" will be used to refer to the subset of reef fishes found on coral reefs. The ETP region has been separated for millions of years from the Indo-West Pacific by the 4000–7000 km wide East Pacific Barrier and from the Caribbean by the Isthmus of Panama (Glynn and Ault 2000; Robertson et al. 2004; Robertson and Allen 2008). This isolation of the ETP is reflected in its high level of endemism, 16 % of genera and 70 % of species of marine fishes are endemic to the region (Robertson and Allen 2008). This is also generally the case for reef fishes, 73 % of which are endemic to the ETP (Table 7.12). However, among the coral reef fish component the level of endemism is lower (55 % of 115 reef fishes associated with corals). This reduction in regional endemism can be partially explained by faunal enrichment due to immigration from the Indo-West Pacific region: nearly 43 % of the 115 reef-exclusive fishes found on coral reefs are presumed to be Indo-West Pacific immigrants (Robertson et al. 2004). However, only 31 of these species are considered to have resident populations in the ETP (Robertson and Allen 2008). In contrast, faunal enrichment from the Indo-West Pacific is low among the rocky-reef exclusive component, although there is some peripheral enrichment from the adjacent Californian (17 species) and Peruvian (20 species) warm temperate regions, which lack coral reefs (Table 7.12). Therefore, discounting the vagrant Indo-West Pacific species, the entire coral reef fish fauna in the ETP is just below 100 species.

Total fish species richness estimates on coral reefs based on visual censuses generally agree with the notion that the ETP coral reef fish fauna is "depauperate" (Ekman 1953; Sale 1980; Thresher 1991). For example, while only 71 species were observed during 99 visual censuses made during one year at a single ~10 ha coral reef at Gorgona Island, Colombia, a log-normal abundance-distribution model fitted to these data predicted the occurrence of 102 species on this reef (Zapata and Morales 1997). Dominici-Arosemena and Wolff (2006) observed a total of 88 species at four coral reef sites in the Gulf of Chiriquí, two dominated by massive corals and two by branching corals. Benfield et al. (2008) observed a total of 79 species on 45 coral reefs and coral communities at Las Perlas Archipelago in Panama. Alvarez-Filip et al. (2006) observed a total of 62 species on the coral reef at Cabo Pulmo, Gulf of California. Estimates based on visual counts, however, generally underestimate the cryptic fish fauna, which can represent a relatively large proportion of the total

Fig. 7.7 Common echinoderms of ETP coral reefs. **a** *Phataria unifascialis,* Islas Tortuga, Costa Rica; **b** *Acanthaster planci*, Isla Espíritu Santo, Mexico; **c** *Holothuria (Halodeima) kefersteini*, Galápagos Islands, Ecuador; **d** *Isostichopus fuscus*, Golfo Dulce, Costa Rica; **e** *Tripneustes depressus*, Galápagos Islands, Ecuador; **f** *Diadema mexicanum*, Bahía Culebra, Costa Rica. Photographs by Cindy Fernández-García (**a**), Juan José Alvarado (**b**, **f**), Ángel Chiriboga (**c**, **e**), and Jeffrey Sibaja-Cordero (**d**)

Table 7.11 Numbers of ETP echinoderm species associated with coral reefs, by class for each country or region

	Country-continental mainland							Oceanic islands			
	MEX	SAL	NIC	CRC	PAN	COL	ECU	REV	COC	MAL	GAL
Asteroidea	13	4	2	6	9	13	5	5	8	10	10
Ophiuroidea	25	11	8	17	21	16	5	4	9	7	17
Echinoidea	13	6	7	10	9	11	11	5	11	10	11
Holothuroidea	28	13	5	18	24	15	10	11	15	7	20
Total spp. associated with reefs	79	34	22	51	63	55	31	25	43	34	58
Total echinoderm spp. in region	274	57	38	105	267	133	104	43	123	81	216

MEX Mexico; *SAL* El Salvador; *NIC* Nicaragua; *CRC* Costa Rica; *PAN* Panama; *COL* Colombia; *ECU* Ecuador; *REV* Revillagigedo Archipelago; *COC* Isla del Coco; *MAL* Isla Malpelo; *GAL* Galápagos Islands
Sources Pérez-Ruzafa et al. (2013), Solís-Marín et al. (2013a)

Table 7.12 Number of ETP fish species per family that occur exclusively on coral and rocky reefs

Family	Coral and rocky reefs			Rocky reefs only			Total
	ETP	IP	Other	ETP	IP	Other	
Acanthuridae[a]		10		2			12
Antennariidae	2	2					4
Aplodactylidae						1	1
Apogonidae[a]	5					1	6
Aulostomidae		1					1
Balistidae	2	2	2				6
Blenniidae[a]	3			6		3	12
Bythitidae				13			13
Carapidae					1		1
Carcharhinidae		1					1
Chaenopsidae				23			23
Chaetodontidae[a]	1	5		2			8
Cirrhitidae	1	2					3
Gobiesocidae				28		1	29
Gobiidae	4			19		1	24
Haemulidae	1			3		2	6
Holocentridae[a]	4[b]	2					6
Kyphosidae				4		4	8
Labridae[a]	6[b]	2		3		4	15
Labrisomidae				40		2	42
Lutjanidae	1			1			2
Microdesmidae				3			3
Monacanthidae		1	1				2
Muraenidae	5	11		4		3	23
Myrocongridae				1			1
Ophidiidae	1						1
Oplegnathidae						1	1
Ostraciidae		1					1
Pomacanthidae	4[b]						4
Pomacentridae[a]	8[b]			10		4	22

(continued)

Table 7.12 (continued)

Family	Coral and rocky reefs			Rocky reefs only			Total
	ETP	IP	Other	ETP	IP	Other	
Priacanthidae		1		1		2	4
Scaridae[a]		1		3	1		5
Sciaenidae				4		3	7
Scorpaenidae	1	1		3			5
Serranidae[a]	11[b]			9		8	28
Syngnathidae		1		1			2
Tetraodontidae	1	4					5
Tripterygiidae	2			13			15
Zanclidae		1					1
Total	63	49	3	196	2	40	353

Species are divided in two groups: those that occur both on coral and rocky reefs (except for six species occurring only on coral reefs, indicated by a superscript) and those restricted to rocky reefs. In each habitat, species are subdivided into biogeographic categories as follows: ETP, endemic to the ETP; IP, Indo-West Pacific immigrant; Other, other regions including Californian, Circumtropical, E to W migrant, Peruvian, and Western Atlantic/Eastern Pacific

[a]Bellwood (1996, 1997) coral reef fish consensus families (i.e., families present on coral reefs everywhere). Other consensus families (Carangidae and Mullidae) also occur on ETP reefs but are not exclusive reef inhabitants

[b]Species occurring exclusively on coral reefs are: *Myripristis gildi* (Holocentridae), *Thalassoma robertsoni* (Labridae), *Holacanthus limbaughi* (Pomacanthidae), *Stegastes baldwini* (Pomacentridae), *Epinephelus clippertonensis* and *Pseudogramma axelrodi* (Serranidae)

Sources Robertson et al. (2004), and Robertson and Allen (2008)

fish fauna in particular microhabitats (e.g., up to 36 % on coral rubble; Glynn 2006). Comparisons of visual counts and collections using rotenone and clove oil made on isolated live coral colonies at Gorgona Island, indicated that visual counts underestimated overall species richness by 28–36 % and number of individuals by 16–35 % (Alzate et al. 2014).

In spite of the dominance of rocky-reef habitats and the depauperate nature of the ETP coral reef fish fauna, its taxonomic structure does not differ significantly from that of more diverse fish assemblages in coral-rich regions (Robertson 1998). Secondary colonization of rocky reef habitats and subsequent adaptation by a fish fauna that originally evolved on coral reefs may explain this pattern (Dominici-Arosemena and Wolff 2006). Alternatively, rocky-reef habitats may offer sufficient resources to fulfill the requirements of coral reef fishes. Whatever the cause, all of the 11 families of fishes that are characteristic of coral reef fish assemblages everywhere (referred to as coral-reef-fish consensus families: Bellwood 1996, 1997) have representatives in the ETP (Table 7.12). However, the most species-rich families of ETP reef fishes are either restricted to rocky habitats, for example, weed blennies (family Labrisomidae with 42 species), clingfishes (Gobiesocidae, 29 spp.) and tube blennies (Chaenopsidae, 23 spp.), or have few representatives on coral reefs, e.g. gobies (Gobiidae) with four species on coral reefs out of 24 species on reefs). Other families have variable representation on coral reefs. For instance, 11 species of grouper (Serranidae) occur on coral reefs out of 28 species occurring on any type of reef. Similarly, 16 out of 23 moray eels (Muraenidae), 8 out of 22 damselfishes (Pomacentridae), and 8 out of 15 wrasses (Labridae) occur on coral reefs (Table 7.12). Some families relatively rich in coral reef species are composed of a large proportion of Indo-West Pacific immigrants, most notably the surgeonfishes (Acanthuridae) and the moray eels (Muraenidae). Other families, while poor in species, have a large Indo-West Pacific component, e.g., the hawkfishes (Cirrhitidae), and yet others are represented by a single Indo-West Pacific immigrant species, e.g., Aulostomidae, Ostraciidae and Zanclidae (Robertson and Allen 2008).

While fish families typical of coral reefs everywhere (i.e., Bellwood's consensus families) have representatives in the ETP, few of those species are either numerically dominant or ecologically important (Fig. 7.8). For instance, two of the five most abundant species (the snapper *Lutjanus viridis* and the hawkfish *Cirrhitichthys oxycephalus*; Zapata and Morales 1997) and two of the most ecologically important species on a coral reef at Gorgona Island, Colombia (the pufferfish *Arothron meleagris* and the triggerfish *Pseudobalistes naufragium*; Guzmán 1988; Palacios et al. 2014), do not belong to Bellwood's consensus families (Bellwood 1996, 1997). Only in some cases, such as in the damselfishes (Pomacentridae), can species in Bellwood's consensus families be either very abundant (e.g., *Chromis atrilobata*; Zapata and Morales 1997) or ecologically important (e.g.,

Fig. 7.8 Some numerically dominant (*left column*) and ecologically important (*right column*) fishes from Gorgona Island, Colombia coral reefs. **a** Scissortail damselfish, *Chromis atrilobata* (Pomacentridae); **b** Guineafowl puffer, *Arothron meleagris* (Tetraodontidae); **c** Cortez rainbow wrasse, *Thalassoma lucasanum* (Labridae); **d** Blunthead triggerfish, *Pseudobalistes naufragium* (Balistidae); **e** Blue and gold snapper, *Lutjanus viridis*, (Lutjanidae); **f** Acapulco damselfish, *Stegastes acapulcoensis* (Pomacentridae). All photographs by Fernando A. Zapata

Stegastes acapulcoensis; Wellington 1982a, b). Rarely, a single species from a coral reef fish consensus family is both numerically dominant and ecologically important in an ETP coral reef, although it might be argued that abundant species are always ecologically important (e.g., the wrasse *Thalassoma lucasanum*).

A noteworthy feature of the ETP reef fish fauna is the relatively distinctive composition of the oceanic island assemblages. The five oceanic islands or archipelagos in the region (Galápagos, Malpelo, Coco, Clipperton and Revillagigedo islands) are sufficiently isolated to exhibit both a significant degree of local endemism and a reduction in species richness. Only 64 % of the exclusively reef-associated fish species occur on islands and, individually, each of the islands or archipelagos exhibit between 1 and 4.5 % local endemism. An additional 5 % of the reef fishes are endemic to more than one oceanic island (Robertson and Allen 2008). Overall, 18 % of the reef-associated fish species are endemic to ETP oceanic islands and including other environments the level of endemism is 22 % on oceanic islands. Such a level of endemism led Robertson and Cramer (2009), in their re-evaluation of fish faunal relationships in the ETP, to propose the existence of an Oceanic Island Province. In contrast to the full set of reef species, those found exclusively on coral reefs exhibit different patterns. For instance, the reduction in species richness is minimal (only 3 out of 115 species do not occur on oceanic island coral reefs), but local endemism on oceanic islands is lower than in the full set of reef fishes (<1 % on each island except at Clipperton Atoll, where it is 4 %). However, endemism on multiple oceanic islands is higher (9 %) than in the entire reef fish fauna.

Latitudinal variation in species richness has been previously examined for the continental shoreline component of the endemic ETP fish fauna (Mora and Robertson 2005). This analysis found a major species richness peak between Panama and Costa Rica and a second, smaller one, in the lower Gulf of California, Mexico. However, when considering only species that occur exclusively on either coral or rocky reefs (including those on islands), the major peak in species richness is located on the south-eastern tip of the Baja California Peninsula, with secondary high-diversity areas along the south-western coast of the Mexican mainland and in the Costa Rica and Panama area. When considering the subset of species found on coral reefs only, the peak in species richness is located in the Gulf of Chiriquí, Panama, where coral reefs are most abundant (Maté 2003; Robertson and Allen 2006).

7.4 Reef-Associated Communities and Habitats

Several interdependent communities and habitat niches make up coral reef ecosystems. In this section we discuss two communities that are integral parts of ETP coral reefs: zooplankton and the cryptofauna. Zooplankton inhabits coral reef structures closely (emergent or demersal zooplankton, see below) as well as the overlying water column, and is a necessary food source for corals and other reefs organisms. Numerous reef species also contribute importantly to plankton communities during early stages of dispersal and immigration (settlement events). Some species of zooplankton have been found only in reef environments, suggesting that reefs may harbor a distinct zooplankton assemblage whose functional and energetic roles are still poorly understood. The reef cryptofauna is composed of a diverse assemblage of organisms that dwell within reef frameworks and sediments, and are therefore rarely visible on the reef's surface. These assemblages are just beginning to be studied in the ETP. In addition, we summarize present knowledge on soft benthic environments, rhodolith beds, and mesophotic environments neighboring coral reefs. Most of the sediment in reef-associated soft benthic environments in the ETP comes from freshwater run-off and from the coral reefs themselves, mainly as a product of bioerosion. The organisms living in these sediments are important sources of food for some reef fishes. Rhodolith beds are understudied habitats commonly associated with ETP coral reef ecosystems, and are important nursery grounds for several reef species. Much like rhodolith beds, mesophotic environments are very poorly known, and contrary to other ocean regions, there is little evidence of a strong connection between shallow and deep reef habitats.

7.4.1 Reef Zooplankton

Eastern tropical Pacific coral reefs are typically species poor relative to other reef regions, and some do not occur in nutrient poor waters. Although the roles of zooplankton in coral reef communities are not yet well known (Roman et al. 1990), zooplankton represents an important trophic link between primary producers and higher trophic levels, including scleractinian corals and fishes (Porter 1974; Robichaux et al. 1981; Hamner et al. 1988; Sorokin 1990; Sedberry and Cuellar 1993).

Emery (1968) established the term "emergent zooplankton," to refer to the zooplankton living over and within reef substrates. This plankton remains associated with the benthos during the day and migrates into the water column at night (Ohlhorst 1982). A complex and dynamic mix of pelagic zooplankton, eggs and larvae may also be found in the water above reef habitats (Carleton et al. 2001). The abundance and seasonality of coral reef zooplankton have been linked to various environmental conditions (Glynn 1973; McWilliam et al. 1981), reef morphology (Lefevre 1984), and habitat patchiness (Greenblatt 1982). The presence of some groups has been correlated with differences in substrate composition (Alldredge and King 1977; Porter and Porter 1977; Birkeland and Smally 1981), lunar cycles (Alldredge and King 1980), and predation intensity (Holzman et al. 2005; Motro et al. 2005; Yahel et al. 2005).

Information concerning the horizontal distribution of zooplankton is scarce (Morales and Murillo 1996). Some studies, however, have demonstrated the importance of the net import of zooplankton to reef ecosystems (Ferraris 1982). Others have described the differences in abundance, composition and behavior between the diverse community of resident, reef-dwelling zooplankton and the community occupying waters outside of reefs (Echelman and Fishelson 1990; Carleton 1993). Nevertheless, quantitative data on variations in the abundance, type, and distribution of coral reef zooplankton on spatial scales of centimeters to meters and temporal scales of one to several hours are rare (Jacoby and Greenwood 1989; Yahel et al. 2005). Early studies concerned with the abundance, diversity, flux and fate of zooplankton as it approaches and crosses a windward coral reef suffered from the limitations of traditional sampling procedures in reef environments and a poor understanding of fine scale physical oceanographic processes near reefs (Carleton 1993). More recent methods used for the study of zooplankton associated with coral reefs, including nets, benthic traps, core samplers, bags, video footage and suction devices (Heidelberg et al. 2004), have allowed for a better understanding of zooplankton dynamics. Although each method provides information about a specific subset of the zooplankton community, such as the pelagic or demersal component (Alldredge and King 1977), each has limitations (reviewed by Sorokin 1990).

Compared to studies on the Great Barrier Reef and in the Caribbean Sea (Heidelberg et al. 2004), information on zooplankton associated with coral reef ecosystems in the ETP is limited (Table 7.13). Studies in Pacific Panama have quantified the important role of zooplankton as a source of food for corals. This work suggested that coral feeding is facultative and that rates can vary widely (Palardy et al. 2005, 2006). Wellington (1982a) concluded that corals were largely phototrophic, but some species under shading could compensate for the reduction in phototrophic nutrition by feeding on plankton. Grottoli and Wellington (1999) determined the effects of light and zooplankton on skeletal $\delta^{13}C$ values in the ETP corals *Pavona clavus* and *Pavona gigantea*. They found that decreases in light or increases in zooplankton abundances resulted in decreases in skeletal $\delta^{13}C$ levels, thus establishing a preliminary basis for reconstructing past environmental changes in irradiance and zooplankton prevalence using coral records.

Work off the Pacific coast of Colombia, including areas encompassing coral reefs, has examined the taxonomy, spatial distribution, and abundance of ichthyoplankton, plankton productivity and composition, as well as the abundance and diversity of hyperiid amphipods (Giraldo et al. 2006; Escarria et al. 2007; Valencia and Giraldo 2009). In this region, copepods (56.8 % of spp.) and appendicularians (16.9 % of spp.) were the dominant taxa. Also collected were 35 species and 14 families of fish larvae, as well as 20 species of amphipods.

Several studies are available on the reef zooplankton from Costa Rica's Pacific coast. Guzmán and Obando (1988) reported on the composition of zooplankton from coral reefs at Isla del Caño. In this study, copepods were the dominant group, accounting for 40–70 % of the total zooplankton abundance. These workers also found that the abundances of other groups changed at night, in some cases reaching higher levels than the copepods. A detailed study of the isopod *Excorallana tricornis occidentalis* from the same samples indicated a nocturnal migration into the water column and changes in density throughout the year (Guzmán et al. 1988). In coral reef studies at Bahía Culebra, Costa Rica, copepods were numerically dominant among the zooplankton sampled with traps over colonies of *Pavona clavus* and *Pavona gigantea* (Morales-Ramírez, unpub. data). In this study a new monstrilloid copepod was discovered, namely *Cymbasoma concepcione* (Suárez-Morales and Morales-Ramírez 2003).

The first synthesis of zooplankton at Isla del Coco was published by Morales-Ramírez (2008). He reported 136 species, including the species of ichthyoplankton previously reported by Fernández-Leiva (1996). Three coral reefs at Isla del Coco were sampled between 2009 and 2012, and copepods were again the dominant group, with abundances accounting for between 56 and 75 % of the total number of individuals collected, including two new species of monstrilloid copepods (Suárez-Morales and Morales-Ramírez 2009). Other groups such as chaetognaths and larvaceans could, depending on water temperature and food availability, exhibit abundances similar to those of copepods (Esquivel-Garrote 2015). The emergent zooplankton associated with coral reefs was captured with nets and traps during vertical migration to the surface in the evening, and a change in the population structure of zooplankton was observed on these reefs according to the time of day (Esquivel-Garrote 2015).

Studies on the ecology and relationships of the demersal zooplankton among corals and reef communities in the ETP are few compared to other regions; therefore, more research is needed on the diversity, distribution, composition, abundance and trophic role of zooplankton on coral reefs in the ETP region.

7.4.2 Cryptofauna

The term cryptofauna refers to animals that shelter within reef frameworks (Ginsburg 1983; Kobluk 1988; Glynn and Enochs 2011) (Fig. 7.9). Their shelters are structurally diverse, ranging from large framework cavities to thin cracks under loose rubble fragments, and from voids between coral

Table 7.13 Coral reef zooplankton studies in the ETP

Site	Country	Sampling method	Depth (m)	Dominant groups	Study objective
Isla del Caño[1]	Costa Rica	Net (280 μm)	Surface	Copepoda	Diversity and abundance of zooplankton
				Mysidacea	
				Crab zoea	
Isla del Coco[2]	Costa Rica	Net (200 μm-500 μm)	Surface	Copepoda	Diversity and abundance of zooplankton
				Chaetognatha	
				Appendicularia	
Isla Contadora[3]	Panama	Bucket	1–6	Isopoda	Corals feeding on zooplankton
				Crab Zoea	
				Amphipoda	
Isla Contadora[4]	Panama	Net (50 μm)	0–6	Copepoda	Corals feeding on zooplankton
				Isopoda	
				Amphipoda	
Isla Gorgona[5]	Colombia	Net (250 μm)	0–50	Ichthyoplankton	Ichthyoplankton taxa
Isla Gorgona[6]	Colombia	Net (250 μm)	0–50	Copepoda	Plankton productivity
				Appendicularia	
Punta Cruces, Cabo Marzo[7]	Colombia	Net (250 μm)	0–50	Amphipoda	Abundance and diversity of amphipods

Sources [1]Guzmán and Obando (1987), [2]Esquivel-Garrote (2015), [3]Palardy et al. (2005), [4]Palardy et al. (2006), [5]Escarria et al. (2007), [6]Giraldo et al. (2006), [7]Valencia and Giraldo (2009)

Fig. 7.9 Collage of cryptofauna extracted from coral reef frameworks and rubble in the Gulf of Chiriquí, Panama. Photographs by Ian C. Enochs

branches to burrows into mounding coral colonies. Compared to the reef surface the cryptic habitat generally experiences reduced irradiance and water flow, which can influence the distribution of both photosynthetic flora (Kobluk 1988) and sessile suspension feeding fauna alike (Buss and Jackson 1981; Richter et al. 2001). This can, in turn, affect the distribution of a suite of motile cryptic herbivores and invertivores, respectively. The cryptic habitat is also unique in that it confers upon its occupants a degree of protection from large surface predators (Bakus 1966), though many of the cryptofauna may be predators themselves (Glynn 2006, 2013; Enochs 2012).

The animals that occupy the cryptic reef habitat are incredibly diverse, accounting for the majority of reef metazoan biodiversity (Reaka-Kudla 1997). Despite this, the cryptofauna community is understudied relative to surface reef fauna. This holds true for the ETP, though structurally complex pocilloporid reefs and associated rubble fields provide ideal habitats for a diverse cryptofauna assemblage. Investigation of ETP cryptic reef fauna, ecology and biodiversity has focused primarily on associates of living *Pocillopora damicornis* (e.g., Abele 1976; Alvarado and Vargas-Castillo 2012). Recent work (Glynn 2006, 2011; Enochs et al. 2011; Enochs 2012; Enochs and Manzello 2012a), however, has expanded investigation to the multi-phyletic assemblage associated with dead coral, reef frameworks, and rubble. Cryptic communities associated with each of these habitats have been shown to be diverse. For example, Abele (1979) collected more than 55 species of decapods from living *Pocillopora* colonies in the Pearl Islands, Panama and 37 species from colonies at Uva Island, Panama. Glynn (2006) also collected 23 species of cryptic fishes from Uva Island. Alvarado and Vargas-Castillo (2012) identified 35 species from 448 cryptic individuals collected from *P. damicornis* at Playa Blanca, Bahía Culebra, Costa Rica. Enochs and Manzello (2012a) collected more than 289 cryptic species from live and dead coral at Playa Larga, Isla Contadora, in the Gulf of Panama.

7.4.3 Soft Benthic Environments

Coral reefs are one of the main sources of carbonate sediments in soft sandy benthic environments in the ETP. These sediments are the product of erosive processes, both by abiotic factors and by eroding invertebrates (e.g., echinoids, annelids, molluscs and crustaceans) and fishes (e.g., Scaridae, Tetraodontidae, and Balistidae) (Glynn 1997). The sand produced creates intertidal and subtidal soft benthic environments (adjacent to or within the coral reef), where a characteristic benthic fauna lives (Westheide 1991; Sibaja-Cordero et al. 2013). Few studies have been carried out on the benthic fauna of sandy bottoms on or near coral reefs of the ETP.

The meiofauna (metazoans in the size range 63–500 μm) of the Galápagos Islands was studied at 35 intertidal sites composed of carbonate sands (Ax and Schmidt 1973; Schmidt 1978; Westheide 1991). Westheide (1991) recorded 390 species of meiofauna from the archipelago. While the Platyhelminthes were highly species rich (79 spp.) and abundant (50 % of indiv.), other groups, including polychaetes (100 spp.), molluscs and peracarids, were also prominent. Poorly known groups were the Gastrotricha (23 spp.), Acoelomorpha (16 spp.), Nemertea (7 spp.), Gnathostomulida (4 spp.) and Kinorhyncha (1 sp.). Meiofaunal abundances were found to be highest in the warmer equatorial months (Westheide 1991).

Guzmán et al. (1987) studied the dynamics of meiofaunal groups in the sediments at Isla del Caño, Costa Rica. In this study the most abundant groups were the Foraminifera (21.2 %), Copepoda (19.7 %), Nematoda (19.1 %) and Gastropoda (16.5 %). Community composition shifted with season, likely due to the input of organic matter during the rainy months. The highest diversity was observed in heterogeneous carbonate sediments. No other studies on meiofauna are presently known from coral reefs in the ETP.

Studies of reef sediment-dwelling macrofauna (animals >500 μm) are also limited. Examples include studies on the taxonomy of polychaetes from the sublittoral sandy substrates of the coral reefs at Mali Rock and Granito de Oro, Coiba Island, Panama (Aguado and López 2003; Aguado and San Martín 2004, 2006; López et al. 2005). This fauna includes nine species of the family Paraonidae, including one new species (Aguado and López 2003), four species of the family Pisionidae, including one new species (Aguado and San Martín 2004), 19 species of the family Syllidae, including a probable new species (Aguado and San Martín 2006), and nine species of the family Capitellidae, including one new species (López et al. 2005).

Other studies of the macrofauna associated with coral colonies have been performed in the Las Perlas Archipelago, Panama. For example, Mair et al. (2009) reported 201 invertebrate taxa from the sand and sandy-shell bottoms (<30 m depth) near coral colonies. The crustacean families Ampeliscidae, Anthuridae, Ogyridae, Processidae, Pasiphaeidae, and Squillidae were prominent with higher densities in mixed sediments rather than in other benthic substrates. The annelids of the families Nepthyidae, Lubrinereidae, Orbiniidae and Goniadidae, and bivalves of the family Tellinidae were dominant in calcareous sands (Mair et al. 2009).

Valencia et al. (2014) sampled the soft bottom benthos (10 m depth) at Gorgona Island, Colombia, and collected 73 species belonging to 58 families, primarily polychaetes and crustaceans. The sediments contained more carbonates (>50 %) at the stations near the dead Remanso coral reef and the Playa Blanca live reef. Other stations near the live coral reefs of Azufrada, Farrallones and Planchón contained less than 9 % carbonates. The species composition of the benthic fauna indicated that stations in close proximity to coral reefs were different from those located elsewhere. Communities near the dead reef were unique with coarse sediments dominated by both suspension (Ampeliscidae, amphipods) and deposit (Spionidae, polychaete worms) feeders. A previous study by Cortés FA et al. (2012) demonstrated similar patterns with high densities and species richness at El Remanso, Gorgona Island.

Sibaja-Cordero (2012) recorded 267 taxa in sediments primarily produced by the erosion of reef carbonates (<80 m depth; Sibaja-Cordero et al. 2014) at Isla del Coco, Costa Rica. These macrofauna were evidently dependent on detrital trophic inputs from nearby coral reefs and insular terrestrial sources. The stations with sand substrates at the coral reefs of Chatham Bay revealed 18–34 taxa, at Weston Bay 24–26 taxa, and at Wafer Bay 7–31 taxa. Community composition changed with depth, with shallow reef substrates dominated by surface deposit and suspension feeders, such as the polychaete *Magelona californica* and the bivalve *Gouldia californica*, respectively (Sibaja-Cordero 2012). Isla del Coco has a marked thermocline around 50 m depth; at this depth Sibaja-Cordero et al. (2012) observed a reduction in the diversity of polychaetes and other changes in the species composition of the sediments.

Finally, the macrofauna associated with the carbonate sediments of Isla del Coco demonstrated a high species richness of molluscs compared with other soft bottom habitats near the Costa Rican mainland. Sibaja-Cordero et al. (2013) reported 40 species of molluscs from Cocos Island, including 11 new records and one new species of Caecidae (*Caecum cocoensei* Sibaja-Cordero et al. 2014). A total of about 180 species of macrobenthos inhabits the sediments at Isla del Coco (Sibaja-Cordero 2012).

7.4.4 Rhodolith Beds

Rhodolith or mäerl beds (Fig. 7.10a, b) are associated with rocky (Foster 2001; Foster et al. 2007) and coral reefs (Littler and Littler 2008) in the ETP. These communities are made up of aggregations of free-living, non-geniculate coralline red algae (Woelkerling 1988; Foster et al. 2013). Individual rhodoliths usually grow around a rock, coral, shell or any other element that serves as a nucleus for their development. They are composed of erect calcified branches (Fig. 7.10b) that usually grow in a consistent number of concentric cellular layers (McConnnico et al. 2014). Rhodoliths display a wide variety of growth forms ranging from fruticose to lumpy and are discoid or irregularly shaped, averaging approximately six cm in diameter (Riosmena-Rodriguez et al. 1999). Individual rhodoliths, as well as rhodolith communities, can be composed of a single species or a combination of species (Yabur-Pacheco et al. 2007; Steller et al. 2009; Riosmena-Rodríguez et al. 2010; Hernández-Kantún et al. 2014).

Rhodolith beds have been known in the ETP (Fig. 7.10a) since the pioneering taxonomic studies of Dawson (1960) in the Gulf of California. In recent years, rhodolith beds have been observed in other regions of the ETP including Panama (Littler and Littler 2008; Reijmer et al. 2012), Costa Rica (Solano-Barquero 2011; Robinson et al. 2013), the Pacific coast of Baja California (Ávila and Riosmena-Rodríguez 2011; Riosmena-Rodríguez et al. 2012; Robinson et al. 2013; Hinojosa-Arango et al. 2014), the Galápagos Islands (Halfar and Riegl 2013), Nicaragua (Robinson et al. 2013), and the tropical Mexican Pacific coast (Peralta-García and Rosas-Alquicira 2014).

Rhodolith beds are ecologically important because their complex three-dimensional structure provides microhabitats and shelter that sustain a rich diversity of algae, invertebrates, and fishes, some of which are commercially important or endemic (Steller et al. 2003; Hinojosa-Arango and Riosmena-Rodríguez 2004; Solano-Barquero 2011). Reyes-Bonilla et al. (1997) reported the presence of five species of unattached hermatypic corals, and James et al. (2006) described free-living bryozoans within rhodolith communities in the Gulf of California. New records (Hernández-Kantún et al. 2010; Solano-Barquero 2011) and new species of exclusively rhodolith-associated organisms (Cho and Riosmena-Rodríguez 2008) have been reported. Solano-Barquero (2011) has documented the presence of 145 morphospecies of algae and animals associated with rhodolith beds at Isla del Coco National Park, Costa Rica. Rhodolith beds also provide a habitat for the recruitment of hundreds of species (Steller and Cáceres 2009; Amado-Filho et al. 2010; Solano-Barquero 2011), as evidenced from diverse larvae collected in these communities by Riosmena-Rodríguez and Medina (2010).

7.4.5 Mesophotic Environments

The mesophotic or 'twilight zone' has been defined as water depths between 30–40 and 150 m (Menza et al. 2008; Kahng et al. 2014). Mesophotic coral reefs (MCR) or mesophotic coral ecosystems occur throughout the tropics (Hinderstein et al. 2010; Kahng et al. 2010, 2014; Bridge et al. 2013). In the ETP, however, MCRs or prominent zooxanthellate coral communities have not been observed at coastal sites. The deepest ETP zooxanthellate corals typically occur between 10 to 30 m depth, near the thermocline where they can form refuge populations (see Chap. 17, Smith et al.). This absence of MCRs is not due to a lack of exploration since deep areas in the Galápagos Islands (Cairns 1991a), Costa Rica (Cortés and Blum 2008; Starr et al. 2012) and Panama (Cunningham et al. 2013) have been surveyed. In the Galápagos Islands and at Isla del Coco, Costa Rica, only azooxanthellate ahermatypic corals have been observed in deeper habitats (Cairns 1991a; Cortés and Blum 2008). Nonetheless, the occurrence of abundant coral populations at some oceanic island sites, such as Clipperton Atoll (Glynn et al. 1996) and Easter Island (Glynn et al. 2003), to depths of 30–60 m, are worth further scrutiny. At mesophotic depths in the Galápagos Islands, kelp forests

Fig. 7.10 **a** Rhodolith bed at 25 m depth, Isla del Coco, Costa Rica, in 2012; **b** closeup view of rhodoliths, Bahía Magdalena, Mexico. Photographs by Cindy Fernández-García (**a**) and Rafael Riosmena-Rodríguez (**b**)

(*Eisenia galapagensis*) have been documented at some sites (Graham et al. 2007), but have not been found in other areas of the ETP. Recently, a mesophotic octocoral (*Eugorgia siedenburgae*), not associated with euphotic coral reefs, was described from Panama (Breedy and Guzmán 2013).

In other coral reef regions there is a gradual transition of species assemblages from shallow to deep reefs (Kahng et al. 2014), but not in the ETP. At Isla del Coco, Sibaja-Cordero et al. (2012) reported an abrupt separation of the deep (>50 m) and shallow fauna. These workers examined polychaete diversity in soft sedimentary enviroments and demonstrated a difference in species composition and even of families above and below 50 m depth. The possible difference between the ETP mesophotic area and other regions is the characteristic permanent shallow thermocline at roughly 50 m depth (Fiedler and Talley 2006; see Chap. 3 Fiedler and Lavín). This shallow thermocline often results in water-quality conditions at mesophotic depths being much different than those in shallow water. In the ETP, mesophotic depths often have marked declines in temperature and pH, while nutrients are substantially elevated (D'Croz and O'Dea 2007; Manzello et al. 2008; Manzello 2010a). More studies are necessary to better document and understand these regional differences, and to guage the level of connectivity between shallow and deep habitats in the ETP.

7.5 Factors Affecting Biodiversity

7.5.1 Biological Effects

7.5.1.1 Live Coral

In a study of fish communities spanning three decades and highly dynamic fluctuations in benthic coral cover at Uva Island, Gulf of Chiriquí, Panama, Glynn et al. (2014) recorded a parabolic relationship between percent live coral and the species richness of reef-associated fishes. The highest species richness occurred at roughly 20 % live coral cover. This relationship between fish species richness and benthic cover is likely related to the diversity of available habitat and nutritional resources. Dominici-Arosemena and Wolff (2006) analyzed fish communities across a suite of rocky and coral reef habitats, also in the Panamanian Gulf of Chiriquí. Similar to the Glynn et al. (2014) study, they recorded a correlation between fish community richness and substrate heterogeneity, but did not find a significant relationship between coral cover and species richness. Live coral, therefore, provides important shelter and food for a suite of species such as corallivores. Other species utilizing different feeding strategies (e.g., herbivores, non-corallivorous invertivores) are more dependent on other benthic communities (e.g., algae, sponges) and are more often present when their food sources are as well. It should be noted that while high coral cover is not necessary for and may even inhibit highly diverse reef fish communities, it is ultimately responsible for habitat formation and its structure, independent of whether it is alive or dead, and is therefore vital for supporting diverse reef flora and fauna.

Similar to fish communities, there are numerous cryptic invertebrates that are obligate associates of live coral. Perhaps the most conspicuous and well studied of these are the coral guard crabs (*Trapezia* spp.) and the snapping shrimp (*Alpheus lottini*) that associate with live pocilloporid corals. These species are known to defend aggressively their host corals from corallivores such as the crown-of-thorns seastar (*Acanthaster planci*) (Glynn 1976). These crustacean "guards" live in male/female pairs and actively exclude conspecific crabs (Castro 1978). The interactions of crabs and shrimps are complex, and evidence suggests that the association on a coral colony is orchestrated by the shrimps' adoption of the crab's behaviors, eventually resulting in a stable yet territorially-established community (Vannini 1985). Their aggressive behavior can also extend toward other cryptic coral inhabitants (Preston 1973; Abele and Patton 1976; Lassig 1977), though congeneric aggression may be less pronounced and less important in structuring species' distributions (Gotelli and Abele 1983). Nevertheless, it is possible that these behaviors may alter localized diversity through territorial exclusion.

The coral itself is perhaps of equal or greater importance in restricting the biodiversity of its cryptic associates. Corals employ a variety of defensive mechanisms (e.g., nematocysts, mucus) that make them inhospitable to some taxa that would otherwise shelter within their branches (Kirsteuer 1969; Lang and Chornesky 1990). Furthermore, the live coral habitat can be relatively homogeneous, providing relatively few unique niches and thereby limiting diversity (Tews et al. 2004). Enochs and Manzello (2012a) collected more than 289 species of cryptic fauna belonging to six phyla from live and dead coral substrates on a pocilloporid reef in the Gulf of Panama. They employed individual-based rarefaction to standardize sampling effort and fitted six different models to their data. While the community richness estimates of these models varied greatly, they consistently predicted more species associated with dead vs live coral substrates. Many of the species collected on live coral in their study were known obligate commensals, specially adapted to tolerate the defensive characteristics of their hosts and capitalize from the food sources they provide (Patton 1974, 1994). In a similar study conducted at Uva Island, in the Panamanian Gulf of Chiriquí, Enochs and Hockensmith (2009) experimentally induced coral mortality and monitored cryptofauna settlement and colonization relative to live corals. Communities settling on dead coral substrates were less similar than those settling on live corals, at both one and two years after mortality, indicating potentially higher species diversity in dead coral habitats (Enochs and Hockensmith 2009).

At greater distances from coral, the influence of live coral on cryptic community biodiversity becomes more tenuous. To investigate this relationship, Enochs et al. (2011) created cryptic reef habitats from baskets of dead reef framework and covered them with either live or dead coral substrates. They found no significant influence on community richness, diversity, or on the multivariate species assemblages of the underlying experimental habitat. By contrast, in other reef regions a significant positive correlation between coral cover and the generic richness of the more conspicuous benthic invertivore fauna has been established (Idjadi and Edmunds 2006). More investigation is necessary to determine to what extent and at what scales these relationships hold in the ETP.

7.5.1.2 Predation

In addition to influencing actual species numbers, predation plays a role in the behavior and physical location of the majority of the metazoan species on ETP reefs. While coral reefs are known for their great biodiversity and have often been described as the "rainforests of the sea," swimming over the surface of a pocilloporid reef will reveal at first glance fewer species than might be expected. Those species that are conspicuous often possess highly evolved anti-predatory traits such as venom, repellants, and hard skeletal features. This pattern highlights the high level of predation on reef surfaces, a factor that also strongly influences behavior (Castro 1978), and has driven many of the more "palatable" species into the cryptic recesses of reef frameworks and sediments (Bakus 1966). Indeed, the number of species (>289) of small invertebrates and fishes collected by Enochs and Manzello (2012a) from a reef framework on one reef at Isla Contadora in the Gulf of Panama is likely a gross underestimation of true metazoan biodiversity but an order of magnitude or more greater than that readily recognizable on the reef surface.

The influence of predation on species diversity is complex. Given especially high predation levels, or exceptionally vulnerable species, predation can lead to species elimination and decreased diversity. Glynn (1985) described a scenario where El Niño-related mortality of pocilloporid corals would make previously inaccessible and more desirable coral prey available for *Acanthaster* predators. Glynn hypothesized that *Acanthaster* would quickly consume these species and thereby reduce coral species diversity in that area. At lower abundances, however, *Acanthaster* has been hypothesized to increase the diversity of its coral prey. Porter (1972) measured coral diversity at two reefs with very low (1 indiv. per 50,000 m^2, Islas Secas) and moderate (1 indiv. per 50 m^2, Isla Uva) abundances of *A. planci*. He found that the moderate levels of *Acanthaster* were correlated with higher coral diversity and that while all coral species were present at both sites, the site with more predation had greater evenness; a higher relative abundance of the rarer species. While apparently contradictory, the aforementioned studies and hypotheses are both in keeping with the intermediate disturbance hypothesis (Connell 1978). High abundances of *Acanthaster* can cause widespread mortality and ultimately reduce species numbers whereas the absence of disturbance from *Acanthaster* can result in homogeneous stands of fast-growing and competitively dominant coral species such as *Pocillopora*. Infrequent and patchy consumption of coral provides substrate for less common and slower-growing species to become established and grow, thereby increasing site biodiversity.

7.5.1.3 Competition

Competition among species is a ubiquitous feature of reefs (e.g., Lang and Chornesky 1990) and manifests from the common utilization of resources such as food and shelter. For instance, the damselfish *Stegastes acapulcoensis* aggressively defends territories that it establishes in protected shallow reef environments (Wellington 1982b). Once established, these fish will chase away other fishes and exclude sea urchins from the immediate vicinity (Glynn 1988; Eakin 1996). Additionally, *S. acapulcoensis* will preferentially bite at flat coral colonies as it expands and maintains the algal lawn from which it feeds. This can eventually lead to colony mortality and can restrict the distribution of pavonid corals to deeper regions where the damselfish are less abundant (Wellington 1982b). In this manner, territorial exclusion and mortality can lower biodiversity at small scales, among and within zones on a given reef. By contrast, studies from the western Pacific have shown that damselfishes can increase the diversity of algae within their territories (Hixon and Brostoff 1981; Sammarco 1983) and enhance cryptofauna populations (Klumpp et al. 1988).

We have already mentioned the aggressive territorial nature of some species of live coral associates, and this strong affinity for their coral hosts may be rooted in both nutritional requirements (Stimson 1990) and escape from predation (Castro 1978). Huber and Coles (1986), however, presented evidence that the aggressive behavior of trapeziid crabs often results in space occupation greatly in excess of their nutritional demands. As such, the subtle influences of competition on ETP reef biodiversity may be inextricably linked to other ecological relationships and difficult to identify. Further research is therefore needed to examine the degree to which the utilization of a common resource structures species distributions and biodiversity on reefs in the ETP.

7.5.2 Physical Effects

7.5.2.1 Substrate Structure

Substrate structure is likely one of the most studied physical influences on reef community biodiversity, and this holds true in the ETP. Nowhere is this relationship more apparent than among the cryptofauna (>2 mm) that associate directly with their substrate. To investigate experimentally the relationship between substrate porosity and cryptic community diversity, Enochs et al. (2011) created artificial reef frameworks from pocilloporid rubble. Replicate framework habitats were created from two different sizes of rubble, resulting in two treatments: low-porosity/small-rubble and high-porosity/larger-rubble. Cryptofauna community richness was higher in low porosity replicates where higher numbers of individuals were collected, likely due to the greater density of substrate in the same volume of habitat. Comparison of diversity indices (Fisher's α, Shannon's H′), however, revealed a more diverse community associated with high porosity frameworks. These habitats likely contained a greater variety of shelters, more diverse sessile food sources, and allowed organisms with larger body sizes to settle within them.

Enochs and Manzello (2012a) collected naturally occurring cryptofauna from the pocilloporid skeletal habitat along a continuum of erosion and degradation, from intact solid framework to thin veneers of rubble. In order to standardize measurements across these various substrates and the differing sampling procedures, the authors used both rarefaction and non-parametric richness estimators to calculate total community richness at infinite sampling effort. Undisturbed framework sheltered the least species rich community, had the most homogeneous living space, and was covered with a relatively uninterrupted layer of live coral. Areas with higher levels of erosion and lower coral cover had incrementally more species of cryptofauna associated with them, likely due to more diverse and heterogeneous habitat.

Glynn (2006) examined fish communities, both cryptic and surface dwelling, and recorded more species associated with artificial substrates of greater structural complexity and higher relief. This study represents an analysis of communities associated with more extreme levels of habitat erosion than that in the aforementioned studies (Enochs et al. 2011; Enochs and Manzello 2012a). Simulated reef structures were relatively low-relief and composed of fragments of coral rubble, whereas the degraded structure was flat, two-dimensional sand and small rubble fragments. As such, Glynn (2006) picks up where the Enochs and Manzello studies leave off, and ultimately lends insight into the complete loss of reef habitat and the coincident decline in community diversity (Enochs and Manzello 2012b).

Relationships between substrate structure and biodiversity also extend to communities that are not exclusively cryptic in nature. For example, Dominici-Arosemena and Wolff (2006) found a positive correlation between the species richness of reef-associated fishes and the benthic complexity of their coral and rocky reef habitats. These workers also considered aspects of the habitat's porosity and found a significant correlation between richness and the number, variability, and size of holes in the substrates. While it is likely that some of the fish species they considered exhibit cryptic tendencies, it is equally likely that numerous other species associate with these habitat characteristics because of more subtle and indirect relationships, for example pursuing cryptic prey.

7.5.2.2 Other Environmental Influences

The interaction of current flow and coral reef biodiversity is poorly understood. This relationship has been investigated in the Gulf of Panama using artificial reef frameworks deployed at two sites of different current speed and exposure (Enochs et al. 2011). Cryptofauna communities at the more exposed, high-flow site were more species rich and had greater biodiversity (Fisher's α, Shannon's H′) than those in the protected low-flow site. This could be due to a variety of factors, including access to a greater supply of pelagic larvae and the settlement of rare species (Palardy and Witman 2011) or to a greater diversity of prey items available to sessile filter-feeding species (Leviten and Kohn 1980).

Depth is correlated with coral species' distributions on ETP reefs, with relatively homogeneous stands of pocilloporid corals in shallow areas, giving way to more massive species in deeper regions (Wellington 1982a; Guzmán and Cortés 1989). This may, in part, be related to differential dependence on autotrophy versus heterotrophy among species, and the capability of massive coral morphologies to rely on particulate feeding to offset reduced light at depth (Wellington 1982a). It is cautioned, however, that nutritional requirements cannot explain this zonation alone and that both physical (e.g., wave action, tidal exposure, salinity) and biological factors (e.g., competition, predation) may also play a role (Guzmán and Cortés 1989). As previously mentioned, Wellington (1982b) has shown that the behavior of fishes plays a part in structuring this gradient, where massive species in shallow areas are more susceptible to damselfish-inflicted mortality and *Pocillopora* colonies in deeper waters are subjected to a greater degree of predation from corallivorous pufferfishes and grazing parrotfishes. Differences in communities associated with depth and reef zonation have also been observed in other ETP taxa. For example, Gotelli and Abele (1983), investigating crustacean-associates of living *Pocillopora*, observed statistically different species/area relationships between communities present on shallow reef flats versus the deeper reef flanks. These workers concluded, however, that this relationship was likely driven by the density of individuals, which was higher on the reef flat than in deeper reef zones.

7.5.2.3 Disturbance

Abele (1979) examined decapod associates of *Pocillopora damicornis* in the dynamic upwelling Gulf of Panama and the more stable nonupwelling Gulf of Chiriquí. Abele found higher species richness (55 vs. 37 species) in the thermally fluctuating gulf, largely due to the presence of generalists. By contrast, there were greater numbers of obligate coral associates in the thermally constant environment. Abele hypothesized that the higher biodiversity in the periodically upwelling Gulf of Panama was due to mortality of territorial and obligate symbionts, caused by low temperature exposure and subsequent disturbance of their coral food sources. This allowed for the subsequent colonization of opportunistic, non-obligate commensals and thereby increased regional diversity.

If environmental disturbances are too extreme or too frequent they can result in localized or even global extinction if the distribution of the species is limited to the extent of the disturbance. This may have been the case with *Millepora boschmai* in the ETP, which was reported to have been rendered extinct by the 1982–83 ENSO event at the time of its taxonomic description in 1991 (de Weerdt and Glynn 1991). Subsequent searches within its originally reported distribution revealed several colonies (Glynn and Feingold 1992), which were eventually killed in the subsequent 1997–98 ENSO (Glynn 2000). While specimens have since been discovered in Indonesia (Razak and Hoeksema 2003), no living colonies have been documented from the ETP, thereby supporting the suggestion that this represents a disturbance-induced regional extinction.

Enochs and Manzello (2012b) synthesized findings from previously published work (Enochs 2012; Enochs and Manzello 2012a) and created a conceptual model for how the species richness and trophic potential of cryptofauna would respond to coral mortality and erosion, charting community response to reef disturbance and habitat loss. They predicted increased cryptofauna richness following the disruption and partial mortality of previously undisturbed homogeneous stands of *Pocillopora*. Richness is expected to rise further still as dead frameworks are eroded, providing diverse habitats and substrates for benthic communities to colonize and develop. During this time, however, community abundances and trophic potential will likely decrease as important live coral food sources are lost and the quantity of habitat space declines. Ultimately, collapse in both richness and abundance is posited as reef habitat degrades into sand. The effective management of live coral populations is therefore critical for maintaining reef habitat and biodiversity alike (Enochs and Manzello 2012b).

7.5.3 Chemical Effects

The ocean's carbonate chemistry experiences changes due to a combination of biological and physical processes, as well as human-induced ocean acidification (OA). Any alteration of the balance of this system will impact marine biogeochemical cycles, ecosystems and their constituent organisms (Kleypas et al. 2006; Fabry et al. 2008; Doney et al. 2009; Feely et al. 2009; Kleypas and Yates 2009). Of the affected taxa, calcifying organisms are particularly sensitive to changes in pH and carbonate saturation state (Ω) (Kleypas et al. 2006; Kuffner et al. 2008; Dove et al. 2013; van Woesik et al. 2013). Besides the scleractinian corals, some calcifying reef organisms previously mentioned in this chapter (e.g., coralline algae and rhodoliths, polychaete worms, molluscs, crustaceans and echinoderms) can also be negatively impacted to different degrees by changes in seawater carbonate chemistry (Kleypas and Yates 2009; Ragazzola et al. 2012; Wittmann and Pörtner 2013; Fabricius et al. 2015; Gómez et al. 2015). Thus, community composition and diversity could well experience marked changes due to future OA conditions.

Due to the occurrence of seasonal upwelling events in the ETP (McCreary et al. 1989; Millero 2007), coral reefs from this region are naturally exposed to low-pH and low-Ω conditions (Manzello 2008, 2010a, b; Manzello et al. 2008; Rixen et al. 2012; see Chap. 18, Manzello et al.). This results in a potential disadvantage to calcification and, ultimately, for reef accretion as well. Coral reefs are widely recognized as the most threatened ecosystems by OA (Kleypas and Yates 2009) and it is expected that under the ETP's unusual biogeochemical conditions that reefs in this region will be strongly impacted by OA (Kleypas et al. 2006; Manzello et al. 2008; Veron 2008; Manzello 2010a). For example, accelerated carbonate dissolution combined with a reduction in calcification could preclude coral reef accretion within the next few decades (Hoegh-Guldberg 1999; Feely et al. 2004; Hoegh-Guldberg et al. 2007; Silverman et al. 2009). Furthermore, reef ecosystems and associated biota may experience the deleterious impacts of OA via numerous other pathways. For example, increased levels of CO_2 have been shown to cause branching corals to lose their zooxanthellae (Kaniewska et al. 2012), resulting in coral bleaching that could lead to massive mortality events (Anthony et al. 2008). By means of such effects, unfavorable carbonate chemistry could lead to a loss in the complexity of reef structures and a reduction in associated biodiversity, potentially leading to a decline in reef resilience (Kleypas and Yates 2009; Kroeker et al. 2011). Sensitivity and tolerance of marine organisms to OA will differ from one group to the next (Fabry et al. 2008; Gattuso et al. 2011; Kroeker et al. 2013), and these differences have the potential to trigger phase shifts in coral ecosystems (Hall-Spencer et al. 2008; Fabricius et al. 2011). Therefore, coral reefs as we know them today may disappear, leading to macroalgae- or octocoral-dominated communities instead of the current scleractinian coral ecosystems (Veron 2008; Silverman et al. 2009; Gómez et al. 2014; Johnson et al. 2014; see Chap. 11, Fong et al.).

In summary, OA could impact coral reef biodiversity in the ETP, not only by reducing the complexity of reef structures, but also by changing their community composition, including reef benthos and all other associated species. Published studies on this topic in the ETP are scarce, and the best-studied areas are in Panama and the Galápagos Islands (Manzello 2008, 2010a; Manzello et al. 2008), with an additional contribution from the Pacific coast of Costa Rica (Rixen et al. 2012). Nevertheless, there are several ongoing projects in the ETP, and it is expected that forthcoming findings will improve our knowledge of this issue.

7.6 ETP Reefs as a Model for Studying Biodiversity

The ETP is one of the most isolated ocean regions (Glynn and Ault 2000; Cortés 2011), separated from the Central Pacific by the East Pacific Barrier (Grigg and Hey 1992), and from the Caribbean after the uplift of the Central American Isthmus (Coates et al. 1992). The closure of the Central American Seaway resulted in extreme conditions with ENSO disturbance events and cold episodes due to seasonal upwelling (Cortés 1997, 2011). In addition, sea level fluctuations during glacial-interglacial periods, low pH, and cooling during the Little Ice Age have limited reef growth in the ETP (Glynn et al. 1983; Manzello 2010b; Cortés 2011). The contemporary ETP coral fauna derives mostly from the Central and Western Pacific (Cortés 1986; Glynn and Ault 2000; see Chap. 5, Glynn et al.). Coral reefs in the ETP are small, isolated, and constructed by a few coral species, and could be considered as the minimum expression of a structural reef (Cortés 1997), with some reefs constructed solely of one or two species (Cortés 2011). These reefs, however, harbor diverse and complex communities of associated organisms, have a long growth history, and occur adjacent to diverse and interdependent communities and habitats (Guzmán and Cortés 1993; Cortés et al. 1994; Glynn 2004; Cortés 2011).

ETP coral reefs are ideal systems for investigating biological and environmental factors influencing coral reef biodiversity. Relative to reef ecosystems in the Western Pacific and Caribbean, they are less diverse, with fewer described species of fishes, corals, molluscs, and crustaceans (Roberts et al. 2002). Given the enormity and complexity of highly diverse reef communities, this comparatively simple system can facilitate work that would not be feasible at other localities (e.g., Enochs and Manzello 2012a). Furthermore, there is a long history of taxonomic work in the ETP, for example, the expeditions of the US Fish Commission, New York Zoological Society, California Academy of Sciences, and the Allan Hancock Foundation (see Cortés 2008, for an example of the history of research at one ETP oceanic island). Additionally, there are taxonomic studies available for the identification of many taxa (e.g., Keen 1971; Wells 1983; Hickman 1998, 1999, 2008; Hickman and Zimmerman 2000; Breedy and Guzman 2002, 2007, 2011; Carballo et al. 2004; Allen et al. 2005; Robertson and Allen 2008; Breedy et al. 2009b; Littler and Littler 2010; Breedy and Cortés 2014). While excellent taxonomic work is currently ongoing in the region (e.g., Wehrtmann and Cortés 2009; Breedy and coworkers; Carballo and collaborators) and the biodiversity of some localities has received an exceptional amount of attention (e.g., the Galápagos Islands by C. P. Hickman), still other areas of the ETP remain open for further exploration and taxonomic attention.

Facilitating the investigation of factors influencing reef biodiversity is the morphology of many ETP reefs, namely those formed by pocilloporid corals. These reefs develop distinct zones that can be mapped and subsequently sampled in a stratified random manner (Eakin 1996; Enochs 2012). Substrates within these zones are relatively homogeneous compared to structurally diverse Caribbean and Western Pacific coral communities, allowing replicate sampling and zone-specific inference. Living pocilloporid corals and dead framework are relatively homogeneous but structurally complex, providing abundant shelter for diverse assemblages of associated cryptofauna.

Pocilloporid corals are present across the Pacific and have been used as model systems (especially *P. damicornis*) to study the ecology and diversity of coral-associated fauna (e.g., Grassle 1973; Kropp and Birkeland 1981; Black and Prince 1983; Gotelli and Abele 1983; Chang et al. 1987). Though especially prominent in the ETP, large clonal stands of *Pocillopora* and the resulting reef frameworks are known from other parts of the world, e.g., western Arabian Sea (Glynn 1993), Gulf of Aden (Kemp and Benzoni 1999), and western Australia (Miller and Ayre 2004). For these reasons research on the biodiversity of ETP reef communities is not limited in its applicability to the ETP; it may also be applied in some instances to systems across large geographic scales.

Finally, ETP reefs are distributed across pronounced gradients in nutrients, temperature, and carbonate chemistry (D'Croz and O'Dea 2007; Manzello et al. 2008; Cortés 2011; see Chap. 5, Glynn et al.). These naturally occurring conditions are closely related to predicted extreme conditions under climate change scenarios and can inform scientists on how anthropogenic stressors will likely structure the reefs of the future. Combined with the relatively depauperate nature of the reef-associated biota and the spatially homogeneous yet structurally complex framework habitat, ETP coral reefs are ideal natural laboratories for studying reef biodiversity and the factors that affect it.

7.7 Research Needs and Closing Remarks

While there are areas in the ETP that are presently considered to be more biotically diverse than others, more studies are necessary as the local species richness of select taxa is likely correlated with sampling effort (Cruz et al. 2003; Wehrtmann et al. 2009). The detailed study of particular groups can increase the known diversity of given sites significantly. For example, previously only nine species in six families of polychaetes were known from Isla del Coco (Dean et al. 2012). As a result of further sampling and analysis, at least 106 species in 35 families are known to occur there. Of these, 92 are new records for Costa Rica and 40 are new records for Central America (Dean et al. 2012). Even more species have been identified since (J.A. Sibaja-Cordero pers. comm. 2014). Additional inventories and compilations such as those conducted in Costa Rica (Wehrtmann and Cortés 2009) and Oaxaca, Mexico (Bastida-Zavala et al. 2013) are needed for a better understanding of the richness of the ETP region and to develop a basis to gauge expected changes in biodiversity. Furthermore, regional workshops should be instituted and supported to improve taxonomic capabilities in the ETP region.

Recent studies of taxonomically cryptic or sibling taxa have demonstrated that species richness is presently greatly underestimated (Knowlton 1993). Not recognizing the full diversity of a taxon can have important implications in the understanding of ecological function, biogeography, resilience, and conservation (Bickford et al. 2007). This issue is especially apparent in the ETP as evidenced by a recent discovery of cryptic species in the genus *Porites*, one of the principal reef-building coral genera (Boulay et al. 2014). What was formerly known as *P. lobata* is actually two species, *P. lobata* and *P. evermanni*, with similar colony morphology, but different reproductive modes and responses to bleaching stress events (Boulay et al. 2014).

Another area in need of further research is the biogeography of ETP marine organisms. As mentioned in the introduction to this chapter, several taxa have been studied, revealing a strong Indo-Pacific affinity. For example, in fishes, a relatively well-studied group, biogeographic inferences can be made with a high degree of confidence (Robertson et al. 2004; Robertson and Cramer 2009). This sort of analysis is not yet possible with other groups. Vega et al. (2012), for example, attempted to determine the biogeographic affinities of ETP sponges, but concluded that more studies are needed to establish species' distributions before geographic relationships can be determined. Biogeographic studies combined with genetic analyses (Hellberg 2009) will help determine the connectivity of populations, information which is critically important in management and conservation. In a detailed analysis of invertebrates from the Galápagos Islands, Hickman (2009) found that some groups are under-represented with respect to continental areas, while others are over represented. He attributed these differences to the relative isolation of the islands, taxon-specific dispersal capabilities, the spatial distribution of particular habitats, and patterns of ocean circulation. Additional research of this genre is needed to understand the distribution and ecology of the contemporary ETP biota.

In the future, biogeographic analyses of the ETP marine fauna should include Rapa Nui (Easter Island) and Salas y Gómez Island, which exist at the southeastern-most limits of coral reef distribution in the Pacific. Most ETP studies exclude these islands. In an updated review of 88 published articles and 10 web databases, Fernández et al. (2014) reported the occurrence at Rapa Nui of 964 species, including 143 algal species (56 % red algae), 605 marine invertebrate species (68 % crustaceans and molluscs) and 216 fish species (6 % chondrichthyans). Rapa Nui is characterized by a high degree of endemism. For example, >30 % of sponges and molluscs as well as >20 % of reef fishes are endemic to the region (Randall and Cea 2011), and examples of restricted ranges in other groups are reported as well (Fernández et al. 2014). This synthesis by Fernández et al. (2014) is an excellent starting point for biogeographic comparisons with the core ETP. A recent study by Veron et al. (2015) reinforces the validity of a connection between Rapa Nui and the remainder of the ETP. In their analysis of the affinity of zooxanthellate corals by ecoregions, Rapa Nui is closer to the core ETP region than is Clipperton Atoll, despite its closer geographic proximity.

Coral reefs at Rapa Nui are constructed predominantly by only two species (*Pocillopora verrucosa* and *Porites lobata*), which comprise nearly 80 % of the total benthic cover (Wells 1972; Glynn et al. 2003, 2007; Wieters et al. 2014). Wieters et al. (2014) reported consistent increases in coral abundance in shallow habitats at Rapa Nui over the past 15 years, since the ENSO bleaching/mortality event in 2000 (Wellington et al. 2001). Top-down effects on this recovery are unlikely because the naturally depauperate fish assemblages and high fishing pressure likely contribute to low fish abundances, small body sizes, and a skewed trophic structure (Friedlander et al. 2013). It is hypothesized that the reported higher abundance of sea urchins (*Diadema savignyi*) increases the likelihood of coral larval settlement, colonization and re-growth. The coral reefs at Rapa Nui are therefore likely resilient systems, potentially capable of maintaining normal ecosystem function in spite of perturbation (Wieters et al. 2014).

Coral reefs across the ETP are impacted by a variety of natural disturbances and anthropogenic stressors (see Chap. 8, Glynn et al.; Chap. 20, Cortés and Reyes-Bonilla). Such conditions are expected to negatively impact the biodiversity of ETP coral reefs. Accordingly, monitoring programs must be implemented to establish present-day baselines and to track changes in biodiversity over time. Presently, only zooxanthellate corals, and select fishes and macro-invertebrates are monitored in the ETP and, while commendable, this approach is lacking in scope (Cortés et al. 2010; Zapata et al. 2010; Glynn et al. 2014).

In closing we cite Rogers (2013), who has captured the essence of coral reef vitality and the possibility of long-term persistence: "The biodiversity of coral reefs is the basis for their remarkable beauty and for the benefits they provide to society. The extraordinary complexity of these ecosystems makes it both more difficult to predict their future and more likely they will have a future."

Acknowledgements We thank Peter W. Glynn and Derek P. Manzello for the invitation to contribute this chapter to the eastern Pacific coral reef book. J. Cortés thanks the Vicerrectoría de Investigación, Universidad de Costa Rica, for the internship received (2014) to work on this and other manuscripts. J.C. also acknowledges the extra help by Luis Hernández in reviewing the complete text, and helping with tables and figures. F.A. Zapata thanks D.R. Robertson, Smithsonian Tropical Research Institute (STRI) for providing access to his Online Information System on the Shore Fishes of the Tropical Eastern Pacific during a time when the website was closed for maintenance. K.L. Kaiser, P. Medina-Rosas and A. Hermosillo thank D.R. Robertson for allowing us to participate in the Expeditions aboard the *RV Urracá* to Panama, Costa Rica, El Salvador, Isla del Coco, Costa Rica, and Clipperton Atoll, sponsored by STRI, and, to Clipperton Atoll by the National Geographic Society (Grant NGS 5831–96); also our thanks to the Santa Barbara Museum of Natural History (SBMNH), Malacology Department, for use of their facilities. A special thanks to Patricia Sadeghian (SBMNH) for her assistance in photography. E. Wieters and A. Pérez-Matus thank the Fondo Nacional de Ciencia y Tecnología, Chile (Fondecyt # 1130167). We thank Peter W. Glynn, Derek P. Manzello and Michael J. Kramer for their detailed reviews and suggestions, and Benjamin Grassian for help with the final figures. Michael P.C. Fuller and Tyler B. Smith offered helpful editorial suggestions.

References

Abbott IA (1966) Elmer Yale Dawson (1918–1966). J Phycol 2:129–132

Abele LG (1972) Comparative habitat diversity and faunal relationships between the Pacific and Caribbean Panamanian decapod Crustacea: a preliminary report, with some remarks on the crustacean fauna of Panamá. Bull Biol Soc Washington 2:125–137

Abele LG (1974) Species diversity of decapod crustaceans in marine habitats. Ecology 55:156–161

Abele LG (1976) Comparative species richness in fluctuating and constant environments: coral-associated decapod crustaceans. Science 192:461–463

Abele LG (1979) The community structure of coral-associated decapod crustaceans in variable environments. In: Livingston RJ (ed) Ecological processes in coastal and marine systems. Plenum Press, New York, pp 265–287

Abele LG, Patton WK (1976) The size of coral heads and the community biology of associated decapod crustaceans. J Biogeogr 3:35–47

Aburto-Oropeza O, Balart EF (2001) Community structure of reef fish in several habitats of a rocky reef in the Gulf of California. PSZNI: Mar Ecol 22:283–305

Acuña FH, Cortés J, Garese A (2012) Occurrence of the sea anemone *Telmatactis panamensis* (Verrill, 1869) (Cnidaria: Anthozoa:

Actiniaria) at Cocos Island National Park. Costa Rica. Rev Biol Trop 60(Suppl 3):201–205

Acuña FH, Garese A, Excoffon AC, Cortés J (2013) New records of sea anemones (Anthozoa: Actiniaria) from Costa Rica. Rev Biol Mar Oceanogr 48:177–184

Aguado MT, López E (2003) Paraonidae (Annelida: Polychaeta) del Parque Nacional de Coiba (Pacífico, Panamá), con la descripción de una nueva especie de *Aricidea* Webster, 1879. Rev Chilena Hist Nat 76:363–370

Aguado MT, San Martín G (2004) Pisionidae (Polychaeta) from Coiba National Park with the description of a new species and two new reports of *Pisione*. J Mar Biol Assoc UK 84:73–79

Aguado MT, San Martín G (2006) Sílidos (Syllidae: Polychaeta) del Parque Nacional de Coiba. Rev Biol Trop 54:725–743

Alldredge A, King JM (1977) Distribution, abundance and substrate preference of demersal reef zooplankton at Lizard Island lagoon, Great Barrier Reef. Mar Biol 41:317–333

Alldredge A, King JM (1980) Effects of moon light on the vertical migration pattern of demersal zooplankton. J Exp Mar Biol Ecol 44:133–156

Allen G, Steene R, Humann P, DeLoach N (2005) Reef fish identification—tropical Pacific. New World Publications, Inc., Jacksonville, Florida, p 480

Alonso D, Ramírez L, Segura-Quintero C, Castillo-Torres P, Diaz JM, Walschburger T (2008) Prioridades de conservación in situ para la biodiversidad marina y costera de la plataforma continental del Caribe y Pacífico colombiano. Santa Marta, Colombia, INVEMAR, TNC, UAESPNN, p 20

Alvarado JJ, Solis-Marin FA, Ahearn C (2010) Echinoderm (Echinodermata) diversity in the Pacific coast of Central America. Mar Biodiv 40:45–56

Alvarado JJ, Aguirre-Rubí J, Ayala A, Buitrago F, Fernández-García C, Reyes-Bonilla H, Sánchez A (2011) Characterization of the coral communities of San Juan del Sur, south Pacific Nicaragua. Bull Mar Sci 87:129–146

Alvarado JJ, Vargas-Castillo R (2012) Invertebrados asociados al coral constructor de arrecifes *Pocillopora damicornis* en Playa Blanca, Bahía Culebra. Costa Rica. Rev Biol Trop 60(Suppl 2):77–92

Alvarado JJ, Baraza E, Sancho-Mejía T (2013) Central American echinoderms: diversity, ecology and future perspectives. In: Alvarado JJ, Solís-Marín FA (eds) Echinoderm research and diversity in Latin American. Springer, Berlin, pp 67–106

Alvarado JJ, Beita A, Mena S, Fernández-García C, Guzmán-Mora AG (2015) Ecosistemas coralinos del Área de Conservación Osa, Costa Rica: estructura y necesidades de conservación. Rev Biol Trop 63 (Suppl 1):219–259

Alvarez-Filip L, Reyes-Bonilla H, Calderón-Aguilera LE (2006) Community structure of fishes in Cabo Pulmo reef, Gulf of California. Mar Ecol 27:253–262

Alzate A, Zapata FA, Giraldo A (2014) A comparison of visual and collection-based methods for assessing community structure of coral reef fishes in the tropical eastern Pacific. Rev Biol Trop 62 (Suppl 1):359–371

Amado-Filho GM, Maneveldt GW, Pereira-Filho GH, Manso RCC, Bahia RG, Barros-Barreto MB, Guimarães SMPB (2010) Seaweed diversity associated with a Brazilian tropical rhodolith bed. Cienc Mar 36:371–391

Amador J, Alfaro EJ, Lizano OG, Magaña V (2006) Atmospheric forcing in the eastern tropical Pacific: a review. Progr Oceanog 69:101–142

Anthony KRN, Kline DI, Diaz-Pulido G, Dove S, Hoegh-Guldberg O (2008) Ocean acidification causes bleaching and productivity loss in coral reef builders. Proc Nat Acad Sci USA 105:17,442–17,446

Ávila E, Riosmena-Rodríguez R (2011) A preliminary evaluation of shallow-water rhodolith beds of Bahía Magdalena, Mexico. Brazilian J Oceanogr 59:365–375

Ávila E, Méndez-Trejo MC, Riosmena-Rodríguez R, López-Vivas JM, Senties A (2012) Epibiotic traits of the invasive red seaweed *Acanthophora spicifera* in La Paz Bay, south Baja California (eastern Pacific). Mar Ecol 33:470–480

Ax P, Schmidt P (1973) Interstitielle fauna von Galapagos. I. Einfiihrung. Mikrof Meeresb 20:1–38

Bailey-Brock JH, Emig CC (2000) Hawaiian Phoronida (Lophophorata) and their distribution in the Pacific region. Pac Sci 54:119–126

Bakus GJ (1966) Some relationships of fishes to benthic organisms on coral reefs. Nature 210:280–284

Barraza E (2000) Comentarios sobre la diversidad de macroinvertebrados marinos de El Salvador. Publicación ocasional Ministerio del Medio Ambiente y Recursos Naturales 2, p 15

Barraza E (2014a) Invertebrados marinos de El Salvador. Ministerio del Medio Ambiente y Recursos Naturales, San Salvador, El Salvador, p 96

Barraza E (2014b) Peces estuarinos y marinos de El Salvador. Ministerio del Medio Ambiente y Recursos Naturales, San Salvador, El Salvador, p 66

Barroso RM, Klautau AM, Solé-Cava Paiva PC (2010) *Eurythoe complanata* (Polychaeta: Amphinomidae), the 'cosmopolitan' fireworm, consists of at least three cryptic species. Mar Biol 157:69–80

Bastida-Zavala JR (1995) Poliquetos (Annelida: Polychaeta) del arrecife coralino de Cabo Pulmo-Los Frailes, B.C.S, Mexico. Rev Zool 6:9–29

Bastida-Zavala JR (2008) Serpulids (Annelida: Polychaeta) from the eastern Pacific, including a brief mention of Hawaiian serpulids. Zootaxa 1722:1–66

Bastida-Zavala JR, García-Madrigal M del S, Rosas-Alquicira EF, López-Pérez RA, Benítez-Villalobos F, Meraz-Hernando JF, Torres-Huerta AM, Montoya-Márquez A, Barrientos-Luján NA (2013) Marine and coastal biodiversity of Oaxaca, Mexico. Check List 9:329–390

Baums IB, Boulay JN, Polato NR, Hellberg ME (2012) No gene flow across the Eastern Pacific Barrier in the reef-building coral *Porites lobata*. Mol Ecol 21:5418–5433

Bautista-Romero J, Reyes-Bonilla H, Lluch-Cota DB, Lluch-Cota SE (1994) Aspectos generales sobre la fauna marina. In: Ortega-Rubio A, Castellanos-Vera A (eds) La Isla Socorro, reserva de la biósfera Archipiélago de Revillagigedo, México. Centro de Investigaciones Biológicas del Noroeste, La Paz, Baja California Sur, México, pp 247–265

Behrens DW, Hermosillo A (2005) Eastern Pacific nudibranchs: a guide to the opisthobranchs from Alaska to Central America. California, Sea Challengers, p 137

Behrens DW, Gosliner TM, Hermosillo A (2009) A new species of dorid nudibranch (Mollusca) from the Revillagigedo Islands of the Mexican Pacific. Proc California Acad Sci, Ser 4, 60:423–429

Bellwood DR (1996) The Eocene fishes of Monte Bolca: the earliest coral reef fish assemblage. Coral Reefs 15:11–19

Bellwood DR (1997) Reef fish biogeography: habitat associations, fossils and phylogenies. In: Proceedings of 8th International Coral Reef Symposium, vol 1, Panamá, pp 379–384

Benavides-Serrato M, Borrero-Pérez GH, Cantera J, Cohen-Renjifo M, Neira R (2013) Echinoderms of Colombia. In: Alvarado JJ, Solís-Marín FA (eds) Echinoderm research and diversity in Latin America. Springer, Berlin, pp 145–182

Benfield S, Baxter L, Guzmán HM, Mair JM (2008) A comparison of coral reef and coral community fish assemblages in Pacific Panama and environmental factors governing their structure. J Mar Biol Assoc UK 88:1331–1341

Bernard FR, McKinnell SM, Jamieson GS (1991) Distribution and zoogeography of the Bivalvia of the eastern Pacific Ocean. Canadian Spec Publ Fish Aquat Sci 112:60

Bickford D, Lohman DJ, Sodhi NS, Ng PKL, Meier R, Winker K, Ingram KK, Das I (2007) Cryptic species as a window on diversity and conservation. Trends Ecol Evol 22:148–155

Birkeland C, Smalley TL (1981) Comparison of demersal plankton from comparable substrates from a high island and an atoll. In: Proceedings of 4th International Coral Reef Symposium, vol 1, Manila, pp 437–442

Black R, Prince J (1983) Fauna associated with the coral *Pocillopora damicornis* at the southern limit of its distribution in Western Australia. J Biogeogr 10:135–152

Blake JA (1991) The polychaete fauna of the Galápagos Islands. In: James MJ (ed) Galápagos marine invertebrates. Topics in Geobiology. Plenum Press, New York, pp 75–96

Boschi E (2000) Species of decapod crustaceans and their distribution in the American marine zoogeographic provinces. Rev Invest Des Pesq 13:1–63

Boulay JN, Hellberg ME, Cortés J, Baums IB (2014) Unrecognized coral species diversity masks differences in functional ecology. Proc Roy Soc B 281:20131580. doi:10.1098/rspb.2013.1580

Breedy O (2001) A new species of *Pacifigorgia* from the eastern Pacific (Coelenterata: Octocorallia: Gorgoniidae). Bull Biol Soc Washington 10:181–187

Breedy O (2009) Octocorals. In: Wehrtmann IS, Cortés J (eds) Marine biodiversity of Costa Rica, Central America. Springer+Business Media, Berlin, Text pp 161–167, Species List: Compact Disc pp 108–111

Breedy O, Cortés J (2008) Octocorals (Coelenterata: Anthozoa: Octocorallia) of Isla del Coco, Costa Rica. Rev Biol Trop 56(Suppl 2):71–77

Breedy O, Cortés J (2011) Morphology and taxonomy of a new species of *Leptogorgia* (Cnidaria: Octocorallia: Gorgoniidae) in Cocos Island National Park, Pacific Costa Rica. Proc Biol Soc Washington 124:62–69

Breedy O, Cortés J (2014) Gorgonias (Anthozoa: Octocorallia: Gorgoniidae) de las aguas someras del Pacífico Norte de Costa Rica. Rev Biol Trop 62(Suppl 4):43–62

Breedy O, Guzman HM (2002) A revision of the genus *Pacifigorgia* (Coelenterata: Octocorallia: Gorgoniidae). Proc Biol Soc Washington 115:782–839

Breedy O, Guzman HM (2003) The genus *Pacifigorgia* (Octocorallia: Gorgonacea) in Costa Rica (Coelenterata: Octocorallia: Gorgoniidae). Zootaxa 281:1–60

Breedy O, Guzman HM (2004) New species of the gorgonian genus *Pacifigorgia* (Coelenterata: Octocorallia: Gorgoniidae) from the Pacific of Panamá. Zootaxa 541:1–15

Breedy O, Guzman HM (2007) A revision of the genus *Leptogorgia* Milne Edwards & Haime, 1857 (Coelenterata: Octocorallia: Gorgoniidae) in the eastern Pacific. Zootaxa 1407:1–90

Breedy O, Guzman HM (2008) *Leptogorgia ignita*, a new shallow-water coral species (Octocorallia: Gorgoniidae) from the tropical eastern Pacific. J Mar Biol Assoc UK 88:893–899

Breedy O, Guzman HM (2011) A revision of the genus *Heterogorgia* Verrill, 1868 (Coelenterata: Octocorallia: Plexauridae. Zootaxa 2995:27–44

Breedy O, Guzman HM (2012) A new species of *Leptogorgia* (Cnidaria: Anthozoa: Octocorallia) from Golfo Dulce, Pacific Ocean, Costa Rica. Zootaxa 3182:65–68

Breedy O, Guzman HM (2013) A new species of the genus *Eugorgia* Verrill, 1868 (Cnidaria: Octocorallia: Gorgoniidae) from mesophotic reefs in the eastern Pacific. Bull Mar Sci 89:735–743

Breedy O, Guzman HM (2014) A new species of alcyonacean octocoral from the Peruvian zoogeographic region. J Mar Biol Assoc UK 94:493–498

Breedy O, Hickman CP Jr, Williams GC (2009a) Octocorals in the Galapagos Islands. Galapagos Res 66:27–31

Breedy O, Guzman HM, Vargas S (2009b) A revision of the genus *Eugorgia* Verrill, 1868 (Coelenterata: Octocorallia: Gorgoniidae). Zootaxa 2151:1–46

Breedy O, Williams G, Guzman HM (2013) Two new species of gorgonian octocorals from the tropical eastern Pacific biogeographic region (Cnidaria, Anthozoa, Gorgoniidae). ZooKeys 350:75–90

Bridge TCL, Hughes TP, Guinotte JM, Bongaerts P (2013) Call to protect all coral reefs. Nature Clim Chan 3:528–530

Briggs JC (1974) Marine zoogeography. McGraw-Hill, New York, p 475

Bula-Meyer G (1995) Macroalgas de la Isla Gorgona (Pacífico Colombiano) con nuevos registros y una explicación de la baja diversidad y biomasa. In: Pinto P (ed) La Isla Gorgona, nuevos estudios biológicos. Editorial Guadalupe Ltda, Bogotá, Colombia, pp 23–45

Buss LW, Jackson JBC (1981) Planktonic food availability and suspension-feeder abundance: evidence of in situ depletion. J Exp Mar Biol Ecol 49:151–161

Bustamante RH, Wellington GM, Branch GM, Edgar GJ, Martinez P, Rivera F, Smith F, Witman J (2002) Outstanding marine features. In: Bensted-Smith R (ed) A biodiversity vision for the Galapagos Islands. Charles Darwin Foundation and World Wildlife Fund, Puerto Ayora. Galápagos, Ecuador, pp 60–71

Caballero-George C, Bolaños J, Ochoa E, Carballo JL, Cruz JA, Arnold E (2010) Protocol to isolate sponge-associated fungi from tropical waters and an examination of their cardioprotective potential. Curr Trends Biotech Pharm 4:881–899

Cairns SD (1986) Stylasteridae (Hydrozoa: Hydroida) of the Galapagos Islands. Smithsonian Contr Zool 426:1–42

Cairns SD (1991a) A revision of the ahermatypic Scleractinia of the Galápagos and Cocos Islands. Smithsonian Contr Zool 504:1–45

Cairns SD (1991b) New records of Stylasteridae (Hydrozoa: Hydroida) from the Galápagos and Cocos Islands. Proc Biol Soc Washington 104:209–228

Calder DR, Mallinson JJ, Collins K, Hickman CP (2003) Additions to the hydroids (Cnidaria) of the Galápagos, with a list of species reported from the islands. J Nat Hist 37:1173–1218

Camacho-García, Y (2009) Benthic opisthobranchs. In: Wehrtmann IS, Cortés J (eds) Marine biodiversity of Costa Rica, Central America. Springer+Business Media, Berlin, Text pp 371–386, Species List: Compact Disc pp 330–347

Capa M, López E (2004) Sabellidae (Annelida: Polychaeta) living in blocks of dead coral in the Coiba National Park, Panamá. J Mar Biol Assoc UK 84:63–72

Carballo JL, Cruz-Barraza JA (2006) The genus *Timea* (Porifera: Hadromerida) in the Eastern Pacific Ocean: revision of type material, new species and intraspecific variability. Cah Biol Mar 47:301–320

Carballo JL, Cruz-Barraza JA (2010) A revision of the genus *Mycale* (Poecilosclerida: Mycalidae) from the Mexican Pacific Ocean. Contrib Zool 79:165–194

Carballo JL, Cruz-Barraza JA, Gómez P (2004) Taxonomy and description of clionaid sponges (Hadromerida, Clionaidae) from the Pacific Ocean of Mexico. Zool J Linn Soc 141:353–397

Carballo JL, Gómez P, Cruz-Barraza JA (2014) Biodiversidad de Porifera en México. Rev Mexicana Biodiv, Supl 85:S143–S153

Carleton JH (1993) Zooplankton and coral reefs: an overview. SPUMS J 23:102–107

Carleton JH, Brinkman R, Doherty PJ (2001) Zooplankton community structure and water flow in the lee of Helix Reef (Great Barrier Reef, Australia). Mar Biol 139:705–717

Carlgren O (1940) Eastern Pacific expeditions of the New York Zoological Society. XIX. Actiniaria from the Gulf of California. Zoologica 25:211–219

Castilla JC, Rozbaczylo N (1987) Invertebrados marinos de Isla de Pascua y Sala y Gómez. In: Castilla JC (ed) Islas oceánicas chilenas: conocimiento científico y necesidades de investigación. Ediciones Universidad Católica de Chile, Santiago, pp 167–189

Castro P (1978) Movements between coral colonies in *Trapezia ferruginea* (Crustacea: Brachyura), an obligate symbiont of scleractinian corals. Mar Biol 46:237–245

Castro P (1982) Notes on symbiotic decapod crustaceans from Gorgona Island, Colombia, with a revision of the eastern Pacific species of *Trapezia* (Brachyura, Xanthidae), symbionts of scleractinian corals. An Inst Invest Mar Punta de Betín 12:9–17

Castroviejo S (1997) Flora y fauna del Parque Nacional de Coiba (Panama): inventario preliminar. AECI, Madrid, p 534

CDB-Panamá (2010) Cuarto Informe Nacional de Panamá ante el Convenio sobre la Diversidad Biológica. Autoridad Nacional del Ambiente, Panamá, República de Panamá, p 110

Chang KH, Chen YS, Chen CP (1987) Xanthid crabs in the corals, *Pocillopora damicornis* and *P. verrucosa* of southern Taiwan. Bull Mar Sci 41:214–220

Charpy L (ed) (2009) Clipperton, environnement et biodiversité d'un microcosme océanique. MNHN, Paris & IRD, Marseille. Patrimoines Naturels 68, p 420

Chávez-Hernández E, Bretado-Aguirre J (1990) Contribución al conocimiento de los gasterópodos macroscópicos litorales de Isla Socorro, Archipiélago de Revillagigedo, México. Mem VIII Simp Int Biol Mar, Universidad Autónoma de Baja California, Ensenada, México, pp 150–155

Chesher RH (1972) The status of knowledge of Panamanian echinoids, 1971, with comments on other echinoderms. Bull Biol Soc Washington 2:139–158

Chiriboga Á, Ruiz D, Banks S (2012) CDF Checklist of Galapagos sponges. In: Bungartz F, Herrera H, Jaramillo P, Tirado N, Jiménez-Uzcátegui G, Ruiz D, Guézou A, Ziemmeck F (eds) Charles Darwin Foundation Galapagos Species Checklist. Charles Darwin Foundation, Puerto Ayora, Galapagos. http://checklists.galadarwin.webfactional.com/marine-invertebrates/porifera/

Cho TO, Riosmena-Rodriguez R (2008) *Ceramium periconicum* sp. nov. (Ceramiaceae, Rhodophyta): a new subtidal species from Baja California Sur, Mexico. Bot Mar 51:307–312

Coates AG, Jackson JBC, Collins LS, Cronin TM, Dowsett HJ, Bybell LM, Jung P, Obando JA (1992) Closure of the Isthmus of Panama: the near-shore marine records of Costa Rica and western Panama. Geol Soc Amer Bull 104:14–28

Coe WR (1940) Revision of the nemertean fauna of the Pacific coasts of North, Central, and Northern South America. Allan Hancock Pac Exp 2:247–323

Connell JH (1978) Diversity in tropical rain forests and coral reefs. Science 199:1302–1310

Connell SD, Foster MS, Airoldi L (2014) What are algal turfs? Toward a better description of turfs. Mar Ecol Prog Ser 495:299–307

Coppard S, Alvarado JJ (2013) Echinoderm diversity in Panama: 146 years of research across the isthmus. In: Alvarado JJ, Solís-Marín FA (eds) Echinoderm research and diversity in Latin America. Springer, Berlin, pp 107–144

Cortés FA, Solano OD, Ruiz-López JA (2012) Variación espacio-temporal de la fauna macrobentónica asociada a fondos blandos y su relación con factores ambientales en el Parque Nacional Natural Gorgona, Pacífico colombiano. Bol Invest Mar Cost 41:323–353

Cortés J (1986) Biogeografía de corales hermatípicos: el istmo centroamericano. An Inst Cienc Mar Limnol, UNAM 13:297–304

Cortés J (1997) Biology and geology of eastern Pacific coral reefs. Coral Reefs 16:S39–S46

Cortés J (2008) Historia de la investigación marina de la Isla del Coco, Costa Rica. Rev Biol Trop 56(Suppl 2):1–18

Cortés J (2009) Stony corals. In: Wehrtmann IS, Cortés J (eds) Marine biodiversity of Costa Rica, Central America. Springer+Business Media, Berlin, Text pp 169–173, Species List: Compact Disc pp 112–118

Cortés J (2011) Eastern Tropical Pacific coral reefs. In: Hopley D (ed) The encyclopedia of modern coral reefs: structure, form and process. Springer, Berlin, pp 351–358

Cortés J (2012) Marine biodiversity of an eastern tropical Pacific oceanic island, Isla del Coco. Costa Rica. Rev Biol Trop 60(Suppl 3):131–185

Cortés J, Blum S (2008) Life to 450 m depth at Isla del Coco. Costa Rica. Rev Biol Trop 56(Suppl 2):189–206

Cortés J, Jiménez C (2003) Corals and coral reefs of the Pacific of Costa Rica: history, research and status. In: Cortés J (ed) Latin American coral reefs. Elsevier Science B.V, Amsterdam, pp 361–385

Cortés J, Macintyre IG, Glynn PW (1994) Holocene growth history of an eastern Pacific fringing reef, Punta Islotes, Costa Rica. Coral Reefs 13:65–73

Cortés J, der Hal N van, Soest RWM van (2009) Sponges. In: Wehrtmann IS, Cortés J (eds) Marine biodiversity of Costa Rica, Central America. Springer+Business Media, Berlin, Text pp 137–142, Species List: Compact Disc pp 83–93

Cortés J, Jiménez CE, Fonseca AC, Alvarado JJ (2010) Status and conservation of coral reefs in Costa Rica. Rev Biol Trop 58(Suppl 1):33–50

Cortés J, Vargas-Castillo R, Nivia-Ruiz J (2012) Marine biodiversity of Bahía Culebra, Guanacaste, Costa Rica: published records. Rev Biol Trop 60(Suppl 2):39–71

Cortés-Núñez J, Sánchez-Noguera C, Breedy-Shadid O, Pacheco-Solano C (2012) Los Ecosistemas marinos del Refugio de Vida Silvestre La Flor y del Corredor Biológico Paso del Istmo, Rivas, Nicaragua. Proyecto Investigación Marino-Costero: Asegurando las bases científicas y educativas para la protección y manejo de tortugas marinas en los RVS La Flor y Río Escalante-Chococente. CIMAR, Paso Pacífico, DANIDA, Managua, Nicaragua, p 52

Cruz M, Gabor N, Mora E, Jiménez R, Mair J (2003) The known and unknown about marine biodiversity in Ecuador (continental and insular). Gayana 67:232–260

Cruz-Barraza JA, Vega C, Carballo JL (2014) Taxonomy of family Plakinidae (Porifera: Homoscleromorpha) from eastern Pacific coral reefs, through morphology and cox1 and cob mtDNA data. Zool J Linn Soc 171:254–276

Cunningham S, Guzman HM, Bates R (2013) The morphology and structure of the Hannibal Bank fisheries management zone, Pacific Panama using acoustic seabed mapping. Rev Biol Trop 61:1967–1979

Cutler EB (1994) The Sipuncula. Their systematics, biology, and evolution. Comstock Publishing Association, Ithaca, New York, p 480

Cutler NJ, Cutler EB, Vargas JA (1992) Peanut worms (Phylum Sipuncula) from Costa Rica. Rev Biol Trop 40:153–158

Daly M, Fautin DG (2004) *Anthopleura mariscali*, a new species of sea anemone (Cnidaria: Anthozoa: Actiniaria) from the Galápagos Islands. Zootaxa 416:1–8

Dawson EY (1960) Marine red algae of Pacific Mexico. Part 3. Cryptonemiales. Corallinaceae subf. Melobesioideae. Pac Nat 2:3–125

D'Croz L, O'Dea A (2007) Variability in upwelling along the Pacific shelf of Panama and implications for the distribution of nutrients and chlorophyll. Estuar Coast Shelf Sci 73:325–340

de Weerdt WH, Glynn PW (1991) A new and presumably now extinct species of *Millepora* (Hydrozoa) in the eastern Pacific. Zool Med Leiden 65:267–276

Dean HK (2001) Marine biodiversity of Costa Rica: the phyla Sipuncula and Echiura. Rev Biol Trop 49(Suppl 2):85–90

Dean HK (2004) Marine biodiversity of Costa Rica: Class Polychaeta (Annelida). Rev Biol Trop 52(Suppl 2):131–181

Dean HK (2009) Polychaetes and echiurans. In: Wehrtmann IS, Cortés J (eds) Marine biodiversity of Costa Rica, Central America. Springer+Business Media, Berlin, Text pp 181–191, Species List: Compact Disc pp 122–159

Dean HK, Sibaja-Cordero JA, Cortés J, Vargas R, Kawauchi GY (2010a) Sipunculans and echiurans of Isla del Coco (Cocos Island), Costa Rica. Zootaxa 2557:60–68

Dean HK, Sibaja-Cordero JA, Cortés J (2010b) Occurrence of the phoronid *Phoronopsis albomaculata* in Cocos Island, Costa Rica. Pac Sci 64:459–462

Dean HK, Sibaja-Cordero JA, Cortés J (2012) Polychaetes (Annelida: Polychaeta) of Cocos Island National Park, Pacific Costa Rica. Pac Sci 66:347–386

Díaz JM, Acero A (2003) Marine biodiversity in Colombia: achievements, status of knowledge, and challenges. Gayana 67:261–274

Diaz MC, van Soest RWM, Guzmán HM (2005) *Aplysina chiriquiensis*, a new pedunculate sponge from the Gulf of Chiriquí, Panamá, Eastern Pacific (Aplysinidae, Verongida). Zootaxa 1012:1–12

DiSalvo L, Randall J, Cea A (1988) Ecological reconnaissance of the Easter Island sublittoral marine environment. Natl Geogr Res 4:451–473

Dominici-Arosemena A, Wolff M (2006) Reef fish community structure in the tropical eastern Pacific (Panamá): living on a relatively stable rocky reef environment. Helgol Mar Res 60:287–305

Dominici-Arosemena A, Brugnoli-Olivera E, Cortés-Núñez J, Molina-Ureña H, Quesada-Alpízar M (2005) Community structure of eastern Pacific reef fishes (Gulf of Papagayo, Costa Rica). Tecnociencia 7:19–41

Doney SC, Fabry VJ, Feely RA, Kleypas JA (2009) Ocean acidification: the other CO_2 problem. Ann Rev Mar Sci 1:169–192

Dove SG, Kline DI, Pantos O, Angly FE, Tyson GW, Hoegh-Guldberg O (2013) Future reef decalcification under a business-as-usual CO_2 emission scenario. Proc Nat Acad Sci USA 110:15,342–15,347

Eakin CM (1996) Where have all the carbonates gone? A model comparison of calcium carbonate budgets before and after the 1982-1983 El Niño at Uva Island in the eastern Pacific. Coral Reefs 15:109–119

Earle SA (1972) A review of the marine plants of Panamá. Bull Biol Soc Washington 2:69–97

Echelman T, Fishelson L (1990) Surface zooplankton dynamics and community structure in the Gulf of Aqaba (Eilat), Red Sea. Mar Biol 107:179–190

Edgar GJ, Banks SA, Bessudo S, Cortés J, Guzmán HM, Henderson S, Martinez C, Rivera F, Soler G, Ruiz D, Zapata FA (2011) Variation in reef fish and invertebrate communities with level of protection from fishing across the eastern tropical Pacific seascape. Global Ecol Biogeogr 20:730–743

Ekman S (1953) Zoogeography of the Sea. Sidgwick and Jackson, London, p 418

Emerson WK (1995) A zoogeographic summary of the marine mollusks of the Revillagigedo Islands (tropical eastern Pacific ocean). Festivus 27:3–18

Emery AR (1968) Preliminary observations on coral reef plankton. Limnol Oceanogr 13:293–303

Emig CC (1982) Nouvelles localisations de phoronidiens. Téthys 10:287–290

Emig CC (2009) Phoronids. In: Wehrtmann IS, Cortés J (eds) Marine biodiversity of Costa Rica, Central America. Springer+Business Media, Berlin, Text pp 409– 411, Species list: Compact Disc pp 383–384

Enochs IC (2012) Motile cryptofauna associated with live and dead coral substrates: implications for coral mortality and framework erosion. Mar Biol 159:709–722

Enochs IC, Hockensmith G (2009) The effects of coral mortality on the community composition of cryptic metazoans associated with *Pocillopora damicornis*. In: Proceedings of 11th International Coral Reef Symposium, Ft. Lauderdale, Florida, pp 1375–1379

Enochs IC, Manzello DP (2012a) Species richness of motile cryptofauna across a gradient of reef framework erosion. Coral Reefs 31:653–661

Enochs IC, Manzello DP (2012b) Responses of cryptofaunal species richness and trophic potential to coral reef habitat degradation. Diversity 4:94–104

Enochs IC, Toth LT, Brandtneris VW, Afflerbach JA, Manzello DP (2011) Environmental determinants of motile cryptofauna on an eastern Pacific coral reef. Mar Ecol Prog Ser 438:105–118

Escarria E, Beltrán-León BS, Giraldo A, Zapata F (2007) Ichthyoplankton in the National Park Isla Gorgona (Pacific Ocean of Colombia) during September 2005. Inves Mar 35:127–133

Eschmeyer WN (2014) Catalog of fishes: genera, species, references. http://research.calacademy.org/research/ichthyology/catalog/fishcatmain.asp. Accessed 5 Jul 2014

Esquivel-Garrote O (2015) Abundancia, diversidad, biomasa y composición del zooplancton, asociado a los parches de arrecife coralino en el Parque Nacional Isla del Coco (PNIC), Pacífico de Costa Rica. Thesis, Universidad de Costa Rica, San Pedro, San José, Costa Rica, p 70

Excoffon AC, Acuña FH, Cortés J (2009) The sea anemone *Nemanthus californicus* (Cnidaria, Actiniaria, Nemanthidae) from Costa Rica: re-description and first record outside the type locality. Mar Biodiv Rec 2:e142. doi:10.1017/S1755267209990601

Fabricius KE, Langdon C, Uthicke S, Humphrey C, Noonan S, De'ath G, Okazaki R, Muehllehner N, Glas MS, Lough JM (2011) Losers and winners in coral reefs acclimatized to elevated carbon dioxide concentrations. Nat Clim Chang 1:165–169

Fabricius KE, Kluibenschedl A, Harrington L, Noonan S, De'ath G (2015) In situ changes of tropical crustose coralline algae along carbon dioxide gradients. Sci Rep 5:9537. doi:10.1038/srep09537

Fabry VJ, Seibel BA, Feely RA, Orr JC (2008) Impacts of ocean acidification on marine fauna and ecosystem processes. ICES J Mar Sci 65:414–432

Fauchald K, Jumars PA (1979) The diet of worms: a study of polychete feeding guilds. Ocean Mar Biol Ann Rev 17:193–284

Fautin DG, Hickman CP Jr, Daly M, Molodtsova T (2007) Shallow-water sea anemones (Cnidaria: Anthozoa: Actiniaria) and tube anemones (Cnidaria: Anthozoa: Ceriantharia) of the Galápagos Islands. Pac Sci 61:549–573

Feely RA, Sabine CL, Lee K, Berelson W, Kleypas J, Fabry VJ, Millero FJ (2004) Impact of anthropogenic CO_2 on the $CaCO_3$ system in the oceans. Science 305:362–366

Feely RA, Doney SC, Cooley SR (2009) Ocean acidification: present conditions and future changes in a high-CO_2 world. Oceanography 22:36–47

Fernández C (2008) Flora marina del Parque Nacional Isla del Coco, Costa Rica. Pacífico Tropical Oriental. Rev Biol Trop 56(Suppl 2):57–69

Fernández C, Cortés J (2005) *Caulerpa sertularioides*, a green alga spreading aggressively over coral reef communities in Culebra Bay, North Pacific of Costa Rica. Coral Reefs 24:10

Fernández M, Pappalardo P, Rodríguez-Ruiz MC, Castilla JC (2014) Synthesis of the state of knowledge about species richness of macroalgae, macroinvertebrates and fishes in coastal and oceanic waters of Easter and Salas y Gómez islands. Latin Amer J Aquat Res 42:760–802

Fernández-García C, Riosmena-Rodríguez R, Wysor B, Tejada O, Cortés J (2011) Checklist of the Pacific marine macroalgae of Central America. Bot Mar 54:53–73

Fernández-García C, Cortés J, Alvarado JJ, Nivia-Ruiz J (2012) Physical factors contributing to the benthic dominance of the alga *Caulerpa sertularioides* (Caulerpaceae, Chlorophyta) in the upwelling Bahía Culebra, north Pacific of Costa Rica. Rev Biol Trop 60 (Suppl 2):93–107

Fernández-Leiva S (1996) Taxonomía del ictioplancton en la Isla del Coco. Thesis, Universidad de Costa Rica, San Pedro, San José, Costa Rica, Costa Rica, p 175

Ferraris JD (1982) Surface zooplankton at Carrie Bow Cay Belize. In: Rützler K, McIntyre IG (eds) The Atlantic barrier reef ecosystem at Carrie Bow cay Belize I. Structure and communities. Smithsonian Institution, Washington DC, pp 143–151

Ferreira AJ (1983) The chiton fauna of the Revillagigedo Archipelago, Mexico. Veliger 25:307–322

Fiedler PC, Lavín MF (2006) Introduction: a review of eastern tropical Pacific oceanography. Prog Oceanogr 69:94–100

Fiedler PC, Talley LD (2006) Hydrography of the eastern tropical Pacific: a review. Prog Oceanogr 69:143–180

Fiege D, Ten Hove HA (1999) Redescription of *Spirobranchus gaymardi* (Quatrefages, 1866) (Polychaeta: Serpulidae) from the Indo-Pacific with remarks on the *Spirobranchus giganteus* complex. Zool J Linn Soc 126:355–364

Fisher WK (1946) Echiuroid worms of the North Pacific Ocean. Proc US Nat Mus 96:215–292

Flot JF, Adjeroud M (2009) Les coraux de Clipperton. In: L Charpy (ed) Clipperton, environnement et biodiversité d'un microcosme océanique. Patrim Nat 68:155–162

Fong P, Smith TB, Wartian MJ (2006) Epiphytic cyanobacteria maintain shifts to macroalgal dominance on coral reefs following ENSO disturbance. Ecology 87:1162–1168

Fonseca ACE, Dean HK, Cortés J (2006) Non-colonial coral macro-borers as indicators of coral reef status in the south Pacific of Costa Rica. Rev Biol Trop 54:101–115

Foster MS (2001) Rhodoliths: between rocks and soft places. J Phycol 37:659–667

Foster MS, McConnico LM, Lundsten L, Wadsworth T, Kimbal T, Brooks LB., Medina-López M, Riosmena-Rodríguez R, Hernández-Carmona G, Vásquez-Elizando RM, Johnson S, Steller DL (2007) Diversity and natural history of a *Lithothamnion muelleri-Sargassum horridum* community in the Gulf of California. Cienc Mar 33:367–384

Foster MS, Amado-Filho GM, Kamenos NA, Riosmena-Rodríguez R, Steller DL (2013) Rhodoliths and rhodolith beds. Smithsonian Contr Mar Sci 39:143–155

Fourriére M, Reyes-Bonilla H, Rodríguez-Zaragoza FA, Crane N (2014) Fishes of Clipperton Atoll, Eastern Pacific: checklist, endemism, and analysis of completeness of the inventory. Pac Sci 68:375–395

Fraser CM (1938a) Hydroids of the 1934 Hancock Pacific expeditions. Allan Hancock Pac Exped 4:1–105

Fraser CM (1938b) Hydroids of the 1936 and 1937 Allan Hancock Pacific Expeditions. Allan Hancock Pac Exped 4:107–127

Fraser CM (1938c) Hydroids of the 1932, 1933, 1935, and 1938 Allan Hancock Pacific expeditions. Allan Hancock Pac Exped 4:129–153

Fraser CM (1948) Hydroids of the Allan Hancock Pacific expeditions since March, 1938. Allan Hancock Pac Exped 4:179–335

Friedlander AM, Ballesteros E, Beets J, Berkenpas E, Gaymer CF, Gorny M, Sala E (2013) Effects of isolation and fishing on the marine ecosystems of Easter Island and Salas y Gómez Island, Chile. Aquat Conserv 23:515–531

Garese A, Guzman HM, Acuña FH (2009) Sea anemones (Cnidaria, Anthozoa: Actiniaria and Corallimorpharia) from Panama. Rev Biol Mar Oceanogr 44:791–802

Garske LE (2002) Macroalgas marinas. In: Danulat E, Edgar GJ (eds) Reserva Marina de Galápagos. Línea Base de la Biodiversidad, Fundación Charles Darwin/Servicio del Parque Nacional Galápagos, Santa Cruz, Galápagos, Ecuador, pp 419–439

Garth JS (1974) On the occurrence in the eastern tropical Pacific of Indo-west Pacific decapod crustaceans commensal with reef-building corals. In: Proceedings of 2nd International Coral Reef Symposium, vol 1, Brisbane, pp 397–404

Garth JS (1992) The brachyuran crabs of the Revillagigedo Islands, Colima, Mexico, with remarks on insular endemism in the Eastern Tropical Pacific. Proc San Diego Soc Nat Hist 24:1–6

Gattuso JP, Bijma J, Gehlen M, Riebesell U, Turley C (2011) Ocean acidification: knowns, unknowns and perspectives. In: Gattuso JP, Hanson L (eds) Ocean acidification. Oxford University Press, Oxford, p 326

Geiger DL (2012) Monograph of the little slit shells. Vol 1. Introduction, Scissurellidae. pp 1–728. Vol 2. Anatomidae, Larocheidae, Depressizonidae, Sutilizonidae, Temnocinclidae. pp. 729–1291. Santa Barbara Museum of Natural History Monographs Number 7. Santa Barbara Museum of Natural History: Santa Barbara, California

Ginsburg RN (1983) Geological and biological roles of cavities in coral reefs. In: Barnes D (ed) Perspectives on coral reefs. Australian Institute of Marine Science, Townsville, QLD, Australia, pp 148–153

Giraldo A, Valencia B, Ramírez D (2006) Productividad planctónica y condiciones oceanográficas locales en la Isla Gorgona durante julio 2006. Bol Invest Mar Cost 40:185–201

Glynn PW (1972a) Observations on the ecology of the Caribbean and Pacific coasts of Panamá. Bull Biol Soc Washington 2:13–29

Glynn PW (1972b) Pacific coral reefs of Panamá: structure, distribution and predators. Geol Rundsch 61:483–519

Glynn PW (1973) Ecology of a Caribbean coral reef. The *Porites* reef-flat biotope: part 2. Plankton community with evidence for depletion. Mar Biol 22:1–21

Glynn PW (1976) Some physical and biological determinants of coral community structure in the eastern Pacific. Ecol Monogr 46:431–456

Glynn PW (1984) An amphinomid worm predator of the crown-of-thorns sea star and general predation on asteroids in eastern and western Pacific coral reefs. Bull Mar Sci 35:54–71

Glynn PW (1985) Corallivore population sizes and feeding effects following El Niño (1982–1983) associated coral mortality in Panama. In: Proceedings of 5th International Coral Reef Congress, vol 4, Tahiti, pp 183–188

Glynn PW (1988) El Niño warning, coral mortality and reef framework destruction by echinoid bioerosion in the eastern Pacific. Galaxea 7:129–160

Glynn PW (1993) Monsoonal upwelling and episodic *Acanthaster* predation as probable controls of coral reef distribution and community structure in Oman, Indian Ocean. Atoll Res Bull 379:1–66

Glynn PW (1997) Bioerosion and coral reef growth: a dynamic balance. In: Birkeland C (ed) Life and death of coral reefs. Chapman and Hall, New York, pp 69–98

Glynn PW (2000) El Niño-Southern Oscillation mass mortalities of reef corals: a model of high temperature marine extinctions? In: Insalaco E, Skelton PW, Palmer TJ (eds) Carbonate platform systems: components and interactions. Geological Society, London, pp 117–133

Glynn PW (2004) High complexity food webs in low diversity eastern Pacific reef-coral communities. Ecosystems 7:358–367

Glynn PW (2006) Fish utilization of simulated coral reef frameworks versus eroded rubble substrates off Panama, eastern Pacific. Proceedings of 10th International Coral Reef Symposium, vol 1, Okinawa, pp 250–256

Glynn PW (2011) In tandem reef coral and cryptic metazoan declines and extinctions. Bull Mar Sci 87:1–28

Glynn PW (2013) Fine-scale interspecific interactions on coral reefs: functional roles of small and cryptic metazoans. Smithsonian Contr Mar Sci 39:229–248

Glynn PW, Ault JS (2000) A biogeographic analysis and review of the far eastern Pacific coral reef region. Coral Reefs 19:1–23

Glynn PW, de Weerdt WH (1991) Elimination of two reef-building hydrocorals following the 1982–83 El Niño warming event. Science 253:69–71

Glynn PW, Enochs IC (2011) Invertebrates and their roles in coral reef ecosystems. In: Dubinsky Z, Stambler N (eds) Coral reefs: an ecosystem in transition. Springer, Berlin, pp 273–325

Glynn PW, Feingold JS (1992) Hydrocoral species not extinct. Science 257:1845

Glynn, PW, Maté JL (1997) Field guide to the Pacific coral reefs of Panamá. Proceedings of 8th International Coral Reef Symposium, vol 1, Panamá, pp 145–166

Glynn PW, Druffel EM, Dunbar RB (1983) A dead Central American coral reef tract: possible link with the Little Ice Age. J Mar Res 41:605–637

Glynn PW, Veron JEN, Wellington GM (1996) Clipperton Atoll (eastern Pacific): oceanography, geomorphology, reef-building coral ecology and biogeography. Coral Reefs 15:71–99

Glynn PW, Wellington GM, Wieters EA, Navarrete SA (2003) Reef-building coral communities of Easter Island (Rapa Nui), Chile. In: Cortés J (ed) Latin American coral reefs. Elsevier Science, Amsterdam, pp 474–494

Glynn PW, Wellington GM, Riegl B, Olson DB, Borneman E, Wieters EA (2007) Diversity and biogeography of the scleractinian coral fauna of Easter Island (Rapa Nui). Pac Sci 61:67–90

Glynn PW, Enochs IC, McCosker JE, Graefe AN (2008) First record of a pearlfish, *Carapus mourlani*, inhabiting the aplysiid opisthobranch mollusc *Dolabella auricularia*. Pac Sci 62:593–602

Glynn PW, Enochs IC, Afflerbach JA, Brandtneris VW, Serafy JE (2014) Eastern Pacific reef fish responses to coral recovery following El Niño disturbances. Mar Ecol Prog Ser 495:233–247

Gómez CE, Paul VJ, Ritson-Williams R, Muehllehner N, Langdon C, Sánchez JA (2015) Responses of the tropical gorgonian coral *Eunicea fusca* to ocean acidification conditions. Coral Reefs 34:451–460

González-González J, Gold-Morgan M, León-Tejera H, Candelaria C, León-Alvarez D, Serviere Zaragoza E, Fragoso D (1996) Catálogo onomástico (nomenclátor) y bibliografía indexada de las algas bentónicas marinas de México. UNAM, Cuad Inst Biol 29:1–492

González-Nakawaga O, Sánchez-Nava S (1986) Nota de moluscos como fauna de acompañamiento de crustáceos de la Isla Clarión, México. Inv Oceanogr BC 3:153–182

Gosliner T (1991) The opisthobranch gastropod fauna of the Galápagos Islands. In: James MJ (ed) Galápagos marine invertebrates: taxonomy, biogeography and evolution in Darwin's Islands. Plenum Press, New York, pp 281–305

Gotelli NJ, Abele LG (1983) Community patterns of coral-associated decapods. Mar Ecol 13:131–139

Graham MH, Kinlan BP, Druehl LD, Garske LE, Banks S (2007) Deep-water kelp refugia as potential hotspots of tropical marine diversity and productivity. Proc Nat Acad Sci USA 104:16,576–16,580

Grassle JF (1973) Variety in coral reef communities. In: Jones OA, Endean R (eds) Biology and geology of coral reefs II. Biology 1. Academic Press, New York, pp 247–270

Greenblatt PR (1982) Small-scale horizontal distributions of zooplankton taxa. Mar Biol 67:97–111

Grigg RW, Hey R (1992) Paleoceanography of the tropical eastern Pacific Ocean. Science 255:172–178

Grottoli AG, Wellington GM (1999) Effect of light and zooplankton on skeletal $\delta^{13}C$ values in the eastern Pacific corals *Pavona clavus* and *Pavona gigantea*. Coral Reefs 18:29–41

Guzmán HM (1988) Distribución y abundancia de organismos coralívoros en los arrecifes coralinos de la Isla del Caño, Costa Rica. Rev Biol Trop 36:191–207

Guzman HM, Breedy O (2008) *Leptogorgia christiae* (Octocorallia: Gorgoniidae) a new shallow gorgonian from Pacific Panama. J Mar Biol Assoc UK 88:719–722

Guzman HM, Breedy O (2012) *Pacifigorgia marviva* (Anthozoa: Octocorallia) a new species from Coiba National Park, Pacific Panama. J Mar Biol Assoc UK 92:693–698

Guzmán HM, Cortés J (1989) Coral reef community structure at Caño Island. Pacific Costa Rica. PSZNI: Mar Ecol 10:23–41

Guzmán HM, Cortés J (1993) Los arrecifes coralinos del Pacífico Oriental Tropical: revisión y perspectivas. Rev Biol Trop 41:535–557

Guzmán HM, Obando VL (1988) Diversidad y abundancia diaria y estacional del zooplancton marino de la Isla del Caño, Costa Rica. Rev Biol Trop 36:139–150

Guzmán HM, Obando VL, Cortés J (1987) Meiofauna associated with a Pacific coral reef in Costa Rica. Coral Reefs 6:107–112

Guzmán HM, Obando VL, Brusca RC, Delaney PM (1988) Aspects of the population biology of the marine isopod *Excorallana tricornis occidentalis* Richardson, 1905 (Crustacea: Isopoda: Corallanidae) at Caño Island, Pacific Costa Rica. Bull Mar Sci 43:77–87

Guzman HM, Guevara CA, Breedy O (2004) Distribution, diversity, and conservation of coral reefs and coral communities in the largest marine protected area of Pacific Panama (Coiba Island). Environ Conserv 31:111–121

Guzman HM, Benfield S, Breedy O, Mair JM (2008) Broadening reef protection across the marine conservation corridor of the eastern tropical Pacific: distribution and diversity of reefs in Las Perlas Archipelago, Panama. Environ Conserv 35:46–54

Halfar J, Riegl B (2013) From coral framework to rhodolith bed: sedimentary footprint of the 1982/1983 ENSO in Galápagos. Coral Reefs 32:985

Hall-Spencer JM, Rodolfo-Metalpa R, Martin S, Ransome E, Fine M, Turner SM, Rowley SJ, Tedesco D, Buia MC (2008) Volcanic carbon dioxide vents show ecosystem effects of ocean acidification. Nature 454:96–99

Hamner WM, Jones MS, Carleton JH, Hauri IR, Williams DM (1988) Zooplankton, planktivorous fishes and water currents on a windward reef face, Great Barrier Reef, Australia. Bull Mar Sci 42:459–479

Hebbeln D, Cortés J (2001) Sedimentation in a tropical fjord: Golfo Dulce, Costa Rica. Geo-Marine Lett 20:142–148

Heidelberg KB, Sebens KP, Purcell JE (2004) Composition and sources of near reef zooplankton on a Jamaican forereef along with implications for coral feeding. Coral Reefs 23:263–276

Hellberg ME (2009) Gene flow and isolation among populations of marine animals. Ann Rev Ecol Evol Syst 40:291–310

Hendrickx ME (1993) Crustáceos decápodos del Pacífico mexicano. In: Salazar-Vallejo SI, González NE (eds) Biodiversidad marina y costera de México. Comisión Nacional para el uso y conocimiento de la Biodiversidad y CIRQRO, México, pp 271–318

Hermosillo A (2009a) A note on the opisthobranch mollusks of Rocas Alijos. Festivus 41:118–119

Hermosillo A (2009b) The opisthobranch fauna of Islas Tres Marías, Mexican Pacific. Festivus 41:3–9

Hermosillo A, Camacho-García YE (2006) A note on the opisthobranchs of Parque Nacional de Coiba, Panama (tropical eastern Pacific). Festivus 38:95–98

Hermosillo A, Gosliner TM (2008) The opisthobranch fauna of Archipiélago de Revillagigedo, Mexican Pacific. Festivus 40:25–34

Hermosillo A, Valdés A (2004) Two new dorids (Mollusca: Opisthobranchia) of Bahía de Banderas and La Paz, México. Proc California Acad Sci 55:550–560

Hermosillo A, Behrens DW, Ríos-Jara E (2006) Opistobranquios de México. Guía de babosas marinas del Pacífico, Golfo de California y las islas oceánicas. Dirección de Artes Escénicas y Literatura, Universidad de Guadalajara-CONABIO, México, p 143

Hernández L, Balart EF, Reyes-Bonilla H (2009) Checklist of reef decapod crustaceans (Crustacea: Decapoda) in the southern Gulf of California, Mexico. Zootaxa 2119:39–50

Hernández L, Ramírez-Ortiz G, Reyes-Bonilla H (2013) Coral-associated decapods (Crustacea) from the Mexican tropical Pacific coast. Zootaxa 3609:451–464

Hernández MA (1992) Inventario preliminar de moluscos marinos de la costa de El Salvador. Secretaría Ejecutiva del Medio Ambiente, San Salvador, El Salvador, p 172

Hernández-Kantún J, Riosmena-Rodríguez R, López-Vivas JM, Pacheco-Ruíz I (2010) Range extension for *Kallymenia* spp. (Kallymeniaceae: Rhodophyta) associated with rhodolith beds, new records from the Gulf of California, México. Mar Biodiv Rec 3:e84 doi:10.1017/S1755267210000783

Hernández-Kantún JJ, Riosmena-Rodríguez R, Adey WH, Rindi F (2014) Analysis of the cox 2-3 spacer region for population diversity and taxonomic implications in rhodolith-forming species (Rhodophyta, Corallinales). Phytotaxa 190:331–354

Hetzinger S, Halfar J, Riegl B, Godinez-Orta L (2006) Sedimentology and acoustic mapping of modern rhodolith beds on a non-tropical carbonate shelf (Gulf of California, Mexico). J Sediment Res 76:670–682

Hickman CP (1998) A field guide to sea stars and other echinoderms of Galápagos. Sugar Spring Press, Lexington, Virginia, p 83

Hickman CP (1999) A field guide to marine molluscs of Galápagos. Sugar Spring Press, Lexington, Virginia, p 150

Hickman CP (2008) A field guide to corals and other radiates of Galápagos. Sugar Spring Press, Lexington, Virginia, p 172

Hickman CP (2009) Evolutionary responses of marine invertebrates to insular isolation in Galapagos. Galapagos Res 66:32–42

Hickman CP, Zimmerman TL (2000) A field guide to crustaceans of Galápagos. Sugar Spring Press, Lexington, Virginia p156

Hinderstein LM, Marr JCA, Martinez FA, Dowgiallo MJ, Puglise KA, Pyle RL, Zawada DG, Appeldoorn R (2010) Theme section on "mesophotic coral ecosystems: characterization, ecology, and management". Coral Reefs 29:247–251

Hinojosa-Arango G, Riosmena-Rodríguez R (2004) Influence of rhodolith-forming species and growth-form on the associated fauna of rhodolith beds in the Central-West Gulf of California. Mexico. PSZNI: Mar Ecol 25:109–127

Hinojosa-Arango G, Rioja-Nieto R, Suárez-Castillo AN, Riosmena-Rodríguez R (2014) Using GIS methods to evaluate rhodolith and *Sargassum* beds as Critical Habitats for commercially important marine species in Bahía Concepción, B.C.S., México. Cryptogamie, Algologie 35:49–65

Hixon MA, Brostoff WN (1981) Fish grazing and community structure of Hawaiian reef algae. Proceedings of 4th International Coral Reef Symposium, vol 2, Manila, pp 508–514

Hoegh-Guldberg O (1999) Climate change, coral bleaching and the future of the world's coral reefs. Mar Freshw Res 50:839–866

Hoegh-Guldberg O, Mumby PJ, Hooten AJ, Steneck RS, Greenfield P, Gomez E, Harvell CD, Sale PF, Edwards AJ, Caldeira K, Knowlton N, Eakin CM, Iglesias-Prieto R, Muthiga N, Bradburry RH, Dubi A, Hatziolos ME (2007) Coral reefs under rapid climate change and ocean acidification. Science 318:1737–1742

Holguín-Quiñones O (1994) Comunidades bentónicas marinas. In: Ortega-Rubio A, Castellanos-Vera A (eds) La Isla Socorro, Reserva de la Biósfera Archipiélago de Revillagigedo. Publ Esp 8, CIBNOR, La Paz, México, pp 225–245

Holguín-Quiñones O, Mille-Pagaza S, Pérez-Chi A (1992) Resultado de las campañas de muestreo de 1991 para el estudio del bentos marino de Isla Socorro, Revillagigedo, Colima, México. Zool Inf 24:1–20

Holzman R, Reidenbach MA, Monismith SG, Koseff JR, Genin A (2005) Near-bottom depletion of zooplankton over a coral reef II: relationships with zooplankton swimming ability. Coral Reefs 24:87–94

Hondt, J-L d' (2009) Les bryozoaires. In: Charpy L (ed) Clipperton, environnement et biodiversité d'un microcosme océanique. Patrim Nat 68:271–276

Hooper JNA, Kennedy JA, Quinn RJ (2002) Biodiversity "hotspots", patterns of richness and endemism, and taxonomic affinities of tropical Australian sponges (Porifera). Biodiver Conserv 11:851–885

Huber ME, Coles SL (1986) Resource utilization and competition among the five Hawaiian species of *Trapezia* (Crustacea, Brachyura). Mar Ecol Prog Ser 30:21–31

Hyman LH (1953) Some polyclad flatworms from the Galapagos Islands. Allan Hancock Pac Exp 15:183–210

Idjadi JA, Edmunds PJ (2006) Scleractinian corals as facilitators for other invertebrates on a Caribbean reef. Mar Ecol Prog Ser 319:117–127

Jacoby CA, Greenwood JG (1989) Emergent zooplankton in Moreton Bay, Queensland, Australia: seasonal, lunar, and diel patterns in emergence and distribution with respect to substrata. Mar Ecol Prog Ser 51:131–154

James DW, Foster MS, O'Sullivan J (2006) Bryoliths (Bryozoa) in the Gulf of California. Pac Sci 60:117–124

James MJ (ed) (1991) Galápagos marine invertebrates: taxonomy, biogeography, and evolution in Darwin's Islands. Plenum Press, New York, p 474

Johnson MD, Price NN, Smith JE (2014) Contrasting effects of ocean acidification on tropical fleshy and calcareous algae. PeerJ 2:e411. doi:10.7717/peerj.411

Jones ML (1972) The Panamic biota: some observations prior to a sea-level canal. Bull Biol Soc Washington 2:270

Jumars PA, Dorgan KM, Lindsay SM (2015) Diet of worms emended: an update of polychaete feeding guilds. Ann Rev Mar Sci 7:1–20

Kahng SE, Garcia-Sais JR, Spalding HL, Brokovich E, Wagner D, Weil E, Hinderstein L, Toonen RJ (2010) A review of community ecology of mesophotic coral ecosystems. Coral Reefs 29:255–275

Kahng SE, Copus JM, Wagner D (2014) Recent advances in the ecology of mesophotic coral ecosystems (MCEs). Curr Opin Environ Sust 7:72–81

Kaiser KL (1997) The Recent molluscan marine fauna of the Islas Galápagos. Festivus 24:1–67

Kaiser KL (2006) Additional mollusk species from Rocas Alijos. México. Festivus 38:22

Kaiser KL (2007) The Recent molluscan fauna of Île Clipperton (tropical Eastern Pacific). Festivus 39:1–162

Kaiser, KL (2009) Les mollusques. In: Charpy L (ed) Clipperton, environment et biodiversité d'un microcosme océanique. Patrim Nat 68:217–234

Kaiser KL, Bryce CW (2001) The Recent molluscan marine fauna of Isla de Malpelo, Colombia. Festivus 33:1–149

Kajihara H, Yoshida R, Uyeno D (2012) First record of *Baseodiscus mexicanus* (Bürger, 1893) (Nemertea: Heteronemertea) from Japanese waters. Check List 8:754–755

Kaniewska P, Campbell PR, Kline DI, Rodriguez-Lanetty M, Miller DJ, Dove S, Hoegh-Guldberg O (2012) Major cellular and physiological impacts of ocean acidification on a reef building coral. PLoS ONE 7(4):e34659. doi:10.1371/journal.pone.0034659

Keen MA (1971) Sea shells of Tropical West America: Marine Molluscs from Baja California to Peru. Stanford University Press, Stanford, California, p 1064

Kemp JM, Benzoni F (1999) Monospecific coral areas on the northern shore of the Gulf of Aden, Yemen. Coral Reefs 18:280

Kerstitch A, Bertsch H (2007) Sea of Cortez marine invertebrates. A guide for the Pacific Coast, México to Perú, 2nd ed. Sea Challengers, Monterey, California, p 124

Kessler WS (2006) The circulation of the eastern tropical Pacific: a review. Prog Oceanogr 69:181–217

Kirsteuer E (1969) Quantitative and qualitative aspects of the nemertean fauna in tropical coral reefs. In: Proceedings of 1st International Coral Reef Symposium, vol 1, Mandapam Camp, pp 363–371

Kleypas JA, Yates KK (2009) Coral reefs and ocean acidification. Oceanography 22:108–117

Kleypas JA, Feely RA, Fabry VJ, Langdon C, Sabine CL, Robbins L (2006) Impacts of ocean acidification on coral reefs and other marine calcifiers: a guide for future research. Report of a Workshop held, St. Petersburg, Florida, NSF, NOAA, USGS, 18–20 Apr 2005

Klumpp DW, McKinnon AD, Mundy CN (1988) Motile cryptofauna of a coral reef: abundance, distribution and trophic potential. Mar Ecol Prog Ser 45:95–108

Knowlton N (1993) Sibling species in the sea. Annu Rev Ecol Syst 24:189–216

Kobluk DR (1988) Cryptic faunas in reefs: ecology and geologic importance. Palaios 3:379–390

Kroeker KJ, Micheli F, Gambi MC, Martz TR (2011) Divergent ecosystem responses within a benthic marine community to ocean acidification. Proc Nat Acad Sci USA 108:14,515–14,520

Kroeker KJ, Kordas RL, Crim R, Hendriks IE, Ramajos L, Singh GS, Duarte CM, Gattuso JP (2013) Impacts of ocean acidification on marine organisms: quantifying sensitivities and interactions with warming. Glob Chan Biol 19:1884–1896

Kropp RK, Birkeland C (1981) Comparison of Crustacea associates on *Pocillopora verrucosa* from a high island and an atoll. In: Proceedings of 4th International Coral Reef Symposium, vol 1, Manila, pp 627–632

Kuffner IB, Andersson AJ, Jokiel PL, Rodgers KS, Mackenzie FT (2008) Decreased abundance of crustose coralline algae due to ocean acidification. Nat Geosci 1:114–117

Landa-Jaime V, Michel-Morfín E, Arciniega-Flores J, Castillo-Vargasmachuca S, Saucedo-Lozano M (2013) Mollusks associated to the coral reefs of Tenacatita, Jalisco, in the central Mexican Pacific. Rev Mexicana Biodiv 84:1121–1136

Lang JC, Chornesky EA (1990) Competition between scleractinian reef corals: a review of mechanisms and effects. In: Dubinsky Z (ed) Coral reefs, ecosystems of the world, vol 25. Elsevier, Amsterdam, pp 209–252

Lassig BR (1977) Communication and coexistence in a coral community. Mar Biol 42:85–92

Laubenfels MW de (1936) A comparison of the shallow-water sponges near the Pacific end of the Panama Canal with those at the Caribbean end. Proc US Natl Mus 83:441–466

Lefevre M (1984) Répartition de la biomasse zooplanctonique autour de l'île de Moorea, Polynésie Française. J Rech Oceanogr 9:20–22

Lemaitre R, Álvarez-León R (1992) Crustáceos decápodos del Pacífico colombiano: Lista de especies y consideraciones zoogeográficas. An Inst Inv Mar Punta Betín 21:33–76

Lendenfeld R von (1910) Reports on the scientific results of the expedition to the Eastern Tropical Pacific in charge of Alexander Agassiz, by the U.S. Fish Commission Steamer "Albatross" from 1904–1905. The sponges I. The Geodiidae. Mem Mus Comp Zoöl, Harvard Coll 41:1–259

Lessios HA, Kessing BD, Pearse JS (2001) Population structure and speciation in tropical seas: global phylogeography of the sea urchin *Diadema*. Evolution 55:955–975

Leviten PJ, Kohn AJ (1980) Microhabitat resource use, activity patterns, and episodic catastrophe: *Conus* on tropical intertidal reef rock benches. Ecol Monogr 50:55–75

Lindner A, Cairns SD, Guzmán HM (2004) *Distichopora robusta* sp. nov., the first shallow-water stylasterid (Cnidaria: Hydrozoa: Stylasteridae) from the tropical eastern Pacific. J Mar Biol Assoc UK 84:943–947

Littler DS, Littler MM (2010) Marine plants of Pacific Panama. Smithsonian Tropical Research Institute, Smithsonian Institution. http://biogeodb.stri.si.edu/pacificalgae: Updated on Jan 2010

Littler MM, Littler DS (2008) Coralline algal rhodoliths form extensive benthic communities in the Gulf of Chiriqui, Pacific Panama. Coral Reefs 27:553

Lizano OG (2008) Dinámica de aguas alrededor de la Isla del Coco, Costa Rica. Rev Biol Trop 56(Suppl 2):31–48

López E, Capa M, Aguado MT, Cladera P (2005) Capitellidae (Annelida: Polychaeta) from Coiba National Park, Pacific coast of Panamá with a new species of the genus *Amastigos*. Cah Biol Mar 46:57–67

Mair JM, Cunningham SL, Sibaja-Cordero JA, Guzman HM, Arroyo MF, Merino D, Vargas R (2009) Mapping benthic faunal communities in the shallow and deep sediments of Las Perlas Archipelago, Pacific Panama. Mar Poll Bull 58:375–383

Manzello DP (2008) Reef development and resilience to acute (El Niño warming) and chronic (high-CO_2) disturbances in eastern tropical Pacific: a real-world climate change model. In: Proceedings of 11th International Coral Reef Symposium, Ft Lauderdale, pp 1299–1303

Manzello DP (2010a) Ocean acidification hotspots: spatiotemporal dynamics of the seawater CO_2 system of eastern Pacific coral reefs. Limnol Oceanogr 55:239–248

Manzello DP (2010b) Coral growth with thermal stress and ocean acidification: lessons from the eastern tropical Pacific. Coral Reefs 29:749–758

Manzello DP, Kleypas JA, Budd DA, Eakin CM, Glynn PW, Langdon C (2008) Poorly cemented coral reefs of the eastern tropical Pacific: possible insights into reef development in a high-CO_2 world. Proc Nat Acad Sci USA 105:10,450–10,455

Maté JL (2003) Corals and coral reefs of the Pacific coast of Panamá. In: Cortés J (ed) Latin American coral reefs. Elsevier, Amsterdam, pp 387–417

McConnico LA, Foster MS, Steller DL, Riosmena-Rodríguez R (2014) Population biology of a long-lived rhodolith: the consequences of becoming old and large. Mar Ecol Prog Ser 504:109–118

McCreary JP, Lee HS, Enfield DB (1989) The response of the coastal ocean to strong offshore winds: with application to circulation in the gulfs of Tehuantepec and Papagayo. J Mar Res 47:81–109

McLean JH, Coan EV (1996) Marine mollusks of Rocas Alijos. In: Schmieder RW (ed) Rocas Alijos. Cordell Expeditions. Kluwer Acad Publ, Dordrecht, Netherlands, pp 305–318

McWilliam PS, Sale PF, Anderson DT (1981) Seasonal changes in resident zooplankton sampled by emergence traps in One Tree Lagoon, Great Barrier Reef. J Exp Mar Biol Ecol 52:185–203

Menza C, Kendall M, Hile S (2008) The deeper we go the less we know. Rev Biol Trop 56(Suppl. 1):11–24

Mille-Pagaza SR, Pérez-Chi A, Holguín-Quiñones O (1994) Fauna malacológica bentónica del litoral de isla Socorro, Revillagigedo, México. Cienc Mar 20:467–486

Millen S, Hermosillo A (2012) Three new species of aeolid nudibranchs (Opisthobranchia) from the Pacific coast of Mexico, Panama, and the Indopacific, with a redescription and redesignation of a fourth species. Veliger 51:145–164

Miller KJ, Ayre DJ (2004) The role of sexual and asexual reproduction in structuring high latitude populations of the reef coral *Pocillopora damicornis*. Heredity 92:557–568

Millero FJ (2007) The marine inorganic carbon cycle. Chem Rev 107:308–341

Miloslavich P, Klein E, Díaz JM, Hernández CE, Bigatti G, Campos L, Artigas F, Castillo J, Penchaszadeh PE, Neill PE, Carranza A, Retana MV, Díaz de Astarloa JM, Lewis M, Yorio P, Piriz ML, Rodríguez D, Yoneshigue-Valentin Y, Gamboa L, Martín A (2011) Marine biodiversity in the Atlantic and Pacific coasts of South America: knowledge and gaps. PLoS ONE 6:e14631. doi:10.1371/journal.pone.0014631

Mora C, Robertson DR (2005) Causes of latitudinal gradients in species richness: a test with fishes of the tropical eastern Pacific. Ecology 86:1771–1782

Morales A, Murillo MM (1996) Distribution, abundance and composition of coral reef zooplankton, Cahuita National Park, Limon, Costa Rica. Rev Biol Trop 44:619–630

Morales-Ramírez A (2008) Caracterización cualitativa del zooplancton del Área de Conservación Marina Isla del Coco (ACMIC), Océano Pacífico de Costa Rica. Rev Biol Trop 56(Suppl 2):159–169

Morales-Ramírez A (2011) La diversidad marina del Golfo Dulce, Pacífico sur de Costa Rica: amenazas a su conservación. Biocenosis 24:9–20

Motro R, Ayalon I, Genin A (2005) Near-bottom depletion of zooplankton over coral reefs: III. Vertical gradient of predation pressure. Coral Reefs 24:95–98

Müller-Gelinek H, Salazar-Orquera M (1996) Algas marinas del Ecuador. Comisión Asesora Ambiental, Instituto Nacional de Pesca, República de Ecuador, p 187

Murillo-Muñoz M, Peña-Salamanca EJ (2014) Benthic marine algae of Gorgona Island, Colombian Pacific coast. Rev Biol Trop 62(Suppl 1):27–41

Newman LJ, Cannon LRG (2003) Marine flatworms: the world of polyclads. CSIRO Publishing, Melbourne, p 112

Ohlhorst SL (1982) Diel migration patterns of demersal reef zooplankton. J Exp Mar Biol Ecol 60:1–15

Pacheco C (2012) Estudio taxonómico de las esponjas perforadoras (Porifera, Demospongiae) de arrecifes de coral del Pacífico de Costa Rica. Thesis, Universidad de Costa Rica. San Pedro, Costa Rica, p 69

Palacios MM, Muñoz CG, Zapata FA (2014) Fish corallivory on a pocilloporid reef and experimental coral responses to predation. Coral Reefs 33:625–636

Palardy JE, Witman JD (2011) Water flow drives biodiversity by mediating rarity in marine benthic communities. Ecol Lett 14:63–68

Palardy JE, Grottoli AG, Matthews KA (2005) Effect of upwelling, depth, morphology, and polyp size on feeding in three species of Panamanian corals. Mar Ecol Prog Ser 300:79–89

Palardy JE, Grottoli AG, Matthews KA (2006) Effect of naturally changing zooplankton concentrations on feeding rates of two coral species in the eastern Pacific. J Exp Mar Biol Ecol 331:99–107

Patton WK (1974) Community structure among the animals inhabiting the coral *Pocillopora damicornis* at Heron Island, Australia. In: Vernberg W (ed) Symbiosis in the sea. University of South Carolina Press, Columbia, South Carolina, pp 219–243

Patton WK (1994) Distribution and ecology of animals associated with branching corals (*Acropora* spp.) from the Great Barrier Reef, Australia. Bull Mar Sci 55:193–211

Payri C, Menou J-L, N'Yeurt A (2009) La flore marine du complexe récifal et quelques aspects de la biodiversité et de la géomorphologie de l'île. In: Charpy L (ed) Clipperton, environnement et biodiversité d'un microcosme océanique. Patrim Nat 68:129–141

Pedroche FF, Silva PC, Aguilar-Rosas LE, Dreckmann KM, Aguilar-Rosas R (2005) Catálogo de las algas marinas bentónicas del Pacífico de México. I. Chlorophycota. Universidad Autónoma de Baja California, Baja California, México, p 135

Pedroche FF, Silva PC, Aguilar-Rosas LE, Dreckmann KM, Aguilar-Rosas R (2008) Catálogo de las algas marinas bentónicas del Pacífico de México. UAM, Estado de México, México, II. Phaeophycota, p 146

Peña-Salamanca EJ (2008) Dinámica espacial y temporal de la biomasa algal asociada a las raíces de mangle en la Bahía de Buenaventura, costa Pacífica de Colombia. Bol Invest Mar Cost 37:55–70

Pepe PJ (1985) Littoral endolithic fauna of the Central American Isthmus. Rev Biol Trop 33:191–194

Peralta-García EC, Rosas-Alquicira EF (2014) Range extension and morphological characterization of rhodolith-forming species (Corallinales, Rhodophyta) from shallow water in the Mexican South Pacific. Helgol Mar Res 68:503–513

Pérez-Ruzafa A et al (2013) Chapter 16: Latin America echinoderm biodiversity and biogeography: patterns and affinities. In: Alvarado JJ, Solís-Marín FA (eds) Echinoderm research and diversity in Latin America. Springer, Berlin, Heidelberg, pp 511–542

Porter JW (1972) Predation by *Acanthaster* and its effect on coral species diversity. Am Nat 106:487–492

Porter JW (1974) Zooplankton feeding by the Caribbean reef-building coral *Montastraea cavernosa*. In: Proceedings of 2nd International Coral Reef Symposium, vol 1, Brisbane, pp 111–125

Porter JM, Porter KG (1977) Quantitative sampling of demersal zooplankton migrating from different coral reef substrates. Limnol Oceanogr 22:553–556

Poupin J, Bouchard J-M, Albenga L, Cleva R, Hermoso-Salazar M, Solis-Weiss V (2009) Les crustacés décapodes et stomatopodes, inventaire, écologia et zoogéographie. In: Charpy L (ed) Clipperton, environnement et biodiversité d'un microcosme océanique. Patrim Nat 68:163–216

Preston E (1973) A computer simulation of competition among five sympatric congeneric species of xanthid crabs. Ecology 54:469–483

Ragazzola F, Foster LC, Form A, Anderson PSL, Hansteen TH, Fietzke J (2012) Ocean acidification weakens the structural integrity of coralline algae. Glob Chan Biol 18:2804–2812

Ramírez-Luna S, de la Cruz G, Barrientos-Luján N (2002) Variación espacio temporal de Porcellanidae, Majoidae y Xanthoidae asociados a los corales del género *Pocillopora* en bahías de Huatulco, México. In: Hendrickx M (ed) Contribuciones al estudio de los Crustáceos del Pacífico Este, 1:233-254. UNAM, ICMyL, Mazatlán, México

Randall JE, Cea A (2011) Shore fishes of Easter Island. University of Hawaii Press, Honolulu, Hawai'i, p 164

Razak TB, Hoeksema BW (2003) The hydrocoral genus *Millepora* (Hydrozoa: Capitata: Milleporidae) in Indonesia. Zool Verh Leiden 345:313–336

Reaka-Kudla ML (1997) The global biodiversity of coral reefs. In: Reaka-Kudla ML, Wilson DE, Wilson EO (eds) Biodiversity II: understanding and protecting our biological resources. Joseph Henry Press, Washington, DC, pp 83–108

Reijmer JJG, Bauch T, Schäfer P (2012) Carbonate facies patterns in surface sediments of upwelling and non-upwelling shelf environments (Panama, East Pacific). Sedimentology 59:32–56

Reimer JD, Sinniger F, Hickman CP Jr (2008) Zoanthid diversity (Anthozoa: Hexacorallia) in the Galapagos Islands: a molecular examination. Coral Reefs 27:641–654

Reyes-Bonilla H, Riosmena-Rodríguez Foster MS (1997) Hermatyic corals associated with rhodolith beds in the Gulf of California, Mexico. Pac Sci 51:328–337

Reyes-Bonilla H, Walther-Mendoza M, Ramírez-Ortiz G (2012) Biodiversidad y turismo ecológico. In: Ibáñez-Pérez RM, Ivanova-Boncheva A (eds) Medio ambiente y política turística en México. Instituto Nacional de Ecología, México, pp 135–148

Richter C, Wunsch M, Rasheed M, Kötter I, Badran MI (2001) Endoscopic exploration of Red Sea coral reefs reveals dense populations of cavity-dwelling sponges. Nature 413:726–730

Riosmena-Rodríguez R, Medina-López M (2010) The role of rhodolith beds in the recruitment of invertebrate species from the southwestern Gulf of California, Mexico. In: Israel A, Einav R, Seckbach J (eds) Seaweeds and their role in globally changing environments. Springer, Berlin, pp 127–138

Riosmena-Rodríguez R, Woelkerling WJ, Foster MS (1999) Taxonomic reassessment of rhodolith-forming species of *Lithophyllum* (Corallinales, Rhodophyta) in the Gulf of California, Mexico. Phycologia 38:401–417

Riosmena-Rodríguez R, Steller DL, Hinojosa-Arango G, Foster MS (2010) Reefs that rock and roll: biology and conservation of rhodolith beds in the Gulf of California. In: Brusca RC (ed) The Gulf of California biodiversity and conservation. University of Arizona Press and the Arizona-Sonora Desert Museum, Tucson, Arizona, pp 49–71

Riosmena-Rodríguez R, López-Calderón JM, Mariano-Meléndez E, Sánchez-Rodríguez A, Fernández-García C (2012) Size and distribution of rhodolith beds in the Loreto Marine Park: their role in coastal processes. J Coast Res 28:255–260

Rivera F, Martínez P (2011) Guía fotográfica de corales y octocorales: Parque Nacional Machalilla y Reserva de Producción Faunística Marino Costera Puntilla de Santa Elena. NAZCA, Instituto de Investigaciones Marinas. Informe Técnico, Conservación Internacional, Ecuador, Ecuador, p 78

Rixen T, Jiménez C, Cortés J (2012) Impact of upwelling events on the sea water chemistry in the Gulf of Papagayo (Culebra Bay). Costa Rica. Rev Biol Trop 60(Suppl 2):187–195

Roberts CM, McClean CJ, Veron JEN, Hawkins JP, Allen GR, McAllister DE, Mittermeier CG, Schueler FW, Spalding M, Wells F, Vynne C, Werner TB (2002) Marine biodiversity hotspots and conservation priorities for tropical reefs. Science 295:1280–1284

Robertson DR (1998) Do coral-reef fish faunas have a distinctive taxonomic structure? Coral Reefs 17:179–186

Robertson DR, Allen GR (2006) Shorefishes of the tropical eastern Pacific: an information system. Version 2.0. Smithsonian Tropical Research Institute, Balboa, Panama

Robertson DR, Allen GR (2008) Shorefishes of the Tropical Eastern Pacific online information system. Version 1.0.4.53. Smithsonian Tropical Research Institute, Balboa, Panama (available at: www.stri.org/sftep)

Robertson DR, Cramer KL (2009) Shore fishes and biogeographic subdivisions of the tropical eastern Pacific. Mar Ecol Prog Ser 380:1–17

Robertson DR, Grove JS, McCosker JE (2004) Tropical transpacific shorefishes. Pac Sci 58:507–565

Robichaux DM, Cohen AC, Reaka MJ, Allen D (1981) Experiments with zooplankton on coral reefs, or, will the real demersal please come up? PSZNI: Mar Ecol 2:77–94

Robins CR (1972) The state of knowledge of the coastal fish fauna of the Panamic region prior to the construction of an interoceanic sea-level canal. Bull Biol Soc Washington 2:159–166

Robinson NM, Hansen GI, Fernández-García C, Riosmena-Rodríguez R (2013) A taxonomic and distributional study of the rhodolith-forming species *Lithothamnion muelleri* (Corallinales, Rhodophyta) in the Eastern Pacific Ocean. Algae 28:63–71

Rodríguez-Sevilla L, Vargas R, Cortés J (2009) Benthic, shelled Gastropods. In: Werhtmann IS, Cortés J (eds) Marine biodiversity of Costa Rica, Central America. Springer+Business Media, Berlin, pp 333–356

Rogers CS (2013) Coral reef resilience through biodiversity. ISRN Oceanogr 2013, Article ID 739034. doi:10.5402/2013/739034

Roman MR, Furnas MJ, Mullin MM (1990) Zooplankton abundance and grazing at Davies Reef, Great Barrier Reef. Mar Biol 105:73–82

Ruggiero MA, Gordon DP, Orrell TM, Bailly N, Bourgoin T, Brusca RC, Cavalier-Smith T, Guiry MD, Kirk PM (2015) A higher level classification of all living organisms. PLoS ONE 10(4): e0119248

Rusch DB et al (2007) The Sorcerer II Global Ocean Sampling Expedition: northwest Atlantic through eastern tropical Pacific. PLoS Biol 5(3):e77. doi:10.1371/journal.pbio.0050077

Russell DJ (1992) The ecological invasion of Hawaiian reefs by two marine red algae, *Acanthophora spicifera* (Vahl) Børg. and *Hypnea musciformis* (Wulfen) J. Ag., and their association with two native species, *Laurencia nidifica* J.Ag. and *Hypnea cervicornis* J. Ag. ICES Marine Science Symposiom, vol. 194, pp 110–125

Sachet MH (1962) Flora and vegetation of Clipperton Island. Proc California Acad Sci 31:249–307

Salazar-Vallejo SI, Londoño-Mesa MH (2004) Lista de especies y bibliografía de poliquetos (Polychaeta) del Pacífico Oriental Tropical. An Inst Biol UNAM Ser Zool 75:9–97

Sale PF (1980) The ecology of fishes on coral reefs. Oceanogr Mar Biol Ann Rev 18:367–421

Sammarco PW (1983) Effects of fish grazing and damselfish territoriality on coral reef algae. I. Algal community structure. Mar Ecol Prog Ser 13:1–14

Sánchez JA, Gómez CE, Escobar D, Dueñas LF (2011) Diversidad, abundancia y amenazas de los octocorales de la Isla Malpelo, Pacífico Oriental Tropical, Colombia. Bol Inv Mar Cost 40(Supl Esp):139–154

Sánchez JA, Ardila NE, Andrade J, Dueñas LF, Navas R, Ballesteros D (2014) Octocoral densities and mortalities in Gorgona Island, Colombia. Tropical Eastern Pacific. Rev Biol Trop 62(Suppl 2):209–219

Schmidt P (1978) Interstitielle fauna von Galapagos. XXI. Lebensraum, Umweltfaktoren. Gesamtfauna. Mikrof Meeresb 68:1–52

Schnetter R, Bula-Meyer G (1982) Algas marinas del litoral Pacífico de Colombia. Gantner Verlag, Berlin, p 287

Sedberry GR, Cuellar N (1993) Planktonic and benthic feeding by the reef-associated vermilion snapper, *Rhomboplites aurorubens* (Teleostei, Lutjanidae). US Fish Bull 91:699–709

Segovia-Prado JV, Navarrete-Calero MT (2007) Biodiversidad a nivel de ecosistema en parches de corales hermatípicos (*Porites lobata*, *Pocillopora* sp.) en la zona intermareal de la Playa los Cóbanos, Departamento de Sonsonate, El Salvador. Thesis, Universidad de El Salvador, San Salvador, El Salvador, p 68

Segovia-Prado JV (2012) Comunidades de octocorales (Cnidaria: Octocorallia) en El Salvador, América Central. MSc thesis, Universidad de Costa Rica, San Pedro, San José, Costa Rica, p 126

Serviere-Zaragoza E, Riosmena-Rodríguez R, León-Tejera H, González-González J (2007) Distribución espacial de macroalgas marinas en las Islas Revillagigedo, México. Ciencia y Mar 11:3–13

Sibaja-Cordero JA (2012) Composición de la macrofauna bentónica en substratos sedimentarios del Parque Nacional Isla del Coco. PhD Dissertation, Universidad de Vigo, Vigo, España, Costa Rica, p 306

Sibaja-Cordero JA, Cortés J, Dean HK (2012) Depth diversity profile of polychaete worms in Bahía Chatham, Parque Nacional Isla del Coco, Pacific of Costa Rica. Rev Biol Trop 60(Suppl 3):293–301
Sibaja-Cordero JA, García-Méndez K, Troncoso JS (2013) Additions to the mollusk checklist of Cocos Island National Park, Costa Rica (Eastern Tropical Pacific). Iberus 31:127–163
Sibaja-Cordero JA, Troncoso JS, Gómez-Ramírez E (2014) Recent surface marine sediments of Cocos Island in Costa Rica. Int J Sed Res 29:59–72
Silverman J, Lazar B, Cao L, Caldeira K, Erez J (2009) Coral reefs may start dissolving when atmospheric CO_2 doubles. Geophys Res Lett 36:L05606. doi:10.1029/2008GL036282
Small M (1998) Ten days in May: a trip to Las Islas Tres Marías. Festivus 30:75–82
Smith TB, Fong P, Kennison R, Smith J (2010) Spatial refuges and associational defenses promote harmful blooms of the alga *Caulerpa sertularioides* onto coral reefs. Oecologia 164:1039–1048
Solano-Barquero A (2011) Macrofauna asociada a rodolitos en el Parque Nacional Isla del Coco. Thesis, Universidad de Costa Rica, San Pedro, San José, Costa Rica, Costa Rica, p 50
Solís-Marín FA, Laguarda-Figueras A (2009) Les échinodermes. In: Charpy L (ed) Clipperton, environnement et biodiversité d'un microcosme océanique. Patrim Nat 68:235–247
Solís-Marín FA et al. (2013a) Appendix. In: Alvarado JJ, Solís-Marín FA (eds) Echinoderm research and diversity in Latin America. Springer, Berlin, pp 543–654
Solís-Marín FA, Honey-Escandón MBI, Herrero-Perezrul MD, Benitez-Villalobos F, Díaz-Martínez JP, Buitrón-Sánchez BE, Palleiro-Nayar JS, Durán-González A (2013b) Echinoderms from Mexico: Biodiversity, distribution and current state of knowledge. In: Alvarado JJ, Solís-Marín FA (eds) Echinoderm Research and Diversity in Latin America. Springer, Berlin, pp 11–65
Solís-Weiss V, Hernández-Alcántara P (2009) Les annélides polychètes de la cryptofaune benthique assciées au corail mort. In: Charpy L (ed) Clipperton, environnement et biodiversité d'un microcosme océanique. Patrim Nat 68:249–270
Sonnenholzner J, Brandt M, Francisco V, Hearn A, Luzuriaga M, Guarderas P, Navarro JC (2013) Echinoderms of Ecuador. In: Alvarado JJ, Solís-Marín FA (eds) Echinoderm research and diversity in Latin America. Springer, Berlin, pp 183–233
Sorokin YI (1990) Plankton in the reef ecosystems. In: Dubinsky Z (ed) Ecosystems of the world, vol 25., Coral Reefs. Elsevier, Amsterdam, pp 291–327
Spalding MD, Fox HE, Allen GR, Davidson N, Ferdaña ZA, Finlayson M, Halpern BS, Jorge MA, Lombana A, Lourie SA, Martin KD, McManus E, Molnar J, Recchia CA, Robertson J (2007) Marine ecoregions of the world: a bioregionalization of coastal and shelf areas. Bioscience 57:573–583
Starr RM, Cortés J, Barnes CL, Green K, Breedy O (2012) Characterization of deepwater invertebrates at Coco Island National Park and Las Gemelas Seamounts. Costa Rica. Rev Biol Trop 60 (Suppl 3):303–319
Steller DL, Cáceres C (2009) Coralline algal rhodoliths enhance larval settlement and early growth of the Pacific calico scallop *Argopecten ventricosus*. Mar Ecol Prog Ser 396:49–60
Steller DL, Riosmena-Rodríguez R, Foster MS, Roberts C (2003) Rhodolith bed diversity in the Gulf of California: the importance of rhodolith structure and consequences of anthropogenic disturbances. Aquat Conserv Mar Freshw Ecosyst 13:S5–S20
Steller DL, Foster MS, Riosmena-Rodríguez R (2009) Living rhodolith bed ecosystems in the Gulf of California. In: Johnson JM, Ledesma-Vazquez J (eds) Atlas of coastal ecosystems in the Gulf of California: past and present. University of Arizona Press, Tucson, Arizona, pp 72–82
Stimson J (1990) Stimulation of fat-body production in the polyps of the coral *Pocillopora damicornis* by the presence of mutualistic crabs of the genus *Trapezia*. Mar Biol 106:211–218
Strong AM, Hanna GD (1930) Marine Mollusca of the Revillagigedo Islands, Mexico. Proc California Acad Sci 19:7–12
Suárez-Morales E, Morales-Ramírez A (2003) A new species of *Cymbasoma* (Crustacea: Copepoda: Monstrilloida) from the Pacific coast of Costa Rica, Central America. Proc Biol Soc Washington 116:206–214
Suárez-Morales E, Morales-Ramírez A (2009) New species of Monstrilloida (Crustacea: Copepoda) from the Eastern Tropical Pacific. J Nat Hist 43:1257–1271
Taylor WR (1945) Pacific marine algae of the Allan Hancock Pacific Expeditions to the Galapagos Islands. Allan Hancock Pac Exped 12:1–528
Tews J, Brose U, Grimm V, Tielborger K, Wichmann MC, Schwager M, Jeltsch F (2004) Animal species diversity driven by habitat heterogeneity/diversity: the importance of keystone structures. J Biogeogr 31:79–92
Thiel M, Kruse I (2001) Status of the Nemertea as predators in marine ecosystems. Hydrobiologia 456:21–32
Thresher RE (1991) Geographic variability in the ecology of coral reef fishes: evidence, evolution, and possible implications. In: Sale PF (ed) The ecology of fishes on coral reefs. Academic Press Inc, San Diego, pp 401–436
Treadwell AL (1941) Eastern Pacific expeditions of the New York Zoological Society. XXIII. Polychaetous annelids from the west coast of Mexico and Central America. Zoologica 26:17–24
Valencia B, Giraldo A (2009) Hipéridos (Crustacea: Amphipoda) en el sector norte del Pacífico oriental tropical Colombiano. Lat Am J Aquat Res 37:265–273
Valencia B, Herrera L, Giraldo A (2014) Estructura de la comunidad y distribución vertical de la macrofauna de fondos blandos en Isla Gorgona, Pacifico Colombiano. Rev Biol Trop 62(Suppl 1):169–188
Vannini M (1985) A shrimp that speaks crab-ese. J Crust Biol 5:160–167
van Soest RWM, Kaiser KL, van Syoc R (2011) Sponges from Clipperton Island, East Pacific. Zootaxa 2839:1–46
van Woesik R, van Woesik K, van Woesik L, van Woesik S (2013) Effects of ocean acidification on the dissolution rates of reef-coral skeletons. PeerJ 1:e208. doi:10.7717/peerj.208
Vargas R, Wehrtmann IS (2009) Decapod crustaceans. In: Wehrtmann IS, Cortés J (eds) Marine biodiversity of Costa Rica, Central America. Springer+Business Media, Berlin, Text pp 209–228, Species List: Compact Disc pp 167–198
Vargas S, Guzman HM, Breedy O (2008) Distribution patterns of the genus *Pacifigorgia* (Octocorallia Gorgoniidae): track analysis and parsimony analysis of endemicity. J Biogeogr 35:241–247
Vega AJ, González A (2001) Moluscos del Pacífico veragüense, parte I (Clases: Bivalvia y Poliplacophora). Tecnociencia 3:89–107
Vega AJ, González A (2002) Moluscos del Pacífico veragüense, parte II (Gasterópoda). Tecnociencia 4:23–45
Vega C, Hernández-Guerrero J, Cruz-Barraza JA (2012) Biogeografía de esponjas marinas (Phylum Porifera); Estudios en el Pacífico Oriental. CICIMAR Oceánides 27:35–50
Veron JEN (2008) Mass extinctions and ocean acidification: biological constraints on geological dilemmas. Coral Reefs 27:459–472
Veron J, Stafford-Smith M, DeVantier L, Turak E (2015) Overview of distribution patterns of zooxanthellate Scleractinia. Front Mar Sci 1:81. doi:10.3389/fmars.2014.00081
Verrill AE (1868–70) Notes on Radiata in the museum of Yale College, Number 6: Review of the corals and polyps of the West Coast of America. Trans Connecticut Acad Arts Sci, 2nd ed., vol 1, pp 377–558

Villalobos A (1960) Notas acerca del aspecto hidrobiológico de la parte Sur de la Isla. In: Adem J, Cobo E, Blásquez L, Miranda F, Villalobos A, Herrera T, Villa B, Vásquez L (eds) La Isla Socorro. Monogr Inst Geofís 2:155–180

von Prahl H, Guhl F, Grögl M (1978) Crustáceos decápodos comensales del coral *Pocillopora damicornis* L. en la Isla de Gorgona, Colombia. An Inst Inv Mar Punta Betín 10:81–93

Wehrtmann IS, Cortés J (eds) (2009) Marine biodiversity of Costa Rica, Central America. Springer+Business Media, Berlin. Text 538 p, List of species in Compact Disk p 500

Wehrtmann IS, Cortés J, Echeverría-Sáenz S (2009) Marine biodiversity of Costa Rica: perspectives and conclusions. In: Wehrtmann IS, Cortés J (eds) Marine biodiversity of Costa Rica, Central America. Springer+Business Media, Berlin, pp 521–533

Wellington GM (1982a) An experimental analysis of the effects of light and zooplankton on coral zonation. Oecologia 52:311–320

Wellington GM (1982b) Depth zonation of corals in the Gulf of Panama: control and facilitation by resident reef fishes. Ecol Monogr 52:223–241

Wellington GM, Glynn PW, Strong AE, Navarrete SA, Wieters E, Hubbard D (2001) Crisis on coral reefs linked to climate change. EOS, Trans Amer Geophys Union 82(1):5

Wells JW (1972) Notes on Indo-Pacific scleractinian corals. Part 8. Scleractinian corals from Easter Island. Pac Sci 26:183–190

Wells JW (1983) Annotated list of the scleractinian corals of the Galápagos. In: Glynn PW, Wellington GM (eds) Corals and coral reefs of the Galápagos Islands. Univ California Press, Berkeley, pp 212–295

Westheide W (1974) Interstitielle fauna von Galapagos XI. Pisionidae, Hesionidae, Pilargidae, Syllidae (Polychaeta). Mikr Meeresb 44:1–146

Westheide W (1977) Interstitielle fauna von Galapagos XVIII. Nereidae, Eunicidae, Dorvilleidae (Polychaeta). Mikr Meeresb 63:357–393

Westheide W (1991) The meiofauna of the Galapagos: a review. In: James MJ (ed) Galapagos marine invertebrates: taxonomy, biogeography, and evolution in Darwin's Islands. Plenum, New York, pp 37–73

Wicksten MK, Hendrickx ME (1992) Checklist of penaeoid and caridean shrimps (Decapoda: Penaeoidea, Caridea) from the Eastern Tropical Pacific. Proc San Diego Soc Nat Hist 9:1–11

Wieters EA, Medrano A, Pérez-Matus A (2014) Functional community structure of shallow hard bottom communities at Easter Island (Rapa Nui). Lat Am J Aquat Res 42:827–844

Williams G, Breedy O (2004) The Panamic genus *Pacifigorgia* (Octocorallia: Gorgoniidae) in the Galápagos Archipelago. Proc California Acad Sci 55:54–87

Williamson SJ, Rusch DB, Yooseph S, Halpern AL, Heidelberg KB, Glass JI, Andrews-Pfannkoch C, Fadrosh D, Miller CS, Sutton G, Frazier M, Venter JC (2008) The Sorcerer II global ocean sampling expedition: metagenomic characterization of viruses within aquatic microbial samples. PLoS ONE 3(1):e1456. doi:10.1371/journal.pone.0001456

Wittmann AC, Pörtner HO (2013) Sensitivities of extant animal taxa to ocean acidification. Nature Clim Change. doi:10.1038/nclimate1982

Wilson HV (1904) The sponges. Reports on an exploration off the west coasts of Mexico, Central and South America, and off the Galapagos Islands. Mem Mus Comp Zoöl, Harvard Coll 30:1–164

Woelkerling WJ (1988) The coralline red algae: an analysis of the genera and subfamilies of nongeniculate Corallinaceae. British Museum (Natural History)/Oxford University Press, London/Oxford, p 268

Wulff J (1997) Causes and consequences of differences in sponge diversity and abundance between the Caribbean and Eastern Pacific of Panama. In: Proceedings of 8th International Coral Reef Symoposium, vol 2, Panamá, pp 1377–1382

Wysor B (2004) An annotated list of marine Chlorophyta from the Pacific coast of the Republic of Panama with a comparison to Caribbean Panama species. Nova Hedwigia 78:209–241

Yabur-Pacheco R, Riosmena-Rodríguez R (2007) Rhodolith bed composition in the Southwestern Gulf of California, México. Publ Seto Mar Biol Lab Spec Publ Ser 8:37–47

Yahel R, Yahel G, Genin A (2005) Near-bottom depletion of zooplankton over coral reefs: I. Diurnal dynamics and size distribution. Coral Reefs 24:75–85

Zapata FA, Morales YA (1997) Spatial and temporal patterns of fish diversity in a coral reef at Gorgona Island, Colombia. In: Proceedings of 8th International Coral Reef Symposium, vol 1, Panama, pp 1029–1034

Zapata FA, Robertson DR (2007) How many species of shore fishes are there in the tropical Eastern Pacific? J Biogeogr 34:38–51

Zapata FA, Rodríguez-Ramírez A, Caro-Zambrano C, Garzón-Ferreira J (2010) Mid-term coral-algal dynamics and conservation status of a Gorgona Island (tropical Eastern Pacific) coral reef. Rev Biol Trop 58(Suppl 1):81–94

El Niño-Southern Oscillation: Effects on Eastern Pacific Coral Reefs and Associated Biota

8

Peter W. Glynn, Alissa B. Mones, Guillermo P. Podestá, Angela Colbert, and Mitchell W. Colgan

Abstract

The sudden and sporadic occurrence of anomalous conditions associated with El Niño-Southern Oscillation (ENSO) can precipitate diverse, immediate and long-term effects on eastern Pacific reef-building corals and associated organisms. ENSO is manifested in two complementary phases, namely warm (El Niño) and cool (La Niña) events, which have contrasting and potentially negative effects on coral reef ecosystems. Of the two distinct types of El Niño activity—Eastern-Pacific (EP) and Central-Pacific (CP)—the former exhibits maximum SST anomalies and related climate and weather impacts that affect eastern Pacific coral reefs. Relevant ENSO conditions, with direct or indirect effects, include (a) high and low sea temperature extremes, (b) thermocline and nutricline depths, (c) high and low sea level stands, (d) storm activity, (e) rainfall patterns (and terrestrial runoff), and (f) deviances in surface current direction, velocity and spatial extent. The first sign of ENSO stress to zooxanthellate corals is tissue blanching or bleaching, which may occur during periods of elevated (El Niño) or depressed (La Niña) thermal anomalies. Earlier studies of thermally induced coral bleaching in the Galápagos Islands and Panama are updated to 2012 with attention to anomalous warm and cool events that are stressful to reef-building corals. When thermal conditions normalize, surviving reef-building corals typically re-gain their usual complement of zooxanthellae (symbiotic dinoflagellates) and pigmentation. Long-term ecological effects from extreme ENSO activity, particularly during El Niño sea warming events, may occur over months or years following initial impacts. Such effects can markedly reduce coral cover, cause local species disappearances and significantly change the abundances of coral-associated taxa. The bioerosion of dead

P.W. Glynn (✉)
Department of Marine Biology and Ecology, Rosenstiel School of Marine and Atmospheric Science, University of Miami, 4600 Rickenbacker Causeway, Miami, FL 33149, USA
e-mail: pglynn@rsmas.miami.edu

A.B. Mones
Marine and Atmospheric Science Program, University of Miami, 1365 Memorial Drive, 210 Ungar Building, Coral Gables, FL 33124-4250, USA
e-mail: mones2@illinois.edu

A.B. Mones
College of Veterinary Medicine, University of Illinois at Urbana-Champaign, 3505 Veterinary Medicine Basic Science Building, 2001 South Lincoln Avenue, Urbana, IL 61802, USA

G.P. Podestá · A. Colbert
Department of Ocean Sciences, Rosenstiel School of Marine and Atmospheric Science, University of Miami, 4600 Rickenbacker Causeway, Miami, FL 33149, USA
e-mail: gpodesta@rsmas.miami.edu

A. Colbert
e-mail: acolbert@frostscience.org

A. Colbert
Patricia and Phillip Frost Museum of Science, 3280 South Miami Ave, Miami, FL 33129, USA

M.W. Colgan
Department of Geology and Environmental Geosciences, College of Charleston, SSMB 224 • 202 Calhoun Street, Charleston, SC 29401, USA
e-mail: ColganM@cofc.edu

P.W. Glynn et al. (eds.), *Coral Reefs of the Eastern Tropical Pacific*, Coral Reefs of the World 8, DOI 10.1007/978-94-017-7499-4_8

corals and carbonate frameworks can eliminate essential habitat space for a multitude of species if coral recruitment is suppressed and recovery unduly prolonged. Coral reef recovery and resiliency are examined in the context of recent ENSO disturbances. Since the last very strong 1997–98 El Niño coral reef bleaching event, live coral cover has increased significantly on many but not all equatorial eastern Pacific (EEP) reefs.

Keywords
Coral bleaching • Thermal thresholds • Bleaching indices • Panama • Galápagos

8.1 Introduction

The very strong 1982–83 El Niño warming event was touted as the most severe of the century and was the first reported to have had widespread impacts on coral reefs (Glynn 1984, 1988a, 1990a). High levels of coral bleaching and mortality were documented in the eastern Pacific (Glynn 1983a, 1990b; von Prahl 1983, 1985, 1986a; Cortés et al. 1984; Robinson 1985; Guzmán et al. 1987), and at several localities in the Indo-Pacific and Caribbean regions (Brown 1987; Coffroth et al. 1990; Williams and Bunkley-Williams 1990). Extensive coral bleaching was first observed in early 1982 on the Great Barrier Reef, and although coincident with sea warming temperature anomalies and increased solar radiation, this event was not causally linked to the 1982–83 El Niño disturbance in the eastern Pacific (Fisk and Done 1985; Harriott 1985; Oliver 1985). Previous records of thermally related coral bleaching and mortality exist, one in the Florida Keys as early as 1911 (Mayer 1914). These events, however, were often local, restricted to particular biotic zones, temporally infrequent and sometimes associated with other stress conditions, such as spring tidal exposures and lowered salinities caused by torrential rains and river discharge (Goreau 1964; Egaña and DiSalvo 1982). Oliver et al. (2009) noted that there are a total of 26 records of coral bleaching before the first well documented 1982–83 event.

What sets the 1982–83 coral bleaching event apart from these earlier reports are the severity and scale of the disturbance. Coral bleaching and mortality coincident with sea warming in 1982–83 occurred regionally over extensive areas, inter-regionally in the eastern Pacific, the south, central and western Pacific, Indian Ocean and Arabian Sea, and the greater Caribbean region (Glynn 1984, 1993, 1996; Brown 1987; Coffroth et al. 1990; Baker et al. 2008). Zooxanthellate corals in all reef zones, from 1 to 20 m depth, experienced whole to partial colony bleaching and mortality. In the equatorial eastern Pacific, mean coral mortality was correlated with elevated sea surface temperature anomalies and rates of increase, and ranged from 51 % in Costa Rica to 97 % in the Galápagos Islands (Glynn et al. 1988).

Coral reef bleaching and mortality have continued globally since 1982–83 with the number of events spiking in the 1980s (Ogden and Wicklund 1988; Williams and Bunkley-Williams 1990; Wilkinson 1992; Ginsburg 1993; Glynn 1993; Goreau and Hayes 1994). Significant coral bleaching and mortality occurred world-wide during the 1986–87 El Niño and through the 1980s. Wilkinson (1998) observed that coral reef bleaching in 1997–98 was the most geographically widespread ever reported and probably the most severe in recorded history. This impact to coral reefs coincided with the very strong 1997–98 El Niño event, immediately switching over to the strong 1999–2001 La Niña event (Fig. 8.1). Coral bleaching and mortality, frequently in conjunction with high temperature anomalies, have continued worldwide to 2007 (Baker et al. 2008) and more recently. Catastrophic coral mortality occurred in the eastern Caribbean in 2005, not concurrently with an El Niño event, but

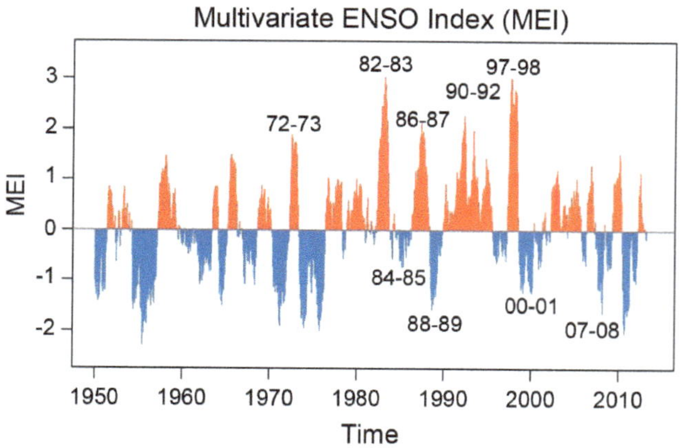

Fig. 8.1 Pacific-wide thermal history, 1950–2012, based on the multivariate ENSO index (MEI). The MEI is composed of six main observed variables over the tropical Pacific: sea level pressure, zonal and meridional components of the surface winds, sea surface temperature, surface air temperature, and sky cloud cover (www.esrl.noaa.gov/psd/enso/mei/). Three of the five major El Niño events identified by year were of the EP or eastern Pacific type (1972–73, 1982–83, 1997–98) and two of the CP or central Pacific type (1986–87, 1991–94) (Wang et al. Chap. 4). Four La Niña events discussed in this chapter are also identified

Table 8.1 Variety of immediate and long-term ENSO disturbances affecting reef-building corals and coral reef communities. Modified and updated after Glynn 2000

Stressors	Impacts[a]	Locations	Sources
Temperature			
High	Bleaching, mortality (EN)	Eastern Pacific, Indonesia, French Polynesia, Japan, Tokelau Iss	Glynn (1990), Brown (1987)
		Several regions globally	Wilkinson (1998)
Low	Bleaching, mortality (LN)	Panama	Glynn and D'Croz (1990)
		Galápagos Iss	Banks et al. (2009)
		East Kalimantan	Wilkinson (1998)
UV radiation	Bleaching, mortality (EN)	Caribbean	Gleason and Wellington (1993)
Storm damage	Scouring, dislodgment of corals, burial by sediments (EN)	Galápagos Iss	Robinson (1985)
		Huatulco, Mexico	Lirman et al. (2001)
		Tuamotu Archipelago	Laboute (1985) Harmelin-Vivien and Laboute (1986)
Subaerial exposure	Bleaching, mortality (LN)	Panama	Eakin et al. (1989)
	Reduced circulation (EN)	Tokelau Iss	Glynn (1984)
Nutrients High	Overgrowth of corals by filamentous and macroalgae (LN)	Panama	Glynn and Maté (1997)
			Smith et al. (2010)
Low	Trophic deprivation (barnacles, oysters) fish larvae (EN)	Equatorial eastern Pacific	Robinson (1985)
			Robertson et al. (2004)
Dinoflagellate Blooms	Asphyxiation, toxicity (LN)	Costa Rica, Panama	Guzmán et al. (1990)
		Colombia	Vargas-Ángel (1996)
		W Sumatra IOD[b]	Abram et al. (2003)
Calcification	Lowered coral growth (EN)	Colombia, equatorial eastern Pacific	von Prahl 1986b Wellington and Dunbar (1995)
	Lowered calcification, cementation (LN)	Panama	Eakin (1996) Manzello et al. (2008) Manzello (2010)
Bioerosion	Higher rates following coral bleaching and death (EN)	Equatorial eastern Pacific	Glynn (1988b) Scott et al. (1988) Eakin (1996) Reaka-Kudla et al. (1996)
		Northeastern Brazil	Attrill et al. (2004)
Coral sexual reproduction	Reduced/absent (EN); continuing/enhanced reproduction (moderate EN); lower recruitment (post-EN)	Equatorial eastern Pacific	Glynn et al. (1994), (1996) Colley et al. (2006)
		Florida keys	Szmant and Gassman (1990)
Predator concentration	Accelerated mortality of surviving corals (post-EN)	Panama	Glynn (1985)
Predator-prey spatial relationships	Disruption of prey refugia (post-EN)	Panama	Glynn (1985)
Diseases	Bleaching, coral tissue debilitation (EN, LN)	Galápagos Iss	Vera and Banks (2009)
	El Niño warm phase	Indian Ocean	McClanahan (2004) Wilkinson et al. (1999)

[a]Impacts occurring during El Niño (EN) and La Niña (LN) phases of ENSO event
[b]IOD, upwelling caused by Indian Ocean Dipole

during the warmest recorded year in Planet Earth's Holocene history (Wilkinson and Souter 2008; Eakin et al. 2010).

High sea temperatures associated with ENSO activity have been experimentally linked to coral bleaching by Glynn and D'Croz (1990) in the eastern Pacific, and Glynn et al. (1993) and Berkelmans (2002, 2009) in the western Pacific. With the increased frequency, severity and widespread extent of sea warming impacts during the 1980s and 1990s, expert opinions have changed on the threats and causes of coral reef decline, namely with an emphasis on direct anthropogenic pressures to global climate change (contrast Wilkinson and Buddemeier 1994; Wilkinson 2000). Colgan (1990) hypothesized that high coral mortality resulting from recurrent ENSO activity would depress coral reef growth, with bioerosion gaining ascendency and the eventual loss of eastern Pacific reef structures. A recent paleoecologic study in Panama has provided compelling evidence of coral reef collapse over a 2500 year period, due to increased variability and intensification of ENSO activity (Toth et al. 2012).

A brief overview of ENSO oceanographic and meteorological conditions that can affect reef-building corals and associated biota is presented in the following section. Examples of ENSO-related physical and biological disturbances reported to affect coral reef communities are noted in Table 8.1. These known potential stressors are examined here in Sect. 8.2 chiefly in relation to the eastern tropical Pacific (ETP), but with reference to other coral reef regions where they may also have an impact. Biotic responses are reviewed in Sect. 8.3 with emphasis on ETP studies, and from observations in other regions because of the global reach of ENSO conditions.

8.2 ENSO Conditions

El Niño-Southern Oscillation is comprised of two phases, namely El Niño, characterized by sea warming and high rainfall in parts of the eastern Pacific, and La Niña, a period of anomalously cool sea temperatures and low rainfall (Philander 1985, 1990). La Niña, the cool phase, is also sometimes referred to as anti-El Niño or El Viejo (Feingold 2011). The ENSO two-phase phenomenon is accompanied by several additional atmospheric and oceanographic conditions that can have varying effects on corals. Geographically, ENSO spans the entire tropical Pacific Ocean (Enfield 1989; Cane 1986; Philander 1990), and can have effects as far afield as the Atlantic and Indian Oceans. El Niño mean frequency is 3.8 ± 1.8 (SD) year and strong to very strong events occur every 9.9 ± 4.2 year (Enfield and Cid 1991). A 62 year record of ENSO activity across the Pacific basin, with La Niña events dominant from 1950 to 1976 and El Niño events more frequent thereafter (1977–2012), is denoted in the Multivariate ENSO indices (MEI) produced by NOAA (Fig. 8.1).

The earliest ENSO stressors are predominantly physical, leading to short-term (days, months) effects/responses to corals, and are often followed by cascading biotic effects that can cause lasting, long-term (years, decades) ecosystem-level perturbations. A variety of impacts to coral reefs, with emphasis on eastern Pacific reefs, have been noted in Glynn (1988, 2000a, 2001), Reyes-Bonilla et al. (2002), Cortés (2003), Wellington and Glynn (2007) and Baker et al. (2008). Below we first consider potentially stressful physical conditions that typically accompany ENSO, followed by short-term and then longer-term effects and responses.

8.2.1 Temperature Extremes

8.2.1.1 Eastern Tropical Pacific

Before the extreme high sea temperatures that accompanied the 1982–83 El Niño event, low temperatures were generally understood to limit coral reef development in the eastern Pacific. Several renowned scientists in the 19th Century championed this notion (e.g., Couthouy 1842; Darwin 1842; Dana 1843; Verrill 1866; see Chap. 1, Glynn). Low temperature effects in limiting eastern Pacific reef development also have figured importantly in studies up to the early 1980s (Glynn et al. 1972, 1983; Glynn and Stewart 1973; Glynn and Wellington 1983; Wellington and Glynn 1983). Experimental work on the sensitivity of corals to high temperatures in the central and south Pacific regions demonstrated the significance of the upper thermal tolerance limits of zooxanthellate corals (Jokiel and Coles 1977; Coles and Jokiel 1978). Since the large scale coral bleaching/mortality events of 1982–83, additional disturbances associated with ENSO activity have been documented globally (e.g., Wellington and Glynn 2007; Baker et al. 2008; van Oppen and Lough 2009). AVHRR (Advanced Very High Resolution Radiometer) sensors on NOAA satellites reliably record SST (sea surface temperature) (Wellington et al. 2001), and these data are used to monitor ENSO conditions in the tropical Pacific. Since satellite images capture SST values across the tropical Pacific, the asymmetrical thermal signatures between El Niño warm and La Niña cold occurrences can be discerned (Fig. 8.2). Because of the pivotal role of ENSO-related sea temperature variability, we examine in detail the warm and cold thermal conditions that can stress corals and lead to their death.

8.2.1.2 Local Signatures of ENSO-Related SST Variability

Long-term data sets from Galápagos (Academy Bay) and Panama (Naos Island) allowed Podestá and Glynn (1997, 2001) to determine stressful thermal metrics that caused coral bleaching in these areas during periods of extreme high and low SSTs. These are updated here to include data through most of 2012.

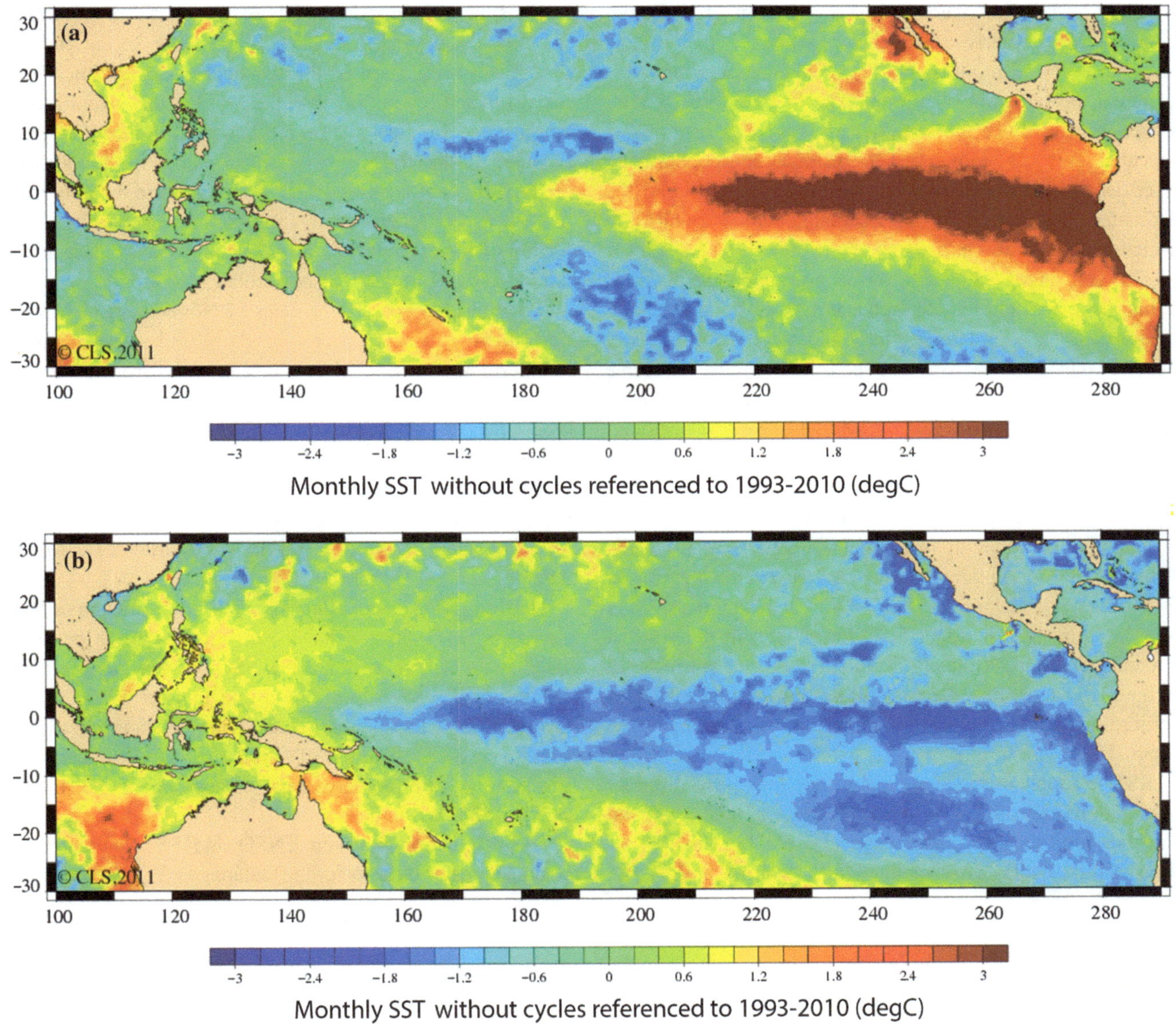

Fig. 8.2 El Niño warm thermal signature, December 1997 **a** La Niña cool thermal signature, December 2010, **b** Satellite images show monthly mean SST (°C) anomalies across the tropical Pacific Ocean as referenced to 1993–2010. From NOAA's AVHRR satellite sensor. Image produced from data stored at the Centre National d'Etudes Spatiales (CNES) and Collecte de Localisation Satellite (CLS). Map generated at http://bulletin.aviso.oceanobs.com/html/produits/indic/enso/welcome_uk.php3

In the Galápagos Islands, daily SST measurements were made at the Charles Darwin Research Station in Academy Bay (Bahía Academia) on the island of Santa Cruz, one of the main islands in the archipelago (see Glynn [1994, Fig. 2], for detailed map). Original daily SST records available encompassed the period 1 January 1965 to 30 September 2012 (approximately 48 years). A total of 241 daily SSTs are missing, with the length of data gaps ranging from 1 to 91 days. Short gaps (1–3 days) were filled by repeating the last SST value before each gap. This procedure still left 197 unavailable values.

In the Gulf of Panama, SST was measured daily at the Smithsonian Tropical Research Institute, Naos Island, near the Pacific entrance to the Panama Canal. Daily SST observations were available for the period 1 January 1970 to 21 November 2012 (about 43 years). The total number of missing daily SSTs was 458, with the data gaps ranging from 1 day to a very wide gap of 319 missing values that encompassed most of 1999 (1 January to 15 November). As for the Galápagos data, short gaps (1–3 days) were filled by repeating the last SST value before each gap. After filling in missing values, 336 values remained missing.

Daily SST values were treated as the sum of two components: (1) a yearly seasonal component, a pattern that repeated every 12 months, and (2) a residual component that included the remaining variation. The first step in the analysis was to estimate the seasonal SST cycle, a pattern that repeated every 12 months. The residual component—defined as the difference between the observed SST and the estimated seasonal cycle—is associated with variability at scales from days to years.

The Galápagos SST cycle shows no obvious semiannual component, and the annual signal is clearly dominant. The cycle peaks between late February and early April; the cool part of the cycle occurs between mid-August and late September. In Panama, the seasonal SST cycle shows only small changes between early June and early December. A marked cooling begins in December that is tied to the seasonal intensification of the northeast trade winds. The prevailing northerly winds displace the Gulf of Panama waters offshore, inducing upwelling of deeper, cooler waters. Although the amplitudes of the seasonal cycles are similar for Panama and Galápagos (4.2°−4.4 °C), the annual median SST in Galápagos is about 5 °C lower than in the Gulf of Panama. The approximately 5-month season of warm SSTs in the Galápagos, however, is still within the range of temperatures supporting coral reproductive activity (see Chap. 15, Glynn et al.).

SST residuals of the same magnitude may have different implications depending on whether they coincide with the warm or cool portions of the seasonal cycle. For this reason, the seasonal cycles are separated into "warm" and "cool" portions, respectively defined as seasonal SSTs being above or below the overall median of SST values. In the Galápagos, the warm portion of the seasonal cycle (during which the estimated cycle is above 23.3 °C, the overall SST median) encompasses the period between 18 December and 17 June of the following calendar year (Fig. 8.3a). Conversely, the cool portion of the seasonal cycle extends from 18 June to 17 December. In the Gulf of Panama, the warm portion of the cycle (SST above overall median of 28.1 °C) extends from 9 May to 16 November; correspondingly, the cool period encompasses the dates 17 November to 8 May of the following calendar year (Fig. 8.3b). Individual warm or cool periods were labeled by the calendar year in which that period begins. For example, the Galápagos warm period of 1982 extending from December 1982 to 17 June 1983, was denoted as "WP82". All subsequent calculations are based on full seasonal cycles. For this reason, the data finally employed encompass the period 18 June 1965 to 17 June 2012 for Galápagos, and 9 May 1970 to 8 May 2012 for the Gulf of Panama.

The long, consistent runs of positive and negative SST anomalies in Galápagos have been recognized to be associated with warm and cold ENSO events, respectively (Dunbar et al. 1994; Wellington and Dunbar 1995). Large positive MEI values are associated with the occurrence of El Niño events and warm SST anomalies in the Galápagos (Fig. 8.4). In turn, negative MEI values correspond to cool La Niña SST anomalies.

The association between ENSO and SST anomalies in Panama is not as clear as for the Galápagos (Fig. 8.5) and has

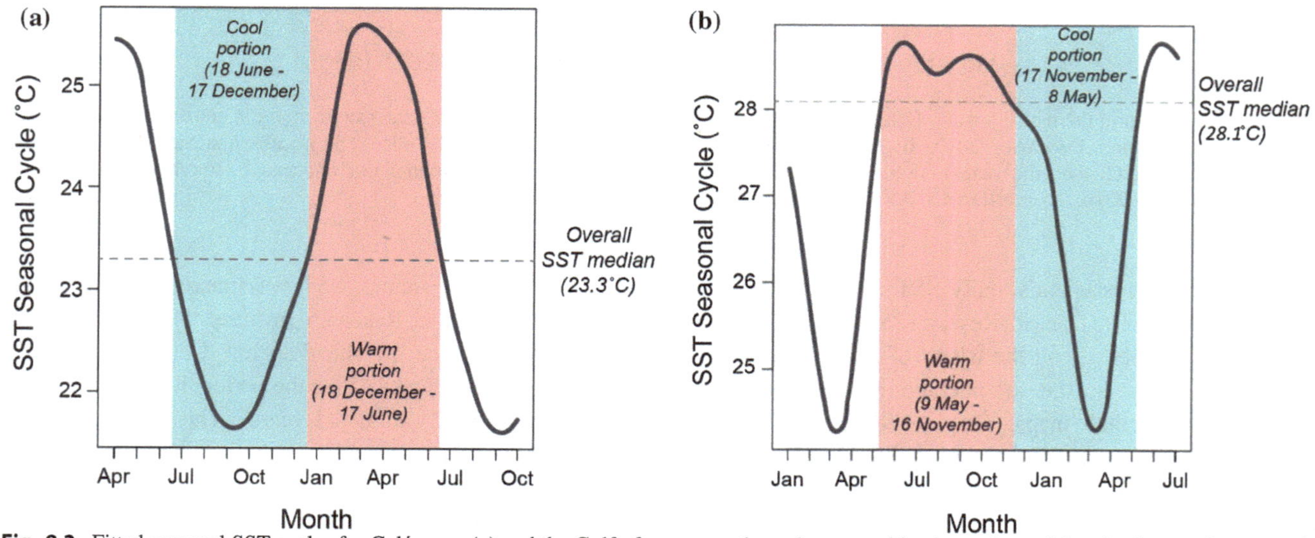

Fig. 8.3 Fitted seasonal SST cycles for Galápagos (**a**) and the Gulf of Panama (**b**). In previous work a non-parametric smoother (loess) was used to estimate the seasonal cycle. Here a generalized additive model (GAM) was employed, where smoothing splines were used instead of linear coefficients for covariates (Hastie and Tibshirani 1990). In this case, seasonal SST is expressed as a smooth function of day of the year, through a combination of model selection and automatic smoothing parameter selection using penalized cyclic splines (Wood 2006). This analysis was conducted using the gam modeling function in package "mgcv" (Wood 2006) of R environment for statistical computing (R Development Core Team 2009)

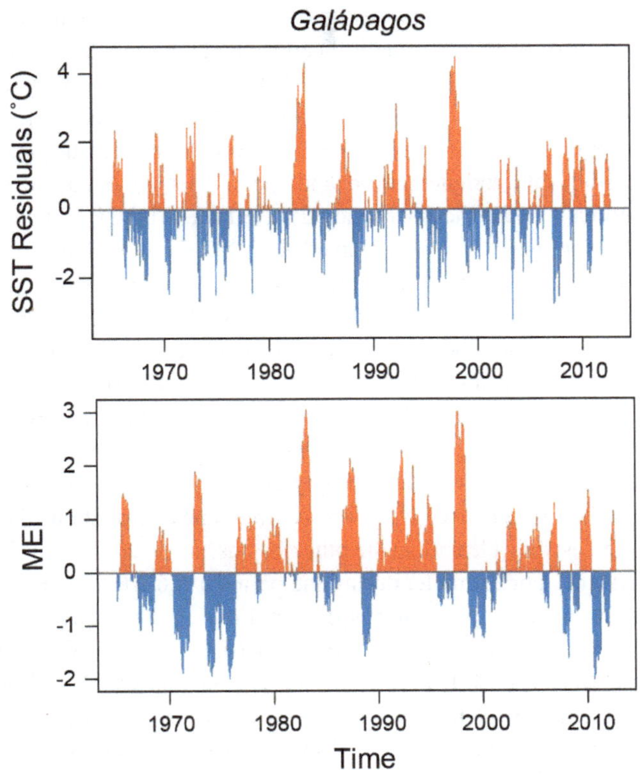

Fig. 8.4 SST anomalies (residuals from the seasonal cycle) and Multivariate ENSO Index values, Galápagos. ENSO state is monitored with the use of a Multivariate ENSO Index, a measure of the ENSO signal expressed as the composite of six observed variables over the tropical Pacific: sea-level pressure, zonal and meridional components of the surface wind, sea surface temperature, surface air temperature, and total cloudiness fraction of sky (Wolter and Timlin 1998, 2011). The MEI is computed separately for each of the twelve sliding bi-monthly seasons (Dec/Jan, Jan/Feb, etc.). Here, MEI values were assigned to the second month in the bi-monthly window. By comparing the monthly-averaged SST residuals with monthly values of Multivariate ENSO Index values, associations between ENSO and local SST anomalies in Galápagos and Panama become apparent (see also Fig. 8.5)

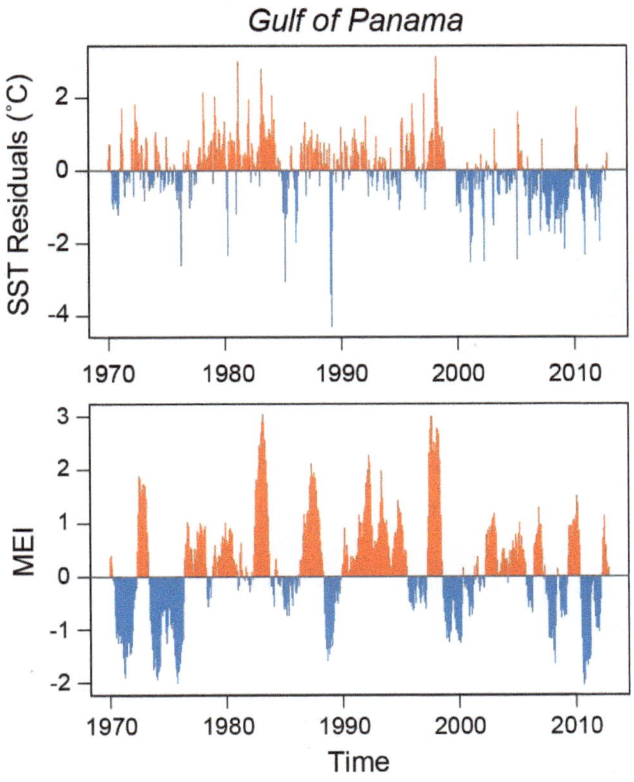

Fig. 8.5 SST anomalies (residuals from the seasonal cycle) and Multivariate ENSO Index values, Panama

been the subject of some debate (Kwiecinski and Chial 1987; Glynn 1990). Previous work has shown that ENSO effects (at least as manifested in SST) were only apparent in the Gulf of Panama during strong events (Podestá and Glynn 1997).

The difference in local SST anomalies as a function of the Pacific-wide state of the ENSO phenomenon is illustrated by boxplots of monthly SST residuals partitioned by positive or negative values of the MEI. Figure 8.6a shows a clear difference in the distribution of SST residuals in Galápagos, where SST anomalies are mostly positive for positive values of the MEI; for negative MEI values, over 75 % of monthly residuals are negative (the box is below the 0 °C reference line). In contrast, the corresponding boxplots for Panama (Fig. 8.6b) show a much less clear ENSO signal on local SST anomalies. Although there is a tendency for warm SST anomalies to correspond to positive MEI, the distributions of SST residuals for positive and negative MEI values show much more overlap than for Galápagos. The difference in local SST linkages to ENSO patterns is also confirmed by the cross-correlation of local monthly SST anomalies and MEI values: the 0-lag (i.e., co-temporal) correlation is 0.662 for Galápagos, but only about half as high (0.369) for Panama.

The relationship between ENSO-related SST variability and thermally-induced coral bleaching events at these locations is discussed in further detail in Sect. 8.3.1.1 (Thermally-Induced Coral Bleaching Mortality).

8.2.2 Sea Level

Fluctuations in sea level height are associated with ENSO activity, a response to altered wind fields and the thermal expansion/retraction of sea water (Wyrtki 1973, 1985a; Cane 1983; Hansen 1990). During El Niño events sea level in the South Pacific may drop by as much as 40 cm below the norm and rise by a comparable amount in the equatorial eastern Pacific (Wyrtki 1985b). These dynamic, complimentary sea level fluctuations across the tropical Pacific can be observed in satellite images captured during the El Niño event of December 1997 when there was a sea level drop of 30 cm in the western Pacific and a rise of 30 cm in the eastern Pacific (Fig. 8.7a). Extreme sea level lowering has

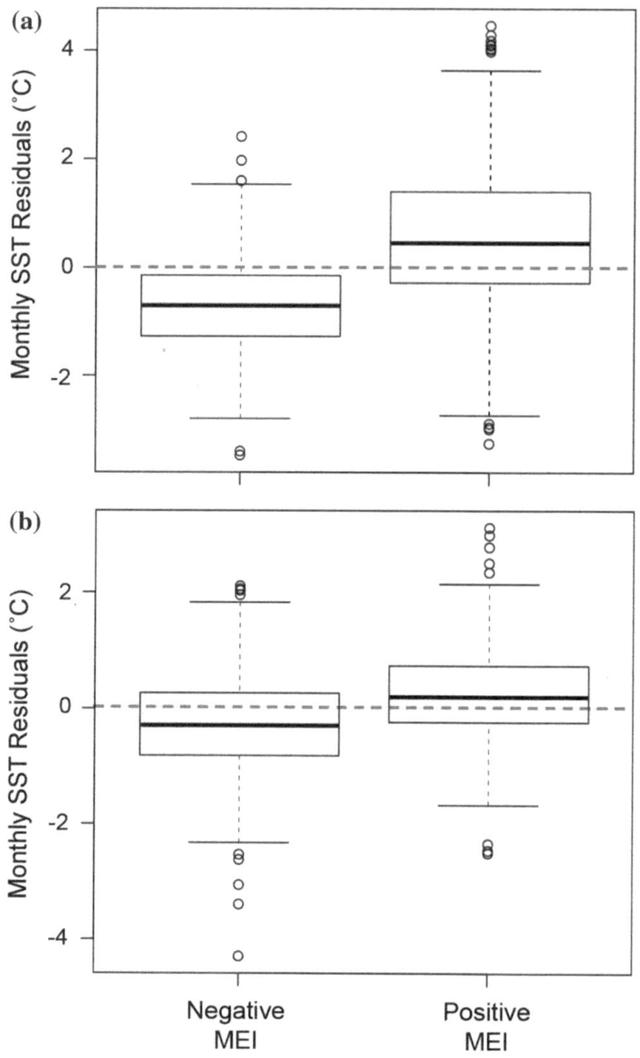

Fig. 8.6 Boxplots of monthly SST anomalies partitioned by negative and positive values of the Multivariate ENSO Index (MEI) for **a** Galápagos and **b** Panama

occurred in the eastern Pacific during La Niña events (Fig. 8.7b), and at other times. When La Niña low sea level stands coincide with spring low tides, extreme reef flat exposures can cause catastrophic mortalities of corals and associated species (e.g., Eakin and Glynn 1996; Cortés and Jiménez 2003). This effect is amplified during mid-day exposures when solar heating and desiccation are pronounced, and during storms with high precipitation (day or night) leading to osmotic shock.

8.2.3 Storm Activity

Storm activity is greatest between about 10° and 30°N latitude where tropical cyclones (TCs) are spawned across the eastern Pacific warm pool (see Chap. 3, Fiedler and Lavín). From the longitudinal locations of maximum SST anomalies and associated atmospheric heating, El Niño events can be classified as eastern Pacific (EP) and central Pacific (CP) warming events. The eastern Pacific El Niño has maximum SST anomalies centered over the eastern tropical Pacific cold tongue region and the CP El Niño has SST anomalies near the International Date Line. The CP El Niño also has been variously named the Date Line El Niño, El Niño Modoki, and the Warm Pool El Niño (see Chap. 4, Wang et al.). Previous work has suggested differences in the formation and location of tropical cyclones (TCs) during these two types of El Niño events (Kim et al. 2012); however, due to a limited subset of years to include in observational analysis, we focus on results where these two types are not distinguished and simply classified as El Niño.

The impact of El Niño on TC frequency throughout the basin is less clear than in other basins, such as the North Atlantic. For example, during the period 1966–2000, a maximum of 26 tropical storms occurred during a prolonged El Niño event in 1992 (Amador et al. 2006). Conversely, Collins and Mason (2000) found that the region east of 116°W is favorable for tropical cyclogenesis regardless of El Niño--Southern Oscillation phase. However, when examining other metrics of TC activity, including intensity and genesis location with corresponding tracks, a stronger relationship emerges. An increase in the frequency of intense hurricanes is observed during El Niño years (Gray and Sheaffer 1991; Whitney and Hobgood 1997). Furthermore, a westward shift in genesis location during El Niño years leads to more storms that travel along the western portion of the eastern North Pacific and away from the Mexican coastline (Chu and Wang 1997; Irwin and Davis 1999; Chu 2004; Camargo et al. 2007).

A 41-year record (1970–2010) of eastern north Pacific storms is examined from the International Best Track Archive for Climate Stewardship (IBTrACS) database (Knapp et al. 2010). An analysis of storm frequency failed to reveal any difference in storm genesis frequency in the eastern development region between El Niño and non-El Niño (neutral) years. El Niño and non-El Niño years demonstrated seasonal mean frequencies of 16.7 and 16.8 tropical storms, respectively. During La Niña years, however, storm frequency was lower with a mean of 14.3 tropical cyclones per season. None of these storm frequencies is found to be statistically significant from the total mean of 16.0 storms per season. Figures 8.8 and 8.9 show the resulting mean storm genesis frequency and track frequency for El Niño (a), La Niña (b), and neutral (c) years. In agreement with previous studies, there is a westward shift in genesis during El Niño that leads to more tracks propagating away from the coastline (Figs. 8.8a, 8.9a). During La Niña and neutral years, however, there are more storms that form near the Mexican coastline and proceed to travel along it (Figs. 8.8b, c and 8.9b, c).

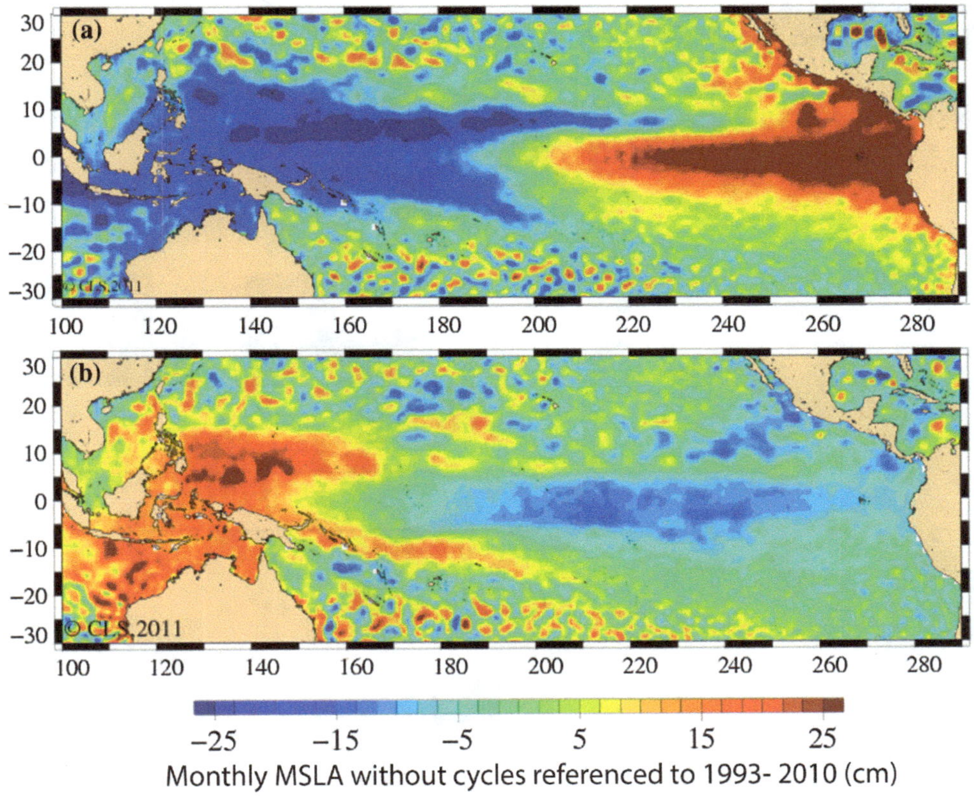

Fig. 8.7 Monthly mean sea surface height (cm) across the tropical Pacific Ocean during an El Niño event in December 1997 (**a**), and a La Niña event in December 2010 (**b**). *Source* TOPEX/Poseidon satellite altimetry data from Centre National d'Etudes Spatiales (CNES) and Collecte de Localisation Satellite (CLS). http://bulletin.aviso.oceanobs.com/html/produits/indic/enso/welcome_uk.php3

Vulnerable reef areas occur along coastal Mexico, and the oceanic Revillagigedo Archipelago and Clipperton Atoll. Cyclones generate storm surge and strong wave action that can damage coral communities and reef structures. Damage can also occur at lower (equatorial) latitudes due to contrary swell direction generated during El Niño storm conditions (Robinson 1985).

8.2.4 Solar Radiation

High levels of irradiance can cause oxidative stress and damage to photosynthetic systems, leading to coral bleaching and death (Shick et al. 1996; Lesser 1997). Highly dependent on geographic location, four factors can contribute to increased penetration of solar radiation, including potentially damaging ultraviolet (UV) wave lengths, during ENSO events: (a) drought conditions with lowered cloud cover, (b) decreased terrestrial runoff and sedimentation, (c) calm sea conditions, and (d) depressed plankton production lowering water column turbidity. While such conditions have been implicated in coral bleaching during El Niño activity in the Caribbean (Gleason and Wellington 1993) and Indo-Pacific (Brown 1997), they have not been systematically linked to El Niño disturbances in the eastern Pacific. El Niño events in the eastern Pacific, particularly at low latitudinal locations, lead to increases in rainfall, cloudiness, and runoff, which decrease the depth of UV radiation penetration of the water column. A widespread but non-lethal bleaching event observed in the southern Gulf of California in 2006, during a non-El Niño period, was possibly a stress response to increased solar radiation (LaJeunesse et al. 2007). The water column was exceptionally clear during this disturbance, prompting the suggestion that anomalously high light levels may have caused the coral bleaching. However, since UV radiation perturbation of eastern Pacific marine ecosystems during El Niño events is unlikely, it will not be discussed in the Biotic Responses section below.

8.2.5 Thermocline/Nutricline Depth

An El Niño event depresses the thermocline, minimizing nutrient replenishment to the photic zone (Fig. 8.10). This in turn reduces primary productivity with negative effects cascading through all trophic levels (Cane 1983, 1986; Barber and Chavez 1986; Barber and Kogelschatz 1990; McPhaden 1999). In the equatorial eastern Pacific Ocean, the increased depth of the thermocline, the boundary between warm and cold sea water at 20 °C, nearly equalizes the thermocline depth across the tropical Pacific (Fig. 8.10). During La Niña the thermocline depth shallows, enabling increased upwelling of nutrients and a significant increase in primary productivity (Fig. 8.11). These contrasting ENSO conditions were evident in low surface water

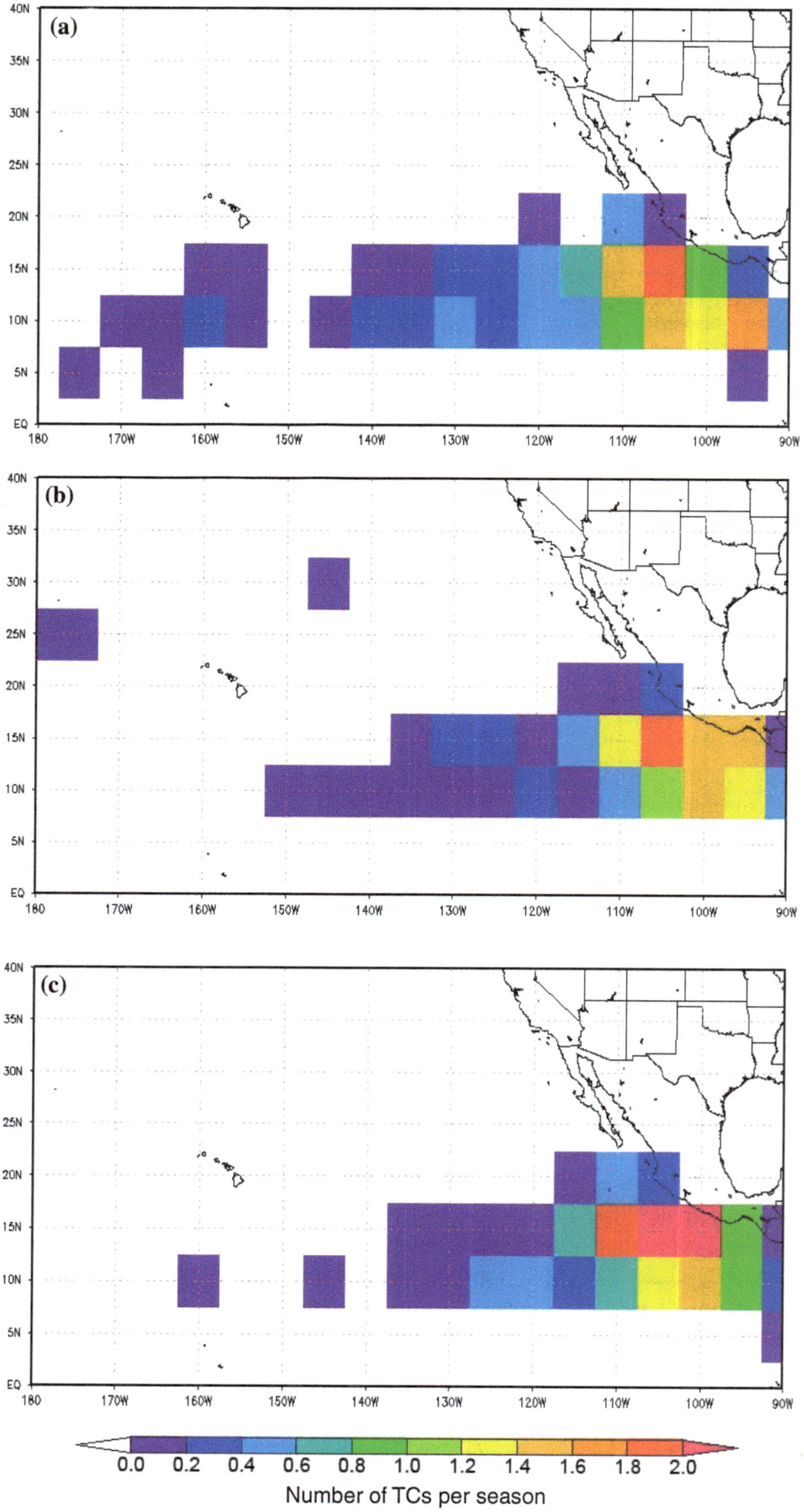

Fig. 8.8 Mean genesis frequency of tropical cyclones per season during **a** El Niño, **b** La Niña, and **c** neutral years. Grid boxes are 5° × 5°

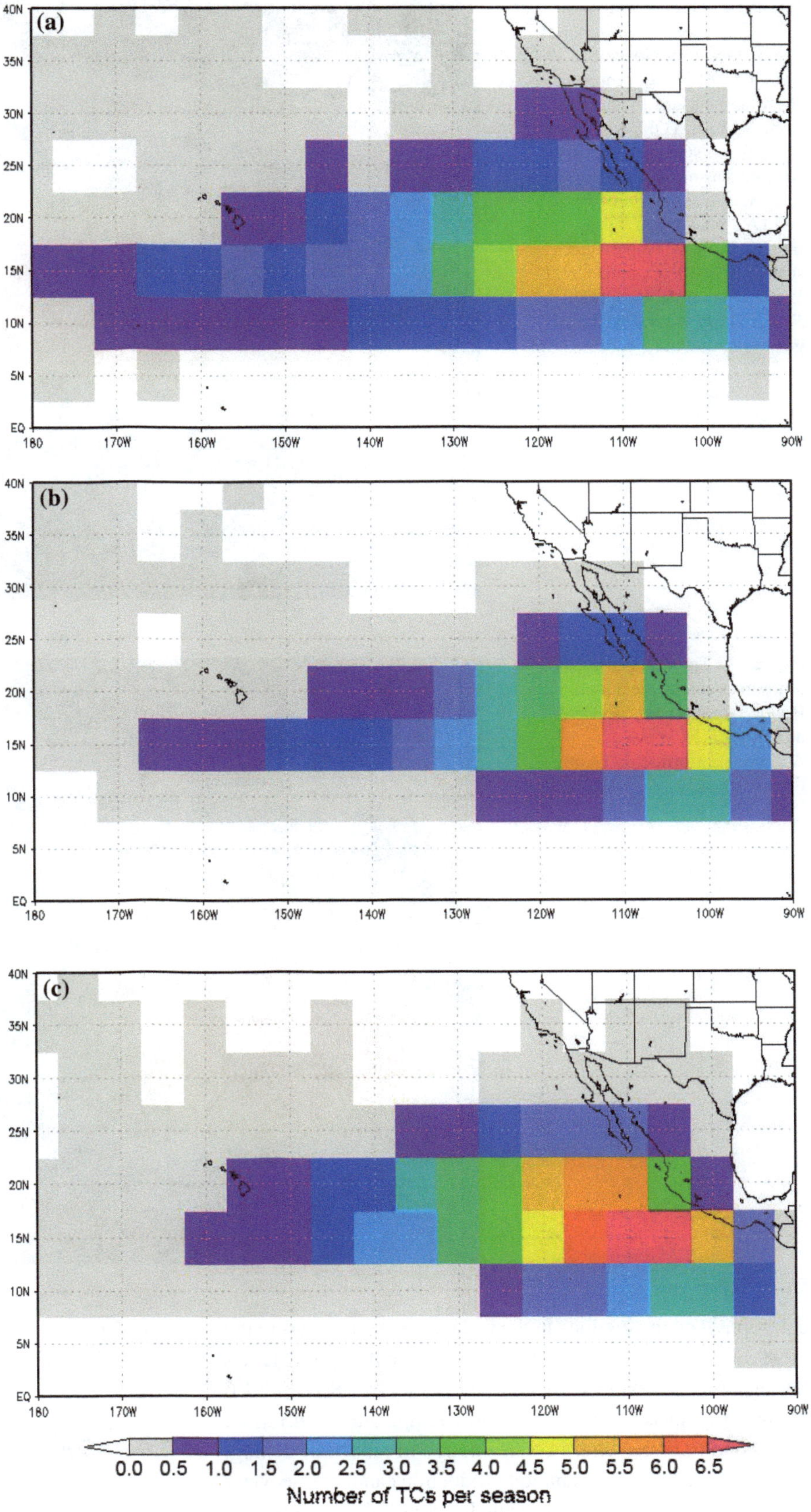

Fig. 8.9 Mean track frequency of tropical cyclones per season during (**a**) El Niño, (**b**) La Niña, and (**c**) neutral years. Grid boxes are 5° × 5°, where a given storm is only counted once per grid box for the duration of its lifetime

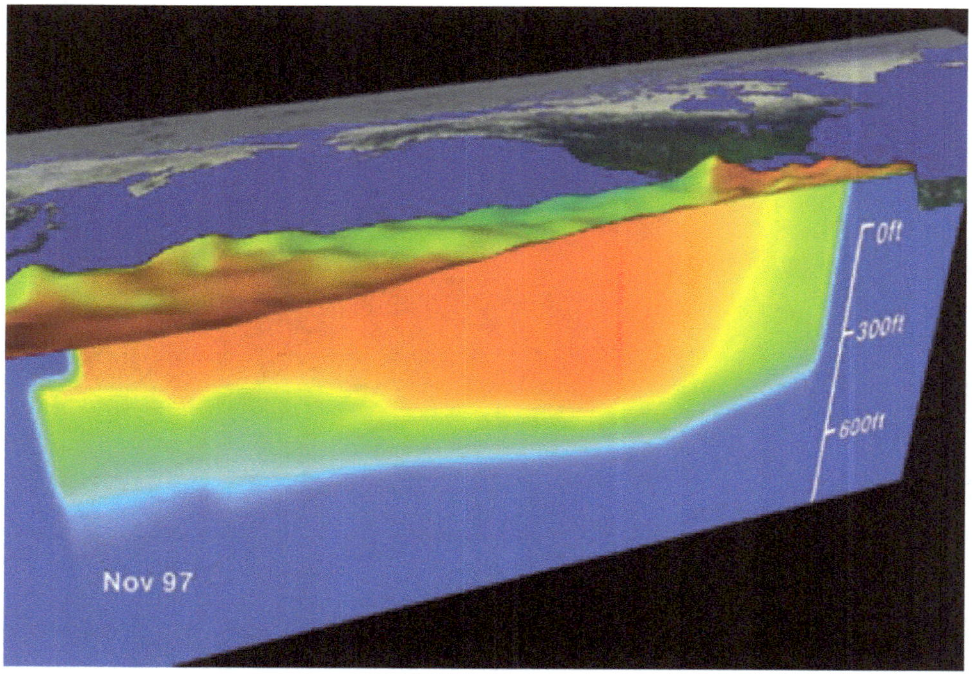

Fig. 8.10 Three-dimensional relief map across the equatorial Pacific Ocean in November 1997 during an El Niño event. Image shows sea surface topography from NASA's TOPEX satellite, SSTs from NOAA's AVHRR satellite sensor, and subsurface temperature as measured by NOAA's network of TAO moored buoys. Temperature scale: *red* = 30 °C; *blue* = 8 °C. Water column depth in feet. *Source* http://www.pmel.noaa.gov/tao/elnino/nino_profiles.html

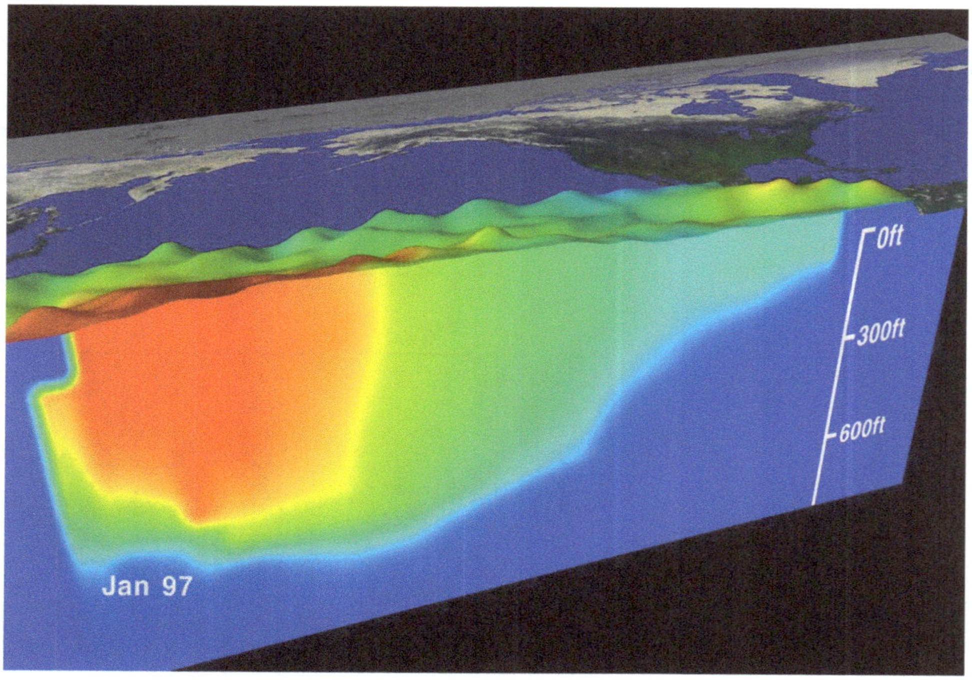

Fig. 8.11 Three-dimensional relief map across the equatorial Pacific Ocean in January 1997 during a La Niña event. Additional information as in Fig. 8.10

chlorophyll concentrations in the Galápagos Islands during the 1998 El Niño event in March, and high chlorophyll levels during a La Niña event 4 months later in July (see Sect. 8.3.1.6 below). As noted above, extreme low temperatures during La Niña conditions can stress corals enough to induce bleaching, however, the increased plankton production can increase water column turbidity reducing potentially stressful UV radiation. Also, if these La Niña nutrient pulses are sufficiently intense they can stimulate toxic phytoplankton blooms that have the potential to affect negatively the greater marine coastal environment (Guzmán et al. 1990).

8.2.6 Surface Currents/Water Masses

Upper ocean current variability in relation to El Niño-La Niña cycles has been extensively sampled and described in recent years. In general, the westward flowing NEC and SEC weaken during El Niño because of weakened easterly trade winds. Also the North Equatorial Countercurrent (NECC) typically increases in velocity, volume transport and duration (e.g., Firing and Lukas 1983; Wyrtki 1985; Mangum et al. 1986; Kessler and Taft 1987; Hansen and Herman 1989; Glynn et al. 1996; Johnson et al. 2000). Picaut et al. (1996) have reported that NECC transport can increase by about 25 %. This is potentially important in terms of larval dispersal and biogeographic connectivity between marine biotas of the central and eastern Pacific regions (Glynn and Ault 2000; see Chap. 16, Lessios and Baums). The deeper flowing Equatorial Undercurrent (EUC), 50–100 m depth, weakens or disappears during the height of El Niño, as observed for example during the events of 1982–83, 1986–87, 1991–92, and 1997–98 (Johnson et al. 2002).

Several investigators have shown evidence of increased penetration of tropical waters northward along the coasts of Central America and Mexico during El Niño events (Kessler 2006). This would allow the transport of planktonic larvae northward along the Central American coast via the Costa Rica Coastal Current, and as far north as the mouth of the Gulf of California via the West Mexican Current. In the southern hemisphere, reversal of current direction during El Niño allows a southerly transport of tropical and subtropical species, via the paths of the South Equatorial Current, and the Peru Oceanic and Peru Coastal Currents, into Peru and as far south as the northern Chilean coast (Arntz et al. 1985; Arntz and Tarazona 1990).

8.3 Biotic Responses

8.3.1 Immediate/Short-Term Responses

It is not possible to establish a well-defined time line separating immediate and short-term direct biotic responses from longer-term indirect effects stemming from ENSO disturbances. Thermal stress from an ENSO disturbance can cause certain coral species' populations to bleach and die during or soon after the perturbation, but the full measure of such losses may not be evident for months or years following. Also, El Niño and La Niña phases generate different kinds of physical stresses, which directly elicit biotic responses (e.g., thermal stressors causing coral bleaching, and nutrient enrichment accelerating macroalgal growth). Adding to the difficulties of establishing a well-defined time line is that a new La Niña event can interrupt coral recovery from a previous El Niño disturbance and vice versa. The overlapping biotic responses to these different physical stresses can have cascading effects on reef organisms, further degrading reef ecosystems. In addition, long after the initial disturbance, indirect effects (such as the loss of obligate metazoan symbionts) may either lower the health or defense of a coral host. These multiple stresses elevate the risk of disease, predation, and bioerosion. These longer-term responses considered below (Sect. 8.3.2) are indirect responses that have consequences for coral community structure, biodiversity, and reef framework integrity, and can continue for years or decades slowing or preventing reef community recovery. With respect to these complicating factors, we consider immediate and short-term responses that directly stress the coral community for up to one year following the beginning of the disturbance.

8.3.1.1 Thermally-Induced Coral Bleaching/Mortality

It has long been recognized that zooxanthellate corals are sensitive to sea temperature extremes (Darwin 1842; Dana 1843), and early studies emphasized that low temperatures stressed corals (Mayer 1916; Vaughan 1919) with negative effects on reproduction (Yonge 1940) and food capture (Mayer 1916; Edmondson 1928). See also Table 15.9 in Chap. 15 (Coral Reproduction in the Eastern Pacific) for various negative effects on reproduction caused by elevated sea temperatures. More recently Coles and Jokiel in the 1970s (see Jokiel and Coles 1990) underlined the pivotal role that high temperature stress (thermal pollution) plays in causing coral bleaching and dysfunctional physiological effects. A large body of evidence from field and laboratory studies demonstrates that high and low temperature extremes cause coral bleaching in the eastern Pacific (Glynn and d'Croz 1990; Maté 1997; D'Croz et al. 2001; Hueerkamp et al. 2001; Wellington and Glynn 2007; Baker et al. 2008). Temperature stress also plays a pivotal role in the identity and density of the algal endosymbiont communities associated with corals during stressful periods (Baker et al. 2008; Cunning and Baker 2013; see also Chap. 13, Baker et al.). Although bleaching in the eastern Pacific is often associated with ENSO thermal extremes it is always necessary to consider other potential causative agents such as, for example, sudden declines in salinity (Cea Egaña and DiSalvo 1982) or increasing sedimentation (Cortés 1990; Jiménez 2001).

El Niño Warm Bleaching

Coral bleaching, due primarily to the loss of endosymbiotic algae (*Symbiodinium*), can be initiated by a variety of stressors with elevated temperatures often playing a dominant role (e.g., Coffroth et al. 1990; Glynn 1993; Brown 1997; Hoegh-Guldberg 1999; see Chap. 13, Baker et al.). The 1982–83 very strong El Niño warming episode resulted in regional coral bleaching and was the first widespread mortality event linked to elevated sea temperatures (e.g., von Prahl 1986a; Williams and Bunkley-Williams 1990; Goreau and Hayes

1994; Brown 1997; Wellington and Glynn 2007). Podestá and Glynn (1997, 2001) documented the relationship between ENSO-related elevations in SST and bleaching events in the Galápagos Islands and Gulf of Panama, and these studies are updated and expanded upon in the following sections.

The maximum SST experienced by corals and the combined effect of magnitude and temporal persistence of SST anomalies ('degree days index') are important quantities associated with coral bleaching (Podestá and Glynn 1997). Sea surface thermal metrics based on these variables have been developed to explore high temperature stress thresholds that correspond to the occurrence of coral bleaching (Podestá and Glynn 1997, 2001; Berkelmans 2009).

The maximum SST experienced by corals is one of the most important quantities in relation to bleaching. Berkelmans (2009) used statistical techniques (classification trees) to determine the relative influence of SST, solar and UV radiation, and a number of other potential weather variables in explaining past coral bleaching events at Magnetic Island, Australia. Results from the Magnetic Island study showed that sea surface temperature is the most important factor separating bleaching and non-bleaching events. Corals bleached when the monthly mean temperature was >31.46 °C (see Fig. 1 in Berkelmans 2002).

Unless SSTs are very high, brief periods of high temperatures are unlikely to cause significant bleaching (Glynn 1993); therefore we discuss monthly SST means rather than daily values. Daily SSTs were first averaged by month; then the highest monthly mean SST within a year was identified. The highest monthly mean SSTs for each year are shown for Galápagos (Fig. 8.12a) and Panama (Fig. 8.12b). In Galápagos, the four highest monthly SST values correspond to years in which coral bleaching was detected. High monthly mean SSTs occurred in March 1983 (28.6 °C), April 1987 (29.0 °C), April 1992 (28.4 °C) and March 1998 (28.5 °C) (Fig. 8.12a). The next highest monthly SST for which bleaching was *not* observed was 27.7 °C in April 1972. So, it would appear that in Galápagos the threshold for bleaching based on warmest monthly mean SST may lie somewhere between 27.7 °C and 28.4 °C. In Panama, only one high monthly SST was associated with bleaching, namely 30.1 °C in June 1983 (Fig. 12b). Similar values (30.1 °C) were observed in both May and June 1972, yet apparently there was no bleaching in that year. Therefore, no potential threshold for bleaching could be ascertained, as was accomplished for Galápagos. Possible reasons for the differences between 1972 and 1983 are discussed below.

Previous work has reported that SSTs averaged over periods shorter than a month (e.g., 3–6 days) can predict occurrence of bleaching (Berkelmans et al. 2004). Therefore such a metric also was explored in this work. Daily SSTs were averaged over rolling windows of 3, 6 and 15 days (to test various averaging spans shorter than one month). The

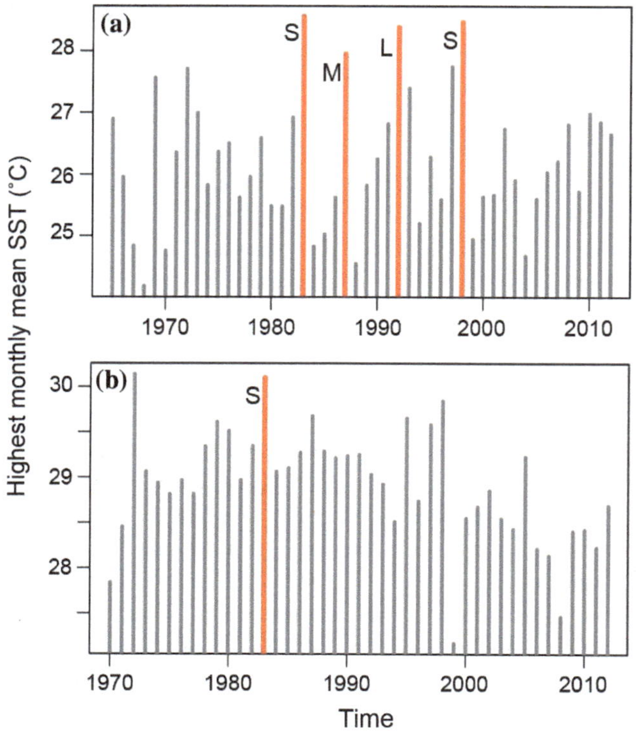

Fig. 8.12 Highest monthly mean SSTs on each calendar year for **a** Galápagos and **b** Gulf of Panama. Bars highlighted in red correspond to years in which coral bleaching occurred in each location. The letters next to the colored bars indicate "strong" (S), "mild" (M) and "light" (L) bleaching

maximum average SST on any warm cycle was identified. In Galápagos, the four bleaching events (1972, 1983, 1987 and 1998) showed the top average SSTs for windows of 6 and 15 days; for a window of 3 days, 1972 and 1993 preceded 1987. In Panama, the top three years with highest average SSTs were consistent for 3-, 6- and 15-day windows: 1972, 1983 and 1998. Again, 1972 was a non-bleaching year, despite showing slightly higher average SSTs than 1983 (31.0 vs. 30.8 °C for 6-day windows).

Another common metric used to predict coral bleaching is the HotSpot index, developed by NOAA in the mid-1990s. This index is based on the positive anomaly of temperatures that exceed the maximum monthly mean SST in a year. As such, it is very close to the highest monthly mean SST within a year discussed above. Nevertheless, it should be noted that HotSpot index assumes that the highest SSTs occur in the same month each year, whereas the results shown here suggest that its occurrence can fluctuate: in Galápagos highest SSTs have occurred in either March or April.

A "degree days" (DD) index is a convenient way of summarizing the thermal stress on coral reefs for a given year. The index used was similar to the "degree-heating weeks" described by Gleeson (1994), and Gleeson and Strong (1995), although computed here in a slightly different manner.

The WDD index describes persistent SST anomalies during the warm portion of the SST annual cycle (Fig. 8.13 a, b). This index reflects previous findings that high SST values during the warm phase of the year are particularly relevant to bleaching. To compute the WDD, the "warm phase" was determined for each area studied (see Sect. 8.2.1.2). Then, the WDD for each year was calculated by summing over the duration of the warm phase the differences between the observed SST values and the overall SST median. To avoid biased WDD values associated with data gaps, the WDD value was defined as missing for years in which the number of missing daily values during the warm phase was >9.

As was found for the highest monthly mean SSTs (Fig. 8.12), in Galápagos the four events for which coral bleaching was detected occurred in periods with the highest WDD values (Fig. 8.13a). The next WDD value corresponds to the warm period from December 1972 to June 1973 (which coincided with an El Niño event), but no bleaching was detected in that year. In Panama, the only bleaching event occurred in May-November 1983 (Fig. 8.13b). No bleaching occurred in May-November 1972 or 1998, even though degree-day values were relatively high. Although no bleaching occurred in the Gulf of Panama during this time, bleaching was observed in the Gulf of Chiriquí, where seasonal upwelling does not occur. This suggests that Gulf of Panama corals were less impacted by the strong 1997–1998 ENSO event due to upwelling offsetting potential stress resulting from elevated SSTs.

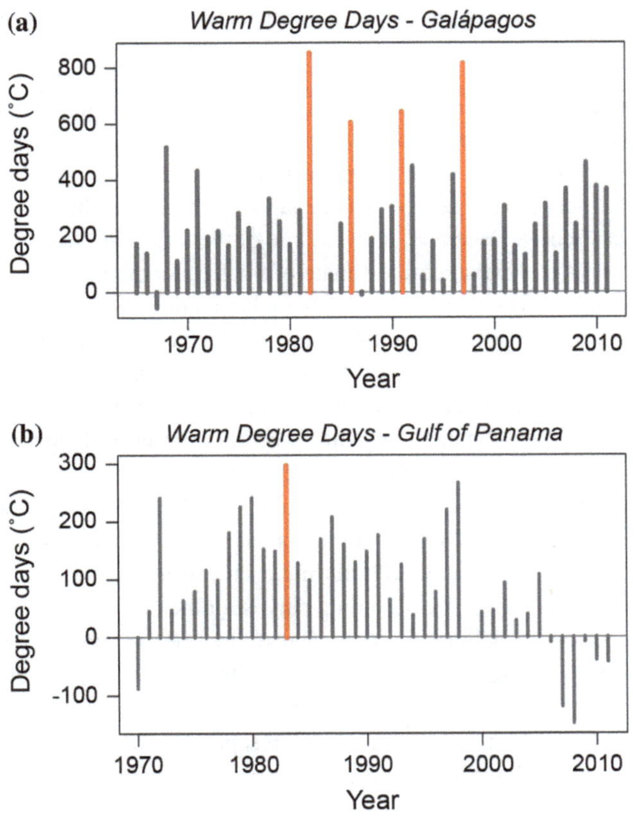

Fig. 8.13 Degree days for the warm portion of the seasonal SST cycle, Galápagos (**a**) and Gulf of Panama (**b**). In Galápagos the warm portion of the seasonal cycle goes from 18 December of a year to 17 June of the following calendar year. Note that, as calendar year boundaries are crossed by this definition, the label shown in Galápagos figure corresponds to the year in which the warm phase started. That is, the bar corresponding to "1970" corresponds to the period 18 December 1970 to 17 June 1971. In Panama, the warm cycle encompasses the period from 9 May to 16 November

Some interesting spatial differences in thermal stressors were observed during the 1982–83 and 1997–98 events. In Panama in 1982–83, the Gulf of Panama experienced a greater warming anomaly and rate of warming than in the Gulf of Chiriquí and this was reflected in higher coral mortalities at the former locality (Glynn et al. 1988). During the 1997–98 El Niño Fong and Glynn (2001) and Glynn et al. (2001) observed a reversed trend with higher elevated temperature anomalies and accelerated warming in the Gulf of Chiriquí compared to the Gulf of Panama. This resulted in high coral mortality in the former gulf and no observed mortality in the latter. At Bahía Banderas, Mexico coral communities experienced overall mortality rates of 96 % in 1997–98, comparable with the high coral losses in the Galápagos Islands in 1982–83 (Carriquiry et al. 2001). Carriquiry and Reyes-Bonilla (1997) noted that the Mexican reefs, surveyed in 1991, showed no evidence of prior ENSO-induced damage such as that observed elsewhere during the 1982–83 El Niño event.

The bleaching event of 1983 in Panama (Figs. 8.13b, 8.14) may be used to illustrate how bleaching may result from a combination of relative (anomalies) and absolute temperatures. Even though strong SST anomalies (ranging from 2 to 4 °C) occurred in March–April 1983, the actual SST experienced by corals during those months was moderated by the peak of the cool phase (i.e., the lowest annual SSTs, associated with seasonal upwelling). In contrast, even though SST anomalies were somewhat lower in April-May 1983 (order of 2 °C), those months coincided with the rapid transition towards the warm portion of the seasonal SST cycle. Bleaching was first reported in June 1983, once seasonal warming combined with positive anomalies to yield SSTs around 30 °C. In other words, in this case the bleaching apparently was not determined by the strength of the SST anomalies but by reaching a threshold SST value around 30 °C. SST anomalies remained positive during the rest of 1983 and strengthened again in February–March 1984, but no bleaching was reported in that year because, again, the strong anomalies were moderated by the cool seasonal upwelling temperatures.

The limitations of using any single metric to predict bleaching are illustrated by the differences between the 1983 bleaching event in Panama and conditions experienced in

1972, when no bleaching was reported (Figs. 8.14 and 8.15). Highest monthly SSTs were very similar in both years: in May and June 1973 monthly mean SSTs were 30.1 °C, the same value reached in June 1983. The WDD value for May-November 1983 was 297 °C, higher than the 241 °C value for 1972 (Fig. 8.13b), suggesting this metric contributes to the separation of bleaching and non-bleaching conditions. Nevertheless, a difference not shown by Figs. 8.12b and 8.13b was the fact that anomalies during the cool phase were much higher in 1983 than in 1972.

The degree-days (DD) index attempts to combine exposure time and magnitude of temperature effect into a single number. Nevertheless, some studies have correctly pointed out that implicit in the DD calculation is the assumption that exposure time and temperature can be combined linearly: for instance, the DD index assumes that x days at y °C above threshold temperature would have the same effect as y days at x °C above threshold temperature (Berkelmans 2002). To address the concerns about the DD index, Berkelmans (2002) proposed the construction of "predicted bleaching curves," that essentially are frequency distributions of binned SST values (in this case, daily observations) for each year or seasonal cycle. By identifying the curves corresponding to years or cycles when bleaching occurred and did not occur, one can identify thresholds for each SST bin; the combination of those thresholds defines the "bleaching curve"; for a graphical example see Fig. 4 in Berkelmans (2002). Similar

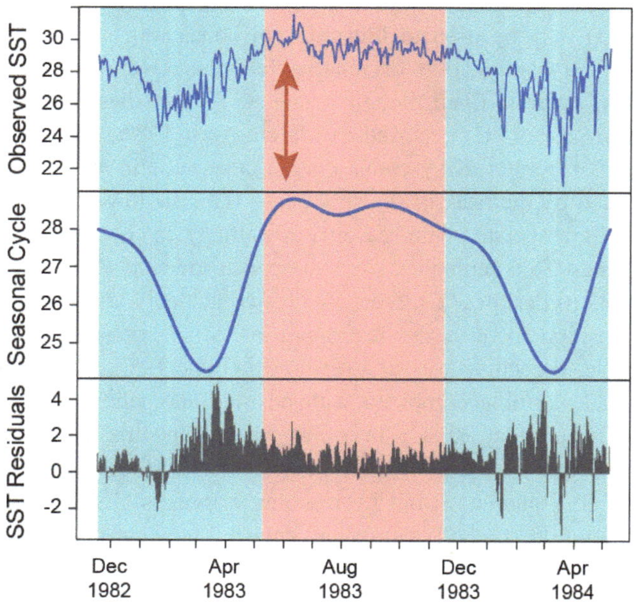

Fig. 8.14 Observed SST, fitted seasonal SST cycle and SST anomalies in the Gulf of Panama, November 1982 to May 1984. The arrow indicates the time when coral bleaching was first detected (around June 1983). The timing of bleaching coincided with rapid warming as the SST cycle entered its warm phase

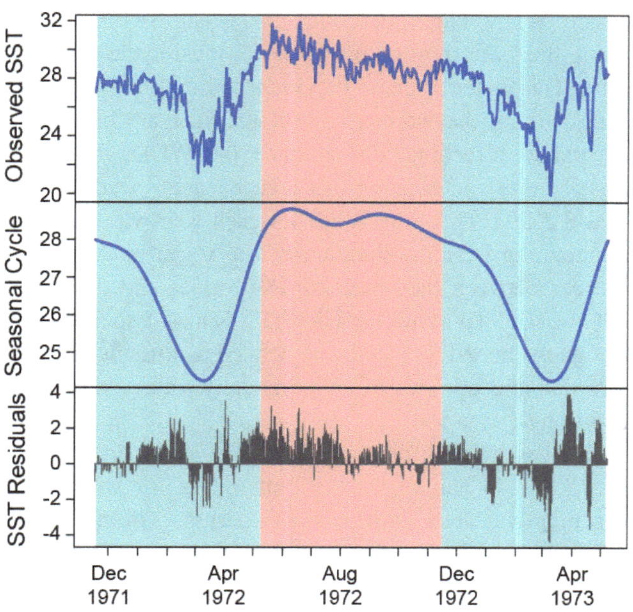

Fig. 8.15 Observed SST, fitted seasonal SST cycle and SST anomalies in the Gulf of Panama, November 1971 to May 1973. No bleaching was reported this year, although large SST anomalies occurred in March–April 1972

Berkelmans-style "bleaching curves" were constructed for the two locations studied. This is the first attempt to build these curves for reefs in the eastern tropical Pacific.

For each warm cycle (in Galápagos, for example, the period between 18 December and 17 June of the following calendar year) a non-parametric estimate of the frequency distribution of daily SSTs was obtained using a kernel density estimation procedure (Bowman and Azzalini 1997). The kernel density estimates are related to Berkelman's histograms (values binned by SST intervals) but are smooth and continuous. Densities were estimated with the "sm" package in R (Bowman and Azzalini 2013). The resultant time-temperature curves, therefore, are the upper end values (SST $\geq$ 25.5 °C for Galápagos and 28.5 °C for Panama) of smooth frequency distributions of daily SSTs during each warm period.

In Galápagos, the bleaching curves for the four cycles with reported bleaching (colored lines) are clearly above the non-bleaching cycles (shown in grey) (Fig. 8.16a). That is, the bleaching curves that aim to combine experienced temperature and exposure time are successful in separating bleaching and non-bleaching conditions (although other metrics did as well). This illustrates how persistent elevated SSTs create conditions conducive to bleaching at this location.

Bleaching curves were also constructed for the Gulf of Panama, where a single bleaching event was recorded (1983) (Fig. 8.16b). Unlike the bleaching curves for the Galápagos Islands site, which show a clear divergence between bleaching and non-bleaching years, there is one curve in Panama with high SSTs during which no bleaching was

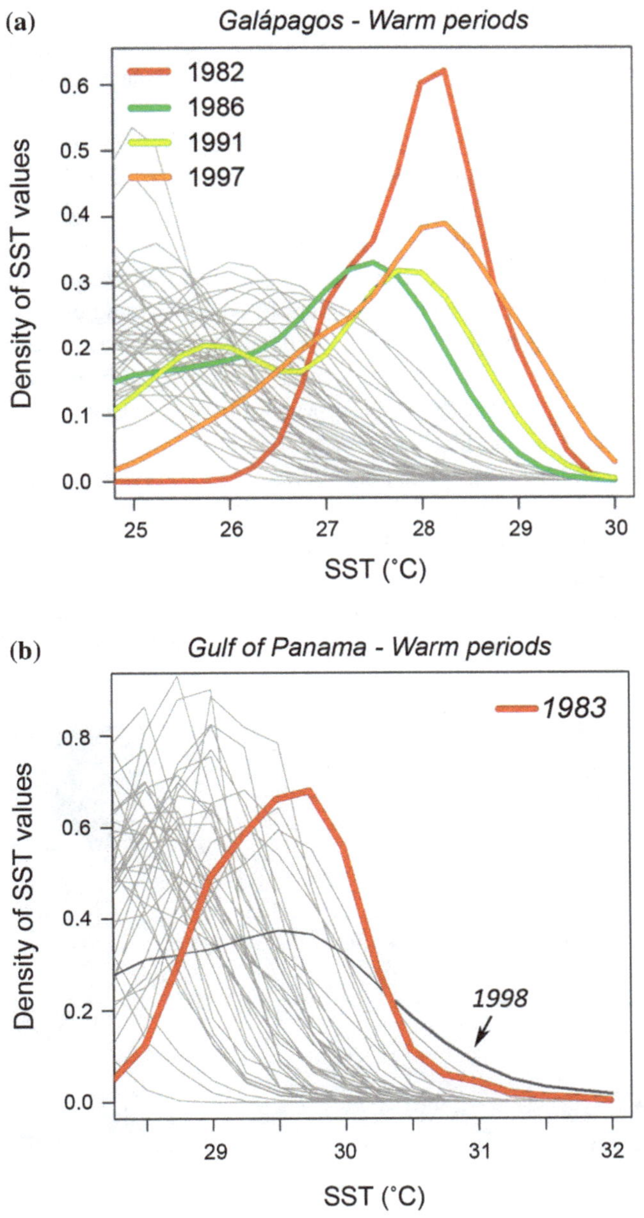

Fig. 8.16 Bleaching curves for the warm portion of SST seasonal cycle in Galápagos (**a**) and Panama (**b**)

observed (1998). This can be attributed to seasonal upwelling mitigating the effects of extreme high temperature stress on corals at this location (Glynn et al. 2001). This result illustrates the need to understand local processes that can modulate or modify the influence of large-scale sea temperature fluctuations on corals.

In summary, both metrics (degree-days index and bleaching curve) used to explore impacts on corals of warm sea temperature anomalies attempt to combine the effects of both exposure time and magnitude of the anomalies. Perhaps a future approach would be to describe impacts on corals as the joint distribution of duration and magnitude of the anomalies via multivariate distributions based on copulas, a technique that recently has begun to be increasingly used for hydrological analyses, characterization of droughts, etc. (Chen et al. 2013).

La Niña Cool Bleaching

Extreme low temperatures resulting from atmospheric winter cooling or upwelling also cause coral bleaching and mortality (e.g., Glynn 1990; Coles and Fadlallah 1991; Lirman et al. 2011). The first La Niña event that was quantified in terms of coral bleaching and mortality occurred in mid-March 1985, 16 months following the very strong 1982–83 El Niño warming event (Glynn and D'Croz 1990). A low temperature of 17.0 °C was recorded at 7 m depth on the Saboga Island, Pearl Islands (Panama) coral reef. Richmond (1990) also reported extreme low temperatures on a coral reef in the Pearl Islands in March 1985, with minimum lows of 14.2 °C recorded on two occasions. Bleaching of *Pocillopora damicornis* was widespread across the Saboga reef with numerous colonies sloughing partially bleached tissues. In contrast to El Niño high temperature bleaching/mortality when coral colonies typically appear stark white (Fig. 8.17a), the affected tissues during the cold event were faintly pigmented, displaying a pinkish-brown color (Fig. 8.17b). Of over 100 colonies surveyed, 27.2 % were bleached with tissue sloughing; 10.4 % were dead. This same coral population experienced 68.5 % mortality during the 1982–83 El Niño warm bleaching event.

Three additional cold water bleaching events occurred in the Pearl Islands during the upwelling seasons in 1989, 2000, and 2001 (Podestá and Glynn 1997; Glynn et al. 2001; Maté 2003; Glynn and Fong 2006). These extreme cold water events followed the El Niño events of 1986–87 and 1997–98 (Fig. 8.1). Low temperatures of 15.4 °C and 14.7 °C were recorded on the Saboga Island coral reef in February and March 2000 and 2001, respectively. In March 2000, reef temperatures were below 20 °C for several weeks before the annual minimum temperature occurred (Fig. 8.18). Unfortunately, coral mortality was not assessed immediately following these events. However, an indication of the severity of the 2000–2001 cold period can be inferred from a coral recovery study conducted by Glynn and Fong (2006). They reported highly significant declines in the colony sizes of *Pocillopora elegans* (50 % mean decrease) and the live tissue cover of *Pavona clavus* colonies (25 % mean decrease). These losses were due primarily to the shedding of dead branches from *P. elegans* colonies and tissue-free patches resulting from partial colony mortality in *P. clavus*.

Cold water coral bleaching in the Galápagos Islands, located at the southern-most, low thermal limit of eastern Pacific zooxanthellate corals, was reported on two occasions since 1975. Bleached *Pocillopora* was first observed (Glynn pers. obs.) but not reported during the La Niña event of

1988. Mild bleaching was later observed in 2007 at Wenman and Darwin Islands in the northern-most sector of the Galápagos Archipelago. This occurred during a 12 °C sudden drop in temperature (28°–16 °C at 15 m depth over a 6-day period) from the end of February to early March (Banks et al. 2009; Glynn et al. 2009). Bleaching was most pronounced below 15 m depth in *Porites lobata* and *Pavona clavus*, corals with massive colony morphologies (Table 8.2). At 18–34 m depth, nearly all colonies of both massive and branching species were pale or completely bleached. The mixed layer thermocline in the northern islands, generally located at ~50 m depth, ascended to shallower depths apparently due to internal wave forcing.

A "cool degree days" (CDD) index was computed here to focus on potential bleaching effects of persistent cold SST anomalies. The CDD was computed similarly as the WDD (see Section "El Niño warm bleaching"), but by summing over the duration of the cool phase of the seasonal cycle the differences between observed SST values and the overall SST median. The "cool phase" was defined as the time of year during which the estimated annual cycle is below the overall SST median. Figure 8.19 illustrates the cool degree days calculated for the bleaching events in the Galápagos Islands (a) and Panama (b), respectively.

The cool degree days value at Academy Bay (southern Galápagos Islands, ~300 km SE of Wenman and Darwin Islands) exceeded −500 °C in 2007 (Fig. 8.19a), a reflection of the prolonged period of stressful low temperatures in 2007. Corals island-wide were subjected to an abrupt temperature change from a moderate El Niño warming event to a strong La Niña cold event at the height of the warm season (Banks et al. 2009). A second abrupt cold shock also occurred in May 2007. From May 2007 to January 2008, a widespread "cold shock" bleaching of *Porites lobata* and *Pocillopora* spp. occurred across the entire Galápagos Archipelago (Fig. 8.20).

To explore cool thermal stress on corals, Berkelmans-style bleaching curves were estimated for each cold period in the Galápagos Islands and Panama. The curves were built in the same way as their warm period

Fig. 8.17 ENSO-induced thermal bleaching in *Pocillopora damicornis*. **a** El Niño warm-water bleaching, Uva reef, 3 m depth, 22 March 1983; **b** La Niña cold-water bleaching, Saboga Island reef, 3 m depth, 19 March 1985

Fig. 8.18 March 2000 monthly mean SST (°C), tropical Pacific, referenced to 1998–2010 mean temperatures. *Source* Fig. 8.2 caption

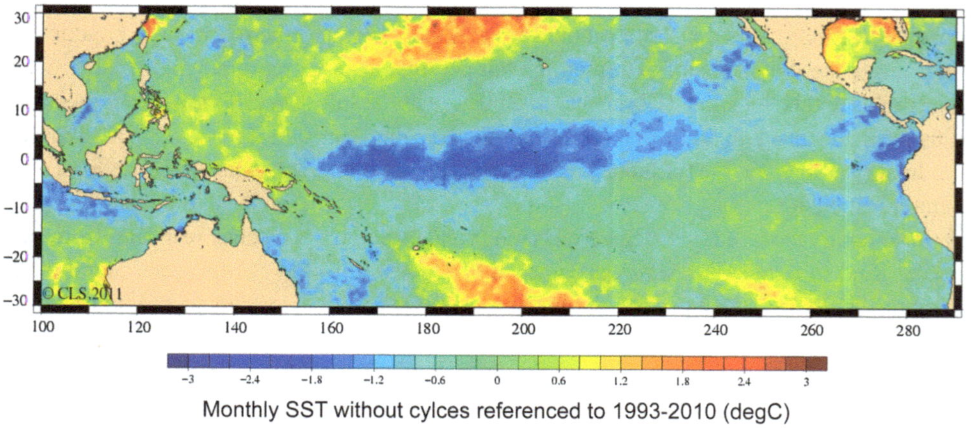

Table 8.2 Initiation of severe 2007 cold-water bleaching event, central and northern-most Galápagos Islands

Location	Date	Depth	Species	Condition		
				Normal	Pale	Bleached
Marchena Is.	3 March	4–8	*Pocillopora elegans*	5	0	0
			Porites lobata	5	0	0
			Pavona clavus	5	0	0
Wenman Is.	4 March	9–14	*Pocillopora elegans*	5	0	0
			Porites lobata	29	0	0
			Pavona clavus	5	0	0
	8 March	12–16	*Pocillopora eydouxi*	4	0	0
		15–18	*Porites lobata*	0	0	2
		8–16	*Pavona clavus*	0	4	2
Darwin I	7 March	9–10	*Porites lobata*	46	1	1
		16–18		2	0	8
		11–20	*Pocillopora* spp.	14	0	0

counterparts (Section "El Niño warm bleaching"), but using daily SSTs observed during the cold period of each seasonal cycle. In this case, the focus is on the lower end of the estimated densities of SST values.

In Galápagos, cold water bleaching was observed in 1988 and 2007, coinciding with periods of persistent low SST anomalies (Glynn et al. 2009). The bleaching curves (Fig. 8.21a) clearly show that these two years had a higher frequency of occurrence of low SST values (<19 °C) than all other years. Indeed, these two periods coincided with persistent low SST anomalies (Banks et al. 2009; Glynn et al. 2009). In 1988, cold SST anomalies first developed around February (actually, during the warm seasonal phase), peaked at the beginning of the 1988 cold period (around July) and persisted through April 1989 (i.e., the following warm period) (Fig. 8.22). Similarly, cold anomalies in 2007 started in mid-March (during the warm seasonal phase) and continued through the cold period of 2007, and into the following warm period, finally dissipating around February 2008 (Figs. 8.23, 8.24).

In the Gulf of Panama, four cold periods were associated with cold bleaching: 1985, 1989, 2000 and 2001. For three of these events (1985, 2000 and 2001), the density curves show relatively high frequencies of low SSTs (<22 °C) (Fig. 8.21b). Densities for some years (1984, 1988 and 2005) lie above the curves for the bleaching events below 19 °C, yet apparently these years are not associated with cold bleaching. Because the differences are observed at the lowest SST values, where densities are lowest, perhaps the frequency of occurrence of low SSTs is not sufficient to cause bleaching, whereas between 20 and 23 °C (where more days occur), the frequency is highest for the years with reported bleaching. It is unclear why 1989 was tied to cold bleaching, as the density of cool SSTs lies in the middle of

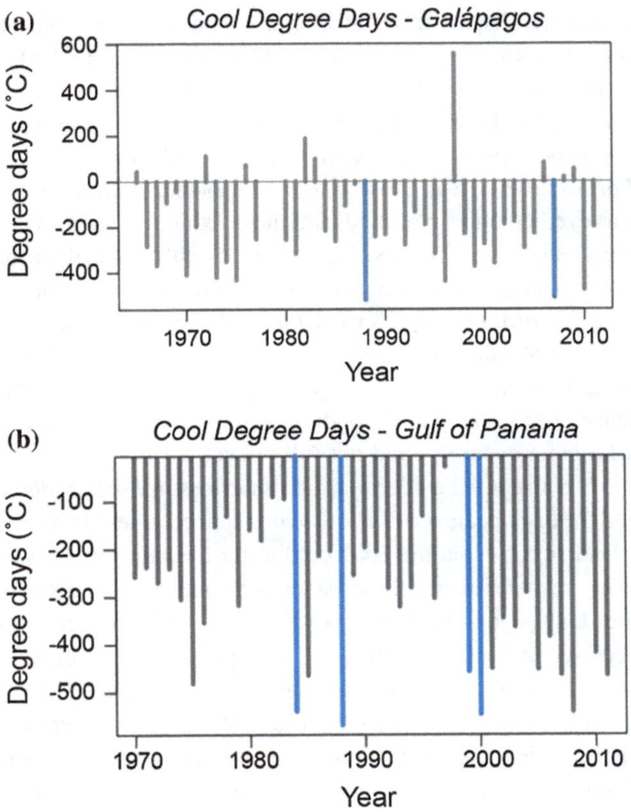

Fig. 8.19 Degree days for the cool portion of the seasonal SST cycle, Galápagos (**a**) and Gulf of Panama (**b**). In Galápagos the cool portion of the seasonal cycle extends from 18 June to 17 December. In Panama, the warm cycle encompasses the period from 17 November to 8 May of the next calendar year. Note that, as calendar year boundaries are crossed by this definition, the label shown in the Panama figure corresponds to the year in which the warm phase began. That is, the bar corresponding to "1970" corresponds to the period 17 November 1970 to 8 May 1971

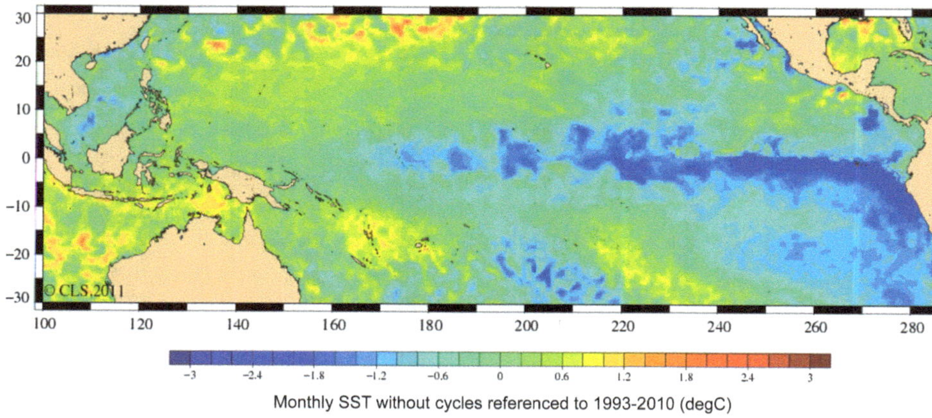

Fig. 8.20 December 2007 monthly mean SST (°C), tropical Pacific referenced to 1993–2010 mean temperatures. *Source* Fig. 8.2 caption

all non-bleaching lines. A run of SST anomalies ~2 °C (data not shown) occurs in January–March 1990, just before the lowest seasonal SSTs. It is not clear if the bleaching could be due to these warm anomalies.

Elsewhere in the equatorial eastern Pacific a cold water bleaching episode in 2009 was reported at Malpelo Island, Colombia (Zapata et al. 2011), and data from an AVHRR sensor captured an image of the SST cooling around Malpelo Island and other areas in the EEP in March 2009 (Fig. 8.25). Four of 10 species bleached, including *Gardineroseris planulata*, *Pocillopora* spp. and *Porites lobata*. *Porites lobata* was most heavily impacted, with 72 % of surveyed colonies bleached, and these exhibiting up to 87 % bleached tissue surfaces. This stress event occurred in the dry season (January–March) with a minimum temperature of 15.8 °C recorded at 20 m depth, the lowest temperature reported at Malpelo Island in 21 years. Zapata et al. (2011) suggested this bleaching response was due to the concurrence of seasonally cold conditions in the Panama Bight and a La Niña event that did not fully develop.

Circumstantial evidence from tissue loss in *Porites lobata* and the disappearance of *Pocillopora* spp. colonies at Onslow Island suggest that the 2007 cold water bleaching event was likely responsible for these losses at a southern island site in the Galápagos Archipelago. Extreme low temperatures occurred in September 2007 at 2 m depth with a minimum of 16.8 °C recorded, the lowest temperature in 17 years. Cold water bleaching was not observed (the study sites were not visited at this time), but *P. lobata* colonies lost 19 % live tissue (Paul 2012) and *Pocillopora* spp. colony abundances declined from 154 to 11 colonies during this period (2007–2009; see Baker et al. 2008; Feingold and Glynn 2014).

Coral communities and small patch reefs in the southern-most region of the Gulf of California also occur near the lower thermal tolerance limits of zooxanthellate corals in northern Mexico. As in the southern Galápagos Islands, pocilloporid species in northern Mexico are predominant and contribute toward incipient reef building. Cold water bleaching, however, has occurred more frequently in Mexico than in the Galápagos Islands with five bleaching/mortality events reported during the last 26 years (Fig. 8.26). Up to 90–95 % mortality occurred in *Pocillopora* spp. at some surveyed sites in 1988 (Wilson 1990) and 2008 (Hernández et al. 2010; LaJeunesse et al. 2010). Four of the five cold-water bleaching events occurred during negative phases of the Oceanic ENSO index. During the last five years monthly mean temperatures below 19 °C have resulted in coral bleaching.

Interspecific Differences to Thermal Stressors

Severe bleaching often leads to partial or whole coral colony mortality. Colonies that experience partial mortality commonly recover and continue growing, often re-sheeting dead patches (Glynn and Fong 2006). In some instances, however, when whole colony mortality is locally intense, notable changes in the relative abundances of species can occur. Bleaching-induced disturbances often lead to differential mortalities, favoring certain species and/or colony morphologies over others. Branching species generally demonstrate higher tissue mortality than massive species (Cortés et al. 1984; Glynn 1990; Hueerkamp et al. 2001), but this is not always the case (Vargas-Ángel et al. 2001).

Quantitative assessments of coral populations at several eastern Pacific sites following the 1982–83 El Niño event demonstrated marked post-disturbance changes, ranging from extreme reductions in species abundances to local and regional disappearances (Glynn 1997). On pocilloporid patch reefs in the southern Galápagos Islands, coral mortalities reached 98–100 % with the disappearance of virtually all frame-building species (Glynn 1994). Local extirpations of *Porites panamensis* and *Leptoseris papyracea* were documented in Panama, and of *Diaseris distorta* and *Cycloseris curvata* in the Galápagos Islands. Presumed eastern Pacific regional extinctions of *Porites rus* (Costa Rica) and *Millepora platyphylla* (Panama), both species observed bleached in 1983, were reported following the El

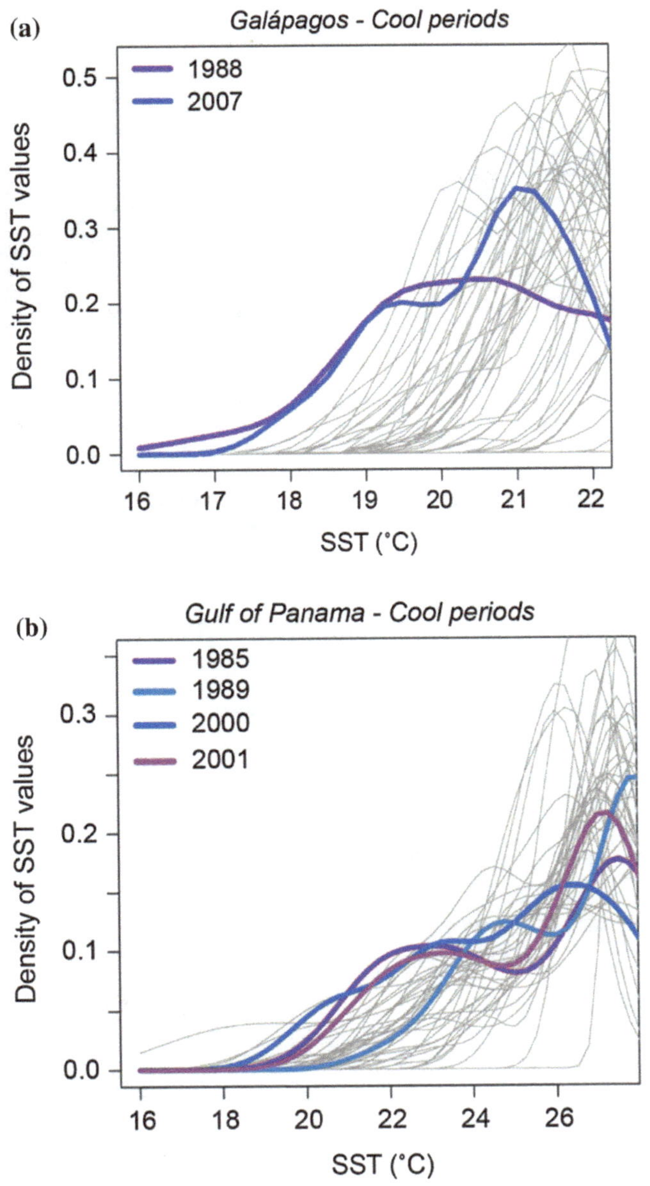

Fig. 8.21 Bleaching curves for the cool portion of the SST seasonal cycle in Galápagos (**a**) and Panama (**b**)

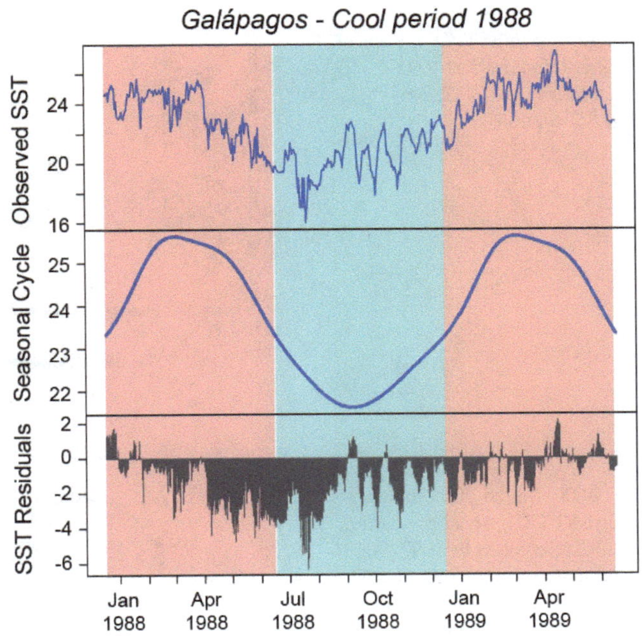

Fig. 8.22 Observed SSTs, estimated SST seasonal cycle and SST residuals in the Galápagos Islands, January 1988–June 1989

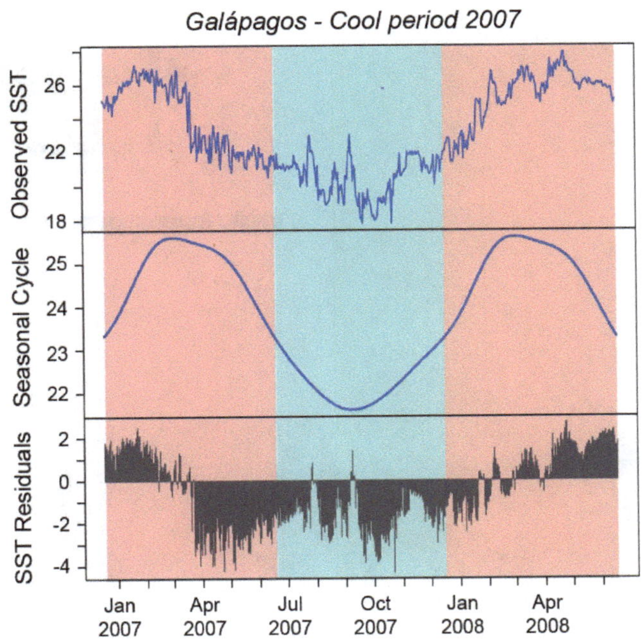

Fig. 8.23 Observed SSTs, estimated SST seasonal cycle and SST residuals in the Galápagos Islands, January 2007–June 2008

Niño warming event (Cortés et al. 1984; Glynn and de Weerdt 1991). It is likely that *Millepora boschmai*, formerly restricted to the Gulf of Chiriquí, is also now regionally extinct (Glynn 2011). No living colonies have been observed during the past decade, until as recently as 2015.

The 1997–98 very strong El Niño event was comparable in magnitude and duration to the 1982–83 record-setting event (Enfield 2001). Its impact on eastern Pacific corals and coral reefs, however, was less than in 1982–83 (Glynn 2000b). Nonetheless, some sites and species not affected earlier were impacted by the more recent event, and some earlier negative effects were exacerbated in 1997–98. For example, only a single known colony of *Gardineroseris planulata* that survived the 1982–83 bleaching event in the southern Galápagos Islands disappeared following the 1997–98 El Niño event, thus resulting in a local extinction (Glynn 1994; Glynn et al. 2001). Coral mortality was low for most species at Culebra Bay, Costa Rica during the 1997–98 El Niño event, however, a relatively deep (7–15 m) 2.5 ha

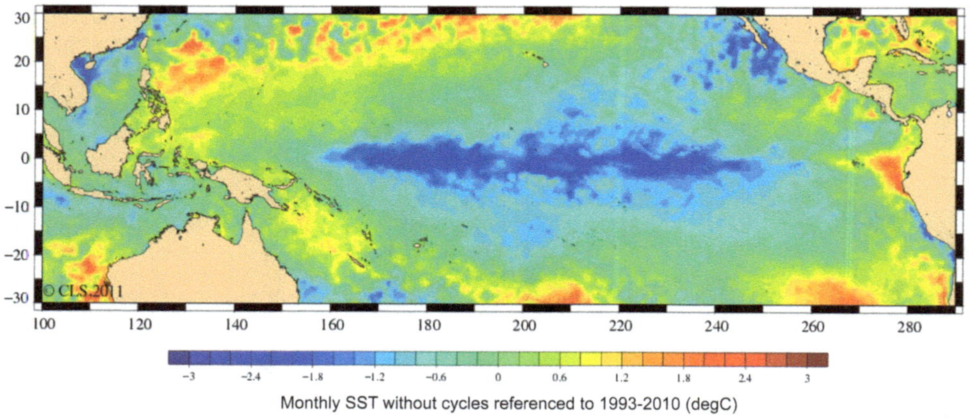

Fig. 8.24 February 2008 monthly mean SST (°C), tropical Pacific, referenced to 1993–2010 mean temperatures. *Source* Fig. 8.2 caption

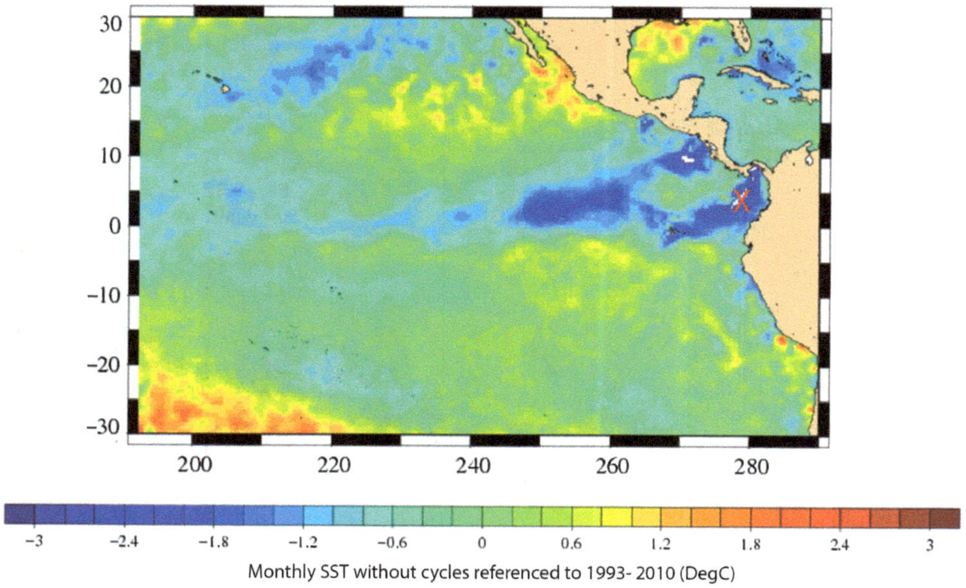

Fig. 8.25 March 2009 monthly mean SST (°C), equatorial eastern Pacific, referenced to 1993–2010 mean temperatures. Red 'X' off Colombian coast marks position of Malpelo Island. *Source* Fig. 8.2 caption

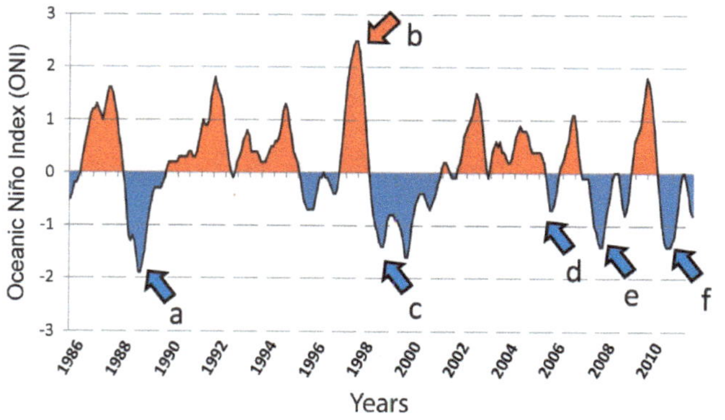

Fig. 8.26 Cold-water bleaching events (**a**, **c–f**) observed in the Gulf of California, Mexico during negative ONI anomalies over a 26-year period. A warm-water bleaching event (**b**) was observed during the very strong 1997–98 El Niño event. ONI is based on SST anomalies in the central equatorial Pacific, in the Niño 3.4 region (120°W–160°E). Modified after Paz-García et al. (2012)

patch of *Leptoseris papyracea* experienced especially high mortality, with a 94.7 % decrease of live cover (Jiménez et al. 2001). Since this incipient, monospecific reef was discovered as recently as the mid-1990s, the effects of the 1982–83 El Niño disturbance on this coral formation are unknown.

8.3.1.2 Coral Symbionts

The loss of endosymbiotic dinoflagellates (*Symbiodinium*) from gastrodermal tissues is usually the first sign of a stress response in zooxanthellate corals. This subject is considered in Chap. 13 by Baker et al. In addition, new molecular sequencing and gene function methods are revealing taxonomically-rich and specialized endolithic and surface inhabiting microbial populations that can initiate metabolic changes in corals. These organisms may offer stress resistance and protection against coral pathogens but are outside the scope of this review (e.g., Knowlton and Rohwer 2003; Thurber et al. 2009; Rypien et al. 2010).

Corals also host a diverse assemblage of metazoan symbionts that are affected during bleaching events. The abundances of obligate crustacean symbionts have been observed to decline rapidly in bleached, moribund and dead pocilloporid corals during El Niño disturbances. *Trapezia* (4 species) and *Alpheus lottini* demonstrated marked declines in abundance in their *Pocillopora* hosts during the 1982–83 and 1997–98 El Niño events in the eastern Pacific (Glynn et al. 1985a; Prahl 1985; Iglesias-Prieto et al. 2003). These crustaceans feed on coral mucus and entrapped detritus, lipid bodies produced by corals, and occasionally on coral tissues (Glynn et al. 1985a; Stimson 1990). Lacking this food source, which is diminished during El Niño bleaching events, the crustaceans die in situ or vacate their hosts, and when exposed (host-free) become more vulnerable to predation (Castro 1978).

The loss of metazoan symbionts can negatively affect coral colonies that survive bleaching events. For example, it has been demonstrated in laboratory and field studies that obligate crustacean symbionts enhance coral health, growth, competitive ability, and defense against corallivores (Glynn 1983b; Pratchett 2001; Stewart et al. 2006). Thus, if pre-bleaching symbiont assemblages are not reestablished in surviving corals they would likely experience elevated mortality rates from a variety of post-bleaching causes.

8.3.1.3 Corallivores

In general, the initial stressors of ENSO-induced coral bleaching/mortality do not adversely affect many corallivores. *Jenneria pustulata*, an ovulid gastropod mollusc, is an obligate corallivore with a strong dietary preference for *Pocillopora* spp. (Achurra Cárdenas and Valdés Araúz 1980). This mollusc was the only coral macropredator at the Uva Island reef that was observed to die in large numbers during the 1982–83 El Niño event. Numerous empty and undamaged *J. pustulata* shells were found in and around dead *Pocillopora* spp. colonies that had died during the warming episode (Glynn et al. 1985a). Since *Jenneria* has limited mobility and dies quickly if deprived of coral tissues, within a few days to 1–2 weeks (Oramas 1979), it is likely that the observed predator-prey death assemblages resulted from starvation. Mean population densities assessed at the Uva Island reef in 1982 and 1984, before and after coral prey mortality respectively, were 23.7 ind m^{-2} and 1.7 ind m^{-2} (Glynn 1985a).

The obligate corallivore sea star *Acanthaster planci* did not experience population declines in Panama during the 1982–83 El Niño. Sea star densities ranged from 5 to 15 ind ha^{-1} in different reef censuses during and following the coral mortality period (see Chap. 10, Enochs and Glynn). A notable decline in *A. planci* abundance on the Uva Island reef did not occur until 1990, and the sea star has been absent or present at low densities (0–1 ind ha^{-1}), for some unknown reason(s), to 2011 (when last surveyed). Sea stars that were present on the Uva reef after 1983 had a greater impact on the coral community because they attacked lower abundances of corals that survived the bleaching event (Glynn 1985a, b). A spatial effect also was observed—some corals that were highly preferred by *Acanthaster* but unavailable before the disturbance event were now exposed to predation. The elimination of large tracts of live pocilloporid corals that formerly surrounded and thus created a refuge for some preferred massive and encrusting species (e.g., *Gardineroseris planulata, Pavona varians*), could then be traversed by *Acanthaster* and thus allowed entry into pre-disturbance refugia. *Acanthaster* generally avoids contact with live *Pocillopora* spp. colonies because of their defensive cnidae and protective crustacean guards (Glynn 1976, 1985b).

The abundance of *Arothron meleagris*, a facultative pufferfish corallivore that feeds predominantly on the branch tips of *Pocillopora* spp. in Panama, also was unaffected by the 1982–83 El Niño event. Mean annual population densities on the Uva reef in 1981, 1983, and 1984 were respectively 5.0, 8.2, and 5.7 ind 1000 m^{-2}, which were not statistically different (Glynn 1985a). These fish concentrated their feeding on the remaining healthy pocilloporid corals, and avoided bleached and moribund colonies. Guzmán and Robertson (1989) found that the abundances of *A. meleagris* at different eastern Pacific reef sites affected during the 1982–83 El Niño did not decline as did their usual pocilloporid prey. These workers demonstrated that the pufferfish at Caño Island (Costa Rica) switched its dietary preferences to other less affected corals, e.g. *Porites lobata* and *Psammocora stellata*, and even to coralline algae. A similar dietary switching response was reported for a butterfly fish corallivore on the Great Barrier Reef (Pratchett et al. 2004). After a mass bleaching/mortality event in Australia that severely reduced the numbers of preferred *Acropora* spp. prey, *Chaetodon lunulatus* began feeding on a higher proportion of surviving non-acroporid corals, which did not experience a decrease in abundance.

Damselfishes in Panama did not undergo significant changes in abundance during and following El Niño 1982–83 warming (Wellington and Victor 1985), but they often expanded their territories onto stressed and moribund coral colonies (Glynn 1990). *Stegastes acapulcoensis* expanded its territories onto moribund and dead coral surfaces, nipping and killing coral tissues in the process. Species of massive corals that suffered partial mortality were often subject to additional tissue loss by *S. acapulcoensis* that established territories on damaged colonies (Fig. 8.27). Coral tissues are removed and ingested during the expansion of algal lawns, but territorial damselfishes are herbivores, consuming

Fig. 8.27 *Stegastes acapulcoensis* in algal garden territory on *Gardineroseris planulata* at Uva Island coral reef, Gulf of Chiriquí, Panama (4 m depth, 13 August 1985). Brown and green patches of filamentous algae are visible on coral surfaces that bleached during the 1982–83 El Niño event. Damselfish nipped at bleached and moribund coral tissues thereby extending their territories

predominantly microfilamentous algae (Wellington 1982; see Chap. 10, Enochs and Glynn). A similar interaction resulting in additional post-El Niño coral tissue loss occurred in the Galápagos Islands. The damselfish species involved in the Galápagos Islands were *Stegastes arcifrons* and *Stegastes beebei*, and the coral species affected were the same as those in Panama (Glynn 1988b).

Under certain conditions, for example at reef sites with abundant or outbreak sea urchin populations, damselfishes can actually reduce framework breakdown by interfering with sea urchin bioerosion. This is accomplished by the ability of damselfishes to remove sea urchins from their territories (Eakin 1987, 1988). A damselfish typically grasps a spine of an intruding sea urchin and then swims with it beyond the algal lawn, often over a distance of 0.5 m or more, where it is then released. The territorial defensive behavior of *Stegastes acapulcoensis* in Panama involves "body brushing", the nipping of sea urchin spines, and "tail beating" (Glynn 1988b). Surface-eroding sea urchins increased in abundance at numerous reef sites after 1982–83. In Panama, *Diadema mexicanum* increased from a mean density of 3 ind m^{-2} before 1983 to 80 ind m^{-2} after 1983, due to high recruitment. In the Galápagos Islands, *Eucidaris galapagensis* demonstrated a pre- to post-El Niño increase of 5–30 ind m^{-2}, a result of the redistribution of sea urchins onto dead coral frameworks. In Panama, estimated bioerosion rates of coral substrates with territorial damselfish present (*S. acapulcoensis*) were substantially lower than in areas without damselfish, namely 3.7–14.8 versus 9.5–28.5 gm $CaCO_3$ m^{-2} d^{-1}, respectively (Glynn 1988b). A similar damselfish-sea urchin interaction, involving ecologically equivalent species, also occurred in the Galápagos Islands. At the Onslow Island reef, estimated bioerosion rates on substrates with and without damselfishes (*Stegastes beebei*, *Stegastes arcifrons*) were, respectively, 19.8–39.6 gm $CaCO_3$ m^{-2} d^{-1}, versus 49.5–99.0 gm $CaCO_3$ m^{-2} d^{-1}. Damselfishes can also limit the entry of *Acanthaster* corallivores into their territories, but this behavior did not appear to have an important effect in limiting post El Niño coral mortality in the eastern Pacific (Glynn and Colgan 1988).

8.3.1.4 Sea Level

The first documented reef flat exposure mortalities of corals in the eastern Pacific occurred in Panama during the 1974–75 La Niña event (Glynn 1976). Widespread coral bleaching and mortality were observed along the Pacific Panamanian coast, affecting shallow reef flats in the Gulf of Panama and the Gulf of Chiriquí. The reef flat exposures occurred during mid-day hours in the dry season (February, March) when cloud cover is normally low (Glynn 1977). At the Saboga Island reef (Pearl Islands, Gulf of Panama), reef flat exposures ranged between −60 to −80 cm (relative to MLWS sea level elevations) for 8–10 consecutive days during each of two spring tidal cycles. At −80 cm elevations, reef flat exposures in both Panamanian gulfs lasted from 2.2 to 2.4 h. Pocilloporid coral mortality was extensive, due primarily to desiccation and actinic damage. As the succeeding tide rose, flooding the reef flat after each exposure event, mucus release and tissue sloughing from affected corals were notable. Diverse reef-associated fishes moved onto the reef flat with the rising tide to feed on suspended particulate organic matter, dead algal-coated coral skeletons, and moribund and dead metazoans. Quantitative sampling on two reefs in the Gulf of Chiriquí after the exposures revealed that *Pocillopora* spp., the predominant reef flat corals, experienced overall mortalities of 61 % and 63–70 %.

A later extreme reef flat exposure, with accompanying coral bleaching and mortality, occurred in Panama in 1989, during the 1988–89 La Niña event (Eakin et al. 1989). This exposure resulted in 97 % partial to complete coral mortality on the Uva Island reef flat. A multitude of fish and invertebrate species that were stranded in isolated superheated pools also died in large numbers (Fig. 8.28a, b). Eakin and co-workers suggested that the high mortality of corals in 1989 could have resulted from an interactive effect of exposure and cold water stress due to thermocline shoaling. Summarizing low tidal exposure events in Costa Rica, Cortés and Jiménez (2003) noted mid-day coral mortalities in three different reef areas, all occurring during La Niña years. Subaerial reef flat exposures have been observed along the Colombian coast as well and may have long-term effects in increasing dead coral substrates for algal colonization and growth, increased herbivory, and bioerosion (Zapata and Vargas-Ángel 2003). Extreme reef flat exposures during darkness may also contribute to high coral mortality if

Fig. 8.28 Coral bleaching and mortality associated with mid-day, extreme, low water exposures at Uva Island coral reef (10 March 1993). **a** Panoramic view of exposed ~2 ha reef flat. **b** Exposed reef flat showing isolated pools that experienced intense heating. White stake in center foreground is 30 cm in height. Photographs courtesy of C. M. Eakin

accompanied by heavy rainfall. Such an event occurred at the Uva Island reef in mid-May 1999, causing extensive mortality to five reef-building coral species (Maté 2003).

Reef flat exposure events resulting in extensive mortality have been reported in the South Pacific region as well, during the early phases of El Nino activity. Wyrtki (1985a, b) documented spatial fluctuations in sea level elevations in response to changing wind fields across the equatorial Pacific Ocean during the progression of the 1982–83 El Niño event. In December 1982, monthly mean sea level anomalies of −20 to −30 cm were centered around the Marshall and Mariana Islands. In early 1983, corals and associated organisms in shallow reef habitats on Moorea Island (French Polynesia) and at Nukunonu Atoll (Tokelau Islands) experienced high mortalities in conjunction with anomalously low sea level exposures (Glynn 1984). Yamaguchi's (1975) early comprehensive assessment of reef flat mortalities in Guam (Mariana Islands) was correctly related to sea level fluctuations during the 1972–73 El Niño event. Virtually all reef flat macrobiota at Pago Bay succumbed during low sea level stands (−44 cm below MSL) that persisted for more than two weeks. Extreme stressful conditions—water temperatures as high as 37.7 °C and dissolved O_2 levels <1.00 ml l^{-1}—were recorded in isolated reef flat pools. Recovery to pre-exposure conditions was incomplete after 3 years, with many pre-existing taxa still absent. Extreme sea level drops also occurred at Kwajelein Atoll (Marshall Islands) and Truk Island (Caroline Islands) during this period, but effects to shallow reef biota were unreported.

8.3.1.5 Storm Effects

Robinson (1985) offered the first observations of El Niño-related storm conditions affecting corals in the eastern tropical Pacific. He recounted the damage caused by a reversal of swell direction at Onslow Island, Floreana Island (Galápagos Islands) in January 1983. The usual swell direction is from the S-SE, but during the early phase of El Niño this reversed to the NW, and impacted shallow coral communities. Large colonies of *Porites lobata* and *Pavona* spp. were dislodged from the bottom and deposited along the shore above the high tide line. *Pocillopora* colonies were also broken free and fragmented, along with associated species such as the echinoid *Eucidaris galapagensis*, and deposited on the shore. When the Onslow Island patch reef was resurveyed in May 1983, much of the pre-existing pocilloporid cover had disappeared, likely a result of strong wave action, scouring, and transport away from the former reef site. A similar series of events likely destroyed a pocilloporid reef at Pinta Island (central Galápagos Islands) in 1983, creating a strandline coral rubble berm (Wellington and Glynn 2007).

Hurricane impacts off the SW Mexican coast during the very strong 1997–98 El Niño event caused substantial damage to coral reefs (Lirman et al. 2001). Within a two month period late in 1997, three major storms with maximum sustained winds of 205 km h^{-1} (TC 'Nora'), 110 km h^{-1} (TC 'Olaf') and 215 km h^{-1} (TC 'Pauline'), skirted the Huatulco reef tract. Coral damage, principally affecting *Pocillopora* spp., was spatially variable, but at several sites resulting in the destruction (separation) of reef frameworks (Fig. 8.29), dislodgment and scattering of colonies, and extensive colony fragmentation and abrasion. Of six reefs surveyed before and after the storms, three experienced significant declines in live coral cover and three demonstrated increases in coral cover, however, these latter differences were not statistically significant. The increases in coral cover were due to branch fragmentation. Live coral recovery could be accelerated if fragmented branches were deposited on suitable substrates that would increase survivorship. An analysis of reef orientation and coral damage in relation to the predominant direction of storm-generated swells did not reveal any discernible patterns.

Fig. 8.29 Upended pocilloporid reef frame blocks (~1 m in length along vertical growth axis) dislodged from Chachacual reef during tropical storm activity across the Huatulco reef tract in 1997 (2–3 m depth, 8 December 1997)

Storm damage also affected coral reefs in the central and south Pacific during the 1982–83 El Niño event. At that time the region of tropical storm genesis shifted far to the east (Rasmusson and Wallace 1983), and as a result French Polynesia was battered by hurricanes. Five hurricanes and one tropical storm moved across French Polynesia over a 4–month period. The damage to coral communities on atolls in the Tuamotu Archipelago and the high island of Moorea in the Society Islands was well documented (Laboute 1985; Harmelin-Vivien and Laboute 1986). Storm-generated waves caused severe damage on steep atoll slopes with 100 % coral destruction from about 25 m to at least 100 m depth. Corals were broken and dislodged, and transported to deep waters by slumping. Associated epifauna and cryptic species also disappeared during the loss of coral habitat. The chief damage on the gently sloping Moorea shelf occurred between sea level and about 20 m depth. Dislodged corals at the shallower depths of all surveyed sites were transported to the island's exposed boulder ramparts.

Recent model analysis studies project an increase in the frequency and intensity of central Pacific (CP) warming El Niño years compared to EP warming El Niño years under increasing greenhouse gases (Yeh et al. 2009; Kim and Yu 2012). However, although ENSO is projected to change under climate change conditions, there is still uncertainty in the projections (Collins et al. 2010).

8.3.1.6 Nutrients

Sea-viewing Wide Field-of-view Sensor (SeaWiFS) satellite data portray chlorophyll concentrations, which serve as a proxy for primary production (O'Reilly et al. 1998). Low surface chlorophyll concentrations during the 1998 El Niño event corresponded with decreased productivity in waters off the western sector of the Galápagos Islands. The decrease in productivity during an El Niño event is in contrast to the elevated productivity of a La Niña event (Fig. 8.30). Suspension-feeding barnacles and oysters died in large numbers in the Galápagos Islands during the 1982–83 El Niño disturbance. This mortality may have resulted in part from low primary and secondary production, and heavy rains and freshwater runoff during low tides (Robinson 1985). Three species of *Lithophaga*, boring bivalves inhabiting massive corals such as *Porites lobata*, suffered high mortalities in the equatorial eastern Pacific in 1982–83 (Scott et al. 1988). These bivalves are suspension feeders, and exhibited increasing site-specific mortality rates along an El Niño gradient of increasing coral bleaching/mortality (Glynn et al. 1988), namely 16 % bivalve mortality in Costa Rica, 29 % in Panama, and 72 % in the Galápagos Islands (Scott et al. 1988).

Most likely a result of La Niña nutricline shoaling in 1985, dinoflagellate blooms occurred on coral reefs in Costa Rica and Panama, discoloring surface waters from red to red-brown and yellow, and producing a viscous foam (Guzmán et al. 1988). Water samples collected at Caño Island, Costa Rica were dominated by *Cochlodinium catennatum* and *Gonyaulax monilata* with cell counts of $8.3 \times 10^9\ l^{-1}$ and $3.0 \times 10^6\ l^{-1}$ respectively. Coral mortality was greatest in shallow reef zones (1 to 3 m depth), and reached as high as 100 % in *Pocillopora* spp. and 95 % in *Tubastraea coccinea* on some parts of the Caño Island

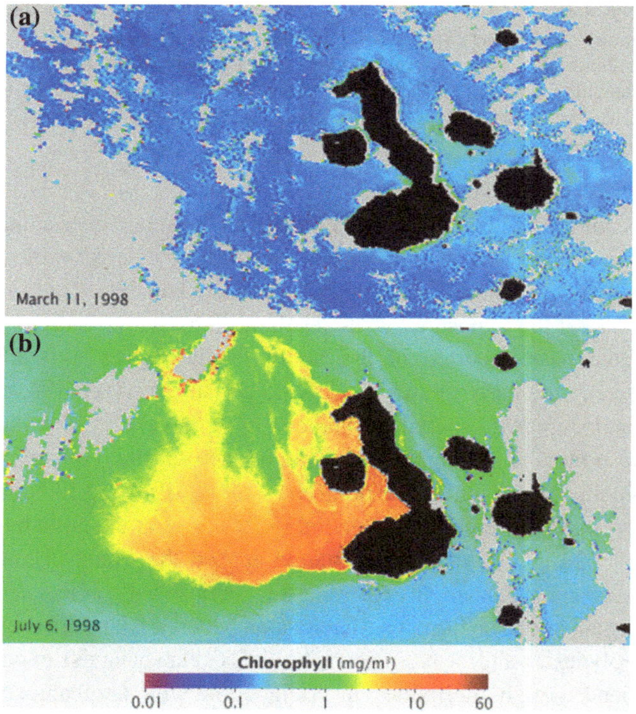

Fig. 8.30 NASA's SeaWiFS maps of chlorophyll concentrations (mg m^{-3}) during **a** El Niño (11 March 1998) and **b** La Niña (6 July 1998) events around the Galápagos Islands. *Source* http://earthobservatory.nasa.gov/blogs/fromthefield/2009/07/16/oceanographer-gene-feldman-is-going-home-for-the-first-time/

reef. Pocilloporid mortality on the Uva Island reef was considerably less (12.8 %), but also confined to shallow zones exposed to discolored, viscous water. Numerous non-coral species also experienced high mortalities coincident with the dinoflagellate blooms, including reef fishes (scarids, balistids, acanthurids, pomacentrids, tetraodontids) and invertebrates (hermit and brachyuran crabs, gastropods). A combination of factors was probably involved in causing the death of reef metazoans: (a) toxicity due to *Gonyaulax monilata* (= *Alexandrium monilatum*), (b) oxygen depletion, and (c) coating of epidermal respiratory and food-collecting tissues with mucus. Corals and fishes also experienced catastrophic mortalities in southwestern Sumatra during the Indian Ocean Dipole event in 1997 when strong upwelling and phytoplankton blooms occurred (Abram et al. 2003). Geochemical records from corals indicated a link with Indonesian wildfires that caused iron fertilization, thus promoting massive phytoplankton production. Low oxygen levels and asphyxiation were the likely causes of death.

Elevated nutrient pulses, such as those occurring during thermocline shoaling and upwelling, stimulate the growth of macroalgae and suspension-feeding taxa that often exhibit a competitive advantage over corals (Highsmith 1980). Algal overgrowth of corals was first observed on the Uva Island reef during 1988–89. Continuous carpets of *Caulerpa sertularioides*, with intermixed and fewer *Caulerpa racemosa* thalli, overgrew and smothered low-profile colonies of *Psammocora stellata* across the seaward reef slope to 10–12 m depth, causing 100 % coral mortality (Glynn and Maté 1997). On the reef proper, between 2 and 6 m depth, *C. racemosa* occasionally over-topped *Pocillopora* colonies, but the occurrence of this alga was short-lived (a few weeks) and usually did not kill the erect coral colonies (Fig. 8.31). Smith et al. (2010) concluded that *Caulerpa* blooms on the Uva reef in 1989 and 1999, were initiated by nutrient pulses but likely also were influenced by two interacting factors, namely a cryptic habitat niche (on the undersides of coral rubble) and an associational mutualism with a toxic epiphytic cyanobacterium. Both of these factors limited herbivory and allowed for the local persistence of algae and an immediate growth response to increasing nutrient levels (see Chap. 11, Fong et al.). Macroalgae also increased in abundance, and coral cover decreased during upwelling pulses in the Pearl Islands (Wartian 2006). This inverse relationship was supported experimentally by Muthukrisnan (2013), who demonstrated a partial shift in dominance from coral to algal cover with the addition of nutrients.

Fig. 8.31 *Caulerpa racemosa* surrounding and overgrowing *Pocillopora elegans* following La Niña conditions (upwelling) at Uva Island coral reef, Gulf of Chiriquí, Panama (4 m depth, 13 August 1985)

8.3.1.7 Dispersal and Range Extensions

Dana's (1975) long distance dispersal hypothesis posits the transport of central Pacific planktonic larvae to the eastern Pacific via the North Equatorial Countercurrent. Richmond (1990) proposed that this mode of dispersal would be enhanced during El Niño events when easterly flow of equatorial currents is greatly accelerated along with increased volume transport.

New records of Indo-Pacific species in the eastern tropical Pacific during and shortly after El Niño activity offer indirect evidence in support of the dispersal hypothesis. The first recorded occurrences of central and western Pacific species in the equatorial eastern Pacific after recent El Niño events included several molluscs (Emerson 1991; Finet 1991; Kay 1991; Emerson and Chaney 1995), echinoderms (Lessios et al. 1996, 1998), and fishes (Grove 1984, 1989; Merlen 1988; McCosker and Humann 1996; Robertson and Allen 1996; Grove and Lavenberg 1997). Increased abundances and new records of Indo-Pacific reef fishes, including several acanthurid, labrid and chaetodontid species, appeared throughout the Galápagos Islands during the 1982–83 El Niño event (Grove 1984, 1985; Merlen 1988; Edgar et al. 2004). Many of the presumed migrants possess long-lived (teleplanic) pelagic larvae that have been found at various distances across the eastern Pacific barrier (EPB), supporting the feasibility of crossing this broad expanse of open ocean (Scheltema 1988). Robertson et al. (2004), however, have cautioned that the coincidence of El Niño activity and first occurrences of central Pacific species in the eastern Pacific may be an artifact of a concentration of effort to document new records during these events. Robertson et al. (2004), and McCosker and Rosenblatt (2010) also suggested that certain oceanographic conditions during El Niño are not conducive to the long-distance dispersal of fish larvae. For example, a deepening of the thermocline lowers productivity and potential food sources, and higher temperatures could shorten pelagic larval duration and lead to smaller size at metamorphosis.

Within the eastern Pacific, the effects of ENSO-induced changes in circulation on reef fish abundances and distributions are more tractable and certain. The combination of widespread ocean warming, and reversals and other alterations in surface currents during El Niño events corresponds with more effective dispersal and recruitment of certain fish groups within the eastern Pacific region. This has been demonstrated among some labrid and pomacentrid fishes with long larval lives (Wellington and Victor 1992; Meekan et al. 2001). Range extensions have been documented to the northern (Baja California) and southern (Galápagos Islands) limits of tropical species' distributions in the eastern Pacific region. In the Galápagos Islands, where observations were made before, during and following ENSO activity, inter-island changes in fish populations were well documented. Several coral-associated species normally confined to the warm waters of the northern Galápagos Islands (Darwin and Wolf) extended their ranges throughout the archipelago, including cool water zones, during the influx of warm El Niño waters (Edgar et al. 2004). In addition, some damselfishes showed enhanced recruitment during the 1982–83 El Niño (Grove 1984; Grove et al. 1986), and year classes with age structures corresponding to the timing of El Niño events (Meekan et al. 2001). Ruttenberg (2000) noted as well a close correspondence between recruitment pulses of a reef-associated goatfish and parrotfish in the Galápagos Islands during the 1997–98 El Niño event. Finally, the advection of warm waters further south, with a surface temperature increase of up to 11 °C, resulted in the appearance of vagrant subtropical fishes along the coasts of Peru and Chile (Arntz and Tarazona 1990).

An influx of warm, surface waters towards northern Mexico and southern California during recent El Niño events has led to numerous range extensions of Panamanian fishes to higher latitudes. Most of these records, however, are ephemeral with the disappearance of tropical species introductions upon the return of more usual, cooler conditions. At such times the expansion of the NECC, an increase in the northerly flow of the CRCC and the West Mexican Current (WMC), and a more pronounced westerly flow of the California Current, offer potential conduits of transport from Cocos Island, Clipperton Atoll, and the Revillagigedo Islands towards northern Mexico and California (Norton et al. 1985).

Victor et al. (2001) documented the range extensions of two labrid species from the Revillagigedo Archipelago and/or Clipperton Atoll to the southern Gulf of California and tip of Baja California during the 1997–98 ENSO. A suggested hypothetical route involved the easterly transport of planktonic larvae from Clipperton via the NECC, and then north via the CRCC and WMC. These introductions involved massive settlements that lasted for weeks. Range extensions of Panamic fishes, including species in 29 families, also were reported as far north as southern California (Lea and Rosenblatt 2000). Most records coincided closely with the 1997–98 El Niño warm water event when mean annual SST anomalies reached 2 °C. The tropical migrants arrived off California as planktonic larvae, juveniles or adults with flotsam, and as free-swimming adults in the northward-flowing warm water masses.

8.3.1.8 Colonization of Opportunists

Dead coral surfaces are quickly colonized by micro-filamentous algae and micro-boring algae, fungi, foraminifera and other prokaryotic and eukaryotic microorganisms (Tribollet and Golubic 2011). Within a year after the 1982–83 El Niño event, dead coral surfaces in Costa Rica, Panama, and the Galápagos Islands were often covered with turf algae and/or crustose coralline algae (Cortés et al. 1984; Glynn 1984, 1990; Robinson 1985). After 6–9 months on the Uva Island reef, much of the substratum was covered with 1–3 cm thick carpets of red algae (mainly *Gelidiopsis intricata* intermixed with lesser amounts of *Hypnaea pannosa* and *Amphiroa beauvosii*). The settlement of opportunistic metazoans, such as sponges, barnacles, molluscs, and tunicates, often occurs later in succession, over monthly-to-yearly time scales, and is considered below under 'Long-term effects'.

8.3.2 Long-Term Effects

8.3.2.1 Damselfish Interactions

Dead coral surfaces resulting from ENSO disturbances potentially provide space for the colonization of territorial damselfishes. Damselfishes can exercise both negative and positive effects on corals. The colonization of damselfish onto dead patches of massive corals that experience partial mortality can cause additional tissue damage, impede recovery by asexual re-sheeting, and perhaps prevent the settlement of coral larvae. At the Uva Island reef, *Stegastes acapulcoensis* colonized dead patches on *Gardineroseris planulata* and *Pavona* spp., causing increased tissue mortality by expanding their algal lawn territories (Fig. 8.27; Glynn 1990). Even though 'free' space became available immediately after coral tissue death, new damselfish lawns were not observed until nearly two years after the bleaching event. Once established, however, damselfish territories persisted for several years.

Some damselfish species can exercise a positive effect by limiting the access of corallivores into their territories, which often include surviving and recovering coral patches. For example, damselfishes have been observed to prevent the entry of *Acanthaster* into or near their territories in Panama and Tahiti, thus providing a refuge for corals (Glynn and Colgan 1988). This has been documented by behavioral observations in which, for example, defending damselfish will nip and tail- brush intruding corallivores. A comparison of small-scale differences in coral assemblages inside versus outside defended territories has demonstrated respectively significantly (a) more abundant preferred seastar prey, and

(b) higher coral species diversity. Damselfishes may limit bioerosion as well, by preventing the entry into their territories of grazing fishes and sea urchins (Eakin 1992; see Sect. 8.3.2.2).

Damselfish recruitment in relation to El Niño events has exhibited different responses, both spatially and temporally. In the Pearl Islands, Panama Wellington and Victor (1985) found that post-El Niño 1982–83 damselfish abundances were not significantly different from pre-El Niño abundances. They concluded that damselfish population abundances were controlled more by recruitment success than by the availability of newly created space. However, in the Galápagos Islands a massive influx of *Stegastes acapulcoensis* occurred during the 1997–98 El Niño and earlier warming events (Grove and Lavenberg 1997; Meekan et al. 2001; Victor et al. 2001). This damselfish species occurs commonly in coral communities along the eastern Pacific mainland, but only sporadically in the Galápagos Islands. Its more recent occurrence in the Galápagos Islands has persisted through 2000 at least, indicating that it may have become established at this offshore location.

8.3.2.2 Bioerosion

The skeletons of living corals are subject to bioerosion, the mechanical or chemical breakdown of calcium carbonate by biologic agents, but generally the pace of erosion accelerates in dead corals (Highsmith 1981; Hutchins 2011; Tribollet and Golubic 2011). Here we consider bioerosion in response to ENSO activity and chiefly with reference to alterations in echinoid abundances, preeminent external agents of carbonate degradation. Alvarado et al. (Chap. 12) should be consulted for an in-depth treatment of bioerosion in the eastern Pacific region, including the diverse taxa involved, differences in external and internal bioerosion, and geographic patterns.

Aside from the high and unprecedented coral mortality that occurred during the 1982–83 El Niño disturbance, an additional change that soon followed was a notable increase in echinoid abundances on numerous coral reefs throughout the equatorial eastern Pacific region. These relatively sudden increases in sea urchin population densities occurred about a year after coral mortality, and continued in many areas until the early to mid-1990s. In Panama (Uva Island, Gulf of Chiriquí), *Diadema mexicanum*, a wide-ranging diadematid bioeroder, experienced increases in mean population densities from <10 ind m^{-2} pre-El Niño up to 60–80 ind m^{-2} post-El Niño (Glynn 1988; Eakin 2001). These high densities were centered around dead pocilloporid frameworks in forereef and reef base zones (Fig. 8.32a). The feeding activities of the sea urchins were concentrated largely on the filamentous and crustose coralline algae coating dead coral surfaces. The persistent deep-grazing of the large *Diadema* populations, continuing for several years, resulted in the removal of epilithic fleshy and calcareous algae as well as the erosion of coral skeletons and reef framework structures. The mean bioerosion rate of *Diadema* in Panama was 0.2 gm dry weight $CaCO_3$ ind^{-1} day^{-1}. This resulted in severe bioerosion and reef framework degradation (see Chaps. 12, Alvarado et al., and 18, Manzello et al.). Many reef sites in the equatorial eastern Pacific experienced significant structural losses, e.g., Panama, including the Pearl Islands in the Gulf of Panama (Glynn 1990), Cocos Island, Costa Rica (Guzmán and Cortés 1992; Macintyre et al. 1992), Colombia (Toro-Farmer 1998), mainland Ecuador, and the northern Galápagos Islands (Glynn 1994, 2003).

It was hypothesized that these sea urchin outbreaks resulted from high recruitment during La Niña 1985 upwelling, and increased productivity (contributing to added algal resources) and availability of sheltering habitat space (Glynn 1988; Eakin 1996). The ultimate outcome of high *Diadema* abundances was self-limiting, however—the topographic complexity of reef substrates was reduced by over-grazing, thus eliminating shelter sites and algal food resources, which resulted in significant population declines (Eakin 2001). Alvarado (2013) provides comprehensive and up-dated information on *Diadema* abundances and bioerosion rates across the entire eastern Pacific region.

A similar pattern of coral reef degradation and loss was documented in the southern and central Galápagos Islands following the 1982–83 El Niño event. It began with major coral bleaching and mortality, followed shortly thereafter with large aggregations of sea urchins that over-grazed dead corals, reducing corals and framework structures to rubble and sand-sized sediments. *Eucidaris galapagensis*, a cidarid sea urchin endemic to the Galápagos Islands and Cocos Island, was chiefly responsible for the coral erosion in the central and southern Galápagos Islands (Fig. 8.32b; see also Glynn et al. 2015). This echinoid species had a mean bioerosion rate of 2.0 gm $CaCO_3$ (DW, dry weight) ind^{-1} day^{-1}, an order of magnitude higher than *Diadema mexicanum* (Glynn 1988). Moreover, *E. galapagensis* population densities commonly reached 25–50 ind m^{-2} on dead carbonate substrates. The rapidity of framework loss by *Eucidaris* is evident from the bioerosion of an incipient *Pavona* spp. reef at Champion Island. This 4–5 m thick patch reef (pre-1982–83) was reduced to a rubble pile of <1 m in just 8 years after the bleaching/mortality El Niño event (Fig. 8.32b; and Fig. 26 in Glynn and Wellington 1983).

Eucidaris was known to limit reef growth in the Galápagos Islands before the 1982–83 El Niño event, but its activities were largely confined to reef margins where live coral cover was patchy and discontinuous (Glynn et al. 1979). Following the post-El Niño increase in dead coral, *Eucidaris* underwent a re-distribution, forming dense aggregations around dead coral substrates. Increases in sea urchin densities have also been attributed to the overfishing

Fig. 8.32 **a** *Diadema mexicanum* grazing on dead, algal-encrusted pocilloporid reef frame, Uva Island reef, seaward slope, 6 m depth, 13 August 1985. **b** *Eucidaris galapagensis* grazing on dead, algal-encrusted colony of *Pavona clavus*, Champion Island, 6 m depth, 18 March 1991

of the predators of *E. galapagensis* at sites experiencing high fishing pressure (Sonnenholzner et al. 2009; Edgar et al. 2010).

8.3.2.3 Species Extirpations

A local or regional species extinction may occur suddenly after an ENSO disturbance, but this is nearly impossible to document without thorough and sustained searches of all suitable habitats. With the disappearance of a species and the passage of time, one cannot be certain when the extinction actually occurred. Several azooxanthellate and zooxanthellate corals have suffered dramatic declines in abundances in the eastern Pacific following El Niño impacts (Glynn 1997, 2011; Jiménez 1998; Jiménez et al. 2001; Edgar et al. 2010). The disappearances of four zooxanthellate species (one scleractinian coral and three hydrocorals), all sharing extremely restricted distributions, were closely correlated with high SST anomalies during the 1982–83 El Niño event. Bleached coral tissues observed during the disturbance indicated that these species were stressed directly by anomalously high temperatures. *Porites rus* was known only at Sámara, Costa Rica and was not found there again after the El Niño event (Cortés and Murillo 1985; Guzmán and Cortés 1993; Glynn 1997). Three hydrocoral species bleached and died on the Uva Island reef and in surrounding areas in the Gulf of Chiriquí, Panama in 1983. These species were known only from the Gulf of Chiriquí in the eastern Pacific region. One of these species, *Millepora intricata*, was absent from shallow reef zones for 3–4 years and then re-appeared, presumably by recruitment from surviving source populations in deeper waters (Smith et al. 2014). Comprehensive post-El Niño surveys over a 7-year period failed to reveal live colonies of *Millepora boschmai* or *Millepora platyphylla*, suggesting the disappearance of these species from the eastern tropical Pacific (Glynn and deWeerdt 1991). However, five live colonies of *M. boschmai* were re-discovered at Uva Island, ~4 km distant from the reef, and three colonies at Coiba Island, about 25 km south of Uva Island (Glynn and Feingold 1992; Glynn 2011). All eight of these live colonies disappeared sometime from 2001 to 2010, and have not been observed since, suggesting a regional extinction. Live colonies of *M. platyphylla* have not been reported anywhere in the eastern Pacific since 1983. Both species still occur in the Indo-Pacific region, however, west of Ekman's Barrier (Glynn 2011).

Several other zooxanthellate corals experienced marked declines in abundance following the 1982–83 El Niño event, e.g. *Pocillopora capitata*, *Porites panamensis*, *Psammocora stellata*, *Gardineroseris planulata* (see Glynn 1997), *Leptoseris papyracea* (in Cortés and Jiménez 2003), and *Pocillopora elegans* and *Cycloseris curvata* (in Edgar et al. 2010). The severe reduction in abundance of two species, *L. papyracea* and *P. elegans*, resulted in the cessation of carbonate skeletal accumulation and framework construction respectively. This, combined with bioerosion, led to the degradation and loss of a unique leptoserid mound-like reef in Culebra Bay, Costa Rica (Jiménez et al. 2001), and the loss of all known pocilloporid patch reefs in the central and southern Galápagos Islands (Glynn 1994, Glynn et al. 2015). *Leptoseris papyracea* colonized a reef area (Matapalo) in Costa Rica after the 1997–98 mortality event, but then disappeared a few years later, due possibly to coastal development. Three small colonies of *Acropora valida* were established as new distributional records in the eastern Pacific in September 1983 (von Prahl and Mejía 1985). Age estimates indicate they settled 5–10 years prior to their discovery (Guzmán 1988). They disappeared soon after the 1982–83 El Niño event, but apparently in a healthy, non-bleached state. Therefore, the reason(s) for their disappearance cannot be directly attributed to El Niño stressors.

Invertebrate and fish corallivores, obligate coral symbionts, and species that utilize corals for shelter or settlement and nesting sites can experience severe declines in abundances following coral mortality events (Pratchett et al. 2008; Enochs and Hockensmith 2009; Glynn 2011; Stella et al. 2011). The death of pocilloporid corals that provide diverse services to many species can precipitate massive mortalities of associated species. In the eastern Pacific, numerous obligate crustacean and mollusc symbionts die with the demise of their pocilloporid hosts (Glynn et al. 1985a; Glynn et al. 1988; Iglesias-Prieto et al. 2003). Similar crustacean symbiont mortalities have occurred in pocilloporid host corals in the western Pacific (Okinawa) in the wake of El Niño bleaching events (Tsuchiya et al. 1992; Tsuchiya 1999). However, eastern Pacific pocilloporid reefs support diverse motile cryptofaunal communities of high biomass that are not dependent on live corals (Enochs et al. 2011; Enochs 2012). In fact, the species richness and abundances of the motile cryptofauna were found to increase across a gradient of pocilloporid framework erosion and degradation (Enochs and Manzello 2012a, b). These cryptofaunal populations may persist long after coral mortality, providing a food source for reef-associated fish omnivores and carnivores.

Some facultative fish corallivores in the eastern Pacific, for example monacanthids, pomacanthids, tetraodontids and chaetodontids (see Chap. 10, Enochs and Glynn; Glynn 2008; Glynn et al. 2014), may be affected by declines in live coral cover. Obligate coral fish associates are absent from eastern Pacific reefs, but where present in the Indo-Pacific experience significant negative effects, even local extinctions (e.g., Jones et al. 2004; Munday 2004; Pratchett et al. 2006).

Long-term quantitative data on the abundances of reef-associated invertebrates and protozoans in Brazil have revealed marked population responses of numerous taxa corresponding to ENSO episodes. Notable declines in abundances of hundreds of species and some local extinctions have been documented for cnidarians (Kelmo et al. 2003), bryozoans (Kelmo et al. 2004), ascidians (Kelmo et al. 2006), and foraminiferans (Kelmo and Hallock 2013). Initial mortalities often corresponded closely with elevated temperatures, solar insolation and sediment loads from rainfall and runoff, but continued mortality of some species, for up to two years after El Niño impacts, suggests other factors could be involved.

8.3.2.4 Coral Diseases

Beyond bleaching, coral diseases and parasitism have not been considered to be important agents of mortality in the eastern Pacific (e.g., Glynn et al. 1985b; see regional treatment chapters in Cortés 2003). The first account of apparent but not confirmed coral diseases, "white plague" in *Porites lobata* and "white band" in *Pocillopora eydouxi* at Malpelo Island, was not reported until as recently as 1999 (Garzón--Ferreira and Pinzón 1999). A later survey of coral diseases in the Galápagos Islands revealed unexpected high levels of tissue anomalies compared with areas in the Caribbean, Florida Keys, and Hawaii (Vera and Banks 2009). Of eight coral communities sampled in the north-central and northern islands in 2005, 2006 and 2007, potential disease symptoms exhibited a prevalence rate of 23.9 %. *Porites lobata* was the most affected species with 35 % of colonies showing signs of parasitism or illness. The most common anomaly was presumed trematodiasis. The general appearance and color of the tissue anomalies of Galápagos corals are consistent with a digenean trematode parasite of *Porites* that occurs commonly in the central and western Pacific (Aeby 1998). The suspected parasitic condition in the Galápagos Islands, however, is in need of verification because the pinkish discoloration, a sign by which the parasite is recognized, can result from several causes (Benzoni et al. 2010).

Field and laboratory studies support a causative connection between temperature stress and compromised coral resistance that would facilitate the invasion of opportunistic infections (Kushmaro et al. 1996; Harvell et al. 1999, 2001; Hughes et al. 2003; Lesser et al. 2007; Rosenberg and Kushmaro 2011). Vera and Banks (2009) concluded that the high incidence of health anomalies could be attributed to recent warm and cold-water shock bleaching events, and possible compounding effects of anthropogenic disturbances.

8.4 Fate of Eastern Pacific Coral Reefs

8.4.1 Recovery

Coral reef recovery, by one measure the return of live coral cover to pre-ENSO (pre-1982) levels, has occurred in some but not all eastern Pacific areas (Guzmán and Cortés 2007; Wellington and Glynn 2007; Baker et al. 2008; Glynn et al. 2009). In addition to coral cover, however, it is also necessary to consider coral species richness and topographic complexity, essential components of reef-coral communities. Here we briefly note examples of reef recovery in the EEP, and areas that have not demonstrated recovery (Sect. 8.4.2 below). The extent of El Niño 1982–83 coral bleaching and mortality, and subsequent recovery along the Mexican coast, is largely unknown (Reyes-Bonilla 2003). Since the 1997–98 El Niño warming event several reef sites along the Mexican coast are now under study. With an increasing number of monitored sites, it should be possible in the near future to undertake a critical comparison of the impacts and recovery potential of ENSO warm and cool coral bleaching disturbances across all eastern Pacific latitudes.

Two general assessments of reef recovery following ENSO disturbances have revealed increases in live coral cover at 4 of 12 sites (Wellington and Glynn 2007), and 9 of 20 sites (Baker et al. 2008). These sites are not all identical and cannot be compared directly, but nonetheless document recovery in several areas. Reefs consisting predominantly of branching (*Pocillopora*) or massive (*Porites, Pavona*) coral genera have demonstrated comparable levels of recovery across the EEP region. Generally, the same reef-building species present before ENSO disturbances are preeminent in recovering communities. However, at Cocos Island, Costa Rica *Leptoseris scabra*, formerly a rare species there, has contributed substantially toward recovery at some monitored sites (Guzmán and Cortés 2007). Recent surveys in the northern-most Galápagos Islands also are revealing rapid recovery of the Darwin Island coral reef and the coral communities at Wenman Island. Prominent patches of pocilloporid corals on the Darwin Island reef in the mid-1970s, however, are now absent with *Porites lobata* presently the predominant species (compare Glynn and Wellington 1983; Glynn et al. 2009, 2015). *Pocillopora* spp. cover at Onslow Island in the southern Galápagos Islands also has declined dramatically; however, *Porites* and *Psammocora* patches at this site have fully recovered to pre-1982 levels (Paul 2012; Feingold and Glynn 2014). Presently in the Galápagos Islands the only known patches of *Pocillopora* with juxtaposed colonies, a pre-condition necessary for framework development, occur in the Concha y Perla lagoon at Villamil, Isabela Island (Feingold and Glynn 2014; Baums et al. 2014). Lacking pre-1982–83 and 1997–98 data at this site, it is not possible to know if this population is recovering from surviving remnant corals or is the result of new recruitment and growth.

Numerous large, massive colonies of *Porites* and *Pavona* in the EEP suffered partial mortality during the 1982–83 El Niño, but have recovered by asexual regeneration or de novo sexual recruitment onto dead colony surfaces (see Chap. 15, Glynn et al.). Many of these colonies have maintained their integrity, with limited bioerosion, and still contribute importantly to the structural complexity of reefs. Pocilloporid patches have begun to re-appear in some areas in Panama, particularly on stable substrates such as dead, rigid pocilloporid frameworks. On the Uva Island forereef slope, *Pocillopora* cover has increased from 0 to ~35 % (pre-El Niño live cover) over a 27 year period (Glynn et al. 2014). At this site framework development has been re-established and, as of 2012, has attained a vertical height of ~0.5 m in some areas.

8.4.2 Absence of Recovery

Over one-half of monitored EEP reef sites have demonstrated declining coral cover or no change since the early 1980s. As noted earlier, local coral extirpations and structural losses have occurred at several former reef sites. Most notable in this regard has been the high mortality of reef-building corals in the southern and central Galápagos Islands in 1983, followed by elevated echinoid bioerosion and eventual loss of virtually all reef formations within a 10-year period (Glynn 1994; Wellington and Glynn 2007; Baker et al. 2009). Nearly 80 % (13 of 17) of monitored south/central Galápagos coral communities have demonstrated no recovery (Glynn 2003; Glynn et al. 2015). Originally (mid-1970s) most of the pocilloporid patch reefs consisted of frameworks with modal thicknesses of 0.6 and 1.0 m. These frameworks were constructed of vertically oriented and interlocking colony branches that exhibited continuous, un-interrupted growth. Hundreds of large massive colonies (*Porites*, *Pavona*), many with skeletal growth axes exceeding 100 and 200 cm, have also been eroded and reduced to rubble and finer sediments (Glynn 1990).

Only a single site in the southern islands, the former Onslow Island (Devil's Crown) pocilloporid reef, began to undergo recovery, but this was short-lived. This site, consisting of a stable basalt/carbonate substrate and subject to strong current flow, experienced a steady increase in the sexual recruitment of *Pocillopora*, from 4 colonies in 1996 to 167 colonies in 2007 (Feingold in Baker et al. 2008; Feingold and Glynn 2014). A sudden decline to 12 colonies occurred sometime after 2007, likely in response to an island-wide La Niña bleaching event (Figs. 8.19a, 8.20, 8.21a; >400 cool degree days).

8.4.3 Long-Term Outlook

To gain some perspective of the temporal scale, it is instructive to consider the time during which coral growth and framework accumulation occurred before and up to the 1982–83 El Niño disturbance. This has been estimated for the predominant branching and massive reef-building species in the EEP (Glynn 1990). Age estimates of *Pocillopora* reef frame blocks 1–2 m in thickness in Panama and the Galápagos Islands commonly ranged between 100 and 200 years. Maximum reef frame thicknesses and age estimates in Galápagos and Panama were 2.7 m = 482 − 551 years, and 2.5 m = 338 years respectively. In the Galápagos Islands, large massive colonies (*Porites lobata, Pavona clavus*) were about 1 m in height with estimated ages of 100 years. A few of the largest colonies ranged between 2 and 5 m in height with estimated ages between 300 and 400 years. This suggests that if pocilloporid frameworks and massive corals become established relatively quickly after a disturbance, and then grow without interruption, it would be possible for modest coral formations to develop in a few hundreds of years. However, as noted earlier in this review, rapid coral recovery and uninterrupted growth are not commonplace in many EEP areas.

Several ENSO-related stressors have been shown to cause various levels of disturbance in the eastern Pacific coral reef region, often with interacting effects over diverse spatial and time scales. Therefore, to a large extent the structure, dynamics, and persistence of coral communities and reefs could be expected to depend on future ENSO activity. If ENSO activity intensifies this could lead to an onset of temperature extremes and greater inter-annual variability of high and low thermal stressors. In addition, global greenhouse warming could, by intensifying the alongshore wind stress on the ocean surface, lead to acceleration of coastal upwelling (Bakun 1990; Sydeman et al. 2014). Increased upwelling would contribute to temperature variability and ocean acidification, which could increase coral bleaching and mortality and depress calcification and reef building (Manzello 2010; see Chap. 18, Manzello et al.). Such a scenario is supported, at least on a decadal time scale, by 155 years of instrumental and coral isotope records. These records indicate an abrupt change in the tropical Pacific in 1976 towards warmer and wetter conditions and a higher frequency of interannual variability in more recent years (Fig. 8.1, Urban et al. 2000; Fiedler 2002). Such conditions would be expected to have a major impact on reef growth and the persistence of coral formations.

A recent study has noted that over one-half of the 860 known coral-associated invertebrate species have an obligate dependence on live corals (Stella et al. 2011). If symbiont-rich corals that are vulnerable to bleaching, e.g. branching pocilloporid species, disappear this could lead to a cascading effect of species extinctions. In addition to direct effects of coral host extinctions, longer-term reef framework loss through storm action and bioerosion would deplete structural and micro-habitat resources leading to large-scale extinctions of cryptic species (Glynn 2011). Diverse communities of cryptic invertebrate species also occupy dead coral rubble habitats, however, and therefore would not suffer losses, but in some cases might actually increase in abundance (Enochs and Manzello 2012). In like manner, cryptic fish communities associated with degraded corals, consisting mostly of small carnivores, omnivores, and herbivores (e.g., gobies, blennies, wrasses, eels, and damselfishes) do not undergo changes in abundance following bleaching disturbances (Bellwood et al. 2006; Glynn 2006; Feary et al. 2007). In the long term, however, many of these species would also likely decrease in abundance given the absence of corals and a dwindling supply of carbonate rubble.

Is it possible that heightened ENSO conditions in the past have interfered with eastern Pacific reef growth (Colgan 1990; Glynn and Colgan 1992)? A Holocene paleoecologic analysis of Panamanian pocilloporid reefs offers evidence of significant declines in reef-coral framework construction and overall reef degradation over a ~2500 year period, from 4200 to 1800 cal year BP (Toth et al. 2012). The proposed causes of this reef-building hiatus were (a) increased intensity, frequency and variability of ENSO with concomitant extreme temperatures, (b) elevated turbidity and reduced light from increased precipitation, and (c) lethal subaerial exposure events. This depositional hiatus is supported by climatic reconstructions of several ENSO-related effects. Similar contemporaneous hiatuses in reef accretion in Costa Rica, Australia and Japan suggest a millennial-scale, basin-wide setback in Pacific coral reef growth. If recent climate change trends continue, it is possible that increasing ENSO activity could again interfere with coral reef growth in the eastern Pacific and even in other parts of the Pacific Ocean.

As Enfield (2003), Collins et al. (2010) and Wang et al. (Chap. 4) have cautioned, there is a great deal of uncertainty on how global warming might influence the frequency and/or severity of the ENSO phenomenon. At the very least, however, it is highly probable that El Niño and La Niña events will continue into the future, with potentially strong direct and indirect impacts on eastern Pacific coral reefs.

Acknowledgments Much of the research reviewed in the EEP was supported by the Smithsonian Tropical Research Institute and Scholarly Studies Program, Smithsonian Institution, and the US National Science Foundation, Biological Oceanography Program, grant OCE-0526361 and earlier awards. Draft copies of the manuscript were greatly improved by help from Rafael Araujo, Stuart Banks, Beltran Espinosa, Peggy Fong, Benjamin Grassian, Carlos Jiménez, Frank Kelmo, Joshua Levy, Tyler Smith, and Brian Soden. Permission to use Fig. 8.18 was kindly approved by David A. Paz-García. Julia Cole, Stuart Banks, and Juan Maté assisted in the acquisition of temperature data.

References

Abram NJ, Gagan MK, McCulloch MT, Chappell J, Hantoro WS (2003) Coral reef death during the 1997 Indian Ocean dipole linked to Indonesian wildfires. Science 301:952–955

Achura Cárdenas JB, Valdéz Araúz E (1980) Estudio sobre los efectos predatorios de los moluscos *Coralliophila abbreviata* y *Jenneria pustulata* de ambos litorales del Istmo de Panamá. Thesis, University of Panamá, Panamá, p 66

Aeby GS (1998) A digenean metacercaria from the reef coral, *Porites compressa*, experimentally identified as *Podocotyloides stenometra*. J Parasitol 84:1259–1261

Amador JA, Alfaro EJ, Lizano OG, Magaña VO (2006) Atmospheric forcing of the eastern tropical Pacific: a review. Prog Oceanogr 69:101–142

Arntz WE, Tarazona J (1990) Effects of El Niño 1982–83 on benthos, fish and fisheries off the South American Pacific coast. In: Glynn PW (ed) Global ecological consequences of the 1982–83 El Niño-Southern Oscillation. Elsevier Oceanogr Ser 52, Amsterdam, pp 323–360

Arntz WE, Landa A, Tarazona J (eds) (1985) El Niño, su impacto en la fauna marina. Bol Inst Mar Perú—Callao (special vol), p 222

Attrill MJ, Kelmo F, Jones MB (2004) Impact of the 1997–98 El Niño event on the coral-associated echinoderm assemblage from northern Bahia, northeastern Brazil. Inter-Res Clim Res 26:151–158

Baker AC, Glynn PW, Riegl B (2008) Climate change and coral reef bleaching: an ecological assessment of long-term impacts, recovery trends and future outlook. Estuar Coast Shelf Sci 80:435–471

Bakun A (1990) Global climate change and intensification of coastal upwelling. Science 247:198–201

Banks S, Vera M, Chiriboga Á (2009) Characterizing the last remaining reefs: establishing reference points to assess long term change in Galápagos zooxanthellate coral communities. Galapagos Res 66:43–64
Barber RT, Chavez FP (1986) Ocean variability in relation to living resources during the 1982/83 El Niño. Nature 319:279–285
Barber RT, Kogelschatz JE (1990) Nutrients and productivity during the 1982/83 El Niño. In: Glynn PW (ed) Global ecological consequences of the 1982–83 El Niño-Southern Oscillation. Elsevier Oceanogr Ser 52, Amsterdam, pp 21–53
Baums IB, Devlin-Durante M, Laing BAA, Feingold J, Smith T, Bruckner A, Monteiro J (2014) Marginal coral populations: the densest known aggregation of *Pocillopora* in the Galápagos Archipelago is of asexual origin. Front Mar Sci. doi:10.3389/fmars.2014.00059
Bellwood DR, Hoey AS, Ackerman JL, Depczyinski M (2006) Coral bleaching, reef fish community phase shifts and the resilience of coral reefs. Global Change Biol 12:1587–1594
Benzoni F, Galli P, Pichon M (2010) Pink spots on *Porites*: not always a coral disease. Coral Reefs 29:153
Berkelmans R (2002) Time-integrated thermal bleaching thresholds of reefs and their variation on the Great Barrier Reef. Mar Ecol Prog Ser 229:73–82
Berkelmans R (2009) Bleaching and mortality thresholds: how much is too much? In: van Oppen MJH, Lough JM (eds) Coral bleaching. Patterns, processes, causes and consequences. Ecol Stud, Springer-Verlag, Berlin, pp 103–119
Bowman AW, Azzalini A (1997) Applied smoothing techniques for data analysis. Clarendon Press, Oxford, p 193
Bowman AW, Azzalini A (2013) R package 'sm': nonparametric smoothing methods (version 2.2–5). URL http://www.stats.gla.ac.uk/~adrian/sm, http://azzalini.stat.unipd.it/Book_sm
Brown BE (1987) Worldwide death of corals—natural cyclical events or man-made pollution? Mar Poll Bull 18:9–13
Brown BE (1997) Coral bleaching: causes and consequences. Coral Reefs 16(suppl):S129–S138
Camargo SJ, Emanuel KA, Sobel AH (2007) Use of a genesis potential index to diagnose ENSO effects on tropical cyclone genesis. J Climate 20:4819–4834
Cane MA (1983) Oceanographic events during El Niño. Science 222:1189–1195
Cane MA (1986) El Niño. Annu Rev Planet Sci 4:43–70
Carriquiry JD, Reyes-Bonilla H (1997) Community structure and geographic distribution of the coral reefs of Nayarit, Mexican Pacific. Cienc Mar 23:227–248
Carriquiry JD, Cupul-Magaña AL, Rodríguez-Zaragosa F, Medina-Rosas P (2001) Coral bleaching and mortality in the Mexican Pacific during the 1997–98 El Niño and prediction from a remote sensing approach. Bull Mar Sci 69:237–249
Castro P (1978) Movements between coral colonies in *Trapezia ferruginea* (Crustacea: Brachyura), an obligate symbiont of scleractinian corals. Mar Biol 46:237–245
Cea Egaña A, DiSalvo LH (1982) Mass expulsion of zooxanthellae by Easter Island corals. Pac Sci 36:61–63
Chen L, Singh V, Guo S, Mishra A, Guo J (2013) Drought analysis using copulas. J Hydrol Eng 18:797–808
Chu PS (2004) ENSO and tropical cyclone activity. In: Murnane RJ, Liu K-B (eds) Hurricanes and typhoons, past, present and future. Columbia Univ Press, New York, pp 297–332
Chu PS, Wang J (1997) Tropical cyclone occurrences in the vicinity of Hawaii: are the differences between El Niño and non-El Niño years significant? J Climate 10:2683–2689
Coffroth MA, Lasker HR, Oliver JK (1990) Coral mortality outside of the eastern Pacific during 1982–83: relationship to El Niño. In: Glynn PW (ed) Global ecological consequences of the 1982–83 El Niño-Southern Oscillation. Elsevier Oceanogr Ser 52, Amsterdam, pp 141–182
Coles SL, Fadlallah YH (1991) Reef coral survival and mortality at low temperatures in the Arabian Gulf: new species-specific lower temperature limits. Coral Reefs 9:231–237
Colgan MW (1990) El Niño and the history of eastern Pacific reef building. In: Glynn PW (ed) Global ecological consequences of the 1982–83 El Niño-Southern Oscillation. Elsevier Oceanogr Ser 52, Amsterdam, pp 183–232
Colley SB, Glynn PW, May AS, Maté JL (2006) Species-dependent reproductive responses of eastern Pacific corals to the 1997–1998 ENSO event. Proceedings of 10th International Coral Reef Symposium, vol 1, Okinawa, pp 61–70
Collins JM, Mason IM (2000) Local environmental conditions related to seasonal tropical cyclone activity in the northeast Pacific basin. Geo Res Lett 27:3881–3884
Collins M, An SI, Cai W, Ganachaud A, Guilyardi E, Jin F-F, Jochum M, Lengaigne M, Power S, Timmermann A, Vecchi G, Wittenberg A (2010) The impact of global warming on the tropical Pacific Ocean and El Niño. Nat Geosci 3:391–397
Cortés J (1990) The coral reefs of Golfo Dulce, Costa Rica: distribution and community structure. Atoll Res Bull 344:1–37
Cortés J (1997) Biology and geology of eastern Pacific coral reefs. Coral Reefs 16(suppl):539–546
Cortés J (ed) (2003) Latin American coral reefs. Elsevier, Amsterdam, p 497
Cortés J, Murillo MM (1985) Comunidades coralinas y arrecifes del Pacífico de Costa Rica. Rev Biol Trop 33:197–202
Cortés J, Guzmán HM (1998) Organismos de los arrecifes coralinos de Costa Rica: descripción, distribución geográfica e historia natural de los corales zooxantelados (Anthozoa: Scleractinia) del Pacífico. Rev Biol Trop 46:55–92
Cortés J, Jiménez C (2003) Corals and coral reefs of the Pacific coast of Costa Rica: history, research and status. In: Cortés J (ed) Latin American coral reefs. Elsevier, Amsterdam, pp 361–385
Cortés J, Murillo MM, Guzmán HM, Acuña J (1984) Pérdida de zooxantelas y muerte de corales y otros organismos arrecifales en el Caribe y Pacífico de Costa Rica. Rev Biol Trop 32:227–231
Cunning R, Baker AC (2013) Excess symbionts increase the susceptibility of reef corals to bleaching. Nat Clim Change doi:10.1038/nclimate 1711
Dana JD (1843) On the temperature limiting the distribution of corals. Amer J Sci 45:130–131
Dana TF (1975) Development of contemporary eastern Pacific coral reefs. Mar Biol 33:355–374
Darwin C (1842) The structure and distribution of coral reefs. Reprinted 1962, Univ Calif Press, Berkeley, p 214
D'Croz L, Maté JL, Oke JE (2001) Responses to elevated sea water temperature and UV radiation in the coral *Porites lobata* from upwelling and non-upwelling environments on the Pacific coast of Panama. Bull Mar Sci 69:203–214
Dunbar RB, Wellington GM, Colgan MW, Glynn PW (1994) Eastern Pacific sea surface temperature since 1600 A.D.: The δO^{18} record of climate variability in Galápagos corals. Paleoceanography 9:291–315
Eakin CM (1987) Damselfishes and their algal lawns: a case of plural mutualism. Symbiosis 4:275–288
Eakin CM (1988) Avoidance of damselfish lawns by the sea urchin *Diadema mexicanum* at Uva Island, Panama. Proceedings of 6th International Coral Reef Symposium, vol 2, Townsville, 21–26
Eakin CM (1992) Post-El Niño Panamanian reefs: less accretion, more erosion and damselfish protection. Proceedings of the 7th International Coral Reef Symposium, vol 1, Guam, pp 387–396
Eakin CM (1996) Where have all the carbonates gone? A model comparison of calcium carbonate budgets before and after the

1982–83 El Niño at Uva Island in the eastern Pacific. Coral Reefs 15:109–119
Eakin CM, Glynn PW (1996) Low tidal exposure and reef mortalities in the eastern Pacific. Coral Reefs 15:120
Eakin CM (2001) A tale of two ENSO events: carbonate budgets and the influence of two warming disturbances and intervening variability, Uva Island, Panama. Bull Mar Sci 69:171–186
Eakin CM, Smith DB, Glynn PW, D'Croz L, Gil J (1989) Extreme tidal exposures, cool upwelling and coral mortality in the eastern Pacific (Panamá). Ass Mar Labs Carib 22:29
Eakin CM, Morgan JA, Heron SF, Smith TB, Liu G, Alvarez-Filip L, Baca B, Bartels E, Bastidas C, Bouchon C, Brandt M, Bruckner AW, Bunkley-Williams L+54 others (2010) Caribbean corals in crisis: record thermal stress, bleaching, and mortality in 2005. PLoS ONE 5(11):e13969. doi:10.1371/journal.pone.0013969
Edgar GJ, Banks S, Fariña JM, Calvopiña M, Martínez C (2004) Regional biogeography of shallow reef fish and macro-invertebrate communities in the Galápagos Archipelago. J Biogeogr 31:1107–1124
Edgar GJ, Banks SA, Brandt M, Bustamante RH, Chiriboga A, Earle SA, Garske LE, Glynn PW, Grove JS, Henderson S, Hickman CP, Miller KA, Rivera F, Wellington GM (2010) El Niño, grazers and fisheries interact to greatly elevate extinction risk for Galápagos marine species. Global Change Biol 16:2876–2890
Edmondson CH (1928) The ecology of an Hawaiian coral reef. Bull Bernice P Bishop Mus 45:1–64
Egaña AC, DiSalvo LH (1982) Mass expulsion of zooxanthellae by Easter Island corals. Pacific Sci 36:61–63
Emerson WK (1991) First records for *Cymatium mundum* (Gould) in the eastern Pacific Ocean, with comments on the zoogeography of the tropical trans-Pacific tonnacean and non-tonnacean prosobranch gastropods with Indo-Pacific faunal affinities in west American waters. The Nautilus 105:62–80
Emerson WK, Chaney HW (1995) A zoogeographic review of the Cypraeidae (Mollusca: Gastropoda) occurring in the eastern Pacific Ocean. Veliger 38:8–21
Enfield DB (2001) Evolution and historical perspective of the 1997-1998 El Niño-Southern Oscillation event. Bull Mar Sci 69:7–25
Enfield DB (2003) The "El Niño" FAQ http://www.aoml.noaa.gov/general/enso_faq
Enfield DB, Cid L (1991) Low-frequency changes in El Niño-Southern Oscillation. J Climate 4:1137–1146
Enochs IC (2012) Motile cryptofauna associated with live and dead coral substrates: implications for coral mortality and framework erosion. Mar Biol 159:709–722
Enochs IC, Hockensmith G (2009) Effects of coral mortality on the community composition of cryptic metazoans associated with *Pocillopora damicornis*. Proceedings of 11th International Coral Reef Symposium, vol 26, Ft Lauderdale, pp 1368–1372
Enochs IC, Manzello DP (2012a) Species richness of motile cryptofauna across a gradient of reef framework erosion. Coral Reefs 31:653–661
Enochs IC, Manzello DP (2012b) Responses of cryptofaunal species richness and trophic potential to coral reef habitat degradation. Diversity 4:94–104
Enochs IC, Toth LT, Brandtneris VW, Afflerbach JC, Manzello DP (2011) Environmental determinants of motile cryptofauna on an eastern Pacific coral reef. Mar Ecol Prog Ser 438:105–118
Feary DA, Almany GR, Jones GP, McCormick MI (2007) Coral degradation and the structure of tropical reef fish communities. Mar Ecol Prog Ser 333:243–248
Feingold JS (2011) El Niño, La Niña and ENSO. In: Hopley D (ed) Encyclopedia of Modern coral reefs. Springer, Berlin, part 5, pp 365–368
Feingold JS, Glynn PW (2014) Coral research in the Galápagos Islands, Ecuador. In: Vinueza L, Denkinger J (eds) The Galapagos marine reserve: a dynamic socio-ecological system. Springer, Berlin, pp 3–22
Fiedler PC (2002) Environmental change in the eastern tropical Pacific Ocean: review of ENSO and decadal variability. Mar Ecol Prog Ser 244:265–283
Finet Y (1991) The marine mollusks of the Galápagos Islands. In: James MJ (ed) Galápagos marine invertebrates: taxonomy, biogeography, and evolution in Darwin's Islands. Plenum Press, New York, London, pp 253–280
Firing E, Lukas R (1983) El Niño at the Equator and 159° W. Trop Ocean-Atmos Newsl 21:9–11
Fisk DA, Done TJ (1985) Taxonomic and bathymetric patterns of bleaching in corals, Myrmidon Reef (Queensland). Proceedings of 5th International Coral Reef Congress, vol 6, Tahiti, pp 149–154
Fong P, Glynn PW (2001) Population abundance and size-structure of an eastern tropical Pacific reef coral after the 1997–98 ENSO: a simulation model predicts field measures. Bull Mar Sci 69:187–202
Garzón-Ferreira J, Pinzón J (1999) Evaluación rápida de estructura y salud de las formaciones coralinas de la Isla de Malpelo (Pacífico colombiano). Bol Inves Mar Cost INVEMAR 28:137–154
Ginsburg RN (ed) (1993) Global aspects of coral reefs: health, hazards and history. 7–11 June 1993, Univ Miami, Coral Gables, p 420
Gleason DF, Wellington GM (1993) Ultraviolet radiation and coral bleaching. Nature 365:836–838
Gleeson MW (1994) Correlation of coral bleaching events and remotely-sensed sea surface temperatures. Trident Scholar Proj Rep, US Naval Acad, Annapolis, MD, p 215
Gleeson MW, Strong AE (1995) Applying MCSST to coral reef bleaching. Adv Space Res 16(10):151–154
Glynn PW (1976) Some physical and biological determinants of coral community structure in the Eastern Pacific. Ecol Monogr 46:431–456
Glynn PW (1977) Coral growth in upwelling and nonupwelling areas off the Pacific coast of Panamá. J Mar Res 35:567–585
Glynn PW (1983a) Extensive 'bleaching' and death of reef corals on the Pacific coast of Panama. Environ Conserv 10:149–154
Glynn PW (1983b) Crustacean symbionts and the defense of corals: coevolution on the reef? In: Nitecki MH (ed) Coevolution. Univ Chicago Press, Chicago, pp 111–178
Glynn PW (1984) Widespread coral mortality and the 1982–83 El Niño warming event. Environ Conserv 11:133–146
Glynn PW (1985a) Corallivore population sizes and feeding effects following El Niño (1982–83) associated coral mortality in Panama. Proceedings of 5th International Coral Reef Congress, vol 4, Tahiti, pp 183–188
Glynn PW (1985b) El Niño-associated disturbance to coral reefs and post disturbance mortality by *Acanthaster planci*. Mar Ecol Prog Ser 26:295–300
Glynn PW (1988a) El Niño-Southern Oscillation 1982–1983: nearshore population, community and ecosystem responses. Annu Rev Ecol Syst 19:309–345
Glynn PW (1988b) El Niño warming, coral mortality and reef framework destruction by echinoid bioerosion in the eastern Pacific. Galaxea 7:129–160
Glynn PW (ed) (1990a) Global ecological consequences of the 1982–83 El Niño-Southern Oscillation. Elsevier Oceanogr Ser 52, Amsterdam, p 563
Glynn PW (1990b) Coral mortality and disturbances to coral reefs in the tropical eastern Pacific. In: PW Glynn (ed) Global ecological consequences of the 1982–83 El Niño-Southern Oscillation. Elsevier Oceanogr Ser 52, pp 55–126
Glynn PW (1993) Coral reef bleaching: ecological perspectives. Coral Reefs 12:1–17

Glynn PW (1994) Status of coral reefs in the Galápagos Islands: natural versus anthropogenic impacts. Mar Poll Bull 29:131–140

Glynn PW (1996) Coral reef bleaching: facts, hypotheses, and implications. Glob Change Biol 2:495–509

Glynn PW (1997) Eastern Pacific coral reef biogeography and faunal flux: Durham's dilemma revisited. Proceedings of 8th International Coral Reef Symposium, vol 1, Panama, pp 371–378

Glynn PW (2000a) El Niño-Southern Oscillation mass mortalities of reef corals: a model of high temperature marine extinctions? In: Insalaco E, Skelton PW, Palmer TJ (eds) Carbonate platform systems: components and interactions. Geol Soc, London, Spec Pubs 178:117–133

Glynn PW (2000b) Effects of the 1997–98 El Niño-Southern Oscillation on eastern Pacific corals and coral reefs: an overview. Proceedings of 9th International Coral Reef Symposium, vol 2, Bali, pp 1169–1174

Glynn PW (ed) (2001) A collection of studies on the effects of the 1997–98 El Niño-Southern Oscillation event on corals and coral reefs in the eastern tropical Pacific. Bull Mar Sci 69:1–288

Glynn PW (2003) Coral communities and coral reefs of Ecuador. In: Cortés J (ed) Latin American coral reefs. Elsevier, Amsterdam, pp 449–472

Glynn PW (2006) Fish utilization of simulated coral reef frameworks versus eroded rubble substrates off Panamá, eastern Pacific. Proceedings of 10th International Coral Reef Symposium, vol 1, Okinawa, pp 250–256

Glynn PW (2008) Food-web structure and dynamics of eastern tropical Pacific coral reefs: Panamá and Galápagos Islands. In: McClanahan TR, Branch GM (eds) Food webs and the dynamics of marine reefs. Oxford Univ Press, Oxford, pp 185–208

Glynn PW (2011) In tandem reef coral and cryptic metazoan declines and extinctions. Bull Mar Sci 87:767–794

Glynn PW, Wellington GM (1983) Corals and coral reefs of the Galápagos Islands. Univ California Press, Berkeley, p 330

Glynn PW, D'Croz L (1990) Experimental evidence for high temperature stress as the cause of El Niño-coincident mortality. Coral Reefs 8:181–191

Glynn PW, Colgan MW (1988) Defense of corals and enhancement of coral diversity by territorial damselfishes. Proceedings of 6th International Coral Reef Symposium, vol 2, Townsville, pp 157–163

Glynn PW, Colgan MW (1992) Sporadic disturbances in fluctuating coral reef environments: El Niño and coral reef development in the eastern Pacific. Amer Zool 32:707–718

Glynn PW, Cortés J, Guzmán HM, Richmond RH (1988) El Niño (1982–83) associated coral mortality and relationship to sea surface temperature deviations in the tropical eastern Pacific. Proceedings of 6th International Coral Reef Symposium, vol 3, Townsville, pp 237–243

Glynn PW, de Weerdt WH (1991) Elimination of two reef-building hydrocorals following the 1982–83 El Niño warming event. Science 253:69–71

Glynn PW, Feingold JS (1992) Hydrocoral species not extinct. Science 257:1845

Glynn PW, Fong P (2006) Patterns of reef coral recovery by the regrowth of surviving tissues following the 1997–98 El Niño warming and 2000, 2001 upwelling cool events in Panamá, eastern Pacific. Proceedings of 10th International Coral Reef Symposium, vol 1, Okinawa, pp 624-630

Glynn PW, Maté JL (1997) Field guide to the Pacific coral reefs of Panama. Proceedings of 8th International Coral Reef Symposium, vol 1, Panama, pp 145–166

Glynn PW, Perez M, Gilchrist SL (1985a) Lipid decline in stressed corals and their crustacean symbionts. Biol Bull 168:276–284

Glynn PW, Peters EC, Muscatine L (1985b) Coral tissue microstructure and necrosis: relation to catastrophic coral mortality in Panama. Dis Aquat Org 1:29–37

Glynn PW, Veron JEN, Wellington GM (1996) Clipperton Atoll (eastern Pacific): oceanography, geomorphology, reef-building coral ecology and biogeography. Coral Reefs 15:71–99

Glynn PW, Wellington GM, Birkeland C (1979) Coral reef growth in the Galápagos: limitation by sea urchins. Science 203:47–49

Glynn PW, Maté JL, Baker AC, Calderón MO (2001) Coral bleaching and mortality in Panama and Ecuador during the 1997–1998 El Niño-Southern Oscillation event: spatial/temporal patterns and comparisons with the 1982–1983 event. Bull Mar Sci 69:79–109

Glynn PW, Riegl B, Correa AMS, Baums IB (2009) Rapid recovery of a coral reef at Darwin Island, Galápagos Islands. Galápagos Res 66:6–13

Glynn PW, Imai R, Sakai K, Nakano Y, Yamazato K (1993) Experimental responses of Okinawan (Ryukyu Islands, Japan) reef corals to high sea temperature and UV radiation. Proceedings of 7th International Coral Reef Symposium, vol 1, Guam, pp 27–37

Glynn PW, Riegl B, Purkis S, Kerr JM, Smith TB (2015) Coral reef recovery in the Galápagos Islands: the northernmost islands (Darwin and Wenman). Coral Reefs. doi:10.1007/s00338-015-1280-4

Glynn PW, Colley SB, Maté JL, Baums IB, Feingold JS, Cortés J, Guzmán HM, Afflerbach JC, Brandtneris VM, Ault JS (2012) Reef coral reproduction in the equatorial eastern Pacific: Costa Rica, Panamá, and the Galápagos Islands (Ecuador). VII *Siderastreidae*, *Psammocora stellata* and *Psammocora profundacella*. Mar Biol 159:1917–1932

Glynn PW, Enochs IC, Afflerbach JA, Brandtneris VW, Serafy JE (2014) Eastern Pacific reef fish responses to coral recovery following El Niño disturbances. Mar Ecol Prog Ser 495:233–247

Goreau TF (1964) Mass expulsion of zooxanthellae from Jamican reef communities after Hurricane Flora. Science 145:383–386

Goreau TJ, Hayes RL (1994) Coral bleaching and ocean 'hot spots'. Ambio 23:176–180

Gray WM, Sheaffer JD (1991) El Niño and QBO influences on tropical cyclone activity. In: Glantz MH, Katz RW, Nicholls N (eds) Teleconnections linking worldwide climate anomalies. Cambridge Univ Press, Cambridge, pp 257–284

Grove JS (1984) Influence of the 1982–83 El Niño on the ichthyofauna of the Galápagos Archipelago. Trop Ocean Atmos Newsl 28:18–19

Grove JS (1985) Influence of the 1982–1983 El Niño event upon the ichthyofauna of the Galápagos Archipelago. In: Robinson G, del Pino EM (eds) El Niño in the Galápagos Islands: the 1982–1983 event. Publication of the Charles Darwin Foundation for the Galápagos Islands, Quito, Ecuador, pp 191–198

Grove JS, Lavenberg RJ (1997) The fishes of the Galápagos Islands. Stanford Univ Press, Stanford, p 912

Grove JS, Gerzon D, Saa D, Strang M (1986) Distribución y ecología de la familia Pomacentridae (Pisces) en las Islas Galápagos. Rev Biol Trop 34:127–140

Guzmán HM (1986) Estructura de la comunidad arrecifal de la Isla del Caño, Costa Rica, y el efecto de perturbaciones naturales severas. MSc thesis, Univ Costa Rica, San Pedro, Costa Rica, p 179

Guzmán HM (1988) Primer informe de un coral acropórido para el Pacífico oriental: crítica a Prahl y Mejía. Rev Biol Trop 36: 163–166

Guzmán HM, Cortés J (1992) Cocos Island (Pacific of Costa Rica) coral reefs after the 1982–83 El Niño disturbance. Rev Biol Trop 40:309–324

Guzmán HM, Cortés J (1993) Arrecifes coralinos del Pacífico oriental tropical: revisión y perspectivas. Rev Biol Trop 41:535–557

Guzmán HM, Robertson DR (1989) Population and feeding responses of the corallivorous pufferfish *Arothron meleagris* to coral mortality in the eastern Pacific. Mar Ecol Prog Ser 55:121–131

Guzmán HM, Cortés J (2007) Reef recovery 20 years after the 1982–1983 El Niño massive mortality. Mar Biol 151:401–411

Guzmán HM, Cortés J, Richmond RH, Glynn PW (1987) Efectos del fenómeno de El Niño-Oscilación Sureña 1982–83 en los arrecifes coralinos de la Isla del Caño, Costa Rica. Rev Biol Trop 35:325–332

Guzmán HM, Cortés J, Glynn PW, Richmond RH (1990) Coral mortality associated with dinoflagellate blooms in the eastern Pacific (Costa Rica and Panama). Mar Ecol Prog Ser 60:299–303

Hansen DV (1990) Physical aspects of the El Niño event of 1982–1983. In: Glynn PW (ed) Global ecological consequences of the 1982–83 El Niño-Southern Oscillation. Elsevier Oceanogr Ser 52, Amsterdam, pp 1–20

Hansen DV, Herman A (1989) Evolution of isotherm depth anomalies in the eastern tropical Pacific Ocean during the El Niño event of 1982–1983. J Geophys Res C10:14461–14473

Harriott VJ (1985) Mortality rates of scleractinian corals before and during a mass bleaching event. Mar Ecol Prog Ser 21:81–88

Harmelin-Vivien ML, Laboute P (1986) Catastrophic impact of hurricanes on atoll outer reef slopes in the Tuamotu (French Polynesia). Coral Reefs 5:55–62

Harvell CD, Kim K, Quirolo C, Weir J, Smith G (2001) Coral bleaching and disease: contributions to 1998 mass mortality in *Briareum asbestinum* (Octocorallia, Gorgonacea). Hydrobiologia 460:97–104

Harvell CD, Kim K, Burkholder JM, Colwell RR, Epstein PR, Grimes J, Hofmann EE, Lipp EK, Osterhaus ADME, Overstreet R, Porter JW, Smith GW, Vasta GR (1999) Emerging marine diseases—climate links and anthropogenic factors. Science 285:1505–1510

Hastie T, Tibshirani R (1990) Generalized additive models. Chapman and Hall/CRC, vol 43

Hernández L, Reyes-Bonilla H, Balart EF (2010) Effect of coral bleaching induced by low temperature on reef-associated decapod crustaceans of the southwestern Gulf of California. Rev Mexicana Biodivers 81:S113–S119

Highsmith RC (1980) Geographic pattern of coral bioerosion: a productivity hypothesis. J Exp Mar Biol Ecol 46:177–196

Highsmith RC (1981) Coral bioerosion at Enewetak: agents and dynamics. Int Revue ges Hydrobiol 66:335–375

Hoegh-Guldberg O (1999) Coral bleaching, climate change and the future of the world's coral reefs. Mar Freshwater Res 50:839–866

Hueerkamp C, Glynn PW, D'Croz L, Maté JL, Colley SB (2001) Bleaching and recovery of five eastern Pacific corals in an El Niño-related temperature experiment. Bull Mar Sci 69:215–236

Hughes T, Baird A, Bellwood AD, Card DM, Connolly MS, Folke SC, Grosber CR, Hoegh-Guldberg RO, Jackson OJ, Kleypas JJ, Lough JJ, Marshall JP, Nyström PM, Palumbi MS, Pandolfi SJ, Rosen B, Roughgarden J (2003) Climate change, human impacts and the resilience of coral reefs. Science 301:929–933

Hutchins PA (2011) Bioerosion. In: Gabioch G, Davies P, Done TJ, Guschler E, Macintyre IG, Wood R, Woodroffe CD, Hopley D (eds) Encyclopedia of modern coral reefs: structure, form and process. Encyclopedia Earth Sci Ser, Springer, Berlin, pp 139–156

Iglesias-Prieto R, Reyes-Bonilla H, Riosmena-Rodríguez R (2003) Effects of the 1997-1998 ENSO on coral reef communities in the Gulf of California, Mexico. Geofis Int 42:467–471

Irwin RP, Davis R (1999) The relationship between the Southern Oscillation index and tropical cyclone tracks in the Eastern North Pacific. Geophys Res Lett 26:2251–2254

Jiménez C (1998) Arrecifes y comunidades coralinas de Bahía Culebra, Pacífico norte de Costa Rica (Golfo de Papagayo). MSc Thesis, Univ Costa Rica, San Pedro, Costa Rica, p 218

Jiménez CE (2001) Arrecifes y ambientes coralinos de Bahía Culebra, Pacífico de Costa Rica: aspectos biológicos, económicos, recreactivos y de manejo. Rev Biol Trop 49(suppl 2):215–231

Jiménez C, Cortés J, León A, Ruíz E (2001) Coral bleaching and mortality associated with the 1997–98 El Niño in an upwelling environment in the eastern Pacific (Gulf of Papagayo, Costa Rica). Bull Mar Sci 69:151–169

Johnson GC, McPhaden MJ, Rowe GD (2000) Upper equatorial Pacific Ocean current and salinity variability during the 1996-1998 El Niño-La Niña cycle. J Geophys Res 105(C1):1037–1053

Johnson GC, Sloyan BM, Kessler WS, McTaggart KE (2002) Direct measurements of upper ocean currents and water properties across the tropical Pacific during the 1990s. Prog Oceanogr 52:31–61

Jokiel PL, Coles SL (1990) Response of Hawaiian and other Indo-Pacific reef corals to elevated temperature. Coral Reefs 8:155–162

Jones GP, McCormick MI, Srinivasan M, Eagle JV (2004) Coral decline threatens fish biodiversity in marine reserves. Proc Natl Acad Sci USA 101:8251–8253

Kay EA (1991) The marine mollusks of the Galápagos: determinants of insular marine faunas. In: James MJ (ed) Galápagos marine invertebrates: taxonomy, biogeography, and evolution in Darwin's Islands. Plenum Press, New York, London, pp 235–252

Kelmo F, Hallock P (2013) Responses of foraminiferal assemblages to ENSO climate patterns on bank reefs of northern Bahia, Brazil: a 17-year record. Ecol Ind 30:148–157

Kelmo F, Attrill MJ, Jones MB (2003) Effects of the 1997–98 El Niño on the cnidarian community of a high turbidity coral reef system (Northern Bahia, Brazil). Coral Reefs 22:541–550

Kelmo F, Attrill MJ, Gomes RCT, Jones MB (2004) El Niño induced local extinction of coral reef bryozoan species from northern Bahía, Brazil. Biol Conserv 118:609–617

Kelmo F, Attrill MJ, Jones MB (2006) Mass mortality of coral reef ascidians following the 1997/1998 El Niño event. Hydrobiologia 555:231–240

Kessler WS (2006) The circulation of the eastern tropical Pacific: a review. Prog Oceanogr 69:181–217

Kessler WS, Taft BA (1987) Dynamic heights and zonal geostrophic transports in the central tropical Pacific during 1979-84. J Phys Oceanogr 17:97–122

Kim ST, Yu J-Y (2012) The two types of ENSO in CMIP5 models. Geophys Res Lett 39:L11704. doi:10.1029/2012GL052006

Knapp KR, Kruk MC, Levinson DH, Diamond HJ, Neumann CJ (2010) The International Best Track Archive for Climate Stewardship (IBTrACS): unifying tropical cyclone best track data. Bull Amer Meteor Soc 91:363–376

Knowlton N, Rohwer F (2003) Multispecies microbial mutualisms on coral reefs: the host as a habitat. Amer Nat 162:S51–S62

Kushmaro A, Loya Y, Fine M, Rosenberg E (1996) Bacterial infection and coral bleaching. Nature 380:396

Kwiecinski B, Chial B (1987) Manifestations of El Niño in the Gulf of Panamá. Trop Ocean-Atmos News, Coral Gables 42:7–9

Laboute P (1985) Evaluation of damage done by the cyclones of 1982–1983 to the outer slopes of the Tikehau and Takapoto Atolls (Tuamotu Archipelago). Proceedings of 5th International Coral Reef Congress, vol 3, Tahiti, pp 323–329

LaJeunesse TC, Reyes-Bonilla H, Warner ME (2007) Spring "bleaching" among *Pocillopora* in the Sea of Cortez, Eastern Pacific. Coral Reefs 26:265–270

LaJeunesse TC, Smith R, Walther M, Pinzón J, Pettay DT, McGinley M, Aschaffenburg M, Medina-Rosas P, Cupul-Magaña AL, López-Pérez A, Reyes-Bonilla H, Warner ME (2010) Host-symbiont recombination versus natural selection in the response of coral dinoflagellate symbioses to environmental disturbance. Proc Royal Soc B 277:2925–2934

Lesser MP (1997) Oxidative stress causes coral bleaching during exposure to elevated temperatures. Coral Reefs 16:187–192

Lesser MP, Bythell JC, Gates RD, Johnstone RW, Hoegh-Guldberg O (2007) Are infectious diseases really killing corals? Alternative interpretations of the experimental and ecological data. J Exp Mar Biol Ecol 346:36–44

Lessios HA, Kessing BD, Wellington GM, Graybeal A (1996) Indo-Pacific echinoids in the tropical eastern Pacific. Coral Reefs 15:133–142

Lessios HA, Kessing BD, Robertson DR (1998) Massive gene flow across the world's most potent marine biogeographic barrier. Proc R Soc Lond B 265:583–588

Lirman D, Glynn PW, Baker AC, Leyte Morales GE (2001) Combined effects of three sequential storms on the Huatulco coral reef tract, Mexico. Bull Mar Sci 69:267–278

Lirman D, Schopmeyer S, Manzello D, Gramer LJ, Precht WF, Muller-Karger F, Banks K, Barnes B, Bartels E, Bourque A, Byrne J, Donahue S, Duquesnel J, Fisher L, Gilliam D, Hendee J, Johnson M, Maxwell K, McDevitt E, Monty J, Rueda D, Ruzicka R, Thanner S (2011) Severe 2010 cold-water event caused unprecedented mortality to corals of the Florida reef tract and reversed previous survivorship patterns. PLoS ONE 8:e23047. doi: 10.1371/journal pone.0023047

Macintyre IG, Glynn PW, Cortés J (1992) Holocene reef history in the eastern Pacific: mainland Costa Rica, Caño Island, Cocos Island, and Galápagos Islands. Proceedings of 7th International Coral Reef Symposium, vol 2, Guam, pp 1174–1184

Mangum LJ, Hayes SP, Toole JM (1986) Eastern Pacific Ocean circulation near the onset of the 1982–1983 El Niño. J Geophys Res 91:8428–8436

Manzello DP (2010) Coral growth with thermal stress and ocean acidification: lessons from the eastern tropical Pacific. Coral Reefs 29:749–758

Manzello DP, Kleypas JA, Budd DA, Eakin CM, Glynn PW, Langdon C (2008) Poorly cemented coral reefs of the eastern tropical Pacific: possible insights into reef development in a high-CO_2 world. Proc Nat Acad Sci USA 105:10,450–10,455

Maté-T JL (1997) Experimental responses of Panamanian reef corals to high temperature and nutrients. Proceedings of 8th International Coral Reef Symposium, vol 1, Panamá, pp 515–520

Maté JL (2003) Corals and coral reefs of the Pacific coast of Panamá. In: Cortés J (ed) Latin American coral reefs. Elsevier Science BV, Amsterdam, pp 387–417

Mayer AG (1914) The effects of temperature upon marine animals. Carnegie Inst Washington Pub Papers, Mar Lab Tortugas 6:3–24

Mayer AG (1916) The lower temperatures at which reef-corals lose their ability to capture food. Carnegie Inst Wash Yearbook 14:212

McClanahan T (2004) Coral bleaching, diseases and mortality in the western Indian Ocean. In: Rosenberg E, Loya Y (eds) Coral health and disease. Springer, Berlin, pp 157–176

McCosker JE, Humann PH (1996) New records of Galápagos fishes. Noticias de Galápagos 56:18–22

McCosker JE, Rosenblatt RH (2010) The fishes of the Galápagos Archipelago: an update. Proc Calif Acad Sci 61(suppl II):167–195

McPhaden MJ (1999) Genesis and evolution of the 1997–98 El Niño. Science 283:950–954

Meekan MG, Ackerman JL, Wellington GM (2001) Demography and age structures of coral reef damselfishes in the tropical eastern Pacific Ocean. Mar Ecol Prog Ser 212:223–232

Merlen G (1988) Two new butterfly fish records from Wenman Island. Noticias Galápagos 46:8–9

Munday PL (2004) Habitat loss, resource specialization, and extinction on coral reefs. Glob Change Biol 10:1642–1647

Muthukrishnan R (2013) An integrated empirical and modeling approach to evaluate determinants of community structure and alternate stable states dynamics on tropical reefs. PhD dissert, Univ Calif, Los Angeles, p 115

Norton J, McLain D, Brainard R, Husby D (1985) The 1982–83 El Niño event off Baja and Alta California and its ocean climate context. In: Wooster WS, Fluharty DL (eds) El Niño North: Niño effects in the eastern subarctic Pacific Ocean. Univ Washington, Washington Sea Grant Prog, pp 44–72

Ogden J, Wicklund R (1988) Mass bleaching of coral reefs in the Caribbean: a research strategy. National Undersea Research Program, Research Report 88-2, p 51

Oliver J (1985) Recurrent seasonal bleaching and mortality of corals on the Great Barrier Reef. Proceedings of 5th International Coral Reef Congress, vol 4, Tahiti, pp 201–206

Oliver JK, Berkelmans R, Eakin CM (2009) Coral bleaching in space and time. In: van Oppen MJH, Lough JM (eds) Coral bleaching: patterns, processes, causes and consequences. Springer, Berlin, pp 21–39

Oramas-N FA (1979) Estudio ecológico del *Jenneria pustulata* (Mollusca, Gastropoda). Thesis, Univ Panamá, Panamá, p 100

O'Reilly JE, Maritorena S, Mitchell BG, Siegel DA, Carder KL, Garver SA, Kahru M, McClain C (1998) Ocean color chlorophyll algorithms for SeaWiFS. J Geophys Res 103:24, 937–24 ,953

Paul NC (2012) Variable recovery of the massive coral, *Porites lobata*, in response to El Niño-Southern Oscillation events at Devil's Crown, Galápagos Islands, Ecuador. MSc thesis, Nova Southeastern Univ Oceanographic Center, Dania Beach, Florida, p 90

Paz-García DA, Balart EF, García-de-Léon FJ (2012) Cold water bleaching of *Pocillopora* in the Gulf of California. Proceedings of 12th International Coral Reef Symposium, vol 9A-10, Cairns, pp 1–6

Philander SGH (1985) El Niño and La Niña. J Atmos Sci 42:2652–2662

Philander SGH (1990) El Niño, La Niña, and the Southern Oscillation. Academic Press, San Diego, p 304

Picaut J, Ioualalen M, Menkes C, Delcroix T, McPhaden MJ (1996) Mechanism of the zonal displacements of the Pacific warm pools: implications for ENSO. Science 274:1486–1489

Podestá GP, Glynn PW (1997) Sea surface temperature variability in Panama and Galápagos: extreme temperatures causing coral bleaching. J Geophys Res 102:15, 749–15, 759

Podestá GP, Glynn PW (2001) The 1997–98 El Niño event in Panama and Galápagos: an update of thermal stress indices relative to coral bleaching. Bull Mar Sci 69:43–59

Pratchett MS (2001) Influence of coral symbionts on feeding preferences of crown-of-thorns starfish *Acanthaster planci* in the western Pacific. Mar Ecol Prog Ser 214:111–119

Pratchett MS, Wilson SK, Berumen ML, McCormick MI (2004) Sublethal effects of coral bleaching on an obligate coral feeding butterflyfish. Coral Reefs 23:352–356

Pratchett MS, Wilson SK, Baird AH (2006) Declines in the abundance of *Chaetodon* butterfly fishes following extensive coral depletion. J Fish Biol 69:1269–1280

Pratchett MS, Munday PL, Wilson SK, Graham NAJ, Cinner JE, Bellwood DR, Jones GP, Polunin NVC, McClanahan TR (2008) Effects of climate-induced coral bleaching on coral-reef fishes – ecological and economic consequences. Oceanogr Mar Biol: Annu Rev 46:251–296

R Development Core Team (2009) R: a language and environment for statistical computing. R Foundation for Statistical Computing, Vienna, Austria, p 409

Rasmusson EM, Wallace JM (1983) Meteorological aspects of the El Niño/Southern Oscillation. Science 222:1195–1202

Reaka-Kudla ML, Feingold JS, Glynn PW (1996) Experimental studies of rapid bioerosion of coral reefs in the Galápagos Islands. Coral Reefs 15:101–107

Reyes-Bonilla H (2003) Coral reefs of the Pacific coast of México. In: Cortés J (ed) Latin American coral reefs. Elsevier, Amsterdam, pp 331–349

Reyes-Bonilla H, Carriquiry JD, Leyte-Morales GE, Cupul-Magaña AL (2002) Effects of the El Niño-Southern Oscillation and the anti-El

Niño event (1997–1999) on coral reefs of the western coast of México. Coral Reefs 21:368–372
Richmond RH (1990) The effects of the El Niño/Southern Oscillation on the dispersal of corals and other marine organisms. In: Glynn PW (ed) Global ecological consequences of the 1982–83 El Niño-Southern Oscillation. Elsevier Oceanogr Ser 52, Elsevier, Amsterdam, pp 127–140
Robertson DR, Grove JS, McCosker JE (2004) Tropical transpacific shore fishes. Pac Sci 58:507–565
Robinson G (1985) Influence of the 1982–83 El Niño on Galápagos marine life. In: Robinson G, del Pino EM (eds) El Niño in the Galápagos Islands: the 1982–1983 event. Publication of the Charles Darwin Foundation for the Galápagos Islands, Quito, Ecuador, pp 153–190
Rosenberg E, Kushmaro A (2011) Microbial diseases of corals: pathology and ecology. In: Dubinsky Z, Stambler N (eds) Coral reefs: an ecosystem in transition. Springer, Dordrecht, pp 451–464
Ruttenberg B (2000) An unusual pulse of recruitment of two reef fishes in the Galápagos Islands coincident with the 1997–1998 El Niño. Bull Mar Sci 67:869–874
Rypien KL, Ward JR, Azam F (2010) Antagonistic interactions among coral-associated bacteria. Environ Microbiol 12:28–39
Scott PJB, Risk MJ, Carriquiry JD (1988) El Niño, bioerosion, and the survival of east Pacific reefs. Proceedings of 6th International Coral Reef Symposium, vol 2, Townsville, pp 517–520
Sheltema RS (1988) Initial evidence for the transport of teleplanic larvae of benthic invertebrates across the east Pacific barrier. Biol Bull 174:145–152
Shick JM, Lesser MP, Jokiel PJ (1996) Effects of ultraviolet radiation on corals and other coral reef organisms. Glob Change Biol 2:527–545
Smith TB, Fong P, Kennison R, Smith J (2010) Spatial refuges and associational defenses promote harmful blooms of the alga *Caulerpa sertularioides* onto coral reefs. Oecologia 164:1039–1048
Smith TB, Glynn PW, Maté JL, Toth LT, Gyory J (2014) A depth refugium from catastrophic coral bleaching prevents regional extinction. Ecology 95:1663–1673
Sonnenholzner JI, Ladah LB, Lafferty KD (2009) Cascading effects of fishing on Galápagos rocky reef communities: reanalysis using corrected data. Mar Ecol Prog Ser 375:209–218
Stella JS, Pratchett MS, Hutchings PA, Jones GP (2011) Coral-associated invertebrates: diversity, ecological importance and vulnerability to disturbance. Oceanogr Mar Biol Annu Rev 49:43–104
Stewart HL, Holbrook SJ, Schmitt RJ, Brooks AJ (2006) Symbiotic crabs maintain coral health by cleaning sediments. Coral Reefs 25:609–615
Stimson JS (1990) Stimulation of fat body production in the polyps of the coral *Pocillopora damicornis* by the presence of mutualistic crabs of the genus *Trapezia*. Mar Biol 106:211–218
Sydeman WJ, García-Reyes M, Schoeman DS, Rykaczewski RR, Thompson SA, Black BA, Bograd SJ (2014) Climate change and wind intensification in coastal upwelling ecosystems. Science 345:77–80
Thurber RV, Wilner-Hall D, Rodriguez-Mueller B, Desnues C, Edwards RA, Angly F, Dinsdale E, Kelly L, Rohwer F (2009) Metagenomic analysis of stressed coral holobionts. Environ Microbiol 11:2148–2163
Toth LT, Aronson RB, Vollmer SV, Hobbs JW, Urrego DH, Hai C, Enochs IC, Combosch DJ, van Woesik R, Macintyre IG (2012) ENSO drove 2500-year collapse of eastern Pacific coral reefs. Science 337:81–84
Tribollet A, Golubic S (2011) Reef bioerosion: agents and processes. In: Dubinsky Z, Stambler N (eds) Coral reefs: an ecosystem in transition. Springer, Dordrecht, pp 435–449
Tsuchiya M (1999) Effect of mass coral bleaching on the community structure of small animals associated with the hermatypic coral *Pocillopora damicornis*. Galaxea, Jap Coral Reef Soc 1:65–72
Tsuchiya M, Yamauchi Y, Moretzsohn F, Tsukiji M (1992) Species composition and some population traits of obligate symbiotic xanthid crabs, *Trapezia* and *Tetralia*, associated with bleached corals. Proceedings of 7th International Coral Reef Symposium, vol 1, Guam, pp 56–63
Urban FE, Cole JE, Overpeck JT (2000) Influence of mean climate change on climate variability from a 155-year tropical Pacific coral record. Nature 407:989–993
Van Oppen MJH, Lough JM (2009) Coral bleaching: patterns, processes, causes and consequences. Springer-Verlag, Berlin, p 178
Vargas-Ángel B (1996) Distribution and community structure of the reef corals of Ensenada de Utría, Pacific coast of Colombia. Rev Biol Trop 44:643–651
Vargas-Ángel B, Zapata FA, Hernández H, Jiménez JM (2001) Coral and coral reef responses to the 1997–98 El Niño event on the Pacific coast of Colombia. Bull Mar Sci 69:111–132
Vaughan TW (1919) Corals and the formation of coral reefs. Smith Inst Annu Rep 1917:189–276
Vera M, Banks S (2009) Health status of the coral communities of the northern Galápagos Islands Darwin, Wolf and Marchena. Galápagos Res 66:65–74
Verrill AE (1866) On the polyps and corals of Panama with descriptions of new species. Proc Boston Soc Nat Hist 10:323–333
Victor BC, Wellington GM, Robertson DR, Ruttenberg BI (2001) The effect of the El Niño-Southern Oscillation event on the distribution of reef-associated labrid fishes in the eastern Pacific Ocean. Bull Mar Sci 69:279–288
von Prahl H (1983) Blanqueo masivo y muerte de corales en la Isla de Gorgona, Pacífico colombiano. Cespedesia 12:125–129
von Prahl H (1985) Blanqueo y muerte de corales hermatípicos en el Pacífico colombiano atribuidos al fenómeno de El Niño 1982–83. Boletín ERFEN 12:22–24
von Prahl H (1986a) Mortandad de corales hermatípicos en América a consecuencia del fenómeno El Niño 1982–1983. Boletín ERFEN 17:14–15
von Prahl H (1986b) Crecimiento del coral *Pocillopora damicornis* durante y después del fenómeno El Niño 1982–1983 en la Isla de Gorgona, Colombia. Boletín ERFEN 18:11–12
von Prahl H, Mejía A (1985) Primer informe de un coral acropórido *Acropora valida* (Dana 1846) (Scleractinia: Astrocoeniida: Acroporidae) para el Pacífico Americano. Rev Biol Trop 33:39–43
Wartian MJ (2006) Determinants of community structure and resilience on eastern tropical Pacific coral reefs. PhD disser, Univ California, Los Angeles, p 247
Wellington GM (1982) Depth zonation of corals in the Gulf of Panama: control and facilitation by resident reef fishes. Ecol Monogr 52:223–241
Wellington GM, Dunbar RB (1995) Stable isotopic signature of El Niño-Southern Oscillation events in eastern tropical Pacific reef corals. Coral Reefs 14:5–25
Wellington GM, Glynn PW (2007) Responses of coral reefs to El Niño-Southern Oscillation sea-warming events. In: Aronson RB (ed) Geological approaches to coral reef ecology, Ecol Stud 192:342–385
Wellington GM, Victor BC (1985) El Niño mass coral mortality: a test of resource limitation in a coral reef damselfish population. Oecologia 68:15–19
Wellington GM, Strong AE, Merlen G (2001) Sea surface temperature variation in the Galápagos Archipelago: a comparison between AVHRR nighttime satellite data and *in situ* instrumentation (1982–1998). Bull Mar Sci 69:27–42

Whitney LD, Hobgood J (1997) The relationship between sea surface temperatures and maximum intensities of tropical cyclones in the eastern North Pacific Ocean. J Climate 10:2921–2930

Wilkinson CR (1992) Coral reefs of the world are facing widespread devastation: can we prevent this through sustainable management practices? Proceedings of 7th International Coral Reef Symposium, vol 1, Guam, pp 11–21

Wilkinson CR compiler (1998) The 1997–1998 mass bleaching event around the world. In: CR Wilkinson (ed), Status of coral reefs of the world:1998. Global Coral Reef Monitoring Network, Australian Institute of Marine Science, Cape Ferguson, Queensland, p 184

Wilkinson CR (2000) Executive summary. In: Wilkinson CR (ed) Status of coral reefs of the world: 2000. Global Coral Reef Monitoring Network, Australian Institute of Marine Science, Cape Ferguson, Queensland, pp 7–19

Wilkinson CR, Buddemeier RW (1994) Global climate change and coral reefs: implications for people and reefs. Rept UNEP-IOC-ASPEI-IUCN Global Task Team on Coral Reefs. IUCN, Gland, p 124

Wilkinson CR, Linden O, Cesar H, Hodgson G, Rubens J, Strong AE (1999) Ecological and socioeconomic impacts of 1998 coral mortality in the Indian Ocean: an ENSO impact and a warning of future change? Ambio 28:188–196

Williams EH Jr, Bunkley-Williams L (1990) The world-wide coral reef bleaching cycle and related sources of coral mortality. Atoll Res Bull 335:1–71

Wilson EC (1990) Mass mortality of the reef coral *Pocillopora* on the south coast of Baja California Sur, Mexico. Bull So California Acad Sci 89:39–41

Wolter K, Timlin MS (1998) Measuring the strength of ENSO events: How does 1997/98 rank? Weather 53(9):315–324

Wolter K, Timlin MS (2011) El Niño/Southern Oscillation behavior since 1871 as diagnosed in an extended multivariate ENSO index (MEI.ext). Int J Climate 31(7):1074–1087

Wood SN (2006) Generalized additive models: an introduction with R/. Chapman & Hall/CRC, Boca Raton, FL, p 392

Wyrtki K (1973) Teleconnections in the equatorial Pacific Ocean. Science 180:66–68

Wyrtki K (1985a) Sea level fluctuations in the Pacific during the 1982–83 El Niño. Geophys Res Lett 12:125–128

Wyrtki K (1985b) Pacific-wide sea level fluctuations during the 1982–83 El Niño. In: Robinson G, del Pino EM (eds) El Niño in the Galápagos Islands: the 1982–1983 event. Pub Charles Darwin Found Galápagos Iss, Quito, Ecuador, pp 29–48

Yamaguchi M (1975) Sea level fluctuations and mass mortalities of reef animals in Guam, Mariana Islands. Micronesica 11:227–243

Yeh S-W, Kug J-S, Dewitte B, Kwon M-H, Kirtman BP, Jin F-F (2009) El Niño in a changing climate. Nature 461:511–514

Yonge CM (1940) The biology of reef-building corals. Gt Barrier Reef Exped 1928–29, Sci Rep 1:353-391

Zapata FA, Vargas-Ángel B (2003) Corals and coral reefs of the Pacific coast of Colombia. In: J Cortés (ed), Latin American coral reefs. Amsterdam, Elsevier pp 419–447

Zapata FA, Jaramillo-González J, Navas-Camacho R (2011) Extensive bleaching of the coral *Porites lobata* at Malpelo Island, Colombia, during a cold water episode in 2009. Bol Invest Mar Cost 40 (Spec Suppl):185–193

Trophodynamics of Eastern Pacific Coral Reefs

9

Ian C. Enochs and Peter W. Glynn

Abstract

Trophic interactions on eastern Pacific coral reefs are complex and highly dynamic, ever changing due to numerous biological and physical factors. In this chapter, we first address the sources of energy at the base of food webs, i.e. photosynthetic carbon fixation by benthic algae and endosymbiotic zooxanthellae, secondarily derived organic deposits, detritus, and fecal matter, as well as demersal (within reef) and allochthonous plankton food sources. Next we consider consumers, covering the major reef trophic guilds in the eastern Pacific—suspension feeders, deposit and detritus feeders, herbivores, carnivores (predators and carnivorous grazers), as well as scavengers. The diversity and relative abundance of consumer taxa are described and considered in terms of their ecological roles in community processes. The complex interplay of these guilds is examined through food webs constructed for Panama, Cabo Pulmo reef in the Gulf of California, Mexico, and the Floreana Island rocky reef in the Galápagos Islands. Finally, the effects of physical and biotic perturbations on food webs, interactions, indirect effects, and trophic cascades conclude this review.

Keywords

Food webs • Biomass • Trophic interactions • Feeding guilds • Cascades

9.1 Introduction

Trophodynamics, the ecological processes involved in consumer-resource interactions, plays a fundamental and pervasive role in coral reef ecosystems. In addition to sustaining species populations, trophic processes have a profound influence on the evolution of reef species (e.g., Bakus 1969, 1981; Steneck 1983; Vermeij 1987), and the structuring and dynamics of coral communities (e.g., Wellington 1982; Birkeland and Lucas 1990; Carpenter 1997; Pennings 1997; Karlson 1999; Knowlton and Jackson 2001; Glynn and Enochs 2011).

Trophic studies begin with tracing the flow of nutrients and energy among the diverse organisms making up coral communities and coral reef ecosystems. This genre of study involves constructing food webs, which are highly complex in species-rich coral reefs. Not only does the exceedingly high species richness add to the complexity of consumer interactions, but also nearly all species feed on multiple prey items and diverse organic matter resources, frequently at multiple trophic levels. In addition, because of the complicated structure of reefs and the cryptic nature of the majority of reef organisms, many of the species involved in trophic processes are unrecognized and taxonomically unknown. It

I.C. Enochs
Atlantic Oceanographic and Meteorological Laboratories (AOML) NOAA, 4301 Rickenbacker Causeway, Miami, FL 33149, USA

I.C. Enochs (✉)
Cooperative Institue of Marine and Atmospheric Studies, Rosenstiel School of Marine and Atmospheric Science, 4600 Rickenbacker Causeway, Miami, FL 33149, USA
e-mail: ian.enochs@NOAA.gov

P.W. Glynn
Department of Marine Biology and Ecology, Rosenstiel School of Marine and Atmospheric Science, 4600 Rickenbacker Causeway, Miami, FL 33149, USA
e-mail: pglynn@rsmas.miami.edu

P.W. Glynn et al. (eds.), *Coral Reefs of the Eastern Tropical Pacific*, Coral Reefs of the World 8, DOI 10.1007/978-94-017-7499-4_9

is probably safe to assert that no definitive food web exists for any coral reef community, and this is no less true for the eastern tropical Pacific (ETP).

Just as important as sunlight (radiant energy) in fueling primary production, mineral nutrient supplies are necessary for the synthesis of organic molecules, energy flow and hence biomass accumulation and production. Nutrients can enter coral communities physically via river outflow, from terrestrial runoff, wind-driven currents, upwelling, and also biologically by nitrogen fixation and autochthonous recycling (Hatcher 1997; Atkinson 2011; Suzuki and Casareto 2011). Even oceanic reefs in oligotrophic waters receive nutrients in sufficient amounts, for example via trade wind-driven currents and the mass island effect, to support one of planet Earth's most productive ecosystems.

As observed on coral reefs worldwide, the feeding activities of fishes and invertebrates in the ETP figure prominently in the ecological processes structuring coral communities. Consumer guilds can exercise important top-down control on the rates of secondary production and reef growth. For example, pufferfish corallivores, where abundant, can reduce coral carbonate production by up to 16 % (Palacios et al. 2014). Additionally, the feeding activities of pufferfish can promote coral fragmentation and asexual propagation, thus contributing to the expansion of coral cover. Secondary and tertiary invertivores can also influence bottom-up control by preying on and limiting the abundances of scavengers and detritivores, such as polychaete worms and sea cucumbers, involved in the regeneration of nutrients necessary for primary production.

Recent studies have offered evidence of the pivotal roles of particular consumer species or guilds in affecting prey abundances across trophic levels, creating indirect effects or trophic cascades. For example, the removal of macropredators (e.g. sharks and groupers) by overfishing can allow the release of predation pressure on their prey, thus causing increases in the abundance of their carnivorous prey that could lead to increased predation at lower trophic levels with notable effects on coral community structure. At the landscape level, it is now understood that significant trophic subsidies often cross diverse ecosystem boundaries (Polis et al. 2004), a condition recognized early by the pioneering studies of Sargent and Austin (1949) and Odum and Odum (1955) in the Marshall Islands. Wind-driven currents, internal waves, upwelling and other physical forces have been shown to convey important trophic resources, such as plankton and particulate organic matter, to coral reefs (Leichter et al. 1996; Witman et al. 2004).

9.2 Relevance/Significance

Coral reefs are one of the most productive and species-rich ecosystems globally. As such, they provide essential goods and services—nourishment, coastal protection, tourism, biomedical resources, research and a pedagogical resource—to the great majority of peoples inhabiting tropical coastal zones. When perturbed, healthy and actively accreting reefs can decline or be replaced by other community types (e.g., phase shifts), depending in many instances on the presence and abundance of particular consumers and their feeding intensity and prey choices. Phase shifts can occur from direct and indirect trophic interactions (as noted above) that can lead to the decimation of coral populations and their replacement by fleshy and filamentous algae or other non-reef-building benthos. Coral reef decline in many parts of tropical seas has been attributed to overfishing and the loss of top-level predators (e.g., Jackson et al. 2001; McClanahan 2008; Fong and Paul 2011). Fish predators, for example, naturally exert control over corals indirectly by limiting corallivore activity or the population growth of competing benthos. Various stressors and perturbations, such as overfishing and eutrophication are now compounded by global-scale climate change effects that can exacerbate existing disturbances (Hughes et al. 2007). Regrettably, reef formations that have developed over centuries or millennia can be degraded in decades or less by disturbances that initiate changes in trophic structure and function.

9.3 Energy Sources

Coral reef ecosystems thrive in nutrient-poor, unproductive waters, often producing high concentrations of biomass relative to surrounding systems. This is accomplished through a variety of interesting mechanisms, involving the efficient capture, utilization, and recycling of nutrients. The importance of these processes is often quite evident, especially in remote offshore reefs like those at Clipperton Atoll. Nonetheless, there are notable exceptions to the proverbial reef-oasis concept in the eastern Pacific, such as strongly upwelling regions at Cabo Pulmo (Gulf of California), in the Gulf of Panama, and in the northern and central Galápagos Islands. Regardless, many of the same ecosystem processes that promote high biomass in remote nutrient-poor areas are also present in dynamic upwelling regions.

At the very base of an eastern Pacific reef food web, as with all but a few aphotic ecosystems, are primary producers that photosynthetically fix carbon. On coral reefs, this group of organisms includes benthic and planktonic algae, cyanobacteria, as well as endosymbiotic zooxanthellae associated with various cnidarians, most notably corals (Fig. 9.1). Primary production is dynamic and dependent on a variety of factors including light and nutrients. As previously noted, the ETP coral reefs that exist in upwelling regions (e.g., Gulf of Tehuantepec, Mexico; Gulf of

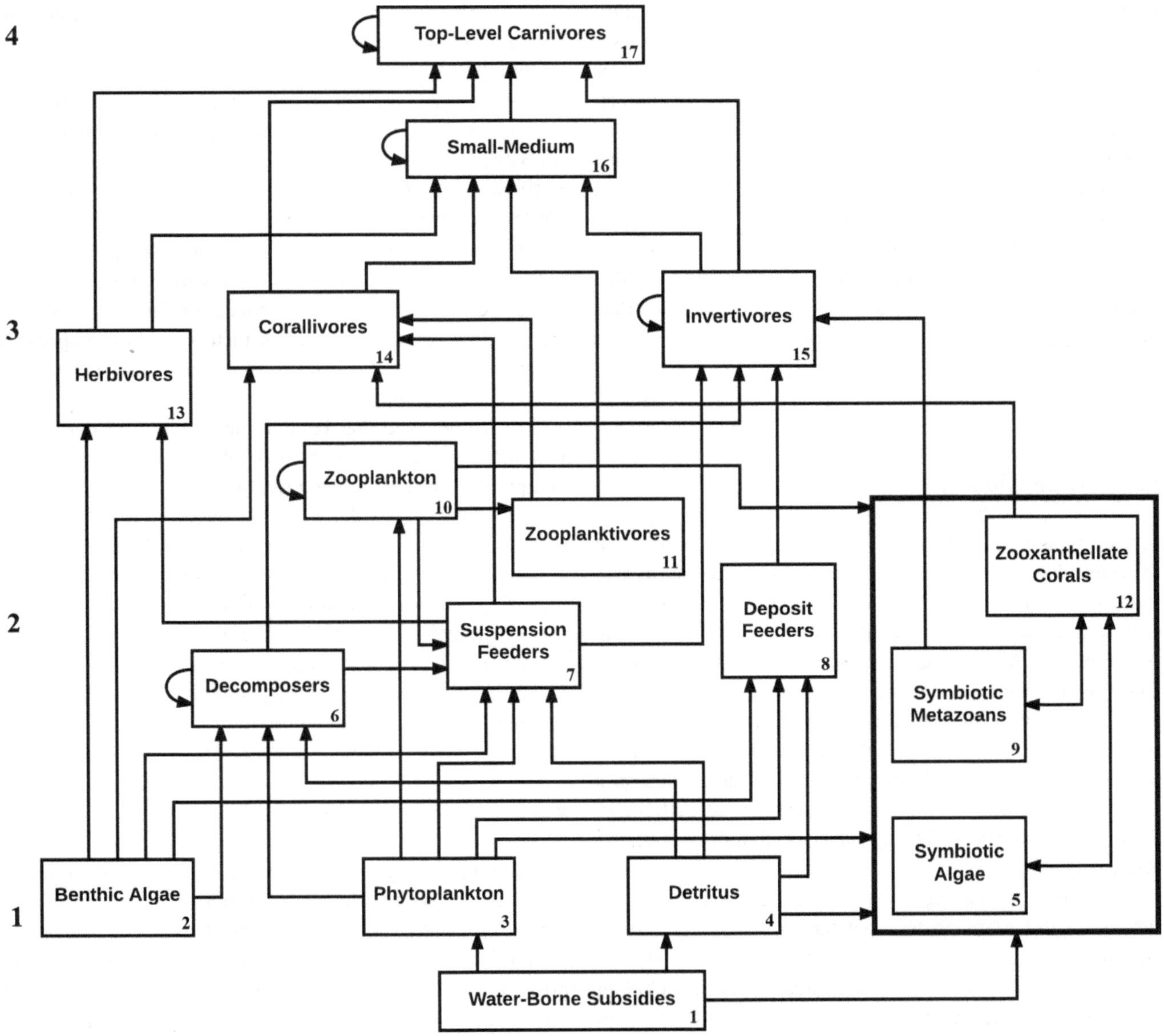

Fig. 9.1 Schematic food-web diagram of the Uva Island coral reef, Gulf of Chiriquí, Panama. Trophic level assignments follow Opitz (1996). *1* water-borne subsidies include nutrients, and dissolved and particulate organic molecules delivered to the reef by tidal and wind-driven currents, upwelling, and internal waves. Zooxanthellate coral holobionts and associated symbiotic metazoans are noted within the bold frame. Representative guild membership: *2* benthic algae (algal turf, endolithic algae, crustose coralline algae, macroalgae); *3* phytoplankton (femtoplankton to mesoplankton); *4* detritus (particulate organic matter, mucus); *5* endosymbiotic algae (*Symbiodinium* clades A, C, D); *6* decomposers (microorganisms, meiofauna and microfauna); *7* suspension and filter feeders (sponges, cnidarians, polychaete worms, bryozoans, lithophagine bivalves, barnacles); *8* deposit feeders (echiurans, sea cucumbers, shrimp, lobsters); *9* symbiotic metazoans (crabs, shrimps, copepods); *10* zooplankton (cnidarians, crustaceans, diverse larval stages); *11* zooplanktivores (cnidarians, ctenophores, chaetognaths, damselfishes, creole fish, chaetodontids, pomacanthids, juvenile parrotfishes, manta rays); *12* zooxanthellate corals (scleractinian corals and hydrocorals); *13* herbivores (gastropods, echinoids, damselfishes, pomacanthids, parrotfishes, kyphosids, acanthurids, turtles; *14* corallivores (hermit crabs, gastropod molluscs, sea stars, pufferfishes, parrotfishes, see Chap. 10, Enochs and Glynn); *15* invertivores (polychaete worms, octopuses, diverse crustaceans, hawkfish, jacks, triggerfishes, wrasses, serranids; *16* small-medium carnivores (polychaete worms, diverse crustaceans, pufferfishes, balistids, eels, snappers, wrasses, groupers, jacks, sharks; *17* top-level carnivores (sharks, barracuda, wahoo, snappers, groupers, mackerels, belonids). Modified after Glynn (2004)

Papagayo, Costa Rica; Gulf of Panama, Panama; Galápagos Islands, Ecuador) may be periodically exposed to elevated nutrients. Populations of primary producers, including benthic algae and phytoplankton, respond to these conditions, causing cascading effects and implications for higher trophic levels.

While we consider reef ecosystems to be functional units, they are by no means isolated systems. Connectivity is a ubiquitous characteristic of the marine realm and organic material may be introduced from numerous outside sources. This may occur from planktonic inputs, both autotrophic and heterotrophic, that are advected to a reef via currents and tides. It may also occur when transient fishes, invertebrates (e.g., cephalopods), birds, mammals (e.g., sea lions), or reptiles (e.g., sea snakes, crocodiles, iguanas) cross from one ecosystem into another and are subsequently consumed, die, or defecate. While this is likely of lesser importance than other energy sources, in some regions and for some species this pathway can potentially be quite important (Rothans and Miller 1991).

9.3.1 Benthic Algae

Benthic algae are an important source of primary production on eastern Pacific coral reefs. Aspects of their taxonomy and population dynamics are reviewed in Chap. 11 (Fong et al.). With respect to trophic interactions, diverse invertebrate and fish herbivores directly consume algae within the eukaryotic Chlorophyta, Rhodophyta, and Phaeophyceae, as well as the prokaryotic Cyanophyta; with growth forms ranging from filamentous turfs to more developed branching and foliose morphologies. Benthic algae function in diverse roles on a reef, ranging from N_2 fixation (Charpy et al. 2010), to substrate stabilization (Wulff 1984), and habitat creation (Kramer et al. 2012).

Crustose coralline algae, which is abundant on some eastern Pacific reefs, is less palatable and thought to be indicative of reef systems dominated by herbivory (Littler and Littler 1984). Many coral reef algal species, including cyanobacteria, have evolved chemical compounds that render them unpalatable to herbivores (Fong and Paul 2011; Charpy et al. 2012; see Chap. 11, Fong et al.). Regardless, specialized consumers exist that are able to consume many of these taxa, and even sequester their toxic metabolites as an anti-predatory defensive mechanism.

Algae, as primary producers, are important in that they capture and utilize nutrients from the water column. In experiments performed in the Panamanian Gulf of Chiriquí, the red alga *Acanthophora spicifera* was found to increase its growth under high nutrient concentrations, particularly phosphorus (Fong et al. 2006). This nutrient-enriched alga was preferentially grazed upon by herbivores, effectively keeping algal biomass in check (Fong et al. 2006). This suggests a dynamic pathway whereby nutrient enrichment may rapidly lead to the nutritional benefit of herbivores, and underscores the importance of top-down control on algal biomass. Indeed, in the Gulf of Panama, the seasonal upwelling of nutrient-rich waters has been experimentally shown to increase algal biomass in the absence of herbivory (Smith 2006). While this may not be immediately evident in heavily grazed shallow reefs (Smith 2008), deeper reef margins and rubble zones may experience algal blooms, probably due to lower herbivory and periodically high nutrient levels (Glynn and Maté 1997; Smith et al. 2010). Little is known concerning how these blooms translate into higher trophic levels. However, considering the strong top-down control of algae by herbivores, it is likely that enhanced algal biomass influences trophic structure and energy flow, unless concomitant factors such as temperature or epiphytic cyanobacteria depress herbivory (Fong et al. 2006; Smith 2008).

Finally we note that benthic algae may be important in reef food webs for reasons that fall outside of direct consumption of primary production by herbivores. The epilithic algal matrix (EAM), is of great importance for non-herbivorous taxa in that it produces and retains detritus, which in turn is fed upon by diverse detritivorous taxa, including crustaceans and fishes (Depczynski and Bellwood 2003; Kramer et al. 2013). The magnitude of this food source and the extent to which algae-based detritus supports higher trophic levels is presently unknown in the ETP and is in need of further investigation.

9.3.2 Zooxanthellae

Endosymbiotic algae, namely zooxanthellae (*Symbiodinium* spp.), reside within the gastrodermal cells of hermatypic corals (Stambler 2011). These photosynthetic symbionts obtain nutrients from both the water column and their coral hosts, and in turn fix carbon that can be used by the coral. Often, this production is sufficiently high to nearly satisfy the metabolic requirements of the zooxanthellae and the coral host, resulting in an almost fully autotrophic holobiont. While these algae are microscopic, their high abundance can account for a substantial proportion of total algal biomass and production where coral cover is high (e.g., Odum and Odum 1955; Sorokin 1990). At the ecosystem level, their production has been consistently credited for the persistence of coral reefs in generally oligotrophic waters (e.g., Muscatine and Porter 1977; Muscatine 1990).

Coral bleaching, or the disruption of the coral-algal symbiosis, can have strong implications for the metabolic function of a coral and can often lead to the host's mortality (Baker et al. 2008). Given that diverse corallivores directly consume coral tissue and metabolic byproducts (see Chap. 10, Enochs and Glynn), perturbation of zooxanthellae production can have cascading effects for higher trophic levels. For instance, Glynn et al. (2014) observed that certain fish species experienced population fluctuations correlated with ENSO-related coral mortality and postulated that this was due to both direct and indirect trophic linkages to coral/algal production. Enochs (2012) showed higher biomass of cryptofauna associated directly with live vs. dead coral, and again linked this to a dietary dependence on the coral/algal production.

Symbiodinium communities associated with corals in the eastern Pacific can be quite diverse, and some clades are known to have different thermo-tolerances, potentially conferring an advantage to their host during warming events (see Chap.13, Baker et al.). Evidence suggests, however, that those clades that are more thermo-tolerant (e.g., clade D) do not necessarily confer the same metabolic advantage as their less-tolerant counterparts (e.g., clade C) (Little et al. 2004). Since the trophic potential of *Symbiodinium* in the ETP is dependent on both their high production and persistence, understanding the complexities of community diversity and dynamics is important for understanding larger-scale ecosystem trophodynamics.

9.3.3 Phytoplankton

Phytoplankton, while not consumed directly by corals, is a food source for numerous species of zooplankton, as well as reef-dwelling benthic invertebrates (Ferrier-Pagès et al. 2011). Many of the suspension feeding fauna are sessile and cryptic, including sponges and to a lesser extent, ascidians, bivalves, bryozoans, and polychaetous annelids (Richter et al. 2001). Water circulating through the reef cavities where these taxa reside has been shown to experience significant chlorophyll depletion and evidence suggests selective capture of picoplankton-sized food sources (Richter and Wunsch 1999; Richter et al. 2001).

With respect to reef-scale trophodynamics, gross production from phytoplankton is likely less than that of benthic algae and zooxanthellae (e.g., Sorokin 1990). This is not to say that it is unimportant in some systems. For instance, Ayukai (1995) calculated capture rates of phyto-and microbial plankton as water moved from the reef front to the reef flat at a site on the Great Barrier Reef (GBR). Rates of plankton removal were as high as 0.09 g C m^{-2} d^{-1}, which is nearly equal to mean net production (0.1 g C m^{-2} d^{-1}) but much less than gross production (3 g C m^{-2} d^{-1}) as calculated by Crossland et al. (1991). In the Red Sea, Richter et al. (2001) have measured phytoplankton capture by sessile reef cryptofauna to account for 0.89 g C m^{-2} d^{-1}, and estimated that it equated to 22 % of the gross community metabolism. Much of this was likely due to extensive networks of cavities within the reef framework, providing several times the surface area of exposed reef surfaces and two orders of magnitude more total biomass than in communities on exposed surfaces (Richter et al. 2001).

While phytoplankton capture rates have not been measured within frameworks in the ETP, it is noteworthy that pocilloporid reef structures are highly porous and complex (Enochs and Manzello 2012). Enochs (2012) has demonstrated high biomass and a diverse suspension-feeding fauna within these cavities, and Wulff (1997) has highlighted the prevalence of the cryptic strategy among sponges in the ETP. It is therefore possible that, given their surface area and complexity, these frameworks act as a biological catalyst to increase the capture rate of phytoplankton similar to that reported by Richter et al. (2001).

With respect to the quantity of phytoplankton available to ETP coral reefs, plankton productivity is variable and dependent on location and season. A review of the spatial and temporal dynamics of phytoplankton productivity in the ETP can be found in Pennington et al. (2006), and earlier studies include those by Blackburn et al. (1970) and Feldman et al. (1984), among others. D'Croz and O'Dea (2007) examined seasonal fluctuations in nutrients and chlorophyll along the Pacific coast of Panama. They concluded that nitrate was limiting for phytoplankton in this region, and found that upwelling of nutrient-rich waters during the dry season in the Gulf of Panama was accompanied by elevated concentrations of chlorophyll (1.5 mg m^{-3}). These plankton- and nutrient-rich waters are advected onto reefs in the Pearl Islands and elsewhere, likely causing seasonal fluctuations in food web structure.

Finally, it is noted that phytoplankton is not always beneficial for reef trophodynamics in the ETP. Guzmán et al. (1990) reported harmful dinoflagellate blooms at reefs in both Panama and Costa Rica. These blooms were potentially the result of shoaling of the thermocline and influx of nutrient-rich waters into generally non-upwelling reef environments. In both cases, and to a greater degree in Costa Rica, blooms were accompanied by widespread mortality of corals and fishes. While the indirect effects on higher-level consumers were not quantified in this study, it is likely that food webs were significantly altered.

9.3.4 Zooplankton

Coral reef zooplankton exists as both resident populations that emerge from reef frameworks at night (Alldredge and King 1977; Porter and Porter 1977) or as more transient populations that are advected onto reefs from other habitats (allochthonous). In this latter case, zooplankton represents a means by which new organic matter can be brought to and incorporated in a reef ecosystem. Resident zooplankton taxa include polychaetes and crustaceans, in particular members of the Ostracoda, Copepoda, Mysidacea, Tanaidacea, Isopoda, Amphipoda, and Caridea. Free-spawning polychaetes, such as species in the families Syllidae, Nereidae and Eunicidae, can become highly abundant around full moon phases after sunset. Transient taxa include meroplankton, as well as holoplankton within Ostracoda, Copepoda, Euphausiacea and Chaetognatha (Hobson and Chess 1986). Zooplankton are captured and consumed by diverse benthic invertebrates, as well as fishes that are adapted to feed in the water column at night (Hobson 1991). Zooplankton, both resident and transient, can also be an important food source for corals (reviewed in Ferrier-Pagès et al. 2011). During times of impaired autotrophy (e.g., coral bleaching) some coral species may be able to increase feeding rates and supplement their metabolic requirements by means of zooplanktivory (Grottoli et al. 2006). Not all species are necessarily capable of this compensatory feeding mode, however, and evidence suggests that the common ETP massive coral *Porites lobata* does not exhibit this heterotrophic resilience-enabling behavior (Grottoli et al. 2006).

Coral reef zooplankton abundances fluctuate greatly in the ETP, dependent on both biological and environmental cycles. For instance, at Isla Contadora in the Gulf of Panama, total zooplankton abundance can vary by up to 50 % according to the lunar cycle, and the relative abundance of various taxa can change as well (Palardy et al. 2006). The feeding rate of animals (e.g., corals) that consume plankton can be significantly correlated with the fluctuating abundances of their food source (Palardy et al. 2006), but can also depend on other environmental conditions such as water temperature and depth (Palardy et al. 2005). The relative metabolic contribution of various planktonic food sources can therefore vary greatly over time and by extension, the importance of zooplankton within reef ecosystem food webs is dynamic.

9.4 Organic Deposits and Detritus

Dead and discarded organic matter, along with diverse associated microbes, is collectively known as detritus (reviewed in Wilson et al. 2003). The efficient recycling of plant and animal matter into more nutritionally-accessible forms is paramount to a functioning coral reef ecosystem. On coral reefs, detritus can be found suspended in the water column (Sorokin 1990) and accumulates in the benthos, within sediments (Lopez and Levinton 1987), cavities in reef frameworks, and within the EAM (Wilson et al. 2003). Larger detrital deposits may be found in areas of restricted flushing, where organic matter can accumulate and slowly decompose. For example, the back reef at Uva Island, Panama is sheltered with limited water movement, and decomposing matter from both marine and terrestrial sources accumulates in this habitat.

Detrital sources on a reef are numerous. As previously mentioned, isolated macroalgae and the EAM both produce and accumulate detritus, which in turn is consumed by numerous detritivores and herbivores alike. This detrital source is reviewed by Wilson et al. (2003), and it is estimated that 10–80 % of the net production of algae ends up as detritus (Hatcher 1997). Metabolic byproducts, such as feces, are also an important source of detritus. In some situations this matter may provide a novel source of nutrients and organic carbon, rather than intra-ecosystem waste recycling. For example, in St. Croix Meyer and Schultz (1985) observed grunts (Haemulidae) that fed in seagrass beds at night and defecated on reefs during the day. Similar inter-ecosystem migrations likely occur on eastern Pacific reefs and abundant coprophagous invertebrate communities have been documented to exist on eastern Pacific rocky reefs (Rothans and Miller 1991). New detritus may also enter an eastern Pacific reef ecosystem from terrestrial sources. Many near-shore reefs in Panama and elsewhere, especially those in close proximity to streams and rivers, can accumulate plant matter, including leaves, branches, and tree trunks. This organic matter is decomposed in the reef environment and eventually incorporated into the ecosystem.

Another major source of detritus on a reef is coral mucus, which is produced as a means of feeding as well as a mechanism to protect against physical, biological, and chemical stress (Brown and Bythell 2005). Rates of production are highly variable and are dependent on environmental conditions as well as coral species (Richman et al. 1975). Consumption of coral mucus provides an alternate pathway to corallivory by which photosynthetically fixed carbon from zooxanthellae is transformed into higher trophic levels (Wild et al. 2004). Additionally, coral mucus may be nutritionally enriched by adhering organic matter and microorganisms (Wild et al. 2004). While the magnitude and trophic importance of this detrital source is not completely known (e.g., Coffroth 1990), in the case of the western Pacific staghorn coral *Acropora acuminata*, it has been estimated that mucus accounts for 40 % (370 μg C mg^{-1} protein) of net carbon production (Crossland et al. 1980).

Metazoans that consume detritus include suspension and deposit feeders as well as herbivores. Prominent examples of detritivores living on eastern Pacific coral reefs include crustacean associates of pocilloporid corals such as *Trapezia* spp., holothurians such as *Euapta godeffroyi*, fishes in the families Blenniidae and Gobiidae, sediment-dwelling meiofauna, as well as numerous polychaetes, ophiuroids, hermit crabs, and gastropods.

At the ecosystem level, it is very difficult to quantify the standing stock of detritus, its turnover, and assimilation by higher trophic levels. Investigation of these topics is still lacking in the eastern Pacific. However, from other reef systems and modeling efforts, we know that detritus is of great importance. For instance, in Opitz's (1996) 50-box Ecopath model of a Caribbean reef, detritus represented the single greatest amount of "biomass" per square meter (2000 g), even higher than the biomass of benthic producers (1300 g m^{-2}) and plankton (40 g m^{-2}) combined. Studies and models of this type are needed in the ETP as strong detrital pathways and active recycling of this organic matter are closely linked to ecosystem stability in spite of perturbation (Vasconcellos et al. 1997).

9.5 Consumers

In order to address the diverse diets and feeding methods of coral reef consumers, we divide our analysis among trophic guilds. Each encompasses a group of animals that utilize similar food sources in a similar manner (Simberloff and Dayan 1991). While useful in a broader analysis of trophic interactions, we acknowledge that these groupings are not finite, and possess a high degree of overlap. Here we address herbivores and carnivores, with the latter separated into both piscivore and invertivore predators, as well as grazers. Next we discuss scavengers, which opportunistically consume a variety of food sources, and may often be considered omnivores. We introduce suspension feeders that capture nutrients from the water, often in an unselective or semi-selective manner. While these animals may not be clearly classified as carnivores or herbivores, they utilize similar feeding behaviors and as such merit separation. Finally, we review the detritivores that consume decomposing organic matter and waste, as well as associated bacteria. These animals are instrumental in recycling organic matter within reef ecosystems, and as such are an important component of reef food webs.

9.5.1 Herbivores

A comprehensive review of ETP reef herbivores can be found in Chap. 11 (Fong et al.). Briefly, taxa belonging to the herbivore trophic group are diverse, including fishes such as scarids, acanthurids, and pomacentrids, as well as epibenthic and cryptic invertebrates, such as echinoids, opisthobranchs, gastropods, brachyuran crabs, amphipods, and others. The biomass of fish herbivores is high (Smith 2006), and apparently quite stable in the ETP despite major disturbances to ecosystem structure and function (Glynn et al. 2014). This guild persistently exerts top-down control over algal standing stocks despite environmental fluctuations, such as nutrient pulses, that could otherwise enhance algal biomass (Smith 2006).

Smith (2006) calculated the biomass of fish herbivores in the Panamanian Gulfs of Chiriquí and Panama. Mean herbivore biomass was significantly higher in the Gulf of Chiriquí (55.3 g m^{-2} ± 5.85 SE) than in the Gulf of Panama (28.6 g m^{-2} ± 3.27 SE). In both gulfs, site-attached herbivores, primarily damselfishes, accounted for a relatively small fraction of total herbivore biomass (1.7 and 16.8 %, Chiriquí and Panama respectively). Roving herbivores, by contrast, accounted for the majority of the fish herbivore biomass, which was primarily due to fishes in the family Scaridae. Especially prominent fish herbivores were *Scarus ghobban* and *Scarus rubroviolaceus* in both gulfs, as well as *Acanthurus xanthopterus* (family Acanthuridae) in the Gulf of Panama. With respect to invertebrate herbivores, Enochs et al. (2011) found that herbivorous taxa accounted for 17.3 % of the biomass of the motile cryptofauna occupying dead coral substrates in the Gulf of Panama, a much higher proportion than that associated with living pocilloporid corals. In the same region, experimental manipulation of dead coral habitats revealed that the prominence of the herbivore trophic guild among the motile cryptofauna was significantly influenced by flow, with higher flow environments supporting communities with 10.1 to 19.1 % herbivore biomass.

While the body size and biomass of cryptic micrograzers (e.g., harpacticoid copepods) may be small (e.g., 0.48 g m^{-2} ± 0.05 SE, Orpheus Island, GBR, Kramer et al. 2012) relative to large herbivorous fishes, these crustaceans have comparatively shorter generation times (<1 month). Coupled with high consumption rates, both by benthic invertivores and incidentally by herbivorous fishes, micrograzers may represent an important trophic connection between the EAM and fishes (Kramer et al. 2013). For instance, at Orpheus Island, GBR, Kramer et al. (2013) calculated that herbivorous fishes consumed 5.3 % of the harpacticoid copepods associated with algal turfs daily (12,000 ind m^{-2} d^{-1}), demonstrating remarkable turnover and recovery for this community.

Alteration of herbivore populations has been shown, in turn, to influence the abundance of algae in the ETP. On rocky reefs in the Galápagos, Sonnenholzner et al. (2009) has shown that sites where predators of *Eucidaris galapagensis* are removed by fishing, resulting higher *E. galapagensis* densities are correlated with lower abundances of non-coralline algae. It is possible that if ETP reef ecosystems experience elevated fishing pressure of important fish herbivores, such as parrotfishes, algae standing stocks could increase. Given the prominence and persistence of herbivores on ETP reefs, their role in coral reef food webs should not be neglected. The Sonnenholzner et al. (2009) and

Kramer et al. (2013) studies underscore the potential for connectedness among trophic groups within ETP reef ecosystems, and highlights the importance of herbivores as an integral link between benthic productivity and high-level consumers.

9.5.2 Carnivores

ETP reef carnivores vary in size from itinerant reef sharks to meiofaunal nematodes living between sand grains. These animals utilize a myriad of feeding strategies ranging from predation to parasitism, and exhibit a diversity of behaviors from direct capture to grazing. Carnivore diets are diverse as well, including invertivores and piscivores, as well as more-specialized consumers such as corallivores. Some species of suspension feeders that mainly consume zooplankton can be thought of as carnivorous, however we address these taxa separately due to their unique feeding mechanisms and behaviors. A thorough description of all species, strategies, and behaviors is outside the scope of this chapter. We focus instead on notable and better-documented predators and grazers. Predators are carnivores that consume more motile taxa, often completely or to the point of prey mortality. Carnivorous grazers, by contrast, consume sessile colonial taxa and in a manner that is often not lethal to the entire animal or colony (partial predation).

Many species of fishes are conspicuous predators within ETP reef ecosystems, and include large piscivores such as whitetip reef sharks (*Triaenodon obesus*) that have been observed breaking apart coral frameworks while hunting (Jiménez 1997). Similarly, fish invertivores such as *Pseudobalistes naufragium* are known to fragment corals in search of their endolithic bivalve prey (*Lithophaga* spp., Guzmán 1988; see Chaps. 12 and 15, Alvarado et al., Glynn et al. respectively). Other large transient fish predators include members of the Carangidae, Belonidae, Lutjanidae, Scombridae, Serranidae, and Sphyraenidae. Notable reef predators with more resident behaviors include fish species within the families Apogonidae, Balistidae, Chaetodontidae, Cirrhitidae, Gobiidae, Haemulidae, Holocentridae, Labridae, Lutjanidae, Monacanthidae, Mullidae, Muraenidae, Pomacentridae, Scorpaenidae, Serranidae, Tetraodontidae, and Zanclidae. Many species within these families, such as *Arothron meleagris* (Tetraodontidae) can also be categorized as carnivorous grazers as they consume sessile colonial invertebrates such as corals.

In a study of 126 fish species from 44 families living on rocky reefs in the Panamanian Gulf of Chiriquí, Dominici-Arosemana and Wolff (2006) found that planktivorous pomacentrids and labrids, along with benthic-feeding haemulid invertivores, were the most abundant families. They also noted that the planktivore, carnivore, mobile-invertebrate feeder, and piscivore feeding guilds were the most abundant. Glynn et al. (2014) have shown that abundances of many fish predators, including both corallivores and invertivores, are strongly correlated with coral cover. Mass coral mortality events, such as the 1982–83 El Niño bleaching disturbance, therefore have widespread ramifications for reef community structure, influencing not only corallivores but also those predators that consume invertebrates that shelter within and feed upon corals.

Non-piscine carnivores living on reef surfaces include other chordates such as grazing turtles (*Lepidochelys olivacea, Caretta caretta*), which can destructively forage for sponges within pocilloporid frameworks (Glynn 2004). Other charismatic invertebrate predators include cephalopods such as octopuses and squids that prey on surface-dwelling animals. Finally, numerous examples of invertebrate predators and grazers that prey on epibenthic animals are found among Echinodermata (e.g., *Acanthaster planci*), Gastropoda (e.g., *Conus* spp.), Crustacea (e.g., stomatopods, lobsters), Polychaeta (e.g., amphinomid worms), and Platyhelminthes (e.g., polyclad flatworms), among others.

Intense levels of predation on reef surfaces have likely contributed to the prevalence of cryptic behaviors among potential prey species in the ETP (Bakus 1966; Dee et al. 2012). However, while pocilloporid reef frameworks do provide shelter from nektonic and other predators, many of the cryptic species that live within them are carnivores (Enochs 2012). Glynn (2006) simulated cryptic reef habitats using netted bags of pocilloporid rubble deployed at Uva Island Reef, Panama. The majority of fishes found to shelter within these habitats exhibited predatory feeding behaviors. Similarly, fishes that recruited to dead *Pocillopora* colonies at Gorgona Island, Colombia were dominated by carnivorous species (Lozano et al. 2003). Cryptic predators are diverse, and include numerous invertebrates such as stomatopods, portunid crabs, isopods, octopods, gastropods, as well as polychaete and polyclad worms, among others (Glynn and Enochs 2011). Examples of carnivorous cryptofauna species living among pocilloporid reef frameworks can be found in Enochs et al. (2011). The high abundances and great diversity of these cryptic carnivores have potential implications for vulnerable juvenile organisms recruiting to reef habitats (Glynn 2006, 2013). Additionally, the multitude of these higher level trophic groups residing within the interstices of pocilloporid reefs points to the presence of complex cryptic food webs.

Corallivory, a form of carnivorous grazing, is discussed separately and in greater detail in Chap. 10 (Enochs and Glynn). Diverse organisms are adapted to consume coral tissues in the ETP, ranging from fishes, to crustaceans, echinoderms, and molluscs. The biomass of cryptofaunal

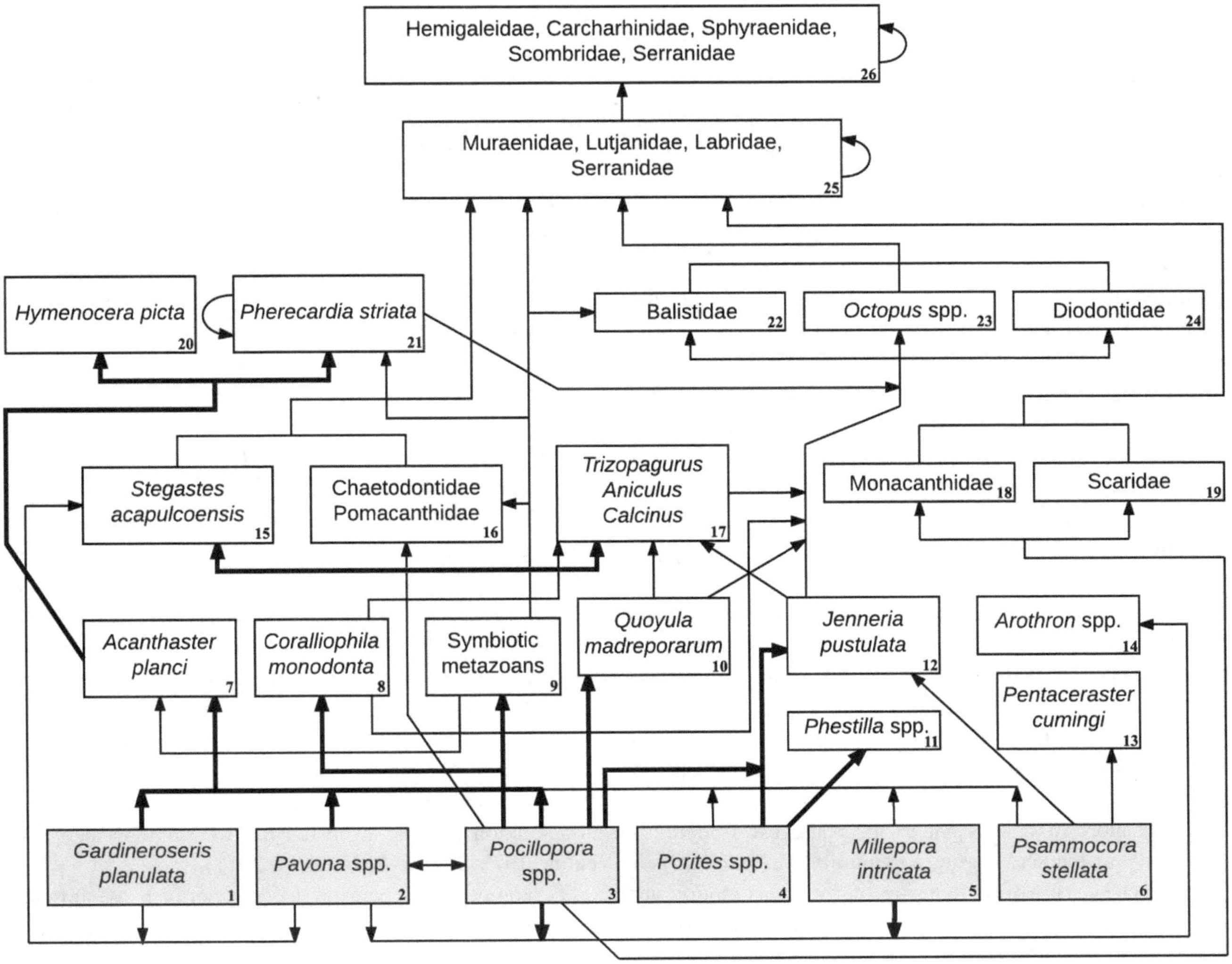

Fig. 9.2 Schematic diagram of the corallivore trophic subweb illustrating the chief Uva Island zooxanthellate stony corals (taxa 1–6), their known consumers and predators. Thick arrows denote strong trophic pathways with frequent prey consumption. Species identities: *1 Gardineroseris planulata*; *2 Pavona* spp. (*Pavona clavus*, *Pavona gigantea*, *Pavona varians*, *Pavona chiriquiensis*); *3 Pocillopora* spp. (*Pocillopora damicornis*, *Pocillopora elegans*, *Pocillopora eydouxi*); *4 Porites lobata*, *Porites panamensis*; *5 Millepora intricata*; *6 Psammocora stellata*; *7 Acanthaster planci*; *8* Symbiotic metazoans (*Trapezia* spp., *Alpheus lottini*, *Thor amboinensis*, *Harpiliopsis depressus*); *9 Coralliophila monodonta*; *10 Quoyula madreporarum*; *11 Phestilla* spp. (*Phestilla lugubris*, *Phestilla melanobrachia*); *12 Jenneria pustulata*; *13 Pentaceraster cumingi*; *14 Arothron* spp. (*Arothron meleagris*, *Arothron hispidus*); *15 Stegastes acapulcoensis*; *16* Chaetodontidae (*Johnrandallia nigrirostris*, *Chaetodon humeralis*), Pomacanthidae (*Holacanthus passer*, *Pomacanthus zonipectus*); *17* Hermit crabs (*Trizopagurus magnificus*, *Aniculus elegans*, *Calcinus obscurus*); *18* Monacanthidae (*Aluterus scriptus*, *Cantherhinus dumerilli*);*19* Scaridae (*Scarus perrico*, *Scarus rubroviolaceus*); *20 Hymenocera picta*; *21 Pherecardia striata*; *22* Balistidae (*Sufflamen verres*, *Pseudobalistes naufragium*); *23 Octopus* spp.; *24* Diodontidae (*Diodon holocanthus*, *Diodon hystrix*); *25* Muraenidae (*Gymnothorax castaneus*, *Gymnothorax dovii*), Lutjanidae (*Lutjanus viridis*, *Lutjanus argentiventris*, *Lutjanus novemfasciatus*), Labridae (*Bodianus diplotaenia*, *Novaculichthys taeniourus*, *Thalassoma grammaticum*), and Serranidae (*Epinephelus panamensis*, *Epinephelus labriformis*); *26* Hemigaleidae (*Triaenodon obesus*), Carcharhinidae (*Carcharhinus leucas*), Sphyraenidae (*Sphyraena ensis*), Scombridae (*Acanthocybium solandri*), and Serranidae (*Mycteroperca xenarcha*). Modified after Glynn (2004)

communities directly associated with live coral is significantly greater than that associated with dead substrates, thereby demonstrating the capacity of coral food sources to support abundant and vibrant reef communities. For this reason, it is not surprising that corals and corallivores form an integral component of ETP food webs that ultimately transforms the primary production of microscopic zooxanthellae into high-level piscivores (Fig. 9.2; Glynn 2004). Consequently, perturbations that affect coral health and abundance can have a far-reaching influence throughout the ecosystem, affecting trophic groups not directly dependent on corals for food (Glynn et al. 2014).

9.5.3 Scavengers

Scavengers are a poorly studied group of consumers on ETP coral reefs. Given the relative paucity of allochthonous energy sources available to coral reef systems, this guild performs an important role in the efficient consumption and recycling of its food sources. For many consumers, scavenging is an opportunistic feeding strategy; predatory, herbivorous, and detritivorous feeding behaviors may be exhibited as well. Records of scavenging behavior are rare from the ETP. Enochs (2012) noted that a large proportion of the cryptofauna collected from dead *Pocillopora* substrates in the Pearl Islands, Panama were opportunistic omnivores, many utilizing scavenging as a feeding strategy. Glynn (2013) deployed traps baited with dead fish to assess the diversity and abundance of scavengers at Uva Island, Panama. Prominent among the captured specimens were the amphinomid polychaete *Pherecardia striata*, cirolanid isopods, ostracods, hermit crabs, and gastropods. Submersible night-vision cameras, deployed in conjunction with the scavenger traps at the Uva Island reef have revealed that muraenid eels are also common and active nocturnal scavengers (pers. obs., I.C.E. and P.W.G.).

Similar to the scavenging taxa collected by Glynn (2013), cirolanids were strongly represented among the 108 species collected by Keable (1995) using comparable baited traps on Lizard Island, GBR. This pattern may suggest a degree of uniformity in the scavenging community across large geographic scales. However, focusing on the consumption and decomposition of *Acanthaster* following injury and mortality, Glynn (1984) observed that the sea stars were consumed nearly twice as fast in eastern compared with western Pacific study sites (4 vs. 8 days for complete consumption, Panama vs. Guam, Micronesia). In Panama, a diverse array of species were attracted to and scavenged upon the dead *Acanthaster*, namely amphinomid polychaete spp., *Diadema mexicanum*, *Hymenocera picta*, *Trizopagurus magnificus*, *Eupomacentrus acapulcoensis*, *Thalassoma lucasanum*, *Holacanthus passer*, *Sufflamen verres*, *Arothron meleagris*, and *Arothron hispidus*. In Guam, however, only a single species of butterfly fish (*Chaetodon citrinellus*) was observed to feed on several sea star carcasses at three different reef sites. It is possible that the complex structure of pocilloporid reefs in the ETP, the chief type of coral formation in this region, may shelter abundant populations of small cryptic scavengers. Indeed, in Panama Glynn (1984) reported densities of *Pherecardia striata* as high as 380 ind m^{-2}, and Enochs et al. (2011) found exceptionally high abundances of the amphinomid worm associated with low porosity pocilloporid substrates in low flow habitats.

9.5.4 Suspension Feeders

The term suspension feeder refers to those animals that capture either all or part of their nutriment from matter suspended in the water column. Food sources can include both phyto- and zooplankton (Sects. 9.3.3 and 9.3.4, respectively), as well as suspended organic matter. Because many suspension feeders live at the interface of the water column and benthos, those taxa that feed on organic matter deposits and detritus may exhibit suspension feeding behaviors as well, thereby blurring the line between the two feeding modes (Riisgård and Kamermans 2001). While the term filter feeder is often used interchangeably with suspension feeder, we prefer the latter term as “filtering” is not always the mechanism employed to collect suspended matter. Methods of capture can, in fact, be quite diverse. On ETP reefs, examples include direct capture by fishes preying on zooplankton, both ciliary and muscular flushing of water across complex feeding structures and specialized cells, as well as entrapment of particles in mucus. A review of active suspension feeding mechanisms can be found in Riisgård and Larsen (2000). Often a single taxon can employ multiple suspension feeding mechanisms, each of which may be more suitable for capturing a specific type of organism or size-class of plankton. For instance, corals are known to use tentacles, nematocysts, and mucus to capture zooplankton ranging from 0.4 μm to 2 mm in size (Muscatine 1973; Ferrier-Pagès et al. 2011).

Ecologically, suspension feeders perform the important function of capturing nutrients and organic matter from the water column. While coral reefs are generally recognized to exist in oligotrophic waters, both resident and transient plankton populations may at times be abundant in the ETP, and important in reef ecosystem trophodynamics (see Sects. 9.3.3, 9.3.4). Efficient capturing of plankton can enhance local biomass by acquiring matter that is otherwise new to the ecosystem, and also minimizing the leaking of biomass that could potentially be lost.

Corals are the most studied of the reef-associated suspension feeding fauna in the ETP. They often rely on heterotrophy in addition to photosynthetically fixed carbon from symbiotic zooxanthellae (Sect. 9.3.2). During times of stress and compromised autotrophy, such as during periods of warming and reduced pH, some species of coral may further supplement their metabolic requirements with plankton, reflecting trophic plasticity and a pathway towards resilience (Grottoli et al. 2006; Edmunds 2011). The degree to which this applies to ETP corals is not well understood, and it is likely that not all corals are capable of this compensatory trophic switching. Certainly feeding rates differ amongst coral species, with the massive corals *Pavona gigantea* and *Pavona clavus* consuming more zooplankton per unit surface area than branching *Pocillopora damicornis* (Palardy et al. 2005). Of these species, *P. clavus* may exhibit the greatest potential to employ suspension feeding as a resilience mechanism. Wellington (1982) demonstrated that this coral can offset depressed growth (calcification) due to shading via the consumption of plankton.

While corals are known to consume particulate organic matter and absorb dissolved organic matter, their consumption of plankton is limited to zooplankton (reviewed in Ferrier-Pagès et al. 2011). In the ETP, zooplanktivory by corals is strongly correlated with zooplankton abundance, specifically that in the 200–400 μm size class (Palardy et al. 2006). Plankton captured by *Pocillopora damicornis* and *Pavona gigantea* was dominated by isopods, amphipods, and crustacean zoea larvae (Palardy et al. 2006). Many of the plankton species were slow swimming meroplankton, possibly advected onto the reef from surrounding ecosystems.

In addition to plankton abundance, feeding rates of ETP corals may be dependent on aspects of their physical environment. Palardy et al. (2005) observed that suspension feeding was higher (169 %) for three species of corals (*Pocillopora damicornis, Pavona clavus*, and *Pavona gigantea*) during the warmer non-upwelling season. Feeding rates also increased as a function of depth, and there was an apparent interaction with season, with deeper (6 m) corals feeding 18.7 % more than shallow (1 m) corals during cool upwelling periods, and 35 % more during warm non-upwelling periods. These differences likely reflect a variety of different influences, ranging from depressed rates of polyp contraction and nematocyst function in cooler waters, to trophic plasticity and an increased reliance on heterotrophy when autotrophy is limited by low light levels (Palardy et al. 2005).

In addition to corals and cnidarians, sessile suspension feeding reef fauna in the ETP belong to Porifera, Annelida, Arthropoda, Bryozoa, and Mollusca. The Pearl Islands in the Gulf of Panama, which support numerous coral reef formations, are so named because of the once abundant pearl-bearing bivalve *Pinctada mazatlanica*. This species is a suspension feeder that often colonizes reef substrates and is found throughout the ETP. It has a rich fisheries history in the Gulf of Panama, with corresponding dynamic population fluctuations, dating back to the 1500s (Cipriani et al. 2008). In addition to the large epibenthic bivalves, lithophage bivalves bore into coral rock and suspension feed by circulating water through small key hole-like openings on the surface of the coral (see Fig. 12.4 in Chap. 12, Alvarado et al.). They can be present in high abundances on ETP reefs but populations fluctuate and are dependent on the availability of suitable coral substrate (Scott et al. 1988; Kleemann 2013; see Chap. 12, Alvarado et al.). Despite their cryptic nature, lithophage bivalves are frequently consumed by the triggerfish *Pseudobalistes naufragium*, which breaks apart coral rock to expose them (Guzmán 1988; see Fig. 15. 9, Chap. 15, Glynn et al.). Other notable sessile reef-dwelling suspension feeders include sponges which, relative to other regions, are not abundant on exposed reef surfaces in the ETP (Wulff 1997). Instead, cryptic and boring species permeate reef carbonates and cavities, and are present in high abundances (Carballo et al. 2008; Bautista-Guerrero et al. 2014; see Chap. 12, Alvarado et al.). From studies in other localities, it is known that these cryptic suspension feeders efficiently capture transient plankton (e.g., Richter and Wunsch 1999; Richter et al. 2001). Despite their cryptic nature, ETP reef sponges are in turn consumed by fishes (Wulff 1997) and turtles (Glynn 2008).

Fishes are prominent zooplanktivores on ETP reefs. A review of reef fishes that prey on zooplankton can be found in Hobson (1991). This feeding strategy is utilized by numerous species, and notable families include Apogonidae, Balistidae, Holocentridae, Labridae, Pomacentridae, and Priacanthidae. On reefs in the Gulf of Chiriquí, fish planktivores were found to be the most abundant trophic guild, accounting for more than 50 % of individuals at the majority of study sites (Dominici-Arosemena and Wolff 2006). Especially common species include the apogonids *Apogon dovii* and *Apogon pacificus*, the pomacentrid *Chromis atrilobata*, the serranid *Paranthias colonus*, and the labrid *Thalassoma lucasanum* (Dominici-Arosemana and Wolff 2006).

Fish zooplanktivores feed on crustaceans (e.g., euphausids, mysids, copepods), larvaceans, and eggs, extending their protruding jaws to create suction and help grasp prey. This behavior can also be classified as carnivorous and predatory. However, the small planktonic and suspended nature of the prey also merits classification as suspension feeding. Zooplanktivory is common among both diurnal and nocturnally active fishes in the ETP. Those that feed during the day, such as *Chromis atrilobata*, often do so in large schools that may reduce predation by piscivores. Nocturnal species, such as the holocentrid *Myripristis* sp., have large eyes that are more efficient in the location and capture of plankton under low-light conditions (Hobson 1991). Whereas many nocturnal planktivores may avoid strong currents to better feed on resident holoplankton, diurnal species may be found in areas with stronger water movement, where transient plankters are more abundant (Hobson 1991). Consequently, diurnal planktivorous fishes are often strong swimmers, and Dominici-Arosemena and Wolff (2006) have proposed that the apparent lower abundance of these taxa in southern Baja California relative to other ETP regions is possibly due to the lower tidal amplitude in this region, and therefore less tidal flushing.

Many of the plankton-feeding fishes exhibit cryptic tendencies, especially nocturnal species that can shelter within reef frameworks during the day. Suspension feeding is also utilized by diverse cryptic invertebrates. From 3.9 to 14.5 % of the biomass of motile cryptofauna colonizing cages of *Pocillopora* rubble in the Gulf of Panama was classified as belonging to suspension feeders by Enochs et al. (2011). Animals using this feeding strategy were especially prominent among high-flow, high-porosity substrates, and flow was found to significantly influence the proportion of suspension feeding biomass, presumably due to greater access to planktonic food sources or to flow enhanced recruitment (Palardy and Witman 2011).

Motile suspension feeding cryptofauna include ophiuroids, crustaceans, molluscs, and annelids. As with many cryptic invertebrates, the feeding behaviors of these taxa are poorly known in the ETP, and it is often necessary to infer their diets from information on related taxa from other regions (Enochs et al. 2011). It is likely that further investigation will reveal that the suspension feeding strategy is utilized by additional taxa, and that feeding behaviors are complex and not necessarily easy to clearly delineate (Riisgård and Kamermans 2001).

Six of the ten cryptic ophiuroid species collected by Enochs et al. (2011) from pocilloporid reefs in Panama were classified as primary suspension feeders, with the remaining species utilizing multiple strategies that potentially included suspension feeding. At night on reefs off Saboga Island, Gulf of Panama, numerous brittle stars have been observed to gather among the distal branches of pocilloporid corals and extend their arms into the water column. This behavior is similar to feeding observations of *Ophiocoma* spp. observed at other localities (e.g., Enewetak Atoll and Kenya), and it is likely that these individuals were suspension feeding (Chartock 1983; Oak and Scheibling 2006).

Suspension feeding crustaceans include macrozooplankton as well as many epibenthic and cryptic species. Crabs in the family Cryptochiridae are known to form gall-like structures in the skeletons of pocilloporid corals and likely feed on passing plankton (Garth 1974; Glynn and Enochs 2011). Porcellanid crabs, such as those in the genus *Petrolisthes*, are common on ETP reefs (Haig 1960). They are primarily suspension feeders, and utilize setae-fringed thoracic appendages to generate currents from which they extract particulate food. Barnacles, including both endo- and epilithic species, are also active suspension feeders, and are important prey for gastropods and fishes (Witman et al. 2010). Numerous other motile suspension feeders live on ETP reefs, including gastropods and annelids, among others. Further information is needed to evaluate their contribution to reef trophodynamics.

9.5.5 Detritivores

Diverse detritivores occupy reef frameworks and corals in the ETP. Analysis of the trophic group composition of the cryptofauna inhabiting an eastern Pacific pocilloporid coral reef revealed a large biomass of deposit feeders living among both live and dead coral habitats (Enochs 2012). Within each substrate, their biomass density was second only to opportunistic omnivores, which included many taxa exhibiting detritivory. Cryptic detritivores included terebellid polychaetes, echiurans (*Anelassorhynchus* sp.), tanaids, as well as holothurians. Rothans and Miller (1991) deployed traps baited with planktivorous fish feces on rocky reefs off Catalina Island, California and off St. Croix in the Caribbean. Numerous invertebrates were attracted to the traps containing feces, including many genera that are common on ETP reefs. The majority of these were shrimps (e.g., *Lysmata, Palaemonella, Periclimenes, Thor, Alpheus*) and hermit crabs (*Pylopagurus, Paguristes, Iridopagurus,* and *Pagurus*), but other crustaceans and gastropods were also collected. Working in Palau, Robertson (1982) observed coprophagy among 45 species of reef fishes, including members of the families Acanthuridae, Balistidae, Kyphosidae, Pomacentridae, Siganidae, and Scaridae. Robertson noted that this feeding behavior usually occurred in addition to other feeding modes. The feces of carnivorous planktivores and piscivores were often consumed by herbivorous taxa, thereby complicating the directionality of energy flow and adding a level of complexity to food webs not often considered. While not necessarily coprophagous, Depczynski and Bellwood (2003) examined cryptic fish communities from the GBR and found that detritivores accounted for 63 % of species and 39 % of individuals. Together, the aforementioned studies underscore the prevalence of the detritivore community, and its potentially important role in ETP reef food webs should not be overlooked (Glynn 2004).

Crustacean associates of *Pocillopora* (e.g., trapeziid crabs, caridean shrimps) obtain nutriment from detritial sources by either direct consumption of mucus or removal of entrapped organic matter (e.g., Knudsen 1967; Patton 1976). These coral-generated food sources are likely responsible for the high cryptic biomass associated with live vs. dead *Pocillopora* (Enochs 2012). The effects of these taxa on their coral hosts are reviewed in Chap. 10 (Enochs and Glynn). Numerous other mucus-feeding taxa have been described from other regions, including fishes (e.g., Benson and Muscatine 1974) and zooplankton. McCloskey (1970) noted that mucus and coral-derived detritus can accumulate on and around the bases of branching *Oculina* in Atlantic temperate waters. These deposits corresponded to high abundances of deposit feeders, presumably consuming these food sources. Pocilloporid corals in the ETP have a similar branching morphology and their mucus production likely experiences a parallel fate. Indeed, communities associated with the dead coral substrates around these colony bases may be more species rich than those on the live corals themselves (Enochs and Manzello 2012).

Wild et al. (2004) estimated that from 56–80 % of cast-off coral mucus dissolves in the water column, where it is consumed by bacterioplankton. At Heron Island, Australia the authors estimate that 50 % of both the dissolved and undissolved mucus finds its way to the reef lagoon, where it accumulates along with detritus from other sources such as algae (Koop and Larkum 1987). This organic matter settles into sediments and is consumed by benthic fauna and bacteria. In the ETP, these organic-rich reef sediments contain diverse populations of meiofauna (Guzmán et al. 1987). In Costa Rica, this community was found to include numerous individuals (up to 575 ind. per 10 cm^2) belonging to Foraminifera, Copepoda, Nematoda, Gastropoda, and Polychaeta, among others.

Sediment-dwelling detritivorous macrofauna feed on the meiofauna, bacteria, and organic deposits. These so called deposit feeders are some of the most well-known detritivores (reviewed in Lopez and Levinten 1987). Their members include gastropods, bivalves, annelids, crustaceans, and echinoderms. Because these taxa feed on a relatively nutrient-poor food source, they often consume many times their body weight in sediments daily, sorting through the sediment and selecting particles that are of greater nutritional quality (Hammond 1983; Lopez and Levinton 1987). A prominent ETP reef dwelling deposit feeder is the holothurian *Euapta godeffroyi*, which feeds by extending its tentacles onto sediments (Glynn 2008; Granja Fernández et al. 2013). This species is found throughout the eastern and western Pacific and is known from a variety of sedimentary habitats, including reefs (Granja Fernández et al. 2013). Like other deposit-feeding holothurians from other reef regions (e.g., Hammond 1983), *E. godeffroyi* is nocturnally active and remains hidden within reef frameworks during the day.

The ecology of other deposit-feeding detritivores, their prevalence, and influence on reef trophodynamics is poorly known in the ETP. From other reef regions we have estimates of the standing stock of sediment-dwelling bacteria and microalgae, as well as abundances of the deposit-feeding taxa that feed on them (e.g., the gastropod *Rhinoclavis aspera*, Hansen and Skilleter 1994). We know that bacterial production can fluctuate over seasonal cycles, and that experimentally elevated abundances of deposit feeders can reduce the abundances of their bacterial food source, though this last relationship is potentially complex, possibly involving the enhancement of predatory meiofauna populations via the bioturbation of macro-deposit feeders (Hansen and Skilleter 1994). Further research is necessary to tease apart these pathways and to quantify the trophic importance of deposit-feeding detritivores on ETP coral reefs.

9.6 Structure of Eastern Tropical Pacific Food Webs

A comprehensive bio-regional analysis of the ETP coral reef invertebrate faunas and their feeding habits is not presently available. However, many of the functionally important producer and invertebrate consumer species are shared among the ETP coral provinces (e.g., Jones 1972; James 1991; Hickman 1998–2008; Cortés 2003; Alvarado and Cortés 2005; Wehrtman and Cortés 2009; see Chap. 7, Cortés et al.). Alert to possible local differences in predator-prey interactions, it is useful to compare trophic patterns across the ETP. An early attempt to assess the effects of corallivore taxa on coral community development in upwelling and non-upwelling environments was initiated by Glynn and Wellington (1983, Table 25), but only seven species were examined. While the ETP invertebrate corallivore fauna is species rich compared with other coral reef regions (see Chap. 10, Enochs and Glynn), population outbreaks and resulting marked coral declines are uncommon. *Acanthaster planci* is absent or rare at most coral sites, and present at moderate densities in only a few localities (southern Gulf of California, Revillagigedo Islands, and Gulf of Chiriquí, Panama). Outbreaks of the corallivorous gastropod *Jenneria pustulata* are infrequent and highly localized (Glynn 1985; Paz-García et al. 2012). High abundances of *Eucidaris galapagensis*, an effective facultative corallivore, are restricted to the central and southern Galápagos Islands (Glynn et al. 2015).

With the exception of relatively few endemic species, the coral reef fish fauna of the three major ETP provinces recognized by Robertson and Cramer (2009)—Panamic, Cortez, and Oceanic Islands—are similar, sharing many of the same species in the major functional groups (Robertson and Allen 2008). Trophic assignments differ somewhat among studies and localities, but herbivore, detritivore, invertivore, corallivore, piscivore, and plankton-feeding guilds are always represented (e.g., Jennings et al. 1994; Zapata and Morales 1997; Edgar et al. 2002, 2004; Dominici-Arosemena et al. 2005; Espinoza and Salas 2005; Dominici-Arosemana and Wolff 2006; Alvarado et al. 2011; Arburto-Oropeza et al. 2011; López-Pérez et al. 2013; Glynn et al. 2014). Dominici-Arosemena and Wolff (2006) identified 25 abundant species in eight trophic groups in Panama (Gulf of Chiriquí) with mobile invertivores, planktivores, and higher level carnivores making up the most prominent guilds. The reef-fish fauna of Rapa Nui (Easter Island) and Salas y Gómez Islands, with 77 % and 73 % endemicity respectively, diverges substantially from other ETP localities in terms of shared species (Friedlander et al. 2013). However, four major functional groups—apex predators, herbivores, invertivores, and planktivores—figure prominently in the structuring of Rapa Nui coral communities. Notable differences in apex predator abundances, due to fishing pressure, were hypothesized to drive inter-island differences in planktivore, invertivore, and herbivore guild biomass (Friedlander et al. 2013).

Relatively few food web studies have been reported on eastern Pacific coral reefs. One of the first focused on a pocilloporid patch reef at Señorita Island in the Pearl Islands, Panama, and the macro-corallivores that prey on reef-building *Pocillopora* corals (Glynn et al. 1972). This study reported that an ovulid snail, two hermit crab species, and a pufferfish consumed about 1/3 of the annual coral carbonate production. Gross coral production at the Señorita patch reef, calcareous skeleton + tissue, was estimated to equal 20.5 tonnes ha^{-1} $year^{-1}$; consumption by the four corallivores amounted to 6.7 tonnes ha^{-1} $year^{-1}$. While coral erosion was substantial, this patch reef still demonstrated positive growth.

A community-wide conceptual food web proposed for the Uva Island (Gulf of Chiriquí, Panama) coral reef revealed numerous consumer groups and the complexity of trophic interactions on an ETP reef (Fig. 9.1, Glynn 2004, 2008). This trophic web consisted of 17 functional groups with 94 links, and a substantial contribution of water-borne subsidies. The coral holobionts and associated metazoan symbionts are an essential component of this schema with links to plankton food sources and consumer invertivore and corallivore guilds. The considerable number of corallivores that prey on all coral species is now well documented (Fig. 9.2; see Chap. 10, Enochs and Glynn). There are 14 known corallivore guilds (single species or species complexes), including gastropod molluscs, crustaceans, asteroids and six fish families, that consume the tissues, secretions, and skeletons of the 12 Uva reef coral species. Eleven links (thick lines) involve heavy consumption of corals. *Pocillopora* species in particular, with nine links to corallivores, provide nutritional sustenance to a host of consumers, which in turn are potential prey of secondary and tertiary consumers. A minimum of 105 trophic links constitute this corallivore subweb.

Adding to the sessile cryptic fauna studies of Jackson and Winston (1982) and Jackson (1984), it is now recognized that the motile cryptic faunas are also species rich and abundant (e.g., Keable 1995; Gischler and Ginsburg 1996; Depczynski and Bellwood 2003; Glynn and Enochs 2011; Enochs and Manzello 2012). The prevalent cavity-rich pocilloporid reefs of the ETP shelter numerous cryptic species that belong to all major trophic functional groups (Enochs and Hockensmith 2009; Enochs et al. 2011; Enochs 2012; Enochs and Manzello 2012; Glynn 2006, 2013). More than 121 identified motile species representative of six phyla and 83 families were collected from experimental pocilloporid substrates in Panama with representative members of carnivore, herbivore, scavenger, detritus, and suspension-feeding guilds (Fig. 9.3, Enochs et al. 2011; Enochs 2012). Diverse scavenger species are prominent with 14 links to all cryptic guilds, frequently recipients of dead and dying organisms, and also serving as prey. Several fish macropredators are able to gain access to cryptic invertebrate scavengers (and other cryptos) by rooting into carbonate rubble (e.g., *Gnathanodon speciosus* and *Mulloidichthys dentatus*) and breaking apart isolated coral colonies and framework structures (e.g. *Bodianus diplotaenia* and balistids).

Two Ecopath-modelled trophic webs with a zooxanthellate coral guild are available from studies in the ETP, for the Cabo Pulmo reef in the Gulf of California, Mexico (Fig. 9.4), and a subtidal rocky reef at Floreana Island, Galápagos Islands (Fig. 9.5). Considering first the Cabo Pulmo reef trophic web, 34 guilds ranging from one to about 4.4 trophic levels were identified, and 12 of these consisted of single species. Zooxanthellate corals exhibited the highest live tissue biomass with phytoplankton, detritus, zooplankton, benthic invertebrates, and bivalve molluscs contributing importantly to the community. Energy sources from detritus and zooplankton to corals were high (blue arrows) and corals were observed as prey (red arrows) of four invertebrate (echinoderms, crustaceans, gastropods) functional groups and five species of fishes. Pufferfishes (*Arothron* spp.) are also important corallivores present on the Cabo Pulmo reef (Reyes-Bonilla and Calderón-Aguilera 1999), and although they are not shown, they should be considered in this trophic web.

The Galápagos food web model is based on the biota in most of the western, northern, and eastern coasts of Floreana Island from 0 to 20 m depth, a 6.11 km^2 area (Okey et al. 2004). A total of 42 living groups was recognized, and these were apportioned among 24 functional groups (including detritus) in two simplified food web diagrams according to high and low levels of biomass flow (Fig. 9.5a, b; Bustamante et al. 2008). The number of trophic links depicting these high and low schema were 74 and 83 respectively. Live coral biomass at Floreana Island was about 18 % lower than at Cabo Pulmo, namely 0.91 tonnes (wet weight) ha^{-1}, also with a decidedly low productivity of 0.082 tonnes ha^{-1} $year^{-1}$. Seven functional groups demonstrated higher biomass values than corals at Floreana: macro-invertebrate grazers (18), micro-gastropod grazers (15), vertebrate grazers (19), pelagic planktivorous fishes (7), filter and suspension feeders (20), detritus (24), and macroalgae and microphytobenthos (23). Other notable differences were

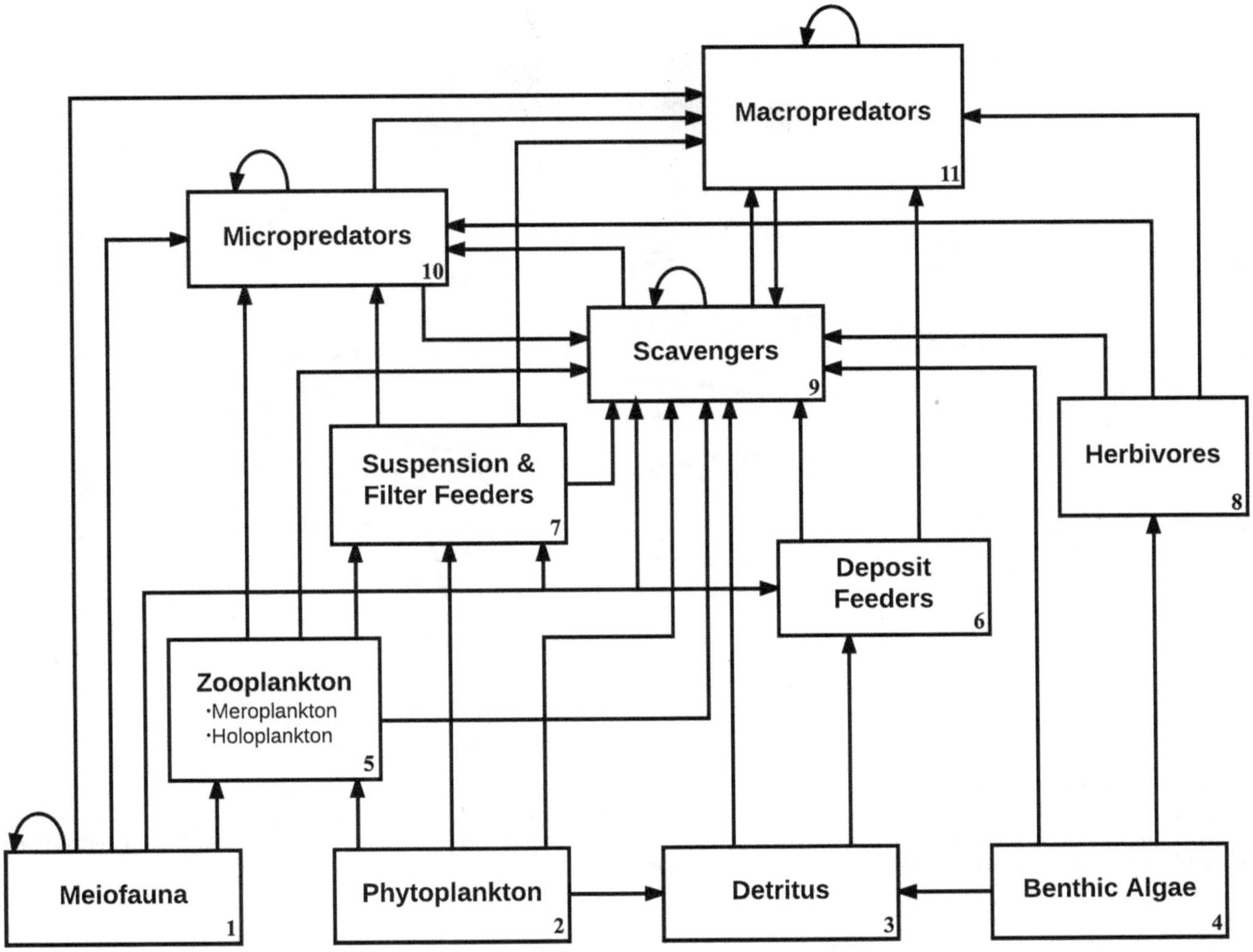

Fig. 9.3 Schematic trophic subweb of cryptic species in the pocilloporid coral framework of the Uva Island reef, Panama. *1* Meiofauna (foraminifers, copepods, nematods, gastropods, polychaetes, bivalves); *2* phytoplankton; *3* detritus; *4* benthic algae (microfilamentous and crustose coralline algae); *5* zooplankton (copepods, fish larvae, ostracods, mysids); *6* Deposit feeders (holothurians, polychaetes, sipunculids, echiurans, *Hipponix grayanus*); *7* Suspension and filter feeders (sponges, hydroids, corals, bryozoans, bivalves, vermetids, ophiuroids, serpulid and sabellid polychaetes, porcellanid crabs, barnacles); *8* Herbivores (diverse gastropods, crustaceans, *Dolabella*, sea urchins); *9* Scavengers (*Pherecardia striata*, gastropods, cirolanid isopods, ostracods, hermit crabs, holothurians; *10* Micropredators (polyclad flatworms, isopods, *Trapezia*, nemerteans, polychaetes, pygmy octopuses); *11* Macropredators (cone and ovulid gastropods), stomatopods, *Trapezia* spp., Muraenidae (moray eels), Ophichthidae (snake eels), Ophidiidae (brotulids), Carapidae (pearlfishes), Bythitidae (cuskeels), Antennariidae (frogfishes), Holocentridae (squirrelfishes), Scorpaenidae (scorpionfishes), Serranidae (juvenile groupers), Grammistidae (soapfishes), Apogonidae (cardinalfishes), Lutjanidae (juvenile snappers), Labrisomidae (blennies). Taxa in 1 and 5 after Guzmán et al. 1987, and Guzmán and Obando 1988, Caño Island coral reef, Costa Rica. Modified after Glynn (2008)

evident in ecotrophic efficiency, with low coral biomass passing to higher trophic levels at Cabo Pulmo, and the relatively high contribution of corals (24 vs. 3.5 %) to community biomass structure at Cabo Pulmo (Table 9.1). There are three factors contributing to the paucity of scleractinian corals at Floreana: (a) a high rate of coral mortality during the 1982–83 El Niño warming event with little to no recovery (Glynn 1994; Glynn et al. 2015), (b) the coastal shelf is a dominantly basalt reef/sand habitat with only patchy coral development, and (c) upwelling favoring phytoplankton and benthic algal production (Okey et al. 2004).

9.7 Perturbations of Feeding Group Populations

Chief among eastern tropical Pacific disturbances affecting nutrient sources and consumer populations are upwelling events and ENSO activity (as noted above). The over-exploitation of reef resources by fishers can also have far-reaching effects on coral community trophic structure. Such perturbations may affect multiple trophic levels, generating a cascading effect that extends throughout the food web. A trophic cascade was initially defined by Carpenter

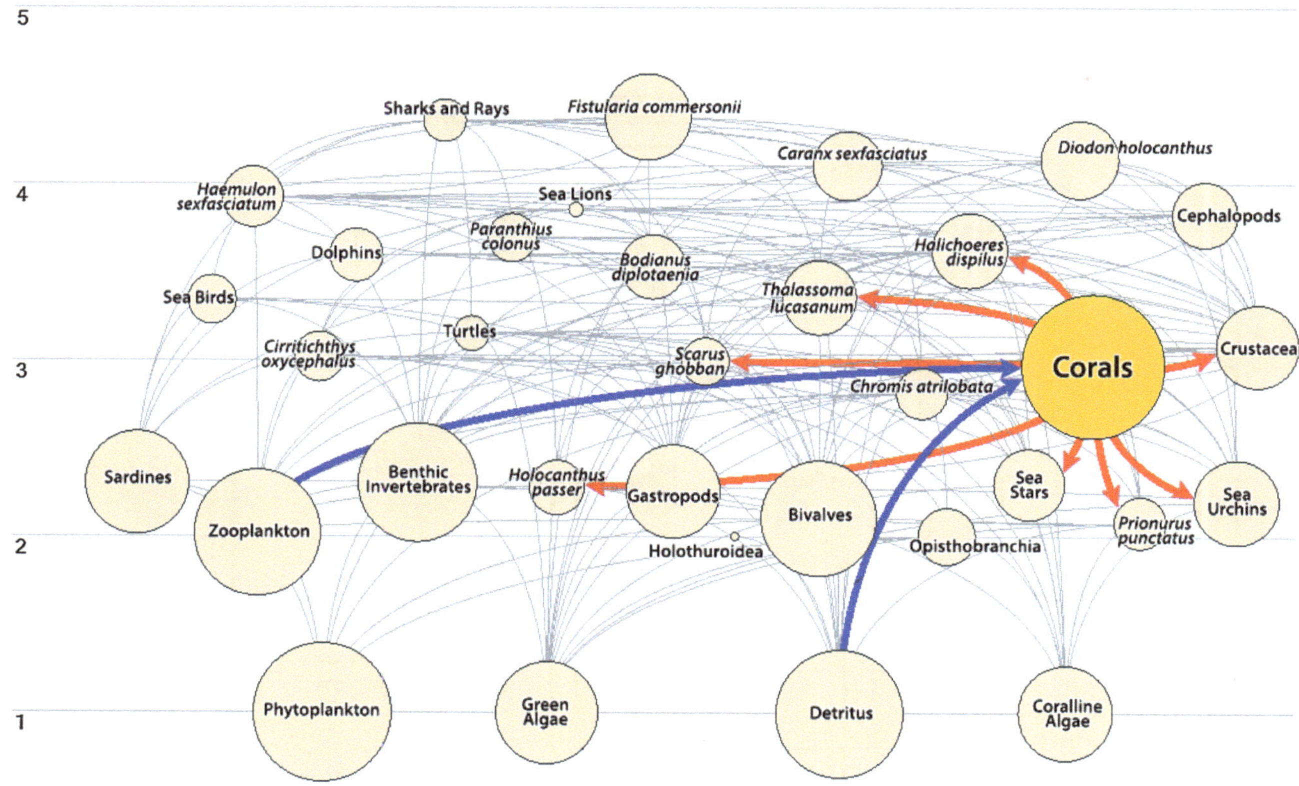

Fig. 9.4 Ecopath flow diagram for the Cabo Pulmo coral reef, Baja California, Mexico (modified after Reyes-Bonilla et al. 2014). Each functional group or species is positioned vertically according to its calculated trophic level (ordinate scale) and the sizes of circles represent the relative biomasses. Blue and red arrows denote energy flow to and from zooxanthellate corals respectively, rendered from predator/prey diet composition values. See also Reyes-Bonilla (1993)

et al. (1985) as an indirect interaction chain that is triggered by an apex predator's effect on plant abundance. This has subsequently been extended to any biotic or physical perturbation affecting top- or bottom-trophic level population abundances and interactions with often far-reaching effects on community structure and productivity.

9.7.1 Direct Effects

Predator-prey interactions, including parasitism, have been abundantly documented on coral reefs, and represent an obvious means of energy transfer to all consumer levels. Direct effects of predatory sea stars (*Acanthaster planci*) have been studied in Panama relative to changes in coral prey abundances, community structure and species diversity (Porter 1972, 1974; Glynn 1976), and reef growth (Glynn et al. 1979). In the early 1970s, when *Acanthaster* was abundant in the non-upwelling Gulf of Chiriquí, Panama, (25–35 ind ha^{-1} at Uva Island), live coral cover demonstrated significant declines. Porter (1972, 1974) reported a diversifying effect in coral community structure when *Acanthaster* exhibited a feeding preference for the most abundant coral taxa (*Pocillopora* spp.). Glynn (1976), however, found that *Acanthaster* does often feed on *Pocillopora* spp. colonies, but usually on broken and loose branches. Large intact colonies of *Pocillopora* host crustacean (shrimp and crabs) guards that defend colonies from attack whereas broken branches are usually devoid of guards and are thus vulnerable to *Acanthaster* predation. The end result is a low diversity coral community composed of mostly pocilloporid species. In recent years, *Acanthaster* has been on the decline in Panama (Fong and Glynn 2000) and now, as of 2015, exercises little influence on community structure.

Arothron meleagris, a ubiquitous ETP pufferfish corallivore, can attain very high densities, up to 320 ind ha^{-1} (Guzmán and Robertson 1989). At Gorgona Island, Colombia it was reported to consume approximately 16 % of the annual pocilloporid carbonate production of 56 tonnes ha^{-1} year^{-1} (Palacios et al. 2014). Its feeding responses are complex; it may benefit reef recovery by switching its diet from coral to coralline algae following mass mortality events, and by promoting asexual propagation of clipped but uningested surviving branches (Palacios et al. 2014). Other direct consumer interactions in the ETP are numerous, including for example apex piscivores (Fig. 9.1), corallivores (Fig. 9.2, see Chap. 10, Enochs and Glynn), and

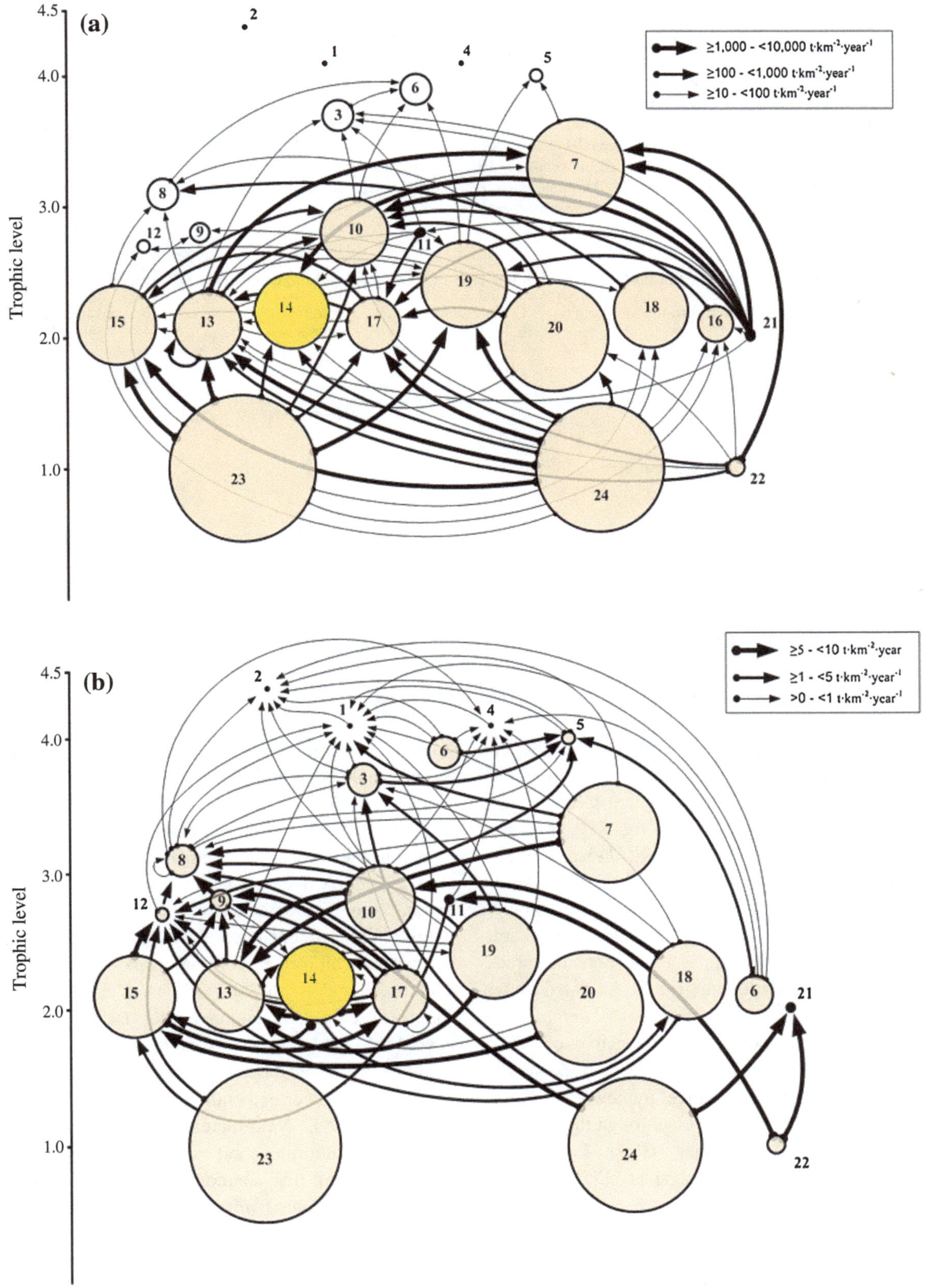
(a)
≥1,000 - <10,000 t·km⁻²·year⁻¹
≥100 - <1,000 t·km⁻²·year⁻¹
≥10 - <100 t·km⁻²·year⁻¹
Trophic level
4.5
4.0
3.0
2.0
1.0
(b)
≥5 - <10 t·km⁻²·year
≥1 - <5 t·km⁻²·year⁻¹
>0 - <1 t·km⁻²·year⁻¹

◀ **Fig. 9.5** Simplified Ecopath flow diagram for the Floreana Island rocky reef ecosystem, Galápagos Islands, Ecuador (modified after Bustamante et al. 2008). Circles denote functional groups and numbers the taxa involved (see below); sizes indicate a proportional scaling of the relative biomasses of each group. Arrows indicate direction of biomass flow from prey/biomass to predator/consumer. Panel **'a'** depicts 74 links for biomass flows that range between 10 and 10,000 mt km^{-2} $year^{-1}$; panel **'b'** depicts 83 links that range between 0.1–10 mt km^{-2} $year^{-1}$. Numbers are: *1* pelagic sharks, *2* dolphins, *3* large predatory reef fishes, *4* seabirds, *5* sea lions, *6* pelagic predatory fishes, *7* pelagic planktivorous fishes, *8* large benthic predatory fishes, *9* macro-invertebrate carnivores, *10* small omnivorous reef fishes, *11* carnivorous zooplankton, *12* macro-crustacean carnivores, *13* micro-crustacean carnivores (e.g., *Trapezia*), *14* zooxanthellate scleractinian corals (*yellow*), *15* micro-gastropod grazers (e.g., *16* detritivorous fishes, *17* benthic invertebrate detritivores, *18* macro-invertebrate grazers [*Eucidaris*, *Pentaceraster*]), *19* vertebrate grazers, *20* filter and suspension feeders, *21* herbivorous zooplankton, *22* phytoplankton, *23* benthic macroalgae and microphytobenthos, *24* detritus

Table 9.1 Ecopath II model parameters and outputs, Cabo Pulmo (Gulf of California) and Floreana Island (Galápagos Islands)

Location	Trophic Level	Biomass (t km^{-2})	Production/biomass y^{-1}	Consumption/biomass y^{-1}	Ecotrophic efficiency	Production/consumption	Biomass (%)
Cabo Pulmo	2.97	111.9	1.09	3.15	0.20	0.346	24.0
Floreana Is.	2.20	91.2	1.09	15.0	0.90	0.073	3.5

Courtesy of Luis E. Calderón-Aguilera. See Calderón-Aguilera et al. (2007)

diverse micro- and macro-herbivores (Fig. 9.1, see Chap. 11, Fong et al.).

9.7.2 Indirect Trophic Cascades

Indirect trophic interactions, whereby predators or parasites have effects often far removed from direct consumption, are ubiquitous in coral reef food webs (Pennings 1997). Several coral reef species are involved in indirect effects or trophic cascades, including polychaete worms, crustaceans, echinoids, and fishes whose interactions in the ETP have been extensively studied. In Panama, a worm (*Pherecardia striata*) and *Hymenocera picta*, the harlequin shrimp, prey on *Acanthaster* and thus limit the sea star's predation on corals (Glynn 1977, 1984). The pocilloporid crustacean guards, *Alpheus lottini* and *Trapezia* spp. ward off potential predators from their coral hosts (Glynn 1983). As noted above, this reduces the direct predatory effects of *Acanthaster* on specific coral taxa. The potentially destructive effects of *Diadema mexicanum* and *Eucidaris galapagensis*—predation on corals and the bioerosion of reef frameworks—can be reduced by the removal of sea urchins through the territorial defensive behavior of damselfishes (Glynn 1988; Eakin 1992; Irving and Witman 2009; Edgar et al. 2010; Glynn et al. 2015). A somewhat more involved concatenation of conditions was found to control the depth zonation and types of coral communities in the Pearl Islands, Panama (Wellington 1982). The interactions in this case center on the presence of a damselfish (*Stegastes acapulcoensis*) and its ability to prevent the feeding of herbivorous fishes and corallivores on pocilloporid corals within its territories. Shallow-occurring (0–6 m) pocilloporid corals provide shelter sites for *Stegastes*, which lowers predation of the damselfish by piscivores and allows for their increased abundances. Deeper-occurring (6–10 m) agariciid corals (*Pavona gigantea*) lack shelter sites, thus increasing the exposure and vulnerability of damselfish to predation.

The sea urchin outbreaks on coral reefs in Panama (*Diadema*) and Galápagos (*Eucidaris*) following the 1982–83 El Niño event caused extensive bioerosion of dead carbonate structures (Glynn 1988, 1994; Eakin 1996; Reaka-Kudla et al. 1996). In Panama, *Diadema* abundances gradually declined as coral recovery progressed. Perhaps this decline was related to the reduced algal food source on live corals. In the central and southern Galápagos Islands, however, *Eucidaris* abundances have remained high, and their grazing on carbonate-encrusted algae (endolithic and micro-epilithic filamentous algae and macroalgae) has eroded virtually all former reef frameworks (Glynn 1994, 2003; Wellington and Glynn 2007; Glynn et al. 2015). Areas that supported coral communities prior to 1982–83 are now devoid of corals and consist predominantly of crustose coralline algae and high abundances of *Eucidaris* ('urchin barrens'), and other epibenthos.

Due to post El Niño increases in a competitively robust sea anemone in former coral and algal patches, it is necessary to be aware of this anemone's influence on community recovery. Anemone (*Aiptasia* sp.) barrens, low diversity patches possibly created by a combination of elevated temperature stress during El Niño events and the depletion of sea cucumbers from overfishing, have been reported at several sites in the Galápagos Islands (Okey et al. 2003, Okey 2004). *Isostichopus fuscus*, a suspected holothurian predator of *Aiptasia* (Bermeo-Sarmiento 1995), has been severely reduced in numbers in recent years. Like *Isostichopus*,

Aiptasia is also an effective predator, consuming the recruiting propagules of corals, other benthic metazoans, and algae. If *Aiptasia* is now free of sea cucumber predation at former coral sites (e.g., at Bartolomé and Onslow islands), it could prevent the recruitment of coral propagules, thus replacing pre-disturbance coral communities with "anemone barrens". It is not known if this secondarily derived community state will replace the pre-disturbance coral communities (Okey et al. 2003; Okey 2004).

Several studies in the Galápagos Islands have suggested that persistently high abundances of *Eucidaris* are a result of indirect effects of overfishing (e.g., Ruttenberg 2001; Okey et al. 2004; Sonnenholzner et al. 2009; Edgar et al. 2010). None of these studies, however, is based on manipulative field experiments where predation was actually observed. The "evidence" for top-down control by fishes is based solely on inverse correlations between fishes and sea urchins. The over-exploitation by artisanal fishers of the hogfish *Bodianus diplotaenia* and lobsters (*Panulirus* spp. and *Scyllarides astori*) was hypothesized to allow *Eucidaris* to escape predation from these predators (Sonnenholzner et al. 2009). In a 30-year Ecosim modelling study, Okey et al. (2004) showed that the removal of sharks from the Floreana rocky reef ecosystem could lead to an increase in meso-predator abundances, a small decline in *Eucidaris* numbers, and a small increase in stony corals. Edgar et al. (2010) suggested that the heavily grazed macroalgal and coral habitats frequented by *Eucidaris* magnified El Niño impacts, resulting in local and regional declines in biodiversity and elevated risk of extinction of numerous taxa. These studies were broadened by Edgar et al. (2011) to examine fishing-induced trophic cascades in a range of protected and unprotected habitats across the equatorial eastern Pacific seascape. These workers found that poorly enforced MPAs generate food webs more similar to those present in fished areas than in well-protected MPAs. Finally, differences in community structure and types driven by over-exploitation of fish piscivore and herbivore guilds have been reported among the two major islands in the Rapa Nui region (Friedlander et al. 2013), lending support for the generality of this trophic response in the eastern Pacific.

9.8 Conclusions and Synthesis

Coral reefs in the ETP have remarkably complex food webs, despite relatively depauperate flora and fauna compared to other reef regions. Elevated production from corals/zooxanthellae (e.g., Palacios et al. 2014; Rico-Esenaro et al. 2014), benthic algae (Smith 2006; see Chap. 11, Fong et al.), and plankton (Palardy et al. 2006) is quickly captured by specialized fauna. These corallivores (see Chap. 10, Enochs and Glynn), herbivores (Smith 2006; see Chap. 11, Fong et al.), and suspension feeders (Palardy et al. 2005) are in turn consumed by abundant piscivores and invertivores (Dominici-Arosemena and Wolff 2006). Evidence of efficiency and recycling is present at all stages of the food web, from mucus (Enochs 2012) and deposit feeders (Granja-Fernández et al. 2013) as well as meiofauna (Guzmán et al. 1987), to coprophagous taxa (Rothans and Miller 1991) and scavengers (Glynn 1984, 2013).

To date, the bulk of our understanding of ETP reef trophodynamics pertains primarily to corals and corallivores. Feeding rates and abundances of prominent corallivorous taxa have been quantified for numerous reefs throughout the ETP. Dynamic fluctuations in coral cover due to ENSO-associated bleaching and mortality events have revealed concomitant fluctuations in corallivore populations, primarily fishes. While preliminary evidence suggests that coral mortality will have far-reaching ramifications for trophic groups that do not necessarily depend on coral for a food source (e.g., Glynn et al. 2014), more experimental and modeling work are needed to determine how alteration to the corallivore sub-web affects the structure of the entire reef ecosystem.

To a lesser extent the role of herbivores, their biomass and consumption is known, but only for select regions, and for select taxa. Information on algal production is similarly limited, and only available for certain reefs. Evidence suggests, however, that it can potentially be dynamic, especially if herbivory is limited and nutrients are elevated. While plankton productivity in the ETP is well-studied (especially in non-reef regions), our understanding of the capacity of reef ecosystems to capture and incorporate this resource is limited. Considering the importance of this trophic pathway in reefs in other areas of the world (e.g., Richter et al. 2001) and the periodic spikes in productivity that occur in select regions of the ETP due to upwelling, research into this topic is needed. Similarly, our understanding of detritivores and recycling in ETP food webs is presently incomplete. Evidence from the GBR suggests that coral mucus and benthic algae provide large quantities of detritus that support numerous higher-level consumers (Wild et al. 2004; Kramer et al. 2012). The magnitude of these sources is unquantified in the ETP and the pathways by which this organic matter is incorporated back into the ecosystem is presently poorly described.

While a relatively large amount of attention has been directed toward the cryptic metazoan symbionts associated with live corals, the majority of studies focus on biodiversity and territorial behaviors rather than the consumption of coral/zooxanthellae productivity and translation of this biomass to the remainder of the food web. Even more poorly described are the cryptofauna that associate with the more heterogeneous dead coral substrates and reef frameworks. These taxa must be studied in greater detail as both food sources for fishes and as consumers that are potentially important herbivores, carnivores, suspension feeders, detritivores, and scavengers. Finally, further information is needed on the structure of the cryptic subweb, the degree of biomass retention within framework cavities, and the extent to which it is connected to epibenthic and nektonic communities.

All of these knowledge gaps should be filled so that more comprehensive trophic models can be constructed. With these models, it will be possible to examine the far reaching ramifications of stress on reef structure and function, as well as to more carefully manage reef resources in an era of climate change. Experimental manipulations to test hypotheses of top down control of prey species and community structure in the ETP are lacking and sorely needed to move beyond correlation to causation. While there are numerous trophic groups and relationships that are in need of further study in the ETP, reefs in this region are perhaps perfect systems for future investigation of trophodynamics. Though ETP food webs are complex, their reduced biodiversity compared with western Pacific and Atlantic systems likely renders them easier to model. Dynamic fluctuation in nutrients due to upwelling will provide insight into eutrophication and its influence on reef food web structure. Coral mortality due to ENSO-related bleaching can help to predict how climate change-associated alteration of coral cover may influence ecosystem function. Finally, since reef ecosystems in the ETP have displayed an ability to recover from mass coral mortality, ETP reef food webs may reveal recovery mechanisms important for preserving the reefs of the future.

Acknowledgments This chapter benefitted greatly from the data and discussions offered by Rodrigo H. Bustamante, Luis E. Calderón-Aguilera, Francesca Forrestal, Gordon Hendler, Thomas A. Okey, Héctor Reyes-Bonilla, and Jon Witman. Special thanks are due Benjamin Grassian for modifying and preparing the figures, as well as providing valuable insight into the topics discussed herein.

References

Alldredge AL, King JM (1977) Distribution, abundance, and substrate preferences of demersal reef zooplankton at Lizard Island Lagoon, Great Barrier Reef. Mar Biol 41:317–333

Alvarado JJ, Cortés J (2005) Research on echinoderms in Latin America: Int J Trop Biol Conserv 53(suppl 3):1–387

Alvarado JJ, Ayala A, Álvarez del Castillo-Cárdenas PA, Fernández C, Aguirre-Rubí J, Buitrago F, Reyes-Bonilla H (2011) Coral communities of San Juan del Sur, Pacific Nicaragua. Bull Mar Sci 87:129–146

Arburto-Oropeza O, Erisman B, Galland GR, Mascareñas-Osorio I, Sala E, Ezcurra E (2011) Large recovery of fish biomass in a no-take marine reserve. PLoS ONE 6(8):e23601. doi:10.1371/journal.pone.0023601

Atkinson MJ (2011) Biogeochemistry of nutrients. In: Dubinsky Z, Stambler N (eds) Coral reefs: an ecosystem in transition. Springer, Dordrecht, pp 199–206

Ayukai T (1995) Retention of phytoplankton and planktonic microbes on coral reefs within the Great Barrier Reef, Australia. Coral Reefs 14:141–147

Baker AC, Glynn PW, Riegl B (2008) Climate change and coral reef bleaching: an ecological assessment of long-term impacts, recovery trends and future outlook. Estuar Coast Shelf Sci 80:435–471

Bakus GJ (1966) Some relationships of fishes to benthic organisms on coral reefs. Nature 210:280–284

Bakus GJ (1969) Energetics and feeding in shallow marine waters. Int Rev Gen Exp Zool 4:275–369

Bakus GJ (1981) Chemical defense mechanisms on the Great Barrier Reef, Australia. Science 211:497–499

Bautista-Guerrero E, Carballo JL, Maldonado M (2014) Abundance and reproductive patterns of the excavating sponge *Cliona vermifera*: a threat to Pacific coral reefs? Coral Reefs 33:259–266

Benson AA, Muscatine L (1974) Wax in coral mucus: energy transfer from corals to reef fishes. Limnol Oceanogr 19:810–814

Bermeo-Sarmiento JD (1995) Informe sobre la evaluación del recurso pepinos de mar *Isostichopus fuscus* (Ludwig 1875) en áreas interiores del Canal Bolívar en las Islas Galápagos. Informe, Instituto Nacional de Pesca, p 11

Birkeland C, Lucas JS (1990) *Acanthaster planci*: major management problem of coral reefs. CRC Press, Boca Raton, Florida p 257

Blackburn M, Laurs R, Owen R, Zeitzschel B (1970) Seasonal and areal changes in standing stocks of phytoplankton, zooplankton and micronekton in the eastern tropical Pacific. Mar Biol 7:14–31

Brown B, Bythell J (2005) Perspectives on mucus secretion in reef corals. Mar Ecol Prog Ser 296:291–309

Bustamante RH, Okey TA, Banks S (2008) Biodiversity and food web structure of a Galápagos shallow rocky-reef ecosystem. In: McClanahan TR, Branch GM (eds) Food webs and the dynamics of marine reefs. Oxford Univ Press, Oxford, pp 135–161

Calderón-Aguilera LE, Reyes Bonilla H, Carriquiry JD (2007) El papel de los arrecifes coralinos en el flujo de carbono en el océano: estudios en el PacÍfico mexicano. In: Hernández de la Torre B, Gaxiola Castro G (eds) Carbono en ecosistemas acuáticos de México. México, SEMARNAT-INE-CICESE. pp 215–226 ISBN:978-968-817-855-3

Carballo JL, Bautista-Guerrero E, Leyte-Morales GE (2008) Boring sponges and the modeling of coral reefs in the Eastern Pacific Ocean. Mar Ecol Prog Ser 356:113–122

Carpenter RC (1997) Invertebrate predators and grazers. In: Birkeland C (ed) Life and death of coral reefs. Chapman & Hall, New York, pp 198–229

Carpenter SR, Kitchell JF, Hodgson JR (1985) Cascading trophic interactions and lake productivity. Bioscience 35:634–639

Charpy L, Palinka KA, Casareto B, Langlade MJ, Suzuki Y, Abed RMM, Golubic S (2010) Dinitrogen-fixing cyanobacteria in microbial mats of two shallow coral reef ecosystems. Microb Ecol 59:174–186

Charpy L, Casareto BE, Langlade MJ, Suzuki Y (2012) Cyanobacteria in coral reef ecosystems: a review. J Mar Biol. doi:10.1155/2012/259571

Chartock MA (1983) Habitat and feeding observations on species of *Ophiocoma* (Ophiocomidae) at Enewetak. Micronesica 19:131–149

Cipriani R, Guzmán HM, López M (2008) Harvest history and current densities of the pearl oyster *Pinctada mazatlanica* (Bivalvia: Pteriidae) in Las Perlas and Coiba Archipelagos, Panama. J Shellfish Res 27:691–700

Coffroth MA (1990) Mucous sheet formation on poritid corals: an evaluation of coral mucus as a nutrient source on reefs. Mar Biol 105:39–49

Cortés J (ed) (2003) Latin American coral reefs. Elsevier, Amsterdam p 497

Crossland CJ, Barnes DJ, Borowitzka MA (1980) Diurnal lipid and mucus production in the staghorn coral *Acropora acuminata*. Mar Biol 60:81–90

Crossland CJ, Hatcher BG, Smith SV (1991) Role of coral reefs in global ocean production. Coral Reefs 10:55–64

D'Croz L, O'Dea A (2007) Variability in upwelling along the Pacific shelf of Panama and implications for the distribution of nutrients and chlorophyll. Estuar Coast Shelf Sci 73:325–340

Dee LE, Witman JD, Brandt M (2012) Refugia and top-down control of the pencil urchin *Eucidaris galapagensis* in the Galápagos Marine Reserve. J Exp Mar Biol Ecol 416:135–143

Depczynski M, Bellwood DR (2003) The role of cryptobenthic reef fishes in coral reef trophodynamics. Mar Ecol Prog Ser 256:183–191

Dominici-Arosemena A, Wolff M (2005) Reef fish community structure in Bocas del Toro (Caribbean, Panamá): gradients in habitat complexity and exposure. Carib J Sci 41:613–637

Dominici-Arosemena A, Wolff M (2006) Reef fish community structure in the Tropical Eastern Pacific (Panamá): living on a relatively stable rocky reef environment. Helgol Mar Res 60:287–305

Eakin CM (1992) Post-El Niño Panamanian reefs: less accretion, more erosion and damselfish protection. In: Proceedings of 7th International Coral Reef Symposium, vol 1, pp 387–396

Eakin CM (1996) Where have all the carbonates gone? A model comparison of calcium carbonate budgets before and after the 1982–1983 El Niño at Uva Island in the eastern Pacific. Coral Reefs 15:109–119

Edgar GJ, Fariña JM, Calvopiña M, Martínez C, Banks S (2002) Comunidades submareales rocosas II: peces y macroinvertebrados móviles. In: Danulat E, Edgar GJ (eds) Reserva marina de Galápagos. Línea base de la biodiversidad, Fundación Charles Darwin/Servicio Parque Nacional Galápagos, Santa Cruz, Galápagos, Ecuador pp 68–97

Edgar GJ, Banks S, Fariña JM, Calvopiña M, Martínez C (2004) Regional biogeography of shallow reef fish and macro-invertebrate communities in the Galápagos Archipelago. J Biogeogr 31:1107–1124

Edgar GJ, Banks SA, Bessudo S, Cortés J, Guzmán HM, Henderson S, Martinez C, Rivera F, Soler G, Ruiz D, Zapata FA (2011) Variation in reef fish and invertebrate communities with level of protection from fishing across the eastern tropical Pacific seascape. Global Ecol Biogeogr 20:730–743

Edgar GJ, Banks SA, Brandt M, Bustamante RH, Chiriboga Á, Earle SA, Garske LE, Glynn PW, Grove JS, Henderson S, Hickman CP, Miller KA, Rivera F, Wellington GM (2010) El Niño, fisheries and animal grazers interact to magnify extinction risk for marine species in Galápagos. Global Change Biol 16:2876–2890

Edmunds PJ (2011) Zooplanktivory ameliorates the effects of ocean acidification on the reef coral *Porites* spp. Limnol Oceanogr 56:2402–2410

Enochs IC (2012) Motile cryptofauna associated with live and dead coral substrates: implications for coral mortality and framework erosion. Mar Biol 159:709–722

Enochs IC, Hockensmith G (2009) Effects of coral mortality on the community composition of cryptic metazoans associated with *Pocillopora damicornis*. In: Proceedings of 11th International Coral Reef Symposium, vol 2, Ft Lauderdale, Florida, pp 1368–1372

Enochs IC, Manzello DP (2012) Responses of cryptofaunal species richness and trophic potential to coral reef habitat degradation. Diversity 4:94–104

Enochs IC, Toth LTL, Brandtneris V, Afflerbach J, Manzello DP (2011) Environmental determinants of motile cryptofauna on an eastern Pacific coral reef. Mar Ecol Prog Ser 438:105–118

Espinoza M, Salas E (2005) Estructura de las comunidades de peces de arrecifes en las Islas Catalinas y Playa Ocotal, Pacífico norte de Costa Rica. Rev Biol Trop 53:523–526

Feldman G, Clark D, Halpern D (1984) Satellite color observations of the phytoplankton distribution in the eastern equatorial Pacific during the 1982–1983 El Niño. Science 226:1069–1071

Ferrier-Pagès C, Hoogenboom M, Houlbrèque F (2011) The role of plankton in coral trophodynamics. In: Dubinsky Z, Stambler N (eds) Coral reefs: an ecosystem in transition. Springer, New York, pp 215–230

Fong P, Glynn PW (2000) A regional model to predict coral population dynamics in response to El Niño-Southern Oscillation. Ecol Appl 10:842–854

Fong P, Smith TB, Wartian MJ (2006) Epiphytic cyanobacteria maintain shifts to macroalgal dominance on coral reefs following ENSO disturbance. Ecology 87:1162–1168

Fong P, Paul VJ (2011) Coral reef algae. In: Dubinsky Z, Stambler N (eds) Coral reefs: an ecosystem in transition. Springer, New York, pp 241–272

Friedlander AM, Ballesteros E, Beets J, Berkenpas E, Gaymer CF, Gorny M, Sala E (2013) Effects of isolation and fishing on the marine ecosystems of Easter Island and Salas y Gómez, Chile. Aquatic Conserv Mar Freshw Ecosyst 23:515–531

Garth JS (1974) On the occurrence in the eastern tropical Pacific of Indo-west Pacific decapod crustaceans commensal with reef-building corals. In: Proceedings of 2nd International Coral Reef Symposium, vol 1, Australia, pp 397–404

Gischler E, Ginsburg RN (1996) Cavity dwellers (coelobites) under coral rubble in southern Belize barrier and atoll reefs. Bull Mar Sci 58:570–589

Glynn PW (1976) Some physical and biological determinants of coral community structure in the eastern Pacific. Ecol Monogr 46:431–456

Glynn PW (1977) Interactions between *Acanthaster* and *Hymenocera* in the field and laboratory. In: Proceedings of 3rd International Coral Reef Symposium, vol 1, Miami, pp 209–215

Glynn PW (1983) Crustacean symbionts and the defense of corals: coevolution on the reef? In: Nitecki MH (ed) Coevolution. Univ Chicago Press, Chicago, pp 111–178

Glynn PW (1984) An amphinomid worm predator of the crown-of-thorns sea star and general predation on asteroids in eastern and western Pacific coral reefs. Bull Mar Sci 35:54–71

Glynn PW (1985) Corallivore population sizes and feeding effects following El Niño (1982-83) associated coral mortality in Panama. In: Proceedings of 5th International Coral Reef Congress, vol 4, Tahiti, pp 183–188

Glynn PW (1988) El Niño warming, coral mortality and reef framework destruction by echinoid bioerosion in the eastern Pacific. Galaxea 7:129–160

Glynn PW (1994) State of coral reefs in the Galápagos Islands: natural vs anthropogenic impacts. Mar Pollut Bull 29:131–140

Glynn PW (2003) Coral communities and coral reefs of Ecuador. In: Cortés J (ed) Latin American coral reefs. Elsevier, Amsterdam, pp 449–472

Glynn PW (2004) High complexity food webs in low-diversity eastern Pacific reef-coral communities. Ecosystems 7:358–367

Glynn PW (2006) Fish utilization of simulated coral reef frameworks versus eroded rubble substrates off Panama, eastern Pacific. In: Proceedings of 10th International Coral Reef Symposium, vol 1, Okinawa, pp 250–256

Glynn PW (2008) Food-web structure and dynamics of eastern tropical Pacific coral reefs: Panamá and Galápagos Islands. In: McClanahan TR, Branch GM (eds) Food webs and the dynamics of marine reefs. Oxford Univ Press, Oxford, pp 185–208

Glynn PW (2013) Fine-scale interspecific interactions on coral reefs: functional roles of small and cryptic metazoans. In: Lang MA,

Marinelli RL, Roberts SJ, Taylor PR (eds) Research and discoveries: the revolution of science through scuba. Smith Contribution Marine Science, vol 39, pp 229–248

Glynn PW, Wellington GM (1983) Corals and coral reefs of the Galápagos Islands. Univ Calif Press, Berkeley, p 330

Glynn PW, Maté JL (1997) Field guide to the Pacific coral reefs of Panama. In: Proceedings of 8th International Coral Reef Symposium, vol 1, Panama, pp 145–166

Glynn PW, Enochs IC (2011) Invertebrates and their roles in coral reef ecosystems. In: Dubinsky Z, Stambler N (eds) Coral reefs: an ecosystem in transition. Springer, New York, pp 273–326

Glynn PW, Stewart RH, McCosker JE (1972) Pacific coral reefs of Panamá: structure, distribution and predators. Geol Rundsch 61:483–519

Glynn PW, Wellington GM, Birkeland C (1979) Coral reef growth in the Galápagos: limitation by sea urchins. Science 203:47–49

Glynn PW, Enochs IC, Afflerbach JA, Brandtneris VW, Serafy JE (2014) Eastern Pacific reef fish response to coral recovery following El Niño disturbances. Mar Ecol Prog Ser 495:233–247

Glynn PW, Riegl B, Purkis S, Kerr JM, Smith TB (2015) Coral reef recovery in the Galápagos Islands: the northern-most islands (Darwin and Wenman). Coral Reefs. doi:10.1007/s00338-015-1280-4

Granja-Fernández R, Antonio-Pérez V, López-Pérez RA (2013) *Euapta godeffroyi* (Holothuroidea: Synaptidae): filling the distribution gap between Mexico and Costa Rica, eastern tropical Pacific. Hidrobiológica 23:130–132

Grottoli AG, Rodrigues LJ, Palardy JE (2006) Heterotrophic plasticity and resilience in bleached corals. Nature 440:1186–1189

Guzmán HM (1988) Distribución y abundancia de organismos coralívoros en los arrecifes coralinos de la Isla del Caño, Costa Rica. Rev Biol Trop 36:191–207

Guzmán HM, Cortés J, Glynn PW, Richmond RH (1990) Coral mortality associated with dinoflagellate blooms in the eastern Pacific (Costa Rica and Panama). Mar Ecol Prog Ser 60:299–303

Guzmán HM, Obando VL (1988) Diversidad y abundancia diaria y estacional del zooplankton marino de la Isla del Caño, Costa Rica. Rev Biol Trop 36:139–150

Guzmán HM, Obando VL, Cortés J (1987) Meiofauna associated with a Pacific coral reef in Costa Rica. Coral Reefs 6:107–112

Guzmán HM, Robertson DR (1989) Population and feeding responses of the corallivorous pufferfish *Arothron meleagris* to coral mortality in the eastern Pacific. Mar Ecol Prog Ser 55:121–131

Hammond LS (1983) Nutrition of deposit-feeding holothuroids and echinoids (Echinodermata) from a shallow reef lagoon, Discovery Bay, Jamaica. Mar Ecol Prog Ser 10:297–305

Haig J (1960) The Porcellanidae (Crustacea Anomura) of the eastern Pacific. Allan Hancock Pacific Exped 24:1–440

Hansen JA, Skilleter GA (1994) Effects of the gastropod *Rhinoclavis aspera* (Linnaeus, 1758) on microbial biomass and productivity in coral-reef-flat sediments. Aust J Mar Freshwater Res 45:569–584

Hatcher BG (1997) Organic production and decomposition. In: Birkeland C (ed) Life and death of coral reefs. Chapman and Hall, New York, pp 140–174

Hickman CP Jr (1998–2008) Galápagos marine life series (four field guides treating invertebrate taxa). Sugar Spring Press, Lexington, Virginia

Hobson ES (1991) Trophic relationships of fishes specialized to feed on zooplankters above coral reefs. In: Sale P (ed) The ecology of fishes on coral reefs. Academic Press, San Diego, pp 69–95

Hobson ES, Chess JR (1986) Diel movements of resident and transient zooplankters above lagoon reefs at Enewetak Atoll, Marshall Islands. Pac Sci 40:7–26

Hughes TP, Rodrigues MJ, Bellwood DR, Ceccarelli D, Hoegh-Guldberg O, McCook L, Moltschaniwskyj N, Pratchett MS, Steneck RS, Willis B (2007) Phase shifts, herbivory, and the resilience of coral reefs to climate change. Current Biol 17:360–365

Irving A, Witman JD (2009) Positive effects of damselfish override negative effects of urchins to prevent an algal habitat switch. J Ecol 97:337–347

Jackson JBC (1984) Ecology of cryptic coral reef communities. III. Abundance and aggregation of encrusting organisms with particular reference to cheilostome Bryozoa. J Exp Mar Biol Ecol 75:37–57

Jackson JBC, Winston JE (1982) Ecology of cryptic coral reef communities. I. Distribution and abundance of major groups of encrusting organisms. J Exp Mar Biol Ecol 57:135–147

Jackson JBC, Kirby MX, Berger WH, Bjorndal KA, Botsford LW, Bourque BJ, Lange CB, Lenihan HS, Pandolfi JS, Peterson CH, Steneck RS, Tegner MJ, Warner RR (2001) Historical overfishing and the recent collapse of coastal ecosystems. Science 293:629–638

James MJ (ed) (1991) Galápagos marine invertebrates. Taxonomy, biogeography, and evolution in Darwin's Islands. Springer, New York p 474

Jennings S, Brierly AS, Walker JW (1994) The inshore fish assemblages of the Galapagos Archipelago. Biol Conserv 70:49–57

Jiménez C (1997) Coral colony fragmentation by whitetip reef sharks at Coiba Island National Park, Panama. Rev Biol Trop 44,45:698–700

Jones ML, ed (1972) The Panamic biota: some observations prior to a sea-level canal. Bull Biol Soc Wash 2:270

Karlson RH (1999) Dynamics of coral communities. Chapter 6 Consumer-resource interactions, pp 118–139. Kluwer Academic Pub, Dordrecht p 250

Keable SJ (1995) Structure of the marine invertebrate scavenging guild of a tropical reef ecosystem: field studies at Lizard Island, Queensland, Australia. J Nat Hist 29:27–45

Kleemann K (2013) Fast and massive settlement of boring bivalves on coral slabs at Taboga Islands, eastern Pacific, Panama. Boll Malacol 49:104–113

Knowlton N, Jackson JBC (2001) The ecology of coral reefs. In: Bertness MD, Gaines SD, Hay ME (eds) Marine community ecology. Sinauer Assoc Inc, Pub, Sunderland, Massachusetts, pp 395–422

Knudsen JW (1967) *Trapezia* and *Tetralia* (Decapoda, Brachyura, Xanthidae) as obligate ectoparasites of pocilloporid and acroporid corals. Pac Sci 21:51–57

Koop K, Larkum AWD (1987) Deposition of organic material in a coral reef lagoon, One Tree Island, Great Barrier Reef. Estuar Coast Shelf Sci 25:1–9

Kramer MJ, Bellwood DR, Bellwood O (2012) Cryptofauna of the epilithic algal matrix on an inshore coral reef, Great Barrier Reef. Coral Reefs 31:1007–1015

Kramer MJ, Bellwood O, Bellwood DR (2013) The trophic importance of algal turfs for coral reef fishes: the crustacean link. Coral Reefs 32:575–583

Leichter JJ, Wing SR, Miller SL, Denny MW (1996) Pulsed delivery of subthermocline water to Conch Reef (Florida Keys) by internal tidal bores. Limnol Oceanogr 41:1490–1501

Little AF, van Oppen MJH, Willis BL (2004) Flexibility in algal endosymbioses shapes growth in reef corals. Science 304:1492–1494

Littler MM and Littler DS (1984) A relative-dominance model for biotic reefs. Advances in reef sciences. In: Proceedings of Joint Meeting of Atlantic Reef Communication and International Society Reef Studies. Miami, Florida pp 1–2

Lopez GRG, Levinton JS (1987) Ecology of deposit-feeding animals in marine sediments. Q Rev Biol 62:235–260

López-Pérez RA, Calderon-Aguilera LE, Zepeta-Vilchis RC, López--Pérez Maldonado I, López-Ortiz AM (2013) Species composition, habitat configuration and seasonal changes of coral reef fish assemblages in western Mexico. J Appl Ichthyol 29:437–448

Lozano S, Zapata A, Zapata F (2003) Short-term temporal patterns of early recruitment of coral reef fishes in the tropical eastern Pacific. Mar Biol 142:399–409
McClanahan TR (2008) Food-web structure and dynamics of east African coral reefs. In: McClanahan TR, Branch GM (eds) Food webs and the dynamics of marine reefs. Oxford Univ Press, Oxford pp 162–184
McCloskey LR (1970) The dynamics of the community associated with a marine scleractinian coral. Int Rev Gesamten Hydrobiol Hydrogr 55:13–81
Meyer JL, Schultz ET (1985) Migrating haemulid fishes as a source of nutrients and organic matter on coral reefs. Limnol Oceanogr 30:146–156
Muscatine L (1973) Nutrition of corals. In: Jones OA, Endean R (eds) Biology and geology of coral reefs. Academic Press, New York, pp 77–115
Muscatine L (1990) The role of symbiotic algae in carbon and energy flux in reef corals. In: Dubinsky Z (ed) Ecosystems of the world 25, coral reefs. Elsevier, Amsterdam, pp 75–87
Muscatine L, Porter J (1977) Reef corals: mutualistic symbioses adapted to nutrient-poor environments. Bioscience 27:454–460
Oak T, Scheibling RE (2006) Tidal activity pattern and feeding behaviour of the ophiuroid *Ophiocoma scolopendrina* on a Kenyan reef flat. Coral Reefs 25:213–222
Odum HH, Odum EP (1955) Trophic structure and productivity of a windward coral reef community on Eniwetok Atoll. Ecol Monogr 25:291–320
Okey TA (2004) Shifted community states in four marine ecosystems: some potential mechanisms. Ph.D. Dissertation, Univ British Columbia p 173
Okey TA, Shepherd SA, Martínez PC (2003) A new record of anemone barrens in the Galápagos. Noticias Galápagos 62:17–20
Okey TA, Banks S, Born AF, Bustamante RH, Calvopiña M, Edgar GJ, Espinoza E, Fariña JM, Garske LE, Reck GK, Salazar S, Shepherd S, Toral-Granda V, Wallem P (2004) A trophic model of a Galápagos subtidal rocky reef for evaluating fisheries and conservation strategies. Ecol Model 172:383–401
Opitz S (1996) Trophic interactions in Caribbean coral reefs. ICLARM Tech Rep 43, Makati City, Philippines p 341
Palacios MM, Muñoz CG, Zapata FA (2014) Fish corallivory on a pocilloporid reef and experimental coral responses to predation. Coral Reefs. doi:10.1007/s00338-014-1173-y
Palardy JE, Witman JD (2011) Water flow drives biodiversity by mediating rarity in marine benthic communities. Ecol Lett 14:63–68
Palardy JE, Grottoli AG, Matthews KA (2005) Effects of upwelling, depth, morphology and polyp size on feeding in three species of Panamanian corals. Mar Ecol Prog Ser 300:79–89
Palardy JE, Grottoli AG, Matthews KA (2006) Effect of naturally changing zooplankton concentrations on feeding rates of two coral species in the Eastern Pacific. J Exp Mar Biol Ecol 331:99–107
Patton WK (1976) Animal associates of living reef corals. In: Jones OA, Endean R (eds) Biology and geology of coral reefs vol 3, Biol 2. Academic Press, New York, pp 1–36
Paz-García DA, Aldana-Moreno A, Cabral-Tena RA, Balart EF (2012) High predation by the corallivore sea snail *Jenneria pustulata* in a high-latitude reef in the Gulf of California. Mar Biodivers Rec 5: e94. doi:10.1017/S1755267212000826
Pennings SC (1997) Indirect interactions on coral reefs. In: Birkeland C (ed) Life and death of coral reefs. Chapman & Hall, New York, pp 249–272
Pennington TJ, Mahoney KL, Kuwahara VS, Kolber DD, Calienes R, Chavez FP (2006) Primary production in the eastern tropical Pacific: a review. Prog Oceanogr 69:285–317
Polis GA, Power ME, Huxel GR (2004) Food webs at the landscape level. Univ Chicago Press, Chicago p 548
Porter JW (1972) Predation by *Acanthaster* and its effect on coral species diversity. Am Nat 106:487–492
Porter JW (1974) Community structure of coral reefs on opposite sides of the Isthmus of Panama. Science 186:543–545
Porter JW, Porter KG (1977) Quantitative sampling of demersal plankton migrating from different coral reef substrates. Limnol Oceanogr 22:553–556
Reaka-Kudla ML, Feingold JS, Glynn PW (1996) Experimental studies of rapid bioerosion of coral reefs in the Galápagos Islands. Coral Reefs 15:101–107
Reyes-Bonilla H (1993) Estructura de la comunidad, influencia de la depredación y biología poblacional de corales hermatípicos en el arrecife de Cabo Pulmo, BCS. MSc thesis, Centro de Investigación Científica y Enseñanza Superior de Ensenada, p 169
Reyes-Bonilla H, Calderón-Aguilera LE (1999) Population density, distribution and consumption rates of three corallivores at Cabo Pulmo reef, Gulf of California, Mexico. PSZNI: Mar Ecol 20:347–357
Reyes Bonilla H, Álvarez del Castillo Cárdenas PA, Calderón Aguilera LE, Erosa Ricárdez CE, Fernández Rivera Melo FJ, Frausto TC, Luna Salguero BM, Moreno Sánchez XG, Mozqueda Torres MC, Norzagaray López CO, Petatán Ramírez D (2014) Servicios ambientales de arrecifes coralinos: el caso del Parque Nacional Cabo Pulmo, B.C.S. In: Urciaga García JI (ed) Desarrollo regional en Baja California Sur. Una perspectiva de los servicios ecosistémicos. UABCS, La Paz pp 38–63
Richman S, Loya Y, Slobodkin LB (1975) The rate of mucus production by corals and its assimilation by coral reef copepod *Acartia negligens*. Limnol Oceanogr 20:918–923
Richter C, Wunsch M (1999) Cavity-dwelling suspension feeders in coral reefs—a new link in reef trophodynamics. Mar Ecol Prog Ser 188:105–116
Richter C, Wunsch M, Rasheed M, Kötter I, Badran MI (2001) Endoscopic exploration of Red Sea coral reefs reveals dense populations of cavity-dwelling sponges. Nature 413:726–730
Rico-Esenaro SD, Signoret Poillon M, Aldeco J, Reyes-Bonilla H (2014) Metabolic balance of the polyp-algae mutualistic symbiosis in the hermatypic coral *Porites panamensis* in La Paz, Baja California Sur. México. CICIMAR Oceánides 29(1):1–10
Riisgård HU, Larsen PS (2000) Comparative ecophysiology of active zoobenthic filter feeding, essence of current knowledge. J Sea Res 44:169–193
Riisgård HU, Kamermans P (2001) Switching between deposit and suspension feeding in coastal zoobenthos. In: Riese K (ed) Ecological comparisons of sedimentary shores. Springer, Part I, pp 73–101
Roberston DR (1982) Fish feces as fish food on a Pacific coral reef. Mar Ecol Prog Ser 7:253–265
Robertson DR, Allen GR (2008) Shorefishes of the tropical eastern Pacific online information system. Smith Trop Res Inst, Balboa, Panama (available at: www.stri.org/sftep)
Robertson DR, Cramer KL (2009) Shore fishes and biogeographic subdivisions of the tropical eastern Pacific. Mar Ecol Prog Ser 380:1–17
Rothans TC, Miller AC (1991) A link between biologically imported particulate organic nutrients and the detritus food web in reef communities. Mar Biol 110:145–150
Ruttenberg B (2001) Effects of artisanal fishing on marine communities in the Galápagos Islands. Conserv Biol 15:1691–1699
Sargent MC, Austin TS (1949) Organic productivity of an atoll. Trans Amer Geophys Union 30:245–249
Scott PJB, Risk MJ, Carriquiry JD (1988) El Niño, bioerosion and the survival of east Pacific reefs. In: Proceedings of 6th International Coral Reef Symposium, vol 2, Townsville, pp 517–520
Simberloff D, Dayan T (1991) The guild concept and the structure of ecological communities. Ann Rev Ecol Syst 22:115–143

Smith TB (2006) Benthic algal production in an upwelling coral reef environment (Gulf of Panamá): evidence for control by fish herbivores. In: Proceedings of 10th International Coral Reef Symposium, vol 3, Okinawa, pp 854–861

Smith TB (2008) Temperature effects on herbivory for an Indo-Pacific parrotfish in Panamá: implications for coral-algal competition. Coral Reefs 27:397–405

Smith TB, Fong P, Kennison R, Smith J (2010) Spatial refuges and associational defenses promote harmful alga *Caulerpa sertularioides* onto coral reefs. Oecologia 164:1039–1048

Sonnenholzner JI, Ladah LB, Lafferty KD (2009) Cascading effects of fishing on Galápagos rocky reef communities: reanalysis using corrected data. Mar Ecol Prog Ser 375:209–218

Sorokin YI (1990) Aspects of trophic relations, productivity and energy balance in coral-reef ecosystems. In: Dubinsky Z (ed) Coral Reefs. Elsevier, Amsterdam, pp 401–410

Stambler N (2011) Zooxanthellae: the yellow symbionts inside animals. In: Dubinsky Z, Stambler N (eds) Coral reefs: an ecosystem in transition. Springer, Dordrecht, pp 87–106

Steneck RS (1983) Escalating herbivory and resulting adaptive trends in calcareous algal crusts. Paleobiology 9:44–61

Suzuki Y, Casareto BE (2011) The role of dissolved organic nitrogen (DON) in coral biology and reef ecology. In: Dubinsky Z, Stambler N (eds) Coral reefs: an ecosystem in transition. Springer, Dordrecht, pp 207–214

Vasconcellos M, Mackinson S, Sloman K, Pauly D (1997) The stability of trophic mass-balance models of marine ecosystems: a comparative analysis. Ecol Modeling 100:125–134

Vermeij GJ (1987) Evolution and escalation: an ecological history of life. Princeton Univ Press, Princeton 527

Wehrtman IS, Cortés J (2009) Marine biodiversity of Costa Rica. Central America, Springer, Berlin p 538

Wellington GM (1982) Depth zonation of corals in the Gulf of Panama: control and facilitation by resident reef fishes. Ecol Monogr 52:223–241

Wellington GM, Glynn PW (2007) Responses of coral reefs to El Niño-Southern Oscillation sea-warming events. In: Aronson RB (ed) Geological approaches to coral reef ecology. Ecol Stud 192. Springer, New York, pp 342–385

Wild C, Rasheed M, Werner U, Franke U, Johnstone R, Huettel M (2004) Degradation and mineralization of coral mucus in reef environments. Mar Ecol Prog Ser 267:159–171

Wilson S, Bellwood DR Choat JH, Furnas MJ (2003) Detritus in the epilithic algal matrix and its use by coral reef fishes. Oceanogr Mar Biol Annu Rev 41:279–309

Witman JD, Patterson MR, Genovese SJ (2004) Benthic-pelagic linkages in subtidal communities: influence of food subsidy by internal waves. In: Polis GA, Power ME, Huxel GR (eds) Food webs at the landscape level. Univ Chicago Press, Chicago, pp 133–153

Witman JD, Brandt M, Smith F (2010) Coupling between subtidal prey and consumers along a mesoscale upwelling gradient in the Galápagos Islands. Ecol Monogr 80:153–177

Wulff JL (1984) Sponge-mediated coral reef growth and rejuvenation. Coral Reefs 3:157–163

Wulff JL (1997) Causes and consequences of differences in sponge diversity and abundance between the Caribbean and eastern Pacific of Panama. In: Proceedings of 8th International Coral Reef Symposium, vol 2, Panama, pp 1377–1382

Zapata FA, Morales YA (1997) Spatial and temporal patterns of fish diversity in a coral reef at Gorgona Island, Colombia. In: Proceedings of 8th International Coral Reef Symposium, vol 1, Panama, pp 1029–1034

Corallivory in the Eastern Pacific

10

Ian C. Enochs and Peter W. Glynn

Abstract

Eastern Pacific reef ecosystems are home to a diverse assemblage of corallivorous fishes and invertebrates. It is therefore not surprising that there is a rich history of research on corallivores in the eastern Pacific. In fact, much of what is known today on the topic of corallivory has built upon studies from the eastern Pacific region. Here we review the progression of our understanding of eastern Pacific corallivory and corallivores. We discuss the behavior and ecology of these specialized consumers, dividing our analysis into the larger conspicuous taxa such as the crown-of-thorns sea star (*Acanthaster planci*) and the guineafowl puffer (*Arothron meleagris*), as well as into the smaller cryptic species such as the pustulate egg shell (*Jenneria pustulata*) and coral crustacean guards (*Trapezia* spp., *Alpheus lottini*). The majority of species that consume coral tissues are facultative corallivores, feeding on corals only incidentally. Both the negative and positive interactions of corallivores to their prey/hosts are reviewed. We address detrimental direct consumption of coral and how it can ultimately influence growth form, species distributions, population structure, and the asexual reproduction of corals. We examine the cleaning behavior of some corallivorous species as well as their territorial tendencies, which may potentially lead to the exclusion of more lethal coral predators. Despite the high diversity of corallivore taxa, no population outbreaks have been observed in the eastern Pacific; coral colony growth rates and reef accretion proceed apace. Finally, we explore the far-reaching implications of the corallivore feeding strategy, touching on the connections that ultimately link coral biomass with higher trophic levels and the rest of the reef ecosystem.

Keywords

Corallivory • *Acanthaster* • *Arothron meleagris* • Food webs • Coral mortality

10.1 Introduction

Where there are reef-building corals, there are nearly always animals present that feed on them. The coral-consuming guild is known collectively as corallivores. This is, admittedly, a broad categorization, encompassing species that not only feed on the tissues of adult coral colonies, but also on metabolic byproducts such as mucus, fat bodies, and larvae. Taxa exhibiting corallivorous behavior are phyletically diverse, including species belonging to Platyhelminthes, Annelida, Arthropoda, Chordata, Echinodermata, and Mollusca. They exhibit a wind range of morphologies and feeding behaviors, from jawed fishes that scrape away

I.C. Enochs (✉)
Cooperative Institute of Marine and Atmospheric Studies, Rosenstiel School of Marine and Atmospheric Science, University of Miami, 4600 Rickenbacker Causeway, Miami, FL 33149, USA
e-mail: ian.enochs@noaa.gov

I.C. Enochs
Atlantic Oceanographic and Meteorological Laboratory (AOML), Ocean Chemistry and Ecosystem Division (OCED), National Oceanic and Atmospheric Administration (NOAA), 4301 Rickenbacker Cswy, Miami, FL 33149, USA

P.W. Glynn
Department of Marine Biology and Ecology, Rosenstiel School of Marine and Atmospheric Science, University of Miami, 4600 Rickenbacker Causeway, Miami, FL 33149, USA
e-mail: pglynn@rsmas.miami.edu

P.W. Glynn et al. (eds.), *Coral Reefs of the Eastern Tropical Pacific*, Coral Reefs of the World 8, DOI 10.1007/978-94-017-7499-4_10

tissues and skeleton (e.g., Scaridae, Tetraodontidae) to those that remove individual coral polyps (e.g., Chaetodontidae), from sea stars that evert their stomach and digest underlying coral tissues (e.g., Acanthasteridae) to crabs that use setose appendages to scrape the surface of living corals to remove mucus, tissues, and organic deposits (e.g., Trapeziidae). While many of these animals naturally occur at low densities, dynamic fluctuations in population size may result in "outbreaks," potentially leading to devastating consequences for coral cover and reef health (e.g., Colgan 1987). As such, corallivores are of paramount importance to reef ecology and conservation. Reviews of corallivorous reef taxa, their behavior and feeding mechanisms, as well as the impacts that they have on corals and coral reef ecosystems can be found in Robertson (1970), Endean (1976), Glynn (1982, 1983a, 1988, 1990), Carpenter (1997), Cole et al. (2008), Rotjan and Lewis (2008), and in Glynn and Enochs (2011). Here we focus primarily on the corallivores of the eastern Pacific, including literature on pan-Pacific species from other regions when relevant.

The corallivore feeding strategy is especially prominent in the eastern Pacific, where the proportion of described corallivorous species (from Rotjan and Lewis 2008) to known coral species is much higher than all other reef regions (Fig. 10.1). While a more species rich corallivorous fauna exists in other areas (e.g., Indo-Asian Archipelago, 73 spp.), these regions also contain a greater number of coral species. Therefore, the biogeography of corallivores and corals in the eastern Pacific is unique, in that species feeding on corals have been favored, while corals themselves have not. This trend may be influenced by the exceptionally depauperate nature of the eastern Pacific coral fauna. It is of note, however, that the eastern Pacific contains more corallivorous species than nine of the 16 reef regions from which data are presented, suggesting that this pattern is not due to the lower numbers of coral species alone. We hypothesize that the prominence of fast growing (Guzmán and Cortés 1989a) nutritionally rich *Pocillopora* spp. (Stimson 1990), that contribute importantly to the coral communities and reefs in this region, have conferred a competitive advantage to species utilizing corallivory as a feeding method. Whereas various hydrographic (e.g., upwelling) and climatic (ENSO) phases have limited the diversity of corals in the eastern tropical Pacific, the persistence, and in some cases proliferation, of *Pocillopora* have provided an abundant food source for corallivores and contributed to their prevalence.

Given the richness of corallivore species, the remote, understudied nature of many eastern Pacific coral reef sites,

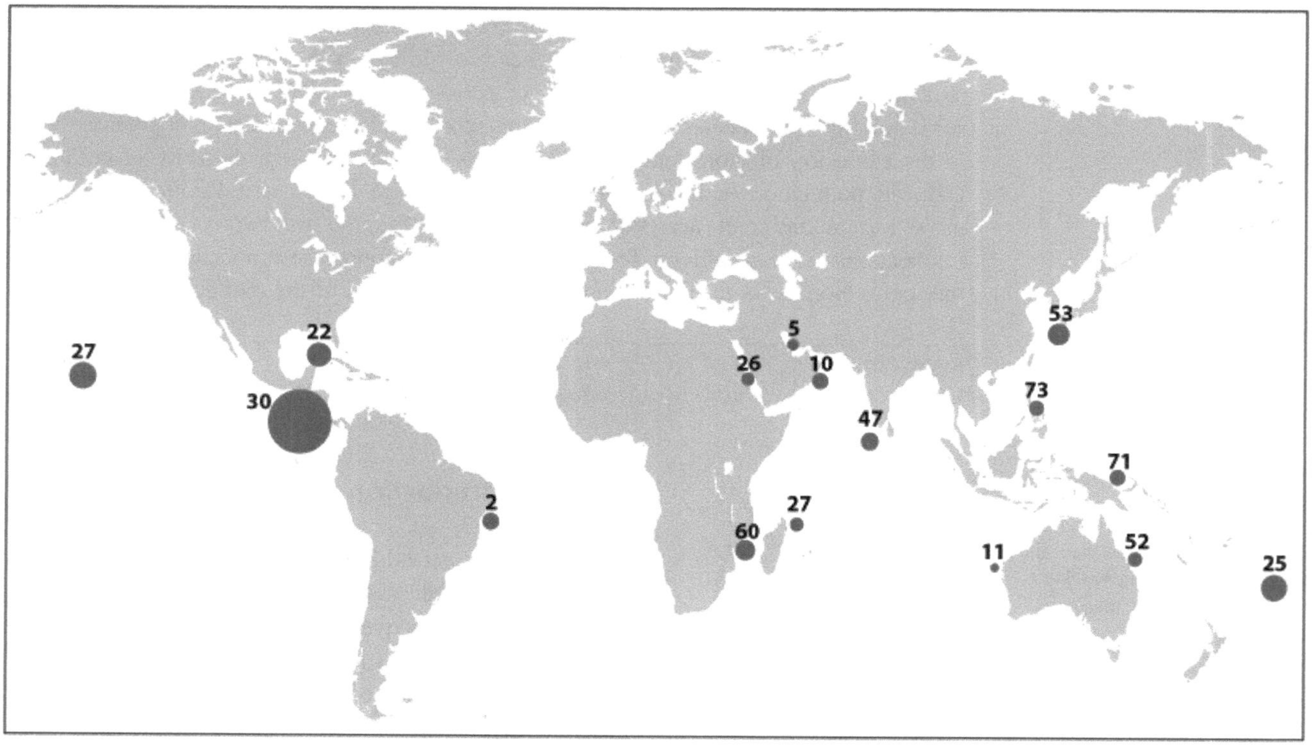

Fig. 10.1 Prominence of corallivore feeding strategy among global coral reef regions. Numbers are known corallivore species from Rotjan and Lewis (2008). Circle areas represent the number of corallivore species divided by the number of coral species in each region (Veron 1995). Note that this review (Table 10.1) includes several corallivore species not included in the analysis of Rotjan and Lewis. These numbers would further inflate this ratio yet are not included here due to the bias of our study towards the eastern Pacific fauna

and the cryptic behaviors of many small inconspicuous corallivorous taxa, it is unsurprising that information as to the diversity and distribution of corallivores within the eastern Pacific are incomplete. Herein an attempt is made to compile the disparate sources containing species distributions throughout the 14 eastern Pacific biogeographic regions recognized by Glynn and Ault (2000), as well as Rapa Nui (Table 10.1). From this compilation, it is apparent that Guatemala, Nicaragua, and Rapa Nui are in need of further study, while the Gulf of California, Panama, and the Galápagos Islands have received a greater amount of attention in the literature. Fishes account for 25 of the 47 corallivorous species presently known from the eastern Pacific, with the remaining invertebrate species divided among Arthropoda, Echinodermata, and Mollusca. The known distribution of fishes throughout the eastern Pacific is therefore likely more complete than for the invertebrates. In contrast to other reef regions, currently no corallivorous annelids are known to exist in the eastern Pacific.

Study of corallivores in the eastern Pacific likely began with the conspicuous and often destructive crown-of-thorns sea star (*Acanthaster planci*). A reference and photograph of this sea star exists in the popular American author John Steinbeck's "Sea of Cortez: A Leisurely Journal of Travel and Research" published with Edward Ricketts in 1941. The genus was subsequently reviewed by Madsen (1955) and Caso (1962), who focussed on specimens from the eastern Pacific (Mexico). These papers were followed by Chesher's (1969) review of the corallivorous, and potentially destructive nature of *Acanthaster* in Guam and Palau. Following this seminal work, the 1970s saw an explosion of interest into the ecology and corallivorous nature of *Acanthaster*, with many notable papers focusing on populations within the eastern Pacific (Dana and Wolfson 1970; Glynn et al. 1972; Porter 1972; Glynn 1973, 1974). Investigation of *Acanthaster* has steadily continued since and is discussed in more detail below. Since the 1970s, reef ecologists have expanded their scope to include numerous other invertebrate and fish corallivores and much of this work has been conducted on eastern Pacific reef systems (e.g., Robertson 1970; Glynn et al. 1972; Glynn 1983a; Gilchrist 1985; Reyes-Bonilla and Calderon-Aguilera 1999).

In this chapter we first begin with a review of the biology and ecology of the more conspicuous and well-studied corallivorous species, including numerous families of fishes as well as the invertebrates *Acanthaster planci* and *Eucidaris galapagensis*. We follow with a review of cryptic and lesser known corallivores, including species within the phyla Mollusca and Arthropoda. Finally, we address the various ecological roles of corallivores, highlighting the numerous positive influences that corallivory can have on corals and coral reef ecosystems in the eastern Pacific.

10.2 Conspicuous Corallivores

10.2.1 *Acanthaster planci*

Acanthaster planci is remarkably well adapted to feed on the tissues of all growth forms of virtually all zooxanthellate coral species (Moran 1986; Birkeland and Lucas 1990). Although *Acanthaster* was collected in the eastern Pacific as early as the Nineteenth Century (Caso 1962), its ecological importance did not become realized until the early 1970s, following Chesher's (1969) alarming report of coral devastation in the western Pacific. The sea star's feeding behavior and effects on corals in the Gulf of California (Mexico) were first reported by Dana and Wolfson (1970), and Barham et al. (1973), and in Panama by Glynn (1972) and Porter (1972).

In early systematic studies, two taxa of *Acanthaster* were recognized in the eastern Pacific, *Acanthaster ellisii* (Gray) and the subspecies *Acanthaster ellisii pseudoplanci* (Caso). On the basis of gross morphology and meristic data, Glynn (1974) concluded that eastern Pacific populations of *Acanthaster* should be assigned to the species *planci*, contrary to the studies of Madsen (1955), Caso (1962) and Barham et al. (1973). Recent molecular evidence supports the occurrence of *A. planci* in the eastern tropical Pacific, and demonstrates significant genetic divergence of *Acanthaster* within the Indo-Pacific region. Eastern Pacific *A. planci* has been shown to belong to one of four clades broadly distributed across the central and western Pacific Ocean (Vogler et al. 2008). *Acanthaster* in the three remaining clades demonstrates distinct distributions in the Indian Ocean and Red Sea.

Relatively high numbers of *Acanthaster* have been observed in the lower half of the Gulf of California, in the Gulf of Chiriquí in western Panama, at Costa Rican mainland sites, and also at Mexico's Revillagigedo Islands. *Acanthaster* has been reported to be rare at the offshore island sites of Clipperton (Glynn et al. 1996), Malpelo Island (Narváez and Zapata 2010), and the Galápagos Islands (Hickman 1998) with only one or a few individuals reported. A new distributional record was made recently of two individuals on coral reefs at Gorgona Island, southern Colombia (M.M. Palacios, pers. comm.). This extends the southern occurrence of *Acanthaster* on the eastern Pacific continental shelf by five degrees latitude, from 8°N in Panama to 3°N in Colombia (Glynn 1974). The absence of *Acanthaster* from eastern Pacific upwelling centers (Glynn 1974) is likely due to low sea temperatures that inhibit larval development (Henderson and Lucas 1971; Yamaguchi 1973). Larval developmental stages are also sensitive to small changes in salinity (Henderson and Lucas 1971; Lucas 1973) and this may limit the sea star's presence in other

Table 10.1 Recorded distribution of corallivores in the eastern tropical Pacific

	GOC	MXM	REV	CLP	GUA	ELS	NIC	CRC	COC	PAN	COL	MAL	ECD	GAL	RPN	Sources
Arthropoda																
Alpheus lottini	1	1	1	1				1		1	1	1		1	1	1,2,3,4,5,6,7
Aniculus elegans	1	1				1		1		1			1	1		1,3,5,6,7,8,9,10,11
Calcinus obscurus						1	1	1		1	1		1	1		1,4,8,10
Trapezia corralina	1			1						1				1		1,3,4,7,12
T. cymodoce	1		1							1	1				1	1,4
T. digitalis	1	1	1	1						1	1	1	1	1		1,2,3,4,5,7,12
T. ferruginea	1	1	1	1					1	1		1	1	1	1	1,2,3,4,5,7,12
Trizopagurus magnificus	1	1				1		1		1			1	1		1,3,4,5,7,8,9,10,13
Echinodermata																
Acanthaster planci	1		1	1				1	1	1	1	1		1		1,2,3,4,7,8,9,11,14,15,16,17,18,19
Diadema mexicanum	1	1		1		1	1	1	1	1	1	1		1		1,2,4,8,9,14,15,18,19,20,21
Eucidaris galapagensis	1	1	1	1		1		1	1	1	1	1		1		1,2,3,4,7,8,9,11,15,16,18,19,20
Nidorellia armata	1	1				1	1	1		1	1	1	1	1		1,2,3,4,7,11,18,19,20
Pentaceraster cumingi	1	1						1		1			1	1		1,2,4,8,18,19,20
Pharia pyramidata	1	1	1			1		1		1	1	1	1	1		1,2,3,4,7,11,18,19,20
Mollusca																
Calliostoma mcleani	1							1		1	1					2,4,11,22
Coralliophila monodonta	1	1		1			1	1		1	1			1		2,4,9,13,14,23,24
C. violacea			1	1					1					1	1	1,14,22,23,25
Jenneria pustulata	1	1		1		1		1		1	1		1	1		1,2,3,4,7,8,9,11,13,14,22,23,24
Latiaxis hindsii	1									1		1	1	1		2,3,4,13,22
Muricopsis zeteki	1	1						1	1	1	1		1	1		1,2,3,4,13,22,24
Phestilla spp.[a]	1			1				1	1	1						2,4,8,9,11
Reliquiaecava robillardi				1												14

(continued)

Table 10.1 (continued)

	GOC	MXM	REV	CLP	GUA	ELS	NIC	CRC	COC	PAN	COL	MAL	ECD	GAL	RPN	Sources
Chordata																
Balistidae																
Balistes polylepis	1	1	1	1	1	1	1	1	1	1	1	1	1	1		2,4,26,27,28,29
Melichthys niger	1		1	1	1	1	1	1	1	1	1	1		1		2,4,14,26,27,28,29
Sufflamen verres	1	1	1	1	1	1	1	1	1	1	1	1	1	1		2,3,4,8,9,26,27,28,29
Chaetodontidae																
Chaetodon humeralis	1	1	1		1	1	1	1	1	1	1	1	1	1		2,4,8,26,27,28,29
Forcipiger flavissimus	1		1	1					1			1	1	1	1	2,4,26,27,28,29
Johnrandallia nigrirostris	1	1	1	1	1	1	1	1	1	1	1	1	1	1		2,4,8,26,27,28,29
Prognathodes falcifer	1		1											1		2,4,28,29
Monacanthidae																
Aluterus scriptus	1	1	1	1	1	1	1	1	1	1	1	1	1	1	1	2,4,8,26,27,28,29
Cantherhines dumerilii	1	1	1	1					1	1	1	1		1	1	2,4,8,25,26,27,28,29
Pomacanthidae																
Holacanthus passer	1	1	1		1	1	1	1	1	1	1	1	1	1		2,4,8,26,27,28,29
Pomacanthus zonipectus	1	1	1	1	1	1	1	1	1	1	1	1	1	1		2,4,8,26,27,28,29
Pomacentridae																
Microspathodon dorsalis	1	1	1		1	1	1	1	1	1	1	1	1	1		2,4,9,26,27,28,29
Stegastes acapulcoensis	1	1	1		1	1	1	1	1	1	1	1	1	1		2,4,8,9,26,27,28,29
S. arcifrons								1	1		1	1	1	1		2,4,26,28,29
Scaridae																
Calotomus carolinus	1		1						1					1		2,4,27,28,29
Scarus ghobban	1	1			1	1	1	1	1	1	1	1	1	1		2,3,4,8,11,26,27,28,29
S. perrico	1	1	1		1	1	1	1	1	1	1	1	1	1		2,3,4,11,26,27,28,29
S. rubroviolaceus	1	1	1	1				1	1	1	1	1	1	1		2,3,4,8,26,27,28,29
Tetraodontidae																

(continued)

Table 10.1 (continued)

	GOC	MXM	REV	CLP	GUA	ELS	NIC	CRC	COC	PAN	COL	MAL	ECD	GAL	RPN	Sources
Arothron hispidus	1	1	1	1	1	1	1	1	1	1	1	1	1	1		2,3,4,7,8,9,11,26,27,28,29
A. meleagris	1	1	1	1	1	1	1	1	1	1	1	1	1	1	1	2,3,4,7,8,9,11,14,16,25,26,27,28,29
Canthigaster amboinensis									1	1				1		2,11,28,29
C. janthinoptera										1	1			1		2,4,28,29
C. punctatissima	1	1	1	1		1	1	1	1	1	1	1	1	1		2,4,26,27,28,29
C. valentini									1					1		2,28,29
Zanclidae																
Zanclus cornutus	1	1	1	1	1	1	1	1	1	1	1	1	1	1	1	2,4, 26,27,28,29
Total	40	30	28	25	15	24	20	33	30	40	33	29	29	43	9	

[a]Two species are known to graze on corals in the eastern Pacific—*Phestilla lugubris* and *Phestilla melanobrachia*

Reef Regions: GOC—Gulf of California, including all islands, MXM—Mexican mainland south of Mazatlán, REV—Revillagigedo Islands, CLP—Clipperton Atoll, GUA—Guatemala, ELS—El Salvador, NIC—Nicaragua, CRC—Costa Rican mainland, COC—Cocos Island, PAN—Panamanian mainland, COL—Colombian mainland, MAL—Malpelo Island, ECD—Ecuadorian mainland, GAL—Galápagos Islands, RPN—Rapa Nui

Sources (1) NMNH Invertebrate Collections http://collections.nmnh.si.edu/search/iz/; (2) OBIS http://www.iobis.org; (3) Glynn (1983a); (4) GBIF http://data.gbif.org/occurrences/; (5) Hickman and Zimmerman (2000); (6) Wicksten and Hendrickx (2003); (7) Glynn (1983b); (8) Glynn (2008); (9) Guzmán (1988a); (10) Moran (1984); (11) Glynn (1982); (12) Garth (1974); (13) Guzmán (1988b); (14) Glynn et al. (1996); (15) Lessios (2005); (16) Reyes-Bonilla and Calderón-Aguilera (1999); (17) Narváez and Zapata (2010); (18) Alvarado et al. (2010a); (19) Hickman (1998); (20) Alvarado and Fernández (2005); (21) Alvarado et al. (2010b); (22) Hickman and Finet (1999); (23) Robertson (1970); (24) Zamorano et al. (2008); (25) Glynn et al. (2003); (26) NMNH Vertebrate Collections Fishes http://collections.mnh.si.edu/search/fishes/; (27) Allen and Robertson (1994); (28) Froese and Pauly (2011) http://www.fishbase.org; (29) Robertson (2009)

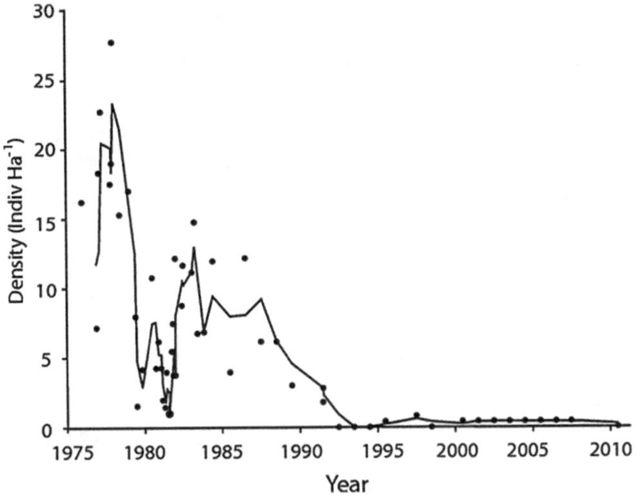

Fig. 10.2 Population density of *Acanthaster planci* at Uva Island coral reef, 1976–2010. Moving average curve is based on two census points

eastern Pacific areas. High abundances of 27 individuals (1 ind per 117 m^2) were reported for one site in the Gulf of California (Dana and Wolfson 1970) and 36 individuals (1 ind per 143 m^2) at the Uva Island reef in Panama. Comparing feeding rates and coral growth at sea star population densities of 26–65 ind ha^{-1}, Glynn (1973) concluded that the Uva reef in Panama could still exhibit positive growth. At a simulated outbreak density of 260 ind ha^{-1}, however, the Uva coral reef would cease to grow. The maximum numbers of *Acanthaster* observed at the ~2.5 ha Uva reef, 36 individuals in September 1972, has demonstrated an irregular decline to only 1–2 individuals over a 22 year period (Fong and Glynn 1998). Updated censuses (n = 12) at the Uva reef from 1995 to 2010 revealed a mean abundance of 0.9 individuals ha^{-1}(Fig. 10.2).

Observations on the feeding behavior of *Acanthaster* have demonstrated distinct prey preferences. Since *Acanthaster* was often found feeding on *Pocillopora* spp., fast-growing and competitively dominant eastern Pacific taxa, Porter (1972, 1974) proposed that the selective removal of these species would increase coral community diversity. Glynn's (1974, 1976) observations and field experiments showed that *Acanthaster* preyed predominantly on small pocilloporid colonies and the dislocated broken branches of larger colonies. These small and fragmented coral prey were defended from *Acanthaster* attacks less intensely by symbiotic crustacean guards than were large intact colonies. When *Acanthaster* was offered a choice between pocilloporid and certain non-pocilloporid species the latter were preferred, especially if the former were well defended by their crustacean guards. Examples of species with a high electivity index were *Gardineroseris planulata* and *Pavona* spp. (Glynn 1974). In the eastern Pacific these guards are trapeziid crabs, four species in the genus *Trapezia* (Castro 1982, 1996), and the alpheid shrimp *Alpheus lottini* (Kim and Abele 1988). Thus, Glynn (1974, 1976) argued that the selective removal of relatively rare coral species and their replacement by *Pocillopora* spp. would tend to depress coral community diversity.

Other effects observed in Panama in areas with and without *Acanthaster* were colony size differences and the abundances of preferred prey (Glynn 1987). Six coral species demonstrated significantly smaller colony sizes where *Acanthaster* was present than in habitats without the corallivore. Fong and Glynn (1998) also demonstrated shrinkage of the size classes of *Gardineroseris planulata*, a preferred prey species, at increasing numbers of *Acanthaster*. Additionally, in four of five species the frequency of dead colonies in habitats exposed to *Acanthaster* predation was significantly higher than at sites without the corallivore. A final example of how pocilloporid crustacean guards can affect coral longevity was observed during the 1982–83 El Niño disturbance at the Uva Island reef in Panama. Elevated sea temperatures caused the bleaching and mortality of continuous stands of crustacean guard-defended *Pocillopora* surrounding massive colonies of *Gardineroseris*. Before the bleaching and death of *Pocillopora* spp. these corals formed a protective barrier around *Gardineroseris*, preventing *Acanthaster* from feeding on this species. With the elimination of the protective barrier, *Acanthaster* then gained access to the refuge and began preying on *Gardineroseris*.

Because of concerns over the relatively recent *Acanthaster* outbreaks in the Indo-Pacific region, particularly beginning in the 1960s, some workers have searched for evidence of earlier sea star abundances and have proposed that such outbreaks are not unprecedented. Sediment coring studies have found *Acanthaster* skeletal remains concentrated at particular depth levels in reef sediments of 3000 (Frankel 1977, 1978) to 7000 year BP (Walbran et al. 1989a, b), suggesting cyclic abundances and probable past sea star outbreaks similar to those of today. The evidence supporting this work has been criticized, however, with potential difficulties arising from bioturbation and size-dependent preservation that could distort the sedimentary record (Moran et al. 1986). In spite of this criticism, Birkeland and Lucas (1990) concluded that the occurrence of skeletal elements in sediments probably at least indicates the presence of *Acanthaster* in the past.

To investigate the possibility of the occurrence of *Acanthaster* in the eastern Pacific in the recent past, sediment samples at Uva Island were cored and analyzed quantitatively. These samples were collected from fore reef habitats where *Acanthaster* was foraging. Approximately 1 kg samples of sediment were collected and analyzed at ~30 cm depth intervals from four sites. Each sample was examined microscopically with reference to a known collection of *Acanthaster* skeletal elements. Push-core samples revealed the presence of skeletal elements throughout the sediment pile to a maximum depth of 260–270 cm. Carbon-14 dating of coral and crustose

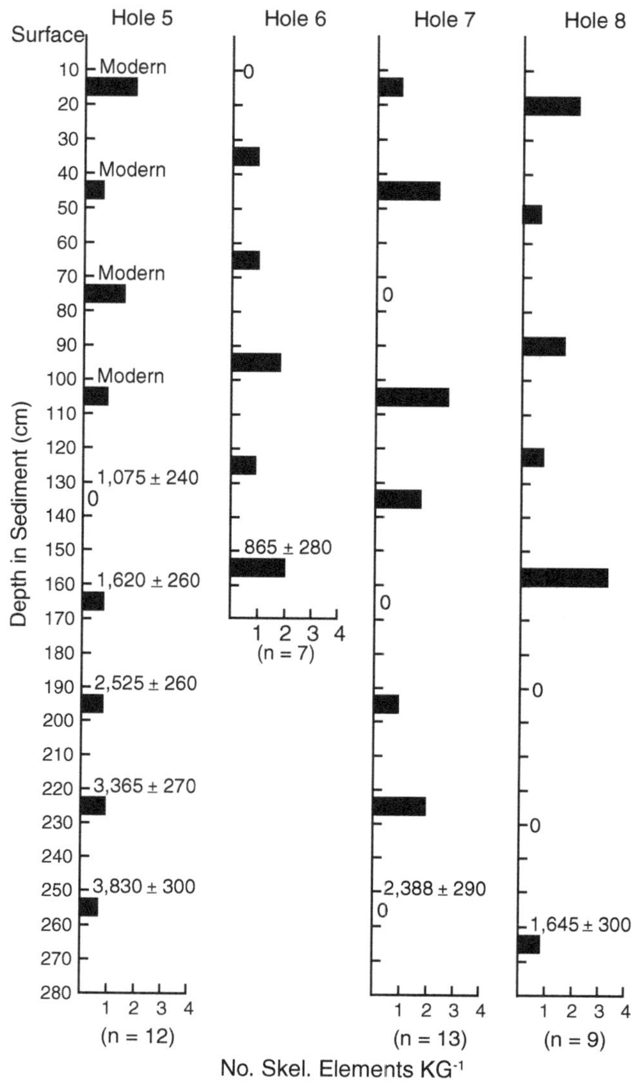

Fig. 10.3 Recovery of *Acanthaster planci* skeletal elements from push core samples collected in seaward reef-front zone sediments, 3 m isobath, Uva Island reef, 8 August 1985. Maximum penetration was 260–270 cm with samples collected at approximately 30 cm intervals. Approximately 1 kg samples, passed through 0.461 mm mesh sieve were sorted under 5 X magnification and compared with a reference set of known *A. planci* skeletal elements. Radiocarbon dating performed at the National Ocean Sciences Accelerator Mass Spectrometry Facility (NOSAMS), Woods Hole Oceanographic Institution (Druffel 1995). Radiocarbon dates were calibrated using Calib 6.0 software (Stuiver and Reimer 1993; Stuiver et al. 2005)

coralline algae at the same sediment horizons with *Acanthaster* skeletal ossicles suggests a long-term occurrence of the sea star on the Uva reef of nearly 4000 year BP (Fig. 10.3).

A comparison of the numerical densities of skeletal elements in surface sediment samples from Panama, with relatively low population densities of *Acanthaster*, and from Oman where sea star outbreaks occurred in 1978–80 (Glynn 1987, 1993) revealed notable differences. Skeletal elements were absent from over one-half of samples (~58 %) from Panama whereas only a single sample from Oman (~11 %, 1 of 9 total) did not contain *Acanthaster* skeletal remains (Table 10.2). In addition, the mean number of skeletal elements per sample in Panama was only about one-tenth of that in Oman. The high skeletal densities in Oman are in line with those reported by Walbran et al. (1989a, b) for the John Brewer reef in Australia, which experienced two major episodes of *Acanthaster* predation since 1962. These results provide indirect evidence supporting low, but sustained population densities of *Acanthaster* on the Uva Island reef.

Table 10.2 *Acanthaster planci* skeletal elements recovered from ~1 kg surface samples collected at Uva Island coral reef, Gulf of Chiriquí, Panama, and Gulf of Oman, Sultanate of Oman

Reef site	Dates collected	Number of samples	Samples without elements	Average elements per sample
Panama				
Uva Island	Jul'84; Feb, Aug '85	40	23 (57.5 %)	0.65 ± 0.14
Oman				
Daymaniyat Iss	Oct '85	4	1 (25.0 %)	11.5
Banda Jissah	Oct '85	5	0 (0 %)	2.8
		9	1 (11.1 %)	6.7 ± 2.9

10.2.2 *Eucidaris galapagensis*

The first eastern Pacific record of the sea urchin *Eucidaris* feeding on live coral tissues was made in 1975 in the Galápagos Islands by Glynn et al. (1979). *Eucidaris galapagensis* is a facultative corallivore, and was observed grazing on live *Pocillopora* spp. and massive species of *Porites* and *Pavona*. Most of its feeding activities, however, involve the grazing of micro-filamentous algae, both epibenthic and endolithic, on dead coral skeletons and basalt rock. It is an effective grazer, capable of rapid bioerosion when present in large numbers (Glynn et al. 1979; Glynn 1994; Reaka-Kudla et al. 1996; Glynn et al. 2015). *Eucidaris* efficiently abrades coral and rock substrates employing Aristotle's lantern, a complex feeding apparatus of plates, muscles and ligaments that support five sharp calcareous teeth.

Recognizing the correct identity of *Eucidaris* in the eastern Pacific has created some confusion. For many years, the Galápagos species of *Eucidaris* was assigned to *thouarsii*, which occurs commonly along mainland shores from the Gulf of California to southern Ecuador. *Eucidaris thouarsii*, unlike *E. galapagensis*, is relatively small, a herbivore/omnivore in feeding habit, and seldom attains high abundances. In a phylogeographic study, Lessios et al. (1999) found that populations of *Eucidaris* inhabiting off-shore islands (Galápagos Islands, Isla del Coco and

Clipperton Atoll) comprised a clade sufficiently distinct from mainland populations to justify separation and assignment to *E. galapagensis*, a resurrected species from the older taxonomic literature. Although Kroh (2010) has reduced *E. galapagensis* to subspecies rank, namely *E. thouarsii galapagensis*, we prefer to retain the full species binomen.

Eucidaris galapagensis can reach high abundances in Galápagos coral communities with population densities of 20–30 ind m^{-2} not uncommon, and maximum densities as high as 50–90 ind m^{-2} reported (Glynn et al. 1979; Glynn and Wellington 1983; Glynn 1988). High densities were observed at the Devil's Crown patch reef (Floreana Island) in areas of moderate to high coral cover, whether the coral was alive or dead. *Eucidaris* is a large species, with test diameters of 5 to 6 cm in adults. At the Devil's Crown study reef, about 36 % of *Eucidaris* were feeding on live corals on the reef crest and 52 % at the reef edge. Pocilloporid skeletal grains made up 30–40 % and coralline algae 40–50 % by weight of the gut contents of sea urchins feeding on the reef.

Recent studies in the Galápagos Islands suggest possible negative interactions between El Niño disturbances and anthropogenic activities that affect sea urchin abundances. A strong relationship between *E. galapagensis* abundances (and at least two other sea urchin species) and fishing pressure has been observed in the southern archipelago. Sea urchin numbers are generally higher where their natural predators (e.g., spiny and slipper lobsters, and hogfish) have been over-harvested (Sonnenholzner et al. 2009; Edgar et al. 2010) or in rubble refugia (Dee et al. 2012). This has resulted in intense sea urchin grazing and a shift from macroalgal and coral communities to "urchin barrens" dominated by crustose coralline algae. This phase shift has been implicated in causing declines and local extinctions of species associated with the more biologically-structured habitats (Edgar et al. 2010). If sea urchin overgrazing continues it is likely that coral recruitment would be severely limited, resulting in a prolonged delay of coral community recovery.

10.2.3 Other Echinoderms

Three additional species of echinoderms, all asteroids, are known to prey on corals in the eastern Pacific. *Pharia pyramidata* everts its stomach to feed on pocilloporid corals, in a manner similar to *Acanthaster* (Dana and Wolfson 1970). This feeding behavior has been observed in the Gulf of California and mainland Ecuador, but not in the Galápagos Islands (Dana and Wolfson 1970; Glynn 1983a). By contrast, *Nidorellia armata* is known to feed on corals in the Galápagos, but not in Panama where it instead subsists on a diet of sponges and sometimes algae (Chesher 1972; Glynn 1983a). In the Galápagos, *N. armata* consumes the tissues of *Pavona clavus* (0.5–1 m diam.), on which it leaves small (2–5 cm) feeding scars (Glynn 1983a). Experiments on the exclusion of the above species from damselfish territories in the Galápagos suggest that the order in which they are preferentially ejected by damselfish correlates with their corallivory potential. The corallivorous sea urchin *Eucidaris galapagensis* was removed most actively, followed by *N. armata*, and finally *P. pyramidata*, which is not known to exhibit corallivorous tendencies in the Galápagos (Glynn 1983a). A third species of asteroid, *Pentaceraster cumingi*, has been recorded in large groups among fungiid coral communities in the Galápagos (Glynn 2003). While these sea stars feed on fungiids such as *Diaseris distorta*, they more frequently consume algae, sponges, sea urchins, and even colonies of *Psammocora stellata* (Glynn 2003). It should be noted that none of the aforementioned asteroid species, with the exception of *Acanthaster*, is considered to be an obligate corallivore. This dietary diversity, in conjunction with relatively low population densities, usually limits the destructive potential of these sea stars on eastern Pacific coral reefs.

10.2.4 Fishes

No fewer than 25 fish species have been observed to prey consistently, or at least incidentally, on live corals in the eastern Pacific (Table 10.1). Only a few of these species are obligate corallivores with corals making up the bulk of their food intake. Most are facultative corallivores, feeding predominantly on algae, other non-scleractinian invertebrates, and only incidentally on corals. Some species appear to concentrate more on coral grazing as juveniles than as adults. Feeding behavior varies greatly among fish taxa, from browsers that consume live coral tissue only, to scrapers that remove live tissues and some skeletal carbonate, and excavators that remove live tissues as well as a major portion of the skeleton (Rotjan and Lewis 2008). Remarkably, as early as the 1920s William Beebe, an intrepid pioneering naturalist *cum* marine biologist, performed observations on the feeding behavior of eastern Pacific reef fishes. This was accomplished by hardhat diving in Darwin Bay at Tower Island, Galápagos Islands (Beebe 1924, 1926). Beebe coined the term "coral grazers" in an early attempt to relate mouth structure and dentition to coral feeding, herbivory and invertivory in several fish families.

10.2.4.1 Tetraodontidae

Several species of puffers (Tetraodontidae) and butterflyfishes (Chaetodontidae) are often the most conspicuous corallivores in eastern Pacific coral communities. One of the most important corallivores in terms of biomass consumed is the guineafowl puffer, *Arothron meleagris* (Fig. 10.4). This species often concentrates its feeding activities on the

Fig. 10.4 *Arothron meleagris* feeding on *Pocillopora damicornis* at Uva Island reef, Panamá (1 m depth, March 2004, photo by T.B. Smith)

branch-tips of *Pocillopora* spp. and *Psammocora* spp., biting off 1 to 2 cm-long fragments with its sharp beak-like teeth. Analysis of the stomach contents of 14 individuals of *A. meleagris* in Panama revealed that the branch tips of *Pocillopora* spp. made up 91.3 % of the dry mass (Glynn et al. 1972). It also feeds on small nodular species such as *Porites panamensis*, and massive corals in the genera *Porites*, *Pavona*, and *Gardineroseris*. Grazing is usually concentrated along the margins, on nodules or ridges of large massive species where sufficient leverage can be achieved. Population densities are reviewed in Palacios et al. (2014) and can range from 0 to 320 ind ha^{-1} in Costa Rica, Panama, Colombia and the Galápagos Islands (Guzmán and Robertson 1989; Guzmán and López 1991).

Recent work by Palacios et al. (2014) at Gorgona Island, Colombia has shown that the feeding activities of *Arothron meleagris* can result in the removal of nearly 16 % of the annual calcium carbonate production of pocilloporid coral prey. These authors observed that the majority of predation events were sublethal and that colonies that were preyed upon had calcification rates comparable to those that were shielded from predators. Exposure of *Pocillopora* spp. to *A. meleagris* bites, however, significantly reduced linear extension and over time resulted in a change in colony morphology that favored a more robust and "stubby" form (see Chap. 12, Alvarado et al.).

Although *Arothron meleagris* consumes large amounts of coral, it may not contribute directly to higher trophic level predators because of its toxicity and apparent avoidance by piscivores. It is possible that its distinctive color phases have an aposomatic function, deterring potential predators. Nuñez-Vázquez et al. (2000) have found relatively high concentrations of tetrodotoxin in the tissues of *A. meleagris* in the Gulf of California. The pufferfish deposits large accumulations of skeletal fragments in its feces, and this likely attracts micro-scavengers that would play a role in nutrient recycling. Another potentially important ecological role could be served in the dispersal of zooxanthellae. Parker (1984) found that the zooxanthellae of *Aiptasia* ingested by *A. meleagris* in Hawaii remained viable after defecation, and could infect aposymbiotic *Aiptasia*. Whether this type of transmittal can occur in eastern Pacific puffers feeding on zooxanthellate scleractinian corals needs to be investigated.

Arothron hispidus is a facultative corallivore, feeding commonly on other non-scleractinian taxa such as sponges, hydroids, polychaete worms, bryozoans, crustaceans and molluscs (Glynn et al. 1972). When feeding on corals it often concentrates on *Pocillopora* spp., but also has been observed feeding on *Porites lobata* in the Galápagos Islands (Isabela Island, F. Rivera, pers. comm.). Its abundance at Caño Island (Costa Rica), ~8 ind ha^{-1}, is similar to *A. meleagris* at that site, however, it varies greatly in numbers in other eastern Pacific areas (Guzmán 1988b).

At least three species of small monogeneric pufferfishes have been observed in eastern Pacific coral communities, however, they all have varied diets including, in addition to coral polyps, filamentous algae, sponges, polychaete worms, and crustaceans. *Canthigaster punctatissima* is endemic to the eastern Pacific, occurring in coral communities from the Gulf of California to the Galápagos Islands. Two additional species, *Canthigaster amboinensis* and *Canthigaster janthinoptera*, occur sporadically in the eastern Pacific, perhaps a result of dispersal from the central Pacific region during ENSO events.

10.2.4.2 Chaetodontidae

Of 128 fish corallivores listed by Cole et al. (2008), no fewer than 54 % belong to the family Chaetodontidae, commonly known as butterflyfishes. Butterflyfishes have small protractile mouths with bristle-like dentition, which are well suited for grazing on corals (Motta 1988), but only 52 % of butterflyfish species do actually feed on corals (Cole and Pratchett 2014). In all, there are 24 species of butterflyfishes that are obligate hard coral-feeders (Pratchett 2014), which feed almost exclusively (>80 %) on scleractinian corals, and a further 65 species that only occasionally graze on corals. Obligate coral feeding butterflyfishes are numerically dominant on many Indo-Pacific reefs (Cole and Pratchett 2014), but are often under-represented at marginal reef locations (Pratchett et al. 2013). Similarly, there are no obligate corallivores in the eastern Pacific, and even facultative coral-feeding species are scarce; only three species of butterflyfishes are routinely found in the eastern Pacific (*Chaetodon humeralis*, *Johnrandallia nigrirostris*, and *Forcipiger flavissimus*) feed on corals, but all these species feed predominantly on non-coral prey including algae and mobile benthic invertebrates. In Panama, *C. humeralis* and *J. nigrirostris* are often observed nipping on *Pocillopora* spp. (Glynn 2008). *Johnrandallia nigrirostris* also occupies 'cleaning stations' where it visits large fish clients to feed on ectoparasites, tissues and mucus. At least six other facultative coral-feeding butterflyfishes (*Chaetodon auriga*, *Chaetodon kleini*, *Chaetodon lunula*) occur sporadically in the eastern Pacific, usually during or shortly following El Niño events (Allen and Robertson 1994).

While there are marked interspecific differences in dietary preferences, most coral-feeding butterflyfishes feed preferentially on *Acropora* and/or *Pocillopora* corals (Pratchett 2014). In locations where butterflyfishes are abundant, strong selective feeding may exert a strong influence on the abundance and fitness of preferred corals (Cox 1986; Cole et al. 2012). In the eastern Pacific, however, coral-feeding butterflyfishes are relatively rare, and also feed on a diversity of different prey. In the Gulf of Chiriquí (Panama) two chaetodontid species are ranked among the top 20 in abundance, but at densities much lower than in the western Pacific; *Chaetodon humeralis* and *Johnrandallia nigrirostris* occur at densities of only 0.4 and 0.9 ind per 1000 m^2 respectively (Dominici-Arosemena and Wolff 2006; Glynn et al. 2014). Considering the broad diets (i.e. extending well beyond corals) and low abundances of eastern Pacific butterflyfishes, their impacts on coral communities are expected to be negligible. There is, however, a need for explicit research on the diet composition, and especially relative consumption of different coral species, for coral-feeding butterflyfishes in the eastern Pacific.

10.2.4.3 Scaridae

Four species of parrotfishes have been implicated in excavating corals, namely *Scarus ghobban*, *Scarus perrico*, *Scarus rubroviolaceous* and *Calotomus carolinus*. These parrotfishes feed on various species of *Pocillopora*, *Pavona* and *Porites* (Table 10.1). Particularly conspicuous are the bite marks on massive corals after a feeding bout (Fig. 10.5). The bite marks

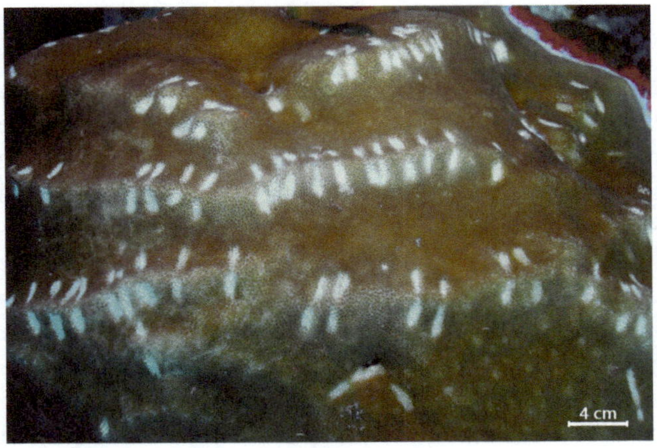

Fig. 10.5 Probable parrotfish bite marks on *Porites lobata*, Darwin Island reef, Galápagos Islands (13 m depth, 4 June 2012, photo by J.S. Feingold)

Fig. 10.6 Bite marks on *Porites lobata*, Darwin Island reef, Galápagos Islands (12 m depth, 3 June 2012, photo by V.W. Brandtneris). Star-shaped pattern is likely due to parrotfish targeting raised surfaces surrounding lithophagine bivalve bore holes

are typically 1–2 cm in length and a few mm deep. They are most often scattered over the surface of massive corals, a condition known as 'spot biting' (Bruckner et al. 2000) but may also target particular skeletal features (Fig. 10.6). On the Uva Island reef in Panama, schools of *S. rubroviolaceus* have been observed spot biting *Gardineroseris planulata*; this is a new corallivore/coral feeding record.

Rotjan and Lewis (2008) have argued that the negative effects of parrotfishes grazing on live corals have been underestimated, especially in recent decades with rapidly declining coral populations. This view was challenged by Mumby (2009) who concluded that parrotfishes on Caribbean coral reefs have a net positive effect by limiting algal growth that would interfere competitively with coral recruits. These ecological interactions are in need of further study in the eastern Pacific.

10.2.4.4 Balistidae

Four triggerfish species have been implicated in feeding on corals in the eastern Pacific (Glynn et al. 1972; Guzmán and Cortés 1989b). This number can be reduced to three triggerfish species since Glynn et al. (1972) mistakenly identified *Balistes polylepis* for *Pseudobalistes naufragium*; the latter was observed massively excavating *Pavona varians*. Actually, *P. naufragium* was breaking apart the coral to expose endolithic bivalves, the prey objects of this destructive behavior. *Porites* spp. and other massive corals with bivalves also are subject to significant breakage by *P. naufragium*, but without ingestion of live coral tissues (Guzmán and Cortés 1989b).

Branching and massive coral fragments contribute importantly to the gut contents of some individuals of *Sufflamen verres*, suggesting a concerted targeting of corals

(Glynn et al. 1972). The gut contents of other triggerfishes, however, reveal few coral fragments, perhaps a result of incidental ingestion while extracting coral-associated invertebrates. *Melichthys niger*, a tropical cosmopolite triggerfish, has been classified as a facultative corallivore on the basis of small coral fragments (mean < 1 % by volume) present in the guts of 17 individuals examined (Randall 1967). While this species has been observed feeding in small groups on large colonies of *Pavona gigantea* (>1.5 m high) at Gorgona Island, Colombia (M. Palacios pers. comm.), we are unaware of any other records of this feeding interaction from other regions in the eastern Pacific.

10.2.4.5 Monacanthidae

Two species of filefishes occasionally observed on eastern Pacific reefs are facultative corallivores: *Aluterus scriptus* and *Cantherhines dumerilii*. Randall (2005) noted that *A. scriptus* feeds on *Millepora*, and *C. dumerilii* on *Acropora, Pocillopora, Montipora* and *Leptoseris* in the South Pacific. Both species prey on a variety of taxa in addition to corals. Glynn (2008) indicated that *Pocillopora* spp. are consumed by these fishes in the eastern Pacific.

10.2.4.6 Pomacentridae

The high percentage of acroporid corals in the guts of some Indo-Pacific damselfishes, up to 96 %, suggests a dominantly corallivorous feeding mode (Rotjan and Lewis 2008). Three species that bite and kill corals, not listed by Rotjan and Lewis (2008), are *Stegastes acapulcoensis*, *Stegastes arcifrons*, and *Microspathodon dorsalis*. All are widely distributed across the eastern tropical Pacific (Wellington 1982; Guzmán 1988a; Guzmán and Cortés 1989b; Dominici-Arosemena and Wolf 2006; Glynn 2008; Table 10.1). In the Pearl Islands, Panama, Wellington (1982) observed persistent biting of *Pavona gigantea* by *S. acapulcoensis* at shallow depths, which resulted in extensive mortality and an influential factor in controlling coral zonation. It is not known if the damselfish gains a nutritional benefit from this behavior. Numerous circular bite marks on *P. lobata* (Fig. 10.7), adjacent to an algal lawn defended by *S. acapulcoensis*, are an example of the lesions caused by a browser and suggest directed coral feeding and a nutritional benefit. The damselfish is the suspected predator, but it was not observed biting the coral.

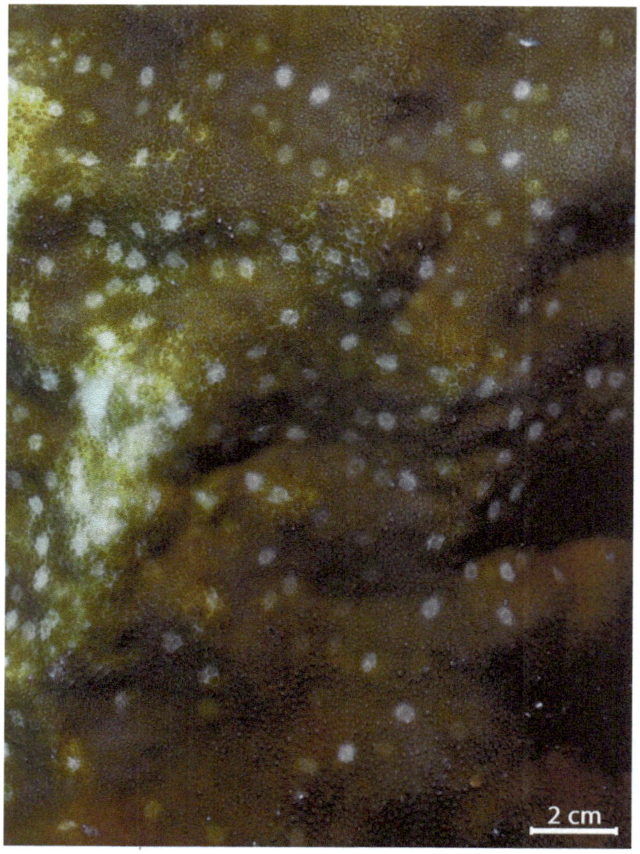

Fig. 10.7 Possible *Stegastes arcifrons* bite marks on *Porites lobata*, Darwin Island reef, Galápagos Islands (13 m depth, 3 June 2012, photo by J.S. Feingold). Circular marks 0.5–1.0 cm in diameter

10.2.4.7 Zanclidae, Pomacanthidae

Three species occasionally observed grazing on corals are the Moorish Idol *Zanclus cornutus* (Zanclidae) and two angelfishes, *Holacanthus passer* and *Pomacanthus zonipectus* (Pomacanthidae). *Zanclus cornutus* was recently observed browsing on *P. lobata* polyps at Darwin Island, Galápagos Islands (J.S. Feingold, pers. comm.). The fish left circular, 3–5 mm diameter lesions with no apparent damage to the underlying skeleton. McClanahan et al. (2005) observed *Zanclus canescens* and other fishes (Chaetodontidae, Tetraodontidae, Scaridae) preying on corals in the Indian Ocean after bleaching, concluding this would interefere with recovery. Coral browsing by angelfishes may be more common in juvenile than adult life history stages.

10.3 Cryptic Corallivores

Many of the species known to exhibit corallivorous feeding behaviors in the eastern Pacific have cryptic tendencies. The prevalence of this behavior is especially telling considering that these species remain hidden from easy observation and study, and are poorly known relative to epibenthic and nektonic species. The potential that numerous additional cryptic corallivores exist is therefore high; presently unknown due to minute, obscured, and often brief feeding interactions. Those eastern Pacific cryptic corallivores that are known are diverse, both taxonomically (Table 10.1) and in their feeding behaviors. Literature concerning these species is sparse, yet perhaps less so in the eastern Pacific relative to other reef regions. Below we discuss many of the more inconspicuous taxa belonging to the phyla Mollusca and Arthropoda.

10.3.1 *Coralliophila monodonta*

Several studies refer to the corallivorous activity of both *Quoyula madreporarum* (Glynn et al. 1972; Glynn 2004) and *Quoyula monodonta* (e.g., Guzmán 1988b) in the eastern Pacific. Robertson (1970), however, referenced the host-dependent morphological plasticity demonstrated by Maes (1967) and concluded that *Q. madreporarum* is synonymous with the earlier named taxon *Q. monodonta*. More recently, the genus *Quoyula* has been considered a junior synonym of *Coralliophila* and the correct name is therefore *Coralliophila monodonta* (Oliverio 2008).

Coralliophila monodonta is known from reefs throughout the Pacific and almost exclusively associates with corals in the family Pocilloporidae (e.g., Robertson 1970; Coles 1980; Black and Prince 1983). Demond (1957), however, reports several instances of these gastropods associating with both acroporid and poritid corals in Micronesia; this family-specific affinity, therefore, cannot be considered obligate.

In the eastern Pacific, *Coralliophila monodonta* generally occupies the sheltered bases of *Pocillopora* colonies (Guzmán 1988b), perhaps to escape from predatory hermit crabs, octopuses, and various invertivore fishes (Glynn 2004). These gastropods create small feeding scars on the coral surface immediately under their shells and remain relatively sedentary at these sites (Robertson 1970; Glynn et al. 1972). Activity, however, increases during the night and many individuals have been collected that were not associated with scars (Guzmán 1988b).

Densities of *C. monodonta* have been recorded as high as 29 individuals in a 1/8 m^2 quadrat on a *Pocillopora* reef at Chapera Island, Pearl Islands, Panama (Glynn et al. 1972). Guzmán (1986) reported a mean density of 10.6 *C. monodonta* per m^2 on a Caño Island reef, Costa Rica, however these measurements were made shortly after the 1982–83 El Niño-related and 1985 red tide-coincident coral die-offs. Later, at the same locality, Guzmán (1988a) recorded *C. monodonta* on 72 % of 245 *Pocillopora* colonies and a single 27 cm diameter colony was found to host 17 snails. Guzmán (1988a) recorded a mean feeding rate of 0.64 cm^2 coral tissue ind^{-1} d^{-1} on a Costa Rican reef which, given observed densities of *C. monodonta*, could result in the complete elimination of coral tissue from the base of a 10 cm diameter *Pocillopora* colony within 30 days. Guzmán concluded that this corallivorous activity could have widespread ramifications for the reproduction and erosion of pocilloporid corals as well as the distribution and community structure of coral communities on eastern Pacific reefs.

10.3.2 *Jenneria pustulata*

The pustulate egg shell, *Jenneria pustulata*, is well documented throughout the eastern Pacific (Table 10.1), in part due to its conspicuous and uniquely orange-spotted shell (Bertsch 1984). It is the sole species within the genus *Jenneria* and it occurs in eastern Pacific and Hawaiian waters (Robertson 1970). While present in the Galápagos (Glynn 1983a), it was absent during early surveys of Clipperton (Glynn et al. 1996). This gastropod is primarily found associated with pocilloporid corals across all reef zones (Glynn et al. 1982). It remains cryptic within coral colonies during the day and moves outward to feed at night (Glynn et al. 1972). High risk from predation may be responsible for this behavior as indicated by an abundance of broken *J. pustulata* shells within reef sediments (Glynn et al. 1972). Indeed, in a laboratory setting, Oramas (1979) observed that the feeding of *Jenneria* was restricted to the basal branches of pocilloporid corals in the presence of a potential porcupine fish predator (*Diodon holocanthus*).

Jenneria feeds primarily on pocilloporid corals (Glynn et al. 1982; Navas-Camacho et al. 2010), though it has also been observed to consume several other eastern Pacific coral species (e.g., *Pavona* spp., *Porites lobata*, *Porites panamensis* and *Psammocora* sp., Achurra-Cárdenas and Valdés-Araúz 1980; Paz-García et al. 2012). Under laboratory conditions, *J. pustulata* has been noted to feed on other non-eastern Pacific coral species including *Porites* sp., *Phyllangia americana*, *Siderastrea siderea*, and *Madracis mirabilis* (D'Asaro 1969; Glynn 1985). *Jenneria* feeds on all surfaces of its pocilloporid hosts, removing 0.80 g skeleton and tissue a day on average (Glynn et al. 1972). Feeding scars appear similar to white band disease and consist of areas of white, denuded coral skeleton (Navas-Camacho et al. 2010; Paz-García et al. 2012).

Jenneria is often found in clusters as high as 50–100 individuals per colony of *Pocillopora* (Panama, Glynn 2004), 15–40 per colony of *Porites* (Gulf of California, Paz-García et al. 2012), and 24 ind m^{-2} on reefs around Gorgona Island, Colombia (Glynn et al. 1982). Mean densities, however, were much lower in the Panamanian Pearl Islands at 1.8 ind m^{-2} (Glynn et al. 1972). At this last density, and at the aforementioned feeding rate reported by Glynn et al. (1972), *Jenneria* was responsible for an estimated 5.31 metric tons of coral mortality ha^{-1} $year^{-1}$ (skeleton and tissue), accounting for 79 % of the total coral mass removal considered by Glynn et al. (1972) in a corallivore-coral model.

Densities of *Jenneria pustulata* have been shown to be significantly affected by bleaching events associated with El Niño thermal anomalies. Glynn (1985) observed that after the 1983 El Niño-associated bleaching event, *J. pustulata* population densities dropped from 23.7 to 1.7 ind m^{-2} and dead, intact shells were found in 15 % of quadrats at the Panamanian study sites. *Jenneria* is known to starve within weeks if it has limited or no access to coral food, and it is likely that coral mortality played a part in its rapid die-off (Glynn 1985). Laboratory experiments, however, indicate that elevated water temperatures may also have direct detrimental impacts, as *J. pustulata* held in aquaria at 31.8 ° C died even when food was present (Glynn 1985).

10.3.3 Other Molluscan Corallivores

Latiaxis hindsii and *Muricopsis* (*Babelomurex*) *zeteki* are not as well-known as the aforementioned molluscan corallivore species, and are muricid gastropods that have been observed to consume coral tissues in the Galápagos Islands (Glynn and Wellington 1983). While documented as present at numerous additional localities (Table 10.1), observations on the corallivorous activities of these species are limited. In the Galápagos, both *L. hindsii* and *M. zeteki* have been observed feeding on pocilloporid corals in groups of two to five individuals (Glynn and Wellington 1983). In each instance where a colony was observed to be infested by these gastropods, tissue damage was not severe, accounting for less than 10 % of the total colony surface area.

An aeolid nudibranch in the genus *Phestilla* was first reported feeding on coral in the eastern Pacific by Highsmith (pers. comm.) who observed it grazing at night on *Porites lobata* in Panama (Glynn 1982). This species was described as *Phestilla panamica* (Rudman 1982) and more recently has been considered as *Phestilla lugubris* (Behrens and Hermosillo 2005; Camacho-Garcia et al. 2005). More recently an additional species—*Phestilla melanobrachia*—was reported to feed on *Tubastrea*. Terrence Gosliner (pers. comm.) observed *P. lugubris* feeding on *Porites* sp. (probably *Porites lobata*) in the Gulf of Chiriquí, Panama, and at Caño Island, Costa Rica. *Phestilla lugubris* also was observed laying eggs on *Porites* in Panama (T. Gosliner, pers. comm.). *Phestilla lugubris* is wide ranging throughout the eastern tropical Pacific and Indo-Pacific regions while *P. melanobrachia* is widespread throughout the Indo-Pacific (Gosliner et al. 2008). Within the genus *Phestilla*, strong species-specific feeding preferences have been reported with a single nudibranch species only consuming coral prey in a particular family or genus (e.g., *Phestilla sibogae* consumes *Porites* spp.; Harris 1975; Gosliner et al. 2008). Whether *Phestilla* consumes other coral taxa in the eastern Pacific is in need of study. *Phestilla* spp. are better known throughout the western Pacific and have been the focus of a study in Hawaii (Rotjan and Lewis 2008). In experimental tanks in Hawaii, *Phestilla* has been observed to infest and ultimately kill *Porites compressa*. Evidence of infestations of this magnitude are lacking from the wild, leading some to conclude that naturally occurring populations are controlled by fish and crustacean predators (Gochfeld and Aeby 1997). Regardless, single individuals may consume an average of 3.1 cm^2 of tissue per day (Haramaty 1991). The nudibranch utilizes small radular teeth to remove both polyps and coenosarc, indiscriminately scraping the soft tissues into its mouth (Rudman 1981). There is, however, some evidence that food selection occurs within the guts of the nudibranchs, with different species deriving nutriment from different parts of their coral prey (Rudman 1981). Live zooxanthellae have been observed within *Phestilla* species, and measurements of O_2 consumption by animals held in illuminated and dark aquaria revealed a reduction in net oxygen consumption in the light, thereby suggesting that the zooxanthellae were photosynthetically active and capable of conferring energetic benefits to their corallivore hosts. Finally, it should be noted that while *Phestilla* spp. are likely not capable of significant damage to eastern Pacific corals and reefs through their corallivorous activities, recent evidence suggests that they are a potential vector for transmitting coral disease (Dalton and Godwin 2006). These small, cryptic, and poorly known corallivores should therefore not be dismissed as unimportant to reef ecosystem health, and further investigation is necessary.

10.3.4 Hermit Crabs

Three species of diogeniid hermit crabs (*Trizopagurus magnificus*, *Calcinus obscurus*, and *Aniculus elegans*) have been observed to exhibit corallivorous feeding behaviors in the eastern Pacific. All are known from coral and rocky habitats at shallow depths (<40 m), however *A. elegans* has also been found to occupy sand and gravel habitats, and *C. obscurus* algal substrates (Ayón-Parente and Hendrickx 2010). *Trizopagurus magnificus* and *A. elegans* occur in the Sea of Cortez, Mexican, and Panamic provinces as well as the offshore Malpelo and Galápagos islands. *Calcinus obscurus*, however, is known solely from Panamic waters (Ayón-Parente and Hendrickx 2010).

Both *Trizopagurus magnificus* and *Aniculus elegans* are known to feed primarily at night; scraping the branch tips of pocilloporid corals, removing both skeletal material and tissues, and producing sediments (Glynn et al. 1972). Gilchrist (1985) described the feeding behaviors of *T. magnificus* and *Calcinus obscurus* in detail, noting size-dependent feeding behavior in *T. magnificus*. Larger individuals of this species scraped the surface of coral tissues with their major chelae, transferring organic matter to their maxillipeds and then to their mouths. Smaller individuals, however, picked at the surface of the coral with their maxillipeds. *Aniculus elegans* exhibited a scraping behavior similar to larger *T. magnificus*, though this feeding method was used regardless of size-class and was displayed subsequent to the snipping of the coral's surface tissues with their chelae.

In addition to damage inflicted by the feeding behavior of these species, both were observed to abrade coral tissues as they moved their shell across the live colony surfaces. In cases where this behavior was especially concentrated, such as when a crab tried to free itself when wedged among branches, colonies exhibited pronounced stress responses such as mucus production. Gilchrist (1985) also noted the sheltering behavior of hermit crab corallivores, which take refuge among the branches and dead bases of *Pocillopora* colonies, exhibiting limited mobility. Larger *Trizopagurus magnificus*, however, were observed to move between individual colonies more frequently, perhaps due to a reduced susceptibility to invertivore fishes.

At Isla del Caño, Costa Rica, densities of *Trizopagurus magnificus* and *Aniculus elegans* ranged from 0–24 and 0–28 ind m^{-2}, respectively (Guzmán 1988a). Glynn et al. (1972) recorded mean densities of *T. magnificus* in the Pearl Islands as high as 27.5 ind m^{-2}, and densities of *A. elegans* at 0.02 ind m^{-2}. In this same study, *T. magnificus* was found to consume 10.3 mg mean coral mass (tissue and skeleton) per day while large individuals of *A. elegans* removed a mean of 1.24 g of coral per day. At these rates and densities, hermit crabs were estimated to remove 1.12 metric tons of coral material per hectare of reef per year (Glynn 1974).

10.3.5 *Trapezia* spp.

The coral guard crabs *Trapezia* spp. as well as the alpheid shrimp *Alpheus lottini* are among the few corallivorous species that have not received the, albeit anthropocentric and over-simplified, distinction of harmful to corals and coral reefs. These obligate coral associates are known to reduce mortality and predation pressure (see below), yet still utilize their coral hosts as a food source.

Trapeziid crabs are primarily found associated with corals in the family Pocilloporidae and are widely distributed across the Indian and Pacific Oceans. Adults generally live in male-female pairs and occupy the living inter-branch spaces of their coral hosts, while juveniles are known to have an intra-colony distribution confined to the more-protected dead base of the coral. Numerous studies have investigated various aspects of the ecology of these crab associates, including resource partitioning and competition (Preston 1973; Huber and Coles 1986), seasonal reproduction patterns (Gotelli et al. 1985), inter-colony movements (Castro 1978) as well as their responses to coral bleaching and host death (Glynn et al. 1985; Caley et al. 2001; Stella et al. 2011).

Patton (1974) reviewed early studies discussing the diet of *Trapezia* spp. and lists invertebrates, detritus, and sediments, as well as coral mucus and tissues as food sources. Knudsen (1967) presented detailed descriptions of the feeding behaviors of *Trapezia ferruginea* and *Trapezia areolata* collected from Enewetok Atoll. Crabs were observed to introduce the tips of their walking legs into living polyps and rapidly scrape away surface materials. Substances adhering to the brush-like setae of their legs, including mucus, were removed by the mouthparts and consumed. Additionally, Knudsen (1967) described scraping of polyps and inter-polyp tissues (coenosarc), and noted food transfer from the crab's chelae.

Experimental evidence for the metabolic reliance of *Trapezia* on host corals can be found in Glynn et al. (1985). Heterosexual pairs of *Trapezia* were maintained on healthy, bleached, and dead coral colonies. Significant decreases in crab lipid content were observed on metabolically impaired (bleached and dead) corals versus those associated with healthy live coral colonies. Furthermore, emigration of crabs from their hosts and crab mortality were observed. Later studies by Stimson (1990) revealed that trapeziid crabs feed upon lipid-rich fat bodies produced by host *Pocillopora damicornis* colonies. This food source is available in abundance only when the crabs are present, suggesting that the crabs directly stimulate their hosts to produce it.

The most conclusive evidence for *Trapezia* corallivory is that of Rinkevich et al. (1991), who incubated *Stylophora pistillata* colonies with radioactive ^{14}C and then transplanted *Trapezia cymodoce* to each coral colony. These workers observed that compounds made by the photosynthetic activity of zooxanthellae were transferred to their coral hosts and finally into the tissues of *T. cymodoce*. The translocation of radioactive carbon from coral to crab associate was accomplished primarily through tissue consumption rather than mucus. Rates of ^{14}C accumulation corresponded to roughly 140 cm^2 of coral tissue surface area per month, roughly the amount of tissue on coral branches 40–45 cm in length.

10.3.6 *Alpheus lottini* and Other Caridean Shrimps

Alpheus lottini is well known throughout the eastern Pacific where it is found in heterosexual pairs, exclusively associated with pocilloporid corals. In other Indo-Pacific localities, however, it is also known to inhabit coral colonies within the genera *Seriatopora* and *Stylophora* (Castro 1971). The shrimp is omnivorous and has been observed to consume eggs (Coutière 1899), small invertebrates, algae, coral tissues (Barry 1965 in Castro 1971), and coral mucus (Barry 1965 in Castro 1971; Patton 1974). The diverse food sources of *A. lottini* may be explained by its seemingly indiscriminate feeding behavior, which involves brushing setose appendages across the surface of its coral host and removing the collected scrapings with its mouthparts (Patton 1974).

There are several other species of caridean shrimps known to associate with *Pocillopora* spp. in the eastern Pacific and it is likely that many of these rely, at least in part, on the metabolic byproducts of their hosts for nutriment. For example, *Thor amboinensis* is a small hyppolitid shrimp known to associate with a variety of invertebrate hosts across the Pacific, including in Panamanian waters (Patton 1974; Abele and Patton 1976). Gut content analysis of those collected from coral colonies has indicated coral mucus as a probable food source though individuals collected from sea anemones have been observed to utilize suspension feeding (Patton 1974). Similarly, shrimps belonging to the genus *Harpiliopsis* are known to associate with a diverse array of coral genera, including *Pocillopora* in the eastern Pacific (Castro 1971; Abele and Patton 1976). Knowledge of their feeding habits is limited to the gut contents of four individuals of *Harpiliopsis depressus* examined in the 1960s, which revealed a diet of coral mucus, zooxanthellae, and algal spores (Barry 1965 in Castro 1971). While the feeding habits of the aforementioned caridean shrimps have not been observed in as great of detail as *A. lottini*, since they consume coral products they may therefore be considered corallivorous.

10.4 Ecological Role of Corallivores

The positive impacts of corallivores on reef ecosystem health and function are often overlooked and underappreciated. While it is true that population explosions or "outbreaks" of corallivorous species such as the crown-of-thorns sea star can have devastating consequences for corals and reef communities (e.g., Chesher 1969; Birkeland and Lucas 1990), corallivory in the eastern Pacific is integral to ecosystem function. Corallivores are known to ameliorate the deleterious consequences of bleaching, they deter more efficient predators, aid in coral asexual reproduction, and finally they form a trophic pathway that supports diverse and abundant invertivorous and piscivorous species (see Chap. 9, Enochs and Glynn). Several of these positive impacts are noted by Rotjan and Lewis (2008), Glynn (2013), and Castro (2015).

The prevalence of these positive impacts within eastern Pacific reefs, and the importance of coral food sources in reef ecosystem trophodynamics, is likely related to the unique biology and ecology of pocilloporid corals. Species within this genus form many of the reef frameworks within the region. They are fast growing and branching, providing an abundant and accessible food source that regenerates relatively rapidly. Furthermore, their morphology provides an ideal shelter for cryptic associates, and their tissues and metabolic by products are an especially nutritious food source for their occupants (Stimson 1990). Together, these characteristics contribute to not just the enhancement of corallivorous populations, but to the corals' persistence despite the abundant and diverse species that feed on them.

10.4.1 Keystone Predators

Do any eastern Pacific corallivores exhibit keystone species characteristics, i.e. relative to a low or moderate abundance can they exert disproportionately strong effects on coral community structure? Among invertebrate corallivores, *Acanthaster planci* is a likely candidate, but perhaps only locally because this sea star is generally in low abundance except at some sites in the Gulf of California, the Revillagigedo Islands, and the Gulf of Chiriquí, Panama. Where *A. planci* is abundant in some areas (Panama), preferred prey species are small and exhibit high proportions of pre-existing dead skeleton (Glynn 1987). One could argue that the crustacean symbionts of pocilloporid corals, *Trapezia* spp. and *Alpheus lottini* would qualify as keystone corallivores because they enhance the survival of their hosts by warding off potential predators, prevent the settlement of alien taxa, and enhance boundary layer circulation and gas exchange. The facultative corallivore *Eucidaris galapagensis* plays a keystone role in the Galápagos Islands because of its pronounced excavating feeding behavior, which affects mature live corals, coral recruits, and the integrity of coral skeletons and carbonate framework structures. The degradation and loss of habitat structure is critical for a multitude of cryptic organisms (Glynn 2011; Enochs 2012; Enochs and Manzello 2012).

Among the facultative fish corallivores, the ubiquitous territorial damselfishes have been suggested to play keystone roles on coral reefs (Williams 1980; Hixon and Brostoff 1983). Gochfeld (2010) has offered data to support the role of damselfishes in affecting the relative abundances of corals. She found that *Pocillopora damicornis* colonies inside damselfish territories were protected from butterflyfish predation, and persisted for several years. Colonies experimentally removed and placed outside of territories were attacked and heavily grazed by butterflyfishes. Other studies in the central and western Pacific have demonstrated that several territorial damselfish species also protect acroporid corals, resulting in a local diversifying effect (e.g., Glynn and Colgan 1988; Done et al. 1991; Jones et al. 2006). Wellington's (1982) demonstration of *Stegastes acapulcoensis* controlling the zonation of branching and massive corals in the Pearl Islands (Panama) is a compelling example of the functional role of damselfish in shaping coral community structure and regulating the depth distribution of community types.

It is highly likely that the feeding behavior of the guineafowl puffer, *Arothron meleagris*, which characteristically nips the branch-tips of *Pocillopora* spp., can promote the lateral expansion of these coral communities by asexual propagation. Some coral branch segments that are broken but not ingested may survive and continue to grow if deposited on suitable substrates, e.g. course sediments at reef margins with ample sunlight. Several eastern Pacific localities in Mexico, Panama, and the Galápagos Islands have exhibited high abundances of pocilloporid asexual recruits, and a high proportion of these have probably been generated by pufferfish feeding (see Tables 15.12 and 15.13 in Chap. 15, Glynn et al.). One of the highest population densities of *A. meleagris* reported in the eastern Pacific, 171 ind ha^{-1} on a reef at Gorgona Island (Colombia), has not depressed coral colony mass, calcification or reef persistence and growth at this site (Palacios et al. 2014).

10.4.2 Cleaning

In a three month study conducted by Glynn (1983b) on the Pacific coast of Panama, corallivorous crab and shrimp symbionts (see Sects. 10.3.5 and 10.3.6), were found to increase mucus production, skeletal extension, and ultimately the survivorship of their pocilloporid coral hosts. Glynn postulated that these positive effects were conferred to the host due to the symbionts directly removing harmful substances settling on the coral's surface (e.g., sediment and colonizing organisms such as algae and invertebrates) as well as their stimulation of the coral's natural mucosal cleansing. Stewart et al. (2006) observed a similar phenomenon in Moorea. When *Trapezia* crabs were experimentally removed from *Acropora* and *Pocillopora* colonies in the field, the corals accumulated sediments, bleached, and ultimately died.

10.4.3 Defense from Harmful Species

Symbiotic crustaceans (trapeziid crabs and alpheid shrimps) are known to defend their coral hosts from *A. planci*, emerging from the protective interstices of the coral's branches and aggressively snipping with their chelae the tube feet and spines of the attacking asteroid (e.g., Weber and Woodhead 1970; Glynn 1976, 1983c). This defensive behavior is triggered, in large part, by chemicals naturally produced by *Acanthaster* that are detected at close range by the crustacean guards (Glynn 1980). In the western Pacific, Pratchett (2001) observed that the defensive behavior of trapeziid crabs is of sufficient magnitude to influence the hierarchical feeding preferences of *Acanthaster*, with the sea star preferentially preying upon less protected corals harboring weaker symbionts.

The strong defensive behavior of these symbiotic crabs and shrimps presents an apparent paradox in that they are able to coexist despite strongly territorial tendencies. It has been shown that the competitively inferior shrimp mimics the mannerisms of the trapeziid crabs, communicating with behaviors that are not used with conspecific shrimp (Vannini 1985). This display can result in the appeasement of the crabs' aggressive tendencies and facilitate the coexistence of the different species.

The positive nature of these corallivorous symbionts in ameliorating the deleterious effects of interspecific interactions is not limited to the exclusion of predatory *Acanthaster*. In Moorea, and throughout much of the Indo-Pacific, the sessile vermetid gastropod *Dendropoma maximum* is known to inhabit live coral colonies. While *Dendropoma* is not a corallivore, it extends mucus nets during feeding and is known to negatively influence coral growth and colony morphology (e.g., Shima et al. 2010). When present, trapeziid crabs were found to reduce the negative effects that *Dendropoma* had on coral growth rate (Stier et al. 2010). The authors of the latter study hypothesized that this was due to mucus removal by the crab, either inadvertently or through feeding and coral cleansing activities. It is likely that similar interactions occur in the eastern Pacific but remain unstudied.

10.4.4 Asexual Coral Reproduction

Many fast growing branching coral species fragment naturally due to physical and biological disturbances (e.g., Tunnicliffe 1979; Highsmith 1982). If tissue mortality is not severe following fragmentation, the resulting pieces may grow into new colonies. This form of asexual reproduction results in localized scattering, but when combined with long-distance dispersal of sexually produced offspring, it can be a highly effective life-history strategy in marginal reef environments with patchy habitable substrates (Bothwell 1981). While clonal reproduction is of limited importance to populations of *Pocillopora damicornis* in some localities (e.g., Miller and Ayre 2004), in many areas throughout its geographical range it remains one of the dominant forms of reproduction (Stoddart 1984; Adjeroud and Tsuchiya 1999; Whitaker 2006). This is true in the eastern Pacific, where *Pocillopora* forms large contiguous stands and reef structures (Richmond 1987).

The pufferfish *Arothron meleagris* is known to break off and ingest the tips of *Pocillopora damicornis* branches up to

3 cm in length (Guzmán 1988a). In some situations, where the resulting fragments are not consumed, these branch tips may grow into new coral colonies. This process, combined with physical disturbance by wave action, occurs on the reef slopes of eastern Pacific pocilloporid reefs. The resulting loose coral fragments spread out into adjacent rubble plains, where they eventually take hold and extend the consolidated reef framework seaward (Highsmith 1982). The extent that this process functions in the asexual propagation of *Pocillopora* remains unquantified, and investigation into the number of fragments created and their subsequent mortality rate is needed. However, given that *A. meleagris* can exist in high abundances (up to 171 ind ha^{-1}, Palacios et al. 2014) and consume less than 50 % of the live coral they remove (Glynn et al. 1972), the potential exists for this to be an ecologically important means of reproduction.

10.4.5 Trophic Contribution

In addition to the aforementioned positive influences on their coral food sources, corallivores in turn provide an important food source for the taxa that consume them, transferring energy from coral suspension feeding and zooxanthellae primary production to higher trophic level predators. In the eastern Pacific, this may occur when various piscivorous (e.g., Sphyraenidae, Scombridae, Carcharhinidae) or fish invertivores (e.g., Balistidae, Tetraodontidae) consume corallivore prey. Additionally, small cryptic invertebrates such as octopuses consume corallivorous gastropods and the caridean shrimp and polychaete worms (*Hymenocera picta*, *Pherecardia striata*, respectively) that feed on *Acanthaster planci* (Glynn 1981, 2004). Finally, it is likely that feces produced by corallivorous taxa may be consumed by detritivorous fauna, creating a pathway through which biological materials are recycled on a reef (e.g., Rothans and Miller 1991).

Enochs (2012) sampled motile cryptofauna associated with live and dead coral substrates from a pocilloporid reef on the Pacific coast of Panama. Live coral substrates were found to support significantly more cryptofauna biomass per unit volume than dead coral, suggesting that the trophic benefits of coral may extend to entire communities, beyond those species that directly consume them. In actuality, the number of feeding trophic interactions involving corals and corallivores is high, even in depauperate eastern Pacific systems (Glynn 2004). Quantitative analyses of complex coral reef food webs are rare (e.g., Opitz 1996; Reyes-Bonilla et al. 2014), especially for the multispecies assemblages and corallivore sub-webs found on eastern Pacific reefs.

10.5 Synthesis and Conclusions

In light of recent reviews (Cole et al. 2008; Rotjan and Lewis 2008) and the elevated levels of corallivory summarized here for the eastern Pacific, the pivotal role of coral consumption and the influence of the coral feeding guild to reef trophodynamics and community structure are further reinforced. In spite of a diverse and abundant guild of eastern Pacific corallivores and their prominence relative to the diversity of coral species prey, no records exist of local or large-scale outbreaks that have resulted in massive coral mortality as commonly observed in the Indo-Pacific region. This apparent exception to the rule may provide insight into natural ecological control of predator populations. It may signify that eastern Pacific reef ecosystems have uniquely evolved to support balanced but thriving corallivore populations or it may simply suggest that we have just not yet observed a predator outbreak event.

Regardless of the reasoning, we know that populations of corallivores are inexorably linked to the availability of their prey. Widespread coral mortality that resulted from the 1982–83 El Niño-related thermal anomalies and recovery in the years following have been shown to be strongly correlated with the abundance of corallivorous fishes (Glynn et al. 2014). Similar patterns have also been observed with coral cover and invertivore fishes, many of which feed on cryptic corallivore prey sheltering within the branches of pocilloporid corals (Enochs 2012; Glynn et al. 2014). These correlations point to the importance of the relationships between corals, corallivores, and higher level predators in the eastern Pacific. They indicate strong interconnectivity, a dependence on coral production and ultimately the consumption thereof. These trophic interactions provide a mechanism whereby climate-related coral mortality can have far reaching ramifications for numerous dependent taxa. These relationships/mechanisms are doubtlessly applicable to numerous other reef systems but are more difficult to discern due to higher species and habitat diversity. In their relative simplicity, eastern Pacific reefs and coral/corallivore interactions can therefore serve as informative systems with implications for ecosystem persistence, meriting further investigation.

Acknowledgments Thanks are due Michael J. Risk and Ellen R.M. Druffel for help with the radiometric dating. Aníbal Velarde performed and supervised the sorting of *Acanthaster* skeletal ossicles. G. Kolodziej provided help with manuscript preparation. We are grateful for information and editorial comments provided by T.M. Gosliner, M.M. Palacios, M.S. Pratchett, and an anonymous reviewer who helped to improve this chapter.

References

Abele LG, Patton WK (1976) The size of coral heads and the community biology of associated decapod crustaceans. J Biogeogr 3(1):35–47

Achurra-Cárdenas JB, Valdés-Araúz E (1980) Estudio sobre los efectos predatorios de los moluscos *Coralliophila abbreviata* y *Jenneria pustulata* de ambos litorales del Istmo de Panama. BSc thesis, Universidad de Panamá, Panamá

Adjeroud M, Tsuchiya M (1999) Genetic variation and clonal structure in the scleractinian coral *Pocillopora damicornis* in the Ryukyu Archipelago, southern Japan. Mar Biol 134(4):753–759

Allen GR, Robertson DR (1994) Fishes of the tropical eastern Pacific. Crawford House Press, Bathurst, p 332

Allen GR, Robertson DR (1997) An annotated checklist of the fishes of Clipperton Atoll, tropical eastern Pacific. Rev Biol Trop 45(2): 813–843

Alvarado JJ, Fernández C (2005) Equinodermos del Parque Nacional Marino Ballena, Pacífico, Costa Rica. Rev Biol Trop 53(3):275–284

Alvarado JJ, Solís-Marín FA, Ahearn CG (2010a) Echinoderm (Echinodermata) diversity in the Pacific coast of Central America. Mar Biodiv 40:45–56

Alvarado JJ, Reyes-Bonilla H, Buitrago F, Aguirre-Rubí J, Álvarez del Castillo Cardenas PA (2010b) Coral reefs of the Pacific coast of Nicaragua. Coral Reefs 29(1):201

Ayón-Parente M, Hendrickx ME (2010) Species richness and distribution of hermit crabs of the family Diogenidae (Crustacea: Decapoda: Anomura) in the eastern Pacific. Nauplius 18(1):1–12

Barham EG, Gowdy RW, Wolfson FH (1973) *Acanthaster* (Echinodermata, Asteroidea) in the Gulf of California. Fish Bull 71:927–942

Barry CK (1965) Ecological study of the decapod crustaceans commensal with branching coral *Pocillopora meandrina* var. *nobilis* Verrill. University of Hawai'i, Honolulu

Beebe W (1924) Galápagos: world's end. G.P. Putnam's Sons, New York

Beebe W (1926) The Arcturus adventure. G.P. Putnam's Sons, New York

Behrens DW, Hermosillo A (2005) Eastern Pacific nudibranchs: a guide to the opisthobranchs from Alaska to Central America. Sea Challengers, Global Interprint Inc, Santa Rosa, CA, p 137

Bertsch H (1984) *Jenneria pustulata*, the pustulate "cowrie". Opisthobranch 16(2):10

Birkeland C, Lucas JS (1990) *Acanthaster planci*: major management problem of coral reefs. CRC Press, Boca Raton

Black R, Prince J (1983) Fauna associated with the coral *Pocillopora damicornis* at the southern limit of its distribution in Western Australia. J Biogeogr 10(2):135–152

Bothwell AM (1981) Fragmentation, a means of asexual reproduction and dispersal in the coral genus *Acropora* (Scleractinia: Astrocoeniida: Acroporidae)—a preliminary report. In: Proceedings of 4th International Coral Reef Symposium, vol 2, pp 137–144

Bruckner AW, Bruckner RJ, Sollins P (2000) Parrotfish predation on live coral: "spot biting" and "focused biting". Coral Reefs 19(1):50

Caley MJ, Buckley KA, Jones GP (2001) Separating ecological effects of habitat fragmentation, degradation, and loss on coral commensals. Ecology 82(12):3435–3448

Camacho-Garcia Y, Gosliner TM, Valdés Á (2005) Field guide to the sea slugs of the tropical eastern Pacific. California Academy of Sciences, San Francisco, p 134

Carpenter RC (1997) Invertebrate predators and grazers. In: Birkeland C (ed) Life and death of coral reefs. Chapman and Hall, New York, pp 198–229

Caso ME (1962) Estudios sobre astéridos de México. Observaciones sobre especies pacíficas del género *Acanthaster* y descripción de una subespecie nueva, *Acanthaster ellisii pseudoplanci*. An Inst Biol UNAM 32(1–2):313–331

Castro P (1971) The natantian shrimps (Crustacea, Decapoda) associated with invertebrates in Hawaii. Pac Sci 25:395–403

Castro P (1978) Movements between coral colonies in *Trapezia ferruginea* (Crustacea: Brachyura), an obligate symbiont of scleractinian corals. Mar Biol 46(3):237–245

Castro P (1982) Notes on symbiotic decapod crustaceans from Gorgona Island, Colombia, with a revision of the eastern Pacific species of *Trapezia* (Brachyura, Xanthidae), symbionts of scleractinian corals. An Inst Inv Mar Punta de Betín 12:9–17

Castro P (1996) The eastern Pacific species of *Trapezia* (Crustacea: Brachyura: Trapeziidae) sibling species symbiotic with scleractinian corals. Bull Mar Sci 58:531–554

Castro P (2015) Symbiotic Brachyura. In: Castro P, Davie PJF, Guinot D, Schram FR, Vaupel Klein JC von (eds) Traité de Zoologie, The Crustacea, vol 9, Part C-I Decapoda: Brachyura (Part 1) pp 543–581

Chesher RH (1969) Destruction of Pacific corals by the sea star *Acanthaster planci*. Science 165(3890):280–283

Chesher RH (1972) The status of knowledge of Panamanian echinoids, 1971, with comments on other echinoderms. Bull Biol Soc Wash 2:139–158

Cole AJ, Pratchett MS (2014) Diversity in diet and feeding behaviour of butterflyfishes; reliance on reef corals versus reef habitats. In: Pratchett MS, Berumen ML, Kapoor B (eds) Biology of butterflyfishes. CRC Press, Boca Raton, pp 107–139

Cole AJ, Pratchett MS, Jones GP (2008) Diversity and functional importance of coral-feeding fishes on tropical coral reefs. Fish Fisheries 9:286–307

Cole AJ, Lawton RJ, Wilson SK, Pratchett MS (2012) Consumption of tabular acroporid corals by reef fishes: a comparison with plant–herbivore interactions. Funct Ecol 26:307–316

Coles SL (1980) Species diversity of decapods associated with living and dead reef coral *Pocillopora meandrina*. Mar Ecol Prog Ser 2:281–291

Colgan MW (1987) Coral reef recovery on Guam (Micronesia) after catastrophic predation by *Acanthaster planci*. Ecology 68(6):1592–1605

Coutière MH (1899) Les Alpheidae, morphologie externe et interne, formes larvaires, bionomie. Masson, Paris

Cox EP (1986) The effects of a selective corallivore on growth rates and competition for space between two species of Hawaiian corals. J Exper Mar Biol Ecol 101:161–174

D'Asaro CN (1969) The egg capsules of *Jenneria pustulata* (Lightfoot, 1786) with notes on spawning in the laboratory. Veliger 11(3):182–184

Dalton SJ, Godwin S (2006) Progressive coral tissue mortality following predation by a corallivorous nudibranch (*Phestilla* sp.). Coral Reefs 25(4):529

Dana T, Wolfson A (1970) Eastern Pacific crown-of-thorns starfish populations in the lower Gulf of California. Trans San Diego Soc Nat Hist 16:83–90

Dee LE, Witman JD, Brandt M (2012) Refugia and top-down control of the pencil urchin *Eucidaris galapagensis* in the Galapagos Marine Reserve. J Exp Mar Biol Ecol 416:135–143

Demond J (1957) Micronesian reef-associated gastropods. Pac Sci 11:275–341

Dominici-Arosemena A, Wolff M (2006) Reef fish community structure in the tropical eastern Pacific (Panamá): living on a relatively stable rocky reef environment. Helgol Mar Res 60:287–305

Done TJ, Dayton PK, Dayton AE, Steger R (1991) Regional and local variability in recovery of shallow coral communities: Moorea, French Polynesia and central Great Barrier Reef. Coral Reefs 9:183–192

Druffel E (1995) Pacific bomb radiocarbon coral data. In: IGBP PAGES/World Data Center-A for Paleoclimatology. NOAA/NGDC Paleoclimatology Program, Boulder

Edgar GJ, Banks SA, Brandt M, Bustamante RH, Chiriboga A, Earle SA, Garske LE, Glynn PW, Grove JS, Henderson S, Hickman CP, Miller KA, Rivera F, Wellington GM (2010) El Niño, grazers and fisheries interact to greatly elevate extinction risk for Galápagos marine species. Global Change Biol 16:2876–2890

Endean R (1976) Destruction and recovery of coral reef communities. In: Jones OA, Endean R (eds) Biology and geology of coral reefs, vol III., Biology 2 Academic Press, New York, pp 215–254

Enochs IC (2012) Motile cryptofauna associated with live and dead coral substrates: implications for coral mortality and framework erosion. Mar Biol 159:709–722

Enochs IC, Manzello DP (2012) Species richness of motile cryptofauna across a gradient of reef framework erosion. Coral Reefs 31:653–661

Fong P, Glynn PW (1998) A dynamic size-structured population model: does disturbance control size structure of a population of the massive coral *Gardineroseris planulata* in the eastern Pacific. Mar Biol 130:663–674

Frankel E (1977) Previous *Acanthaster* aggregations in the Great Barrier Reef. In: Proceedings of 3rd International Coral Reef Symposium, vol 1, Miami, pp 201–208

Frankel E (1978) Evidence from the Great Barrier Reef of ancient *Acanthaster* aggregations. Atoll Res Bull 220:75–93

Froese R, Pauly D (2011) FishBase. Online database. www.fishbase.org

Garth JS (1974) On the occurrence in the eastern tropical Pacific of Indo-west Pacific decapod crustaceans commensal with reef-building corals. In: Proceedings of 2nd International Coral Reef Symposium, vol 1, Brisbane, pp 397–404

Gilchrist SL (1985) Hermit crab corallivore activity. In: Proceedings of 5th International Coral Reef Symposium, vol 5, Tahiti, pp 211–214

Global Biodiversity Information Facility. Online database. http://www.gbif.org

Glynn PW (1972) Observations on the ecology of the Caribbean and Pacific coasts of Panama. In: Jones ML (ed) The Panamic biota: some observations prior to a sea-level canal. Bull Biol Soc Wash 2:13–30

Glynn PW (1973) *Acanthaster*: effect on coral reef growth in Panama. Science 180:504–506

Glynn PW (1974) The impact of *Acanthaster* on corals and coral reefs in the eastern Pacific. Environ Conserv 1(4):295–304

Glynn PW (1976) Some physical and biological determinants of coral community structure in the eastern Pacific. Ecol Monogr 46(4):431–456

Glynn PW (1980) Defense by symbiotic Crustacea of host corals elicited by chemical cues from predator. Oecologia 47(3):287–290

Glynn PW (1981) *Acanthaster* population regulation by a shrimp and a worm. In: Proceedings of 4th International Coral Reef Symposium, vol 2, Manila, pp 607–612

Glynn PW (1982) Coral communities and their modifications relative to past and prospective Central American seaways. Adv Mar Biol 19:91–132

Glynn PW (1983a) Biological interactions. In: Glynn PW, Wellington GM, Wells JW (eds) Corals and coral reefs of the Galápagos Islands. University of California Press, Los Angeles, pp 127–160

Glynn PW (1983b) Increased survivorship in corals harboring crustacean symbionts. Mar Biol Lett 4(2):105–111

Glynn PW (1983c) Crustacean symbionts and the defense of corals: coevolution on the reef? In: Nitecki MH (ed) Coevolution. University of Chicago Press, Chicago, pp 111–178

Glynn PW (1985) Corallivore population sizes and feeding effects following El Niño (1982–83)-associated coral mortality in Panamá. In: Proceedings of 5th International Coral Reef Symposium, vol 4, Tahiti, pp 183-188

Glynn PW (1987) Some ecological consequences of coral-crustacean guard mutualisms in the Indian and Pacific Oceans. Symbiosis 4:301–324

Glynn PW (1988) El Niño warming, coral mortality and reef framework destruction by echinoid bioerosion in the eastern Pacific. Galaxea 7:129–160

Glynn PW (1990) Feeding ecology of selected coral-reef macroconsumers: patterns and effects on coral community structure. In: Dubinsky Z (ed) Ecosystems of the world, vol 25. Elsevier, Amsterdam, pp 365–400

Glynn PW (1993) Monsoonal upwelling and episodic *Acanthaster* predation as probable controls of coral reef distribution and community structure in Oman, Indian Ocean. Atoll Res Bull 379:1–66

Glynn PW (1994) State of coral reefs in the Galápagos Islands: natural vs anthropogenic impacts. Mar Poll Bull 29:131–140

Glynn PW (2003) Coral communities and coral reefs of Ecuador. In: Cortés J (ed) Latin American coral reefs. Elsevier Science, Amsterdam, pp 449–472

Glynn PW (2004) High complexity food webs in low-diversity eastern Pacific reef-coral communities. Ecosystems 7(4):358–367

Glynn PW (2008) Food-web structure and dynamics of eastern tropical Pacific coral reefs: Panamá and Galápagos Islands. In: McClanahan TR, Branch GM (eds) Food webs and the dynamics of marine reefs. Oxford University Press, Oxford, pp 185–208

Glynn PW (2011) In tandem reef coral and cryptic metazoan declines and extinctions. Bull Mar Sci 87:767–794

Glynn PW (2013) Fine-scale interspecific interactions on coral reefs: functional roles of small and cryptic metazoans. Smith Contrib Mar Sci 39:229–248

Glynn PW, Ault JS (2000) A biogeographic analysis and review of the far eastern Pacific coral reef region. Coral Reefs 19(1):1–23

Glynn PW, Colgan MW (1988) Defense of corals and enhancement of coral diversity by territorial damselfishes. In: Proceedings of 6th International Coral Reef Symposium, vol 2, Townsville, pp 157–163

Glynn PW, Enochs IC (2011) Invertebrates and their roles in coral reef ecosystems. In: Dubinsky Z, Stambler N (eds) Coral reefs: an ecosystem in transition. Springer, New York, pp 273–326

Glynn PW, Wellington GM (1983) Corals and coral reefs of the Galápagos Islands. University of California Press, Berkeley

Glynn PW, Stewart RH, McCosker JE (1972) Pacific coral reefs of Panama: structure, distribution and predators. Geol Rundsch, Int J Earth Sci 61(2):483–519

Glynn PW, Wellington GM, Birkeland C (1979) Coral reef growth in the Galapagos: limitation by sea urchins. Science 203:47–49

Glynn PW, von Prahl H, Guhl F (1982) Coral reefs of Gorgona Island, Colombia, with special reference to corallivores and their influence on community structure and reef development. An Inst Invest Mar Punta de Betin 12:185–214

Glynn PW, Perez M, Gilchrist SL (1985) Lipid decline in stressed corals and their crustacean symbionts. Biol Bull 168(2):276–284

Glynn PW, Veron JEN, Wellington GM (1996) Clipperton Atoll (eastern Pacific): oceanography, geomorphology, reef-building coral ecology and biogeography. Coral Reefs 15:71–99

Glynn PW, Wellington GM, Wieters EA, Navarrete SA (2003) Reef-building coral communities of Easter Island (Rapa Nui). In: Cortés J (ed) Latin American coral reefs. Elsevier Science BV, Amsterdam, pp 473–494

Glynn PW, Enochs IC, Afflerbach JA, Brandtneris VW, Serafy JE (2014) Eastern Pacific reef fish responses to coral recovery following El Niño disturbances. Mar Ecol Prog Ser 495:233–247

Glynn PW, Riegl B, Purkis S, Kerr JM, Smith TB (2015) Coral reef recovery in the Galápagos Islands: the northernmost islands (Darwin and Wenman). Coral Reefs. doi:10.1007/s00338-015-1280-4

Gochfeld DJ (2010) Territorial damselfishes facilitate survival of corals by providing an associational defense against predators. Mar Ecol Prog Ser 398:137–148

Gochfeld DJ, Aeby GS (1997) Control of populations of the coral-feeding nudibranch *Phestilla sibogae* by fish and crustacean predators. Mar Biol 130(1):63–69

Gosliner TM, Behrens DW, Valdés Á (2008) Indo-Pacific nudibranchs and sea slugs: a field guide to the world's most diverse fauna. Sea Challengers/California Academy of Sciences, Gig Harbor/San Francisco, p 426

Gotelli NJ, Gilchrist SL, Abele LG (1985) Population biology of *Trapezia* spp. and other coral-associated decapods. Mar Ecol Prog Ser 21(1):89–98

Guzmán HM (1986) Estructura de la comunidad arrecifal de la Isla del Caño, Costa Rica y el efecto de perturbaciones naturales severas. MS thesis, University of Costa Rica, San Pedro, Costa Rica

Guzmán HM (1988a) Distribución y abundancia de organismos coralívoros en los arrecifes coralinos de la Isla del Caño, Costa Rica. Rev Biol Trop 36(2):191–207

Guzmán HM (1988b) Feeding activity of the corallivorous gastropod *Quoyula monodonta* (Blainville). Rev Biol Trop 36(2A):209–212

Guzmán HM, Cortés J (1989a) Growth rates of eight species of scleractinian corals in the eastern Pacific (Costa Rica). Bull Mar Sci 44(3):1186–1194

Guzmán HM, Cortés J (1989b) Coral reef community structure at Caño Island, Pacific Costa Rica. Pubb Staz Zool Napoli I: Mar Ecol 10:23–41

Guzmán HM, López JD (1991) Diet of the corallivorous pufferfish *Arothron meleagris* (Pisces: Tetraodontidae) at Gorgona Island, Colombia. Rev Biol Trop 39:203–206

Guzmán HM, Robertson DR (1989) Population and feeding responses of the corallivorous pufferfish *Arothron meleagris* to coral mortality in the eastern Pacific. Mar Ecol Prog Ser 55:121–131

Haramaty L (1991) Reproduction effort in the nudibranch *Phestilla sibogae*: calorimetric analysis of food and eggs. Pac Sci 45(3):257–262

Harris LG (1975) Studies on the life history of two coral-eating nudibranchs of the genus *Phestilla*. Biol Bull 149(3):539–550

Henderson JA, Lucas JS (1971) Larval development and metamorphosis of *Acanthaster planci* (Asteroidea). Nature 232:655–657

Hickman CP Jr (1998) A field guide to the sea stars and other echinoderms of Galápagos., Galápagos marine life series, Sugar Spring Press, Lexington, p 83

Hickman CP Jr, Finet Y (1999) A field guide to marine molluscs of Galápagos: an illustrated guidebook to the common intertidal and shallow-water snails, bivalves, and chitons of the Galápagos Islands. Sugar Spring Press, Lexington, p 150

Hickman CP Jr, Zimmerman TL (2000) A field guide to crustaceans of Galápagos: an illustrated guidebook to the barnacles, shrimps, lobsters, and crabs of the Galápagos Islands., Galápagos marine life series, Sugar Spring Press, Lexington, p 156

Highsmith RC (1982) Reproduction by fragmentation in corals. Mar Ecol Prog Ser 7(2):207–226

Hixon MA, Brostoff WN (1983) Damselfish as keystone species in reverse: intermediate disturbance and diversity of reef algae. Science 220:511–513

Huber ME, Coles SL (1986) Resource utilization and competition among the five Hawaiian species of *Trapezia* (Crustacea, Brachyura). Mar Ecol Prog Ser 30:21–31

Intergovernmental Oceanographic Commission (IOC) of UNESCO. The Ocean Biogeographic Information System. Online database. http://www.iobis.org

Jones GP, Santana L, McCook LJ, McCormick MI (2006) Resource use and impact of three herbivorous damselfishes on coral reef communities. Mar Ecol Prog Ser 328:215–224

Kim W, Abele LG (1988) The snapping shrimp genus *Alpheus* from the eastern Pacific (Decapoda: Caridea: Alpheidae). Smith Contrib Zool 454:1–119

Knudsen JW (1967) *Trapezia* and *Tetralia* (Decapoda, Brachyura, Xanthidae) as obligate ectoparasites of pocilloporid and acroporid corals. Pac Sci 21:51–57

Kroh A (2010) *Eucidaris thouarsii galapagensis* Döderlein, 1887. In: Kroh A, Mooi R (eds) World Echinoidea database. World register of marine species. http://www.marinespecies.org/echinoidea/aphia.php?p=taxdetails&id=513653

Lessios HA (2005) Echinoids of the Pacific waters of Panama: status of knowledge and new records. Rev Biol Trop (Int J Trop Biol) 53 (Suppl 3):147–170

Lessios HA, Kessing BD, Robertson DR, Paulay G (1999) Phylogeography of the pantropical sea urchin *Eucidaris* in relation to land barriers and ocean currents. Evolution 53:806–817

Lucas JS (1973) Reproductive and larval biology of *Acanthaster planci* (L.) in Great Barrier Reef waters. Micronesica 9:197–203

Madsen FJ (1955) A note on the sea-star genus *Acanthaster*. Vidensk Medd fra Dansk naturh Foren 117:179–192

Maes VO (1967) The littoral marine mollusks of Cocos-Keeling Islands (Indian Ocean). P Acad Nat Sci Phila 119(4):93–217

McClanahan TR, Maina J, Starger CJ, Herron-Perez P, Dusek E (2005) Detriments to post-bleaching recovery of corals. Coral Reefs 24:230–246

Miller KJ, Ayre DJ (2004) The role of sexual and asexual reproduction in structuring high latitude populations of the reef coral *Pocillopora damicornis*. Heredity 92(6):557–568

Moran DA (1984) Additions to the known anomuran fauna of El Salvador, Central America (Crustacea: Decapoda). J Crustacean Biol 4(1):72–84

Moran PJ (1986) The *Acanthaster* phenomenon. Oceanogr Mar Biol Annu Rev 24:379–480

Moran PJ, Reichelt RE, Bradbury RH (1986) An assessment of the geological evidence for previous *Acanthaster* outbreaks. Coral Reefs 4:235–238

Motta PJ (1988) Functional morphology of the feeding apparatus of ten species of Pacific butterflyfishes (Perciformes, Chaetodontidae): an ecomorphological approach. Environ Biol Fishes 22:39–67

Mumby PJ (2009) Herbivory versus corallivory: are parrotfish good or bad for Caribbean coral reefs? Coral Reefs 28:683–690

Narváez K, Zapata FA (2010) First record and impact of the crown-of-thorns starfish, *Acanthaster planci* (Spinulosida: Acanthasteridae) on corals of Malpelo Island, Colombian Pacific. Rev Biol Trop 58(supp 1):139–143

National Museum of Natural History, Smithsonian Institution. NMNH Invertebrate Zoology Collections. Online database. http://collections.nmnh.si.edu/search/iz/

National Museum of Natural History, Smithsonian Institution: NMNH Vertebrate Zoology Fishes Collections. Online database. http://collections.nmnh.si.edu/search/fishes/

Navas-Camacho R, Rodríguez-Ramírez A, Reyes-Nivia MC (2010) Agents of coral mortality on reef formations of the Colombian Pacific. Rev Biol Trop 58(supp 1):133–138

Nuñez-Vázquez EJ, Yotsu-Yamashita M, Sierra-Beltrán AP, Yasumoto T, Ochoa JL (2000) Toxicities and distribution of tetrodotoxin in the tissues of puffer fish found in the coast of the Baja California Peninsula, Mexico. Toxicon 38:729–734

Oliverio M (2008) Coralliophilinae (Neogastropoda: Muricidae) from the southwest Pacific. In: Héros V, Cowie RH, Bouchet P (eds) Tropical deep-sea benthos: Mémoires du Muséum National d'Histoire Naturelle. MNHN, Paris, pp 481–585

Opitz S (1996) Trophic interactions in Caribbean coral reefs. ICLARM, Manila

Oramas NFA (1979) Estudio ecológico del *Jenneria pustulata* (Mollusca, Gastropoda). BSc thesis, Escuela de Biología, University of Panama, Panamá

Palacios MM, Muñoz CG, Zapata FA (2014) Fish corallivory on a pocilloporid reef and experimental coral responses to predation. Coral Reefs. doi:10.1007/s00338-014-1173-y

Parker GM (1984) Dispersal of zooxanthellae on coral reefs by predators on cnidarians. Biol Bull 167:159–167

Patton WK (1974) Community structure among the animals inhabiting the coral *Pocillopora damicornis* at Heron Island, Australia. In: Vernberg WB (ed) Symbiosis in the sea 2. University of South Carolina Press, Columbia, pp 219–243

Paz-García DA, Aldana-Moreno A, Cabral-Tena RA (2012) High predation by the corallivore sea snail *Jenneria pustulata* in a high-latitude reef in the Gulf of California. Mar Biodiv Rec 5:e94

Porter JW (1972) Predation by *Acanthaster* and its effect on coral species diversity. Am Nat 106(950):487–492

Porter JW (1974) Community structure of coral reefs on opposite sides of the Isthmus of Panama. Science 186:543–545

Pratchett MS (2001) Influence of coral symbionts on feeding preferences of crown-of-thorns starfish *Acanthaster planci* in the western Pacific. Mar Ecol Prog Ser 214:111–119

Pratchett MS (2014) Feeding preferences and dietary specialization among obligate coral-feeding butterflyfishes. In: Pratchett MS, Berumen ML, Kapoor B (eds) Biology of butterflyfishes. CRC Press, Boca Raton, pp 140–179

Pratchett MS, Hoey AS, Feary DA, Bauman AG, Burt JA, Riegl B (2013) Functional composition of *Chaetodon* butterflyfishes at a peripheral and extreme coral reef location, the Persian Gulf. Mar Poll Bull 72:333–341

Preston EM (1973) A Computer simulation of competition among five sympatric congeneric species of xanthid crabs. Ecology 54(3):469–483

Randall JE (1967) Food habits of reef fishes of the West Indies. Stud Trop Oceanogr 5:665–847

Randall JE (2005) Reef and shore fishes of the South Pacific, New Caledonia to Tahiti and the Pitcairn Islands. University of Hawai'i Press, Honolulu

Reaka-Kudla ML, Feingold JS, Glynn PW (1996) Experimental studies of rapid bioerosion of coral reefs in the Galapagos Islands. Coral Reefs 15:101–107

Reyes-Bonilla H, Calderon-Aguilera LE (1999) Population density, distribution and consumption rates of three corallivores at Cabo Pulmo Reef, Gulf of California, Mexico. Mar Ecol-Evol Persp 20 (3–4):347–357

Reyes-Bonilla H, Alvarez del Castillo Cárdenas PA, Calderón Aguilera LE, Erosa Ricárdez CE, Fernández Rivera Melo FJ, Frausto TC, Luna Salguero BM, Moreno Sánchez XG, Mozqueda Torres MC, Norzagaray López CO, Petatán Ramírez D (2014) Servicios ambientales de arrecifes coralinos: el caso del Parque Nacional Cabo Pulmo, B.C.S. In: Urciaga García JI (ed) Desarrollo regional en Baja California Sur. Una perspectiva de los servicios ecosistémicos. UABCS, La Paz, Mexico, pp 38–63

Richmond RH (1987) Energetic relationships and biogeographical differences among fecundity, growth and reproduction in the reef coral *Pocillopora damicornis*. Bull Mar Sci 41(2):594–604

Rinkevich B, Wolodarsky Z, Loya Y (1991) Coral-crab association: a compact domain of a multilevel trophic system. Hydrobiologia 216–217(1):279–284

Robertson R (1970) Review of the predators and parasites of stony corals, with special reference to symbiotic prosobranch gastropods. Pac Sci 24:43–54

Robertson DR (2009) Shorefishes of the tropical eastern Pacific online information system. Smith Contr Mar Sci 197–208. Online database. http://biogeodb.stri.edu/sftep/

Rothans TC, Miller AC (1991) A link between biologically imported particulate organic nutrients and the detritus food web in reef communities. Mar Biol 110:145–150

Rotjan RD, Lewis SM (2008) Impact of coral predators on tropical reefs. Mar Ecol Prog Ser 367:73–91

Rudman WB (1981) Further studies on the anatomy and ecology of opisthobranch molluscs feeding on the scleractinian coral *Porites*. Zool J Linn Soc Lond 71(4):373–412

Rudman WB (1982) A new species of *Phestilla*: the first record of a corallivorous aeolid nudibranch from tropical America. J Zool 198:465–471

Shima JS, Osenberg CW, Stier AC (2010) The vermetid gastropod *Dendropoma maximum* reduces coral growth and survival. Biol Lett 6(6):815–818

Sonnenholzner JI, Ladah LB, Lafferty KD (2009) Cascading effects of fishing on Galapagos rocky reef communities: reanalysis using corrected data. Mar Ecol Prog Ser 375:209–218

Steinbeck J, Ricketts EF (1941) Sea of Cortez: a leisurely journal of travel and research. Penguin Group USA Inc., New York

Stella JS, Munday PL, Jones GP (2011) Effects of coral bleaching on the obligate coral-dwelling crab *Trapezia cymodoce*. Coral Reefs 30:719–727

Stewart HL, Holbrook SJ, Schmitt RJ, Brooks AJ (2006) Symbiotic crabs maintain coral health by clearing sediments. Coral Reefs 25 (4):609–615

Stier AC, McKeon CS, Osenberg CW, Shima JS (2010) Guard crabs alleviate deleterious effects of vermetid snails on a branching coral. Coral Reefs 29(4):1019–1022

Stimson J (1990) Stimulation of fat-body production in the polyps of the coral *Pocillopora damicornis* by the presence of mutualistic crabs of the genus *Trapezia*. Mar Biol 106:211–218

Stoddart JA (1984) Genetic differentiation amongst populations of the coral *Pocillopora damicornis* off southwestern Australia. Coral Reefs 3(3):149–156

Stuiver M, Reimer PJ, Reimer RW (2005) CALIB 5.0. http://calib.qub.ac.uk/calib/

Stuivier M, Reimer PJ (1993) Extended ^{14}C database and revised CALIB radiocarbon calibration program. Radiocarbon 35:215–230

Tunnicliffe V (1979) The role of boring sponges in coral fracture. In: Levi C, Boury-Esnault N (eds) Collogues Internationaux du Centre National de la Recherche Scientifique, vol 291. CNRS, Paris, pp 309–315

Vannini M (1985) A shrimp that speaks crab-ese. J Crust Biol 5 (1):160–167

Veron JEN (1995) Corals in space and time. The biogeography and evolution of the Scleractinia. UNSW Press, Sydney, Australia, p 321

Vogler C, Benzie J, Lessios H, Barber P, Wörheide G (2008) A threat to coral reefs multiplied? Four species of crown-of-thorns starfish. Biol Lett 4:696–699

Walbran PD, Henderson RA, Jull AJT, Head MJ (1989a) Evidence from sediments of long-term *Acanthaster planci* predation on corals of the Great Barrier Reef. Science 245:847–850

Walbran PD, Henderson RA, Faithful JW, Polach HA, Sparks RJ, Wallace G, Lowe DC (1989b) Crown-of-thorns starfish outbreaks on the Great Barrier Reef: a geological perspective based upon the sediment record. Coral Reefs 8:67–78

Weber JN, Woodhead PMJ (1970) Ecological studies of the coral predator *Acanthaster planci* in the south Pacific. Mar Biol 6(1): 12–17

Wellington GM (1982) Depth zonation of corals in the Gulf of Panama: control and facilitation by resident reef fishes. Ecol Monogr 52:223–241

Whitaker K (2006) Genetic evidence for mixed modes of reproduction in the coral *Pocillopora damicornis* and its effect on population structure. Mar Ecol Prog Ser 306:115–124

Wicksten MK, Hendrickx ME (2003) An updated checklist of benthic marine and brackish water shrimps (Decapoda: Penaeoidea,

Stenopodidea, Caridea) from the eastern tropical Pacific. In: Hendrickx ME (ed) Contributions to the study of eastern Pacific crustaceans, vol 2. Instituto de Ciencias del Mar y Limnología. Universidad Nacional Autónoma de México, Mexico, DF, pp 49–76

Williams AH (1980) The threespot damselfish: a noncarnivorous keystone species. The Amer Nat 116:138–142

Yamaguchi M (1973) Early life-histories of coral reef asteroids, with special reference to *Acanthaster planci* (L.). In: Jones OA, Endean R (eds) Biology and geology of coral reefs, vol 2. Biology I. Academic Press, London, pp 369–387

Zamorano P, Barrientos-Luján NA, Ramírez-Luna S (2008) Malacofauna del infralitoral rocoso de Agua Blanca, Santa Elena Cozoaltepec, Oaxaca. Ciencia y Mar 12:19–33. http://www.umar.mx/revista/36/malacofauna.pdf

Algal Dynamics: Alternate Stable States of Reefs in the Eastern Tropical Pacific

11

Peggy Fong, Tyler B. Smith, and Ranjan Muthukrishnan

Abstract

Algae from two kingdoms and four major phyla form productive benthic communities that play key ecological roles on coral reefs of the eastern tropical Pacific (ETP). The diversity of algae that comprise these communities is being actively explored throughout the region, with new species still being discovered. Due to its physical setting, there are several distinct algal habitats in the ETP, including reef crests dominated by crustose coralline algae and algal turfs, reef flats that support a higher abundance of foliose macroalgae, unconsolidated accumulations of dead coral rubble that may support rhodoliths and cryptic macroalgae, and deeper-water refuges from herbivory dominated by more of the palatable forms of macroalgae rarely found on reefs. Past work has identified a number of ecological processes that control algal populations and communities at play in the ETP, but research elucidating their relative roles is still limited in this region. Recent work suggests that these reefs are supported by allochthonous nutrient supplies such as upwelling and thermocline shoaling, though reduced light during these events may limit productivity to shallow reef zones. Though at least partially sheltered from major disturbances that structure other reef systems, such as hurricanes and outbreaks of crown-of-thorns sea stars, reefs of the ETP are strongly affected by sea surface temperature extremes associated with El-Niño-Southern Oscillation (ENSO) leading to major coral mortality with the opening of space for colonization by algae. Biotic processes that control algal communities include herbivory and competition with coral, both of which limit algal proliferation and may enhance recovery after disturbance. However, facilitative interactions within algal communities, such as reduced herbivory and amelioration of nutrient limitation by internal recycling that act as positive feedbacks within areas dominated by algae, may stabilize algal communities after disturbance. There is some empirical and modeling evidence that suggests these processes may drive reefs of the ETP to display alternative stable states. This evidence includes a bifurcated pattern of recovery after ENSO disturbance, the patchy nature of this recovery within certain reefs, and the existence of stabilizing feedbacks that support resilience of both coral and algal states. Further, a simulation model (cellular automaton) of ETP reefs

P. Fong (✉) · R. Muthukrishnan
Department of Ecology and Evolutionary Biology, University of California Los Angeles, 621 Young Drive South, Los Angeles, CA 90095-1606, USA
e-mail: pfong@biology.ucla.edu

R. Muthukrishnan
e-mail: mrunjj@gmail.com

T.B. Smith
Center for Marine and Environmental Studies, University of the Virgin Islands, #2 John Brewers Bay, St. Thomas, VI 00802, USA
e-mail: tsmith@uvi.edu

P.W. Glynn et al. (eds.), *Coral Reefs of the Eastern Tropical Pacific*, Coral Reefs of the World 8, DOI 10.1007/978-94-017-7499-4_11

parameterized with experimentally-derived feedback relationships suggests these reefs exist under environmental conditions that produce bistability. However, much research is still needed to fully understand the physical and ecological processes that structure algal communities, and how these controls may shift with anthropogenic impacts.

Keywords
Diversity • Phase shifts • Community stability • Ecological processes • Algal habitats

11.1 Importance of Coral Reef Algae in the Eastern Tropical Pacific

The importance of algae on reefs of the eastern tropical Pacific (ETP) has not always been recognized. Earle (1972) noted that the first botanist to study Pacific Panama in 1909–1910, Marshall A. Howe, claimed "a marine flora, in the ordinary sense, was in that region, almost non-existent"; that the shores were "nearly destitute of plant life—at least of conspicuous species" (Fig. 11.1a). Howe suggested the intense tropical sun combined with the wide range of tides might have produced unfavorable conditions. As recently as 1945, despite several research cruises and expeditions in the region, published lists of marine algae of Pacific Panama only consisted of 24 species (Earle 1972). Earle (1972) believed that the impression of the region, that it lacked a diversity of macroalgae, was a misconception based largely on the paucity of algae in the intertidal zone and the lack of substantial beach wrack—something she attributed to the strength of herbivory. It was only with the advent of the use of SCUBA diving in research in the late 1950s and early 1960s that the importance of benthic algae in the ETP began to be revealed, and the extent of its distribution and abundance documented. For example, in cruises from Mexico to Panama in the late 1950s and aided by early use of SCUBA, Dawson (1960) recorded 58 new species of algae in the ETP. The documentation of species' range expansion and endemic species is still an ongoing process. However, we now know that marine algae of many taxonomic and morphological forms occur across all marine habitats in the ETP, and can be found on coral reefs from the upper limits of the intertidal zone to depths of more than 200 m (Fig. 11.1b–f).

Benthic marine macroalgae are key members of all coral reef communities that provide vital ecological functions such as stabilization of reef structures, production of calcareous sediments, nutrient retention and recycling, primary productivity, habitat for burrowing species and meiofauna, and trophic support. Macroalgae—defined here as all algae that are not microscopic and are therefore visible to the human eye—occur on all tropical reefs in an amazing display of diversity, abundance, and morphological and functional forms, and provide a unique set of equally diverse ecological functions and ecosystem services. However, the ETP region is different among reef systems due to its physical and chemical setting (see Chap. 3, Fiedler and Lavín), and this uniqueness is reflected in the roles of the algal community. For example, on most coral reefs crustose coralline red algae (CCA) make an essential contribution to reef structural strength by forming calcite "cement" (Littler and Littler 1988, 1995) and often form intertidal ridges at reef crests that protect back reef zones from the full force of oceanic waves. However, while CCA commonly dominates algal communities on ETP reefs (e.g., Guzmán and Cortés 2001; Edgar et al. 2011), its ability to provide structural strength may be greatly diminished due to low pH environments that result in relatively weak calcification, though this varies along a pH gradient within the region (see Chap. 18, Manzello et al.). Similarly, on most tropical reefs calcifying green algae are important sources of the carbonate sands that form key reef habitats. For example, one pan-tropical genus of siphonaceous (multi-nucleate but single-celled) green algae, *Halimeda*, forms rigid calcareous plates strengthened by aragonite that ultimately comprise a large portion of the sandy sediments in back reefs and lagoons (Macintyre et al. 1987; Hillis 1997). However, in the ETP members of this calcified algal genus are not abundant (Earle 1972; Wysor 2004; Fernández-García et al. 2011), and *Halimeda discoidea*, the only know species, lacks heavily calcified plates. Also missing in the ETP are expansive stretches of shallow sandy areas that support seagrass beds and the associated algal community—perhaps due to the lack of a sufficient supply of carbonate sands.

Tropical reefs have been called "oases of productivity in a nutrient desert" and algae play key roles in supporting this high productivity. Many processes contribute to this productivity, including (1) high advection rates supplying low concentrations of nutrients yet high volumes of oceanic water, (2) algae efficiently taking up even these low concentrations of nutrients, (3) extensive nitrogen fixation by cyanobacteria and bacteria, and (4) a spatially rugose set of algal habitats that dramatically increase the retention and recycling capability of the ecosystem (Littler and Littler 1988). A highly diverse assemblage of tropical reef algae is responsible for a large amount of the nutrient retention and recycling, contributing to the high level of primary

Fig. 11.1 Communities of algae of the ETP. **a** Rocky intertidal zones are typically sparse of algal communities, leading to the early belief algae were not diverse or abundant in the ETP (Naos Island, Panamá, April 7, 2005). Subsequent work using SCUBA dispelled this misconception, with abundant communities found in different depth zones including **b** the reef flat, **c** the reef crest, **d** deeper massive coral communities, **e** the reef slope where coral transitions to algal dominance, and **f** throughout all zones where it is often found in epiphytic relationships. All photos were taken in the Panamanian Pacific; **b** and **c** Dog Island, Panamá, March 12, 2003, **d** Uva Island, March 14, 2005, **e** Uva Island, March 8, 2003, **f** Canales de Tierra Island, March 14, 2003. All photos in chapter are credited to T.B. Smith unless otherwise noted

productivity typical of coral reefs and providing trophic support for consumers (see Chap. 9, Enochs and Glynn).

Although tropical reefs generally occur in highly oligotrophic oceanic water, many ETP reefs experience conditions of elevated nutrients and productivity driven by the regionally shallow thermocline that positions elevated nutrient supplies in shallow, sunlit waters near coral reefs (Fiedler and Talley 2006). Processes driving translocation of deep nutrients to reef environments include equatorial upwelling (the western Galápagos Archipelago), seasonal wind jet-Ekman upwelling (the Gulf of Tehuantepec, the Costa Rican Dome, the Gulf of Papagayo, and the Gulf of Panama), coastal upwelling (Ecuadorian coastline), and internal waves and tidal bores (e.g., the Gulf of Chiriquí; see Chap. 3, Fiedler and Lavín). Where these processes are particularly intense, such as in the western Galápagos and the Ecuadorian coast, coral reef development can be limited by low temperatures or competition with macroalgae (Glynn and Wellington 1983). Seasonally upwelling environments experience 3–4 month intrusions of nutrient-rich waters that supply inorganic nutrients (up to 15 μM NO_3 and 1.2 μM PO_4) to the upper water column, which in turn increases phytoplankton abundance (up to 1.5 mg·m^{-3}) (D'Croz and O'Dea 2007), decreases water clarity, and can support seasonal increases in benthic macroalgal cover (Wartian 2006) and productivity (Smith 2005, 2006).

11.2 Diversity

11.2.1 Taxonomy

Marine macroalgae are a functional rather than phylogenetic group with members from two Kingdoms and at least four phyla including the Cyanobacteria (prokaryotic blue-green algae, sometimes termed Cyanophyta), Rhodophyta (red algae), Chlorophyta (green algae), and Heterokontophyta (taxa including brown algae, Class Phaeophyceae) (Lee 2008). They encompass a diversity of forms ranging from simple chains of prokaryotic cells to upright foliose forms with complex internal structures analogous to vascular plants.

Cyanobacteria on tropical reefs have adapted to inhabit multiple locations, including taxa that form mats along reef margins or on coral (Smith et al. 2009), are epiphytic on algae or various reef metazoans (Paul et al. 2005; Ritson-Williams et al. 2005; Fong et al. 2006; Smith et al. 2010), rapidly colonize open space after disturbances, and bloom in response to nutrient enrichment (Miller et al. 1999; Littler et al. 2006; Ahern et al. 2007; Paerl et al. 2008; Arthur et al. 2009; Burkepile and Hay 2009). Cyanobacteria seem to be increasing in abundance on many ETP reefs, both as free-living organisms and as epiphytes (Fig. 11.2a; Fong et al. 2006; Smith et al. 2010); their worldwide increase in abundance has been linked to climate change (Paerl and Paul 2012), though this has not been studied for ETP reefs. Because they are chemically defended against herbivory and many species fix atmospheric nitrogen, they may not be strongly regulated by herbivores or nutrient supply, unlike

Fig. 11.2 Algae in four major Divisions occur on ETP reefs, including **a** Cyanophyta (blue-green algae/cyanobacteria), **b** Chlorophyta (green algae), **c** Heterokontophyta (including brown algae, Class Phaeophyceae), and Rhodophyta (red algae) with three functional forms, including (**d**) fleshy algae (**e**), crustose coralline algae (**f**), and upright articulated coralline algae. All photos from the Panamanian Pacific: **a** Saboga Island, October 7, 2002, **b**, **d**, and **e** Dog Island, March 12, 2003, **c** Uva Island, March 14, 2004, **f** Uva Island, September 18, 2002

the vast majority of other types of algae (but see Miller et al. 1999; Burkepile and Hay 2009).

The **Rhodophyta** (red algae) is the most speciose group across Pacific Central America (Bernecker and Wehrtmann 2009), and this pattern is likely to hold for the entire ETP. The most common forms of red algae include the crustose members of the family Corallinaceae as well as a high diversity of small, filamentous turf-forming species. Both groups are ubiquitous across reef zones, with crustose taxa covering 50–80 % of the benthos in some reef zones (Fig. 11.2e) (Guzmán and Cortés 1989; Edgar et al. 2011). However, in areas where herbivory is reduced and nutrient supplies enhanced, some genera of upright and branching calcifying forms, such as *Galaxaura* and branching or flattened foliose red algae in the genera *Acanthophora*, *Laurencia*, and *Halymenia*, can be quite abundant (Fig. 11.2d, f). These genera can also be found in off-reef habitats where structural complexity is low and shelter for herbivorous fishes sparse.

Chlorophyta is the second most speciose division of algae in the ETP (Wysor 2004; Bernecker and Wehrtmann 2009) with a breadth of ecological strategies ranging from fast-growing, extremely productive, and highly palatable species to slow-growing persistent forms with both chemical and physical defenses against herbivores. While some green algae such as *Caulerpa* can be found in high abundances and can even form harmful algal blooms on ETP reefs (Fernández and Cortés 2005; Smith et al. 2010), some genera that are common in other reef systems, such as *Udotea, Penicillus,* and *Halimeda,* are missing or rare on ETP reefs (Wysor 2004; Bernecker and Wehrtmann 2009). In many regions of the ETP soft sediment habitats are limited and this may, in part, explain the scarcity of members of these genera.

Brown algae in the class Phaeophyceae (**Heterokontophyta**) have overall low diversity and abundance in tropical areas, including regions of the ETP (Wysor 2004; Bernecker and Wehrtmann 2009). However, some genera of complex and structurally robust forms, such as *Sargassum*, dominate in high-energy reef zones, though they are less common in the ETP than on reefs in the Indo-Pacific or the Caribbean. Other groups of fast-growing and more opportunistic genera, such as *Dictyota* and *Padina*, may occur on ETP reefs, but are usually restricted to refuges from herbivory (Fig. 11.2c).

The most recent effort to document the diversity of marine algae in the ETP is a species checklist for Pacific Central America, from Guatemala to Panama (Fernández-García et al. 2011). A total of 379 species were listed (Fernández-García et al. 2011) with red algae comprising ~66 %, followed by green algae with ~21 %, and brown algae with ~12 %. Costa Rica has the greatest diversity of marine macroalgae (216 species), then Panamá (174), El Salvador (146), Nicaragua (24), and Guatemala (16), with no published records for Honduras. The authors suggested that geographical variation in number of species was likely due to differences in geomorphology, habitat heterogeneity among national coastal-marine zones, and unbalanced research efforts. They stress that our knowledge of macroalgal diversity in this region is constrained by a short history of investigation and a limited number of taxonomists. New records of species were found in collections in Pacific Costa Rica as recently as 2009 (Bernecker and Wehrtmann 2009), suggesting our underestimation of algal diversity may be extensive.

11.2.2 Functional Forms and the Relative Dominance Model

Although to a phycologist all algae that can be seen with the naked eye are considered macroalgae, coral reef scientists have developed two complementary grouping methods, functional forms and the relative dominance model, that are commonly used during benthic surveys. Confusion can arise with these systems because the term macroalgae is often only used for algae that are large and foliose, implying that some types of smaller filamentous turfs or crustose benthic algae are not macroalgae. However, field ecologists often must rely on these grouping systems when surveying algal communities because they are typically extremely diverse (http://www.algaebase.org) and field identification is difficult or impossible for the vast majority of species (Abbott 1999; Lee 2008). Thus, to avoid confusion in this chapter, hereafter we will always discriminate between foliose macroalgae, algal turfs (diminutive, often filamentous forms), and CCA. We will use the term foliose macroalgae for those plants that are in the functional groups containing algae that are large, upright, and have complex morphology.

The rationale underlying the use of functional form groups is that algae of similar morphological form may function more similarly than those that are morphologically disparate yet more closely related phylogenetically. Steneck and Watling (1982) formed seven functional groups for macroalgae based on susceptibility to grazing by gastropods. Littler and Littler (1984) based their functional form groups on a broader set of characteristics including nutrient uptake rates, productivity, turnover rates, and resistance to herbivory. The six functional form groups proposed by Littler and Littler (1984) are sheet-like, filamentous, coarsely-branched, thick-leathery, jointed-calcareous, and crustose forms (Table 11.1). These groups are listed from fast-growing opportunistic species that respond quickly to increases in available nutrients and are most susceptible to herbivory (sheet-like) to the slowest-growing persistent species that are not readily consumed (crustose). While functional form groupings have been used extensively over the last two decades, there is a burgeoning recognition that macroalgae do not always fit neatly into discrete categories, especially in areas subject to multiple human impacts (e.g., Fong and Fong 2014).

Table 11.1 Functional-form groups of predominant macroalgae with their characteristics and representative taxa

	Functional-form group	External morphology	Comparative anatomy	Thallus size/texture	Example genera
1	Sheet-like algae	Flattened or thin tubular (foliose)	One to several cell layers thick	Soft, flexible	*Ulva* *Halymenia* *Dictyota*
2	Filamentous algae	Delicately branched	Uniseriate, multiseriate, or lightly corticated	Soft, flexible	*Chaetomorpha* *Cladophora* *Gelidium*
3	Coarsely branched algae	Terete, upright, thicker branches	Corticated	Wiry to fleshy	*Acanthophora* *Laurencia*
4	Thick leathery macrophytes	Thick blades and branches	Differentiated, heavily corticated, thick walled	Leathery-rubbery	*Sargassum*
5	Jointed calcareous algae	Articulated, calcareous, upright	Calcified genicula, flexible, intergenicula	Stony	*Galaxaura* *Amphiroa*
6	Crustose algae	Epilithic, prostrate, encrusting	Calcified, heterotrichous	Stony and tough	*Porolithon* *Hydrolithon*

Adapted from Littler and Littler (1984)

The Relative Dominance Model is a broader grouping of major space-holding organisms on coral reefs (Littler and Littler 1984). This model predicts which of four groups will be more abundant under different combinations of nutrient supply and herbivory pressure (Fig. 11.3): under low nutrients and high herbivory stony corals predominate, under low nutrients and low herbivory algal turfs predominate, under high nutrients and high herbivory CCA predominate, and under high nutrients and low herbivory frondose/foliose macroalgae (this group is often referred to simply as macroalgae in other references) predominate.

While algal turfs are often spatially dominated by filamentous red algae, they also can include filamentous green algae and cyanobacteria, as well as cropped bases of larger algae. Turfs globally are characterized by extremely high rates of primary productivity, though biomass is usually very low, suggesting an opportunistic life-history strategy where success is a result of growing slightly faster than the rate at which they are consumed by herbivores (Carpenter 1986; Duffy and Hay 2001). Overall, algal turfs dominate reefs of

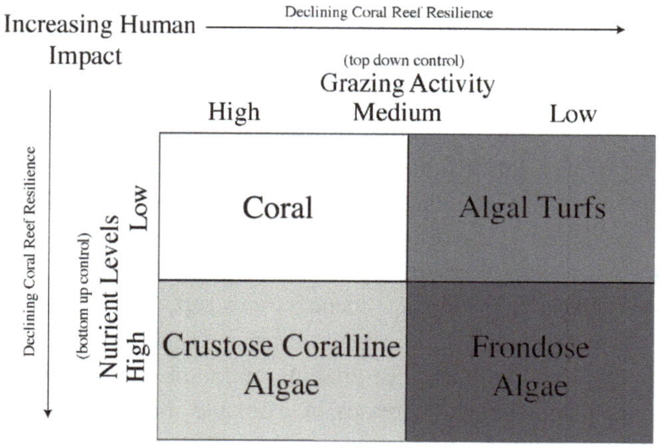

Fig. 11.3 The Relative Dominance Model proposed by Littler and Littler (1984) groups coral reef benthic communities into four categories predicted to predominate under different combinations of herbivory and nutrient supply. Note that the category "frondose algae" in Littler and Littler (1984) is the same group here termed "foliose macroalgae"

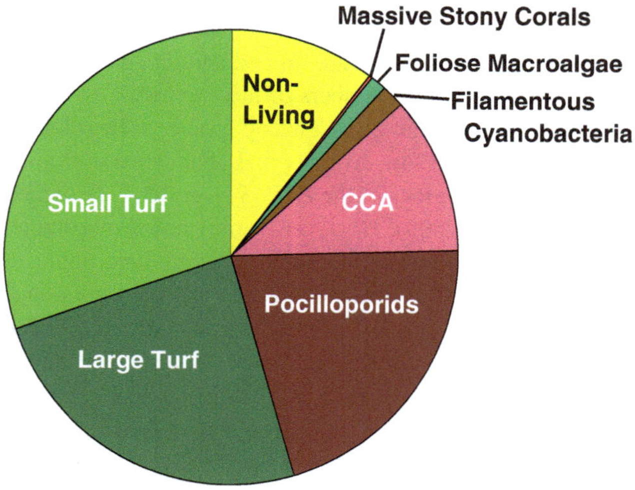

Fig. 11.4 An example of benthic coral reef composition from the ETP. Relative coverage of sessile epibenthic biota and non-living substrate (e.g., rock, sand) from six selected reefs in the Gulf of Chiriquí (3) and Gulf of Panamá (3), Panamá. Coverage estimates are derived from benthic diver estimated or photo quadrats (28 × 21 cm) taken every 5 m along a randomly laid 25 m linear transect (n = 173) across fore reef, crest, reef flat, and back reef habitats from 2002 to 2004 (Smith 2005). Turfs were composed of filamentous algae and separated into small turf algae (<1 cm in height) and large turf algae (>1 cm)

the ETP, occupying over half of all available space on the reef (Fig. 11.4).

The CCA are also members of the Rhodophyta and occur ubiquitously on coral reefs. Productivity levels, however, are low relative to other algal groups, suggesting a "persister" life history strategy. In general, CCA are both abundant and widely distributed in the ETP, with maximum abundance in shallow turbulent areas. The mean abundance of CCA from reef surveys in Panama (11.7 %, Fig. 11.4) is just below the range seen in offshore coral reefs of the Great Barrier Reef, but greater than sediment impacted nearshore reefs (Fabricius and De'ath 2001).

11.3 Algal Habitats Unique to the ETP

Coral reefs of the ETP exhibit typical reef zonation with reef bases, slopes, crests, flats, and back reef habitats (Glynn and Maté 1997), though they lack intertidal algal ridges typical of reefs in high energy environments. On ETP reefs, corals reach their highest cover on the reef crests and the lowest on reef flats, while algal turfs predominate in algal communities across all shallow reef zones, followed by CCA (Fig. 11.5). Cyanobacteria and foliose macroalgae comprise only a small portion of the algal community of shallow reef zones. Along with these typical reef zones, there are several unique habitats on ETP reefs including crests comprised of living pocilloporid framework, large areas of loose dead coral rubble, and deeper algal-dominated areas.

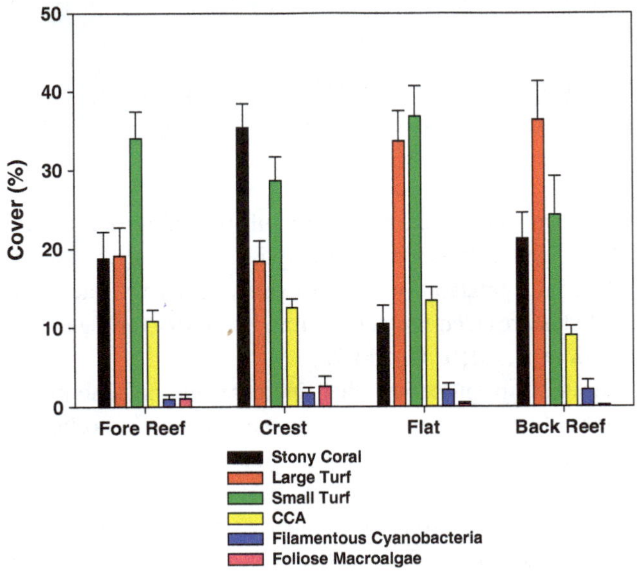

Fig. 11.5 The benthic composition of different coral reef habitats in shallow reef zones (<7 m depth) in the Panamanian Pacific. Methods as in Fig. 11.4

11.3.1 Reef Crests and Flats

Reef crests and flats on coral reefs in the ETP with true framework construction can be highly productive habitats (Smith 2005). La Niña conditions result in extreme low tides and coral mortality that open up space for algae (see Chap. 8, Glynn et al.). While reef flats are typically dominated spatially by CCA and algal turfs (Figs. 11.5 and 11.6a), the reef crest often supports a slightly higher abundance of foliose macroalgae (Fig. 11.6b). Part of this difference may be due to refuges for macroalgae from intense grazing within the living pocilloporid framework of the crest and the high abundance of territorial damselfishes where grazing by other herbivores is reduced. Overall, however, these zones are subject to very high grazing and typically support among the highest biomass of roving herbivorous fishes and sea urchins, although there is much reef-to-reef variability (Smith 2005). Many afternoons the reef flat is seen to effervesce with gas, presumably photosynthetically-generated oxygen (Fig. 11.6c), and suggesting very high rates of photosynthesis.

11.3.2 The Unique Rubble Habitat

Shallow coral reefs of the ETP are typically dominated by branching *Pocillopora* spp. and have a long history of widespread coral mortality during severe ENSO events. Therefore, it is not surprising that dead pocilloporid frameworks and rubble aggregations of dead pocilloporid branches are the dominant habitat supporting macroalgal communities. In particular, rubble fields are often present around living frameworks of *Pocillopora* spp. (Enochs 2012) and support extensive and productive algal communities as well as diverse cryptic metazoan populations (Fig. 11.6d). These substrates are unique in that they are often unconsolidated, and therefore can move under the influence of swell or other disturbances. However, since pocilloporid reefs of the ETP tend to form in the lee of islands, much movement of rubble is likely driven by bioturbation from fishes, particularly large balistids, carangids, and parrotfishes (Glynn 1974; Glynn et al. 2014; see Chap. 12, Alvarado et al.). This provides a shifting three-dimensional substrate for algal turfs and CCA and can be the nucleus for the formation of dense deposits of rhodoliths. Sides and undersides of rubble are temporarily protected from intense grazing and large algal thalli may be concentrated around the fringes of rubble fragments (Fig. 11.6e). The area of these rubble habitats usually comprises a large percentage of the reef (Enochs 2012) and is likely one of the major sources of primary productivity. However, it is unknown how the instability of these rubble fragments affects the overall productivity of this habitat.

Fig. 11.6 Habitats dominated by algae in the ETP. **a** Reef flat dominated by CCA and algal turf, **b** reef crest with frondose algae amidst branching coral, **c** reef flat with photosynthetically-produced effervescence (center of picture) **d** "rubble fields", that are dominated by **e** large algal thalli under and on sides of rubble, and **f** deeper water algal communities on sand or rubble plain adjacent to reefs. All photos are from the Panamanian Pacific; **a** Contadora Island, March 15, 2009, **b** Uva Island, March 14, 2004, **c** Uva Island, March 6, 2003, **d** Uva Island, March 10, 2003, **e** Saboga Island, October 7, 2002, **f** Coibita Island, March 4, 2003

11.3.3 Near-Reef and Deeper-Water Algal Refuges

Deeper water areas (>12 m depth) around the talus slope of reefs and adjacent sand and rubble plains of ETP reefs support extensive communities of large filamentous cyanobacteria and foliose macroalgae. These areas experience a unique set of bio-physical properties, such as higher nutrients and lower herbivory, that support algal proliferation. Lower herbivory is likely due to the relatively flat topography of these areas providing a refuge from herbivores as there are limited sheltering opportunities for herbivorous fishes and sea urchins. One off-reef algal community dominated by *Caulerpa sertularioides* was shown to experience reduced herbivory compared to areas on the reef. However, herbivory still played a role in regulating algal abundance as epiphytization by cyanobacteria reduced the palatability or detectability of macroalgae. Thus, epiphytes provided protection from herbivory for algal thalli (Smith et al. 2010; Fig. 11.7a).

Bottom-up processes that increase the availability of water column inorganic nutrients may also be important for deep water and near-reef algal communities. The thermocline is often present at depths below about 12 m and is associated with a strong nutricline (D'Croz and O'Dea 2007; Smith et al. 2014). Near-reef areas are often influenced by fluctuations of the shallow thermocline through internal waves that can advect subthermocline/nutricline waters up reef slopes, potentially enriching reef areas (Smith et al. 2010). In addition, the processes that maintain near-reef

(a)

(b)

Fig. 11.7 Deeper water algal communities, such as the one dominated by **a** *Caulerpa sertularioides* are thought to proliferate due to reduced herbivory, increased nutrients, and an association with an epiphytic cyanobacterium that provides an associational defense. **b** These processes that maintain near-reef populations may also be responsible for promoting blooms of macroalgae on reefs, especially during La Niña years. Both photos taken at Uva Island, Panama; **a** September 14, 2002, 6 m depth, *photo credit* T.B. Smith, **b** March 14, 1990, 2 m depth, *photo credit* JS Feingold

populations of bloom-forming macroalgae, such as *Caulerpa sertularioides*, may also be responsible for promoting blooms of macroalgae on reefs (Fig. 11.7b; Glynn and Maté 1997; Fernández and Cortés 2005) when stimulated by large-scale nutrient pulses, such as occur during La Niña phases of ENSO.

Physical processes can also limit the maximum depth of algal community development. In much of the neritic ETP, areas deeper than ~25 m have strongly attenuated light, and mean oxygen concentrations can be less than 3 ml l^{-1} (Prince and Goodyear 2006; Smith et al. 2014). Given these constraints, the biomass of algae appears to peak between 10 and 20 m in off-reef habitats. This biomass includes genera of larger red algae (*Halymenia, Acanthophora*), as well as green (*Chaetophora, Halimeda*) brown (*Colpomenia, Padina*), and a variety of long (>1 cm) filamentous algae.

11.4 Ecological Processes Controlling Algal Populations and Communities

The distribution and abundance of coral reef algae in the ETP are controlled by the same ecological processes that control other primary producers: abiotic limits include availability of suitable substrates, light quantity and quality, nutrients, and disturbance. In addition, biotic processes may strongly control biomass accumulation, including herbivory, competition, and facilitation.

11.4.1 Abiotic Factors Limiting Settlement and Growth

11.4.1.1 Suitable Substrate

While marine algae require "hard" substrata to settle upon, the definition of hard is very broad; hard substrates can be abiotic or living surfaces, such as rock, coral rubble, shells of animals alive or dead, seagrass blades, mangrove roots, and other algae. When an alga settles and grows on another living organism, it is called epiphytic, and this type of association is quite common on tropical reefs. The effects of epiphytes on their hosts can range from strongly negative to strongly positive. Negative effects may include competition for light and nutrients, accumulation of sediments, increased drag, and reduced gas exchange while positive effects can be reduced herbivory if the epiphyte has defenses (Fong et al. 2006; Smith et al. 2010), and potentially increased nutrient supply if the epiphyte is a nitrogen fixer. Some benthic organisms, like many coralline algae, can shed outer layers of cells where epiphytes recruit, effectively acting as an anti-fouling mechanism. Others, like corals, produce and slough mucus off their surfaces to rid themselves of algal settlers. Chemical defenses are also employed by marine algae and invertebrates to deter algal settlers and other fouling organisms (Lane and Kubanek 2008; Chadwick and Morrow 2010). Thus, not all hard substrate is equally "available" for algal recruitment.

Algae rapidly and efficiently colonize almost any substrate space that has been newly opened by disturbances and where enough light penetrates. Several workers have hypothesized that the widespread and global occurrence of phase shifts from coral to algal dominance is largely a result of algae opportunistically settling on open space resulting from coral mortality (e.g., Bruno et al. 2009). While some types of algae in the ETP, such as *Caulerpa* with its root-like rhizomes, have been observed to directly over grow live coral (Fig. 11.7b); how frequently and under what conditions algae can grow over live coral is highly debated. It is known that algae can rapidly colonize dead coral skeletons following episodes of coral bleaching and mortality (e.g., Baker et al. 2008) as has been observed in the ETP after massive coral mortality following ENSO-associated warming. However, in the ETP corals can recover after mortality quite rapidly by literally re-sheeting the algal-covered skeleton (Glynn and Fong 2006), suggesting that certain corals can be better competitors for space than at least some forms of algae over the longer-term.

11.4.1.2 Light

Like all primary producers, reef algae depend on light to photosynthesize. Although tropical waters are generally clear with high light penetration, in the ETP many areas are subject to upwelling (see Chap. 3, Fiedler and Lavín) with increased nutrient supplies resulting in increased phytoplankton and decreased light availability (D'Croz and O'Dea 2009). Given the range of upwelling systems in the ETP, light penetration is continuously, episodically, or seasonally reduced, depending on the area. In addition, neritic areas subject to high terrestrial runoff also experience high turbidity due to shading and light scattering by water column particulates. Preliminary work suggests that reduced light and increased nutrients, which interact, should both have effects on algal productivity, diversity, and abundance (Smith et al. 2014; see Chap. 17, Smith et al.); however, these relationships have not been thoroughly characterized across reefs and reef zones.

Both the quality and quantity of light available for photosynthesis change with depth, though these changes vary dramatically with different water characteristics. Light attenuation is increased by suspended material such as sediment, detrital particles, dissolved organic matter, and biota, including phytoplankton. The net result is most red light is absorbed in the first few meters of depth while blue and green light penetrates the deepest (Fig. 11.8). Different algal taxa have unique adaptations to changes in quality and quantity of light that may affect their depth distribution. Algae in different divisions contain unique pigments (light absorbing compounds) that have different action spectra. Peaks in this spectrum show which wavelengths of light are most effective in fueling photosynthesis for each algal pigment.

Light availability can be affected over short time scales by changes in cloud cover, variation in shading by the macroalgal community during wind or wave events, and changes in light attenuation with sediment resuspension. In some areas of the ETP, light availability can change over longer time scales between upwelling and non-upwelling seasons. In response, algae can adapt to changing light conditions across temporal scales that vary from minutes to a few months. These adaptations include increasing pigment content, which increases light utilization efficiency and allows macroalgae to

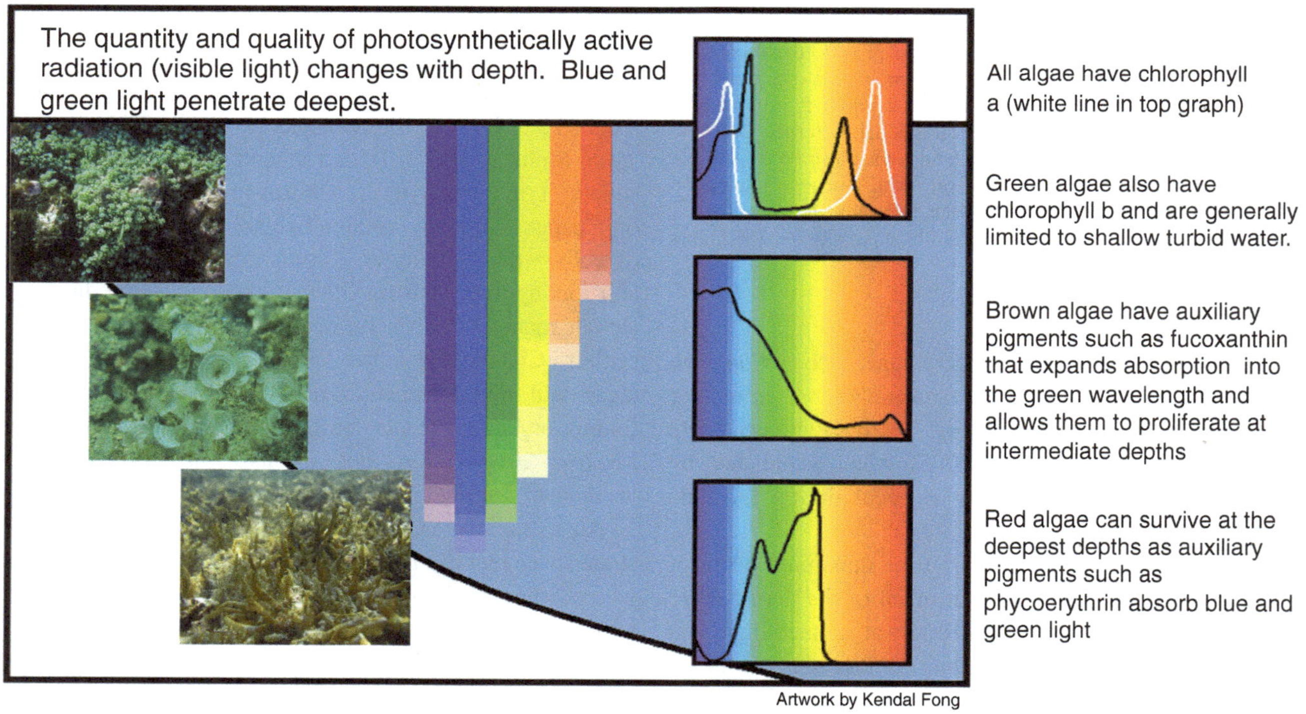

Fig. 11.8 Absorption spectrum of major pigments for three divisions of eukaryotic marine algae. Note that blue-green algae share pigments with red algae. While pigments set the lower depth limit of algae, they do not restrict algae from shallower depths. Thus, red algae are often both the shallowest and deepest algae found on coral reefs

photosynthesize more effectively at low light levels. There are energetic costs to the production and maintenance of pigments, however, so at high irradiances pigment concentrations are typically lower. At high irradiances or under high UV stress, photoinhibition may occur, and damage to algal photosystems during such periods may not be reversible.

11.4.1.3 Nutrients

Six decades ago in a seminal paper, Odum and Odum (1955) estimated that supplies of allochthonous nutrients to coral reefs of Eniwetok Atoll via advection of oceanic water were not sufficient to sustain the high levels of productivity characteristic of these systems. This finding motivated 50 years of research into how high levels of productivity are supported on coral reefs. This research was focused on the relative importance of allochthonous versus autochthonous sources of nutrients, the role of nutrient limitation and key limiting resources, and the role of benthic algae in nutrient cycling, including uptake, storage, retention, and recycling of water column nutrients. Each of these topics has received some attention in the ETP. Unsurprisingly, nutrient cycling and productivity in the ETP are likely higher than in other coral reef systems such as Eniwetok Atoll, due to a shallow thermocline where any vertical advection of water may provide nutrients from deeper sources.

Allochthonous Versus Autochthonous Sources—Upwelling, Thermocline Shoaling, and Internal Recycling

On coral reefs worldwide, the primary source of allochthonous or "new" nutrients depends on a variety of factors including: (1) the geographic location and oceanographic setting of the reef system, (2) the type of coral reef (fringing, barrier, or atoll), and (3) the type of terrestrial system to which it is coupled (Fong and Paul 2011). Many global factors that influence nutrient supplies depend upon the geographic location of a reef in relation to currents (Garzón-Ferreira et al. 2004), atmospheric aerosol/particle deposition (Muhs et al. 2007), and degree of isolation from major continental influences. System-specific factors include the type of reef and the type of associated terrestrial system, if any. These factors impact the connectivity of the reef system to oceanic versus terrestrial sources of nutrients and will therefore greatly influence their relative importance. Further, recent work across marine systems has suggested that sources of new nitrogen to all marine systems are rapidly increasing globally as a result of anthropogenic alterations of the global nitrogen cycle (for reviews see Vitousek et al. 1997; Downing et al. 1999), though the impacts of these increases to coral reefs are highly debated (for a review see Fong 2008).

It is likely that productivity in many areas of the ETP is strongly controlled by allochthonous nutrients derived from natural oceanic sources. Because many ETP reefs occur well offshore, and are therefore relatively remote from continental terrestrial influences, they are likely to experience less terrestrially-derived allochthonous nutrient sources. Instead, the relatively shallow thermocline of the ETP brings nutrient-enriched water close to the surface and the likelihood that upwelling fuels productivity in this region. Evidence to support this claim includes a correlation between upwelled nutrients and increased phytoplankton abundance (D'Croz and O'Dea 2007) and greater cover of foliose macroalgae and filamentous cyanobacteria on Saboga reef, Panama during upwelling (Smith 2005). Furthermore, in situ experiments in the Gulf of Panama showed increased algal biomass accumulation in herbivore exclosures (Smith 2005) or with elevated nutrient or sediment levels (Muthukrishnan 2013) during upwelling.

Autochthonous or in situ nutrient supplies to coral reef algae come from sources including nitrogen fixation, recycling from other biota, regeneration from deposition of organic matter in any associated sediment areas, and regeneration of nutrients from primary producers and consumers during decomposition (Fong 2008; see Chap. 9, Enochs and Glynn). Nitrogen-fixing benthic cyanobacteria, especially those in the genus *Lyngbya*, are becoming increasingly common on coral reefs worldwide, presumably due to their ability to proliferate under stressful environmental conditions, including warm temperatures, and their strong chemical defenses against herbivores (Paul 2008). Nitrogen fixation on coral reefs may be very high, especially when filamentous cyanobacteria colonize newly opened substrates following disturbances (Larkum 1988). However, although abundant communities of cyanobacteria have been observed on ETP reefs (Fong et al. 2006; Smith et al. 2010), rates of N-fixation have not been studied. Other sources of autochthonous nutrients on reefs include nutrient-rich waste products of invertebrates (e.g., Williams and Carpenter 1988), and fishes (Burkepile et al. 2013; Allgeier et al. 2014), flocculent material settling on the surfaces of algal thalli (Schaffelke 1999), and regeneration of nutrients during decomposition (Stimson and Larned 2000). These sources, however, have also not been quantified on ETP reefs, so their importance remains unknown.

The Role of Nutrient Limitation and Key Limiting Resources

The extent to which nutrient limitation of coral reef algae shapes the benthic structure of coral reefs is an extremely controversial issue across all reef systems (e.g., McCook 1999; Szmant 2002). Some workers hypothesize that nutrients are of key importance (e.g., Lapointe et al. 2005), while others suggest that the typically high productivity of tropical systems, even in areas characterized by low water column nutrient concentrations, suggests that algae may be rarely, if ever, limited by nutrients (Hatcher 1988; McCook 1999). In the ETP, nutrients were found to strongly limit growth of two species of algae that form blooms on the Uva Island reef, Gulf of Chiriquí, Panama. *Caulerpa sertularioides* forms persistent and large patches in off-reef habitats where they are subjected to less intense herbivory pressure due to lack of shelter for herbivorous fishes and provision of an associational defense by epiphytic cyanobacteria (Smith et al. 2010). These patches proliferate onto reefs and kill corals during La Niña conditions, presumably in response to increased allochthonous nutrients supplied to the reefs from thermocline shoaling. *Caulerpa sertularioides* showed a strong growth response to nutrient addition in in situ experiments repeated both on the Uva Island reef and in *C. sertularioides* patches off reef (Fig. 11.9a). Enrichment led to increases in growth responses up to an order of magnitude above ambient conditions, demonstrating strong nutrient limitation. *Acanthophora spicifera* formed a persistent bloom for at least seven years on Uva Island reef; this bloom also was protected from herbivory by epiphytic cyanobacteria and was strongly nutrient limited. When protected from herbivory, experimental nutrient addition in situ stimulated growth of *A. spicifera* across all reef zones, including slope, base, crest and reef flat, with increases ranging from ~30 to 150 % in four days (Fig. 11.9b). Herbivores, however, also targeted enriched thalli; when the same algal treatments were deployed without cages, consumption increased from 30 to 100 % in 24 h. Thus, increased herbivory over-compensated for the increased growth, masking evidence for nutrient-limitation for this species.

Nutrients also were found to limit non bloom-forming algae on some ETP reefs (Wartian 2006; Muthukrishnan and Fong 2014). Experimental addition of nutrients was found to increase cover of algae on a reef in the non-upwelling Gulf of Chiriquí (Wartian 2006) whereas in the upwelling Gulf of Panama similar experiments had mixed results (Wartian 2006; Muthukrishnan and Fong 2014). In addition, overall increases in algal cover in the Gulf of Chiriquí were modest, less than 10 % over 6 months compared to about 25 % increases in algal cover during upwelling in the Gulf of Panama during the same time period. In the Gulf of Chiriquí, increases in algal cover were due to increased algal turf and small filamentous forms of algae. On Contadora reef in the Gulf of Panama, the pattern was more complicated. Wartian's earlier experiment (Wartian 2006) documented benthic shifts during upwelling with increases in foliose or corticated macroalgae overgrowing coral; these increases were replaced by turf algae and CCA after upwelling ended. In contrast, Muthukrishnan and Fong (2014) showed turf

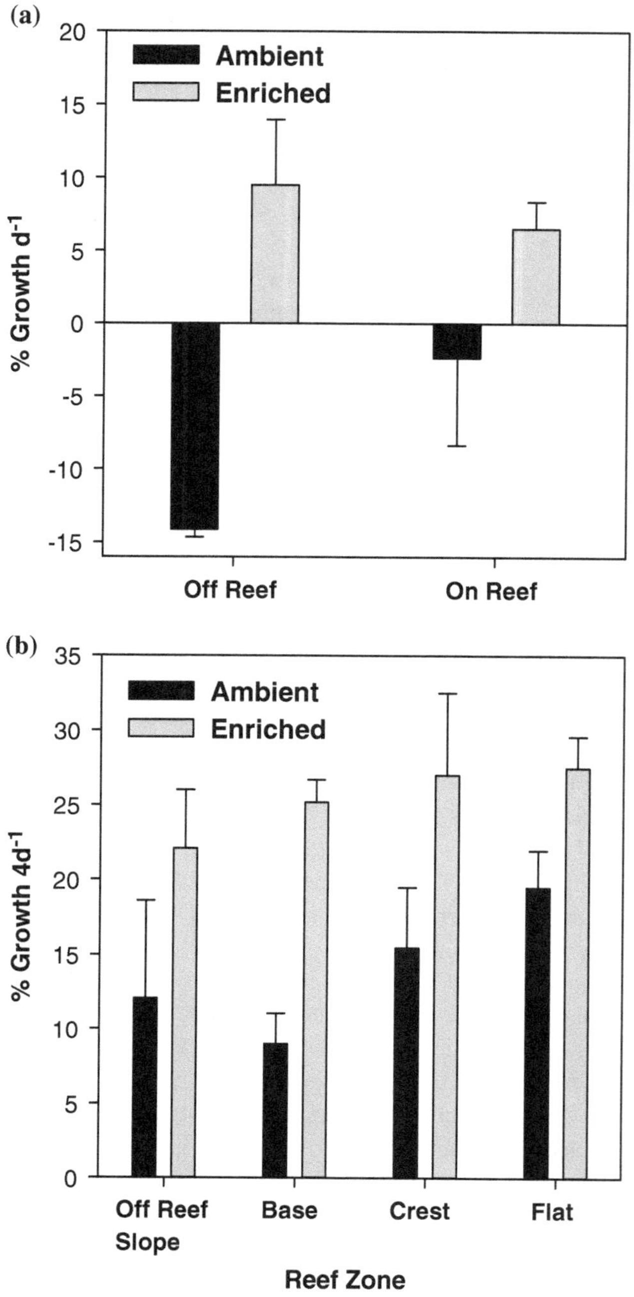

Fig. 11.9 Nutrients strongly limited growth (mean ± SE) of **a** *Caulerpa sertularioides* measured both on the reef (3 m depth) and in a deeper off reef refuge (6 m depth), and **b** *Acanthophora spicifera* across four reef zones: off reef slope (4–15 m depth), base (2 m depth), crest (1 m depth), and reef flat (1 m depth). The negative growth of *C. sertularioides* when not supplied with nutrients in the off reef location was attributed to dissociation of algal thalli from sediment nutrient sources (Smith et al. 2010). Both studies were conducted on Uva Island reef: **a** March 2005, **b** March 2002

algae responded to nutrient enrichment and directly replaced coral during the upwelling season. While both experiments showed that experimental nutrient enrichment could influence the benthic community, it is clear that the large-scale increases in nutrient supply during upwelling were more efficacious in alleviating nutrient limitation than small-scale experimental additions.

Although more data are needed to understand the role of nutrient limitation in the ETP, it is important to continue efforts to unravel the complexity of this process. While ETP reefs have not historically been impacted by terrestrial nutrient sources, the global rise in anthropogenic nutrient enrichment has increased the likelihood this may become a concern in the future (see Chap. 20, Cortés and Reyes Bonilla).

For tropical marine algae on coral reefs worldwide, developing generalizations about the relative importance of nitrogen (N) and phosphorus (P) in limiting growth have remained elusive. There is no consensus for tropical systems, unlike temperate systems where N limitation is the paradigm (e.g., Fujita et al. 1989; Thybo-Christesen et al. 1993; Rivers and Peckol 1995; Taylor et al. 1995; Gallegos and Jordan 1997; Sfriso and Marcomini 1997). Rather, while some tropical studies found that P could limit productivity and growth more frequently than N (e.g., Lapointe 1987, 1989; Lapointe et al. 1992) other studies have found growth limitation by both N and P (Lapointe 1987; Schaffelke and Klumpp 1998; Fong et al. 2003; Paerl et al. 2008), and one study (Larned 1998) found N to be strongly limiting across many dominant species of algae.

In the ETP, results of factorial enrichment experiments testing for N and/or P limitation for macroalgae followed the global pattern of mixed results (Fig. 11.10 a, b). In these bioassays, algae were collected from the study site and known masses placed in experimental units where N and P were added alone and in combination, and growth responses quantified. When addition of a nutrient increased growth, it was considered to be in limiting supply. These studies were conducted on *Acanthophora spificera* and *Caulerpa sertularioides,* both bloom-forming species on Uva Island reef. *Acanthophora* was strongly limited by P and secondarily by N (Fong et al. 2006), while *C. sertularioides* was limited by both nutrients, with N showing the greatest stimulation of growth (Smith et al. 2010). These experiments, however, were conducted at different times and with algae from different locations, so results cannot be directly compared (Keeley et al. 2015). Other studies have found relative nutrient limitation to vary among species from the same location and over time (for a review see Fong and Paul 2011). Thus, much work is needed to understand the role of key limiting resources in the ETP.

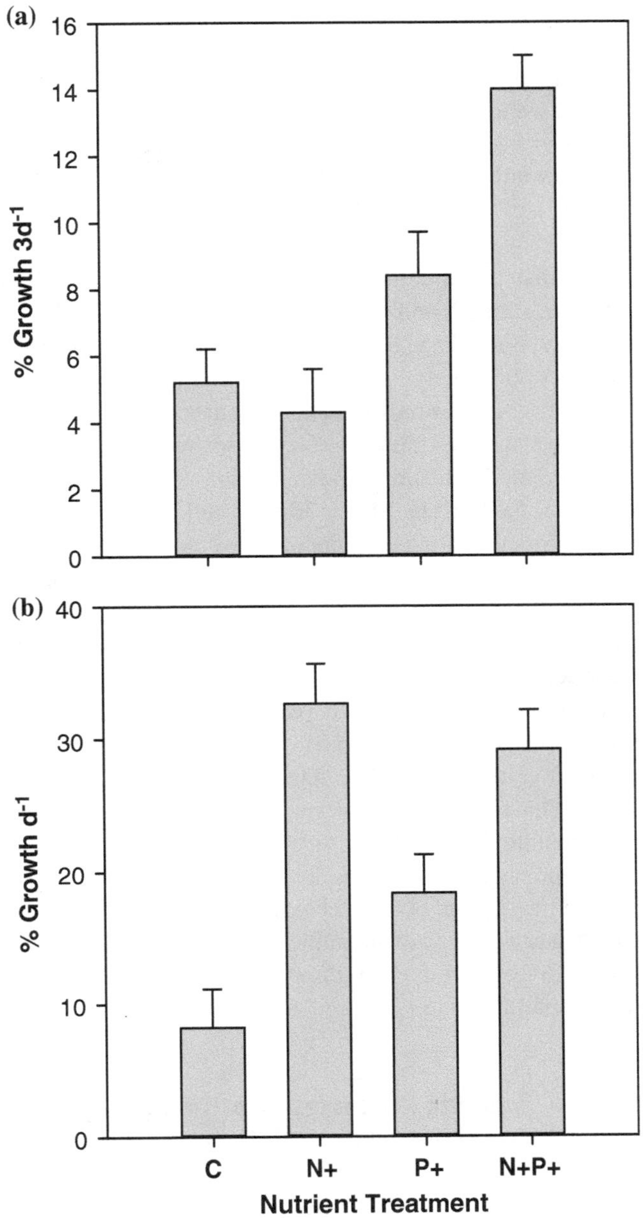

Fig. 11.10 The relative importance of nitrogen versus phosphorus limitation on growth (mean ± SE) for two species of algae that form blooms on ETP reefs, **a** *Acanthophora spicifera*, and **b** *Caulerpa sertularioides*. Treatments are ambient seawater control (C), addition of 20 μM N (N+), addition of 2 μ P (P+), and addition of both 20 μM N and 2 μ P (N+P+). Both studies were conducted in Panama; **a** Uva Island, February 2001, **b** Cavada Island, May 2000

11.4.1.4 Disturbance

The importance of disturbance in structuring coral reef communities has long been the focus of intensive research, dating back to a seminal paper relating hurricane disturbance as a diversifying agent of tropical forests and coral communities in Australia (Connell 1978). While disturbance due to hurricanes and tropical storms has been widely documented as an important force that affects many coral reef communities, including algae, much of the ETP is out of the path of major hurricanes (see Chap. 3, Fiedler and Lavín). The exceptions include Mexico, Clipperton Atoll, and the Revillagigedo Islands; also, Easter Island and the western Galápagos Islands can be subject to high wave energy. Similarly, the ETP is relatively safe from outbreaks of the crown-of-thorns sea star,

Acanthaster. *Acanthaster planci* has been found on ETP reefs in thermally stable regions, such as Uva Island in the Gulf of Chiriquí, and in the Gulf of California (Glynn 1972). Thus, tropical cyclones and predator outbreaks, known to be very important elsewhere in the Pacific and Indian Oceans, are relatively unimportant in the ETP.

The key disturbance affecting reefs in the ETP is increased sea surface temperatures associated with ENSO events that cause coral bleaching, and, if the event is of sufficient strength and duration, results in widespread coral mortality (see Chap. 8, Glynn et al.). The effect on algae appears to be largely indirect as coral mortality opens space for rapid colonization by algae. In other systems, benthic algae and corals exhibit a variety of interactions, including competition, facilitation, and inhibition (for a review see Fong and Paul 2011). This is likely true in the ETP, where long-term monitoring has documented that recovery after ENSO disturbances can be very rapid, but also very patchy (Wellington and Glynn 2007; Baker et al. 2008). In many areas of the ETP, remnant coral patches expand and recover (re-sheet) old algal-covered skeletons very rapidly, almost regaining original cover in <6 years (Glynn and Fong 2006). This suggests that corals, under certain conditions, are at a competitive advantage over algae for space in many areas. We also have evidence, however, that in these same regions, algal communities exhibit positive feedbacks that perpetuate algal dominance (Fong et al. 2006; Smith et al. 2010). The relative importance of these ecological processes controlling maintenance of algal communities after disturbance has only begun to be investigated through field experiments and simulation modeling (see Sect. 11.5).

11.4.2 Biotic Processes Controlling Algal Biomass Accumulation

11.4.2.1 Herbivory Limits Algal Proliferation

Historically, herbivory has been considered the primary force limiting the proliferation of algae on coral reefs (e.g., Ogden and Lobel 1978; Hay 1991; Hay and Steinberg 1992; Paul et al. 2001; Hughes et al. 2007; Amsler 2008; Bellwood and Fulton 2008). Coral reef herbivores are capable of removing almost 100 % of the biomass produced daily by marine algae in certain reef habitats (Carpenter 1986; Hay 1991; Hay and Steinberg 1992; Choat and Clements 1998). However, because of the prevalence of overfishing on modern reefs the strength of herbivory may be reduced globally (e.g., Jackson et al. 2001). Populations and communities of herbivorous fishes have been reduced in abundance and their size structures shifted toward smaller size classes as a result of "fishing down the food web" (Pauly et al. 1998). Overall, it appears that the strong limitation by herbivory that existed in the past may be restricted to remote reefs or those in well-enforced marine protected areas (Jackson 2008).

Despite global overfishing, some portions of the ETP retain relatively intact herbivore populations and communities and therefore support high herbivory pressure. Invertebrate herbivores, particularly diadematid sea urchins, are quite common on reefs of the ETP and can provide intense focal herbivory in areas where they occur (Eakin 1996). Long-term survey data showed that, despite high inter-annual variability, over the past 30 years there was no overall decline in communities of herbivorous fishes on Uva Island reef, Gulf of Chiriquí, Panama (Glynn et al. 2014). Bioassays measuring the strength of herbivory on this same reef found rapid rates of herbivory on the palatable red alga, *Acanthophora spicifera*, with up to 75 % biomass consumption in 24 h (Fig. 11.11a) (Fong et al. 2006). Similarly, herbivory was strong even on the chemically defended alga *Caulerpa sertularioides*, with up to 60 % biomass removal in 24 h (Fig. 11.11b; Smith et al. 2010). Factors that may contribute to sustaining this strong top-down control on the Uva Island reef include its protected status as part of a marine reserve, distance from mainland Panama, and absence of a human population. In contrast, surveys of herbivorous fishes on reefs in the Pearl Islands, Gulf of Panama, Panama suggest larger size classes have been reduced, either by natural disturbance (low temperature mortality) or fishing, since size frequency distributions have been skewed toward smaller size classes (Smith 2005). Reductions in herbivorous fishes may have occurred because protection status is recent, a management plan has not yet been approved, and existing laws are poorly enforced (J. Maté, personal communication). However, herbivory remains intense on the Contadora Island reef in the Gulf of Panama. Bioassays similar to those employed on Uva Island showed up to 90 % removal in 24 h of a palatable red alga in the genus *Laurencia* (Muthukrishnan 2013). In addition, proliferation of *Laurencia* and other foliose macroalgae only occurred when protected from reef herbivores, even during the upwelling period (Smith 2006). Thus, while the current status of fish communities suggests human impacts, the force of herbivory appears to be somewhat resilient, at least to this level of impact.

Diversity of Micro-herbivores and Macro-herbivores and Their Functional Roles

Coral reef algae are consumed by a wide variety of invertebrates and vertebrates, and these consumers have different impacts on algal communities depending on their functional roles. Because the herbivore communities of the ETP are less diverse than in the western Pacific and Atlantic regions, there may be limited functional redundancy in feeding guilds, which

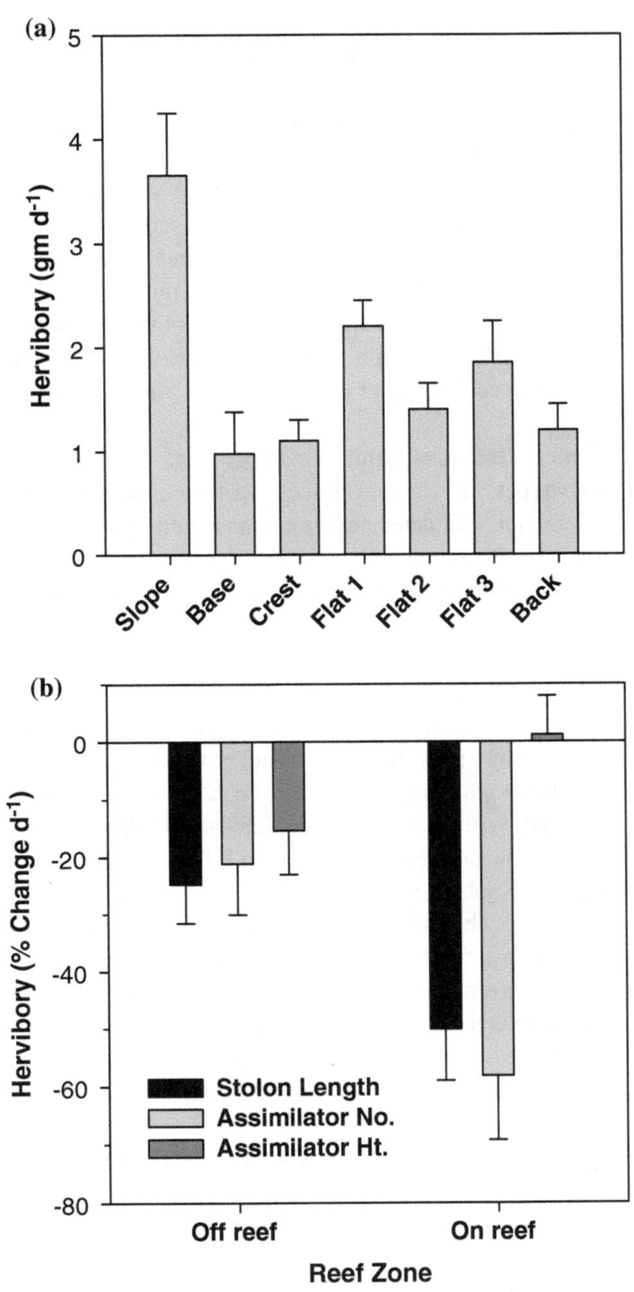

Fig. 11.11 Mean herbivory rates (±SE) on two species of algae **a** *Acanthophora spicifera* and **b** *Caulerpa sertularioides*. In panel **a** consumption rates are reported with an initial biomass of 5 g while in panel **b** herbivory is calculated as percent loss in biomass over time. Note that in panel **a** the reef slope was a shallow, more topographically complex area than most slope habitats. Both studies were conducted at Uva Island, Panama; **a** May 2000, **b** March 2005

may lead to reduced impacts on algal communities if herbivore population densities are reduced. On many ETP reefs, however, herbivorous fishes are very abundant and include blennies (Blenniidae), damselfishes (Pomacentridae), chubs (Kyphosidae), surgeonfishes (Acanthuridae), and parrotfishes (Scaridae), whereas the Indo-Pacific rabbitfishes (Syganidae) are absent. Although the impacts of blennies and damselfishes can be quite localized, the impact of certain damselfishes such as the Acapulco Gregory (*Stegastes acapulcoensis*) can be inordinately important due to their ability to deter large grazing herbivores and to damage corals (Wellington 1982). Three of the six parrotfish species known from the ETP are large-bodied (>50 cm total length) and include *Scarus ghobban*, *Scarus rubroviolaceus*, and *Scarus perrico* (endemic). These species are likely to be important in grazing as scrapers and excavators, and should have large impacts on reef algae (Ong 2007; Plass-Johnson et al. 2013). Browsers that graze foliose macroalgae, such as the small parrotfish *Nicholsina denticulata* (endemic) (Streelman et al. 2002), are relatively rare, perhaps because macroalgae are rare. Surgeonfishes are locally important on some reefs and in some habitats, and include *Acanthurus xanthopterus*, *Acanthurus triostegus*, *Ctenochaetus marginatus*, *Prionurus laticlavis* (endemic), and *Prionurus punctatus* (endemic) (Smith 2005).

Total biomass of herbivorous fishes in the ETP can be relatively high and is generally not limited by the unique oceanography and habitats of the ETP; the exception includes some areas of the Galápagos Islands where reef-associated species are likely limited by the consistent influence of cold water. The mean biomass of all herbivorous fishes on reefs in Panama ranges from 28.6 g m^{-2} ± 3.3 SE (Gulf of Panama) to 55.3 g m^{-2} ± 5.8 SE (Gulf of Chiriquí) (Smith 2005), which is equal to or greater than total herbivore biomass found in a compilation of studies across the Indo-Pacific region (29.0 g m^{-2} ± 4.1 SE) and Caribbean (9.25 g m^{-2} ± 0.8 SE) (Roff and Mumby 2012). Over 85 % of fish herbivore biomass is composed of large-bodied scarids that likely strongly inhibit the proliferation of algal biomass on ETP reefs. Furthermore, in the study by Roff and Mumby (2012) such an exceptionally high biomass of herbivorous fishes was associated with macroalgal cover at or near zero, which is consistent with typical cover on ETP reefs.

The influence of fish herbivores is reduced with declines in temperature on both temporal and spatial scales. The grazing rate of young blue-chin parrotfish, *Scarus ghobban*, rapidly decreased four-fold in the naturally occurring range from high ambient temperatures (29.4 °C) to upwelling temperatures (21.2 °C) in the Gulf of Panama (Smith 2008). Tropical herbivorous fishes are not generally found in sub-tropical or temperate waters, such as in the western Atlantic (Floeter et al. 2005). A similar pattern appears across the ETP, where the influence of tropical fish herbivores is reduced in cooler areas, such as higher latitudes and consistent upwelling centers. In the Galápagos Islands, the abundance of herbivorous fishes decreases in the western and southern parts of the archipelago where sub-tropical and temperate-like cool temperatures are common (Smith, unpublished data). In areas where herbivorous fish abundances are reduced, herbivorous invertebrates tend to become relatively more common (Fig. 11.12).

Herbivorous macroinvertebrates that play strong ecological roles in maintaining low abundances of algae on ETP coral reefs, as elsewhere, are dominated by sea urchins (Echinoidea). Reef sea urchins in the ETP commonly include *Diadema mexicanum*, *Eucidaris thouarsii*, *Eucidaris galapagensis*, *Arbacia stellata*, *Echinometra vanbrunti*, and *Lytechinus semituberculatus*. On reef substrates, *D. mexicanum* and *Eucidaris* spp. are numerically dominant, with a shift in relative dominance from *D. mexicanum* in warmer regions to *Eucidaris* spp. in cooler regions (Fig. 11.12). In warmer areas, the reef-wide densities of *D. mexicanum* are typically less than 1 ind m^{-2}; however, within the crest and fore reef zones sea urchin densities can exceed 18 ind m^{-2} (Eakin 1996), values that are among the highest recorded globally (O'Leary and McClanahan 2010; see Chap. 12, Alvarado et al.). Populations of *E. galapagensis* in the central and southern Galápagos Islands have much higher densities in and around reefs and have been known to have extreme population fluctuations, with an outbreak after the 1982–83 El Niño that decimated carbonate reef structures following coral mortality (Reaka-Kudla et al. 1996; Glynn et al. 2015). Other species that largely account for the balance of sea urchins in the Galápagos in more temperate conditions include *Toxopneustes depressus* and *Lytechinus semituberculatus*.

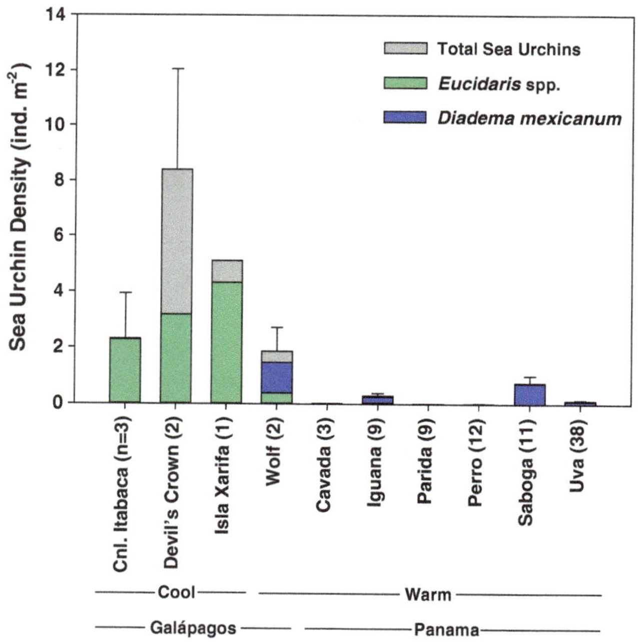

Fig. 11.12 Sea urchin mean population densities (±SE) in coral communities and on coral reefs in the ETP, 2001–2002. Densities were highest in cooler environments of the Galápagos Islands. Numbers in parentheses with localities are the numbers of 25 × 2 m belt transects sampled

Relatively little is known about the impacts of herbivorous micro-invertebrates on reef algae in the ETP or elsewhere (but see Carpenter 1986 for a Caribbean study). A variety of micro-invertebrate crustacean and gastropod molluscs are known to be herbivores and their impact on algal communities deserves further scrutiny. While overall algal control on ETP reefs may fall to large herbivorous fishes and sea urchins, micro-invertebrates with specialist diets could have important roles by affecting specific algal populations, such as ascoglossan molluscs that specialize on *Caulerpa* spp. (Gavagnin et al. 1994), and have frequently been observed feeding at coral reef study sites in Panama.

Chemical Defenses Limit Herbivory

Many types of algae produce structural and chemical defenses that can deter herbivores and thereby reduce consumption (Duffy and Hay 1990; Wahl and Hay 1995). Structural defenses, such as calcification and toughness (e.g., Paul and Hay 1986; Steneck 1988; Duffy and Hay 1990), and chemical defenses (for reviews see Paul et al. 2001, 2007; Amsler 2008; Paul and Ritson-Williams 2008) are common, and multiple mechanisms of defense may function together (Hay 1997; Paul 1997; Paul et al. 2001). It is likely that benthic community structure on ETP reefs is strongly influenced by the anti-herbivory defenses of algae. Most of the foliose macroalgae that occur on ETP reefs, and are not restricted to spatial refuges, contain structural and/or chemical defenses that allow them to persist in the presence of abundant and diverse herbivores. Studies from other reefs subject to overfishing suggest that, once overall herbivory is reduced, selective grazing will remove only the most palatable species, shifting algal community dominance to unpalatable algae (Fong and Paul 2011).

11.4.2.2 Competition Between Corals and Algae

Mechanisms of competition between algae and corals include pre-emption of space, shading, allelopathy, attraction of settling larvae to ephemeral algal surfaces, abrasion, basal encroachment and increased sedimentation (Chadwick and Morrow 2010). According to many reviews, the nature and importance of competitive interactions between corals and macroalgae cannot be generalized as they vary greatly across different species of macroalgae (for reviews see McCook et al. 2001; Ritson-Williams et al. 2009; Chadwick and Morrow 2010; Fong and Paul 2011). For the ETP, we would add that the strength and direction of competitive interactions appear also to be dependent on the type of coral. For example, when *Caulerpa* formed blooms on the Uva Island reef during a La Niña year, it was able to overgrow and kill *Psammocora stellata*, a low-profile species that dominates rubble fields, but *Pocillopora* spp., which are higher profile corals, were not completely overgrown and suffered less mortality (Glynn and Maté 1997).

Thus, in the ETP, both coral and algal morphology and their placement with respect to each other may be important in determining the outcome of competition between these taxa.

Several other factors are known to mediate competitive interactions between algae and corals on other reefs. Some types of foliose macroalgae may function as a reservoir for coral pathogens (Nugues et al. 2004; Ritson-Williams et al. 2009; Chadwick and Morrow 2010), reduce water flow and therefore increase coral mortality through increased allelochemical concentrations (Bak and Borsboom 1984), and enhance microbial activity by stimulating the microbial community with algal exudates (Smith et al. 2006; Vermeij et al. 2009). Some foliose macroalgae compete with juvenile corals or recruits by creating a less suitable habitat. In contrast, some crustose coralline algae provide settlement cues for coral larvae, reinforcing the importance of algal identity in competitive interactions. Higher nutrient availability can interact with other processes producing indirect benefits for algae. Algae with greater nutrient availability may be able to allocate more resources to structural or chemical defenses, thus reducing consumption by herbivores and supporting their persistence and expansion (Stiger et al. 2004; S.J. Bittick, unpublished data, but see Boyer et al. 2004; Chan et al. 2012 for examples of herbivores preferentially consuming nutrient-enriched algae). Nutrient enrichment has also been linked to increased rates of coral disease and mortality (Bruno et al. 2003), creating space for algal expansion and reducing coral competition. However, none of these factors mediating coral/algal competition have been studied on ETP reefs.

11.4.2.3 Facilitative Interactions Can Stabilize Algal Communities

In the ETP, facilitative interactions within algal communities have been documented to help sustain certain algae or even aid in their proliferation on coral reefs. These include feedbacks that intensify herbivory within coral-dominated communities and reduce it in algal-dominated communities as well as processes that alleviate nutrient limitation in algal-dominated but not coral-dominated communities.

The strength of herbivory on ETP reefs, although overall quite intense, has been shown to be affected by the spatially patchy nature of the coral- and algal-dominated communities. Herbivory is much stronger in coral- versus algal-dominated areas of reefs (Fig. 11.13). One explanation for the pattern of higher herbivory in coral areas is that the complex structure of coral communities provides a refuge from predation for herbivores (Friedlander and Parrish 1998; Lee 2006). Increased abundances of herbivores taking shelter in this structurally heterogeneous habitat leads to higher local herbivory pressure and stronger top down control on algae (Graham and Nash 2013; Muthukrishnan 2013). Another possible explanation is that some algal communities may cause reductions in herbivory rates. For example, the presence of unpalatable algae within a community can diminish herbivore activity, providing an associational refuge to other more palatable algae living in close association (Hay 1986; Littler et al. 1986; Bittick et al. 2010; Pereira et al. 2010). Similarly, the association of palatable algae with unpalatable epiphytes, such as cyanobacteria, can

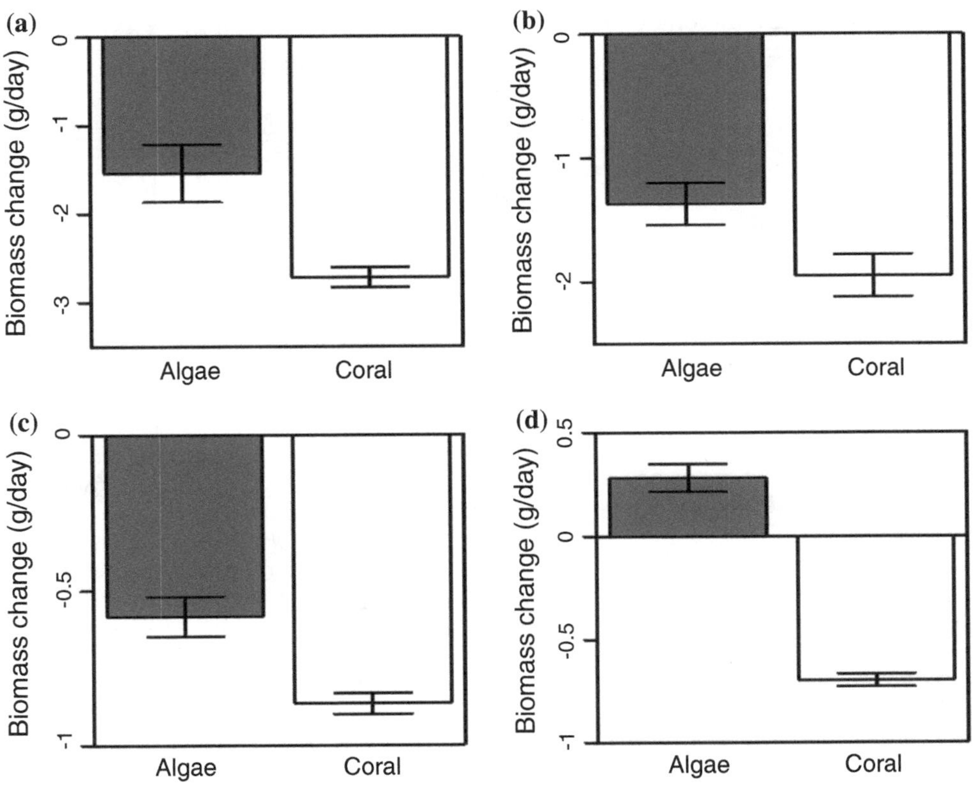

Fig. 11.13 The strength of herbivory is consistently higher within coral-dominated compared to algal-dominated communities at multiple reefs and across multiple algal species. Panels show average herbivory (mean ± SE) over 24 h in different experiments, **a** with *Laurencia* sp. at Contadora Island in 2009, n = 10, **b** and **c** with *Acanthophora spicifera* at Uva Island in 2010, n = 6 and n = 5 respectively, and **d** with *Boodlea* sp. also at Uva Island in 2010, n = 5

offer protection from herbivory and allow algae to persist even in the presence of strong top-down grazing pressure; herbivory on both a palatable red alga and a defended green alga was greatly reduced when epiphytized by cyanobacteria (Fig. 11.14) (Fong et al. 2006; Smith et al. 2010). Both increases in herbivory driven by corals and reductions in herbivory caused by algae are stabilizing feedbacks that make both communities more resilient to perturbations once a community type has gained dominance.

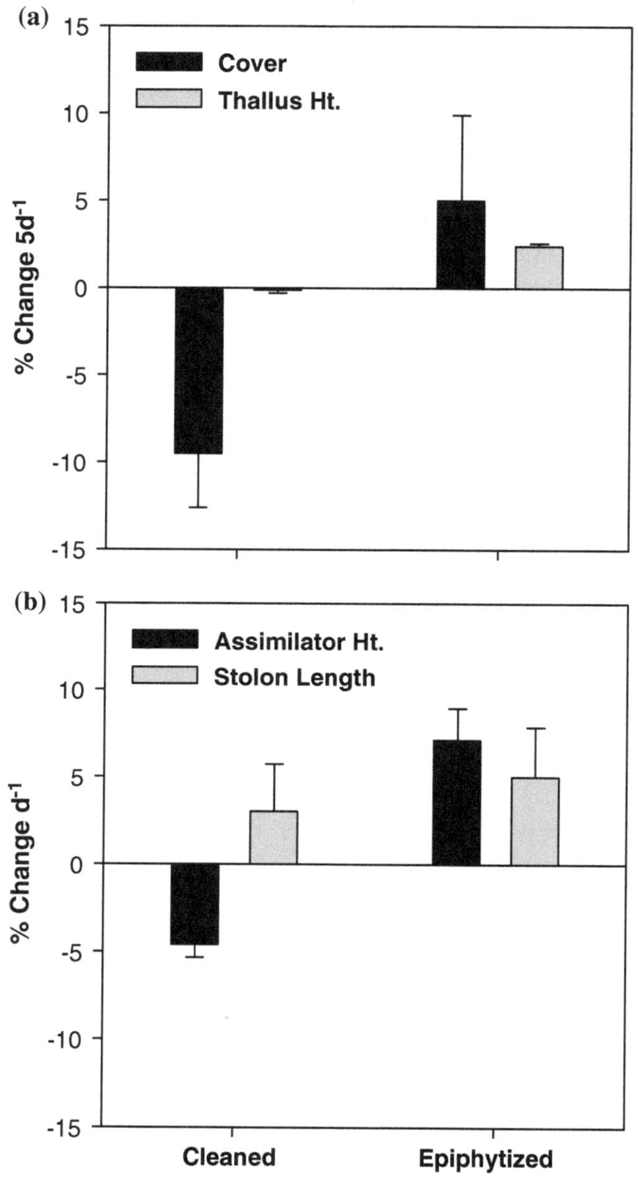

Fig. 11.14 Epiphytes protect **a** *Acanthophora spicifera* (see Fong et al. 2006) and **b** *Caulerpa sertularioides* (see Smith et al. 2010) from herbivory (mean ± SE) on Uva Island reef. Cover in **a** represents substrate coverage

Similarly, differences in nutrient limitation between coral- and algal-dominated patches of reef may provide a second stabilizing feedback for each community type. This mechanism has been identified in the ETP, where growth rates were higher and nutrient limitation lower in algal-dominated relative to coral-dominated patches (Fig. 11.15; Muthukrishnan 2013). One possible explanation for this result is that some types of algae have been shown to enhance local nutrient levels by taking up dissolved inorganic nitrogen and subsequently releasing dissolved organic nitrogen, making it available to other algae (Fong et al. 2004). By increasing N processing and retention, algae can increase the N available (Fong et al. 2004), particularly in systems with pulses of nutrient input such as coral reefs. This acts to ameliorate nutrient limitation and produces a positive interaction between algae. In addition, cyanobacteria can increase nitrogen availability through nitrogen fixation (Howarth 1988), and in certain cases algal-cyanobacterial symbiosis can increase the efficiency of nitrogen fixation (Gordon et al. 1994). Reduced nutrient limitation results in faster growth that allows algae to fill empty habitats more quickly, thus reducing the ability of corals to recover or recolonize after a disturbance (Benayahu and Loya 1977), which may be particularly relevant in a system like the ETP with frequent disturbances.

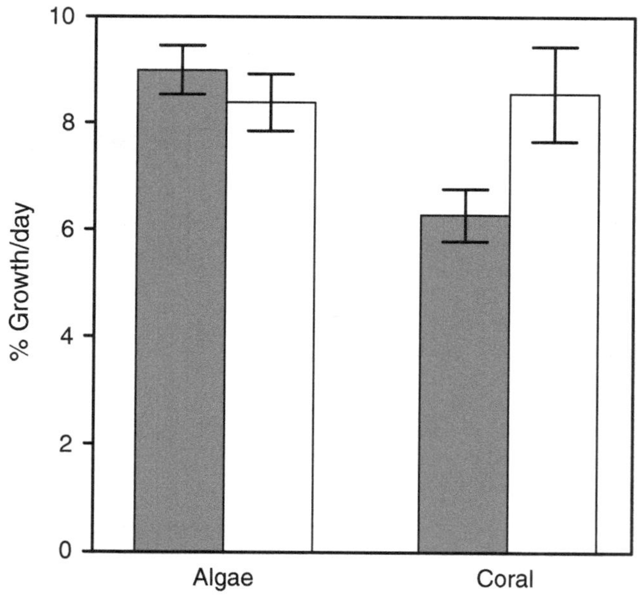

Fig. 11.15 Rates of *Laurencia* sp. growth (mean ± SE) in coral-dominated and algal-dominated communities at Contadora Island in 2009 (n = 15). *White bars* represent nutrient-enriched conditions while *dark bars* are ambient nutrient conditions. Nutrients stimulated growth (indicating nutrient limitation) only in the coral-dominated state

11.5 Phase Shifts, Alternative Stable States, and the Stability of Coral and Algal-Dominated ETP Communities

Tropical reefs worldwide have undergone declines in coral abundance (Gardner et al. 2003; Pandolfi et al. 2003; Bruno and Selig 2007; Carpenter et al. 2008), and some regions of the ETP mirror this global pattern (e.g., Glynn 1990; Baker et al. 2008). These declines are often associated with transitions to reefs increasingly dominated by non-structure forming taxa, such as foliose macroalgae, taller algal turfs, or bioeroding sea urchins (Hughes et al. 2007; Maliao et al. 2008). Coral declines have been linked to stressors such as ENSO, especially in the ETP (Glynn 1990; Wellington and Glynn 2007), and in some cases outside of the ETP, to anthropogenic stressors such as overfishing (Jackson et al. 2001; Edgar et al. 2010), nutrient enrichment (Lapointe 1997; Stimson et al. 2001), and global climate change (Hoegh-Guldberg et al. 2007). Interest in preventing continued loss of coral reefs and in strategies for restoration of degraded reefs has focused attention on the nature of these often catastrophic and seemingly irreversible transitions between community states. Because shifts from coral-dominated to algal-dominated reefs are often sudden and occur with little warning there has been a particular effort to apply threshold models to better understand their dynamics (e.g., Mumby et al. 2007; Hughes et al. 2010; Muthukrishnan 2013).

11.5.1 Alternate Stable States: A Theoretical Framework

Threshold transitions (sensu Suding and Hobbs 2009, but also referred to as regime shifts or critical transitions), include two general categories of community shifts, namely phase shifts (or discontinuous thresholds) and alternate stable states (or hysteresis thresholds), both of which contrast with smooth community transitions (Fig. 11.16). In all three cases community composition changes along a gradient of environmental conditions, often from a community that is considered pristine to one that is degraded. When a community undergoes a smooth transition, there is a gradual and continuous loss of dominant members of one community with replacement by those of the alternate community. Thus mixed coral/algal communities should be common, with their relative abundances predictable based on the present environmental condition. In contrast, for systems that undergo phase shifts, the coral community is initially resilient to degrading environmental conditions before transitioning to algal-dominance at a critical threshold. These systems have a single threshold or tipping point separating environmental conditions that produce very different community states, and there is only a very narrow range of conditions that produce mixed communities. Thus, these communities should be strongly dominated by either corals or algae, with few mixed states. Another key characteristic of phase shift systems is that shifts occur at the same threshold independent of the direction of change,

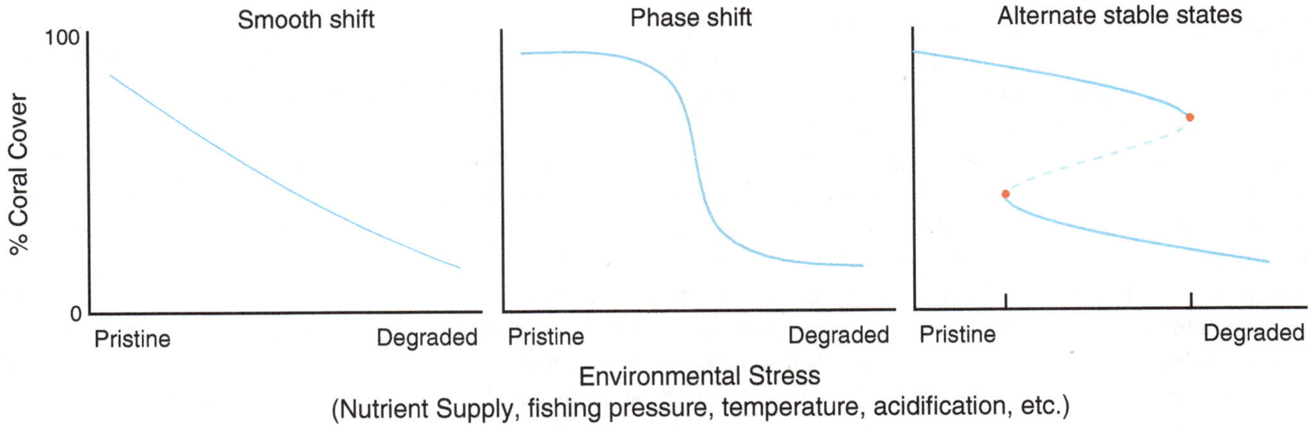

Fig. 11.16 Conceptual models of three types of transitions between community states along a gradient of environmental conditions. Both phase shifts and alternative stable states (ASS) are threshold models that predict rapid and dramatic changes at critical tipping points. For smooth and phase shift responses there is only one possible community state for any environmental condition, while under ASS there is a range of conditions where multiple community states are possible. Thus for ASS both current environmental conditions and the recent history of the community are necessary to predict the current community state. *Dashed line* in ASS graph represents the unstable equilibrium between community states, and *red dots* indicate the environmental thresholds where community shifts occur

degradation (corals to algae) or restoration (algae to corals) (Scheffer and Carpenter 2003). For both of these types of shifts, if the community is disturbed it should always recover to the original state as long as there is not a persistent shift in environmental conditions due to the disturbance.

In contrast, while a system with alternate stable states (ASS) undergoes similar rapid transitions between community states as for phase shifts, there are two different environmental thresholds depending on the direction of change (Fig. 11.16). As for phase shifts, the coral community is resilient to degradation before the critical transition. However, if environmental conditions improve, the algal community is also resilient to recovering conditions and requires a much more "pristine" condition to transition back to a coral-dominated community (Scheffer and Carpenter 2003). This path dependency, or hysteresis, creates a situation where two community types can potentially persist under the same range of environmental conditions. Thus, with ASS the current community state is determined by both the current environmental conditions and the history of that community. Ecosystems with environmental conditions between the two thresholds will display bistability with two separate community states that could potentially persist. Thus, historical community composition is the main determinant of the long-term community state rather than environmental conditions, unless some external perturbation forces the system to flip to the alternate state.

Mechanisms that control community structure also determine whether communities will undergo phase shifts or ASS. For example, competition alone (either direct or indirect) can produce phase shifts if there is a switch in the competitive dominance of community members at some point along an environmental gradient (Terborgh 1971). For communities to exist as ASS, however, stabilizing feedback mechanisms are required that make communities resilient to future environmental variability once the community is established (Scheffer 2009). Stabilizing feedbacks are produced when the presence of key community members increase that population or assemblage's growth rate in a density-dependent manner. For example, a stabilizing feedback thought to be essential for maintaining dominance of coral communities arises from the high topographic complexity produced by corals, which provides shelter for herbivorous fishes. Increased abundances of fish herbivores limit algal proliferation and in turn help maintain coral dominance. Because positive feedback mechanisms drive community composition towards coral- or algal-dominated states, mixed coral and algal communities are rare. Theory also predicts that ASS communities will recover quickly from small disturbances but undergo rapid transitions if large disturbances force them beyond the unstable equilibrium that separates basins of attraction (blue broken line in the ASS panel of Fig. 11.16; Scheffer 2009).

Tropical reefs have been used as a classic example of a system with ASS dynamics, though such a designation remains contentious (for a review see Dudgeon et al. 2010). Initial expectations of ASS were driven by the observation that two major community states appear to dominate tropical reefs: those with high cover of live corals, high structural complexity, high biodiversity, and complex food webs, and those with high cover of algae, lower faunal diversity and less complex foodwebs (Done 1992; Hughes 1994; Pandolfi et al. 2005; Hughes et al. 2010). Other community types, however, have been observed to dominate tropical reefs in certain cases (Norstrom et al. 2009), including corallimorpharians (Chadwick-Furman and Spiegel 2000; Work et al. 2008), soft corals (Linden et al. 2002; Stobart et al. 2005) or sponges (Aronson et al. 2002; Ward-Paige et al. 2005), suggesting the possibility of more than two possible states. It also has been relatively well established that tropical reef communities are strongly influenced by environmental drivers that are affected by anthropogenic activities such as increased nutrient supply (Lapointe 1997; McClanahan et al. 2003), reduced herbivorous fish and invertebrate abundances (Jackson et al. 2001; Burkepile and Hay 2009; Edgar et al. 2010; Rasher et al. 2012; Roff and Mumby 2012), increased sediment loading (Umar et al. 1998; Birrell et al. 2005; Fabricius 2005; Weber et al. 2012), increased ENSO frequency (Toth et al. 2012), and climate change (Hughes et al. 2003; Hoegh-Guldberg et al. 2007; Pandolfi et al. 2011). What remains uncertain, however, is whether coral-dominated and algal-dominated states represent extremes along a continuous environmental gradient, a discontinuous phase shift, or alternate stable states (Bruno et al. 2009; Dudgeon et al. 2010; Żychaluk et al. 2012).

11.5.2 Alternate Stable States and ETP Coral Reefs: Empirical Evidence

Interest in threshold shifts and the potential for ASS has been particularly strong in ETP reefs because of the existing spatial pattern of sharp transitions between community states. Many ETP reefs persist with distinct patches of high percentage live coral cover adjacent to patches of coral rubble covered by turf algae with little live coral (Enochs 2012; Muthukrishnan 2013). In addition, major blooms of algae such as *Caulerpa*, *Acanthophora*, or large mat-forming filamentous green algae (putatively *Boodlea*) cover large areas of some reefs and have persisted for at least 10 years (Fong et al. 2006; Smith et al. 2010; Fig. 11.17). The presence of sharp spatial boundaries provides support for threshold transitions of some sort (Heffernan 2008; Scheffer 2009), though not necessarily ASS. Phase shift dynamics can produce sharp boundaries between community states in a landscape with heterogeneous environmental conditions (Fig. 11.18). However, those shifts will still follow the

Fig. 11.17 The persistence of macroalgal blooms on Uva Island reef has renewed interest in determining whether these reefs exist as alternative stable states. Blooms include, among others, a large filamentous green algal mat composed, putatively, of **a** *Boodlea* sp., and **b** *Caulerpa sertularioides*. Both photos from Uva Island, Panama; **a** March 4, 2003, 5 m depth, **b** March 13, 2005, 10 m depth

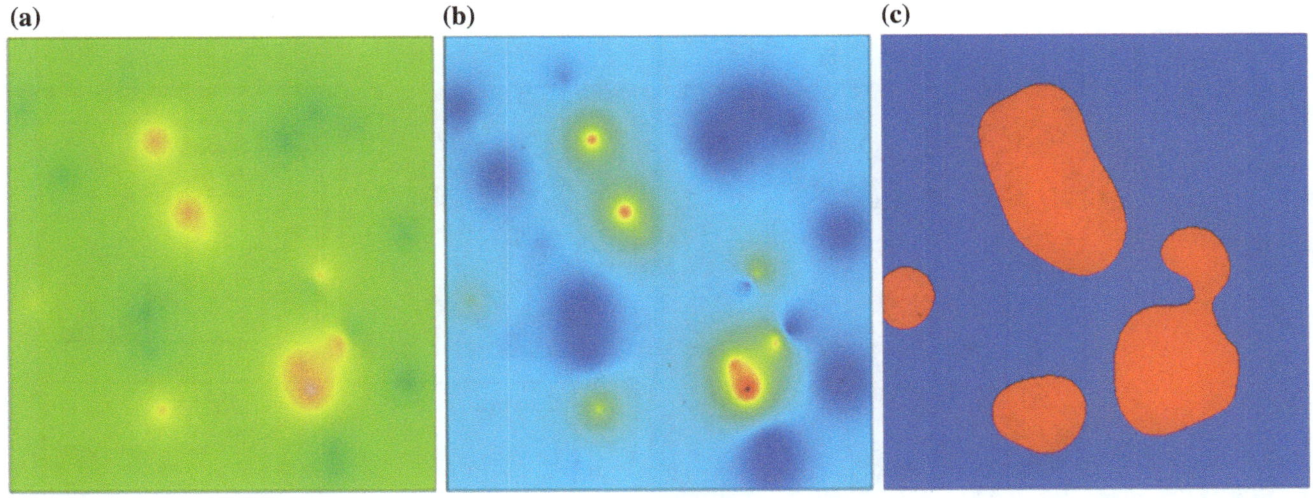

Fig. 11.18 Expectations for the spatial patterning of community composition on a reef with **a** a heterogeneous environment. **b** The community in this same heterogeneous environment when the pattern of community change has a smooth relationship to environmental conditions, and **c** the community in this same heterogeneous environment when the pattern of change is phase-shift dynamics. In panel **a** colors represent levels of a hypothetical environmental gradient, and in **b** and **c** colors represent community state with *blue* indicating algal-dominated, *red* indicating coral-dominated, and other colors representing different relative abundances in mixed communities

underlying spatial patterns of the environmental gradients, with the community shifts at the key environmental thresholds. In the ETP, patches of different community states are often on the scale of meters to tens of meters (Enochs 2012); this scale is likely too small to match the scale of many underlying environmental gradients (such as overfishing). This difference in scales makes it unlikely that phase shift dynamics drive the spatial structure of ETP reef communities. In contrast, facilitative mechanisms that allow organisms to modify the environment locally and stabilize communities can produce patchy, sharply segregated spatial distributions (Rietkerk et al. 2004; Rietkerk and van de Koppel 2008) in systems that are under a single environmental regime. This pattern arises because feedbacks create a discontinuous gradient in local environmental conditions experienced by the community even in the presence of uniform underlying conditions, and transitions can occur on smaller spatial scales than might be expected for natural environmental variability. The presence of repeated sharp transitions between community states in the ETP suggests that community composition is not controlled by a single threshold response to environmental conditions, but rather is driven by positive feedbacks forcing areas toward the alternate community states.

A second line of evidence for ASS in the ETP comes from recurrent ENSO-related disturbances that may have caused reefs to switch between basins of attraction for coral and algal communities. High temperature anomalies during severe ENSO events led to massive coral mortality and rapid changes in community composition (Glynn 1990; Glynn and D'Croz 1990). Recovery from the severe coral mortality caused during the 1982–83 ENSO has shown high variability with some sites recovering readily while others have experienced almost no recovery (e.g., Baker et al. 2008; Fig. 11.19). This bifurcation in recovery trajectories provides another line of evidence supporting the presence of multiple attractors. This pattern also suggests that the reefs that have shown little recovery may be locked into an algal-dominated state via stabilizing feedbacks and are unlikely to return to their initial coral-dominated state. Alternative explanations, however, include high rates of bioerosion and substrate instability on non-recovering reefs slowing recovery in these systems.

Identifying mechanisms that could produce ASS dynamics rather than the patterns that arise from them offers a complementary strategy for evaluating ASS in natural ecosystems. In particular, positive feedback mechanisms are a critical aspect of systems with ASS and can be an important indicator of their presence (Scheffer 2009; Scheffer et al. 2012). Positive feedbacks can arise in any system where community members modify environmental conditions or change access to certain resources in a way that is self-beneficial. Such processes are common in many ecosystems (Stachowicz 2001; Bruno et al. 2003) including coral reefs (Bruno et al. 2003; Nugues et al. 2004; Lee 2006; Smith et al. 2006; Hoey and Bellwood 2011). In the ETP an increase in herbivory rates in coral-dominated habitats and amelioration of nutrient limitation in algal dominated habitats have been identified (Figs. 11.13 and 11.15) indicating at least two positive feedback mechanisms are present. By limiting algal accumulation in the coral-dominated state and enhancing algal growth in the algal-dominated state, these mechanisms help create the basins of attraction for the separate community states making them resilient to external perturbations or stochastic variability.

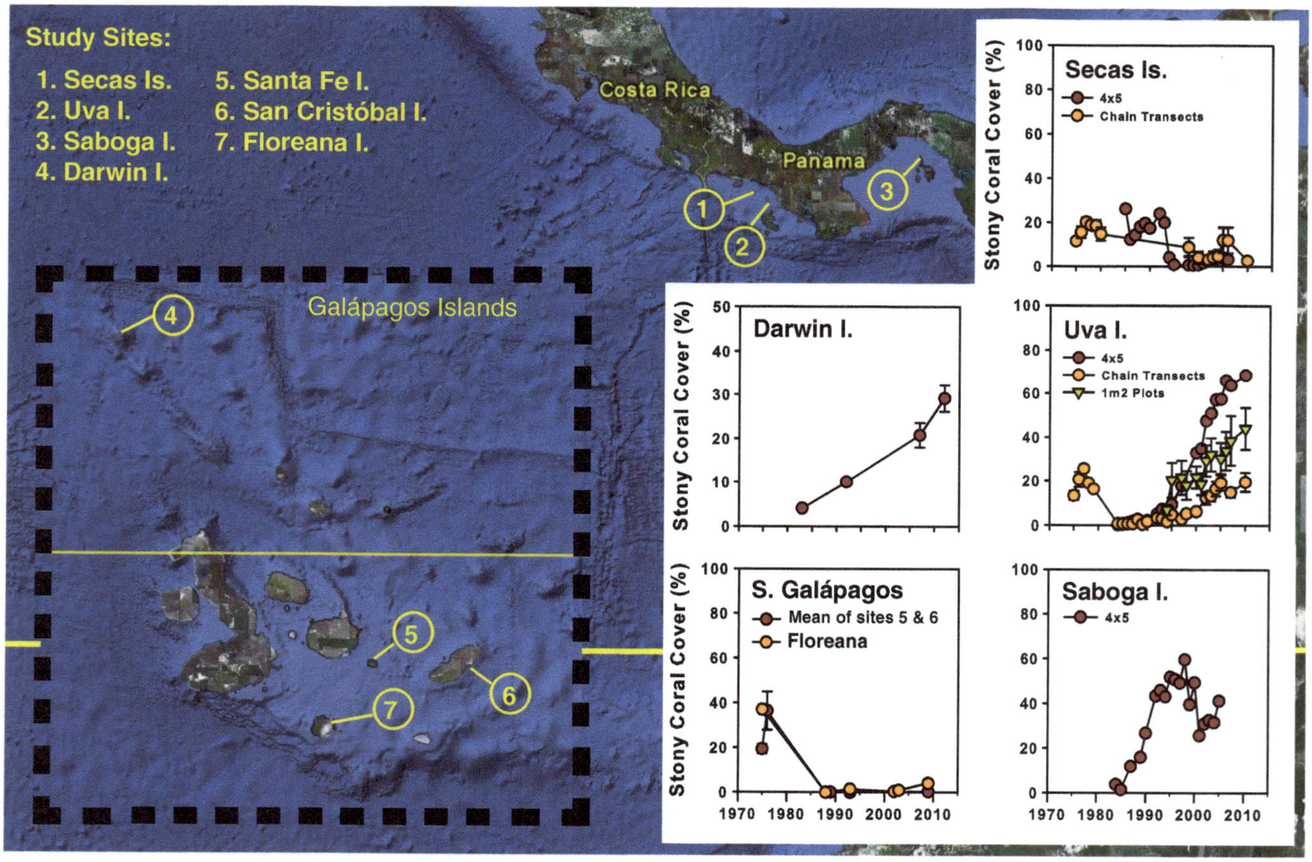

Fig. 11.19 Long-term recovery trajectories for several reefs of the ETP. Recovery of coral cover (mean ± SE) seems to be either vigorous or is lacking entirely, which may be driven by positive feedbacks. When data sources are not indicated, data are from random chain transects, otherwise data are from 4 × 5 m recruitment plots, chain transects, and 1 m^2 plots (last three methods involved repeated sampling at permanently marked sites)

11.5.3 Alternate Stable States and ETP Coral Reefs: Modeling Evidence

Well-established empirical methods to evaluate ASS (Connell and Sousa 1983; Petraitis and Latham 1999) are difficult to employ in the ETP, and coral reefs generally, because they require large destructive perturbations that would be logistically difficult and ethically questionable to conduct. In addition, the appropriate time scale to evaluate stability could be decades for an ecosystem largely dominated by long-lived clonal species as the classic requirement to prove stability (Connell and Sousa 1983) is turnover of all individuals (including clones). An alternate strategy is to evaluate ASS based on the analysis of models developed and parameterized with empirical data on competition and positive feedbacks in ETP reefs (Muthukrishnan 2013).

While the presence of positive feedbacks is a necessary condition for ASS, whether they are sufficiently strong to produce ASS dynamics remains the more challenging question (Scheffer 2009). Analysis of simulation models parameterized from empirical data offers one strategy to address that question. Muthukrishnan (2013) developed a cellular automaton model that allowed for patches of reef to be occupied by coral or different types of algae and for transitions among these space-occupiers to be controlled by processes such as coral or algal growth, recruitment, and herbivory (Fig. 11.20). The model included parameters for environmental conditions such as external nutrient supply and abundance of consumers, and for strengths of different feedback processes that locally altered nutrient supply and herbivory to allow evaluation of dynamics under different environmental and feedback conditions. Those local environmental conditions were then used to determine probabilities for each transition among community types. Because local transition probabilities were able to change through

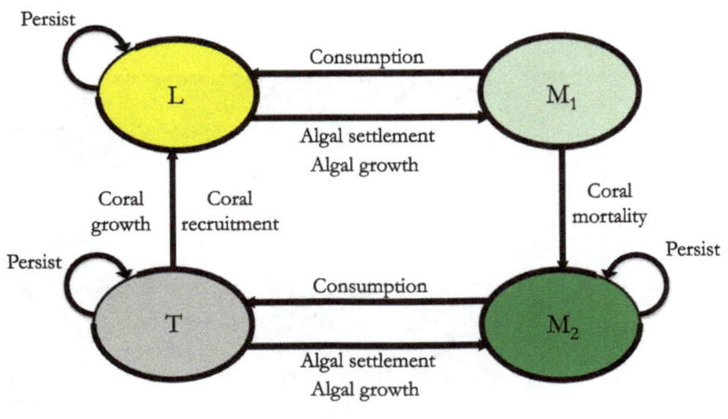

Fig. 11.20 A schematic of the simulation model in Muthukrishnan (2013). The four compartments represent the different community members that can occupy individual cells in the cellular automaton: live coral (L), mixed turf and crustose coralline algae (T), macroalgae over live coral framework (M_1), and macroalgae over dead coral framework (M_2). *Lines* indicate different processes that cause cells to shift from one state to another

time in response to local community composition, relatively complex and realistic patterns were able to arise from fairly simple rule sets, making cellular automata modeling an excellent simulation strategy for real systems (Wootton 2001). General simulations showed that strong competition for space drove threshold dynamics under almost all circumstances (Fig. 11.21). Furthermore, alternate stable states readily appeared in simulations that included positive feedbacks, with the range of environmental conditions that showed bistability driven by the strength of those feedbacks. Similarly, simulations also showed that the strength of positive feedbacks was the determining factor between a system with phase shifts or ASS.

Further simulations parameterized specifically from reefs at Isla Contadora, Gulf of Panama, showed that ETP reefs exist under the specific range of model conditions that produced ASS. Simulations using those empirical parameters showed bistability with mixed communities of corals and algae being unstable and the system attracted to either a coral-dominated or algal-dominated state (Fig. 11.22). The final community state of individual reefs is strongly predicted by the initial composition indicating a strong sensitivity to initial conditions, a sign of ASS. This suggests that with feedbacks of the strengths measured on ETP reefs the ability of the present community to modify local environmental conditions is critical to determining community trajectories. These results also indicate that, at a certain threshold community composition, the density-dependent feedbacks that support one community state are overwhelmed by the processes that support the other, creating the separate basins of attraction for the two states (or the unstable equilibrium in Fig. 11.16). This could in turn produce the bifurcated recovery trajectories seen after major disturbances in the ETP (Fig. 11.19).

This modeling approach has shown that environmental conditions and positive feedbacks observed in the ETP can produce bistability (Figs. 11.21 and 11.22) and that positive feedbacks provide a plausible mechanism for the spatial structure of sharp boundaries between community states observed there (Muthukrishnan 2013). At the same time, other work from the ETP has argued against ASS suggesting that major shifts in ETP community structure over geological time can be attributed to changes in ENSO variability rather than stochastic transitions between states (Toth et al. 2012). Even if historical transitions were driven by shifts in ENSO frequency and magnitude, if the historical environmental regimes that led to different community states reflected environmental conditions outside the bistable range (as shown in Fig. 11.16) it remains possible that ETP reefs currently display ASS. It is possible that there could be an intermediate set of conditions (incorporating ENSO conditions as well as other environmental parameters) where both states could persist. The combination of the present day

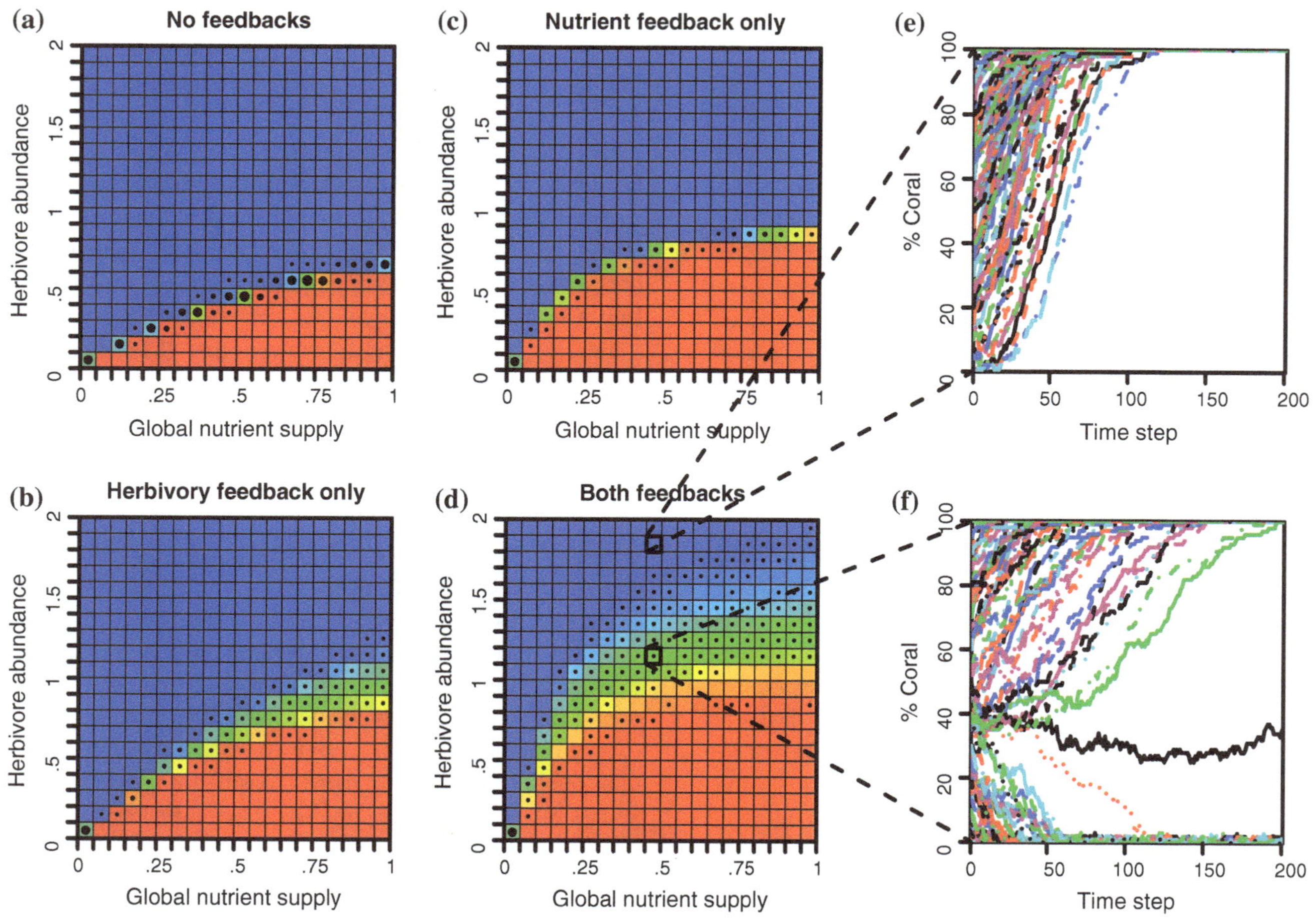

Fig. 11.21 Evaluation of bistability under a range of global environmental conditions (herbivore abundance and ambient nutrient supply; from Muthukrishnan (2013). Simulations were initialized with a range of initial coral cover levels and the number of simulations that ended as either coral or algal dominated (above or below 50 % coral) was measured. Simulations were run for four feedback scenarios: **a** no feedbacks, **b** herbivory feedback only, **c** nutrient feedback only, and **d** both herbivory and nutrient feedbacks. *Dark blue squares* indicate conditions where all simulations ended in a coral-dominated state, *red squares* where all simulations ended in an algal-dominated state, and *intermediate colors* denote split outcomes. In addition, the size of the *black dot* in each cell represents the proportion of runs that ended with intermediate (25–75 %) coral cover. Thus, only *intermediate-colored squares with small or no dots* represent bistability as opposed to stochastic noise. Tracks of individual simulation runs are shown for two parameter sets with both feedbacks that lead to **e** a coral-dominated state, and **f** a bi-stable system

Fig. 11.22 Reefs simulated using empirically derived parameters for Isla Contadora, Pearl Islands (Panama) in the ETP show the presence of multiple attractors for community composition producing bistability (adapted from Muthukrishnan 2013)

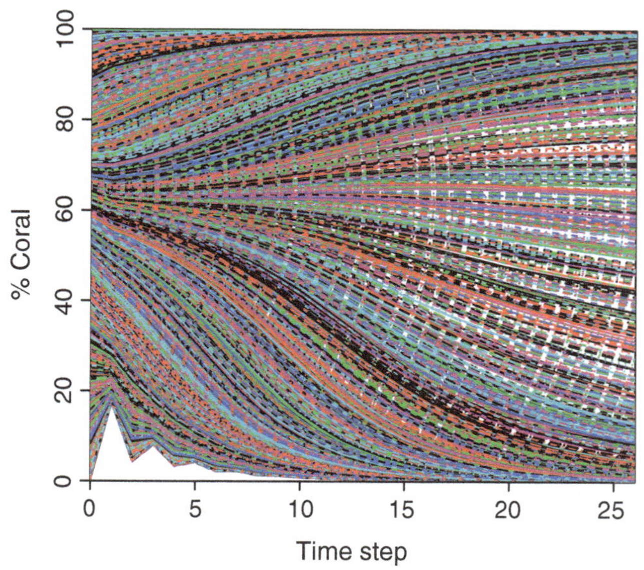

results (Muthukrishnan 2013) and the historical patterns (Toth et al. 2012; see Chap. 6, Toth et al.) raises the possibility that ASS dynamics in the ETP are a relatively recent circumstance only occurring in response to current anthropogenic modification of the environment, a scenario that has been suggested for a number of other ecosystems (Suding et al. 2004; Suding and Hobbs 2009).

Acknowledgments We thank Peter W. Glynn for his continual mentorship, friendship, and intellectual stimulation that have kept us working together in the eastern tropical Pacific for so many years. And, of course, we greatly appreciate his sense of humor that transformed it from work to fun, and our group from colleagues into family. The National Science Foundation (Biological Oceanography) supported our work through grants to P.W.G. from 1984 to 2010. We are also grateful for the support of UCLA's Department of Ecology and Evolutionary Biology (P.F.), the University of the Virgin Islands' Center for Marine and Environmental Studies (T.S.), and NSF's GRFP (R.M.) for support. Our colleague Juan Maté efficiently facilitated our work in Panama, and Jordan Fong verified all literature citations.

References

Abbott IA (1999) Marine red algae of the Hawaiian Islands. Bishop Museum Press, Hawaii

Ahern KS, Udy JW, Ahern CR (2007) Nutrient additions generate prolific growth of *Lyngbya majuscula* (cyanobacteria) in field and bioassay experiments. Harmful Algae 6:134–151

Allgeier JE, Layman CA, Mumby PJ, Rosemond AD (2014) Consistent nutrient storage and supply mediated by diverse fish communities in coral reef systems. Global Change Biol 20:2459–2472

Allgeier CD (ed) (2008) Algal chemical ecology. Springer, Berlin, pp 91–103

Aronson RB, Precht WF, Toscano MA, Koltes KH (2002) The 1998 bleaching event and its aftermath on a coral reef in Belize. Mar Biol 141:435–447

Arthur KE, Paul VJ, Paerl HW, O'Neil JM, Joyner J, Meickle T (2009) Effects of nutrient enrichment of the cyanobacterium *Lyngbya* sp. on growth, secondary metabolite concentration and feeding by the specialist grazer *Stylocheilus striatus*. Mar Ecol Prog Ser 394:101–110

Bak RPM, Borsboom JLA (1984) Allelopathic interaction between a reef coelenterate and benthic algae. Oecologia 63:194–198

Baker AC, Glynn PW, Riegl B (2008) Climate change and coral reef bleaching: an ecological assessment of long-term impacts, recovery trends, and future outlook. Estuar Coastal Shelf Sci 80:435–471

Bellwood DR, Fulton CJ (2008) Sediment-mediated suppression of herbivory on coral reefs: decreasing resilience to rising sea levels and climate change? Limnol Oceanogr 53:2695–2701

Benayahu Y, Loya Y (1977) Space partitioning by stony corals, soft corals, and benthic algae on the coral reefs of northern Gulf of Eilat (Red Sea). Helgoländer 30:361–382

Bernecker A, Wehrtmann IS (2009) New records of benthic marine algae and Cyanobacteria for Costa Rica, and a comparison with other Central American countries. Helgol Mar Res 63:219–229

Birrell CL, McCook LJ, Willis BL (2005) Effects of algal turfs and sediment on coral settlement. Mar Pollut Bull 51:408–414

Bittick SJ, Bilotti ND, Peterson HA, Stewart HL (2010) *Turbinaria ornata* as an herbivory refuge for associate algae. Mar Biol 157:317–323

Boyer KE, Fong P, Armitage AR, Cohen RA (2004) Elevated nutrient content of macroalgae increases rates of herbivory in coral, seagrass, and mangrove habitats. Coral Reefs 23:530–538

Bruno JF, Selig ER (2007) Regional decline of coral cover in the Indo-Pacific: timing, extent, and subregional comparisons. PLoS ONE 2:e711

Bruno JF, Petes LE, Harvell CD, Hettinger A (2003) Nutrient enrichment can increase the severity of coral diseases. Ecol Lett 6:1056–1061

Bruno JF, Sweatman H, Precht WF, Selig ER, Schutte VGW (2009) Assessing evidence of phase shifts from coral to macroalgal dominance on coral reefs. Ecology 90:1478–1484

Burkepile DE, Hay ME (2009) Nutrient versus herbivore control of macroalgal community development and coral growth on a Caribbean reef. Mar Ecol Prog Ser 389:71–84

Burkepile DE, Allgeier JE, Shantz AA, Pritchard CE, Lemoine NP, Bhatti LH, Layman CA (2013) Nutrient supply from fishes facilitates macroalgae and suppresses corals in a Caribbean coral reef ecosytem. Sci Rep 3:1493. doi:10.1038/srep01493

Carpenter RC (1986) Partitioning herbivory and its effects on coral reef algal communities. Ecol Monogr 56:345–363

Carpenter KE, Abrar M, Aeby G, Aronson RB, Banks S, Bruckner A, Chiriboga Á, Cortés J, Delbeek JC, DeVantier L, Edgar GJ, Edwards AJ, Fenner D, Guzmán HM, Hoeksema BW, Hodgson G, Johan O, Licuana WY, Livingstone SR, Lovell ER, Moore JA, Obura DO, Ochavillo D, Polidoro BA, Precht WF, Quibilan MC, Reboton C, Richards ZT, Rogers AD, Sanciangco J, Sheppard A, Sheppard C, Smith J, Stuart S, Turak E, Verson JEN, Wallace C, Weil E, Wood E (2008) One-third of reef-building corals face elevated extinction risk from climate change and local impacts. Science 321:560–563

Chadwick-Furman NE, Spiegel M (2000) Abundance and clonal replication in the tropical corallimorpharian *Rhodactis rhodostoma*. Invert Biol 119:351–360

Chadwick NE, Morrow KM (2010) Competition among sessile organisms on coral reefs. In: Dubinsky Z, Stambler N (eds) Coral reefs: an ecosystem in transition. Springer, Dordrecht, pp 347–371

Chan AY, Lubarsky K, Judy KN, Fong P (2012) Nutrient addition increases consumption rates of tropical algae with different initial palatabilities. Mar Ecol Prog Ser 465:25–31

Choat JH, Clements KD (1998) Vertebrate herbivores in marine and terrestrial environments: a nutritional ecology perspective. Annu Rev Ecol Syst 29:375–403

Connell JH (1978) Diversity in tropical rain forests and coral reefs. Science 199:1302–1310

Connell JH, Sousa WP (1983) On the evidence needed to judge ecological stability or persistence. Am Nat 121:789–824

D'Croz L, O'Dea A (2007) Variability in upwelling along the Pacific shelf of Panamá and implications for the distribution of nutrients and chlorophyll. Estuar Coast Shelf Sci 73:325–340

D'Croz L, O'Dea A (2009) Nutrient and chlorophyll dynamics in Pacific Central America (Panamá). In: Proceedings of Smithsonian Marine Science Symposium, vol 38, pp 335–344

Dawson EY (1960) New records of marine algae from Pacific Mexico and Central America. Pac Nat 1:31–52

Done TJ (1992) Phase shifts in coral reef communities and their ecological significance. Hydrobiologia 247:121–132

Downing JA, McClain M, Twilley RW, Melack JM, Elser J, Rabalais NN, Lewis WM, Turner RE, Corredor J, Soto D, Yanez-Arancibia A, Kopaska JA, Howarth RW (1999) The impact of accelerating land-use change on the N-cycle of tropical aquatic ecosystems: current conditions and projected changes. Biogeochemistry 46:109–148

Dudgeon SR, Aronson RB, Bruno JF, Precht WF (2010) Phase shifts and stable states on coral reefs. Mar Ecol Prog Ser 413:201–216

Duffy JE, Hay ME (1990) Seaweed adaptations to herbivory. Bioscience 40:368–375

Duffy JE, Hay ME (2001) The ecology and evolution of marine consumer-prey interactions. In: Bertness MD, Gaines SD, Hay ME (eds) Marine community ecology. Sinauer Associates Inc., Sunderland, pp 131–157

Eakin CM (1996) Where have all the carbonates gone? A model comparison of calcium carbonate budgets before and after the 1982–1983 El Niño at Uva Island in the eastern Pacific. Coral Reefs 15:109–119

Earle SA (1972) A review of the marine plants of Panamá. In: Jones ML (ed) The Panamic biota: some observations prior to a sea-level canal. Bull Biol Soc Wash 2:69–87

Edgar GJ, Banks SA, Brandt M, Bustamante RH, Chiriboga Á, Earle SA, Garkse LE, Glynn PW, Grove JS, Henderson S, Hickman CP, Miller KA, Rivera F, Wellington GM (2010) El Niño, grazers and fisheries interact to greatly elevate extinction risk for Galápagos marine species. Global Change Biol 16:2876–2890

Edgar GJ, Banks SA, Bessudo S, Cortés J, Guzmán HM, Henderson S, Martinez C, Rivera F, Soler G, Ruiz D, Zapata FA (2011) Variation in reef fish and invertebrate communities with level of protection from fishing across the Eastern Tropical Pacific seascape. Global Ecol Biogeogr 20:730–743

Enochs IC (2012) Motile cryptofauna associated with live and dead coral substrates: implications for coral mortality and framework erosion. Mar Biol 159:709–722

Fabricius KE (2005) Effects of terrestrial runoff on the ecology of corals and coral reefs: review and synthesis. Mar Pollut Bull 50:125–146

Fabricius KE, De'ath G (2001) Biodiversity on the Great Barrier Reef: large-scale patterns and turbidity-related local loss of soft coral taxa. In: Wolanski E (ed) Oceanographic processes of coral reefs: physical and biological links in the Great Barrier Reef, vol 1. CRC Press, London, pp 127–144

Fernández C, Cortés J (2005) *Caulerpa sertularioides*, a green alga spreading aggressively over coral reef communities in Culebra Bay, North Pacific of Costa Rica. Coral Reefs 24:10

Fernández-García C, Riosmena-Rodrígues R, Wysor B, Tejada OL, Cortés J (2011) Checklist of the Pacific marine macroalgae of Central America. Bot Mar 54:53–73

Fiedler PC, Talley LD (2006) Hydrography of the eastern tropical Pacific: a review. Prog Oceanogr 69:143–180

Floeter SR, Behrens MD, Ferreira CEL, Paddack MJ, Horn MH (2005) Geographical gradients of marine herbivorous fishes: patterns and processes. Mar Biol 147:1435–1447

Fong P (2008) Macroalgal dominated ecosystems. In: Capone DG, Carpenter EJ (eds) Nitrogen in the marine enviroment. Springer, New York, pp 917–948

Fong CR, Fong P (2014) Why species matter: an experimental assessment of assumption and predictive ability of two functional-group models. Ecology 95:2055–2061

Fong P, Paul VJ (2011) Coral reef algae. In: Dubinsky Z, Stambler N (eds) Coral reefs: an ecosystem in transition. Springer, Dordrecht, pp 241–272

Fong P, Boyer KE, Kamer K, Boyle KA (2003) Influence of initial tissue nutrient status of tropical marine algae on response to nitrogen and phosphorus additions. Mar Ecol Prog Ser 262:111–123

Fong P, Fong JJ, Fong CR (2004) Growth, nutrient storage, and release of dissolved organic nitrogen by *Enteromorpha intestinalis* in response to pulses of nitrogen and phosphorus. Aquat Bot 78:83–95

Fong P, Smith TB, Wartian MJ (2006) Epiphytic cyanobacteria maintain shifts to macroalgal dominance on coral reefs following ENSO disturbance. Ecology 87:1162–1168

Friedlander AM, Parrish JD (1998) Habitat characteristics affecting fish assemblages on a Hawaiian coral reef. J Exp Mar Biol Ecol 224: 1–30

Fujita RM, Wheeler PA, Edwards RL (1989) Assessment of macroalgal nitrogen limitation in a seasonal upwelling region. Mar Ecol Progr Ser 53:293–303

Gallegos CL, Jordan TE (1997) Seasonal progression of factors limiting phytoplankton pigment biomass in the Rhode River estuary, Maryland (USA): II. Modeling N versus P limitation. Mar Ecol Progr Ser 161:199–212

Gardner TA, Cote IM, Gill JA, Grant A, Watkinson AR (2003) Long-term region-wide declines in Caribbean corals. Science 301:958–960

Garzón-Ferreira J, Cortés J, Croquer A, Guzmán H, Leao Z, Rodríguez-Ramírez A (2004) Southern tropical America: coral reef status and consolidation as GCRMN regional node. In: Wilkinson C (ed) Status of coral reefs of the world vol 2, pp 509–522

Gavagnin M, Marin A, Castelluccio F, Villani G, Cimino G (1994) Defensive relationships between *Caulerpa prolifera* and its shelled sacoglossan predators. J Exp Mar Biol Ecol 175:197–210

Glynn PW (1972) Observations on the ecology of the Caribbean and Pacific coasts of Panamá. In: Jones ML (ed) The Panamic biota: some observations prior to a sea-level canal. Bull Biol Soc Wash 2:13–30

Glynn PW (1974) Rolling stones among the Scleractinia: mobile coralliths in the Gulf of Panamá. In: Proceedings of 2nd International Coral Reef Symposium, vol 2, Brisbane, pp 183–200

Glynn PW (1990) Coral mortality and disturbances to coral reefs in the tropical eastern Pacific. In: Glynn PW (ed) Global ecological consequences of the 1982–83 El Niño-Southern Oscillation, vol 52. Elsevier oceanography series, pp 55–126

Glynn PW, D'Croz L (1990) Experimental evidence for high temperature stress as the cause of El Niño-coincident coral mortality. Coral Reefs 8:181–191

Glynn PW, Fong P (2006) Patterns of reef coral recovery by the regrowth of surviving tissues following the 1997–98 El Niño warming and 2000, 2001 upwelling events in Panamá, eastern Pacific. In: Proceedings of 10th International Coral Reef Symposium, Okinawa, pp 624–630

Glynn PW, Maté JL (1997) Field guide to the Pacific coral reefs of Panamá. In: Proceedings of 8th International Coral Reef Symposium, vol 1, Panama, pp 145–166

Glynn PW, Wellington GM (1983) Corals and coral reefs of the Galápagos Islands. University of California Press, Berkeley, p 330

Glynn PW, Enochs IC, Afflerbach JA, Brandtneris VW, Serafy JE (2014) Eastern Pacific reef fish responses to coral recovery following El Niño disturbances. Mar Ecol Prog Ser 495:233–247

Glynn PW, Riegl BM, Purkis S, Kerr J, Smith T (2015) Coral reef recovery in the Galápagos Islands: the northern-most islands (Darwin and Wenman). Coral Reefs. doi:10.1007/s00338-015-1280-4

Gordon N, Angel D, Neori A, Kress N, Kimor B (1994) Heterotrophic dinoflagellates with symbiotic cyanobacteria and nitrogen limitation in the Gulf of Aqaba. Israel Oceanogr Limnol Res, Haifa, IS 83–88

Graham NAJ, Nash KL (2013) The importance of structural complexity in coral reef ecosystems. Coral Reefs 32:315–326

Guzmán HM, Cortés J (1989) Coral reef community structure at Caño Island. Pacific Costa Rica. PSZNI: Mar Ecol 10:23–41

Guzmán HM, Cortés J (2001) Changes in reef community structure after fifteen years of natural disturbances in the eastern Pacific (Costa Rica). Bull Mar Sci 69:133–149

Hatcher BG (1988) Coral reef primary productivity: a beggar's banquet. Trends Ecol Evol 3:106–111

Hay ME (1986) Associational plant defenses and the maintenance of species diversity: turning competitors into accomplices. Am Nat 128:617–641

Hay ME (1991) Fish-seaweed interactions on coral reefs: effects of herbivorous fishes and adaptations of their prey. In: Sale PF (ed) The ecology of fishes on coral reefs. Academic Press, New York, pp 96–119

Hay ME (1997) The ecology and evolution of seaweed-herbivore interactions on coral reefs. Coral Reefs 16:67–76

Hay ME, Steinberg PD (1992) The chemical ecology of plant-herbivore interactions in marine versus terrestrial communities. In: Rosenthal GA, Berenbaum MR (eds) Herbivores: their interactions with secondary plant metabolites, vol II. Ecological and evolutionary processes. Academic Press, pp 371–413

Heffernan JB (2008) Wetlands as an alternative stable state in desert streams. Ecology 89:1261–1271

Hillis L (1997) Coralgal reefs from a calcareous green alga perspective, and a first carbonate budget. In: Proceedings of 8th International Coral Reef Symposium, vol 1, Panama, pp 761–766

Hoegh-Guldberg O, Mumby PJ, Hooten AJ, Steneck RS, Greenfield P, Gomez E, Harvell CD, Sale PF, Edwards AJ, Caldeira K, Knowlton N, Eakin CM, Iglesias-Prieto R, Muthiga N, Bradbury RH, Dubi A, Hatziolos ME (2007) Coral reefs under rapid climate change and ocean acidification. Science 318:1737–1742

Hoey AS, Bellwood DR (2011) Suppression of herbivory by macroalgal density: a critical feedback on coral reefs? Ecol Lett 14:267–273

Howarth RW (1988) Nutrient limitation of net primary production in marine ecosystems. Ann Rev Ecol Syst 19:89–110

Hughes TP (1994) Catastrophes, phase shifts, and large-scale degradation of a Caribbean coral reef. Science 265:1547–1551

Hughes TP, Baird AH, Bellwood DR, Card M, Connolly SR, Folke C, Grosberg R, Hoegh-Guldberg O, Jackson JBC, Kleypas J, Lough JM, Marshall P, Nyström M, Palumbi SR, Pandolfi JM, Rosen B, Roughgarden J (2003) Climate change, human impacts, and the resilience of coral reefs. Science 301:929–933

Hughes TP, Rodrigues MJ, Bellwood DR, Ceccarelli D, Hoegh-Guldberg O, McCook LJ, Moltschaniwskyj N, Pratchett MS, Steneck RS, Willis B (2007) Phase shifts, herbivory, and the resilience of coral reefs to climate change. Curr Biol 17:360–365

Hughes TP, Graham NAJ, Jackson JBC, Mumby PJ, Steneck RS (2010) Rising to the challenge of sustaining coral reef resilience. Trends Ecol Evol 25:633–642

Jackson JBC (2008) Ecological extinction and evolution in the brave new ocean. Proc Nat Acad Sci USA 105:11458–11465

Jackson JBC, Kirby MX, Berger WH, Bjorndal KA, Botsford LW, Bourque BJ, Bradbury RH, Cooke R, Erlandson J, Estes JA, Hughes TP, Kidwell S, Lange CB, Lenihan HS, Pandolfi JM, Peterson CH (2001) Historical overfishing and the recent collapse of coastal ecosystems. Science 293:629–638

Keeley KN, Stroh JD, Tran DSC, Fong CR, Fong P (2015) Location, location, location: small shifts in collection site result in large intraspecific differences in macroalgal palatability. Coral Reefs. doi:10.007/s00338-015-1274-2

Lane AL, Kubanek J (2008) Secondary metabolite defense against pathogens and biofoulers. In: Amsler CD (ed) Algal chemical ecology. Springer, Berlin, pp 229–243

Lapointe BE (1987) Phosphorus- and nitrogen-limited photosynthesis and growth of *Gracilaria tikvahiae* (Rhodophyceae) in the Florida Keys: an experimental field study. Mar Biol 93:561–568

Lapointe BE (1989) Macroalgal production and nutrient relations in oligotrophic areas of Florida Bay. Bull Mar Sci 44:312–323

Lapointe BE (1997) Nutrient thresholds for bottom-up control of macroalgal blooms on coral reefs in Jamaica and southeast Florida. Limnol Oceanogr 42:1119–1131

Lapointe BE, Littler MM, Littler DS (1992) Nutrient availability to marine macroalgae in siliciclastic versus carbonate-rich coastal waters. Estuaries 15:75–82

Lapointe BE, Barile PJ, Littler MM, Littler DS (2005) Macroalgal blooms on southeast Florida coral reefs II. Cross-shelf discrimination of nitrogen sources indicates widespread assimilation of sewage nitrogen. Harmful Algae 4:1106–1122

Larkum AWD (1988) High rates of nitrogen fixation on coral skeletons after predation by the crown of thorns starfish *Acanthaster planci*. Mar Biol 97:503–506

Larned ST (1998) Nitrogen- versus phosphorus-limited growth and sources of nutrients for coral reef macroalgae. Mar Biol 132:409–421

Lee SC (2006) Habitat complexity and consumer-mediated positive feedbacks on a Caribbean coral reef. Oikos 112:442–447

Lee RE (2008) Phycology, 4th edn. Cambridge University Press, Cambridge, p 547

Lindén O, Souter D, Wilhelmsson D, Obura D (2002) Coral reef degradation in the Indian Ocean status report 2002. Kalmar, Sweden, Coral Reef Degradation in the Indian Ocean, p 283

Littler MM, Littler DS (1984) A relative-dominance model for biotic reefs. In: Proceedings of the joint meeting of the Atlantic Reef Committee Society of reef studies, Miami, Florida

Littler MM, Littler DS (1988) Structure and role of algae in tropical reef communities. In: Lembi CA, Waaland JR (eds) Algae and human affairs. Cambridge University Press, Cambridge, pp 29–56

Littler MM, Littler DS (1995) Selective herbivore increases biomass of its prey: a chiton-coralline reef-building association. Ecology 76:1666–1681

Littler MM, Littler DS, Blair SM, Norris JN (1986) Deep-water plant communities from an uncharted seamount off San Salvador Island, Bahamas: distribution, abundance, and primary productivity. Deep-Sea Res 33:881–892

Littler MM, Littler DS, Lapointe BE, Bariles PJ (2006) Toxic cyanobacteria (blue-green algae) associated with groundwater conduits in the Bahamas. Coral Reefs 25:186

Macintyre IG, Graus RR, Reinthal PN, Littler MM, Littler DS (1987) The barrier reef sediment apron: Tobacco Reef, Belize. Coral Reefs 6:1–12

Maliao RJ, Turingan RG, Lin J (2008) Phase-shift in coral reef communities in the Florida Keys National Marine Sanctuary (FKNMS), USA. Mar Biol 154:841–853

McClanahan TR, Sala E, Stickels PA, Cokos BA, Baker AC, Starger CJ, Jones SHI (2003) Interaction between nutrients and herbivory in controlling algal communities and coral condition on Glover's Reef, Belize. Mar Ecol Prog Ser 261:135–147

McCook LJ (1999) Macroalgae, nutrients and phase shifts on coral reefs: scientific issues and management consequences for the Great Barrier Reef. Coral Reefs 18:357–367

McCook LJ, Jompa J, Diaz-Palido G (2001) Competition between corals and algae on coral reefs: a review of evidence and mechanisms. Coral Reefs 19:400–417

Miller MW, Hay ME, Miller SL, Malone D, Sotka EE, Szmant AM (1999) Effects of nutrients versus herbivores on reef algae: a new method for manipulating nutrients on coral reefs. Limnol Oceanogr 44:1847–1861

Muhs DR, Budahn JR, Prospero JM, Carey SN (2007) Geochemical evidence of African dust inputs to soils of western Atlantic islands: Barbados, the Bahamas, and Florida. J Geophys Res Part F Earth Surf 112:1–26

Mumby PJ, Hastings A, Edwards HJ (2007) Thresholds and the resilience of Caribbean coral reefs. Nature 450:98–101

Muthukrishnan R (2013) An integrated empirical and modeling approach to evaluate determinants of community structure and alternate stable states dynamics on tropical reefs. PhD dissertation, University California, Los Angeles, p 129

Muthukrishnana R, Fong P (2014) Multiple anthropogenic stressors exert complex interactive effects on a coral reef. Coral Reefs. doi:10.1007/s00338-014-1199-1

Norstrom AV, Nystrom M, Lokrantz J, Folke C (2009) Alternative states on coral reefs: beyond coral-macroalgal phase shifts. Mar Ecol Prog Ser 376:295–306

Nugues MM, Smith GW, Hooidonk RJ (2004) Algal contact as a trigger for coral disease. Ecol Lett 7:919–923

O'Leary JK, McClanahan TR (2010) Trophic cascades result in large-scale coralline algae loss through differential grazer effects. Ecology 91:3584–3597

Odum HT, Odum EP (1955) Trophic structure and productivity of a windward coral reef community on Eniwetok Atoll. Ecol Monogr 25:291–320

Ogden JC, Lobel PS (1978) The role of herbivorous fishes and urchins in coral reef communities. Environ Biol Fish 3:49–63

Ong L (2007) The ecological importance of parrotfish as bioeroders and sediment producers and their conservation within small marine protected areas. PhD dissertation, University of Hawai'i, Manoa, p 133

Paerl HW, Paul VJ (2012) Climate change: links to global expansion of harmful cyanobacteria. Water Res 46:1349–1363

Paerl HW, Joyner JA, Joyner AR, Arthur KE, Paul VJ, O'Neil JM, Heil CA (2008) Co-occurrence of dinoflagellate and cyanobacterial harmful algal blooms in southwest Florida coastal waters: dual nutrient (N and P) input controls. Mar Ecol Prog Ser 371:143–153

Pandolfi JM, Bradbury RH, Sala E, Hughes TP, Bjorndal KA, Cooke RG, McArdle D, McClenechan L, Newman MJH, Paredes G, Warner RR, Jackson JBC (2003) Global trajectories of the long-term decline of coral reef ecosystems. Science 301:955–958

Pandolfi JM, Jackson JBC, Baron N, Bradbury RH, Guzman HM, Hughes TP, Kappel CV, Micheli F, Ogden JC, Possingham HP, Sala E (2005) Ecology—are US coral reefs on the slippery slope to slime? Science 307:1725–1726

Pandolfi JM, Connolly SR, Marshall DJ, Cohen AL (2011) Projecting coral reef futures under global warming and ocean acidification. Science 333:418–422

Paul VJ (1997) Secondary metabolites and calcium carbonate as defenses of calcareous algae on coral reefs. In: Proceedings of 8th International Coral Reef Symposium, vol 1, Panama, pp 707–712

Paul VJ (2008) Global warming and cyanobacterial harmful algal blooms. In: Hudnell HK et al (eds) Cyanobacterial harmful algal blooms: state of the science and research needs, federal responses & HAB control. Adv Exp Med Biol:239–257. Springer, Heidelberg

Paul VJ, Hay ME (1986) Seaweed susceptibility to herbivory: chemical and morphological correlates. Mar Ecol Prog Ser 33:255–264

Paul VJ, Ritson-Williams R (2008) Marine chemical ecology. Nat Prod Rep 25:662–695

Paul VJ, Cruz-Rivera E, Thacker RW (2001) Chemical mediation of macroalgal-herbivore interactions: ecological and evolutionary perspectives. In: McClintock J, Baker B (eds) Marine chemical ecology. CRC Press, LLC, pp 227–265

Paul VJ, Thacker R, Banks K, Golubic S (2005) Benthic cyanobacterial bloom impacts on the reefs of South Florida (Broward County, USA). Coral Reefs 24:693–697

Paul VJ, Arthur KE, Ritson-Williams R, Ross C, Sharp K (2007) Chemical defenses: from compounds to communities. Biol Bull 213:226–251

Pauly D, Christensen V, Dalsgaard J, Froese R, Torres F Jr (1998) Fishing down marine food webs. Science 279:860–863

Pereira RC, Bianco EM, Bueno LB, de Oliveira MAL, Pamplona OS, da Gama BAP (2010) Associational defense against herbivory between brown seaweeds. Phycologia 49:424–428

Petraitis PS, Latham RE (1999) The importance of scale in testing the origins of alternative community states. Ecology 80:429–442

Plass-Johnson JG, McQuaid CD, Hill JM (2013) Stable isotope analysis indicates a lack of inter- and intra-specific dietary redundancy among ecologically important coral reef fishes. Coral Reefs 32:429–440

Prince ED, Goodyear CP (2006) Hypoxia-based habitat compression of tropical pelagic fishes. Fish Oceanogr 15:451–464

Rasher DB, Engel S, Bonito V, Fraser GJ, Montoya JP, Hay ME (2012) Effects of herbivory, nutrients, and reef protection on algal proliferation and coral growth on a tropical reef. Oecologia 169:187–198

Reaka-Kudla ML, Feingold JS, Glynn PW (1996) Experimental studies of rapid bioerosion of coral reefs in the Galápagos Islands. Coral Reefs 15:101–107

Rietkerk M, van de Koppel J (2008) Regular pattern formation in real ecosystems. Trends Ecol Evol 23:169–175

Rietkerk M, Dekker SC, de Ruiter PC, van de Koppel J (2004) Self-organized patchiness and catastrophic shifts in ecosystems. Science 305:1926–1929

Ritson-Williams R, Paul VJ, Bonito V (2005) Marine benthic cyanobacteria overgrow coral reef organisms. Coral Reefs 24:629

Ritson-Williams R, Arnold S, Fogarty N, Steneck R, Vermeij M, Paul VJ (2009) New perspectives on ecological mechanisms affecting coral recruitment on reefs. Smith Contrib Mar Sci 38:437–457

Rivers JS, Peckol P (1995) Interaction effects of nitrogen and dissolved inorganic carbon on photosynthesis, growth, and ammonium uptake of the macroalgae *Cladophora vagabunda* and *Gracilaria tikvahiae*. Mar Biol 121:747–753

Roff G, Mumby PJ (2012) Global disparity in the resilience of coral reefs. Trends Ecol Evol 27:404–413

Schaffelke B (1999) Particulate organic matter as an alternative nutrient source for tropical *Sargassum* species (Fucales, Phaeophyceae). J Phycol 35:1150–1157

Schaffelke B, Klumpp DW (1998) Nutrient-limited growth of the coral reef macroalga *Sargassum baccularia* and experimental growth enhancement by nutrient addition in continuous flow culture. Mar Ecol Prog Ser 164:199–211

Scheffer M (2009) Alternative stable states and regime shifts in ecosystems. In: Levin S (ed) Princeton guide to ecology. Princeton University Press, Princeton, pp 395–406

Scheffer M, Carpenter SR (2003) Catastrophic regime shifts in ecosystems: linking theory to observation. Trends Ecol Evol 18:648–656

Scheffer M, Carpenter SR, Lenton TM, Bascompte J, Brock W, Dakos V, van de Koppel J, van de Leemput IA, Levin SA, van Nes EH, Pascual M, Vandermeer J (2012) Anticipating critical transitions. Science 338:344–348

Sfriso A, Marcomini A (1997) Macrophyte production in a shallow coastal lagoon. Part I: coupling with chemico-physical parameters and nutrient concentrations in waters. Mar Environ Res 44:351–375

Smith JE, Shaw M, Edwards RA, Obura D, Pantos O, Sala E, Sandin SA, Smriga S, Hatay M, Rohwer FL (2006) Indirect effects of algae on coral: algae-mediated, microbe-induced coral mortality. Ecol lett 9(7):835–845

Smith JE, Kuwabara J, Coney J, Flanagan K, Beets J, Brown D, Stanton F, Takabayashi M, duPlesses S, Griesemer BK, Barnes S, Turner J (2009) An unusual cyanobacterial bloom in Hawaii. Coral Reefs 27:851

Smith TB (2005) The dynamics of coral reef algae in an upwelling system. PhD dissertation, University of Miami, Coral Gables, p 156

Smith TB (2006) Benthic algal production in an upwelling coral reef environment (Gulf of Panamá): evidence for control by fish herbivores. In: Proceedings of 10th International Coral Reef Symposium, vol 3, Okinawa, pp 854–861

Smith TB (2008) Temperature effects on herbivory for an Indo-Pacific parrotfish in Panamá: implications for coral–algal competition. Coral Reefs 27:397–405

Smith TB, Fong P, Kennison R, Smith J (2010) Spatial refuges and associational defenses promote harmful blooms of the alga *Caulerpa sertularioides* onto coral reefs. Oecologia 164:1039–1048

Smith TB, Glynn PW, Maté JL, Toth LT, Gyory J (2014) A depth refugium from catastrophic coral bleaching prevents regional extinction. Ecology 95:1663–1673

Stachowicz JJ (2001) Mutualism, facilitation, and the structure of ecological communities. Bioscience 51:235–246

Steneck RS (1988) Herbivory on coral reefs: a synthesis. In: Proceedings of 6th International Coral Reef Symposium, vol 1, Townsville, pp 37–49

Steneck RS, Watling L (1982) Feeding capabilities and limitation of herbivorous mollusks: a functional group approach. Mar Biol 68:299–319

Stiger V, Deslandes E, Payri CE (2004) Phenolic contents of two brown algae, *Turbinaria ornata* and *Sargassum mangarevense* on Tahiti (French Polynesia): interspecific, ontogenic and spatio-temporal variations. Bot Mar 47:402–409

Stimson J, Larned ST (2000) Nitrogen efflux from the sediments of a subtropical bay and the potential contribution to macroalgal nutrient requirements. J Exp Mar Biol Ecol 252:159–180

Stimson J, Larned ST, Conklin E (2001) Effects of herbivory, nutrient levels, and introduced algae on the distribution and abundance of the invasive macroalga *Dictyosphaeria cavernosa* in Kaneohe Bay, Hawaii. Coral Reefs 19:343–357

Streelman JT, Alfaro M, Westneat MW, Bellwood DR, Karl SA (2002) Evolutionary history of the parrotfishes: biogeography, ecomorphology, and comparitive diversity. Evolution 56:961–971

Stobart B, Teleki K, Buckley R, Downing N, Callow M (2005) Coral recovery at Aldabra Atoll, Seychelles: five years after the 1998 bleaching event. Philos T R Soc A: Math Phys Eng Sci 363:251–255

Suding KN, Hobbs RJ (2009) Threshold models in restoration and conservation: a developing framework. Trends Ecol Evol 24:271–279

Suding KN, Gross KL, Houseman GR (2004) Alternative states and positive feedbacks in restoration ecology. Trends Ecol Evol 19: 46–53

Szmant AM (2002) Nutrient enrichment on coral reefs: is it a major cause of coral reef decline? Estuaries 25:743–766

Taylor DI, Nixon SW, Granger SL, Buckley BA, McMahon JP, Lin HJ (1995) Responses of coastal lagoon plant communities to different forms of nutrient enrichment: a mesocosm experiment. Aquat Bot 52:19–34

Terborgh J (1971) Distribution on environmental gradients: theory and a preliminary interpretation of distributional patterns in the avifauna of the Cordillera Vilcabamba, Peru. Ecology 52:23–40

Thybo-Christesen M, Rasmussen MB, Blackburn TH (1993) Nutrient fluxes and growth of *Cladophora sericea* in a shallow Danish bay. Mar Ecol Progr Ser 100:273–281

Toth LT, Aronson RB, Vollmer SV, Hobbs JW, Urrego DH, Cheng H, Enochs IC, Combosch DJ, van Woesik R, Macintyre IG (2012) ENSO drove 2500-year collapse of eastern Pacific coral reefs. Science 337:81–84

Umar MJ, McCook LJ, Price IR (1998) Effects of sediment deposition on the seaweed *Sargassum* on a fringing coral reef. Coral Reefs 17:169–177

Vermeij MJA, Smith JE, Smith CM, Thurber RV, Sandin SA (2009) Survival and settlement success of coral planulae: independent and synergistic effects of macroalgae and microbes. Oecologia 159:325–336

Vitousek PM, Aber JD, Howarth RW, Likens GE, Matson PA, Schindler DW, Schlesinger WH, Tilman DG (1997) Human alteration of the global nitrogen cycle: sources and consequences. Ecol Appl 7:737–750

Wahl M, Hay ME (1995) Associational resistance and shared doom: effects of epibiosis on herbivory. Oecologia 102:329–340

Ward-Paige CA, Risk MJ, Sherwood OA, Jaap WC (2005) Clionid sponge surveys on the Florida Reef Tract suggest land-based nutrient inputs. Mar Pollut Bull 51:570–579

Wartian MJ (2006) Determinants of community structure and resilience on tropical Eastern Pacific coral reefs. PhD dissertation, University of California, Los Angeles, p 150

Weber M, de Beer D, Lott C, Polerecky L, Kohls K, Abed RMM, Ferdelman TG, Fabricius KE (2012) Mechanisms of damage to corals exposed to sedimentation. Proc Nat Acad Sci 109:E1558–E1567

Wellington GM (1982) Depth zonation of corals in the Gulf of Panamá: control and facilitation by resident reef fishes. Ecol Monogr 52:223–241

Wellington GM, Glynn PW (2007) Responses of coral reefs to El Niño-Southern Oscillation sea-warming events. In: Aronson RB (ed) Geological approaches to coral reef ecology. Ecol Stud 192:342–385. Springer, New York

Williams SL, Carpenter RC (1988) Nitrogen-limited primary productivity of coral reef algal turfs: potential contribution of ammonium excreted *by Diadema antillarum*. Mar Ecol Prog Ser 47:145–152

Wootton JT (2001) Local interactions predict large-scale pattern in empirically derived cellular automata. Nature 413:841–844

Work TM, Aeby GS, Maragos JE (2008) Phase shift from a coral to a corallimorph-dominated reef associated with a shipwreck on Palmyra Atoll. PLoS ONE 3:e2989

Wysor B (2004) An annotated list of marine Chlorophyta from the Pacific coast of the Republic of Panamá with a comparison to Caribbean Panamá species. Nova Hedwigia 78:1–2

Żychaluk K, Bruno JF, Clancy D, McClanahan TR, Spencer M (2012) Data-driven models for regional coral-reef dynamics. Ecol Lett 15:151–158

Coral Reef Bioerosion in the Eastern Tropical Pacific

12

Juan José Alvarado, Benjamin Grassian, Jaime Ricardo Cantera-Kintz, José Luis Carballo, and Edgardo Londoño-Cruz

In loving memory of our colleague Ana Cecilia Fonseca Escalante (1970–2013)
Dreaming about boring a tunnel to escape from prison?
Dreaming about boring a reef and discover its formation history?
Dreaming about boring a being to possess it?
Dreaming about boring an oil pit that attaches us to money? Or
Dreaming about boring a skeleton to keep enclosed?
It is finally just dreaming about "boring to cope with boredom."
Revista de Biología Tropical 2006 54:101–115

Abstract

Bioerosion, the weakening and erosion of hard substrates by boring, etching, and grazing organisms, is a major structuring force on coral reefs of the Eastern Tropical Pacific (ETP). Bioerosional processes are the main source of reef erosion, and facilitate recycling of reefal carbonate. In healthy reefs, a dynamic balance exists between destructive (i.e. bioerosion) and constructive (i.e. bioaccretion) processes, allowing for maintenance and growth of reef frameworks. In changing environments, however, bioerosion rates can exceed those of coral calcification, leading to reduced reef development and the destruction of reef frameworks. In the ETP, high rates of bioerosion are promoted by nutrient-rich upwelling and high primary productivity conditions, recurrent coral bleaching and mortality events, and a chemical environment characterized by high-pCO_2 and low aragonite saturation state. Here we examine bioerosion in ETP coral habitats and the variable roles of reef-dwelling bioeroder taxa: microbial euendoliths (microendoliths), sponges, polychaetes, sipunculans, crustaceans, molluscs, echinoids, and reef fishes. Among these agents of bioerosion, sponges, sipunculans, bivalves, and echinoderms have been relatively well studied in this

J.J. Alvarado (✉)
Centro de Investigación en Ciencias del Mar y Limnología, Escuela de Biología, Universidad de Costa Rica, 2060-1000 San José, Costa Rica
e-mail: juan.alvarado@ucr.ac.cr

B. Grassian
Rosenstiel School of Marine and Atmospheric Science, Marine Biology and Ecology, University of Miami, 4600 Rickenbacker Causeway, Miami, FL 33149, USA
e-mail: bgrassian@gmail.com

B. Grassian
Graduate School of Oceanography, University of Rhode Island, 215 S Ferry Rd, Narragansett, RI 02882, USA

J.R. Cantera-Kintz · E. Londoño-Cruz
Departamento de Biología, Universidad del Valle, Calle 100, no 13-00, Cali, Colombia
e-mail: jaime.cantera@correounivalle.edu.co

E. Londoño-Cruz
e-mail: edgardo.londono@correounivalle.edu.co

J.L. Carballo
Instituto de Ciencias del Mar y Limnologia (estación Mazatlán), Universidad Nacional Autónoma de Mexico, Mazatlán, Mexico
e-mail: jlcarballo@ola.icmyl.unam.mx

P.W. Glynn et al. (eds.), *Coral Reefs of the Eastern Tropical Pacific*, Coral Reefs of the World 8, DOI 10.1007/978-94-017-7499-4_12

region, while information is currently lacking or limited for microendolith assemblages, polychaetes and reef fishes. The frequency of coral invasion by clionaid sponges (e.g., *Cliona vermifera* and *Thoosa mismalolli*) is variable between ETP coral habitats. Dense boring sponge assemblages can lead to high rates of carbonate losses exceeding those of bioaccretion. Boring bivalves (i.e., species of *Lithophaga* and *Gastrochaena*) are very abundant on many actively accreting reefs and are generally more prominent contributors to reef erosion in the ETP than in other regions. Sea urchins are by far the most destructive grazers of coral substrates in habitats where abundant. Following ENSO-associated coral mortality events, intense bioerosion by sea urchins has impeded coral recovery and compromised reef health at many eastern Pacific sites. This chapter reviews factors important in ETP bioerosion, and current knowledge of bioeroder populations in the region.

Keywords
Carbonate degradation • Echinoids • Boring bivalves • Bioeroder diversity • Grazing and boring

12.1 Introduction

Coral reefs are ecologic and geomorphic formations constructed by calcifying organisms. The biological origin of these ecosystems presents a unique scenario, in that biotic processes not only drive ecological functioning but also the geological construction of these habitats. The degradation of coral habitats is predominantly of a biological derivation as well. Coral reefs are so often introduced as the ocean's "rainforests"—marine ecosystems supporting enormous biodiversity and biological activity. Among this biodiversity, many reef-dwelling organisms are bioeroders responsible for the erosion of coral structures. The occurrence and prevalence of reefal bioerosion may appear outwardly as an unsustainable ecological feedback. Bioerosional processes, however, are natural phenomena intrinsic to the functioning of these ecosystems, present on healthy reefs at rates approaching carbonate production (Hutchings 1986, 2011; Glynn 1997). The development of coral habitats relies on the maintenance of a dynamic equilibrium between destructive and constructive modifications of reefal carbonate. In scenarios where destructive processes gain favor over constructive ones, bioerosion can threaten coral reef structures and ecological functioning.

Currently, global climate change and increasing anthropogenic activities are testing the resilience of reef ecosystems. In response to shifting environmental conditions and heightened coral mortality, it is becoming progressively evident that on impacted habitats enhanced bioerosional processes can outperform constructive ones. Bioerosion appears to be playing important roles in the transformation of impacted reef habitats away from coral-dominated states (Perry et al. 2008). For effective conservation of coral reefs, a comprehensive understanding of bioerosional processes must be established; and to understand these processes is to better grasp the biotic and ecological dynamics driving reefal carbonate budgets, and how these dynamics change on impacted reef environments.

Bioerosion is the removal and breakdown of geological materials by the actions of living organisms (Neumann 1966; Hutchings 1986; Tribollet et al. 2011). Depending on the organisms involved, bioerosion is performed through chemical means (e.g., etching), mechanical means (e.g., abrasion), or by a combination of both mechanisms (Neumann 1966; Warme 1975; Spencer 1992; Glynn 1997). For some organisms, the mechanisms of bioerosion are still not fully understood (Hutchings 2011). The agents of reef bioerosion can be broadly classified into either those performing internal or external erosion of substrates—endoliths and epiliths respectively. Endoliths (i.e., algae and microendoliths, sponges, polychaete worms, sipunculan worms, acrothoracican crustaceans, and molluscs) are cryptic organisms that excavate carbonate substrates internally and often become permanent residents within coral skeletons. Epilithic bioeroders (i.e., crustaceans, echinoids and fishes) are external grazers that erode coral structures via foraging behavior. Among and between these populations, numerous functional interactions provide feedback that drive the pace and pattern of the overall bioerosion process (Chazottes et al. 1995; Tribollet et al. 2002, 2005). Bioerosion is a dynamic and collectively complex process, varying across reef zones, among coral habitats, and over time (Eakin 2001; Perry et al. 2008; Hutchings 2011).

The impacts of bioeroder activities on corals and coral reefs are functionally diverse. Carbonate excavation by bioeroding organisms creates and elaborates upon the three-dimensional habitat used by a wide range of reefal organisms. Nestling invertebrates constitute a large component of reef communities, and these cryptic populations (chasmoliths) occupy habitats provisioned by bioerosional activites. Sediments generated by bioerosional processes contribute to reef cementation, produce lagoonal sediments, and facilitate the recycling of carbonate materials (Hutchings 2011). The weakening and fragmentation of coral skeletons by bioerosion promotes asexual dispersal of coral colonies (Tunnicliffe 1979; Highsmith 1982; Guzman and López 1991). Herbivorous grazing by epilithic bioeroders, while destructive, can limit algal competition (Tribollet et al. 2011). Bioerosion is inherently a destructive process, however, and when not balanced by reef accretion impedes the development, maintenance, and functioning of coral ecosystems. Bioerosional processes can compromise the structural integrity of reef frameworks and, when extreme, reduce the topographic complexity of coral habitats or can engender their destruction altogether (Tunnicliffe 1979; Scoffin et al. 1980; Glynn 1997; Enochs and Manzello 2012).

This chapter will review environmental and ecological conditions influencing bioerosional processes in the ETP, followed by a summary of bioeroder populations in coral habitats, emphasizing current trends in their occurrence and contributions to reef erosion. Bioerosional dynamics in the ETP are assessed focally within this regional scope, and discussed in a global context to establish relevant comparisons. Among the agents of bioerosion: microendoliths, sponges, polychaetes, sipunculans, crustaceans, bivalve molluscs, echinoids, and fishes; not all groups have been well studied in ETP coral habitats (i.e., microendolith, polychaete, and fish bioerosion currently lack or have limited investigation). The scope of these taxa-specific discussions reflects current knowledge of the agents of bioerosion in coral habitats as well as the apparent significance of their contributions to ETP bioerosional processes.

12.2 Bioerosion in the Eastern Tropical Pacific

Bioeroder activites exert enormous influence on coral habitats in the ETP, perhaps uniquely so. For example, reefal bioerosion rates in the Galápagos Islands and Panama are among the highest reported worldwide to date (Glynn 1988; Eakin 1996; Reaka-Kudla et al. 1996; Glynn et al. 2015). In the ETP, bioerosional processes are promoted by a suite of environmental conditions that impede reef development and are responsible for the relatively ephemeral existence of this region's reefs over geological time (Colgan 1990; Toth et al. 2012): productive waters and seasonal upwelling; a shallow and fluctuating thermocline; surface waters characterized by elevated pCO_2, relatively low pH and low aragonite saturation state (Ω); poor reef cementation; and the considerable influence of ENSO thermal shock events. Intense bioerosion regimes promoted by these conditions may, in turn, contribute to the poor development of coral reefs in the ETP.

In general, bioerosional processes are promoted by conditions (1) causing coral death and thereby enlarging opportunity for bioeroder recruitment, or (2) conditions providing growth advantage to bioeroders over calcifying organisms (Glynn and Manzello in press). The eastern Pacific environment is replete with conditions stimulating bioeroder activites by these means. Nutrient-loading of ETP surface waters, promoting bioeroder food chains, may be principal among these influences. Predominant bioeroders (i.e., boring bivalves and sponges) are suspension-feeders and a direct correlation between primary productivity, the abundance of these taxa, and bioerosion rates has been well-demonstrated globally (Highsmith 1980; Hallock 1988; Hutchings et al. 2005). Coinciding with this trend, boring bivalves occur in higher densities in the ETP and are responsible for more carbonate excavation than found in regions elsewhere (Highsmith 1980; Scott and Risk 1988; Kleemann 1990). Eutrophic conditions in the ETP increase the ecological competiveness of algae and macroborers relative to hermatypic corals, thereby increasing bioerosion rate (Highsmith 1980; Hallock 1988).

Corals in the ETP are often subjected to severe episodic bleaching and mortality events associated with ENSO thermal perturbation. Coral communities impacted by these events are susceptible to increased bioerosion regimes, and enhanced bioeroder recruitment and bioerosion rates often accompany scenarios where there is a sudden reduction in live coral cover (Glynn 1984, 1990; Eakin 1992). ENSO-induced coral mortality events help drive bioeroder abundances and biogeographic trends in the ETP. Current distributions of boring sponges in Mexican reefs largely coincide with variable impact of ENSO events from reef to reef (Carballo et al. 2013). The rapid degradation of reef frameworks by echinoids in Panama and the Galápagos Islands after the severe 1982–83 El Niño event is a well-documented phenomenon (Glynn 1988, 1990, 1994; Eakin 1992, 1996; Wellington and Glynn 2007) and demonstrates the potential for rapid shifts in reef states via sudden promotion of bioerosional processes (Perry et al. 2008). Changes in coral reef community structure can also lead to or prolong outbreaks of high bioeroder population densities. Trophic-cascades triggered by over-harvesting of sea urchin predators (e.g., lobsters, fishes) in the Galápagos Islands has been proposed as a link to high echinoid abundance and bioerosion rate currently persisting on certain coral communites impacted by ENSO events years prior (Sonnenholzer et al. 2009, 2011; Edgar et al. 2011; Glynn et al. 2015).

The effects of ocean acidification and accelerating oceanic uptake of atmospheric CO_2 on coral reef maintenance and growth are key issues for the future of these ecosystems. Demonstration of low pH and low Ω waters acting as physiochemical impediment to reef deposition is becoming increasingly robust (Orr et al. 2005; Kleypas et al. 2006; Kroecker et al. 2010; Manzello 2010a). Ocean acidification and lowered Ω enhance bioerosion phenomena as well via thermodynamic facilitation of chemical erosion mechanisms employed by predominant bioeroders, e.g. microbial euendoliths, sponges, and bivalve molluscs (Lazar and Loya 1991; Tribollet et al. 2009; Wisshak et al. 2012, 2013), while simultaneously impairing calcifying processes (Manzello et al. 2010b; Wisshak et al. 2012; DeCarlo et al. 2015). Upwelling of high pCO_2 subthermocline waters in the ETP creates low pH and low Ω surface waters throughout most of the region (Millero 2007; Manzello et al. 2008; Manzello 2010b). Although the direct interaction between ocean acidification and bioerosion phenomena is only recently becoming elucidated, ETP reefs provide a contemporary model for the interface between these processes. A high-CO_2 physical environment contributes to poor reef development and cementation in the ETP (Manzello et al. 2008, 2010a, b), and also increases bioerosion rates (Tribollet et al. 2009; Wisshak et al. 2012, 2013; DeCarlo et al. 2015). Coupling of high-CO_2 conditions with elevated nutrients in the ETP appears to accelerate synergistically reefal bioerosion (DeCarlo et al. 2015).

While it is becoming increasingly important to assess the threats bioerosional processes impose on impacted coral reefs, these activities also incite ecological benefits for coral communities. The weakening and fragmentation of coral skeletons by bioeroders stimulate propagation of coral colonies and are important initiators of asexual coral reproduction (Highsmith 1982). In the ETP, characterized by ephemeral reefs (Glynn and Macintyre 1977; Colgan 1990; Cortés 1997; Toth et al. 2012), rapid recycling of carbonate materials, and the predominance of fast-growing pocilloporid corals that reproduce chiefly through asexual means (Glynn and Macintyre 1977; Richmond 1985, 1987; Glynn et al. 1991), the disperal of corals by bioerosion-induced colony fragmentation may contribute to the persistence of ETP corals in unfavorable environments (Scott et al. 1988; Scott and Risk 1988). In particular, the high abundance of boring bivalves that by their activities lower coral skeleton breaking stability (Scott and Risk 1988), and persistent excavation by reef fishes (e.g., tetraodontids and balistids) appear to be significant initiators of coral reproduction in the ETP (Guzman 1988; Guzman and López 1991; Palacios et al. 2014; see Chap. 15, Glynn et al.). In this way, the strong influence of bioerosional processes in ETP coral habitats likely helps to define both the transience and persistence of these structures over geological time—contributing to their rapid destruction and continuous rebuilding.

12.3 Microendolith Communities

A diverse assemblage of endolithic microorganisms inhabits skeletons of live and dead coral colonies. Certain cyanobacteria, chlorophyte and rhodophyte algae, fungi, and foraminifera are endolithic bioeroders of coral, chemically dissolving carbonate substratum as they invade within. Microendoliths are among the first recognized bioeroders in the fossil record, with microendolith borings identified in ooid grains of the late Precambrian, 570–700 million year (Campbell 1982). There is a relative paucity of research addressing microendoliths in comparison to studies of macroborers and grazers, and the role of micro-organismal populations in bioerosion of modern corals is less clear (Tribollet 2008). Reports suggest microendolith boring accelerates the overall bioerosion processes by structural weakening of the carbonate substratum. In some reef environments these assemblages perform high rates of carbonate erosion themselves (Tudhope and Risk 1985). In the ETP, there has not yet been an assessment of microendolith bioerosion rates or taxonomy. A brief inclusion is made here to outline functional interactions between microendolith bioerosion and other bioerosional agents demonstrated in coral habitats outside the ETP (e.g.. French Polynesia, Great Barrier Reef)—these broad roles are more than likely to be conserved in eastern Pacific reef environments.

Microendoliths are present in both living and dead corals, but pioneer assemblages that recruit following coral mortality are taxonomically dissimilar from those that colonize and keep pace with live, calcifying corals (Hutchings 2011). Chlorophytes of the genus *Ostreobium* are reported to be ubiquitous in the skeletons of live scleractinian corals (Bentis et al. 2000; Gutner-Hoch and Fine 2011), with certain cyanobacteria (i.e. *Plectonema* spp.) and heterotrophic fungi being cosmopolitan as well.

Microendoliths can be highly abundant in dead coral skeletons (Le Campion-Alsumard et al. 1995). Microendoliths colonize dead coral substrata well before macroborer taxa and pioneer epibenthic communities can be established within days of coral death (Tribollet and Golubic 2005). Initial surface invasion by pioneer microborers is followed by a distinct succession of microendolith assemblages that bore deeper into the limestone substrate. Mature communities taxonomically similar to assemblages found in live corals are established at roughly six months after substrate exposure in tropical waters, and are dominated by low-light specialists and heterotrophic fungi (Hutchings 2011; Tribollet and Golubic 2011). Most notably, there is a suggestion that rapid post-mortem invasion by microendoliths may condition newly available reef rock and render it suitable for recruitment of the larger bioeroding community (Hutchings 1986, 2011).

Several studies have demonstrated a salient functional interconnectedness between microendolith and grazer activities (Chazotte et al. 1995; Zubia et al. 2001; Tribollet and Golubic 2005). Initially, colonization of dead corals by microendolith assemblages attracts grazers, (many of which forage on algal mats), and thereby mediates high rates of epilithic bioerosion. The larger-scale excavation by grazers in turn accommodates for the progression of microendolith invasion, a process that otherwise would be self-stabilizing. As grazers remove epilithic material and continuously reduce shading, phototrophic microendoliths that have light-limited distributions are able to penetrate deeper into coral rock. In this way, reinforcing interactions between microendolith and grazer activity provide a synergistic expansion of the bioeroder-substrate interface and the overall bioerosion process (Chazotte et al. 1995).

It is well known that grazing species (i.e., sea urchins) are among the most destructive bioeroders in ETP reefs (Glynn 1988; Eakin 1996). Given our current understanding of interaction between microendolith assemblages and epilithic agents, ETP microendolith assemblages are near-assured to have substantial involvement in bioerosional processes. It therefore becomes necessary for microendoliths to be assessed by future studies attempting to explain the organismal and functional dynamics driving ETP bioerosion processes.

12.4 Excavating Sponges

Sponges are the most common macroendolith present in reef ecosystems worldwide and are responsible for the highest rates of internal coral destruction than any other endolithic borer (Goreau and Hartman 1963; MacGeachy 1977; Perry 1998; Schönberg and Ortiz 2008; Carballo et al. 2010a). Boring sponges are generally cryptic in nature and have great taxonomic and ecological complexity, thereby exacerbating difficulties in sampling and identification. This has led to frequent omission of boring sponges in community-level reef assessments, despite their well-established importance to reefal bioerosion (Schönberg 2008).

Sponge assemblages have been relatively well-studied in the ETP, especially in the Mexican Pacific region (e.g., Carballo et al. 2004, 2008; Nava and Carballo 2013). Assessment of boring sponges is robust on Mexican reefs, both ecologically and geographically. Sponge diversity, abundance, and prevalence within corals have been quantified at several of the most prominent reef sites in Mexico: Huatulco, Guerrero, Bahía Banderas, Baja California, and islands such as Isabel, Marietas, Marías and Socorro (Carballo et al. 2013). Reef-dwelling sponges were assessed earlier in Costa Rica (Guzman 1988; Scott et al. 1988) and Panama (Wulff 1997), and provide variable degrees of additional ecological and biological insight (see Chap. 5 for site locations).

The early surveys of sponges in Panama established an initial characterization of ETP sponges by drawing a comparison with sponge assemblages inhabiting Caribbean reefs (Wulff 1997). In contrast to Caribbean species, sponges in the ETP are most often small, cryptic and, except for a few massive, exposed species (Cruz-Barraza and Carballo 2008; Fig. 12.1a, b), principally coral endoliths. These findings were later corroborated by studies on Mexican reefs, as a high abundance and diversity of boring sponges compared to encrusting (Fig. 12.1c–e) and non-boring massive sponges (Fig. 12.1f) are consistently reported (Carballo et al. 2004, 2008). This characterization is relatively unique to the ETP and should be seen as having much relevance for understanding biological and environmental factors influencing sponges in this region. While massive sponges do not contribute directly to bioerosion, these species establish spatial competition with corals and can induce coral mortality as a result.

The high proportion of cryptic azoothanthellate sponges on ETP reefs is thought to be caused by environmentally-induced selectivity in larval recruitment (Wulff 1997). Sponge larvae, and the larvae of sessile macrofauna in general, demonstrate a strong preference for recruitment to cryptic, aphotic reef surfaces in the ETP (Zea 1993). Birkeland (1977) hypothesized that the enhanced accumulation of epibenthic biomass, resulting from high nutrient loading in ETP waters, promotes selective recruitment to cryptic spaces by decreasing survival of larvae settling on exposed substrates. In addition, grazing pressure by opportunistic predators may further restrict ETP sponges to cryptic spaces. Reef fishes abundant in the ETP, such as *Arothron hispidus*, are observed to readily consume cryptic sponges that become exposed and lack inherent defenses against predation (Wulff 1997).

A high diversity of cryptic boring sponges is found within coral bases and reef frameworks on ETP reefs. Twenty-two species have been described in the region to date, a majority belonging to the genus *Cliona* (Carballo et al. 2010a) (Table 12.1). Among ETP sponges, clionaid species are dominant, both in incidence and impact. *Cliona vermifera*, specifically, is the most abundant coral-excavating sponge in the ETP, and the prevailing significance of *C. vermifera* is conserved among all assessments from Mexican and Costa Rican reefs (Guzman 1988; Carballo et al. 2008; Bautista-Guerrero et al. 2014). *Cliona vermifera* is a highly adaptable species that demonstrates considerable resilience to environmental stress and an ability to thrive in a wide range of microhabitats (Bautista-Guerrero et al. 2014). *Thoosa mismalolli* is also found in high and consistent abundance in Mexico (Carballo et al. 2004, 2008), and is common on Costa Rican reefs as well (Pacheco-Solano

Fig. 12.1 ETP boring sponges. **a** *Amphimedon texotli* living between colonies of pocilloporid corals. **b** *Chalinula nematifera*, an encrusting sponge able to overgrow live corals of the genus *Pocillopora*. **c** *Cliona raromicrosclera* overgrowing corals of the genus *Porites*. **d** *Cliona raromicrosclera* to *left*, and *Cliona californiana* to *right*, both encrusting (beta) growth forms. **e** *Cliona vallartense* in beta stage growing on dead coral. **f** Large, massive (gamma stage) specimen of *C. californiana*. All photographs from J.L. Carballo except **c** and **d** by E. Avila

2011). Both *C. vermifera* and *T. mismalolli* are known to be highly destructive coral-excavating sponges and are capable of producing large galleries in coral skeletons, especially those of branching *Pocillopora* spp. (Nava and Carballo 2008; Carballo et al. 2013; Bautista-Guerrero et al. 2014).

Endolithic sponges excavate distinct systems of chambers and branching galleries within coral skeletons (Fig. 12.2a–h). The boring process occurs at a cellular level by the combined chemical and mechanical actions of amebocytic etching cells (Rützler and Rieger 1973; Pomponi 1977). Along the internal substrate interface, etching cells are flattened against the substratum and pseudopodial structures are extended into the surrounding limestone. The penetrating filopodia then ramify and coalesce forming "pseudopodial baskets," thereby cutting small, lenticular chips (15–100 μm in diameter) from the substrate. Cutting by filopodia is accomplished by enzymatic dissolution of both the inorganic calcium carbonate and the organic matrix of the coral skeleton. Once this dissolution occurs, the etching cells and calcareous chips are transported away from the substrate interface and expelled through the sponge's aquiferous system (Rützler and Rieger 1973; Pomponi 1977).

This boring action can perform high rates of sediment production, and excavating sponge activities are known to contribute up to 30–40 % of the total sediment deposited on some Caribbean reefs (Goreau and Hartman 1963; Neumann 1966; Fütterer 1974). On reefs of the Mexican Pacific, calcareous chips characteristic of the sponge boring mechanism can comprise up to 15 % of the fine sediments present (Ovalle 2011). Until recently, it has been estimated that dissolved substrate represents only a small fraction (2–3 %) of the total skeletal material eroded by sponge boring (Rützler and Rieger 1973; Rützler 1975). Recent estimates, however, report a much larger fraction of material removed by chemical erosion than previously realized (Nava and Carballo 2008), with one study finding chemical dissolution by *Pione* cf. *vastifica* removes three times more carbonate than is produced as sediment (Zundelevich et al. 2007).

There is only one study that has quantified bioerosion and sediment production rates for boring sponges in the ETP (Nava and Carballo 2008). Rates for both chemical and mechanical bioerosion of living *Pocillopora verrucosa* by the boring sponges *Cliona vermifera* and *Cliona flavifodina* were determined at Isla Isabel, Mexico. Chemical dissolution accounted for a significantly larger fraction of eroded skeletal material than the 2–3 % initially proposed and mean rates of dissolution were 1.2 and 0.5 kg $CaCO_3$ m^{-2} $year^{-1}$ for *C. vermifera* and *C. flavifodina* respectively. The highest reported carbonate dissolution rate for *C. vermifera* was 2.4 kg $CaCO_3$ m^{-2} $year^{-1}$. Sediment production rates were higher than those of dissolution, with means of 3.3 and 4.6 kg $CaCO_3$ m^{-2} $year^{-1}$ calculated for *C. vermifera* and *C. flavifodina* respectively. Hence, a summation of chemical dissolution and sediment production gives mean bioerosion rates of 4.5 for *C. vermifera* and 5.1 kg $CaCO_3$ m^{-2} $year^{-1}$ for *C. flavifodina* (Nava and Carballo 2008), among the highest recorded for ETP endoliths. These bioerosion rates are within the range of those calculated for clionaids of the Caribbean, the Indian Ocean and Red Sea on coral substrates, though well below the highest reported rates—i.e. *Pione lampa* in the Caribbean (Rützler 1975), and *Cliona albimarginata* in the eastern Indian Ocean (Calcinai et al. 2007; see Fig. 12.3).

Of the 22 boring sponge species observed on ETP reefs, none has demonstrated a capacity to penetrate corals through soft tissue (Carballo et al. 2013). Boring sponges invade the skeletons of dead corals (Carballo et al. 2013; Nava and Carballo 2013) and live coral colonies having surfaces that lack soft tissues (Highsmith 1980, 1983; Nava and Carballo 2008). This dead skeletal requirement provides an explanation for why, on Mexican reefs, higher densities of boring sponges are found in corals on reef margins and structural depressions than densities found on reef platforms with high live cover (Carballo et al. 2008). It is on reef margins and depressions where boring sponge recruits can access the bases of coral colonies and other structures not covered by living tissue.

A recent study undertaken on two reefs from the Mexican Pacific found high sedimentation rates to restrict boring sponge distributions (Nava and Carballo 2013). Sediment movement associated with water flow causes abrasion or clogging impeding physiological functions, and accumulated sediments can also prevent larval settlement as has been reported for free-living sponges (Hutchings et al. 2005). Boring sponges such as *Cliona carteri* and *Cliona viridis* exhibit their lowest abundance near sites with high sedimentation (Muricy 1991). On some Oaxacan coral reefs on the Mexican Pacific coast, the highest abundance of boring sponges was recorded on reefs with a low sedimentation rate (<0.2 kg m^{-2} $days^{-1}$) (Carballo et al. 2008). *Cliona vermifera,* one of the most abundant and widely distributed boring sponges on eastern Pacific coral reefs, is considered to have a low tolerance to sediment deposition that restricts the species' distribution (Carballo et al. 1994, 2008). In the ETP, only *Siphonodictyon* and *Spheciospongia* species are considered tolerant of high sedimentation, since their aquiferous system is not easily overloaded (Rützler 1971). The abundance of *Siphonodictyon crypticum* on Mexican coral reefs under high sedimentation supports this relationship (Carballo et al. 2007, 2008).

Substantial increases in sponge abundance have been repeatedly reported for reefs having undergone bleaching events or reductions in live coral cover (Rützler 2002; Schönberg and Ortiz 2009), and this correlation is well documented in the ETP (Carballo et al. 2013; Nava and Carballo 2013). Reefs in the Mexican Pacific that have been

Fig. 12.2 Sponge excavation. **a** Chambers excavated by *Cliona californiana* in a bivalve shell after removing superficial layer. Cross-sections of coral skeletons showing chambers formed by **b** *Thoosa purpurea*; **c** *Pione carpenteri*; **d** *Cliona tropicalis*; **e** *Cliona mucronata*; and **f** *Cliona vermifera* (the latter boring into the coral *Pavona gigantea*). **g** Sediment sample containing a sponge chip (*arrow*) with characteristically faceted surface. **h** Sponge chips inside tissue before being expelled from oscula. Photographs by J.L. Carballo

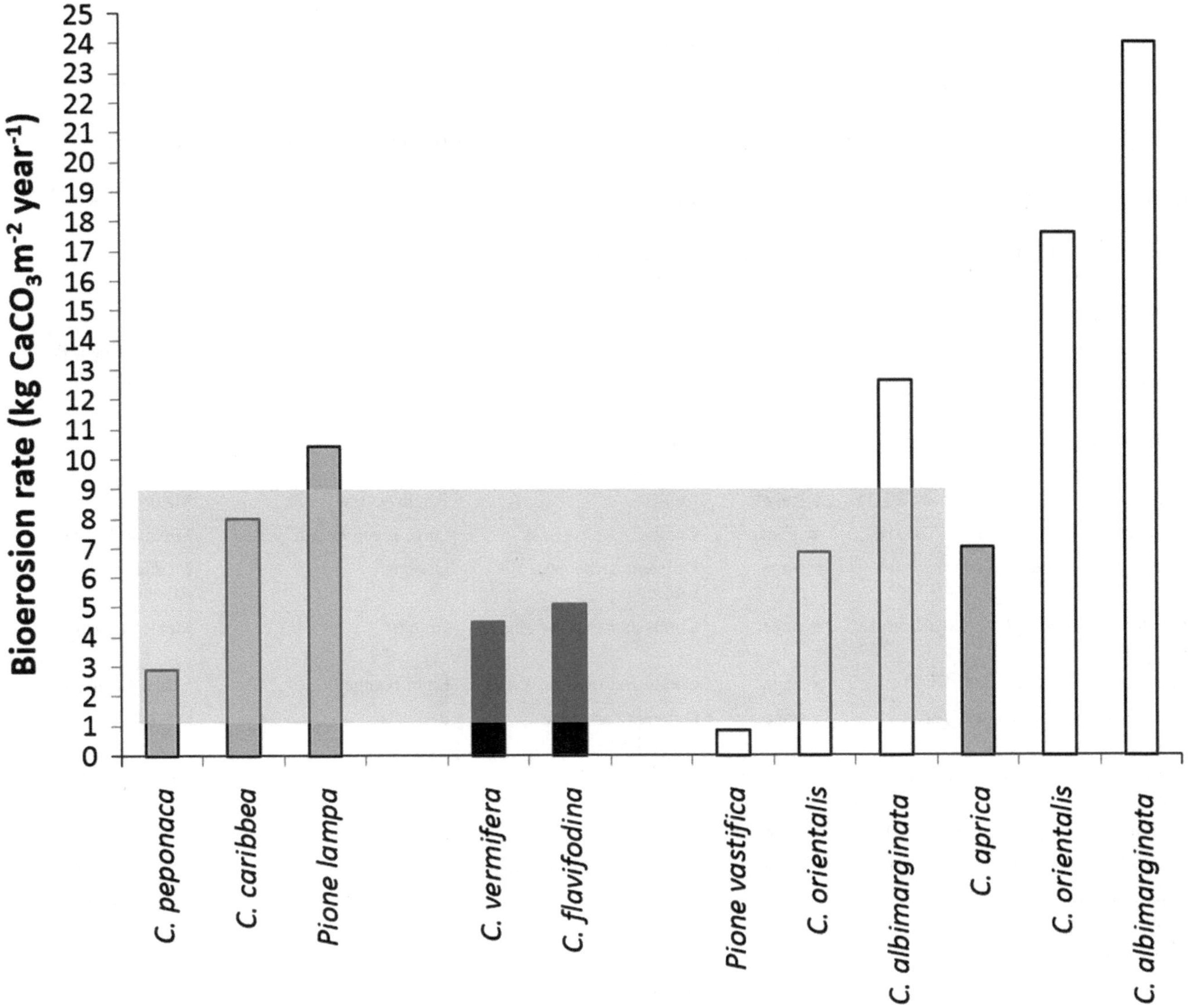

Fig. 12.3 Bioerosion rates (kg $CaCO_3$ m^{-2} $year^{-1}$) of boring sponges of the genera *Cliona* and *Pione* in different oceans: grey bars represent sponge bioerosion rates from the Caribbean, black bars from the east Pacific Ocean, and white bars from the Indian Ocean and Red Sea. Shadowed rectangular area inside plot represents calcification rates in coral communities, which have been estimated between 1.1 and 9 kg $CaCO_3$ m^{-2} $year^{-1}$ (Stearn et al. 1977; Kinsey 1985; Silverman et al. 2007). Bars inside shadowed area represent bioerosion rates on coral substrata; bars outside represent rates on mollusc shells in coral reef environments. *Sources C. peponaca* from Bak (1976), *C. caribbaea* from Acker and Risk (1985), *C. orientalis* from Schönberg (2002), *C. vermifera* and *C. flavifodina* from Nava and Carballo (2008), *C. aprica* and *P. lampa* from Rützler (1975), *C. albimarginata* from Calcinai et al. (2007); *Pione* cf. *vastifica* from Zundelevich et al. (2007). Information on *P. lampa* from Neumann (1996) is not included (see text)

impacted by ENSO-associated bleaching events are currently heavily infested by boring sponges (Carballo et al. 2013). Forty six percent of coral samples on Mexican reefs are invaded by boring sponges, and reef frameworks are especially infested (56 % mean invasion in sampled frameworks). Furthermore, boring sponge abundances and assemblage diversity are highest at sites where coral communities have been most affected by ENSO bleaching events, such as: Carayeros, San Juanitas, and Islas Marías reefs in Nayarit, Cabo Pulmo in the Gulf of California, and La Entrega in Oaxaca (Table 12.2; see Chap. 5 for site locations).

Certain clionaid sponges are highly resilient to anomalous temperature shifts (Shirley and Rützler 2010; Duckworth and Peterson 2013; Bautista-Guerrero et al. 2014), especially when compared to tolerance thresholds found for corals (Glynn et al. 2001; Hueerkamp et al. 2001). *Cliona celata* explants exposed to high water temperatures of up to 31 °C, demonstrated near unaltered growth, survival, and boring rates (Duckworth and Peterson 2013). In another study, *C. celata* explants were found to undergo no significant loss of sponge pigment in response to thermal and salinity stress (Shirley and Rützler 2010). Particularly relevant, a study of

Table 12.1 General characteristics and distribution of coral boring sponges from the Mexican Pacific Ocean

Species	Form[a]	Substrate	Live color	Distribution
Cliona amplicavata (Rützler, 1974)	α stage	Corals and mollusc shells	Bright yellow	Mediterranean, Caribbean, Pacific
C. californiana (de Laubenfels, 1930)	α, β, γ stage	Different carbonate substrates	Golden yellow, light pink, reddish brown	east Pacific
C. *euryphylla* (Topsent, 1887)	α stage	Different carbonate substrates	From yellow to light orange	Atlantic, Pacific
C. *flavifodina* (Rützler, 1974)	α, β stage	Different carbonate substrates	Bright yellow, purple brown	Caribbean, east Pacific
C. medinae (Cruz-Barraza et al., 2011)	α stage	Corals	Yellow-orange	Mexican Pacific
C. *mucronata* (Sollas, 1878)	α stage	Corals	Red to red-orange	Indian Ocean, Pacific
C. *pocillopora* (Bautista-Guerrero et al., 2006)	α stage	Corals	Light brown	Mexican Pacific
C. *raromicrosclera* (Dickinson, 1945)	α, β stage	Different carbonate substrates	Orange, brown	Mexican Pacific
C. *tropicalis* Cruz (Barraza et al., 2011)	α stage	Corals	Bright yellow	Mexican Pacific
C. *vallartense* (Carballo et al., 2004)	α, β stage	Corals	Olive green to pale yellow	Mexican Pacific
C. *vermifera* (Hancock, 1867)	α stage	Different carbonate substrates	Orange	Mediterranean, Atlantic, Pacific
Cliothosa tylostrongylata (Cruz-Barraza et al., 2011)	α stage	Corals and mollusc shells	Orange	Mexican Pacific
Pione carpenteri (Hancock, 1867)	α stage	Corals and mollusc shells	Light orange	Indian Ocean, Pacific
P. mazatlanensis (Hancock, 1867)	α stage	Different carbonate substrates	Red	Mexican Pacific
Spheciospongia incrustans (Carballo et al., 2004)	Encrusting	Corals and calcareous rocks	Purplish	Mexican Pacific
Thoosa calpulli (Carballo et al., 2004)	α stage	Corals	Light brown	Mexican Pacific
T. mismalolli (Carballo et al., 2004)	α stage	Corals and mollusc shells	Light brown	Mexican Pacific
T. purpurea (Cruz Barraza et al., 2011)	α stage	Corals	Bright purple	Mexican Pacific
Siphonodictyum crypticum (Carballo et al., 2007)	Fistulae	Corals	Whitish, translucent	Mexican Pacific

[a]α, β, and γ stages represent cryptic, encrusting, and massive growth forms respectively

Cliona vermifera at Isabel Island, Mexico over a four-year period reported anomalous temperature rises that were detrimental to corals had no negative effect on *C. vermifera* abundances or reproduction in the same study area (Bautista-Guerrero et al. 2014). In fact, reproductive potential and developmental rates increased with water temperature, being highest during the warmest months and during periods of anomalous temperature elevation.

The resilience demonstrated by boring sponge species abundant in the ETP to thermal shock supports ecological projections that sponges will become an increasing threat to coral and coral reef health. The studies available so far suggest that boring sponges are competitively superior to corals in stressful environments that render corals beyond their survival threshold (Rützler 2002). In the ETP, corals are repeatedly affected by episodes of ENSO-associated thermal stress, persistently compromising coral health and recovery on many reefs (Glynn et al. 2001; Baker et al. 2008). It is these reefs impacted by coral bleaching and mortality events that are most vulnerable to invasion by boring sponges. More generally, coral reef calcification rates at community level have been estimated between 1.1 and 9 kg $CaCO_3$ m^{-2} $year^{-1}$ globally (Stearn et al. 1977; Kinsey 1985; Silverman et al. 2007). Bioerosion rates recorded for a single sponge species can reach similar values (Fig. 12.3). The current degradation of eastern Pacific reefs is likely to increase with the progression of temperature rise and ocean acidification (Glynn et al. 2001; Cortés and Jiménez 2003; Manzello et al. 2008; Carballo et al. 2013). It is suggested, concurrently, that sponge bioerosion will increase as coral health declines and the availability of reef substrates for bioeroders expands (Schönberg and Ortiz 2009; Carballo et al. 2010b). Thus, sponge bioerosion is, and will remain, a key issue for the future health of coral reef ecosystems.

Table 12.2 Variation of coral cover and boring sponge abundance on Mexican coral reefs (with respect to ENSO events)

State	Site	Coral coverage[a]						Framework extension (ha)	Frequency of sponge invasion (%)		No Species
		1991	1998	2001	2003	2005	2010		2005–2006	2009–2010	
Oaxaca	La Entraga		76		41			7.5	50	55	5
	Isla Cacaluta		90		61			1.7	41	*	9
	San Agustín		73		64			2.5	31	47	7
Manzanillo	Isla Ixtapa						24	6.9	*	14	6
	Playa Manzanillo						36	0.5	*	15	6
	Playa Blanca					13	12		38	41	13
Nayarit	Careyeros	44	22	3	1	0	0	83	77	**	12
	Antiguo Corral del Risco	38	33	3		7.5	6.5	0.001	41	52	12
	Guayabitos					0	0		55	57	7
	Bahía Tiburones				8.6	6.7	6.5	0.2	47	55	12
	Maria Cleofas					49			*	60	13
	San Juanito		40	42					*	63	8
	Maria Madre Sur		45	40					*	61	10
	San Lorenzo		40			47		0.3	32	52	12
Baja	San Gabriel					99	80		22	28	10
California	Cabo Pulmo	30	29	30	12		18	150	56	58	11
	Caleritas						74		37	32	8

Adapted from Carballo et al. (2013), see original publication for references on coral coverage data

[a]Little information exists for coral habitats prior to the 1997/98 coral bleaching event. Mean frequency of sponge invasion represents average percent invasion in 450 coral samples per site, comprised of coral rubble, small fragments of coral reef framework, and complete branches of live coral colonies in equal proportions (150 samples per category per site)

*Not sampled

**Sampled, but not in coral substrate

12.5 Polychaete Worms

Polychaete worms that bore into reef rock are regarded as dominant macroborers during initial stages of bioerosion, and are important bioeroders in various reef environments worldwide (Hutchings 1981, 2008; Davies and Hutchings 1983). On coral reefs of Australia and French Polynesia, boring polychaetes are found in enormously high densities and their contribution to reef erosion is considerable (Kiene and Hutchings 1994; Hutchings and Peyrot-Clausade 1989, 2002; Hutchings 2008). To date, there are few ETP studies that describe the abundance or distribution of polychaete borers and fewer that address their significance as agents of reef bioerosion. Assessments of bioeroding communities in the ETP that include polychaete borers have generally found their contributions to reef erosion marginal, especially in comparison to other macroborer taxa (i.e., bivalves and sponges). High densities of boring polychaetes have not been reported in ETP corals. However, colonization of reef frameworks by polychaetes and other boring organisms is highly variable in time and space (Hutchings 1981; Kiene and Hutchings 1994). It is thus likely that existing assessments of boring polychaetes in the ETP are collectively inadequate to understand their roles as bioeroders in the region.

Species belonging to six polychaete families recognized as endolithic bioeroders are reported in the ETP: Eunicidae, Flabelligeridae, Sabellidae, Cirratulidae, Spionidae, and Dorvilleidae (Fauchald1977; Blake 1991; Hutchings et al. 1992; Fonseca et al. 2006). The various boring species belonging to these families penetrate up to 10 cm into the interiors of carbonate substrates, excavating circular galleries from 0.5 to 2 mm in diameter (Glynn 1997). While there have been no studies describing boring mechanisms employed by polychaetes in the ETP, some authors (Haigler 1969; Hein and Risk 1975; Hutchings 2008) have described the mechanisms for eunicid and spionid borers elsewhere.

Eunicids excavate sinuous, anastomosing passages in carbonate substrates mechanically via the chewing action of a spade-like lower mandible. Spionid borers produce "U-shaped" burrows of constant diameter (0.3–0.5 mm) excavated by chemical dissolution, with abrasion by chaetae likely contributing minimally to substrate removal (Haigler 1969; Hein and Risk 1975).

A survey of infaunal associates and macroborers of *Pocillopora damicornis* colonies at Playa Blanca, Gorgona Island (see Chap. 5 for site location) identified fifteen polychaete species, of which five were suspected borers belonging to the families Eunicidae, Spionidae, and Flabelligeridae (Cantera et al. 2003). *Eunice* spp. and *Flabelligera* spp. were the most abundant boring polychaetes on the reef, accounting for 5.5 and 5.7 % of the total cryptofauna respectively. *Eunice* spp. occurred mainly on the reef platform and crest within the bases of dead *P. damicornis* colonies. This was characteristic of the overall distribution of polychaete borers on the reef—concentrated in dead coral colonies in the reef zone having the highest coral cover. *Flabelligera* spp. were more evenly distributed across reef zones than other boring species, and inhabited both live and dead coral colonies. Relative to other cryptofauna (i.e., bivalves and sipunculans), the impact of boring polychaete infauna was minimal at Gorgona Island (Cantera et al. 2003).

A subsequent study of polychaete macroborer assemblages in *Porites lobata*-dominated reefs of Costa Rica found similar trends in their distributions and relative abundance (Fonseca et al. 2006). Among the six families of polychaete borers identified, eunicid species were again found dominant. Higher polychaete densities were reported in coral colonies with 50–100 % tissue mortality compared to polychaete densities found in less degraded colonies (combined mean polychaete densities of 34.4 and 5.0 ind cm^{-2} for the high and low coral mortality levels respectively). While this trend was statistically nonsignificant, a qualitative assessment demonstrates higher polychaete abundances in more degraded coral colonies across all three reefs surveyed. As was observed at Gorgona Island, boring polychaete abundances were a fraction of those recorded for other macroborer taxa, with endolithic bivalve and sipunculan populations dominating on the three surveyed reefs.

A few additional studies provide brief insight into ETP polychaete bioeroders. An assessment of cryptofaunal assemblages in Panama (Enochs 2012) found polychaetes to comprise a fraction percent of total infauna biomass, and were more abundant on dead rather than live coral colonies. A study of coral bioerosion in the Galápagos Islands (Reaka-Kudla et al. 1996) reported polychaetes and bivalves as the most abundant infaunal borers present in experimental blocks of *Porites lobata*. The contribution of endolithic agents to total carbonate loss, however, was again marginal, with rates of internal bioerosion approaching an order of magnitude below rates calculated for external bioerosion (2.4 and 21.3 kg m^{-2} respectively) by sea urchin grazers. The observation of high boring polychaete abundances on experimental carbonate blocks may support the importance placed on these bioeroders as opportunistic colonizers of newly available substrates, as is observed in French Polynesia and on the Great Barrier Reef (Osorno et al. 2005; Hutchings 1981, 2008). Given that there are no data on in situ macroborer abundances or recruitment patterns in the Galápagos for comparison, it is possible that the relative dominance of polychaete assemblages on newly exposed substrates also occurs in Galápagos coral habitats.

12.6 Crustaceans

There are three groups of crustaceans that erode coral skeletons and have relevance to ETP reef bioerosion: hermit crabs, ghost shrimps, and barnacles. Crustacean bioeroders have received little attention and are generally considered a fringe, inconsequential group. To assign no importance to crustacean bioeroders may be a premature conclusion, however, with certain ETP species, (specifically pagurid crabs), exhibiting a capacity for coral damage that is not insignificant in early studies. These findings should be provided contemporary inquiry and expansion.

12.6.1 Pagurid Crabs

There are two known species of coral-associated hermit crabs that abrade and remove skeletal material when foraging on live coral colonies. *Trizopagurus magnificus* and *Aniculus elegans* are found on reefs throughout the ETP often sheltering among branching pocilloporid corals and are especially common on *Pocillopora damicornis* (Gilchrist 1985). These pagurid crabs produce large amounts of fine and coarse-grained calcareous sediments when foraging on soft tissues of coral prey (Glynn et al. 1972).

Feeding rates and fractional mass components of targeted *Pocillopora* colonies were assessed experimentally in the 1970s using *Trizopagurus magnificus* and *Aniculus elegans* in Panama (Glynn et al. 1972). Inorganic skeletal material represented more than 90 % of the total coral mass abraded by pagurid foraging, and mean sediment production rates of 9.6 and 1160 mg dry wt day^{-1} of carbonate were removed per individual for *T. magnificus* and *A. elegans* respectively. Given these rates, corallivorous predation by hermit crabs may have greater implications for bioerosion processes than for trophically-driven removal of coral tissues. A subsequent experiment in Panama reported an average of 3.6 cm^2 coral area damaged by individual *T. magnificus* every two days (Gilchrist 1985). Furthermore, crabs from Pacific waters are persistent coral inhabitants compared to transient coral occupation by Caribbean species. As a result, non-feeding damage to corals by Pacific crabs is greater, with sustained contact and shell movement causing further surface abrasion and local polyp death (Gilchrist 1985).

12.6.2 Burrowing Shrimps

Burrowing thalassinidean shrimps in the genus *Pomatogebia* (formerly included in *Upogebia*) are specialized for boring into scleractinian coral skeletons. Two species are found in the ETP, *Pomatogebia rugosa* and *Pomatogebia cocosia* (Williams and Ngoc-Ho 1990). Male/female pairs excavate complex, ramifying burrows in massive corals. *Pomatogebia rugosoa* tunnels are roughly 2.5 mm in diameter (Fonseca and Cortés 1998). With the exception of *Alpheus simus* infestation in some Caribbean reefs (Cortés 1985), boring shrimps are found to be minimal contributors to reef erosion. Sampling of thalassinidean shrimps in ETP reefs commonly report them absent or at very low densities, and excavating reef rock at near negligible rates—though reports of thalassinidean invasion on Costa Rican reefs are highly variable (Cortés 1991; Fonseca and Cortés 1998; Fonseca et al. 2006).

12.6.3 Acrothoracican Barnacles

Burrowing acrothoracican barnacles infest a wide variety of skeletal materials including carbonate reef rock (Warme 1975). Acrothoracican excavation is accomplished largely by mechanical abrasion via the action of chitinous teeth along the carapace. The initial penetration of carbonate substrates by acrothoracican recruits is thought to be performed by chemical dissolution, as suggested by experimental observations and the absence of abrading structures in larvae (Tomlinson 1969). Reports of acrothoracicans on reef substrates are rare (Zullo 1991), with sampling difficulties likely contributing to this paucity. Reef-dwelling barnacles in the ETP are poorly understood and little is known about this group's significance to reef bioerosion. At Gorgona Island, barnacles were early colonizers of experimental units of *Pocillopora* substrates and highly abundant during initial exposure periods (Londoño-Cruz et al. 2003). Pioneer borings by acrothoracicans may be important for facilitating subsequent macroborer recruitment. Barnacles belonging to the Family Lithotryidae perform very high bioerosion rates in certain shallow, near-shore reef environments (Scoffin et al. 1980), but lithotryid species are not reported in ETP coral habitats (Tomlinson 1969; Zullo 1991).

12.7 Sipuncula

Species in several Sipuncula (peanut worms) genera bore into coral skeletons. As endolithic bioeroders, sipunculan worms share a few apparent characteristics with boring polychaetes. Both 'worm' taxa are exceptionally common marine cryptofauna, but occur in highly variable densities and distributions that shift across a broad range of spatial-temporal scales. As a result, there is no general agreement on the overall significance of sipunculans as bioeroders and, as is the case with polychaetes, it is more necessary to assess regional impacts or roles within coral habitats. Additionally, both sipunculans and polychaetes are known to have distinct roles within bioeroder community succession, and temporal trends of substrate colonization have been foci in the study of both groups. While polychaetes are well-recognized as early colonizers of newly available coral substrates, longer-lived sipunculan borers are predominant in the latter stages of the bioerosion process (Hutchings et al. 1992). Despite the morphological and ecological similarities between these endolithic worms, these assemblages demonstrate contrasting requirements for coral invasion and the role sipunculans play in bioerosional processes is generally less well understood.

Sipunculan burrows are found in dead corals or in the dead parts of live colonies (Williams and Margolis 1974). Sipunculan recruits excavate a single flask-shaped burrow that is inhabited individually and permanently. Sipunculan burrows are highly variable and species-specific, ranging from straight to sinuous, narrow to broad, and from near surface to penetrating centimeters deep into the coral skeleton (Glynn 1997). There is substantial evidence that sipunculan boring is accomplished by chemical and mechanical means in combination, though the exact manner is not known. It is likely that acidic or chelating fluids secreted by epidermal glands soften and dissolve the substrate, while cuticular papillae and a calcareous shield on the posterior trunk abrade concurrently (Rice 1969; Rice and Macintyre 1972; Williams and Margolis 1974). Sipunculan species reported on ETP corals are small (Fonseca and Cortés 1998; Londõno-Cruz et al. 2003), limiting their contribution to reefal bioerosion.

In Costa Rica, there has been considerable effort to describe sipunculan diversity since the early 1990s, and more than 20 species have been described to date (Cutler et al. 1992; Dean et al. 2010). Sipunculans that excavate calcareous substrata belong to the families Aspidosiphonidae and Phascolosomatidae (Cutler 1994). Of these families, aspidosiphonid species are larger contributors to coral erosion in the eastern Pacific, with sipunculans in the genus *Aspidosiphon* being particularly abundant (Cantera et al. 2003; Fonseca et al. 2006). Coral-excavating sipunculans are highly abundant on certain Costa Rican reefs. In Golfo Dulce, 92 and 83 % of *Porites lobata* colonies were infested by the sipunculan *Aspidosiphon elegans* at the Sándalo and Punta Islotes reefs respectively (see Chap. 5 for site locations), and *A. elegans* densities reached up to 300 ind m^{-3} (Fonseca and Cortés 1998). A subsequent study found sipunculan infauna to comprise more than 50 % of non-colonial macroborers surveyed at Platanillo (Caño Island) and Golfo Dulce, and were reported to have a mean density of 131.4 ind m^{-2} across sites—sipunculan densities were more than three-fold higher than those found for bivalves, the second most abundant borers (Fonseca et al. 2006).

In the same study, reefal abundance of sipunculan infauna was suggested as a potential bio-indicator of overall reef health (Fonseca et al. 2006). Sipunculan distributions, specifically those of aspidosiphonids, demonstrated apparent and contrasting relationships with both overall reef health and individual coral colony mortality. Mean sipunculan densities were inversely related to the intensity of site degradation, with sipunculans being more abundant at reefs having higher coral cover (coral cover was highest at Platanillo, lowest at Sándalo and mean sipunculan densities were 131.2, 52.6, and 28.6 ind m^{-2} at Platanillo, Punta Islotes and Sándalo respectively) (see Chap. 5 for site locations). With respect to individual host corals, however, sipunculan densities were significantly higher in corals having more than 50 % tissue mortality (Fonseca et al. 2006). Distributional separation between live and dead host coral was more pronounced in healthier reefs where sipunculan densities and relative abundance were highest (i.e. at Platanillo, mean sipunculan densities were 261.4 and 1.1 ind m^{-2} in corals with greater and less than 50 % tissue mortality respectively). These findings support the notion that sipunculans are dead coral borers abundant in healthy reefs. In extensively degraded reef sites sipunculan populations are diminished and other macroborers (i.e. bivalves) dominate.

At Gorgona Island Colombia, sipunculan borers are reported to have a small impact on reefal bioerosion compared to populations observed on Costa Rican reefs (Cantera et al. 2003). Assessment of internal bioerosion at Gorgona found sipunculans to be among the least important macroborers of experimental units of *Pocillopora* branches, with small-sized individuals recruited in low abundances (Londõno-Cruz et al. 2003). While temporal patterns in bioeroder recruitment reported in this study were not consistent with findings elsewhere (Hutchings et al. 1992), there was apparent increase in sipunculan abundance over time in the majority of experimental units. These findings, however, were not considered supportive of sipuncalans as late invaders, due to recruits being observed after only 6 months of exposure (Londoño-Cruz et al. 2003). The rate of total internal bioerosion, however, increased over time more rapidly than is typical for non-ETP reefs (Davies and Hutchings 1983; Kiene and Hutchings 1994). It is possible that the nutrient-rich ETP environment accommodates for more rapid substrate conditioning as a result of bioerosion processes operating more quickly, thereby establishing unique or condensed temporal patterns in macroborer recruitment (Highsmith 1980; Reaka-Kudla et al. 1996).

At the Playa Blanca fringing reef off the southern coast of Gorgona Island, six sipunculan species were identified in a survey of *Pocillopora damicornis* cryptofauna (Cantera et al. 2003). *Aspidosiphon steenstrupii*, the most abundant sipunculan species, occurred in relatively moderate numbers among the infaunal community. Sipunculan distributions exhibited strong zonal preferences and *A. steenstrupii* occurred almost excusively in the reef flat crest and reef front where coral cover was highest. More sipunculans were found in dead *P. damicornis* than in live colonies at the reef flat-crest (45 and 9 individuals extracted respectively), while sipunculans were found to be evenly distributed among dead and live corals at the reef front where coral cover was lower (44 and 56 individuals in dead and live coral respectively). These trends coincide with distributions reported for sipunculan populations in Costa Rican habitats by Fonseca et al. (2006).

12.8 Bivalve Molluscs

In the ETP, boring bivalves are often the most abundant and destructive endolithic bioeroders of coral skeletons. While gastropod and polyplacophoran molluscs can be important epilithic bioeroders in regions elsewhere (see Table 4-2 in Glynn 1997), these taxa do not directly or substantially contribute to carbonate excavation in ETP coral habitats. Bivalve assemblages in the ETP, however, demonstrate a much greater capacity for coral erosion than is found for taxonomically similar assemblages in regions elsewhere (Kleemann 1986, 1990, 2013; Scott et al. 1988). Endolithic bivalves, *Lithophaga* spp. in particular, often reach unusually high abundances on eastern Pacific reefs and densities of several hundred boring bivalves per m^2 of coral surface area are commonly reported (Table 12.3). Assessments of bioeroder communities at sites throughout the ETP frequently report bivalve excavating activity to have the most significant impact on coral erosion, and bivalve molluscs consistently dominate macroborer communities (Cantera et al. 2003; Londoño-Cruz et al. 2003; Fonseca et al. 2006).

The high abundance, growth potential, and capacity for carbonate erosion demonstrated by ETP bivalve assemblages are environmentally conditioned, with genetic or biological distinctions appearing less influential (Kleemann 1990). Nutrient loading and high primary productivity in ETP

Table 12.3 Bivalve densities reported in the eastern Pacific

Site	Host coral	Total	Density (ind m^{-2})		Reference
			Live host	Dead host	
Panama	All genera		3060	76[a]	Scott et al. (1988)
	Porites	35.2	18.7 (14.0)[b]	16.5 (21.2)	Highsmith (1980)[c]
	Pavona	20.9	5.1 (4.4)	15.8 (16.5)	Highsmith (1980)
Galápagos	All genera		480	381	Scott et al. (1988)
	Porites	13.8	12.3	1.5	Highsmith (1980)
	Pavona	31	4.9 (4.6)	26.1 (26.4)	Highsmith (1980)
Costa Rica	All genera		4810	3920	Fonseca et al. (2006)[d,e]
Caño Island			1870	2021	Fonseca et al. (2006)
		2529	2110	2920	Scott et al. (1988)
Punta Islotes		4350	5580	3120	Fonseca et al. (2006)
Sándalo		6220	6730	5710	Fonseca et al. (2006)

[a]Values represent ind kg^{-1}
[b]Values represent mean number of bivalves per coral colony
[c]Values in parentheses are adjusted to show individuals encircled by coral growth after settlement and moved from live to dead category
[d]Value converted from ind 100 cm^{-2}
[e]Live versus dead substrate category reflects coral colonies with 0–50 % and 50–100 % tissue mortality respectively

waters endow considerable growth advantage to bivalve endoliths (Highsmith 1982; Kleemann 1986, 1990). The abundance of filter-feeding bivalves in reef environments worldwide corresponds with a relative gradient in primary production, peaking in the ETP (Highsmith 1980). Nevertheless, the ecological success and dominance of bivalves among ETP macroborers may not be conditioned by trophic factors alone, with reefal (e.g., live cover, zonation, disturbance) and taxonomic composition (i.e., bivalves and coral substrate species) likely having additional influence.

Two groups of boring bivalves are found in ETP coral skeletons: bivalves belonging to the genus *Gastrochaena* in the family Gastrochaenidae and species in the genus *Lithophaga* (*Leiosolenus*) belonging to the family Mytilidae. Lithophagid bivalves are well-established as being the most important infaunal coral borers in the ETP in regards to both abundance and boring rates (Scott et al. 1988; Reaka-Kudla et al. 1996). The borings performed by less abundant and smaller-sized gastrochaenid bivalves are reported to have a large impact on pocilloporid corals at certain sites, however (Cantera et al. 2003; Londoño-Cruz et al. 2003).

Lithophaginid and gastrochaenid bivalves reside exclusively in calcareous substrates, with the possible exception of at least one lithophagid species (i.e., *Lithophaga plumula*) observed boring into mudstone and quartzose sandstone by Warme (1975), substrates that are largely non-calcareous in composition. Boring bivalves that inhabit coral skeletons gain shelter from predators while retaining ability to filter-feed via an apical siphonal opening. Endolithic bivalves are prominently invaders of dead coral colonies (Kleemann 1986; Enochs 2012). Within the lithophaginids, however, there are a few species, i.e. *Lithophaga laevigata*, whose larvae can penetrate soft coral tissues and invade live corals (Kleemann 1977, 1980; Hutchings 1986; Guzman 1988). Boring bivalves are generally found to have strict requirements with respect to host coral species (Kleemann 1980, 1986, 1992, 1995, 2008; Kleeman and Hoeksema 2002). In the eastern Pacific, however, boring bivalves are atypically abundant in live coral skeletons (Kleeman 2013), and species considered to invade exclusively dead coral colonies (i.e., *Lithophaga aristata* and *Lithophaga attenuata*) are occasionally reported to inhabit live coral colonies as well (Highsmith 1980; Scott et al. 1988; Cantera et al. 2003; Kleemann 2013)—though such reports may owe in part to lateral overgrowth of coral tissues around borehole openings following invasion (Highsmith 1980; Kleemann, personal communication). *L. laevigata* is reported to be an obligate associate of live *Porites* and *Pavona* massive corals, not found in dead coral colonies, and declining in abundance with mortality of coral tissue (Guzman 1988). Trends reported for bivalve species invading dead versus live coral colonies are complex and demonstrate highly variable preferences with respect to bivalve species and reef zone (Table 12.4).

It is generally understood that the recruitment of bivalve larvae and subsequent invasion of carbonate rock occurs only after long periods of substrate exposure and preconditioning by pioneer boring assemblages (mainly microendoliths and polychaete worms), as is observed in many reefs outside the ETP (Kiene and Hutchings 1994; Chazottes et al. 1995; Hutchings 2011). Numerous studies of bioeroder community succession in a wide range of reef environments

Table 12.4 Abundance of infaunal bivalves sampled in three live and three dead *Pocillopora damicornis* colonies at major reef zones, Playa Blanca fringing reef, Gorgona Island, Colombia

Species	Backreef		Reef flat crest		Reef front		Outer slope		Totals		
	Live	Dead	Live	Dead	Live	Dead	Live	Dead	Live	Dead	All
Lithophaga aristata	0	3 (100)	10 (19)	43 (81)	4 (20)	16 (80)	3 (10)	27 (90)	17 (16)	89 (83)	106
L. plumula	2 (33)	6 (67)	12 (29)	29 (71)	8 (57)	6 (43)	3 (43)	4 (57)	25 (36)	45 (64)	70
L. hancocki	2 (22)	7 (78)	17 (7)	243 (93)	53 (82)	12 (18)	2 (2)	99 (98)	74 (17)	361 (83)	435
L. calyculata	4 (24)	13 (76)	21 (14)	125 (86)	50 (65)	27 (35)	9 (23)	30 (77)	84 (30)	195 (70)	279
L. hastasia	0	0	2 (100)	0	0	0	0	0	2 (100)	0	2
Gastrochaena ovata	2 (40)	3 (60)	7 (10)	61 (90)	21 (64)	12 (36)	4 (17)	19 (83)	34 (26)	95 (74)	129
All species	10 (30)	32 (70)	69 (12)	501 (88)	136 (65)	73 (35)	21 (10)	179 (90)	236 (23)	785 (77)	1021

Values in parentheses represent percent abundance in live or dead coral. Adapted from Cantera et al. (2003)

support the later-stage settlement of bivalve assemblages: boring bivalves were absent for the first 1–4 years of substrate exposure on the Great Barrier Reef (Kiene and Hutchings 1994; Tribollet et al. 2002; Osorno et al. 2005), and after two years at Moorea, French Polynesia and Aqaba, Red Sea (Chazottes et al. 1995; Hassan 1988). In the eastern Pacific, however, the settlement of bivalves on newly available substrates is much faster and can occur within days of coral death (Kleemann 2013). Macroborer community succession occurs more rapidly in the ETP, with temporal and successional trends deviating from trends demonstrated in reef habitats elsewhere (Kiene and Hutchings 1994; Londoño-Cruz 2001; Londoño-Cruz et al. 2003). In coral habitats in the Galápagos Islands, the profusion of vacant lithophage borings may have aided recruitment of *Eucidaris* and a massive increase in echinoid abundance following the 1982–83 El Nino event (Reaka-Kudla et al. 1996; Glynn et al. 2015; see Sect. 12.9.2.1).

Carbonate excavation by bivalves has historically been regarded as a predominantly mechanical process which, within the Mytilacea, may be assisted chemically (Yonge 1955; Ansel and Nair 1969; Hutchings 1986; Morton 1990). All other boring bivalves were considered to use exclusively mechanical means, i.e. abrasion (Ansel and Nair 1969; Pojeta and Palmer 1976). The controversial view that bivalves can bore by chemical means (corrosion) found little support in the literature despite many early studies dealing with boring bivalves in general (Cailliaud 1850; Deshayes 1850) and *Gastrochaena* in particular (Cailliaud 1843). It is now known that chelating agents secreted from pallial glands soften and dissolve calcium carbonate, initiating the boring process (Jaccarini et al. 1968; Morton and Scott 1980). Loosened small calcareous fragments are then removed by ciliary action of the mantle and exhalent siphon, producing fine-grain sediment (Kleemann 1973, 1974). Mechanical abrasion contributes to boring in some gastrochaenids, i.e. *Spengleria* spp. (Carter 1978; Hutchings 1986; Kleemann 1996). Coral excavation by lithophagid and gastrochaenid species produces vase-shaped cavities penetrating 1 to >15 cm into the coral skeleton (Glynn 1997). Characteristically, lithophagid borehole openings usually resemble old-fashioned keyholes marking the surface of invaded host corals (Fig. 12.4b). Bivalves that invade dead corals may prevent lateral overgrowth of this siphonal opening by the secretion of chelating agents during the boring process (Morton and Scott 1980; Kleemann 2013).

In coral reefs at Costa Rica, Panama, and the Galápagos Islands, heavy infestation by associated lithophagids [i.e., *Lithophaga* (*Leiosolenus*) *laevigata*] is exceptionally common in massive *Porites lobata* corals (Fig. 12.4a–c), a predominant reef-building species (Kleemann 1982, 2013; Scott et al. 1988). Structurally, *Porites lobata* has the lowest compressive and bending strength of tested coral species (Scott and Risk 1988). The strongest un-bored *P. lobata* skeleton demonstrates bending strength an order of magnitude below that of un-bored *Acropora cervicornis* skeleton of approximately the same diameter. The high skeletal porosity and arrangement of septa and theca are thought to most affect the unusually low strength of *P. lobata* skeletons. Boring by lithophagid bivalves greatly reduces the breaking stability of reef-building *P. lobata* (Scott and Risk 1988) and corals in general.

Coral breakage generated or so predisposed by bivalve excavation promotes asexual dispersal and the recruitment of remnant coral colonies following major disturbances, such as mortality events that accompany ENSO phenomena (Scott et al. 1988). Endolithic bivalves also mediate coral colony fragmentation indirectly by attracting grazing predators (Guzman 1986; Scott and Risk 1988; see Chaps. 10 and 15, Enochs and Glynn, and Glynn et al. respectively). Triggerfishes (e.g. *Pseudobalistes naufragium*) in particular break apart large portions of *Porites lobata* in their foraging activities. After accessing bivalve prey, triggerfish leave behind uningested coral fragments, and a significant proportion of these coral fragments survive (Guzman 1986). Thus, a paradoxical couplet of destructive and propagating

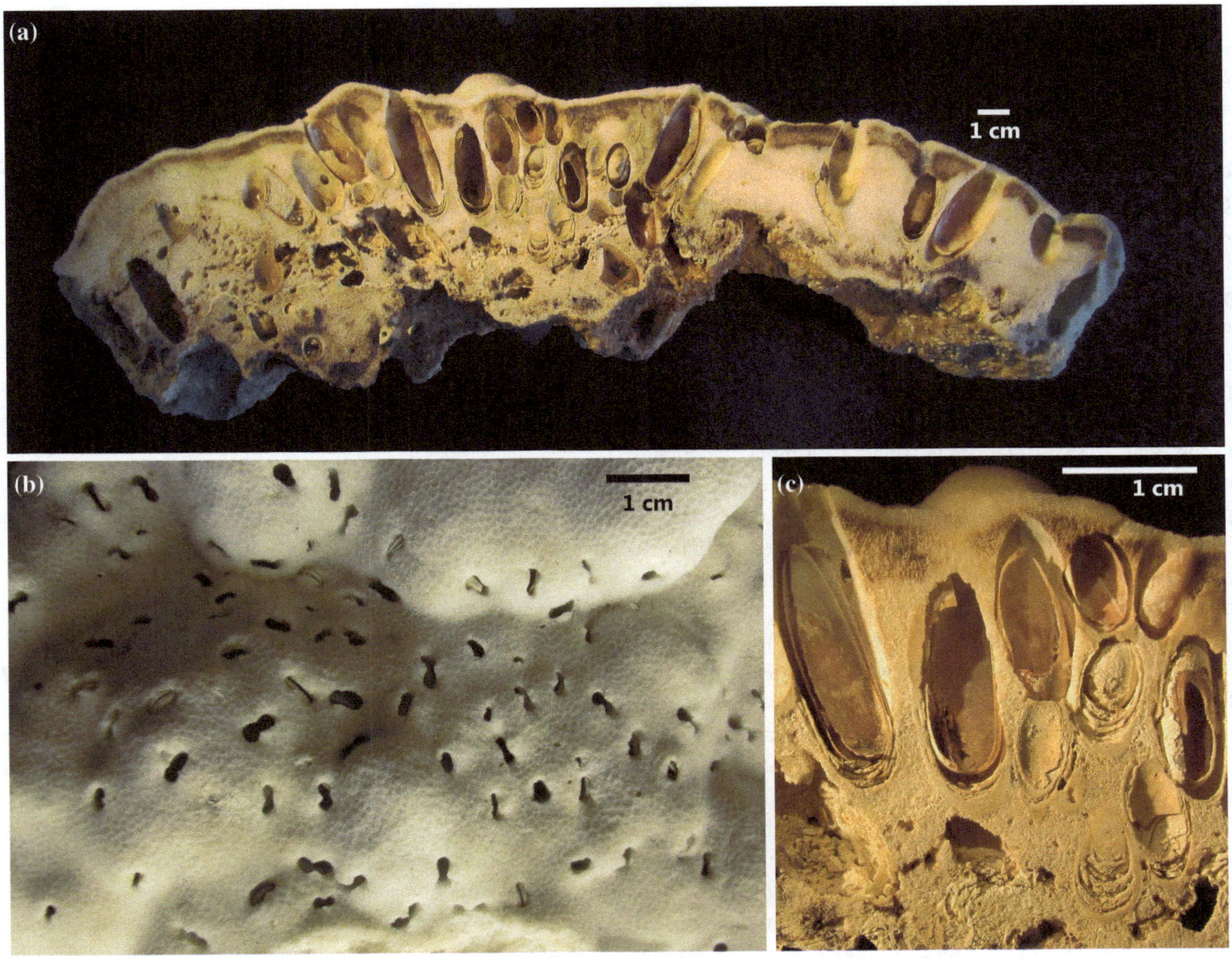

Fig. 12.4 *Lithophaga laevigata* excavations in live *Porites lobata*. **a** Boring bivalves revealed in a cross-section of coral skeleton, **b** coral surface with characteristic *L. laevigata* siphonal apertures, **c** magnified cross-section. Photographs by B. Grassian

reefal processes is driven by the activities of boring bivalves and their predators. Although boring bivalves perform extensive coral erosion, they also may play an important role in establishing coral populations in unfavorable ETP environments, and promote reef recovery after ENSO-associated mortality events (Scott and Risk 1988; Scott et al. 1988).

Boring bivalve assemblages have been assessed in most ETP coral environments, including: Panama, Galápagos Islands, Costa Rica, and Gorgona Island, Colombia (Kleemann 1986, 1990; Scott et al. 1988; Cortés 1991; Cantera et al. 2003; Londoño-Cruz et al. 2003; Fonseca et al. 2006). In all of these assessments, bivalve assemblages are highly abundant and large contributors to bioerosional processes (Table 12.3). Boring bivalves typically dominate macroborer communities and have been reported in extraordinarily high densities on ETP reefs. As example, one study reports *Lithophaga* population densities reaching upwards of 10,000 ind m^{-2}, with means of 3060 and 1870 ind m^{-2} in Panama and Costa Rica, respectively (Scott et al. 1988). These high densities are supported by subsequent surveys (Fonseca et al. 2006). An assessment of macroborer communities at Gorgona Island, Colombia reports boring bivalves to comprise 67.3 % of total macroborer abundance (Cantera et al. 2003). The importance of lithophagid bivalves in bioeroder communities throughout the ETP has been demonstrated with both definite consistency and significance.

Of ETP gastrochaenid bivalves, *Gastrochaena ovata* is found in highest abundance and is common in the dead bases of *Pocillopora damicornis* corals at Gorgona Island, Colombia (Cantera et al. 2001, 2003; Londoño-Cruz 2001). *Gastrochaena ovata* is observed less frequently in tissue-covered parts of branching corals as well, and is not reported to inhabit massive coral species (Cantera et al. 2001, 2003). Burrows excavated by *G. ovata* are cylindrical, deep and typically much larger than the bivalve occupant (Londoño-Cruz et al. 2003; Kleemann 2008). Boring rates for *G. ovata* are unknown but Guzman and Cortés (1993) reported a rate for Costa Rican *Gastrochaena rugulosa* of 0.5 g $CaCO_3$ m^{-2} $days^{-1}$ on experimental plates of *Porites lobata*. (The calculation for this rate, based on percent of skeletal removal, was not explicitly noted.)

Several studies have calculated lithophagid boring rates on ETP reefs, though most provide a combined rate for macroborer communities dominated by bivalve assemblages. A study investigating bioerosional processes on reefs impacted by the 1982–83 El Niño event reported 9 kg m^{-2} $year^{-1}$ of carbonate removed by infaunal bioeroders (Scott et al. 1988). The authors attribute a majority of sediment production to *Lithophaga* boring, corroborated by the preponderance of very fine to silt-sized grains resembling sediments produced by *Lithophaga* in laboratory experiments (Scott et al. 1988). At Gorgona Island, Colombia, mean rates of total internal bioerosion on experimental units of dead pocilloporid branches ranged from 0.69 ± 0.03 (SEM) to 2.58 ± 0.59 (SEM) kg m^{-2} $year^{-1}$. These rates include the entire endolithic guild, but *Lithophaga* spp. again dominated and were responsible for a large majority of the observed carbonate removal (Londoño-Cruz et al. 2003).

A comparative assessment between functionally similar ETP and Caribbean bivalve species, as well as observations of bivalve transplants demonstrates considerable enhancement in bivalve growth rates and boring potential conditioned in eastern Pacific waters (Kleemann 1990, 2013). This study finds: Pacific *Lithophaga attenuata* bores twice as fast and grows two and a half times faster than similar-sized Caribbean *Lithophaga corrugata*; Pacific *Lithophaga plumula* bores twice as fast and has four-fold higher growth rates than similar-sized Caribbean *Lithophaga teres*; Pacific *Lithophaga aristata* bores three times faster and has four-fold higher growth rates than Caribbean *Lithophaga malaccana* (Kleemann 1990). Furthermore, Caribbean bivalve species transplanted to Urabá Island, Panama demonstrate substantially increased boring and growth rates approaching those observed for similar ETP species (rates for Caribbean transplants were only 10–20 % less than similar Pacific species). Although the transplanting of non-native species would no longer be practiced in field-based experiments today, these early findings are quite important and establish a strong linkage between the marine environment and bioerosion potential of bivalve macroborers (Highsmith 1980; Hallock 1988; Kleemann 1990).

12.9 Echinoderms

Among the echinoderms, echinoids are the only group that performs reefal bioerosion, although *Acanthaster* predation sets the stage for bioerosion by other organisms (Glynn 1973; Glynn and Manzello, in press). In certain regions (e.g., ETP, Caribbean, and French Polynesia), sea urchins are

Fig. 12.5 *Pocillopora* framework impacted by bioerosion activity of *Diadema mexicanum* in Bahía Culebra, Costa Rica (see Alvarado et al. 2012)

common, conspicuous inhabitants of coral reefs, and can be responsible for considerable carbonate erosion (Scoffin et al. 1980; Reaka-Kudla et al. 1996; Pari et al. 1998). In the ETP specifically, abrupt increases in sea urchin population densities have led to large-scale modification of reef structure and functioning at certain sites—phenomena often initiated by sudden declines in reef condition (Birkeland 1989). Following ENSO thermal shock events, echinoid populations have rapidly expanded in eastern Pacific coral habitats affected by mass bleaching and coral mortality, establishing intense regimes of bioerosion and framework degradation (Fig. 12.5). The recovery of impacted coral habitats can be significantly influenced by attendant changes in sea urchin abundance (Glynn 1988; Eakin 1996; Glynn et al. 2015).

Sea urchins possess a highly developed jaw apparatus, the Aristotle's lantern; a flexible and protractile masticatory organ consisting of five radially arranged calcified teeth, which are harder than the surfaces they scrape (Glynn 1997). When feeding, sea urchins employ Aristotle's lantern to graze on epilithic algae, including those on the exposed skeletons of dead corals. During the foraging process, small chips of coral are scraped off and ingested with the primary algal food source. The carbonate material is then transformed into a paste from which the ruptured algal cells and sap are absorbed. Fecal pellets consist almost entirely of triturated calcium carbonate, which is deposited back on the reef (Hutchings 2011).

Grazing typically occurs at night when the echinoids leave their shelters to feed on exposed surfaces. Foraging is primarily on dead corals, though echinoids occasionally attack live corals as well (Bak and van Eys 1975; Glynn et al. 1979). Echinoid populations can perform substantial bioerosion at low to moderate densities (0.5–5 ind m^{-2}), but at high densities (>5 ind m^{-2}) degradation of reef substrata rivals that of the most destructive endolith assemblages (Hutchings 1986; Glynn 1997). Where abundant, sea urchins are the most destructive grazers on coral reefs (Hutchings 1986), and are responsible for 80 % of total bioerosion at certain sites (Scoffin et al. 1980). The magnitude of echinoid bioerosion depends on species, mean test size, and population density (Bak 1994)—factors concurrently influenced by overall reef health, the availability of coral substrates, and oceanographic conditions affecting larval recruitment and growth.

12.9.1 Echinoid Bioerosion World-Wide

Sea urchins are extremely abundant on certain reefs: the highest population densities reported are on degraded reefs in French Polynesia (210 ± 60 SD ind m^{-2}) and the Galápagos Islands (60 ± 5 SEM ind m^{-2}) (Glynn et al. 1979; Reaka-Kudla et al. 1996; Pari et al. 1998). In these cases, coral reef frameworks can be altogether destroyed by sea urchin bioerosion. This situation commonly occurs on highly fished or over-fished reefs (Glynn 1988; Reaka-Kudla et al. 1996; Peyrot-Clausade et al. 2000; Tribollet and Golubic 2011), where echinoid populations are unfettered ostensibly by predation pressure (Sonnenholzer et al. 2009, 2011; Edgar et al. 2011), and execute unusually high bioerosion rates (McClanahan et al. 1994). It is hypothesized that on reefs where fishes maintain abundance, sea urchin populations are biologically controlled and usually do not threaten reef condition. Large increases in echinoid abundance can act as an alert signal to stakeholders and/or managers, indicating that either significant over-fishing may be occurring or changing environmental conditions (e.g., water quality and terrestrial run-off) are leading to declining coral health, and remedial action should be taken (Sonnenholzner et al. 2009, 2011; Hutchings 2011).

Damselfish populations also effect echinoid distributions and bioerosion rates, however not by predation (Eakin 1988). Damselfishes deter and clear benthic grazers from small, well-defined patches on the reef. The territorial behavior of damselfishes causes substantial localized reduction in sea urchin densities within defended algal lawns (Eakin 1987, 1988, 1991; Glynn 1988). In this way, spatial competition by damselfishes protects exposed coral substrates from echinoid bioerosion.

12.9.2 Echinoid Bioerosion in the ETP

The first reference to sea urchin erosion in the ETP was noted by E.W. Craig for *Echinometra vanbrunti* and *Eucidaris thouarsii* in Guaymas, Mexico (Fewkes 1890). More recent studies have shown that *Eucidaris galapagensis, Eucidaris thouarsii, Centrostephanus coronatus* and *Diadema mexicanum* are the main external bioeroders in the ETP (Glynn et al. 1979; Eakin 1988; Glynn 1988; Colgan 1990; Guzman and Cortés 1992, 2007; Reaka-Kudla et al. 1996; Reyes-Bonilla and Calderón-Aguilar 1999; Toro-Farmer et al. 2004; Herrera-Escalante et al. 2005). The occurrence, relative abundance, and role of these ETP species vary across smaller, local scales. The impact of echinoid bioerosion on coral reef carbonate dynamics is greater in the ETP than on reefs elsewhere (Ogden 1977; Eakin 1996; Glynn et al. 2015; Table 12.5). The intensity of sea urchin bioerosion is generally responsive to that of ENSO-associated coral mortality events, and these phenomena together establish an erratic regime of reef degradation varying in severity across relatively small spatio-temporal scales. The following sections assess, within the ETP region, variable relationships between echinoid bioerosion and ENSO-associated reef disturbance (see Chap. 5 for site locations).

12.9.2.1 Galápagos Islands

Glynn et al. (1979) demonstrated that bioerosion by the sea urchin *Eucidaris galapagensis* in the Galápagos Islands prevented establishment of reef frameworks, composed mainly of pocilloporid corals. Intensive grazing by *E. galapagensis* on reef-building *Pocillopora damicornis, Pocillopora elegans*, and *Pocillopora capitata* corals, resulted in zero net carbonate production when coral cover was equal to or less than 30 %. The impact of echinoid bioerosion on Galápagos reefs increased following the dramatic 1982–83 El Niño event. Among ETP coral habitats, those in the Galápagos Islands experienced the largest decrease in coral cover as a result of the 1982–83 El Niño, with a mean coral mortality of 97 % estimated from 14 surveyed reefs (Glynn 1988). Following coral mortality, the high numbers of *E. galapagensis* that were mainly found on reef margins aggregated on the dead coral framework, resulting in a rapid loss of reef structure (Table 12.5; Fig. 12.6).

A decade later, Reaka-Kudla et al. (1996) showed that *Eucidaris galapagensis* abundance and bioerosion rates remained high at Champion Island in the southern Galápagos, with *E. galapagensis* responsible for 67–75 % of total reefal bioerosion. High echinoid population densities still persist in most coral communities in the central and southern Galápagos Islands, continuing for more than two decades after the initial population expansion (i.e., mean *Eucidaris galapagensis* population densities of 43.4 ind m^{-2} at Champion Island between 1985 and 2009). The indirect release of predation pressure via over-exploitation of piscine (i.e., the hogfish *Bodianus diplotaenia*) and crustacean predators of echinoids (i.e., *Panulirus* spp. and *Scyllarides astori* lobsters) is strongly implicated as being causal to the maintenance of high echinoid densities on these reefs (Ruttenberg 2001; Okey et al. 2004; Sonnenholzner et al. 2009; Edgar et al. 2011). Additionally, nutrient-rich

Table 12.5 Bioerosion rates of different sea urchin species at coral reef localities around the world

Species	Locality	Density (ind m^{-2})	Test diam (cm)	Bioerosion rate (kg m^{-2} $year^{-1}$)	Source
Eucidaris galapagensis	Galápagos	10–50	4.9–5.5	20–40	Glynn (1988)
		10–60	–	22.8	Reaka-Kudla et al. (1996)
Eucidaris thouarsii	Mexico	0.2	4.3	0.7	Reyes-Bonilla and Calderón-Aguilera (1999)
Diadema antillarum	Barbados	–	–	9.7	Hunter (1977)
		23	0.5–6.0	9.0	Stearn et al. 1977
		23	1.0–6.0	9.7	Scoffin et al. (1980)
	St. Croix, US Virgin Islands	9	6.0	4.6	Ogden (1977)
	Venezuela	4	–	6.0	Weil 1980 in Griffin et al. (2003)
		3		5.9	
		2.6		3.8	
	Curaçao	12	–	2.9	Bak et al. (1984)
Diadema mexicanum	Uva Is, Panama	3–150	1.8–2.7	5.0–20.0	Glynn (1988)
		1.78	–	0.1	Eakin (1996)
		0.06		0.0	
		18.76		1.0	
		48.26		4.4	
	Saboga Is, Panama	27.2	3.4–3.5	13–17	Glynn (1988)
	Huatulco, Mexico	1.0	4.0	0.9	Herrera-Escalante et al. (2005)
		1.3	3.4	0.9	
		2.9	3.0	2.0	
		6.8	2.9	3.3	
		0.26	4.72	0.3	Benítez-Villalobos et al. (2008)
		0.36	4.83	0.4	
		0.48	4.41	0.4	
		4.17	4.71	4.0	
Diadema savignyi	French Polynesia	4.87	5.5–5.9	3.9	Bak (1990)
	Kenya	0–0.1	5.7–8.3	0.1–1.2	Carreiro-Silva and McClanahan (2001)
Diadema setosum	Gulf of Eilat	0.1–6.4	3.0	0.0–0.7	Mokady et al. (1996)
	Kenya	0.01–0.4	5.3–7.8	0.65	Carreiro-Silva and McClanahan (2001)
	Thailand	0.4–11.8	2.0–7.0	1.64–5.5	Ruengsawang and Yeemin (2000)
Echinothrix diadema	French Polynesia	0.65	8.8	0.80	Bak (1990)
	Kenya	0.02–0.07	9.1–12.0	2.00	Carreiro-Silva and McClanahan (2001)
Centrostephanus coronatus	Colombia	6.86	0.91–2.58	0.2	Toro-Farmer et al. (2004)
Echinostrephus aciculatus	Marshall Islands	0.50	2.4	0.1	Russo (1980)
		1.05	1.8	0.06	
		0.66	1.68	0.03	

(continued)

Table 12.5 (continued)

Species	Locality	Density (ind m^{-2})	Test diam (cm)	Bioerosion rate (kg m^{-2} $year^{-1}$)	Source
Echinometra sp. A	Fiji	1.08	38	0.04	Appana and Vuki (2006)
		0.62	37	0.03	
		0.39	39	0.03	
		0.36	41	0.04	
Echinometra mathaei	Marshall Islands	0.16	2.2	0.01	Russo (1980)
		6.50	1.9	0.26	
		1.75	1.95	0.07	
	La Réunion	49.1	–	8.3	Conand et al. (1997)
		12.4		2.9	
		2.5		0.4	
		3.8–73.6	–	0.4–8.3	Peyrot-Clausade et al. (2000)
	French Polynesia	7.38	1.5–1.9	0.372	Bak (1990)
			2–2.4		
			2.5–2.9		
		4.8	–	0.89	Pari et al. (1998)
		210		6.87	
		8		0.63	
	Kenya	1.7–14.2	3.8	0.4–3.6	McClanahan and Muthiga (1988)
		0.03–5.6	2.6–4.9	0.15	Carreiro-Silva and McClanahan (2001)
	Okinawa, Japan	10	–	0.2	Hibino and van Woesik (2000)
	Gulf of Eilat	3.7–10.5	2.3	0.6–1.6	Mokady et al. (1996)
	Kuwait	30	3.7	0.33	Downing and El-Zahr (1987)
Echinometra lucunter	St. Croix, US Virgin Islands	100	–	3.9	Ogden (1977)
	Bermuda	25	–	7.0	Hunt 1969 in Russo (1980)
Echinometra viridis	Puerto Rico	0.8–62.6	2–2.5	1.1–4.1	Griffin et al. (2003)
Toxopneustes roseus	Panama	–	5.1–7.5	–	Glynn (1988)
Several species	Belize[a]	17.6	2.75	1.2	Brown-Saracino et al. (2007)
		6.5	2.87	0.6	
		0.8	3.41	0.2	
		12.5	2.44	0.8	
		17.3	2.86	1.4	
		32.8	2.37	1.2	
		30	2.11	1.1	
		25	2.56	1.3	
		17.8	1.64	0.3	
		31.6	1.40	0.3	
		40.0	1.33	0.3	
	French Polynesia[b]	7.12–10.0	–	0.6–7.5	Peyrot-Clausade et al. (2000)

[a]*Echinometra viridis, Echinometra lucunter, Diadema antillarum, Eucidaris tribuloides, Lytechinus variegatus, Lytechinus williamsi*
[b]*Echinometra* types A and B, *Echinometra diadema, Echinothrix calamaris, Diadema savignyi*

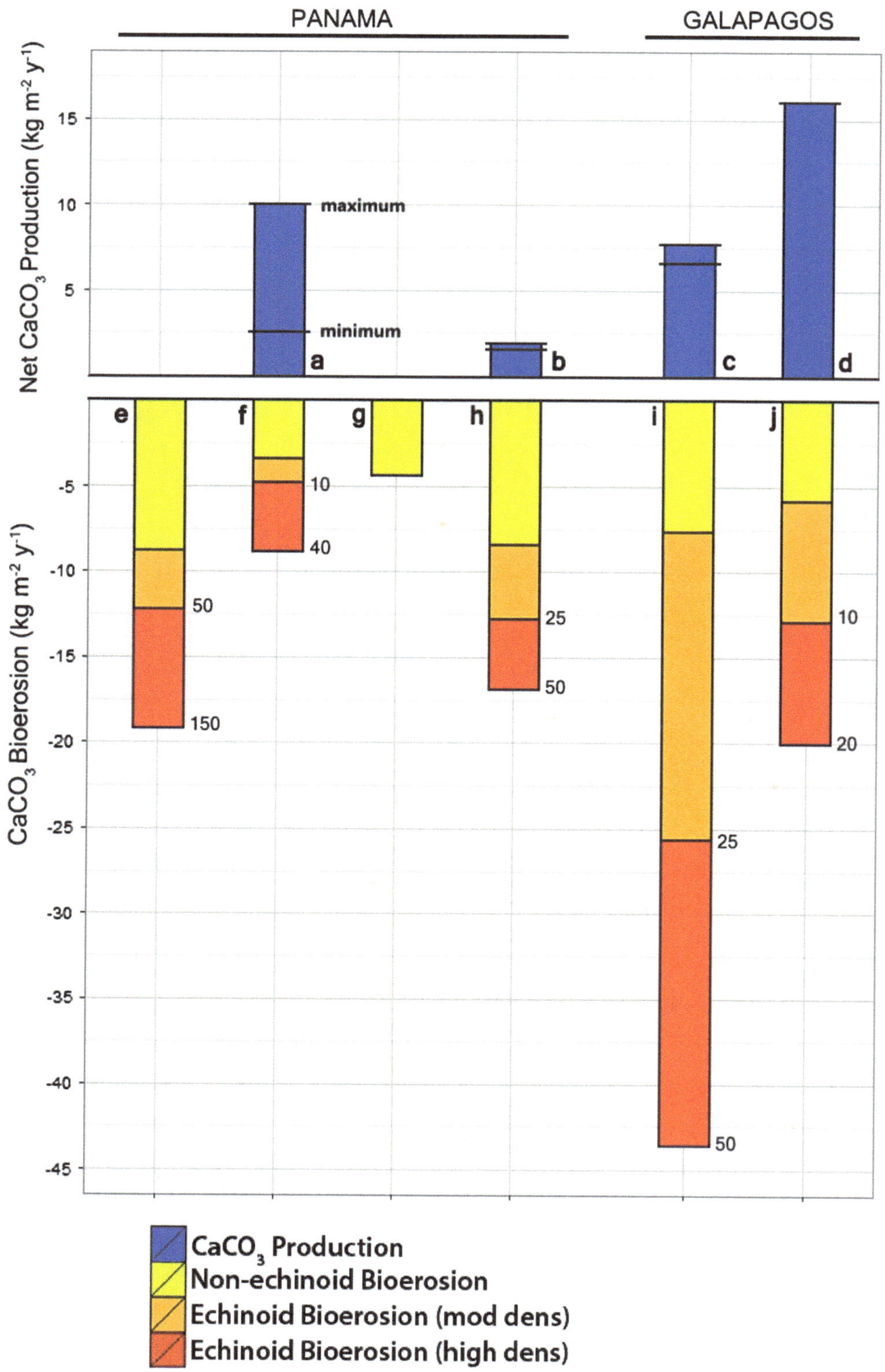

Fig. 12.6 Net carbonate production pre-1982 (*above*) and carbonate bioerosion post-1983 (*below*) on coral reefs in Panama and the Galápagos Islands. Letters denote geographic areas and habitat conditions, and numbers densities (ind m^{-2}) of sea urchins. Production (**a**–**d**): **a**—Gulf of Chiriquí, Uva Island; **b**—Gulf of Panama, Saboga Island; **c**—Floreana Island, Onslow Island reef, 70 % coral cover, maximum and minimum production levels at *Eucidaris* densities of 3 and 34 ind m^{-2} respectively; **d**—Fernandina Island, *Eucidaris* absent; horizontal lines mark ranges of production values (after Glynn 1988). Bioerosion (**e**–**j**): **e**–**g** Gulf of Chiriquí, Uva Island; **e**—lower seaward slope, damselfish absent; **f**—lower seaward slope, damselfish present; **g**—upper seaward slope, live pocilloporid substratum; **h**—Gulf of Panama, Saboga Island, lower seaward slope; **i**, **j** Floreana Island, Onslow Island reef; **i**—damselfish absent; **j**—damselfish present

upwelling and high primary production in the southern Galápagos provisions a larger food source for echinoid recruits and adults. Coral habitats in the central and southern Galápagos Islands have undergone little or no coral recovery due to intense and persistent bioerosion by *Eucidaris galapagensis* (Glynn et al. 2015).

In comparison, coral habitats of the northern Galápagos Islands (i.e., the Wellington Reef at Darwin Island and Wenman Island coral community), have demonstrated periods of robust coral recovery since the 1982–83 El Niño, in spite of subsequent ENSO thermal shock events (Glynn et al. 2015). The Darwin and Wenman coral communities are remote and protected habitats where minimal near-shore fishing pressure is exerted, and sea urchin abundances remain low at these sites. It then appears that large differences in the recovery of Galápagos coral reefs, occurring across a small latitudinal scale, results from differences in local echinoid abundances which in turn are controlled by nutrient upwelling and possibly fishing-induced trophic cascades (Glynn et al. 2015).

12.9.2.2 Gorgona Island, Colombia

At Gorgona Island, Colombia, Toro-Farmer et al. (2004) calculated the bioerosional impacts of the sea urchin *Centrostephanus coronatus* at Playa Blanca fringing reef, reporting small bioerosion rates compared to those calculated for *Eucidaris* and *Diadema* populations at ETP sites elsewhere. Significant statistical differences in *C. coronatus* bioerosion rates with both test size and reef zone were demonstated: mean bioerosion rates were highest in the crest-platform (0.103 kg $CaCO_3$ m^{-2} $year^{-1}$), medium in the back reef (0.071 kg $CaCO_3$ m^{-2} $year^{-1}$), and lowest in the reef front (0.052 kg $CaCO_3$ m^{-2} $year^{-1}$). Using coral cover, growth rate and skeletal density data for *Pocillopora*, a carbonate accretion rate of 5.02 kg $CaCO_3$ m^{-2} $year^{-1}$ was calculated for the Playa Blanca reef. Thus, with respect to the overall carbonate budget, *C. coronatus* demonstrated only minor effects. Nevertheless, this study provides estimates for a single echinoid species. *Diadema mexicanum* present at higher densities (14 ind m^{-2}) than *C. coronatus* on Colombian reefs, likely effects greater bioerosion rates (Toro-Farmer 1998). Further complicating these findings, *D. mexicanum* juveniles are easily confused with *C. coronatus*, and therefore some individuals identified as *C. coronatus* may have actually been juvenile *D. mexicanum*.

12.9.2.3 Uva Reef, Panama

In the Pacific of Panama, mass coral bleaching and mortality caused by the severe 1982–83 El Niño event precipitated a rapid increase in *Diadema mexicanum* densities and bioerosion rates. The 1982–83 El Niño was responsible for a 50 % reduction in live coral cover at reefs in the Gulf of Chiriquí (Glynn 1985). Within years, the opportunistic expansion of *D. mexicanum* populations led to bioerosion rates that exceeded reefal calcification and rendered the overall carbonate budget in deficit. Glynn (1988) pointed out that the continuing recruitment success of echinoid larvae onto the Uva Island reef contributed to high *D. mexicanum* densities being maintained for years following the initial El Niño disturbance. The exposure of dead coral substratum also had the compounding effect of promoting an increase in algal biomass, expanding the echinoid food source. It is interesting to note that no significant fishing pressure had been exerted in near-shore Panamanian waters since the early 1970s, and stable populations of fish predators were well-established at the Uva reef during the period of study. These observations suggest that echinoid abundances increased in response to ENSO-associated coral mortality and oceanographic conditions that favored a strong settlement pulse of *Diadema*, and not changes in predator-prey dynamics (Glynn 1988).

Prior to the 1982–83 El Niño, *Diadema mexicanum* population densities on the lower reef slope were between 2 and 5 ind m^{-2} at Uva reef (Glynn 1988). These densities persisted on deep reef areas with high coral cover from 1978 through 1983. By mid-1984, *D. mexicanum* densities had already approximately doubled, and from 1985 to early 1989 *D. mexicanum* densities fluctuated between 60 and 100 ind m^{-2}—enormously elevated from years prior to El Niño (Glynn 1988, 1990). The occurrence of these high echinoid densities over dead coral substrates was spatially heterogeneous, however, and damselfish lawns induced patchy echinoid distributions (Eakin 1987, 1988, 1991; Glynn 1990). At the Uva reef, vertical coral framework erosion was more than three-fold lower in defended damselfish lawns (0.63 cm $year^{-1}$) than in non-lawn substrates (2.2 cm $year^{-1}$) (Eakin 1992). Eakin (1992) noted that without the protective influence of the damselfish algal lawn symbiosis, the carbonate budget of Uva reef would be more easily shifted towards deficit—especially during periods when reefs are rendered vulnerable to increased bioerosional processes via extensive coral mortality, i.e. following ENSO thermal shock events. Prior to 1982–83, damselfish protected, in total, less than 800 kg $CaCO_3$ $year^{-1}$ from being eroded, whereas after the El Niño event this value rose to 2000 kg $CaCO_3$ $year^{-1}$.

Grazing by *Diadema mexicanum* on Panamanian reefs contributed 78 % or >10 (to 20) kg m^{-2} $year^{-1}$ of local bioerosion (Glynn 1988; Eakin 1991), more than doubling the total impact of non-echinoid grazers (Fig. 12.6). Eakin (1996) developed a box model for estimating $CaCO_3$ fluxes at Uva Island after the 1982–83 El Niño disturbance. Based on budget calculations, the 2.5 ha reef underwent a net loss of $CaCO_3$ structures due to intense bioerosion at the reef base, where *D. mexicanum* densities were highest (mean density of 48.3 ind m^{-2}). El Niño had shifted the carbonate budget at Uva Reef from a net deposition of 8600 kg $CaCO_3$ $year^{-1}$ before 1982 to a net loss of 4800 kg $CaCO_3$ $year^{-1}$ after the event (Eakin 2001). Echinoid bioerosion of the reef base was estimated at 4.4 kg m^{-2} $CaCO_3$ $year^{-1}$—higher than accretion rates calculated in the most actively developing reef zones. The Eakin model implied that most material eroded by echinoids is defecated directly onto the reef framework, with roughly 43 % retention of echinoid produced sediments (Eakin 2001).

Perry et al. (2008) developed a ternary approach to assess temporal linkages between ecological and carbonate production states (as dictated by corals and secondary carbonate producers) and the geomorphic dynamics of reef landforms. Shift in the Uva Reef carbonate budget and ecological status after the 1982–83 El Niño was used by Perry et al. (2008) to illustrate the potential for substantial changeover in reef states following major disturbance and that such phase-shifts can occur during ecological timescales. The lasting impacts of rapid erosion of reef structures over geological timescales are not certain (Perry et al. 2008).

Since Eakin's (1996) model, several changes have occurred at Uva Island, including minor bleaching events during the 1990s and a widespread El Niño event in 1997–98. High densities of *Diadema mexicanum* at the reef base and forereef persisted until the mid-1990s (~40–60 ind m^{-2}) when they began to decline to near 1974 levels in the year 2000 (~10 ind m^{-2}). The increase of *D. mexicanum* density during the 1980s and 1990s is attributed to both habitat expansion via coral mortality, and increased sea urchin recruitment owing to high oceanic productivity during the subsequent La Niña period (Glynn 1988; Eakin 1996, 2001). Declines in *D. mexicanum* densities coincided with declining topographic complexity of the reef habitat, suggesting a sea urchin population collapse regulated by their increased bioerosional activities (Eakin 2001). Reduced shelter availability may have rendered *D. mexicanum* vulnerable to predators such as pufferfishes, wrasses and balistids, possibly leading to the population collapse by 2000. No other causes of the *D. mexicanum* population collapse (e.g., disease) were evident (Eakin 2001).

12.9.2.4 Costa Rica

The impact of the 1982–83 El Niño event varied among the three most prominent Costa Rican reef sites: coral mortality was highest at Cocos Island, moderate at Caño Island, and lowest in Culebra Bay. Increases in *Diadema mexicanum* populations at these sites correlated with coral mortality intensity, i.e. greatest at oceanic Cocos Island and least in Culebra Bay (Alvarado et al. 2012). At Caño Island, Costa Rica, Guzman (1988) found low densities of *D. mexicanum* eroding the bases of massive corals. At Isla del Coco, however, *D. mexicanum* grazing activities caused high levels of bioerosion and reduced coral recruitment (Guzman and Cortés 1992). After the 1982–83 El Niño event, Guzman and Cortés (1992) predicted that coral recovery at Cocos Island would take centuries due to high bioeroder densities, and low coral cover and reproduction. Guzman and Cortés (1992, 2007) indicated that part of the erosion of reef structures at Isla del Coco was due to the bioerosive action of *D. mexicanum*. In 2002, there was a five-fold increase of coral cover and a large reduction in sea urchin abundance, and *D. mexicanum* was no longer playing an important role in reefal bioerosion (Guzman and Cortés 2007). At these lower densities, echinoids reduce spatial competition by algae and assist coral recruitment, as has been observed on Caribbean reefs (Sammarco et al. 1974; Sammarco 1980, 1982a, b; Mumby et al. 2006). The impact of *D. mexicanum* at Cocos Island transitioned from impeding reef recovery via high bioerosion intensity from 1987 to 2002, to promoting reef recovery by performing a key herbivorous role at low densities from 2002 to 2009 (Alvarado et al. 2012).

The recovery from an erosional reef state at Isla del Coco following the 1982–83 El Niño took roughly 25 years. Conservation practices, aided by the distance of Cocos Island from mainland (500 km) Costa Rica, prevented further reef decline and allowed for fish populations to recover by 2002. Sea urchin abundances were then reduced and buffered by fish predators (Alvarado et al. 2012). In contrast, at Culebra Bay, on the Costa Rican mainland, *D. mexicanum* bioerosion rates have increased several orders of magnitude since 2009, in step with coral cover declines (Fig. 12.5).

12.9.2.5 Mexico

Coral reefs in the Gulf of California, Mexico were generally less impacted by the 1982–83 El Niño and subsequent ENSO events than were reefs closer to the equator (e.g., Galápagos Islands and Panama). Assessments of Mexican reefs since the early 1990s have generally reported small echinoid populations that result in much lower rates of carbonate removal than echinoids at highly impacted reef sites. Between 1991 and 1992, Reyes-Bonilla and Calderón-Aguilera (1999) surveyed *Eucidaris thouarsii* populations at the Cabo Pulmo reef in the Gulf of California, Mexico, reporting a low mean population density (0.67 ind m^{-2}). Bioerosional activities by *E. thouarsii* demonstrated a minimal impact on coral carbonates, resulting in a mean bioerosion rate roughly twenty times lower than rates reported for *Eucidaris galapagensis* in the Galápagos Islands (Table 12.3). Together with *Arothron meleagris*, *E. thouarsii* reduced the coral standing crop at Cabo Pulmo by <4 % (performing a combined bioerosion rate of 0.002 kg $CaCO_3$ m^{-2} $year^{-1}$), and had little influence on scleractinian coral abundance (Reyes-Bonilla and Calderón-Aguilera 1999).

Herrera-Escalante et al. (2005) assessed carbonate removal resulting from *Diadema mexicanum* foraging and spine abrasion at Bahías de Huatulco, Oaxaca, Mexico. Reefs at Huatulco are well-developed with high coral cover persisting through ENSO events. As was the case with *Eucidaris thouarsii* at Cabo Pulmo, *D. mexicanum* populations did not demonstrate a large overall impact on corals and coral reef structures at Oaxaca. *Diadema mexicanum* abundance and bioerosion rates varied among reef sites, however, and was responsible for carbonate losses ranging from 5 to 27 % of local carbonate production. Higher densities and bioerosion rates were reported at La Entrega (6.8 ind m^{-2} and 3.3 kg $CaCO_3$ m^{-2} $year^{-1}$ respectively) compared with the other reefs at Bahías de Huatulco. The coral reef at La Entrega has been more affected by ENSO events than reefs at neighboring sites. The 1987 ENSO event caused coral mortality in deep reef areas of La Entrega, where *D. mexicanum* was found in highest abundance. The high abundances of *D. mexicanum* on La Entrega reef was atypical for coral habitats in the Gulf of California, where most reefs had been well protected from ENSO-associated disturbances. These findings further demonstrate the variable response of echinoid populations to ENSO thermal shock events across relatively small spatial scales.

Benitez-Villalobos et al. (2008) reevaluated echinoid populations on the Huatulco reefs and found a further reduction in the bioerosion intensity at all sites sampled previously by Herrera-Escalante et al. (2005). This later study reported a large reduction in *D. mexicanum* densities and bioerosion rates. A slight increase in mean test size at all sites was found as well, demonstrating population size to be a larger driver of echinoid bioerosion intensity than is the mean size of individuals.

12.9.3 Global Comparison of Echinoid Bioerosion Rates

Comparing echinoid bioerosion rates in reef provinces around the world, the highest values are found in the ETP and far surpass the highest rates reported in the Caribbean or Indo-Pacific (Table 12.5). Within the ETP, sea urchin densities and bioerosion rates are highest in the Galápagos Islands and Panama, and lowest in the Gulf of California, Mexico. Coral communities in the Galápagos Islands and Panama have been severely impacted by ENSO thermal shock events that have led to extensive coral cover losses from echinoids not observed in Mexican habitats (Glynn 1990; Reyes-Bonilla 2003). Records of *Eucidaris galapagensis* and *Diadema mexicanum* bioerosion in the Galápagos Islands and Panama demonstrate bioerosional regimes unmatched in intensity worldwide.

12.10 Fishes

Three fish guilds erode or break coral substrates as a result of foraging activities: (1) herbivorous fishes that abrade and remove substrate, (2) fish corallivores that denude and fragment corals, and (3) fish invertivores that break coral skeletons when searching for associated infauna.

12.10.1 Herbivorous Fishes: Family Scaridae

Parrotfishes are key herbivores and occasional corallivores that cause bioerosion while scraping or biting into coral skeletons with strong dental plates adapted for substrate excavation. When population densities are high, grazing parrotfishes can cause extensive reef erosion and produce large amounts of fine sediments, though relatively few species possess the capacity to do so (Bellwood and Choat 1990). On ETP reefs, the common scarid species (e.g., *Scarus ghobban* and *Scarus rubroviolaceus*) scrape and abrade coral skeletons while foraging, which is generally less destructive than excavating mechanisms. According to Bellwood and Choat (1990), *S. ghobban* does not consistently produce feeding scars and only exceptionally large individuals do so, though scarring does become more common on internally eroded or soft substrates. Bioerosion by scraping species can be important on certain protected reefs, however, where there is minimal fishing pressure and high scarid densities (Bruggemann et al. 1996; Jiménez 1999; Peyrot-Clausade et al. 2000).

On coral reefs at Gorgona Island, Colombia, both *Scarus ghobban* and *Scarus rubroviolaceus* are dominant herbivorous fishes (Zapata and Morales 1997). A large *S. ghobban* population (1544 ind ha^{-1}) was responsible for losses of 1.59 ± 0.76 (SEM) kg $CaCO_3$ m^{-2} $year^{-1}$ at Gorgona Island, a relatively high bioerosion rate on the reef, and compared with rates calculated for substrate scrapers elsewhere (Jiménez 1999; Cantera et al. 2001). *Scarus ghobban* bioerosion rates varied among reef zones, correlating with their abundance and distance from the shoreline: highest on the reef flat, but otherwise decreasing progressively from the backreef to reef slope (Jiménez 1999). Across reef zones, *Scarus ghobban* foraged mainly on coral rubble, but on the reef flat foraged equally on dead branching corals and rubble substrates. When feeding on coral colonies, parrotfishes will break off branches and convert intact substrates to rubble. If coral branches are ingested, the pharyngeal mill, a triturating structure unique to parrotfishes, further reduces coral skeletons into sediments (Glynn 1997). When foraging on rubble substrates, the change in substrate composition is much less severe (Jiménez 1999).

12.10.2 Fish Corallivores: Family Tetraodontidae

Pufferfishes (Family Tetraodontidae) are dominant corallivores in the ETP that prey upon branching corals, and massive species to a lesser extent (Glynn and Wellington 1983). Pufferfishes scrape into and bite off the tips of coral branches, removing coral skeleton and leaving small areas denuded and exposed. In spite of persistent grazing pressure, ETP pocilloporid communities are largely resilient to tissue loss and skeletal erosion inflicted by pufferfish corallivory. Following *Arothron meleagris* foraging, pocilloporid corals can regenerate tissues quickly (approximately two weeks, and two months for superficial damage and deeper lesions respectively), limiting opportunity for bioeroder recruitment (Neudecker 1979; Palacios et al. 2014).

While pufferfishes can perform considerable damage to individual coral colonies, their reef-wide impact on bioerosional processes appears to be minor (Glynn 1997). An early study in Panama reported a small *Arothron meleagris* population (40 ind ha^{-1}) that removed only 34 g of coral mass (soft tissue and skeleton) m^{-2} $year^{-1}$ (Glynn et al. 1972). At Cabo Pulmo, Mexico, *A. meleagris* populations have a low impact on coral health and reef bioerosion, in total removing 16.4 g $CaCO_3$ m^{-2} $year^{-1}$ (Reyes-Bonilla and Calderón-Aguilera 1999). *Arothron meleagris* is not abundant at Cabo Pulmo reefs (mean densities are 39 ind ha^{-1}) and coral tissues were found to be a small component of these fish's diets—11.8 and 4.8 % for *Pocillopora* spp. and *Porites* spp. respectively (Reyes-Bonilla and Calderón-Aguilera 1999; Moreno et al. 2009).

At Gorgona Island, Colombia, an early study reported that corallivorous foraging by pufferfishes did not adversely affect overall coral health and was ecologically beneficial; *Arothron meleagris*-mediated coral colony fragmentation may be aiding reef recovery and coral colony dispersal at this site (Guzman and López 1991). Pocilloporid corals often reproduce asexually (Glynn et al. 1991; see Chap. 15, Glynn et al.) and, as applies to all bioerosional processes inducing coral fragmentation, pufferfish foraging may enhance coral propagation in the ETP. Observations on a Panamanian reef demonstrated that more than 50 % of the coral mass lost due to pufferfish grazing represents fragments broken away from colonies and not ingested (Glynn et al. 1972). While pufferfishes mainly remove the tips of coral branches, the large number of fragments generated likely offsets the lowered survival probability attributable to their typically small-size (≤ 1 cm).

More recently, relatively high *Arothron meleagris* bioerosion rates have been reported at Gorgona Island, on La Azufrada reef. A large *A. meleagris* population (172 ind ha^{-1}) exerts a standing bite density of 366.2 ± 329.7 (SD) bites m^{-2}, and removes 0.89 kg $CaCO_3$ m^{-2} $year^{-1}$ (Palacios et al. 2014). This carbonate 'removal' rate may be misleading, however, due to a novel finding: coral nubbins eroded by *A. meleagris* demonstrated similar rates of mass growth as found for nubbins protected from pufferfish foraging. Bioerosional pufferfish foraging stimulates skeletal deposition that compensates for carbonate losses. By way of this enhanced accretion, pufferfish (i.e., *Arothron meleagris*) also alter the morphology of pocilloporid colonies via reorienting corallum growth (Palacios et al. 2014). Pufferfish bioerosion reduces linear growth rates of pocilloporid nubbins and induces girth-wise carbonate deposition. Pocilloporid nubbins thus become shorter and thicker than nubbins protected from pufferfish corallivory (Fig. 12.7). The bioerosional activities of pufferfish thereby can change topographies of coral habitats, and engender variation among pocilloporid colonies locally (Palacios et al. 2014).

The electivity of pufferfish for coral prey, and subsequently bioerosional impact, is however variable. Guzman and Robertson (1989) investigated the relationship between *Arothron meleagris* diet composition and changes in coral cover. After a red tide-induced coral mortality event at Caño Island, Costa Rica, *A. meleagris* fed mainly on coralline algae and *Psammocora* corals (40.6 and 23.7 % of diet composition respectively). Two years after this event, *A. meleagris* electivity was appreciably altered, with *Porites lobata* representing 53.7 % of the diet composition. *Arothron meleagris* coral electivities also varied among ETP reef sites: on the Uva Island reef (Panama), pocilloporid corals comprised 96.8–100 % of pufferfish diets, but only 63.3 % of the live coral cover, while at Caño Island (Costa Rica) *P. lobata* comprised the large majority (75.3 %) of *A. meleagris* diet composition at $\sim$90 % live coral substrate cover (Guzman and Robertson 1989). Feeding preferences of *A. meleagris* are flexible and variable across ETP reef sites, but do not consistently relate to the relative availability of potential coral species prey (Guzman and Robertson 1989). These findings illustrate pufferfish as opportunistic predators and bioeroders of live corals, having complex feeding habits. Consequently, the impacts of pufferfish bioerosion are probably highly variable from reef to reef, although it appears unlikely these populations typically play large roles in reefal carbonate budget losses—especially provided recent evidence that pufferfish predation stimulates skeletal deposition and asexual propagation in pocilloporid corals (Palacios et al. 2014). The ability for recovery and maintenance of carbonate production in branching corals may also help explain why pufferfish predation demonstrates negligible impact on ETP reef development and accretion, despite their dominance among the region's corallivores and negative impacts observed for corallivores on other tropical reefs (Guzman and Robertson 1989; Rotjan and Lewis 2008; see Chap. 10, Enochs and Glynn et al.).

12.10.3 Fish Invertivores and Other Predators

Predators of coral-associated invertebrates (e.g., bivalves, gastropods, echinoids, and crustaceans) can cause extensive damage to coral skeletons while foraging. Among fish invertivores in the ETP, triggerfishes (Family Balistidae) are frequent contributors to reef bioerosion. While extracting invertebrates from coral skeletons, mainly lithophagid bivalves, triggerfishes excavate carbonate material and expel sediments through the gill chambers, as well as causing coral breakage (Francisco 2000; Cantera et al. 2001). Thus, feedback exists whereby coral endoliths attract predators that perform further carbonate erosion. Bivalve distributions dictate which live and dead coral colonies are targets of triggerfish bioerosion, predominantly dead pocilloporid and live massive corals such as *Porites* spp. and *Pavona* spp. (Glynn et al. 1972; Francisco 2000; Cantera et al. 2001). Alternatively, endolith abundances are reduced and limited by triggerfish predation.

On reefs at Gorgona Island, Colombia, the balistids *Sufflamen verres* and *Pseudobalistes naufragium* forage predominantly on dead pocilloporid corals and coral debris (Cantera et al. 2001). *Pseudobalistes naufragium* performs considerable damage to live massive corals when searching for associated lithophagid bivalves, as has been reported on reefs in Panama and at Gorgona Island (Glynn et al. 1972; Francisco 2000), and many balistid feeding scars are often apparent on the surfaces of live *Pavona varians, Gardineroseris planulata* and *Porites lobata* colonies. Live coral patches damaged by triggerfishes are readily colonized by algae and diverse endolith taxa (i.e., sipunculans, polychaetes, acrothoracican cirripeds, and bivalves) and are then vulnerable to further bioerosion by epilithic grazers (Cantera et al. 2001).

Triggerfish bioerosion rates have been calculated at the Playa Blanca reef at Gorgona Island, and rates among species correspond to their local abundance. The most abundant triggerfish, *Sufflamen verres* (270 ± 351 [SD] ind ha^{-1}), removes up to 1.13 kg $CaCO_3$ m^{-2} $year^{-1}$; a moderately abundant *P. naufragium* population (30 ± 47 [SD] ind ha^{-1}) erodes 0.62 kg $CaCO_3$ m^{-2} $year^{-1}$; and infrequently encountered *Balistes polylepis* was thought to perform minimal bioerosion, and was not assessed (Francisco 2000; Cantera et al. 2001). There are currently no additional data on triggerfish bioerosion from reefs elsewhere in the ETP for comparison.

White-tip reef sharks, *Triaenodon obesus* (Family Carcharhinidae) at times cause extensive damage to coral colonies when pursuing balistids and other fishes (Jiménez 1996–1997; Glynn 2008). At Coiba Island, Panama, white-tip sharks forage aggressively in pocilloporid recesses while chasing reef fishes, detaching and fragmenting coral colonies in the process (Jiménez 1996–1997). In an hour-long observation of three *T. obesus* hunting two *Sufflamen verres* juveniles sheltering among pocilloporid colonies, the sharks produced more than 60 coral fragments. Coral fragments resulting from this event were large (mean size 11.2 ± 4.6 [SD] cm) and their expected survival rate was estimated at >90 % (Guzman 1991; Jiménez 1996–1997). On ETP reefs, sea turtles also perform similarly-derived coral destruction while foraging (Glynn 2004, 2008). These periodic bioerosion events performed by large reef predators perhaps contribute more importantly to coral propagation than to the conversion of coral substrates into sediments. Furthermore, coral colony fragmentation mediated by triggerfishes also strongly promotes asexual propagation of coral colonies (see Chaps. 10 and 15, Enochs and Glynn, and Glynn et al. respectively).

12.11 Summary and Conclusions

Bioerosion is a highly diverse and complex process that affects the development of coral formations and exerts great influence on coral habitats of the ETP. The ETP physical environment is enormously conducive to bioerosion processes. Eutrophic conditions throughout most of the ETP enhance the ecological competiveness of macroborers and increase overall bioerosion rates (Highsmith 1980; DeCarlo et al. 2015). A high-CO_2, low Ω physical environment not only contributes to marginal reef development and cementation in the ETP (Manzello et al. 2008), but also magnifies bioerosion rates (Tribollet et al. 2009; Wisshak et al. 2012; DeCarlo et al. 2015). Persistent interannual ENSO-induced coral mortality events are major drivers of ETP bioerosional dynamics and can bring about intense bioerosion regimes (Scott et al. 1988; Eakin 2001). ENSO thermal shock events promote and help shape bioeroder abundances and bioerosion rates. While heightened bioerosion has impeded the recovery of impacted coral habitats (Glynn et al. 2015), such activities are an important initiator of coral colony dispersal and may be responsible for the high occurrence of asexually derived coral communities, and enhance coral reef resistance to ENSO disturbances in the ETP (Highsmith 1982; Scott and Risk 1988; Scott et al. 1988; Guzman and López 1991).

Echinoid populations in particular have performed extensive bioerosion in many eastern Pacific coral habitats. The recovery of coral reefs affected by El Niño events can be significantly influenced by attendant changes in sea urchin abundances (Glynn 1988; Eakin 1996; Alvarado et al. 2012; Glynn et al. 2015). Destructive grazing by sea urchins has led to large-scale modification of reef structure and functioning at certain sites (Birkeland 1989). Coral communities of the Galápagos Islands and Panama have been especially affected by echinoid bioerosion (Eakin 2001; Glynn et al. 2015). This situation is likely exacerbated by cascading effects from over-fishing (Glynn 1988; Reaka-Kudla et al. 1996; Peyrot-Clausade et al. 2000; Tribollet and Golubic 2011), causing echinoid population densities to be unfettered by predation pressure (Sonnenholzer et al. 2009; Edgar et al. 2011; Glynn et al. 2015).

There has not yet been assessment of carbonate erosion by microendolith assemblages in the ETP. The activites of microendolith assemblages function in tandem with those of epilithic grazers (Chazotte et al. 1995; Zubia et al. 2001; Tribollet and Golubic 2005). It therefore is necessary to investigate microendolith bioerosion to better understand the ecological dynamics driving bioerosional activites of echinoids and fishes. However, irrespective of microendolith studies, bioerosion by fishes—parrotfishes, pufferfishes, and triggerfishes—has received relatively poor study to date in respect to their likely contributions. Scarids are reported to perform high bioerosion rates on protected reefs where populations are abundant (Jiménez 1999; Cantera et al. 2001). Recent studies have demonstrated the persistent influence of pufferfish bioerosion on Colombian reefs (Jiménez 1999), and an induction of developmental and morphological changes in coral colonies resulting from their activities (Palacios et al. 2014). Triggerfishes perform significant carbonate excavation while foraging on endolithic prey (e.g., bivalves) and promote coral colony dispersal through asexual fragmentation (Francisco 2000; Cantera et al. 2001). Triggerfishes as prey further generate bioerosion and coral colony fragmentation via periodic highly destructive foraging performed by shark predators (Jiménez 1996–1997).

The high abundance and diversity of endolithic sponges compared to encrusting and non-boring massive sponges are a strong characteristic of sponge assemblages in ETP coral habitats (Carballo et al. 2004, 2008), and may be due to eutrophic conditions causing selection for larval recruitment to cryptic aphotic spaces, as well as strong grazing pressure exerted by opportunistic fishes (e.g., *Arothron hispidus*; Birkeland 1977; Wulff 1997). Sponge assemblages have variable biogeographic importance in the ETP, and high sedimentation restricts the distribution of dominant clionaid species (Hutchings et al. 2005; Carballo et al. 2008; Nava and Carballo 2013). Many ETP boring sponges, however, demonstrate exceptional resiliency to temperature perturbation (Shirley and Rützler 2010; Duckworth and Peterson 2013; Bautista-Guerrero et al. 2014). This adaptive physiology renders boring sponges a threat to reefs having undergone

Fig. 12.7 *Arothron meleagris* bioerosion and induction of morphological change in pocilloporid corals at Gorgona Island, Colombia, adapted from Palacios et al. (2014). **a** Pufferfish bite marks on a *Pocillopora* sp. colony; **b** *Arothron meleagris* feeding on *Pocillopora* sp.; **c** excavation of a *Pocillopora* sp. skeletal protuberance by *A. meleagris*; **d** corallum morphology of pocilloporid branch exposed to pufferfish corallivory; **e** corallum morphology of pocilloporid branch protected from fish corallivory

coral bleaching and mortality due to environmental stressors such as ENSO thermal shock (Carballo et al. 2013).

The profusion and predominance of lithophagine bivalves in dead coral colonies and live massive corals is an important characteristic of ETP reef bioerosion. Boring bivalve populations reach exceptionally high densities in the ETP not found in other regions (Scott et al. 1988). Lithophagine bivalves are dominant endolithic bioeroders throughout the ETP that weaken coral skeletons and are responsible for substantial carbonate losses (Scott and Risk 1988; Scott et al. 1988). The predominance of lithophagids among ETP endolith assemblages may engender a more limited contribution by other co-occurring coral borers (Scott et al. 1988; Reaka-Kudla et al. 1996; Londoño-Cruz et al. 2003).

The role polychaetes play in ETP coral bioerosion has not yet been made clear. Given current assessments, polychaete populations appear to be less abundant and less significant members of ETP bioeroder communities compared to the importance of their activities in coral habitats elsewhere (Hutchings 2008). Polychaete populations can be locally abundant (Reaka-Kudla et al. 1996), and resolving their contribution to bioerosion processes requires further study. Sipunculan worms can be found in very high densities locally and are dominant coral endoliths on Costa Rican reefs, though appear to be less significant bioeroders on Colombian reefs. Both sipunculan and polychaete assemblages demonstrate highly variable abundances and strong zonal preferences (Cantera et al. 2003; Fonseca et al. 2006). Measurements of bioerosion rates for either assemblage where abundant have yet to be performed.

Functional interaction among bioeroder recruits and temporal trends in bioeroder community succession have not been fully described for ETP coral habitats. It does appear characteristics of bioeroder community succession in the ETP deviates from regions elsewhere, and there is a strong indication that bioeroder recruitment, substrate conditioning, and community succession are especially rapid (Londoño--Cruz et al. 2003). Rapid and massive lithophagine bivalve recruitment on newly exposed substrates has been well-illustrated (Kleemann 2013), and the pace of boring bivalve activities likely helps to shape endolith community maturation. Sipunculan borers, also thought to be late-stage invaders, are present in ETP coral skeletons soon after exposure. Conditioning of newly exposed substrates by polychaetes may be less important for the maturation of bioerosion processes in the ETP, while acrothoracican crustaceans' pioneering behavior may play larger roles (Londoño-Cruz et al. 2003). There is thus considerable evidence that temporal and functional patterns of bioeroder community succession in ETP coral habitats deviate from generalities well-described elsewhere, but these trends have not yet been resolved.

Acknowledgments J.J. Alvarado acknowledges several sources for their financial support: Vicerectoría de Investigación, Universidad de Costa Rica, Ministerio de Ciencia y Tecnología de Costa Rica (MICIT), Consejo Nacional para Investigaciones Científicas y Tecnológicas de Costa Rica (CONICIT), Consejo Nacional de Ciencia y Tecnología de Mexico (CONACYT). J.L. Carballo acknowledges the CONACYT SEP for funding project 2008-102239. We are grateful for the comments offered by J. Cortés and two anonymous reviewers that improved this chapter. B. Grassian acknowledges P. Hutchings and K. Kleemann for providing comments that substantially improved the quality of this chapter, D. F. McNeill for preparation of specimens for photography, and M. Palacios and colleagues for their contribution of Fig. 12.7.

References

Acker KL, Risk MJ (1985) Substrate destruction and sediment production by the boring sponge *Cliona caribbaea* on Grand Cayman Island. J Sed Petrol 56:705–711

Alvarado JJ, Cortés J, Reyes-Bonilla H (2012) Reconstruction of *Diadema mexicanum* A. Agassiz, 1863 bioerosion impact on three Costa Rican Pacific coral reefs. Rev Biol Trop 60(Suppl 2):121–132

Ansell AD, Nair NB (1969) The mechanisms of boring in *Martesia striata* Linné (Bivalvia: Pholadidae) and *Xylophaga dorsalis* Turton (Bivalvia: Xylophaginidae). Proc Royal Soc Lond Ser B Biol Sci 174(1034):123–133

Appana SD, Vuki VC (2006) Bioerosion patterns and abundance and spatial distribution of *Echinometra* sp. A (green white tip) ecomorph on Nukubuco reef, Fiji. In: Proceedings of 10th International Coral Reef Symposium, Okinawa, pp 938–945

Bak RPM (1976) The growth of coral colonies and the importance of crustose coralline algae and burrowing sponges in relation with carbonate accumulation. Neth J Sea Res 10:285–337

Bak RPM (1990) Patterns of echinoid bioerosion in two Pacific coral reef lagoons. Mar Ecol Prog Ser 66:267–272

Bak RPM (1994) Sea urchin bioerosion on coral reefs: place in the carbonate budget and relevant variables. Coral Reefs 13:99–103

Bak RPM, van Eys G (1975) Predation of the sea urchin *Diadema antillarum* Philippi on living coral. Oecologia 20:111–115

Bak RPM, Carpay MJE, de Ruyter van Steveninck ED (1984) Densities of the sea urchin *Diadema antillarum* before and after mass mortalities on the coral reefs of Curaçao. Mar Ecol Prog Ser 17:105–108

Baker AC, Glynn PW, Riegl B (2008) Climate change and coral reef bleaching: an ecological assessment of long-term impacts, recovery trends and future outlook. Estuar Coast Shelf S 80(4):435–471

Bautista-Guerrero E, Carballo JL, Maldonado M (2014) Abundance and reproductive patterns of the excavating sponge *Cliona vermifera*: a threat to Pacific coral reefs? Coral Reefs 33(1):259–266

Bellwood DR, Choat JH (1990) A functional analysis of grazing in parrotfishes (Family Scaridae)—The ecological implications. Env Biol Fish 28:189–214

Benítez-Villalobos F, Domínguez-Gómez MT, López-Pérez RA (2008) Temporal variation of the sea urchin *Diadema mexicanum* population density at Bahías de Huatulco, Western Mexico. Rev Biol Trop 56(Suppl 3):255–263

Bentis CJ, Kaufman L, Golubic S (2000) Endolithic fungi in reef-building corals (Order: Scleractinia) are common, cosmopolitan, and potentially pathogenic. Biol Bull 198:254–260

Birkeland C (1977) The importance of rate of biomass accumulation in early successional stages of benthic communities to the survival of coral recruits. In: Proceedings of 3rd International Coral Reef Symposium, vol 1, Miami, pp 15–21

Birkeland C (1989) The influence of echinoderms on coral-reef communities. In: Jangoux M, Lawrence JM (eds) Echinoderm Studies 3. Balkema, Rotterdam, pp 1–79

Blake JA (1991) The polychaete fauna of the Galápagos Islands. In: James MJ (ed) Galápagos marine invertebrates. Taxonomy, biogeography, and evolution in Darwin's Islands. Plenum Press, New York, pp 75–96

Brown-Saracino J, Peckol P, Curran HA, Robbart ML (2007) Spatial variation in sea urchins, fish predators, and bioerosion rates on coral reefs of Belize. Coral Reefs 26:71–78

Bruggemann JH, van Kessel AM, van Rooij JM, Breeman AM (1996) Bioerosion and sediment ingestion by the Caribbean parrotfish *Scarus vetula* and *Sparisoma viride*: implications of fish size, feeding mode and habitat use. Mar Ecol Prog Ser 134:59–71

Cailliaud F (1843) Notice sur le genre Gatrochène. Mag Zool 1843:1–10

Cailliaud F (1850) Nouvelles obvservations au sujet de la perforation des pierres par les mollusques. J Conchol 1:363

Calcinai B, Azzini F, Bavestrello G, Gaggero L, Cerrano C (2007) Excavating rates and boring pattern of *Cliona albimarginata* (Porifera: Clionaidae) in different substrata. In: Custódio MR, Lôbo-Hajdu G, Hajdu E, Muricy G (eds) Porifera research: biodiversity, innovation and sustainability. Série Livros 28. Museu Nacional, Rio de Janeiro, pp 203–207

Campbell SE (1982) Precambrian endoliths discovered. Nature 299:429–431

Cantera JR, Zapata F, Forero P, Francisco V, Jiménez JM, Londoño E, Narvaez K, Neira R, Orozco CA, Toro-Farmer G (2001) Organismos bioerosionadores en arrecifes de Isla Gorgona. In: Barrios LM, López-Victoria M (eds) Gorgona marina: contribución al conocimiento de una isla única, Serie Publicaciones Especiales 7. INVEMAR, Magdalena, pp 51–64

Cantera JR, Orozco C, Londoño-Cruz E, Toro G (2003) Abundance and distribution patterns of infaunal associates and macroborers of the branched coral in Gorgona Island. Bull Mar Sci 72:207–219

Carballo JL, Sánchez-Moyano JE, García Gómez JC (1994) Taxonomic and ecological remarks on boring sponges (Clionidae) from the Straits of Gibraltar (southern Spain): tentative bioindicators? Zool J Lin Soc 112:407–424

Carballo JL, Cruz-Barraza JA, Gómez P (2004) Taxonomy and description of clionaid sponges (Hadromerida, Clionaidae) from the Pacific Ocean of Mexico. Zool J Lin Soc 141:353–397

Carballo JL, Hepburn L, Nava H, Cruz-Barraza JA, Bautista-Guerrero E (2007) Coral boring *Aka*-species (Porifera: Phloeodictyidae) from Mexico with description of *Aka cryptica* sp. nov. J Mar Biol Assoc UK 87:1477–1484

Carballo JL, Bautista-Guerrero E, Leyte-Morales GE (2008) Boring sponges and the modelling of coral reefs in the East Pacific Ocean. Mar Ecol Prog Ser 356:113–122

Carballo JL, Cruz-Barraza JA, Nava H, Bautista E (2010a) Esponjas perforadoras de sustratos calcáreos: importancia en los ecosistemas arrecifales del Pacífico este. Rev Biol Trop 58:1591–1592

Carballo JL, Bautista-Guerrero E, Nava H, Cruz Barraza JA (2010b) Cambio climático y ecosistemas costeros. Bases fundamentales para la conservación de los arrecifes de coral del Pacífico Este. In: Hernández-Zanuy AC, Alcolado PM (eds) La biodiversidad en ecosistemas marinos y costeros del litoral de Iberoamérica y el cambio climático: I. Red Cyted Biodivmar, Havana, pp 183–193

Carballo JL, Bautista-Guerrero E, Nava H, Cruz-Barraza JA, Chávez JA (2013) Boring sponges, an increasing threat for coral reefs affected by bleaching events. Ecol Evol 4:872–886

Carreiro-Silva M, McClanahan TR (2001) Echinoid bioerosion and hervibory on Kenyan coral reefs: the role of protection from fishing. J Exp Mar Biol Ecol 262:133–153

Carter JG (1978) Ecology and evolution of the Gastrochaenacea (Mollusca, Bivalvia) with notes on the evolution of the endolithic habitat. Peabod Mus Nat Hist Yale Univ Bull 41:48–51

Chazottes V, Le Campion-Alsumard T, Peyrot-Clausade M (1995) Bioerosion rates on coral reefs: interactions between macroborers, microborers and grazers (Moorea, French Polynesia). Palaeogeogr Palaeocl 113(2):189–198

Colgan M (1990) El Niño and the history of Eastern Pacific reef building. In: Glynn PW (ed) Global ecological consequences of the 1982–83 El Niño-Southern Oscillation. Elsevier, Amsterdam, pp 183–232

Conand C, Chabanet P, Cuet P, Letourneur Y (1997) The carbonate budget of a fringing reef in La Réunion Island (Indian Ocean): sea urchin and fish bioerosion and net calcification. In: Proceedings of 8th International Coral Reef Symposium, vol 1, Panama, pp 953–958

Cortés J (1985) Preliminary observations of *Alpheus simus* Guérin--Méneville, 1856 (Crustacea: Alpheidae): a little-known Caribbean bioeroder. In: Proceedings of 5th International Coral Reef Congress, vol 5, Tahiti, pp 351–353

Cortés J (1991) Los arrecifes coralinos del Golfo Dulce, Costa Rica: aspectos geológicos. Rev Geol Amer Central 13:15–24

Cortés J (1997) Biology and geology of eastern Pacific coral reefs. Coral Reefs 16(Suppl):S39–S46

Cortés J, Jiménez C (2003) Corals and coral reefs of the Pacific of Costa Rica: history, research and status. In: Cortés J (ed) Latin American coral reefs. Elsevier Science, Amsterdam, pp 361–385

Cruz-Barraza JA, Carballo JL (2008) Taxonomy of sponges associated with corals from the Mexican Pacific Ocean. Zool Stud 47:741–758

Cutler EB (ed) (1994) The Sipuncula: their systematics, biology, and evolution. Cornell University Press, Ithaca, p 453

Cutler NJ, Cutler EB, Vargas JA (1992) Peanut worms (phylum Sipuncula) from Costa Rica. Gusanos maní (phylum Sipuncula) de Costa Rica. Rev Biol Trop 40:153–158

Davies PJ, Hutchings PA (1983) Initial colonization, erosion and accretion on coral substrate: experimental results, Lizard Island, Great Barrier Reef. Coral Reefs 2:27–35

Dean HK, Sibaja-Cordero JA, Cortés J (2010) Sipunculans and echiurans of Isla del Coco (Cocos Island), Costa Rica. Zootaxa 2557:60–68

DeCarlo TM, Cohen AL, Barkley HC, Cobban Q, Young C, Shamberger KE, Brainard RE, Golbuu Y (2015) Coral macrobioerosion is accelerated by ocean acidification and nutrients. Geology 43:7–10

Deshayes GP (1850) Traité élémentaire de conchyliologie: avec les applications de cette science à la géologie. Tome premier. Seconde partie. V Mason, Paris, p 824

Downing N, El-Zahr CR (1987) Gut evacuation and filling rates in the rock-boring sea urchin, *Echinometra mathaei*. Bull Mar Sci 41:579–584

Duckworth AR, Peterson BJ (2013) Effects of seawater temperature and pH on the boring rates of the sponge *Cliona celata* in scallop shells. Mar Biol 160:27–35

Eakin CM (1987) Damselfishes and their algal lawns: a case of plural mutualism. Symbiosis 4(1–3):275–288

Eakin CM (1988) Avoidance of damselfish lawns by the sea urchin *Diadema mexicanum* at Uva Island, Panama. In: Proceedings of 6th International Coral Reef Symposium, vol 2, Townsville, pp 21–26

Eakin CM (1991) The damselfish-algal lawn symbiosis and its influence on the bioerosion of an El Niño impacted coral reef, Uva Island, Pacific Panama. PhD Dissert, Univ Miami, Coral Gables, Florida, p 158

Eakin CM (1992) Post-El Niño Panamanian reefs: less accretion, more erosion and damselfish protection. In: Proceedings of 7th International Coral Reef Symposium, vol 1, Guam, pp 387–396
Eakin CM (1996) Where have all the carbonates gone? A model comparison of calcium carbonate budgets befote and after the 1982–1983 El Niño at Uva Island in the Eastern Pacific. Coral Reefs 15:109–119
Eakin CM (2001) A tale of two ENSO events: carbonate budgets and the influence of two warming disturbances and intervening variability, Uva Island, Panama. Bull Mar Sci 69:171–186
Edgar GJ, Banks SA, Bessudo S, Cortés J, Guzman HM, Henderson S, Martinez C, Rivera F, Soler G, Ruiz D, Zapata FA (2011) Variation in reef fish and invertebrate communities with level of protection from fishing across the eastern tropical Pacific seascape. Glob Ecol Biogeogr 20(5):730–743
Enochs IC (2012) Motile cryptofauna associated with live and dead coral substrates: implications for coral mortality and framework erosion. Mar Biol 159:709–722
Enochs IC, Manzello DP (2012) Species richness of motile cryptofauna across a gradient of reef framework erosion. Coral Reefs 31:653–661
Fauchald K (1977) The polychaete worms. Definitions and keys to the orders, families and genera. Nat Hist Mus LA County, Los Angeles, pp 1–188
Fewkes W (1890) Sea-urchin excavations at Guaymas, Mexico. Am Nat 24:478–480
Fonseca AC, Cortés J (1998) Coral borers of the eastern Pacific: *Aspidosiphon (A.) elegans* (Sipuncula: Aspidosiphonidae) and *Pomatogebia rugosa* (Crustacea: Upogebiidae). Pac Sci 52:170–175
Fonseca AC, Dean HK, Cortés J (2006) Non-colonial coral macro-borers as indicators of coral reef status in the south Pacific of Costa Rica. Rev Biol Trop 54:101–115
Francisco V (2000) Determinación de las tasas de bioerosión por peces balístidos en un arrecife coralino del Pacífico colombiano. MSc Thesis, Universidad del Valle, Cali, Colombia, p 65
Fütterer DK (1974) Significance of the boring sponge *Cliona* for the origin of fine grained material of carbonate sediments. J Sed Petrol 44:79–84
Gilchrist S (1985) Hermit crab corallivore activity. In: Proceedings of 5th International Coral Reef Symposium, vol 5, Tahiti, pp 211–214
Glynn PW (1973) *Acanthaster*: effect on coral reef growth in Panama. Science 180(4085):504–506
Glynn PW (1984) Widespread coral mortality and the 1982/83 El Niño warming event. Environ Conserv 11(2):133–146
Glynn PW (1985) Corallivore population sizes and feeding effects following El Niño (1982–1983) associated coral mortality in Panama. In: Proceedings of 5th International Coral Reef Symposium, vol 4, Tahiti, pp 183–188
Glynn PW (1988) El Niño warming, coral mortality and reef framework destruction by echinoid bioerosion in the eastern Pacific. Galaxea 7:129–160
Glynn PW (1990) Coral mortality and disturbances to coral reefs in the tropical eastern Pacific. In: Glynn PW (ed) Global ecological consequences of the 1982–83 El Niño-Southern Oscillation. Elsevier, Amsterdam, pp 55–126
Glynn PW (1994) State of coral reefs in the Galápagos Islands: natural vs anthropogenic impacts. Mar Pollut Bull 29(1):131–140
Glynn PW (1997) Bioerosion and coral-reef growth: a dynamic balance. In: Birkeland C (ed) Life and death of coral reefs. Chapman & Hall, New York, pp 68–95
Glynn PW (2004) High complexity food webs in low-diversity Eastern Pacific reef-coral communities. Ecosystems 7:358–367
Glynn PW (2008) Food-web structure and dynamics of eastern tropical Pacific coral reefs: Panamá and Galápagos Islands. In: McClanahan TR, Branch GM (eds) Food webs and the dynamics of marine reefs. Oxford University Press, Oxford, pp 185–208
Glynn PW, Macintyre IG (1977) Growth rate and age of coral reefs on the Pacific coast of Panama. In: Proceedings of 3rd International Coral Reef Symposium, vol 2, Miami, pp 251–259
Glynn PW, Wellington GM (1983) Corals and coral reefs of the Galápagos Islands (with an annotated list of the scleractinian corals of the Galápagos by JW Wells). Univ California Press, Berkeley, p 330
Glynn PW, Stewart RH, McCosker JE (1972) Pacific coral reefs of Panama: structure, distribution and predators. Geol Rundsch 61 (2):483–519
Glynn PW, Wellington GM, Birkeland C (1979) Coral reef growth in the Galápagos: limitation by sea urchins. Science 203:47–49
Glynn PW, Gassman NJ, Eakin CM, Cortés J, Smith DB, Guzman HM (1991) Reef coral reproduction in the eastern Pacific: Costa Rica, Panama and Galápagos Islands (Ecuador). I. Pocilloporidae. Mar Biol 109:355–368
Glynn PW, Maté JL, Baker AC, Calderón MO (2001) Coral bleaching and mortality in Panama and Ecuador during the 1997–1998 El Niño-Southern Oscillation event: spatial/temporal patterns and comparisons with the 1982–1983 event. Bull Mar Sci 69(1):79–109
Glynn PW, Riegl BM, Purkis S, Kerr J, Smith T (2015) Coral reef recovery in the Galápagos Islands: the northern-most islands (Darwin and Wenman). Coral Reefs 34(2):421–436
Goreau TF, Hartman WD (1963) Boring sponges as controlling factors in the formation and maintenance of coral reefs. Am Assoc Advan Sci Publ 75:25–54
Griffin SP, García RP, Weil E (2003) Bioerosion in coral reef communities in southwest Puerto Rico by the sea urchin *Echinometra viridis*. Mar Biol 143:79–84
Gutner-Hoch E, Fine M (2011) Genotypic diversity and distribution of *Ostreobium quekettii* within scleractinian corals. Coral Reefs 30:643–650
Guzman HM (1986) Estructura de la comunidad arrecifal de la Isla del Caño, Costa Rica, y el efecto de perturbaciones naturales severas. MS thesis, Universidad de Costa Rica, San José, Costa Rica, p 179
Guzman HM (1988) Distribución y abundancia de organismos coralívoros en los arrecifes coralinos de la Isla del Caño, Costa Rica. Rev Biol Trop 36(2):191–207
Guzman HM (1991) Restoration of coral reefs in Pacific Costa Rica. Conser Biol 5:189–195
Guzman HM, Cortés J (1992) Isla del Coco (Pacific of Costa Rica) coral reefs after the 1982–83 El Niño disturbance. Rev Biol Trop 40:309–324
Guzman HM, Cortés J (1993) Arrecifes coralinos del Pacífico Oriental Tropical: revisión y perspectivas. Rev Biol Trop 41:535–557
Guzman HM, Cortés J (2007) Reef recovery 20 years after the 1982–1983 El Niño massive mortality. Mar Biol 151:401–411
Guzman HM, López JV (1991) Diet of the corallivorous pufferfish *Arothron meleagris* (Pisces: Tetraodontidae) at Gorgona Island, Colombia. Rev Biol Trop 39:203–206
Guzman HM, Robertson DR (1989) Population and feeding responses of the corallivorous pufferfish *Arothron meleagris* to coral mortality in the Eastern Pacific. Mar Ecol Prog Ser 55:121–131
Haigler SM (1969) Boring mechanism of *Polydora websteri* inhabiting *Crassostrea virginica*. Am Zool 9:821–828
Hallock P (1988) The role of nutrient availability in bioerosion: consequences to carbonate buildups. Palaeogeogr Palaeoclimat Palaeoecol 62:275–291

Hassan M (1988) Modification of carbonate substrata by bioerosion and bioaccretion on coral reefs of the Red Sea. Shaker Verlag, Aachen, p 124
Hein FJ, Risk MJ (1975) Bioerosion of coral heads: inner patch reefs, Florida Reef Tract. Bull Mar Sci 25:133–138
Herrera-Escalante T, López-Pérez RA, Leyte-Morales GE (2005) Bioerosion caused by the sea urchin *Diadema mexicanum* (Echinodermata: Echinoidea) at Bahías de Huatulco, Western Mexico. Rev Biol Trop 53(Suppl 3):263–273
Hibino K, Van Woesik R (2000) Spatial differences and seasonal changes of net carbonate accumulation on some coral reefs of the Ryukyu Islands, Japan. J Exp Mar Biol Ecol 252:1–14
Highsmith RC (1980) Geographic patterns of coral bioerosion: a productivity hypothesis. J Exp Mar Biol Ecol 46:177–196
Highsmith RC (1982) Reproduction by fragmentation in corals. Mar Ecol Prog Ser 7:207–226
Highsmith RC, Lueptow RL, Schonberg SC (1983) Growth and bioerosion of three massive corals on the Belize barrier reef. Mar Ecol Prog Ser 13(2–3):261–271
Hunter IG (1977) Sediment production by *Diadema antillarum* on a Barbados fringing reef. In: Proceedings of 3th International Coral Reef Symposium, vol 2, Miami, pp 105–109
Hueerkamp C, Glynn PW, D'Croz L, Maté JL, Colley SB (2001) Bleaching and recovery of five eastern Pacific corals in an El Niño-related temperature experiment. B Mar Sci 69(1):215–236.
Hutchings PA (1981) Polychaete recruitment on to dead coral substrates at Lizard Island, Great Barrier Reef, Australia. Bull Mar Sci 31:410–423
Hutchings PA (1986) Biological destruction of coral reefs. Coral Reefs 4:239–252
Hutchings PA (2008) Role of polychaetes in bioerosion of coral substrates. In: Wisshak M, Tapanila L (eds) Current developments in bioerosion. Springer, Berlin, pp 249–264
Hutchings PA (2011) Bioerosion. In: Hopley D (ed) Encyclopedia of modern coral reefs: structure, form, and process. Springer, Berlin, pp 139–156
Hutchings PA, Peyrot-Clausade M (1989) Macro-boring communities of *Porites*. A biological comparison. In: Proceedings of 6th International Coral Reef Symposium, vol 3, Townsville, pp 263–268
Hutchings PA, Peyrot-Clausade M (2002) The distribution and abundance of boring species of polychaetes and sipunculans in coral substrates in French Polynesia. J Exp Mar Biol Ecol 269 (1):101–121
Hutchings PA, Kiene WE, Cummingham RB, Donnelly C (1992) Spatial and temporal patterns of non-colonial boring organisms: (polychaetes, sipunculans and bivalve molluscs) in *Porites* at Lizard Island, Great Barrier Reef. Coral Reefs 11:23–31
Hutchings PA, Peyrot-Clausade M, Osorno A (2005) Influence of land runoff on rates and agents of bioerosion of coral substrates. Mar Poll Bull 51:438–447
Jaccarini V, Bannister WH, Micallef H (1968) The pallial glands and rock boring in *Lithophaga lithophaga* (Lamellibranchia, Mytilidae). J Zool 154(4):397–401
Jiménez C (1996–1997) Coral colony fragmentation by whitetip reef sharks at Coiba Island National Park, Panamá. Rev Biol Trop 45 (1B):698–700
Jiménez JM (1999) *Scarus ghobban* (Pisces: Scaridae) en Isla Gorgona (Pacífico Oriental Tropical): abundancia, comportamiento alimenticio y papel en la bioerosión de arrecifes. BSc thesis, Univ del Valle, Cali, Colombia, p 74
Kiene WE, Hutchings PA (1994) Bioerosion experiments at Lizard Island, Great Barrier Reef. Coral Reefs 13:91–98
Kinsey DW (1985) Metabolism, calcification and carbon production. I. Systems level studies. In: Proceedings of 5th International Coral Reef Congress, vol 4, Tahiti, pp 505–526
Kleemann K (1973) *Lithophaga lithophaga* (L.) (Bivalvia) in different limestones. Malacologia 14:345–347
Kleemann K (1974) Beitrag zur Kenntnis des Verhaltens von *Lithophaga lithophaga* (L.) (Bivalvia) im Bohrloch. Sitzubgsberichte Österr Akad Wiss (math-naturw Kl, Abt. I) 182:201–210
Kleemann KH (1977) A new species of *Lithophaga* (Bivalvia) from the Great Barrier Reef, Australia. Veliger 20(2):151–154
Kleemann KH (1980) Boring bivalves and their host corals from the Great Barrier Reef. J Moll Stud 46:13–54
Kleemann KH (1982) Ätzmuscheln im Ghetto? *Lithophaga* (Bivalvia) aus dem Leithakalk (Mittel-Miozän: Badenian) von Müllendorf im Wiener Becken, Österreich. Beitr Paläonto Österr 9:211–231
Kleemann K (1986) Lithophagines (Bivalvia) from the Caribbean and the Eastern Pacific. In: Proceedings of 8th International Coral Reef Congress, Budapest, pp 113–118
Kleemann K (1990) Boring and growth in chemically boring bivalves from the Caribbean, Eastern Pacific and Australia's Great Barrier Reef. Senck Marit 21:101–154
Kleemann K (1992) Coral communities and coral-bivalve associations in the northern Red Sea at Safaga, Egypt. Facies 26:1–10
Kleemann K (1995) Associations of coral and boring bivalves: Lizard Island (Great Barrier Reef, Australia) versus Safaga (N Red Sea). Beitr Paläont 20:31–39
Kleemann K (1996) Biocorrosion by bivalves. PSZNI Mar Ecol 17(1–3):145–158
Kleemann K (2008) *Lithophaga (Leiosolenus) purpurea* (Bivalvia: Mytilidae): one species becomes three. Club Conchylia Informationen 39(3/4):32–45
Kleemann K (2013) Fast and massive settlement of boring bivalves on coral slabs at Taboga Islands, Eastern Pacific, Panama. Boll Malacol 49:104–113
Kleemann K, Hoeksema BW (2002) *Lithophaga* (Bivalvia, Mytilidae), including a new species, boring into mushroom corals (Scleractinia, Fungiidae) off South Sulawesi, Indonesia. Basteria 66:11–24
Kleypas JA, Feely RA, Fabry VJ, Langdon C, Sabine C, Robbins LL (2006) Impacts of ocean acidification on coral reefs and other marine calcifiers: a guide for future research. Report of a workshop held 18–20 April 2005, St. Petersburg, FL, sponsored by NSF, NOAA, and USGS, p 88
Kroeker KJ, Kordas RL, Crim RN, Singh GG (2010) Meta-analysis reveals negative yet variable effects of ocean acidification on marine organisms. Ecol Lett 13:1419–1434
Lazar B, Loya Y (1991) Bioerosion of coral reefs—A chemical approach. Limnol Oceanogr 36:377–383
Le Campion-Alsumard T, Golubic S, Hutchings P (1995) Microbial endoliths in skeletons of live and dead corals: *Porites lobata* (Moorea, French Polynesia). Mar Ecol Prog Ser 117:149–157
Londoño-Cruz E (2001) Proceso de bioerosión por organismos perforadores en un ecosistema arrecifal del Pacífico Oriental Tropical (Isla Gorgona, Colombia). MSc Thesis, Universidad del Valle, Cali, Colombia, p 56
Londoño-Cruz E, Cantera JR, Toro-Farmer G, Orozco C (2003) Internal bioerosion by macroborers in *Pocillopora* spp. in the Tropical Eastern Pacific. Mar Ecol Prog Ser 265:289–295
MacGeachy JK (1977) Factors controlling sponge boring in Barbados reef corals. In: Proceedings of 3rd International Coral Reef Symposium, vol 2, Miami, pp 477–483
Manzello DP (2010a) Coral growth with thermal stress and ocean acidification: lessons from the eastern tropical Pacific. Coral Reefs 29:749–758

Manzello DP (2010b) Ocean acidification hotspots: spatiotemporal dynamics of the seawater CO_2 system of eastern Pacific coral reefs. Limnol Oceanogr 55:239–248

Manzello DP, Kleypas JA, Budd DA, Eakin CM, Glynn PW, Langdon C (2008) Poorly cemented coral reefs of the eastern tropical Pacific: possible insights into reef development in a high-CO_2 world. Proc Natl Acad Sci USA 105(30):10450–10455

McClanahan TR, Muthiga NA (1988) Changes in Kenyan coral reef community structure and function due to exploitation. Hydrobiologia 166:269–276

McClanahan TR, Nugues M, Mwachireya S (1994) Assaying fish and sea urchin herbivory and competition in Kenyan coral reef lagoons. J Exp Mar Biol Ecol 184:237–254

Millero FJ (2007) The marine inorganic carbon cycle. Chem Rev 107 (2):308–341

Mokady O, Lazar B, Loya Y (1996) Echinoid bioerosion as a major structuring force of Red Sea coral reefs. Biol Bull 190:367–372

Moreno XG, Abitia LA, Favila A, Gutiérrez FJ, Palacios DS (2009) Ecología trófica del pez *Arothron meleagris* (Tetraodontiformes: Tetraodontidae) en el arrecife de Los Frailes, Baja California Sur, México. Rev Biol Trop 57:113–123

Morton B (1990) Corals and their bivalve borers: the evolution of a symbiosis. In: Morton B (ed) The Bivalvia: proceedings of a memorial symposium in honour of Sir Charles Maurice Yonge (1899–1986) at the 9th international malacological congress, 1986, Edinburgh, Scotland, UK. Hong Kong University Press, Hong Kong, pp 11–46

Morton B, Scott PJB (1980) Morphological and functional specializations of the shell, musculature and pallial glands in the Lithophaginae (Mollusca: Bivalvia). J Zool 192(2):179–203

Mumby PJ, Hedley JD, Zychaluk K, Harborne AR, Blackwell PG (2006) Revisiting the catastrophic die-off of the urchin *Diadema antillarum* on Caribbean coral reefs: fresh insights on resilience from a simulation model. Ecol Mod 196:131–148

Muricy G (1991) Structure des peuplements de spongiaires autour de l'égout de Cortiou (Marseille, France). Vie Milieu 41(4):205–221

Nava H, Carballo JL (2008) Chemical and mechanical bioerosion of boring sponges from Mexican Pacific coral reefs. J Exp Biol 211 (17):2827–2831

Nava H, Carballo JL (2013) Environmental factors shaping boring sponge assemblages at Mexican Pacific coral reefs. Mar Ecol 34(3):269–279

Neudecker S (1979) Effects of grazing and browsing fishes on the zonation of corals in Guam. Ecology 60(4):666–672

Neumann AC (1966) Observations on coastal erosion in Bermuda and measurements of the boring rate of the sponge, *Cliona lampa*. Limnol Oceanogr 11:92–108

Ogden JC (1977) Carbonate-sediment production by parrot fish and sea urchins on Caribbean reefs. In: Frost SH, Weiss MP, Saunders JB (eds) Reefs and related carbonates-ecology and sedimentology. Am Assoc Petrol Geol Stud Geol 4:281–288

Okey TA, Banks S, Born AF, Bustamante RH, Calvopiña M, Edgar GJ, Espinoza E, Fariña JM, Garske LE, Reck GK, Salazar S (2004) A trophic model of a Galápagos subtidal rocky reef for evaluating fisheries and conservation strategies. Ecol Model 172(2):383–401

Orr JC, Fabry VJ, Aumont O, Bopp L, Doney SC, Feely RA et al (2005) Anthropogenic ocean acidification over the twenty-first century and its impact on calcifying organisms. Nature 437:681–686

Osorno A, Peyrot-Clausade M, Hutchings PA (2005) Patterns and rates of erosion in dead *Porites* across the Great Barrier Reef (Australia) after 2 years and 4 years of exposure. Coral Reefs 24(2):292–303

Ovalle H (2011) Cuantificación de la erosión por esponjas perforadoras en comunidades arrecifales del Pacífico Mexicano. Licentiate thesis, Facultad de Ciencias del Mar, Univ Autónoma de Sinaloa, Sinaloa, Mexico

Pacheco-Solano C (2011) Estudio taxonómico de las esponjas perforadoras (Porifera, Demospongiae) de arrecifes de coral del Pacífico de Costa Rica. Licentiate thesis, Escuela de Biología, Univ de Costa Rica, San José, Costa Rica

Palacios MM, Muñoz CG, Zapata FA (2014) Fish corallivory on a pocilloporid reef and experimental coral responses to predation. Coral Reefs 33:625–636

Pari N, Peyrot-Clausade M, LeCampion-Alsumard T, Hutchings P, Chazottes V, Golubic S, Le Campion J, Fontaine MF (1998) Bioerosion of experimental substrates on high islands and on atoll lagoons (French Polynesia) after two years of exposure. Mar Ecol Prog Ser 166:119–130

Perry CT (1998) Macroborers within coral framework at Discovery Bay, north Jamaica: species distribution and abundance, and effects on coral preservation. Coral Reefs 17(2):277–287

Perry CT, Spencer T, Kench PS (2008) Carbonate budgets and reef production states: a geomorphic perspective on the ecological phase-shift concept. Coral Reefs 27:853–866

Peyrot-Clausade M, Chabanet P, Conand C, Fontaine MF, Letourner Y, Harmelin-Vivien H (2000) Sea urchin and fish bioerosion on La Réunion and Moorea reefs. Bull Mar Sci 66:477–485

Pojeta J Jr, Palmer TJ (1976) The origin of rock boring in mytilacean pelecypods. Alcheringa 1(2):167–179

Pomponi SA (1977) Etching cells of boring sponges: an ultrastructural analysis. In: Proceedings of 3th International Coral Reef Symposium, vol 2, Miami, pp 485–490

Reaka-Kudla ML, Feingold JS, Glynn PW (1996) Experimental studies of rapid bioerosion of coral reefs in the Galápagos Islands. Coral Reefs 15:101–107

Reyes-Bonilla H (2003) Coral reefs of the Pacific coast of Mexico. In: Cortés J (ed) Latin American coral reefs. Elsevier, Amsterdam, pp 331–349

Reyes-Bonilla H, Calderón-Aguilera LE (1999) Population density, distribution and consumption rates of three corallivores at Cabo Pulmo reef, Gulf of California. PZNSI Mar Ecol 20:347–357

Rice ME (1969) Possible boring structures of sipunculids. Am Zool 9(3):803–812

Rice ME, Macintyre IG (1972) A preliminary study of sipunculan burrows in rock thin sections. Carib J Sci 12(1–2):41–43

Richmond RH (1985) Variations in the population biology of *Pocillopora damicornis* across the Pacific. In: Proceedings of 5th International Coral Reef Congress, vol 6, Tahiti, pp 101–106

Richmond RH (1987) Energetic relationships and biogeographical differences among fecundity, growth, and reproduction in the reef coral, *Pocillopora damicornis*. Bull Mar Sci 41:595–604

Rotjan RD, Lewis SM (2008) Impact of coral predators on tropical reefs. Mar Ecol Prog Ser 367:73–91

Ruengsawang N, Yeemin T (2000) Bioerosion caused by grazing activities on coral communities in the Gulf of Thailand. In: Proceedings of 9th International Coral Reef Symposium, vol 1, Bali, pp 289–294

Russo AR (1980) Bioerosion by two rock boring echinoids (*Echinometra mathaei* and *Echinostrephus aciculatus*) on Enewetak Atoll, Marshall Islands. J Mar Res 38:99–110

Ruttenberg BI (2001) Effects of artisanal fishing on marine communities in the Galapagos Islands. Conserv Biol 15(6):1691–1699

Rützler K (1971) Bredin-Archbold-Smithsonian Biological Survey of Dominica: burrowing sponges, genus *Siphonodictyon* Bergquist, from the Caribbean. Smith Contr Zool 77:1–37

Rützler K (1975) The role of burrowing sponges in bioerosion. Oecologia 19:203–216

Rützler K (2002) Impact of crustose clionid sponges on Caribbean reef corals. Acta Geol Hisp 37:61–72

Rützler K, Rieger G (1973) Sponge burrowing: fine structure of *Cliona lampa* penetrating calcareous substrata. Mar Biol 21:144–162

Sammarco PW (1980) *Diadema* and its relationship to coral spat mortality: grazing, competition and biological disturbance. J Exp Mar Biol Ecol 45:245–472

Sammarco PW (1982a) Echinoid grazing as a structuring force in coral communities: whole reef manipulations. J Exp Mar Biol Ecol 61:31–55

Sammarco PW (1982b) Effects of grazing by *Diadema antillarum* Phillipi (Echinodermata: Echinoidea) on algal diversity and community structure. J Exp Mar Biol Ecol 65:83–105

Sammarco PW, Levington JS, Ogden JC (1974) Grazing and control of coral reef community structure by *Diadema antillarum* Phillipi (Echinodermata: Echinoidea): a preliminary study. J Mar Res 32:47–53

Schoënberg CH (2002) Substrate effects on the bioeroding demosponge *Cliona orientalis*. 1. Bioerosion rates. Mar Ecol 23(4):313–326

Schönberg CHL (2008) A history of sponge erosion: from past myths and hypotheses to recent approaches. In: Wisshak M, Tapanila L (eds) Current developments in bioerosion. Springer, Berlin, pp 165–202

Schönberg CHL, Ortiz JC (2009) Is sponge bioerosion increasing? In: Proceedings of 11th International Coral Reef Symposium, Ft. Lauderdale, pp 520–523

Scoffin TP, Stearn CW, Boucher D, Frydl P, Hawkins CM, Hunter JG, MacGeachy JK (1980) Calcium carbonate budget of fringing reef of the west coast of Barbados. Part II. Erosion, sediments and internal structure. Bull Mar Sci 30:475–508

Scott PJB, Risk MJ (1988) The effects of *Lithophaga* (Bivalvia: Mytilidae) boreholes on the strength of the coral *Porites lobata*. Coral Reefs 7:145–151

Scott PJB, Risk MJ, Carriquiry JD (1988) El Niño, bioerosion and the survival of east Pacific reefs. In: Proceedings of 6th International Coral Reef Symposium, vol 2, Townsville, pp 517–520

Shirley TC, Rützler K (2010) Effects of heat and salinity stress on the sponge *Cliona celata*. Int J Biol 2:3–16

Silverman J, Lazar B, Erez J (2007) Effect of aragonite saturation, temperature, and nutrients on the community calcification rate of a coral reef. J Geophys Res 112. doi:10.1029/2006JC003770

Sonnenholzner JI, Ladah LB, Lafferty KD (2009) Cascading effects of fishing on Galapagos rocky reef communities: reanalysis using corrected data. Mar Ecol Prog Ser 375:209–218

Sonnenholzner JI, Lafferty KD, Ladah LB (2011) Food webs and fishing affect parasitism of the sea urchin *Eucidaris galapagensis* in the Galápagos. Ecology 92:2276–2284

Spencer T (1992) Bioerosion and biogeomorphology. In: John DM, Hawkins SJ, Price JH (eds) Plant-animal interactions in the marine benthos. Clarendon Press, Oxford, pp 493–509

Stearn CW, Scoffin TP, Martindale W (1977) Calcium carbonate budget of a fringing reef on the west coast of Barbados Part I—Zonation and productivity. Bull Mar Sci 27:479–510

Tomlinson JT (1969) The burrowing barnacles (Cirripedia: order Acrothoracica). Bull US Natl Mus 296:1–162

Toro-Farmer GA (1998) Estimación de la intensidad bioerosionadora de los erizos (Echinodermata) en un arrecife coralino de la isla Gorgona-Pacífico colombiano. BSc thesis, Fac Cien, Univ Valle, Cali, Colombia

Toro-Farmer G, Cantera JR, Londoño-Cruz E, Orozco C, Neira R (2004) Patrones de distribución y tasas de bioerosión del erizo *Centrostephanus coronatus* (Diadematoida: Diadematidae), en el arrecife de Playa Blanca, Pacífico colombiano. Rev Biol Trop 52:67–76

Toth LT, Aronson RB, Vollmer SV, Hobbs JW, Urrego DH, Cheng H, Enochs IC, Combosch DJ, van Woesik R, Macintyre IG (2012) ENSO drove 2500-year collapse of eastern Pacific coral reefs. Science 337:81–84

Tribollet A (2008) Dissolution of dead corals by euendolithic microorganisms across the northern Great Barrier Reef (Australia). Microbial Ecol 55(4):569–580

Tribollet A, Golubic S (2005) Cross-shelf differences in the pattern and pace of bioerosion of experimental carbonate substrates exposed for 3 years on the northern Great Barrier Reef, Australia. Coral Reefs 24(3):422–434

Tribollet A, Golubic S (2011) Reef bioerosion: agents and processes. In: Dubinsky Z, Stambler N (eds) Coral Reefs: an ecosystem in transition. Springer, Berlin, pp 435–449

Tribollet A, Decherf G, Hutchings P, Peyrot-Clausade M (2002) Large-scale spatial variability in bioerosion of experimental coral substrates on the Great Barrier Reef (Australia): importance of microborers. Coral Reefs 21(4):424–432

Tribollet A, Godinot C, Atkinson, M Langdon C (2009) Effects of elevated PCO_2 on dissolution of coral carbonates by microbial euendoliths. Global Biogeochem Cy 23 GB3008. doi:10.1029/2008GB003286

Tribollet A, Radtke G, Golubic S (2011) Bioerosion. In: Reitner J, Thiel V (eds) Encyclopedia of geobiology. Springer, Berlin, pp 117–134

Tudhope AW, Risk MJ (1985) Rate of dissolution of carbonate sediments by microboring organisms, Davies Reef, Australia. J Sediment Res 55(3):440–447

Tunnicliffe V (1979) The role of boring sponges in coral fracture. In: Lévi C, Boury-Esnault N (eds) Biologie des spongiaires, vol 291. Coll Int CNRS, Paris, pp 309–315

Warme JE (1975) Borings as trace fossils, and the processes of marine bioerosion. In: Frey RW (ed) The study of trace fossils. Springer, Berlin, pp 181–227

Wellington GM, Glynn PW (2007) Responses of coral reefs to El Niño-Southern Oscillation sea-warming events. In: Aronson RB (ed) Geological approaches to coral reef ecology. Ecological studies 192. Springer, New York, pp 342–385

Williams JA, Margolis SV (1974) Sipunculid burrows in coral reefs: evidence for chemical and mechanical excavation. Pac Sci 28 (4):357–359

Williams AB, Ngoc-Ho N (1990) *Pomatogebia*, a new genus of thalassinidean shrimps from western hemisphere tropics (Crustacea: Upogebiidae). P Biol Soc Wash 103(3):614–616

Wisshak M, Schönberg CHL, Form A, Freiwald A (2012) Ocean acidification accelerates reef bioerosion. PLoS ONE 7(9):e45124

Wisshak M, Schönberg CH, Form AU, Freiwald A (2013) Effects of ocean acidification and global warming on reef bioerosion—Lessons from a clionaid sponge. Aquat Biol 19:111–127

Wulff JL (1997) Causes and consequences of differences in sponge diversity and abundance between the Caribbean and eastern Pacific of Panama. In: Proceedings of 8th International Coral Reef Symposium, vol 2, Panama, pp 1377–1382

Yonge CM (1955) Adaptation to rock boring in *Botula* and *Lithophaga* (Lamellibranchia, Mytilidae) with a discussion on the evolution of this habit. Q J Microsc Sci 3(35):383–410

Zapata FA, Morales YA (1997) Spatial and temporal patterns of fish diversity in a coral reef at Gorgona Island, Colombia. In: Proceedings of 8th International Coral Reef Symposium, vol 1, Panama, pp 1029–1034

Zea S (1993) Recruitment of demosponges (Porifera, Demospongiae) in rocky and coral reef habitats of Santa Marta, Colombian Caribbean. Mar Ecol 14(1):1–21

Zubia M, Peyrot-Clausade M (2001) Internal bioerosion of *Acropora formosa* in Réunion (Indian Ocean): microborer and macroborer activities. Oceanol Acta 24(3):251–262

Zullo VA (1991) Zoogeography of the shallow-water cirriped fauna of the Galápagos Islands and adjacent regions in the tropical eastern Pacific. In: James MJ (ed) Galápagos Marine Invertebrates. Taxonomy, biogeography, and evolution in Darwin's Islands. Plenum Press, New York, pp 173–192

Zundelevich A, Lazar B, Ilan M (2007) Chemical versus mechanical bioerosion of coral reefs by boring sponges—Lessons from *Pione* cf. *vastifica*. J Exp Biol 210:91–96

Diversity, Distribution and Stability of *Symbiodinium* in Reef Corals of the Eastern Tropical Pacific

13

Andrew C. Baker, Adrienne M.S. Correa, and Ross Cunning

Abstract

Coral reefs of the eastern tropical Pacific (ETP) are unique in being the only reef region in the world that has experienced multiple episodes of mass coral bleaching while also being exposed to chronically depressed aragonite saturation states as a result of regional upwelling. These characteristics make them ideal case studies for the continued effects of climate change on coral reefs, and in particular for the responses of reef corals and their dinoflagellate algal symbionts (*Symbiodinium* spp.) to combined climate stressors. As a result, the diversity, distribution and stability of *Symbiodinium* in ETP corals have been studied since the mid-1990s, shortly after contemporary molecular methods to identify *Symbiodinium* were first developed. ETP reefs have been instrumental in the discovery that certain members of *Symbiodinium* in clade D impart bleaching resistance to their coral hosts. In the ETP, clade D is represented by a single symbiont type (D1, also referred to as *S. glynni*), which has been shown to be tolerant of both episodic El Niño-driven high temperature stress (e.g., 1997 in the Gulf of Chiriquí, Panama), and low temperature stress during an unusually cold winter (e.g., 2008 in Baja California, Mexico). Virtually all studies in the region have focused on corals in the genus *Pocillopora*, which is both the dominant reef-building genus and the most symbiotically diverse, being the only coral genus in the ETP that routinely hosts heat tolerant D1 symbionts at high abundance. There is debate over the mechanisms by which D1 becomes dominant on pocilloporid reefs, with evidence for both differential mortality of corals, and dynamic change in symbiont communities in response to thermal history and/or disturbance. The relative importance of these two mechanisms is likely to be critical in determining reef survival trajectories over the coming century, as is the degree to which non-pocilloporid corals in the region can associate with, or become dominated by, D1. Dynamic change in symbiont communities may allow corals to survive recurrent bleaching events if stepwise increases in the abundance of D1 following each bleaching event allow these thermotolerant symbionts to accumulate. However, controlled experiments and continued monitoring are required to resolve the debate over symbiont stability versus lability, and there may be significant differences in coral response across the region. Finally, understanding whether different

A.C. Baker (✉) · A.M.S. Correa · R. Cunning
Department of Marine Biology and Ecology, Rosenstiel School of Marine and Atmospheric Science, University of Miami, 4600 Rickenbacker Causeway, Miami, FL 33149, USA
e-mail: abaker@rsmas.miami.edu

A.M.S. Correa
e-mail: adymscorrea@gmail.com

R. Cunning
e-mail: ross.cunning@gmail.com

A.C. Baker
Marine Program, Wildlife Conservation Society, 2300 Southern Boulevard, Bronx, NY 10460, USA

A.M.S. Correa
Biosciences at Rice, Rice University, 6100 Main Street, Houston, TX 77005, USA

R. Cunning
Hawaii Institute of Marine Biology, University of Hawaii, PO Box 1346, Kaneohe, HI 96744, USA

P.W. Glynn et al. (eds.), *Coral Reefs of the Eastern Tropical Pacific*, Coral Reefs of the World 8, DOI 10.1007/978-94-017-7499-4_13

Symbiodinium can alter coral response to high CO_2 or depressed aragonite saturation state (Ω_{arag}) remains a priority research area for the region, especially given the potential for strong interactions between Ω_{arag}, coral growth, and symbiont community structure.

Keywords
Symbiosis • Mutualism • Bleaching • Climate change • Microbial ecology

13.1 Introduction

13.1.1 The ETP as a Case Study for Studying Coral-Algal Symbiosis Under Climate Change

Coral reefs of the eastern tropical Pacific (ETP) are unique case studies for the long-term ecological effects of thermal variability on coral reefs, and particularly for the impacts of repetitive high temperature stress on reef coral symbioses. Due to the influence of the El Niño-Southern Oscillation (ENSO) on these reefs, they have experienced multiple episodes of severe coral "bleaching" (stress-induced expulsion of algal symbionts), which have led to coral mortality events followed by sustained periods of recovery. In addition, this region includes corals: (1) at the northern limit of their latitudinal distribution, with relatively cool winter temperatures and lower irradiance (Gulf of California, Mexico), (2) in nutrient-rich upwelling environments (Gulf of Panama, Panama), (3) in relatively warm, stable conditions (Gulf of Chiriquí, Panama), and (4) exposed to some of the lowest equatorial temperatures in the world (Galápagos, Ecuador). All of these circumstances result in a living laboratory for the investigation of temperature regime, and thermal extremes, on coral-algal symbioses, and represent unique systems for understanding the effects of climate change on coral reefs.

In addition, many coral reefs in the ETP, by virtue of their exposure to upwelling conditions, also experience chronically depressed aragonite saturation state (Ω_{arag}). These conditions reduce calcification rates, resulting in slower coral growth and decreased reef formation. These conditions, combined with the thermal stress history of the ETP, make these reefs ideal "space-for-time" case studies for coral reefs under projected climate change scenarios (Manzello et al. 2008). Although the implications of depressed Ω_{arag} for *Symbiodinium* diversity have not been explicitly investigated in the ETP, these conditions nevertheless underscore the relevance of these reefs for studying future climate change scenarios.

13.1.2 Marginal Environmental Conditions in the ETP

Prevailing environmental conditions in the eastern Pacific are marginal for reef development because of the: (1) narrow continental shelf on the western side of the Americas; (2) strong, permanent shallow thermocline; (3) relatively frequent occurrence of aerial exposure or mortality caused by extreme low tides, ENSO-related sea level drops and/or tectonic uplifts (Maté 2003); and (4) existence of dry and wet seasons (Dana 1975; Cortés 2003). During the wet season, many areas experience high cloud cover and intense rainfall, and the resulting terrestrial run-off leads to high turbidity, low light, and hyposaline conditions. In some areas, such as the Gulf of Panama (Panama), the Revillagigedo Islands, the Gulf of Tehuantepec (Mexico), and the northern coast of Costa Rica (Cortés and Jiménez 2003), high atmospheric pressure during the dry season pushes coastal surface waters offshore. This results in seasonal upwelling, which bathes associated reefs in relatively cold, nutrient-rich waters for part of the year. These conditions challenge coral reef growth and survival in several ways. First, temperatures may be relatively low compared to the average tolerance of coral-algal symbioses, sometimes resulting in cold-water bleaching events (Fig. 13.1). Second, coral animals are most competitive against filter feeders and macroalgae under low nutrient conditions; the nutrient-rich conditions that accompany upwelling can increase the competitive interactions experienced by corals on the benthos (Glynn 1996). Third, upwelled waters are relatively pCO_2-enriched and low in pH and aragonite saturation state (Ω_{arag}), making calcification more difficult on ETP reefs than in other areas, such as the Caribbean. The low Ω_{arag} and high nutrient concentrations associated with upwelled waters limit early marine cementation of reef framework and/or stimulate bioerosion (Manzello et al. 2008). Consequently, ETP reefs can be viewed as proxies for reefs of the future under various ocean acidification (OA) scenarios predicted to occur in the world's oceans due to increasing atmospheric pCO_2 (IPCC 2007).

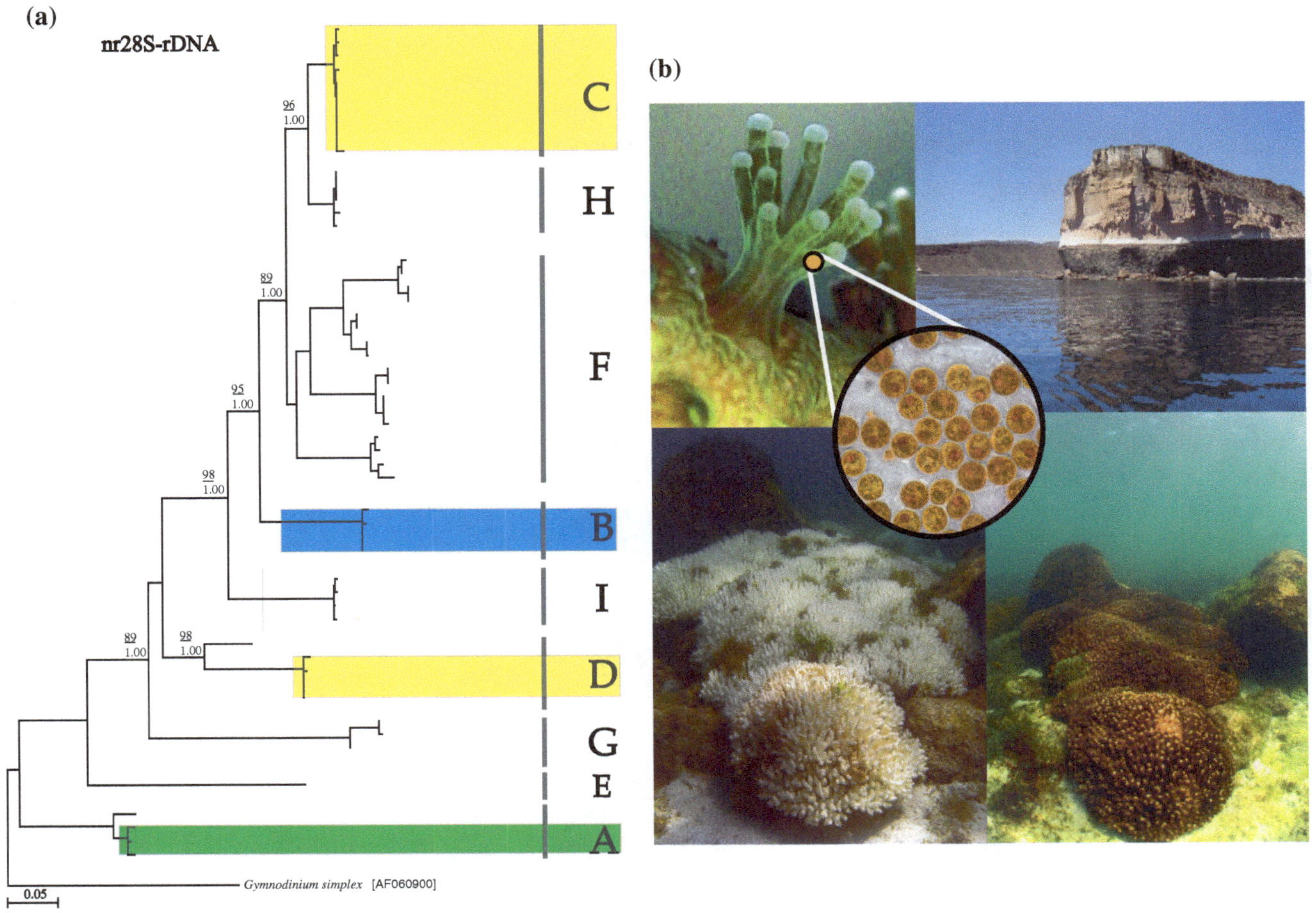

Fig. 13.1 Diversity of algal symbionts (*Symbiodinium* spp.) in reef corals of the ETP and their importance in understanding coral reef ecology in the region. **a** Molecular phylogeny of *Symbiodinium* based on nuclear 28S (large subunit) ribosomal DNA (nr28S-rDNA). Shown are 9 clades (A-I), of which four have been found in scleractinian corals of the ETP. Members of clades C and D (*yellow*) are the most common symbionts in the region, with *Symbiodinium* D1 frequently occurring in *Pocillopora*. Members of clades A (*green*) and B (*blue*) have been found at low abundance in ETP scleractinians using high resolution molecular techniques, but B (*blue*) has also been found at high abundance in *Pocillopora* recovering from cold-water bleaching. Modified from Pochon and Gates (2010). **b** Spatial relationships in coral-algal symbiosis at the cellular, polyp, and reef-scale. Pocilloporid reef frameworks, shown here bleaching in response to cold temperatures, can show dramatic patterns of bleaching in response to temperature extremes. From McGinley et al. (2012)

Upwelling also affects sea surface temperatures (SSTs) across the ETP region. In areas that experience seasonal upwelling, temperature can drop by as much as 10 °C from the wet to the dry season (Manzello et al. 2008; see Chap. 8, Glynn et al.). In contrast, areas of the ETP that do not experience upwelling are more thermally stable throughout the year, although shoaling thermoclines and other effects may also result from seasonal weather patterns. In addition to this thermal variability over seasonal timescales, some regions of the ETP, including the west coast of the Baja California Peninsula (Mexico), Panama, and Colombia, are also subject to the El Niño-Southern Oscillation (ENSO). During intense ENSO events, SST temperatures can increase by 2–3 °C and upwelling is reduced, resulting in significant thermal stress and potentially catastrophic mass bleaching events (Glynn 1984). Mass bleaching in the ETP was associated with ENSO events occurring in 1982–83 (Glynn and Colgan 1992), and 1997–98, with coral mortality following the 1982-83 event being significantly higher than the 1997–98 event (Glynn et al. 2001a, b). This can likely be attributed to both differential survival and reproduction of temperature tolerant genotypes during the 1982–83 event, as well as symbiont community shifts within individual colonies in favor of thermally tolerant members of *Symbiodinium* clade D (Baker et al. 2004; LaJeunesse et al. 2010a, see also Sect. 13.4). Although ENSO events may ultimately lead to the collapse of ETP reefs under predicted climate scenarios for the current century (Toth et al. 2012), the unusual environmental and hydrodynamic settings that characterize this region provide a unique opportunity to examine the prevalence of thermally tolerant symbionts before, during, and after environmental stress events, and how these patterns are affected by differences in aragonite saturation state between reefs.

13.1.3 Isolation of ETP Reefs and Its Implications for the Study of *Symbiodinium*

In addition to the difficult environmental conditions encountered by corals in the ETP, they also face significant challenges colonizing the region from elsewhere. Coral reefs of the ETP are depauperate and have been isolated from the Caribbean Sea since the closure of the Isthmus of Panama about 3.7–3.0 Ma (Duque-Caro 1990; Coates and Obando 1996). Although occupying the same ocean basin as the center of global coral diversity (the 'Coral Triangle'), migration from the Indo-west Pacific requires extreme larval dispersal (5000–8000 km, Dana 1975; Grigg and Hey 1992). The eastern Pacific 'barrier' is formed by strong currents running from east to west between the Coral Triangle and the ETP (see Chap. 3, Fiedler and Lavín) and, if not 'impassable' (Baums et al. 2012; see Chap. 16, Lessios and Baums), has long been considered the world's most effective marine barrier to larval dispersal (Ekman 1953). This isolation has led to a distinct eastern Pacific reef-building coral fauna, with some endemism (e.g., *Pavona chiriquiensis*, Glynn et al. 2001a; see also Chap. 2, López-Pérez). Thirty-six stony coral species are found in the ETP (Glynn and Ault 2000; see Chap. 5, Glynn et al.), which contrasts starkly with the hundreds of species that can be found in the Indo-west Pacific. *Pocillopora*, the main reef-building genus of the region, typically dominates shallow-water ETP reefs. In some areas, reef framework development is minimal and only coral communities exist, often dominated by massive colonies in the genera *Pavona* and *Porites*. ETP coral species exist at the limits of their respective geographic ranges, which may subject these populations to reduced gene flow, and provide opportunities for speciation. The relatively depauperate nature of these reefs from both a coral host and an algal symbiont perspective also provides an opportunity to understand symbiont specificity and co-evolution (Pinzón and LaJeunesse 2011), and to study niche differentiation (Iglesias-Prieto et al. 2004), as well as ecological interactions and responses among corals and their *Symbiodinium* communities (Glynn et al. 2001a, b; Baker et al. 2004; LaJeunesse et al. 2008).

13.2 *Symbiodinium* Diversity in the ETP

13.2.1 Overview of Diversity in *Symbiodinium*

Modern investigations of diversity in *Symbiodinium* use molecular genetic methods to survey *Symbiodinium* and document its distribution and dynamics. Currently, the genus *Symbiodinium* consists of nine major clades (A-I, Pochon and Gates 2010), most of which are characterized by significant additional diversity at finer taxonomic scales (LaJeunesse 2005), leading to the description of a variety of new species (e.g., LaJeunesse et al. 2010a). Although no formal taxonomic revisions of the genus have yet been undertaken, the diversity present within this genus strongly supports its subdivision into multiple new genera.

Early investigations used variation in the small subunit ribosomal DNA (rDNA) to classify *Symbiodinium* into clades, and distinguish some variation within these clades (Rowan and Powers 1991a, b). These methods were then used to study the distribution and dynamics of *Symbiodinium* in Caribbean *Montastraea (Orbicella)* (Rowan and Knowlton 1995; Rowan et al. 1997). Shortly afterwards, using large subunit (LSU) rDNA, these methods were applied to ETP reefs to study symbiont diversity, distribution, and dynamics with a higher degree of taxonomic resolution (Baker and Rowan 1997; Baker 1999; Glynn et al. 2001a, b). Due to the relatively rapid adoption of these methods in the region, ETP reefs are represented by some of the longest running datasets on *Symbiodinium* in existence (dating to 1995 in Panama). Subsequent investigations of symbiont diversity applied even greater taxonomic resolution by analyzing the Internal Transcribed Spacer-2 (ITS-2) region of rDNA (LaJeunesse et al. 2008, 2010a, b), and more recent studies are beginning to reveal even finer-scale genetic variation among *Symbiodinium* types through the use of the non-coding region of the psbA minicircle in *Symbiodinium* chloroplasts ($psbA^{ncr}$), which resolves more diversity than the ITS-2 (Pinzón and LaJeunesse 2011), while also reducing the problems associated with the latter marker's high copy number and intragenomic variation (LaJeunesse and Thornhill 2011). These methods have greatly increased our understanding of *Symbiodinium* distributions in the ETP and elsewhere, but they do preclude comparisons with earlier datasets for all but the highest (typically clade-level) diagnoses. Additionally, microsatellite markers can further differentiate among *Symbiodinium* genotypes, with ~20–30 distinct *Symbiodinium* D1 genotypes identified at individual reefs in the Gulf of California (Pettay et al. 2011). Coral colonies can host single or multiple *Symbiodinium* genotypes simultaneously, and these associations can change over time. Particular *Symbiodinium* D1 genotypes may also be adapted to specific local environmental regimes (Pettay and LaJeunesse 2013). Parsing out this fine-scale genetic diversity is an important ongoing research objective in understanding *Symbiodinium* diversity and ecology.

Recently, higher detection resolution using quantitative PCR (qPCR) of both ITS-2 (LaJeunesse et al. 2010a) and the actin gene (Cunning and Baker 2013) has been applied to *Symbiodinium* communities in ETP hosts. Although these methods have greatly increased our ability to detect, distinguish, and quantify different *Symbiodinium*, they are also subject to constraints that can limit data interpretation (see Sect. 13.2.2). Next-generation sequencing may further advance these objectives; high-throughput ITS amplicon sequencing has been applied to resolve fungal community diversity (Jumpponen and Jones 2009), and is beginning to be applied to *Symbiodinium* (Kenkel et al. 2013; Green et al. 2014), although no studies to date have utilized these approaches to characterize *Symbiodinium* communities in the ETP.

Globally, scleractinian corals have been observed to host *Symbiodinium* in clades A, B, C, D, and less commonly, clades F and G (Baker 2003; van Oppen et al. 2009; Franklin et al. 2012). In the Pacific, most stony corals tend to host members of clades C and/or D (Baker and Rowan 1997; LaJeunesse et al. 2003), with clades A, B, F and G being only rarely found (e.g., Darius et al. 1998, 2000; Rodriguez-Lanetty et al. 2002; LaJeunesse et al. 2010b). *Symbiodinium* diversity in the ETP has mainly been assessed along the coast of Panama (Glynn et al. 2001a, b; Baker et al. 2004), in the Gulf of California, Mexico (LaJeunesse et al. 2008) and in the Galápagos Islands, Ecuador (Baker 1999; Pinzón and LaJeunesse 2011), as well as on Clipperton Atoll (Pinzón and LaJeunesse 2011, Pettay and LaJeunesse 2013). In these regions, *Symbiodinium* in clade C most commonly dominates scleractinian coral colonies, except in the genus *Pocillopora,* which commonly also hosts *Symbiodinium* in clade D, often at high abundance. Despite relatively higher diversity in clade D in the Indo-west Pacific, clade D in the ETP appears to be solely represented by a single "type" of *Symbiodinium*, D1, also referred to as *S. glynni* (LaJeunesse et al. 2010a). *Symbiodinium* B1 has also been observed in recovering branch tips of *Pocillopora* (LaJeunesse et al. 2010a). Using rt-PCR assays, Silverstein et al. (2012) detected members of clades A and B at low abundance in a variety of scleractinian corals from Panama and the Galápagos, but made no attempt to detect low abundance *Symbiodinium* in clades F and G, which have not yet been reported for the ETP. Although the dominant symbiont(s) within coral species are predictable to some extent, based on coral species, geographic location, and environment (see Chap. 14, Pinzón), the relative abundance of different symbionts can shift in response to seasonal cues or environmental disturbances, both acute and chronic (see Sect. 13.3.2). Since *Symbiodinium* genotypes can influence colony health and survival, these processes are key to understanding the potential trajectories of ETP reefs over the coming century.

13.2.2 Methodological Constraints in Detecting *Symbiodinium*

Symbiodinium diversity has been assessed from corals in the ETP using a variety of molecular methods including Restriction Fragment Length Polymorphisms (RFLP), Denaturing Gradient Gel Electrophoresis (DGGE), direct sequencing, and real-time PCR (rt-PCR) (Table 13.1). Markers that have been used in conjunction with these tools are multi-copy and include various regions of the ribosomal RNA gene, a non-coding region of the chloroplast psbA gene, as well as microsatellite loci, and the actin gene. These molecular approaches have been pivotal in understanding coral-algal symbioses because they revealed high genetic diversity within *Symbiodinium* (Stat et al. 2012), and previously unrecognized complexity in coral-algal associations. The sensitivity and taxonomic resolution of molecular assays applied to *Symbiodinium* communities have increased with time. Given this, comparisons of studies employing molecular assays of varied resolution (e.g., targeting clades, types, or sub-types) should be made conservatively and explicitly detail the approaches and markers involved. High-resolution approaches consistently report one or more symbiont variants present at low abundance within at least some surveyed host individuals, in addition to numerically dominant symbiont types (e.g., Mieog et al. 2007; Correa et al. 2009; McGinley et al. 2012; Silverstein et al. 2012; Cunning and Baker 2013; Thornhill et al. 2013). These detections must be interpreted carefully, given that *Symbiodinium* variants likely differ in terms of their impact (e.g., amount of photosynthates released, residence time) on the holobiont (Hill and Hill 2012), and that some symbionts detected at low abundance may represent ephemeral endosymbioses, transients in the coral gastrovascular cavity, or colony surface contaminants (Silverstein et al. 2012). Nevertheless, to obtain a complete understanding of coral-algal associations, both the numerically dominant symbionts, and those found at low abundance, should be characterized. Future research will confirm which (if any) low abundance symbionts impact holobiont physiology and ecology, and the time scales and contexts over which this may occur. Some of the major limitations in the detection of *Symbiodinium* using molecular techniques are briefly discussed below; for comprehensive discussion of technical details, the reader is referred to the individual papers cited.

All of the molecular approaches that have been applied to coral-algal symbioses in the ETP (Table 13.1) are potentially susceptible to PCR bias. By preferentially amplifying some *Symbiodinium* genetic sequence variants over others, PCR bias can skew assessments of the relative abundance of different symbiont types within individual hosts. Additionally, if some *Symbiodinium* types (e.g., based on ITS-2) contain a high copy number for a given gene relative to other types, this could also produce a bias by artificially inflating detections of a given symbiont variant.

Community fingerprinting techniques, such as RFLP and DGGE, which utilize conventional PCR amplicons, require caution in assessing the diversity of coral-algal symbioses because they do not necessarily detect community members present at low relative abundances (e.g., <5–20 % of the

Table 13.1 Summary of the molecular tools and markers that have been used to assess *Symbiodinium* diversity in the eastern tropical Pacific (ETP)

Molecular approach/marker	Host taxa	Country	Citation
RFLP of LSU 28S rRNA	*Gardineroseris, Pocillopora, Pavona, Porites, Psammocora*	Panama, Galápagos (Ecuador)	Glynn et al. (2001a)
	Pocillopora damicornis, Pocillopora elegans	Panama	Baker et al. (2004)
DGGE of ITS-2	*Pocillopora elegans (verrucosa), Pavona gigantea*	Mexico	Iglesias-Prieto et al. (2004)
	Pocillopora	Mexico	LaJeunesse et al. (2007)
	Pavona, Pocillopora, Porites, Psammocora	Mexico	LaJeunesse et al. (2008)
	Porites lobata	Galápagos (Ecuador)	Glynn et al. (2009)
	Pocillopora	Mexico, Clipperton Atoll, Galápagos (Ecuador), Panama	LaJeunesse et al. (2010a); Pinzón and LaJeunesse (2011)
	Pocillopora	Mexico	McGinley et al. (2012)
psbA non-coding region	*Pocillopora*	Mexico, Clipperton Atoll, Galápagos (Ecuador), Panama, Gulf of California	Pinzón and LaJeunesse (2011)
Microsatellites	*Pocillopora*	Mexico, Clipperton Atoll, Galápagos (Ecuador), Panama	LaJeunesse et al. (2010a)
	Pocillopora	Mexico	Pettay et al. (2011)
	Pocillopora	Mexico, Clipperton Atoll, Galápagos (Ecuador), Panama, Gulf of California	Pettay and LaJeunesse (2013)
rt-PCR of Domain 2 of the LSU 28S rRNA[a]	*Gardineroseris, Pavona clavus, Pavona gigantea, Pavona varians, Pocillopora damicornis, Porites panamensis, Psammocora superficialis*	Panama, Galápagos (Ecuador)	Silverstein et al. (2012)
	Pocillopora	Mexico	McGinley et al. (2012)
rt-PCR of actin[a]	*Pocillopora damicornis*	Panama	Cunning and Baker (2013)
	Pocillopora	Panama, Galápagos (Ecuador)	Cunning et al. (2013)

[a]Indicates a high-sensitivity molecular approach

total community, Lien et al. 2007; Loram et al. 2007). In a relatively extreme case, DGGE was shown to detect *Symbiodinium* ITS-2 type D1 only when it comprised at least 10–30 % of the total community (LaJeunesse et al. 2008). In scleractinian corals, which typically host 1–2 million *Symbiodinium* cells per cm^2 of coral surface tissue (Drew 1972), diverse symbionts might therefore be present at densities well in excess of 100,000 cells per cm^2, yet remain undetected using these approaches.

The problem of failed detections is further exemplified, for the ETP in particular, in comparisons of RFLP and rt-PCR data from colonies of *Pocillopora*. Correa (2009) showed that 96 % (n = 204 of 213) of samples contained both C and D symbionts based on rt-PCR, but these mixed communities were detected in only 12 % (n = 26 of 213) of samples analyzed using RFLPs. This problem occurred even in samples that contained relatively even abundances of C and D symbionts (e.g., each clade comprised ~30 % or more of the total community). The overall RFLP type II error rate (178 errors in 204 assays) suggests that earlier studies (Glynn et al. 2001a, b; Baker et al. 2004) identifying *Symbiodinium* based on LSU rDNA may have underestimated the presence of mixed symbiont clades by nearly 90 % (Correa 2009). This suggests that: (1) RFLP type II errors in published datasets are likely more pervasive than previously inferred (e.g., Baker and Romanski 2007); and (2) some abundant symbiont variants have likely been 'hidden in plain sight' by early RFLP analysis, at least in ETP *Pocillopora*.

Real-time PCR provides a 1000–10,000-fold increase in detection sensitivity over conventional PCR methods (Mieog et al. 2007). Such high-resolution techniques present their own caveats in assessing *Symbiodinium* diversity *in hospite*, however. For example, rt-PCR can potentially detect *Symbiodinium* cells present as surface contaminants on colonies or within the guts of sampled hosts, calling into question the biological relevance of high C_T value amplifications. Silverstein et al. (2012) applied a *Symbiodinium*-specific rt-PCR assay (Correa et al. 2009) to the azooxanthellate coral *Tubastraea coccinea*; only 2 % (1 out of 52) of the analyzed samples harbored sufficient surface contaminants or recently ingested cells to produce a positive detection for *Symbiodinium*. Although this 'biological negative control' does not exactly estimate false positive *Symbiodinium* detections for zooxanthellate corals, it does suggest that the vast majority of positive *Symbiodinium* detections using rt-PCR represent *Symbiodinium* that are not surface or gut contaminants. Nevertheless, careful use of biological and technical negative controls is important to establish appropriate detection thresholds using rt-PCR. Another constraint of rt-PCR is that detection is limited to the specific symbiont taxa queried by user-developed primers and probes, which are based on known diversity and have, to date, targeted relatively coarse levels of taxonomic resolution (i.e., clades, Correa et al. 2009; Cunning and Baker 2013). Clade-level detection is unable to resolve sub-clade diversity, which may correlate with important functional variation (see Sect. 13.2.3). Fortunately, rt-PCR assays can be developed to detect and quantify any target sequence variant (e.g., Cunning and Baker 2013), facilitating its application at any scale of taxonomic resolution.

Another advantage of rt-PCR analysis is its ability to quantify the abundance and/or proportion of symbiont types in mixed communities, including those that contain low abundance symbionts, enabling more detailed studies of dynamism and shuffling in symbiont communities over time. Absolute quantitation of different *Symbiodinium* variants remains difficult due to the multi-copy nature of the markers currently in use (Mieog et al. 2007) and unsatisfactory parameters for standardizing symbiont numbers (e.g., per cm^2 or ng DNA). These issues have been addressed via the development of markers with lower copy number (e.g., actin) normalized to host cell numbers (Mieog et al. 2009; Cunning and Baker 2013). However, the copy number of these markers may still vary among taxa, and must therefore be quantified for each *Symbiodinium* type studied. Normalizing symbionts to host cells (e.g., S/H cell ratio in Mieog et al. 2009) provides a metric of symbiont abundance (i.e., density) that relative ratios do not, but these metrics must still be interpreted carefully due to the dynamic nature of host cells (Cunning and Baker 2014).

The most informative snapshot of a *Symbiodinium* community may be obtained by utilizing a combination of these approaches to attain both high sensitivity detection and fine-scale taxonomic resolution, along with some quantitative information. Although this may be achieved by rt-PCR assays that target specific *Symbiodinium* types, next-generation sequencing is poised to provide another step forward in allowing investigators to comprehensively and quantitatively assess *Symbiodinium* community structure (Kenkel et al. 2013; Green et al. 2014), but this has yet to be applied to the ETP.

13.2.3 Functional Differences in *Symbiodinium*

Different *Symbiodinium* taxa can have distinct physiological traits (Iglesias-Prieto and Trench 1994, 1997; Warner et al. 1999; Savage et al. 2002; Goulet et al. 2005; Loram et al. 2007), causing them to be differentially suited to varied environmental conditions. Based on these differences, they are often predictably distributed along gradients of light and temperature across reefs (Fabricius et al. 2004; Iglesias-Prieto et al. 2004) and even within single coral colonies (Rowan et al. 1997; Kemp et al. 2008). This intraspecific (and intracolonial) symbiont diversity can influence holobiont physiology in a number of ways.

One of the clearest differences among symbionts, at least from the perspective of their coral hosts, is that some *Symbiodinium*, particularly members of clade D, are heat-tolerant (Rowan 2004) and confer increased resistance to thermal bleaching (Rowan et al. 1997; Glynn et al. 2001b; Berkelmans and van Oppen 2006; LaJeunesse et al. 2008). In the ETP, *Pocillopora* colonies dominated by D1 symbionts are resistant to heat-induced bleaching compared to those dominated by clade C (Glynn et al. 2001b). D1 symbionts also protect corals from bleaching events triggered by high irradiance (LaJeunesse et al. 2007) and cold stress (LaJeunesse et al. 2010a; McGinley et al. 2012). The higher incidence of bleaching in shallow water, and the higher irradiance at these depths, may thus explain the depth gradient in D1 characterizing Panamanian *Pocillopora* (Fig. 13.2, see also Sect. 13.3.2). These thermal differences may also occur at finer taxonomic scales. Sampayo et al. (2008) showed that *Stylophora pistillata* hosting C78 and C8/a on the southern Great Barrier Reef (GBR) was more thermally tolerant than those hosting C79 and C35/a, suggesting that fine-scale differences in symbiont type might similarly affect physiological outcomes in ETP corals. Different multilocus genotypes within *Symbiodinium* D1 also show distinct geographic partitioning in the ETP even though their hosts are widely distributed, suggesting that certain symbiont clones may be functionally distinct (Pettay and LaJeunesse 2013). Other more subtle aspects of coral physiology may also be impacted by symbiont type. DeSalvo et al. (2010) showed that gene expression profiles in the host are highly dependent on the genetic identity of the symbiont, suggesting that symbionts may modulate host physiology in complex ways.

In addition to bleaching susceptibility, other aspects of coral fitness may be dependent on the algal taxa they host, including the transfer of photosynthetically fixed carbon to the host (Loram et al. 2007; Cantin et al. 2009), and coral growth rates.

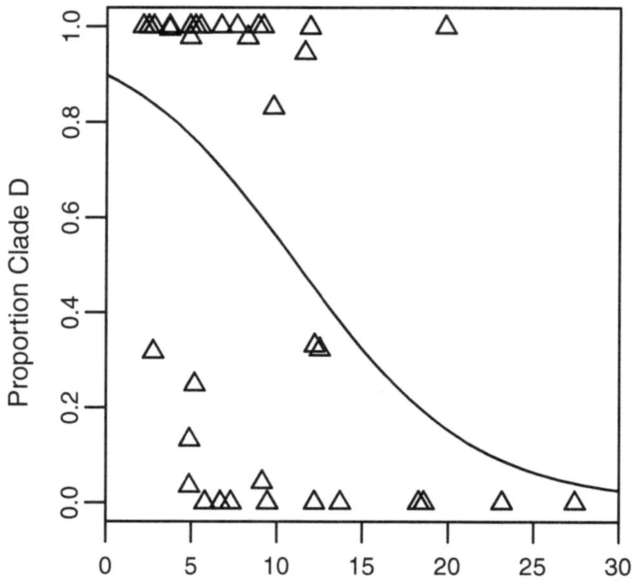

Fig. 13.2 Proportion of clade D in *Pocillopora* colonies with clade C and/or D *Symbiodinium* across a depth gradient at Uva Island, Panama (n = 44). A quasibinomial generalized linear model fit indicates that the proportion of clade D declines with depth (p = 0.00693)

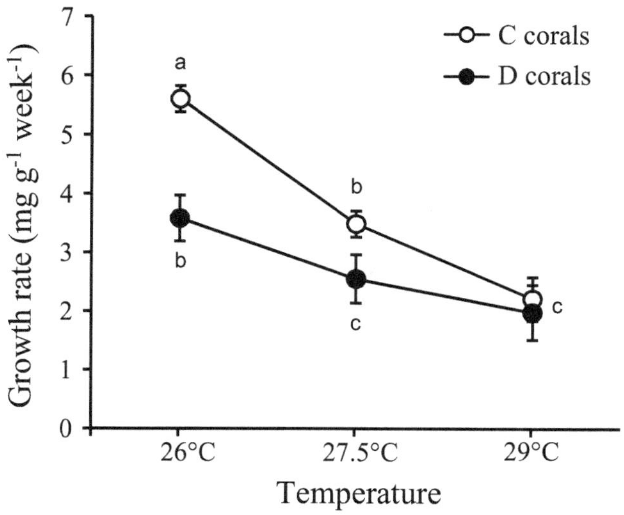

Fig. 13.3 Growth rates of *Pocillopora damicornis* fragments dominated by either *Symbiodinium* in clade C and D corals grown at different temperatures over 55 weeks. Sample sizes for 26, 27.5, and 29 °C treatments were n = 177, 164, and 62 for clade C fragments, and n = 72, 66, and 50 for clade D fragments. *Error bars* represent SEM. Group means that do not share a letter are significantly different ($p < 0.05$). From Cunning et al. (2014)

Little et al. (2004) found that juvenile *Acropora millepora* and *Acropora tenuis* on the GBR that were infected by *Symbiodinium* in clade D grew 2–3 times more slowly than juveniles infected with members of clade C, as measured by the rate of addition of new polyps over the first 6 months (Little et al. 2004). This growth differential was also found in adult corals of *A. millepora* containing clade D, which grew 29 % more slowly than conspecifics with clade C in controlled experimental conditions, and 38 % more slowly in the field (Jones and Berkelmans 2010). However, studies of *Pocillopora* in the ETP found that these tradeoffs were temperature-dependent, with increasing temperatures lessening, and eventually eliminating, the growth reduction among corals hosting clade D (Fig. 13.3, Cunning et al. 2014). These findings suggest a more nuanced view in which different symbionts vary in their environmental optima, which in turn influences rates of carbon translocation and the eventual growth of coral hosts (Cunning et al. 2015).

13.3 Distribution of *Symbiodinium* in the ETP

13.3.1 Host Systematic Distribution

Most coral species in the ETP are dominated by closely related *Symbiodinium* in clade C that include ITS-2 types C1c, C1b-c, C1f, C1d and C1ee (Table 13.2; LaJeunesse et al. 2008). In addition, some coral species can be found dominated by variants of C66 or C75 (*Porites panamensis*) or D1 (*Pocillopora* spp.). Only one non-pocilloporid has been found containing high abundance of D1, a single colony of *Porites lobata* collected shortly after the 1997–98 bleaching event at Uva Island (Baker 1999). Pocilloporid corals in the ETP can be dominated by a variety of *Symbiodinium* in clade C (including types C1b-c, C1d, and C1ee), D1 (LaJeunesse et al. 2008, 2010a, b; Cunning et al. 2013), and even B1, albeit temporarily (LaJeunesse et al. 2010a). Mixtures of D1 with clade C-types have been detected by real-time PCR in over half of examined *Pocillopora* colonies, but typically involve dominance of one symbiont type with low background levels of one or more additional types (Fig. 13.4; Correa 2009; McGinley et al. 2012; Cunning and Baker 2013). Although the proportions of *Pocillopora* dominated by members of clade C or D may vary among locations, both holobiont combinations are common throughout the region, and readily co-occur across reefscapes (Fig. 13.5; Glynn et al. 2001b; LaJeunesse et al. 2008). Although species boundaries within the genus *Pocillopora* are not clearly established (see Chap. 14, Pinzón), there may be some patterns in the distribution of *Symbiodinium* types among host taxa. Pinzón and LaJeunesse (2011) found that different lineages of *Pocillopora* hosted distinct types of *Symbiodinium* within clade C, with "type 1" *Pocillopora* hosting *Symbiodinium* C1b-c, "type 2" hosting C1ee, and "type 3" hosting C1d. In addition, they suggested that only type 1 *Pocillopora* was able to associate with the stress tolerant *Symbiodinium* D1 based on DGGE fingerprinting of the ITS-2 region. However, additional sampling in a subsequent study

Table 13.2 Diversity of *Symbiodinium* (based on conventional and high-resolution molecular approaches) in scleractinian corals in the ETP

Species	Location	Clades	Type(s)	References
Gardineroseris planulata	PAN	A, C, D		Silverstein et al. (2012)
Pavona gigantea	PAN	A, B, C, D		Silverstein et al. (2012)
	GAL	C, D		Silverstein et al. (2012)
	GOC	C	C1c	LaJeunesse et al. (2008)
Pavona clavus	PAN	A, B, C, D		Silverstein et al. (2012)
	GAL	A, C, D		Silverstein et al. (2012)
	GOC	C	C1c	LaJeunesse et al. (2008)
Pavona varians	PAN	C, D		Silverstein et al. (2012)
Pocillopora spp.[a]	GAL	C, D	C1b-c, D1	Silverstein et al. (2012), Cunning et al. (2013)
	PAN	C, D	C1b-c, C1d, D1	LaJeunesse et al. (2010a, b), Cunning and Baker (2013), Cunning et al. (2013)
	MEX	D	D1	Baker (1999), LaJeunesse et al. (2010a)
	GOC	B, C, D	B1, C1b-c, D1	LaJeunesse et al. (2008, 2010)
	PAN	C, D	C1b-c, C1d, D1	LaJeunesse et al. (2010a, b)
	RVG	C	C1b-c	LaJeunesse et al. (2010a, b)
	CLP	C, D	C1b-c, C1ee, D1	LaJeunesse et al. (2010a, b)
Porites panamensis	PAN	A, B, C, D		Silverstein et al. (2012)
	GOC	C	C1, C66, C66a, C66b, C75	LaJeunesse et al. (2008)
Porites lobata	PAN	C, D		Glynn et al. (2001a), Baker (1999)
Psammocora superficialis	PAN	C, D		Silverstein et al. (2012)
	GOC	C	C1f	LaJeunesse et al. (2008)
P. brighami	GOC	C	C1f	LaJeunesse et al. (2008)
P. profundacella	GOC	C	C1f	LaJeunesse et al. (2008)
P. stellata	GOC	C	C1f	LaJeunesse et al. (2008)

GAL Galápagos, Ecuador; *GOC* Gulf of California, Mexico; *MEX* Mainland Mexico; *PAN* Panama; *RVG* Revillagigedo Islands; *CLP* Clipperton Atoll

[a]Due to taxonomic uncertainty in this genus, all *Pocillopora* species are grouped by genus

revealed that both type 1 and 3 *Pocillopora* in Panama and the Galápagos routinely associated with *Symbiodinium* in clade D (based on rt-PCR of the actin gene, Cunning et al. 2013), suggesting that certain host-symbiont combinations that were not previously identified may have restricted or patchy distributions. Branch tips of *Pocillopora* in the ETP have also been found to be dominated by *Symbiodinium* B1 (a common symbiont of the anemone *Aiptasia*) during the early stages of recovery from cold water bleaching. However, these symbionts were no longer detected in the corals once fully recovered (LaJeunesse et al. 2010a), suggesting these symbionts may have opportunistically colonized bleaching tissue. Nevertheless, these various results indicate that *Pocillopora* in the ETP can form associations with diverse symbionts in at least three clades of *Symbiodinium*.

Other coral species in the ETP have not been studied as thoroughly as *Pocillopora*, but studies using high-resolution rt-PCR indicate that they also associate with multiple *Symbiodinium* clades (Table 13.2), although they are rarely (if ever) dominated by symbionts other than clade C. Surveys of symbiont diversity in other scleractinian species on Panamanian reefs have detected both clade C and clade D *Symbiodinium* in association with *Gardineroseris planulata*, *Pavona clavus*, *Pavona gigantea*, and *Porites lobata, Porites panamensis, Pavona varians,* and *Psammocora superficialis* (Silverstein et al. 2012; Cunning et al., unpub data). Members of clade A *Symbiodinium* have also been found in Panamanian *P. gigantea, P. clavus, G. planulata,* and *P. panamensis*, and clade B has been found in *P. gigantea, P. clavus,* and *P. panamensis*. Together, these findings suggest that the total diversity of *Symbiodinium* present in corals of the ETP (and elsewhere) extends beyond the dominant or most abundant symbionts typically characterized to date, and likely includes members of one or more additional clades present at low densities.

In addition to the symbionts of scleractinian corals on ETP reefs, additional zooxanthellate non-scleractinians also host a variety of *Symbiodinium* that add to the total pool of symbionts

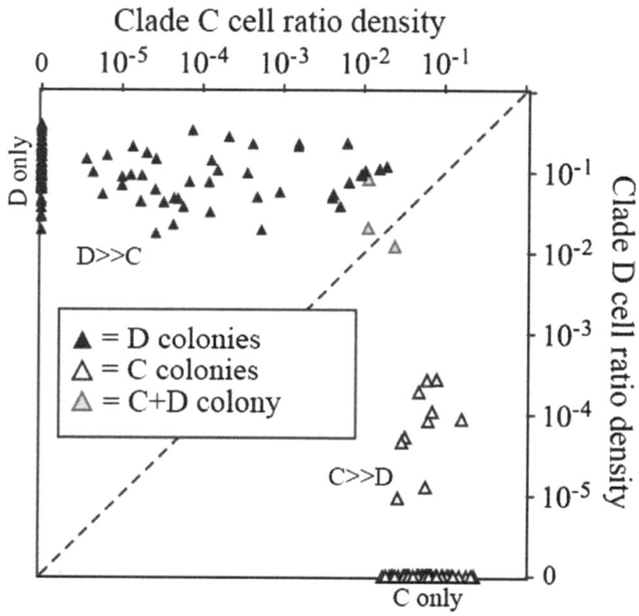

Fig. 13.4 Community structure of *Symbiodinium* in *Pocillopora* from Panama. Cell ratio densities (symbiont to host cell ratios) of clade C and D symbionts show a high incidence of mixed communities (64.1 % of colonies) that are heavily dominated by one clade, with background populations of a second clade. *Open triangles* represent colonies categorized as C-dominated (99.6–100 % clade C), and filled triangles are D-dominated (87.6–100 % clade D). *Dashed line* represents equal amounts of clades C and D in a sample. From Cunning and Baker (2013)

that occur in the region. As mentioned previously, the anemone *Aiptasia* hosts *Symbiodinium* B1, while anemones in the genus *Isoaulactinia* (=*Bunodactis*) and the zoanthid *Zoanthus pacificus* have both been found to host C66 (which is also found in the scleractinian *Porites panamensis*). *Zoanthus pacificus* has also been found hosting C29 and A12 (LaJeunesse et al. 2008). In contrast, zoanthids in the genera *Palythoa* and *Protopalythoa* host C1 and C1o, respectively (LaJeunesse et al. 2008), while the hydrozoan fire coral *Millepora intricata* hosts members of clade A (Baker 1999).

13.3.2 Environmental Control of *Symbiodinium* Distribution

ETP reefs show high variability in both temperature and water clarity, as well as large diurnal changes in irradiance due to extreme tidal ranges in some locations, such as Pacific Panama. While *Pocillopora* dominated by D1 or C1b-c co-occur throughout the ETP (with the exception of mainland Mexico, including Banderas Bay and Oaxaca, where only D1 has been reported, Baker 1999; LaJeunesse et al. 2010a), their relative dominance does appear to be related to environmental conditions at each location. High temperatures have been associated with clade D *Symbiodinium* (including D1 in the ETP), and areas that routinely experience high temperatures show high relative abundance of these symbionts (Baker 2003; Baker et al. 2004; Fabricius et al. 2004; Berkelmans and van Oppen 2006). Therefore, while some members of clade C may tolerate high seasonal temperature variability (e.g., C1b-c, LaJeunesse et al. 2010a), their sensitivity to extreme high temperatures, combined with the high thermal tolerance of D1, has resulted in dominance by D1 in areas that experienced severe thermal stress during El Niño bleaching events (LaJeunesse et al. 2010a; Cunning et al. 2013).

Depth-related trends in the distribution of *Symbiodinium* types in the Gulf of California (Iglesias-Prieto et al. 2004) suggest that stress-tolerant *Symbiodinium* D1 perform better in shallow, high light environments, whereas the more sensitive C1c performed better in deeper, low light environments. This niche differentiation between symbiont types was thought to drive the vertical distribution of two different host species at this site: *Pocillopora verrucosa*, which associated with clade D, was restricted to shallower habitats, while *Pavona gigantea*, which associated with clade C, was restricted to deeper habitats. Thus, host specificity may limit the vertical distribution of less flexible associations to the niche space occupied by the symbiont. However, D1 is also abundant in areas with relatively high turbidity and therefore lower light conditions (LaJeunesse et al. 2010a).

To date, intraspecific depth-related trends in *Symbiodinium* diversity have not been investigated in detail in the ETP, particularly for coral taxa that routinely host multiple symbiont types (such as *Pocillopora*). To remedy this, we analyzed 44 samples of *Pocillopora damicornis* across a depth gradient at Uva Island, Panama, using real-time PCR (Cunning and Baker 2013) to quantify the abundance of clades C and D *Symbiodinium*. We found that *P. damicornis* was more commonly dominated by clade D in shallow water compared to deeper water (Fig. 13.3). These patterns confirm intraspecific depth zonation in coral-algal symbiosis for ETP coral hosts (Rowan and Knowlton 1995) and suggest that abiotic factors, especially light and temperature, are important axes for niche diversification between different symbionts, influencing their distribution across reefscapes. However, differences in disturbance history, such as higher rates of bleaching and mortality at shallower depths, likely also contribute to the depth zonation of symbionts in *P. damicornis*.

13.3.3 Biogeographic and Taxonomic Gaps in Our Understanding of *Symbiodinium* Distribution in the ETP

Due to tectonic and hydrodynamic factors, it was originally thought that coral reefs were not present in the ETP (Darwin 1842; Dana 1843; see Chap. 1, Glynn). The discovery of structural coral reefs in Panama (Glynn 1972; Glynn et al.

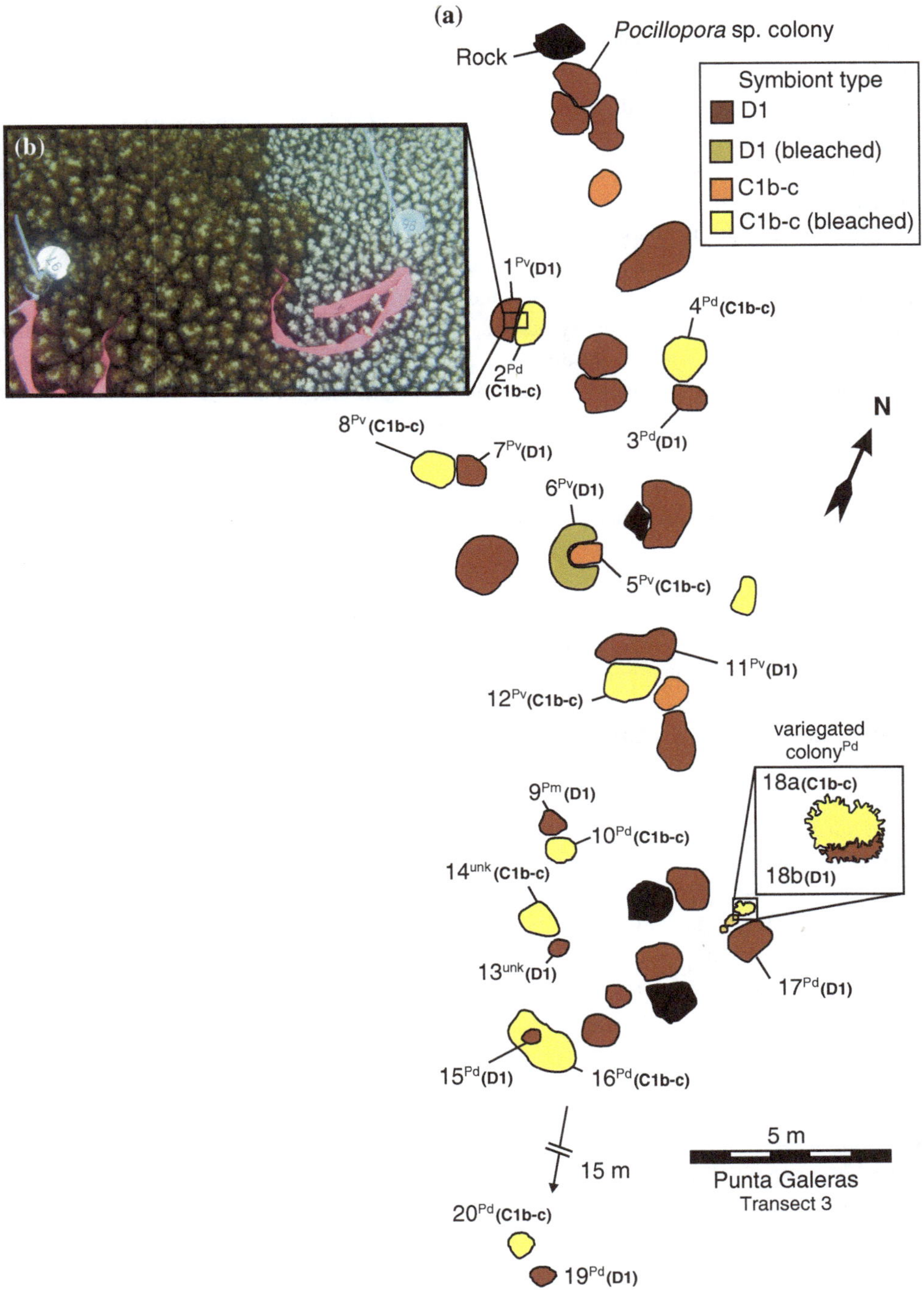

Fig. 13.5 Reefscape distribution of *Symbiodinium* in *Pocillopora* at 2 m depth at Punta Galeras, Baja California, Mexico, and mosaic patterns of bleaching resulting from this distribution in response to spring bleaching in 2006. **a** Survey of 36 *Pocillopora* spp. colonies along a 25 m transect. Relative distances from each other, colony size, bleached or pigmented, and resident *Symbiodinium* taxon are given. Only those colonies surveyed are depicted. *P. damicornis* (*Pd*), *P. verrucosa* (*Pv*), *P. meandrina* (*Pm*); **b** a colony of *P. damicornis* is bleached while adjacent *P. verrucosa* exhibits "healthy" pigmentation. From LaJeunesse et al. (2007)

1972; Porter 1972) marked the beginning of a period of scientific exploration in the region. However, despite sustained efforts to document and characterize ETP reefs (many of which are documented in this volume; see Chap. 5, Glynn et al.; Chap. 6, Toth et al.), studies of *Symbiodinium* diversity in the ETP have largely focused on the Gulf of California (Mexico), the Galápagos (Ecuador), and two areas in Panama (the Gulfs of Chiriquí and Panama). Almost all knowledge of ETP *Symbiodinium* distributions is based on studies from these areas (but see Table 13.2). Yet, well-developed reefs also exist off the Pacific coasts of Ecuador, Costa Rica, southern Mexico and Colombia (Cortés 2003). The least known reefs in the ETP, however, are arguably those within the "Pacific central American faunal gap" (Springer 1958), which includes Nicaragua, El Salvador, and Guatemala. Although coral framework was thought to be largely absent from the gap (e.g., Ryan and Zapata 2003), coral communities and developed reefs built mainly by *Pavona gigantea*, *Gardineroseris planulata*, and *Pocillopora elegans* were recently reported from Nicaragua (Alvarado et al. 2010). Additional data on *Symbiodinium* diversity should be obtained for these areas, as they may play an important role in the connectivity of ETP coral meta-populations in coastal Mexico, offshore areas such as the Revillagigedo Islands, and Central and South America (i.e., Costa Rica to Ecuador; see Chap. 16, Lessios and Baums). Additionally, coral-algal associations in the faunal gap may provide insights regarding how, and to what extent, these symbioses cope with naturally marginal environmental conditions, such as high turbidity in central Mexico (LaJeunesse et al. 2010a). Such associations may be useful proxies for the potential responses of conspecifics to analogous anthropogenic disturbances, such as eutrophication.

We are not aware of any data on *Symbiodinium* diversity or distribution from Guatemala, El Salvador, Nicaragua, Costa Rica, or Colombia. *Symbiodinium* diversity is also relatively unknown from ETP oceanographic islands, particularly Cocos Island, and Salas y Gómez and Easter (Rapa Nui) Islands (Chile), although limited *Symbiodinium* surveys have been conducted on pocilloporid hosts of the Revillagigedo Islands (LaJeunesse et al. 2010a) and Clipperton Atoll (LaJeunesse et al. 2010a; Pinzón and LaJeunesse 2011). Rapid surveys of these coral-dominated ecosystems, including documentation of coral-algal associations, should be performed in these relatively undescribed regions to generate baseline data necessary for habitat management, valuation, and protection. As one of the first shallow platforms in the ETP encountered by the North Equatorial Countercurrent (Glynn 1996), a higher (albeit still modest) proportion of coral-algal associations at Cocos Island may represent long-distance colonizers (Cortés and Jiménez 2003). Thus, *Symbiodinium* surveys from Cocos Island reefs are of particular interest.

Taxonomic gaps also exist in our knowledge of ETP *Symbiodinium* distributions. By far, most information regarding symbiont distributions comes from the dominant reef-building stony coral genus in the region, *Pocillopora. Symbiodinium* diversity within a few other ETP branching corals (e.g., *Psammocora*) has been documented, and some data are available for massive and encrusting coral species common to Panama, the Galápagos, and the Gulf of California (e.g., *Porites lobata*, *Pavona gigantea*, *Pavona clavus*, Baker 1999; Iglesias-Prieto et al. 2004; LaJeunesse et al. 2008; Glynn et al. 2009; see Table 13.2). Under-sampled ETP sites, however, contain some host species that are absent or rare elsewhere. These host species constitute another *Symbiodinium* frontier in the ETP. Host sampling gaps include *Leptoseris papyracea* (Costa Rica, Colombia, Ecuador), *Leptoseris scabra* (Ecuador), *Pavona maldivensis* (Colombia, Ecuador), *Porites paschalensis* (Easter Island or Rapa Nui), and even, potentially, *Porites rus* (Costa Rica) and *Acropora valida* (Colombia) (Cortés and Jiménez 2003; Glynn 2003; Glynn et al. 2003; Zapata and Vargas-Ángel 2003).

13.4 Stability Versus Change in the Distribution of *Symbiodinium* in the ETP

Because ETP reefs have been exposed to repetitive heat stress and experience gradients in pH across the region, they are excellent case studies for understanding the mechanisms by which corals might adapt or acclimatize to combined temperature and acidification stressors. In particular, since laboratory experiments cannot capture the long-term timescales over which projected changes are likely to occur, these real-world changes are more appropriate for identifying potential compensatory mechanisms. Symbiont community changes have been proposed as one way by which corals might respond to changing environments, including those occurring as a result of climate change (Buddemeier and Fautin 1993). These changes might occur as a result of dynamic change in community structure within individual colonies in response to changing environmental conditions, or as a result of differential mortality of hosts containing unsuitable symbionts. Baker et al. (2004) documented shifts to favor symbionts in clade D following large scale bleaching of *Pocillopora* during the 1997–98 El Niño warming event at Uva Island (Gulf of Chiriquí, Panama), and suggested that both symbiont "shuffling" as well as differential mortality, might account for these changes.

In contrast, in a study of cold-water bleaching and mortality, LaJeunesse et al. (2010a) found that relatively few of the surviving colonies experienced directional change in their symbiont communities, suggesting that differential mortality, rather than dynamic change on the colony level, resulted in the observed *Symbiodinium* distributions across the ETP. This was also suggested by McGinley et al. (2012), who used high resolution rt-PCR to study changes in the abundance of background symbionts and found that changes in symbiont dominance occurred in only 3 % of colonies following the same cold water bleaching event, suggesting that changes in symbiont communities were rare. Given these different findings, it seems likely that both mechanisms—natural selection (differential mortality and reproduction) of coral colonies with different symbionts, and dynamic change in symbiont communities within colonies—are likely to be important in understanding the distribution and stability of symbionts in reef corals over time and space. Responses may be expected to vary depending on whether bleaching was caused by high or low temperatures, with different responses across the ETP as a result.

13.5 The Future of Coral-Algal Symbiosis in the ETP

13.5.1 Interactions Between High Temperature and High CO_2

ETP reefs are model systems for the effects of periodic high temperature stress under conditions of high CO_2. Chronically depressed aragonite saturation states at some sites in the ETP are expected to decrease calcification rates (Langdon and Atkinson 2005; Manzello et al. 2008), resulting in slower coral growth and lower reef resilience (Anthony et al. 2011). At the same time, symbiont shifts to favor thermotolerant symbionts (*Symbiodinium* D1) may also lead to slower growth as a result of tradeoffs (see Sect. 13.2.3), although these tradeoffs may not be present at higher temperatures (Cunning et al. 2014). Nevertheless, investigating whether there is an interaction between ocean acidification (OA), symbiont type, and coral growth rate should be a high research priority, especially in the ETP. These interactions have not yet been explored in detail.

Studies of isolated *Symbiodinium* in culture have revealed significant variation among different types in their growth rates and photosynthetic capacities (Brading et al. 2011), but studies of how corals of the same species respond to OA if they have different symbiont types have not yet been undertaken in a controlled experiment. At three natural CO_2 seeps in Papua New Guinea, no difference in the dominant symbiont type was documented in six coral species at the seep sites versus nearby control sites (Noonan et al. 2013). This suggests that high CO_2 does not select for particular symbionts, and that, consequently, corals are unlikely to adapt or acclimatize to high CO_2 environments by shifts in the composition of their symbiont communities. However, it also implies that high CO_2 will not impede ongoing shifts to *Symbiodinium* D1 (due to increased warming and recurrent bleaching events) if these shifts are neutral from the point of view of high CO_2.

The few studies undertaken to date suggest that OA effects on *Symbiodinium* distributions may not be dramatic, but controlled experiments are needed to properly test this idea. In addition, the interaction between high CO_2 and high temperature in causing bleaching (Anthony et al. 2008) requires further research, especially in the ETP. It is possible that, instead of favoring particular symbiont types, high CO_2 may increase the per-cell productivity of symbiont communities, resulting in higher symbiont productivity and lower overall skeletal density in the coral host. If this is the case, OA might actually increase the thermal tolerance of coral hosts because corals with relatively fewer symbionts may show less severe bleaching in response to heat stress (Cunning and Baker 2013). Similarly, the effect of high pCO_2 on the rate of coral recovery from bleaching also represents a potential research question of particular relevance to the ETP.

13.5.2 Response of Reef Coral Symbioses to Climate Change: The Ratchet Hypothesis

Reef corals in the ETP represent an excellent "test bed" for how corals will respond to the combined effects of climate change, having already experienced multiple mass bleaching episodes as a result of recurring El Niño events, and also being subject to depressed aragonite saturation states as a result of regional upwelling conditions. Reefs throughout the region suffered severe mortality (52–97 %) following the 1982–83 El Niño event, but considerably less mortality (0–26.2 %) following the 1997–98 El Niño, despite the fact that the magnitude and duration of the two events were very similar (Enfield 2001; Glynn et al. 2001b). This suggests that the corals that survived and propagated on these reefs in the intervening ~15 years were considerably more thermotolerant than those that dominated reefs prior to 1982 (Baker 2002). This is probably a result of both differential mortality of thermally sensitive coral-symbiont combinations and increases in the abundance of thermotolerant symbionts within surviving colonies, with the relative importance of both of these mechanisms depending on site, species, stress exposure, and environmental history. Maynard et al. (2008) reached similar conclusions regarding the responses of *Pocillopora damicornis*, *Acropora* spp., and *Porites* spp. to two major thermal anomalies on the Great Barrier Reef (GBR). These corals experienced a major bleaching event in 1998, but during a severe thermal anomaly in 2002 bleached 30–100 % less than predicted. The mechanisms responsible for acclimatization/adaptation on the GBR were not directly observed, but Maynard et al. (2008) inferred that several mechanisms must have been acting in addition to differential mortality, because rates of mortality in 1998 were not well correlated with increases in thermal tolerance (measured as the difference between predicted and observed bleaching).

From a research perspective, what makes corals unusual is their ability to engage in mutualisms with diverse symbionts, and not the fact that they can experience differential mortality. Consequently, if dynamism in these symbiont communities is at least partially responsible for increased thermotolerance, then understanding how changes in symbiont communities occur, and the degree to which they are reversible, represent opportunities for insight into how corals might behave differently from other organisms, an element which is critical to projecting future survival trajectories for coral reefs. The apparent increase in the thermal tolerance of

ETP corals between 1983 and 1997 suggests that, if mass bleaching resulted in corals that recovered with a higher abundance of thermotolerant symbionts, then these symbionts (*Symbiodinium* D1) must have remained in coral tissues at sufficient abundance over ~15 years to have had a measureable effect on the bleaching susceptibility and/or survivorship of corals in 1997–98. This suggests that, as bleaching events become more frequent and more severe, the abundance of *Symbiodinium* D1 may be subject to stepwise increases, with rapid increases in abundance following bleaching, and much slower declines when conditions return to normal. This "ratchet" mechanism may be one way in which coral communities might make the transition to thermotolerance without suffering the high levels of mortality which might be expected under scenarios of differential mortality and natural selection. Testing the ratchet hypothesis is a priority for further progress in this field, and additional studies are required that apply high-resolution methods to quantify change and persistence in symbiont communities before, during, and after a bleaching event. Furthermore, symbiont communities in the field should also be monitored over time using high-resolution methods, and any changes related to disturbance history and environmental conditions. Such studies are already underway in the ETP and continuing these efforts should be a priority research area for coral reef scientists interested in projecting reef futures for the coming century.

Acknowledgements We dedicate this chapter to Peter Glynn for first introducing us to coral reef ecosystems of the eastern tropical Pacific, and for his guidance and friendship over many years. A.B. also thanks N. Knowlton and R. Rowan for their long-lasting impact as mentors in Panama. We thank I. Enochs, P. Fong, J. Jara, D. Manzello, J. Maté, and T. Smith for field support, K. Dziedzic for laboratory assistance, and J. Maté for long-term permit and logistical support. This work was supported by grants from the National Science Foundation (BIO-OCE 0099301, 0526361, and 0547169), the Wildlife Conservation Society (Tiffany & Co. Foundation), and the Lenfest Ocean Program. It was also supported by a University of Miami Fellowship and NSF Graduate Research Fellowship (to R.C.), and a Columbia University Graduate Fellowship (to A.M.S.C.). No corals were harmed during the writing of this chapter.

References

Alvarado JJ, Reyes-Bonilla H, Buitrago F, Aguirre-Rubí J, Álvarez del Castillo Cardenas PA (2010) Coral reefs of the Pacific coast of Nicaragua. Coral Reefs 29:201

Anthony KRN, Kline DI, Diaz-Pulido G, Dove S, Hoegh-Guldberg O (2008) Ocean acidification causes bleaching and productivity loss in coral reef builders. Proc Natl Acad Sci USA 105:17442–17446

Anthony KRN, Maynard JA, Diaz-Pulido G, Mumby PJ, Marshall PA, Cao L, Hoegh-Guldberg O (2011) Ocean acidification and warming will lower coral reef resilience. Glob Change Biol 17(5):1798–1808

Baker AC (1999) Symbiosis ecology of reef-building corals. Ph.D. dissert, Univ Miami, Coral Gables, Miami, p 120

Baker AC (2003) Flexibility and specificity in coral-algal symbiosis: diversity, ecology, and biogeography of *Symbiodinium*. Annu Rev Ecol Evol Syst 34:661–689

Baker AC, Romanski AM (2007) Multiple symbiotic partnerships are common in scleractinian corals, but not in octocorals: comment on Goulet (2006). Mar Ecol Prog Ser 335:237–242

Baker AC, Rowan R (1997) Diversity of symbiotic dinoflagellates (zooxanthellae) in scleractinian corals of the Caribbean and eastern Pacific. In: Proceedings of 8th International Coral Reef Symposium, vol 1, Panama, pp 1301–1306

Baker AC (2002) Is bleaching really adaptive? Nature 415:601–602

Baker AC, Starger CJ, McClanahan TR, Glynn PW (2004) Corals' adaptive response to climate change. Nature 430:741

Baums IB, Boulay JN, Polato NR, Hellberg ME (2012) No gene flow across the Eastern Pacific Barrier in the reef-building coral *Porites lobata*. Mol Ecol 21:5418–5433

Berkelmans R, van Oppen MJH (2006) The role of zooxanthellae in the thermal tolerance of corals: a 'nugget of hope' for coral reefs in an era of climate change. Proc R Soc B 273:2305–2312

Brading P, Warner ME, Davey P, Smith DJ, Achterberg EP et al (2011) Differential effects of ocean acidification on growth and photosynthesis among phylotypes of *Symbiodinium* (Dinophyceae). Limnol Oceanogr 56:927–938

Buddemeier RW, Fautin DG (1993) Coral bleaching as an adaptive mechanism. Bioscience 43:320–326

Cantin NE, van Oppen MJH, Willis BL, Mieog JC, Negri AP (2009) Juvenile corals can acquire more carbon from high-performance algal symbionts. Coral Reefs 28:405–414

Coates AG, Obando JA (1996) The geologic evolution of the Central American Isthmus. In: Jackson JBC, Budd AF, Coates AG (eds) Evolution and environment in tropical America. University of Chicago Press, Chicago, pp 21–56

Correa AMS (2009) Molecular ecology of endosymbiotic dinoflagellates (genus *Symbiodinium*) in scleractinian corals. Ph.D. dissert, Columbia University, New York, p 235

Correa AMS, McDonald MD, Baker AC (2009) Development of clade-specific *Symbiodinium* primers for quantitative PCR (qPCR) and their application to detecting clade D symbionts in Caribbean corals. Mar Biol 156:2403–2411

Cortés J (ed) (2003) Latin American coral reefs. Elsevier Science BV, Amsterdam p 497

Cortés J, Jiménez C (2003) Corals and coral reefs of the Pacific of Costa Rica: history, research and status. In: Cortes J (ed) Latin American coral reefs. Elsevier Science BV, Amsterdam, pp 361–385

Cunning R, Baker AC (2013) Excess algal symbionts increase the susceptibility of reef corals to bleaching. Nat Clim Change 3:259–262

Cunning R, Baker AC (2014) Not just who, but how many: the importance of partner abundance in reef coral symbioses. Front Microbiol 5:400

Cunning R, Glynn PW, Baker AC (2013) Flexible associations between *Pocillopora* corals and *Symbiodinium* limit utility of symbiosis ecology in defining species. Coral Reefs. doi:10.1007/s00338-013-1036-y

Cunning R, Gillette P, Capo T, Galvez K, Baker AC (2014) Growth tradeoffs associated with thermotolerant symbionts in the coral *Pocillopora damicornis* are lost in warmer oceans. Coral Reefs. doi:10.1007/s00338-014-1216-4

Cunning R, Vaughan N, Gillette P, Capo T, Maté J, Baker AC (2015) Dynamic regulation of partner abundance mediates response of reef coral symbioses to environmental change. Ecology 96:1411–1420

Dana JD (1843) On the temperature limiting the distribution of corals. Am J Sci Arts 45:130–131

Dana T (1975) Development of contemporary eastern Pacific coral reefs. Mar Biol 33:355–374

Darius HT, Dauga C, Grimont PAD, Chungue E, Martin PMV (1998) Diversity in symbiotic dinoflagellates (Pyrrhophyta) from seven scleractinian coral species: restriction enzyme analysis of small subunit ribosomal RNA genes. J Eukaryot Microbiol 45:619–627

Darius HT, Martin PMV, Grimont PAD, Dauga C (2000) Small subunit rDNA sequence analysis of symbiotic dinoflagellates from seven scleractinian corals in a Tahitian lagoon. J Phycol 36:951–959

Darwin C (1842) The structure and distribution of coral reefs. Smith Elder & Co., London, p 214

DeSalvo MK, Sunagawa S, Fisher PL, Voolstra CR, Iglesias-Prieto R, Medina M (2010) Coral host transtriptomic states are correlated with *Symbiodinium* genotypes. Mol Ecol 19:1174–1186

Drew EA (1972) The biology and physiology of algal-invertebrate symbiosis. II. The density of algal cells in a number of hermatypic corals and alcyonarians from various depths. J Exp Mar Biol Ecol 9:71–75

Duque-Caro H (1990) Neogene stratigraphy, paleoceanography and paleobiogeography in northwest South America and evolution of the Panama Seaway. Palaeogeogr Palaeocl 77:203–234

Ekman S (1953) Zoogeography of the sea, 1967 edn. Sidgwick & Jackson Ltd., London 417

Enfield DB (2001) Evolution and historical perspective of the 1997–1998 El Niño-Southern oscillation event. Bull Mar Sci 69:7–25

Fabricius KE, Mieog JC, Colin PL, Idip D, van Oppen MJH (2004) Identity and diversity of coral endosymbionts (zooxanthellae) from three Palauan reefs with contrasting bleaching, temperature and shading histories. Mol Ecol 13:2445–2458

Franklin EC, Stat M, Pochon X, Putnam H, Gates RD (2012) GeoSymbio: a hybrid, cloud-based web application of global geospatial bioinformatics and ecoinformatics for *Symbiodinium*-host symbioses. Mol Ecol Resour 12:369–373

Glynn PW (1972) Observations on the ecology of the Caribbean and Pacific coasts of Panama. In: Jones ML (ed) The Panamic biota: some observations prior to a sea-level canal, vol 2, pp 13–30 (Bull Biol Soc Wash)

Glynn PW (1984) Widespread coral mortality and the 1982–83 El Niño warming event. Environ Conserv 11:133–146

Glynn PW (ed) (1996) Coral reefs of the eastern Pacific. Coral Reefs 15:69

Glynn PW (2003) Coral communities and coral reefs of Ecuador. In: Cortés J (ed) Latin American coral reefs. Elsevier Sci BV, Amsterdam, pp 449–472

Glynn PW, Ault JS (2000) A biogeographic analysis and review of the far eastern Pacific coral reef region. Coral Reefs 19:1–23

Glynn PW, Colgan MW (1992) Sporadic disturbances in fluctuating coral reef environments—El Niño and coral reef development in the eastern Pacific. Am Zool 32:707–718

Glynn PW, Stewart RH, McCosker JE (1972) Pacific coral reefs of Panama: structure, distribution and predators. Geol Rundsch 61:483–519

Glynn PW, Maté JL, Baker AC, Calderón MO (2001a) Coral bleaching and mortality in Panama and Ecuador during the 1997–1998 El Niño-Southern Oscillation event: spatial/temporal patterns and comparisons with the 1982–1983 event. Bull Mar Sci 69:79–109

Glynn PW, Maté JL, Stemann TA (2001b) *Pavona chiriquiensis*, a new species of zooxanthellate scleractinian coral (Cnidaria: Anthozoa: Agariciidae) from the eastern tropical Pacific. Bull Biol Soc Wash 10:210–225

Glynn PW, Wellington GM, Wieters EA, Navarrete SA (2003) Reef-building coral communities of Easter Island (Rapa Nui), Chile. In: Cortés J (ed) Latin American coral reefs. Elsevier Sci BV, Amsterdam, pp 473–494

Glynn PW, Riegl B, Correa AMS, Baums IB (2009) Rapid recovery of a coral reef at Darwin Island, Galápagos Islands. Galápagos Res 66:6–13

Goulet TL, Cook CB, Goulet D (2005) Effect of short-term exposure to elevated temperatures and light levels on photosynthesis of different host-symbiont combinations in the *Aiptasia pallida/Symbiodinium* symbiosis. Limnol Oceanogr 50:1490–1498

Green E, Davies SW, Matz MV, Medina M (2014) Next-generation sequencing reveals cryptic *Symbiodinium* diversity within *Orbicella faveolata* and *Orbicella franksi* at the Flower Garden Banks, Gulf of Mexico. PeerJ PrePrints 2:e246v1

Grigg RW, Hey R (1992) Paleoceanography of the tropical eastern Pacific Ocean. Science 255:172–178

Hill M, Hill A (2012) The magnesium inhibition and arrested phagosome hypotheses: new perspectives on the evolution and ecology of *Symbiodinium* symbioses. Biol Rev 87(804–821):804. doi:10.1111/j.1469-185X.2012.00223.x

Iglesias-Prieto R, Trench RK (1994) Acclimation and adaptation to irradiance in symbiotic dinoflagellates. I. Responses of the photosynthetic unit to changes in photon flux density. Mar Ecol Prog Ser 113:163–175

Iglesias-Prieto R, Trench RK (1997) Acclimation and adaptation to irradiance in symbiotic dinoflagellates. II. Response of chlorophyll-protein complexes to different photon-flux densities. Mar Biol 130:23–33

Iglesias-Prieto R, Beltrán VH, LaJeunesse TC, Reyes-Bonilla H, Thomé PE (2004) Different algal symbionts explain the vertical distribution of dominant reef corals in the eastern Pacific. Proc Roy Soc Lond Ser B 271:1757–1763

IPCC (2007) The physical science basis. In: Solomon S et al (eds) Contribution of working group 1 to the fourth assessment report of the Intergovernmental Panel on climate change. Cambridge University Press, Cambridge, UK, New York

Jones A, Berkelmans R (2010) Potential costs of acclimatization to a warmer climate: growth of a reef coral with heat tolerant vs. sensitive symbiont types. PLoS One 5:1–9

Jumpponen A, Jones KL (2009) Massively parallel 454 sequencing indicates hyperdiverse fungal communities in temperate *Quercus macrocarpa* phyllosphere. New Phytol 184:438–448

Kemp DW, Fitt WK, Schmidt GW (2008) A microsampling method for genotyping coral symbionts. Coral Reefs 27:289–293

Kenkel CD, Goodbody-Gringley G, Caillaud D, Davies SW, Bartels E, Matz MV (2013) Evidence for a host role in thermotolerance divergence between populations of the mustard hill coral (*Porites astreoides*) from different reef environments. Mol Ecol 22:4335–4348

LaJeunesse TC (2005) "Species" radiations of symbiotic dinoflagellates in the Atlantic and Indo-Pacific since the Miocene-Pliocene transition. Mol Biol Evol 22:570–581

LaJeunesse TC, Thornhill DJ (2011) Improved resolution of reef-coral endosymbiont (*Symbiodinium*) species diversity, ecology, and evolution through *psbA* non-coding region genotyping. PLoS One 6:e29013

LaJeunesse TC, Loh WKW, van Woesik R, Hoegh-Guldberg O, Schmidt GW, Fitt WK (2003) Low symbiont diversity in southern Great Barrier Reef corals relative to those of the Caribbean. Limnol Oceanogr 48:2046–2054

LaJeunesse TC, Reyes-Bonilla H, Warner ME (2007) Spring "bleaching" among *Pocillopora* in the Sea of Cortez, eastern Pacific. Coral Reefs 26:265–270

LaJeunesse TC, Reyes-Bonilla H, Warner ME, Wills M, Schmidt GW, Fitt WK (2008) Specificity and stability in high latitude eastern Pacific coral-algal symbioses. Limnol Oceanogr 53:719–727

LaJeunesse TC, Pettay T, Phongsuwan N, Brown B, Obura D, Hoegh-Guldberg O, Fitt WK (2010a) Long-standing environmental conditions, geographic isolation and host-symbiont specificity influence the relative ecological dominance and genetic diversification of coral endosymbionts in the genus *Symbiodinium*. J Biogeogr 37:785–800

LaJeunesse TC, Smith R, Walther M, Pinzón J, Pettay DT, McGinley M et al (2010b) Host-symbiont recombination versus natural selection in the response of coral-dinoflagellate symbioses to environmental disturbance. Proc Roy Soc B-Biol Sci 277:2925–2934

Langdon C, Atkinson MJ (2005) Effects of elevated pCO_2 on photosynthesis and calcification of corals and interactions with seasonal change in temperature/irradiance and nutrient enrichment. J Geophys Res 110:C09S07

Lien Y-T, Nakano Y, Plathong S, Fukami H, Wang J-T, Chen CA (2007) Occurrence of the putatively heat-tolerant *Symbiodinium* phylotype D in high-latitudinal outlying coral communities. Coral Reefs 26:35–44

Little A, van Oppen MJH, Willis BL (2004) Flexibility in algal endosymbioses shapes growth in reef corals. Science 304:1492

Loram JE, Boonham N, O'Toole P, Trapido-Rosenthal HG, Douglas AE (2007) Molecular quantification of symbiotic dinoflagellate algae of the genus *Symbiodinium*. Biol Bull 212:259–268

Manzello DP, Kleypas JA, Budd DA, Eakin CM, Glynn PW, Langdon C (2008) Poorly cemented coral reefs of the eastern tropical Pacific: possible insights into reef development in a high-CO_2 world. Proc Nat Acad Sci USA 105:10450–10455

Maté JL (2003) Corals and coral reefs of the Pacific coast of Panama. In: Cortés J (ed) Latin American coral reefs. Elsevier Science BV, Amsterdam, pp 387–417

Maynard JA, Anthony KRN, Marshall PA, Masiri I (2008) Major bleaching events can lead to increased thermal tolerance in corals. Mar Biol 155:173–182

McGinley MP, Aschaffenburg MD, Pettay DT, Smith RT, LaJeunesse TC, Warner ME (2012) *Symbiodinium* spp. in colonies of eastern Pacific *Pocillopora* spp. are highly stable despite the prevalence of low-abundance background populations. Mar Ecol Prog Ser 462:1–7

Mieog JC, van Oppen MJH, Cantin NC, Stam WT, Olsen JL (2007) Real-time PCR reveals a high incidence of *Symbiodinium* clade D at low levels in four scleractinian corals across the Great Barrier Reef: implications for symbiont shuffling. Coral Reefs 26:449–457

Mieog JC, van Oppen MJH, Berkelmans R, Stam WT, Olsen JL (2009) Quantification of algal endosymbionts (*Symbiodinium*) in coral tissue using real-time PCR. Mol Ecol Resour 9:74–82

Noonan SHC, Fabricius KE, Humphrey C (2013) *Symbiodinium* community composition in scleractinian corals is not affected by life-long exposure to elevated carbon dioxide. PLoS One 8(5): e63985. doi:10.1371/journal.pone.0063985

Pettay DT, LaJeunesse TC (2013) Long-range dispersal and high-latitude environments influence the population structure of a 'stress-tolerant' dinoflagellate endosymbiont. PLoS One 8:e79208

Pettay DT, Wham DC, Pinzón JH, LaJeunesse TC (2011) Genotypic diversity and spatial-temporal distribution of *Symbiodinium* clones in an abundant reef coral. Mol Ecol 20:5197–5212

Pinzón JH, LaJeunesse TC (2011) Species delimitation of common reef corals in the genus *Pocillopora* using nucleotide sequence phylogenies, population genetics and symbiosis ecology. Mol Ecol 20:311–325

Pochon X, Gates RD (2010) A new *Symbiodinium* clade (Dinophyceae) from soritid foraminifera in Hawai'i. Mol Phylogenet Evol 56: 492–497

Porter JW (1972) Ecology and species diversity of coral reefs on opposite sides of the Isthmus of Panamá. In Jones ML (ed) The Panamic biota: some observations prior to a sea level canal. Bull Biol Soc Wash 2:89–116

Rodriguez-Lanetty M, Cha H, Song JI (2002) Genetic diversity of symbiotic dinoflagellates associated with anthozoans from Korean waters. In: Proceedings of 9th International Coral Reef Symposium, vol 1, Bali, pp 163–166

Rowan R (2004) Coral bleaching: thermal adaptation in reef coral symbionts. Nature 430:742

Rowan R, Powers DA (1991a) A molecular genetic classification of zooxanthellae and the evolution of animal-algal symbiosis. Science 251:1348–1351

Rowan R, Powers DA (1991b) Molecular genetic identification of symbiotic dinoflagellates (zooxanthellae). Mar Ecol Prog Ser 71:65–73

Rowan R, Knowlton N (1995) Intraspecific diversity and ecological zonation in coral-algal symbiosis. Proc Nat Acad Sci USA 92:2850–2853

Rowan R, Knowlton N, Baker AC, Jara J (1997) Landscape ecology of algal symbiont communities explains variation in episodes of coral bleaching. Nature 388:265–269

Ryan J, Zapata Y (2003) Nicaragua's coral reefs: status, health and management strategies. In: Cortés J (ed) Latin American coral reefs. Elsevier Sci BV, Amsterdam, pp 203–222

Sampayo EM, Ridgway T, Bongaerts P, Hoegh-Guldberg O (2008) Bleaching susceptibility and mortality of corals are determined by fine-scale differences in symbiont type. Proc Nat Acad Sci USA 105:10444–10449

Savage A, Trapido-Rosenthal HG, Douglas AE (2002) On the functional significance of molecular variation in *Symbiodinium*, the symbiotic algae of Cnidaria: photosynthetic response to irradiance. Mar Ecol Prog Ser 244:27–37

Silverstein RN, Correa AMS, Baker AC (2012) Specificity is rarely absolute in coral-algal symbiosis: implications for coral response to climate change. Proc Roy Soc B Biol Sci 279:2609–2618

Springer VG (1958) Systematics and zoogeography of the clinid fishes of the subtribe Labrisomini Hubbs. PhD dissert, University of Texas, Austin, Texas, p 226

Stat M, Baker AC, Bourne D, Correa AMS, Forsman Z, Huggett M, Pochon X, Skillings D, Toonen R, van Oppen MJH, Gates RD (2012) Molecular delineation of species in the coral holobiont. Adv Mar Biol 63:1–65

Thornhill DJ, Xiang Y, Pettay DT, Zhong M, Santos SR (2013) Population genetic data of a model symbiotic cnidarian system reveal remarkable symbiotic specificity and vectored introductions across ocean basins. Mol Ecol 22:4499–4515

Toth LT, Aronson RB, Vollmer SV, Hobbs JW, Urrego DH, Cheng H et al (2012) ENSO drove 2500-year collapse of eastern Pacific coral reefs. Science 337:81–84

van Oppen MJH, Baker AC, Coffroth MA, Willis BL (2009) Bleaching resistance and the role of algal endosymbionts. In: van Oppen MJH, Lough JM (eds) Coral bleaching. Springer, Berlin, Heidelberg, pp 83–102

Warner ME, Fitt WK, Schmidt GW (1999) Damage to photosystem II in symbiotic dinoflagellates: a determinant of coral bleaching. Proc Nat Acad Sci USA 96:8007–8012

Zapata FA, Vargas-Ángel B (2003) Corals and coral reefs of the Pacific coast of Colombia. In: Cortés J (ed) Latin American coral reefs. Elsevier Sci BV, Amsterdam, pp 419–447

Insights from the Application of Genetics on *Pocillopora-Symbiodinium* Associations in the Eastern Tropical Pacific

14

Jorge H. Pinzón C.

Abstract

As one of the most widely distributed and most studied scleractinian genera in the world, *Pocillopora* encompasses an important group of corals. In the eastern tropical Pacific, *Pocillopora* species thrive and are the major reef-building scleractinian taxon, even though conditions are considered suboptimal for coral growth and reef development. Early observations on reproduction and species distributions appear to be complicated by high phenotypic diversity and often inaccurate species identifications. New genetic-based evidence reorganizes species classifications within *Pocillopora* by delimiting boundaries to genetic recombination. Such improvements toward a natural and accurate taxonomy have further revealed important patterns in *Symbiodinium* diversity and distribution associated with *Pocillopora* in the eastern Pacific. Here, I review work on genetic connectivity and symbiosis ecology that may explain physiological, ecological and evolutionary characteristics that account for the differential success of this coral genus in the marginal eastern tropical Pacific.

Keywords

Symbiosis • Coevolution • Coral-algae interactions • Genetic connectivity • Diversity

14.1 Introduction

Corals in the genus *Pocillopora* are widely distributed throughout the Indo-Pacific region. Yet, they are sufficiently abundant to build reefs only in places where environmental factors appear to limit coral diversity and carbonate accretion, including Western Australia (Veron and Marsh 1988), the Arabian Gulf (Burchard 1979; Riegl et al. 2012), the Gulf of Oman (Glynn 1993), and the eastern Pacific (Glynn 1976; Cortés 1997; see Chap. 6, Toth et al.). Relatively cold, turbid, seasonal conditions, as well as aperiodic El Niño-Southern Oscillation events in the eastern Pacific (Veron 1995; Toth et al. 2012; see Chap. 6, Toth et al.; see Chap. 8, Glynn et al.), combined with the region's isolation from the central Indo-Pacific, severely limit scleractinian diversity (Richmond 1987a). However, shallow reef coral communities are widespread throughout the eastern tropical Pacific (ETP) and are built largely by the branching colonies of *Pocillopora* spp. that dominate these communities (Fig. 14.1) (Glynn and Ault 2000; Cortés and Jiménez 2003; Glynn 2003; Maté 2003; Reyes-Bonilla 2003; Reyes-Bonilla and Barraza 2003; Zapata and Vargas-Ángel 2003; see Chap. 6, Toth et al.). The ecological abundance and wide geographic distributional range of *Pocillopora* in the eastern Pacific have attracted the interest of many researchers, making this taxon one of the most studied scleractinian reef corals in the world (Richmond 1981, 1987a, b; Richmond and Jokiel 1984; Cortés 1997; Combosch et al. 2008; Traylor-Knowles et al. 2011; Pinzón et al. 2012; Vidal-Dupiol et al. 2013).

Research findings on eastern Pacific *Pocillopora* over the past several decades have reported old and new morphological species (Cantera et al. 1989; Glynn 1999; Reyes-Bonilla 2002; Veron 2002), extended known geographic distributions of the species (Glynn and Ault 2000), examined their associations with endosymbiotic dinoflagellates (Baker et al. 2004; LaJeunesse et al. 2008; Pettay et al. 2011; Pettay and

J.H. Pinzón C. (✉)
Department of Biology, University of Texas at Arlington, Arlington, TX, USA
e-mail: Jorge.Pinzon@UTSouthwestern.edu

P.W. Glynn et al. (eds.), *Coral Reefs of the Eastern Tropical Pacific*, Coral Reefs of the World 8, DOI 10.1007/978-94-017-7499-4_14

Fig. 14.1 Shallow water communities across the eastern tropical Pacific are dominated by colonies of the scleractinian genus *Pocillopora*. Communities of these corals are shown from **a** Taboga Island in the Gulf of Panama (picture taken in January 2009 at ~3 m) and **b** Punta Galeras in La Paz, Mexico (July 2009, ~3 m). **c** Skeletons of dead *Pocillopora* build much of the reef framework in this region and serve as substrate for new colonies (Gulf of Panama, January 2009 ~3 m)

LaJeunesse 2013), and measured their response to physiological stress brought on by El Niño-Southern Oscillation events (Eakin 2001; Iglesias-Prieto et al. 2003). Many of these investigations relied on assessment of visual characters to identify species by morphology, despite the complications created by phenotypic variation (Veron and Pichon 1976; Cantera et al. 1989; Cairns 1999) and transitional morphologies between several species (Schmidt-Roach et al. 2014). Recently, genetic markers applied to the analysis of seven to eight morphospecies resolved only three biological species in the ETP (Flot et al. 2008, 2010; Pinzón and LaJeunesse 2011; Pinzón et al. 2013a). In light of this new context, reinterpretations of past observations and experimental results appear to improve our understanding of the physiology and ecology of *Pocillopora* in the region (Pettay et al. 2011; Pinzón and LaJeunesse 2011; Pinzón et al. 2012, 2013a).

Many workers have speculated why *Pocillopora*, a group that is relatively susceptible to stress, is common in areas with sub-optimal conditions for coral growth and reef development (Veron 1995; Glynn and Colley 2008). While other corals also inhabit ETP reefs, *Pocillopora* species have the ability to disperse over long distances by means of larvae that can persist in the plankton for weeks or months (Richmond 1987a) and colony rafting on floating debris (Jokiel 1984), and exhibit rapid colony growth (Richmond 1987b; Guzmán and Cortés 1989) and frequent asexual propagation (Stoddart 1983; Richmond 1987b; Combosch et al. 2008; Yeoh and Dai 2010; Pinzón et al. 2012). An additional reason why eastern Pacific *Pocillopora* appear to endure environmental stressors better than in other regions (where colonies are stress-sensitive and have a lower threshold for coral bleaching—the dissociation of the coral-algal association rendering the coral colony white or bleached) is the presence of a particularly stress-tolerant symbiont. In this region, both host and symbiont are subject to selection depending on their physiological characteristics and resistance to variations in habitat conditions (LaJeunesse et al. 2009, 2010b; Pettay et al. 2011; Pettay and LaJeunesse 2013). In this chapter, I review and analyze published findings that may further explain why *Pocillopora* spp. are ecologically dominant relative to other scleractinians in the eastern Pacific. I present a brief summary of the recent failure to delimit *Pocillopora* morphospecies with genetic markers. This work has motivated a reassessment of the species diversity in the eastern tropical Pacific region and its relationship to species found in other regions. Finally, I discuss ecological and physiological attributes related to both the coral host and the associated symbionts that help explain the success of *Pocillopora* in the ETP.

14.2 Eastern Pacific *Pocillopora* Species Diversity

The genus *Pocillopora* as a group has historically challenged taxonomists (Veron and Pichon 1976). Colonies in this genus are described as branching, and only showing crustose and massive forms under extreme environmental conditions with small skeletal protuberances or verrucae, and poorly developed calices (Veron and Pichon 1976; Veron 2002). Until recently, species within *Pocillopora* were classified and identified using morphological traits, including general colony appearance, branch morphology (i.e. width, length and shape), verrucae morphology (i.e. shape, size and density), and micro-structures within calices (Veron and Pichon 1976; Glynn 1999; Veron 2000; Schmidt-Roach et al. 2014). However, there is no clear consensus among taxonomists as to which, if any, of these morphological features accurately delineate species (Budd 1990). This has resulted in uncertainty regarding the actual number of valid species (Cairns 1999; Veron 2000) and inconsistencies in the identification of individual colonies (Forsman et al. 2013).

The emergence of genetic markers in the Scleractinia has facilitated the identification of reproductively isolated groups, providing a more objective delineation of species boundaries (Stefani et al. 2008, 2011). In several cases, the integration of genetic markers with morphological studies has improved species resolutions (Forsman et al. 2005; Stefani et al. 2008; Benzoni et al. 2010; Forsman et al. 2010) and increased our understanding of the evolutionary history of Scleractinia (Fukami 2008; Budd et al. 2010). Genetic assessment of the genus *Pocillopora* appears more definitive in the identification and classification of species, as defined by the Biological Species Concept (Pinzón and LaJeunesse 2011; Pinzón et al. 2013a); and with this resolution comes a better understanding of their ecology and evolution upon which to pursue well-informed research questions.

Genetic markers used in the study of *Pocillopora* were first developed after sequencing the mitochondrial genomes of two morphospecies, *P. damicornis* and *P. eydouxi* (Flot and Tillier 2007). Comparative analyses of these mitochondrial genomes revealed a highly variable region, denoted as the Open Reading Frame (ORF). The use of this nucleotide sequence in phylogenetics has helped to distinguish genetically distinct *Pocillopora* lineages (Flot et al. 2008, 2010; Souter 2010; Pinzón and LaJeunesse 2011; Schmidt-Roach et al. 2012; Pinzón et al. 2013a). Furthermore, other mitochondrial (a control region) and nuclear (internal transcribed spacer region—ITS, the ATP synthase β—ATPSβ and the Heat Shock Protein 70—HSP70) (Flot et al. 2008, 2010; Schmidt-Roach et al. 2012, 2014) markers, including population genetic markers (microsatellites) (Pinzón and LaJeunesse 2011; Pinzón et al. 2013a) have provided similar results as those delineations obtained by means of ORF analyses. Lastly, ecological (symbiotic associations) and biogeographic data support the delineations of *Pocillopora* spp. (Pinzón and LaJeunesse 2011).

Phylogenetic patterns correspond with morphology in some geographic locales (Flot et al. 2008); but at other locations, morphology and genetics can be incongruent (Flot et al. 2010; Forsman et al. 2013; Pinzón et al. 2013a). Correspondence between genetics and morphology is restricted to locations where conditions appear "ideal" for coral development, such as Hawai'i and the Great Barrier Reef. The disparity between genetic characters and morphological features appears to be the most extreme in the ETP, especially in the Gulf of California where a single genetically-defined lineage encompassed six morphospecies (Pinzón and LaJeunesse 2011; Pinzón et al. 2012, 2013a).

Progress in coral systematics and evolution now relies on genetic-based identification and classification (Fukami et al. 2004; Fukami 2008; Huang et al. 2011; Pinzón and LaJeunesse 2011; Budd et al. 2012). A global genetic analysis recently differentiated the genus *Pocillopora* into at least nine separately evolving lineages (types 1 to 9; Fig. 14.2) based on samples obtained across the Indo-Pacific, Arabian Gulf, and Red Sea (Pinzón et al. 2013a). Only three of these lineages are known to occur in the ETP (Pinzón and LaJeunesse 2011).

Genetic boundaries between *Pocillopora* spp. are maintained among populations distributed over large geographic distances (Pinzón et al. 2013a). In the ETP, *Pocillopora* was believed to be composed of approximately seven morphologically defined species, but genetic data delineate only three distinct groups: *Pocillopora* types 1, 2 and 3 (Fig. 14.2) (Pinzón and LaJeunesse 2011). Genetic delineations were corroborated by geographic distribution analyses and the identification of a specific algal species from *Symbiodinium* clade C associated exclusively with each *Pocillopora* lineage (*Pocillopora* type 1 with C1b-c, *P.* type 2 with C1ee, and *P.* type 3 with C1d). Additional support for this conclusion was presented in an independent analysis of coral colonies from Clipperton Atoll (Flot et al. 2010).

Two studies with samples collected in the ETP disagree with the findings presented here and with those in Pinzón and LaJeunesse (2011). The first analyzed the population structure of *Pocillopora damicornis* across various locations in Panama (Combosch and Vollmer 2011). These results suggested that there were significant levels of population genetic structure with restricted gene flow among and within the surveyed regions. This might be an indication of localized differential adaptation and population differentiation, and suggests that gene flow among *P. damicornis* populations in the ETP is limited, even at scales involving 10 s of kilometers (Combosch and Vollmer 2011). This work did not include a method to phylogenetically delimit species, and based specimen identifications and subsequent data analysis on morphological features. It is possible that Combosch and Vollmer (2011) in actuality sampled colonies from different lineages and therefore their reported patterns of allele diversity and differences in relative frequency resulted from differences in the relative number of colonies from a sampling location that represents genetically distinct species.

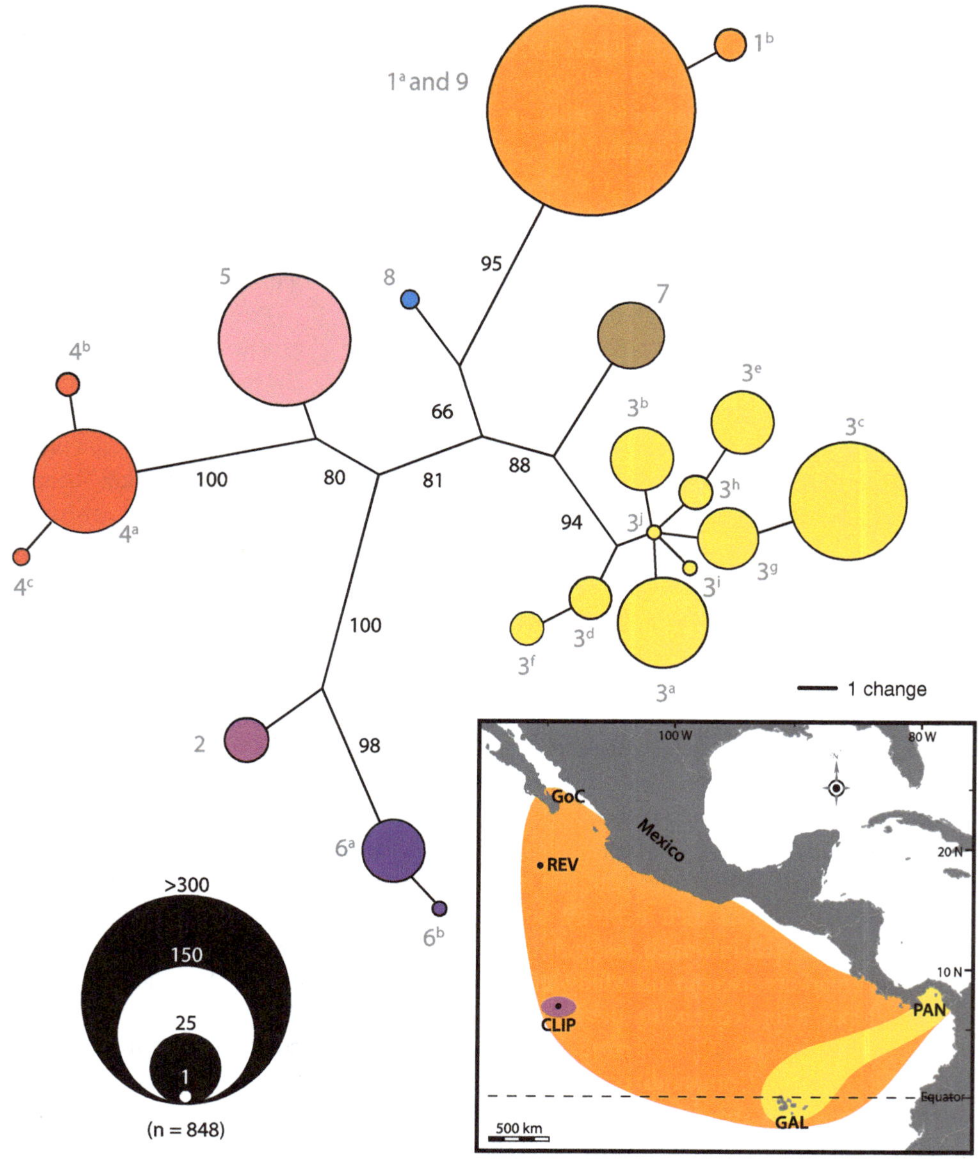

Fig. 14.2 Unrooted maximum parsimony tree showing phylogenetic relations between the genetically-delineated types of *Pocillopora* using the mitochondrial ORF region (redrawn from Pinzón et al. 2013a). Map denotes distribution of three *Pocillopora* types (1, 2 and 3) inhabiting the eastern tropical Pacific (redrawn from Pinzón and LaJeunesse 2011). Lineage designations are labeled with light grey numbers, and the size and color of each circle correspond to species-lineages and sample size. Numbers listed above branches represent bootstrap support values, on 1000 replicates, between nodes. Black and white colored circles correspond to sample size. Color shading on map matches that in tree and corresponds to known distributions of types 1, 2 and 3. (*GoC* Gulf of California, *REV* Revillagigedo Islands, *CLIP* Clipperton Atoll, *GAL* Galápagos, *PAN* Panama)

In the second study, the disagreement is more on the methodology than with the identity of the *Pocillopora* species. This study argues that the use of *Symbiodinium* identity, namely the presence or absence of a stress tolerant taxon, to delineate species of *Pocillopora* is not accurate (Cunning et al. 2013), as it found both *Symbiodinium* clades C and D in the two analyzed host species. The disagreement is based on the method used to identify the symbiont species. On one hand, the method (actin targeted quantitative PCR—qPCR) used in Cunning et al. (2013) differentiates clade diversity, while the methods (Denature Gradient Gel Electrophoresis of the ITS2 rDNA region and sequencing of the non-coding region of the psbA minicircle—*psbA*^{ncr}) used by Pinzón and LaJeunesse (2011) each resolves within clade diversity, that is *Symbiodinium* types. The latter approach clearly provides better definition and accuracy when it comes to identify *Symbiodinium* species. In fact, Pinzón and LaJeunesse (2011) found that each *Pocillopora* lineage harbors a particular species of *Symbiodinium* clade C (*Pocillopora* type 1 with C1b-c, *P.* type 2 with C1ee, and *P.* type 3 with C1d). The thermally tolerant type (*Symbiodinium glynni*—D1) was not used to delineate host species as it was common in *Pocillopora* type 1, but was also found in the other two *Pocillopora* lineages (Pinzón and LaJeunesse 2011). Regardless of the different approaches and the argument concerning the symbiont's identity, the delineation of *Pocillopora* in Panama, from both studies (Pinzón and LaJeunesse 2011; Cunning et al. 2013) suggests the presence of only two host lineages at this locality (*Pocillopora* type 1 and type 3).

Pocillopora type 1 is the most widely-distributed and abundant species in the ETP. Colonies of this lineage are found from the Gulf of California to Panama and at several offshore islands along this broad latitudinal gradient (Fig. 14.2). Type 2 appeared to be endemic to Clipperton Atoll, but also may occur at Malpelo and Cocos Islands, and possibly in some reef communities from the Central Pacific (Pinzón and LaJeunesse 2011; Pinzón et al. 2013a). Finally, *Pocillopora* type 3 is found in the southern ETP (Panama and the Galápagos Islands). This type also occurs worldwide with genetically interconnected populations (Pinzón et al. 2013a). More extensive sampling across the eastern Pacific region may eventually find it at other localities as well (Fig. 14.2).

A recent attempt to review the taxonomy of the "*Pocillopora* complex" (Schmidt-Roach et al. 2014) partially disagrees with the findings presented here and with Pinzón et al. (2013a). In Schmidt-Roach et al. (2014), an assessment of the diversity of *Pocillopora* from Australia is presented. The main conclusion of this work is that the use of morphological features, in combination with genetic analyses, resolves species in the genus. However, this work is geographically restricted and appears to have several contradictions; for example, the main result is the description of eight species, but these are grouped into five randomly-defined genetic clades (Schmidt-Roach et al. 2014). This contradiction is explained by the fact that there are morphological "transitions" between all species in the genus, leading to taxonomic confusion. In fact, the study presents a reassignment analysis from which one of the proposed morphospecies has been removed because it possesses many morphological features that overlap with other morphospecies (Schmidt-Roach et al. 2014). Despite this clear disconnection between morphology and genetics, the species descriptions presented are largely based on morphological characters and continue to force traditional morphology-based names of the species (Schmidt-Roach et al. 2014). The exception was *Pocillopora bairdi*, a putatively newly described species delineated mainly by genetics, and whose ORF sequence appeared central to all other "clade 2" (i.e. type 3 in Pinzón et al. 2013a) sequences. Thus, *P. bairdi* appears to be a variant of the pocilloporid "clade" that exhibits high levels of genetic and morphological variation (Pinzón et al. 2013a; Schmidt-Roach et al. 2014).

This latest attempt to combine morphology and genetics in *Pocillopora* species' delimitations highlights the confusion created by forcing morphologically based nomenclature on a group whose genetics is not concordant with traditional binomial nomenclature. In the future, such a situation could be avoided by utilizing binomial names that do not reflect previous, and misleading, conceptions caused by morphology having been a poor proxy for species differentiations in *Pocillopora*. Genetic tools have been shown to delineate species accurately in the genus, and the type-based nomenclature used by Flot et al. (2010), Pinzón and LaJeunesse (2011) and Pinzón et al. (2013a) is less confusing while generating the most parsimonious conclusions about the ecology and evolution of *Pocillopora* (see below). Additionally, molecular tools are now part of the day-to-day endeavors in many research laboratories since it has become less expensive to sequence genes from several samples than, for example, acquiring an electron microscope for micro-skeletal measurements. In addition to direct sequencing, other genetic methods have been explored to identify *Pocillopora* types in the eastern Pacific (Cunning et al. 2013) and Australia (Torda et al. 2013), and these might work on a larger geographic scale. More reliable and inexpensive genetic approaches provide better resolution of species diversity than traditional morphological identifications, at least among *Pocillopora*.

14.3 Assessing Symbiont Species Diversity

Symbiosis is a common type of interaction among marine organisms, with most associations formed between invertebrates and various taxa of microorganisms (Bucher et al. 2016). Such relationships are important for the survival of individuals, and in some cases the productivity of ecosystems (Muscatine 1990; Friedrich et al. 1999; see Chap. 9, Enochs and Glynn). The key factor in the structure of coral reefs is the mutualism between scleractinian corals and the dinoflagellate algae in the genus *Symbiodinium*. In this relationship, the coral host provides an ideal environment and inorganic nutrients for the algae, which provide energy for the coral (Muscatine 1967; Pearse and Muscatine 1971; Muscatine and Porter 1977; Trench 1979; Sutton and Hoegh-Guldberg 1990; Swanson and Hoegh-Guldberg 1998; Stat et al. 2006).

The genus *Symbiodinium* comprises numerous evolutionarily divergent lineages or clades, with some of these containing tens and potentially 100s of diverging sub-lineages distinguished ecologically by differences in host species association, depth, and geographic range, and differences in physiology (Berkelmans and van Oppen 2006; Rodriguez-Lanetty et al. 2006; Sampayo et al. 2008; Hannes et al. 2009). The ETP is populated with a diverse array of coral-*Symbiodinium* associations (see Chap. 13, Baker et al.), among them the *Pocillopora–Symbiodinium* symbioses represent the most common and chief reef-building associations in the region; therefore, study of these symbioses is imperative for understanding the dynamics of ETP coral reef ecosystems.

Pocilloporids are among the corals that transfer their endosymbiotic dinoflagellates from the maternal parent to developing ova (Hirose et al. 2008). This explains, in part, the relatively high specificity these corals display for certain symbiont species (LaJeunesse et al. 2003, 2004; Sampayo et al. 2007). While most *Pocillopora* symbionts currently lack formal taxonomic binomial names, they do represent distinct genetic and ecological entities. At coarse levels of genetic resolution, two clades of *Symbiodinium*, Clade C and Clade D, predominate at varying frequencies among *Pocillopora* populations across the eastern Pacific, relating to regional temperatures, water turbidity, and bleaching-mortality history (Glynn et al. 2001; Baker et al. 2004; LaJeunesse et al. 2007, 2008, 2010b; Cunning et al. 2013).

ETP *Pocillopora-Symbiodinium* combinations involve a broadly occurring representative of Clade D (LaJeunesse et al. 2010a; Pinzón and LaJeunesse 2011; Cunning et al. 2013). Genetic markers, as well as host specific ecology, distinguish this *Symbiodinium* species from other members of clade D that occur elsewhere in the world (LaJeunesse et al. 2014). Given this finding, it has tentatively been designated *Symbiodinium "glynni"* (*nomen nudum*, pending formal description). While *S. "glynni"* may exist at low abundance background populations in other eastern Pacific corals (LaJeunesse et al. 2007), it is known only to occur in a stable state and at high abundances in *Pocillopora* type 1, and at a lesser prevalence among *Pocillopora* type 3 colonies (Cunning et al. 2013). The presence of this symbiont clearly affects the physiology of the symbiosis in response to stress created by natural temperature anomalies lasting several weeks or more (Glynn et al. 2001; LaJeunesse et al. 2007, 2010b) and during experimental exposures to high temperatures (McGinley et al. 2012). The presence or absence of *S. "glynni"* ultimately influences differential survival of colonies in a population (Sampayo et al. 2008; LaJeunesse et al. 2010b), and may explain why colonies with this symbiont are abundant in areas where seasonal temperatures are variable and water conditions more turbid (LaJeunesse et al. 2010a; Cunning et al. 2013).

In the ETP, high-resolution genetic markers identified distinct lineages of Clade C *Symbiodinium* that associate specifically with each of the three *Pocillopora* types. Sequences of the non-coding region of the psbA minicircle (*psbA*^{ncr}), which can evolve at rates 10 to 20 times faster than ITS rDNA, resolve Clade C lineages that have been evolutionarily distinct for several millions of years and more (LaJeunesse and Thornhill 2011; Thornhill et al. 2014). The phylogenetic comparison among closely related Clade C *Symbiodinium* from other scleractinians in the ETP reveals that *Pocillopora*-specific lineages are monophyletic (Fig. 14.3), and that many of the other ETP scleractinian genera associate with distinct *Symbiodinium* C lineages as well (LaJeunesse et al. 2008). The host-specific distributions of these *Symbiodinium* suggest that they are distinct species that may exhibit differences in physiology, and whether they respond differently to thermal stress requires further experimental research. Pinzón and LaJeunesse (2011) suggested that the ecological dominance of *Pocillopora* type 1 was explained by its propensity to harbor *S. "glynni"* (Clade D), but the clade C symbiont (C1b-c) may also impart a level of thermal tolerance that is greater than the symbionts associated with *Pocillopora* types 2 and 3 (C1ee and C1b, respectively).

Clade C symbionts of eastern Pacific *Pocillopora* spp. also associate with these corals in the central and western Pacific. The increased phylogenetic resolution provided by *psbA*^{ncr} indicates that type *C1b-c* in the eastern Pacific is simply a genetic variant of type *C1c* present in western and central Pacific regions. Together, these populations comprise a distinct and undescribed species of *Symbiodinium* (Thornhill et al. 2014) that occurs with *Pocillopora* type 1 across the Pacific.

Changes in host-symbiont combinations that occur over large geographic scales may be influenced by either environmental gradients or genetic drift in isolation. For example, *Pocillopora* type 3 on O'ahu, Hawai'i, is synonymous with *P. molokensis*, and harbors a distinct symbiont type *C34* (Pinzón et al. 2013a). The number of colonies examined, however, were few and collected at considerable depth (>20 m; LaJeunesse et al. 2004). Likewise, *Pocillopora* type 5 (the canonical *P. damicornis*), which is not known to occur in the eastern Pacific (LaJeunesse et al. 2004; Pinzón and LaJeunesse 2011; Cunning et al. 2013), harbored *C1d* at sites around O'ahu, Hawai'i (LaJeunesse et al. 2004; Pinzón and LaJeunesse 2011: unpub. observ.). This indicates that while *C1d* is *Pocillopora* specific, it can associate with more than one species, depending on geographical and historical factors. Phylogeographic partitions based on *psbA*^{ncr} haplotypes discerned Hawai'i *C1d* populations from those in the eastern Pacific. However, genotypes from Palau also occurred in each grouping. The lack of clear genetic divergence among *psbA*^{ncr} haplotypes within each *Symbiodinium* sp. further suggests that populations of *Pocillopora* are genetically connected across the Pacific basin, which likely engenders greater connectivity among populations of their symbionts (Pettay and LaJeunesse 2013).

The developmental biology of *Pocillopora*, which involves translocation of symbionts to the developing ovum, may considerably enhance the dispersal capability of their associated *Symbiodinium*. Population genetic markers applied to the analyses of both host and symbiont indicate successful co-dispersal across large expanses of the eastern Pacific (Pettay and LaJeunesse 2013). The only location where strong genetic structure exists for *S. "glynni"* (but not for populations of the host) is at the boundary between the Gulf of California and populations along the central Mexican coast. The Gulf of California represents the northern most range of *Pocillopora*, where seasonal environmental conditions (e.g., light and temperature) are extreme relative to southern locations (Pettay and LaJeunesse 2013). Based on available evidence, it is reasonable to propose that environmental selection has created this population genetic break and those populations of *Pocillopora* type 1 associated with locally adapted genotypes of *Symbiodinium "glynni"*.

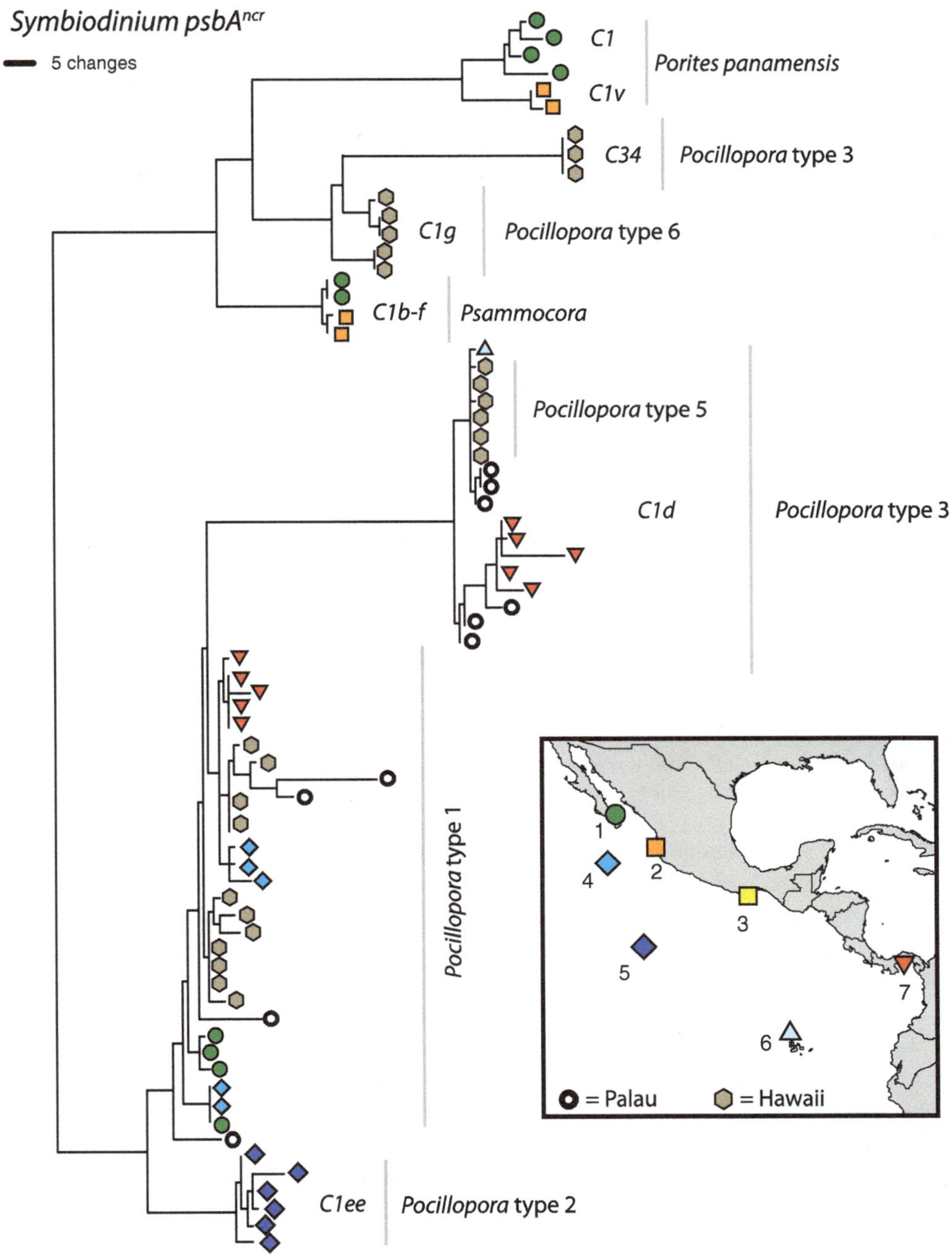

Fig. 14.3 Maximum parsimony phylogenetic reconstruction of Clade C *Symbiodinium* associated with eastern Pacific Scleractinia based on *psbA^ncr* haplotype sequences. Geographic origin of each sample is indicated by symbols that correspond to its location on the map (inset; *1* Gulf of California, *2* Banderas Bay, *3* Oaxaca, *4* Revillagigedo Islands, *5* Clipperton Atoll, *6* Galápagos, *7* Panama). The ITS2-DGGE fingerprint type is indicated to the right of each lineage indicating that nuclear and chloroplast DNA loci are concordant. The sequence similarity break evident between lineages suggests that most corals represent distinct species (data provided by Todd C. LaJeunesse. Alignment in nexus format can be found at the University of Texas Research Commons Library—https://uta-ir.tdl.org/uta-ir/handle/10106/24302)

14.4 Factors Driving *Pocillopora's* Success in the Eastern Tropical Pacific

14.4.1 *Pocillopora* Shows High Connectivity Within the Eastern Pacific and Across the Indo-Pacific

Gene flow is a homogenizing force of evolution (Slatkin 1987), and varying rates can influence the local adaptation of populations to different environments that may exist across a species' geographic range (Bridle et al. 2010). For species with broad distributions, the influence of gene flow and selection pressure impact the adaptation of populations across environmental clines. In some cases, there is sufficient genetic variation among individuals that all populations are adapted to the full range of environmental conditions found throughout the species' distribution. In other examples, geographically discreet populations contain individuals adapted to local environmental conditions (Kirkpatrick and Barton 1997; Bridle et al. 2010).

A wide expanse of deep water between the central and eastern Pacific (~5000 km), known as the eastern Pacific Barrier, poses a significant challenge for the dispersal of most marine species (Grigg and Hey 1992; Lessios and Robertson

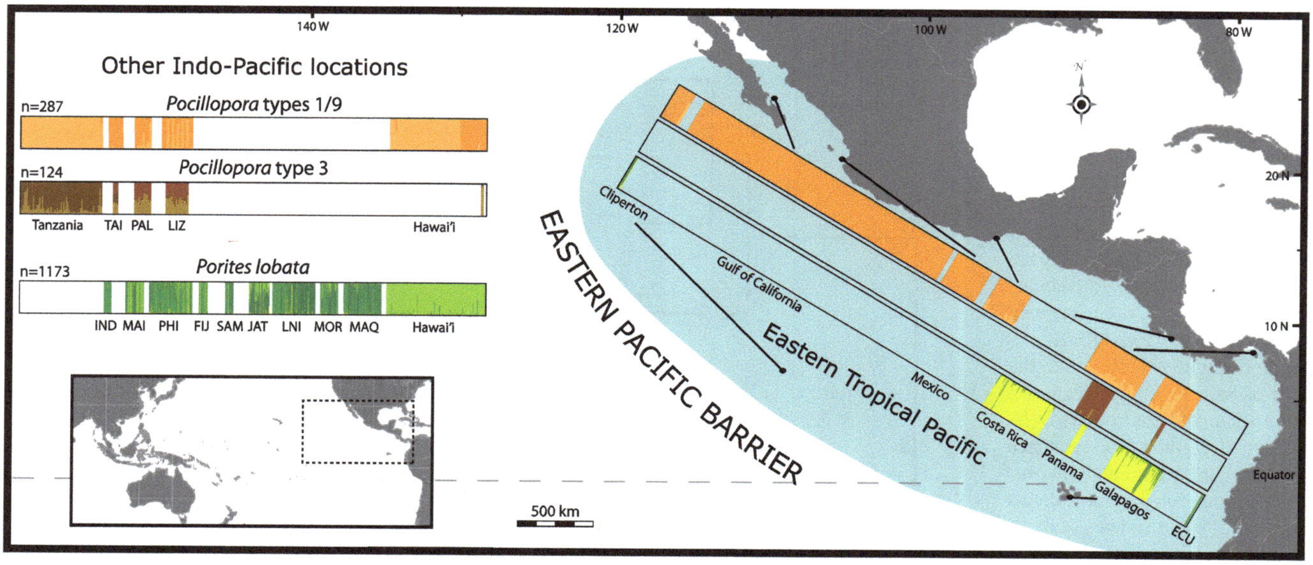

Fig. 14.4 Map of the eastern tropical Pacific and a portion of the Central Pacific showing the population structure of *Pocillopora* and *Porites* species. Clustering was determined on Structure (version 3.2) using 7 microsatellite loci for *Pocillopora* types 1/9 (*orange tones*) and *Pocillopora* type 3 (*brown tones*), and 12 for *Porites lobata* (*yellow* and *green tones*). These scleractinian corals showed no population differentiation within the ETP, having single, region-wide, well-connected populations. On a larger scale, however, the *Pocillopora* species, contrary to *P. lobata*, remain genetically connected to populations from other locations in the central Pacific and elsewhere in the Indo-Pacific region. (*TAI* Thailand, *PAL* Palau, *LIZ* Lizard Island, *IND* Indonesia, *MAI* Marshall Islands, *PHI* Phoenix Islands, *FIJ* Fiji, *SAM* Samoa, *JAT* Johnston Atoll, *LAI* Line Islands, *MOR* Moorea, *MAQ* Marquesas Islands, *ECU* Ecuador). *Pocillopora* (Pinzón and Lajeunesse 2010; Pinzón et al. 2013a, b) and *Porites* (Baums et al. 2012a, b) data downloaded from Dryad Digital Repository

2006; see Chap. 16, Lessios and Baums). This barrier consequently generates various levels of isolation across different species and is an ideal region to study genetic connectivity, or the lack thereof, in marine species. Population genetic studies of the corals *Porites lobata* (Baums et al. 2012a), *Pocillopora* type 1, and *Pocillopora* type 3 (Pinzón and LaJeunesse 2011; Pinzón et al. 2013a), show contrasting patterns in genetic structure and relative isolation across the same Indo-Pacific region. *Porites lobata* and *Pocillopora* type 1 are distributed widely throughout the Indo-Pacific and co-occur in the ETP from the equatorial Galápagos Islands to the high northern latitude of Banderas Bay, Mexico (Pinzón and LaJeunesse 2011; Baums et al. 2012a). Data from both species in the ETP suggest high gene flow at regional scales (Pinzón and LaJeunesse 2011; Baums et al. 2012a; Pinzón et al. 2013a) (Fig. 14.4). Other aspects (such as migration from the Central Pacific) could generate similar patterns, however this appears unlikely, at least for *Porites lobata* (Baums et al. 2012a). Over larger scales, encompassing the thousands of kilometers separating the central Pacific (Hawai'i) from the eastern Pacific, *Pocillopora* type 1 populations appear well connected and show no perceptible differentiation between these two regions (Pinzón et al. 2013a). In contrast, populations of *Porites lobata* are genetically isolated (Baums et al. 2012a) (Fig. 14.4). Wide interconnected populations appear to be common among *Pocillopora* species with geographically wide distributions. *Pocillopora* type 3, for example, is less abundant than *Pocillopora* type 1 and *Porites lobata* in the ETP (Pinzón and LaJeunesse 2011; Baums et al. 2012a; Pinzón et al. 2013a), but is also widely distributed and shows intermediate levels of genetic connectivity throughout the Indo-Pacific region (Pinzón et al. 2013a).

Gene flow and genetic variation affect the adaptive potential of all animals, including corals. In the ETP, *Porites lobata* has relatively lower abundances with populations apparently more isolated (Baums et al. 2012a; see Chap. 16, Lessios and Baums); in contrast, *Pocillopora* types 1 and 3 have larger population sizes and greater gene flow between populations based on previously analyzed data (Pinzón and LaJeunesse 2011; Pinzón et al. 2013a). Individuals of *Pocillopora* spp. are exposed to diverse conditions across their distributional range, affecting growth rates and morphology, and local distributions among reefs. Population connectivity, combined with biological and ecological adaptations, suggests that these species are well adapted to each location, while preserving a single widely distributed genetically connected population.

14.4.2 Is Phenotypic Variation Enhanced by Inter- or Intraspecific Competition?

Among the many factors that can influence diversification, competition plays an important role (Pfennig and Pfennig 2012; Robinson and Pfennig 2013). Natural selection favors

Hawaii
Pocillopora type 1/9
(P. meandrina morphospecies)
(a)
(b)
pd
pm
pd
Pocillopora type 5
(P. damicornis morphospecies)
(c)
Eastern tropical Pacific - Pocillopora type 1
1.0
1.3
2.3
2.8
Depth (m)*
b
v
Depth ~3 m
One genet, four ramets
different morphologies
One genet, one ramet
different morphologies

◀ **Fig. 14.5** High levels of phenotypic plasticity in *Pocillopora* spp. are common across different locations. In Hawai'i (*upper panel*), where morphology and genetics have shown some agreement in the delineation of species, **a**, **b** colonies of *P. meandrina* (=*Pocillopora* type 1/9) develop morphological features characteristic of **c** *P. damicornis* (= *P.* type 5) after being exposed to environmental conditions of deeper habitats (i.e., less water movement and lower light levels—Forsman, unpub. data). In the ETP (lower panel) due to high levels of plasticity, colony morphology is less informative for the identity of species than in Hawai'i. **d** Ramets of the same genet of *Pocillopora* type 1 from the Gulf of California show different colony morphologies, across relatively short distances (less than 90 m), as a response to deep, and possibly wave action, differences. **e** Additionally, different areas of a single *Pocillopora* type 1 colony developed morphologies characteristic of different *Pocillopora* morphospecies; notice spacing between branches (b) and presence of verrucae (v) on left side of colony. All five colonies depicted in the lower panel were identified using mitochondrial ORF and genotyped with 7 microsatellite loci. (pm = *Pocillopora meandrina*, pd = *Pocillopora damicornis*, * = depth not to scale)

individuals with phenotypes that utilize resources unexploited by competing species (Darwin 1859; Robinson and Pfennig 2013), and/or individuals or populations within the same species (Dayan and Simberloff 2005; Rueffler et al. 2006; Johansson and Richter-Boix 2013). In areas where competition is weak, specialized phenotypes are less likely to be eliminated and capable of utilizing all available resources. In that context, the reduced scleractinian diversity in the ETP is ideally suited to offer an environment of reduced competition where dominant organisms, such as *Pocillopora* species, exploit resources otherwise utilized by competitors. Niche expansion may have resulted in character release, an evolutionary phenomenon suggested to have occurred within the genus *Montastraea* (=*Orbicella*) in the Caribbean (Pandolfi et al. 2002).

High phenotypic plasticity among *Pocillopora* spp. affects morphologically-based species' delineations, but this could be a key feature of their success in the ETP. In Hawai'i, colonies of *Pocillopora* type 9 (*P. meandrina*) develop branches similar to the genetically distinct *Pocillopora* type 5 (*P. damicornis*) after being transplanted to other than usual environments (Forsman unpub; Fig. 14.5 upper panel). Coral species diversity in Hawai'i may increase competition while limiting phenotypic divergence between and within species. In other words, phenotypic variation might be constrained by competition between different species. However, intermediate morphologies among distinct species are common, suggesting that species of *Pocillopora* may contain substantial genetic variation that can account for broad phenotypic variation in some populations/locations. In the ETP, the lack of species diversity suggests that competition for space and other resources is relaxed. Inter- and intraspecific competition in *Pocillopora* allows individuals with different phenotypes to broaden the range of resources. Colonies and ramets from the same genet are known to develop distinctive morphologies that may be related to environmental conditions, such as a constant rate of water flow, wave energy, and depth (Fig. 14.5, lower panel).

14.5 Conclusions

Species of *Pocillopora* in general display high phenotypic plasticity, increased reproductive rates, long-distance dispersal, and high fragmentation rates across several locations. Populations in the ETP show particular *Pocillopora*-*Symbiodinium* associations, high levels of gene flow, and high genetic variation. The success of *Pocillopora* in the ETP is likely a result of the combination and synergy between the general aspects of the species in the genus coupled with the particular characteristics of the populations inhabiting the region.

Studies of natural populations distributed through heterogeneous environments can help discern the ecological and/or genetic factors that promote or hinder evolutionary adaptation. Likewise, similarities and differences among genetic lineages can provide insight on the history of a "species" as well as its potential survival in the future (Dionne et al. 2008). Thus, further research is needed to assess how variable environments and lack of competition affect morphological variation and plasticity among *Pocillopora* populations in the ETP, as these traits may influence the success of corals across the Indo-Pacific region in generations to come. Ultimately the survival of any organism depends on its interactions with the surrounding physical environment and other organisms, as well as its capacity for coping with stressful conditions. Therefore physiological and ecological studies on populations that appear more resilient to environmental stress, such as *Pocillopora* in the ETP, may offer important insights towards understanding how corals respond to adverse environmental change.

Acknowledgements I would like to express my gratitude to Todd C. LaJeunesse for the *Symbiodinium* data provided and help with the manuscript. Scott Santos and Dan Thornhill provided comments during the review process. Zac Forsman provided the photographs of *Pocillopora* from Hawai'i. Finally, Peter Glynn, Waleska Castro and Whitney Mann offered helpful comments on early versions of this chapter.

References

Baker AC, Starger CJ, McClanahan T, Glynn PW (2004) Corals' adaptive response to climate change. Nature 430:410

Baums IB, Boulay JN, Polato NR, Hellberg ME (2012a) No gene flow across the Eastern Pacific Barrier in the reef-building coral *Porites lobata*. Mol Ecol 21:5418–5433

Baums IB, Boulay JN, Polato NR, Hellberg ME (2012b) Data from: No gene flow across the Eastern Pacific Barrier in the reef-building coral *Porites lobata*. doi:10.5061/dryad.7gp1f

Benzoni F, Stefani F, Pichon M, Galli P (2010) The name game: morpho-molecular species boundaries in the genus *Psammocora* (Cnidaria, Scleractinia). Zool J Linn Soc 160:421–456

Berkelmans R, van Oppen MJ (2006) The role of zooxanthellae in the thermal tolerance of corals: a 'nugget of hope' for coral reefs in an era of climate change. Proc R Soc Lond, Ser B: Biol Sci 273:2305–2312

Bridle JR, Polechová J, Kawata M, Butlin RK (2010) Why is adaptation prevented at ecological margins? New insights from individual-based simulations. Ecol Lett 13:485–494

Bucher M, Wolfowicz I, Voss PA, Hambleton EA, Guse A (2016) Development and symbiosis establishment in the cnidarian endosymbiosis model *Aiptasia* sp. Sci Rep 6:19867. doi:10.1038/srep19867

Budd AF (1990) Longterm patterns of morphological variation within and among species of reef-corals and their relationship to sexual reproduction. Syst Biol 15:150–165

Budd AF, Romano S, Smith N, Barbeitos M (2010) Rethinking the phylogeny of scleractinian corals: a review of morphological and molecular data. Integr Comp Biol 50:411–427

Budd AF, Fukami H, Smith ND, Knowlton N (2012) Taxonomic classification of the reef coral family Mussidae (Cnidaria: Anthozoa: Scleractinia). Mol Ecol Resour 166:465–529

Burchard JE (1979) Coral fauna of the western Arabian Gulf. Arabian American Oil Company, Dhahran, Saudi Arabia, p 129

Cairns SD (1999) Species richness of recent Scleractinia. Atoll Res Bull 459:1–45

Cantera JR, von Prahl H, Escobar JC, Peña E (1989) Sistemática de los corales del género *Pocillopora* del pacífico colombiano utilizando taxonomía numérica. Rev Biol Trop 37:23–28

Combosch DJ, Vollmer SV (2011) Population genetics of an ecosystem-defining reef coral *Pocillopora damicornis* in the tropical eastern Pacific. PLoS ONE 6:e21200

Combosch DJ, Guzmán HM, Schuhmacher H, Vollmer SV (2008) Interspecific hybridization and restricted trans-Pacific gene flow in the tropical eastern Pacific *Pocillopora*. Mol Ecol 17:1304–1312

Cortés J (1997) Biology and geology of eastern Pacific coral reefs. Coral Reefs 16:S39–S46

Cortés J, Jiménez C (2003) Corals and coral reefs of the Pacific of Costa Rica: history, research and status. In: Cortes J (ed) Latin American coral reefs. Elsevier Science, Amsterdam, pp 361–386

Cunning R, Glynn PW, Baker AC (2013) Flexible associations between *Pocillopora* corals and *Symbiodinium* limit utility of symbiosis ecology in defining species. Coral Reefs 1–7

Darwin C (1859) On the origin of species by means of natural selection, or the preservation of favoured races in the struggle for life. John Murray, London

Dayan T, Simberloff D (2005) Ecological and community-wide character displacement: the next generation. Ecol Lett 8:875–894

Dionne M, Caron F, Dodson JJ, Bernatchez L (2008) Landscape genetics and hierarchical genetic structure in Atlantic salmon: the interaction of gene flow and local adaptation. Mol Ecol 17:2382–2396

Eakin CM (2001) A tale of two ENSO events: carbonate budgets and the influence of two warming disturbances and intervening variability, Uva Island, Panama. Bull Mar Sci 69:171–186

Flot J-F, Tillier S (2007) The mitochondrial genome of *Pocillopora* (Cnidaria: Scleractinia) contains two variable regions: the putative D-loop and a novel ORF of unknown function. Gene 401:80–87

Flot J-F, Magalon H, Cruaud C, Couloux A, Tillier S (2008) Patterns of genetic structure among Hawaiian corals of the genus *Pocillopora* yield clusters of individuals that are compatible with morphology. C R Biol 331:239–247

Flot J-F, Couloux A, Tillier S (2010) Haplowebs as a graphical tool for delimiting species: a revival of Doyle's "field for recombination" approach and its application to the coral genus *Pocillopora* in Clipperton. BMC Evol Biol 10:372. doi:10.1186/1471-2148-10-372

Forsman ZH, Guzmán HM, Chen CA, Fox GE, Wellington GM (2005) An ITS region phylogeny of *Siderastrea* (Cnidaria: Anthozoa): is *S. glynni* endangered or introduced? Coral Reefs 24:343–347

Forsman ZH, Martinez JA, Maragos JE, Toonen RJ (2010) Resurrection of *Porites hawaiiensis* Vaughan, 1907; a Hawaiian coral obscured by small size, cryptic habitat, and confused taxonomy. Zootaxa 2624:67–68

Forsman ZH, Johnston E, Brooks A, Adam TC, Toonen RJ (2013) Genetic evidence for regional isolation of *Pocillopora* corals from Moorea. Oceanography 26:153–155

Friedrich A, Merkert H, Fendert T, Hacker J, Proksch P, Hentschel U (1999) Microbial diversity in the marine sponge *Aplysina cavernicola* (formerly *Verongia cavernicola*) analyzed by fluorescence in situ hybridization (FISH). Mar Biol 134:461–470

Fukami H (2008) Short review: molecular phylogenetics analyses of reef corals. Galaxea 10:47–55

Fukami H, Budd AF, Paulay G, Sole-Cava A, Chen CA, Iwao K, Knowlton N (2004) Conventional taxonomy obscures deep divergence between Pacific and Atlantic corals. Nature 427:832–835

Glynn PW (1976) Some physical and biological determinants of coral community structure in the eastern Pacific. Ecol Monogr 46:431–456

Glynn PW (1993) Monsoonal upwelling and episodic *Acanthaster* predation as probable controls of coral reef distribution and community structure in Oman, Indian Ocean. Atoll Res Bull 379:1–86

Glynn PW (1999) *Pocillopora inflata*, a new species of scleractinian coral (Cnidaria:Anthozoa) from the eastern Pacific. Pac Sci 53:168–180

Glynn PW (2003) Coral communities and coral reefs of Ecuador. In: Cortes J (ed) Latin American coral reefs. Elsevier Science, Amsterdam, pp 449–472

Glynn PW, Ault J (2000) A biogeographic analysis and review of the far eastern Pacific coral reef region. Coral Reefs 19:1–23

Glynn PW, Colley SB (2008) Survival of brooding and broadcasting reef corals following large scale disturbances: is there any hope for broadcasting species during global warming? In: Proceedings of 11th International Coral Reef Symposium, vol 1, Ft Lauderdale, pp 361–365

Glynn PW, Maté JL, Baker AC, Calderón MO (2001) Coral bleaching and mortality in Panama and Ecuador during the 1997-1998 El Niño-Southern Oscillation event: spatial/temporal patterns and comparisons with the 1982-1983 event. Bull Mar Sci 69:79–109

Grigg RW, Hey R (1992) Paleoceanography of the Tropical Eastern Pacific Ocean. Science 255:172–178

Guzmán HM, Cortés J (1989) Growth rates of eight species of scleractinian corals in the eastern Pacific (Costa Rica). Bull Mar Sci 44:1186–1194

Hannes AR, Barbeitos M, Coffroth MA (2009) Stability of symbiotic dinoflagellate type in the octocoral *Briareum asbestinum*. Mar Ecol Prog Ser 391:65–72

Hirose M, Reimer JD, Hidaka M, Suda S (2008) Phylogenetic analyses of potentially free-living *Symbiodinium* spp. isolated from coral reef sand in Okinawa. Japan. Mar Biol 155:105–112

Huang D, Licuanan WY, Baird AH, Fukami H (2011) Cleaning up the "Bigmessidae": molecular phylogeny of scleractinian corals from

Faviidae, Merulinidae Pectiniidae and Trachyphylliidae. BMC Evol Biol 11:37
Iglesias-Prieto R, Reyes-Bonilla H, Riosmena Rodríguez R (2003) Effects of 1997–1998 ENSO on coral reef communities in the Gulf of California, Mexico. Geofis Int 42:467–472
Johansson F, Richter-Boix A (2013) Within-population developmental and morphological plasticity is mirrored in between-population differences: linking plasticity and diversity. Evol Biol 1–10
Jokiel PL (1984) Long distance dispersal of reef corals by rafting. Coral Reefs 3:113–116
Kirkpatrick M, Barton NH (1997) Evolution of a species' range. The Amer Nat 150:1–23
LaJeunesse TC, Thornhill DJ (2011) Improved resolution of reef-coral endosymbiont (*Symbiodinium*) species diversity, ecology, and evolution through psbA non-coding region genotyping. PLoS ONE 6:e29013
LaJeunesse TC, Loh W, van Woesik R, Hoegh-Guldberg O, Schmidt GW, Fitt WK (2003) Low symbiont diversity in southern Great Barrier Reef corals, relative to those of the Caribbean. Limnol Oceanogr 48:2046–2054
LaJeunesse TC, Bhagooli R, Hidaka M, de Vantier L, Done T, Schmidt GW, Fitt WK, Hoegh-Guldberg O (2004) Closely related *Symbiodinium* spp. differ in relative dominance in coral reef host communities across environmental, latitudinal and biogeographic gradients. Mar Ecol Prog Ser 284:147–161
LaJeunesse TC, Reyes-Bonilla H, Warner ME (2007) Spring "bleaching" among *Pocillopora* in the Sea of Cortez. Eastern Pacific. Coral Reefs 26:265
LaJeunesse TC, Reyes-Bonilla H, Warner ME, Wills M, Schmidt GW, Fitt WK (2008) Specificity and stability in high latitude eastern Pacific coral–algal symbioses. Limnol Oceanogr 53:719–727
LaJeunesse TC, Loh W, Trench RK (2009) Do introduced endosymbiotic dinoflagellates 'take' to new hosts? Biol Invasions 11:995–1003
LaJeunesse TC, Pettay D, Sampayo EM (2010a) Long-standing environmental conditions, geographic isolation and host–symbiont specificity influence the relative ecological dominance and genetic diversification of coral endosymbionts in the genus *Symbiodinium*. J Biogeogr 37:785–800
LaJeunesse TC, Smith R, Pinzón JH, Pettay DT, McGinley M, Medina-Rosas P, Cupul-Magaña AL, Pérez A, Reyes-Bonilla H, Warner ME (2010b) Host-symbiont recombination vs. natural selection in the response of coral-dinoflagellate symbioses to environmental disturbance. Proc R Soc Lond, Ser B: Biol Sci 277:2925–2934
LaJeunesse T, Wham D, Pettay D, Parkinson J, Keshavmurthy S, Chen C (2014) Ecologically differentiated, stress tolerant endosymbionts in the dinoflagellate genus *Symbiodinium* Clade D are different species. Phycologia 53:305–319
Lessios H, Robertson D (2006) Crossing the impassable: genetic connections in 20 reef fishes across the eastern Pacific barrier. Proc R Soc Lond, Ser B: Biol Sci 273:2201–2208
Maté JL (2003) Corals and coral reefs of the Pacific coast of Panamá. In: Cortes J (ed) Latin American coral reefs. Elsevier Science, Amsterdam, pp 387–418
McGinley MP, Aschaffenburg MD, Pettay DT, Smith RT, LaJeunesse TC, Warner ME (2012) *Symbiodinium* spp. in colonies of eastern Pacific *Pocillopora* spp. are highly stable despite the prevalence of low-abundance background populations. Mar Ecol Prog Ser 462:1–7
Muscatine L (1967) Glycerol excretion by symbiotic algae from corals and *Tridacna* and its control by the host. Science 156:516–519
Muscatine L (1990) The role of symbiotic algae in carbon and energy flux in reef corals. In: Dubinsky Z (ed) Ecosystems of the World 25. Elsevier, Amsterdam, pp 75–87
Muscatine L, Porter J (1977) Reef corals: mutualistic symbioses adapted to nutrient-poor environments. Bioscience 27:454–460
Pandolfi JM, Lovelock CE, Budd AF (2002) Character release following extinction in a Caribbean reef coral species complex. Evolution 56:479–501
Pearse VB, Muscatine L (1971) Role of symbiotic algae (zooxanthellae) in coral calcification. Biol Bull (Woods Hole) 141:350–363
Pettay DT, LaJeunesse T (2013) Long-range dispersal and high-latitude environments influence the population structure of a "stress-tolerant" dinoflagellate endosymbiont. PLoS ONE. doi:10.1371/journal.pone.0079208
Pettay DT, Wham D, Pinzón JH, LaJeunesse TC (2011) Genotypic diversity and spatial-temporal distribution of *Symbiodinium* clones in an abundant reef coral. Mol Ecol 20:5197–5212
Pfennig DW, Pfennig KS (2012) Development and evolution of character displacement. Ann N Y Acad Sci 1256:89–107
Pinzón JH, Lajeunesse TC (2010) Data from: species delimitation of common reef corals in the genus *Pocillopora* using nucleotide sequence phylogenies, population genetics, and symbiosis ecology. doi:10.5061/dryad.7908
Pinzón JH, LaJeunesse TC (2011) Species delimitation of common reef corals in the genus *Pocillopora* using nucleotide sequence phylogenies, population genetics and symbiosis ecology. Mol Ecol 20:311–325
Pinzón JH, Reyes-Bonilla H, Baums I, LaJeunesse T (2012) Contrasting clonal structure of a single population of *Pocillopora* (Scleractinia) at two nearby sites in the Gulf of California. Coral Reefs 31:765–777
Pinzón JH, Sampayo EM, Cox EF, Chauka LJ, Chen CA, Voolstra CR, LaJeunesse TC (2013a) Blind to morphology: genetics identifies several widespread ecologically common species and few endemics among Indo-Pacific cauliflower corals (*Pocillopora*, Scleractinia). J Biogeogr 40:1595–1608
Pinzón JH, Sampayo EM, Cox EF, Chauka LJ, Chen CA, Voolstra CR, LaJeunesse TC (2013b) Data from: Blind to morphology: genetics identifies several widespread ecologically common species and few endemics among Indo-Pacific cauliflower corals (*Pocillopora*, Scleractinia). Dryad digital repository. doi:10.5061/dryad.jd512
Reyes-Bonilla H (2002) Checklist of valid names and synonyms of stony corals (Anthozoa: Scleractinia) from the eastern Pacific. J Nat Hist 36:1–13
Reyes-Bonilla H (2003) Coral reefs of the Pacific coast of Mexico. In: Cortés J (ed) Latin American coral reefs. Elsevier Science, pp 331–349
Reyes-Bonilla H, Barraza JE (2003) Corals and associated marine communities from El Salvador. In: Cortés J (ed) Latin American coral reefs. Elsevier Science, Amsterdam, pp 351–360
Richmond RH (1981) Energetic considerations in the dispersal of *Pocillopora damicornis* (Linnaeus) planulae. In: Proceedings of 4th International Coral Reef Symposium, vol 1, Manila, pp 153–156
Richmond RH (1987a) Energetics, competency, and long-distance dispersal of planula larvae of the coral *Pocillopora damicornis*. Mar Biol 93:527–533
Richmond RH (1987b) Energetic relationships and biogeographical differences among fecundity, growth and reproduction in the reef coral *Pocillopora damicornis*. Bull Mar Sci 41:594–604
Richmond RH, Jokiel PL (1984) Lunar periodicity in larva release in the reef coral *Pocillopora damicornis* at Enewetak and Hawaii. Bull Mar Sci 34:280–287
Riegl BM, Purkis SJ eds (2012) Coral reefs of the Gulf: adaptation to climate extremes. Coral reefs of the world 3, Springer p 379
Robinson BW, Pfennig DW (2013) Inducible competitors and adaptive diversification. Curr Zool 59:537–552
Rodriguez-Lanetty M, Wood-Charlson E, Hollingsworth L, Krupp D, Weis V (2006) Temporal and spatial infection dynamics indicate

recognition events in the early hours of a dinoflagellate/coral symbiosis. Mar Biol 149:713–719
Rueffler C, Van Dooren TJM, Leimar O, Abrams PA (2006) Disruptive selection and then what? Trends Ecol Evol 21:238–245
Sampayo EM, Franceschinis L, Hoegh-Guldberg O, Dove S (2007) Niche partitioning of closely related symbiotic dinoflagellates. Mol Ecol 16:3721–3733
Sampayo EM, Ridgway T, Bongaerts P, Hoegh-Guldberg O (2008) Bleaching susceptibility and mortality of corals are determined by fine-scale differences in symbiont type. Proc Natl Acad Sci USA 105:10,444–10,449
Schmidt-Roach S, Lundgren P, Miller KJ, Gerlach G, Noreen AME, Andreakis N (2012) Assessing hidden species diversity in the coral *Pocillopora damicornis* from eastern Australia. Coral Reefs 1–12
Schmidt-Roach S, Miller KJ, Lundgren P, Andreakis N (2014) With eyes wide open: a revision of species within and closely related to the *Pocillopora damicornis* species complex (Scleractinia; Pocilloporidae) using morphology and genetics. Zool J Linn Soc 170:1–33
Slatkin M (1987) Gene flow and the geographic structure of natural populations. Science 236:787–792
Souter P (2010) Hidden genetic diversity in a key model species of coral. Mar Biol. doi:10.1007/s00227-009-1370-3
Stat M, Carter D, Hoegh-Guldberg O (2006) The evolutionary history of *Symbiodinium* and scleractinian hosts—Symbiosis, diversity, and the effect of climate change. Perspect Plant Ecol Evol Syst 8:23–43
Stefani F, Benzoni F, Pichon M, Cancelliere C, Galli P (2008) A multidisciplinary approach to the definition of species boundaries in branching species of the coral genus *Psammocora* (Cnidaria, Scleractinia). Zool Scr 37:71–91
Stefani F, Benzoni F, Yang SY, Pichon M, Galli P, Chen C (2011) Comparison of morphological and genetic analyses reveals cryptic divergence and morphological plasticity in *Stylophora* (Cnidaria, Scleractinia). Coral Reefs 30:1033–1049
Stoddart JA (1983) Asexual production of planulae in the coral *Pocillopora damicornis*. Mar Biol 76:279–284
Sutton D, Hoegh-Guldberg O (1990) Host-zooxanthella interactions in four temperate marine invertebrate symbioses: assessment of effect of host extracts on symbionts. Biol Bull (Woods Hole) 178:175–186
Swanson R, Hoegh-Guldberg O (1998) Amino acid synthesis in the symbiotic sea anemone *Aiptasia pulchella*. Mar Biol 131:83–93
Thornhill DJ, Lewis AM, Wham DC, LaJeunesse TC (2014) Host-specialist lineages dominate the adaptive radiation of reef coral endosymbionts. Evolution 68:352–367
Torda G, Schmidt-Roach S, Peplow L, Lundgren P, van Oppen MJH (2013) A rapid genetic assay for the identification of the most common *Pocillopora damicornis* genetic lineages on the Great Barrier Reef. PLoS ONE. doi:10.1371/journal.pone.0058447
Toth LT, Aronson RB, Vollmer SV, Hobbs JW, Urrego DH, Cheng H, Enochs IC, Combosch DJ, van Woesik R, Macintyre IG (2012) ENSO drove 2500-year collapse of eastern Pacific coral reefs. Science 337:81–84
Traylor-Knowles N, Granger BR, Lubinski TJ, Parikh JR, Garamszegi S, Xia Y, Marto JA, Kaufman L, Finnerty JR (2011) Production of a reference transcriptome and transcriptomic database (PocilloporaBase) for the cauliflower coral *Pocillopora damicornis*. BMC Genomics 12:585
Trench RK (1979) The cell biology of plant-animal symbiosis. Annu Rev Plant Physiol 30:485–531
Veron JEN (1995) Corals in space and time. The biogeography and evolution of the Scleractinia. Comstock/Cornell, Ithaca and London, p 321
Veron JE (2000) Corals of the world, 3 vols. Australian Institute of Marine Science, Townsville, pp 429, 463, 490
Veron JE (2002) New species described in corals of the world. Australian Institute of Marine Science, Townsville, Australia
Veron JEN, Marsh LM (1988) Hermatypic corals of Western Australia: records and annotated species list. Rec West Aust Mus 29:1–136
Veron JE, Pichon M (1976) Scleractinia of eastern Australia. Part I. Families Thamnasteriidae, Astrocoeniidae, Pocilloporidae. Australian Institute of Marine Science, Monograph Series, vol 1, p 86
Vidal-Dupiol J, Zoccola D, Tambutté E, Grunau C, Cosseau C, Smith KM, Freitag M, Dheilly NM, Allemand D, Tambutte S (2013) Genes related to ion-transport and energy production are upregulated in response to CO_2-driven pH decrease in corals: New insights from transcriptome analysis. PLoS ONE 8:e58652. doi:10.1371/journal.pone.0058652
Yeoh S-R, Dai C-F (2010) The production of sexual and asexual larvae within single broods of the scleractinian coral, *Pocillopora damicornis*. Mar Biol 157:351–359
Zapata FA, Vargas-Ángel B (2003) Corals and coral reefs of the Pacific coast of Colombia. In: Cortés J (ed) Latin American coral reefs. Elsevier Science, Amsterdam, pp 419–448

Coral Reproduction in the Eastern Pacific

15

Peter W. Glynn, Susan B. Colley, Eugenio Carpizo-Ituarte, and Robert H. Richmond

Abstract

Sexual reproductive activity has been demonstrated in all reef-building (zooxanthellate) scleractinian corals examined from Mexico to the equatorial eastern Pacific (Galápagos Islands). Eleven of 13 species spawn gametes, six are gonochoric, three hermaphroditic, and four exhibit significant mixed sexuality (both gonochoric and hermaphroditic). Four or 30.1 %, two species each of *Pocillopora* and *Porites*, produce autotrophic ova. *Porites panamensis* is the only known zooxanthellate brooder. Also sexually active are the azooxanthellate scleractinian *Tubastraea coccinea* and the zooxanthellate hydrocoral *Millepora intricata*. Reproductive structures, sex ratios, age at sexual maturity, sexuality, and developmental mode have been determined from largely histological evidence. Agariciid corals, comprising more than one-third of investigated species, exhibit predominantly mixed sexual systems with sequential cosexual hermaphroditic cycles in four species. Mixed sexuality is also minimally exhibited in populations of two dominantly gonochoric species. Several eastern Pacific corals spawn mostly on lunar day 17 and 1–2 days following; however, multispecific spawning has not been observed probably because of seasonal, diel, and variable timing in spawning behavior. Factors contributing to the high fecundity of eastern Pacific corals include (1) seasonally prolonged reproductive activity, (2) small size of mature gametes allowing for production of high numbers, (3) split spawning with bimonthly gamete production in some species, (4) alternation of sex maturation in gamete development, and possibly (5) their low latitudinal location under relatively constant and high thermal conditions. Coral community persistence, reef growth

P.W. Glynn (✉)
Department of Marine Biology and Ecology, Rosenstiel School of Marine and Atmospheric Science, University of Miami, 4600 Rickenbacker Causeway, Miami, FL, 33149 USA
e-mail: pglynn@rsmas.miami.edu

S.B. Colley
Stanley S. Scott Cancer Center, LSU Health Science Center New Orleans, 1700 Tulane Avenue Room 914, New Orleans, LA, 70112 USA
e-mail: scolle6@tigers.lsu.edu

E. Carpizo-Ituarte
Laboratorio de Ecología y Biología Del Desarrollo, Instituto de Investigaciones Oceanológicas, Universidad Autónoma de Baja California, Baja California, Mexico
e-mail: ecarpizo@gmail.com

R.H. Richmond
Kewalo Marine Laboratory, University of Hawaii at Manoa, 41 Ahui Street, Honolulu, HI, 96813USA
e-mail: Richmond@hawaii.edu

P.W. Glynn et al. (eds.), *Coral Reefs of the Eastern Tropical Pacific*,
Coral Reefs of the World 8, DOI 10.1007/978-94-017-7499-4_15

and recovery are highly dependent on both sexual and asexual reproductive processes. Asexual fragmentation by physical and biotic causes is particularly important, especially for branching pocilloporid species and the fungiid coral *Diaseris distorta*. Asexual propagation in massive and encrusting poritid and agariciid species is also common-place, often the result of bioerosion and colony breakage by foraging reef fishes. Some research areas in need of attention are noted, for example (a) timing of spawning and the behavior of gamete release of several species, (b) life cycles of *Pocillopora* spp. and *Millepora intricata*, and (c) effects of anthropogenic stressors on eastern Pacific coral reproduction and recruitment.

Keywords
Seasonality • Broadcast spawning • Brooders • Fecundity • Gametogenesis

15.1 Introduction

Sexual and asexual reproductive functions are critical life history processes in the maintenance, repopulation and recovery of coral communities. This is especially true for highly dispersed coral populations and for coral reefs subject to frequent disturbances, such as those in the eastern tropical Pacific. The sexual patterns, spawning modes and timing, fecundity, and recruitment are examined in this chapter for several coral species that contribute importantly toward reef-building and coral community structure. In addition, the reproductive biology of eastern Pacific corals will be compared with those of reef corals in other biogeographic regions to underline similarities and differences in relation to environmental conditions.

A central question that emerged early in the study of eastern Pacific coral reefs was how the small and dispersed reefs of this region sustain growth and persist in the face of frequent disturbances (Dana 1975; Richmond 1990). Richmond's (1982, 1985) initial studies in Panama suggested that sexual reproduction was rare or non-existent in *Pocillopora damicornis*, the principal reef-building species in most eastern Pacific areas. This result contributed importantly to the hypothesis that eastern Pacific reefs were populated and maintained primarily by asexual processes such as fragmentation, and by initial and perhaps episodic long distance dispersal of larvae from source populations in the central and western Pacific (Dana 1975; Richmond 1987a, b, 1990).

Studies of the reproductive biology and ecology of eastern Pacific corals have continued in the equatorial eastern Pacific (EEP: Panama, Costa Rica, and the Galápagos Islands) from the 1980s to the present, and in Mexico beginning in the early 2000s. As a result, sexual reproduction has been found to be prevalent among the eastern Pacific genera of major reef-building families (Pocilloporidae, Poritidae, Agariciidae). In addition, less structurally important families, such as the zooxanthellate Siderastreidae (*Psammocora* spp.), Fungiidae (*Diaseris*), and azooxanthellate Dendrophylliidae (*Tubastraea*), have also been shown to exhibit frequent sexual activity. New information on the reproduction of *Millepora intricata*, an abundant zooxanthellate hydrocoral in western Panama (Gulf of Chiriquí), has also shown this species to be sexually active.

Within the present chapter, sexual reproductive processes are first examined, followed by recruitment and the importance of asexual reproduction in conjunction with the effects of physical and biotic (corallivory, fish foraging, bioerosion) factors on coral fragmentation. A discussion of the regionally unique environmental conditions of the eastern Pacific, and potentially stressful anthropogenic effects on coral reproduction, conclude the chapter.

15.2 Patterns of Sexual Reproduction

15.2.1 Gonads, Gametes and Larvae

Coral gonads are not true organs but rather sites of gamete development within or closely associated with the 12 primary mesenteries. Ovaries and testes enlarge, with developing oocytes and spermaries assuming spherical sizes and shapes as gametogenesis proceeds. In order to quantify the pace of gamete maturation, gonads have been assigned to four developmental stages in eastern Pacific studies. Morphological changes during development are briefly summarized for both broadcast spawning and brooding species in Tables 15.1 and 15.2. Color comparisons of gametes that follow are based on histological preparations of Azocarmine G and Aniline Blue (Luna 1968).

Table 15.1 Gonad stages employed in assessing reproductive activity in eastern Pacific broadcast-spawning corals

Gonads	Gametogenesis	Female	Male
Stage I	Interstitial cells enlarging (3–20 μm diameter); light blue, sometimes pink; prominent nuclei; near or entering mesoglea	Interstitial cells enlarging individually; sizes variable (20–65 μm); usually beige	Cluster of interstitial cells move into mesoglea; usually stain gray, blue or beige
Stage II	Oocytes and spermaries enlarging; sizes vary depending on species; no developmentally unique characteristics	Accumulation of cytoplasm, nucleus continues to enlarge, cytoplasm dark pink or blue	Further migration of interstitial cells into spermary; mitosis of spermatocytes begins; stain pink, raspberry, peach
Stage III	Period of most rapid growth	Vitellogenesis increases rapidly; nucleus continues to grow; large clear nucleus generally centrally located; color of ooplasm begins to change, can be species specific culminating in Stage IV color; *Pocillopora* ooplasm peach, *Porites* ooplasm darkens to red or maroon; lipid vesicles in *Porites* and *Pocillopora* become prominent	Mitosis of spermatocytes, tightly packed, regularly shaped, cells often hexagonal from adhesion; gonadal lumen forms, raspberry-like in appearance; spermary generally fuchsia or pink
Stage IV	Period of maturation prior to release	Usually a color change in staining (darkens); red due to Azocarmine (more protein). *Pocillopora* orange, *Pavona* peach or gray blue, *Porites* sometimes maroon; nucleus migrates to periphery of ova, often becomes irregular in shape; changes in nucleus not obvious in *Pavona*, *Psammocora* and *Diaseris*; zooxanthellae begin to enter ova of *Porites* and *Pocillopora*; thickened border is visible just inside ova membrane of *Pavona*, *Porites*, and *Psammocora*; space forms around nucleus that separates it from ooplasm in *Porites lobata* and *Pocillopora* spp; more subtle color change in *Diaseris* (slight), *Pavona* and *Gardineroseris*	Spermatocyte division into spermatozoa (purple), cells without flagella begin to fill lumen, division occurs centrally, flagella form (golden color), spermaries teardrop-shaped, flagella arranged in bouquet (Stages IV & V, Szmant et al. 1985). All species similar in appearance
Spawned	Gametes released into water column	Empty, thin blue sacs of mesoglea, often ruptured	Mesoglea light blue, centralized, torn, usually with few residual spermatozoa; gastrodermis sometimes colorless and puffy

Based on histological preparations and staining with Heidenhain's Aniline Blue and Azocarmine G (Luna 1965, gametocyte description scheme of Szmant 1986)

15.2.1.1 Oogenesis

Histological preparations of *Porites*, *Pocillopora* and *Tubastraea* provided more detail in oogenic development than those of the genera *Pavona*, *Gardineroseris*, *Psammocora* and *Diaseris*. Color changes occurring as development progresses were not as evident in those species of the latter group. Oocytes of this group contained homogeneously fine and grainy ooplasm. This is due in part to a less complex condition of the yolk, which was evident throughout development. Stages I and II oocytes of all species generally displayed similar staining characteristics, however, lipid vesicles appeared at one pole, most notably in both species of *Porites* (*P. lobata*, *P. panamensis*). During Stage III, oocytes generally darkened. *Pocillopora* oocytes transitioned to bright peach, and in *Porites* oocyte gametogenesis progressed into deeper hues of red. In *Tubastraea*, oocytes transitioned from gray (Stage I) to light pink (Stage II) to deep pink as yolk differentiated (Stage III). And in the Agariciidae (*Pavona*, *Gardineroseris*), some darkening was experienced, but could be light blue, light pink or light peach. Ooplasm remained homogeneously granular in this family. *Psammocora* oocytes (two species) approximated the characteristics of the Agariciidae, however, they were often of an orange hue or brick red. Color change was not observed in *Diaseris* oocytes, which remained yellow throughout most of development. Stage IV *Diaseris* oocytes, however, were frequently mottled with raspberry blotches, as were sometimes agariciid oocytes.

Table 15.2 Gonad stages employed in assessing reproductive activity in eastern Pacific brooding corals

Gonads	*Porites panamensis*		*Tubastraea coccinea*	
	Female	Male	Female	Male
Stage I	Generally same appearance as broadcast spawner (*Porites lobata*); oocytes pink to red, 5–10 μm diameter	Red-staining endodermal cells, 2.5–3.0 μm diameter; each clustered near mesoglea	Oocytes migrate into gastrodermis, irregular shape due to migration; pinkish purple, prominent nucleus with scant cytoplasm, generally light pink, can be gray blue; 15–50 μm diameter oocyte: nucleus (1:1–1:2), nucleolus distinct, cherry	Loose cluster of interstitial cells; cluster size ≤ 45 μm, cell size 2–3 μm, enveloped in mesoglea; nucleus prominent; cells faint pink to gray
Stage II	Generally same appearance as broadcast spawner; oocytes pink to red, 20–60 μm; more cytoplasm, larger nucleus	Loose cell bundles, 10–35 μm diameter; pink	Homogeneous cytoplasm finely granular, smooth appearance, edge can be gently ruffled; light pink or gray; prominent nucleus colorless with smooth textureless appearance	Compact sphere, interstitial cells still accumulating from gastrodermis
Stage III	Generally same appearance as broadcast spawner; oocytes pink to red, 60–200 μm; more cytoplasm, larger nucleus, some irregularly shaped oocytes up to 350 μm in diameter	35–170 μm diameter, darken to red, lumen forms centrally	Size increases dramatically, pink, yolk formation beginning, cytoplasm appearance no longer smooth, yolk vesicles up to 3 μm diameter, vesicle contents colorless; nucleus smooth, central, gray to lavender, nucleus: nucleolus ~4:1, oocyte: nucleus 4:1, oocyte surrounded by puffy, empty, colorless endodermal cells, sometimes stain blue	Gray, much smaller than oocytes, honey-combed; regular sized spermatocytes, no lumen observed, clustered, tucked into mesenteries with oocytes
Stage IV	Nucleus darker, moves peripherally, zooxanthellae begin to move into ovum, ovum darkens	Cellular division of spermatocytes; spermatids (1.0–1.5 μm) migrate into lumen then move to periphery during further development	Ova always present in tissues, heterogeneous yolk vesicles 3–10 μm, clumps of 10 vesicles; ova darker pink to maroon, dark red border inside membrane (4–6 μm wide), granular cytoplasm between vesicles, irregular nucleus gray to lavender, migrates peripherally (flattened), ova: nucleus 6.5:1, ovum 300–800 μm; cushiony band of puffy endodermal cells surround ovum	No flagella present in tissues, however heads pointed and flagella very long in 70 % ethanol- preserved spawn, present intermittently in tissues, color dark purple, spermaries 150–200 μm diameter

Based on histological preparations stained with Heidenhain's Aniline Blue and Azocarmine G (Luna 1965, gametocyte description scheme of Szmant 1986)

Both *Psammocora* spp. and *Pavona varians* contained "satellite nucleoli-like structures" (approximately 2–5). These had the same perfectly round, raspberry or dark pink appearance as the prominent nucleolus, however, they were significantly smaller, located in the ooplasm clustered near, but outside, the nucleus and most noticeably in Stage III oocytes. Their function is unknown; these could be remnants of oocyte fusion that has been observed in at least one species of eastern Pacific corals (*Tubastraea coccinea*) where nuclear fragments were present.

Stage IV ovum development had the most obvious changes in cell structure and coloration. A subtle to dramatic color change was represented in Stage IV ova of many species, and the nucleus completely migrated to the ovum membrane. In some species where few changes could be identified, i.e., *Diaseris distorta*, Stage IV ova had an "off center nucleus" that eventually moved to a more peripheral location, but was never observed in preparations to be close to the nuclear membrane. A space often formed between the nucleus and ooplasm in late Stage IV ova of *Porites* and *Pocillopora* species. In *Pocillopora,* Stage IV ova were replete with prominent, colorless lipid vesicles/vacuoles within the peach-colored ooplasm, presumably to aid in buoyancy. This was not apparent in *Porites* spp., which darkened to maroon and contained dense granular material before being released.

Nuclei in *Porites*, *Pocillopora* and *Pavona* became irregular in shape instead of spherical after approaching the cell membrane. Dome, triangular, and cube shapes were observed. Sickle or crescent shapes were in part due to the portion of the nucleus exposed in the histological section as it lay flattened and parallel, adjacent to the oocyte membrane.

A notch or indentation was observed in late Stage IV ova of *Porites* and *Pocillopora*, but was not usually present in ova of other species. This structure had been reported previously in species other than those of the eastern Pacific (Szmant-Froelich et al. 1985).

A border was evident just inside the plasma membrane of developing oocytes of several species. This border was approximately 3 µm wide and was evident in species of *Porites*, *Pavona*, *Tubastraea* and *Psammocora*. It stained more deeply in color than the rest of the ooplasm, but displayed approximately the same hue. This border was most noticeable in *Porites* and *Tubastraea* oocytes and had a granular appearance. In *Tubastraea*, which has very large oocytes (up to 800 µm), the granularity was due to the detection of numerous minute, barely-visible vesicles. Also present in *Pavona* spp., this border had a more subtle, granular appearance. Its function is unknown, but could be associated with a cortical layer of vesicles (Harrsion and Wallace 1990) that bursts, causing an elevation of the membrane to prevent polyspermy at the time of fertilization. In brooders, this border has disappeared in early Stage I planulae.

Porites evermanni is now recognized as occurring commonly in the eastern Pacific, but it reproduces chiefly by asexual fragmentation (Boulay et al. 2013). Due to the high level, year round gametogenesis observed in *P. lobata*, it is likely that few if any *P. evermanni* were confused in the sampling of *P. lobata*.

15.2.1.2 Spermatogenesis

Stage I spermaries in eastern Pacific corals are generally similar; however, there are some differences in staining that are characteristic of certain species. *Porites* Stage I spermaries can be enlarged brown or red cells, while those in the Agariciidae are often gray or bluish. In *Tubastraea,* the spermaries are grayish-pink to light pink and sometimes beige.

Stage II spermaries contain cells that are of the same size, shape and staining properties as those of Stage I. Towards the end of this phase, the spermatocytes become densely packed, and the prominent nucleus seen earlier has disappeared. Stage II spermaries generally take on more rose-colored hues or may darken if previously pink in color. Cells become more regular in size with less detail. Delineation of developmental stages for each species was based on overall gametogenic characteristics. Therefore, stage delineation was determined per species and is not exactly the same for all.

In Stage III, increased volume is due to dense packing and increased number of mitotic products. This results in a honeycomb appearance of the spermary. Cell coloration is similar to that of late Stage II when mitosis often begins. No nucleus is visible in cells that are smooth and without inclusions. Stage III spermaries are generally dark pink, raspberry, or peach colored, but can be light blue or gray if the de-staining of Azocarmine is overly executed. A lumen forming centrally gives the spermary a raspberry-like morphology.

The most consistent characteristic in spermatogenesis of eastern Pacific corals is the color change to deep purple, which occurs in Stage IV. Meiosis occurs during this phase and the color deepens as divisions increase. The lumen becomes filled with division products as meiosis progresses peripherally within the spermary. Spermaries in this transitional phase can be magenta to purple with a variety of cell sizes present. In the eastern Pacific studies, Szmant's (1986) Stages IV and V were combined; therefore both divisions and the formation of spermatozoan tails were included in the same stage (IV). Golden tails typically form at the end of this phase (although not observed in *Tubastraea* histological samples) and purple spermaries assume a teardrop shape with tails aligned at the tapered end, hence the term bouquet. Sperm heads can also occasionally be observed aligned in radial rows.

15.2.1.3 Planula Development

In brooding species, morphological changes during planula development can also be conveniently divided into four stages. Table 15.3 compares planula development in the two eastern Pacific species, *Porites panamensis* (zooxanthellate) and *Tubastraea coccinea* (azooxanthellate). Early embryos or Stage I planulae are released into the gastrovascular cavity or break out of the gastrodermis. This early planula stage typically appears as a bubbly amorphous mass. Blastula or gastrula stages were not observed. A thin mesogleal layer is one of the defining features of Stage II planulae, which appears as a (blue) ring towards the periphery of the larva. Interstitial cell formation initiated earlier, exterior to this (ectoderm formation), progresses in consecutive rows towards the mesoglea

Table 15.3 Planula stages employed in assessing reproductive activity in eastern pacific brooding corals

Planulae	*P. panamensis*	*T. coccinea*
Stage I	Dark red; 240 μm length, 200 μm diameter, red and gold vesicles amorphous, either in gastrodermis or in gastrovascular cavity	Bubbly and amorphous, with bright red and golden vesicles, no longer present in mesoglea, in gastrovascular cavity, devoid of nucleus, dark red border no longer present; yolk migrates and channels into fingerlike projections at peripheral circumference of larva where a row of cells forms, additional cell rows develop progressing interiorly, center of larva more spacious, furrows develop interiorly, interstitial cells forming along furrow
Stage II	First appearance of a thin mesogleal layer, zooxanthellae scattered inside the encircling mesoglea staining similar to Stage I; oral pore, specialization of ectoderm and appearance of nematocytes in late Stage II; 270 μm length, 200 μm diameter	Thin concentric ring of mesoglea forms peripherally (50 μm from edge), rows of interstitial cells exterior to mesogleal ring are differentiating into ciliated, gland, mucous and cnidoblast cell types, migrating yolk vesicles diminish in size as they approach periphery
Stage III	First appearance of mesenteries, 325 μm length, 150 μm diameter	Oral pore forms; interior of larva clears, colorless; a temporary space forms centrally, then colorless gastrodermal cells begin to form a central network peripherally to the mesoglea, few vesicles remain in this area
Stage IV	Larvae white when released in aquaria, planula length 400–600 μm, 200–300 μm diameter (estimated) 329–512[a]	Mesenteries with mesenterial filaments forming from grainy gray/blue material aggregating linearly through the gastrodermis and eventually encircling oral pore; larvae are bright orange when released; planula length highly variable, ranging from 0.5–1.0 mm in live larvae, 300–500 μm diameter

Based on histological preparations and staining with Heidenhain's Analine Blue and Azocarmine G (Luna 1965) and stage IV live larvae. *Porites panamensis* chiefly from Smith (1991); *Tubastraea coccinea* from Glynn et al. (2008)
[a]Range of planula diameters sampled at La Entrega Bay, Huatulco, Mexico (Rodriguez-Troncoso et al. 2011)

and differentiation of cell types begins. Stage III planulae are generally elongate with an oral pore opening into the gastrovascular cavity. Differentiation of the ectodermal cell types continues. Formation of Stage IV planulae results in the possession of two or more mesenteries with mesenterial filaments that can be well developed. Cells of the gastrodermis are formed. The planulae of *Porites panamensis* enlarge as development proceeds, from 200–300 μm in length in early stages to 400–600 μm in length in Stage IV. *Tubastraea* planulae were 0.5 to 1.5 mm in length when released. Additional developmental changes in yolk color and disposition, appearance of mesoglea and mesenteries, and cellular differentiation are noted in Table 15.3.

15.2.1.4 Histological Documentation

To substantiate the prevalence of sexual reproduction in eastern Pacific corals, following is a collection of photomicrographs of the reproductive products of 11 coral species from Mexico to equatorial locations. [Sampling sites and some relevant oceanographic conditions are noted in Fig. 15.1a, b. The reader is referred to Fiedler and Lavín (Chap. 3) for more details.] Three abundant species in Banderas Bay, Mexico demonstrate the hermaphroditic condition (*Pocillopora damicornis*, Fig. 15.2a, b), the gonochoric condition with mature ovaries and a spermary with the bouquet arrangement (*Porites panamensis*, Fig. 15.2c, d), and Stage IV ova and Stage III spermaries with centrally-located lumina (*Pavona gigantea*, Fig. 15.2e, f) of a species that demonstrates a mixed sexual pattern. Zooxanthellae are visible within a mature ovum of *P. panamensis* (Fig. 15.2c).

Stages I-IV ovaries are present in *Porites panamensis* from Uva Island, Panama (Fig. 15.3a–c). Advanced vitellogenesis is evident in Stage IV ova (Fig. 15.3b, c). A mature ovum is readily identified by its crescent-shaped nucleus that has migrated to the peripheral cell membrane (Fig. 15.3c). Numerous zooxanthellae are also visible in mature ova. All spermary stages (I-IV) are present in Fig. 15.3d–f, including the centrally located lumen in Stage III and a presumed empty spermary ('x') after spawning. Early Stage I larvae are typically irregularly shaped and contain abundant lipid stores (Fig. 15.4a). A thin mesogleal layer divides the ciliated ectoderm and gastrodermis in Stage II larvae (Fig. 15.4b). An oral pore, cnidocytes and mesenteries are present in Stage III planulae (Fig. 15.4c). Mesenterial filaments make their appearance in Stage IV planulae (Fig. 15.4d). Zooxanthellae are present in all larval stages with a tendency to aggregate peripherally in the gastrodermis, just inside the mesoglea, as development proceeds.

The disposition of gonads in advanced stages of development in seven broadcast spawning species is illustrated in Figs. 15.5a–f and 15.6a, b. Numerous Stage IV spermaries are visible in the cross-section of most of the 12 mesenteries surrounding the gastrovascular cavity in many *Porites lobata* polyps (Fig. 15.5a). Of the agariciid species, Stage IV ova are arranged longitudinally along the mesenteries of

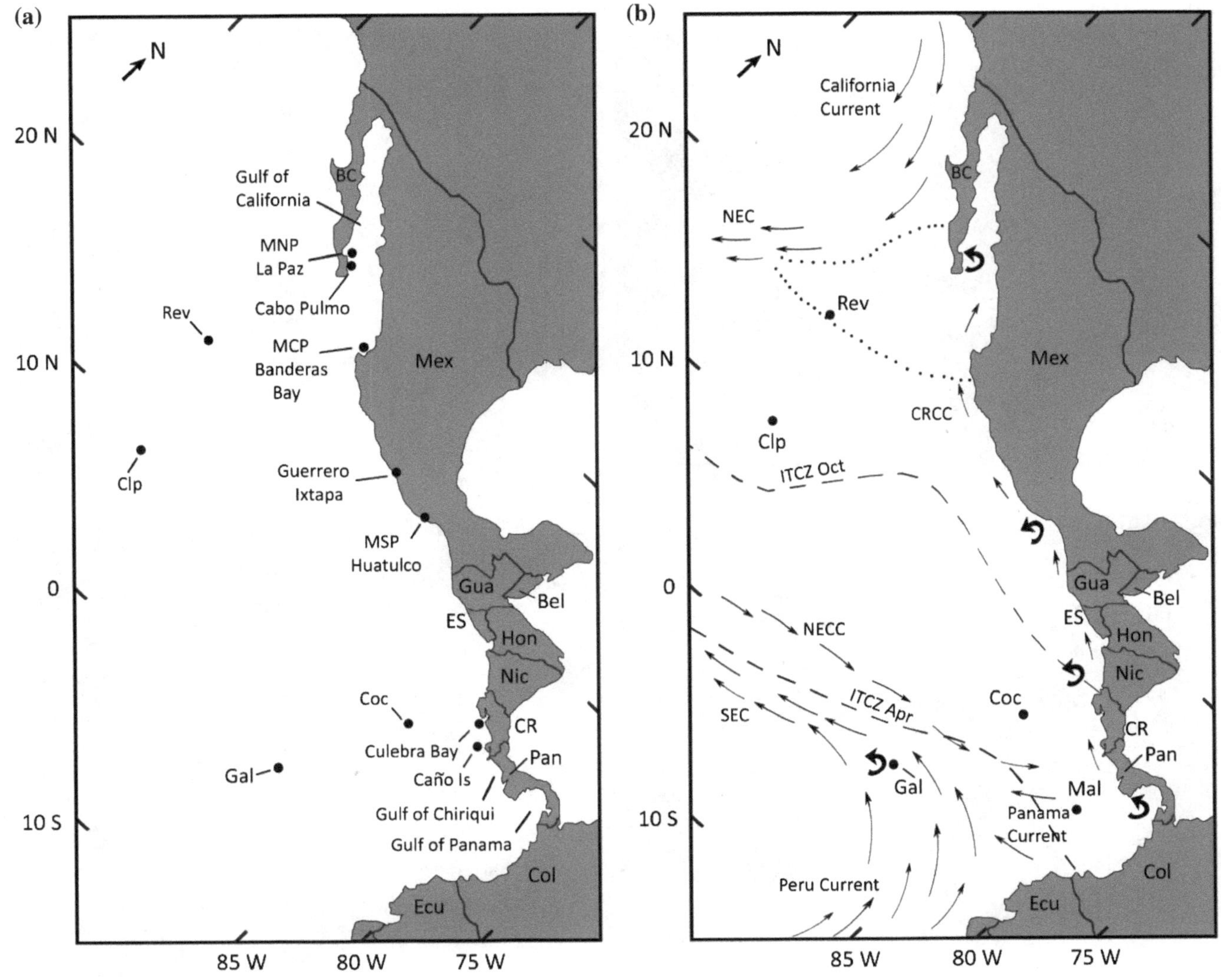

Fig. 15.1 **a** Locations of reproductive sampling sites and **b** oceanographic conditions across the eastern tropical Pacific. **a** MNP, MCP, MSP, Mexico northern (25°N), central (20°N), and southern (15°N) Pacific; MSP includes Entrega Bay, Montosa and Riscalillo sites; Gulf of Chiriquí, Panama includes Uva I and Bahía Honda sites; Gulf of Panama includes Taboga I and Saboga I sites; Galápagos Is include Santa Cruz I, Itabaca Canal, Floreana I and Playa La Picona. *BC* Baja California; *ES* El Salvador; *CR* Costa Rica. Oceanic islands: *Rev* Revillagigedo Is; *Clp* Clipperton Atoll; *Coc* Isla del Coco; *Mal* Malpelo I; *Gal* Galápagos Is. Equatorial eastern Pacific (EEP) sites, study areas between 11°N and 2°S, include Costa Rica, Panama, and Galápagos Islands. **b** Dotted lines off Mexican coast denote limits of frontal boundary between California Current and tropical surface water. *CRCC* Costa Rican Coastal Current; *dashed lines* denote positions of Intertropical Convergence Zone (ITCZ) in October and April. Panama Current flows seasonally from the Panama Bight towards the Galápagos Islands from January to mid–April. *Bold arrows* denote upwelling centers at Cabo Pulmo, Baja California, Gulf of Tehuantepec (Mexico), Gulf of Papagayo (Costa Rica/Nicaragua), Gulf of Panama (Panama), and western Galápagos Islands

Pavona gigantea (Fig. 15.5b). Two clusters of Stage III ovaries are present at the bases of mesenteries in *Pavona clavus* (Fig. 15.5c). Stage IV spermaries, several with bouquet arrangements of spermatozoa, are shown for *Pavona chiriquiensis* (Fig. 15.5d) and *Pavona varians* (Fig. 15.5e, f). Bouquet structures are also shown for *Gardineroseris planulata* (Fig. 15.6a) and *Psammocora stellata* (Fig. 15.6b) with free spermatozoa in the former section. An advanced planula of azooxanthellate *Tubastraea coccinea*, in transition between Stages III and IV, shows clear colorless areas that are gastrodermal anlage and gray granular strings that will form mesenteries (Fig. 15.6c). Asynchronous spermatocyte development (Stages II-IV) and a single Stage II oocyte (with nucleus and nucleolus) illustrate the hermaphroditic condition of *T. coccinea* (Fig. 15.6d).

Sexual reproduction in the hydrozoan coral *Millepora intricata* begins in the sessile polypoid generation by the budding off of male and female planktonic medusae (Soong and Cho 1998; Lewis 2006). Evidence of sexual reproduction was difficult to detect in *M. intricata*, which is restricted in

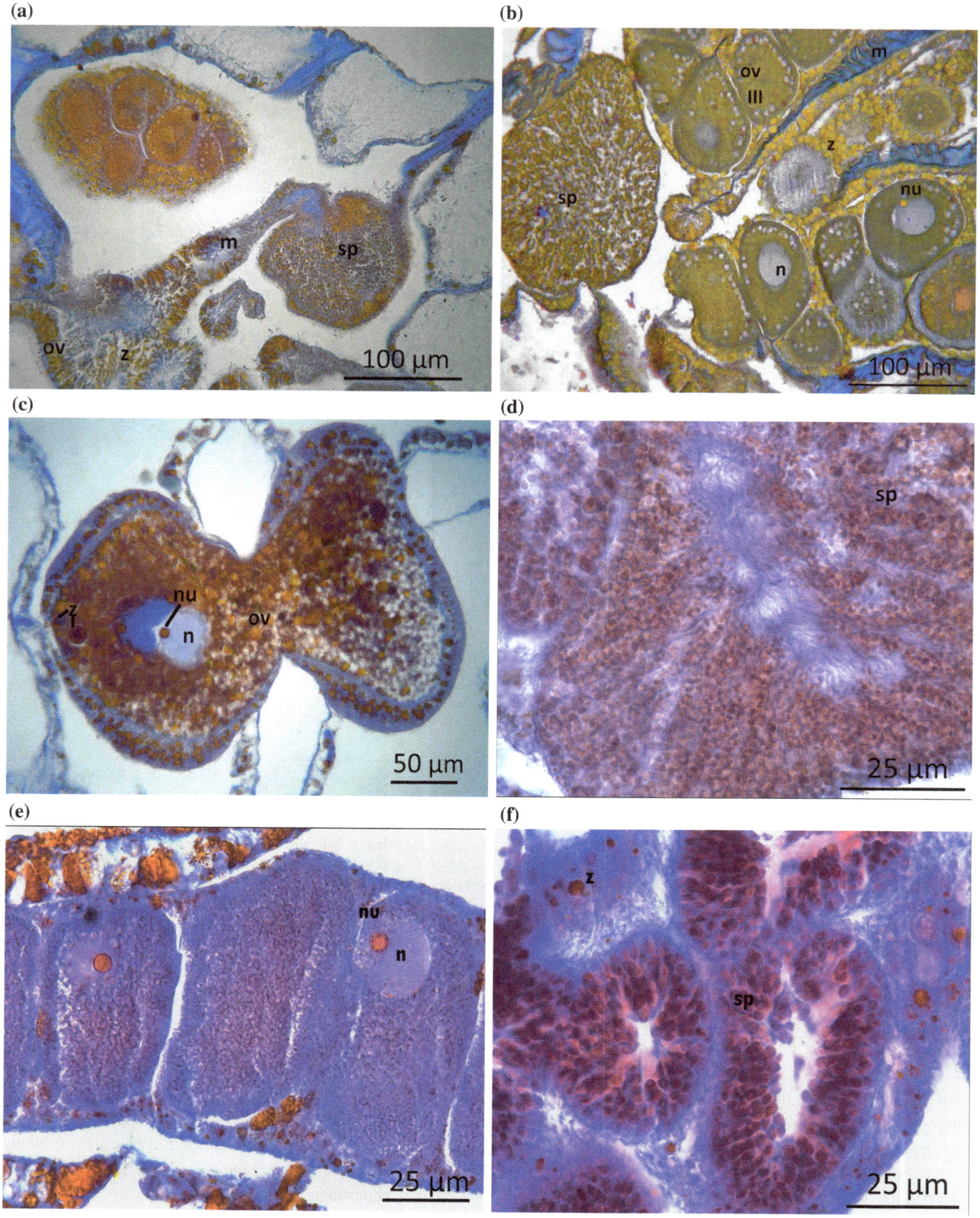

Fig. 15.2 Photomicrographs of gonads from corals sampled at Banderas Bay, Mexico. **a**, **b** *Pocillopora damicornis* simultaneous hermaphrodites with ovaries and spermaries. **c** *Porites panamensis* Stage IV ovum with zooxanthellae. **d** *Porites panamensis* Stage IV spermary. **e** *Pavona gigantea* Stage III oocytes. **f** *Pavona gigantea* Stage III spermaries. Staining: **a**, **c–f**, Azocarmine G and Toluidine Blue (Humason 1967); **b**, Mallory Heidenhain. *ov* ovary, *sp* spermary, *n* nucleus, *nu* nucleolus, *z* zooxanthellae

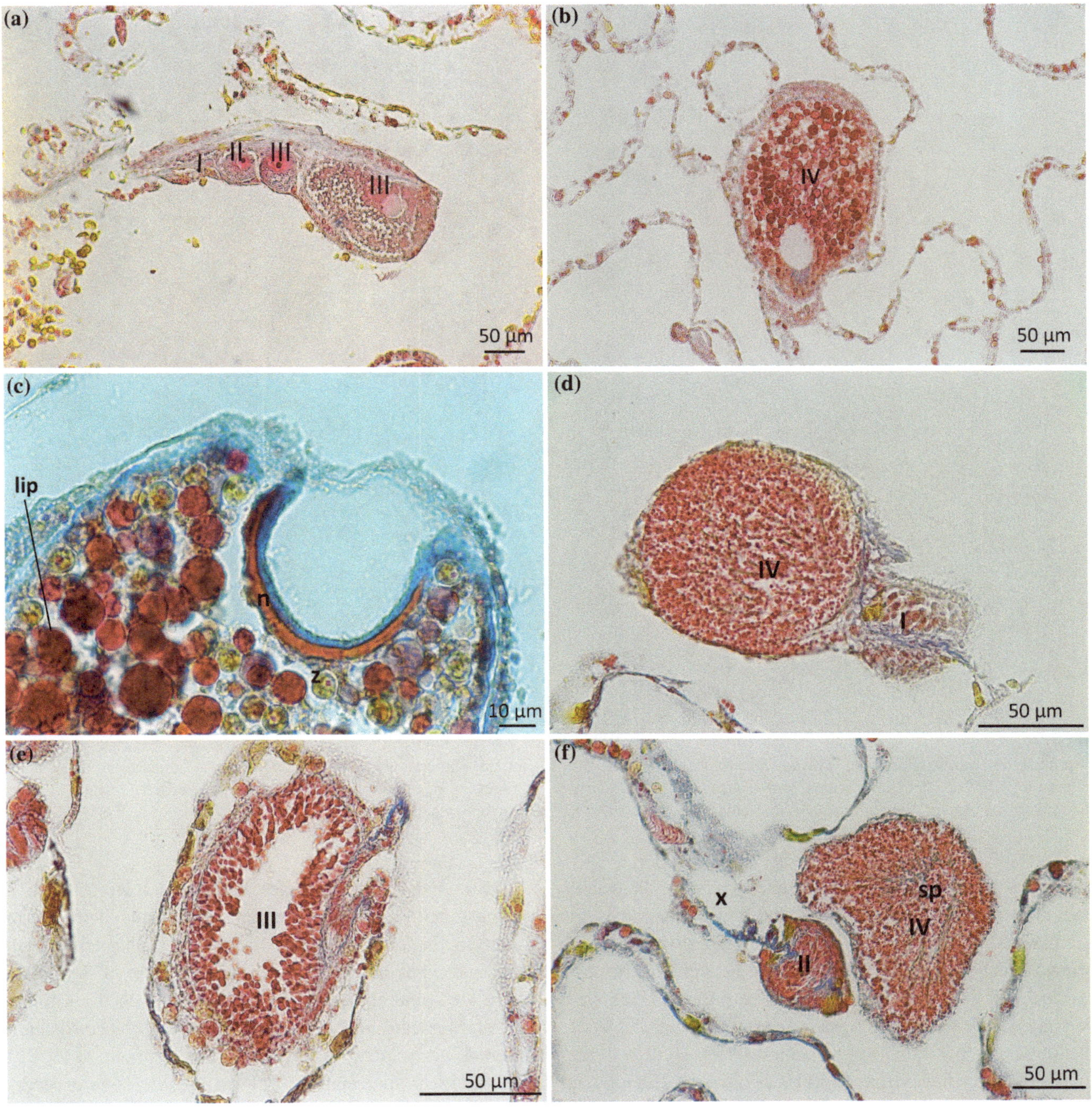

Fig. 15.3 Photomicrographs of gonads from *Porites panamensis*, Uva Island study reef, Gulf of Chiriquí, Panama. **a** Ovary with Stages I, II and III oocytes. **b** Stage IV ovum. **c** Stage IV ovum. **d** Spermary with Stages I and IV spermatocytes. **e** Stage III spermary. **f** Stages II and IV spermaries, '**x**' denotes probable location of spent spermary. Staining: Heidenhain's Aniline-blue (Luna 1968) and Azocarmine G. *sp* spermary, *n* nucleus, *l* lipid vesicle, *z* zooxanthellae

distribution to the Gulf of Chiriquí (Panama) in the eastern Pacific. Histological preparations, however, did demonstrate gametogenesis. A developmentally advanced medusa with two oocytes, one large with a nucleus present and a small portion of one near the ring canal, is shown in Fig. 15.6e. Zooxanthellae are visible in the large oocyte. Two immature medusae within their respective ampullae are shown in Fig. 15.6f. The medusa in the lower left corner contains two developing oocytes. These specimens were collected in January, but due to sparse data, the timing and seasonality of gamete production was not delineated for this species.

15.2.1.5 Sizes of Sexual Products

Critical metrics in assessing the fecundity of corals are gonad, gamete and planula sizes, and their rates of growth and production during the reproductive period. The size ranges of

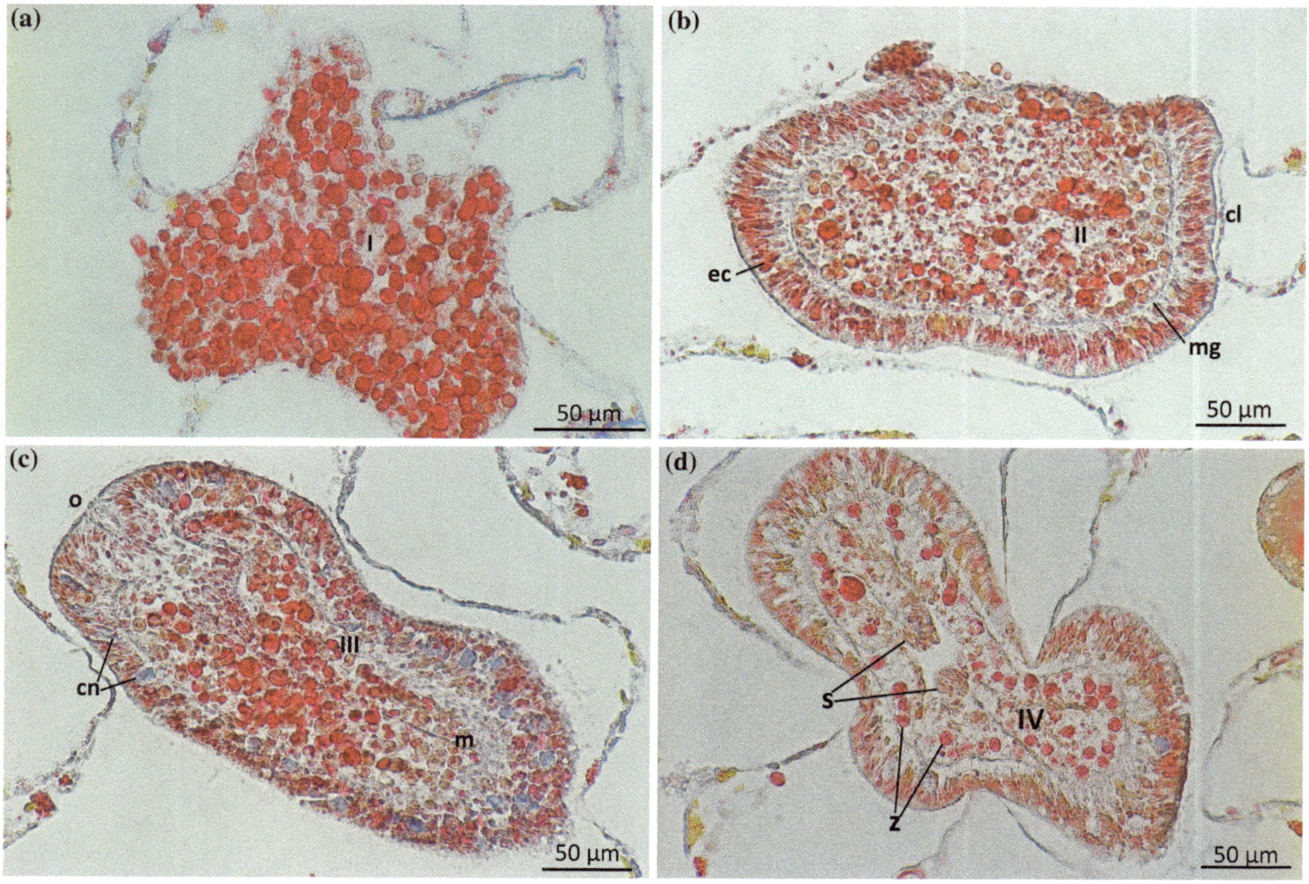

Fig. 15.4 Photomicrographs of *Porites panamensis* planulae, Uva Island reef, Gulf of Chiriquí, Panamá. **a–d** Stages I-IV planula larvae respectively. Staining: Heidenhain's Aniline-blue and Azocarmine G. Roman numerals denote planula stages. *o* oral pore, *cl* cilia, *cn* cnidocytes, *ec* ectoderm, *m* mesentery, *mg* mesoglea, *s* septal filament, *z* zooxanthellae

oocytes and spermaries (minimum and maximum diameters), and the mean mature ova sizes and planula volumes are summarized in Tables 15.4 and 15.5, respectively, for the principal reef-building and reef-associated eastern Pacific coral species. It is important to note that size measurements were derived largely from histologically processed samples. These measurements do not take into account shrinkage of cells and tissues that may range from 15 to 30 % (Harriott 1983; Glynn et al. 2008). This issue, and adjustments to compensate for it, is considered later under Fecundity (Sect. 15.4).

Mature ova (Stage IV) in broadcast spawning species commonly ranged from about 50–150 µm in diameter (Table 15.5). Ova in brooding corals were considerably larger, ranging from 300 to 800 µm in diameter. The planulae of brooding species (*Porites panamensis* and *Tubastraea coccinea*) were considerably larger than the ova of spawning species. This is a general trend across different coral reef regions (Harrison and Wallace 1990). Mean ovum volume measurements of broadcast spawning species were similar among species and across localities. Mature spermaries (Stage IV) ranged from about 50–250 µm in diameter.

Ovum sizes of nine species at four localities in the EEP demonstrated relatively similar dimensions with mean diameters that ranged from 90 to 117 µm (Table 15.5). Only *Tubastraea* has larger ova, beyond the range of the others. There are some apparent size differences with the sampled ova of Mexican corals, but whether these are consistently different will require further study. For example, at all EEP sites the mean diameters of *Pavona gigantea* ova exceeded 100 µm whereas at Banderas Bay, Mexico mean ovum diameter was 71 µm with a maximum diameter of 107 µm (Carpizo-Ituarte et al. 2011). *Porites panamensis* ova in Mexico, ~140 µm at Banderas Bay and 132–287 µm at Huatulco (unpub. data), overlap the EEP population at Uva Island, Panama with diameters of 60–250 µm (Smith 1991). Planula diameters of *P. panamensis* in southern Mexico (Huatulco), however, that ranged from 329 to 512 µm were notably larger than those reported from Panama, 170–330 µm (Uva Island) and 174 µm (Gulf of Panama, Table 15.5). Maximum diameters of Stage IV ova of *Pocillopora verrucosa* and *Pocillopora meandrina* from Isla Gaviota, Bahía de La Paz, Mexico were 116 µm

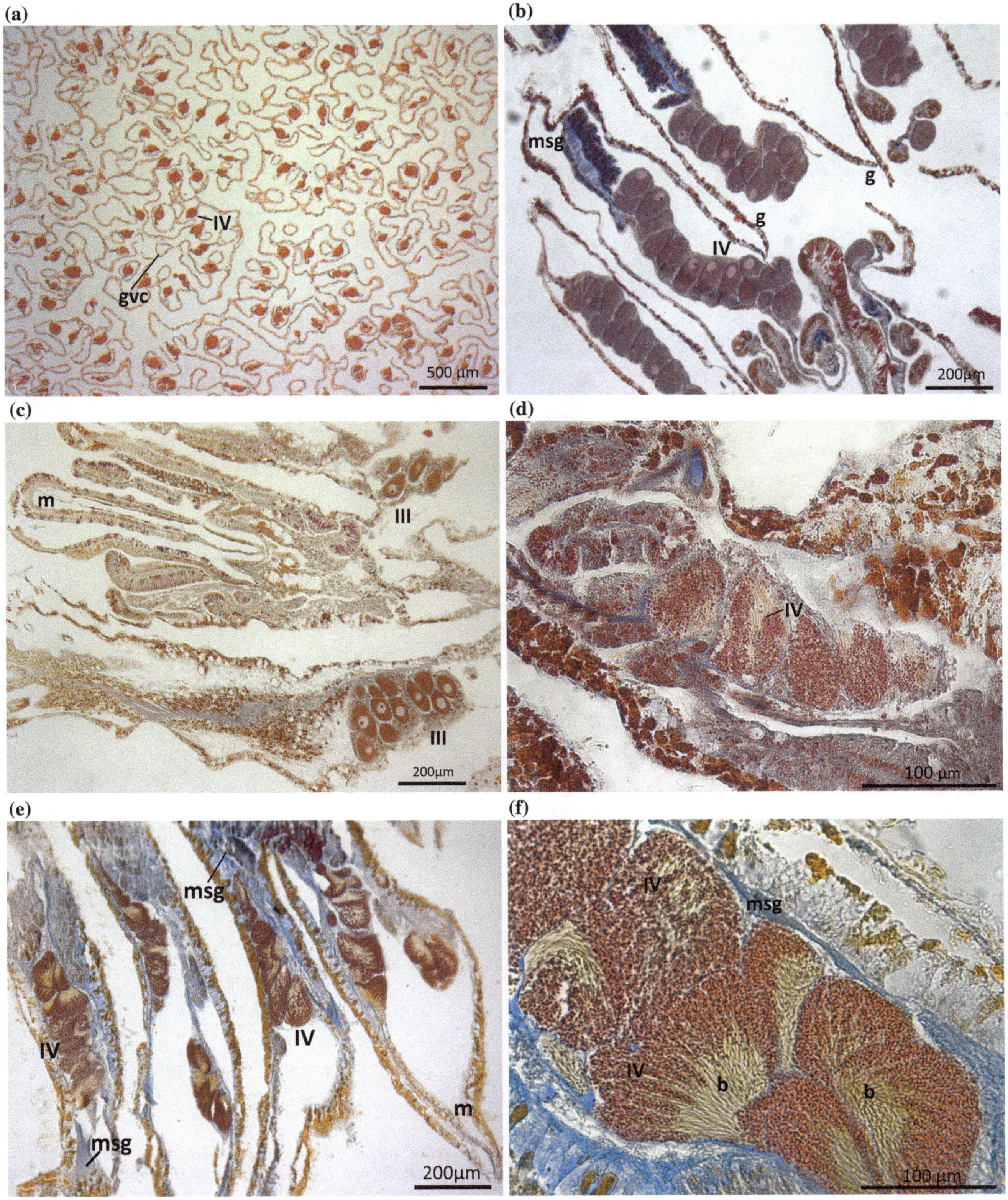

Fig. 15.5 Photomicrographs of gonads. **a** Stage IV spermaries and associated mesenteries in *Porites lobata* (Santa Cruz I, Galápagos Is). **b** *Pavona gigantea* with several Stage IV ova (Itabaca Canal, Galápagos Is). **c** *Pavona clavus* (Saboga I, Gulf of Panama) with two clusters of Stage III oocytes. **d** *Pavona chiriquiensis* (Uva I, Gulf of Chiriquí) with several Stage IV spermaries. **e** *Pavona varians* with numerous Stage IV spermaries (Uva I, Gulf of Chiriquí). **f** *Pavona varians* with Stage IV spermaries (Uva I, Gulf of Chiriquí). Staining: Heidenhain's Aniline-blue and Azocarmine G. *m* mesentery, *msg* mesoglea, *b* bouquet, *gvc* gastrovascular cavity

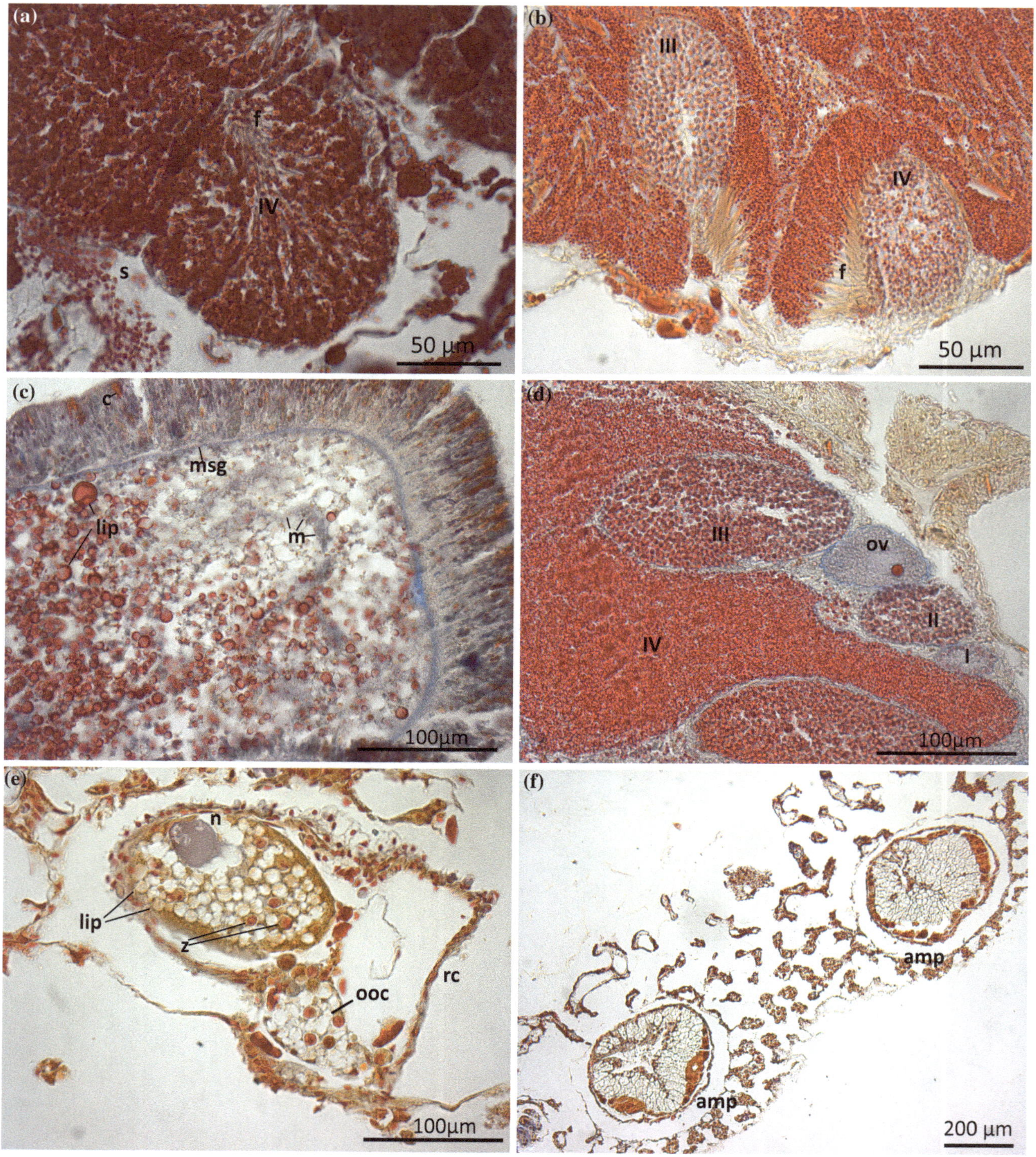

Fig. 15.6 Photomicrographs of reproductive structures of scleractinian corals and hydrocorals sampled at various sites in the equatorial eastern Pacific. **a** *Gardineroseris planulata* (Caño I, Costa Rica) with Stage IV spermaries. **b** *Psammocora stellata* (Floreana I, Galápagos Is) with Stages III and IV spermaries. **c** *Tubastraea coccinea* planula, Stage III (Uva I, Gulf of Chiriquí). **d** *Tubastraea coccinea* (Uva I, Gulf of Chiriquí) with Stages I, II, III, IV spermaries, and Stage II oocyte. **e** *Millepora intricata* (Uva I, Gulf of Chiriquí) with a developing medusa. Parts of two oocytes visible, one with nucleus. **f** *Millepora intricata* (Uva I, Gulf of Chiriquí), two medusae within ampullae. Developing oocytes at periphery. Staining: Heidenhain's Aniline-blue plus Azocarmine G. *s* spermatozoa, *ooc* oocytes, *ov* ovum, *f* sperm flagella, *lip* lipid vesicles, *c* cnidocytes, *msg* mesoglea, *m* mesentery, *z* zooxanthellae, *rc* medusa ring canal, *amp* ampulla

Table 15.4 Minimum and maximum gonad diameters (μm) of 13 eastern Pacific coral species at various stages of development

Gonads	PDAM[a] PELE	PLOB	PGIG	GPLA	PCHR PVAR	PCLA	DDIS	PSTE	PPRO	PPAN	TCOC
Oocyte I	10–20	6–21	5–14	5–14	<15–18	<20	12–20	9–24	9–24	5–10	5–50
Oocyte II	20–40	21–33	10–30	22–35	15–38	20–50	22–62	24–65	24–65	20–60	50–160
Oocyte III	40–80	33–160	30–100	35–110	38–100	50–80	64–88	60–100	60–100	60–350	150–300
Oocyte IV	100	135–170	54–216	62–114	>100	50–120	88–140	90–135	90–135	~300	300–800
Sperm I	–[b]	(~5)[d]	10–15	10–15	(1–5)	(3–5)	15–20	4–24	4–24	20–30 (8–10)	≤45
Sperm II	[c]60–90	27 (10)[d]	15–30	15–30	up to 25	20–35 (5-25)	24–40	24–45	24–45	10–35	~60
Sperm III	110–170	27–260	30–130	30–130	30–120 (>25)	40–200[f]	40–22	39–64	39–64	35–170	~175–200
Sperm IV	180–270	132–230[e]	120–180	50–112	~120	60–200	>84	64–120	64–120	170	150–200

Data are measurements from histological material chiefly from EEP localities in Costa Rica, Panama, and Galápagos Islands (from various publications by Glynn et al. 1991–2012; Smith 1991; Colley et al. 2000)

[a]Species codes: PDAM, *Pocillopora damicornis*; PELE, *Pocillopora elegans*; PLOB, *Porites lobata*; PGIG, *Pavona gigantea*; GPLA, *Gardineroseris planulata*; PCHR, *Pavona chiriquiensis*; PVAR, *Pavona varians*; PCLA, *Pavona clavus*; DDIS, *Diaseris distorta*; PSTE, *Psammocora stellata*; PPRO, *Psammocora profundacella*; PPAN, *Porites panamensis*; TCOC, *Tubastraea coccinea*

[b]Gamete classification scheme of Szmant (1986) employed with exception of Stage IV spermaries; small clusters of interstitial cells located near or entering mesoglea, sizes not stated

[c]Spermary II diameters from Mexican samples (Carpizo-Ituarte et al. 2011)

[d]Number of cells noted in parentheses

[e]The 132 diameter value is the mean of 20 measurements from Caño Island

[f]Spermary III and IV values for *Pavona clavus* are from gonads sampled in the Galápagos Islands

Table 15.5 Mean ovum and planula diameters (μm), and ovum and planula volumes (mm^3) for ten taxa at four EEP locations

Species	Location				Authority
	Caño I	Uva I	G Panama[a]	Galápagos Is.	
Porites lobata	168 (2.5×10^{-3})	136 (1.3×10^{-3})	–	51 (7.0×10^{-5})[b]	Glynn et al. (1994)
Porites panamensis[c]	–	60–250	*174 (5.5×10^{-3})*	–	Smith (1991)
		170–330 (2.0×10^{-3})			
Tubastraea coccinea[c]	459 (7.1×10^{-2})	467 (6.9×10^{-2})	494 (8.6×10^{-2})	503 (8.8×10^{-2})	Glynn et al. (2008)
		258 (2.1×10^{-1})		*252 (2.0×10^{-1})*	
Pavona gigantea	104 (6.4×10^{-4})	114 (8.4×10^{-4})	117 (9.0×10^{-4})	104 (6.9×10^{-4})	Glynn et al. (1996)
Gardineroseris planulata	117 (8.8×10^{-4})	107 (7.2×10^{-4})	–	–	Glynn et al. (1996)
Pavona spp[d]	106 (7.6×10^{-4})	103 (7.4×10^{-4})	112 (8.0×10^{-4})	113 (8.2×10^{-4})	Glynn et al. (2000)
Pavona clavus	88 (4.1×10^{-4})	89 (4.6×10^{-4})	84 (3.6×10^{-4})	90 (4.1×10^{-4})	Glynn et al. (2011)
Psammocora stellata	99 (5.6×10^{-4})	92 (4.6×10^{-4})	93 (4.7×10^{-4})	95 (4.8×10^{-4})	Glynn et al. (2012)
Psammocora profundacella	90 (4.1×10^{-4})	94 (4.8×10^{-4})	–	95 (4.8×10^{-4})	Glynn et al. (2012)
Diaseris distorta	–	–	–	104 (6.2×10^{-4})	Colley et al. (2000)

Volume measurements in parentheses, planula measurements in italics

[a]Combined measurements from Taboga and Saboga Is, Gulf of Panama

[b]Dimensions from a single ovum; only one mature ovum available from 89 colonies examined

[c]Maximum diameter and volume (in parentheses); planula measurements in italics. Planula diameters were not measured directly, but only live, recently released (few hours) and non-swimming larval lengths. A mean length:diameter ratio was established (L:D = 4.3:1) based on larval body proportions in photomicrographs and sketches

[d]Combined measurements from *Pavona varians* and *Pavona chiriquiensis*

(Campos-Vázquez et al. 2014), and comparable to the 100 μm ova of equatorial eastern Pacific *Pocillopora damicornis* and *Pocillopora elegans* of Glynn et al. (1991). Ovum and planula sizes of *Tubastraea coccinea* were markedly larger than those of *P. panamensis* at two localities in Panama. At Uva Island, the azooxanthellate *T. coccinea* planula volume was two orders of magnitude larger than in *P. panamensis* (Table 15.5). If some of these differences can be substantiated it would be of interest to explore the possibility of environmental effects, especially in upwelling and nonupwelling areas.

Ovum sizes in broadcast spawning *Pocillopora* and *Porites* species in the northern Red Sea (Boumeester et al. 2011) and on the Great Barrier Reef (Schmidt-Roach et al. 2012) are within the same size range as comparable eastern Pacific species, i.e. 100–180 μm in diameter. Eastern Pacific *Pavona varians* and *Pavona chiriquiensis*, however, produce mature ova only slightly exceeding 100 μm at all localities. Compared with coral faunas in the Caribbean (Szmant 1986) and Indo-Pacific (Harrison and Wallace 1990; Shlesinger et al. 1998), where egg diameters of diverse genera generally exceed 200–300 μm, eastern Pacific species produce eggs that are small. The planula larvae of *Porites panamensis*, the only zooxanthellate brooder in the eastern Pacific, are also considerably smaller than the larvae of other brooding species from the Caribbean (2–8 times smaller than in *Agaricia*) and Great Barrier Reef (3–9 times smaller than in *Acropora*) listed in Harrison and Wallace (1990).

15.2.1.6 Spermatozoa Morphology

The single study of the ultrastructure of spermatozoa in eastern Pacific corals examined *Pavona gigantea* and *Pocillopora* spp. in Costa Rica (Steiner and Cortés 1996). This study revealed inconsistencies with other suggested sexual patterns. Harrison's survey (1990) and review (2011) have shown that the more primitive conical-shaped sperm heads are mostly associated with gonochoric species, suggesting that gonochorism is an ancestral trait in the Scleractinia. However, *Pavona gigantea* with conical type sperm demonstrated a mixed sexual pattern of gonochoric and hermaphroditic colonies with hermaphrodites predominant at some localities. *Pocillopora damicornis* and *Pocillopora elegans* with bullet-shaped sperm heads and elongated mitochondria are also hermaphrodites but conform closely to acroporid corals. Harrison (1990) suggested that the sperm morphology of hermaphroditic corals may be better suited for swimming, which could reduce rates of self-fertilization in species releasing combined sperm/egg bundles. However, eastern Pacific *Pocillopora* species, although simultaneous hermaphrodites, most likely do not form gamete bundles, therefore this pattern may not apply in this case.

Molecular phylogenetic analyses suggest that the Scleractinia consist of two large clades, the Robust and Complex corals that diverged early in the evolutionary history of the order (e.g., Romano and Palumbi 1996; Fukami et al. 2008). Pocilloporid and acroporid corals are now recognized as members of the Robust and Complex molecular clades, respectively (Kerr 2005; Fukami et al. 2008). This suggests that sperm morphology is more closely linked to sexual pattern than to phylogenetic relationship.

15.2.2 Sexuality and Mode of Development

Sexuality and mode of development are summarized for 13 scleractinian corals in six families in relation to colony morphology and presence of zooxanthellae in mature ova (Table 15.6). Six species, in the families Poritidae (*Porites panamensis, Porites lobata*), Siderastreidae (*Psammocora* spp.), Agariciidae (*Pavona clavus*) and Fungiidae (*Diaseris*), are stable gonochoric. Three species were simultaneous hermaphrodites, two in the family Pocilloporidae (*Pocillopora damicornis, P. elegans*) and one in the family Dendrophylliidae (*Tubastraea coccinea*). Mixed breeding systems, with some colonies demonstrating both gonochoric and hermaphroditic sexuality, occurred in all five species in the family Agariciidae (Table 15.6). In addition, *Pocillopora verrucosa* and *Pocillopora meandrina* in Bahía La Paz, Mexico are simultaneous hermaphrodites (Campos-Vázquez et al. 2014), and *Pavona gigantea* at four locations off the Oaxaca coast in the southern Mexican Pacific were found to have mixed breeding systems (Santiago-Valentín et al. 2015).

It is unclear if *Tubastraea* displays a mixed breeding system or is simultaneously hermaphroditic. Each polyp section is dominated by the intense staining and characterisitic large sizes of Stages III and IV oocytes. This makes earlier stages of female gametes and male gametes difficult to detect and less numerous in the section viewed. It appears that this species is simultaneously hermaphroditic with spermaries appearing at certain times of the year, while oocytes are present year round. The life histories of the simultaneous hermaphroditic genera seem to be very different, since one of them, *Tubastraea*, is a brooder with spermaries identified in only a few polyps, while *Pocillopora* spp. contain approximately the same number of oocytes and spermaries in many polyps, and are broadcast spawners.

In four of the mixed breeding agariciid species, the proportion of hermaphroditic colonies was relatively high, ranging from 16.0 % (*Gardineroseris planulata*) to 70.7 % (*Pavona chiriquiensis*). Sequential cosexual hermaphroditism, or alternating maturation of different sex gametogenesis, has been established in these species (Glynn et al. 1996, 2000). Within this hermaphroditic condition, a cosexual sequential colony alternates maturation of each gender, and therefore, alternates between release of male gametes during one spawning episode and then mature female gametes during the next spawning event. Therefore, cosexual sequential hermaphroditic colonies function gonochorically during breeding season periods.

Table 15.6 Sexuality and mode of development in 13 scleractinian coral species in six EEP families in relation to colony morphology and occurrence of zooxanthellae in mature ova

Species	Gender	Mode	Hermaphroditic development	Morphology	Zooxanthellae[d]
Pocilloporidae					
Pocillopora damicornis	H	SG	SM	branching	+
Pocillopora elegans	H	SG	SM	branching	+
Poritidae					
Porites lobata	G[a]	SG	N/A	massive	+
Porites panamensis	G	BP	N/A	encrusting	+
Agariciidae					
Pavona gigantea	M (64.5 %)	SG	SC	massive	–
Pavona clavus	G[b] (3.3 %)	SG	SC	massive	–
Pavona varians	M (33.6 %)	SG	SC	encrusting	–
Pavona chiriquiensis	M (70.7 %)	SG	SC	encrusting	–
Gardineroseris planulata	M (16.0 %)	SG	SC	massive	–
Siderastreidae					
Psammocora stellata	G	SG	N/A	loose nodular, subramose	–
Psammocora profundacella	G	SG	N/A	encrusting[c]	–
Fungiidae					
Diaseris distorta	G	SG	N/A	loose, solitary	–
Dendrophylliidae					
Tubastraea coccinea	H	BP	SM	cryptic	–

Gender: *H* hermaphrodite; *G* gonochoric; *M* mixed (percent hermaphrodites compared with single sex colonies). Mode of larval development: *SG* spawn gametes; *BP* brood planulae. Hermaphroditic development: *SM* simultaneous; *SC* sequential cosexual; *N/A* not applicable
[a]Predominantly gonochoric at four localities; 14 % of 50 colonies from Costa Rica were hermaphroditic (Glynn et al. 1994)
[b]Predominantly gonochoric, hermaphroditic colonies occasionally found
[c]Sometimes cemented to substrate, sessile
[d]Zooxanthellae present in ova and larvae

Two of the stable gonochoric species also displayed a minimal percentage of sequential cosexual polyps. A small proportion (<1 %) of sampled colonies of the agariciid *Pavona clavus* was hermaphroditic over three locations. *Porites lobata* (Poritidae) exhibited 14 % hermaphroditic colonies (n = 50) from Costa Rica and stable gonochorism at other localities (Glynn et al. 1994). These species were classified as stable gonochoric as per Harrison (2011); in cases where hermaphrodites are relatively rare, the species can be considered to exhibit stable gonochorism and are classified as gonochoric (Harrison 2011; sensu Giese and Pearse 1974).

Brooders in the eastern Pacific, *Tubastraea coccinea* and *Pavona panamensis*, have life history strategies that appear to be very different. *Tubastraea coccinea* is a cryptic, hermaphroditic, azooxanthellate ahermatype, while *P. panamensis* is an encrusting, gonochoric, phototrophic species with zooxanthellae present in both ova and planulae. Colonies of *T. coccinea* contain female gametes of several stages year round, but male gametes were detected in only 3.7 % of tissue samples, and spermaries were significantly smaller than the eggs (Table 15.4). This may be due in part to the small size and faint staining characteristics of *Tubastraea* spermaries. An equal sex ratio, approximately 1:1, was evident in gonochoric *P. panamensis* (see Sect. 15.2.3). Spermary maturation in *T. coccinea* generally followed the sequence of planulae development at sampled locations, and sperm release in this species was observed in Panama and the Galápagos Islands (1–5 nights after full moon). Planula release in both brooding species was greatest at full moon with some new moon activity. Planulae of *P. panamensis* are much shorter in length (400–500 μm) than those of *T. coccinea*, which are bright tangerine (same color as adults) and are 0.5–1.5 μm long. Both species seem to settle close to the maternal colonies, however, the dispersal distance of *T. coccinea* may typically be of shorter range. *Tubastraea coccinea* populations can be found clustered in mass on reef frameworks or more generally on non-carbonate rock substrata, commonly under subdued light conditions, while *P. panamensis* can form separate nodular colonies dispersed across reef-rubble substrates. *Porites panamensis* can recruit to basalt outcrops and contribute to reef coral cover as well. *Tubastraea* contributes little to reef structures, although it has invaded and recruited to many coral communities and reefs in the Pacific and Atlantic regions (Fenner 2001; Fenner and

Banks 2004; Figueira de Paula and Creed 2004). In addition, *T. coccinea* thrives in the Galápagos Islands, while the zooxanthellate poritid brooder does not occur at this location, but is endemic only to eastern Pacific mainland localities (Glynn and Ault 2000).

In the Indo-Pacific and Caribbean, many major reef building corals form gamete bundles just prior to spawning. Gamete bundle formation requires that polyps be simultaneously hermaphroditic and that both sex gametes are mature and ready for release concurrently. Although simultaneous hermaphroditic maturation exists in the tissues of eastern Pacific *Pocillopora damicornis* and *Pocillopora elegans* at and near full moon, no spawning activity has been witnessed in this region, and therefore no reports of gamete bundle formation. No evidence of this trait has been detected in hundreds of tissue samples collected around the suspected time of spawning.

In other regions, broadcast spawning species of *Pocillopora* have been observed to release gametes of both sexes in a gamete cloud. *Pocillopora verrucosa*, *P. meandrina*, *P. damicornis*, and *P. eydouxi* on the Great Barrier Reef were observed in aquaria to release gametes separately into the water column (Schmidt-Roach et al. 2012). In Okinawa, Japan and in the central Red Sea, *Pocillopora verrucosa* was observed to spawn clouds of mixed sex gametes in aquaria and in the field, respectively (Boumeester et al. 2011). *Pocillopora eydouxi* (also found in the eastern Pacific) was observed to freely spawn both sexes of gametes during the daytime in Okinawa (Kinzie 1993). Therefore, although *Pocillopora* species are the major reef builders in the eastern Pacific, this genus may likely not bundle gametes.

Hybridization of similar species, which is reported as a result of mass spawning events (Willis et al. 1993, 2006; Richmond 1997), may be possible in eastern Pacific corals. *Pocillopora damicornis* usually dominates reef flat habitats, but two additional species in this genus (*Pocillopora elegans* and *Pocillopora eydouxi*) are also often present in this zone. However, while lunar cycles of gamete maturation in two of these species (*P. elegans* and *P. damicornis*) are very similar, interannual monthly activity may vary. The potential for hybridization also exists within the family Agariciidae, but seasonal trends of each of these species at eastern Pacific locations can be very different, making this less likely, especially due to low population densities and dispersed species' distributions.

Eastern Pacific corals exhibit a preponderance of species with mixed breeding systems. Most agariciid species have developed functional gonochorism within the hermaphroditic mode. Even poritids, whose species generally display stable gonochorism, exhibited some evidence of hermaphroditism (e.g., *Porites lobata* in Costa Rica). Although this breeding strategy is prevalent in the eastern Pacific, mixed breeding systems and sequential cosexual hermaphroditism have rarely been reported in the scleractinian literature. Guest et al. (2012) noted this condition in *Diploastrea heliopora*, a broadcast spawner near Singapore with predominantly gonochoric polyps. The colonies contained male, female and a low proportion of cosexual polyps with the most plausible explanation being that polyps switch sexes with oogenic and spermatogenic cycles occasionally overlapping. This condition occurred in corals on equatorial reefs that were chronically impacted near a large urban center.

Other cases of mixed breeding systems have been reported in different biogeographic regions. These are listed in Harrison (2011) with their respective sources. Caribbean agariciids (two species from Curaçao), *Porites astreoides* in both Puerto Rico and Caribbean Panama, *Monomyces rubrum* (Flabellidae) from South Africa, and members of the Fungiidae, *Fungia scutaria* in the Red Sea, *Heliofungia actiniformis* at Palau and *Sandalolitha robusta* in Hawaii all have displayed mixed breeding characteristics. In addition, Waller et al. (2005), in Guest et al. (2012), discovered all samples of three Atlantic, deep-sea species of *Caryophyllia* (Caryophylliidae) to contain gametes of both sexes, but with only one sex viable at a time (sequential cosexuality).

Therefore, the sequential cosexual strategy may be more common than originally considered. There may be a genetic/systematic component to the development of this type of sexual system or this strategy may develop in species adapted to more equatorial environs, to extreme or marginal habitats or perhaps both. Regardless of the conditions that may induce the development of this reproductive pattern, such corals are common in the eastern Pacific region, and numerous populations with mixed sexual systems have survived here. The agariciids with mixed sexual systems comprise more than a third of EEP species.

15.2.3 Sex Ratios and Sex Allocation

The sex ratios of eastern Pacific coral populations are influenced by fragmentation resulting from a variety of disturbances. Sea surface temperature extremes associated with ENSO events, seasonal upwelling, and variable water temperature result in an ecosystem often at the limits of coral habitation. The sex ratio of simultaneous hermaphrodites, such as species of *Pocillopora*, mature gametes of both sexes at the same time thereby exhibiting colony sex ratios of 1:1. Sequential cosexual hermaphroditism, wherein male and female gamete maturation alternates in each colony, should also display a 1:1 sex ratio. Although the spawning activities and sex ratios of individual colonies have not been tracked in the eastern Pacific, any deviation from 1:1 sex ratios per spawning event would be "balanced" over time for both sexes in alternating cycles of gametogenesis. The sex first matured in each colony may be genetically determined, and

Table 15.7 Sex ratios of 12 zooxanthellate and one azooxanthellate scleractinian corals of varied sexual patterns

Species	Locality	Sexuality[a]	Sex ratio[b]			Chi Square[c]
			Ovaries	Spermaries	♀:♂	
Pocillopora damicornis	Caño I	H	23	22	1.0:1	ns
	Uva I		39	43	0.9:1	ns
	Galápagos Is		10	10	1.0:1	ns
Pocillopora elegans	Caño I	H	62	66	0.9:1	ns
	Uva I		67	61	1.1:1	ns
	Galápagos Is		28	17	1.6:1	ns
Porites lobata	Caño I	G	32	18	1.8:1	ns
	Uva I		30	35	0.9:1	ns
Porites panamensis	Huatulco[d]	G	–	–	0.4:1	–
	Taboga I		31	28	1.1:1	ns
Pavona gigantea	Huatulco[d]	M	–	–	1.0:1	–
	Uva I		18	13	1.4:1	ns
	Taboga I		5	17	0.3:1	$0.02 > p > 0.1$
	Galápagos Is		26	26	1.0:1	ns
Gardineroseris planulata	Uva I	M	19	13	1.5:1	ns
Pavona varians	Caño I	M	41	21	2.0:1	$0.025 > p > 0.01$
	Uva I		16	8	2.0:1	ns
	Taboga I		18	32	0.6:1	$0.05 > p > 0.025$
Pavona chiriquiensis	Uva I	M	10	9	1.1:1	ns
	Galápagos Is		43	29	1.5:1	ns
Pavona clavus	Caño I	G/M[e]	17	4	4.2:1	$0.005 > p > 0.001$
	Uva I		18	9	2.0:1	ns
	Galápagos Is		57	20	2.8:1	$p < 0.001$
Psammocora stellata	Caño I	G	16	17	0.9:1	ns
	Uva I		76	13	5.8:1	$p < 0.001$
	Saboga I		16	3	5.3:1	$0.01 < p < 0.005$
	Taboga I		0	36	0:1	$p < 0.001$
	Galápagos Is		12	30	0.4:1	$0.01 < p < 0.005$
Psammocora profundacella	Uva I	G	28	10	2.8:1	$0.005 < p < 0.001$
	Galápagos Is		19	13	1.5:1	ns
Diaseris distorta	Galápagos Is	G	15	85	0.2:1	$p < 0.001$
Tubastraea coccinea	Caño I	H	20	6	3.3:1	$p < 0.01$
	Uva I		64	10	6.4:1	$p \ll 0.001$
	Saboga I		64	9	7.1:1	$p \ll 0.001$
	Taboga I		64	6	10.3:1	$p \ll 0.001$
	Galápagos Is		154	23	6.7:1	$p \ll 0.001$

[a]*H* hermaphrodite; *G* gonochoric; *M* mixed, with a high proportion of H or G colonies
[b]Ovary and spermary counts are for presence of respective gonad genders in H, G or M colonies
[c]Yates' correction uniformly applied
[d]Huatulco sampling from Rodríguez-Troncoso et al. (2011); '-' indicates not reported
[e]Five hermaphroditic colonies reported, representing <1 % sampled at three locations

therefore the proportion of colonies maturing each sex may not be equal at any given time. Although sex allocation would seem to be approximately equal in simultaneous hermaphrodites (*Pocillopora* spp.), and sequential cosexual hermaphrodites (agariciid species), sex allocation may become more key in gonochoric and brooding species.

Except for *Porites* spp., the sex ratios of species that are gonochoric and/or displayed little sequential cosexual hermaphroditism frequently deviated from 1:1 (Table 15.7). (Only poritid samples of sufficiently large size, n $\geq$ 50 colonies, were tested.) Dominance of a particular sex was not consistent from site to site for any particular species. This could result from frequent disturbances to study (monitored) populations and/or from fragmentation creating clonal populations. For gonochoric species in the eastern Pacific, difficulty in finding a mate could be pronounced, thereby reliance on asexual fragmentation for propagation of these species may be significant.

The burden of generating sufficient energy in gonochoric species appears to be on those colonies that produce eggs, which may require more energy to produce than spermaries (Charnov 1982; Harrison and Wallace 1990). In several EP broadcast spawning species, the sequential hermaphroditic mode of development may allow for efficient allocation of reproductive resources and also offer all colonies of a species the potential to contribute to each spawning event. Two stable gonochoric species, *Porites lobata* and *Pavona clavus*, did display some degree of cosexual hermaphroditism. A low incidence of hermaphroditism (2.7 %) was also detected in a population of *Porites porites* subjected to urban and industrial pollution (eutrophication) in Barbados, West Indies (Tomascik and Sander 1987). Therefore, the reproductive mode in these corals may be more malleable than previously recognized and sequential hermaphroditism in at least some EP coral populations may be a response to local stressors.

Sex ratios, and therefore sex allocation, in eastern Pacific brooding species are very different. The gonochoric brooder, *Porites panamensis* had approximately equal female to male numbers (1.1:1 Taboga Island, Gulf of Panama and 1:0.5 Caño Island). About the same proportion of colonies sampled contained planulae (sporadic sampling, pooled over several years) (Glynn et al. 1994). Smith (1991) reported a 1:1 sex ratio for this species at Uva Island whose population was probably recruited after the1982–1983 El Niño disturbance. The ratio of female to hermaphrodite colonies of *Tubastraea coccinea* ranged from about 3:1 to 10:1 (Table 15.7). Spermaries are sparse and very difficult to observe in tissues of this species, and do not exhibit detectable flagella. Planulae were observed in tissues and in the field generally throughout the year, depending on the location. This genus has been reported to develop oocytes spontaneously within the tissues, which may explain egg maturation throughout the year. Allocation to sperm development appears to be very small in comparison to that of the ova. Mature eggs are 300–800 μm and mature spermaries 150–200 μm.

Siderastrea radians, a gonochoric brooder in the Caribbean, exhibits a significant female bias, which Szmant (1986) attributed to its limited capacity to bear large ova and planulae. If this condition were a genetically determined reproductive trait, then *Porites panamensis* might also be expected to exhibit a preponderance of female colonies, which it does not. Cabral-Tena et al. (2013) have demonstrated a gender bias in *P. panamensis* at some localities, with significantly greater skeletal extension and calcification rates in male compared with female colonies. If relatively high male growth rates confer some advantage during recruitment and early community succession, this might then bias selection toward a greater proportion of male colonies under certain conditions.

Two species sampled at the Taboga Island site (*Pavona gigantea* and *Pavona varians*), located immediately south of the eutrophic waters of Panama Bay (D'Croz and Robertson 1997), exhibited an excess of male colonies. A male bias was also observed in *Porites porites* populations along a eutrophic gradient in Barbados (Tomascik and Sander 1987). It was hypothesized that this pattern in the latter study (Caribbean) was due to increased turbidity and reduced light levels that could depress zooxanthella photosynthesis and thus deprive oogenesis of essential nutrients. Rinkevich (1989) demonstrated experimentally the contribution of zooxanthella photosynthetic products to coral reproduction. Considering the relatively small sample sizes and the high likelihood of a clonal population structure due to asexual fragmentation (see Sect. 15.5 below, Recruitment and Asexual Fragmentation), in this case it is difficult to demonstrate genetic adaptation to local conditions.

True protandrous sex change, and possible protandrous hermaphroditism from initial male function in small corals to female function in larger corals, has been demonstrated for several solitary fungiid species (Loya and Sakai 2008; Loya et al. 2009; Harrison 2011). The possibility of sex change in *Diaseris distorta* was examined in a size series collection (n = 24) in the Galápagos Islands (Colley et al. 2000). Only a single female was found (1190 mm^2, surface area). Since several smaller (238–1088 mm^2) and larger (1260–1980 mm^2) males were collected, and two males were the same size as the female, no relationship between size and sex was evident. However, the skewness of the sex ratio (males dominated the population 5.4:1, $p < 0.001$, chi square) toward males and the strong evidence of fragmentation (only one sexual recruit could be identified) may be keeping individual size small, and therefore, possibly predominantly male, if maleness is related to small size. It is also likely that prolific daughter fragment production is creating a male dominated clonal population. Size series collections of all species are discussed in Sect. 15.2.3 below, Sex Ratios and Sex Allocation.

15.2.4 Age at Sexual Maturity

Size series collections of reproductively active colonies have allowed estimates of the ages to first reproduction of several species. Coral ages can be estimated by relating known species

Table 15.8 Size and age estimates related to reproductive condition

Species	Size (cm)[a]			n	Growth rate (cm year^{-1})	Age (years)		
	Indet[b]	♂	♀			Indet[b]	♂	♀
Porites lobata	3.5 (1.2–4.0)[c]	7.5 (6.9–8.0)	9.8 (6.3–39.9)	25	1.17 (±0.10)[d] 2–3 m 1.05 (± 0.19)[d] 8–10 m	3.0 3.3	6.4 7.1	8.4 9.3
Pavona gigantea	3.1 (2.5–7.5)	17.8 (12.6–35.7)	17.8 (12.0–19.9)	26	0.92 (0.3–1.3)[e,f]	3.4	19.3	19.3
Gardineroseris planulata	2.3 (3.6–8.0)	9.8 (7.5–21.8)	7.5 (7.2–12.5)	30	1.04 (0.07)[g]	2.2	9.4	7.2
Pavona varians	1.5 (1.2–2.3)	7.0 (5.5–8.9)	5.5 (3.2–9.4)	17	0.50 (0.25–6.2)[h]	3.0	14.0	11.0
Pavona chiriquiensis	1.0	2.5 (1.5–2.6)	1.6 (1.5–2.5)	25	3.0[i]	0.3	0.8	0.5
Pavona clavus	8.9 (4.4–13.8)	11.6 (4.5–13.9)	31.4 (21.5–41.8)	30	0.98 (0.2–1.7)[f]	9.1	11.8	32.0

Size series collections during reproductive activity at Uva Island, Gulf of Chiriquí, Panama
[a]Mean vertical growth axis of skeleton
[b]Indeterminate, no reproductive products present
[c]Standard deviation
[d]Standard error, followed by water depth
[e]Depths for all agariciid species 2–5 m
[f]From Manzello (2010), minimum and maximum values in parentheses
[g]High rate of Glynn (1985) used because close to dates of coral collections
[h]From Glynn et al. (2000), deviation from mean are 1st and 3rd quartiles
[i]Maximum growth rate from Glynn et al. (2001a, b)

skeletal growth rates to the greatest growth axes of colonies. There are drawbacks to this method however, e.g., (a) biased sampling of larger more easily encountered colonies, (b) highly variable colony growth rates, and (c) fragmentation and fusion that can obscure the true ages of colonies.

Examination of a size series of *Diaseris distorta* in the Galápagos Islands revealed sexual maturity in small individuals, and likely at an early age (Colley et al. 2000). Both asexual fragments and sexually derived individuals, from 1.8 cm to 2.7 cm in skeletal diameter, were reproductively active, i.e. bearing mature gametes. The ages of these fungiids are not known, but it is possible they approximate two years (Yamashiro and Nishihira 1998). Most adult individuals of *D. distorta* in the study population ranged from 3.7 to 5.0 cm in diameter (Feingold 1995). Colony growth is critical in early benthic stages vis-à-vis interphyletic competition for space.

At Uva Island, Panama the mean size of colonies of six sampled species with gonads varied substantially (Table 15.8). The smallest colonies bearing gonads belonged to the encrusting species *Pavona chiriquiensis* with a mean vertical growth axis of 1.6 cm in females and 2.5 cm in male colonies. Applying a mean skeletal growth rate of 3.0 cm year^{-1}, this extrapolation suggests that both sexes can reach reproductive maturity in less than one year. Sexually mature colonies of *Pavona varians*, also with an encrusting growth habit, were notably larger and possibly much older than *P. chiriquiensis*, i.e., 10 years in age in both sexes. The remaining agariciid species and *Porites lobata* possess large, dome-shaped colonies with reproductive ages similar to *Pavona varians*. The relatively large female colonies of *Pavona clavus* may be a result of biased sampling. Small colonies of this species were unavailable.

Rapid recovery is especially important in environments with high nutrient and light levels that favor opportunistic algae and other epibenthic taxa (Birkeland 1977; Highsmith 1980). Eastern Pacific reef substrates generally consist of low-lying turf algae, due in large measure to intense fish and echinoid grazing (Birkeland 1997; Fong et al. 2006; Wartian 2006). Ten-cm high coral reproductive colonies would offer an escape in size, generally extending above micro-filamentous algal films or turf substrates. Species with very young maturation ages, such as *Pavona chiriquiensis*, could have an advantage in population recovery. Presumably they would have more rapid generation times and hence be better poised to adapt to changing environmental conditions. The large age differences in reaching sexual maturity in eastern Pacific agariciid species would offer an ideal opportunity to test for such a life history difference.

15.3 Timing of Reproduction

Interest in coral reproductive schedules is long-standing, from the first systematic studies conducted during the Great Barrier Reef Expedition (1928–29) to the present (e.g., Fadlallah 1983; Shlesinger and Loya 1985; Szmant 1986; Oliver et al. 1988; Harrison and Wallace 1990; Richmond and Hunter 1990; Richmond 1997; Guest et al. 2005; van Woesik 2009; Harrison 2011). Harrison and Wallace (1990) and Harrison (2011) have offered comprehensive reviews of the environmental factors implicated in regulating the timing of coral reproduction. These range from annual (seasonal) to lunar and diel time-scales, and include such proximate controls as sea temperature, day length, lunar cycles, daily light/dark periods, moon light, calm spells, and tidal height. Of the various factors implicated in regulating annual to daily reproductive processes, it is necessary to caution that the proposed causalities have not been unequivocally demonstrated.

Although lunar cycles of eastern Pacific corals may be similar, the seasonal and diel timing, and differences in annual temperature regimes may prevent multiple species from spawning together. *Pavona varians* and *Pavona chiriquiensis* have spawned monthly in the Gulf of Chiriquí during the dry season (January–April), but do so 12 h apart. *Pavona clavus* is reported to mature gametes monthly at Culebra, Costa Rica (Bezy 2009), but in Panama spawning of this species is restricted to the middle of the wet season, i.e. July–October (Glynn et al. 2011). *Gardineroseris planulata* favors the warmest time of the dry (non-upwelling) season, and *Pavona gigantea* is most active in the Galápagos Islands during the coolest time of year.

Long-term studies have shown that warmer or cooler years result in a variety of breeding outcomes. In the Galápagos Islands, a warm water pulse of about two months duration in 1991 accompanied extended and year round gamete production in *Pavona varians* and *Pavona chiriquiensis* (Colley et al. 2006). Pocilloporid species (*Pocillopora elegans*, *Pocillopora damicornis*) are particularly good examples of variable reproductive activity. For example, mature ova and sperm were usually observed in tissues at full moon, but some collections over a span of 20 years have shown a variety of gametogenic seasonal patterns during several months (Glynn et al. 1991; Colley et al. 2006). This is in contrast to the results of Campos-Vázquez et al. (2014) who found other pocilloporid species, namely *Pocillopora verrucosa* and *Pocillopora meandrina* in Bahía La Paz, Mexico to have one annual gametogenic cycle from approximately May to September (a 16-month study). Commencement and completion of these gametogenic cycles coincided with seawater temperatures between 23–30 °C.

In addition, although ripe gametes have been found historically in both *Pocillopora elegans* and *Pocillopora damicornis* on and just after full moon, field observations have never revealed gamete release in these species. Therefore, it is possible that at least some eastern Pacific species may release gametes on the same lunar days, but generally seasonal months of release can vary between species, due to location, variations in yearly temperature regimes, and presumed sensitivity or insensitivity to temperature (Colley et al. 2006).

Sensitivity to water quality can affect six chemically-mediated steps in the reproductive cycle of broadcast spawning corals: (1) synchronization among conspecific colonies, (2) egg-sperm interactions leading to fertilization, (3) embryological development, (4) larval dispersal and substratum searching, (5) metamorphic induction, and (6) acquisition of zooxanthellae in those species that are lacking vertical, maternal transmission. Pollutants, including freshwater, can result in the failure of one or more of these links, and such interference may result in a cascade effect preventing the replenishment of reef populations (Richmond 1994, 1997).

As the eggs of most coral species float following release, they are subjected to surface waters for varying periods prior to fertilization. As fresh water is less dense than seawater, rainfall and coastal runoff reduce the quality of this upper layer where eggs generally become fertilized. Even moderate reductions in sea surface salinities to about 15 ‰ from 35 or 30 ‰, can result in 50 to 90 % reduction in fertilization rates (Richmond 1994). As eastern Pacific spawning events span the wet season (and peak ENSO thermal anomalies), increasing levels of coastal watershed runoff can lead to reproductive failure following these periods. The addition of sediment and nutrients from runoff (Humphrey et al. 2008) as well as agrochemicals and hydrocarbons further reduce fertilization rates over larger areas. Most eastern Pacific coral species mature gametes during the wet and generally warmer season, which approximates mid–April to mid–December. Eggs of many eastern Pacific coral species are neutral or negatively buoyant as described below (Sect. 15.3.5, Spawning and Gamete Characteristics), and thus may avoid surface water stress effects. Location may also be important since *Psammocora* spp., which mature gametes in nonupwelling Costa Rica in the wet season, are not reproductively active at the same time of year in the nonupwelling Gulf of Chiriquí, Panama.

15.3.1 Seasonal

Sexual reproductive activity is generally greatest seasonally during warm or non-upwelling periods in the three major eastern Pacific upwelling centers (Chávez-Romo and Reyes-Bonilla 2007; Carpizo-Ituarte et al. 2011; Rodríguez-Troncoso et al. 2011; Glynn et al. 2012). The reproductive activity of *Pocillopora damicornis* and *Pavona gigantea* along the Mexican coast is mainly restricted to the summer months (May-August) when sea temperatures are at a maximum (>27–28 °C). This pattern has also been shown in *Pocillopora verrucosa* and *Pocillopora meandrina* (Campos-Vázquez et al. 2014). In central and southern Mexico, *Porites panamensis* is reproductively active from March or May to September, but in seasonally upwelling Gulf of Tehuantepec (Huatulco), *P. panamensis* is most active after upwelling ceases and later, from March until August (Rodríguez-Troncoso et al. 2011). Curiously, a year-long sampling of *P. damicornis* off southern Mexico (Huatulco, 15.7°N), including the height of the warm season

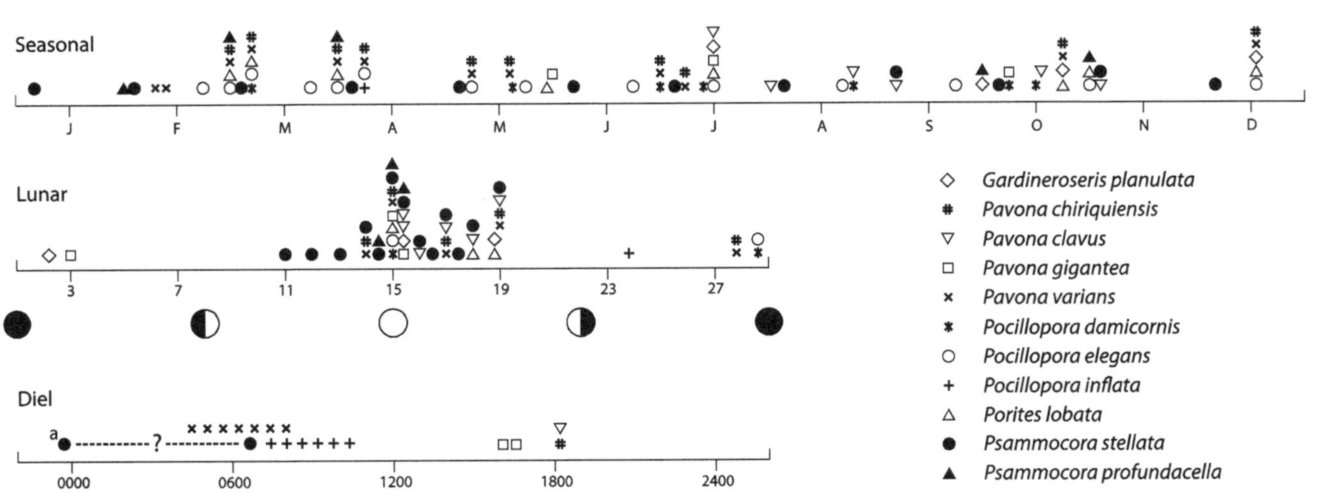

Fig. 15.7 Timing of seasonal, lunar and diel reproductive activities in 11 broadcast spawning EEP zooxanthellate corals. Data are primarily from presence of mature gonads in histological preparations and spawning observations at Uva Island reef, Gulf of Chiriquí, Panama. Lunar and diel spawning times supplemented with observations from Gulf of Panama (*Pocillopora inflata*, *Pavona clavus*) and Galápagos Islands (*Pavona gigantea*). Modified after Glynn et al. (2012)

during 2002–2003, failed to reveal any reproductive activity (Rodríguez-Troncoso 2004).

Further south, near the equator, coral reproduction is also most pronounced during the warmest times of year. At Costa Rica, about 6° latitude farther south, *Pavona clavus* was active monthly in Culebra Bay (10.5°N) from May to November, during the non-upwelling season. In the Gulf of Chiriquí, Panama (8°N), which is subject to seasonal thermocline shoaling, the reproductive season was more restricted to the warmest months. Similarly, in the equatorially-situated Galápagos Islands, reproductive activity was pronounced but limited to the warmest time of year. Spawning was observed every month from August to October in the Gulf of Panama (Glynn et al. 2012) and within this time frame at Culebra Bay, Costa Rica (Bezy 2009). This is at the height of the wet season (non-upwelling) at both localities.

However, there are exceptions to reproduction occurring in the warmest seasonal period. During seasonal upwelling in the Gulf of Panama (January–April) and the cool season in the Galápagos Islands (June–December), the latter region under the influence of the Oceanic Peru Current, coral reproduction in most species ceases or falls off (Colley et al. 2006). In Galápagos waters, where mean temperatures decline rapidly in May and June (from 25 to 22 °C and lower), *Pavona gigantea* alone contained mature gametes during much of the cool season. This suggests unusual local adaptation of this species to low thermal conditions (Glynn et al. 1996). In the Gulf of Panama, most species do not initiate gametogenesis during the upwelling months. *Pavona gigantea*, however, is reproductively active there during both the upwelling and non-upwelling seasons.

In non-upwelling areas in the EEP where mean sea temperatures remain high throughout the year (28–29 °C), gametogenesis is often more prolonged with some species exhibiting gametogenesis year-round (Fig. 15.7). In the Gulf of Chiriquí, Panama, *Pocillopora* spp. were active year around, with *Pocillopora damicornis* exhibiting a decline at the beginning and end of the calendar year. *Pavona chiriquiensis* and *Pavona varians* also show year-round reproductive activity at this locality and in non-upwelling Costa Rica, but are limited to the warm seasons in the Galápagos Islands and the Gulf of Panama.

Porites panamensis, a brooder, reproduces year-round in the EEP at most sites (Smith 1991; Glynn et al. 1994), with reproductively active colonies present in the Gulf of Panama from April–May until November, one month before the onset of upwelling. Planula development occurs in the Gulf of Panama only after upwelling ceases, however, in the Gulf of Chiriquí planulae were released at this and other times of the year (Smith 1991). Therefore, it is most likely that *P. panamensis* has an extended period of larval release in the non-upwelling Gulf of Chiriquí.

Some studies conducted at low latitude reef sites elsewhere, where environmental conditions such as sea temperature and solar radiation are relatively constant year round, reported extended spawning seasons. These extended over several months or during most of the year (Oliver et al. 1988; Richmond and Hunter 1990; Penland et al. 2004; Mangubhai and Harrison 2009) as observed at equatorial locations in the eastern Pacific. Longer spawning seasons are not restricted to lower latitudes, however; for example, in the Solitary Islands, Australia at 30°S latitude, reproductive activity is spread over several months (Wilson and Harrison

Table 15.9 Elevated thermal bleaching effects in relation to coral sexual reproduction

Effects	Species	Location	Years	Authority
Lack of gonadal development Tissue atrophy and necrosis	*Pavona gigantea Psammocora stellata Pocillipora damicornis Porites* spp.	Uva Is, Gulf of Chiriquí Panama	1983	Glynn et al. (1985)
Decline in numbers of reproductive colonies Poor tissue integrity	*Pavona* spp. *Gardineroseris planulata Pocillopora* spp. *Porites* spp.	Gulf of Chiriquí Gulf of Panama Galápagos Is	1997–1998	Colley et al. (2006)
Cessation of gametogenesis	*Pocillopora damicornis*	Banderas Bay, Mexico	2002	Carpizo-Ituarte et al. (2011)
Lack of gonadal development Decline in egg sizes	*Pocillopora damicornis Pavona gigantea*	La Entrega Bay, Oaxaca, Mexico	2002–2003	Rodriguez-Troncoso et al. (2011)
Cessation of gametogenesis Decline in C and N tissue biomass	*Orbicella* (cf *Montastraea) annularis*	Carysfort Reef, Florida	1987–1988	Szmant and Gassman (1990)
Decline in egg numbers and sizes Decline in polyps with gametes	*Acropora* spp. *Montipora* spp. *Porites* spp. *Pocillopora damicornis*	Heron I, Great Barrier Reef	1998–1999	Ward et al. (2000)
Decline of gravid colonies	*Acropora* spp. *Platygyra daedalea, Porites lobata*	Orpheus I, Pelorus I, Great Barrier Reef	1998	Baird and Marshall (2002)
Decline in gametogenesis, spawning colonies, and fertilization rate	*Montipora digitata Pocillopora verrucosa*	Sesoko I, Bise I, Okinawa, Japan	1997–1998	Hirose and Hidaka (2000)
Decline in fertilization success Low sperm motility Decline in concentration of mobile sperm	*Acropora* spp.	Aka I, Okinawa, Japan	1998–1999	Omori et al. (2001)

2003). And at some equatorial locations (e.g. Singapore) coral species spawn seasonally, mainly during a 2-month period (Guest et al. 2005).

A general decline in the reproductive condition of corals has been documented during severe elevated sea temperature bleaching events in several studies from the Caribbean to central/west Pacific regions (Table 15.9). Whole bleached coral colonies or bleached patches on colonies demonstrate a variety of adverse effects that are often manifest several months after recovery. These range from declines in the percentage of reproductive colonies that survived and recovered from bleaching to lowered rates of fertilization, reduced sperm motility, reductions in egg size and numbers (Table 15.9). From Mexico to the EEP, all major reef-building corals have demonstrated varied stages of decline in reproduction associated with ENSO elevated thermal stress (Glynn et al. 1985; Colley et al. 2006; Carpizo-Ituarte et al. 2011; Rodriguez-Troncoso et al. 2011). A study of the effects of bleaching on the Great Barrier Reef not only demonstrated declines in coral reproductive success during and shortly after the stress event, but also suggested that reproduction was impaired one year later and that bleached corals showed an increased susceptibility to future stress (Ward et al. 2000).

Moderate ENSO warming, however, has been observed to increase gonadal activity in certain eastern Pacific localities where low temperatures often depress or inhibit development. Colley et al. (2006) documented that all monitored EEP coral species reproduced during the 1997–98 bleaching event. At Uva Island from 1982 to 1996, sexual recruitment in *Pavona varians* was significantly related to maximum monthly positive sea surface temperature (SST) anomalies that occurred in the year preceding recruitment (Glynn et al. 2000). Recruitment failed when SST anomalies exceeded 1.6–1.9 °C during the severe ENSO events of 1982–1983 and 1997–1998, suggesting an elevated temperature threshold for this species. Sexual reproduction in the Gulf of Panama normally ceases during the upwelling season when mean SSTs decline below 24–25 °C. In 1998, when upwelling was suppressed, *Porites panamensis*, *Porites lobata* and *Pocillopora* spp. demonstrated unseasonable reproductive activity (Colley et al. 2006). Cox (2007) has shown that the sexual cycle of *Montipora capitata* in Hawaii was not disrupted in spite of bleaching. She hypothesized that this species was able to produce sufficient energy stores for reproduction by feeding heterotrophically, a capacity for trophic gains demonstrated by Grottoli et al. (2004, 2006).

15.3.2 Lunar

Compared with coral reef regions with species-rich faunas where multiple and synchronous spawning events are common-place, e.g. in the Caribbean, central and western Pacific, and Indian Ocean (Szmant 1986; Guest et al. 2005; Harrison 2011), lunar spawning schedules in the eastern Pacific are relatively diffuse. Most low latitude eastern Pacific broadcast spawning species produce mature gametes during 3 to 4 day periods before or following full moon (Fig. 15.7). Several species display a post-full moon bias of from one to four days. Six species also show heightened sexual activity (mature gametes) near new moon. *Pocillopora inflata* spawned gametes on lunar day 24, two days after a waning quarter moon. This record is atypical and may be due to stress associated with its collection. The spawning occurred in a 20 l container, albeit with frequent water renewal, about 30 min after collection.

Therefore it may generally be concluded that eastern Pacific corals spawn on or around the 17th lunar day and 1–2 days following (Fig. 15.7). *Pavona varians*, *Pavona chiriquiensis* and *Pavona clavus* begin gamete release on these days (Glynn et al. 2001a, b, 2011). *Pocillopora damicornis* and *Pocillopora elegans* harbor mature gametes on full moon (day 15) and spent gonads were observed in tissue samples 2–3 days later (Glynn et al. 1991). Peak percent colonies of *Porites lobata* with mature gonads occur at full moon and taper off a few days later.

There are at least eight morphospecies of closely related pocilloporid corals that occur on EEP reefs (but see Chap. 14, Pinzón). Only two of these (*Pocillopora elegans* and *Pocillopora damicornis*) have been studied extensively and contain ripe gametes of both sexes around full moon. Although there is evidence of variation in the annual breeding behavior of these *Pocillopora* species, the potential exists that these, and perhaps a third pocilloporid (*Pocillopora eydouxi*), spawn synchronously at least occasionally. Species in the Caribbean reef-building *Orbicella/Montastraea* complex spawn on the same days and times annually during the warmest time of year (Van Veghel 1993). Structurally important *Acropora* species, as well as many other corals, spawn on the Great Barrier Reef in the highly visible mass spawning event, again in the warm season. However, *Pocillopora* species, the major contributor to reef framework in the eastern Pacific appear to be more sensitive to thermal conditions than other species of this region, and thus their timing of gamete release may vary annually (Colley et al. 2006).

15.3.3 Diel

Spawning was observed in only six species, including *Pocillopora inflata*. Spawning in *P. inflata* could have been stimulated prematurely due to collection (Fig. 15.7). Multiple interspecific spawning is unlikely among several of the eastern Pacific species since species-specific spawning activity has been witnessed during daylight hours, soon after sunset, and just before sunrise. *Pavona gigantea* was observed spawning during daylight hours in the Galápagos Islands. Several colonies spawned on two different days, during a flooding spring tide beginning at 1700 and lasting for about 30 min. *Pocillopora inflata* also spawned during daylight hours (0730–1030), but this observation may be atypical as noted above.

Two closely related encrusting species, *Pavona chiriquiensis* and *Pavona varians*, spawn gametes on the same lunar days (2–3 days after full moon) during multiple months in the dry season (Glynn et al. 2000, 2001a, b). Gamete release was also associated with peak high-water stands of spring tides. However, *P. chiriquiensis* spawned about one hour after sunset at Uva Island and *P. varians* began spawning 1–2 h before sunrise at this location with some colonies continuing to spawn until about 0830. No overlap or hybridization would be expected in these species since their diel timing is approximately 12 h out of phase. The spawned gametes observed after 12 h appeared to lose vitality. This timing is well beyond gamete viability and dilution effects of a few hours for corals on the central Great Barrier Reef (Oliver and Babcock 1992).

Histological and field observations indicated that spawning of *Pavona clavus* in the Gulf of Chiriquí was centered around full moon, most frequently on lunar day 17 and near sunset (1800 h) (Glynn et al. 2011). In Culebra Bay, Costa Rica, an upwelling site, *P. clavus* also spawns on the 17th lunar day, 2–3 days after full moon (Bezy 2009).

It is likely that overlap in spawning may occur between species of *Pocillopora*. *Pocillopora damicornis* and *Pocillopora elegans* display very similar gamete development, however, differences in thermal sensitivities may prevent synchronous gametogenic cycles in at least some locations. In Bahía La Paz, Mexico, gamete maturation occurs at the same time of year (September) in *Pocillopora verrucosa* and *Pocillopora meandrina* (Campos-Vázquez et al. 2014), but further study is needed to determine if there is any lunar day synchronization. In the Gulf of Chiriquí, gametogenesis occurs year round in the multi-year pooled sample set, but although gamete maturation occurs at full moon, no spawning activity has ever been observed, even after the completion of numerous monitoring exercises in the field.

As reasoned above, the likelihood of spawning overlap of the majority of studied eastern Pacific corals is slight. Some overlap of *Pavona varians* and *Psammocora stellata* is possible since both species spawn at night around full moon, but the diel timing of *P. stellata* is still unknown. According to the seasonal (June, October), lunar (full and new moon), and diel (1800 h) schedules, *Pavona clavus* and *Pavona chiriquiensis* could conceivably spawn at the same time. But, oocyte maturation peaks in these two species in different seasons. Colonies of both species occurring in close proximity at Uva Island have been monitored to explore this possibility, but simultaneous interspecific spawning has not been observed.

15.3.4 Length of Gametogenic Cycles

The spawning seasons of eastern Pacific species are generally variable in length. The seasonality and duration of breeding at each study site are species or genus specific and also depend on the local temperature regime. However, gametogenic cycles of different species can overlap, and it appears that several eastern Pacific corals can release gametes multiple times during their respective breeding seasons. Although sampled colonies in the EEP were not tagged, and data from all years' collections were recorded but pooled, histological collections suggest that gamete release in many eastern Pacific coral species can occur at least monthly during this time. At least four species at Uva Island (*Pocillopora elegans*, *Porites lobata*, *Pavona varians*, *Pavona chiriquiensis*) produce mature gametes year round (Glynn et al. 2011). Eight of the Uva reef species showed evidence of spawning or loss of mature gametes at or up to four days following full moon, and six of eight species exhibited some evidence of split spawning with possible gamete release occurring a few days around new moon as well.

Observed multiple spawning episodes could indicate population behavior and not necessarily that of individual colonies. That is, individual colonies may not spawn on multiple occasions, but that a population may exhibit multiple spawning behavior due to different colonies spawning at different times. The population of *Pavona clavus* in the Gulf of Panama spawned in August, September and October, however, in one year with 4-month consecutive observations, the sampled colonies spawned every other month. Spawning behavior of *Pavona varians* and *Pavona chiriquiensis* was observed on the Uva Island reef in January and April 1997, and in May of the same year in a flow-through sea water tank at Naos Island (Panama).

Whether the full moon falls at the beginning of a month or near its end can influence in which calendar month lunar synchronized coral species spawn. Since many corals spawn a few days after full moon, if the full moon falls late in the month, spawning can occur within the next calendar month following. For example, in *Orbicella annularis*, mature gametes were present in August–October in 1983, but only in August and September in 1984 (Szmant 1986). This was associated with the full moon falling a week earlier in 1984 than in 1983. There also was evidence that individual colonies spawned more than once. The association of spawning with the lunar period may represent the fundamental timing mechanism used by corals to achieve synchronized spawning within a population, while factors such as temperature and photoperiod, which have important roles in regulating reproductive cycles, may control initiation of gametogenesis (Szmant 1986). Therefore, spawning episodes of different years may not occur in exactly the same set of months. This could artificially "extend" the breeding season viewed in pooled annual data. However, there is ample evidence that at least some species of eastern Pacific corals release gametes monthly over a period of several months and that individual colonies of at least some species spawn monthly as well.

Thus, the exact cycle of gamete development is still somewhat unclear. In support of the monthly developmental cycle, mature gametes were observed in tissues of several species approximately one month after collections showed no gametogenic development. For example, in the Gulf of Panama, gametes were found in tissues at the beginning of April–May just after upwelling ends, and mature gametes were present in tissues approximately a month later. Histological slide preparations have revealed Stage IV eggs and spermaries present in tissues over several months in the same year. And the sexual activity of agariciid corals, of which most species in the eastern Pacific have colonies that display overlapping male/female/male generations, suggest that monthly completion of gamete development is likely.

15.3.5 Spawning and Gamete Characteristics

Similar spawning characteristics were observed in four agariciid species, two encrusting and two with massive colony morphologies. Of the encrusting colonies, gametes from *Pavona varians* were positively buoyant with sperm released in a slow-rising diffuse cloud and ova in mucus-bound strings. The gametes, however, did not rise to the surface, but remained in the water column close to the spawning colonies. Spawning in *Pavona chiriquiensis* (also encrusting) was similar with neutral to negatively buoyant gametes that did not disperse far from the parental colonies. Ova and sperm were emitted slowly from *Pavona gigantea*, a massive species, in mucus-bound filaments that appeared to be nearly neutrally buoyant. Gametes remained suspended in the water column where they were dispersed by currents, imparting a milky appearance to the water in the immediate spawning area. Gametes released by *Pavona clavus* (massive colony morphology) were enveloped in mucus-laden strings that adhered to colony surfaces. Ova of *P. varians* and *P. chiriquiensis* were beige or white and dark green respectively, the latter possibly cryptic after sunset.

Agariciid ova are generally small and contain very fine lipid granules (see Sect. 15.2.1.1 and Fig. 15.5). Although mature agariciid ova generally stain lightly in contrast to the ova of some other eastern Pacific species, the ova of *Psammocora* (2 spp.) and *Diaseris* (1 sp.) also are small and stain lightly. Perhaps a more complex yolk that stains darkly (with Azocarmine), such as that displayed in the broadcast spawners *Porites* and *Pocillopora*, is not necessary or would interfere with the non-buoyant type of dispersal described above. Two species of *Pocillopora* (*P. damicornis* and *P. elegans*) possess ova that are laden with large lipid vesicles (Fig. 15.2). This would suggest positive buoyancy and perhaps contributes to the release of gametes, which would rise to the surface and disperse.

Spawning events in the western Pacific are closely associated with neap tidal cycles in several species and as a result tidal rhythms have been proposed to serve as proximate spawning cues. Neap tidal spawning has been hypothesized to promote the retention of sexual products on natal reefs (see Harrison and Wallace 1990). However, the four *Pavona* species (*P. gigantea, P. varians, P. chiriquiensis, P. clavus*) that were observed spawning in the eastern Pacific did so during the flooding phase of spring tides. Whether this would help to disperse gametes and larvae off reef with a positive effect on survival or some other life history function, is in need of study (Strathmann et al. 2002).

In summary, the asynchronous spawning observed in the eastern Pacific is similar to that reported in the northern Red Sea (Shlesinger and Loya 1985; Shlesinger et al. 1998; but see Hanafy et al. 2010), in Hawaii (Kolinski and Cox 2003), and other areas in the central and western Pacific (Richmond and Hunter 1990). While different coral assemblages demonstrate asynchronous and more protracted spawning across a wide range of latitudes, these patterns are generally more common on reefs in equatorial locations (Baird et al. 2009; Harrison 2011). Several studies have suggested that mass spawning is adaptive because the large volume of gametes released would satiate predators and allow for relatively high survival. This argument can also be inverted for coral faunas that demonstrate irregular spawning schedules. For example, it might be difficult for potential predators to evolve focused feeding behaviors for diverse species that are often relatively dispersed and that release gametes at different seasons, lunar phases and time of day.

15.4 Fecundity

Ovum and larval production rates (fecundity) are essential processes in coral reproduction, and for eastern Pacific corals the importance of sexual reproductive success may vary since several species appear to depend significantly on asexual fragmentation. As noted above, gonad and gamete development may occur up to a certain stage, but then be interrupted by a stress event such as a rapid elevation of sea temperatures. A reduction in fecundity or cessation of gametogenesis and larval development can then greatly reduce or nullify sexual reproduction, dispersal and recruitment.

Fecundity estimates have been determined for some of the more common eastern Pacific corals, but most comprehensively at three of four equatorial localities (Table 15.10). For the majority of the studied zooxanthellate species, thousands of ova cm^{-2} are produced annually at several sites although in a few instances this may vary by an order of magnitude. Most species of *Pavona* and *Psammocora* have demonstrated mean annual fecundity values of 10–20 $\times$ 10^3 ova cm^{-2} at non-upwelling sites (Caño and Uva Islands) and in the Galápagos Islands, the latter region characterized by variable and seasonally cool conditions. The large massive species, *Porites lobata*, *Pavona gigantea*, and *Gardineroseris planulata*, exhibited the lowest fecundities, often producing fewer than 5 $\times$ 10^3 ova cm^{-2} $year^{-1}$. While *Porites lobata* produced only 40 to 100 ova cm^{-2} $year^{-1}$ in the Galápagos Islands, *Pavona gigantea* was highly fecund there, exhibiting between 10 to 31 $\times$ 10^3 ova cm^{-2} $year^{-1}$. Considering all species at all sites, high fecundities were observed at Uva Island and the Galápagos Islands, each represented by three species in the genera *Pavona* and *Psammocora*. *Diaseris distorta*, sampled only where abundantly present in the Galápagos Islands, exhibited the highest fecundity of all studied species, ranging from 32 to 104 $\times$ 10^3 ova cm^{-2} $year^{-1}$, depending on the number of annual gametogenic cycles. The high levels of fecundity are unexpected across sites since temperature regimes differ significantly, being generally stable at Uva Island and highly variable in the Galápagos Islands. Overall, fecundities were lowest in the seasonally upwelling Gulf of Panama where temperatures often reach stressfully low levels during a three and one-half month period at the beginning of each calendar year, and gametogenesis is not initiated until after upwelling ceases.

Eight eastern Pacific species, all broadcast spawners, are clustered around relatively small ovum sizes of ~100 μm, but demonstrate a wide range of fecundities (Fig. 15.8a). Seven of these species (*Pavona*, 4 species; *Psammocora*, 2 species; *Diaseris distorta*) exhibited annual fecundities exceeding 10^4 ova cm^{-2}. Three species, *Pavona gigantea* and *Pavona clavus*, and *D. distorta* exhibited the highest fecundities at some localities, in excess of 3.0 $\times$ 10^4 ova cm^{-2} $year^{-1}$. These occurred in sample sets from the Galápagos Islands. *Pavona varians* and *Pavona chiriquiensis* (*Pavona* spp.) also demonstrated high fecundities, in excess of 10^3 ova cm^{-2} $year^{-1}$. These high fecundities occurred in non-upwelling environments (Caño Island, Uva Island, Galápagos Islands). *Porites lobata*, the seventh broadcast spawner, produced ova sizes at Caño and Uva Islands that were relatively large with fecundity estimates in excess of 10^3 cm^{-2} $year^{-1}$. The Galápagos fecundity estimates for *Porites*

Table 15.10 Fecundity estimates (ova/planulae cm^{-2} $year^{-1}$) of annual egg or planula production in equatorial eastern Pacific

Species	Location				Authority
	Caño I	Uva I	G Panamá[a]	Galápagos Is.	
Porites lobata	2.0–4.0 × 10^3 (1–2)	2.6–5.2 × 10^3 (1–2)	–	40,70,100 (1–3)	Glynn et al. (1994)
Porites panamensis[b]	–	–	720 (6)	–	Glynn et al. (1994)
Pavona gigantea	1.8–7.4 × 10^3 (1–4)	4.9–9.8 × 10^3 (1–2)	0.6–2.4 × 10^3 (1–2)	10.3–30.8 × 10^3 (1–3)	Glynn et al. (1996)
Gardineroseris planulata	0.5–1.0 × 10^3 (1–2)	0.7–1.4 × 10^3 (1–2)	–	–	Glynn et al. (1996)
Pavona spp[c]	19.9–27.9 × 10^3 (5–7)? 39.8 × 10^3 (?10)[d]	14.8–19.8 × 10^3 (6–8)	0.5–9.3 × 10^3 (1–3)	9.0–27.2 × 10^3 (4–11,12)? 58.8 × 10^3 (?24)[d]	Glynn et al. (2000)
Pavona clavus	16.8–26.4 × 10^3 (7–11)	16.0 × 10^3 (3)	6.7–7.8 × 10^3 (3)	21.2–33.2 × 10^3 (7–11)	Glynn et al. (2011)
Psammocora stellata	≥ 18.4 × 10^3 (≥ 7)	12.8 × 10^3 (12)	≥ 13.2 × 10^3 (≥ 8)	≥ 18.2 × 10^3 (≥ 9)	Glynn et al. (2012)
Psammocora profundacella	12.2 × 10^3 (≥ 3)	20.3 × 10^3 (5)	–	16.5 × 10^3 (≥ 6)	Glynn et al. (2012)
Diaseris distorta	–	–	–	31.6–51.8 × 10^3 (4) 63.2–104.0 × 10^3 (8)	Colley et al. (2000)
Tubastrea coccinea[b]	–	5–81[e] (12)	–	330 (12)	Glynn et al. (2008)

Estimated number of spawning cycles per year in parentheses
[a]Includes Saboga and Taboga Is, Gulf of Panama
[b]Planula production
[c]Data are for *Pavona varians* and *Pavona chiriquiensis* combined
[d]Number of cycles could be greater
[e]Potential for possible bi-monthly release

lobata (40–100 ova cm^{-2} $year^{-1}$) are probably uncommonly low due to unfavorable local conditions. As noted above, these values were calculated from a single, presumably mature small ovum. This species was sampled in the Canal de Itabaca (Santa Cruz Island) in a small, shallow embayment with limited circulation. In addition, the colonies were small and generally in poor condition (i.e. with pale/mottled tissue and algal filaments in the gastrovascular cavity). The two brooding species, *Tubastraea coccinea* and *Porites panamensis*, produced planulae that were consistently larger than the ova sizes in broadcast spawning species. This trend is consistent with observations in the Indo-Pacific and Caribbean regions (Harrison and Wallace 1990).

Comparing the size of reproductive products and fecundity of corals in the eastern Pacific with other regions, two major differences stand out. The first is the smaller size of eastern Pacific ova, all with mean diameters <200 µm, and brooder larvae ~300 µm in diameter or less (Fig. 15.8a). Ova sizes of Indo-Pacific and Caribbean species cluster between 200 and 600 µm in diameter (Fig. 15.8b), and the planulae from these regions range between 700 and 1500 µm in diameter (Fig. 15.8a, b). However, ova size of species in the families Poritidae, Agariciidae, Pocilloporidae, and Fungiidae from the Indo-Pacific and Caribbean also tends to be small (Harrison and Wallace 1990). Therefore, this pattern may be phylogenetically determined and species with small ova may have a survival advantage in the eastern Pacific. It is also possible that small ova and planulae are a result of a life history strategy that limits time for lipid storage due to frequent and rapid gametogenesis.

The second difference is the notably high fecundity, 10^3 to 10^5 ova cm^{-2} $year^{-1}$, demonstrated by several eastern Pacific broadcast spawning species. The few observations available indicate that eastern Pacific brooding species are also moderately fecund, but more data are needed to make any meaningful interregional comparisons. Perhaps the development of small ova with less complex yolk allows the generation of more numerous gametes during any given reproductive cycle. Such an opportunistic reproductive trait may be advantageous in an unstable environment such as the eastern Pacific, which is subject to sudden and frequent fluctuations in physico-chemical conditions. Thus, the marginal setting of the eastern Pacific may favor the colonization and persistence of highly fecund coral taxa with small eggs.

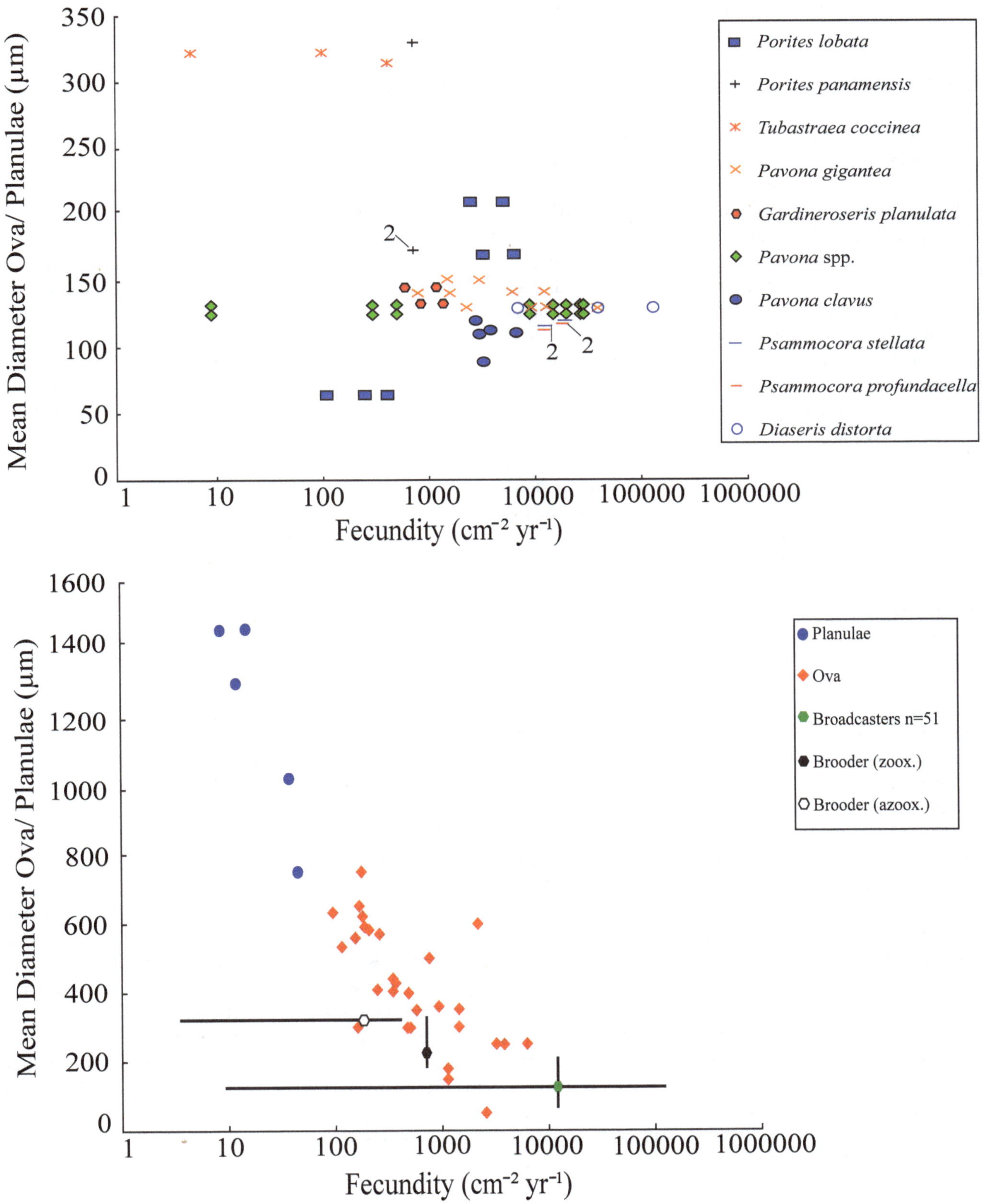

Fig. 15.8 Size of mature ova and planulae versus fecundity of scleractinian corals. **a** Ten equatorial eastern Pacific species sampled at four localities (see Tables 15.1 and 15.6). **b** Indo-Pacific and Caribbean species (n = 32) slightly modified after Harrison and Wallace (1990). Multiple fecundity values plotted for a specific mean ovum or planula size express the range of observed estimates. To assist in comparing two data sets, horizontal and vertical lines in '**b**' denote ranges of ova/planula sizes and fecundities of eastern Pacific species illustrated in '**a**'. Ranges of planula diameters (vertical line) in azooxanthellate brooder, and fecundities (horizontal line) in zooxanthellate brooder were very small, respectively, and therefore are not shown. *Pavona* spp. are *P. varians* and *P. chiriquiensis*; since ova diameters were statistically equal they were averaged; fecundities were also averaged due to similarity of annual spawning cycles (Glynn et al. 2000)

15.5 Recruitment and Asexual Fragmentation

15.5.1 Sexual Recruitment

Coral populations are maintained and replenished primarily by sexual larval recruitment and asexual fragmentation (e.g., Highsmith 1982; Harrison and Wallace 1990; Richmond 1997). The asexual regrowth or budding of patches of tissue resulting from partial mortality during disturbance events, such as bleaching, may also be important in perpetuating colony survival (Fong and Glynn 2001; Glynn and Fong 2006). Sexual recruitment of settling larvae has been studied by the deployment and retrieval of artificial surfaces (settling plates) or the appearance of new juveniles on natural reef substrates. The former method focuses on the number of larvae arriving and settling, whereas the latter focuses on the number of juvenile corals surviving until they are visible to the naked eye (i.e. ≥ 1 cm). Asexual recruitment is inferred from the appearance of colonies according to the criteria noted below in Sect. 15.5.5. The reader is referred to Mundy (2000) for an appraisal of methods employed in sexual recruitment studies, and Baird et al. (2006) for fluorescence, and Edmunds (2010) for aluminum tag detection techniques. All methods described here (Table 15.11) are based on corals counted on natural substrates and artificial surfaces.

Sexually produced planula larvae or genets, which settle onto suitable substrates, are distinguished from ramets that are genetically identical fragments originating from pre-existing colonies. Here a recruit is defined as a new individual introduced to a population whether as a sexual recombinant planula (genet) or an asexual fragment (ramet). Both modes of reproduction offer advantages under different environmental conditions, summarized hypothetically by Pinzón et al. (2012) as follows. Highly clonal species generating ramets would tend to dominate local optimal habitats as asexual reproduction (fragmentation) is important to local persistence. Conversely, sexual recombination increases diversity and would contribute to long-range dispersal and colonization of new habitats and would, therefore, be favored in variable and/or unpredictable environments.

In contrast to central and western Pacific reef areas, notably low rates of sexual recruitment were reported in the first eastern Pacific studies (Birkeland 1977; Wellington 1982; Richmond 1985; Reyes-Bonilla and Calderón-Aguilera 1994). In the Gulf of Panama, no sexual recruitment was observed in the Pearl Islands over an 8-mo period (Wellington 1982) or very low rates of ~1 recruit m^{-2} over 5 years (Birkeland 1977), and 11 colonies of *Pocillopora* sp. after 15 months (Richmond 1985). These low rates of sexual recruitment of the major reef-building species were cited as evidence of the maintenance of eastern Pacific reefs by asexual propagation and long distance dispersal from central/western Pacific populations. In more recent studies in Mexico, low rates of sexual recruitment have also been reported for the major branching and massive reef-building species (Medina-Rosas et al. 2005; López-Pérez et al. 2007). However, one study off the southern Mexican coast of Guerrero reported a low level of sexual recruitment of *Pocillopora damicornis* (García-Ocampo 2005). Recruitment rates at five localities ranged from 0.1 to 0.3 colonies m^{-2} $year^{-1}$ to maximum rates of 0.6–0.8 colonies m^{-2} $years^{-1}$.

Since the early findings from Panama, continuing studies have added to the data base of coral recruitment for the principal eastern Pacific species at several localities. These data are from colonies that recruited to natural and artificial plots monitored over periods that ranged from <1 to 22 years. In the majority of these studies, no distinction was made between sexual and asexual recruitment. The calculated rate estimates do, however, allow for comparisons among species and sites. Inspection of Table 15.11 indicates that recruitment rates are highly variable and generally low for broadcast spawning species, usually with mean colony densities of <1.0 m^{-2} $year^{-1}$. The highest recruitment rates for broadcast spawners (*Pocillopora elegans, Pavona varians, Porites lobata*), ~1.0 colony m^{-2} $year^{-1}$, occurred at non-upwelling sites in Costa Rica and Panama. The sexual recruitment of *P. lobata* was shown by Boulay et al. (2012), utilizing microsatellite markers, to be the predominant mode of recruitment at Isla del Coco, Costa Rica. Although recruitment rates were not reported, the repopulation of *P. lobata* by sexual means has contributed importantly to the recovery of Cocos Island reefs during the last two decades (Guzmán and Cortés 2007). Recruitment of *Porites panamensis*, the single zooxanthellate brooding species, was considerably higher than the broadcast spawners, often exceeding 1.0 colony m^{-2} $year^{-1}$, and as high as 20.4 colonies m^{-2} $year^{-1}$ at one site in Mexico (Montosa Island, Huatulco).

Millepora intricata, a hydrocoral, has only been found in the eastern Pacific in the Gulf of Chiriquí, Panama. At Uva Island colonies are common on the reef slope (2–5 m depth), but are among the first zooxanthellate species to bleach and die during ENSO warming events (Glynn 1990). This species is opportunistic, among the first to recruit to reef slopes after mortality events, presumably from deeper (8–15 m) refuge populations in cooler water. Although colonies of *M. intricata* easily fragment, recolonization of the shallow reef slope presumably occurs via sexual recruits (Smith et al. 2014).

15.5.2 Asexual Fragmentation

Of the different kinds of asexual reproduction in corals that give rise to new individuals or colonies, e.g. budding and

Table 15.11 Colony densities of recently recruited and juvenile coral taxa sampled at northern and EEP reef sites

Taxa	Locality	Depth (m)	Period (year)	Method[a]	Colonies (#)[b]	Density (col m^{-2} year^{-1})			Authority
						Mean[c]	SD	Range	
Psammocora stellata	COSTA RICA								
	Cocos I.	3–24	1987–2002	NS	56	0.098	0.080	0.011–0.233	Guzmán and Cortés (2007)
Psammocora superficialis	COSTA RICA								
	Cocos I.	12–24	1987–2002	NS	67	0.203	–	0.017–0.483	Guzmán and Cortés (2007)
Psammocora spp.	COSTA RICA								
	Caño I.	9–14	1984–99	NS	230	0.575	1.065	0–2.70	Guzmán and Cortés (2001)
Pocillopora elegans	COSTA RICA								
	Caño I.	3–9	1984–99	NS	145	0.362	0.552	0–1.55	Guzmán and Cortés (2001)
		9–14	1984–99	NS	220	1.10	0.79	0.25–2.25	Guzmán and Cortés (2001)
	Cocos I.	3–18	1987–2002	NS	10	0.028	0.011	0.017–0.042	Guzmán and Cortés (2007)
Pocillopora damicornis	COSTA RICA								
	Caño I.	0–3	1984–99	NS	140	0.35	0.829	0–2.25	Guzmán and Cortés (2001)
	PANAMA								
	Contadora I.	0–10	1979	AS	0	0	–	–	Wellington (1982)
Pocillopora eydouxi	COSTA RICA								
	Cocos I.	12–18	1987–2002	NS	2	0.022	–	–	Guzmán and Cortés (2007)
Pocillopora sp.[d]	PANAMA								
	Taboguilla I.	3	1982–84	AS	7	0.99	–	–	Richmond (1985)
		8	1971–75	AS	2	0.53	–	–	Birkeland (1977)
Pavona varians	COSTA RICA								
	Caño I.	5–9	1984–99	NS	27	0.068	0.226	0–0.75	Guzmán and Cortés (2001)
		3–24	1987–2002	NS	168	0.350	0.404	0.022–0.973	Guzmán and Cortés (2007)
	PANAMA								
	Uva I.	2–3	1984–99	NS	95	0.183	0.188	0.050–0.650	Glynn et al. (2000)
		1–7	1988–89	NS	0	0	–	–	Smith (1991)
		4–5	1988	NS	4	0.16	0.62	–	Smith (1991)
		5–7	1989	NS	1	0.04	–	–	Smith (1991)
		4–5	1989	NS	32	1.28	3.59	–	Smith (1991)
Pavona chiriquiensis	COSTA RICA								
	Cocos I.	12–24	1987–2002	NS	41	0.124	–	0.011–0.178	Guzmán and Cortés (2007)

(continued)

Table 15.11 (continued)

Taxa	Locality	Depth (m)	Period (year)	Method[a]	Colonies (#)[b]	Density (col m^{-2} $year^{-1}$)			Authority
						Mean[c]	SD	Range	
Pavona clavus	COSTA RICA								
	Cocos I.	9–24	1987–2002	NS	10	0.030	–	0.011–0.078	Guzmán and Cortés (2007)
	PANAMA								
	Uva I.	4–6	1993–2010	NS	9	0.011	0.013	–	Glynn et al. 2011
		3–5	1993–2010	NS	3	0.004	0.007	–	Glynn et al. (2011)
		2–3	1984–2007	NS	0	0	–	–	Glynn et al. (2011)
		3–6	1994–2007	NS	0	0	–	–	Glynn et al. (2011)
	Saboga I.	2–3	1984–2005	NS	0	0	–	–	Glynn et al. (2011)
	Secas I.	4–5	1984–2006	NS	0	0	–	–	Glynn et al. (2011)
	GALAPAGOS								
	Pta. Estrada	5–8	1985–2001	NS	5	0.042	0.272	–	Glynn et al. (2011)
	Onslow I.	3–5	1986–2001	NS	1	0.010	0.102	–	Glynn et al. (2011)
Pavona gigantea	PANAMA								
	Uva I.	1–7	1988–89	NS	0	0	–	–	Smith (1991)
		5–7	1989	NS	1	0.04	–	–	Smith (1991)
		5–7	1995	NS	57	0.190	–	–	Glynn and Leyte-Morales 1997
	Contadora I.	0–10	1979	AS	0	0	–	–	Wellington (1982)
	MEXICO								
	La Entrega	6–8	1995	NS	74	0.247	–	–	Glynn and Leyte-Morales (1997)
Gardineroseris planulata	COSTA RICA								
	Cocos I.	6–15	1987–2002	NS	1	0.008	–	–	Guzmán and Cortés (2007)
	PANAMA								
	Uva I.	5–7	1988	NS	1	0.100	–	–	Smith (1991)
		5–7	1988–89	NS	0	0	–	–	Smith (1991)
		1–7	1993	NS	24	0.002	–	–	Glynn, unpub. data
		1–7	1994	NS	159	0.013	–	–	Glynn, unpub. data
Leptoseris scabra	COSTA RICA								
	Cocos I.	15–24	1987–2002	NS	3	0.025	–	–	Guzmán and Cortés (2007)

(continued)

Table 15.11 (continued)

Taxa	Locality	Depth (m)	Period (year)	Method[a]	Colonies (#)[b]	Density (col m^{-2} $year^{-1}$)			Authority
						Mean[c]	SD	Range	
Porites lobata	COSTA RICA								
	Caño I.	0–5	1984–99	NS	139	0.350	0.333	0–1.00	Guzmán and Cortés (2001)
	Cocos I.	3–24	1987–2002	NS	690	1.095	0.520	0.22–1.55	Guzmán and Cortés (2007)
	PANAMA	1–7	1988–89	NS	0	0	–	–	Smith (1991)
	Uva I.	5–7	1989	NS	1	0.04	–	–	Smith (1991)
		1–2	1988	NS	10	0.40	0.20	–	Smith (1991)
Porites panamensis	PANAMA								
	Uva I.	5–7	1988	NS	130	13.00	12.28	–	Smith (1991)
		5–7	1989	NS	279	11.16	10.82	–	Smith (1991)
		4–5	1988	NS	6	0.24	0.83	–	Smith (1991)
		4–5	1989	NS	25	1.00	2.12	–	Smith (1991)
		1–5	1988–89	NS	0	0	–	–	Smith (1991)
Porites panamensis	MEXICO								
	La Tijera	2–13	2001–02	AS	44	18.4[f]	–	–	López-Pérez et al. (2007)
	San Agustín I.	2–13	2001–02	AS	16	5.8 [g]	–	–	López-Pérez et al. (2007)
	Jicaral-Chachacual	2–13	2001–02	AS	31	6.6	–	–	López-Pérez et al. (2007)
	Cacaluta I.	2–13	2001–02	AS	7	1.5	–	–	López-Pérez et al. (2007)
	La Entrega	2–13	2001–02	AS	3, 1[e]	0.8	–	–	López-Pérez et al. (2007)
	Montosa I.	2–13	2001–02	AS	96	20.4	–	–	López-Pérez et al. (2007)
	Cabo Pulmo	0–6	1991–92	NS	3.8×10^5	0.26	–	–	Reyes-Bonilla and Calderón-Aguilera (1994)
		3–15	1991–92	NS	3.2×10^5	0.36	–	–	Reyes-Bonilla and Calderón-Aguilera (1994)
		>10	1991-92	NS	4.6×10^5	0.77	–	–	Reyes-Bonilla and Calderón-Aguilera (1994)
Porites sp.[d]	MEXICO								
	Chimo	6–10	1998–99	AS	2	1.08 [h]	–	–	Medina-Rosas et al. (2005)
	Tenacatita	5–7	1998–99	AS	6	4.66 [h]	–	–	Medina-Rosas et al. (2005)
	Additonal Sites (n = 7)	3–13	1998–99	AS	0	0	–	–	Medina-Rosas et al. (2005)

[a]*NS* natural substrate—natural reef surfaces, quadrat and visual transect sampling; *AS* artificial surfaces—blocks, metallic screen, bricks, tiles (ceramic, acrylic)
[b]Total number colonies observed during study period
[c]Mean values calculated on basis of number of surveys and total area sampled
[d]Unidentified as to species
[e]One of four observed recruits was *Pocillopora* sp., unidentified as to species
[f]Time = 1.08 year
[g]Time = 0.91 year
[h]Time = 6 mo

transverse or longitudinal fission, polyp bail-out, detachment of polyp balls, and asexually produced planulae (Harrison and Wallace 1990; Richmond 1997), asexual fragmentation is the predominant process in the eastern Pacific. Several studies have supported the importance of this reproductive mode (e.g., Glynn et al. 1972; Highsmith 1982; Guzmán and Cortés 1989; Colley et al. 2000; Glynn et al. 1996, 2011; Lirman et al. 2001; López-Pérez et al. 2007; Aranceta-Garza et al. 2012).

Although storms can cause fracturing, especially in ramose species (Jiménez 1998), coral fragmentation often has a biotic provenance, i.e. it is initiated or exacerbated by bioerosion (see Chap. 12, Alvarado et al.) and corallivory (see Chap. 10, Enochs and Glynn). Bioerosion by sponges, bivalve molluscs, and other endolithic organisms weakens coral skeletons, rendering them susceptible to physical damage by waves, storm surge and projectiles. Branching species of *Pocillopora* are especially susceptible to breakage by nest-building and foraging fishes (Jiménez 1996–1997; Glynn 2008). It may seem counterintuitive, but large massive colonies of *Porites*, *Pavona*, and *Gardineroseris* are also subject to fragmentation. For example, Highsmith (1982) observed that *Pavona clavus* colonies excavated by burrowing molluscs (*Lithophaga* spp.) resulted in collapse and breakage with the generation of several new colonies. Massive and encrusting colonies, such as *Porites lobata*, *Porites panamensis* and *Pavona varians*, are commonly excavated by the triggerfish *Pseudobalistes naufragium* (Fig. 15.9). The triggerfish, in search of boring bivalves, often produces cm-size fragments that can regenerate, giving rise to new colonies if they remain upright on a suitable substrate (Guzmán 1986; see also Guzmán 1988; Guzmán and Cortés 1989). The combination of a relatively weak skeleton and a high density of lithophagine bivalves, up to 100 ind per 0.01 m^2, render this coral a frequent target of triggerfish attacks (Scott and Risk 1988). A recent study has shown that morphologically similar massive species of *Porites* reproduce mainly asexually (*Porites evermanni*) or sexually (*Porites lobata*), depending on the interaction of triggerfish predators and their endolithic boring bivalve prey (Boulay et al. 2013). The triggerfish preferentially attack *P. evermanni* colonies, which are infested with higher abundances of bivalves than is *P. lobata*.

Fig. 15.9 *Porites lobata* colony presumably attacked by *Pseudobalistes naufragium*. Lithophagine boreholes are visible in two broken sections of the skeleton, top and bottom right side of photograph. Several live fragments lay on bottom to left of colony (arrows). Observed on a patch reef at Bahía Honda, Gulf of Chiriquí, Panamá (5 m depth, 14 March 2003, courtesy Tyler Smith)

Highsmith (1982) has reasoned that some successful corals have evolved a predisposition for colony fragmentation into their life histories. As an example he proposed that pocilloporid corals, which are the predominant reef builders in the eastern Pacific, fragment easily and thus contribute importantly to the lateral expansion of reef frameworks. Fragmented branches, if not stabilized by existing colonies and frameworks, tend to move laterally down slope where they continue to grow and contribute to reef progradation (Fig. 15.10). Additional information on frequent storm-related fragmentation in massive *Porites* colonies can be found in Done and Potts (1992).

A recent study by Baums et al. (2014) indicates that the densest known aggregration of *Pocillopora* in the Galápagos Archipeligo is of asexual origin. A large population of 1614 live *Pocillopora* colonies present on the southern shore of Isabela Island was analyzed for genetic diversity. Multilocus genotyping, capable of resolving individual clones, indicated that this coral stand is monogenotypic, and thus the high density of colonies is a result of asexual reproduction, likely via fragmentation.

Diaseris distorta, free-living fungiid species as adults, exhibits frequent asexual fragmentation, which is aided by the spontaneous formation of sutures and the separation of daughter segments. Morphological evidence indicates this is perhaps the predominant means of propagation in the Galápagos Islands (Colley et al. 2000). This radial asexual fragmentation through natural autonomy is a compelling example of an evolved trait (Yamashiro and Nishihira 1998).

15.5.3 Asexual Recruitment

Data on the abundances of asexually produced recruits in Panama and Mexico indicate that branch fragmentation can potentially contribute substantially to coral community recovery. The asexual recruit densities of *Pocillopora damicornis* in Panama in 1980, increasing from 0.17 to 0.44 colonies m^{-2}, were observed after censuses of 4.5, 8 and 12 months, respectively (Highsmith 1982). This level of fragmentation was attributed to wave action, bioerosion, and excavation by foraging fishes. The high mean pocilloporid

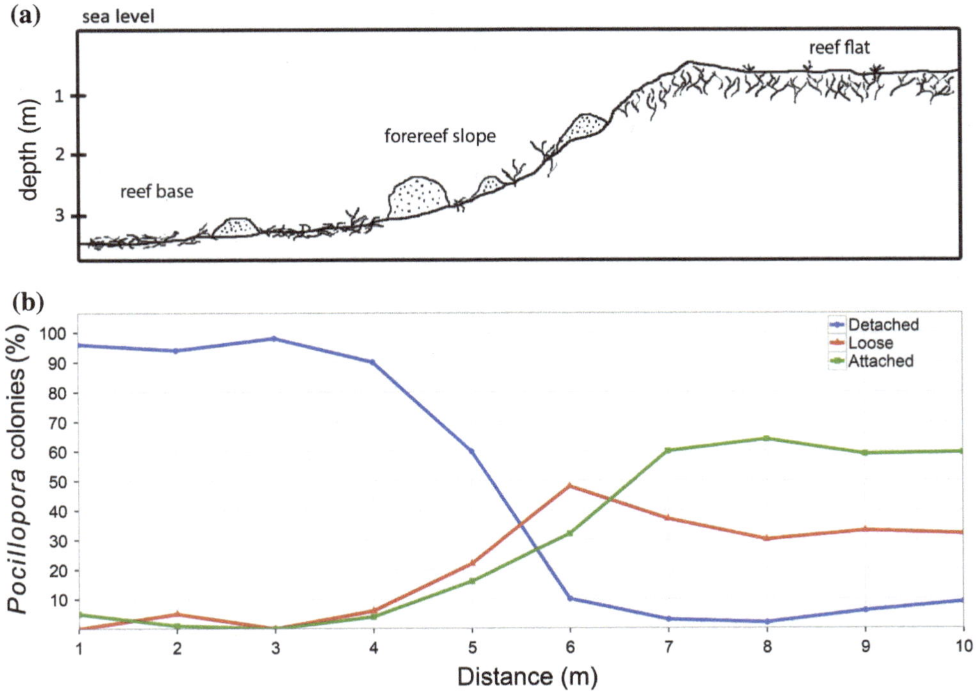

Fig. 15.10 Mean percent attached, detached, and loose colonies of *Pocillopora damicornis* along 9-m long transects (~3-m depth gradient) on three reefs in the Gulf of Chiriquí, Panama. **a** Generalized profiles of reefs sampled showing locations of reef flat, fore reef slope, and reef base. **b** Data from 27, 10-m long transects, with total number of observations n = 1350. Modified after Highsmith (1982)

fragment densities reported by Lirman et al. (2001) in southern Mexico (Huatulco), as high as 26.2 fragments m^{-2} at the Riscalillo sampling site, were observed in 1997, one to two months after three major cyclones passed across the Huatulco reef tract (Table 15.12). These storms caused extensive coral breakage, especially among branching *Pocillopora* spp. The fragment densities observed by López-Pérez et al. (2007), sampled in 2001–2002 along the same Huatulco reef tract, were overall about 2.5 times significantly lower (mean densities = 6.2 vs. 15.7 colonies m^{-2}, $p = 0.012$, t test, 1-tailed) than those assessed in 1997. This significant difference was likely due to the sampling performed soon after the storm impacts in 1997. An expected lasting result of coral communities subject to frequent storms would be the formation of clones and low genetic diversity, a trend demonstrated recently in the lower half of the Gulf of California. A molecular genetic study of branching *Pocillopora verrucosa* colonies revealed a significant relationship between tropical storm activity and the genotypic indexes of richness and diversity (Aranceta-Garza et al. 2012). This analysis offered strong evidence supporting the role of asexual fragmentation in producing clonal populations at localities subject to frequent storms.

The recent study by Baums et al. (2014) at Isabela Island noted above (Sect. 15.5.2) also strongly suggests the asexual origin of this population. All colonies belonged to *Pocillopora* mitochondrial *open reading frame* lineage type *3a*. Typing of additional samples from three other islands indicated that this large stand of colonies was unique in the Galápagos Islands, representative of the only type *3a* population. Colony size distribution, while imperfect, suggested that the population originated from remnant colonies that survived the 1997–98 ENSO event, but may postdate the 1982–83 ENSO.

López-Pérez et al. (2007) underlined the importance of successful and unsuccessful fragmentation in terms of fragment survival. Cementation of pocilloporid fragments can occur rapidly, in 1–2 months, if retained in suitable reef habitats such as stable living or dead reef structures (Lirman et al. 2001). If fragments are deposited in shallow, wave swept zones or in soft sediments, attachment is unlikely and survivorship low (Ayre et al. 1997). Successful cementation was quantified at Huatulco and ranged between 16 and 58 % (López-Pérez et al. 2007). Thus, to determine the efficacy of asexual fragmentation it is necessary to continue monitoring coral survival of past initial fragmentation events.

15.5.4 Other Factors Affecting Recruitment

Recruitment is also chemically mediated in corals. Rather than being a solely stochastic process of random searching and settlement, larvae of some species respond to specific metamorphic inducers, including those associated with certain species of crustose coralline algae, which are hypothesized to enhance survivorship (Golbuu and Richmond 2007). Physical interference with such cues, through sediment and fleshy algal cover, can prevent coral recruitment. Pollutants, including pesticides, may block metamorphic inducers or larval receptors, and prevent settlement and metamorphosis (Peters et al. 1997). However, many eastern Pacific reefs are located at remote island sites; therefore, such chemical pollutants are less

Table 15.12 Asexual pocilloporid and *Gardineroseris planulata* recruits sampled on natural reef substrates in Panama and Mexico

Taxa	Locality	Depth (m)	Period	Colonies	Density (col m^{-2}) Mean (SE)	Authority
Pocillopora damicornis	PANAMA					
	Uva I.	5–6	1980	25	0.444	Highsmith (1982)
		5–6	1980	38	0.380	Highsmith (1982)
		5–6	1980	66	0.440	Highsmith (1982)
Pocillopora spp.	MEXICO					
	San Agustín I.	2–13	2001–02	26[a]	1.3 (2.2)	López-Pérez et al. (2007)
	Jicaral-Chachacual	2–13	2001–02	134[a]	6.7 (7.5)	López-Pérez et al. (2007)
	Cacaluta	2–13	2001–02	106[a]	5.3 (4.7)	López-Pérez et al. (2007)
	La Entrega	2–13	2001–02	262[a]	13.1 (30.3)	López-Pérez et al. (2007)
	Montosa I.	2–13	2001–02	90[a]	4.5 (1.3)	López-Pérez et al. (2007)
	Riscalillo	0–10	1997	1127[a]	26.2 (4.9)	Lirman et al. (2001)
	Mixteca	0–4	1997	513[a]	23.3 (5.3)	Lirman et al. (2001)
	Órgano	0–8	1997	272[a]	20.9 (3.3)	Lirman et al. (2001)
		0–8	1997	406[a]	20.3 (2.4)	Lirman et al. (2001)
	Río Sabroso	0–4	1997	79[a]	9.9 (2.8)	Lirman et al. (2001)
		0–4	1997	298[a]	14.9 (1.6)	Lirman et al. (2001)
	Montosa I.	0–13	1997	540[a]	7.5 (1.3)	Lirman et al. 2001)
	Maguey	0–10	1997	48[a]	4.8 (2.2)	Lirman et al. (2001)
		0–10	1997	348[a]	13.9 (1.6)	Lirman et al. (2001)
Gardineroseris planulata	PANAMA					
	Uva I.	1–7	1993	100	0.008	Glynn, unpub. data
		1–7	1994	111	0.009	Glynn, unpub. data

[a]Colony values were calculated by multiplying mean fragment densities by total area (m^2) sampled

of a concern. Instead, sedimentation during the wet season, especially in areas with significant coastal development and deforestation, can be an obvious contributor to stress.

Fleshy and filamentous algae are often present and abundant on eastern Pacific reefs due to nutrient availability from both upwelling and coastal runoff. Benthic algae are a repository for sediments, which can be re-suspended by water motion. Algae can also draw down oxygen and pH levels during periods of photorespiration at night. As such, the combination of benthic fleshy algae and terrigenous sediments interact to reduce both substrate availability and water quality, affecting coral recruitment and survival.

15.5.5 Comparisons/Discussion

In an attempt to evaluate the relative importance of sexual versus asexual recruitment through fragmentation, colonies of several species in study plots in the EEP were enumerated according to their presumed provenance. The following criteria, employed for live colonies from 1 to 10 cm greatest dimension, were: *sexual recruits*—cemented to substrate, symmetrical colony morphology, no indication of recent or old skeletal fractures; *asexual fragments*—usually loose or weakly cemented to substrate, colony morphology often irregular, recent or old skeletal fractures evident. Some mis-assignments can be expected since these criteria are not mutually exclusive. The magnitude of this type of error has not been reported.

For most species and localities, sexual recruits greatly outnumbered asexual recruits (Table 15.13). In the Galápagos Islands, *Pocillopora* spp. sexual recruits predominated at three of four monitoring sites and made up between about 60 to 100 % of all colonies sampled. Only at Playa La Picona, Floreana Island, were asexual recruits (n = 16) more abundant than sexual recruits (n = 14). This site is often exposed to moderate-to-strong wave action. The majority of sexual recruits were present on the tops and sides of basalt boulders, and asexual recruits were usually unattached on coral rubble/sand substrates. At Cabo Pulmo, Gulf of California, *Pocillopora* fragments were predominant, contributing 59 % to the recruitment pool (Reyes-Bonilla 1993).

Table 15.13 Numbers and colony densities of sexual and asexually-derived coral recruits observed in permanent monitoring plots on Uva Island reef (Panama) and in Galápagos Islands (Ecuador), 1993–1998

Species	Sexual recruits		Asexual recruits		Location[b]	Depth	Year	Total SR/AR[c]
	n	ind. m^{-2}	n	ind. m^{-2}				
Pocillopora spp.[a]	65	0.013	44	0.009	Española I, Gal Gardner Bay	1–10	1995	275/74
	54	0.006	25	0.003			1996	
	57	0.007	0	0			1997	
	99	0.012	5	0.001			1998	
	26	0.006	1	0.002	Santa Cruz I, Gal Caleta Robinson	1–15	1996	37/1
	11	0.003	0	0			1997	
	31	0.004	6	0.001	Santa Fe I, Gal NE anchorage	3–18	1995	86/6
	5	0.001	0	0			1997	
	50	0.007	0	0			1998	
	14	0.004	16	0.005	Floreana I, Gal Playa La Picona	2–6	1995	14/16
Porites lobata	7	0.024	0	0	Uva Reef-South	1–5	1995	7/0
Pavona gigantea	51	0.177	0	0	Uva Reef-North		1995	80/0
	6	0.021	0	0	Uva Reef-South		1998	
	23	0.080	0	0	Uva Reef-North			
Pavona clavus	2	0.007	0	0	Uva Reef-North		1995	6/0
	1	0.004	0	0	Uva Reef-South		1998	
	3	0.010	0	0	Uva Reef-North			
Pavona varians	52	0.181	0	0	Uva Reef-North		1995	132/2
	3	0.010	2	0.007	Uva Reef-South		1998	
	77	0.267	0	0	Uva Reef-North		1998	
Gardineroseris planulata	22	0.002	100	0.008	Uva Reef	2–7	1993	38/100
	6	0.021	0	0	Uva Reef-North		1995	
	1	0.004	0	0	Uva Reef-South		1998	
	5	0.017	0	0	Uva Reef-North		1998	
	4	0.014	0	0	Uva Reef-South			
	1	0.021	4	0.083	Pearl Is Saboga I			1/4
Psammocora profundacella	1	0.004	9	0.031	Uva Reef-North		1995	1/9

[a]Predominantly *Pocillopora elegans* in all surveys
[b]Plot areas sampled: Gardner Bay 1995, 5000 m^2; 1996–1998, 8370 m^2; Caleta Robinson 4100 m^2; NE anchorage (outside bay) 7360 m^2; Playa La Picona 3396 m^2; Uva R (reef) 12,645 m^2; Uva R-N 288 m^2; Uva R-S 289 m^2; Saboga I 48 m^2
[c]The numbers of sexual (SR) and asexual (AR) recruits are summed by species and locality sampled

Only sexual recruits were observed in the massive colonies of *Porites lobata*, *Pavona gigantea*, and *Pavona clavus*. As noted previously, asexual fragmentation in *P. lobata* (and especially in *Porites evermanni*) occurs commonly where this species is present, however, it is relatively rare on the Uva reef. Only seven small colonies (presumably sexual recruits) were found in the monitoring plot at Uva reef-south (Table 15.13). Asexual fragments were encountered in one of three surveys of *Pavona varians* on the Uva Island reef. This species is sometimes targeted by *Pseudobalistes naufragium*, which can severely fragment encrusting colonies such as *P. varians* (Glynn et al. 1972). In four of six surveys, *Gardineroseris planulata* exhibited only sexual recruits; however, asexual recruitment predominated in two surveys. The preponderance of asexual:sexual (100:22) recruits of *G. planulata* on the Uva Island reef in 1993 was ostensibly a result of colony breakage by foraging *Pseudobalistes naufragium*, which were present on the reef at that time. Asexual outnumbered sexual recruits in *Psammocora profundacella* at a north-end sampling site on the Uva reef in 1995. Data are not available for *Porites panamensis* at low latitude sites, but at Cabo Pulmo, Gulf of California, asexual fragmentation was noted in only about 8 % of colonies (Reyes-Bonilla and Calderón-Aguilera 1994).

Molecular genetic methods employing high resolution markers can more accurately distinguish between sexually and asexually derived recruits. Such studies have revealed varying contributions of genets and ramets to the population structure of pocilloporid assemblages. In Panama, *Pocillopora* communities consisted primarily of sexually-recruited colonies (Combosch and Vollmer 2011). Two studies conducted in the Gulf of California have reported dissimilar results. Aranceta-Garza et al. (2012) found that clonal populations of *Pocillopora* were present at sites frequented by storms, whereas Pinzón et al. (2012) reported clonal populations predominant in protected habitats. Since biotic agents (bioerosion, corallivores, foraging fishes) can cause significant colony fragmentation, it is imperative to consider these potential effects.

In a comparative study of the population biology of *Pocillopora damicornis* in the central/western and eastern Pacific, Richmond (1985) observed that 100 % of recruits at Enewetak originated from sexually-produced planula larvae whereas nearly 97 % of recruits in Panama resulted from asexual fragmentation. Other minute recruits of *P. damicornis* in Panama, with 1–12 polyps each, were attributed to 'polyp bail-out' as reported by Sammarco (1982). The nearby presumed donor colonies were under attack by the gastropod corallivore *Jenneria pustulata*. These results indicate marked regional differences in sexual and asexual recruitment among, as well as within, regions for a morphologically similar coral genus.

15.6 Climate Change and Anthropogenic Stressors

Climate change is considered one of the most pressing concerns for the future of coral reefs (e. g., Hughes et al. 2003). Although much discussion in this chapter has included data and other information acquired during a variety of ENSO events, other climate change factors are worthy of mention, although not the major focus of the studies that contributed to this chapter. Elevated seawater temperatures are responsible for mass-bleaching events caused by the loss of symbiotic zooxanthellae, and such stressed corals either die or are energetically compromised with insufficient energetic resources to support normal gamete production (Table 15.9). Elevated levels of pCO_2, such as those predicted to occur by the end of the century, have been found to reduce fertilization, growth and recruitment in colonies of *Acropora palmata* (Albright et al. 2010). While gametogenesis proceeded unabated when ambient seawater pH and aragonite saturation values were experimentally reduced, there is evidence that energy allocated to egg production under such conditions may further reduce calcification and growth rates in gravid colonies (Cohen and Holcomb 2009; Albright 2011).

The energetics of planula larvae in both brooding and spawning species affect dispersal distances and eventual recruitment success (Richmond 1987a, b, 1988). As lipids provide energy reserves, temperature increases associated with climate change are predicted to likewise increase metabolic and energy consumption rates during the planktonic period and reduce overall dispersal potential. This could affect population connectivity among eastern Pacific reefs and recovery trajectories following mortality events.

Reproduction and recruitment can be affected by both natural and anthropogenic stressors, the former often being exacerbated by the latter (e.g., Richmond 1997, 2005; Fabricius 2005; Harrison 2011). The physiological state of a coral will determine the allocation of energy into various functions, including colony growth, tissue repair (when necessary), calcification, and reproduction. Stressed corals may exhibit a reduction in the quantity and quality of gametes, even to the point of reproductive failure under severe conditions, such as elevated temperatures (Table 15.9). In addition to effects on fecundity, reduced water and substratum quality can affect fertilization, embryological development, zooxanthellae community composition (Garren et al. 2006; LaJeunesse et al. 2010), and the metamorphic induction of coral larvae (Richmond 1993).

Eastern Pacific reefs are subject to the common anthropogenic stressors affecting coral reefs elsewhere, including sediment loading and land-based sources of pollution as well as increased competition for space from algae due to overfishing of herbivores. Although the specific effects of these stressors on coral reproduction are unstudied in the eastern Pacific, various potentially disrupting impacts are threatening coral reefs from Mexico (Reyes-Bonilla 2003) to several equatorial eastern Pacific localities (Cortés and Jiménez 2003; Glynn 2003; Maté 2003; Zapata and Vargas-Ángel 2003). Seasonal upwelling is an added source of stress to which reefs from other areas are normally not exposed (Fig. 15.1b), and associated increases in nutrients can alter competition for substrate space in favor of epibenthos other than coral larvae and recruits.

As previously mentioned, high sediment loading onto reefal habitats during the rainy season, due to deforestation of surrounding land forms, decreases visibility and sunlight penetration in many locales. Associated salinity changes are potentially harmful, as well as seasonal low tide exposures that kill tissues on exposed pocilloporid reef flats, leaving behind retracted polyps that can potentially regenerate damaged tissues.

15.7 Conclusions: Reproduction in a Marginal Environment

Environmental physical data, with particular attention to sea temperature and salinity extremes, aragonite saturation state, light penetration, and inorganic nutrient concentration, were analyzed by Kleypas et al. (1999) to identify and classify marginal reef development worldwide. Potentially stressful conditions were listed for several eastern Pacific reef sites and coral communities from the Galápagos Islands to the Gulf of California. Of special note was the low aragonite saturation state of waters across the eastern tropical Pacific region (see also Manzello et al. 2008; Manzello 2010; see Chap. 18, Manzello et al.). The erratic occurrence of ENSO events, warm El Niño and cool La Niña phases (and attendant stressful conditions), also contribute importantly to disturbances in the eastern Pacific (see Chap. 8, Glynn et al.).

Although geographically patchy, reef building has occurred over large parts of the eastern Pacific, at some sites for periods of 100–1000 s of years (see Chap. 6, Toth et al.). It is of interest, therefore, to consider the attributes of coral reproduction that have likely contributed to this success. The following discussion stems largely from Glynn and Ault (2000), who addressed the dispersal potential and regional distribution of corals, and Glynn and Colley (2008), who examined coral survival under diverse conditions of disturbance in the eastern Pacific. These studies were based largely on the possible roles of reproductive traits in larval dispersal and repopulation of devastated coral communities following large scale disturbances.

All eastern Pacific corals with high frame-building potential are regionally widely distributed, and broadcast-spawn gametes: *Pocillopora damicornis*, *Pocillopora elegans*, *Porites lobata*, *Pavona gigantea*, *Pavona clavus*, and *Gardineroseris planulata*. Broadcast spawning species typically produce planktonic larvae capable of long distance dispersal (Harrison and Wallace 1990). Although larval longevity of eastern Pacific corals is not known, recent studies indicate a highly extended and variable duration of planktonic life of western Pacific, non-autotrophic coral larvae (Graham et al. 2008; Connolly and Baird 2010). For example, the larvae of all five species in the Graham et al. (2008) study demonstrated longevities of from 195 to 244 days. The previously reported record, for *Acropora valida*, was 130 days. This may offer a dispersal advantage over brooding corals. Thus, this reproductive mode would offer an advantage to corals subject to local disturbances and high mortality, for example resulting from extreme thermal stress, with distant source populations capable of supplying larvae that would contribute to the recovery of depleted populations. *Porites panamensis*, the sole eastern Pacific zooxanthellate brooder with a relatively narrow geographic range, experienced severe reductions on Panamanian reefs during the 1982–83 El Niño bleaching event (Glynn 1984), and was not observed again on monitored reefs for 3–5 years (Glynn et al. 2001a, b).

Most eastern Pacific corals are reproductively active for several months or in some cases year-round. If acute stress events do not completely inhibit gametogenesis or cause gonad reabsorption, sexual reproduction could possibly resume when conditions improve. Some species have continued to be reproductively active during weak to mild ENSO events when sea temperatures are elevated but do not exceed critical thresholds (Colley et al. 2006). For example, in the Galápagos Islands *Pocillopora damicornis* and *Pocillopora elegans* both demonstrated enhanced gametogenesis during moderate El Niño warming. Also, *Pavona varians* in Panama revealed significant increasing sexual recruitment over a 14 year period with increasing sea temperatures up to a monthly maximum SST anomaly of ~1.6 °C (Glynn et al. 2000).

Extended annual reproductive activity contributes to the high fecundity of eastern Pacific corals. Some other traits also adding to the fecundity of agariciid species (*Pavona*, 3 species; *Gardineroseris planulata*) are the number of monthly spawning cycles. For example, *Pavona* spp. in the Galápagos Islands spawn monthly and may even exhibit split bi-monthly spawning (Glynn et al. 2000). The very small size and prodigious numbers of mature ova in most eastern Pacific species contribute directly and importantly to fecundity. Where estimates of the ages at first reproduction have been reported for eastern Pacific agariciid species, these range from 5 (*Pavona varians*, *Pavona chiriquiensis*) to 20 (*Gardineroseris planulata*) years. These rates cannot be compared with other regions due to the lack of studies on broadcast spawning species in this family.

Perhaps more than in any other coral reef region, corallivore fish foraging and bioturbation activities are at a high level in eastern Pacific coral communities (see Chap. 10, Enochs and Glynn). This leads to frequent asexual colony fragmentation in both branching and massive corals. Where corals occur in favorable environments or under conditions that have normalized after a disturbance, asexual propagation can greatly accelerate local recovery. Unfortunately, information is presently too incomplete to offer any general conclusions on the relative contributions of sexual and asexual reproduction in promoting coral community recovery. It is likely that both reproductive modes are effective, with one or the other predominant depending on the type and severity of disturbance and the particular local conditions prevalent during recovery.

The importance of sexual and asexual reproduction is now clearly recognized for several species within the EEP region. Nonetheless, only 13 of 38 valid zooxanthellate species (Reyes Bonilla 2002) have been studied, and numerous areas within the EEP have not been investigated. Several of the unstudied species, in the genera *Pocillopora*

(6 of 8 species), *Porites* (7 of 9), and the family Agariciidae (8 of 13), contribute significantly to coral communities and reef frameworks in many areas. Since most of our understanding of coral sexuality, mode of development, timing of spawning, and fecundity has been gained from histological studies, numerous questions remain that are best addressed by direct observations in the field.

Both sexual and asexual reproduction contribute importantly to recruitment in the EEP, but their relative contributions with respect to species and environmental conditions need to be clarified and quantified. For example, how might the spatial occurrence of potentially stressful natural impacts affect coral community structure and diversity, reef growth, recruitment, and recovery processes? Coastal Mexico and her offshore islands are subject to frequent hurricane effects; are these critical drivers of asexual propagation compared with other eastern Pacific areas? Upwelling centers occur in several areas off Mexico and in the EEP, and it is known that sexual reproduction in several species is greatly diminished during upwelling activity. Freshwater discharge and sedimentation reach extreme levels off Colombia and northern Ecuador, but no information has been published on how these conditions affect coral reproduction along this coastal corridor. Finally, preliminary evidence shows that the sudden thermal swings that accompany ENSO events are correlated with atypical reproductive activity. Since no two ENSO cycles are identical, spatially or with respect to severity and duration, continuing research is necessary to relate critical stressors to their regional influence on coral reproduction.

While several kinds of anthropogenic stressors have been noted to affect eastern Pacific reefs (Cortés 2003), e.g. deforestation and coastal development, overfishing, and agricultural chemicals, no studies have examined directly how these affect coral reproduction. Past work has been conducted on reefs lacking signs of direct human impact. Therefore, it is urgent that attention be directed toward an assessment of coral reproduction at coastal and island sites in the EEP where human activities are now encroaching on coral populations and coral reefs. The interactive effects of natural and human-induced stresses on coral reproduction are largely unknown. Reproduction is a vital life history function and for coral reef ecosystems to survive into the future it must not be compromised.

Acknowledgments The many individuals who helped with logistics, permitting, collections, histology, and data generation over the years are too numerous to identify here, but are duly acknowledged in the Glynn, Colley et al. publications referenced below. Special thanks go to Bernadette Bezy for sharing information on coral spawning in Costa Rica. This overview benefitted from critiques offered by Joshua S. Feingold and two anonymous reviewers. Information supplied by Peggy Fong, Sascha Steiner and Herman Wirshing is noted with gratitude. We thank Michael C. Schmale for the use of his microscope and digital camera equipment. Joshua Levy and Alissa Mones kindly assisted in producing figures and tables. Principal financial support was provided by grants from the U.S. National Science Foundation (Biological Oceanography Program), the Office of Forestry, Environment and Natural Resources, Bureau for Science and Technology (USAID), the Darwin Initiative (Conservation International), and the National Geographic Society. Essential in-country support has been offered long-term by the Centro de Investigaciones en Ciencias del Mar y Limnología (Costa Rica), the University of Panama, Instituto Nacional de Recursos Naturales, and Smithsonian Tropical Research Institute (Panama), and the Charles Darwin Research Station, Galápagos National Park Service, and Transporte Aereo Militar de Ecuador (Ecuador).

References

Albright R (2011) Reviewing the effects of ocean acidification on sexual reproduction and early life history stages of reef-building corals. J Mar Biol. doi:10.1155/2011/473615

Albright R, Mason B, Miller M, Langdon C (2010) Ocean acidification compromises recruitment success of the threatened Caribbean coral *Acropora palmata*. Proc Nat Acad Sci 107:20,400–20,404

Aranceta-Garza F, Balart EF, Reyes-Bonilla H, Cruz-Hernández P (2012) Effect of tropical storms on sexual and asexual reproduction in coral *Pocillopora verrucosa* subpopulations in the Gulf of California. Coral Reefs 31:1157–1167. doi:10.1007/s00338-012-0941-9

Ayre DJ, Hughes TP, Standish TP (1997) Genetic differentiation, reproductive mode, and gene flow in the brooding coral *Pocillopora damicornis* along the Great Barrier Reef, Australia. Mar Ecol Prog Ser 159:175–187

Baird AH, Marshall PA (2002) Mortality, growth and reproduction in scleractinian corals following bleaching on the Great Barrier Reef. Mar Ecol Prog Ser 237:133–141

Baird AH, Salih A, Trevor-Jones A (2006) Fluorescence census techniques for the early detection of coral recruits. Coral Reefs 25:73–76

Baird AH, Guest JR, Willis BL (2009) Systematic and biogeographical patterns in the reproductive biology of scleractinian corals. Annu Rev Ecol Evol Syst 40:551–571

Baums IB, Durante MD, Laing AA, Feingold J, Smith T, Bruckner A, Monteiro J (2014) Marginal coral populations: the densest known aggregation of *Pocillopora* in the Galápagos Archipeligo is of asexual origin. Front Mar Sci 1:59. doi:10.3389/fmars.2014.00059

Bezy MB (2009) Reproducción sexual y reclutamiento del coral masivo, *Pavona clavus*, en Bahía Culebra, Golfo de Papagayo, Costa Rica. MSc thesis, Univ Costa Rica, San Pedro, Costa Rica, p 151

Birkeland C (1977) The importance of rate of biomass accumulation in early successional stages of benthic communities to the survival of coral recruits. In: Proceedings of 3rd International Coral Reef Symposium, vol 1, Miami, pp 15–21

Birkeland C (1997) Geographic differences in ecological processes on coral reefs. In: Birkeland C (ed) Life and death of coral reefs. Chapman and Hall, New York, pp 273–297

Boulay JN, Cortés J, Nivia-Ruiz J, Baums IB (2012) High genotypic diversity of the reef-building coral *Porites lobata* (Scleractinia: Poritidae) in Isla del Coco National Park, Costa Rica. Rev Biol Trop (suppl 3):279–292

Boulay JN, Hellberg ME, Cortés J, Baums IB (2013) Unrecognized coral species diversity masks differences in functional ecology. Proc R Soc B 281:20131580. doi:10.1098/rspb.2013.1580

Boumeester J, Berumeni ML, Baird AH (2011) Daytime broadcast spawning of *Pocillopora verrucosa* on coral reefs of the central Red Sea, Galaxea. J Coral Reef Stud 13:23–24

Cabral-Tena RA, Reyes-Bonilla H, Lluch-Cota S, Paz-García DA, Calderón-Aguilera LE, Norzagaray-López O, Balart EF (2013) Different calcification rates in males and females of the coral *Porites panamensis* in the Gulf of California. Mar Ecol Prog Ser 476:1–8

Campos-Vázquez RA, Balart EF, Vallejo-Fuerte M, Rodríguez-Jaramillo C (2014). Ciclo reproductivo de los corales *Pocillopora verrucosa* y *Pocillopora meandrina* en Isla Gaviota, Bahía de La Paz, Mexico. Congr Nac Oceanogr June 2014. La Paz, Mexico

Carpizo-Ituarte E, Vizcaíno-Ochoa V, Chi-Barragán G, Tapia-Vázquez O, Cupul-Magaña AL, Medina-Rosas P (2011) Evidence of sexual reproduction in the hermatypic corals *Pocillopora damicornis*, *Porites panamensis*, and *Pavona gigantea* in Banderas Bay, Mexican Pacific. Cienc Mar 37:97–112

Charnov EL (1982) The theory of sex allocation. Princeton monographs in population biology 18. Princeton Univ Press, Princeton NJ, p 355

Chávez-Romo E, Reyes-Bonilla H (2007) Sexual reproduction of the coral *Pocillopora damicornis* in the southern Gulf of California, Mexico. Cienc Mar 33:495–501

Cohen AI, Holcomb M (2009) Why corals care about ocean acidification: uncovering the mechanisms. Oceanogr 22:118–127

Colley SB, Feingold JS, Peña J, Glynn PW (2000) Reproductive ecology of *Diaseris distorta* (Michelin) (Fungiidae) in the Galápagos Islands, Ecuador. In: Proceedings of 9th International Coral Reef Symposium, vol 1, Bali, pp 373–379

Colley SB, Glynn PW, May AS, Maté JL (2006) Species-dependent reproductive responses of eastern Pacific corals to the 1997–1998 ENSO event. In: Proceedings of 10th International Coral Reef Symposium, vol 1, Okinawa, pp 61–70

Combosch DJ, Vollmer SV (2011) Population genetics of an ecosystem-defining reef coral *Pocillopora damicornis* in the tropical eastern Pacific. PLoS ONE 6(8):e21200. doi:10.1371/journal.pone.0021200

Connolly SR, Baird AH (2010) Estimating dispersal potential for marine larvae: dynamic models applied to scleractinian corals. Ecology 91:3572–3583

Cortés J, Jiménez C (2003) Corals and coral reefs of the Pacific of Costa Rica: history, research and status. In: Cortés J (ed) Latin American coral reefs. Elsevier, Amsterdam, pp 361–385

Cox EF (2007) Continuation of sexual reproduction in *Montipora capitata* following bleaching. Coral Reefs 26:721–724

Dana TF (1975) Development of contemporary eastern Pacific coral reefs. Mar Biol 33:355–374

D'Croz L, Robertson DR (1997) Coastal oceanographic conditions affecting coral reefs on both sides of the Isthmus of Panama. In: Proceedings of 8th International Coral Reef Symposium, vol 2, Panama, pp 2053–2058

Done TJ, Potts DC (1992) Influences of habitat and natural disturbances on contributions of massive *Porites* corals to reef communities. Mar Biol 114:479–493

Edmunds PJ (2010) Population biology of *Porites astreoides* and *Diploria strigosa* on a shallow Caribbean reef. Mar Ecol Prog Ser 418:87–104

Fabricius KE (2005) Effects of terrestrial runoff on the ecology of corals and coral reefs: review and synthesis. Mar Poll Bull 50:125–146

Fadlallah YH (1983) Sexual reproduction, development and larval biology in scleractinian corals: a review. Coral Reefs 2:129–150

Feingold JS (1995) Effects of elevated water temperature on coral bleaching and survival during El Niño disturbance events. Ph.D. dissert, Univ of Miami, Coral Gables, Florida p 236

Fenner D (2001) Biogeography of three Caribbean corals (Scleractinia) and the invasion of *Tubastraea coccinea* into the Gulf of Mexico. Bull Mar Sci 69:1175–1189

Fenner D, Banks K (2004) Orange cup coral *Tubastraea coccinea* invades Florida and the Flower Garden Banks, northwestern Gulf of Mexico. Coral Reefs 23:505–507

Figueira de Paula AF, Creed JC (2004) Two species of the coral *Tubastraea* (Cnidaria, Scleractinia) in Brazil: a case of accidental introduction. Bull Mar Sci 74:175–183

Fine D, Tehernov D (2007) Scleractinian coral species survive and recover from decalcification. Science 315:1811(single page)

Fong P, Glynn PW (2001) Population abundance and size-structure of an eastern tropical Pacific reef coral after the 1997–98 ENSO: a simulation model predicts field measures. Bull Mar Sci 69:187–202

Fong P, Smith TB, Wartian MJ (2006) Epiphytic cyanobacteria maintain shift to macroalgal dominance on coral reefs following ENSO disturbance. Ecology 87:1162–1168

Fukami H, Chen CA, Budd AF, Collins A, Wallace C, Chuang Y-Y, Chen C, Dai C-F, Iwao K, Sheppard C, Knowlton N (2008) Mitochondrial and nuclear genes suggest that stony corals are monophyletic but most families of stony corals are not (Order Scleractinia, Class Anthozoa, Phylum Cnidaria). PLoS ONE 3:1–9

García-Ocampo MR (2005) Patrones de reclutamiento de las colonias juveniles de coral del género *Pocillopora* Lamarck 1816 (Anthozoa: Scleractinia) en cinco localidades de Ixtapa-Zihuatanejo, Guerrero, México. Undergrad thesis Mar Biol, Univ del Mar, Pto Ángel, Oaxaca p 56

Garren M, Walsh SM, Caccone A, Knowlton N (2006) Patterns of association between *Symbiodinium* and members of the *Montastraea annularis* species complex on spatial scales ranging from within colonies to between geographic regions. Coral Reefs 25:503–512

Giese AC, Pearse JS (1974) Introduction: general principles. In: Giese AC, Pearse JS (eds) Reproduction of marine invertebrates, vol 1. Academic Press, New York, pp 1–49

Glynn PW (1984) Widespread coral mortality and the 1982–83 El Niño warming event. Environ Conserv 11:133–146

Glynn PW (1985) El Niño-associated disturbance to coral reefs and post disturbance mortality by *Acanthaster planci*. Mar Ecol Prog Ser 26:295–300

Glynn PW (1990) Coral mortality and disturbances to coral reefs in the tropical eastern Pacific. In: Glynn PW (ed) Global ecological consequences of the 1982–83 El Niño-Southern Oscillation. Elsevier Oceanogr Ser 52, Amsterdam, Elsevier, pp 55–126

Glynn PW (2003) Coral communities and coral reefs of Ecuador. In: Cortés J (ed) Latin American coral reefs. Elsevier, Amsterdam, pp 449–472

Glynn PW (2008) Food-web structure and dynamics of eastern tropical Pacific coral reefs: Panamá and Galápagos Islands. In: McClanahan TR, Branch GM (eds) Food webs and the dynamics of marine reefs. Oxford Univ Press, Oxford, pp 185–208

Glynn PW, Ault JS (2000) A biogeographic analysis and review of the far eastern Pacific coral reef region. Coral Reefs 19:1–23

Glynn PW, Colley SB (2008) Survival of brooding and broadcasting reef corals following large scale disturbances: is there any hope for broadcasting species during global warming? In: Proceedings of 11th International Coral Reef Symposium. Ft Lauderdale (session 11), pp 1–5

Glynn PW, Fong P (2006) Patterns of reef coral recovery by the regrowth of surviving tissues following the 1997–98 El Niño warming and 2000, 2001 upwelling cool events in Panamá, eastern Pacific. In: Proceedings of 10th International Coral Reef Symposium, vol 1, Okinawa, pp 624–630

Glynn PW, Stewart RH, McCosker JE (1972) Pacific coral reefs of Panamá: structure, distribution and predators. Geol Rundsch 61:483–519

Glynn PW, Peters EC, Muscatine L (1985) Coral tissue microstructure and necrosis: relation to catastrophic coral mortality in Panamá. Dis Aquat Org 1:29–37

Glynn PW, Gassman NJ, Eakin CM, Cortés J, Smith DB, Guzmán HM (1991) Reef coral reproduction in the eastern Pacific: Costa Rica, Panamá, and Galápagos Islands (Ecuador) I. Pocilloporidae. Mar Biol 109:355–368

Glynn PW, Colley SB, Eakin CM, Smith DB, Cortés J, Gassman NJ, Guzmán HM, Del Rosario JB, Feingold JB (1994) Reef coral reproduction in the eastern Pacific: Costa Rica, Panamá, and Galápagos Islands (Ecuador) II. Poritidae. Mar Biol 118:191–208

Glynn PW, Colley SB, Gassman NJ, Black K, Cortés J, Maté JL (1996) Reef coral reproduction in the eastern Pacific: Costa Rica, Panamá, and Galápagos Islands (Ecuador). III. Agariciidae (*Pavona gigantea* and *Gardineroseris planulata*). Mar Biol 125:579–601

Glynn PW, Colley SB, Ting JH, Maté JL, Guzmán HM (2000) Reef coral reproduction in the eastern Pacific: Costa Rica, Panamá, and Galápagos Islands (Ecuador). IV. Agariciidae, recruitment and recovery of *Pavona varians* and *Pavona* sp a. Mar Biol 136:785–805

Glynn PW, Maté JL, Stemann TA (2001a) *Pavona chiriquiensis*, a new species of zooxanthellate scleractinian coral (Cnidaria: Anthozoa: Agariciidae) from the eastern tropical Pacific. Bull Biol Soc Wash 10:210–225

Glynn PW, Maté JL, Baker AC, Calderón MO (2001b) Coral bleaching and mortality in Panamá and Ecuador during the 1997–1998 El Niño-Southern Oscillation event: spatial/temporal patterns and comparisons with the 1982–1983 event. Bull Mar Sci 69:79–109

Glynn PW, Colley SB, Maté JL, Cortés J, Guzmán HM, Bailey RL, Feingold JS, Enochs IC (2008) Reproductive ecology of the azooxanthellate coral *Tubastraea coccinea* in the equatorial eastern Pacific: Part V. Dendrophylliidae. Mar Biol 153:529–544

Glynn PW, Colley SB, Guzmán HM, Enochs IC, Cortés J, Maté JL, Feingold J (2011) Reef coral reproduction in the eastern Pacific: Costa Rica, Panamá and the Galápagos Islands (Ecuador). VI. Agariciidae, *Pavona clavus*. Mar Biol 158:1601–1617

Glynn PW, Colley SB, Maté JL, Baums IB, Feingold JS, Cortés J, Guzmán HM, Afflerbach JC, Brandtneris VW, Ault JS (2012) Reef coral reproduction in the equatorial eastern Pacific: Costa Rica, Panamá, and the Galápagos Islands (Ecuador). VII Siderastreidae, *Psammocora stellata* and *Psammocora profundacella*. Mar Biol 159:1,917–1,932 doi:10.1007/s00227-012-1979-5

Golbuu Y, Richmond RH (2007) Substratum preferences in planula larvae of two species of scleractinian corals, *Goniastraea retiformis* and *Stylarea punctata*. Mar Biol 152:639–644

Graham EM, Baird AH, Connolly SR (2008) Survival dynamics of scleractinian coral larvae and implications for dispersal. Coral Reefs 27:529–539

Grottoli AG, Rodrigues LJ, Juarez C (2004) Lipids and stable carbon isotopes in two species of Hawaiian corals, *Porites compressa* and *Montipora verrucosa*, following a bleaching event. Mar Biol 145:621–631

Grottoli AG, Rodrigues LJ, Palardy JE (2006) Heterotrophic plasticity and resilience in bleached corals. Nature 440:1186–1189

Guest JR, Baird AH, Goh BPL, Chou LM (2005) Seasonal reproduction in equatorial coral reefs. Inv Repr Dev 48:207–218

Guest JR, Baird AH, Goh BPL, Chou LM (2012) Sexual systems in scleractinian corals: an unusual pattern in the reef-building species *Diploastrea heliopora*. Coral Reefs. doi:10.1007/s00338-012-0881-4

Guzmán HM (1986) Estructura de la comunidad arrecifal de la Isla del Caño, Costa Rica, y el efecto de perturbaciones naturales severas. MSc thesis, Univ Costa Rica, San Pedro, Costa Rica, p 179

Guzmán HM (1988) Distribución y abundancia de organismos coralívoros en los arrecifes coralinos de la Isla del Caño, Costa Rica. Rev Biol Trop 36:191–207

Guzmán HM, Cortés J (1989) Coral reef community structure at Caño Island, Pacific Costa Rica. Publ Staz zool Napoli (I: Mar Ecol) 10:23–41

Guzmán HM, Cortés J (2001) Changes in reef community structure after fifteen years of natural disturbances in the eastern Pacific (Costa Rica). Bull Mar Sci 69:133–149

Guzmán HM, Cortés J (2007) Reef recovery 20 years after the 1982–1983 El Niño massive mortality. Mar Biol 151:401–411

Hanafy MH, Aamer MA, Habib M, Rouphael AB, Baird AH (2010) Synchronous reproduction of corals in the Red Sea. Coral Reefs 29:119–124

Harriott VJ (1983) Reproductive ecology of four scleractinian species at Lizard Island, Great Barrier Reef. Coral Reefs 2:9–18

Harrison PL (1990) Sperm morphology and fertilization strategies in scleractinian corals. Adv Invert Reprod 5:299–304

Harrison PL (2011) Sexual reproduction of scleractinian corals. In: Dubinsky Z, Stambler N (eds) Coral reefs: an ecosystem in transition. Springer, Berlin, pp 59–85

Harrison PL, Wallace CC (1990) Reproduction, dispersal and recruitment of scleractinian corals. In: Dubinsky Z (ed) Coral reefs, ecosystems of the world 25. Elsevier, Amsterdam, pp 133–207

Highsmith RC (1980) Geographic patterns of coral bioerosion: a productivity hypothesis. J Exp Mar Biol Ecol 46:177–196

Highsmith RC (1982) Reproduction by fragmentation in corals. Mar Ecol Prog Ser 7:207–226

Hirose M, Hidaka M (2000) Reduced reproductive success in scleractinian corals that survived the 1998 bleaching in Okinawa. Galaxea JCRS 2:17–21

Hughes TP, Baird AH, Bellwood DR, Card M, Connolly SR, Folke C, Grosberg R, Hoegh-Guldberg O, Jackson JBC, Kleypas J, Lough J, Marshall P, Nystrom M, Palumbi SR, Pandolphi JM, Rosen B, Roughgarden J (2003) Climate change, human impacts, and the resilience of coral reefs. Science 301:929–933

Humason GL (1967) Animal tissue techniques, 2nd edn. WH Freeman and Co, San Francisco, p 569

Humphrey C, Weber M, Lott C, Cooper T, Fabricius K (2008) Effects of suspended sediments, dissolved inorganic nutrients and salinity on fertilization and embryo development in the coral *Acropora millepora* (Ehrenberg, 1834). Coral Reefs 27:837–850

Jiménez CE (1996–1997) Coral colony fragmentation by whitetip reef sharks at Coiba Island National Park, Panamá. Rev Biol Trop 44/45:698–700

Jiménez CE (1998) Arrecifes y comunidades coralinas de Bahía Culebra, Pacífico norte de Costa Rica (Golfo de Papagayo). MSc thesis, Univ Costa Rica, San Pedro, Costa Rica, p 218

Kerr AM (2005) Molecular and morphological supertree of stony corals (Anthozoa: Scleractinia) using matrix representation parsimony. Biol Rev 80:543–558

Kinzie RA III (1993) Spawning in the reef corals *Pocillopora verrucosa* and *P. eydouxi* at Sesoko Island, Okinawa. Galaxea 11 (2):93–105

Kleypas JA, McManus JW, Meñez LAB (1999) Environmental limits to coral reef development: where do we draw the line? Amer Zool 39:146–159

Kolinski SP, Cox EF (2003) An update on modes and timing of gamete and planula release in Hawaiian scleractinian corals with implications for conservation and management. Pac Sci 57:17–27

LaJeunesse TC, Smith R, Walther M, Pinzón J, Pettay DT, McGinley M, Aschaffenburg M, Medina-Rosas P, Cupul-Magaña AL, López Pérez A, Reyes-Bonilla H, Warner ME (2010) Host-symbiont recombination versus natural selection in the response of coral-dinoflagellate symbioses to environmental disturbance. Proc R Soc B 277:2925–2934

Lewis JB (2006) Biology and ecology of the hydrocoral *Millepora* on coral reefs. Adv Mar Biol 50:1–55

Lirman D, Glynn PW, Baker AC, Leyte Morales GE (2001) Combined effects of three sequential storms on the Huatulco coral reef tract, Mexico. Bull Mar Sci 69:267–278

López-Pérez RA, Mora-Pérez MG, Leyte-Morales GE (2007) Coral (Anthozoa: Scleractinia) recruitment at Bahías de Huatulco, western México: implications for coral community structure and dynamics. Pac Sci 61:355–369

Loya Y, Sakai K (2008) Bidirectional sex change in mushroom stony corals. Proc R Soc B 275:2335–2343

Loya Y, Sakai K, Heyward A (2009) Reproductive patterns of fungoid corals in Okinawa, Japan, Galaxea. J Coral Reef Stud 11:119–129

Luna JG (1968) Manual of histological staining methods of the armed forces institute of pathology. McGraw-Hill, New York

Mangubhai S, Harrison PL (2009) Extended breeding seasons and asynchronous spawning among equatorial reef corals in Kenya. Mar Ecol Prog Ser 374:305–310

Manzello DP (2010) Coral growth with thermal stress and ocean acidification: lessons from the eastern tropical Pacific. Coral Reefs 29:749–758

Manzello DP, Kleypas JA, Budd DA, Eakin CM, Glynn PW, Langdon C (2008) Poorly cemented coral reefs of the eastern tropical Pacific: possible insights into reef development in a high-CO_2 world. Proc Nat Acad Sci 105:10,450–10,455

Maté JL (2003) Corals and coral reefs of the Pacific coast of Panamá. In: Cortés J (ed) Latin American coral reefs. Elsevier, Amsterdam, pp 387–417

Medina-Rosas P, Carriquiry JD, Cupul-Magaña AL (2005) Recruitment of *Porites* (Scleractinia) on artificial substrate in reefs affected by the 1997–98 El Niño in Banderas Bay, Mexican Pacific. Cienc Mar 31:1–7

Mundy CN (2000) An appraisal of methods used in coral recruitment studies. Coral Reefs 19:124–131

Oliver J, Babcock R (1992) Aspects of the fertilization ecology of broadcast spawning corals: sperm dilution effects and in situ measurements of fertilization. Biol Bull 183:409–417

Oliver JK, Babcock RC, Harrison PL, Willis BL (1988) Geographic extent of mass coral spawning: clues to ultimate causal factors. In: Proceedings of 6th International Coral Reef Symposium, vol 2, Townsville, pp 803–810

Omori M, Fukami H, Kobinata H, Hatta M (2001) Significant drop of fertilization of *Acropora* corals in 1999: an after effect of heavy coral bleaching? Limnol Oceanogr 46:704–706

Penland L, Kloulechad J, Idip D, van Woesik R (2004) Coral spawning in the western Pacific Ocean is related to solar insolation: evidence of multiple spawning events in Palau. Coral Reefs 23:133–140

Peters EC, Gassman NJ, Firman JC, Richmond RH, Power EA (1997) Ecotoxicology of tropical marine ecosystems. J Environ Tox Chem, Annu Rev 16:12–40

Pinzón JH, Reyes-Bonilla H, Baums IB, LaJeunesse TC (2012) Contrasting clonal structures among *Pocillopora* (Scleractinia) communities at two environmentally distinct sites in the Gulf of California. Coral Reefs. doi:10.1007/s00338-012-0887-4

Reyes-Bonilla H (1993) Estructura de la comunidad, influencia de la depredación y biología poblacional de corales hermatípicos en el arrecife de Cabo Pulmo, B.C.S. MSc thesis, Centro de Investigación Científica y de Educación Superior de Ensenada, Depart Ecología, p 169

Reyes-Bonilla H (2002) Checklist of valid names and synonyms of stony corals (Anthozoa: Scleractinia) from the eastern Pacific. J Nat Hist 36:1–13

Reyes-Bonilla H (2003) Coral reefs of the Pacific coast of Mexico. In: Cortés J (ed) Latin American coral reefs. Elsevier, Amsterdam, pp 331–349

Reyes-Bonilla H, Calderón-Aguilera LE (1994) Parámetros poblacionales de *Porites panamensis* (Anthozoa: Scleractinia), en el arrecife de Cabo Pulmo, México. Rev Biol Trop 42:121–128

Richmond RH (1985) Variations in the population biology of *Pocillopora damicornis* across the Pacific. In: Proceedings of 5th International Coral Reef Congress, vol 6, Tahiti, pp 101–106

Richmond RH (1987a) Energetics, competency, and long-distance dispersal of planula larvae of the coral *Pocillopora damicornis*. Mar Biol 93:527–533

Richmond RH (1987b) Energetic relationships and biogeographical differences among fecundity, growth and reproduction in the reef coral *Pocillopora damicornis*. Bull Mar Sci 41:594–604

Richmond RH (1988) Competency and dispersal potential of planula larvae of a spawning versus a brooding coral. In: Proceedings of 6th International Coral Reef Symposium, vol 2, Townsville, pp 827–831

Richmond RH (1990) The effects of the El Niño/Southern Oscillation on the dispersal of corals and other marine organisms. In: Glynn PW (ed) Global ecological consequences of the 1982–83 El Niño-Southern Oscillation. Elsevier Oceanogr Ser 52, Amsterdam, Elsevier, pp 127–140

Richmond RH (1993) Coral reefs: present problems and future concerns resulting from anthropogenic disturbance. Am Zool 33:524–536

Richmond RH (1994) Effects of coastal runoff on coral reproduction. In: Ginsburg RN (compiler) Proceedings of the colloquium on global aspects of coral reefs: health, hazards and history. Rosenstiel School of Marine and Atmospheric Science, Univ Miami, Miami, pp 360–364

Richmond RH (1997) Reproduction and recruitment in corals: critical links in the persistence of reefs. In: Birkeland C (ed) Life and death of coral reefs. Chapman & Hall, New York, pp 175–197

Richmond RH (2005) Coral reproduction and recruitment as tools for studying the ecotoxicology of coral reef ecosystems. In: Ostrander G (ed) Techniques in aquatic toxicology, vol II. CRC Press, Boca Raton, Florida, pp 331–338

Richmond RH, Hunter CL (1990) Reproduction and recruitment of corals: comparisons among the Caribbean, the tropical Pacific, and the Red Sea. Mar Ecol Prog Ser 60:185–203

Rinkevich B (1989) The contribution of photosynthetic products to coral reproduction. Mar Biol 101:259–263

Rodríguez-Troncoso AP (2004) Caracterización del ciclo reproductivo de *Pocillopora damicornis* (Linnaeus 1758), en el arrecife de la Entrega, Oaxaca, México. Undergrad thesis, Univ del Mar, Pto Ángel, Oaxaca, p 62

Rodríguez-Troncoso AP, Carpizo-Ituarte E, Leyte-Morales GE, Chi-Barragán G, Tapia-Vázquez O (2011) Sexual reproduction of three coral species from the Mexican south Pacific. Mar Biol 158:2673–2683

Romano SL, Palumbi SR (1996) Evolution of scleractinian corals inferred from molecular systematics. Science 271:640–642

Sammarco PW (1982) Polyp bail-out: an escape response to environmental stress and a new means of reproduction in corals. Mar Ecol Prog Ser 10:57–65

Santiago Valentín JD, Rodríguez-Troncoso AP, Carpizo-Ituarte E, Benítez-Villalobos F, Torres-Hernández P, López-Pérez A (2015) Reproductive pattern of the reef building coral *Pavona gigantea* (Scleractinia: Agariciidae) off southwestern Mexico. Cienc Mar 41:233–246

Schmidt-Roach S, Miller KJ, Woolsey E, Gerlach G, Baird AH (2012) Broadcast spawning by *Pocillopora* species on the Great Barrier Reef. PLoS ONE 7(12):e50847. doi:10.1371/journal.pone.0050847

Scott PJB, Risk MJ (1988) The effect of *Lithophaga* (Bivalvia: Mytilidae) boreholes on the strength of the coral *Porites lobata*. Coral Reefs 7:145–151

Shlesinger Y, Loya J (1985) Coral community reproductive patterns: Red Sea versus the Great Barrier Reef. Science 228:1333–1335

Shlesinger Y, Goulet TL, Loya Y (1998) Reproductive patterns of scleractinian corals in the northern Red Sea. Mar Biol 132:691–701

Smith DB (1991) The reproduction and recruitment of *Porites panamensis* Verrill at Uva Island, Pacific Panamá. MS thesis, Univ Miami, Coral Gables, Florida, p 64

Smith TB, Glynn PW, Maté JL, Toth LT, Gyory J (2014) A depth refugium from catastrophic coral bleaching prevents regional extinction. Ecology 95:1663–1673. doi:10.1890/13-0468.1

Soong K, Cho LC (1998) Synchronized release of medusa from three species of hydrozoan fire corals. Coral Reefs 17:145–154

Steiner SCC, Cortés J (1996) Spermatozoan ultrastructure of scleractinian corals from the eastern Pacific: Pocilloporidae and Agariciidae. Coral Reefs 15:143–147

Strathmann RR, Hughes TP, Kuris AM, Lindeman KC, Morgan SG, Pandolfi JM, Warner RR (2002) Evolution of local recruitment and its consequences for marine populations. Bull Mar Sci 70 (suppl):377–396

Szmant-Froelich A, Reutter M, Riggs L (1985) Sexual reproduction of *Favia fragum* (Esper): lunar patterns of gametogenesis, embryogenesis and planulation in Puerto Rico. Bull Mar Sci 37:880–892

Szmant AM (1986) Reproductive ecology of Caribbean reef corals. Coral Reefs 5:43–54

Szmant AM, Gassman NL (1990) The effects of prolonged 'bleaching' on the tissue biomass and reproduction of the reef coral *Montastrea annularis*. Coral Reefs 8:217–224

Tomascik T, Sander F (1987) Effects of eutrophication on reef-building corals III. Reproduction of the reef-building coral *Porites porites*. Mar Biol 94:77–94

van Veghel MLJ (1993) Multiple species spawning on Curaçao reefs. Bull Mar Sci 52:1017–1021

van Woesik R (2009) Calm before the spawn: global coral spawning patterns are explained by regional wind fields. Proc R Soc B 277:715–722

Waller RG, Taylor PA, Gage JD (2005) Sexual reproduction in three hermaphroditic deep-sea *Caryophyllia* species (Anthozoa: Scleractinia) from the NE Atlantic Ocean. Coral Reefs 24:594–602

Ward S, Harrison P, Hoegh-Guldberg O (2000) Coral bleaching reduces reproduction of scleractinian corals and increases susceptibility to future stress. In: Proceedings of 9th International Coral Reef Symposium, vol 2, Bali, pp 1123–1128

Wartian MJ (2006) Determinants of community structure and resilience on tropical eastern Pacific coral reefs. PhD dissert, Univ Calif, Los Angeles, p 247

Wellington GM (1982) Depth zonation of corals in the Gulf of Panama: control and facilitation by resident reef fishes. Ecol Monogr 52:223–241

Willis BL, Babcock RC, Harrison PL, Wallace CC (1993) Experimental evidence of hybridization in reef corals involved in mass spawning events. In: Proceedings of 7th International Coral Reef Symposium, vol 1, Guam, p 504 (abstract)

Willis BL, van Oppen MJH, Miller DJ, Vollmer SV, Ayre DJ (2006) The role of hybridization in the evolution of reef corals. Annu Rev Ecol Evol Syst 37:489–517

Wilson JR, Harrison PL (2003) Spawning patterns of scleractinian corals at the Solitary Islands—a high latitude coral community in eastern Australia. Mar Ecol Prog Ser 260:115–123

Yamashiro H, Nishihira M (1998) Experimental study of growth and asexual reproduction in *Diaseris distorta* (Michelin 1843) a free-living fungoid coral. J Exp Mar Biol Ecol 225:253–267

Zapata FA, Vargas-Ángel B (2003) Corals and coral reefs of the Pacific coast of Colombia. In: Cortés J (ed) Latin American coral reefs. Elsevier, Amsterdam, pp 419–447

Gene Flow in Coral Reef Organisms of the Tropical Eastern Pacific

16

H.A. Lessios and Iliana B. Baums

Abstract

Gene flow can provide cohesion between conspecific populations. In order to obtain an indirect measure of gene flow between coral reef species in the eastern tropical Pacific (ETP) and between these populations and those of the rest of the Pacific we compiled available data from sequences of DNA and microsatellites for corals, gastropods, echinoderms and fishes, and calculated F_{ST} statistics. The ETP consists of a narrow strip of continental shelf along the coast of the Americas and a deeper water gap between the coast and the outer eastern Pacific Islands; a large expanse of deep ocean separates the ETP and the closest islands in the central Pacific. We have, therefore, compared populations in four major directions: (1) between the eastern and the central Pacific, (2) between the coast and the outer islands, (3) among the outer islands, and (4) along the coast and nearshore islands. The available data are biased in favor of showing high levels of gene flow because they contain an excess of transpacific species, which are a minority among ETP biota. Despite this bias, shallow water populations of the ETP are isolated from the rest of the world's oceans. Occasional breaching of the expanse of water between the ETP and the Central Pacific by some species is also possible. Gene flow between the outer eastern Pacific islands and the mainland coast is variable, depending on the species examined. Gene flow among populations at the outer eastern Pacific islands is high except for those at Easter Island (Rapa Nui), in which all but one sampled species show large and significant values of F_{ST} in comparisons with populations from all other islands. Gene flow rates among populations along the ETP coast are high. There is no evident genetic break resulting from the Central American Gap (southern Mexico to the Gulf of Fonseca, Honduras) in any of the sampled species. A trend of isolation by distance along the coast is evident in corals and fishes.

Keywords

Genetic connections • Mitochondrial DNA • Microsatellites • Genetic structure • F_{ST} statistics

16.1 Introduction

Gene flow is the exchange of genes between conspecific populations. It is an important biological process in every species because gene transfer provides cohesion between populations, preventing their independent evolution. The movement of genes over generations assures that new advantageous mutations that arise in one location can eventually spread throughout the species' range. When gene flow becomes restricted populations become "genetically

H.A. Lessios (✉)
Smithsonian Tropical Research Institute, 0843-03092 Balboa, Panama
e-mail: Lessiosh@post.harvard.edu

I.B. Baums
The Pennsylvania State University, 208 Mueller Laboratory, University Park, PA 16802, USA
e-mail: baums@psu.edu

P.W. Glynn et al. (eds.), *Coral Reefs of the Eastern Tropical Pacific*, Coral Reefs of the World 8, DOI 10.1007/978-94-017-7499-4_16

structured", diverging from each other and (if the restrictions become severe and last a long time) possibly turning into separate species. The sharing of genes between populations can also dilute adaptation to local environmental conditions, thus averaging the response to natural selection over the connected populations. It may thus limit the geographic extent over which a species can spread, as peripheral populations are prevented from adapting to conditions at the edge of a species range and then proceed to colonize new, previously unsuitable, areas (Haldane 1956; Case and Taper 2000; Sexton et al. 2009; Dawson et al. 2010).

Genetic exchange is effected by the dispersal of propagules from one population to another, but, as propagules are difficult to follow, most estimates of gene flow do not involve direct observations of such transfers; they rely instead on assessing divergence between populations. This is particularly true for marine organisms in which planktonic larvae maintain genetic connections over very long distances and are virtually impossible to track in the water column. Species with this kind of larvae are expected to show high rates of genetic exchange between populations and thus high genetic homogeneity, although this expectation is not always met. Several studies have indicated that in a number of marine species self-recruitment to the natal population is high, and the dispersal potential expected of their larvae fails to be realized (Jones et al. 1999; Swearer et al. 1999, 2002; Hellberg et al. 2002; Taylor and Hellberg 2003). Such studies have led to doubts that the dogma of near-panmixia in marine organisms is ever realized (Hellberg 2009). Even when local recruitment does occur, however (as it undoubtedly does in most populations), and even if the fraction of the larvae that recruit successfully at a distant locality is small, it may be sufficient to homogenize genetic constitution of populations over large distances.

The position of coral reef organisms of the eastern tropical Pacific (ETP) in the spectrum between panmixia and high genetic structure is the subject of this chapter. We begin by considering geographical and oceanographic features that would be expected to affect rates of contemporary and historical gene flow, then summarize existing data from the literature about genetic connectivity. Not all of the data were gathered to assess population structure, and the organisms for which genetic data exist are far from uniformly spread over taxonomic groups. Nevertheless, a compilation of such data can begin to address the question of whether general patterns exist. In "coral reef organisms" we include all species that are represented by populations resident in coral reefs, even if they are not exclusively found in this particular habitat. As most of these species (even corals) are capable of living outside coral reefs (Robertson 1998; Guzmán et al. 2004), this means that we have attempted to include studies on all organisms that live on hard substrata in the photic zone, except for those limited to the upper intertidal shore line.

16.2 Oceanographic Features of the Tropical Eastern Pacific Relevant to Gene Flow

The ETP (Fig. 16.1) consists of a narrow strip of continental shelf along the coast of the American continent with some islands close to the shore, a deeper water gap between the coast and the outer eastern Pacific Islands (Easter Island, the Galápagos Archipelago, Malpelo Island, Isla del Coco, Clipperton Atoll and the Revillagigedo Archipelago). A large expanse of deep ocean separates the ETP and the closest islands in the central Pacific (the Line Islands, the Marquesas, and Hawaii). Thus, shallow water populations in the ETP can exchange genes in four major directions: (1) between the eastern and the central Pacific, (2) between the mainland coast and the outer islands, (3) among the outer islands, and (4) along the coast and nearshore islands. Potential barriers to gene flow in each direction consist of habitat unsuitable for the establishment of adult populations and of currents that channel the movement of larvae along a particular vector, or that alter environmental conditions as to exceed the tolerances of propagules in a particular area (see Chaps. 3 and 4, Fiedler and Lavín, and Wang et al., respectively). Expanses of deep water that are difficult to cross in a single larval life can be expected to cause the most marked restrictions to the exchange of genes. If impassable, these barriers can prevent the spread of species, thus establishing different biogeographic provinces as determined by patterns of species presence and absence.

Physical barriers that impede gene exchange between populations in the ETP and those in the rest of the world define it as a separate oceanic region. Towards the east, an uninterrupted land bridge has existed for the last 2–3 million years (Coates and Obando 1996). [It has been claimed recently that a nearly complete barrier to water exchange with the Caribbean has existed since the Eocene or early Miocene (Montes et al. 2012) but this claim is incompatible with paleoceanographic (Keigwin 1982; Collins 1996; Haug and Tiedemann 1998; O'Dea et al. 2007), paleontological (Webb 1976; Coates and Obando 1996), and genetic (Lessios 2008) data.] Towards the west lies the widest marine biogeographical barrier on the planet, 4000–7000 km of deep water without any stepping stone habitats on which adults of shallow water marine species can exist (Ekman 1953; Briggs 1974). This ocean configuration, known as the "Eastern Pacific Barrier" (EPB) has been in place for most of the Cenozoic (Grigg and Hey 1992). It is so difficult to cross, that most shallow water benthic genera are represented by different species on its two sides. Had it not been for the small number of "transpacific species", species that span the EPB (Emerson 1978, 1982; Vermeij 1978; Rosenblatt and Waples 1986; Lessios et al. 1996; Robertson et al. 2004; Lessios and Robertson 2006), it would not have figured in a

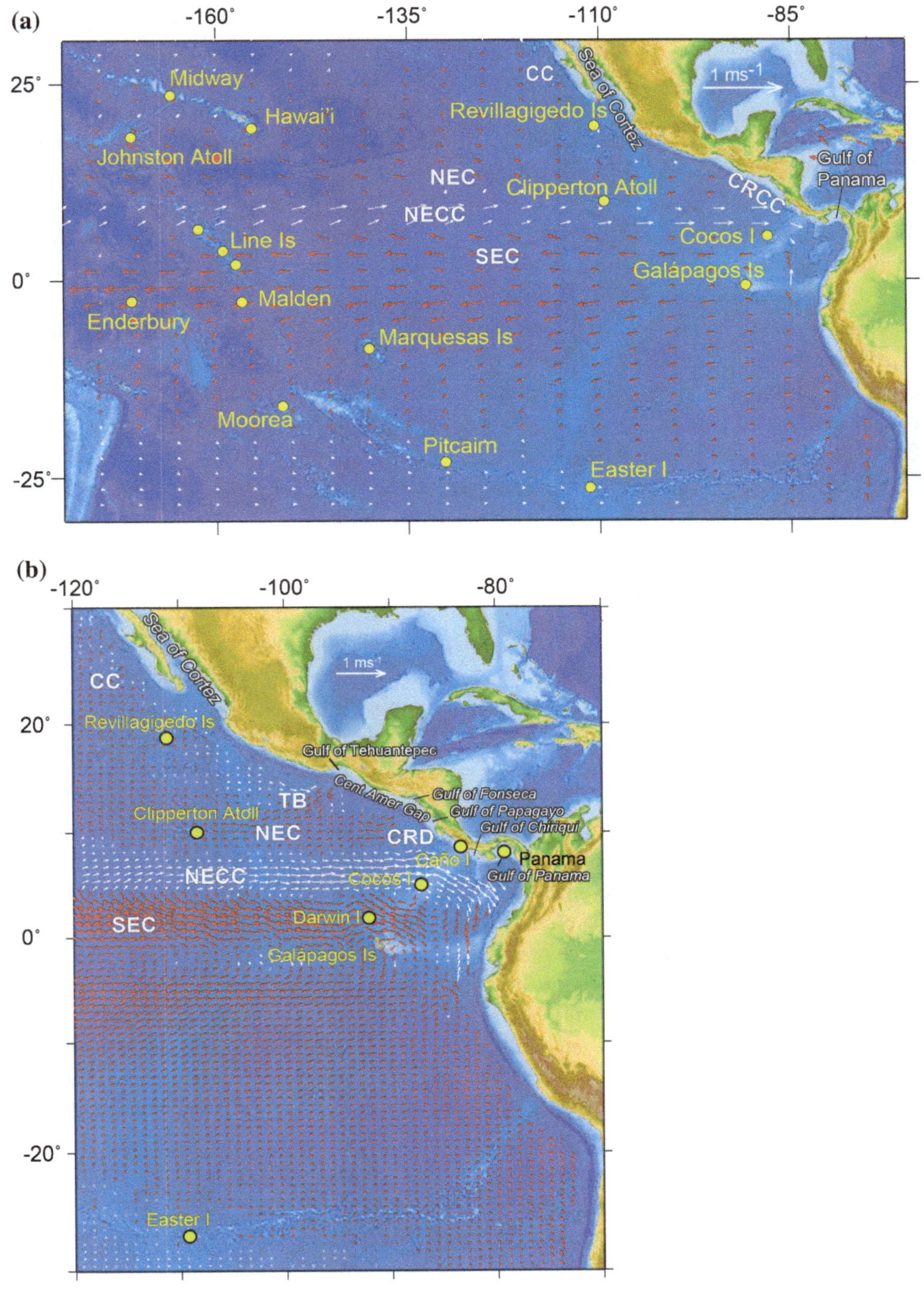

Fig. 16.1 Ten year mean (1993–2003) ocean surface currents in the Central-East Pacific (**a**) and the Eastern Tropical Pacific (**b**). Given are surface current vectors with 1 degree (**a**) or 1/3 degree (**b**) resolution. Basemap generated with ETOPO 2 (USGS). CC = California Current, NEC = North Equatorial Current, NECC = North Equatorial Counter Current, SEC = South Equatorial Current, CRCC = Costa Rica Coastal Current

discussion of conspecific gene flow. The barrier is traversed by the westerly North and South Equatorial Currents and by the easterly North Equatorial Counter Current. The speed of the latter increases greatly during El Niño events, reducing the transit time between the Line Islands and the eastern Pacific to periods that may not exceed the larval duration of a number of organisms (Richmond 1990; Glynn et al. 1996; Glynn and Ault 2000; Robertson et al. 2004). This acceleration of the North Equatorial Counter Current is generally proposed as the conveying mechanism for the dispersal of species from the species-rich central Pacific into the ETP (Dana 1975; Richmond 1990; Robertson et al. 2004) and may also be responsible for recurrent events of gene flow, at least in one direction.

Expanses of deep water narrower than the EPB are also of potential relevance to genetic divergence between conspecific populations of the outer islands of the eastern Pacific and the mainland. Easter Island (Rapa Nui), 4000 km from the coast of Peru, is the most remote of these islands. As its distance from Pitcairn is 1700 km, it can either be considered as the westernmost island of the eastern Pacific or the easternmost island of the central Pacific. Its marine fauna consists of a mixture of endemic, Indo-West-Pacific, transpacific, and cosmopolitan species, but it also contains several species characteristic of the eastern Pacific (Fell 1974; Rehder 1980; Randall 1998; Glynn et al. 2007). The latter may be the result of larval transport in currents that flow predominantly towards the west (Glynn et al. 2007). The next most remote island, Clipperton Atoll, lies inside the tropics, 1000 km from the coast of Mexico and 4000 km from the Marquesas Islands. It is intermittently bathed by either the easterly North Equatorial Counter Current or the westerly North Equatorial Current and may well represent a stepping stone between the central and eastern Pacific (Glynn et al. 1996; see Chap. 5, Glynn et al.). As at Easter Island, the marine fauna of Clipperton is a mixture of Indo-West and east Pacific species, and also endemics (Glynn et al. 1996; Lessios et al. 1996; Robertson and Allen 1996). The Galápagos Islands comprise the only oceanic Archipelago in the Equatorial eastern Pacific. San Cristóbal, the closest island to the mainland, lies 930 km west of the coast of Ecuador. The Galápagos Islands contain the best studied marine biota among those of the outer islands (e.g., Bowman 1966; Glynn and Wellington 1983; James 1991; Grove and Lavenberg 1997). The marine fauna contains numerous species that are also found on the west coast of America, several species endemic to the eastern Pacific outer islands or just to the Galápagos, and a few Indo-Pacific species (Briggs 1974). The Archipelago is influenced by the Peru Oceanic Current flowing towards the Galápagos from the mainland, the North Equatorial Counter Current bringing water from the central Pacific, the South Equatorial Current flowing in the opposite direction, plus a southerly flowing current during the dry season out of the Panama Bight (Ab bott 1966; Glynn and Ault 2000; Kessler 2006). Isla del Coco is situated 690 km northeast of the Galápagos and 500 km west of Costa Rica. Its marine fauna consists mostly of eastern Pacific species, but there is a small number of outer island endemics and of Indo-Pacific species (Hertlein 1963). The four Revillagigedo islands lie approximately 390 km southwest of the southern tip of Baja California. The shallow water marine species on these islands are mostly a subset of those found on the mainland, but some Indo-Pacific species are also present. The corals show affinities with those of Clipperton (Glynn et al. 1996; Glynn and Ault 2000; Ketchum and Bonilla 2001). The Revillagigedo Archipelago is influenced both by the northerly Costa Rica Current and by the southerly California Current (Glynn and Ault 2000; Kessler 2006; see Chap. 3, Fiedler and Lavín).

Marine populations of the coastal ETP are expected to show more genetic connectivity than those of island populations. This coast, however, is not entirely without potential barriers. Hard substrate bottoms are interrupted by river estuaries and mangrove areas, unsuitable as habitats for adults of many coral reef species. A 1000 km stretch from southern Mexico to the Gulf of Fonseca at the Honduras-El Salvador border (the Central American Gap) is devoid of coral reefs or rocky bottoms, except for a small *Pocillopora* reef in Los Cóbanos, El Salvador (Glynn and Ault 2000; Reyes-Bonilla and Barraza 2003; but see Chap. 5, Glynn et al.). Wind-driven upwelling in the Bay of Panama, the Gulf of Papagayo, and at Tehuantepec (Kessler 2006) may also exclude some species with narrow thermal tolerances, such as *Acanthaster planci* (Glynn 1974). Separate biogeographic provinces along the American coast recognized by some authors (Briggs 1974; Veron 1995; Hastings 2000; but see Robertson and Cramer 2009; Briggs and Bowen 2012) attest to the existance of barriers to dispersal that prevent a number of species from spreading along its entire length. Major currents may be of less importance to dispersal along the coast compared to tidal flux and eddies, likely to spread larvae over moderate distances. Nevertheless, larval connectivity might be influenced by the general circulation pattern from south to north between approximately 20°N and 20°S (Kessler 2006; Fiedler and Lavín (Chap. 3)). The northerly Peru Oceanic and Coastal currents reach almost to the Panama Bight; in this area a reversing gyre moves water towards the north or towards the south, depending on the season. From Costa Rica to Baja California, the current flow along the coast is northward, all the way to the entrance of the Sea of Cortez (Gulf of California). The cold California Current flows southward on the west side of Baja California and limits colonization of most tropical species, thus defining the northern termination of the ETP.

16.3 Data on Gene Flow Between Conspecific Populations in the Eastern Pacific

We have attempted to compile all the existing data published until 2013 that can be used to calculate rates of gene flow among ETP populations of coral reef organisms. Most of these data consist of mitochondrial DNA (mtDNA) sequences. The mtDNA of corals, however, evolves at slow rates and does not provide information useful in comparing conspecific populations (Shearer et al. 2002; Hellberg 2006). Population genetic research of corals has, therefore, had to rely on either sequences of ribosomal internal transcribed spacers (ITS) or microsatellites. We have not included data from isozyme studies because the original data are not readily available, and because many of the organisms for which isozyme data exist have been revisited by studies involving DNA, less subject to the problem of "hidden variation". Most of the studies that generated the data were not designed to address the question of gene flow within the eastern Pacific; because of this, the sample sizes for each local population are often smaller than necessary for robust statistical analysis. Nevertheless, sequence data from only a few individuals may present a not too distorted view of gene flow (Pluzhnikov and Donnelly 1996). Some of the published studies have included summary statistics that could be used directly. When they did not, we calculated these statistics from raw data downloaded from GenBank. Some studies had to be omitted because authors have not included locality information with the accessioned sequences, making it impossible to use their data to calculate gene flow between local populations.

In order to quantify degree of divergence relevant to gene flow, we used F_{ST} statistics (Wright 1951). F_{ST} is the ratio of genetic variance between populations over the pooled variance within populations. There are two, qualitatively different, ways of calculating F_{ST} statistics. One, as originally conceived by Wright, is based on the frequency of alleles. The other is relevant only to DNA sequences, and is based both on the frequency of haplotypes and on the number of nucleotide differences between them (Hudson et al. 1992). This measure is sometimes referred to as Φ_{ST}. We have used frequency-based F_{ST} for microsatellite data and Φ_{ST} for sequence data. As most of the published studies have not applied corrections to either of these values, we avoided using them in what we calculated from raw data.

F_{ST} is far from a perfect measure of gene flow. Under an island model (in which migration and genetic drift are at equilibrium, and in which migration rates and effective population sizes are equal between all demes), F_{ST} can be used to estimate the number of propagules dispersed from one population to another per generation (Wright 1951). These conditions, however, are rarely met in any species (Whitlock and McCauley 1999), and much less so in marine populations, which are open to immigration of larvae with potentially different genetic constitutions in each generation (Johnson and Black 1984). An additional problem with F_{ST}, particularly for frequency-based data, is that high within-population heterozygosity can create the appearance of no differentiation between populations, even if they share no alleles (Hedrick 2005; Hellberg 2007, 2009). Φ_{ST} can assume negative values if within-population variation is higher than between-population variation. Such negative values are incompatible with the notion that they estimate gene flow. Finally F_{ST} statistics are incapable of distinguishing between recurrent gene flow at low levels among recently separated populations and higher levels of gene flow between anciently separated ones (Hey and Nielsen 2004; Marko and Hart 2012). Despite these problems, F_{ST} is a useful index of genetic differences between populations over and above local variation, differences that increase with genetic isolation, and, unlike more sophisticated statistics of gene flow, it can be applied to all genetic markers. The latter advantage is the main reason why we use it in this chapter. We have arranged the data in tables organized along four axes in which populations can be compared, i.e. between the central and the eastern Pacific, between the outer (oceanic) islands of the eastern Pacific and the mainland, among the outer islands, and among populations distributed along the coast. This, of course, does not imply that the directions in which genes have traveled over the generations are necessarily limited to these axes.

16.3.1 Gene Flow Between the Central and the Eastern Pacific

As few species are shared between the ETP and the central Pacific, the number of studies comparing conspecific populations in the two areas is necessarily limited. Under the island model of Wright (1951), values of $F_{ST} \geq 0.2$ (if calculated from nuclear markers) or ≥ 0.33 (if calculated from mtDNA), which is haploid and maternally inherited) correspond to an estimate of less than one propagule per generation. Accordingly, F_{ST} values that approach these thresholds are held to indicate genetic exchange that cannot overcome the diversifying effects of genetic drift. By this standard, many of the F_{ST} values in Table 16.1 represent a high degree of divergence, indicative of low rates of gene flow over the enormous distances between shallow water habitats in the two oceans. The patterns of genetic exchange, however, differ between species, as would be expected if the EPB is sporadically breached by larvae carried during periods of acceleration of the North Equatorial Counter Current.

Combosch et al. (2008) compared sequences of the internal transcribed spacer (ITS) from the main reef frame-builder of the eastern Pacific, *Pocillopora damicornis,* with sequences from Hawaii and from the western Pacific. Divergence between populations at Hawaii and at Panama was high (Table 16.1) and approximately equal to divergence between populations from the eastern and the western Pacific. Combosch et al. (2008) attributed part of this divergence to introgression of genes from *P. eydouxi* and *P. elegans* into the genome of ETP populations of *P. damicornis,* facilitated by a shift from a brooding reproductive mode in the central and western Pacific to free-spawning of gametes in the ETP (Glynn et al. 1991). This conclusion has been challenged by Pinzón and LaJeunesse (2011), who regard all three morphospecies as a single genetic entity, and consider evidence of apparent hybridization to be artifacts of cloning and sequencing of ITS, due to its representation by multiple copies in the genome (see below). Despite this disagreement, the F_{ST} value between populations of *Pocillopora* on the two sides of the EPB is not large enough to suggest that what is regarded as *P. damicornis* in Hawaii is a different species than the eastern Pacific form. The planulae

Table 16.1 F_{ST} values and approximate distance by sea (in italics) between populations in the eastern and the central Pacific, sampled by various studies

Organism	Reference	Locus	bp or no of loci		F_{ST}	*Distance (km)*	F_{ST}	*Distance (km)*	F_{ST}	*Distance (km)*	F_{ST}	*Distance (km)*	F_{ST}	*Distance (km)*	F_{ST}	*Distance (km)*	F_{ST}	*Distance (km)*
Corals																		
Pocillopora damicornis	Combosch et al. (2008)	5.8S-ITS2	555 bp	Panama (10)	Hawaii (2)													
					0.419	8400												
Porites lobata	Forsman et al. (2009) *	18S-5.8S-ITS2	673 bp	Galápagos (15)	Tahiti (9)													
					0.070	6600												
				Easter Is. (10)	0.370	4200												
Porites lobata	Baums et al. (2012)	Microsatellites	12 loci		Midway & NW Islands (84)		Hawaii Central (149)		Hawaii Main (53)		Johnston Atoll (58)		Kingman Rf. (22)		Palmyra (19)		Teraina (10)	
				Clipperton (5)	0.380	6760	0.405	6200	0.406	5100	0.191	6500	0.198	5860	0.117	5840	0.073	5700
				Darwin, Gal (46)	0.298	8800	0.324	7900	0.315	7200	0.152	8600	0.165	7800	0.116	7700	0.121	7650
				Wolf, Gal (45)	0.318	8830	0.345	7930	0.342	7230	0.179	8630	0.173	7830	0.124	7730	0.146	7700
				Marchena, Gal (38)	0.319	8890	0.346	7990	0.344	7290	0.130	8690	0.182	7890	0.110	7790	0.142	7850
				S. Galápagos (14)	0.308	8940	0.336	8040	0.335	7340	0.118	8740	0.151	7940	0.088	7840	0.090	8000
				Marino Ballena, CR (32)	0.319	9400	0.350	8700	0.346	7800	0.161	9300	0.155	8700	0.105	8650	0.120	8500
				Is. Caño., CR (79)	0.296	9400	0.323	8700	0.314	7800	0.164	9300	0.154	8700	0.116	8650	0.129	8500
				Golfo Dulce, CR (29)	0.325	9400	0.353	8700	0.350	7800	0.197	9300	0.185	8700	0.150	8650	0.165	8500
				Is. del Coco (55)	0.360	9220	0.388	8500	0.389	7500	0.204	8900	0.189	8200	0.148	8300	0.174	8100
				Panama (17)	0.329	9800	0.357	8900	0.359	8400	0.189	9700	0.175	9000	0.138	8800	0.133	8700
				Ecuador (20)	0.398	10200	0.417	9600	0.442	8500	0.221	9950	0.254	9100	0.190	9080	0.221	8900
				Clipperton (5)	Tabuaeran (7)		Kiritimati (49)		Jarvis (12)		Moorea (50)		Hiva Oa (22)		Motane (82)			
					0.107	5600	0.079	5400	0.110	5740	0.145	5400	0.153	4000	0.151	4000		
				Darwin, Gal (46)	0.075	7500	0.106	7300	0.124	7600	0.129	6600	0.132	5350	0.148	5350		
				Wolf, Gal (45)	0.073	7530	0.127	7270	0.162	7630	0.133	6640	0.124	5500	0.139	5500		
				Marchena, Gal (38)	0.108	7590	0.098	7430	0.098	7690	0.144	6700	0.131	5500	0.138	5500		
				S. Galápagos (14)	0.059	7640	0.078	7500	0.076	7740	0.119	6750	0.134	5525	0.144	5525		
				Marino Ballena, CR (32)	0.061	8400	0.098	8200	0.127	8500	0.126	7800	0.138	6470	0.152	6470		
				Is. Caño., CR (79)	0.063	8400	0.124	8200	0.142	8500	0.133	7800	0.126	6460	0.145	6460		
				Golfo Dulce, CR (29)	0.096	8400	0.156	8200	0.182	8500	0.158	7800	0.159	6460	0.176	6460		
				Is. del Coco (55)	0.084	8000	0.150	7900	0.196	8100	0.161	7300	0.170	6000	0.181	6000		
				Panama (17)	0.065	8600	0.132	8700	0.181	8800	0.134	8200	0.154	6700	0.172	6700		
				Ecuador (20)	0.220	8800	0.206	8600	0.189	8800	0.187	7400	0.215	6600	0.219	6600		

(continued)

Table 16.1 (continued)

Organism	Reference	Locus	bp or no of loci		F_{ST}	*Distance (km)*	F_{ST}	*Distance (km)*	F_{ST}	*Distance (km)*	F_{ST}	*Distance (km)*	F_{ST}	*Distance (km)*	F_{ST}	*Distance (km)*	F_{ST}	*Distance (km)*
Snails																		
Conus ebraeus	Duda and Lessios (2009)	COI	611 bp		Hawaii (17)													
				Clipperton (10)	**0.189**	5100												
				Panama (2)	**0.576**	8400												
Conus chaldaeus	Duda et al. (2012)				Hawaii (29)													
				Clipperton (9)	0.058	5100												
Sea stars																		
Acanthaster planci	Vogler et al. (2008)*	COI	632 bp		Hawaii (2)		Johnston Atol (7)		Kingman Rf. (8)		Moorea (6)							
				Panama (2)	**0.838**	8400	**0.775**	9700	0.095	9000	0.332	8200						
				Is. del Coco (13)	**0.906**	7500	**0.871**	8900	**0.447**	8200	**0.669**	7300						
Sea urchins																		
Echinothrix diadema	Lessios et al. (1998)	CytB	642 bp		Hawaii (10)		Kiritimati (12)											
				Is. del Coco (15)	0.048	7500	−0.032	7900										
				Clipperton (6)	−0.061	5100	−0.004	5400										
Tripneustes depressus/gratilla	Lessios et al. (2003)	COI	640 bp		Marquesas (9)		Kiritimati (10)		Hawaii (10)									
				Clipperton (11)	**0.400**	3900	**0.330**	5400	**0.380**	5000								
				Is. del Coco (10)	**0.410**	6000	**0.440**	7900	**0.560**	7500								
				Panama (4)	**0.380**	6900	**0.400**	8700	**0.550**	8400								
				Galápagos (10)	**0.180**	5400	**0.150**	7400	**0.230**	7400								
				Easter Island (8)	**0.180**	3700	0.080	6000	**0.110**	7300								
Fishes																		
Acanthurus nigricans	Lessios and Robertson (2006)	ATPase8-6	842 bp		Kiritimati (10)		Marquesas (10)											
				Panama (6)	−0.012	8700	0.008	6900										
				Clipperton (6)	−0.064	5400	−0.074	3900										
Acanthurus triostegus	Lessios and Robertson (2006)	ATPase8-6	842 bp		Hawaii (5)		Johnston Atoll (4)		Kiritimati (5)		Marquesas (4)							
				Is. del Coco (4)	**0.883**	7500	**0.845**	8900	0.110	7900	0.060	6000						
				Revillagigedo (5)	**0.941**	4700	**0.915**	6000	**0.537**	5300	**0.706**	4400						
				Panama (5)	**0.898**	8400	**0.866**	9700	**0.316**	8700	0.308	6900						
				Clipperton (5)	**0.905**	5000	**0.873**	6400	**0.328**	5400	0.333	3900						
Arothron meleagris	Lessios and Robertson (2006)	ATPase8-6	842 bp		Hawaii (3)		Kiritimati (5)		Marquesas (5)									
				Panama (5)	**0.547**	8400	**0.625**	8700	**0.531**	6900								
				Clipperton (5)	0.118	5000	0.100	5400	0.091	3900								
				Easter Is. (3)	−0.001	7300	−0.132	6000	−0.010	3700								
Calotomus carolinus	Lessios and Robertson (2006)	ATPase8-6	842 bp	Revillagigedo (5)	Hawaii (4)		Kiritimati (5)		Marquesas (4)									
					−0.073	4700	0.058	5300	−0.040	4400								
Cantherhinus dumerilii	Lessios and Robertson (2006)	ATPase8-6	842 bp		Hawaii (3)		Kiritimati (5)											
				Clipperton (5)	0.189	5000	0.125	5400										
Ctenochaetus marginatus	Lessios and Robertson (2006)	ATPase8-6	842 bp		Kiritimati (5)		Marquesas (5)											
				Panama (5)	0.084	8700	0.034	6900										
				Clipperton (5)	0.050	5400	0.032	3900										

(continued)

Table 16.1 (continued)

Organism	Reference	Locus	bp or no of loci		F_{ST}	*Distance (km)*	F_{ST}	*Distance (km)*	F_{ST}	*Distance (km)*	F_{ST}	*Distance (km)*	F_{ST}	*Distance (km)*	F_{ST}	*Distance (km)*	F_{ST}	*Distance (km)*
Forcipiger flavissimus	Lessios and Robertson (2006)	ATPase8-6	842 bp		Kiritimati (10)		Marquesas (5)											
				Revillagigedo (3)	0.000	*5300*	−0.133	*4400*										
				Clipperton (5)	−0.064	*5400*	0.000	*3900*										
				Easter Is. (5)	0.041	*6000*	0.083	*3700*										
Heteropriacanthus cruentatus	Lessios and Robertson (2006)	ATPase8-6	842 bp		Hawaii (4)		Kiritimati (10)											
				Revillagigedo (4)	**0.082**	*4700*	**0.299**	*5300*										
				Clipperton (7)	**0.276**	*5000*	**0.343**	*5400*										
				Is. del Coco (11)	**0.288**	*7500*	**0.361**	*7900*										
				Panama (2)	0.065	*8400*	0.212	*8700*										
				Galápagos (9)	**0.122**	*7400*	**0.227**	*7400*										
				Easter Is. (2)	−0.264	*7300*	−0.016	*6000*										
Mulloidichthys vanicolensis	Lessios and Robertson (2006)	ATPase8-6	842 bp		Hawaii (10)		Kiritimati (5)		Marquesas (7)		Moorea (5)							
				Clipperton (9)	**0.306**	*5000*	**0.290**	*5400*	**0.225**	*3900*	0.034	*5400*						
				Easter Is. (5)	0.072	*7300*	0.151	*6000*	−0.010	*3700*	−0.066	*4200*						
Myripristis berndti	Lessios and Robertson (2006)	ATPase8-6	842 bp		Hawaii (8)		Kiritimati (11)		Marquesas (9)									
				Clipperton (6)	0.118	*5000*	**0.376**	*5000*	0.089	*3900*								
				Is. del Coco (3)	0.039	*7500*	−0.180	*7900*	0.226	*6000*								
				Panama (12)	0.022	*8400*	**0.241**	*8700*	0.091	*6900*								
Myripristis berndti	Craig et al. (2007)	CytB	700 bp		Hawaii (147)		Kiritimati (12)		Marquesas (18)		Moorea (18)							
				Clipperton (15)	**0.424**	*5000*	0.083	*5400*	0.333	*3900*	0.189	*5400*						
				Panama (19)	**0.393**	*8400*	0.074	*8700*	0.307	*6900*	0.177	*8200*						
Novaculichthys taeniourus	Lessios and Robertson (2006)	ATPase8-6	842 bp		Hawaii (4)		Kiritimati (6)		Marquesas (5)									
				Clipperton (5)	0.069	*5000*	0.117	*5400*	0.125	*3900*								
				Panama (5)	0.144	*8400*	**0.167**	*8700*	**0.167**	*6900*								
Ostracion meleagris	Lessios and Robertson (2006)	ATPase8-6	842 bp		Hawaii (5)		Kiritimati (4)		Marquesas (5)									
				Panama (4)	0.218	*8400*	−0.013	*8700*	−0.061	*6900*								
				Clipperton (5)	0.375	*5000*	0.305	*5400*	0.167	*3900*								
Scarus ghobban	Lessios and Robertson (2006)	ATPase8-6	842 bp		Kiritimati (10)													
				Panama (10)	**0.169**	*8700*												
Scarus rubroviolaceus	Lessios and Robertson (2006)	ATPase8-6	842 bp		Hawaii (5)		Kiritimati (10)		Marquesas (10)									
				Clipperton (6)	−0.128	*5000*	0.268	*5400*	0.156	*3900*								
				Is. del Coco (3)	−0.154	*7500*	**0.411**	*7900*	0.124	*6000*								
				Panama (10)	−0.028	*8400*	**0.261**	*8700*	0.042	*6900*								

(continued)

Table 16.1 (continued)

Organism	Reference	Locus	bp or no of loci		F_{ST}	*Distance (km)*	F_{ST}	*Distance (km)*	F_{ST}	*Distance (km)*	F_{ST}	*Distance (km)*	F_{ST}	*Distance (km)*	F_{ST}	*Distance (km)*	F_{ST}	*Distance (km)*
Scarus rubroviolaceus	Fitzpatrick et al. (2011)	Microsatellites	15 loci		Marquesas (14)		Big Island (42)		Maui (24)		Oahu (44)							
				Clipperton (6)	**0.170**	*3900*	**0.260**	*4970*	**0.270**	*5140*	**0.250**	*5300*						
				Is. del Coco (51)	**0.140**	*6000*	**0.230**	*7480*	**0.230**	*7610*	**0.220**	*7790*						
				Las Perlas (9)	**0.150**	*6900*	**0.250**	*8530*	**0.270**	*8630*	**0.250**	*8800*						
				Panama coast (7)	**0.170**	*6900*	**0.260**	*8580*	**0.260**	*8680*	**0.250**	*8850*						
Sectator ocyurus	Lessios and Robertson (2006)	ATPase8-6	842 bp		Marquesas (10)													
				Panama (11)	**0.120**	*6900*												
Stethojulis bandanensis	Lessios and Robertson (2006)	ATPase8-6	842 bp		Kiritimati (10)													
				Clipperton (7)	0.030	*5400*												
				Is. del Coco (3)	**0.420**	*7900*												
				Panama (4)	**0.190**	*8700*												
Zanclus cornutus	Lessios and Robertson (2006)	ATPase8-6	842 bp		Hawaii (5)		Kiritimati (4)		Marquesas (5)									
				Revillagigedo (3)	−0.079	*4700*	0.071	*5300*	0.036	*4400*								
				Clipperton (3)	−0.151	*5000*	−0.091	*5400*	−0.084	*3900*								
				Panama (4)	−0.099	*8400*	0.018	*8700*	−0.013	*6900*								

Numbers in parentheses next to locality names indicate sample sizes
Asterisks next to references indicate that F_{ST} was calculated by us from the raw data, because it was not shown in the publication. See Fig. 16.1 for a map of locations. 5.8S-ITS2: ribosomal 5.8S and internal transcribed spacer 2, 18S-5.8S-ITS2: ribosomal 18S-5.8S and internal transcribed spacer 2;COI: Cytochrome Oxidase I, ATPase8-6: mitochondrial ATP synthetase8 and 6, CytB: Cytochrome Oxidase B. Values in *bold*: $p < 0.05$ as determined by 1000 reshufflings of individuals between localities

of *P. damicornis* contain phototrophic *Symbiodinium* endosymbionts, which permit them to generate energy during long-distance dispersal (Baird et al. 2009). Baird et al. (2009) have suggested that species with autotrophic larvae were the only ones capable of recolonizing the eastern Pacific after the formation of the Isthmus of Panama, following the extinction of the earlier western American coral fauna. Presumably, such colonization might have occurred during El Niño events, though there may be constraints in the period of development of zooxanthellate planulae at high temperatures (Yakovleva et al. 2009), which may limit the teleplanic advantages conferred by larval autotrophy.

The study of *Porites* by Forsman et al. (2009) included samples of the massive coral *Porites lobata* from Tahiti and the ETP. As expected, little gene flow was evident in ITS between the central Pacific and Easter Island, one of the most isolated islands in the world (Table 16.1). Surprisingly, however, divergence between populations from Tahiti and the Galápagos was very low, although significant. The low F_{ST} value in this comparison may be due to non-equilibrium conditions arising from a recent influx of larvae. In a study of the same species by Baums et al. (2012), based on 12 microsatellites, samples from Moorea were highly divergent from samples from four localities in the Galápagos Archipelago (Table 16.1). The extensive sampling in this study permits firm conclusions regarding gene flow in this widespread, broadcast spawning coral with autotrophic larvae. Gene flow in *P. lobata* is high within the ETP and within the central Pacific but severely restricted across the EPB, despite occasional low F_{ST} values between some populations from the Galápagos and some of the south central Pacific islands. Principal components and Bayesian STRUCTURE (Pritchard et al. 2000) analysis revealed that colonies at Clipperton Atoll group genetically with those from the central, rather than the eastern Pacific (Baums et al. 2012). Populations at Tabuaeran (Fanning Atoll) and Kiritimati were genetically less differentiated than those of other central Pacific islands from a number of locations in the ETP (Table 16.1).

The only genetic comparisons of transpacific molluscs are for two species of *Conus*. Duda and Lessios (2009) found that populations of *C. ebraeus* at Hawaii were highly differentiated in Cytochrome Oxidase I (COI) from two populations in the ETP (Table 16.1). It appears that Clipperton is a stepping stone to the rest of the ETP, because the sample from this atoll consisted of mitochondrial haplotypes also found in either Panama or Hawaii, plus an additional type otherwise only encountered in Okinawa. These three haplotypes were distantly related, suggesting that they did not evolve in situ, and that the Clipperton population is thus the result of haphazard, infrequent influx of larvae. Analysis of Molecular Variance (AMOVA) found high ($\Phi_{ST} = 0.263$) and significant differentiation across the EPB (Duda and Lessios 2009). In contrast to those of *C. ebraeus*, COI haplotypes of *Conus chaldaeus* from Hawaii and Clipperton were very similar (Table 16.1; Duda et al. 2012).

Data from the corallivorous sea star, *Acanthaster planci,* illustrate that different conclusions about gene flow are sometimes drawn when different genetic markers are used. ETP populations of *Acanthaster* were originally described as a separate species, *A. ellisii,* based on morphological differences (Madsen 1955). Sampled by isozymes, their genetic constitution appeared to be similar to those of the central Pacific, which led Nishida and Lucas (1988) to the conclusion that there was only one species, connected by high gene flow across the EPB. When sequences of the mitochondrial gene COI were subsequently obtained from the entire range of *Acanthaster* (Vogler et al. 2008), they showed that mtDNA of the ETP populations belonged to the same lineage as that of populations from the rest of the Pacific. F_{ST} values between eastern and central Pacific localities, however, are very large in seven out of eight comparisons (Table 16.1), suggesting that there is no gene flow across the EPB.

A number of tropicopolitan sea urchin genera show the deepest divergences between central-west and eastern Pacific extant species (Lessios et al. 1999, 2001; McCartney et al. 2000). There are, however, two exceptions. *Echinothrix* is a genus that, until one of its species, *E. diadema*, was observed at Isla del Coco in 1987, was unknown from the ETP (Lessios et al. 1996). Sequencing of Cytochrome B (CytB) found a higher amount of divergence between populations of *E. diadema* at Isla del Coco and at Clipperton, on the same side of the EPB, than between these populations and those at Kiritimati and at Hawaii (Table 16.1). Lack of divergence between ETP and central Pacific populations is more likely to be the result of recent introduction than of recurrent gene flow. Lessios et al. (1996, 1998) suggested that these populations may have become established during the 1982–1983 El Niño, which also introduced a number of central-west Pacific species of fishes into the ETP (Robertson et al. 2004). The second case of a transpacific echinoid is that of *Tripneustes*. Despite doubts by some taxonomic authorities, *T. gratilla* from the Indo-Pacific and *T. depressus* from the ETP were regarded as separate species. Lessios et al. (2003), however, found that these two putative species shared the same mtDNA clade and are, thus, in all probability conspecific. Sequences of bindin, a nuclear gene responsible for sperm-egg recognition, led to the same conclusion (Zigler and Lessios 2003). Gene flow across the EPB in *Tripneustes* was restricted, as evidenced by high pairwise F_{ST} values between five ETP and three central Pacific locations (Table 16.1).

Fishes are the group that contains the highest number of transpacific species (Robertson et al. 2004), and have thus provided the greatest opportunities for assessing gene flow across the EPB. Data from fishes illustrate great diversity of evolutionary histories and gene flow rates, diversity that is consistent with "sweepstakes dispersal" between the oceanic regions on either side of the barrier. An isozyme study by Rosenblatt and Waples (1986) indicated little divergence between ETP and Hawaiian conspecific populations of twelve species. More recently, Lessios and Robertson (2006) compared two mitochondrial genes, ATPase8 and ATPase6, in twenty species considered as transpacific on the basis of morphology. Two of these turned out to be anciently separated between oceanic regions, as indicated by reciprocally monophyletic mtDNA clades, and thus should probably be recognized as separate species. Among the other eighteen, F_{ST} between central and eastern Pacific populations ranged from negative values to +0.94. The highest values, indicative of a complete cessation of gene flow, were those between demes of the surgeon fish *Acanthurus triostegus sandvichensis* from Hawaii and Johnston Atoll, and of *A. triostegus triostegus* from the ETP (Table 16.1), but also between *A. triostegus sandvichensis* and *A. triostegus marquesensis* from the Line Islands and the Marquesas (Lessios and Robertson 2006). It would appear, therefore, that there is genetic isolation between the Hawaiian-Johnston subspecies from the other two subspecies. These comparisons illustrate that populations at Hawaii are often isolated not only from those in the ETP, but also from the rest of the Pacific, as Baums et al. (2012) have also found in *Porites lobata*. Thus, data in Table 16.1 in which the central Pacific is represented only by samples from Hawaii may not be indicative of isolation between oceanic regions.

Divergence between populations of *Acanthurus triostegus* from Kiritimati and the Marquesas and from the ETP was generally high, but at the same time suggestive of low levels of gene flow, also evident in the sharing of the most common haplotypes (Lessios and Robertson 2006). Divergence between ETP and central Pacific localities was occasionally high (but inconsistent between comparisons of different populations) in the glass eye, *Heteropriacanthus cruentatus*, in the goatfish *Mulloidichthys vanicolensis* and in the squirrelfish *Myripristis bendti.* Such inconsistencies may well be the result of small sample sizes, because sampling of *M. bendti* CytB of some of the same localities by Craig et al. (2007) did not always produce similar F_{ST} values. Other species, such as the surgeon fish *Acanthurus nigricans*, the parrotfish *Calotomus carolinus,* the surgeon fish *Ctenochaetus marginatus*, the butterflyfish *Forcipiger flavissimus* and the moorish idol *Zanclus cornutus* showed practically no divergence across the EPB (Table 16.1), indicating that there has been either recent or recurrent gene flow. In an analysis of molecular variation (AMOVA) only two of the 18 transpacific species were found to have significantly higher differentiation between oceanic regions than within regions. F_{ST} and AMOVA cannot distinguish between on-going gene flow and recent isolation. For this reason, Lessios and Robertson (2006) employed "IM", an algorithm of Bayesian estimation based on coalescence (Hey and Nielsen 2004), to deduce time of initial separation and degree and direction of subsequent gene flow. Such analyses do not always produce reliable results when they involve a single locus, but they can provide an indication regarding these parameters. Estimated time of initial separation between central and eastern Pacific conspecific populations ranged from 30,000 to 1,000,000 years ago, times more recent than the 3 million year final closure of the Isthmus of Panama (Coates and Obando 1996). These relatively recent estimates indicate that a vicariance scenario, according to which transpacific species are relicts of circumglobal connections that were severed by the rise of the Isthmus (McCoy and Heck 1976; Heck and McCoy 1978), is unlikely. Direction of gene flow deduced by IM was not always from West to East, as might have been expected if the North Equatorial Counter Current were the only means of conveyance. In eight out of the eighteen cases, migration was actually estimated as being higher in the opposite direction. Reconstruction of ancestral genotypes and comparisons of relative genetic diversity suggested that in at least two cases the original range expansion was from the ETP into the central Pacific. Thus, directions of initial colonization and subsequent gene flow do not always coincide.

In conclusion, marine shallow water biota of the ETP is, indeed, isolated from the rest of the world's oceans, but breaching of the EPB is also possible. Such transpacific migrations of larvae most likely are the result of haphazard combinations of factors favorable for migration, such as timing of spawning relative to current speed intensifications, availability of rafting materials, and good fortune in encountering suitable habitat at the end of the dispersal event.

16.3.2 Gene Flow Between the Outer Eastern Pacific Islands and the Mainland Coast

Compared to the distances between the ETP and the central Pacific, those between the American mainland coast and the outer oceanic islands of the ETP (Revillagigedo, Clipperton, Isla del Coco, the Galápagos and Easter Island) are shorter, but far from negligible. The available data on gene flow across this 520–4000 km oceanic divide (Table 16.2) are biased towards showing high rates of migration because, by and large, they come from transpacific species likely to possess traits conducive towards high rates of dispersal.

Data from corals found on at least one island and on the mainland come from studies on *Pocillopora* and *Porites*. In corals, morphological plasticity (Todd 2008) and gene exchange between distinguishable morphs (Willis et al. 2006)

Table 16.2 F_{ST} values and approximate distance by sea (in italics) between populations on the shore and at the outer islands of the eastern Pacific, sampled by various studies

Organism	Reference	Locus	bp or no of loci		F_{ST}	*Distance (km)*	F_{ST}	*Distance (km)*	F_{ST}	*Distance (km)*	F_{ST}	*Distance (km)*	F_{ST}	*Distance (km)*	F_{ST}	*Distance (km)*
Corals																
Pocillopora type 1	Pinzón and LaJeunesse 2011	Microsatellites	7 loci		Revillagigedo (2)		Clipperton (9)		Galápagos (16)							
				Sea of Cortez (173)	0.121	*700*	0.052	*1600*	0.083	*2200*						
				Banderas Bay, Mex (71)	0.117	*600*	0.021	*1300*	0.089	*2900*						
				Oaxaca, Mexico (15)	0.057	*1700*	0.036	*1700*	0.041	*1800*						
				Panama (22)	0.113	*3800*	0.050	*3400*	0.083	*1500*						
Pocillopora type 3	Pinzón and LaJeunesse (2011)*	Microsatellites	7 loci		Galápagos (3)											
				Panama (26)	**0.239**	*2200*										
Porites lobata	Baums et al. (2012)	Microsatellites	12 loci		Clipperton (5)		Is. delC oco (55)		Darwin, Gal (46)		Wolf, Gal (45)		Marchena, Gal (38)		S. Galápagos (14)	
				Marino Ballena (32)	**0.142**	*2800*	**0.009**	*540*	**0.013**	*1230*	**0.006**	*1240*	**0.046**	*1230*	**0.031**	*1300*
				Is. Caño., CR (79)	**0.173**	*2700*	**0.021**	*520*	**0.018**	*1200*	**0.015**	*1200*	**0.057**	*1200*	**0.057**	*1360*
				Golfo Dulce, CR (29)	**0.190**	*2860*	0.025	*530*	**0.028**	*1230*	**0.014**	*1240*	**0.075**	*1230*	**0.064**	*1300*
				Panama (17)	**0.184**	*3200*	**0.024**	*840*	**0.011**	*1300*	**0.017**	*1300*	**0.082**	*1260*	**0.055**	*1330*
				Ecuador (20)	**0.298**	*3400*	**0.208**	*1000*	**0.145**	*1300*	**0.149**	*1270*	**0.130**	*1100*	**0.150**	*1000*
Sea stars																
Acanthaster planci	Vogler et al. (2008)*	COI	632 bp		Is. del Coco (13)											
				Panama (2)	0.000	*840*										
Sea urchins																
Echinometra vanbrunti	McCartney et al. (2000)*	COI	631		Galápagos (4)											
				Mexico (9)	0.064	*2200*										
				Panama (7)	−0.022	*1500*										
Diadema mexicanum	Lessios et al. (2001)	ATPase8-6	614		Clipperton (12)		Is. del Coco (12)		Galápagos (4)							
				Sea of Cortez (9)	0.000	*1200*	−0.020	*3500*	0.000	*3600*						
				Panama (5)	0.040	*3200*	−0.010	*840*	−0.110	*1500*						
Tripneustes gratilla/depressus	Lessios et al. 2003	COI	640		Clipperton (11)		Is. del Coco (10)		Galápagos (10)		Easter Is.(8)					
				Panama (10)	**0.540**	*3200*	−0.120	*840*	0.170	*1500*	**0.480**	*5100*				
Fishes																
Acanthurus nigricans	Lessios and Robertson (2006)	ATPase8-6	842 bp	Panama (6)	Clipperton (6)											
					−0.082	*3200*										
Acanthurus triostegus	Lessios and Robertson (2006)	ATPase8-6	842 bp		Revillagigedo (5)		Clipperton (5)		Is. del Coco (4)							
				Panama (5)	−0.0006	*3600*	−0.196	*3200*	−0.042	*840*						
Arothron meleagris	Lessios and Robertson (2006)	ATPase8-6	842 bp		Clipperton (5)											
				Panama (5)	0.239	*3200*										
Ctenochaetus marginatus	Lessios and Robertson (2006)	ATPase8-6	842 bp		Clipperton (5)											
				Panama (5)	0.036	*3200*										
Diodon holocanthus	Lessios and Robertson (2006)*	ATPase8-6	842 bp		Revillagigedo (5)		Galápagos (10)		Easter Island (6)							
				Sea or Cortez (20)	0.046	*890*	−0.005	*3700*	**0.442**	*6000*						
				Panama (20)	**0.140**	*3600*	0.036	*1500*	**0.550**	*5100*						

(continued)

Table 16.2 (continued)

Organism	Reference	Locus	bp or no of loci		F_{ST}	*Distance (km)*	F_{ST}	*Distance (km)*	F_{ST}	*Distance (km)*	F_{ST}	*Distance (km)*	F_{ST}	*Distance (km)*	F_{ST}	*Distance (km)*
Epinephelus labriformis	Craig et al. (2006)	CytB	468 bp		Is. del Coco (11)		Galápagos (35)									
				Lower Cape, Baja Calif. (12)	0.064	*3200*	0.064	*3300*								
				Loreto, Baja California (38)	0.045	*3800*	0.048	*3900*								
				Mazatlan (30)	0.050	*3000*	0.052	*3150*								
				Puerto Vallarta (49)	0.098	*2680*	0.036	*2830*								
				Oaxaca (26)	0.040	*1600*	−0.014	*1850*								
				El Salvador (21)	−0.008	*890*	−0.008	*1500*								
				Panama (46)	0.073	*840*	0.012	*1500*								
Heteropriacanthus cruentatus	Lessios and Robertson (2006)	ATPase8-6	842 bp		Revillagigedo (4)		Clipperton (7)		Is. del Coco (11)		Galápagos (9)					
				Panama (2)	−0.191	*3600*	−0.085	*3200*	−0.085	*840*	−0.141	*1500*				
Myripristis berndti	Lessios and Robertson (2006)	ATPase8-6	842 bp		Clipperton (6)		Is. del Coco (3)									
				Panama (12)	−0.049	*3200*	0.234	*840*								
Myripristis berndti	Craig et al. (2007)	CytB	700 bp		Clipperton (15)											
				Panama (19)	−0.061	*3200*										
Novaculichthys taeniourus	Lessios and Robertson (2006)	ATPase8-6	842 bp		Clipperton (5)											
				Panama (5)	0.063	*3200*										
Ophioblennius steindachneri	Muss et al. (2001)	CybB	630 bp		Clipperton (17)		Galápagos (9)									
				Sea of Cortez (11)	**0.396**	*1700*	?	*3200*								
				Panama (13)	?	*3200*	0.000	*1500*								
Ostracion meleagris	Lessios and Robertson (2006)	ATPase8-6	842 bp		Clipperton (5)											
				Panama (4)	−0.122	*3200*										
Scarus rubroviolaceus	Lessios and Robertson (2006)	ATPase8-6	842 bp		Clipperton (6)		Is. del Coco (3)									
				Panama (10)	−0.083	*3200*	−0.135	*840*								
Scarus rubroviolaceus	Fitzpatrick et al. (2011)	Microsatellites	15 loci		Clipperton (6)		Is. del Coco (51)									
				Panama Coast (7)	**0.080**	*3200*	0.000	*840*								
				Is. Perlas (9)	0.020	*3150*	0.000	*790*								
Stethojulis bandanensis	Lessios and Robertson (2006)	ATPase8-6	842 bp		Clipperton (7)		Is. del Coco (3)									
				Panama (4)	**0.590**	*3200*	**0.376**	*840*								
Zanclus cornutus	Lessios and Robertson (2006)	ATPase8-6	842 bp		Revillagigedo (3)		Clipperton (3)									
				Panama (4)	−0.231	*3600*	−0.180	*3200*								

Numbers in parentheses next to locality names indicate sample sizes
Asterisks next to references indicate that F_{ST} was calculated by us from the raw data, because it was not shown in the publication. COI: Cytochrome Oxidase I, ATPase8-6: mitochondrial ATP synthetase8 and 6, CytB: Cytochrome Oxidase B. Values in *bold*: $p < 0.05$ as determined by 1000 reshufflings of individuals between localities. ?: Sequences necessary for the calculations are not in GenBank

contribute to uncertainties as to which populations belong to the same species. This is particularly true for *Pocillopora* in the ETP. Pinzón and LaJeunesse (2011) sampled individuals of this genus that by morphological criteria belonged to *P. damicornis*, *P. verrucosa*, *P. capitata*, *P. meandrina*, and *P. eydouxi*. They determined sequences of the internal transcribed spacer 2 (ITS2) and of an unidentified open reading frame in mitochondrial DNA, and they also genotyped seven microsatellite loci. According to all of these markers, *Pocillopora* in the ETP belong to three distinct clades, independently of morphotype. Microsatellites were the only markers sampled in both the islands and the mainland in *Pocillopora* types 1 and 3. They indicate little differentiation between islands and mainland in *Pocillopora* type 1 (the larger values between Revillagigedo and the mainland are based on only two specimens), but substantial divergence between the Galápagos and Panama in *Pocillopora* type 3 (Table 16.2). *Pocillopora* type 2 is endemic to Clipperton. The population of *Porites lobata* at Clipperton sampled with microsatellites by Baums et al. (2012) was strongly differentiated from populations on the mainland (Table 16.2). Populations from Isla del Coco and the Galápagos maintain high rates of gene flow with populations at the mainland, except for the one at the Ecuadorian coast.

There are high rates of genetic exchange between conspecific populations of echinoderms from the islands and the mainland (Table 16.2). *Acanthaster planci* from the ETP, though genetically differentiated from the same species from the central Pacific, was represented by identical COI haplotypes at Isla del Coco and in the Gulf of Chiriquí, Panama. All three species of sea urchins for which data exist were genetically homogeneous between the islands and the coast with one exception. Paradoxically, this exception consists of the transpacific species *Tripneustes gratilla/depressus*, in which populations at Clipperton and Easter Island maintain very little gene flow with populations along western American shores. Other sea urchin species in which larvae from the islands have not established viable populations on the mainland are *Echinothrix diadema*, which has yet to be observed outside Clipperton and Isla del Coco (Lessios et al. 1996), and *Eucidaris galapagensis* at the islands, which is reciprocally monophyletic in COI with respect to the continental *E. thouarsii* (Lessios et al. 1999). It is doubtful that larvae of these species are incapable of crossing between the islands and the continental shore; a more plausible cause of this pattern is that they fail to become established due to ecological factors.

Practically all the existing data regarding fishes at the oceanic islands of the ETP come from transpacific species, and, as expected, indicate high genetic connectivity with coastal populations (Table 16.2). The grouper *Epinephelus labriformis*, though endemic to the ETP, fits the same pattern. The wrasse *Stethojulis bandanensis*, however, shows a high degree of genetic isolation, both at Clipperton and at Isla del Coco. The cosmopolitan puffer *Diodon holocanthus* is the only fish species sampled from both Easter Island and the mainland; like *Tripneustes*, it illustrates that populations at this remote locality are genetically isolated from populations in the rest of the ETP.

Examination of F_{ST} values from species that have been sampled at more than one island suggests that populations at Easter Island and Clipperton Atoll are less connected to populations along the mainland (Table 16.2). Ideally we would like to test statistically whether there is a general trend showing that populations at geographically more remote islands are also genetically more isolated from conspecific coastal populations. As there are few islands, this cannot be done through correlation of genetic and geographic distances; it can be addressed, instead, by comparison of intraspecific divergence between the residents of each island and those of a common locality on the mainland. The only such comparison that provides a sample size sufficient for statistical testing is between populations from Panama and Clipperton, on the one hand, and between Panama and Isla del Coco, on the other. We compared F_{ST} values of the same nine species sampled in all three localities paired by genetic marker. The results indicate that, as expected from relative geographic distances, genetic isolation of populations at Clipperton from those at Panama was significantly higher than isolation of the same nine species at Isla del Coco (Wilcoxon paired sample test, $p < 0.05$).

16.3.3 Gene Flow Between the Outer Eastern Pacific Islands

Although genes are not necessarily transferred directly between the outer islands (mainland populations may act as stepping stones) it is useful to ask how genetically different island populations are from each other. The general picture in comparisons between island populations of various species (Table 16.3) is almost identical to the one presented by comparisons between the islands and the mainland (Table 16.2). There is high genetic connectivity between conspecific residents of most islands, except for those at Easter Island, in which all but one species show large and significant values of F_{ST} in comparisons with populations from all other islands. Clipperton Atoll shows a mixed pattern. Populations of *Porites lobata*, *Tripneustes gratilla,* the blenny *Ophioblennius steindachneri* and the wrasse *Stethojulis bandanensis* are very different from populations from all other islands. Populations of all other species, however, show high rates of gene flow.

16.3.4 Gene Flow Along the Coast of the Eastern Pacific

We would expect to observe the highest rates of gene flow along the American shores of the ETP. This is generally the case. Pinzón and LaJeunesse (2011) found no significant restrictions of gene flow in microsatellites of *Pocillopora* "type 1" in populations spanning approximately 4200 km from the Sea of Cortez to Panama (Table 16.4). Combosch and Vollmer (2011), on the other hand, reported that their microsatellite data of *Pocillopora damicornis* show significant structure at a much smaller scale on the Panamanian coast. This structure is not evident in F_{ST} values, in which only two comparisons between populations are larger than 0.073 (Table 16.4). It is somewhat more evident in R_{ST} [an F_{ST} equivalent that takes the step-wise mutation pattern expected from microsatellites into account (Slatkin 1995)]. AMOVA analysis found significant, but small variation between individual populations and also between populations that were grouped in three areas along the Panamanian coast. High rates of gene flow were found between microsatellite frequencies of most coastal populations of *Porites lobata* by Baums et al. (2012). One exception was the population at Ecuador, which is different from all populations at Costa Rica, though not from the one at Panama (Table 16.4).

In four species of sea urchins, genetic connectivity in mtDNA along the coast is generally high from Mexico to Panama (Table 16.4). As in *Porites*, the population of the echinoid *Arbacia stellata* in the southernmost periphery of the species is highly differentiated from those in the species' center of distribution.

Among the fishes, those with transpacific ranges, *Diodon holocanthus* and *Scarus rubroviolaceus,* show no genetic structure (Table 16.4). This is hardly surprising in the case of the latter, because both sampling localities of this species are situated close to each other in the Bay of Panama, but in the case of *Diodon* they lie 4000 km apart on either side of the Central American Gap. *Epinephelus labriformis* also shows high genetic connectivity over long spans of the coast from Baja California to Panama. The case of the three species of the grunt *Anisotremus* studied by Bernardi et al. (2008) is somewhat surprising. *Anisotremus interruptus* and *Anisotremus taeniatus*, species usually found only in the proximity of hard bottoms, show no significant structure in either mitochondrial CytB or the nuclear S7 region over 4000 km of coastline. *Anisotremus dovii*, on the other hand, even though it prefers sandy and muddy bottoms (and should thus be more continuously distributed along the coast) appears to experience marked restrictions in gene flow between Mexico and Panama, at least in CytB. Only one out of the total twelve CytB sequences sampled in this species is shared between the two locations, which accounts for the high and significant F_{ST} value.

In contrast to the pattern of high gene flow along most of the tropical west coast of America, genetic connectivity within the Sea of Cortez can be quite low in a number of species. Gene flow in the ovoviviparous sea horse *Hippocampus ingens* was high among all populations sampled from Mexico to Peru by Saarman et al. (2010) except for one; the mitochondrial control region of the population from Guaymas in the Sea of Cortez was highly differentiated from that of every other population (Table 16.4). Sequences of the mitochondrial control region of the scarletfin blenny *Coralliozetus micropes* were reciprocally monophyletic between upper and lower Gulf regions (Riginos 2005), a level of divergence that is reflected in the F_{ST} values in Table 16.4. This is also the case in the Cortez triplefin *Axoclinus nigricaudus* (Table 16.4) and in the Gulf of California endemic sand bass *Paralabrax maculatofasciatus* (Stepien et al. 2001; Riginos 2005), but not in the redside blenny *Malacoctenus hubbsi* (Table 16.4). These genetic breaks were attributed by Riginos (2005) to a hypothetical Pleistocene deep water break of the Baja California Peninsula that bisected habitats of hard bottom fishes, as it did of terrestrial mammals and reptiles. IM analyses yielding similar times of divergence between upper and lower Gulf populations of five fish species supported the hypothesis that a historical barrier was responsible for present-day isolation patterns, but environmental differences might also be responsible (Riginos 2005).

Along a linear coast, such as that of the ETP, one would expect a pattern of isolation by distance (Wright 1943), as one population acts as a stepping stone for the dispersal of genes towards others down the line. To determine whether this was the case, we analyzed the data of all species in Table 16.4 for which more than three localities were sampled. We calculated correlations between F_{ST} values and geographical distance along the coast, using Mantel (1967) tests to estimate probabilities of the correlation coefficient. For this analysis, negative values of F_{ST} were replaced by zero, because negative gene flow has no meaning (Hudson et al. 1992). Combosch and Vollmer (2011) found no correlation between genetic differentiation and distance in *Pocillopora damicornis* over the limited geographic extent of their samples, but our analysis of populations from Mexico to Panama of the data of Pinzón and LaJeunesse (2011) shows that such a correlation does exist in *Pocillopora* type 1 ($r = 0.349$, $p < 0.005$). Baums et al. (2012) found a strong isolation by distance trend in *Porites lobata* in the ETP, including the outer islands. The correlation remains significant when only coastal localities are considered ($r = 0.856$, $0.01 < p < 0.025$). In *Epinephelus labriformis*, most F_{ST} values were negative (and replaced by 0), which resulted in a slight, but still significant correlation with geographic distance ($r = 0.051$, $0.01 < p < 0.025$).

Table 16.3 F_{ST} values and approximate distance by sea (in italics) between populations at the outer islands of the eastern Pacific, sampled by various studies

Organism	Reference	Locus	bp or no of loci		F_{ST}	*Distance (km)*	F_{ST}	*Distance (km)*	F_{ST}	*Distance (km)*	F_{ST}	*Distance (km)*	F_{ST}	*Distance (km)*
Corals														
Pocillopora type 1	Pinzón and LaJeunesse (2011)	Microsatellites	7 loci		Clipperton (9)		Galápagos (16)							
				Revillagigedo (2)	0.079	*950*	0.176	*3000*						
				Clipperton (9)			0.058	*2100*						
Porites lobata	Forsman et al. (2009)*	18S-5.8S-ITS2	673 bp		Easter Island (10)									
				Galápagos (15)	**0.258**	*3500*								
Porites lobata	Baums et al. (2012)	Microsatellites	12 loci		Is. del Coco (55)		Darwin, Gal (46)		Wolf, Gal (45)		Marchena, Gal (38)		S. Galápagos (14)	
				Clipperton (5)	**0.212**	*2500*	**0.134**	*2100*	**0.160**	*2100*	**0.150**	*2350*	**0.089**	*2500*
				Is. del Coco (55)			**0.048**	*700*	**0.032**	*700*	**0.086**	*690*	**0.087**	*815*
				Darwin, Gal (46)					**0.012**	*37*	**0.030**	*217*	**0.007**	*420*
				Wolf, Gal (45)							**0.048**	*180*	**0.045**	*322*
				Marchena, Gal (38)									**0.026**	*197*
Sea urchins														
Echinothrix diadema	Lessios et al. (1998)	CytB	642 bp		Clipperton (6)									
				Is. del Coco (15)	−0.044	*2500*								
Eucidaris galapagensis	Lessios et al. (1999)	COI	640 bp		Galápagos (7)									
				Is. del Coco (6)	−0.060	*750*								
Diadema mexicanum	Lessios et al. (2001)	ATPase8-6	642 bp		Is. del Coco (12)		Galápagos (4)							
				Clipperton (11)	−0.010	*2500*	−0.130	*2300*						
Tripneustes gratilla/depressus	Lessios et al. (2003)	COI	640 bp		Is. del Coco (10)		Galápagos (10)		Easter Island (8)					
				Clipperton (11)	**0.600**	*2500*	0.110	*2300*	**0.400**	*4100*				
				Is. del Coco (10)			0.250	*750*	**0.490**	*4300*				
				Galápagos (10)					**0.190**	*3500*				
Fishes														
Acanthurus triostegus	Lessios and Robertson (2006)	ATPase8-6	842 bp		Clipperton (5)		Is. del Coco (4)							
				Revillagigedo (5)	0.000	*950*	0.358	*3000*						
				Clipperton (5)			−0.042	*2500*						
Arothron meleagris	Lessios and Robertson (2006)	ATPase8-6	842 bp		Easter Is. (3)									
				Clipperton (5)	0.032	*4100*								
Diodon holocanthus	Lessios and Robertson (2006)	ATPase8-6	842 bp		Galápagos (10)		Easter Isl. (6)							
				Revillagigedo (5)	**0.022**	*3000*	**0.325**	*5000*						
				Galápagos (10)			**0.470**	*3500*						
Epinephelus labriformis	Craig et al. (2006)	CytB	468 bp		Galápagos (35)									
				Is. del Coco (11)	0.045	*750*								
Forcipiger flavissimus	Lessios and Robertson (2006)	ATPase8-6	842 bp		Revillagigedo (3)		Easter Is. (5)							
				Clipperton (5)	0.000	*950*	0.250	*4100*						
				Revillagigedo (3)			0.118	*5000*						

(continued)

Table 16.3 (continued)

Organism	Reference	Locus	bp or no of loci		F_{ST}	*Distance (km)*	F_{ST}	*Distance (km)*	F_{ST}	*Distance (km)*	F_{ST}	*Distance (km)*	F_{ST}	*Distance (km)*
Heteropriacanthus cruentatus	Lessios and Robertson (2006)	ATPase8-6	842 bp		Clipperton (7)		Is. del Coco (11)		Galápagos (9)					
				Revillagigedo (4)	−0.075	*950*	−0.117	*3000*	−0.031	*3000*				
				Clipperton (7)			−0.117	*2500*	0.004	*2300*				
				Is. del Coco (11)					0.048	*750*				
Mulloidichthys vanicolensis	Lessios and Robertson (2006)	ATPase8-6	842 bp		Easter Island (5)									
				Clipperton (9)	**0.179**	*4100*								
Myripristis berndti	Lessios and Robertson (2006)	ATPase8-6	842 bp		Is. del Coco (3)									
				Clipperton (6)	0.360	*2500*								
Ophioblennius steindachneri	Muss et al. (2001)	CytB	658 bp		Galápagos (9)									
				Clipperton (17)	**0.379**	*2300*								
Scarus rubroviolaceus	Lessios and Robertson (2006)	ATPase8-6	842 bp		Is. del Coco (3)									
				Clipperton (6)	−0.193	*2500*								
Stethojulis bandanensis	Lessios and Robertson (2006)	ATPase8-6	842 bp		Clipperton (7)									
				Is. del Coco (3)	**0.783**	*2500*								
Zanclus cornutus	Lessios and Robertson (2006)	ATPase8-6	842 bp		Revillagigedo (3)									
				Clipperton (3)	−0.250	*950*								

Numbers in parentheses next to locality names indicate sample sizes
Asterisks next to references indicate that F_{ST} was calculated by us from the raw data, because it was not shown in the publication. 18S-5.8S-ITS2: ribosomal 18S-5.8S and internal transcribed spacer 2, COI: Cytochrome Oxidase I, ATPase8-6: mitochondrial ATP synthetase8 and 6, CytB: Cytochrome Oxidase B. Values in *bold*: $p < 0.05$ as determined by 1000 reshufflings of individuals between localities

Table 16.4 F_{ST} values and approximate distance by sea (in italics) between populations sampled by various studies along the coast of the tropical eastern Pacific

Organism	Reference	Locus	bp or no of loci		F_{ST}	*Distance* (km)	F_{ST}	*Distance* (km)	F_{ST}	*Distance* (km)	F_{ST}	*Distance* (km)	F_{ST}	*Distance* (km)	F_{ST}	*Distance* (km)	F_{ST}	*Distance* (km)	F_{ST}	*Distance* (km)
Corals																				
Pocillopora "type 1"	Pinzón and LaJeunesse (2011)	microsatellites	7 loci		Mexico															
					Banderas Bay (71)		Oaxalaca (15)		Panama (22)											
				Sea of Cortez (173)	0.080	*600*	0.020	*2000*	0.081	*4200*										
				Banderas Bay (71)			0.032	*1600*	0.075	*3500*										
				Oaxaca (15)					0.039	*2000*										
Pocillopora damicornis	Combosch and Vollmer (2011)	microsatellites	6 loci		Gulf of Chiriqui				Azuero Peninsula						Bay of Panama					
					N. of Is. Coiba (21)		B, Damas (30)		Is. C. Afuera (18)		Achotines (30)		Is. Iguana (23)		Taboga (21)		Saboga (22)		Contadora (25)	
				Uva, G. Chiriqui (23)	0.035	*30*	**0.065**	*54*	0.033	*116*	0.015	*250*	0.039	*300*	**0.087**	*420*	0.019	*415*	0.025	*416*
				N. of Is. Coiba (21)			**0.073**	*20*	**0.229**	*100*	0.001	*200*	**0.048**	*240*	**0.072**	*390*	**0.061**	*400*	**0.039**	*401*
				Bahia Damas, Is. Coiba (30)					0.047	*85*	**0.069**	*180*	**0.045**	*230*	**0.065**	*360*	**0.053**	*370*	**0.060**	*371*
				Is. Canal de Afuera. G. of Chiriqui (18)							0.037	*120*	0.025	*160*	**0.064**	*400*	0.015	*320*	**0.040**	*321*
				Achotines, Azuero Pen. (30)									0.029	*41*	**0.071**	*180*	0.040	*185*	0.031	*186*
				Is. Iguana, B. of Panama (23)											**0.069**	*140*	0.022	*150*	**0.033**	*151*
				Taboga, Bay of Panama (21)													**0.046**	*54*	0.030	*57*
				Saboga, Is. Perlas (22)															**0.000**	*1*
Porites lobata	Baums et al. (2012)	microsatellites	12 loci		Costa Rica															
					Is. Caño (79)		Golfo Dulce (29)		Panama (17)		Ecuador (20)									
				Marino Ballena (32)	**0.003**	*119*	0.008	*150*	**0.008**	*320*	**0.154**	*1020*								
				Is. Caño (79)			**0.006**	*242*	**0.015**	*400*	**0.156**	*1150*								
				Golfo Dulce (29)					0.016	*590*	**0.167**	*1000*								
				Panama (17)							**0.001**	770								
Sea urchins																				
Arbacia stellata	Lessios et al. (2012)	COI	660 bp		Panama (9)		Peru (4)													
				El Salvador (11)	0.179	*1300*	0.108	*1700*												
				Panama (9)			**0.373**	*860*												
Diadema mexicanum	Lessios et al. (2001)	ATPase8-6	614 bp		Panama (5)															
				Mexico (9)	0.020	*4000*														
Echinometra vanbrunti	McCartney et al. (2000)*	COI	631 bp		Panama (7)															
				Mexico (5)	−0.095	*4000*														
Eucidaris thouarsi	Lessios et al. (1999)	COI	640 bp		Panama (4)															
				Mexico (6)	−0.030	*4000*														

(continued)

Table 16.4 (continued)

Organism	Reference	Locus	bp or no of loci		F_{ST}	*Distance* (km)	F_{ST}	*Distance* (km)	F_{ST}	*Distance* (km)	F_{ST}	*Distance* (km)	F_{ST}	*Distance* (km)	F_{ST}	*Distance* (km)	F_{ST}	*Distance* (km)	F_{ST}	*Distance* (km)
Fishes																				
Anisotremus dovii	Bernardi et al. (2008)*	S7	531 bp		Panama (4)															
				Mexico (8)	0.166	*4000*														
		CytB	692 bp	Mexico (8)	**0.304**	*4000*														
Anisotremus interruptus	Bernardi et al. (2008)*	S7	529 bp		Costa Rica (5)		Panama (5)													
				Mexico (6)	0.281	*2600*	−0.014	*4000*												
				Costa Rica (5)			0.091	*300*												
		CytB	750 bp		Costa Rica (5)		Panama (6)													
				Mexico (6)	−0.034	*2600*	0.076	*4000*												
				Costa Rica (5)			0.056	*300*												
Anisotremus taeniatus	Bernardi et al. (2008)*	S7	529 bp		Costa Rica (4)		Panama (7)													
				Mexico (2)	0.000	*2600*	−0.316	*4000*												
				Costa Rica (4)			−0.098	*300*												
		CytB	728 bp		Costa Rica (4)		Panama (8)													
				Mexico (2)	−0.264	*2600*	−0.214	*4000*												
				Costa Rica (4)			−0.095	*300*												
Axoclinus nigricaudus	Riginos (2005)*	dloop	206 bp		Bahia Kino (10)															
				Bahia Los Angeles (10)	**0.350**	*154*														
Coralliozetus micropes	Riginos (2005)*	dloop	206 bp		B. S. L. Gonzaga (6)		Isla Monserat (5)		Punta Chivato (5)											
				Bahia Los Angeles (9)	0.030	*260*	**0.689**	*700*	**0.730**	*260*										
				Bahia San Luis Gonzaga (6)			**0.795**	*660*	**0.838**	*460*										
				Isla Monserat (5)					−0.069	*260*										
Diodon holocanthus	Lessios and Robertson (2006)	ATPase8-6	842 bp		Panama (20)															
				Sea of Cortez (20)	−0.004	*4000*														
Epinephelus labriformis	Craig et al. (2006)	CytB	468 bp		Lower Cape (12)		Mazatlan (30)		Puerto Vallarta (49)		Oaxaca (26)		El Salvador (21)		Panama (46)					
				Loreto, Baja California (38)	−0.018	*400*	−0.001	*590*	−0.006	*800*	0.015	*1800*	0.001	*2500*	−0.011	*4000*				
				Lower Cape, Baja California (12)			−0.020	*270*	−0.013	*490*	0.017	*1500*	−0.003	*2400*	−0.004	*3800*				
				Mazatlan (30)					0.000	*300*	0.015	*1400*	0.014	*2300*	0.006	*3700*				
				Puerto Vallarta (49)							0.008	*1200*	0.014	*2000*	−0.011	*3400*				
				Oaxaca (26)									−0.017	*800*	−0.008	*2200*				
				El Salvador (21)											−0.007	*1500*				

(continued)

Table 16.4 (continued)

Organism	Reference	Locus	bp or no of loci		F_{ST}	*Distance* (km)	F_{ST}	*Distance* (km)	F_{ST}	*Distance* (km)	F_{ST}	*Distance* (km)	F_{ST}	*Distance* (km)	F_{ST}	*Distance* (km)	F_{ST}	*Distance* (km)	F_{ST}	*Distance* (km)
Hippocampus ingens	Saarman et al. (2010)	dloop	359 bp		Mexico						Ecuador				Peru					
					Mazatlan (6)		Salina Cruz (5)		Guatemala (30)		Manta/Salango (17)		Guayaquil (6)		Chiclayo (7)		Pisco (4)			
				Sea of Cortez (10)	**0.168**	*600*	**0.134**	*2000*	**0.210**	*2500*	**0.355**	*4300*	**0.263**	*4300*	**0.342**	*5200*	**0.328**	*6000*		
				Mazatlan (6)			−0.099	*1700*	−0.041	*2100*	0.041	*3900*	−0.108	*4200*	−0.024	*4700*	0.002	*5300*		
				Salina Cruz (5)					−0.030	*420*	0.008	*2300*	0.000	*2700*	0.062	*3100*	0.030	*3900*		
				Guatemala (30)							0.022	*2000*	−0.061	*2300*	0.043	*2600*	0.007	*3500*		
				Manta/Salango (17)									0.097	*330*	0.034	*770*	0.028	*1600*		
				Guayaquil (6)											−0.079	*670*	−0.022	*1500*		
				Chiclayo/Sechura (7)													−0.067	*870*		
Malacoctenus hubbsi	Riginos (2005)*	dloop	206 bp		Bahia Kino (10)															
				Bahia Los Angeles (10)	0.080	*154*														
Scarus rubroviolaceus	Fitzpatrick et al. (2011)	microsatellites	15 loci		Perlas Is. (9)															
				Panama coast (7)	0.000	*58*														

Numbers in parentheses next to locality names indicate sample sizes

Asterisks next to references indicate that F_{ST} was calculated by us from the raw data, because it was not shown in the publication. COI: Cytochrome Oxidase I, ATPase8-6: mitochondrial ATP synthetase8 and 6, CytB: Cytochrome Oxidase B, S7: ribosomal protein S7, dloop: mitochondrial control region. Values in *bold*: $p < 0.05$ as determined by 1000 reshufflings of individuals between localities

Saarman et al. (2010) reported no significant correlation between F_{ST} and distance in *Hippocampus ingens*, which would have been surprising given that sea horses are sedentary and have no larval stage. Our re-analysis of their F_{ST} data, however, shows that the expected relationship does, in fact, exist ($r = 0.441$, $0.01 < p < 0.025$). The difference between their analysis and ours is not only that they (presumably) used negative F_{ST} values, but also because some of the geographic distances they listed in their Table 4 cannot be correct because they show Peru situated north of Ecuador. Thus, the available data are consistent with a general trend of isolation by distance along the ETP coast. Such a correlation, however, is not necessarily the result of genes dispersing via stepping stones, because populations on either side of a barrier are also more distant from each other than populations on the same side of the barrier. Riginos and Nachman (2001), using partial Mantel tests, found that genetic divergence of populations of *Axoclinus nigricaudus* in the Sea of Cortez was caused not only by the distance between localities, but also by a genetic break between the upper and central parts of the Gulf of California.

In conclusion, gene flow rates among populations along the ETP coast are high, at least between central Mexico and the Panamanian coast. There is no evident genetic break resulting from the Central American Gap in any of the sampled species. Populations at the northernmost and the southernmost peripheries of the ETP appear to be genetically more isolated, possibly as the result of historical barriers, or possibly due to ecological conditions unfavorable to tropical species. A trend of isolation by distance is evident in corals and fishes.

16.4 General Conclusions and Future Prospects

The compilation of existing data regarding gene flow of coral reef organisms in the ETP suggests that gene flow within this oceanic region is generally high; this is certainly true along the coast, except perhaps for the northernmost and southernmost limits of tropical species ranges. Documented genetic connectivity is also generally high between populations at the outer ETP islands <1000 km offshore and the mainland, and (in some transpacific species) between the residents of islands in the central Pacific and ETP. These generalizations, however, need to be tempered by the realization that all coral reef organisms in which genetic structure has been sampled to date, except for sea horses, possess planktonic larvae. Large genetic differences among closely situated populations of the same morph of the ovoviviparous intertidal isopod *Excirolana braziliensis* were found in both isozymes (Lessios and Weinberg 1993, 1994) and in mtDNA (Sponer and Lessios 2009). The same may well turn out to be true for coral reef organisms with limited means of dispersal. It would be of interest to obtain data from ascidians, bryozoans, and other organisms with abbreviated larval phases to see how they compare with data from organisms with planktonic larvae. It is also important to sample genetically more species of corals, the ecological engineers responsible for creating the habitats in which other coral reef organisms live. The isolation of the ETP from other oceanic regions, the remote location of its outer islands, and the simple spatial arrangement of its coastal populations, can produce interesting contrasts in patterns of gene flow. Such data can address general population genetic theory in addition to producing information regarding the natural history of the organisms in this ocean.

Acknowledgements We thank G. Bernardi, S. Coppard, L. Geyer, A. Hiller, P.W. Glynn, and C. Riginos for comments on the manuscript.

References

Abbott DP (1966) Factors influencing the zoogeographic affinities of the Galápagos. In: Bowman RI (ed) the Galápagos. Univ California Press, Berkeley, pp 108–112

Baird AH, Guest JR, Willis BL (2009) Systematic and biogeographical patterns in the reproductive biology of scleractinian corals. Annu Rev Ecol Evol S 40:551–571

Baums IB, Boulay JN, Polato NR, Hellberg ME (2012) No gene flow across the Eastern Pacific Barrier in the reef-building coral *Porites lobata*. Mol Ecol 21:5418–5433

Bernardi G, Alva-Campbell YR, Gasparini JL, Floeter SR (2008) Molecular ecology, speciation, and evolution of the reef fish genus *Anisotremus*. Mol Phylogenet Evol 48:929–935

Bowman RI (1966) The Galápagos. Univ California Press, Berkeley, California, p 318

Briggs JC (1974) Marine zoogeography. McGraw-Hill, New York, p 475

Briggs JC, Bowen BW (2012) A realignment of marine biogeographic provinces with particular reference to fish distributions. J Biogeogr 39:12–30

Case TJ, Taper ML (2000) Interspecific competition, environmental gradients, gene flow, and the coevolution of species' borders. Am Nat 155:583–605

Coates AG, Obando JA (1996) The geologic evolution of the Central American Isthmus. In: Jackson JBC, Coates AG, Budd A (eds) Evolution and environment in tropical America. Univ Chicago Press, Chicago, pp 21–56

Collins LS (1996) Environmental changes in Caribbean shallow waters relative to the closing tropical American seaway. In: Jackson JBC, Budd AF, Coates AG (eds) Evolution and environment in tropical America. Univ Chicago Press, Chicago, pp 130–167

Combosch DJ, Guzmán HM, Schuhmacher H, Vollmer SV (2008) Interspecific hybridization and restricted trans-Pacific gene flow in the Tropical Eastern Pacific *Pocillopora*. Mol Ecol 17:1304–1312

Combosch DJ, Vollmer SV (2011) Population genetics of an ecosystem-defining reef coral *Pocillopora damicornis* in the Tropical Eastern Pacific. PLoS ONE 6(8):e21200. doi:10.1371/journal.pone.0021200

Craig MT, Hastings PA, Pondella DJ, Robertson DR, Rosales-Casian JA (2006) Phylogeography of the flag cabrilla *Epinephelus labriformis* (Serranidae): implications for the biogeography of the Tropical Eastern Pacific and the early stages of speciation in a marine shore fish. J Biogeogr 33:969–979

Craig MT, Eble JA, Bowen BW, Robertson DR (2007) High genetic connectivity across the Indian and Pacific Oceans in the reef fish *Myripristis berndti* (Holocentridae). Mar Ecol Prog Ser 334:245–254

Dana TF (1975) Development of contemporary eastern Pacific coral reefs. Mar Biol 33:355–374

Dawson MN, Grosberg RK, Stuart YE, Sanford E (2010) Population genetic analysis of a recent range expansion: mechanisms regulating the poleward range limit in the volcano barnacle *Tetraclita rubescens*. Mol Ecol 19:1585–1605

Duda TF, Lessios HA (2009) Connectivity of populations within and between major biogeographic regions of the tropical Pacific in *Conus ebraeus*, a widespread marine gastropod. Coral Reefs 28:651–659

Duda TF, Terbio M, Chen G, Phillips S, Olenzek AM, Chang D, Morris DW (2012) Patterns of population structure and historical demography of *Conus* species in the tropical Pacific. Am Malacol Bull 30:175–187

Ekman S (1953) Zoogeography of the sea. Sidgwick and Jackson Ltd, London, p 417

Emerson WK (1978) Mollusks with Indo-Pacific faunal affinities in the eastern Pacific Ocean. Nautilus 92:91–96

Emerson WK (1982) Zoogeographic implications of the occurrence of Indo-Pacific gastropods on the west American continental borderland. West Soc Malacol Ann Rep 1982:13–14

Fell FJ (1974) The echinoids of Easter Island (Rapa Nui). Pac Sci 28:147–158

Fitzpatrick JM, Carlon DB, Lippe C, Robertson DR (2011) The west Pacific diversity hotspot as a source or sink for new species? Population genetic insights from the Indo-Pacific parrotfish *Scarus rubroviolaceus*. Mol Ecol 20:219–234

Forsman ZH, Barshis DJ, Hunter CL, Toonen RJ (2009) Shape-shifting corals: molecular markers show morphology is evolutionarily plastic in *Porites*. BMC Evol Biol 9:45. doi:10.1186/1471-2148-9-45

Glynn PW (1974) The impact of *Acanthaster* on corals and coral reefs in the eastern Pacific. Environ Conserv 1:295–304

Glynn PW, Wellington GM (1983) Corals and coral reefs of the Galápagos Islands. Univ California Press, Berkeley, p 330

Glynn PW, Ault JS (2000) A biogeographic analysis and review of the far eastern Pacific coral reef region. Coral Reefs 19:1–23

Glynn PW, Gassman NJ, Eakin CM, Cortés J, Smith DB, Guzmán HM (1991) Reef coral reproduction in the eastern Pacific: Costa Rica, Panama, and Galápagos Islands (Ecuador).1. Pocilloporidae. Mar Biol 109:355–368

Glynn PW, Veron JEN, Wellington GM (1996) Clipperton Atoll (eastern Pacific): oceanography, geomorphology, reef-building coral ecology and biogeography. Coral Reefs 15:71–99

Glynn PW, Wellington GM, Riegl B, Olson DB, Borneman E, Wieters EA (2007) Diversity and biogeography of the scleractinian coral fauna of Easter Island (Rapa Nui). Pac Sci 61:67–90

Grigg RW, Hey R (1992) Paleoceanography of the tropical eastern Pacific Ocean. Science 255:172–178

Grove JS, Lavenberg RJ (1997) The fishes of the Galápagos Islands. Stanford Univ Press, Stanford, CA, p 871

Guzmán HM, Guevara CA, Breedy O (2004) Distribution, diversity, and conservation of coral reefs and coral communities in the largest marine protected area of Pacific Panama (Coiba Island). Environ Conserv 31:111–121

Haldane JBS (1956) The relation between density regulation and natural selection. Proc Royal Soc London Ser B-Biol Sci 145:306–308

Hastings PA (2000) Biogeography of the Tropical Eastern Pacific: distribution and phylogeny of chaenopsid fishes. Zool J Linn Soc 128:319–335

Haug GH, Tiedemann R (1998) Effect of the formation of the Isthmus of Panama on Atlantic Ocean thermohaline circulation. Nature 393:673–676

Heck KL, McCoy ED (1978) Long-distance dispersal and the reef-building corals of the eastern Pacific. Mar Biol 48:349–356

Hedrick PW (2005) A standardized genetic differentiation measure. Evolution 59:1633–1638

Hellberg ME (2006) No variation and low synonymous substitution rates in coral mtDNA despite high nuclear variation. BMC Evolutionary Biology 6:24. doi:10.1186/1471-2148-6-24

Hellberg ME (2007) Footprints on water: the genetic wake of dispersal among reefs. Coral Reefs 26:463–473

Hellberg ME (2009) Gene flow and isolation among populations of marine animals. Annu Rev Ecol Evol S 40:291–310

Hellberg ME, Burton RS, Neigel JE, Palumbi SR (2002) Genetic assessment of connectivity among marine populations. Bull Mar Sci 70:273–290

Hertlein LG (1963) Contribution to the biogeography of Cocos Island, including a bibliography. Proc Calif Acad Sci 32:219–289

Hey J, Nielsen R (2004) Multilocus methods for estimating population sizes, migration rates and divergence time, with applications to the divergence of *Drosophila pseudoobscura* and *D. persimilis*. Genetics 167:747–760

Hudson RR, Slatkin M, Maddison WP (1992) Estimation of levels of gene flow from DNA sequence data. Genetics 132:583–589

James MJ (1991) Galápagos marine invertebrates. Taxonomy, biogeography, and evolution in Darwin's islands. Plenum, New York, p 474

Jones GP, Milicich MJ, Emslie MJ, Lunow C (1999) Self-recruitment in a coral reef fish population. Nature 402:802–804

Johnson MS, Black R (1984) Pattern beneath the chaos: the effects of recruitment on genetic patchiness in an intertidal limpet. Evolution 38:1371–1383

Keigwin LD (1982) Isotopic paleoceanography of the Caribbean and east Pacific: role of Panama uplift in Late Neogene time. Science 217:350–353

Kessler WS (2006) The circulation of the eastern tropical Pacific: a review. Prog Oceanogr 69:181–217

Ketchum JT, Bonilla HR (2001) Taxonomy and distribution of the hermatypic corals (Scleractinia) of the Revillagigedo Archipelago, Mexico. Rev Biol Trop 49:803–848

Lessios HA (2008) The great American schism: divergence of marine organisms after the rise of the Central American Isthmus. Annu Rev Ecol Evol S 39:63–91

Lessios HA, Weinberg JR (1993) Migration, gene flow and reproductive isolation between and within morphotypes of the isopod *Excirolana* in two oceans. Heredity 71:561–573

Lessios HA, Weinberg JR (1994) Genetic and morphological divergence among morphotypes of the isopod *Excirolana* on the two sides of the Isthmus of Panama. Evolution 48:530–548

Lessios HA, Robertson DR (2006) Crossing the impassable: genetic connections in 20 reef fishes across the eastern Pacific barrier. Proc R Soc B-Biol Sci 273:2201–2208

Lessios HA, Kessing BD, Wellington GM, Graybeal A (1996) Indo-Pacific echinoids in the tropical eastern Pacific. Coral Reefs 15:133–142

Lessios HA, Kessing BD, Robertson DR (1998) Massive gene flow across the world's most potent marine biogeographic barrier. Proc R Soc Lond Ser B 265:583–588

Lessios HA, Kessing BD, Robertson DR, Paulay G (1999) Phylogeography of the pantropical sea urchin *Eucidaris* in relation to land barriers and ocean currents. Evolution 53:806–817

Lessios HA, Kessing BD, Pearse JS (2001) Population structure and speciation in tropical seas: global phylogeography of the sea urchin *Diadema*. Evolution 55:955–975
Lessios HA, Kane J, Robertson DR (2003) Phylogeography of the pantropical sea urchin *Tripneustes*: contrasting patterns of population structure between oceans. Evolution 57:2026–2036
Lessios HA, Lockhart S, Collin R, Sotil G, Sanchez-Jerez P, Zigler KS, Perez AF, Garrido MJ, Geyer LB, Bernardi G, Vacquier VD, Haroun R, Kessing BD (2012) Phylogeography and bindin evolution in *Arbacia*, a sea urchin genus with an unusual distribution. Mol Ecol 21:130–144
Madsen FJ (1955) A note on the sea star genus *Acanthaster*. Vidensk Medd Dansk Naturhist Foren Kbh 117:179–192
Mantel N (1967) The detection of disease clustering and the generalized regression approach. Cancer Res 27:209–220
Marko PB, Hart MW (2012) Retrospective coalescent methods and the reconstruction of metapopulation histories in the sea. Evol Ecol 26:291–315
McCartney MA, Keller G, Lessios HA (2000) Dispersal barriers in tropical oceans and speciation of Atlantic and eastern Pacific *Echinometra* sea urchins. Mol Ecol 9:1391–1400
McCoy ED, Heck KL Jr (1976) Biogeography of corals, sea grasses, and mangroves: an alternative to the center of origin concept. Syst Zool 25:201–210
Montes C, Cardona A, McFadden R, Moron SE, Silva CA, Restrepo-Moreno S, Ramirez DA, Hoyos N, Wilson J, Farris D, Bayona GA, Jaramillo CA, Valencia V, Bryan J, Flores JA (2012) Evidence for middle Eocene and younger land emergence in central Panama: implications for Isthmus closure. Geol Soc Am Bull 124:780–799
Muss A, Robertson DR, Stepien CA, Wirtz P, Bowen BW (2001) Phylogeography of *Ophioblennius*: the role of ocean currents and geography in reef fish evolution. Evolution 55:561–572
Nishida M, Lucas JS (1988) Genetic differences between geographic populations of the crown-of-thorns starfish throughout the Pacific region. Mar Biol 98:359–368
O'Dea A, Jackson JBC, Fortunato H, Smith JT, D'Croz L, Johnson KG, Todd JA (2007) Environmental change preceded Caribbean extinction by 2 million years. Proc Natl Acad Sci USA 104:5501–5506
Pinzón JH, LaJeunesse TC (2011) Species delimitation of common reef corals in the genus *Pocillopora* using nucleotide sequence phylogenies, population genetics and symbiosis ecology. Mol Ecol 20:311–325
Pluzhnikov A, Donnelly P (1996) Optimal sequencing strategies for surveying molecular genetic diversity. Genetics 144:1247–1262
Pritchard JK, Stephens M, Donnelly P (2000) Inference of population structure using multilocus genotype data. Genetics 155:945–959
Randall JE (1998) Zoogeography of shore fishes of the Indo-Pacific region. Zool Stud 37:227–268
Rehder HA (1980) The marine mollusks of Easter Island (Isla de Pascua) and Sala y Gómez. Smithson Contrib Zool 289:1–167
Reyes-Bonilla H, Barraza JE (2003) Corals and associated marine communities from El Salvador. In: Cortés J (ed) Latin American coral reefs. Elsevier, Amsterdam, pp 351–360
Richmond RH (1990) The effects of the El Niño/Southern Oscillation on the dispersal of corals and other marine organisms. In: Glynn PW (ed) Global ecological consequences of the 1982-83 El Niño-Southern Oscillation. Elsevier, Amsterdam, pp 127–140
Riginos C (2005) Cryptic vicariance in Gulf of California fishes parallels vicariant patterns found in Baja California mammals and reptiles. Evolution 59:2678–2690
Riginos C, Nachman MW (2001) Population subdivision in marine environments: the contributions of biogeography, geographical distance and discontinuous habitat to genetic differentiation in a blennioid fish, *Axoclinus nigricaudus*. Mol Ecol 10:1439–1453
Robertson DR (1998) Do coral-reef fish faunas have a distinctive taxonomic structure? Coral Reefs 17:179–186
Robertson DR, Allen GR (1996) Zoogeography of the shorefish fauna of Clipperton Atoll. Coral Reefs 15:121–131
Robertson DR, Cramer KL (2009) Shore fishes and biogeographic subdivisions of the Tropical Eastern Pacific. Mar Ecol Prog Ser 380:1–17
Robertson DR, Grove JS, McCosker JE (2004) Tropical transpacific shore fishes. Pac Sci 58:507–565
Rosenblatt RH, Waples RS (1986) A genetic comparison of allopatric populations of shore fish species from the eastern and central Pacific Ocean: dispersal or vicariance? Copeia 1986:275–284
Saarman NP, Louie KD, Hamilton H (2010) Genetic differentiation across eastern Pacific oceanographic barriers in the threatened seahorse *Hippocampus ingens*. Conserv Genet 11:1989–2000
Sexton JP, McIntyre PJ, Angert AL, Rice KJ (2009) Evolution and ecology of species range limits. Annu Rev Ecol Evol S 40:415–436
Shearer TL, Van Oppen MJH, Romano SL, Worheide G (2002) Slow mitochondrial DNA sequence evolution in the Anthozoa (Cnidaria). Mol Ecol 11:2475–2487
Slatkin M (1995) A measure of population subdivision based on microsatellite allele frequencies. Genetics 139:457–462
Sponer R, Lessios HA (2009) Mitochondrial phylogeography of the intertidal isopod *Excirolana braziliensis* on the two sides of the Isthmus of Panama. Smith Contr Mar Sci 38:219–228
Stepien CA, Rosenblatt RH, Bargmeyer BA (2001) Phylogeography of the spotted sand bass, *Paralabrax maculatofasciatus*: divergence of Gulf of California and Pacific Coast populations. Evolution 55:1852–1862
Swearer SE, Caselle JE, Lea DW, Warner RR (1999) Larval retention and recruitment in an island population of a coral-reef fish. Nature 402:799–802
Swearer SE, Shima JS, Hellberg ME, Thorrold SR, Jones GP, Robertson DR, Morgan SG, Selkoe KA, Ruiz GM, Warner RR (2002) Evidence of self-recruitment in demersal marine populations. Bull Mar Sci 70:251–271
Taylor MS, Hellberg ME (2003) Genetic evidence for local retention of pelagic larvae in a Caribbean reef fish. Science 299:107–109
Todd PA (2008) Morphological plasticity in scleractinian corals. Biol Rev 83:315–337
Vermeij GJ (1978) Biogeography and adaptation. Harvard Univ Press, Cambridge, Mass, p 332
Veron JEN (1995) Corals in space and time: the biogeography and evolution of the Scleractinia. Ithaca, Comstock/Cornell, p 321
Vogler C, Benzie J, Lessios H, Barber P, Worheide G (2008) A threat to coral reefs multiplied? Four species of Crown-of-thorns Starfish. Biology Lett 4:696–699
Webb SD (1976) Mammalian faunal dynamics of the great American interchange. Paleobiology 2:220–234
Whitlock MC, McCauley DE (1999) Indirect measures of gene flow and migration: $F_{ST} \neq 1/(4Nm+1)$. Heredity 82:117–125
Willis BL, van Oppen MJH, Miller DJ, Vollmer SV, Ayre DJ (2006) The role of hybridization in the evolution of reef corals. Annu Rev Ecol Evol S 37:489–517
Wright S (1943) Isolation by distance. Genetics 28:114–138
Wright S (1951) The genetic structure of populations. Ann Eugen 15:323–354
Yakovleva IM, Baird AH, Yamamoto HH, Bhagooli R, Nonaka M, Hidaka M (2009) Algal symbionts increase oxidative damage and death in coral larvae at high temperatures. Mar Ecol Prog Ser 378:105–112
Zigler KS, Lessios HA (2003) Evolution of bindin in the pantropical sea urchin *Tripneustes*: comparisons to bindin of other genera. Mol Biol Evol 20:220–231

Thermal Refuges and Refugia for Stony Corals in the Eastern Tropical Pacific

17

Tyler B. Smith, Juan L. Maté, and Joanna Gyory

Abstract

Refugia could provide the essential conditions for coral and coral reef persistence in the marginal eastern tropical Pacific Ocean (ETP) by allowing corals to avoid extinction after disturbance events and by maintaining populations that can rapidly recolonize disturbed habitats. The low diversity and restricted ranges of ETP corals make them vulnerable to seawater warming caused by positive phases of the El Niño-Southern Oscillation (El Niño). However, at spatial scales ranging from individual colonies (<10 m) to regions, (>100 km) there are areas or habitats that avoid stress or support corals that are more resistant to stress or with the capacity to recover rapidly. These habitats may form a network of refugia that ensures coral persistence through severe disturbance. This chapter explores evidence for the existence of refugia at multiple spatial and temporal scales across the ETP, with recent examples showing refugia may be common and perhaps even necessary for many species to survive the highly fluctuating environment in this eastern Pacific tropical ocean basin.

Keywords

El Niño-Southern Oscillation • Coral bleaching • Extinction • Resistance • Recovery

17.1 The Concept of Refuges and Refugia

Refuges and refugia are environments that are buffered from acute or chronic disturbance and stress that negatively affect organisms. These environments promote survival of species in an ecosystem that has otherwise become hostile. In this chapter we will follow the convention of Keppel et al. (2012) who distinguished refuges and refugia based on spatial and temporal scales (summarized in Fig. 17.1). At smaller scales, **refuges** (singular: refuge) provide protection from disturbances in three-dimensional space ranging from centimeters to hectares and across time on scales ranging from minutes to years. Refuges can also provide advantages in biotic interactions, such as competition or predation. For example, an associational refuge might deter herbivory on a palatable alga by its close spatial association with a chemically defended plant species. Thus, refuges are related to physiological and ecological spatiotemporal scales. **Refugia** (singular: refugium), however, are "habitats that components of biodiversity retreat to, persist in and can potentially expand from under changing environmental conditions" (Keppel et al. 2012). Thus, refugia are associated with larger scales of kilometers and millennia, and are related to evolutionary spatiotemporal scales.

There are three core processes that control the formation and importance of refuges and refugia, (1) reduction or

T.B. Smith (✉) · J. Gyory
Center for Marine and Environmental Studies, University of the Virgin Islands, #2 John Brewers Bay, St. Thomas, US Virgin Islands, 00802, USA
e-mail: tsmith@uvi.edu

J. Gyory
e-mail: tubularia@gmail.com

J.L. Maté
Smithsonian Tropical Research Institute, Apartado 0843-03092Balboa, Ancón, Republic of Panamá
e-mail: matej@si.edu

J. Gyory
Department of Ecology and Evolutionary Biology, Tulane University, New Orleans, LA 70118, USA

P.W. Glynn et al. (eds.), *Coral Reefs of the Eastern Tropical Pacific*, Coral Reefs of the World 8, DOI 10.1007/978-94-017-7499-4_17

absence of stressful conditions that lead to mortality (**avoidance**), (2) the promotion of conditions that favour **resistance** to disturbance, and (3) enhancement of conditions that lead to faster and more complete **recovery** before the next disturbance. The localized reduction of stressful conditions is the most obvious factor in forming refugia, since an area of space or period of time where the stress does not exist would be de facto a refuge. However, for a refuge to scale to a refugium the avoidance must be "reliable" (West and Salm 2003) and occur frequently enough with a given disturbance to support the persistence of a species. What is less obvious is that there exist conditions that enhance biological resistance to disturbances. An example would be two reef zones where mean thermal stress is equal, yet one area exhibits higher background temperature variability that acclimatizes corals to periods of high temperature stress and thus conditions corals to be more resistant to high mean water temperatures (McClanahan et al. 2007; Davis et al. 2011; Guest et al. 2012). Lastly, enhancement of resistance to pulse disturbances, such as thermal stress, can facilitate robust populations to face the next disturbance, as they may have undergone selection for more resistant coral host and endosymbiont *Symbiodinium* genotypes that might influence greater future resistance (Baker et al. 2004; D'Croz and Maté 2004; Guest et al. 2012; see Chap. 13, Baker et al.). Only resistance, not recovery, is possible for continuous press disturbances (Bender et al. 1984), such as ocean acidification. In press disturbances there is no period during which the population is released from constant pressure to decline in abundance. Therefore, the creation of refugia under press disturbance may demand a greater role of the first two core processes of avoidance and promotion of conditions that enhance resistance.

Refugia for corals are important for preventing extinction and increasing resilience. In the purest sense refugia allow species to avoid local, regional, or global extinction. The focus on extinctions and global scales may be why refugia have received relatively little attention in marine science, since the threat of global extinction for marine species with open populations is widely considered to be comparatively low relative to aquatic and terrestrial species (McKinney 1997). However, we also take a more general view of refuges as demographic buffers that can promote species-level resilience and, perhaps, ecological resilience (Holling 1973) by protecting local, regional, and global sexual and asexual reproductive pools. This view may be more generally applicable to refuges in coral reef regions across the tropics where extinction is not likely (Hughes et al. 2014).

The ETP may be especially reliant on refuges for long-term persistence of coral species and coral ecosystems. Extreme environmental heterogeneity in the ETP ranges on scales from colonies to reefs, habitats, islands, and regions

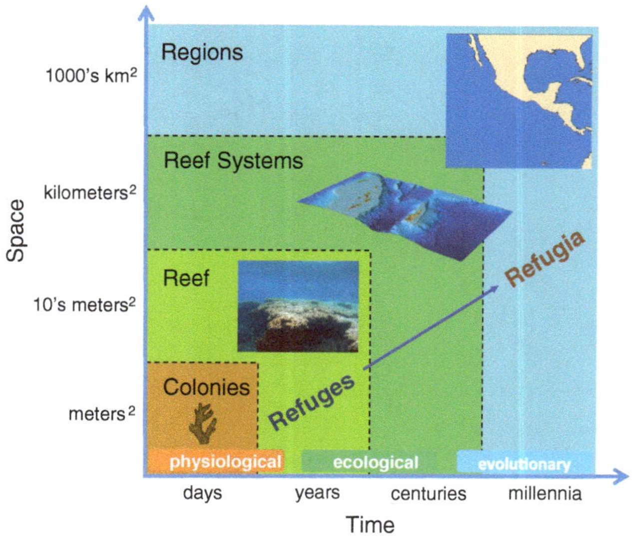

Fig. 17.1 Spatiotemporal scale across which refuges scale to refugia

(Fig. 17.1). While this is certainly not unique for coral reefs anywhere, the ETP is different from most regions due to the presence of mesoscale upwelling, cold currents that impinge from high latitudes, and a shallow upper mixed layer (10–30 m depth). The mixed layer here is defined as the depth from the surface to the depth of active mixing due to turbulence caused largely by local wind forcing, and has been formalized as a change of temperature of 0.2 °C from a near surface value (de Boyer Montégut et al. 2004). In fact, the ETP has some of the shallowest and strongest upper pycnoclines (vertical density gradients) globally, largely because of the shallow thermocline (Fiedler and Talley 2006; see Chap. 3, Fiedler and Lavín). The shallow thermocline and thin upper mixed layer bring cool, nutrient rich water close to the surface and increase the frequency and intensity of upwelling at multiple scales (within reef to regional domains) (D'Croz and O'Dea 2007). Other oceanographic factors associated with the thin upper mixed layer, such as salinity, water column productivity (e.g., concentration and turnover of phytoplankton, zooplankton, and particulate organic matter), and dissolved oxygen create the potential for a mosaic of conditions over relatively small horizontal and vertical areas. Within this diversity of environmental conditions lie multiple refuges that are buffered from acute and chronic disturbance or precondition coral resistance to stress and disturbance.

In our view, refuges in the past have been critical in controlling the modern fauna of ETP coral reefs. The coral fauna of the ETP is more vulnerable to extinctions than most of the other coral reef-supporting areas of the globe because populations are regionally separated from the central and western Pacific Ocean across the Eastern Pacific Barrier (Baums et al. 2012; see Chaps. 5 and 16, Glynn et al., and Lessios and Baums respectively) and within the ETP are sparse and widely separated. Many coastal and oceanic habitats in the ETP are unfavorable for reef and coral growth, restricting available habitat and ranges of occurrence. Examples include, the sedimentary dominated systems of the Pacific Central American Faunal Gap (Springer 1958; Reyes Bonilla and Barraza 2003), areas with a high concentration of dissolved carbon dioxide (Manzello 2010), hard-bottom habitats below the regionally shallow thermocline, and areas of intense upwelling (Glynn 2003). In addition, there is a relatively high degree of endemism within the scleractinian fauna of the ETP (Glynn and Ault 2000). Thus, restricted environments for coral growth could increase the risk of global extinction for ETP endemics and regional extinction among pan Indo-Pacific species in the ETP (Glynn 2011; Harnick et al. 2012). Both prospects reduce ecological redundancy and can render the ecosystem more susceptible to disturbance (McCann 2000; Loreau 2010). In addition, even the regional loss of a species can reduce genetic diversity and the potential for speciation. This may be particularly important in the ETP, where evidence suggests that species have low gene flow with the western Pacific across the Eastern Pacific Barrier, a condition that fosters endemic genotypes (Baums et al. 2012).

In this chapter we focus primarily on refuges and refugia from thermal disturbance, since coral habitats in the ETP have been shown to be particularly susceptible to high and low temperature extremes (Glynn et al. 2001; Glynn and Fong 2006; see Chap. 8, Glynn et al.). However, this discussion could apply equally to other stressors, such as ocean acidity, that are dealt with fully in other chapters (see Chap. 18, Manzello et al.) and will be referenced only briefly in this chapter. In addition, because refuges and refugia are largely ecological and biogeographic topics that are intimately concerned with space and time, we organize our chapter along spatiotemporal scales (low to high). However, we use spatial scales to divide our sections.

17.2 Intra-colony Refuges (Scales Less Than a Meter)

Within a single coral colony there may exist areas with differential resistance to bleaching-related mortality during thermal stress events. Such resistance favours partial colony mortality over whole colony mortality, which can be critical in preserving extant genotypes and jump-starting recovery. Many coral colonies can have morphologies that create heterogeneous environmental conditions, or microhabitats, that respond differently to stressors such as seawater warming (Brown et al. 2002). For example, because light is frequently a co-factor with thermal stress in initiating the bleaching response, factors that reduce light can reduce the severity of bleaching (Lesser 1997). Intra-colony shading is often influenced by colony morphology. Massive, branching, foliose, and encrusting (following substrate contour) colony morphologies often possess shaded surfaces where light stress is reduced. The empirical response of many ETP corals to bleaching (Glynn et al. 2001) shows that shaded areas can confer some resistance to bleaching, even though shaded surfaces may contain different endosymbiont clades that might be more sensitive to thermal stress even at reduced light levels (Baker et al. 2008). In addition, mosaic patterns of bleaching and mortality often occur even when there is no corresponding intra-colony microhabitat identified (Glynn et al. 2001). This suggests differences in symbiont type within colonies, with differing susceptibilities to heat stress (Rowan et al. 1997). Heterogeneous water velocity around corals may also create intra-colony thermal refuges. Higher water velocities around colony surfaces can reduce the height of the free boundary layer that limits diffusion and advective supply, therefore increasing the stripping of oxygen radicals (Nakamura and van Woesik 2001) or increasing the supply of heterotrophic food sources (Sebens et al. 1998). One would expect that more exposed colony surfaces would benefit from increased water velocity, but these areas may also be more exposed to increased light, making the benefit of increased water flow at the colony scale less clear.

Microhabitat refuges, and related partial instead of total colony mortality, can be a potent mechanism leading to recovery of corals after disturbance. Rapid recovery of surviving tissues has been seen in many ETP coral species after massive thermal mortality events (Glynn et al. 2001; Glynn and Fong 2006). In 1997–98 species such as *Gardineroseris planulata*, *Pavona varians*, *Pavona clavus*, *Porites lobata* or *Porites evermanni*, all showed intra-colony variation in bleaching and mortality that led to partial rather than whole colony mortality. In the case of *G. planulata*, upper colony surfaces suffered severe bleaching and mortality, whilst colony sides bleached but survived (Fig. 17.2). In addition, poritid corals with deep tissue reserves in their porous skeletons have been known to be resistant to whole colony mortality as these tissues can survive bleaching events and re-emerge to colony surfaces (Baker et al. 2008; van Woesik et al. 2012a). Furthermore, broadcasting zooxanthellate corals in the ETP, which includes all scleractinian species except *Porites panamensis*, can rapidly recolonize areas through sexual reproduction after thermal disturbances, suggesting reproductive mode may play a role in resilience (Glynn and Colley 2008).

Thus, species-specific traits determine the impact of changing climatic conditions upon coral survival (Loya et al. 2001; van Woesik et al. 2011; Smith et al. 2013) and hence their long-term survival and ability to recover (Darling et al. 2012; van Woesik et al. 2012a). Understanding traits that

Fig. 17.2 Colonies of *Gardineroseris planulata* showing characteristic pattern of upper surface mortality caused by top bleaching during 1997–98 El Niño. Sides of colonies served as intra-colony refuges that allowed colony survival. Recovery is occurring through regrowth onto damaged colony tops (Uva Is Panama, September 16, 2002; ~6 m depth; credit: T Smith)

determine intra-colony refuges will improve predictions of coral community response to future thermal disturbances.

17.3 Intra-reef and Inter-reef Refuges (Scales of 10–1000s of Meters)

The effects of thermal stress at reef and inter-reef scales can be heterogeneous due to interacting bio-physical processes that enhance the three core processes related to avoidance, resistance, and recovery. True reef frameworks include biologically produced carbonate build-ups elevated above the bedrock. Here we use the term "reef" to include coral communities growing on bedrock or incipient carbonate frameworks, since these habitats are likely to surpass the numbers and areal occurrence of true reef frameworks in the ETP (Guzmán et al. 2004). Intra-reef refuges include vertical depth refuges that have unique features and are treated separately in the following section.

As with coral colonies, the reef is a mosaic of heterogeneous physical conditions, some with the ability to change coral exposure or response to disturbance and stress. Also, similar to the scale of colonies, factors that increase the variance of temperature and light conditions (see below) will tend to create more persistent refuges with a greater potential to form refugia, whilst biological interactions that favour recovery, such as decreased competition for substrate, may be inconsistent from one disturbance to the next. The various mechanisms suggested in this section are not unique features of ETP reefs, so this section applies more broadly to intra-reef and inter-reef refugia globally.

Temperature means, maxima, and diel variation can differ greatly within and between reefs on scales of tens of meters to tens of kilometers (Davis et al. 2011; Gorospe and Karl 2011), and during thermal stress events this could lead to both areas of avoidance and areas where tolerance may be increased (Pineda et al. 2013). Areas of avoidance could occur where temperatures are normally lower within and between reefs, such as microhabitats subject to vertical mixing; and the flushing of fore-reef zones or near areas with tidally induced pumping of deep cool water and other localized sources of upwelling (Glynn and Stewart 1973; Riegl and Piller 2003). However, mechanisms that cool these areas would need to be persistent during thermal stress events and maintain temperatures sufficiently below bleaching thresholds to prevent thermal stress. During regional sea warming events that induce bleaching these areas may lose their benign characteristics and experience the regional warming, either because they are exposed to the open ocean where temperatures are elevated or because the thermocline is depressed. Furthermore, in order for areas of thermal stress avoidance to provide long-term refugia at the intra- and inter-reef scale, processes that cause cooling would need to be "reliable" or long-lasting over multiple stress events (West and Salm 2003). For example, Pineda et al. (2013) suggested that a cold-core eddy lowered sea surface temperatures and protected exposed outer shelf reefs in a section of the Red Sea in 2010, providing a spatially explicit refuge from thermal stress. But, cold-core eddies are not normally permanent features and may not always be present during thermal stress events, in which case there would not be a long-term refugium. Lastly, since corals in cooler areas occupy mild thermal environments, they may not have been previously exposed to elevated temperatures, and therefore may show less tolerance (Oliver and Palumbi 2011). Thus, high thermal stress avoidance mechanisms at the reef and inter-reef levels may be relatively rare.

More reliable features in shallow water that would promote avoidance of bleaching conditions and provide potential refugia might be microhabitats shaded by permanent structures and areas subject to strong currents (West and Salm 2003). Light can vary within and among coral reefs due to shading from reef structures, such as erect macroalgae or coral canopies, and nearby terrain such as cliffs and overhanging structures. In addition, water column light attenuation (e.g., turbidity) and cloudiness (Glynn 1996; Brown 1997; Mumby et al. 2002) can reduce irradiance. West and Salm (2003) suggested that shading by subtidal structures and adjacent terrain might be a highly reliable factor that reduces bleaching, whereas turbidity and cloud cover may have low reliability. Water flow that flushes toxins or transports nutrients and food may also differ in its reliability; permanent features, such as tidal and ocean currents, are highly reliable, whereas transient and variable features, such as wave action, tidal range, and wind-driven currents, will be less reliable.

Although long-term avoidance from thermal stress might be rare on the intra- and inter-reef scale, processes that increase tolerance to warming may promote refuges and refugia. For example, slowing of water flows and ponding of water on reef flats, back reef lagoons, and embayments can lead to greater diel warming of water masses interspersed with tidally induced flushing and cooling (Davis et al. 2011; Pineda et al. 2013). This leads to greater variability in daily temperatures, consistent transient exposure to water temperatures greater than the surrounding ocean, and a higher mean temperature. Experimental and empirical evidence show that corals exposed to greater mean, maximum and variable temperatures can be more resistant to bleaching or bleaching-related mortality during heat stress (McClanahan et al. 2007; Bellantuono et al. 2011; Oliver and Palumbi 2011; van Woesik et al. 2012b), and at the inter-reef scale corals exposed to greater temperature variability are more resistant to declining growth during and after heat stress (Castillo et al. 2012). The cause of this effect is not known, but higher temperatures experienced by corals in warmer or variable environments may "frontload" genes for tolerance to high thermal stress (Barshis et al. 2013), with genetic responses that correspond to acclimation and adaptation (Palumbi et al. 2014), or cause shifts in symbiont types (Baker et al. 2008).

Reef flats and back reefs in the ETP may act as refuges from the worst effects of thermally induced coral bleaching mortality. Reef flats experience higher light and more variable temperatures, with higher mean and maximum/minimum values, than surrounding reef areas. These physical regimes might precondition corals for tolerance to conditions that cause bleaching (Fig. 17.3; Davis et al. 2011), potentially forming a refugium for coral species that can develop increased tolerance. Similarly, back reef zones with lagoons or impounded water experience restricted flow that allows greater diel heating and may precondition corals to resist thermal stress (van Woesik et al. 2012b). In addition, lagoons backed by land are often made turbid due to run-off and accumulation of particulate matter (sediments and particulate organic matter) and nutrient enrichment of phytoplankton communities. Turbidity was increased by about 0.2 NTUs in the lagoon at the Uva Island study reef during one of our surveys in 2010, although phytoplankton enrichment (i.e., higher chlorophyll concentration) was not observed during this period (Fig. 17.4).

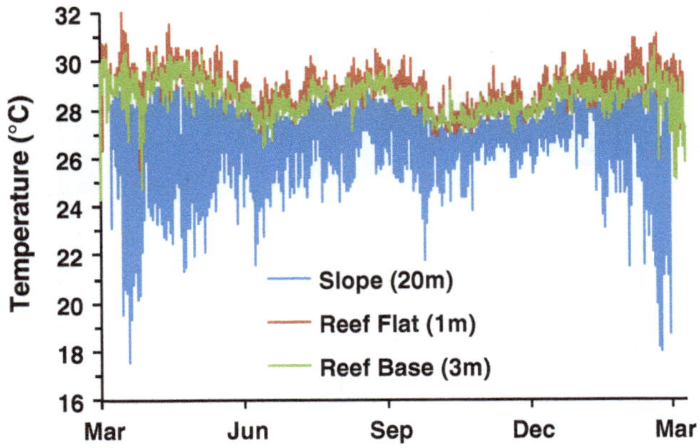

Fig. 17.3 Benthic temperature records at three reef habitats/depths at Uva Is reef, Panama from January 2003 to April 2004. Methods as in Smith et al. (2014)

17.4 Depth Refugia (Vertical Scales of 10s of Meters)

Near-surface conditions in the ETP have the potential to create relatively shallow depth refuges and refugia compared with other tropical regions (Glynn 1996; Riegl and Piller 2003; Bongaerts et al. 2010; Smith et al. 2014). Because in most locations in the ETP the oceanographic conditions limit the formation of true light-dependent deep coral formations (i.e., mesophotic coral reefs; Hinderstein et al. 2010), in this chapter we refer to deeper or deep corals as any colonies deeper than shallow water coral or coral reef formations, typically >5–12 m. In areas without large-scale upwelling, the upper mixed layer depth, from the sea surface to the top of the first thermocline, is between 10 and 30 m (Table 17.1, Fig. 17.4). The typically thin upper mixed layer in the ETP has the potential to lower temperatures at modest depths compared with western ocean basins in the Pacific and Atlantic, increasing the chance to avoid thermal stress. In addition, avoidance of bleaching with depth may be enhanced by light attenuation, which can be substantial in many areas of the ETP because of productive and turbid waters. Furthermore, a prominent and dense chlorophyll maximum layer with high abundance of phytoplankton is associated with the bottom of the upper mixed layer (D'Croz and O'Dea 2007) where vital plant nutrients sequestered below the pycnocline in deeper waters are exposed in the photic zone. The tropical chlorophyll maximum layer may be associated with increased pelagic food (Leichter and

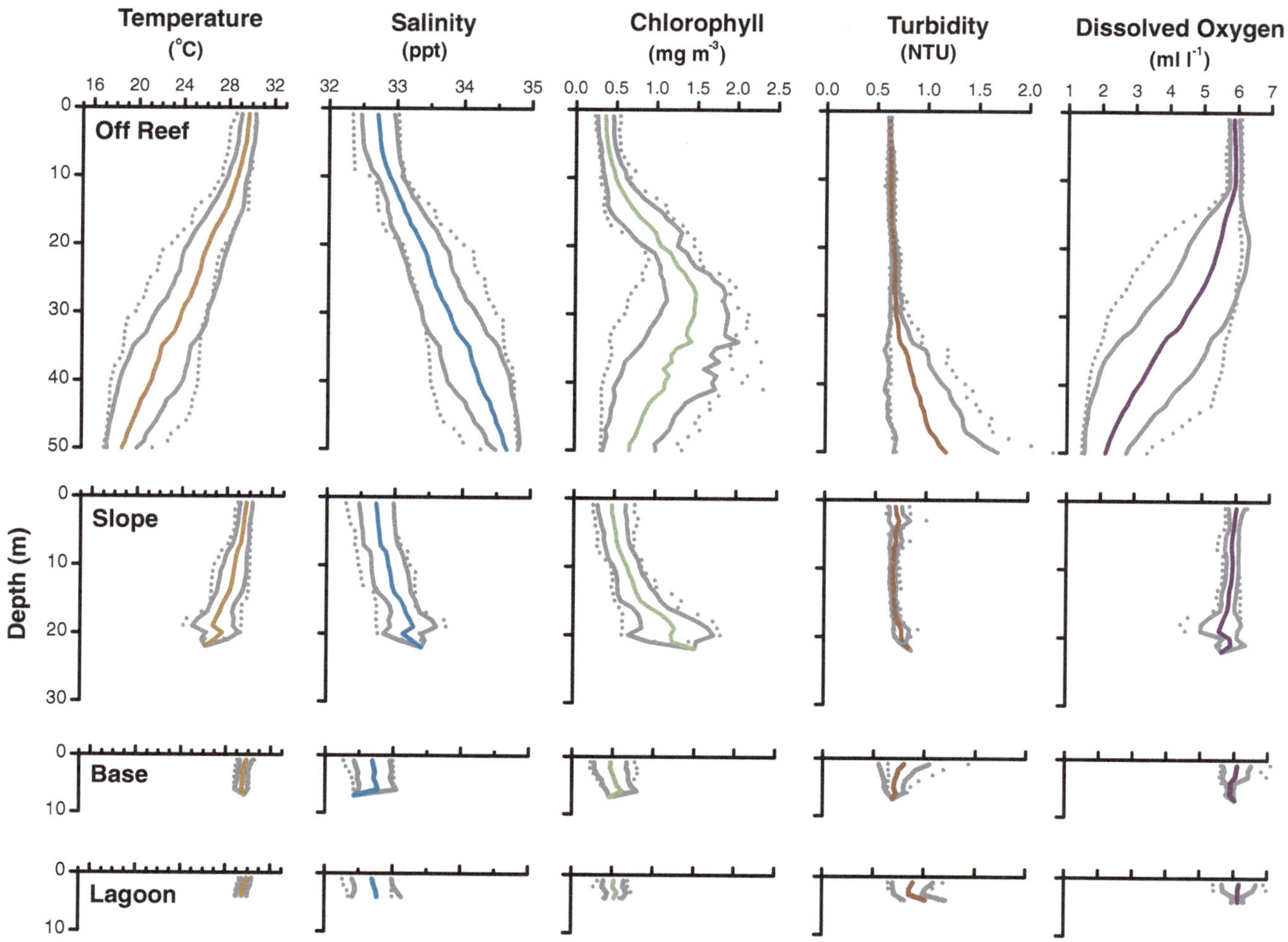

Fig. 17.4 Vertical and horizontal structure of water column bio-physical properties of Uva Island reef system, Gulf of Chiriquí, Republic of Panama, taken from oceanographic CTD casts. Locations of casts are from top to bottom, off-reef (n = 11), reef slope (10), reef base (9), and backreef lagoon (8). Data are means, standard deviation and min/max from 1 m depth bins. Data recorded from March 14 to 22, 2010 with a Seabird 25 sensor package; methods in Smith et al. (2014)

Genovese 2006) that could provide heterotrophic subsidy for bleached corals and improve recovery (Grottoli et al. 2006).

Suitable substrata for the formation of coral reefs is limited in many parts of the ETP. With the exception of oceanic islands, ETP carbonate reef build-ups are typically shallow structures with frameworks forming in less than 10 m depth (Glynn 1976, 2000). Thus, depth refuges in neritic reefs below 10 m will mostly consist of coral communities formed of isolated colonies growing on bedrock or coralliths on unconsolidated substrates. Since coral communities may cover larger areas and host more total hard coral populations than true structural coral reefs (Guzmán et al. 2004), they may be particularly important in determining where refuges or refugia form.

In addition, cooler conditions with depth during periods between El Niño events are necessary but not sufficient evidence to demonstrate that these habitats are refuges or refugia. The thermocline is often deepened by 5–10 m during El Niño conditions in the ETP (Enfield 2001; Fiedler and Talley 2006), which has the potential to erase any thermal buffering. Thus, there may be variability in the cooling provided in depth refuges over El Niño conditions, both from location to location and within a single location over multiple thermal stress events (i.e., reliability), and this may limit the potential for depth refugia. For example, the handful of studies that presented data on temperature with depth during the 1997–98 El Niño in the eastern Pacific showed temperature was isothermal or up to 3.0 °C cooler at 7–30 m depth [Champion Island, Galápagos (Wellington et al. 2001); Devil's Crown, Galápagos (Feingold 2001); Bahía Culebra, Costa Rica (Jiménez et al. 2001)]. An understanding of the processes that control the upper ocean vertical temperature structure during thermal stress events in different regions is lacking both empirical observations and mechanistic studies, and is an area of needed investigation.

Table 17.1 Locations and characteristics of coral community and coral reef environments in eastern tropical Pacific Ocean and their potential to support deep refugia populations

Country	Location	Conditions	Mixed layer depth (m)	Deeper coral species	Max. depth (m)	Source
Chile	Easter Is.	Oligotrophic	80?	Cyva, Lesc, Leso, Poda, Pove, Polo	>50	Glynn et al. (2003); Hubbard and Garcia (2003)
Colombia	Island: Gorgona Is.	Mesotrophic-Eutrophic (upwelling)	7–40	Pagi, Polo, Psspp, Pava	<20	Zapata and Vargas-Ángel (2003); Giraldo et al. (2008); K. Mejía, C. Muñoz, and F. Zapata, pers. comm.
Colombia	Malpelo Is.	Oligotrophic	30	Polo, Poca, Pacl, Pava, Gapl,	>30	Birkeland et al. (1975); Zapata and Vargas-Ángel (2003)
Costa Rica	Island: Caño Is.	Oligotrophic	20*	Polo, Psspp, Poel, Pagi	<22	Guzmán and Cortéz (1989)
Costa Rica	Coastal: Bahía Culebra, Golfo de Papagayo	Eutrophic (upwelling)	0–20	Lepa, Pacl, Pagi, Pava, Poel, Polo, Popa	<20	Jiménez et al. (2001), Cortés and Jiménez (2003)
Costa Rica	Coastal: Bahía Culebra – Pta. Gorda, Península de Nicoya	Mesotrophic	20*	Lepa, Pacl, Pagi, Poel Polo, Popa, Pava	<20	Cortés and Jiménez (2003)
Costa Rica	Cocos Is.	Oligotrophic	13–30	Lesc, Pala, Psam, Gplan, Pvspp	>35	Bakus (1975); Guzmán and Cortés (2007), J. Cortés pers. comm.
Ecuador	Galápagos Iss. North	Oligotrophic-Mesotrophic	15–30	Cyel, Pagi, Pava, Polo, Pacl	>40	Glynn and Wellington (1983); Banks et al. (2009); T. Smith, unpub. data
Ecuador	Galápagos Iss. West: Deep populations not known	Eutrophic (upwelling)	0	Pagi, Polo	5	Glynn and Wellington (1983); T. Smith, unpub. data
Ecuador	Galápagos Iss. East and South	Mesotrophic-Eutrophic (upwelling)	0–25	Cycu, Didi, Psspp, Pagi, Pocl, Polo	25	Glynn and Wellington (1983); T. Smith unpub. data
El Salvador	Coastal: Deep populations not known	Mesotrophic	20*			Reyes-Bonilla and Barraza (2003)
France	Clipperton Atoll	Oligotrophic	30–40*	Polo, Lpap[?]	>50	Glynn et al. (1996)
Mexico	Coastal: Deep populations not known	Mesotrophic-Eutrophic (upwelling)	20–30		<10	Reyes Bonilla (2003), Carriquiry and Reyes-Bonilla (1997)
Mexico	Revillagigedo Iss.	Oligotrophic	20–30*	Polo, Pavona	30	Reyes-Bonilla pers. comm.
Panamá	Island: Secas Iss., Contreras Iss., Coiba Island group	Mesotrophic	12–25	Cycu, Gapl, Miin, Pach, Pacl, Pagi, Polo, Pava, Lapa	27	Maté (2003), Authors, unpub. obs.
Panamá	Island: Pearl Iss., Deep populations not known	Mesotrophic-Eutrophic (upwelling)	0–20		<10	Maté (2003)

Areas are subdivided by country, locations, productivity, depth of mixed layer, species in deep habitats, and depth ranges from 12 m to deepest record of scleractinian coral occurrence
*Mixed layer depths estimated from Fiedler and Talley (2006), Fig. 11. Data are interpolated from predominantly open ocean CTD casts and are approximated near bathymetric highs, such as islands and continental shelves
Species key: *Cycloseris curvata* = Cycu, *Cycloseris vaughani* = Cyva, *Diaseris distorta* = Didi, *Gardineroseris planulata* = Gapl, *Leptoseris papyracea* = Lepa, *Leptoseris scabra* = Lesc, *Leptoseris solida* = Leso, *Millepora intricata* = Miin, *Pavona clavus* = Pacl, *Pavona gigantea* = Pagi, *Pavona varians* = Pava, *Pocillopora capitata* = Poca, *Pocillopora damicornis* = Poda, *Pocillopora elegans* = Poel, *Pocillopora verrucosa* = Pove, *Porites lobata* = Polo, *Porites panamensis* = Popa, *Psammocora* spp. = Psspp

Highly productive eutrophic areas of the ETP are less likely to support depth refuges and refugia because conditions associated with high productivity limit suitable deeper habitats for coral growth. Eutrophic areas of the ETP (Table 19.1) range from turbid inshore locations exposed to terrestrial run-off (e.g., parts of the Gulf of Chiriquí in

Panama, Los Cóbanos in El Salvador), seasonal coastal upwelling regions (i.e., the Gulfs of Tehuantepec, Papagayo, and Panama), and continuous coastal and oceanic upwelling areas (e.g., the western and southern Galápagos, the coasts of Peru and Ecuador). When zooxanthellate corals occur in eutrophic environments, strongly attenuated incident light can reduce the maximum depths of occurrence, such as in Panama (Glynn 1976), and in these areas corals may not exist to depths that interact with the first (shallowest) thermocline. This turbidity can be increased by sediment in terrestrial run-off or upwelling-increased productivity that stimulates blooms of phytoplankton and zooplankton.

In addition, low temperature thermal stress from seasonal and continuous upwelling can limit the occurrence of deeper corals or, in the case of the Peruvian coastline, can prevent the occurrence of any zooxanthellate corals. The 20 °C isotherm often surfaces during upwelling conditions, with even more extreme temperatures in subsurface areas (Glynn and Stewart 1973; Jiménez 2001). Periods of exposure to these temperatures that last longer than a few weeks can inhibit coral growth and lead to coral mortality (Glynn and Stewart 1973; Glynn 1990; Glynn and Fong 2006). Coastal upwelling environments are present in the Gulf of Tehuantepec, Gulf of Papagayo, Gulf of Panama, and off the Ecuadorian mainland, and in these upwelling areas most corals are restricted to shallow depths (<5 m) (Glynn and Leyte-Morales 1997; Glynn 2000, 2003). At Gorgona Island, Colombia the influence of the Gulf of Panama upwelling across the larger Panama Bight likely causes periodic intrusions of cold water to shallow depths (Giraldo et al. 2008), and this may explain why deep corals are not abundant around this nearshore island (Table 19.1; F. Zapata, pers. comm.). Shallow water upwelling refuges and refugia are discussed later in this chapter.

Less productive and turbid oligotrophic areas of the ETP have been shown to support depth refugia in at least one case. In the Contreras Islands a neritic depth refugium has been shown for one species of coral. Populations of the extremely thermally sensitive hydrocoral *Millepora intricata* that are restricted to the Gulf of Chiriquí, Panama survived in depth refuges (>12 m depth) over the two most severe recent El Niño events (1982–83, 1997–98), while the two congeners *Millepora boschmai* and *Millepora platyphylla* without deep populations were driven to regional extinction in the ETP (Smith et al. 2014). Deep populations likely provided planktonic recruits for recolonization of shallow habitats after thermal stress, allowing partial recovery of shallow water populations. In the same study, reef core records indicated that *M. intricata* persisted at two reefs in Panama throughout the last half of the Holocene period (5500 BP), including over a period of heightened ENSO activity that likely generated El Niño conditions that surpassed the severity of those in 1982–83 and 1997–98 (Toth et al. 2012). Smith et al. (2014) concluded that depth refuges in the ETP were necessary for *M. intricata* to avoid regional extinction, and provided evidence of the deep refugia hypothesis (Glynn 1996; Riegl and Piller 2003; Bongaerts et al. 2010). Also, post-El Niño deep refugia hydrocoral populations were not found in structure-forming coral reefs, but were present in coral communities associated with bedrock or in corallith deposits along the deep talus slopes of coral reefs. This further suggests that zooxanthellate coral communities growing outside of true reefs form an important component of depth refugia in the ETP.

Many other areas in the ETP may also support depth refugia for certain coral species, but there are much less data. For this reason we will discuss different oligotrophic areas of the ETP and their potential to support depth refugia, which we hope will encourage further investigations in this region. In neritic environments on the continental shelf of the tropical Pacific Americas, where turbidity and cold water are not limiting, significant coral communities form at locations between 10 and 40 m, and could therefore harbour refuge populations. Deeper scleractinian corals along the continental mainland appear to be sparser than in offshore areas (Table 19.1), but the size of the extensive Central American coastline suggests that the total number of deep coral colonies could be large. Many sections along the continental mainland of the ETP show some deep coral development, particularly in Costa Rica and Panama (Table 19.1). Nearshore deep coral development, however, may be rare or lacking in Mexico and El Salvador (Reyes Bonilla 2003; Reyes Bonilla and Barraza 2003).

There are well-developed deep coral communities present at several neritic islands where waters are clearer than the adjacent continental mainland and deep temperatures are not limiting. Examples include Caño Island (Guzmán and Cortés 1989), Costa Rica; and the Secas Islands, Contreras Islands, and Coiba Island group, Panama (Glynn and Maté 1997). At Caño Island, isolated patches of *Porites lobata, Psammocora* spp., *Pocillopora elegans*, and *Pavona gigantea* occur to 22 m depth although there is a sand margin at a modest 12 m depth near the main reef. Guzmán and Cortés (1989) suggested that deeper corals growing on the reef slope (>5 m depth) had higher survival than shallow corals during the 1982–83 El Niño, which may suggest a depth refuge during this thermal stress event.

In Panama, our recent observations of non-coral, rock outcrops have revealed the presence of zooxanthellate coral communities occurring deeper than 20 m in the Coiba Iss. group, the Secas Iss., and the Contreras Iss.. In the Contreras Iss., along a rock outcrop just north of the Uva Island study reef, abundant isolated colonies of *Gardineroseris planulata*, *Millepora intricata*, *Pavona chiriquiensis*, *Pavona clavus*, *Pavona gigantea*, *Porites lobata*, and *Pavona varians* occur between 18–25 m. Similar communities have been observed around the Coiba Iss. group (Guzmán et al. 2004) and the Secas Iss. (T. Smith, unpub. observ.).

There is great potential for depth refugia at oceanic islands located far from the continent. Offshore islands and archipelagos in the far eastern Pacific exist across an enormous expanse of ocean and a diverse array of abiotic settings (Fiedler 1992; Fiedler and Talley 2006; see Chap. 3, Fiedler and Lavín), ranging from the intense upwelling of the western Galápagos Islands to the oligotrophic waters surrounding Clipperton Atoll. Islands that support light-dependent corals include the Revillagigedo Islands, Clipperton Atoll, Cocos Island, Easter Island (Rapa Nui), Malpelo Island, and the Galápagos Archipelago. Clear oceanic environments with no upwelling to moderate levels of upwelling might support refuges at depths considered mesophotic in the Western Atlantic and Pacific Ocean basins (30–150 m; Hinderstein et al. 2010). Examples include Clipperton Atoll, Cocos Island, Easter Island, Malpelo Island, and the northern Galápagos Islands (Darwin and Wolf Islands) (Table 19.1). Each of these islands supports zooxanthellate corals, including some shallow water species, to depths greater than 30 m (Table 19.1). Perhaps the most spectacular example of deep, light-dependent coral communities is found at Clipperton Atoll, and was well described by Glynn et al. (1996) who sampled high coral cover to 60 m.

There is both direct and indirect evidence that deep coral populations at oceanic islands can endure thermal stress events better than shallow reefs. Cover of *Gardineroseris planulata*, *Pavona* spp., *Pavona varians*, *Porites lobata*, and *Psammocora stellata* on two reefs at Cocos Island was higher in deeper reef areas (>9 m) than shallow areas of the same reefs, where some coral species disappeared after the 1982–83 El Niño event (Guzmán and Cortés 1992). Indeed, some of these colonies may have assisted in recolonization and recovery of shallow water areas (Guzmán and Cortés 2007). There is also direct and indirect evidence from the Galápagos Archipelago that some oceanic islands demonstrated greater coral survival in deeper areas after thermal stress. The best evidence for a depth refuge effect in an oceanic setting comes from Feingold (2001) who showed that fungiid corals at around 33 m did not bleach during the 1997–98 warming event, whereas corals present around 16 m were nearly all bleached. This study attributed the difference to both decreasing irradiance and more variable temperatures at depth caused by the intrusion of cooler water associated with the thermocline.

Coral reefs in the northern Galápagos were heavily impacted by the 1982–83 El Niño event (Glynn et al. 2009), but recent surveys at Darwin and Wolf Islands in 2012 reported extensive deep coral populations in depths greater than 20 m (Glynn et al. 2015; T. Smith, unpub. data). Coral cover peaked at 54 % in 24 m depth in Bahía Tiburón, Wolf Island, whereas coral cover at 10 m was 18.5 %. Deeper coral populations were also observed, but not quantified at Marchena Island. In the southern Galápagos at Floreana Island, cover at 20 m depth was 7.6 % and declined to 0 % at 30 m, showing a more limited potential for a deep refugium possibly related to cooler, more oligotrophic conditions (but see the discussion of fungiid survival above). At other oceanic islands, the evidence for a depth refuge is more equivocal. At Easter Island a 1–2 °C positive temperature anomaly (26–27 °C) caused some bleaching to 30 m (Glynn et al. 2003). Light-dependent corals occur much deeper (to 50+ m) at Easter Island, but their condition was not observed and coral mortality was not quantified (P. Glynn, pers. comm.).

17.4.1 Species-Specific Aspects of Depth Refuges

A prerequisite for a coral species benefitting from a depth refugium is that the species is a depth generalist. For example, it was seen that among milleporid species in the Gulf of Chiriquí that those species without a deep population were driven to regional extinction in the ETP, whereas *Millepora intricata* was able to survive in deep zones over two extreme El Niño disturbances (Smith et al. 2014). Not only does *M. intricata* survive in deep areas but it also seems to be abundant at depth. This is important, since mortality from bleaching did extend into deeper zones, albeit at a reduced rate, and having high abundance at depth helped ensure some survivors for re-expansion. Among the low diversity ETP hermatypic coral faunas there are many species with wide depth ranges that may be able to benefit from depth refuges (Table 19.1).

The exact reason that some species are depth generalists while others are confined to shallower depths has not been well studied. In other systems, heterotrophic plasticity, the ability to shift from autotrophy to heterotrophy to meet colony energetic demands, has been postulated as a mechanism maintaining colony fitness at deeper depths (Lesser et al. 2010). In shallow waters less than 6 m depth in the Gulf of Panama, Palardy et al. (2005) found that the mounding corals *Pavona clavus* and *Pavona gigantea* were more adept at heterotrophic feeding on zooplankton than branching *Pocillopora damicornis* during non-upwelling conditions. The agariciids *P. clavus, P. gigantea,* and *P. varians* all have wide depth ranges across the ETP (Table 19.1) and their survival at greater depths may be facilitated by their heterotrophic food capture abilities.

Among the milleporids, *Millepora intricata* is likely a competent heterotrophic feeder (Lewis 1989) and this may facilitate its ability to occur across a variety of depths. However, under the assumption that all milleporids are efficient at heterotrophic feeding, it is not clear why the two milleporid species that went regionally extinct did not occur deeper than 15 m. They may be more dependent on phototrophy than assumed or the morphologies of *Millepora boschmai* (columnar to upright plates) and *Millepora platyphylla* (upright plates) may be maladapted to persist at low light levels and unidirectional currents of deeper water, as opposed to high light conditions and oscillatory, wave-driven currents of shallow water.

ETP coral species that occur more rarely or not at all in greater depths may have reduced fitness in deeper zones that is related to reliance on phototrophy for energy acquisition. One pattern that supports this hypothesis is that where shallow trending species do have a deep extended range it is in clear oceanic environments (Table 19.1). This includes *Porites lobata*, which may not be able to meet a majority of its energetic needs heterotrophically (Grottoli et al. 2006). This species is confined to substrates less than 20 m depth in more turbid environments, but can occur deeper at locations where the water clarity is higher. In the clear waters of the northern Galápagos Islands and Clipperton Atoll where *P. lobata* does occur below 20 m depth it is much reduced in cover relative to shallow water, unlike species in the genus *Pavona* that have a much more even distribution with depth or even increase in cover below 20 m (Glynn et al. 1996, 2015). This also suggests a greater reliance of *P. lobata* on phototrophy than *Pavona* spp.

Most of the pocilloporid species have narrow and shallow depth ranges (Table 19.1), and reveal markedly reduced fitness in deeper water. Colonies of *Pocillopora damicornis* transplanted from 10 m to 18 m in the Gulf of Chiriquí showed very low survival (6.7 %, n = 15) and negative growth (−28 %, n = 1) over 3.4 years (T. Smith, unpub. data; methods as for *Millepora intricata* in Smith et al. 2014). Reduced fitness at greater depths at the same location where *M. intricata* occurred in a refugium might indicate a decrease in colony energy with decreased photosynthesis, but this was not tested directly.

17.4.2 Vertical Connectivity of Deeper Water Refugia

Very little is known about the connectivity of deeper coral populations via pelagic larval dispersal and other areas (seeding) or among themselves (self-recruitment). While this is difficult to study, larval connectivity has extremely important ramifications for the potential contribution of deeper reef areas to the resilience of the ETP coral fauna after disturbance and its potential to serve as refugia. Deeper coral populations will go extinct if they are only sinks for shallow water larvae and shallow water corals are extirpated. Furthermore, deeper-living corals will never contribute to shallow water recovery if they are only self seeding, which might occur if deep populations are adapted to deeper-occurring habitats. In this case the deeper corals may be a refugium for the species, but not for the full genetic diversity of the species. However, if deeper corals are just part of a network of larval exchange between all habitats, then they may have an important role in coral recovery following disturbance.

In the only demonstrated example of a deep refugium—*Millepora intricata* in the Gulf of Chiriquí—larval contribution from deep to shallow was inferred, suggesting the ability for deep populations to increase resilience (Smith et al. 2014). The inference was made possible because shallow water populations were likely to have been extirpated across most of the range, whereas deeper populations were extant. When shallow water populations reappeared at study sites a few years after a severe El Niño thermal disturbance it seemed likely they recruited from abundant deeper populations within a hundred meters distance, rather than from distant shallow water survivors, if present but not observed. In other species where shallow water populations are not entirely decimated by El Niño disturbances, demonstrating connectivity and its importance for population recovery will require genetic methodologies. This presents its own challenges, such as the development of genetic markers for different species, lack of population structure, and the need for many samples to be compared from each of the habitats. A full evaluation of the hypothesis that deeper areas are refugia from thermal stress events will require these sorts of studies.

17.5 Regional Refugia (Scales of 1000 km and Greater)

Regional refugia occur when large-scale physical oceanographic conditions increase the probability of survival for a large proportion of reef organisms over a large-scale disturbance. The best, and perhaps only, possible example of regional refuges occur in large scale upwelling centers where cooler conditions may act to lower thermal stress during ENSO associated warming events. A regional refugium may also exist, however, as a meta-refuge due to interacting networks of refuges at smaller scales. Thus, there are two types of regional refuges that might occur: (1) an overarching regional refugium where conditions provide avoidance for some of the corals in a region, and (2) a meta-refugium, where regional conditions favour the formation of multiple refuges that combine to increase the potential of a region to maintain biodiversity. We will address overarching and meta-refugia separately.

17.5.1 Overarching Regional Refugia

The most cited overarching regional refugia for corals includes areas where relatively cooler water is present during large-scale high thermal stress events. Conditions for shallow water regional refugia are largely favoured by upwelling of cool water into surface layers (Glynn 1996; West and Salm 2003). Distinct seasonal upwelling occurs at three areas along the Central American coastline (Gulf of Tehuantepec, Gulf of Papagayo, and the Gulf of Panama). Seasonal upwelling in the Gulf of Panama provided a thermal refuge for corals during ENSO related sea warming over the 1997–98 El Niño event (Glynn et al. 2001). In addition, non-seasonal upwelling has been suggested as lowering the incidence of bleaching during the 1997–98 El Niño event in the Gulf of California relative to lower-latitude areas of the ETP (Reyes-Bonilla 2001). More continuous upwelling occurs in the western Galápagos, due to the shoaling of the equatorial undercurrent (Cromwell current), and along the South American mainland coast (Colombia, Ecuador, Peru) from Ekman coastal upwelling.

Scaling upwelling refuges to long-term refugia is compromised if upwelling and thermal stress timing are not coincident (Chollett et al. 2010), or if upwelling prevents acclimation to anomalously warm conditions (D'Croz and Maté 2004). Since high thermal stress created by El Niño conditions can occur throughout the year there is not always synchrony between heat stress and the seasonal upwelling in the Gulf of Panama (Glynn et al. 2001; Chollett et al. 2010). Furthermore, ENSO results in a general deepening of the thermocline (Enfield 2001; Fiedler and Talley 2006) and therefore upwelled waters may still be anomalously warm, even if cooler than the mean temperatures of adjacent areas. Since coral bleaching thresholds are specific to the background climatology (Coles et al. 1976) and corals from the Gulf of Panama have been shown to be more sensitive to thermal stress (D'Croz et al. 2001; D'Croz and Maté 2004), a long-term refugium caused by the Gulf of Panama upwelling may be compromised. The same general principles apply for the upwelling Gulf of Papagayo, whereby there is potentially some benefit from upwelling, but positive thermal anomalies associated with El Niño events do occur and result in moderate coral mortality (Jiménez et al. 2001, 2010). Upwelling as a mechanism creating a thermal refugium has been examined less extensively for coastal upwelling in the Gulf of Tehuantepec.

Positive El Niño-Southern Oscillation anomalies are most pronounced at the equator (+0.5–1.0 °C) and this negates the potential for an upwelling refugium in the Galápagos Archipelago. The corals and coral reefs of the equatorial upwelling Galápagos Archipelago were shown to be very sensitive to thermal stress in the 1982–83 and 1997–98 El Niño events (Glynn 1994; Glynn et al. 2001). Here, heterogeneous thermal zones across the region were uniformly warm, causing anomalies as high as 1–2 °C (Glynn 1990; Wellington et al. 2001). The entire equatorial region is highly susceptible to positive temperature anomalies (+1–2 °C) during El Niño events because the mixed layer depth remains relatively constant, whereas the thermocline is depressed below the mixed layer and cool water is not brought to the surface (Fiedler and Talley 2006).

Other over-arching regional refugia may be found in coastal areas. There is some evidence that neritic regions near the Central American coast may be less sensitive to coral bleaching than offshore regions. There may have been an offshore-to-onshore decline in coral mortality during El Niño events in Ecuador and Panama due to lower SSTs and reduced light penetration and irradiance stress (Glynn et al. 2001). However, it is also possible to expect the same offshore-to-inshore differences due to a reduction in the penetration of Kelvin waves associated with El Niño events by geomorphological and orographic effects in neritic habitats. If true, this may provide additional support for Macintyre's hypothesis (Macintyre et al. 1992) that sheltering from El Niño warming events and cool upwelling conditions could facilitate the accumulation of large coral reef frameworks in some near-shore eastern Pacific areas via avoidance.

17.5.2 Meta-Refugia

Refugia operate on scales of time and space that necessitate protection from multiple disturbances. Different types of refuges across diverse habitats may be protective (e.g., prevent extirpation) in any given disturbance, but perhaps not in subsequent disturbances. In this case no particular refuge may be a refugium because of vulnerability at some point in time and space. However, within a spatial network of habitat patches connected by larval exchange and the possibility of recolonization, different refuges acting in concert could act as a meta-refugium that reduces the risk of regional extinctions. Under this scenario a network of ephemeral refuges permits expansion under classical meta-population dynamics (Hanski 1998). While metapopulation theory is well established, the suggestion of meta-refugia is here used to illustrate the idea that having resistant populations in a diversity of refuges, even if no one refuge can scale to a refugium, may limit the chance of extinction across all interconnected habitat patches and together these multiple patches (refuges) may scale to a refugium.

The prospect of meta-refugia is hypothetical, but some areas of the ETP may be less resistant to long-term changes in ENSO frequency and intensity because of a lack of refuge networks. For example, upwelling areas have low habitat diversity because upwelling limits depth refuges and restricts most coral growth to protected areas in the lee of persistent upwelling streams. Therefore, upwelling areas may be less resistant and resilient in the context of meta-refugia, since all biota may respond equally to large-scale disturbances, such as El Niño seawater warming. Conversely, neritic, non-upwelling areas north of the equator in Costa Rica and Panama may have an increased possibility for meta-refuges since they have the potential for depth refuges, possess a high degree of habitat heterogeneity (turbidity gradients, island and coastal areas, oceanographic variability), and could be well connected. They also have a high abundance of corals, which could be evidence that refugia are present and have been protective in the past, or that refugia are present because of the high abundance of corals. While this may seem circular, large populations increase the number of habitats, including refuges, that are occupied (Hanski 1998), and occupation of more refuge habitats may increase the resilience of metapopulations. Empirically, from the marine fossil record, being widespread and having greater habitat breadth, albeit at a coarser scale than we are examining here, has been shown to decrease the probability of extinction (Harnick et al. 2012).

17.6 Conclusion

Refugia as a means for corals to avoid extinction or to rapidly recolonize susceptible habitats after disturbance seem to be an important mechanism for coral and coral reef persistence in the marginal eastern tropical Pacific Ocean. However, it seems that certain types of refuges may be more likely to scale to a refugium. In particular, conditions that favour partial colony mortality, such as intra-colony environmental heterogeneity, may favour refugia for shallow-water species susceptible to seawater warming, such as *Pocillopora* spp. and *Gardineroseris planulata*. Intra-reef and inter-reef thermal refugia may occur where temperatures are normally elevated, such as lagoons, reef flats, or on reefs that are relatively warm relative to local open ocean conditions. These areas may increase local tolerance through acclimation or adaptation. Avoidance of thermal stress at intra- and inter-reef scales may be comparatively rare, however, since cooling processes are often not reliable during warm anomalies, and corals in cooler environments are susceptible to any change from the highest normal mean thermal conditions. A depth refugium was confirmed for at least one zooxanthellate species, the hydrocoral *Millepora intricata*, and a refuge was shown for a fungiid scleractinian species at one site over the 1997–98 El Niño event. Other depth generalist species, such as *Pavona* spp. may benefit from increased resilience after disturbance because of reduced or no mortality in depth refuges; however, demonstrating depth refugia for other species still needs empirical data through El Niño disturbances. At the scale of regions, seasonal upwelling has been shown to be a refuge over specific events, but whether this scales to a long-term refugium depends critically on whether cooling from upwelling is always timed to avoid periods of thermal stress. Other large-scale upwelling areas, such as the western and southern Galápagos and the South American coast appear vulnerable to the cessation of cooling upwelling during El Niño events. In these areas reduced cooling due to a deepened equatorial thermocline during El Niño and a coral fauna that may be more susceptible to elevated temperatures because of normally cooler temperatures, may compromise regional refuges and refugia. Hypothetically, the existence of meta-refugia may ensure the long-term persistence of coral species and coral reefs in the ETP by supporting corals in diverse types of refuges that together scale to a refugium.

There remains a great deal of work to be done if we are to understand how refugia support corals and other sessile marine species in a warming ocean. Future research in the ETP should target the thermal responses of species in more marginal hermatypic coral habitats, such as deep areas below the thermocline, and corals growing on rock outcrops and unconsolidated substrates. This will help us understand the full demographic ramifications of refugia in maintaining coral populations. In addition, the impacts of chronic stress, such as low aragonite saturation states or low oxygen concentrations, in limiting the extent of coral development and growth are important considerations for understanding the limits to the scale of refugia. Because the ETP has a substantial record of coral ecological research, is prone to repeated thermal disturbances, has a wide diversity of oceanographic conditions, and supports a number of coral species at various stages of rarity and vulnerability to regional extinction, it is an ideal region to continue the study of refugia in the modern ocean.

Acknowledgements We thank Jorge Cortés, Peter Glynn, Katherine Mejía, Carlos Muñoz, Héctor Reyes Bonilla, Bernardo Vargas-Ángel, and Fernando Zapata for personal observations and unpublished data. These workers kindly shared their vast knowledge of corals in the eastern Pacific and helped to make this a more complete chapter. Peter Glynn provided a helpful review of an earlier draft and was particularly important for forming our earlier ideas of reef refugia. Likewise we also thank Iliana Chollet and Aaron O'Dea who provided excellent reviews and provided many helpful comments that are reflected in this chapter. We also thank the editors for their patience and guidance in producing this superb collection of papers on corals and coral reefs of the eastern tropical Pacific Ocean.

References

Baker AC, Glynn PW, Riegl B (2008) Climate change and coral reef bleaching: an ecological assessment of long-term impacts, recovery trends and future outlook. Estuar Coast Shelf S 80:435–471

Baker AC, Starger CJ, McClanahan TR, Glynn PW (2004) Coral reefs: corals' adaptive response to climate change. Nature 430:741

Bakus GJ (1975) Marine zonation and ecology of Cocos Island, off Central America. Atoll Res Bull 179:1–9

Banks S, Vera M, Chiriboga Á (2009) Establishing reference points to assess long-term change in zooxanthellate coral communities of the northern Galápagos coral reefs. Galápagos Res 66:43–64

Barshis DJ, Ladner JT, Oliver TA, Seneca FO, Traylor-Knowles N, Palumbi SR (2013) Genomic basis for coral resilience to climate change. Proc Natl Acad Sci USA 110:1387–1392

Baums IB, Boulay JN, Polato NR, Hellberg ME (2012) No gene flow across the Eastern Pacific Barrier in the reef-building coral *Porites lobata*. Mol Ecol 21:5418–5433

Bellantuono AJ, Hoegh-Guldberg O, Rodriguez-Lanetty M (2011) Resistance to thermal stress in corals without changes in symbiont composition. Proc R Soc B: Biol Sci 279:1100–1107

Bender EA, Case TJ, Gilpin ME (1984) Perturbation experiments in community ecology: theory and practice. Ecology 65:1–13

Birkeland C, Meyer DL, Stames JP, Buford CL (1975) Subtidal communities of Malpelo Island. In: JB Graham (ed) The biological investigation of Malpelo Island, Colombia. Smith Contrib Zool 176:55–68

Bongaerts P, Ridgway T, Sampayo E, Hoegh-Guldberg O (2010) Assessing the 'deep reef refugia' hypothesis: focus on Caribbean reefs. Coral Reefs 29:309–327

Brown BE (1997) Coral bleaching: causes and consequences. Coral Reefs 16:S129–S138

Brown BE, Dunne RP, Goodson MS, Douglas AE (2002) Experience shapes the susceptibility of a reef coral to bleaching. Coral Reefs 21:119–126

Carriquiry JD, Reyes Bonilla H (1997) Community structure and geographic distribution of the coral reefs of Nayarit, Mexican Pacific. Cienc Mar 23:227–248

Castillo KD, Ries JB, Weiss JM, Lima FP (2012) Decline of forereef corals in response to recent warming linked to history of thermal exposure. Nature Clim Change 2:756–760

Chollett I, Mumby P, Cortés J (2010) Upwelling areas do not guarantee refuge for coral reefs in a warming ocean. Mar Ecol Prog Ser 416:47–56

Coles SL, Jokiel PL, Lewis CR (1976) Thermal tolerance in tropical versus subtropical reef corals. Pac Sci 30:159–166

Cortés J, Jiménez C (2003) Corals and coral reefs of the Pacific of Costa Rica: history, research and status. In: Cortés J (ed) Latin American coral reefs. Elsevier, Amsterdam, pp 361–385

D'Croz L, Maté JL (2004) Experimental responses to elevated water temperature in genotypes of the reef coral *Pocillopora damicornis* from upwelling and non-upwelling environments in Panama. Coral Reefs 23:473–483

D'Croz L, O'Dea A (2007) Variability in upwelling along the Pacific shelf of Panama and implications for the distribution of nutrients and chlorophyll. Estuar Coast Shelf S 73:325–340

D'Croz L, Maté J, Oke J (2001) Responses to elevated sea water temperature and UV radiation in the coral *Porites lobata* from upwelling and non-upwelling environments on the Pacific coast of Panama. Bull Mar Sci 69:203–214

Darling ES, Alvarez-Filip L, Oliver TA, McClanahan TR, Côté IM (2012) Evaluating life-history strategies of reef corals from species traits. Ecol Lett 15:1378–1386

Davis KA, Lentz SJ, Pineda J, Farrar JT, Starczak VR, Churchill JH (2011) Observations of the thermal environment on Red Sea platform reefs: a heat budget analysis. Coral Reefs 30:25–36

de Boyer Montégut C, Madec G, Fisher AS, Lazar A, Iudicone D (2004) Mixed layer depth over the global ocean: an examination of profile data and a profile-based climatology. J Geophys Res 109: C12003

Enfield DB (2001) Evolution and historical perspective of the 1997–1998 El Niño-Southern Oscillation event. Bull Mar Sci 69:7–25

Feingold J (2001) Response of three coral communities to the 1997–98 El Niño-Southern Oscillation: Galápagos Islands, Ecuador. Bull Mar Sci 69:61–77

Fiedler P (1992) Seasonal climatologies and variability of eastern tropical Pacific surface waters. NOAA Tech Rept NMFS [NOAA TECH. REP. NMFS], vol 109

Fiedler PC, Talley LD (2006) Hydrography of the eastern tropical Pacific: a review. Prog Oceanogr 69:143–180

Giraldo A, Rodríguez-Rubio E, Zapata FA (2008) Condiciones oceanográficas en Isla Gorgona, Pacífico oriental tropical de Colombia. Lat Am J Aquat Res 36:121–128

Glynn PW (1976) Some physical and biological determinants of coral community structure in the eastern Pacific. Ecol Monogr 46:431–456

Glynn PW (1990) Coral mortality and disturbances to coral reefs in the tropical eastern Pacific. In: Glynn PW (ed) Global ecological consequences of the 1982-83 El Niño-Southern Oscillation. Elsevier Sci Pub BV, Amsterdam, pp 55–126

Glynn PW (1994) State of coral reefs in the Galápagos Islands: natural vs anthropogenic impacts. Mar Poll Bull 29:131–140

Glynn PW (1996) Coral reef bleaching: facts, hypotheses and implications. Glob Change Biol 2:495–509

Glynn PW (2000) Eastern Pacific coral reef ecosystems. In: Seeliger U, Kjerfve B (eds) Coastal marine ecosystems of Latin America. Ecol Stud, vol 144, Springer-Verlag, Berlin pp 281–305

Glynn PW (2003) Coral communities and coral reefs of Ecuador. In: Cortés J (ed) Latin American coral reefs. Elsevier Science BV, Amsterdam, pp 449–472

Glynn PW (2011) In tandem reef coral and cryptic metazoan declines and extinctions. Bull Mar Sci 87:767–794

Glynn PW, Ault JS (2000) A biogeographic analysis and review of the far eastern Pacific coral reef region. Coral Reefs 19:1–23

Glynn PW, Colley SB (2008) Survival of brooding and broadcasting reef corals following large scale disturbances: is there any hope for broadcasting species during global warming? In: Proceedings of 11th International Coral Reef Symposium, Ft Lauderdale, pp 361–365

Glynn PW, Fong P (2006) Patterns of reef coral recovery by the regrowth of surviving tissues following the 1997-98 El Niño warming and 2000, 2001 upwelling cool events in Panamá, eastern Pacific. In: Proceedings of 10th International Coral Reef Symposium, vol 1, Okinawa, pp 624–630

Glynn PW, Leyte-Morales GE (1997) Coral reefs of Huatulco, west México: reef development in upwelling Gulf of Tehuantepec. Rev Biol Trop 45:1033–1047

Glynn PW, Maté JL (1997) Field guide to the coral reefs of Panamá. Proceedings of 8th International Coral Reef Symposium, vol 1, Panama, pp 145–166

Glynn PW, Stewart RH (1973) Distribution of coral reefs in the Pearl Islands (Gulf of Panama) in relation to thermal conditions. Limnol Oceanogr 18:367–379

Glynn PW, Wellington GM (1983) Corals and coral reefs of the Galápagos Islands. Univ California Press, Berkeley, p 330

Glynn PW, Veron JEN, Wellington GM (1996) Clipperton Atoll (eastern Pacific): oceanography, geomorphology, reef-building coral ecology and biogeography. Coral Reefs 15:71–99

Glynn PW, Maté JL, Baker AC, Calderón MO (2001) Coral bleaching and mortality in Panama and Ecuador during the 1997-1998 El Niño-Southern Oscillation event: spatial/temporal patterns and comparisons with the 1982-1983 event. Bull Mar Sci 69:79–109

Glynn PW, Wellington GM, Wieters EA, Navarrete SA (2003) Reef-building coral communities of Easter Island (Rapa Nui), Chile. In: Cortés J (ed) Latin American coral reefs. Elsevier Science, Amsterdam, pp 473–494

Glynn PW, Riegl B, Correa AMS, Baums IB (2009) Rapid recovery of a coral reef at Darwin Island, Galapagos Islands. Galapagos Res 66:6–13

Glynn PW, Riegl B, Purkis S, Kerr J, Smith TB (2015) Coral reef recovery at Darwin, the northern-most of the Galápagos Islands. Coral Reefs 34:421–436

Gorospe KD, Karl SA (2011) Small-scale spatial analysis of *in situ* sea temperature throughout a single coral patch reef. J Mar Biol art ID 719580

Grottoli AG, Rodrigues LJ, Palardy JE (2006) Heterotrophic plasticity and resilience in bleached corals. Nature 440:1186–1189

Guest JR, Baird AH, Maynard JA, Muttaqin E, Edwards AJ, Campbell SJ, Yewdall K, Affendi YA, Chou LM (2012) Contrasting patterns of coral bleaching susceptibility in 2010 suggest an adaptive response to thermal stress. PLoS ONE 7:e33353

Guzmán HM, Cortés J (1989) Coral reef community structure at Caño Island, Pacific Costa Rica. PSZNI: Mar Ecol 10:23–41

Guzmán HM, Cortés J (1992) Cocos Island (Pacific of Costa Rica) coral reefs after the 1982-1983 El Niño disturbance. Rev Biol Trop 40:309–324

Guzmán HM, Cortés J (2007) Reef recovery 20 years after the 1982–1983 El Niño massive mortality. Mar Biol 151:401–411

Guzmán HM, Guevara CA, Breedy O (2004) Distribution, diversity, and conservation of coral reefs and coral communities in the largest marine protected area of Pacific Panama (Coiba Island). Environ Conserv 31:111–121

Hanksi I (1998) Metapopulation dynamics. Nature 396:41–49

Harnik PG, Lotze HK, Anderson SC, Finkel ZV, Finnegan S, Lindberg DR, Liow LH, Lockwood R, McClain CR, McGuire JL, O'Dea A, Pandolfi JM, Simpson C, Tittensor DP (2012) Extinctions in ancient and modern seas. Trends Ecol Evol 27:608–617

Hinderstein L, Marr J, Martinez F, Dowgiallo M, Puglise K, Pyle R, Zawada D, Appeldoorn R (2010) Theme section on mesophotic coral ecosystems: characterization, ecology, and management. Coral Reefs 29:247–251

Holling CS (1973) Resilience and the stability of ecological systems. Annu Rev Ecol Syst 4:1–23

Hubbard DK, Garcia M (2003) The corals and coral reefs of Easter Island—a preliminary look. In: Loret J, Tanacredi JT (eds) Easter Island: scientific exploration into the world's environmental problems in microcosm. Kluwer Academic/Plenum Pub, New York, pp 53–79

Hughes TP, Bellwood DR, Connolly SR, Cornell HV, Karlson RH (2014) Double jeopardy and global extinction risk in corals and reef fishes. Curr Biol 24:2946–2951

Jiménez C (2001) Seawater temperature measured at the surface and at two depths (7 and 12 m) in one coral reef at Culebra Bay, Gulf of Papagayo, Costa Rica. Rev Biol Trop 49(Suppl 2):153–161

Jiménez C, Cortés J, León A, Ruíz E (2001) Coral bleaching and mortality associated with the 1997-98 El Niño in an upwelling environment in the Eastern Pacific (Gulf of Papagayo, Costa Rica). Bull Mar Sci 69:151–169

Jiménez C, Bassey G, Segura Á, Cortés J (2010) Characterization of the coral communities and reefs of two previously undescribed locations in the upwelling region of Gulf of Papagayo (Costa Rica). Rev Cienc Mar Cost 2:95–108

Keppel G, Van Niel KP, Wardell-Johnson GW, Yates CJ, Byrne M, Mucina L, Schut AGT, Hopper SD, Franklin SE (2012) Refugia: identifying and understanding safe havens for biodiversity under climate change. Global Ecol Biogeogr 21:393–404

Leichter JJ, Genovese SJ (2006) Intermittent upwelling and subsidized growth of the scleractinian coral *Madracis mirabilis* on the deep fore-reef of Discovery Bay, Jamaica. Mar Ecol Prog Ser 316:95–103

Lesser MP (1997) Oxidative stress causes coral bleaching during exposure to elevated temperatures. Coral Reefs 16:187–192

Lesser MP, Slattery M, Stat M, Ojimi M, Gates RD, Grottoli A (2010) Photoacclimatization by the coral *Montastraea cavernosa* in the mesophotic zone: light, food, and genetics. Ecology 91:990–1003

Lewis JB (1989) The ecology of *Millepora*. Coral Reefs 8:99–107

Loreau M (2010) Linking biodiversity and ecosystems: towards a unifying ecological theory. Philos T Roy Soc B: Biol Sci 365:49–60

Loya Y, Sakai K, Yamazato K, Nakano Y, Sambali H, van Woesik R (2001) Coral bleaching: the winners and the losers. Ecol Lett 4:122–131

Macintyre IG, Glynn PW, Cortés J (1992) Holocene reef history in the eastern Pacific: mainland Costa Rica, Caño Island, Cocos Island, and Galápagos Islands. In: Proceedings of 7th International Coral Reef Symposium, vol 2, Guam, pp 1174–1184

Manzello DP (2010) Ocean acidification hotspots: spatiotemporal dynamics of the seawater CO_2 system of eastern Pacific coral reefs. Limnol Oceanogr 55:239–248

Maté JL (2003) Corals and coral reefs of the Pacific coast of Panamá. In: Cortés J (ed) Latin American coral reefs. Elsevier, Amsterdam, pp 387–417

McCann KS (2000) The diversity-stability debate. Nature 405:228–233

McClanahan TR, Ateweberhan M, Muhando CA, Maina J, Mohammed MS (2007) Effects of climate and seawater temperature variation on coral bleaching and mortality. Ecol Monogr 77:503–525

McKinney ML (1997) Extinction vulnerability and selectivity: combining ecological and paleontological views. Annu Rev Ecol Syst 28:495–516

Mumby PJ, Chisholm JRM, Edwards AJ, Andrefouet S, Jaubert J (2002) Cloudy weather may have saved Society Island reef corals during the 1998 ENSO event. Mar Ecol Prog Ser 222:209–216

Nakamura T, van Woesik R (2001) Water-flow rates and passive diffusion partially explain differential survival of corals during the 1998 bleaching event. Mar Ecol Prog Ser 212:301–304

Oliver TA, Palumbi SR (2011) Do fluctuating temperature environments elevate coral thermal tolerance? Coral Reefs 30:429–440

Palardy JE, Grottoli AG, Matthews KA (2005) Effects of upwelling, depth, morphology and polyp size on feeding in three species of Panamanian corals. Mar Ecol Prog Ser 300:79–89

Palumbi SR, Barshis DJ, Traylor-Knowles N, Bay RA (2014) Mechanisms of reef coral resistance to future climate change. Science 344:895–898

Pineda J, Starczak VR, Tarrant A, Blythe J, Davis K, Farrar J, Berumen M, da Silva JCB (2013) Two spatial scales in a bleaching event: corals from the mildest and the most extreme thermal environments escape mortality. Limnol Oceanogr 58:1531–1545

Reyes Bonilla H (2001) Effects of the 1997-1998 El Niño-Southern Oscillation on coral communities of the Gulf of California, Mexico. Bull Mar Sci 69:251–266

Reyes Bonilla H (2003) Corals and coral reefs of the Pacific coast of México. In: Cortés J (ed) Latin American coral reefs. Elsevier, Amsterdam, pp 331–350

Reyes Bonilla H, Barraza JE (2003) Corals and associated marine communities from El Salvador. In: Cortés J (ed) Latin American coral reefs. Elsevier, Amsterdam, pp 351–360

Riegl B, Piller WE (2003) Possible refugia for reefs in times of environmental stress. Int J Earth Sci 92:520–531

Rowan R, Knowlton N, Baker A, Jara J (1997) Landscape ecology of algal symbiont communities explains variation in episodes of coral bleaching. Nature 388:265–269

Sebens KP, Grace SP, Helmuth B, Maney EJ Jr, Miles JS (1998) Water flow and prey capture by three scleractinian corals, *Madracis mirabilis*, *Montastrea cavernosa* and *Porites porites*, in a field enclosure. Mar Biol 131:347–360

Smith TB, Glynn PW, Maté JL, Toth LT, Gyory J (2014) A depth refuge from catastrophic coral bleaching mortality twice prevented regional extinction. Ecology 95:1663–1673

Smith TB, Brandt ME, Calnan JM, Nemeth RS, Blondeau J, Kadison E, Taylor M, Rothenberger JP (2013) Convergent mortality responses of Caribbean coral species to seawater warming. Ecosphere 4:art87. http://dx.doi.org/10.1890/ES13-00107.1

Springer VG (1958) Systematics and zoogeography of the clinid fishes of the subtribe Labrisomini Hubbs. Pubs Inst Mar Sci Univ Texas 5:417–492

Toth LT, Aronson RB, Vollmer SV, Hobbs JW, Urrego DH, Cheng H, Enochs IC, Combosch DJ, van Woesik R, Macintyre IG (2012) ENSO drove 2500-year collapse of eastern Pacific coral reefs. Science 337:81–84

van Woesik R, Sakai K, Ganase A, Loya Y (2011) Revisiting the winners and the losers a decade after coral bleaching. Mar Ecol Prog Ser 434:67–76

van Woesik R, Franklin EC, O'Leary J, McClanahan TR, Klaus JS, Budd AF (2012a) Hosts of the Plio-Pleistocene past reflect modern-day coral vulnerability. Proc R Soc B: Biol Sci 279:2448–2456

van Woesik R, Houk P, Isechal AL, Idechong JW, Victor S, Golbuu Y (2012b) Climate-change refugia in the sheltered bays of Palau: analogs of future reefs. Ecol Evol 2:2474–2484

Wellington GM, Strong AE, Merlen G (2001) Sea surface temperature variation in the Galápagos Archipelago: a comparision between AVHRR nighttime satellite data and in situ instrumentation (1982-1998). Bull Mar Sci 69:27–42

West JM, Salm RV (2003) Resistance and resilience to coral bleaching: implications for coral reef conservation and management. Conserv Biol 17:956–967

Zapata FA, Vargas-Ángel B (2003) Corals and coral reefs of the Pacific coast of Colombia. In: Cortés J (ed) Latin American coral reefs. Elsevier Science BV, Amsterdam, pp 419–448

Effects of Global Warming and Ocean Acidification on Carbonate Budgets of Eastern Pacific Coral Reefs

18

Derek P. Manzello, C. Mark Eakin, and Peter W. Glynn

Abstract

Eastern tropical Pacific (ETP) coral reefs provide a real-world example of reef growth, development, structure, and function under the high-pCO_2, low aragonite saturation state (Ω_{arag}) conditions expected for the entire tropical surface ocean with a doubling to tripling of atmospheric CO_2. This provides a unique opportunity to examine various aspects of calcium carbonate ($CaCO_3$) budgets in low-Ω_{arag} conditions in the present day. Unlike anywhere else in the world, the ETP displays a continuum of thermal stress and CO_2 inputs up to levels at which reef building is terminated and reef structures are lost. The response of coral reef $CaCO_3$ budgets to El Niño warming across the ETP shows that reefs can be completely lost after experiencing a 2–3 °C thermal anomaly sustained in excess of two months during the warmest time of the year at Ω_{arag} values expected for the rest of the tropics when atmospheric CO_2 doubles. ETP coral reefs have persisted and shown resilience to this level of thermal stress or acidification when acting alone, but the combination of the two corresponded with the complete elimination of reef framework structures in the southern Galápagos Islands over the decade after the 1982–83 El Niño warming event. Reef carbonate degradation is exacerbated also by diverse agents of bioerosion such as sea urchins, boring bivalves, and excavating sponges, with experimental evidence demonstrating that the latter may even increase their activities during ocean warming and pH decline. This chapter reviews the $CaCO_3$ budget of ETP coral reefs and discusses how a high-CO_2 world may impact the major biotic and abiotic factors responsible for the cycling of carbonate materials. Coral reefs of the ETP serve as a model for conditions that will occur in other regions within a few decades.

Keywords

Climate change • Aragonite saturation state • Calcification • Bioerosion • Thermal stress

D.P. Manzello (✉)
Atlantic Oceanographic and Meteorological Laboratories (AOML), National Oceanic and Atmospheric Administration (NOAA), 4301 Rickenbacker Cswy., Miami, FL 33149, USA
e-mail: derek.manzello@noaa.gov

C. Mark Eakin
Coral Reef Watch, National Environmental Satellite Data and Information Services, National Oceanic and Atmospheric Administration, 1335 East West Hwy., E/RA31, Silver Spring, MD 20910, USA
e-mail: mark.eakin@noaa.gov

P.W. Glynn
Division of Marine Biology and Ecology, Rosenstiel School of Marine and Atmospheric Science, 4600 Rickenbacker Cswy., Miami, FL 33149, USA
e-mail: pglynn@rsmas.miami.edu

P.W. Glynn et al. (eds.), *Coral Reefs of the Eastern Tropical Pacific*,
Coral Reefs of the World 8, DOI 10.1007/978-94-017-7499-4_18

18.1 Introduction

The first scientists to visit the eastern tropical Pacific (ETP), including Charles Darwin, commented on the apparent absence of reef development (Darwin 1842). Structural reefs were later discovered and found to have rapid accretion rates over the past 5600 years, rivaling Holocene reef accretion rates elsewhere (Glynn et al. 1972; Glynn and Macintyre 1977; see Chap. 6, Toth et al.). Despite rapid accretion at a few protected sites, ETP reefs are highly porous, uncemented accumulations of calcium carbonate ($CaCO_3$) relative to those in the Indo-Pacific and Caribbean. They are generally small in areal extent (usually a few hectares), limited to shallow depths (<15 m), patchily distributed, and likely ephemeral on geologic time scales (Glynn et al. 1972; Dana 1975; Manzello 2009). This poor coral reef development in the ETP is thought to be a consequence of multiple stresses that place them at the edge of suitable environmental conditions for reef accretion: colder temperatures and high turbidity from frequent upwelling (Dana 1975) and El Niño-related climate variability (Glynn and Colgan 1992). Recently, it has been appreciated that these reefs are also exposed to chronic high-pCO_2 conditions as a result of upwelling and thermocline shoaling that depresses the saturation state of aragonite (Ω_{arag}) throughout the ETP (Manzello et al. 2008).

18.2 Climate Change

18.2.1 Ocean Warming

Climate change models forced with anthropogenic greenhouse gases predict continued and accelerated global warming due to rising concentrations of CO_2, primarily from the combustion of fossil fuels since the industrial revolution (IPCC 2013). Global mean surface temperature has increased ~0.85 °C during the period 1880–2012 and is expected to increase by 2.6–4.8 °C under Representative Concentration Pathway (RCP) 8.5—the worst case scenario in the latest IPCC report, which has emission rates at or below current levels of atmospheric CO_2 and emission acceleration (IPCC 2013). Proposed increases in temperature are of concern because reef-building corals live at or near their upper thermal limits. Positive ocean temperature anomalies of as little as 1–2 °C sustained for one to two months during the warmest part of the year correlate with mass-coral bleaching and mortality events (Glynn 1993; Goreau and Hayes 1994; Brown 1997; Berkelmans 2002) and such temperature anomalies have been shown experimentally to reproduce this bleaching and mortality (Glynn and D'Croz 1990; Berkelmans and Willis 1999). It is expected that continued warming will increase both the frequency and severity of thermal stress and bleaching events over the next century (Hoegh-Guldberg 1999; Donner et al. 2005, 2009; Logan et al. 2014; van Hooidonk et al. 2014). Thermal stress events and resultant mass coral bleaching have indeed become more severe and widespread over the past 20–30 years (Baker et al. 2008; Eakin et al. 2009), lending support to the notion that reef communities are already being severely impacted by warming (e.g., Eakin et al. 2010).

The response of ETP coral reefs to warming associated with the El Niño-Southern Oscillation (ENSO) is a cornerstone in the understanding of the effects of climate change on coral reefs (e.g., Hughes et al. 2003). The first warning signs of global warming impacts on coral reef ecosystems manifested in the ETP (Glynn 1983, 1984, 1991). While some disregarded these observations when first presented, the deleterious effects of climate change on coral reefs have now been labeled as 'incontrovertible' (Hughes et al. 2003) and are widely accepted as one of the leading threats to the persistence of coral reefs (Kleypas and Eakin 2007). The IPCC (2014) has identified coral reefs as one of the marine ecosystems most vulnerable to damage from climate change.

18.2.2 Ocean Acidification

The oceans have taken up approximately 30 % of the anthropogenic CO_2 released into the atmosphere since the industrial revolution (Sabine et al. 2004; Orr et al. 2005; Le Quéré et al. 2009), causing an acidification of the surface ocean in equilibrium with the lower atmosphere (Caldeira and Wickett 2003). Ocean acidification (OA) decreases the concentration of the carbonate ion in seawater [CO_3^{2-}] and, consequently, decreases the saturation state of carbonate minerals ($\Omega = [CO_3^{2-}][Ca^{2+}]/K'_{sp}$, where K'_{sp} is the solubility product for a carbonate mineral). Acidification is expected to depress coral reef calcification, increase reef dissolution and bioerosion, and reduce reef resilience (Kleypas et al. 1999; Yates and Halley 2006; Tribollet et al. 2009; Anthony et al. 2011)—making coral reefs, along with Arctic ecosystems, one of the two ecosystems considered at greatest risk to ocean acidification (IPCC 2014). Despite these concerns, our understanding of how the combined effects of warming and OA will alter the structure and function of coral reef ecosystems is rudimentary.

Surface waters in many parts of the ETP have lower pH, lower Ω_{arag}, and higher pCO_2 (the partial pressure of CO_2) values relative to other tropical waters because CO_2-enriched deep waters are upwelled to the surface layers along the shallow thermocline (Fig. 18.1) (Takahashi et al. 1997;

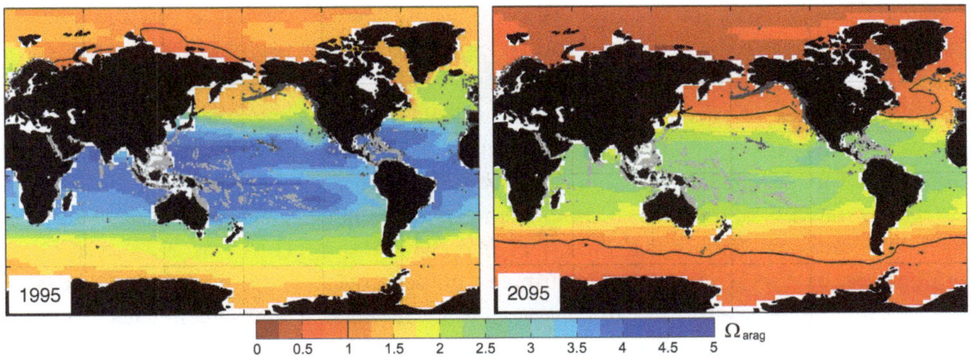

Fig. 18.1 Predicted Ω_{arag} values with increasing atmospheric CO_2 (ppm). Adapted from Feely et al. (2009)

Manzello et al. 2008; see Chap. 3, Fiedler and Lavín). The intensity of this upwelling varies regionally and strongly influences reef development across the ETP (reviewed by Cortés 1997). This unique oceanography of the ETP makes it an exemplary natural laboratory to study the effects of high-pCO_2 conditions on coral reef ecosystem processes and responses to disturbance (Manzello 2009). While other high-pCO_2 tropical reef sites are known in the Pacific (e.g., Jarvis, Line Islands and CO_2 seeps at Maug in the Mariana Islands and Papua New Guinea), these sites are poorly characterized and lack the long-term ecological understanding of the ETP sites (e.g., Fabricius et al. 2011 in contrast with Glynn et al. 2001). Bermuda experiences the lowest Ω_{arag} values in the Atlantic because it is a high-latitude reef system that experiences seasonally low temperatures but the typical values there ($\Omega_{arag} \geq 3$: Bates et al. 2009) are still greater than most ETP reef sites (Manzello 2010a). Some of the lowest known Ω_{arag} values ever documented for any coral reef site in the world are those from the Galápagos Islands (Manzello et al. 2008; Manzello 2010a).

ETP coral reefs provide a real-world example of reef growth, development, structure and function under high-pCO_2, low-Ω_{arag} conditions that encompass the range of expected changes for the entire tropical surface ocean with a doubling to tripling of atmospheric CO_2 (Fig. 18.1) (Manzello et al. 2008; Manzello 2009, 2010a). Reef structural development is highly limited in the marginal low-Ω_{arag} environment of the ETP and, perhaps most striking, ephemeral on geologic timescales (Glynn and Colgan 1992; Macintyre et al. 1992). These high-pCO_2 ETP reefs have been instrumental to understanding the anticipated impacts of OA on coral reefs in the real-world (see discussion of ETP reefs in Cohen and Holcomb 2009; Doney et al. 2009; Hofman et al. 2010; Hoegh-Guldberg 2011; Pandolfi et al. 2011, and others). The naturally high-pCO_2 of the ETP causes reefs in this region to persist near the Ω_{arag} distributional threshold for coral reefs (Kleypas et al. 1999). As a result, these high-pCO_2 reefs are ocean acidification 'hotspots' because they may: 1) display the first negative effects of, and 2) be the most affected by rising CO_2 levels (Manzello et al. 2008; Manzello 2010a). It must be noted that the ETP will experience the most modest oceanic decrease in Ω_{arag} owing to the already very low Ω_{arag} and large natural variability of this region due to effects of upwelling and ENSO on carbonate chemistry (Friedrich et al. 2012). It is possible that the high natural variability and chronically low Ω_{arag} of the ETP could already have led to adaptation to high pCO_2, possibly making resident organisms more tolerant to future acidification. The rate and magnitude of change in Ω_{arag} due to ocean acidification also is important to the impact experienced by organisms. Locations like the Caribbean Sea, despite currently having higher absolute Ω_{arag} values than the ETP, will experience a larger magnitude and rate of change in Ω_{arag} with ocean acidification (Friedrich et al. 2012). Regardless, the ETP will always have the lowest Ω_{arag} values of the tropics. Thus, if there are critical threshold values of Ω_{arag} to reef function, ETP reefs will be the first to cross that value, and may arguably have already done so in certain locations like the southern Galápagos Islands and for certain reef processes.

18.3 Coral Reef Carbonate Budget

18.3.1 $CaCO_3$ Production

There are many calcifying organisms that inhabit coral reefs. Some $CaCO_3$ production contributes to the reef framework (e.g., reef building corals and crustose coralline algae), some to reef sediments [e.g., detrital skeletal material, some articulated calcareous algae (e.g., *Halimeda*)], and some to coating and binding reef materials (e.g., encrusting coralline algae). Corals are generally the largest producers of reefal $CaCO_3$ and serve as the major building blocks of reef framework accumulation (Scoffin et al. 1980; Hubbard et al. 1990; Eakin 1996; Macintyre 1997). An exception to this can be found in environments where corals are unable to live, or where their skeletal remains cannot accumulate into reef framework structures, such as high surge environments like algal ridges (Adey 1978; Grigg 1998). Crustose coralline algae (CCA) also may opportunistically expand to cover space after coral mortality (Eakin 1992) and can indeed cover large areas of reef framework (Eakin 1992, 1996). Frequently, while non-coral $CaCO_3$ production by certain organisms can be quite high, (e.g., *Halimeda*: Rees et al. 2007), the majority of this $CaCO_3$ production generally ends up in the sediment aprons that flank reef frameworks and infilling reef cavities, and does not contribute significantly to the buildup of the rigid, three-dimensional wave resistant structure of the reef unless bound by cements (Orme 1977; Hubbard et al. 1990). CCA, bryozoans, vermetid gastropods, barnacles, and non-boring bivalves can encrust reef frameworks and contribute to the carbonate budget—in some cases becoming major parts of the overall reef carbonate budget. As in most other coral reefs, the carbonate production rates and contribution to reef framework building of these organisms are lower than corals in the ETP (Glynn and Macintyre 1977; Macintyre et al. 1992; Cortés et al. 1994; Eakin 1996).

Another source of $CaCO_3$ production on coral reefs is carbonate cementation. Cementation is the precipitation of $CaCO_3$ that acts to bind framework components and occlude porosity (Perry and Hepburn 2008). The precipitation of much of the aragonite cement in reef environments was originally thought to be inorganic, or abiotic, driven by the carbonate supersaturation of reef waters, but this is still debated (Nothdurft and Webb 2009). The distribution of most other cement types (i.e., high-mg calcite cements) within coral skeletal pores and reef rock suggests that their precipitation is a result of direct microbial precipitation or microbially mediated microenvironments, although broader scale physico-chemical control likely plays an important role (Nothdurft and Webb 2009). For example, endolithic algae are a nearly ubiquitous feature of coral skeletons living directly below the live coral tissue layer that, through photosynthetic CO_2 removal, may facilitate the precipitation of calcium carbonate cements. The high-energy seaward margins of exposed oceanic reefs are usually the most cemented reef formations and cement abundance decreases (often to zero) as water motion decreases across reef crests and into inner shelves and lagoons (James et al. 1976; James and Ginsburg 1979). $CaCO_3$ does precipitate inorganically outside of high-flow areas (e.g., lagoonal environments), but these precipitates usually form unlithified mud that does not bind substrate components (Macintyre and Aronson 2006).

Many processes, particularly those biogeochemical processes that affect reef porewater chemistry, influence reef cementation (James et al. 1976; James and Ginsburg 1979; Macintyre and Marshall 1988; Tribble et al. 1992; Tribble 1993; Rasser and Riegl 2002; Perry and Hepburn 2008). Advection of seawater supersaturated with respect to $CaCO_3$ into reef frameworks is considered a prerequisite for extensive cementation (Buddemeier and Oberdorfer 1986; Tribble et al. 1992). As a result, anthropogenic acidification may reduce future cement precipitation (Andersson et al. 2003) by reducing Ω_{arag} in the oceanic waters bathing reefs. Those cements composed of the more soluble high-mg calcite phase of $CaCO_3$ are expected to be most affected by OA (Andersson et al. 2008). Indeed, $CaCO_3$ cements do not currently precipitate above trace levels in the ETP (Fig. 18.2) (Manzello et al. 2008) and are not found in fossil sediments for the last several thousand years (Glynn and Macintyre 1977). Concomitantly, rates of bioerosion in the ETP are the highest measured anywhere in the world (Manzello et al. 2008; see Chap. 12, Alvarado et al.).

Processes of construction and destruction are often closely balanced on coral reefs (Glynn and Manzello 2015). As such, any disturbance that impairs coral growth can potentially drive coral reefs into a state of degradation. Coral mass mortality

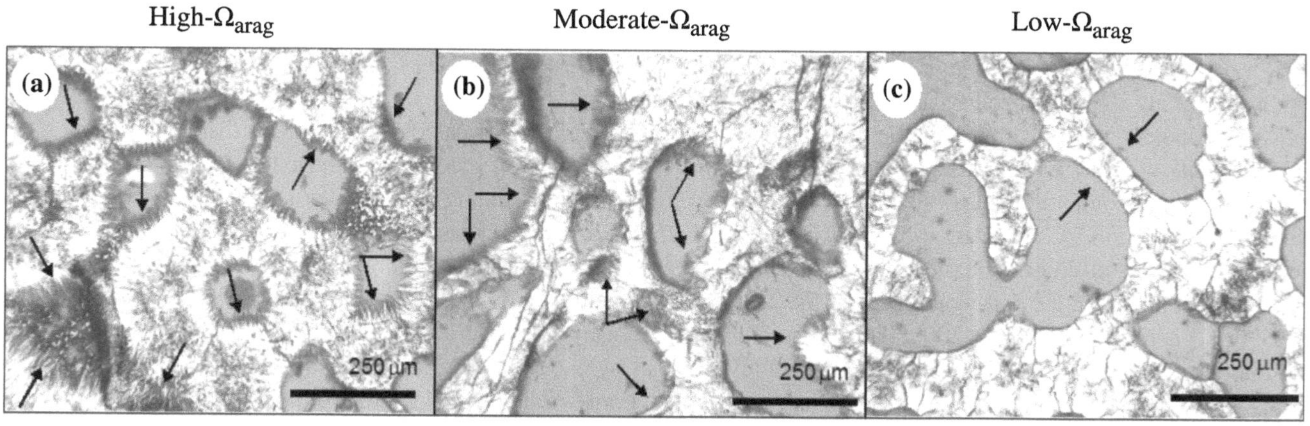

Fig. 18.2 Thin-section photomicrographs of cement distributions. **a** Abundant cementation in the intraskeletal cavities of a coral from high-Ω_{arag} environment in Bahamas ($\Omega_{arag} > 4$). *Arrows point* to examples of aragonite cement crystals. **b** Example of most heavily cemented sample from moderate-Ω_{arag} in Panama. **c** Sample from low-Ω_{arag} Galápagos ($\Omega_{arag} < 3$), in which cement is absent from all intraskeletal pores. From Manzello et al. (2008, © PNAS)

events illustrate a tipping point in the dynamics between reef construction and destruction; as such disturbances are usually followed by a decline in topographic complexity and can even result in the loss of framework structures (Alvarez-Filip et al. 2009; Manzello 2009). Such structural degradation was observed in Panama and the Galápagos Islands after the 1982–83 El Niño event (Glynn 1988; Eakin 2001).

18.3.2 $CaCO_3$ Loss

18.3.2.1 Bioerosion

The breakdown and erosion of coral skeletons and reef framework structures by the activities of organisms is termed bioerosion (Neumann 1966). Bioerosion is a fundamental process that can strongly limit the growth rates and patterns of coral reef development and sustainability if the balance between constructional and destructional processes is altered (Goreau and Hartman 1963; Risk and MacGeachy 1978; Scoffin et al. 1980; Hutchings 1986; Tribollet and Golubic 2005). There are a wide variety of organisms that bioerode coral reef carbonates. These organisms have been characterized according to whether they live externally or internally within calcareous structures, their mechanisms of bioerosion (e.g., grazers, borers, scrapers), and the size of the responsible organism (e.g., macro- vs. microborers; see Chap. 12, Alvarado et al.).

Sea urchins are external bioeroders that play a significant role in the ETP (Glynn et al. 1979, 2015; Glynn 1988; Eakin 1996). The long-spined (*Diadema mexicanum*) and pencil-spined sea urchins (*Eucidaris galapagensis*) are the dominant eroders in terms of $CaCO_3$ degradation per unit area in Panama and the Galápagos Islands, respectively (Glynn et al. 1979; Eakin 1996). The corallivorous pufferfish, *Arothron meleagris*, is also an external bioeroding agent on ETP reefs, but contributes less to the total $CaCO_3$ breakdown on these reefs (Glynn et al. 1972; Guzmán and Robertson 1989; Palacios et al. 2014). Recent research has shown that corallivory by *A. meleagris* can remove up to 16 % of the annual carbonate produced by pocilloporid corals (Palacios et al. 2014). Bioerosion by other fishes on ETP reefs (e.g., scarids, balistids, monacanthids) plays only a minor role in the $CaCO_3$ budget (Eakin 1996).

An entire community of other eroders lives and produces sediment within branching coral frameworks. This community includes internal eroders such as clionaid sponges, endolithic algae, polychaetes, sipunculans, and lithophagine bivalves, and external eroders such as polychaetes, gastropods, and crustaceans. These organisms contributed equally to sediment production as echinoids during erosion experiments in the Galápagos and Panama (Glynn 1988) and were the largest biogenic source of sediment in carbonate budgets in Panama (Eakin 1996).

Among the intraskeletal bioeroders, boring bivalves are recognized as having the greatest impact at several ETP sites (Cantera et al. 2003; Londoño-Cruz et al. 2003; Fonseca et al. 2006; see Chap. 12, Alvarado et al.). *Lithophaga* spp. bivalves can be highly abundant in the ETP, primarily within the skeleton of the coral *Porites lobata*, but also in *Pocillopora* spp. In the ETP, these boring molluscs achieve the highest densities (100s of individuals per m^2 surface area; Scott et al. 1988; Kleemann 1990, 2013) that have ever been documented, where they significantly compromise the integrity and strength of the skeleton they bore into (Scott and Risk 1988) and where feeding on them by triggerfish can cause significant coral fragmentation (Guzmán and Cortés 1989; see Chap. 15, Glynn et al.). Polychaete worms can also be locally abundant within massive pavonid corals in the ETP (Manzello, pers. obs.). Boring sponges (e.g., *Cliona* spp.), which can be dominant agents of $CaCO_3$ breakdown in other geographic areas like the Caribbean (Neumann 1966; Scoffin et al. 1980), appear to play a more limited role in the ETP (Scott et al. 1988; Carballo et al. 2008). This is somewhat paradoxical given that these organisms are thought to be stimulated by elevated nutrient levels (Risk and MacGeachy 1978; Rose and Risk 1985) and upwelling is a source of elevated nutrient concentrations throughout the ETP (D'Croz and O'Dea 2007). In general, there is confusion as to the importance of internal boring organisms as their numbers can dramatically vary in abundance across reef zones and from reef to reef (Macintyre 1984). Recently, the importance of endolithic algae in the process of bioerosion has been recognized (Tribollet et al. 2002; Tribollet and Golubic 2005). The role of these important bioeroders deserves further attention in the ETP, especially given that their rate of dissolution is stimulated by OA (Tribollet et al. 2009; Fang et al. 2013, 2014).

18.3.2.2 Physical Erosion (Wave Energy, Storms)

The majority of coral reefs in the ETP exist in sheltered environments and are not exposed to high levels of wave energy (see Chap. 6, Toth et al.). As such, physical erosion of $CaCO_3$ reef structures is, at best, an episodic occurrence in the ETP that is limited only to the strongest storm events. Furthermore, most ETP reefs occur outside of the tropical cyclone belt, but occasionally tropical storms can affect reef communities (e.g., Lirman et al. 2001). Nevertheless, physical disturbance can be relegated to a minor role in the breakdown and destruction of ETP reef framework structures.

18.3.2.3 Chemical Erosion (Dissolution)

A global model of carbonate production showed that most coral reefs could experience net dissolution with a doubling of atmospheric CO_2 due to expected declines in calcification, even if rates of dissolution do not change with acidification (Silverman et al. 2009). It has been suggested that reef dissolution will increase via the enhancement of pore-water acidity within reef frameworks and sediments as OA progresses (Yates and Halley 2006). Kinsey (1978) pointed out, however, that rates of reef dissolution are erratic, unpredictable, and highly variable. As such, we currently have a very limited understanding of the overall magnitude that purely chemical dissolution plays in reefal $CaCO_3$ in general, and especially among ETP reefs. The dissolution signal measured in bulk seawater overlying coral reefs at night is driven, to some unknown extent, by those organisms that chemically erode carbonate (e.g., *Lithophaga*: Lazar and Loya 1991; *Cliona*: Fang et al. 2013; Wisshak et al. 2013). Preliminary data show that dissolution does occur within pore-waters of the Uva Island reef in Pacific Panama, but it still isn't clear how this relates to other reefs, if the naturally high-pCO_2 stimulates this process, or if it is inherently controlled by the oxidation of organic matter (e.g., Tribble 1993). This is a fruitful area for future research in the ETP and on reefs across the globe.

18.3.3 Carbonate Budgets of the ETP

Rapid growth of pocilloporid corals provides a competitive advantage for space on ecological timescales, while the high skeletal density of these corals imparts a greater preservation potential on geologic timescales. These traits help explain why pocilloporid corals are often the dominant reef builders throughout the ETP (Cortés 1997; Glynn and Maté 1997). The best developed reefs in the ETP occur off the Pacific coast of Panama in the nonupwelling Gulf of Chiriquí (Glynn and Macintyre 1977; Macintyre et al. 1992). Not surprisingly, rates of carbonate production by pocilloporid corals in this region are among the highest reported in the ETP, only exceeded by more recent reports from Mexico. Estimated coral carbonate production rates for the Uva Island reef, in the Gulf of Chiriquí, ranged from 64.4 to 67.2 tonnes ha^{-1} $year^{-1}$ (Table 18.1). The Secas reef, also in the Gulf of Chiriquí, revealed a similar production rate of 67 tonnes ha^{-1} $year^{-1}$. The production rates for corals in the upwelling Gulf of Panama were about 30 % (Señorita reef) and 75 % (Saboga reef) of those reported under nonupwelling conditions in the Gulf of Chiriquí.

Coral carbonate production rates demonstrate high inter-annual variability, which is evident from estimates determined at the Cabo Pulmo reef for three periods from 1987 to 2012 (Table 18.1). The carbonate production rates ranged from low values of 17–39 tonnes ha^{-1} $year^{-1}$ in 2003 to high values of 89.5–203.6 tonnes ha^{-1} $year^{-1}$ in 1987. The very high rates of production in Mexico are based on areas of the reef with very high coral cover and may overestimate whole-reef production (Calderon-Aguilera, pers. commn.). We expect that reef- scale carbonate production tracks SST and Ω_{arag} across the wider ETP, as it does in Panama and Galápagos (Manzello et al. 2008, 2014). Individual coral calcification rates and overall population-level $CaCO_3$ production of the coral species *Porites panamensis* is indeed positively and linearly related to SST in Mexico (Nozagaray-López et al. 2014). Other large differences are related to reductions in live coral cover, a result of severe coral mortality caused by thermal bleaching and physical storm damage. The significant declines in production after 1987 were attributed to the 1997 El Niño warming event and Hurricane Isis in 1998 (Calderón-Aguilera et al. 2007). A large increase in carbonate production by 2012 was accompanied by an increase in live coral cover, a result of community recovery. Three other coral reefs on the Mexican mainland (Chimo, Tenacatita, and La Entrega) demonstrated similar inter-annual differences in carbonate production depending on variations in local live coral cover levels (Table 18.1).

The $CaCO_3$ budgets of ETP reefs are highly dynamic, with many reefs switching from net deposition before to net erosion following the 1982–83 ENSO-related mass mortality of corals (Glynn 1988; Eakin 1996; Reaka-Kudla et al. 1996). Additionally, the lack of cementation has produced reefs consisting of fragile stacks of coral skeletons. In fact, many reef structures in the Galápagos Islands were completely eroded down to the antecedent basalt foundation in less than a decade (Glynn 1994; Manzello 2009). Reef structures have persisted in Panama, despite the switch from net construction to net erosion following two severe ENSO events (Eakin 1996, 2001). Figure 18.3 shows a diagrammatic budget for the solid-phase $CaCO_3$ in the Uva reef system before and after the extensive bleaching caused by the 1982–83 El Niño. The major erosion that followed the 1983 coral bleaching and mortality in Panama did, however, convert a large section of the Uva reef structure to sediment, converting reef framework to a sandy plain (Eakin 2001). It is important to note that each El Niño event has a very different downstream effect. While the 1997–98 event has been argued by some to be the largest El Niño in the historical record (Wolter and

Table 18.1 Coral carbonate production rates across the eastern tropical Pacific

Study Site		Species-composition	Live coral cover (%)	Production[a] metric tons ha^{-1} $year^{-1}$	Authority
Cabo Pulmo Mexico	23°25′N	*Pocillopora, Porites, Pavona, Psammocora*	62	89.5–203.6 (1987)	Reyes-Bonilla (1993)
			12	17–39 (2003)	Calderón-Aguilera et al. (2007)
			25.7	68 (2012)	Reyes-Bonilla et al. (2014)
Chimo Mexico	20°2′N	*Pocillopora*	5.4–6.2	9.2–20.0 (2003)	Calderón-Aguilera et al. (2007)
Tenacatita Mexico	19°16′N	*Pocilliopora*	52.9	88.8–204.0 (2003)	Calderón-Aguilera et al. (2007)
La Entrega Mexico	15°45′N	*Pocillopora*	40.9	65.5–150.5 (2003)	Leyte-Morales (2001)
Saboga Is Panamá	8°38′N	*Pocillopora, Pavona, Porites, Gardineroseris, Psammocora*	45–55	51 (1971)	Glynn (1976) Glynn (1977)
Señorita Is Panamá	8°27′N	*Pocillopora*	25–30	20.5 (1971)	Glynn et al. (1972)
Secas Is Panamá	7°57′N	*Pocillopora*	34	67 (1972)	Glynn et al. (1972) Glynn (1976)
Uva Is Panamá	7°49′N	*Pocillopora,*	36–39	64.4 (pre-1983)	Wellington and Glynn (2007)
		Gardineroseris, Psammocora	25.4	67.2 (post-1983)	Eakin (1996)
Gorgona Is Colombia	2°58′N	*Pocillopora*	49.6	56	Palacios et al. (2014)
Devil's Crown Galápagos	1°12′S	*Pocillopora,*	10	10 (1975–1976)	Glynn et al. (1979)
		Porites, Pavona, Psammocora	50	55	
Floreana Is Galápagos	1°16′–1°19′S	*Pocillopora Porites, Pavona, Psammocora*	0.91	0.082 (2000–2001)	Bustamante et al. (2008) Okey et al. (2004)

[a]Minimum and maximum production values are indicated for Mexican sites. Years sampled are noted in parentheses

Timlin 1998; McPhaden 1999; Wang and Weisburg 2000), warming patterns in the eastern tropical Pacific were quite different than those in 1982–83, resulting in less coral bleaching in Panama (Glynn et al. 2001). The rapid destruction and disappearance of Galápagos reefs is the end result of several cascading factors that include the extreme thermal anomaly with its resultant mass coral bleaching and mortality, less preexisting amounts of reef framework compared to Panama, persistent high abundances and concentration of bioeroding echinoids, and poor sexual recruitment by corals (Glynn 1994). The role of high-pCO_2, low-Ω seawater that minimizes carbonate cement precipitation and facilitates bioerosion in the ETP together add a key piece to the puzzle as to why reefs throughout the ETP are poorly developed and ephemeral on geologic time-scales (Manzello et al. 2008).

18.4 ETP Carbonate Budgets with Warming and Acidification

18.4.1 Effects of Warming/Bleaching on $CaCO_3$ Budget: Acute, Pulse Stressor

To date, thermal stress has been an acute disturbance that has occurred with approximate ENSO periodicity in the ETP. An *acute*, or *pulse* stressor is a single, short-term disturbance that alters the environment (Connell 1997). In contrast, a *chronic*, or *press* stressor is a condition that persistently alters the environment. Acute thermal stress and bleaching are known to reduce coral skeletal extension and calcification (Goreau and Macfarlane 1990; Leder et al. 1991; Manzello et al. 2015a, b). In the best-case scenario, ocean warming will negatively affect $CaCO_3$ production due to

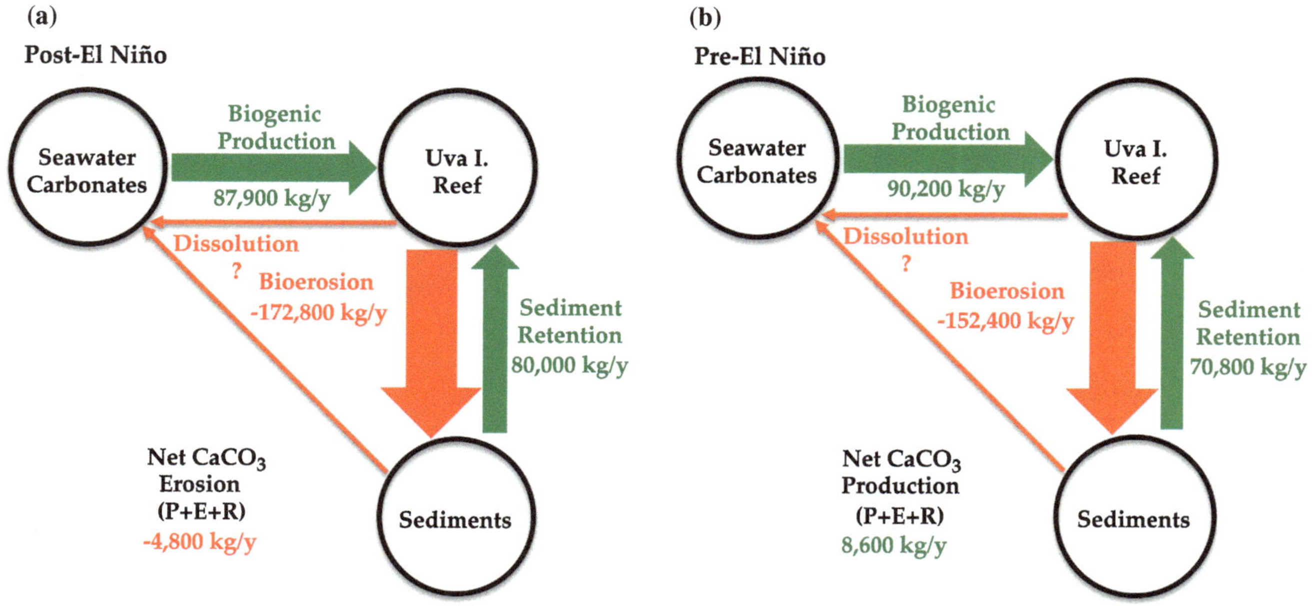

Fig. 18.3 Graphical diagram of $CaCO_3$ fluxes in Uva Island coral reef system, showing flux values from model (rounded to 100 kg/y). **a** Budget after extremely strong 1982–83 El Niño. **b** Budget prior to extremely strong 1982–83 El Niño. Positive fluxes and values in *green*; negative fluxes and values in *red*. Width of *arrows* is proportional to magnitude of fluxes. After Eakin (1996)

depressions in coral growth. If the frequency of thermal stress events increases or if temperatures surpass the optimal range for coral growth, warming would then act like a chronic disturbance, perennially reducing growth. In fact, one potential "nugget of hope" identified for corals is the switching or shuffling of zooxanthellae to clades that are more resistant to thermal stress (e.g., Baker 2001, see Chap. 13, Baker et al.). Thermally-resistant zooxanthellae, such as Clade D *Symbiodinium*, are less productive than other clades (Stat and Gates 2011). Such symbiont switching or shuffling as a response to a warming climate is likely to slow coral growth (Little et al. 2004) and reduce coral fitness (Jones and Berkelmans 2011). Additionally, the growth and calcification of corals and coral reefs will decline due to mortality associated with thermal stress, regardless of acidification.

While most coral species, including the major reef builders, have regained their former abundances in Panama since the 1982–83 El Niño, one species may have been driven to extinction. *Millepora boschmai* is a species of hydrocoral in the ETP that may have already suffered extinction as a result of thermal stress-induced bleaching (Glynn 2011). It was originally believed extinct after the 1982–83 El Niño (Glynn and de Weerdt 1991), but a small population was later located (Glynn and Feingold 1992). This remnant population died during thermal stress from the 1997–98 El Niño event (Glynn et al. 2001). In total, three hydrocorals (*Millepora platyphylla, M. boschmai, M. intricata*) and three scleractinians (*Pavona chiriquiensis, Porites panamensis, Gardineroseris planulata*) experienced presumed extinctions, local extirpations or severe reductions in abundances in the ETP following the 1982–83 and 1997–98 ENSOs (Glynn et al. 2001). Remnant patches of *G. planulata* that survived the 1982–83 ENSO in the Galápagos Islands also disappeared following the 1997–98 event, causing local extinctions (Glynn et al. 2001). While *G. planulata* is a significant reef-builder on some ETP reefs, most of those driven to extinction or extirpation were not. Even so, the potential for the extirpation or extinction of species from ETP reefs points to the impact that high temperatures can have on corals and even the extinction of coral species in general (Brainard et al. 2013).

18.4.2 Effects of Acidification on $CaCO_3$ Budget: Chronic, Press Stressor

Unlike thermal stress, OA is a chronic disturbance. The majority of research indicates that coral calcification will decline as a function of increasing atmospheric CO_2 (e.g., Langdon and Atkinson 2005). Yet the response to high-pCO_2 is not equal for all coral species, with some species being more tolerant to high-pCO_2 conditions than others in both laboratory and field studies (Langdon and Atkinson 2005; Manzello 2010b; Rodolfo-Metalpa et al. 2011; McCulloch et al. 2012). High-pCO_2 and temperature may interact such that the effects of high-pCO_2 do not manifest at low temperatures or the effect of one variable is exacerbated by the other (Reynaud et al. 2003; Anthony et al. 2008; Rodolfo-Metalpa et al. 2011). Growth declines may not begin until a threshold temperature is reached or exceeded. This, or perhaps the ability of some coral species to better up-regulate the pH of their calcifying fluid than others (McCulloch et al. 2012), may explain why declines in massive coral growth have not been observed in some field studies despite rising pCO_2 (Helmle et al. 2011; Cooper et al. 2012). Heterotrophic feeding and elevated nutrients may counteract the effect of declining Ω over some unknown, but limited range providing some degree of optimism as OA progresses (Cohen and Holcomb 2009). The relatively nutrient rich waters of the ETP may foster enhanced heterotrophy or fertilization of endosymbiotic zooxanthellae, enhancing coral calcification in waters that might otherwise have sub-optimal Ω_{arag} for reef growth (Manzello et al. 2014).

In addition to the effects on calcification, there is evidence that reduced pH has a negative effect on coral reproduction, with impacts on fertilization, larval settlement, and growth (Albright et al. 2010) and additionally on larval oxygen consumption and metamorphosis (Nakamura et al. 2011). Impaired recruitment reduces long-term reef growth, especially during recovery from disturbance. Sexual recruitment has been observed to be especially low in the ETP (Glynn et al. 1991, 2011; Glynn and Fong 2006).

Coral growth rates have declined over the past 30 years in the western Pacific, ETP, Indian, and North Atlantic Oceans (Edmunds 2007; Bak et al. 2009; De'ath et al. 2009; Tanzil et al. 2009; Manzello 2010b). *Pocillopora damicornis* has undergone a significant decline in growth at the Uva reef in Panama at a similar rate (approximately 1 % $year^{-1}$) to that documented in two independent studies from the Great Barrier Reef and Thailand (Manzello 2010b). Not all species of coral in Panama, however, have exhibited growth declines (Manzello 2010b).

Selection for thermally tolerant symbionts by ENSO-related bleaching may be causing or contributing to declines in the growth rates of certain species of corals (Manzello 2010b). For instance, pocilloporid corals have shown an increased tolerance to recurrent thermal stress events in Panama because they more frequently host thermo-tolerant symbiotic algae (Glynn et al. 2001; see Chap. 13, Baker et al.). Massive pavonid corals have shown less tolerance to thermal stress presumably because of less flexibility in hosting resistant algal endosymbionts (Glynn et al. 2001). The cost of this increased temperature tolerance in the Pocilloporidae, however, may be a more rapid decline in growth rate than what is anticipated from acidification. Growth rates of *P. damicornis* are correlated with Ω_{arag} across the ETP (Glynn 1977; Manzello 2010b). Conversely, massive pavonid corals have extension rates as high or higher in low-Ω_{arag} environments, which suggests that these species may be more tolerant to acidification than the Pocilloporidae. These differing sensitivities to thermal stress and ocean acidification will be a fundamental determinant of eastern tropical Pacific coral reef community structure with accelerating climate change that has implications to the future of reef communities worldwide.

The biologically-mediated dissolution of $CaCO_3$ by both clionaid sponges and endolithic algae is enhanced by OA (Tribollet et al. 2009; Wisshak et al. 2012; Fang et al. 2013, 2014; Reyes-Nivia et al. 2013; Enochs et al. 2015). Tribollet et al. (2009) found that filaments of the ubiquitous endolithic boring alga, *Ostreobium querkettii*, were able to penetrate deeper into coral rock substrates under high-CO_2 conditions, leading to an increase in $CaCO_3$ dissolution. Reyes-Nivia et al. (2013) observed increases in endolith biomass and respiration during combined exposure to elevated CO_2 and temperature, which led to increases in dissolution. These workers found a significant effect of substrate, as skeletons of the coral *Porites cylindrica* exhibited an increase in the relative abundance of *O. querkettii* within the endolithic community and a greater increase in endolith bioerosion when compared to the denser *Isopora cuneata*. This is intriguing as previous research has indicated that internal bioerosion increases with skeletal density (Highsmith 1981; Schönberg 2002), but the response with climate change and OA may differ and could follow an opposite pattern.

The increase in dissolution by endolithic algae, specifically *Ostreobium* with decreasing seawater pH, may also have implications for carbonate cementation of coral reefs. *Ostreobium* growth initiates endolithically within $CaCO_3$ crystals, but can later become chasmolithic (living inside cavities not of their own making) or epilithic, growing outside of the substrate and extending into the surrounding seawater (Kobluk and Risk 1977a). Following the death of the epilithic algal filaments, they can become completely calcified and this can occur within 2–3 months of substrate infestation (Kobluk and Risk 1977a, b). These calcified filaments can then coalesce to form a cement envelope that was hypothesized to reduce intergranular porosity and bind grains. This process differs from the infilling of vacated microborings by detrital or precipitated micrite (e.g., Bathurst 1966), as only external dead filaments become calcified, whereas those within borings do not.

Given that ocean acidification stimulates the growth of *Ostreobium* (Reyes-Nivia et al. 2013), it could be argued that acidification may promote cementation rather than depress it as has previously been suggested (Manzello et al. 2008) vis-à-vis the micritization process described by Kobluk and Risk (1977a, b). This is an intriguing proposition that deserves further investigation, yet by this rationale it seems that eastern Pacific coral reefs should be heavily cemented. Though the abundance and biomass of endolithic algae have not been directly assessed on ETP reefs, it seems logical to suggest that they most likely achieve high abundances as a result of the combination of low pH seawater, as well as high nutrients. The precipitation of carbonate cements is very low throughout the ETP and microborings were never infilled (Manzello et al. 2008), which is common in other shallow-water carbonate environments. Furthermore, the only limited cements that were observed were aragonitic; no high-mg calcite precipitates or cements were found. This follows the general scheme supporting an overriding thermodynamic control on cement precipitation given that high-mg calcite is more soluble than aragonite and is hypothesized to be a first responder to OA (Andersson et al. 2008). Certain organic coatings/matrices, or the creation of microenvironments due to biological activity, may ultimately be a localized driver of cementation in carbonate depositional environments, including coral reefs (e.g., Reid and Macintyre 1998), but the absence of significant cementation in ETP coral reefs suggests that seawater highly supersaturated with respect to aragonite is a prerequisite for these processes.

Bioerosion by clionaid sponges is also expected to increase in a high-CO_2 world (Wisshak et al. 2012; Fang et al. 2013, 2014; Enochs et al. 2015). Dissolution by the common Caribbean clionaid sponge *Pione lampa* (formerly *Cliona lampa*) is predicted to increase 99 % by the year 2100 due to OA, which is almost two times the anticipated depression in coral calcification (Enochs et al. 2015). Fang et al. (2013) showed that both sponge biomass and bioerosion rate increased in the zooxanthellate Pacific clionaid sponge *Cliona orientalis* with exposure to combined low pH and high temperature conditions. However, these workers found that under experimental conditions mimicking elevated temperatures and CO_2 concentration expected by the end of the century for a business-as-usual emissions scenario, *C. orientalis* bleached and *Symbiodinium* population abundances decreased with increasing CO_2. In spite of this, the bleached sponges under the highest CO_2 and temperature treatments had the greatest rates of bioerosion, even though sponge biomass declined with bleaching and was higher at lower CO_2 levels. In a related study, these workers found that *C. orientalis* consumed more carbon than it produced at high temperatures (Fang et al. 2014). As a result, the stimulation of bioerosion with ocean acidification in this species may be limited ultimately by high temperatures due to bleaching, reductions in biomass, and an overall negative energy balance (Fang et al. 2014). Zooxanthellate excavating sponges may display increased bioerosion rates with acidification to an upper thermal threshold, beyond which rates would likely decline due to bleaching, and cease, if mortality occurs. There could be an optimal pH as well where sponge dissolution peaks as rates of bioerosion decline at the highest CO_2 values for the azooxanthellate sponge *P. lampa* (Enochs et al. 2015).

18.4.3 Interacting Effects of Warming and Acidification

Historical and recent coral growth investigations have revealed the species-specific response of ETP corals to thermal stress and OA is complex. Two massive agariciid coral species (*Pavona clavus* and *Pavona gigantea*) have similar extension rates across natural CO_2 gradients, but are highly susceptible to thermal stress. Branching pocilloporid corals have demonstrated an increased tolerance to recurrent thermal stress events, potentially showing some ability to cope with warming (Glynn et al. 2001; Baker et al. 2004; however, see LaJeunesse et al. 2007, 2008). Pocilloporids have suffered a significant decline in growth rate at the Uva Island reef in Panama, however, possibly due to their increased temperature tolerance from hosting thermally resistant symbionts, OA, or both combined (Manzello 2010b). Despite their sensitivity to thermal stress, the massive pavonid corals rarely suffer complete, whole-colony mortality as tissues usually survive on the sides of the colony not exposed to direct sunlight (Glynn et al. 2001). In summary, the massive pavonids, like many other massive coral taxa, appear to be more tolerant to the combination of OA and ocean warming.

Recent work has shown that warming and acidification interact synergistically to impact corals. Anthony et al. (2008) demonstrated that both the massive coral *Porites lobata* and the branching coral *Acropora intermedia* were more susceptible to bleaching at elevated temperatures when subjected to low pH, compared to susceptibility in waters of contemporary pH levels for the Great Barrier Reef. In fact, they concluded that acidification has a greater impact on coral bleaching and productivity than on calcification. While experiments examining interactions between these factors are still at an early stage, it appears that the effects of warming and acidification may be exacerbated when both occur together. When rising temperatures remain below stressful levels, however, elevated thermal conditions may offset growth depression due to acidification (Lough and Barnes 2000; Cooper et al. 2012).

18.4.4 The Future of $CaCO_3$ Production on ETP Reefs

As stated earlier, ETP reefs provide a real-world example of coral reef growth and development in low-Ω_{arag} waters. This provides a unique opportunity to examine various aspects of $CaCO_3$ budgets under low-Ω_{arag} conditions in the present day. The precipitation of inorganic cements is highly limited in these low-Ω_{arag} reef environments. The thickness of ETP pocilloporid reef frameworks reflects Ω_{arag} in a positive, linear fashion (Fig. 18.4), illustrating the net result of coral calcification ($CaCO_3$ production) minus its loss (erosion) under high-pCO_2. These poorly cemented reef framework components are only held in place by a thin veneer of encrusting organisms, namely crustose coralline algae (CCA) and an organic matrix of sponges and other associated organisms. This point is important given that the geologic record suggests that encrustation by CCA is insignificant and subordinate to cementation in the construction and binding of framework structures (Macintyre 1997). Indeed, bioerosion rates in the Galápagos Islands and Panama are among the highest measured for any reef system to date (Glynn et al. 1979; Glynn 1988; Eakin 1996; Reaka-Kudla et al. 1996; see Chap. 12, Alvarado et al.).

Even with societal or governmental action to begin reducing emissions, in less than 40 years all of the ETP and most of the world's reefs could be at or below Ω_{arag} levels now seen in the Galápagos Islands (van Hooidonk et al. 2014). Without major changes in global fossil fuel use, corals will experience carbonate conditions similar to those of the last great extinction event during the Eocene by the end of this century (Zachos et al. 2005).

Coral reefs are declining globally and the present condition of Caribbean reefs may be the most alarming. There have been multiple interacting disturbances impacting Caribbean coral reefs that have directly caused, or acted to exacerbate large-scale coral mortality. The collapse of Caribbean coral reefs has been attributed to several factors that include: (1) white-band disease, which decimated the acroporid corals that were historically the dominant reef-builders in shallow environments, (2) the basin-wide ecological extinction of the keystone sea urchin herbivore *Diadema antillarum* due to an unidentified pathogen, (3) overfishing, (4) coral bleaching, and (5) land-based sources of pollution (Hughes 1994; Aronson and Precht 2001; Jackson et al. 2001; Eakin et al.

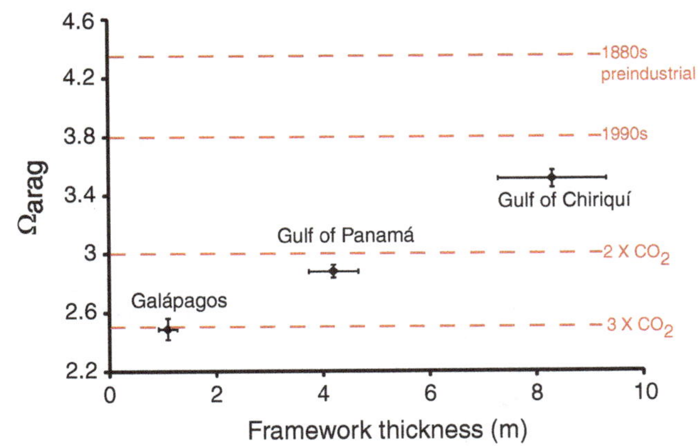

Fig. 18.4 Mean maximum coral reef framework thickness (m, ±SE) in eastern tropical Pacific relative to saturation state of aragonite, Ω_{arag}. Estimated values of Ω_{arag} for average tropical surface ocean as a function of differing levels of atmospheric CO_2 shown in red. From Manzello (2009)

2010). Live coral cover has declined by about 80 % since the 1970s, the remaining reef frameworks are losing architectural complexity, the production of $CaCO_3$ has decreased to half of historical averages, and nearly one-third of sites recently surveyed were net erosional (Gardner et al. 2003; Alvarez-Filip et al. 2009; Perry et al. 2013). Many Caribbean reefs are in *accretionary stasis*, which means that they are at or close to $CaCO_3$ budget neutral; leading to concern about the persistence of architecturally complex reef framework structures (Perry et al. 2013).

Kennedy et al. (2013) simulated carbonate budget dynamics of Caribbean coral reefs based on the expected ecological response to the latest climate projections with the interacting role of local management of fisheries and land-based sources of pollution. A trend towards net erosion was apparent under all increased temperature and CO_2 scenarios, however, local-scale management of fisheries, specifically protection of parrotfishes, delayed the transition to reef loss by a decade. Positive $CaCO_3$ budgets were generated only when local management occurred in concert with aggressive emission reductions to limit global warming to less than 2 °C. Coral calcification declines from warming and acidification were most important in carbonate budget simulations for healthy, coral-dominated reefs. The controls on overall bioerosion rate (e.g., sea urchin population sizes, sponge boring rates, nutrification) became the dominant drivers of the carbonate budget with low coral cover.

Unlike anywhere else in the world, the ETP displays a continuum of thermal stress and CO_2 inputs up to levels at which reef building is terminated and reef structures are lost (Manzello 2009). A thermal anomaly of 2–3 °C for several months killed 95–97 % of zooxanthellate corals in the Galápagos Islands during the 1982–83 ENSO (Glynn 1990). Reef structures were completely lost in <10 years in this natural laboratory (Glynn 1994; Manzello 2009). The carbonate chemistry conditions in the Galápagos Islands provide an analog to what is expected for the rest of the tropical surface ocean with a doubling ($2 \times CO_2$: 560 ppm CO_2, $\Omega_{arag} = 3.0$) and tripling of atmospheric CO_2 ($3 \times CO_2$: 840 ppm CO_2, $\Omega_{arag} = 2.5$) (Manzello et al. 2008). Ω_{arag} values in the southern Galápagos Islands are commonly <3.0 and mean values at one site have been measured as low as 2.3 (Manzello 2010a; Manzello et al. 2014). Conversely, reef frameworks, while fragile, have persisted in Panamá despite net erosion following two severe ENSO events (Eakin 2001). Ω_{arag} values in Panama range from approximately 2.8 to 3.0 in the upwelling Gulf of Panama and from 3.0 to 3.5 in the non-upwelling Gulf of Chiriquí (Manzello 2010a). The thermal anomaly in Panama was 1–2 °C for two months during the 1982–83 ENSO (Podestá and Glynn 2001; see Chap. 8, Glynn et al.). More than 75 % of all zooxanthellate corals died in Panama (Glynn 1990), but sufficient numbers of the predominant reef builders survived, and recovery has been prolific despite another very strong ENSO event in 1997–98 (Glynn and Fong 2006). This suggests that the upper limit for reef growth and persistence may be a 2–3 °C thermal anomaly sustained in excess of two months during the warmest time of the year under more than a doubling of CO_2.

ENSO variability is expected to continue into the future and the frequency of extreme El Niño events is predicted to increase (Cai et al. 2014). The models do project substantial increases in extreme temperatures that include the length, frequency, and/or intensity of warm spells (IPCC 2011). As climate change raises the baseline temperatures upon which extreme events such as El Niño occur, even such events that were not severe in the past become severe in the future. Therefore, there is reason to expect more frequent and severe coral reef bleaching in the ETP in the coming years.

The existence of coral reefs beyond this century is in jeopardy. The concern over the dramatic losses of live coral (Gardner et al. 2003) has quickly evolved to a fear that the underlying framework of coral reefs may erode away in a high-CO_2 world (Hoegh-Guldberg et al. 2007; Manzello et al. 2008; Perry et al. 2013). Coral cover has declined across large geographic scales (Gardner et al. 2003; De'ath et al. 2012), leading to concomitant declines in $CaCO_3$ production. Furthermore, declines in coral growth and calcification have been documented in all tropical seas over the past 30 years (Edmunds 2007; Cooper et al. 2008; Bak et al. 2009; De'ath et al. 2009; Tanzil et al. 2009; Manzello 2010b). The production of $CaCO_3$ on coral reefs has likely declined at a global scale. As a result, the future impacts of OA to the negative side of the coral reef carbonate budget, via stimulation of biologically-mediated dissolution (bioerosion), may be more detrimental than the decline of coral calcification. Indeed, Kennedy et al. (2013) explicitly demonstrated that changes to coral calcification are most significant to the $CaCO_3$ budget when coral cover is high. When coral cover is low, bioerosion becomes the dominant process.

In the ETP, the combination of natural ocean acidification and extreme temperatures during El Niño events has led to the development of fragile, uncemented reefs with a great susceptibility to major losses from thermal stress events. Unfortunately, global ocean acidification and warming will threaten many reefs worldwide with similar conditions within this century and the current trajectories of fossil fuel use will soon drive Ω_{arag} levels in the ETP much lower than those surrounding any coral reefs extant today. At the same time, continued warming threatens the return of mass coral bleaching. Today, coral reefs of the ETP can serve as a model for reef conditions we will likely witness in other regions within a few decades.

Acknowledgements We are grateful for funding from the National Oceanic and Atmospheric Administration's Coral Reef Conservation Program and the National Science Foundation (OCE-00002317 and OCE-0526361 to PW Glynn). The contents of this chapter are solely the opinions of the authors and do not constitute a statement of policy, decision, or position on behalf of NOAA or the U.S. Government.

References

Adey WH (1978) Coral reef morphogenesis: a multidimensional model. Science 202:831–837

Albright R, Mason B, Miller M, Langdon C (2010) Ocean acidification compromises recruitment success of the threatened Caribbean coral *Acropora palmata*. Proc Nat Acad Sci USA 107:20400–20404

Alvarez-Filip L, Dulvy NK, Gill JA, Côté IM, Watkinson AR (2009) Flattening of Caribbean coral reefs: region-wide declines in architectural complexity. Proc R Soc Lond B 276:3019–3025

Andersson AJ, Mackenzie FT, Ver LM (2003) Solution of shallow-water carbonates: an insignificant buffer against rising atmospheric CO_2. Geology 31:513–516

Andersson AJ, Mackenzie FT, Bates NR (2008) Life on the margin: implications of ocean acidification on Mg-calcite, high latitude and cold-water marine calcifiers Mar Ecol Prog Ser 373:265–273

Anthony KRN, Kline DI, Diaz-Pulido G, Dove S, Hoegh-Guldberg O (2008) Ocean acidification causes bleaching and productivity loss in coral reef builders. Proc Nat Acad Sci USA 105:17442–17446

Anthony KRN, Maynard JA, Diaz-Pulido G, Mumby PJ, Cao L, Marshall PA, Hoegh-Guldberg O (2011) Ocean acidification and warming will lower coral reef resilience. Glob Change Biol 17:1798–1808

Aronson RB, Precht WF (2001) White-band disease and the changing face of Caribbean coral reefs. Hydrobiologia 460:25–38

Bak RPM, Nieuwland G, Meesters EH (2009) Coral growth rates revisited after 31 years: what is causing lower extension rates in *Acropora palmata*? Bull Mar Sci 84:287–294

Baker AC (2001) Reef corals bleach to survive change. Nature 411:765–766

Baker AC, Glynn PW, Riegl B (2008) Climate change and coral reef bleaching: an ecological assessment of long-term impacts, recovery trends and future outlook. Estuar Coast Shelf Sci 80:435–471

Baker AC, Starger CJ, McClanahan TR, Glynn PW (2004) Coral reefs: corals' adaptive response to climate change. Nature 430:741

Bates NR, Amat A, Andersson AJ (2009) The interaction of ocean acidification and carbonate chemistry on coral reef calcification: evaluating the carbonate chemistry Coral Reef Ecosystem Feedback (CREF) hypothesis on the Bermuda coral reef. Biogeosci Discuss 6:7627–7672

Bathurst RGC (1966) Boring algae, micrite envelopes, and lithification of molluscan biospartites. Geol J 5:15–32

Berkelmans R (2002) Time-integrated thermal bleaching thresholds of reefs and their variation on the Great Barrier Reef. Mar Ecol Prog Ser 229:73–82

Berkelmans R, Willis B (1999) Seasonal and local spatial patterns in the upper thermal limits of corals on the inshore central Great Barrier Reef. Coral Reefs 18:219–228

Brainard RE, Birkeland C, Eakin CM, McElhany P, Miller MW, Patterson M, Piniak GA, Dunlap MJ, Weijerman M (2013) Incorporating climate change and ocean acidification into extinction risk assessments for 82 coral species. Conserv Biol 27:1169–1178

Brown BE (1997) Coral bleaching: causes and consequences. Coral Reefs 16:S129–S138

Buddemeier RW, Oberdorfer JA (1986) Internal hydrology and geochemistry of coral reefs and atoll islands: key to diagenetic variations. In: Schroeder JH, Purser BH (eds) Reef diagenesis. Springer, Heidelberg, pp 91–111

Bustamante RH, Okey TA, Banks S (2008) Biodiversity and food web structure of a Galápagos shallow rocky-reef ecosystem. In: McClanahan TR, Branch GM (eds) Food webs and the dynamics of marine reefs. Oxford Univ Press, Oxford, pp 135–161

Cai W, Borlace S, Lengaigne M, van Rensch P, Collins M, Vecchi G, Timmermann A, Santoso A, McPhaden MJ, Wu L, England MH, Wang G, Guilyardi E, Jin F-F (2014) Increasing frequency of extreme El Niño events due to greenhouse warming. Nat Clim Change 4:111–116

Caldeira K, Wickett ME (2003) Anthropogenic carbon and ocean pH. Nature 425:365

Calderón-Aguilera LE, Reyes Bonilla H, Carriquiry JD (2007) El papel de los arrecifes coralinos en el flujo de carbono en el océano: estudios en el Pacífico mexicano. In: Hernández de la Torre B, Gaxiola Castro G (eds) Carbono en ecosistemas acuáticos de México. México, SEMARNAT-INE-CICESE pp 215–226. ISBN: 978-968-817-855-3

Cantera JR, Orozco C, Londoño-Cruz E, Toro G (2003) Abundance and distribution patterns of infaunal associates and macroborers of the branched coral in Gorgona Island. Bull Mar Sci 72:207–219

Carballo JL, Bautista-Guerrero E, Leyte-Morales GE (2008) Boring sponges and the modeling of coral reefs in the east Pacific Ocean. Mar Ecol Prog Ser 356:113–122

Cohen AL, Holcomb M (2009) Why corals care about acidification: uncovering the mechanism. Oceanogr 22:118–127

Connell JH (1997) Disturbance and recovery of coral assemblages. In: Proceedings of 8th International Coral Reef Symposium, vol 1, Panama, pp 9–22

Cooper TF, De'ath G, Fabricius K, Lough JM (2008) Declining coral calcification in massive Porites in two nearshore regions of the northern Great Barrier Reef. Glob Change Biol 14:529–538

Cooper TF, O'Leary RA, Lough JA (2012) Growth of Western Australian corals in the Anthropocene. Science 335:593–596

Cortés J (1997) Biology and geology of eastern Pacific coral reefs. Coral Reefs 16:S39–S46

Cortés J, Macintyre IG, Glynn PW (1994) Holocene growth history of an eastern Pacific fringing reef, Punta Islotes, Costa Rica. Coral Reefs 13:65–73

Dana TF (1975) Development of contemporary eastern Pacific coral reefs. Mar Biol 33:355–374

Darwin C (1842) The structure and distribution of coral reefs. Smith, Elder and Co, London, p 214

D'Croz L, O'Dea A (2007) Variability in upwelling along the Pacific shelf of Panamá and implications for the distribution of nutrients and chlorophyll. Est Coast Shelf Sci 73:325–340

De'ath G, Lough JM, Fabricius KE (2009) Declining coral calcification on the Great Barrier Reef. Science 323:116–119

De'ath G, Fabricius KE, Sweatman H, Puotinen M (2012) The 27-year decline of coral cover on the Great Barrier Reef and its causes. Proc Natl Acad Sci USA 109:17995–17999

Doney SC, Fabry VJ, Feely RA, Kleypas JA (2009) Ocean acidification: the other CO_2 problem. Ann Rev Mar Sci 1:169–192

Donner SD, Skirving WJ, Little CM, Hoegh-Guldberg O, Oppenheimer M (2005) Global assessment of coral bleaching and required rates of adaptation under climate change. Glob Change Biol 11:2251–2265

Donner SD, Heron SF, Skirving WJ (2009) Future scenarios: a review of modelling efforts to predict the future of coral reefs in an era of climate change. In: van Oppen MJH, Lough JM (eds) Coral bleaching: patterns, processes, causes and consequences. Springer-Verlag, Berlin, pp 159–173

Eakin CM (1992) Post-El Niño panamanian reefs: less accretion, more erosion and damselfish protection. In: Proceedings of 7th International Coral Reef Symposium, vol 1, Guam, pp 387–396

Eakin CM (1996) Where have all the carbonates gone? A model comparison of calcium carbonate budgets before and after the 1982-1983 El Niño at Uva Island in the eastern Pacific. Coral Reefs 15:109–119

Eakin CM (2001) A tale of two ENSO events: carbonate budgets and the influence of two warming disturbances and intervening variability, Uva Island, Panamá. Bull Mar Sci 69:171–186

Eakin CM, Lough JM, Heron SF (2009) Climate variability and change: monitoring data and evidence for increased coral bleaching stress. In: van Oppen MJH, Lough JM (eds) Coral bleaching: patterns, processes, causes and consequences. Springer-Verlag, Berlin, pp 41–67

Eakin CM, Morgan JA, Heron SF, Smith TB, Liu G et al (2010) Caribbean corals in crisis: record thermal stress, bleaching, and mortality in 2005. PLoS ONE 5(11):e13969. doi:10.1371/journal.pone.0013969

Edmunds PJ (2007) Evidence for a decadal-scale decline in the growth rates of juvenile scleractinian corals. Mar Ecol Prog Ser 341:1–13

Enochs IC, Manzello DP, Carlton RD, Graham DM, Ruzicka R, Collela MA (2015) Ocean acidification enhances the bioerosion of a common coral reef sponge: implications for the persistence of the Florida Reef Tract. Bull Mar Sci 91:271–290

Fabricius KE, Langdon C, Uthicke S, Humphrey C, Noonan S, De'ath G, Okazaki R, Muehllehner N, Glas MS, Lough JM (2011) Losers and winners in coral reefs acclimatized to elevated carbon dioxide concentrations. Nat Clim Change 1:165–169 doi: 10.1038/NCLIMATE1122

Fang JKH, Mello-Athayde MA, Schönberg CHL, Kline DI, Hoegh-Guldberg O, Dove S (2013) Sponge biomass and bioerosion rates increase under ocean warming and acidification. Glob Change Biol 19:3581–3591 doi:10.1111/gcb.12334

Fang JKH, Schönberg CHL, Mello-Athayde MA, Hoegh-Guldberg O, Dove S (2014) Effects of ocean warming and acidification on the energy budget of an excavating sponge. Glob Change Biol 20:1043–1054 doi:10.1111/gcb.12369

Feely RA, Doney SC, Cooley SR (2009) Present conditions and future changes in a high-CO_2 world. Oceanography 22(4):37–47

Fonseca AC, Dean HK, Cortés J (2006) Non-colonial coral macro-borers as indicators of coral reef status in the south Pacific of Costa Rica. Rev Biol Trop 54:101–115

Friedrich T, Timmermann A, Abe-Ouchi A, Bates NR, Chikamoto MO, Church MJ, Dore JE, Gledhill DK, Gonzalez-Davila M, Heinemann M, Ilyina T, Jungclaus JH, McLeod E, Mouchet A, Santana-Casiano JM (2012) Detecting regional anthropogenic trends in ocean acidification against natural variability. Nat Clim Change 2:167–171 doi:10.1038/NCLIMATE1372

Gardner TA, Cote IM, Gill JA, Grant A, Watkinson AR (2003) Long-term region-wide declines in Caribbean corals. Science 301:958–960

Glynn PW (1976) Some physical and biological determinants of coral community structure in the eastern Pacific. Ecol Monogr 46:431–456

Glynn PW (1977) Coral growth in upwelling and nonupwelling areas off the Pacific coast of Panamá. J Mar Res 35:567–585

Glynn PW (1983) Extensive 'bleaching' and death of reef corals on the Pacific coast of Panamá. Environ Conserv 10:149–154

Glynn PW (1984) Widespread coral mortality and the 1982/83 El Niño warming event. Environ Conserv 11:133–146

Glynn PW (1988) El Niño warming, coral mortality and reef framework destruction by echinoid bioerosion in the eastern Pacific. Galaxea 7:129–160

Glynn PW (1990) Coral mortality and disturbance to coral reefs in the eastern tropical Pacific. In: Glynn PW (ed) Global ecological consequences of the 1982-83 El Niño-Southern Oscillation. Elsevier, Amsterdam, pp 55–126

Glynn PW (1991) Coral reef bleaching in the 1980s and possible connections with global warming. Trends Ecol Evol 6:175–179

Glynn PW (1993) Coral reef bleaching: ecological perspectives. Coral Reefs 12:1–17

Glynn PW (1994) State of coral reefs in the Galápagos Islands: natural vs. anthropogenic impacts. Mar Poll Bull 29:131–140

Glynn PW (2011) In tandem reef coral and cryptic metazoan declines and extinctions. Bull Mar Sci 87:767–794

Glynn PW, Macintyre IG (1977) Growth rate and age of coral reefs on the Pacific coast of Panama. In: Proceedings of 3rd International Coral Reef Symposium, vol 2, Miami, pp 251–259

Glynn PW, D'Croz L (1990) Experimental evidence for high temperature stress as the cause of El Niño-coincident coral mortality. Coral Reefs 8:181–191

Glynn PW, de Weerdt (1991) Elimination of two reef-building hydrocorals following the 1982-83 El Niño warming event. Science 253:69–71

Glynn PW, Colgan MW (1992) Sporadic disturbances in fluctuating coral reef environments: El Niño and coral reef development in the eastern Pacific. Am Zool 32:707–718

Glynn PW, Feingold J (1992) Hydrocoral species not extinct. Science 257:1845

Glynn PW, Maté JM (1997) Field guide to the Pacific coral reefs of Panamá. In: Proceedings of 8th International Coral Reef Symposium, vol 1, Panama, pp 145–166

Glynn PW, Fong P (2006) Patterns of reef coral recovery by the regrowth of surviving tissues following the 1997-98 EI Niño warming and 2000, 2001 upwelling events in Panamá, eastern Pacific. In: Proceedings of 10th International Coral Reef Symposium, vol 1, Okinawa, pp 624–630

Glynn PW, Stewart RH, McCosker JE (1972) Pacific coral reefs of Panama: structure, distribution, and predators. Geol Rundsch 61:483–519

Glynn PW, Wellington GM, Birkeland C (1979) Coral reef growth in the Galápagos: limitation by sea urchins. Science 203:47–49

Glynn PW, Maté JM, Baker AC, Calderón MO (2001) Coral bleaching and mortality in Panamá and Ecuador during the 1997-98 El Niño-Southern Oscillation event: spatial/temporal patterns and comparisons with the 1982-83 event. Bull Mar Sci 69:79–110

Glynn PW, Manzello DP (2015) Bioerosion and Coral Reef Growth: A Dynamic Balance. In: Birkeland C (ed) Coral Reefs in the Anthropocene. Springer, Dordrecht. pp. 67–97

Glynn PW, Riegl B, Purkis S, Kerr JM, Smith TB (2015) Coral reef recovery in the Galápagos Islands: the northernnmost islands (Darwin and Wenman). Coral Reefs 34:421–436 doi:10.1007/s00338-015-1280-4

Glynn PW, Gassman NJ, Eakin CM, Cortés J, Smith DB, Guzmán HM (1991) Reef coral reproduction in the eastern Pacific: Costa Rica, Panamá, Galápagos Islands (Ecuador). I. Pocilloporidae. Mar Biol 109:355–368

Glynn PW, Colley SB, Guzmán HM, Enochs IC, Cortés J, Maté JL, Feingold JS (2011) Reef coral reproduction in the eastern Pacific: Costa Rica, Panamá, Galápagos Islands (Ecuador). VI. Agariciidae. Pavona clavus. Mar Biol 158:1601–1617

Goreau TF, Hartman WD (1963) Boring sponges as controlling factors in the formation and maintenance of coral reefs. AAAS Spec Publ 75:25–54

Goreau TF, Macfarlane AH (1990) Reduced growth rate of *Montastraea annularis* following the 1987–1988 coral bleaching event. Coral Reefs 8:211–215

Goreau TJ, Hayes RL (1994) Coral bleaching and ocean Hot Spots. Ambio 23:176–180

Grigg RW (1998) Holocene coral reef accretion in Hawaii: a function of wave exposure and sea level history. Coral Reefs 17:263–272

Guzmán HM, Cortés J (1989) Coral reef community structure at Caño Island. Pacific Costa Rica. PSZNI: Mar Ecol 10:23–41

Guzmán HM, Robertson R (1989) Population and feeding responses of the corallivorous pufferfish *Arothron meleagris* to coral mortality in the eastern Pacific. Mar Ecol Prog Ser 55:121–131

Helmle KP, Dodge RE, Swart PK, Gledhill DK, Eakin CM (2011) Growth rates of Florida corals from 1937 to 1996 and their response to climate change. Nat Comm 2:215

Highsmith RC (1981) Coral bioerosion: damage relative to skeletal density. Am Nat 117:193–198

Hoegh-Guldberg O (1999) Climate change, coral bleaching and the future of the world's coral reefs. Mar Freshwat Res 50:839–866

Hoegh-Guldberg O (2011) The impact of climate change on coral reef ecosystems. In: Dubinsky Z, Stambler N (eds) Coral reefs: an ecosystem in transition. Springer, Berlin, pp 391–403

Hoegh-Guldberg O, Mumby PJ, Hooten AJ, Steneck RS, Greenfield P, Gomez E, Harvell CD, Sale PF, Edwards AJ, Caldeira K, Knowlton N, Eakin CM, Iglesias-Prieto R, Muthiga N, Bradbury RH, Dubi A, Hatziolos ME (2007) Coral reefs under rapid climate change and ocean acidification. Science 318:1737–1742

Hofman GE, Barry JP, Edmunds PJ, Gates RD, Hutchins DA, Klinger T, Sewell MA (2010) The effects of ocean acidification on calcifying organisms in marine ecosystems: an organism-to-ecosystem perspective. Ann Rev Ecol Evol Syst 41:127–147

Hubbard DK, Miller AI, Scutaro D (1990) Production and cycling of calcium carbonate in a shelf-edge reef system (St. Croix, U.S. Virgin Islands): applications to the nature of reef systems in the fossil record. J Sed Petrol 60:335–360

Hughes TP (1994) Catastrophes, phase shifts, and large-scale degradation of a Caribbean coral reef. Science 265:1547–1551

Hughes TP, Baird AH, Bellwood DR, Card M, Connolly SR, Folke C, Grosberg R, Hoegh-Guldberg O, Jackson JBC, Kleypas J, Lough JM, Marshall P, Nystrom M, Palumbi SR, Pandolfi JM, Rosen B, Roughgarden J (2003) Climate change, human impacts, and the resilience of coral reefs. Science 301:929–933

Hutchings PA (1986) Biological destruction of coral reefs. Coral Reefs 4:239–253

IPCC (2011) Summary for policymakers of the *Special Report on Managing the Risks of Extreme Events and Disasters to Advance Climate Change Adaptation* (SREX), p 29

IPCC (2013) Summary for policymakers. In: climate change 2013: the physical science basis. Contribution of Working Group I to the Fifth assessment report of the intergovernmental panel on climate change In: Stocker TF, Qin D, Plattner G-K, Tignor M, Allen SK, Boschung J, Nauels A, Xia Y, Bex V, Midgley PM (eds) Cambridge Univ Press, Cambridge

IPCC (2014) Climate change 2014: impacts, adaptation, and vulnerability. contribution of Working Group II to the Fifth assessment report of the intergovernmental panel on climate change. In: Field CB, Barros VR, Dokken DJ, Mach KJ, Mastrandrea MD, Bilir TE, Chatterjee M, Ebi KL, Estrada YO, Genova RC, Girma B, Kissel ES, Levy AN, MacCracken S, Mastrandrea P, White LL (eds) Cambridge Univ Press, Cambridge

Jackson JBC, Kirby MX, Berger WH, Bjorndal KA, Botsford LW, Bourque BJ, Bradbury RH, Cooke R, Erlandson J, Estes JA, Hughes TP, Kidwell S, Lange CB, Lenihan HS, Pandolfi JM, Peterson CH, Steneck RS, Tegner MJ, Warner RR (2001) Historical overfishing and the recent collapse of coastal ecosystems. Science 293:629–638

James NP, Ginsburg RN, Marzalek DS, Choquette PW (1976) Facies and fabric specificity of early subsea cements in shallow Belize (British Honduras) reefs. J Sed Petrol 46:523–544

James NP, Ginsburg RN (1979) The seaward margin of Belize barrier and atoll reefs. Spec Pub Int Ass Sed Blackwell, Oxford, p 191

Jones AM, Berkelmans R (2011) Tradeoffs to thermal acclimation: energetics and reproduction of a reef coral with heat tolerant *Symbiodinium* type-D. J Mar Bio. doi:10.1155/2011/185890

Kennedy EV, Perry CT, Halloran PR, Iglesias-Prieto R, Schönberg CHL, Wisshak M, Form AU, Carricart-Ganivet JP, Fine M, Eakin CM, Mumby PJ (2013) Avoiding coral reef functional collapse requires local and global action. Current Biol 23:912–918

Kinsey DW (1978) Productivity and calcification estimates using slack-water periods and field enclosures. In: Stoddart DR, Johannes RE (eds) Coral reefs: research methods. UNESCO, pp 439–468

Kleemann K (1990) Boring and growth in chemically boring bivalves from the Caribbean, Eastern Pacific and Australia's Great Barrier Reef. Senck Marit Frankfurt/Main 21(1):101–154

Kleemann K (2013) Fast and massive settlement of boring bivalves on coral slabs at Taboga Islands, Eastern Pacific, Panama. Boll Malacolog 49:104–113

Kleypas JA, Eakin CM (2007) Scientists' perceptions of threats to coral reefs: results of a survey of coral reef researchers. Bull Mar Sci 80:419–436

Kleypas JA, Buddemeier RW, Archer D, Gattuso J-P, Langdon C, Opdyke BN (1999) Geochemical consequences of increased atmospheric carbon dioxide on coral reefs. Science 284:118–120

Kobluk DR, Risk MJ (1977a) Micritization and carbonate-grain binding by endolithic algae. Am Assoc Petrol Geol Bull 61:1069–1082

Kobluk DR, Risk MJ (1977b) Calcification of exposed filaments of endolithic algae, micrite envelope formation and sediment production. J Sed Petrol 47:517–528

Langdon C, Atkinson MJ (2005) Effect of elevated pCO_2 on photosynthesis and calcification of corals and interactions with seasonal change in temperature/irradiance and nutrient enrichment. J Geophys Res 110:1–16

LaJeunesse TC, Reyes-Bonilla H, Warner ME (2007) Spring "bleaching" among *Pocillopora* in the Sea of Cortez, Eastern Pacific. Coral Reefs 26:265–270

LaJeunesse TC, Reyes Bonilla H, Warner ME, Wills M, Schmidt GW, Fitt WK (2008) Specificity and stability in high latitude eastern Pacific coral-algal symbioses. Limnol Oceanogr 53:719–727

Lazar B, Loya Y (1991) Bioerosion of coral reefs-a chemical approach. Limnol Oceanogr 36:377–383

Leder JJ, Szmant AM, Swart PK (1991) The effect of prolonged 'bleaching' on the stable isotope composition and banding patterns in *Montastraea annularis*. Preliminary observations. Coral Reefs 10:19–27

Leyte-Morales GE (2001) Estructura de la comunidad de corales y características geomorfológicas de los arrecifes coralinos de bahías de Huatulco, Oaxaca, México. MSc thesis, UMAR, p 94

Le Quéré C, Raupach MR, Canadell JG, Marland G, Bopp L, Ciais P, Conway TJ, Doney SC, Feely RA, Foster P, Friedlingstein P, Gurney K, Houghton RA, House JI, Huntingford C, Levy PE, Lomas MR, Majkut J, Metzl N, Ometto JP, Peters GP, Prentice IC, Randerson JT, Running SW, Sarmiento JL, Schuster U, Sitch S, Takahashi T, Viovy N, van der Werf GR, Woodward FI (2009) Trends in the sources and sinks of carbon dioxide. Nat Geosci 2:831–836

Lirman D, Glynn PW, Baker AC, Leyte-Morales GE (2001) Combined effects of three sequential storms on the Huatulco coral reef tract, Mexico. Bull Mar Sci 69:267–278

Little AF, van Oppen MJH, Willis BL (2004) Flexibility in algal endosymbioses shapes growth of reef corals. Science 304:1492–1494

Logan CA, Dunne JP, Eakin CM, Donner SD (2014) Incorporating adaptive responses into future projections of coral bleaching. Glob Change Biol 20:125–139

Londoño-Cruz E, Cantera JR, Toro-Farmer G, Orozco C (2003) Internal bioerosion by macroborers in *Pocillopora* spp. in the Tropical Eastern Pacific. Mar Ecol Prog Ser 265:289–295

Lough JM, Barnes DJ (2000) Environmental controls on growth of the massive coral *Porites*. J Exp Mar Biol Ecol 245:225–243

Macintyre IG (1984) Preburial and shallow-subsurface alteration of modern scleractinian corals. In: Oliver WA, Sando WJ, Cairns SD, Coates AG, Macintyre IG, Bayer FM, Sorauf JE (eds) Recent advances in the paleobiology and geology of the Cnidaria. Palaeontogr Americana 54:229–244

Macintyre IG (1997) Reevaluating the role of crustose coralline algae in the construction of coral reefs. In: Proceedings of 8th International Coral Reef Symposium, vol 1, Panama, pp 725–730

Macintyre IG, Marshall JF (1988) Submarine lithification in coral reefs: some facts and misconceptions. In: Proceedings of 6th International Coral Reef Symposium, vol 1, Townsville, pp 263–272

Macintyre IG, Aronson RB (2006) Lithified and unlithified Mg-calcite precipitates in tropical reef environments. J Sed Res 76:81–90

Macintyre IG, Glynn PW, Cortés J (1992) Holocene reef history in the eastern Pacific: mainland Costa Rica, Caño Island, Cocos Island, and Galápagos Islands. In: Proceedings of 7th International Coral Reef Symposium, vol 2, Guam, pp 1174–1178

Manzello DP (2009) Reef development and resilience to acute (El Niño warming) and chronic (high-CO_2) disturbances in the eastern tropical Pacific: a real-world climate change model. In: Proceedings of 11th International Coral Reef Symposium, vol 1, Ft Lauderdale, pp 1299–1304

Manzello DP (2010a) Ocean acidification hotspots: spatiotemporal dynamics of the seawater CO_2 system of eastern Pacific coral reefs. Limnol Oceanogr 55:239–248

Manzello DP (2010b) Coral growth with thermal stress and ocean acidification: lessons from the eastern tropical Pacific. Coral Reefs 29:749–758

Manzello DP, Kleypas JA, Budd DA, Eakin CM, Glynn PW, Langdon C (2008) Poorly cemented coral reefs of the eastern tropical Pacific: possible insights into reef development in a high-CO_2 world. Proc Natl Acad Sci USA 105:10450–10455

Manzello DP, Enochs IC, Bruckner A, Renaud P, Kolodziej G, Budd D, Carlton R, Glynn PW (2014) Galápagos coral reef persistence after ENSO warming across an acidification gradient. Geophys Res Lett 41(24):9001–9008

Manzello DP, Enochs IC, Kolodziej G, Carlton R (2015a) Recent decade of growth and calcification of *Orbicella faveolata* in the Florida keys: An inshore-offshore comparison. Mar Ecol Prog Ser 521:81–89

Manzello DP, Enochs IC, Kolodziej G, Carlton R (2015b) Coral growth patterns of *Montastraea cavernosa* and *Porites astreoides* in the Florida keys: the importance of thermal stress and inimical waters. J Exp Mar Biol Ecol 471:198–207

McCulloch M, Falter J, Trotter J, Montagna P (2012) Coral resilience to ocean acidification and global warming through pH up-regulation. Nat Clim Change 2:623–627

McPhaden M (1999) El Niño: the child prodigy of 1997-98. Nature 398:559–562

Nakamura M, Ohki S, Suzuki A, Sakai K (2011) Coral larvae under ocean acidification: survival, metabolism, and metamorphosis. PLoS ONE 6(1):e14521. doi:10.1371/journal.pone.0014521

Neumann AC (1966) Observations on coastal erosion in Bermuda and measurements of the boring rate of the sponge *Cliona lampa*. Limnol Oceanogr 11:92–108

Nothdurft LD, Webb GE (2009) Earliest diagenesis in scleractinian coral skeletons: implications for palaeoclimate-sensitive geochemical archives. Facies 55:161–201

Nozagaray-López CO, Calderon-Aguilera LE, Hernández-Ayón JM, Reyes-Bonilla H, Carricart-Ganivet JP, Cabral-Tena RA, Balart EF (2014) Low calcification rates and calcium carbonate production in *Porites panamensis* at its northernmost geographic distribution. Mar Ecol. doi:10.1111/maec.12227

Okey TA, Banks S, Born AF, Bustamante RH, Calvopiña M, Edgar GJ, Espinoza E, Fariña JM, Garske LE, Reck GK, Salazar S, Shepherd S, Toral-Granda V, Wallem P (2004) A trophic model of a Galápagos subtidal rocky reef for evaluating fisheries and conservation strategies. Ecol Model 172:383–401

Orme GR (1977) Aspects of sedimentation in the coral reef environment. In: Jones OA, Endean R (eds) Biology and geology of coral reefs, vol 4. Academic Press, New York, pp 129–176

Orr JC, Fabry VJ, Aumont O, Bopp L, Doney SC, Feely RA, Gnanadesikan A, Gruber N, Ishida A, Joos F, Key RM, Lindsay K, Maier-Reimer E, Matear R, Monfray P, Mouchet A, Najjar RG, Plattner G-K, Rodgers KB, Sabine CL, Sarmiento JL, Schlitzer R, Slater RD, Totterdell IJ, Weirig MF, Yamanaka Y, Yool A (2005) Anthropogenic ocean acidification over the twenty-first century and its impact on calcifying organisms. Nature 437:681–686

Palacios MM, Muñoz CG, Zapata FA (2014) Fish corallivory on a pocilloporid reef and experimental coral responses to predation. Coral Reefs 33:625–636 doi:10.1007/s00338-014-1173-y

Pandolfi JM, Connolly SR, Marshall DJ, Cohen AL (2011) Projecting coral reef futures under global warming and ocean acidification. Science 333:418–422

Perry CT, Hepburn LJ (2008) Syn-depositional alteration of coral reef framework through bioerosion, encrustation, and cementation: taphonomic signatures of reef accretion and reef depositional events. Earth Rev Sci 86:106–144

Perry CT, Murphy GN, Kench PS, Smithers SG, Edinger EN, Steneck RS, Mumby PJ (2013) Caribbean-wide decline in carbonate production threatens coral reef growth. Nat Comm 4:1–7

Rasser MW, Riegl B (2002) Holocene coral reef rubble and its binding agents. Coral Reefs 21:57–72

Reaka-Kudla ML, Feingold JS, Glynn PW (1996) Experimental studies of rapid bioerosion of coral reefs in the Galápagos Islands. Coral Reefs 15:109–119

Rees SA, Opdyke BN, Wilson PA, Henstock TJ (2007) Significance of *Halimeda* bioherms to the global carbonate budget based on a geological sediment budget for the Northern Great Barrier Reef, Australia. Coral Reefs 26:177–188

Reid RP, Macintyre IG (1998) Carbonate recrystallization in shallow marine environments: a widespread diagenetic process forming micritized grains. J Sed Res 68:928–946

Reyes-Bonilla H (1993) Estructura de la comunidad, influencia de la depredación y biología poblacional de corales hermatípicos en el arrecife de Cabo Pulmo, BCS. MSc thesis, Centro de Investigación Científica y Enseñanza Superior de Ensenada, p 169

Reyes Bonilla H, Alvarez del Castillo Cárdenas PA, Calderón Aguilera LE, Erosa Ricárdez CE, Fernández Rivera Melo FJ, Frausto TC, Luna Salguero BM, Moreno Sánchez XG, Mozqueda Torres MC, Norzagaray López CO, Petatán Ramírez D (2014) Servicios ambientales de arrecifes coralinos: el caso del Parque Nacional Cabo Pulmo, B.C.S. In: Urciaga García JI (ed) Desarrollo regional en Baja California Sur. Una perspectiva de los servicios

Reyes-Nivia C, Diaz-Pulido G, Kline D, Hoegh-Guldberg O, Dove S (2013) Ocean acidification and warming scenarios increase microbioerosion of coral skeletons. Glob Change Biol 19:1919–1929

Reynaud S, Leclercq N, Romaine-Lioud S, Ferrier-Pages C, Jaubert J, Gattuso J-P (2003) Interacting effects of CO_2 partial pressure and temperature on photosynthesis and calcification in a scleractinian coral. Glob Change Biol 9:1660–1668

Risk MJ, MacGeachy JK (1978) Aspects of bioerosion of modern Caribbean reefs. Rev Biol Trop 2:S85–S105

Rodolfo-Metalpa R, Houlbrèque F, Tambutté E, Boisson F, Baggini C, Patti FP, Jeffree R, Fine M, Foggo A, Gattuso J-P, Hall-Spencer J (2011) Coral and mollusc resistance to ocean acidification adversely affected by warming. Nat Clim Change 1:308–312

Rose CS, Risk MJ (1985) Increase in *Cliona delitrix* infestation of *Montastraea cavernosa* heads on an organically polluted portion of the Grand Cayman fringing reef. PSZNI: Mar Ecol 6:345–363

Sabine CL, Feely RA, Gruber N, Key RM, Lee K, Bullister JL, Wanninkhof R, Wong CS, Wallace DWR, Tilbrook B, Millero FJ, Peng T-H, Kozyr A, Ono T, Rios AF (2004) The oceanic sink for anthropogenic CO_2. Science 305:367–371

Schönberg CHL (2002) Substrate effects on the bioeroding demosponge *Cliona orientalis*. 1. Bioerosion rates. PSZNI: Mar Ecol 23:313–326

Scoffin TP, Stearn CW, Boucher D, Frydl P, Hawkins CM, Hunter IG, MacGeachy JK (1980) Calcium carbonate budget of a fringing reef on the west coast of Barbados. Part II—erosion, sediments and internal structure. Bull Mar Sci 30:475–508

Scott PJB, Risk MJ (1988) The effect of *Lithophaga* (Bivalvia: Mytilidae) boreholes on the strength of the coral *Porites lobata*. Coral Reefs 7:145–151

Scott PJB, Risk MJ, Carriquiry JD (1988) El Niño, bioerosion and the survival of east Pacific coral reefs. In: Proceedings of 6th International Coral Reef Symposium, vol 2, Townsville, pp 517–520

Silverman J, Lazar B, Cao L, Caldeira K, Erez J (2009) Coral reefs may start dissolving when atmospheric CO_2 doubles. Geophys Res Lett 36:L05606. doi:10.1029/2008GL03628

Stat M, Gates RD (2011) Clade D *Symbiodinium* in scleractinian corals: a “nugget” of hope, a selfish opportunist, an ominous sign, or all of the above? J Mar Biol. doi:10.1155/2011/730715

Tanzil JTI, Brown BE, Tudhope AW, Dunne RP (2009) Decline in skeletal growth of the coral *Porites lutea* from the Andaman Sea, South Thailand between 1984 and 2005. Coral Reefs 28:519–528

Takahashi T, Feely RA, Weiss RF, Wanninkhof RH, Chipman DW, Sutherland SC, Takahashi TT (1997) Global air-sea flux of CO2: an estimate based on measurements of sea-air pCO2 difference. Proc Natl Acad Sci USA 94:8292–8299

Tribble GW (1993) Organic matter oxidation and aragonite diagenesis in a coral reef. J Sed Petrol 63:523–527

Tribble GW, Sansone FJ, Buddemeier RW, Li Y-H (1992) Hydraulic exchange between a coral reef and surface sea water. Geol Soc Am Bull 104:1280–1291

Tribollet A, Golubic S (2005) Cross-shelf differences in the pattern and pace of bioerosion of experimental carbonate substrates exposed for 3 years on the northern Great Barrier Reef. Coral Reefs 24:422–434

Tribollet A, Decherf G, Hutchings PA, Peyrot-Clausade M (2002) Large-scale spatial variability in bioerosion of experimental coral substrates on the Great Barrier Reef (Australia): importance of microborers. Coral Reefs 21:424–432

Tribollet A, Godinot C, Atkinson M, Langdon C (2009) Effects of elevated pCO_2 on dissolution of coral carbonates by microbial euendoliths. Glob Biogeochem Cycl 23:GB3008 van Hooidonk R, Maynard J, Manzello DP, Planes S (2014) Opposite latitudinal gradients in projected ocean acidification and bleaching impacts on coral reefs. Glob Change Biol 20:103–112

Wang C, Weisberg RH (2000) The 1997–98 El Niño evolution relative to previous El Niño events. J Climate 13:488–501

Wellington GM, Glynn PW (2007) Responses of coral reefs to El Niño-Southern Oscillation sea-warming events. In: Aronson RB (ed) Geological approaches to coral reef ecology, Ecol Stud 192. Springer, New York, pp 342–385

Wisshak M, Schönberg CHL, Form A, Freiwald A (2012) Ocean acidification accelerates reef bioerosion. PLoS ONE 7:e45124

Wisshak M, Schönberg CHL, Form A, Freiwald A (2013) Effects of ocean acidification and global warming on reef bioerosion—lessons from a clionaid sponge. Aquatic Biol 19:111–127

Wolter K, Timlin MS (1998) Measuring the strength of ENSO events: how does 1997/98 rank? Weather 53:315–324

Yates KK, Halley RB (2006) CO_3^{2-} concentration and pCO_2 thresholds for calcification and dissolution on the Molokai reef flat, Hawaii. Biogeosci Discuss 3:123–154

Zachos JC, Rohl R, Schellenberg SA, Sluijs A, Hodell DA, Kelly DC, Thomas E, Nicolo M, Raffi I, Lourens LJ, McCarren H, Kroon D (2005) Rapid acidification of the ocean during the Paleocene-Eocene thermal maximum. Science 308:1611–1615

Reef-Based Reconstructions of Eastern Pacific Climate Variability

19

Julia Cole and Alexander W. Tudhope

Abstract

In the eastern tropical Pacific, large spatial gradients in climate conditions are associated with oceanic upwelling and the position of the Intertropical Convergence Zone (ITCZ). Dramatic shifts in these systems occur during extremes of the El Niño-Southern Oscillation (ENSO), which also reverberate throughout the global climate system. Motivated by the need to understand ENSO history, and underpinned by ecological research, many of the earliest coral paleoclimate studies originated from the eastern Pacific. Early work in Galápagos, Costa Rica, and Panama highlighted the usefulness of a diverse array of tracers that record past environmental change. Longer records of coral $\delta^{18}O$ have delineated the range of natural variability in ENSO and identified a rich spectrum of tropical Pacific variance that extends into multidecadal and century time scales. Coral-based reconstructions of ocean temperatures generally show strong warming trends, except in Galápagos, where existing records do not span the full 20th century and where strong interannual variability impedes detection of smaller trends. Coral records complement sediment-based and terrestrial records in terms of their length, resolution, and sensitivity. A more complete understanding of variability in the eastern tropical Pacific should emerge from ongoing and future work that integrates reef-based paleorecords with other paleoclimate reconstructions and model simulations. In addition, analysis of additional proxies—e.g. of circulation, pH, and salinity—is recommended to complement temperature reconstructions and provide useful reconstructions of changes in climate, oceanography, and reef stress.

Keywords

Paleoclimate • Coral • Galápagos • El Niño • Sea surface temperature

19.1 Introduction

In the eastern tropical Pacific, climate variations are large and poorly revealed by the limited instrumental datasets. The region integrates climatic influences from higher latitudes and from the equatorial and subsurface oceans, and this variability propagates globally through the El Niño-Southern Oscillation (ENSO) system and related variability. Yet across most of the eastern tropical Pacific, directly observed records of climate are limited to the past several decades. The interannual signal associated with ENSO dynamics is particularly important as a driver for regional variability and as an indicator of impacts throughout the tropics and worldwide (see Chap. 4, Wang et al.). ENSO mechanisms and variations are well known over the past several decades, but instrumental data diminish sharply before the mid-20th century. Thus our knowledge of long-term tropical Pacific variability remains incomplete. The climate systems that

J. Cole (✉)
Department of Geosciences and Atmospheric Sciences, University of Arizona, Tucson, AZ 85721, USA
e-mail: jecole@email.arizona.edu

A.W. Tudhope
School of GeoSciences, Grant Institute, University of Edinburgh, West Mains Road, Edinburgh, EH9 3JW, UK
e-mail: sandy.tudhope@ed.ac.uk

P.W. Glynn et al. (eds.), *Coral Reefs of the Eastern Tropical Pacific*, Coral Reefs of the World 8, DOI 10.1007/978-94-017-7499-4_19

influence this region, particularly ENSO, are poorly understood on multidecade to century time scales. For this reason, the eastern Pacific is a critical site for the development of new paleoclimate and paleoceanographic records. The long-lived coral colonies that grow on eastern Pacific reefs provide important resources for such work. This chapter will summarize climate reconstructions from eastern Pacific coral reefs in the context of both time (linking with longer-term records) and space (linking with coral records from other parts of the tropical oceans).

Because of the robust and influential climate signals that dominate this part of the world, eastern Pacific corals were among the first to be explored for paleoclimatic records. Although motivated by climate reconstruction, coral paleoenvironmental studies in this region have benefited greatly from a foundation of ecological research to identify sites and species for such work. In turn, coral records can help to provide context for environmental variations that stress reef systems, for example warming associated with El Niño events and trends related to anthropogenic influences. Several coral records from eastern Pacific reefs have been published, and these records lay the groundwork for a better understanding of recent changes in the region's environment. The potential for paleorecords to address other ongoing, accelerating changes such as ocean acidification and global warming has not been fully exploited. Important questions remain to be resolved with additional studies, particularly in the detailed characterization of ENSO) variability, changes in ocean chemistry and circulation, and local assessments of environmental variability and change on the reefs.

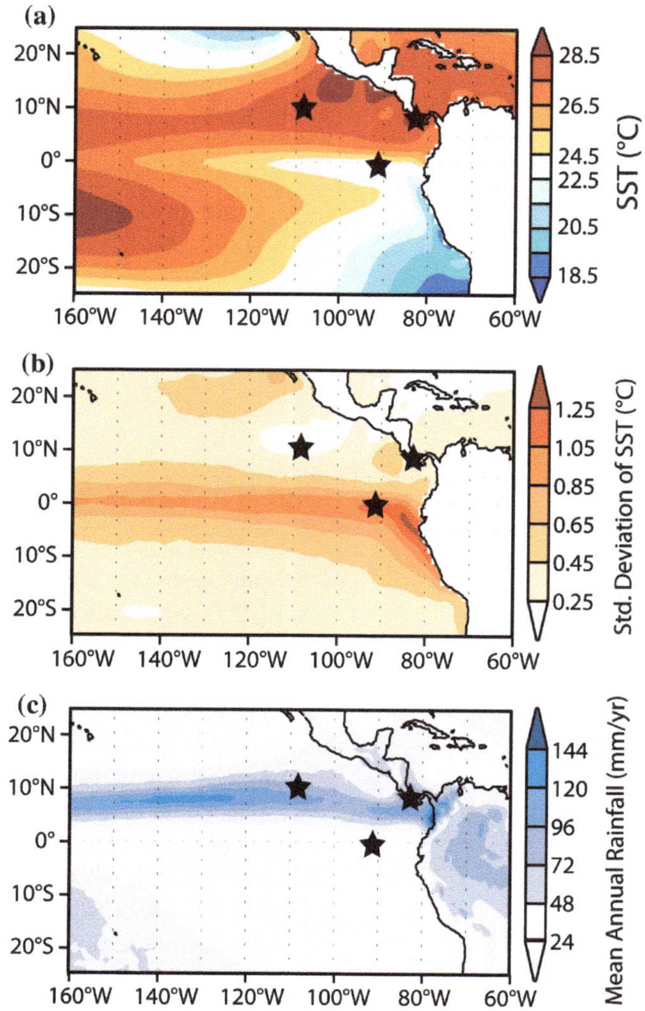

Fig. 19.1 Sea surface temperature mean (**a**) and standard deviation (**b**), and annual mean rainfall (**c**) for the eastern tropical Pacific. SST data are derived from the Reynolds 1° × 1° OISST v2 dataset, and rainfall data are gridded 1° data from TRMM. All data were obtained and plotted using Climate Explorer (http://climexp.knmi.nl). **a** Represents the annual average SST between 1982–2011. In **b**, The standard deviation is calculated on annual means from that interval. The region of high standard deviation corresponds to the strong interannual variability associated with interannual ENSO variability, co-located with "cool tongue" of upwelling water along the equator. **c** Shows annual mean rainfall for 1990–2011; the intertropical convergence zone is clearly seen extending westward from Panama between 5 and 10°N, and it migrates meridionally with the seasons. Stars denote three sites of long coral paleoclimate reconstructions (from W to E: Clipperton, Galápagos, and Chiriquí, Panama)

19.2 Context: Climate Variability and Change in the Eastern Pacific

19.2.1 Modern Context

Unlike most of the tropics, parts of the eastern Pacific experience strong variability on seasonal, interannual, and longer time scales, along with strong spatial gradients in mean conditions (Figs. 19.1 and 19.2). What differentiates this region from other tropical environments is the combination of intense and variable upwelling of cool waters, adjacent to regions that are more typically tropical. These juxtapositions conspire to produce large variations over space and time. The eastern tropical Pacific is a major center of action for the ENSO) system, and the region experiences substantial changes in ocean temperature, chemistry, and circulation, along with rainfall, winds, and runoff, associated with interannual ENSO extremes and longer, decadal to millennial, fluctuations. Thus the eastern Pacific has become a focal region for paleoclimate studies, in large part driven by the need to understand ENSO and longer term variability more clearly. Reef records play a significant part in these and in more localized studies, complemented by sediment-based records that provide a long-term context of past variability. Such studies are increasingly important as researchers strive to clarify how ENSO and the tropical Pacific are responding to increasing greenhouse forcing of climate and a warming world.

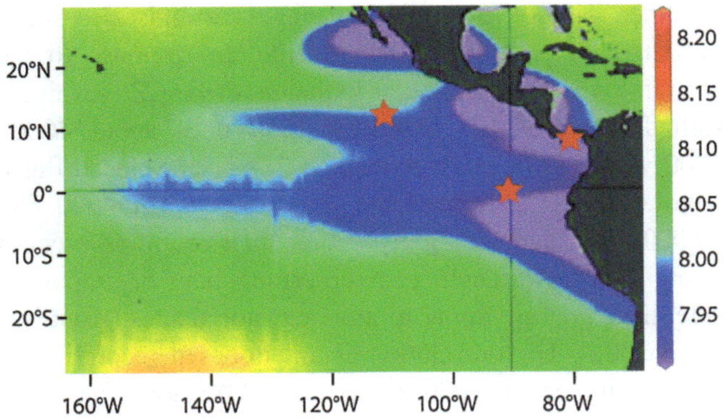

Fig. 19.2 Ocean pH for the upper 50 m in the eastern tropical Pacific, nominally for 1994. Image from Raven et al. (2005). *Stars* show locations of long coral paleoclimate records at (from W to E) Clipperton, Galápagos, and Chiriquí

As the dominant interannual climate signal in the world, ENSO) is of great interest to paleoclimatologists. Particularly for those using reef records, ENSO provides a clear target for interpretation and analysis, as corals are well suited to capture the time scale and location of ENSO variability. With rapid growth rates (1 cm or more per year) and multi-century ages, coral colonies preserve records of both the seasonal-interannual extremes that are an important part of ENSO and the decadal and longer-term changes that are at or beyond the limits of detection from instrumental data. Coral geochemical records from across the tropical Pacific have yielded ENSO reconstructions that correlate strongly with instrumental data and extend the observational record in this poorly instrumented region (e.g., Cole et al. 1993; Dunbar et al. 1994; Guilderson and Schrag 1999; Urban et al. 2000; Linsley et al. 2000; Cobb et al. 2001, 2003; Nurhati et al. 2011).

In addition to the interannual ENSO phenomenon, the Pacific is characterized by coherent variability on decadal-multidecadal time scales (Mantua et al. 1997; Garreaud and Battisti 1999; Deser et al. 2004; see summaries in Deser et al. 2010a, b; Liu 2012; Wang et al. 2012; Chap. 4, Wang et al.). This variability resembles ENSO) in spatial pattern, but it cannot be explained by the primary physical mechanisms of ENSO, which operate mainly on interannual time scales. It is variously termed the Pacific Decadal Oscillation (PDO), Interdecadal Pacific Oscillation (IPO), or Pacific Decadal Variability (PDV), and its time scale ranges from 10 to 50 years. This time scale of variation appears prominently in coral paleoclimate records, which can provide numerous realizations of this phenomenon (e.g., Linsley et al. 2008). In the late 20th century, the strongest El Niño events on record (1982–83 and 1997–98) seemed to be associated with decadal variability, which was in a warm phase from 1976 to 1998. The relationship between PDV and ENSO strength is not yet fully understood, including the degree to which changes in one drive changes in the other. However, we do see that the strength and phase of decadal variability impact extratropical teleconnections of ENSO (e.g., Cole and Cook 1998; Gershunov and Barnett 1998; Brown and Comrie 2004). Finally, recent analyses of changes in the rate of atmospheric warming over the past century suggest that Pacific variability plays an important role: stronger equatorial Pacific winds are associated with a recent slowdown in atmospheric warming, and the phase of the PDO appears to correlate with earlier changes in warming rate as well (Kosaka and Xie 2013; Meehl et al. 2013; England et al. 2014). Eastern Pacific paleorecords have yet to weigh in on this issue, however.

Maps of eastern Pacific climate reveal strong spatial gradients related to upwelling and to oceanic and atmospheric circulation (Figs. 19.1 and 19.2; see Chap. 3, Fiedler and Lavín). In particular, the effect of cool, low-carbonate waters from upwelling is clear in the maps of SST mean and variability, and of pH. Waters are more typically tropical (warm and higher pH) in the western regions of these maps, and SST is warmer north of the equator in the east as well. Salinity is spatially variable, with fresher regions under the intertropical convergence zone (ITCZ) of high precipitation, and particularly in the Gulf of Panama, where the trade winds funnel atmospheric moisture through a topographic low from the warm Caribbean. Features of the ocean circulation include strong equatorial currents and countercurrents and eastern boundary currents from the extratropics (Kessler 2006). The equatorial undercurrent shoals and splits along the western edge of Isabela Island in Galápagos, producing especially cool SST and high chlorophyll in this region (Palacios 2004). Paleorecords can track the strength and position of these features over time, which can change coherently (e.g., as part of ENSO and PDV) or in ways that are unexpected and lend new insights into potential climate configurations.

19.2.2 Long-Term Changes

Marine sediment records of eastern Pacific climate have been used to assess how the region participates in long-term global changes (e.g., glacial cycles). Such records use microfossil-based analyses (geochemical or ecological) to reconstruct past changes in SST, salinity, and/or nutrients. Their relatively coarse resolution (decades-centuries) prevents determination of individual ENSO) events, but mean-state and slowly varying background conditions can be deduced. Although a tendency for cooler conditions during the glacial period is clear in most records, the exact magnitude of cooling and the timing of deglacial transitions remains poorly constrained, as does its connection to high latitude and global drivers (e.g., Lea et al. 2006; Koutavas and Sachs 2008). Also unclear is the response of key oceanographic features; for example, recent studies disagree about whether upwelling in the eastern equatorial Pacific was intensified (Dubois et al. 2009) or weakened (Koutavas et al. 2002; Koutavas and Lynch-Stieglitz 2003) during the last glacial period. A recent study suggests that a weaker SST gradient across the equator corresponded with higher variability during the LGM (Sadekov et al. 2013). In a global summary of glacial SST, the eastern Pacific shows an especially wide range of paleo-SST estimates (MARGO Project Members 2009). Strong seasonal and vertical contrasts likely contribute to this scatter, as different proxies may be sensitive to different seasons or depths. Several studies have noted discrepancies among the different methods of paleoceanographic reconstruction, suggesting that paleotemperature methods could be systematically biased by concurrent changes in oceanic salinity or vertical structure, or by method-specific calibration issues (e.g., Dubois et al. 2009; Kienast et al. 2012). Climate model studies of the last glacial maximum have targeted ENSO and the eastern tropical Pacific for analysis, but results from different modeling groups diverge (DiNezio et al. 2011; Waelbrook et al. 2009).

Holocene climate change, though not as dramatic as glacial-interglacial cycles, has also received increasing scrutiny in the eastern Pacific, particularly for deciphering changes in ENSO) (Fig. 19.3). Several studies have used marine sediment records, despite the limitations imposed by low resolution relative to the interannual timescale of ENSO events. A summary by Koutavas et al. (2006) uses 2 cores each from the western and eastern Pacific to reconstruct the meridional SST gradient across the Pacific. These results suggest that the Holocene was characterized by a stronger zonal SST gradient similar to a La Niña phase, although the cooling in the east is small relative to methodological precision, and reconstructions from the west disagree between cores. Isotope measurements on populations of single foraminiferans (which have a lifespan of a few weeks) further suggest that the interval between ~4000 and 6000 years ago experienced lower SST variability than before or afterwards (Koutavas and Joanides 2012). These results tend to support a picture of weaker ENSO variance in the mid-Holocene that has emerged from other sources, including terrigenous sediments in offshore environments (Rein et al. 2005), lake sediments (Moy et al. 2002; Conroy et al. 2008) and fossil corals from the western Pacific (Tudhope et al. 2001). However, these latter records are all mainly driven by precipitation, and SST changes in the core ENSO region of the eastern equatorial Pacific have not been clearly demonstrated. Central Pacific analyses suggest that although Holocene variability was generally lower than at present, there are inadequate data to resolve significant trend or patterns throughout the past 9000 years (Cobb et al. 2013). Ongoing work on fossil corals in Galápagos should help to resolve this issue (e.g., Tudhope et al. 2008).

Other Holocene tropical changes would also have impacted the eastern tropical Pacific. In response to the changing seasonality of solar radiation, the ITCZ was likely shifted north of its present position until about 4000–6000 years ago (e.g., Haug et al. 2001). Near Soledad Basin (coastal Baja California), the interval between 4000 and 10,000 years ago is slightly cooler than recent (Marchitto et al. 2010). Although local upwelling may be important here, the signal is similar to more equatorial sites (Koutavas et al. 2006). High-resolution sediment-based records from south of the equator (Chazen et al. 2009) suggest an intensification of variability at 5000 years BP and century-scale extremes in which the modern (ENSO-like) relationships among oceanographic variables are not preserved. These data suggest an extratropical source for such low-frequency variability.

Over the past 2000 years, several studies indicate substantial decade-century scale changes occurred in the eastern Pacific climate. Sediment-based records from coastal and island lakes (Moy et al. 2002; Conroy et al. 2009; Sachs et al. 2009) along with marine deposits (Rein et al. 2005) highlight significant variability, but complications with age models, proxy interpretation, spatial variability, and record disagreement have prevented an overall synthesis of recent changes (Fig. 19.3). Because they do not resolve seasonal-interannual variability, these records do not distinguish changes in the strength of interannual variance from mean-state or seasonal shifts. Coral-based paleoclimatology can help to resolve this history, once adequate fossil samples are recovered and analyzed. Corals can be precisely dated and because they resolve monthly information, they can be used to separate intervals of different mean state from intervals of enhanced seasonal or interannual variability.

The presence/absence of reef deposits through time provides another way to understand the influence of climate on reefs. One recent study links diminished reef accretion in Panama between 2 and 3.8 kilo year BP to an intensification of ENSO activity at that time (Toth et al. 2012). Although highly suggestive, this conclusion rests on climate data that remain poorly constrained. Unpublished dates on fossil corals from Galápagos suggest that corals did grow during this interval (Tudhope and Cole, work in progress), although the history of reef accretion in Galápagos remains unknown.

19.3 Coral Paleoclimate Methods: Proof of Concept from Eastern Pacific Samples

Climate reconstructions based on the geochemistry of coral skeletons have been applied in several studies across the eastern tropical Pacific, both in the context of developing and testing particular methods and in the reconstruction of

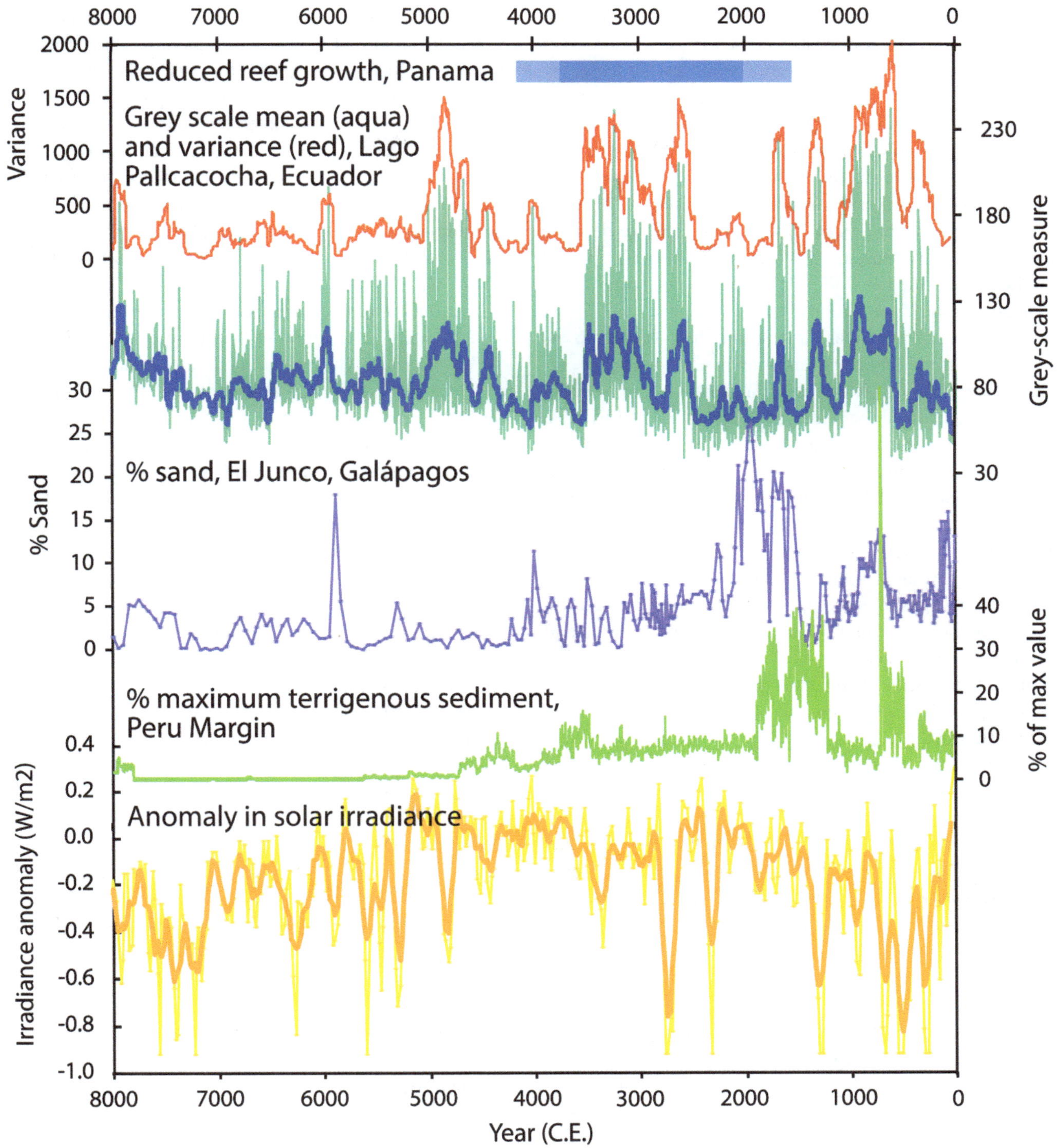

Fig. 19.3 Sediment-based paleoclimate records in the eastern tropical Pacific that are thought to reflect ENSO variability over the past 8000 years. From *top*: color of sediments from Lago Pallcacocha, Ecuador (Moy et al. 2002), presented as 101-year variance (*red*) and mean state (aquamarine; with a 50-year running mean); *bar* indicating interval of reduced reef accretion in Panama (Toth et al. 2012); % sand in core from Lago El Junco, San Cristóbal, Galápagos (Conroy et al. 2008); lithic flux from terrigenous sources, Peru margin (Rein et al. 2004); total solar irradiance anomaly, based on ice core ^{10}Be (Steinhilber et al. 2009). All of the sediment-based records likely reflect the occurrence of strong El Niño events that mobilize coarse sediments. Although such variability seems to increase beginning around 4–5 kilo year BP, other changes in these records do not precisely coincide, perhaps related to age model issues, spatial variability, or an incomplete understanding of the proxies. No clear link to solar forcing is evident

past change over centuries. Some of the earliest demonstrations of coral paleoclimate methods emerged from geochemical analyses of eastern Pacific corals, owing largely to collaborations among ecologists and geoscientists. Among these early pilot studies, a focus on the Galápagos emerges. The support of ecologists in providing samples can be seen in nearly all of these studies. Numerous publications provide general reviews of coral paleoclimate methods (e.g., Cole 2003; Correge 2006; Lough 2004, 2010). Here we describe specifically how eastern Pacific (primarily Galápagos) studies contributed to the development of coral-based paleoclimatology.

Coral radiocarbon was among the first tracers to be exploited for environmental histories in eastern Pacific corals. Using colonies from three Galápagos sites, Druffel (1981) demonstrated that changes in upwelling and El Niño variability are preserved in coral radiocarbon. Also apparent in these data are the Suess effect (declining atmospheric ^{14}C as a consequence of fossil fuel burning) and a substantial contrast with Atlantic corals due to the influence of upwelling (low ^{14}C waters) in Galápagos. Coral radiocarbon has proven to be an important source of information on long-term changes in ocean circulation, including as a tracer to validate global ocean models (Rodgers et al. 2000) and to elucidate potential mechanisms of decadal climate variability in the eastern Pacific (Guilderson and Schrag 1998). Druffel (2002) provides a review of coral-based radiocarbon reconstructions.

Oxygen and carbon isotopes ($\delta^{18}O$ and $\delta\ ^{13}C$) are probably the most commonly measured geochemical parameters in coral paleoclimatology. Oxygen isotopes have become the workhorse tracer of choice for most studies, because they record changes in SST and seawater $\delta^{18}O$ with high reliability. Separating the seawater isotope and temperature signals are typically accomplished using local climate information and relationships, or by using other tracers to constrain these parameters independently. Carbon isotopes exhibit a greater array of dependencies on factors that are not typically well constrained (e.g., the balance between autotrophy and heterotrophy in the coral; Carriquiry et al. 1994; Grottoli and Wellington 1999). Thus $\delta^{13}C$ data are not generally trusted for long-term climate reconstruction, although they do sometimes correlate with climate-related variables.

The groundwork for stable isotopic study of corals was laid in part by studies in the eastern Pacific that compared multiple corals over multiple years. McConnaughey (1989) and Carriquiry et al. (1988) were among the earliest researchers to describe multiyear records from multiple sites. McConnaughey (1989) demonstrated environmental and biological influences on coral isotopes, with three decade-scale records from different islands in the Galápagos, and provided an important foundation to support the reconstruction of SST changes through time using mm-scale sampling of coral skeletons. This work also highlighted important caveats on the choice of sampling methods and potential complications in interpreting coral skeletal $\delta^{13}C$. Follow-on studies included calibration work by Wellington et al. (1996) and multidecade-multicentury records discussed in a later section (Shen et al. 1992; Dunbar et al. 1994). Carriquiry et al. (1988) used five corals from different depths to illustrate how a strong El Niño event was recorded by coral $\delta^{18}O$.

Trace and minor element records from corals provide complementary information to that derived from coral $\delta^{18}O$ and ^{14}C. Because Galápagos corals are among the few to grow in regions of strong upwelling and SST seasonality, researchers have used them to develop tracers of the influence of this cooler, nutrient-rich water. Classic studies (Shen et al. 1987; Lea et al. 1989) used Galápagos corals to define the use of coral Cd/Ca and Ba/Ca as upwelling tracers, based on the nutrient-like (enriched at depth) distribution of Cd and Ba. Later work in Galápagos showed that Mn/Ca tends to vary inversely with those elements, and that volcanism and transport affect coral Mn/Ca as well (Shen et al. 1991; Delaney et al. 1993). Additional elements measured in Galápagos corals (Cu/Ca, Pb/Ca) showed less obvious relationships to environmental variability across seasonal and century time scales (Linn et al. 1990). Short coral records from other parts of the eastern Pacific have refined site-specific interpretations of coral Cd/Ca and Mn/Ca (e.g., Matthews et al. 2008; Carriquiry and Villaescusa 2010). Finally, analyses of Sr/Ca in a Galápagos core showed a strong relationship with local SST across seasonal-interannual scales, in agreement with work from other regions (Schrag 1999). That study provided proof of concept for a new, efficient analytical method (via inductively coupled plasma atomic emission spectrometry, or ICP-AES) that has become the most common method for coral Sr/Ca analysis.

Reef-based records of climate can be derived from skeletal banding properties (e.g., extension rate; see review in Lough 2011), although this method has seen less application in the eastern Pacific than elsewhere. However, many eastern Pacific corals do exhibit exceptionally clear annual bands (e.g., Linsley et al. 1994), probably due to strong seasonal cycles in environmental conditions. This regular banding enables researchers to develop annually precise chronologies. At some sites, interruptions to the banding related to exceptionally stressful events, such as a strong El Niño warming, can be documented (e.g., Dunbar et al. 1994). The annual banding provides clear chronologies that have enabled well-dated geochemical records of past climate variability across the tropical Pacific.

19.4 Coral Records of Eastern Pacific Climate Change Over Decades to Centuries

Coral paleoclimate studies are generally limited by the rarity of large, continuously growing, old coral colonies to support multicentury climate reconstructions. Where such samples can be found, however, coral studies can produce climate records of high fidelity, subseasonal resolution, and multi-century length. Such records have expanded our view of climate variability throughout the warm oceans, but especially in the tropical Indo-Pacific (e.g., Evans et al. 2002; Wilson et al. 2006; Ault et al. 2009).

In the easternmost Pacific, only a few sites have yielded coral-based climate records of several decades or longer (Table 19.1; Figs. 19.4 and 19.5). Clipperton Island (Linsley et al. 1999, 2000) is an open-ocean atoll with a small seasonal cycle in SST ($\sim$2 °C) and El Niño-related warming of generally $\leq$1 °C. The ITCZ brings intense rain to Clipperton seasonally, and interannual changes in ITCZ position and strength may impart interannual variability to coral $\delta^{18}O$. In the Gulf of Chiriquí, Panama (Linsley et al. 1994), the ITCZ and related terrestrial runoff create very large salinity variations that dominate the coral $\delta^{18}O$ signal. At Chiriquí, SST variations are small (1–2 °C) and irregular (Linsley et al. 1994). Other eastern Pacific sites may prove useful in the future (e.g., Costa Rica and the Revillagigedo Islands; Carriquiry et al. 1994; Villaescusa and Carriquiry 2004) if sufficiently long cores can be recovered and analyzed.

In Galápagos, two coral records from different parts of the archipelago both track SST variations primarily. However, one is located in the upwelling zone of the equatorial undercurrent (Urvina Bay; Dunbar et al. 1994) and may be somewhat buffered against surface climate compared to the other from Punta Pitt (Shen et al. 1992; Fig. 19.4), where upwelling is less intense (see Palacios 2004, his Fig. 3). Although the Punta Pitt record is short (47 years) compared to Urvina Bay (375 years), we discuss both here so that 20th century trends may be considered. In earlier analyses, we have patched these together to create a single time series (Ault et al. 2009). Other coral reconstructions from further west in the tropical Pacific complement these interpretations and add context, particularly in the framework of ENSO, decadal variability, and trends.

19.4.1 Interannual Variability and ENSO

Of the three eastern Pacific sites with published century-scale records, Galápagos preserves the clearest ENSO signal. This signal appears clearly in $\delta^{18}O$ data, based mainly on SST changes (although a small salinity imprint may occur due to rainfall and circulation changes; Wellington et al. 1996). It is also apparent in other tracers, particularly those sensitive to upwelling of nutrient-rich waters (e.g., Cd/Ca, Ba/Ca, Mn/Ca, and $\delta^{13}C$). Figure 19.4 shows quarterly measurements of all these tracers in a single coral from Punta Pitt, San Cristóbal (Shen et al. 1992). Clipperton corals also show a moderate ENSO signal in their $\delta^{18}O$, but at Chiriquí, the ENSO signal is weak due to the lack of strong SST variations and to the competing effects of SST and SSS on coral $\delta^{18}O$, which counterbalance each other during ENSO extremes (Linsley et al. 1994, 2000). Galápagos and Clipperton records track known ENSO variations in the 20th century and can be linked to historical and other paleoclimatic records of ENSO in the pre-instrumental period.

Among the clearest lessons to emerge from coral paleoclimatology, both in the eastern Pacific and beyond, is that the strength of interannual ENSO variance has changed significantly during past centuries, and that these changes are not obviously related to external radiative forcing. Two very large El Niño events have occurred in the late 20th century (1982–83 and 1997–98), and it is tempting to speculate that a pattern of stronger events since 1950 is related to global warming. However, strong variability also occurred in the early 20th century. Coral records from Galápagos, Clipperton, and even Chiriquí indicate a lull in the strength of

Table 19.1 Long coral records of climate from the eastern Pacific

Site and original reference	Latitude, longitude	Years (duration, resolution)	Notes (species, chronology, and other details)
Clipperton Atoll Linsley et al. (2000)	10°18′N, 109°13′W	1893–1994 (n = 101; monthly)	*Porites lobata*; well banded; replicated by another shorter core at annual resolution
Gulf of Chiriquí, Panama Linsley et al. (1994)	7°59′N, 82°3′W	1707–1984 (n = 277; monthly)	*Porites lobata*; well banded
Punta Pitt, Galápagos Shen et al. (1992)	0°40′S, 89°10′W	1936–1982 (n = 47; quarterly)	*Pavona gigantea*; one hiatus; well banded
Urvina Bay, Galápagos Dunbar et al. (1994)	0°24′S, 91°14′W	1607–1982 (n = 375; annual)	Combination of cores from *Pavona clavus (1607–1953)* and *P. gigantea (1962–82)*; band quality in long core deteriorates markedly in 20th Century; several hiatuses; age precision ±5 years

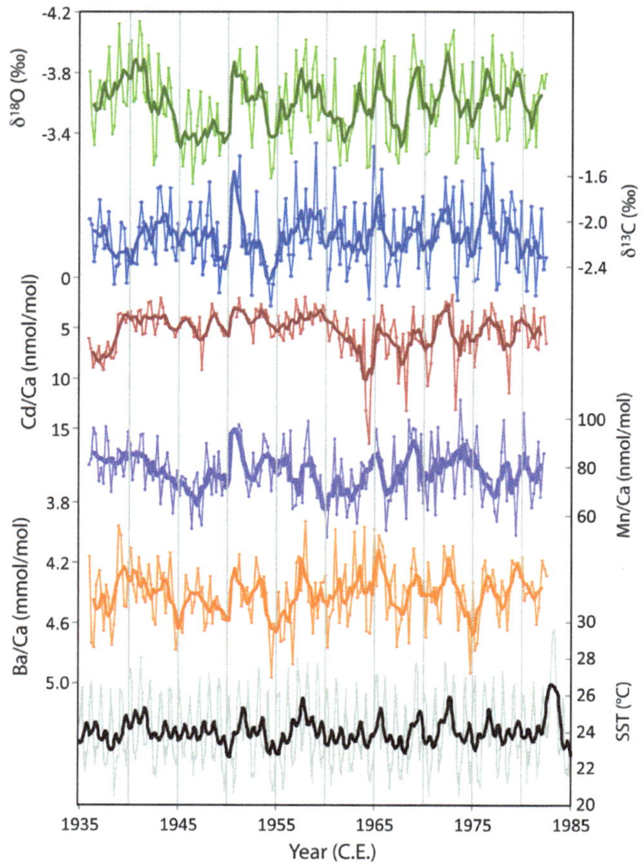

Fig. 19.4 Geochemical records from a Galápagos coral, compared to local SST. From *top*, coral $\delta^{18}O$, $\delta^{13}C$, Cd/Ca, Mn/Ca, and Ba/Ca, followed by SST at the base. All coral data are from a *Pavona gigantea* colony that grew at Punta Pitt, San Cristóbal Is. (Shen et al. 1992); SST data are from the 1 × 1 gridded HadISST product at 0.5°S, 89.5°W (Rayner et al. 2003). Heavy line in all data is a 15-month smooth to highlight interannual ENSO variability (indicated by positive SST deviations). All coral data are plotted so that El Niño events are upwards (in some cases vertical axes have been reversed to achieve this). Like the vast majority of corals in Galápagos, this colony died during the 1982–83 El Niño event

interannual variability during the mid-20th century. This agrees with coral isotope records from Kiribati (Cole et al. 1993; Urban et al. 2000), Nauru (Guilderson and Schrag 1999), and New Guinea (Tudhope et al. 1995). Such a lull is also apparent in instrumental records (Trenberth and Shea 1987). A longer series of fossil coral records from Palmyra Atoll (Cobb et al. 2003) makes clear that the variance in ENSO has changed substantially over the past millennium, without apparent relation to any external cause.

19.4.2 Decadal and Longer-Term Variability

In the tropical Pacific, the lack of long instrumental records poses a grave challenge to resolving decadal to century-scale climate variability. SST data products may reach into the 19th century, but in most of the tropical Pacific, they are not well constrained by actual observations prior to the mid-20th century. Thus they disagree substantially on long-term changes (e.g., Vecchi et al. 2008; Deser et al. 2010a, b). Instrumentally based analyses suggest that multidecadal variability and 20th-century warming are particularly weak in the eastern tropical Pacific (Mestas-Nuñez and Miller 2006). Such variability may appear weak in instrumental data by comparison with higher latitudes. But eastern Pacific coral records provide a broader perspective in which long-term variability plays a more pronounced role (Fig. 19.5). This multidecadal time scale of variability is globally significant.

Decade-century variability in coral $\delta^{18}O$ records is usually pronounced, and may result from SST or from factors that change the seawater isotopic content (specifically hydrological or circulation variations). For example, at Clipperton, fluctuations on a roughly bidecadal time scale are thought to reflect changes in salinity and the equatorial countercurrent strength. In Galápagos, the Urvina Bay record shows substantial variability at periods between 11 and 34 years. At Chiriquí, a similar time scale of variability has been attributed to behavior of the ITCZ. At these longer periods, Chiriquí and Urvina Bay tend to show an anti-correlation that may reflect long-term shifts in the mean position of the ITCZ, influencing sea surface salinity inversely at these sites (Dunbar et al. 1996). Although solar variability may play a role in tropical climate on this time scale (e.g., Mann et al. 2005; Meehl and Arblaster 2009; Meehl et al. 2009), it remains very difficult to disentangle any forced signal from the high background level of apparently unforced decadal variability in Pacific climate records.

Recent studies have synthesized Pacific Ocean SST variability using coral paleorecords and identified significant decadal and longer-term patterns. Using eight Pacific sites spanning 86 years, including Clipperton and Chiriquí, Holland et al. (2007) describe a propagating pattern on decadal time scales. We helped lead a more extensive analysis (Ault et al. 2009) that used 23 Indo-Pacific sites (including Galápagos and Chiriquí) and spanned 150 years. We found substantial decadal variability that showed no evidence of propagation, but that waxes and wanes through time. This variance is tightly confined to a decadal time scale and exhibits an ENSO-like spatial pattern. Like ENSO, it is weak at Chiriquí but strong at Galápagos. Because it was much stronger in the 19th century than in recent decades, this phenomenon has not been well defined by instrumental records. Finally, it is not related to solar forcing: the phasing of the decadal pattern to solar forcing changes phase by 180° over the past century.

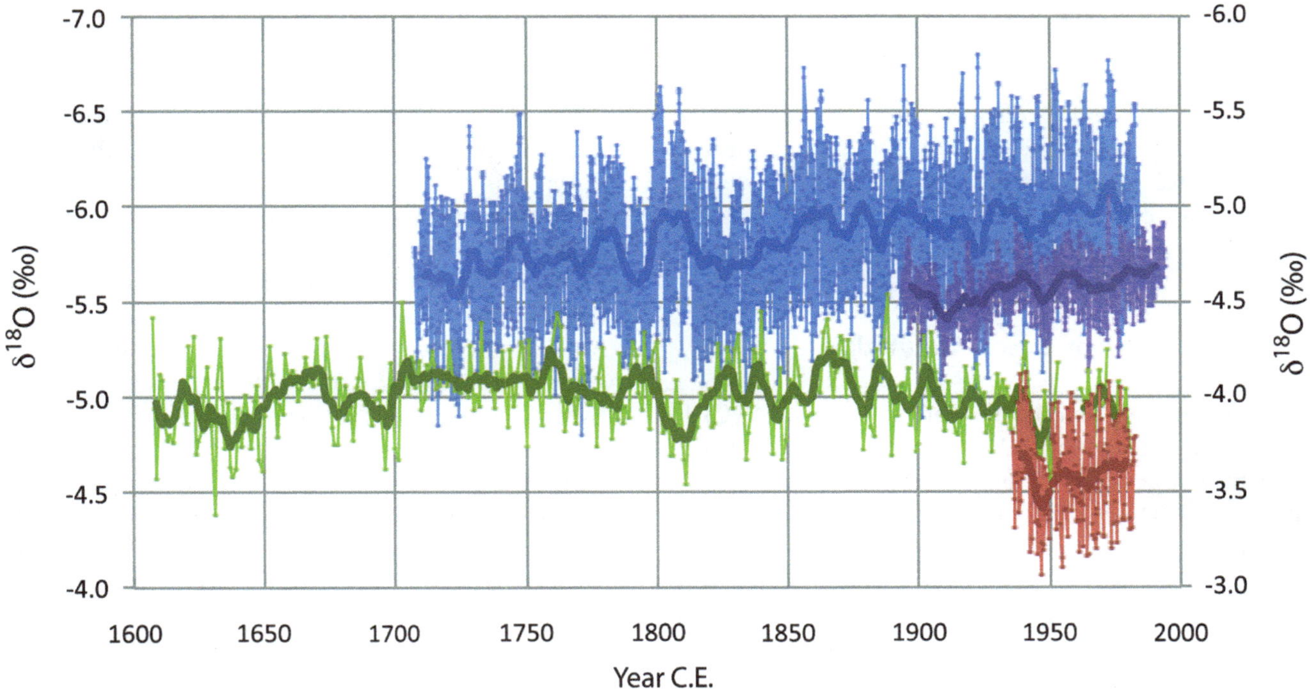

Fig. 19.5 Coral oxygen isotope records from three eastern Pacific sites: Gulf of Chiriquí, Panama (*blue*); Clipperton Atoll (*purple*), and Galápagos (*green*: Urvina Bay, Isabela Is., and *red*: Punta Pitt, San Cristóbal Is.). Site details and references are given in Table 19.1. *Dark line* represents a 7-year running mean through the data at original resolution. Galápagos data are plotted on *right vertical axis*, and Chiriquí and Clipperton are plotted on *left vertical axis*

19.4.3 Recent Trends

Are eastern Pacific corals providing evidence related to recent climate change? Nearly all coral records throughout the tropics demonstrate substantial warming trends over the past 100 years (e.g., Wilson et al. 2006; Thompson et al. 2011) that are unlike earlier century-scale periods (Cobb et al. 2003; Damassa et al. 2006). Yet often the trends are larger than would be expected from locally observed SST. Thompson et al. (2011) argue that a combination of salinity and temperature change can explain these trends. Furthermore, we expect that the sparse nature of instrumental data compromises the early SST record, making "observed" trends difficult to define. Finally, in the eastern tropical Pacific, the internal variability on interannual and decadal time scales is so large that detecting trends is challenging and very sensitive to start and end time, and to seasonal dependencies.

Both Chiriquí and Clipperton coral records exhibit long-term trends. At Chiriquí, the trend persists across the whole 277-year record, suggesting a cause beyond recent warming. Given the importance of hydrological variability in determining this $\delta^{18}O$ record, the cause of the trend is more likely to be related to ITCZ or other hydroclimatic factors. At Clipperton, the overall warming trend seems to occur mainly between 1900 and 1940 and after 1975 (consistent with global temperature trends). According to Linsley et al. (1999), the degree of warming is larger than that seen in a $5 \times 5°$ SST product. The discrepancy may either be due to the inability of such an instrumental dataset to represent reef temperatures accurately, or to the co-occurrence of seawater isotope decreases (increased rainfall) with warming.

In Galápagos, a long-term trend in reconstructed SST is not evident. In the long Urvina Bay record, 20th-century changes are not well represented because the record's quality deteriorates in the 20th century: banding degrades, 10 years are missing between 1954 and 1974, and a splice at 1953 assumes an offset value that could conceal mean-state change. Moreover this site is strongly influenced by the equatorial undercurrent, which could insulate it from surface climate signals. The Punta Pitt record also lacks a significant warming trend and only spans 45 years. Both records end at 1982, predating the strongest part of the global 20th century trend. Interestingly, lake-based data from El Junco (San Cristóbal Is., Galápagos), which are sensitive to precipitation and well correlated with SST, reflect a strong trend to a warmer, wetter climate over the past 200 years (Conroy et al. 2009). Instrumental data are equivocal, and suggest a seasonal dependence to a relatively small trend (e.g., Karnauskas et al. 2009; Wolff 2010). Clearly, Galápagos is less sensitive to the warming trends that are apparent in both instrumental and coral data from other Pacific (and global) sites, likely because of the continuing influence of cool upwelled water. Another potential factor involves the recent predominance of central Pacific warming during El Niño events; this pattern of warming is associated with stable or even cooling SST in the eastern Pacific as a consequence of Kelvin wave dynamics (Dewitte et al. 2012) and may tend to mute a global warming signal (Wolff 2010). New coral samples that we collected in 2010, from Isla Wolf, will help to fill in this picture and place recent changes in context.

19.4.4 Tracers of Ocean Circulation

The age-corrected concentration of radiocarbon ($\delta^{14}C$) in coral skeletons tracks the ^{14}C content of the waters in which the corals grow, which in turn depends on the local circulation regime (Druffel 2002). Coral radiocarbon records from Urvina Bay, Galápagos, provide evidence of seasonal to multidecadal changes in ^{14}C that can be linked to circulation shifts. At Urvina Bay, the radiocarbon tracer reflects changes in the ^{14}C content of Equatorial Undercurrent (EUC) waters that bathe this site. A coral $\delta^{14}C$ record from 1957 to 1982 shows a sudden increase in the seasonal minimum (upwelling season) $\delta^{14}C$ in 1976, coincident with a sudden warming (Guilderson and Schrag 1998). This change in $\delta^{14}C$ has been attributed, via an ocean model, to a decrease in the entrainment of deep subantarctic mode waters (SAMW) into the EUC (Rodgers et al. 2004). Similarly, changes in seasonal minimum $\delta^{14}C$ over the past 400 years are thought to reflect variations in the contribution of high-latitude waters from both northern and southern subpolar latitudes (Druffel et al. 2007, 2014). Such variations may be linked in turn to the strength and position of zonal winds and/or interdecadal fluctuations in the north Pacific (Rodgers et al. 2004; Druffel et al. 2014). Urvina Bay reflects primarily EUC variations, due to its location, but other eastern Pacific sites may be more sensitive to changes among surface (or surface vs upwelling) water masses. Additional coral radiocarbon work is in progress from other eastern Pacific sites.

19.5 Conclusions: What Have We Learned, and What's Next?

Eastern Pacific corals provided some of the earliest proof of concept studies for coral-based paleoclimatology and have yielded three long records that extend before the 20th century. In a region where instrumental data are sparse and climate is spatially heterogeneous, these records have made important contributions to documenting regional climate variability. For example, we have learned that in a location where ENSO-related SST changes are among the strongest, the signal of global warming appears weak. The variability of ENSO changes significantly over recent centuries, with little clear link to radiative forcings. Across the meridional reach of the eastern Pacific ITCZ, coral records suggest a "wobble" over periods of decades, seen as an inverse relation between Galápagos and Chiriquí coral $\delta^{18}O$ that has yet to be well understood. Finally, coral radiocarbon from Galápagos has been used to identify how the contribution of distinct water masses into upwelling waters may have changed in association with Pacific decadal variability. Coral radiocarbon holds promise for understanding past circulation changes and their relation to climate shifts.

And yet only three sites in the eastern Pacific have been developed for records of past climate that extend more than 100 years (additional sites are in preparation; J.D. Carriquiry, personal communication), and the only one of these that is replicated is the shortest, at Clipperton. The Galápagos data, which could be most useful for ENSO, do not include the most recent two large events in 1982–83 and 1997–98 and cannot address recent trends. Tremendous potential exists for developing long records at additional sites, where long coral records may be present, and the need for replication to add confidence is acute. Long coral records from the eastern Pacific could help to resolve questions about long-term trends in zonal SST gradients, spatial variability in ENSO-related SST anomalies, and the spatial footprint of decadal and longer variability. Extensive coral mortality associated with strong El Niño events, for example at Galápagos (e.g., Glynn 1994; Wellington and Glynn 2007), limits the potential in some regions. Yet even where past mortality has been high, some reefs have survived and individual colonies persist (e.g., Glynn 2009; Glynn et al. 2015).

New types of measurement also offer opportunities to expand our reef-based view of past environmental change. At a very basic level, sites with oxygen isotope records would benefit from the paired analysis of Sr/Ca, for example to identify ITCZ-related behavior that affects seawater salinity, for comparison to SST in the $\delta^{18}O$ records cited here. Tracers of upwelling and circulation can lend insight into the dynamic processes that govern climate variability and change, including ENSO, decadal variability, and the response of the eastern tropical Pacific to global warming.

Tracers that have not yet seen wide use may also have potential to provide important information on environmental change. For example, a better understanding of how pH and aragonite saturation change naturally, over decades to centuries, would help elucidate the vulnerability of eastern Pacific reefs to ongoing global pH declines. The boron isotope composition ($\delta^{11}B$) of marine carbonates has seen wide use as a paleo-pH indicator in marine carbonate sediments (as summarized in Pelejero et al. 2010). It may prove increasingly informative in coral skeletons (Honisch et al. 2004; Trotter et al. 2011), although physiological factors and other localized influences on coral skeletal $\delta^{11}B$ could complicate simple paleo-pH interpretations (e.g., Matear and

McNeil 2006; Dissard et al. 2012; McCulloch et al. 2012). Within the eastern Pacific, the Galápagos may be a particularly sensitive "hotspot" for declining pH (Manzello 2010; see Chap. 18, Manzello et al.) and reconstructions of natural changes will be highly informative.

Other environmental tracers hold promise for expanded application in the eastern Pacific. U/Ca has been suggested as a tracer of ocean pH, although here too, work is needed to sort out environmental and vital effects (Shen and Dunbar 1995; Inoue et al. 2011). Coralline P/Ca has been proposed as a direct tracer of environmental phosphate concentrations (LaVigne et al. 2008, 2010) and could be extremely useful in oceanic upwelling regimes. Analysis of other organisms, such as deeper dwelling (non-photosynthetic) corals and other calcifying taxa, could potentially add an important third dimension to studies of past oceanographic change, with reconstructions of conditions in or below the thermocline.

Anticipating the behavior of ENSO as the world warms remains among the biggest challenges for climate science. Even the most advanced climate models disagree on basic metrics of future ENSO variability (Collins et al. 2010). One of the most important outstanding questions remains the relationship between changes in ENSO mean state and variability. By combining sediment-based records of mean state conditions with fossil corals that resolve seasonal-interannual variability, we should be able to make progress on this question. Reef-based records can also help to evaluate climate model simulations. For example, reconstructions of 20th-century trends can be compared with coupled climate models to provide an assessment of how well models simulate observed changes (e.g., Brown et al. 2008; Thompson et al. 2011).

Finally, reef-based records can help determine whether the temporal and spatial patterns of ENSO variability that characterize the last 50 years are typical of the full range of ENSO behavior. Increasingly, paleodata suggest that the instrumental record is providing only one potential realization of the dominant spatial and temporal modes of variability in the tropical Pacific (e.g., Ault et al. 2009; Pak 2012). The tropical Pacific is unlikely to respond to global warming in a strictly El Niño/La Niña-like way (DiNezio et al. 2009), highlighting the importance of providing paleo-observations of alternate patterns. The synthesis of reef-based records with longer sediment records should offer a more complete view of the nature of tropical Pacific variability during times of varying background states and forcings. This new perspective will be critical to understanding how the tropical Pacific could respond as the world warms.

Acknowledgments We thank Peter Glynn, Gerard Wellington, Stuart Banks, Lenín Cruz, Roberto Pepolas, and all our colleagues who have provided encouragement, information and samples that have advanced reef-based paleoclimatic studies in the eastern Pacific. We gratefully acknowledge research support from NSF (JEC: OCE 1416148 and OCE 0957881) and NERC (AWT) that allows us to continue work in Galápagos, and we deeply appreciate the support of the Charles Darwin Research Station and the Galápagos National Park in furthering our research. We thank our colleagues on Galápagos paleoclimate research projects: Colin Chilcott, Meriwether Wilson, Diane Thompson, Jonathan Overpeck, and Jessica Conroy, Jennifer Suarez, and Gloria Jimenez. We also thank the editors for their hard work in organizing the publication of this book. We are grateful to P. Glynn, D. Lea, and J. Carriquiry for comments that have improved the manuscript.

References

Ault TR, Cole JE, Evans MN, Barnett H, Abram NJ, Tudhope AW, Linsley BK (2009) Intensified decadal variability in tropical climate during the late 19th century. Geophys Res Lett 36. doi:10.1029/2008GL036924

Brown DP, Comrie AC (2004) A winter precipitation 'dipole' in the western United States associated with multidecadal ENSO variability. Geophys Res Lett 31. doi:10.1029/2003GL018726

Brown J, Collins M, Tudhope AW, Toniazzo T (2008) Modelling mid-Holocene tropical climate and ENSO variability: towards constraining predictions of future change with palaeo-data. Clim Dyn 30:19–36

Carriquiry JD, Risk MJ, Schwarcz HP (1988) Timing and temperature record from the stable isotopes of the 1982–1983 El Niño warming event in eastern Pacific corals. Palaios 3:359–364

Carriquiry JD, Risk MJ, Schwarcz HP (1994) Stable-Isotope Geochemistry of Corals from Costa-Rica as Proxy Indicator of the El-Nino Southern Oscillation (Enso). Geochim Cosmochim Acta 58:335–351

Carriquiry JD, Villaescusa JA (2010) Coral Cd/Ca and Mn/Ca records of ENSO variability in the Gulf of California. Clim Past 6:401–410

Chazen CR, Altabet MA, Herbert TD (2009) Abrupt mid-Holocene onset of centennial-scale climate variability on the Peru-Chile Margin. Geophys Res Lett. doi:10.1029/2009GL039749

Cobb KM, Charles CD, Cheng H, Edwards RL (2003) El Niño/Southern Oscillation and tropical Pacific climate during the last millennium. Nature 424:271–276

Cobb KM, Westphal N, Sayani HR, Watson JT, Di Lorenzo E, Cheng H, Edwards RL, Charles CD (2013) Highly variable El Niño-Southern Oscillation throughout the Holocene. Science 339:67–70

Cole JE (2003) Holocene coral records: windows on tropical climate variability. In: McKay A, Battarbee R, Birks J, Oldfield F (eds) Global change in the Holocene. Arnold, London, pp 168–184

Cole JE, Cook ER (1998) The changing relationship between ENSO variability and moisture balance in the continental United States. Geophys Res Lett 25:4529–4532

Cole JE, Fairbanks RG, Shen GT (1993) The spectrum of recent variability in the Southern Oscillation: Results from a Tarawa Atoll coral. Science 262:1790–1793

Collins M, Soon-Il A, Wenju C, Genachaud A, Guilyardi E, Fei-Fei J, Jochum M, Lengaigne M, Power S, Timmermann A, Vecchi G, Wittenberg A (2010) The impact of global warming on the tropical Pacific ocean and El Niño. Nat Geosci 3:391–397

Conroy JL, Overpeck JT, Cole JE, Shanahan TM, Steinitz-Kannan M (2008) Holocene changes in eastern tropical Pacific climate inferred from a Galapagos lake sediment record. Quat Sci Rev 27:1166–1180

Conroy JL, Restrepo A, Overpeck JT, Steinitz-Kannan M, Cole JE, Bush MB, Colinvaux PA (2009) Unprecedented recent warming of surface temperatures in the eastern tropical Pacific Ocean. Nat Geosci 2:46–50

Correge T (2006) Sea surface temperature and salinity reconstruction from coral geochemical tracers. Paleogeogr Paleoclimatol Paleoecol 232:408–428

Damassa TD, Cole JE, Barnett H, Ault TR, McClanahan TR (2006) Enhanced multidecadal climate variability in the 17th century from coral isotope records in the western Indian Ocean. Paleoceanography. doi:10.1029/2005PA001217

Delaney ML, Linn LJ, Druffel ERM (1993) Seasonal cycles of manganese and cadmium in coral from the Galápagos Islands. Geochim Cosmochim Acta 57:347–354

Deser C, Phillips AS, Hurrell JW (2004) Pacific interdecadal climate variability: linkages between the tropics and the North Pacific during boreal winter since 1900. J Climate 17:3109–3124

Deser C, Alexander MA, Xie SP, Phillips AS (2010a) Sea surface temperature variability: patterns and mechanisms. Annu Rev Mar Sci 2:115–143

Deser C, Phillips AS, Alexander MA (2010b) Twentieth century tropical sea surface temperature trends revisited. Geophys Res Lett. doi:10.1029/2010GL043321

Dewitte B, Vazquez-Cuervo J, Goubanova K, Illig S, Takahashi K, Cambon G, Purca S, Correa D, Gutierrez D, Sifeddine A, Ortlieb L (2012) Change in El Niño flavours over 1958–2008: implications for the long-term trend of the upwelling off Peru. Deep-Sea Res II 77–80:143–156

DiNezio PN, Clement AC, Vecchi GA, Soden BJ, Kirtman BP (2009) Climate response of the equatorial Pacific to global warming. J Climate 22:4873–4892

DiNezio PN, Clement A, Vecchi GA, Soden B, Broccoli AJ, Otto-Bliesner BL, Braconnot P (2011) The response of the Walker circulation to Last Glacial Maximum forcing: implications for detection in proxies. Paleoceanography 26. doi:10.1029/2010PA002083

Dissard D, Douville E, Reynaud S, Juillet-Leclerc A, Montagna P, Louvat P, McCulloch M (2012) Light and temperature effect on δ11B and B/Ca ratios of the zooxanthellate coral *Acropora* sp.: results from culturing experiments. Biogeosci Discus 9:5969–6014

Druffel EM (1981) Radiocarbon in annual coral rings from the eastern tropical Pacific Ocean. Geophys Res Lett 8:59–62

Druffel ERM (2002) Radiocarbon in corals: records of the carbon cycle, surface circulation, and climate. Oceanography 15:122–127

Druffel ERM, Griffin S, Beaupré SR, Dunbar RB (2007) Oceanic climate and circulation changes during the past four centuries from radiocarbon in corals. Geophys Res Lett. doi:10.1029/2006GL028681

Druffel ERM, Griffin S, Glynn DS, Dunbar RB, Mucciarone DA, Toggweiler JR (2014) Seasonal radiocarbon and oxygen isotopes in a Galapagos coral: calibration with climate indices. Geophys Res Lett 41(14):5099–5105. doi:10.1002/2014GL060504

Dubois N, Kienast M, Normandeau C, Herbert TD (2009) Eastern equatorial Pacific cold tongue during the Last Glacial Maximum as seen from alkenone paleothermometry. Paleoceanography. doi:10.1029/2009PA001781

Dunbar RB, Wellington GM, Colgan MW, Glynn PW (1994) Eastern Pacific sea surface temperature since 1600 A.D.: the $\delta^{18}O$ record of climate variability in Galápagos corals. Paleoceanography 9:291–316

Dunbar R, Linsley B, Wellington GM (1996) Eastern Pacific corals monitor El Niño/Southern Oscillation, precipitation, and sea surface temperature variability over the past 3 centuries. In: Jones PD, Bradley R, Jouzel J (eds) Climatic fluctuations and forcing mechanisms of the last 2000 years. Springer, New York, pp 375–407

England MH, McGregor S, Spence P, Meehl GA, Timmermann A, Cai W, Sen Gupta A, McPhaden MJ, Purich A, Santoso A (2014) Recent intensification of wind-driven circulation in the Pacific and the ongoing warming hiatus. Nature Climate Change 4:222–227

Evans M, Kaplan A, Cane M (2002) Pacific sea surface temperature field reconstruction from coral $\delta^{18}O$ data using reduced space objective analysis. Paleoceanography 17(1):7–1, 7–13

Garreaud RD, Battisti DS (1999) Interannual (ENSO) and interdecadal (ENSO-like) variability in the Southern Hemisphere tropospheric circulation. J Climate 12:2113–2123

Gershunov A, Barnett TP (1998) Interdecadal modulation of ENSO teleconnections. B Am Meteorol Soc 79:2715–2725

Glynn PW (1994) State of coral reefs in the Galápagos Islands: natural vs anthropogenic impacts. Mar Pollut Bull 29:131–140

Glynn PW, Riegl B, Correa AMS, Baums IB (2009) Rapid recovery of a coral reef at Darwin Island, Galápagos Islands. Galapagos Res 66:6–13

Glynn PW, Riegl B, Purkis S, Kerr JM, Smith TB (2015) Coral reef recovery in the Galápagos Islands: the northernmost islands (Darwin and Wenman). Coral Reefs. doi:10.007/s00338-015-1280-4

Grottoli AG, Wellington GM (1999) Effect of light and zooplankton on skeletal delta C-13 values in the eastern Pacific corals *Pavona clavus* and *Pavona gigantea*. Coral Reefs 18:29–41

Guilderson TP, Schrag DP (1998) Abrupt shifts in subsurface temperatures in the tropical Pacific associated with changes in El Niño. Science 281:240–243

Guilderson TP, Schrag DP (1999) Reliability of coral records from the western Pacific warm pool: a comparison using age-optimized records. Paleoceanography 4:457–464

Haug GH, Hughen KA, Sigman DM, Peterson LC, Rohl U (2001) Southward migration of the intertropical convergence zone through the Holocene. Science 293:1304–1308

Holland CL, Scott RB, An S-I, Taylor FW (2007) Propagating decadal sea surface temperature signal identified in modern proxy records of the tropical Pacific. Climate Dynam 28:163–179

Honisch B, Hemming NG, Grottoli AG, Amat A, Hanson GN, et al. (2004) Assessing scleractinian corals as recorders for paleo-pH: Empirical calibration and vital effects. Geochim Cosmochim Acta 68:3675–3685

Inoue M, Suwa R, Suzuki A, Sakai K, Kawahata H (2011) Effects of seawater pH on growth and skeletal U/Ca ratios of *Acropora digitifera* coral polyps. Geophys Res Lett 38(12). doi:10.1029/2011GL047786

Karnauskas KB, Seager R, Kaplan A, Kushnir Y, Cane MA (2009) Observed strengthening of the zonal sea surface temperature gradient across the equatorial Pacific Ocean. J Climate 22: 4316–4321

Kessler WS (2006) The circulation of the eastern tropical Pacific: a review. Prog Oceanogr 69:181–217

Kienast M, MacIntyre G, Dubois N, Higginson S, Normandeau C, Chazen C, Herbert TD (2012) Alkenone unsaturation in surface sediments from the eastern equatorial Pacific: implications for SST reconstructions. Paleoceanography 27(1). doi:10.1029/2011PA002254

Kosaka Y, Xie S-P (2013) Recent global-warming hiatus tied to equatorial Pacific surface cooling. Nature 501:403–407

Koutavas A, Joanides S (2012) El Niño-Southern Oscillation extrema in the Holocene and Last Glacial Maximum. Paleoceanography 27 (4). doi:10.1029/2012PA002378

Koutavas A, Lynch-Stieglitz J (2003) Glacial-interglacial dynamics of the eastern equatorial Pacific cold tongue Intertropical Convergence Zone system reconstructed from oxygen isotope records. Paleoceanography 18(4). doi:10.1029/2003PA000894

Koutavas A, Sachs JP (2008) Northern timing of deglaciation in the eastern equatorial Pacific from alkenone paleothermometry. Paleoceanography 23(4). doi:10.1029/2008PA001593

Koutavas A, Lynch-Stieglitz J, Marchitto TM, Sachs JP (2002) El Niño-like pattern in ice age tropical Pacific sea surface temperature. Science 297:226–230

Koutavas A, Demenocal PB, Olive GC, Lynch-Stieglitz J (2006) Mid-Holocene El Niño-Southern Oscillation (ENSO) attenuation revealed by individual foraminifera in eastern tropical Pacific sediments. Geology 34:993–996

LaVigne M, Field MP, Anagnostou E, Grottoli AG, Wellington GM, Sherrell RM (2008) Skeletal P/Ca tracks upwelling in Gulf of Panama coral: evidence for a new seawater phosphate proxy. Geophys Res Lett 35(5). doi:10.1029/2007GL031926

LaVigne M, Matthews KA, Grottoli AG, Cobb KM, Anagnostou E, Cabioch G, Sherrell RM (2010) Coral skeleton P/Ca proxy for seawater phosphate: multi-colony calibration with a contemporaneous seawater phosphate record. Geochim Cosmochim Acta 74:1282–1293

Lea DW, Boyle EA, Shen GT (1989) Coralline barium records temporal variability in equatorial Pacific upwelling. Nature 340:373–376

Lea DW, Pak DK, Belanger CL, Spero HJ, Hall MA, Shackleton NJ (2006) Paleoclimate history of Galápagos surface waters over the last 135,000 year. Quat Sci Rev 25(11–12):1152–1167

Linn LJ, Delaney ML, Druffel ERM (1990) Trace metals in contemporary and seventeenth-century Galápagos coral: records of seasonal and annual variations. Geochim Cosmochim Acta 54:387–394

Linsley BK, Dunbar RB, Wellington GM, Mucciarone DA (1994) A coral-based reconstruction of Intertropical Convergence Zone variability over Central America since 1707. J Geophys Res 99:9977–9994

Linsley BK, Messier RG, Dunbar RB (1999) Assessing between-colony oxygen isotope variability in the coral *Porites lobata* at Clipperton Atoll. Coral Reefs 18:13–27

Linsley BK, Ren L, Dunbar RB, Howe SS (2000) El Niño-Southern Oscillation (ENSO) and decadal-scale climate variability at 10 N in the eastern Pacific from 1893 to 1994: a coral-based reconstruction from Clipperton Atoll. Paleoceanography 15:322–335

Linsley BP, Zhang A, Kaplan S, Howe S, Wellington GM (2008) Interdecadal-decadal climate variability from multicoral oxygen isotope records in the South Pacific Convergence Zone region since 1650 A.D. Paleoceanography 23. doi:10.1029/2007PA001539

Liu Z (2012) Dynamics of interdecadal climate variability: a historical perspective. J Climate 25:1963–1995

Lough JM (2004) A strategy to improve the contribution of coral data to high-resolution paleoclimatology. Paleogeogr Paleoclimat Paleoecol 204:115–143

Lough JM (2010) Climate records from corals. Wiley Interdiscip Rev-Climate Change 1:318–331

Lough JM, Cooper TF (2011) New insights from coral growth band studies in an era of rapid environmental change. Earth-Sci Rev 108:170–184

Mann ME, Cane MA, Zebiak SE, Clement A (2005) Volcanic and solar forcing of the tropical Pacific over the past 1000 years. J Climate 18:447–456

Mantua NJ, Hare SR, Zhang Y, Wallace JM, Francis RC (1997) A Pacific interdecadal oscillation with impacts on salmon production. B Am Meteorol Soc 78:1069–1079

Manzello D (2010) Ocean acidification hot spots: spatiotemporal dynamics of the seawater CO_2 system of eastern Pacific coral reefs. Limnol Oceanogr 55:239–248

Marchitto TM, Muscheler R, Ortiz JD, Carriquiry JD, van Geen A (2010) Dynamical response of the tropical Pacific Ocean to solar forcing during the early Holocene. Science 330:1378–1381

Matear RJ, McNeil BI (2006) Comment on "Preindustrial to modern interdecadal variability in coral reef pH". Science 314:595–595

Matthews KA, Grottoli AG, McDonough WF, Palardy JE (2008) Upwelling, species, and depth effects on coral skeletal cadmium-to-calcium ratios (Cd/Ca). Geochim Cosmochim Ac 72:4537–4550

McConnaughey TA (1989) C-13 and O-18 isotopic disequilibria in biological carbonates: I. Patterns. Geochim Cosmochim Ac 53: 151–162

McCulloch M, Falter J, Trotter J, Montagna P (2012) Coral resilience to ocean acidification and global warming through pH up-regulation. Nature Climate Change 2:623–627

Meehl GA, Arblaster JM (2009) A lagged warm event-like response to peaks in solar forcing in the Pacific region. J Climate 22:3647–3660

Meehl GA, Arblaster JM, Matthes K, Sassi F, van Loon H (2009) Amplifying the Pacific climate system response to a small 11-year solar cycle forcing. Science 325:1114–1118

Meehl GA, Hu A, Arblaster JM, Fasullo J, Trenberth KE (2013) Externally forced and internally generated decadal climate variability associated with the interdecadal Pacific oscillation. J Climate 26:7298–7310

Mestas-Nuñez AM, Miller AJ (2006) Interdecadal variability and climate change in the eastern tropical Pacific: a review. Prog Oceanogr 69:267–284

Moy CM, Seltzer GO, Rodbell DT, Anderson DM (2002) Variability of El Niño/Southern Oscillation activity at millennial timescales during the Holocene epoch. Nature 420:162–164

Nurhati IS, Cobb KM, Di Lorenzo E (2011) Decadal-Scale SST and Salinity Variations in the Central Tropical Pacific: Signatures of Natural and Anthropogenic Climate Change. J Climate 24:3294–3308

Pak D (2012) Palaeoclimate analogue complexity. Nat Geosci 5:450–451

Palacios DM (2004) Seasonal patterns of sea-surface temperature and ocean color around the Galápagos: regional and local influences. Deep-Sea Res Part II-Top Stud Oceanogr 51:43–57

Pelejero C, Calvo E, Hoegh-Guldberg O (2010) Paleo-perspectives on ocean acidification. Trends Ecol Evol 25(6):332–344

Raven J, Caldeira K, Elderfield H, Hoegh-Guldberg O, Liss P, Riebesell U, Shepherd J, Turley C, Watson A (2005) Ocean acidification due to increasing atmospheric carbon dioxide. The Royal Society. Available online at https://royalsociety.org/policy/publications/2005/ocean-acidification/

Rayner NA, Parker DE, Horton EB, Folland CK, Alexander LV, Rowell DP, Kent EC, Kaplan A (2003) Global analyses of sea surface temperature, sea ice, and night marine air temperature since the late nineteenth century, J Geophys Res-Atmos 108. doi:10.1029/2002JD002670

Rein B, Luckge A, Sirocko F (2004) A major Holocene ENSO anomaly during the Medieval period. Geophys Res Lett 31:17. doi:10.1029/2004GL020161

Rein B, Luckge A, Reinhardt L, Sirocko F, Wolf A, Dullo WC (2005) El Niño variability off Peru during the last 20,000 years. Paleoceanography 20(4). doi:10.1029/2004PA001099

Rodgers KB, Schrag DP, Cane MA, Naik NH (2000) The bomb C-14 transient in the Pacific Ocean. J Geophys Res-Oceans 105 (C4):8489–8512

Rodgers KB, Aumont O, Madec G, Menkes C, Blanke B, Monfray P, Orr JC, Schrag DP (2004) Radiocarbon as a thermocline proxy for the eastern equatorial Pacific. Geophys Res Lett 31(14). doi: 10.1029/2004GL019764

Sachs JP, Sachse D, Smittenberg RH, Zhang ZH, Battisti DS, Golubic S (2009) Southward movement of the intertropical convergence zone, AD 1400-1850. Nat Geosci 2:519–525

Sadekov A, Ganeshram R, Pichevin L, Berdin R, McClymont E, Elderfield H, Tudhope AW (2013) Palaeoclimate reconstructions reveal a strong link between El Niño-Southern Oscillation and tropical Pacific mean state. Nat Commun 4:1–8. doi:10.1038/ncomms3692

Schrag DP (1999) Rapid analysis of high-precision Sr/Ca ratios in scleractinian corals and other marine carbonates. Paleoceanography 14:97–102

Shen GT, Dunbar RB (1995) Environmental controls on uranium in reef corals. Geochim Cosmochim Ac 59:2009–2024

Shen GT, Boyle EA, Lea DW (1987) Cadmium in corals as a tracer of historical upwelling and industrial fallout. Nature 328:794–796

Shen GT, Campbell TM, Dunbar RB, Wellington GM, Colgan MW, Glynn PW (1991) Paleochemistry of manganese in corals from the Galápagos Islands. Coral Reefs 10:91–101

Shen GT, Cole JE, Lea DW, Linn LJ, McConnaughey TA, Fairbanks RG (1992) Surface ocean variability at Galápagos from 1936–1982: calibration of geochemical tracers in corals. Paleoceanography 7:563–588

Steinhilber F, Beer J, Fröhlich C (2009) Total solar irradiance during the Holocene. Geophys Res Lett 36. doi:10.1029/2009GL040142

Thompson DM, Ault TR, Evans MN, Cole JE, Emile-Geay J (2011) Comparison of observed and simulated tropical climate trends using a forward model of coral $\delta^{18}O$. Geophys Res Lett 38(14). doi:10.1029/2011GL048224

Toth LT, Aronson RB, Vollmer SV, Hobbs JW, Urrego DH, Cheng H, Enochs IC, Combosch DJ, van Woesik R, Macintyre IG (2012) ENSO drove 2500-year collapse of eastern Pacific coral reefs. Science 337:81–84

Trenberth KE, Shea DJ (1987) On the evolution of the Southern Oscillation. Mon Weather Rev 115:3078–3096

Trotter J, Montagna P, McCulloch M, Silenzi S, Reynaud S, Mortimer G, Martin S, Ferrier-Pages C, Gattuso JP, Rodolfo-Metapa R (2011) Quantifying the pH 'vital effect' in the temperate zooxanthellate coral *Cladocora caespitosa*: validation of the boron seawater pH proxy. Earth Planet Sc Lett 303:163–173

Tudhope AW, Shimmield GB, Chilcott CP, Jebb M, Fallick AE, Dalgleish AN (1995) Recent changes in climate in the far western equatorial Pacific and their relationship to the southern oscillation: oxygen isotope records from massive corals, Papua New Guinea. Earth Planet Sc Lett 136:575–590

Tudhope AW, Chilcott CP, McCulloch MT, Cook ER, Chappell J, Ellam RM, Lea DW, Lough JM, Shimmield GB (2001) Variability in the El Niño Southern Oscillation through a glacial-interglacial cycle. Science 291:1511–1517

Tudhope A, Cole JE, Chilcott CP, Ellam RM, Brown J, Collins M, Barnett H, Lea DW (2008) Evidence for changes in ENSO in the mid-late Holocene from fossil corals in Galápagos. EOS: Trans AGU 89(53). Fall Meet Suppl, Abstract PP21D-06

Urban FE, Cole JE, Overpeck JT (2000) Modification of tropical Pacific variability by its mean state inferred from a 155 year coral record. Nature 407:989–991

Vecchi GA, Clement A, Soden BJ (2008) Examining the tropical Pacific's response to global warming. EOS: Trans AGU 89(9):81, 83

Villaescusa JA, Carriquiry JD (2004) Calibration of Sr/Ca and Mg/Ca paleothermometers in coral Porites sp from San Benedicto Island, Revillagigedo Archipelago, Mexico. Ciencias Marinas 30:603–618

Waelbrook C, Paul A, Kucera M, Rosell-Melé A, Weinelt M, Schneider R, Mix AC et al (2009) Constraints on the magnitude and patterns of ocean cooling at the Last Glacial Maximum. Nature Geosci 2(2):127–132

Wellington GM, Glynn PW (2007) Responses of coral reefs to El Niño-Southern Oscillation sea-warming events. In: Aronson RB (ed) Geological approaches to coral reef ecology. Springer, Berlin, pp 342–385

Wellington GM, Dunbar RB, Merlen G (1996) Calibration of stable oxygen isotope signatures in Galápagos corals. Paleoceanography 11:467–480

Wilson R, Tudhope A, Brohan P, Briffa K, Osborn T, Tett S (2006) Two-hundred-fifty years of reconstructed and modeled tropical temperatures. J Geophys Res-Oceans. doi:10.1029/2005JC003188

Wolff M (2010) Galapagos does not show recent warming but increased seasonality. Galapagos Res 67:38–44

Human Influences On Eastern Tropical Pacific Coral Communities and Coral Reefs

20

Jorge Cortés and Héctor Reyes-Bonilla

Abstract

Coral reefs world-wide have been impacted by direct and indirect human activity and natural disturbances. This has led to the degradation and disappearance of many reef structures. On a basin-wide scale, the natural impact of El Niño warming has been the main cause of reef decline in the eastern tropical Pacific (ETP). At local scales, human activity has also taken its toll although only limited observations are available on specific impacts to ETP coral reefs. The main direct causes of damage are the extraction of corals and other reef organisms, nonregulated tourist activity, ship groundings, anchor damage, and eutrophication. The main indirect sources of damage to coral reefs are coastal alteration, sedimentation, pollution (including eutrophication), oil pollution, agrochemicals, other pollutants, and plankton blooms. Climate change can impact coral reefs directly (sea warming, sea level rise, ocean acidification, increased storm activity, and possibly stronger and more frequent El Niño events), and indirectly (coastal erosion, increased fresh water runoff and elevated nutrients). Even though human impacts on ETP reefs are low compared to other regions, significant damage has been documented. Since ETP coral reefs are relatively small and few in number, a redoubled effort is necessary for their protection.

Keywords

Coastal alteration • Sedimentation • Pollution • Plankton blooms • Tourism

20.1 Introduction

The direct and indirect impacts by humans on coral reefs are probably the main cause of reef deterioration world-wide. Natural impacts also damage and degrade reefs in the eastern tropical Pacific (ETP); for example, extreme high temperatures (see Glynn et al., Chap. 8), very low temperatures (Glynn and Stewart 1973; Maté 2003; Paz-García et al. 2012), and subaerial exposure during extreme low tides (Glynn 1976; Eakin and Glynn 1996) cause severe physiological stress and, on occasion, mass coral mortality. Nevertheless, many negative effects on reefs have an anthropogenic origin (Chabanet et al. 2005), and regrettably, are increasing in scope and severity. The distinction between natural and human impacts is not always clear-cut; for example, tropical Atlantic cyclones have occurred for millenia, but have demonstrated an increasing 30-year trend in strength coincident with a recent increase in ocean temperatures (Elsner et al. 2008). In this chapter we outline the known or hypothesized causes of coral reef degradation and destruction in the ETP as a result of direct or indirect human activities.

J. Cortés (✉)
Centro de Investigación en Ciencias del Mar y Limnología (CIMAR), Universidad de Costa Rica, San Pedro, 11501-2060, San José, Costa Rica
e-mail: jorge.cortes@ucr.ac.cr

J. Cortés
Escuela de Biología, Universidad de Costa Rica, San Pedro, 11501-2060, San José, Costa Rica

H. Reyes-Bonilla
Universidad Autónoma de Baja California Sur, Departamento Académico de Biología Marina, Carretera al Sur Km 5.5, Col. El Mezquitito, 23080 La Paz, BCS, Mexico
e-mail: hreyes@uabcs.mx

P.W. Glynn et al. (eds.), *Coral Reefs of the Eastern Tropical Pacific*, Coral Reefs of the World 8, DOI 10.1007/978-94-017-7499-4_20

We consider human impacts on coral reefs a result of direct damage (such as the removal of corals and other associated organisms, and diverse physico-chemical disturbances) or indirect damage (for example coastal development and various attendant negative effects such as sedimentation and pollution), with the caveat that these two classes of disturbance are difficult to separate and often interact detrimentally. Anthropogenic climate change stressors can impact corals directly or indirectly, and will be noted accordingly. The objectives of this contribution are to review the information available on human disturbances to coral reefs in the ETP, and to emphasize the need for more quantitative cause and effect information in the western American region. Such knowledge is necessary for the adequate protection, conservation and sustainable use of coral reef resources.

Fig. 20.1 Fishing lines entangled in corals, Bahía Culebra, Guanacaste, Costa Rica. Photograph by Celeste Sánchez Noguera

20.2 Direct Sources of Damage

20.2.1 Extraction

The effects of fisheries on the health of coral reefs have been extensively discussed worldwide, and in general it has been concluded that most stocks are in decline or overexploited (Jackson et al. 2001; Pauly and Zeller 2014; MacNeil et al. 2015). Cheung et al. (2007) have suggested that coral reef fisheries are among the most threatened because of their inherent vulnerability, which stems in large part from the life history traits of reef species. The reef fisheries of the ETP are no exception (Edgar et al. 2010; Friedlander et al. 2013). In the ETP, fishing is mostly artisanal, and is concentrated on species that have limited value in the international market. The majority of fishes, molluscs, crustaceans and echinoderms that comprise the catches are sold for domestic consumption, except for sharks, lobsters and sea cucumbers (Nance et al. 2011; Purcell et al. 2011; Cisneros-Montemayor et al. 2012; Dapp et al. 2013), which bring high prices in Asian markets.

"Fishing down the food web"—with a gradual transition from landings of long-lived, high trophic level piscivores to short-lived, low trophic level invertivores, herbivores and planktivorous fishes (Pauly et al. 1998; Branch et al. 2010)—may have relevance on ETP coral reefs. Sala et al. (2004) have demonstrated that for specific areas in the southern Gulf of California with well-developed coral communities, the trophic level of harvested fishes decreased noticeably over a decade. However, an analysis of larger areas with corals did not support the afore-noted observation (Pérez-España et al. 2006). Therefore, it is unknown if a downward trophic shift occurs only locally or can be considered on a regional scale. This distinction is important for the implementation of remedial actions. In any case, as a consequence of fishing activities in coral reef areas, it is common to observe fishing lines and nets entangled with corals, octocorals and other organisms (Fig. 20.1). Such abandoned fishing gear will continue to injure and kill reef organisms (Macfadyen et al. 2009; CBD/STAP 2012).

Another type of resource exploitation impacting ETP coral habitats is related to the aquarium trade. For example, one known incident involved the removal of 200 colonies of *Tubastraea coccinea* from a single site in Costa Rica (Cortés and Jiménez 2003). There is now a demand for "live rock", which is also extracted and sold to aquarists. This type of harvesting is practised in Mexico (Reyes-Bonilla 2003), Panama (Maté 2003) and Colombia (Zapata and Vargas-Ángel 2003). In western Mexico, two illegal shipments of live coral rock, of three and five tons each, were detained by the authorities (Mexican law prohibits this activity). The corals were about to be transported to Guadalajara and Mexico City to be sold to aquarists who most probably would never know that their live rock purchases were harvested illegally.

The trafficking of reef organisms for souvenirs may also have caused yet undetermined levels of damage to ETP coral communities and reefs (Fig. 20.2). The removal of corals has been widespread, observed in the Galápagos Islands (Glynn 1994, 2003), Mexico (Reyes-Bonilla et al. 2008), El Salvador (Reyes-Bonilla and Barraza 2003), Costa Rica (Cortés and Jiménez 2003), Panama (Maté 2003), Colombia (Zapata and Vargas-Ángel 2003), and Rapa Nui (Easter Island) (Glynn et al. 2003). The concentrated removal of *Pocillopora meandrina* in Costa Rica and *Pocillopora eydouxi* at Rapa Nui has resulted in the local disappearance of these species. Ornamental fisheries in Costa Rica demonstrated a decline in stocks between 1994 and 2000. Alperman (2001) found that populations of the Cortez

Fig. 20.2 Corals for sale in Mazatlán, Mexico. Souvenir shops now have large coral inventories from the Indo-Pacific region to supplement local supplies that have been severely depleted in many areas. Photograph by Cristian Pacheco

Rainbow Wrasse were severly overexploited, and Dominici-Arosemena et al. (2005) reported that many reef fish populations were adversely affected in the Gulf of Papagayo, Costa Rica. Nevertheless, effective management in the central Gulf of California (Mexico), orchestrated by fishers, non-governmental organizations and academics, have allowed improvements in catches and the attainment of stable population levels of several ornamental fish species (Germain et al. 2015). In light of the limited catch information available for aquarium fishes and invertebrates, there is a pressing need to generate accurate assessments of the magnitude of the problem in order to better manage this activity.

20.2.2 Tourism

Tourism is becoming one of the main sources of revenue for Latin America, with the numbers of visitors to ETP coastal areas rapidly rising. In Mexico, for example, tourism development has been especially active in areas with coral habitats (Wilson 2008). This presents new challenges for conservation and management—prior to the 1990s nearly all Pacific American coral reefs experienced only minimal tourist use.

In some areas, divers, both scuba and surface divers, are responsible for much of the observed damage to corals and associated reef organisms. Common sources of injury result from direct contact with corals, colony fragmentation, and the hand-feeding of coral-associated fishes, all activities observed in Mexico, Costa Rica, Panama, and Colombia (Jiménez 1997; Cortés and Jiménez 2003; Maté 2003; Zapata and Vargas-Ángel 2003; Reyes-Bonilla et al. 2012a). Such harmful effects are expected to continue and will require special management actions for their control. Diver briefings on the fragility of coral habitats, and information on the rules and regulations in effect for particular MPAs should help to protect these resources.

Diving carrying capacity (Sorice et al. 2007), a measure of the maximum number of divers that can perform underwater activities without causing damage to coral habitats, was calculated by Reyes-Bonilla et al. (2012b) for the Cabo San Lucas area, one of the most frequently visited sites in western Mexico. In this area, small coral patches occur in the main bay and surroundings (Wilson 2008). The estimated carrying capacity for Cabo San Lucas is 20,400 dives year^{-1}, which is about 30 % higher than the actual number of users, indicating that the present level of activity can be sustained by the local coral resources. A similar analysis by Ríos-Jara et al. (2013) for the Isabel Island Marine Park off the central Mexican coast (Nayarit) also indicated a sustainable level of diving activities. The calculated diving carrying capacity was 8300 dives year^{-1} with an observed rate at this site of 1000 dives year^{-1}.

20.2.3 Ship Groundings and Anchor Damage

Information on the effects of large ship impacts on ETP coral reefs is scarce. One of the most widely publicized groundings occurred in the Galápagos Islands in 2001, when the oil tanker MV *Jessica* spilled over 900,000 L of bunker fuel and diesel oil at San Cristóbal Island (Edgar et al. 2003; Álava et al. 2014). The spill occurred near the west end of the island, several km distant from coral habitats at Punta Pitt on the island's east end. Fortunately, the fuel oil was quickly dispersed by currents and did not greatly harm nearshore biota (Lougheed et al. 2002). Two boat groundings in 1996 and 1997 on the Uva Island (Panama) coral reef caused extensive damage. On both occasions, 40-cm thick sections of the reef framework were destroyed (plained horizontally), and two ~100 year-old colonies of *Gardineroseris* were crushed (Maté 2003).

A noteworthy grounding by the tanker *Lázaro Cárdenas II* impacted a well developed coral reef at La Paz Bay, southwestern Gulf of California in 2004 (Aranceta-Garza et al. 2012). This grounding caused extensive fragmentation of pocilloporid corals over nearly 1000 m^2. In a restoration effort, more than 600 broken pocilloporid fragments of different sizes were collected from the impact zone, cemented to concrete blocks and transported to a slightly deeper sand bottom area about 500 m from the grounding site. This recovery program was initiated in 2005 and was still in progress until 2013, with excellent results. Coral survival on

Fig. 20.3 Anchor damage, Bahía Culebra, Guanacaste, Costa Rica. Photograph by Celeste Sánchez Noguera

the blocks was over 80 %. Storm-generated fragments from the blocks eventually spread to the surrounding area where they now contribute to live cover exceeding 15 %.

Anchor damage is common-place in all countries of the ETP (Glynn 1994; Cortés and Jiménez 2003; Maté 2003; Reyes-Bonilla and Barraza 2003; Zapata and Vargas-Ángel 2003), but no attempts to quantify its destructive effects have been made. Broken branches and displaced colonies can be seen on practically all coral reefs where frequent fishing and tourist visitation occur (Fig. 20.3). This kind of damage can have varying effects, ranging from total colony death to branch survival and regeneration. Smaller broken branches generally demonstrated a lower likelihood of survival than larger branches.

20.2.4 Plankton Blooms

The ETP is a region with normally high concentrations of plankton (Fernández-Álamo and Färber-Lorda 2006; Pennington et al. 2006), and commonly experiences microalgal blooms. Since some algal blooms occur near population centers and agricultural areas, the provenance of nutrient sources initiating the blooms is not always clear.

Phytoplankton blooms on coral reefs in Panama and Costa Rica caused mass mortalities of corals, other invertebrates and some reef fishes in 1985 (Guzmán et al. 1990). Copious amounts of mucus released by dinoflagellates were hypothesized as the responsible agent causing coral death, which resulted in the elimination of some coral species from shallow reef zones. No other similar events have since been reported in the equatorial ETP. Cortés and Jiménez (2003) documented algal blooms at Ballena Marine Park, Costa Rica in 1992, 1994 and 1995, but no significant mortality was noted. Several workers have claimed that zooplankton and phytoplankton blooms occur annually across several ETP sites from the southern Gulf of California (LaJeunesse et al. 2007) to the Gulf of Nicoya, Costa Rica (Víquez and Hargraves 1995; Vargas-Montero et al. 2008), and Panama Bay (Miglietta et al. 2008). These are all sites with modest reef coral habitats (Maté 2003; Reyes-Bonilla and López-Pérez 2009).

Where annual phytoplankton blooms occur, coral communities and associated species have shown only limited harmful effects, if at all (D'Croz and Robertson 1997; Miglietta et al. 2008). Zapata and Vargas-Ángel (2003) described a short-term (one week) coral bleaching event at Utría, Colombia. This occurred during a combined cold water period and phytoplankton (*Gymnodinium* spp.) bloom. It is likely that reef corals in the ETP are repeatedly exposed to phytoplankton blooms but are able to survive these potentially harmful events except when massive. Since 2007, major phytoplankton blooms have impacted coral reefs on the northwestern coast of Costa Rica (C. Jiménez, personal communication, J. Cortés, personal observation). These blooms occurred near human settlements, and were possibly a response to added nutrient inputs as observed in other areas (Heisler et al. 2008).

20.2.5 Climate Change

Climate change projections indicate that coral reef ecosystems will be impacted directly and severly in the coming decades. Among the commonly cited expected stressful conditions are sea warming, sea level rise, increasing storm intensity, ocean acidification, and possibly more frequent and severe El Niño events (Hoegh-Guldberg et al. 2007; Wild et al. 2011). As these changes are expected to occur worldwide, it is safe to assume they will also occur in the ETP. The unique regional conditions of the ETP—cold currents, upwelling centers, ENSO episodes, and narrow continental shelf—will require a specific analysis to better understand the possible consequences.

A straightforward stress to coral reefs will be the projected and unpredictable increase (and decrease, see Chap. 8, Glynn et al.) in ocean temperature. The ETP zooxanthellate coral species have demonstrated a relatively low tolerance to elevated sea temperatures, amply indicated by repeated bleaching events that have occurred across the entire region since the 1980s (Wellington and Glynn 2007; Baker et al. 2008). The loss of algal endosymbionts can cause disruptions at the ecosystem level, including a simplification of the community structure of invertebrates and fishes (Iglesias-Prieto et al. 2003; Glynn et al. 2014). Additionally, thermally-induced bleaching can lead to reduced reef calcification, primary productivity and possibly nutrient regeneration, all resulting from the loss of live coral and increased erosion of reef frameworks (Glynn 2004; Calderón-Aguilera et al. 2007). Although some evidence of possible local adaptation of corals to the new conditions has been forthcoming (Baker et al. 2004; LaJeunesse et al. 2010; Logan et al. 2014), the ETP is a global region that experiences higher historical increases in temperature (Hansen et al. 2010). Since the warming trend is expected to continue, bleaching events will probably occur more often in the near future as the tolerance thresholds of corals are surpassed (Frieler et al. 2012).

Changes in storm regimes are predicted to affect coral communities and reefs due to increased freshwater runoff from rain, and breakage of coral colonies by unusual patterns of wave action (Huntington 2006; Bender et al. 2010). Rising sea level, combined with greater storm activity, will probably result in elevated coastal erosion as well as increased sedimentation (Cortés and Risk 1985; Neumann and Macintyre 1985; Nicholls and Cazenave 2010). Under present-day conditions, hurricanes in the ETP typically occur between 15°N and 24°N latitude (Lirman et al. 2001; Hernández et al. 2008; Aranceta-Garza et al. 2012; see Chap. 3, Fiedler and Lavín). Hurricane damage is not long-lasting except for very strong storms or repeated strikes over short time intervals (Glynn et al. 1998; Calderón-Aguilera et al. 2007). If hurricanes increase in strength in the future, coral reefs may not have the capacity to return to a semblance of their original condition because of continuous fragmentation and subsequent mortality (Knutson et al. 2010). This effect was reported for the Cabo Pulmo reef during the past decade when three cyclones struck this site in a span of four years (Álvarez-Filip and Reyes-Bonilla 2006).

Another major impact of global change on the ecology of coral reefs worldwide would be the reduction in pH ("ocean acidification"), and its corresponding decrease in aragonite saturation state (Ω_{arag}, Doney et al. 2009). This phenomenon has received considerable attention by coral reef researchers due to its actual and potential effects on many reef species and processes, and the fact that Ω_{arag} is expected to decline to levels below those currently experienced by modern coral reefs (Hoegh-Guldberg et al. 2007; Pelejero et al. 2010). Lower pH reduces coral calcification, and causes impaired development of echinoderm and mollusc larvae, interfering with early skeletal development, metamorphosis and ability to settle (Dupont et al. 2010; Nakamura et al. 2011). Also, fishes are affected as cellular changes in acid-base balance block olfactory senses, and consequently the ability of adults and juveniles to detect predators (Dixson et al. 2010; Munday et al. 2013; Heuer and Grosell 2014). Otolith development is affected as well, impacting echolocation and the ability of juvenile fishes to detect suitable habitats (Bignami et al. 2013).

In the ETP, there is naturally low pH and Ω_{arag} due to the upwelling of CO_2-enriched deep waters (Kleypas et al. 1999; Manzello et al. 2008; Rixen et al. 2012). Manzello (2010a) demonstrated that in the Galápagos Islands and Panama, the lowest Ω_{arag} values occur where upwelling is most pronounced as well as in the cool season, while Reyes-Bonilla et al. (2014) reported that the best conditions for carbonate deposition in the southern Gulf of California are between July and November, when $\Omega_{arag} \geq 3.0$. Some species of corals may be more or less resilient to lower pH than others (Manzello 2010b; Cabral-Tena et al. 2013). Rixen et al. (2012), in a study during seasonal upwelling in northern Costa Rica, found conditions unfavorable for coral growth, but a previous study in the same area revealed high skeletal growth rates (Jiménez and Cortés 2003). During upwelling there are more plankton and nutrients available in the water column that may help stimulate coral growth despite depressed pH (e.g., Manzello et al. 2014). More detailed local studies are needed, but it can be hypothesized that in the near future, with even higher atmospheric carbon dioxide concentrations and lower pH conditions, zooxanthellate corals and other calcifying organisms of the west American coast will progressively be subject to habitats of decreasing quality.

Recent tracking of mean global sea level reveals a gradually accelerating 60 mm rise during the last 40 years (Domingues et al. 2008; Cazenave and Llovel 2010). This increasing rate of sea level rise is more than an order of magnitude lower than the vertical skeletal extension rates of most ETP corals (Guzmán and Cortés 1989; Jiménez and Cortés 2003; Reyes-Bonilla 2003), and also lower than known reef accretion rates (Glynn and Macintyre 1977; Glynn et al. 1979; Cortés et al. 1994; Calderón-Aguilera et al. 2007; see Chap. 6, Toth et al.). Therefore, baring any unforeseen rapid acceleration in sea level rise it seems unlikely that this factor alone would interfere with ETP reef growth.

El Niño-Southern Oscillation (ENSO) disturbances to ETP coral reefs are well characterized, as virtually the entire region has been impacted by thermally-induced bleaching and cascading effects (Wellington and Glynn 2007; Baker et al. 2008). This phenomenon is described in much more detail in other chapters of this book (see Chap. 4, Wang et al. and Chap. 8, Glynn et al.); here we note its possible links with predicted changes in sea temperature. Although in the 1990s several workers hypothesized that ENSO might increase in strength under climate change conditions, current understanding is inconclusive (Collins et al. 2010). However, it is possible that ENSO events could increase in duration and variability (Gastineau and Soden 2011), thus exposing reefs to both extreme warming and low thermal anomalies, which also cause coral bleaching (Zapata et al. 2011; Paz-García et al. 2012). In addition, the projected SST

Fig. 20.4 Coastal alteration for construction of luxury homes, Guanacaste, Costa Rica. Photograph by Jorge Cortés

increases, in combination with El Niño anomalies in particular years, would exceed current bleaching thresholds with likely negative consequences. Paleoecologic evidence suggests that from about 4000 to 1500 years ago, during a period of ENSO and ITCZ (Inter-Tropical Convergence Zone) coupling, Panamanian coral reef development collapsed, coincident with stronger, more frequent and higher El Niño variability (Toth et al. 2012).

It has been proposed that upwelling centers might serve as refugia during stressful warming events (Glynn 1996; Riegl 2003; Riegl and Piller 2003; Smith et al. 2014; see Chap. 17, Smith et al.). However, Chollett et al. (2010) cautioned that two conditions must be met for this to be effective, (a) the co-occurrence of warm water stress during upwelling events, and (b) a significant decrease in thermal stress brought on during the upwelling period. In their analysis of four upwelling sites over a 21 year period, including the ETP (Panama), only in two, Colombia (Caribbean) and Oman (Arabian Sea), did the timing of upwelling commonly coincide with the warming interval. In Panama, the average joint occurrence of warming and upwelling was 41 %, compared with 86 % in Oman. Additionally, productive oceanic areas are generally sites of lower pH, so in the long term upwelling zones may not be ideal refugia for corals. In the future, some taxa might adapt (Somero 2010), and that would alleviate perturbations at the ecosystem level, but for long-lived species such as many corals, fishes and other marine vertebrates, it would be challenging to deal with the projected anomalous conditions.

20.3 Indirect Sources of Damage

20.3.1 Coastal Alterations

Changes to the coastal zone from urbanization, agriculture and mariculture, or from construction of tourist and port facilities, have resulted in indirect local impacts to coral reefs (Fig. 20.4). Harmful effects are due mainly to sedimentation, but also to changes in nutrient runoff and local ocean currents. Increasing human population, especially in coastal areas, has resulted in widespread pollution and overexploitation of many marine organisms. The expansion of agriculture has increased the amounts of agrochemicals reaching the ocean, affecting coral reefs in ways that are still not completely understood (Hughes et al. 2003; Pandolfi et al. 2003). These problems have been observed in all ETP countries with coral habitats (Burke and Spalding 2011; see Chap. 5, Glynn et al.), but because of their complexity and potentially large-scale impacts, detailed site-specific information is often absent.

Bahía Culebra on the Pacific coast of Costa Rica has experienced significant land-use changes in shifting from low-scale agriculture to intensive tourism (Sánchez-Noguera 2012). The notable increase in number of visitors has interfered with local fishers and coral reef protection (Jiménez 1997, 2001). In Mexico, tourist developments have also had strong effects on the economy, societal norms and the environment, and not always in a positive way (Wilson 2008; Reyes-Bonilla et al. 2012b).

Construction of an interoceanic canal in Nicaragua began in December 2014 (Laursen 2015). This project has been highly critized because of the environmental damage it will likely cause to forest ecosystems and the water quality of Lake Nicaragua, the largest fresh water body in Central America (Meyer and Huete-Pérez 2014). However, no mention has been made of the probable impact of this project on coral communities and reefs on the Pacific coast, which were recently discovered and described by Alvarado et al. (2011; see Chap. 5, Glynn et al.). The possible canal-related damage to coral reefs on the Caribbean side of Nicaragua is also of great concern. On the basis of known environmental disturbances vis-à-vis the Panama and Suez canals, the main impacts of the Nicaraguan canal may include an increase in sediment loading from dredging and construction (McCullough 1977), oil pollution (Guzmán and Jarvis 1996), and the ocurrence of invasive species on both coasts (Galil et al. 2014).

20.3.2 Sedimentation

One of the main impacts to ETP mainland coral communities and reefs is excessive terrigenous sedimentation, a result of coastal or more inland activities such as deforestation, mining, and poor land management. Loss of vegetation in tropical watersheds can result in high sediment loads in rivers that enter coastal areas with coral habitats (Fig. 20.5). Deforestation occurs from the expansion of urban areas, the building of tourist facilities and construction of ports, but also from logging and land clearing for agriculture, cattle ranches and aquaculture projects. All of these near-coastal alterations can lead to direct or indirect sediment input to reef areas by rivers, coastal runoff, and wind-blown sediments.

Suspended sediments reduce the amount of light penetrating the water column and reaching corals, thus affecting the photosynthetic efficiency of the zooxanthellae. Corals can tolerate certain amounts of suspended sediments, expending energy to shed sediments by ciliary action, mucus production, and by altering colony morphology. All of these processes consume resources that otherwise would be used for growth, wound repair or reproduction. As a result corals degrade slowly, even under low sediment loads, while elevated loads can lead to morbidity and death relatively quickly, i.e. within days to weeks (Cortés and Risk 1985; Edmunds and Davies 1989; Rogers 1990; Fabricius 2005; Erftemeijer et al. 2012).

Sedimentation effects have been reported for many areas in the ETP. In Mexico and Costa Rica, damage to coral reefs has been observed in areas of coastal alteration, particularly in the construction of tourist facilities (Cortés and Jiménez 2003; Reyes-Bonilla 2003). Ochoa-López et al. (1998) reported coral death from sedimentation at Socorro Island (Revillagigedo Archipelago, Mexico), due to overgrazing by feral sheep. At Los Cóbanos, the principal coral habitat in El Salvador, siltation from inappropriate land management poses a threat (Reyes-Bonilla and Barraza 2003; E. Barraza, personal communication 2013). Several areas along the Costa Rican coast are affected by excessive sediment loads: in the north from construction of tourist facilities (Cortés and Murillo 1985; Jiménez 2001; Cortés and Jiménez 2003), along the central coast from a combination of land clearing for agriculture and construction (Alvarado et al. 2005, 2009), and in the south mainly from deforestation for cattle ranching and agriculture (Cortés 1990; Cortés et al. 1994; Cortés and Jiménez 2003). In Panama, sedimentation ostensibly influences the distribution and development of coral reefs, but there are no published accounts documenting this effect (Maté 2003). González and Polo (1982) showed experimentally that in Panama the growth rates of *Pavona gigantea* and *Porites lobata*, two widely distributed and abundant ETP species, were not affected after 103 days of exposure to sediments; *Pavona varians*, however, was greatly affected.

Fig. 20.5 Suspended sediments visible in surface waters, Golfo Dulce, Costa Rica. Area of heavily impacted coral habitats (Cortés 1990; Cortés et al. 1994). Photograph by Silvia Echeverría

Sediment stress also occurs in the southern-most continental limits of ETP coral development where rainfall and freshwater discharge rates are exceptionally high (Glynn and Ault 2000). Forest clearing at Gorgona Island, Colombia, a mid-20th century prison site, resulted in high sediment loads across the largest coral reef of the island (Prahl et al. 1979; Glynn et al. 1982). Suspended sediments from mainland coastal rivers may also reach Gorgona (Zapata and Vargas-Ángel 2003). Prahl and Vargas-Ángel (1990) reported reduced skeletal growth rates of *Pocillopora damicornis* at Ensenada de Utría, a mainland site north of Gorgona Island. Subsequent studies by Vargas-Ángel (1996) reported that severe coastal erosion and sedimentation were affecting the Utría coral reefs, and this condition probably occurred over much of Holocene time (Vargas-Ángel 2001). Also, mainland Ecuadorian reefs are likely regularly impacted by sedimentation and eutrophication while at the oceanic setting of the Galápagos Islands highly turbid conditions are uncommon and ephemeral (Glynn 2003).

20.3.3 Pollution

Three main sources of pollution impacting coral reefs are eutrophication, oil, and agrochemicals; there are additional minor pollutants as well, such as polychlorinated biphenols (PCBs), pharmaceutical and cosmetic products, antibiotics, interalia. More research on the effects of these sources of pollution is needed worldwide, and in the ETP in particular. Here we briefly note what is known, and conclude the section on what has been studied in the ETP.

20.3.3.1 Eutrophication

Nutrient enrichment favors the proliferation of phytoplankton and macroalgae (Heisler et al. 2008). This source of pollution affects corals by light reduction in the water column, interphyletic competition for space, inhibition of coral recruitment, shading by benthic algae, abrasion, and allelopathy (McCook et al. 2001; Jompa and McCook 2003; Arnold et al. 2010; Rasher and Hay 2010; Hoey and Bellwood 2011; Shearer et al. 2012).

Eutrophication-related studies have not been conducted in the ETP, but indirect evidence indicates that nutrient input from land may be impacting coral reefs in some areas. At Bahía Culebra, Costa Rica the proliferation of the alga *Caulerpa sertularioides* has been partly attributed to increases in nutrient runoff during the non-upwelling wet season (Fernández and Cortés 2005; Fernández-García et al. 2012; Fig. 20.6). Also, an increase in frequency, intensity and duration of phytoplankton blooms has coincided with the proliferation of tourist facilities and the lack of waste water treatment (C. Jiménez pers. comm.; J. Cortés pers. obs.). Glynn (2003) noted that eutrophication was not a serious threat in the Galápagos Islands, at least up to the late 1990s.

Eutrophication is apparently not a problem on rocky or coral reefs in western Mexico. In fact, coral habitats are absent from coastal areas with large rivers where high turbidity and frequent fresh water discharge occurs (Reyes-Bonilla 2003). Also, most coastal cities with nearby coral reefs have small human populations; only Acapulco, La Paz, and Puerto Vallarta have more than 200,000 inhabitants each. Finally, the strong mixing of water masses surrounding reefs in the Gulf of California and nearby areas prevents the concentration of nutrients and sediments that could reach detrimental levels at particular sites (Reyes-Bonilla and López-Pérez 2009).

Fig. 20.6 *Caulerpa sertularioides* encroaching on *Pocillopora*, Bahía Culebra, Costa Rica. Photograph by Cindy Fernández

20.3.3.2 Oil Pollution

Two oil spills in El Salvador in 1993 were implicated in the disappearance of a coral community. The first spill resulted in 50–80 % mortality of *Porites lobata* (Molina 1996), and the second was assumed to have killed the few remaining *Pocillopora* colonies at the affected site (E. Barraza, personal communication 2013). In Costa Rica, low levels of oil pollution have been observed, especially in port areas far from coral communities and reefs (Spongberg and Davis 1998; Acuña-González et al. 2004). However, one study carried out in a coral reef area in Costa Rica (Isla del Caño), reported low levels of dissolved hydrocarbons, and rock substrates coated with tar (Acuña and Murillo 1987). D'Croz (1988) reported high levels of oil pollution at Balboa, near the Pacific entrance of the Panama Canal. Guzmán and Holst (1994) suggested that oil pollution may have been responsible for some reef damage in Panama Bay. Further, tar has been observed on beaches at Gorgona Island (Colombia), probably arriving by current transport from Ecuador. These tar deposits, however, had not caused any apparent damage to reef organisms (Zapata and Vargas-Ángel 2003). There have been repeated oil spills in the Galápagos Islands, but their effects have been described for marine invertebrates other than corals (Whelan 1989; Wikelski et al. 2002).

The situation on the western coast of Mexico is not threatening at present due to the absence of oil refineries, and because most transportation of petrochemicals is by train or tanker trucks. There may still exist local threats, however; for example, at Manzanillo, one of the largest Pacific harbors, there are data indicating significant oil input to the ocean due to atmospheric deposition from power plants and mining (Cerón et al. 2012). Also, at other large ports and locations where artisanal fisheries and tourism are prevalent (e.g., Huatulco, Acapulco, and Puerto Vallarta), diffuse contamination caused by gasoline spills from small boats is commonplace, but has not been studied to assess its effects in coral habitats.

20.3.3.3 Agrochemicals

Large coastal zone areas in and around the ETP are, or were until recently, dedicated to agriculture, and in most cases with heavy use of chemicals. Some rivers may contain persistent pesticides (Spongberg and Davis 1998). However, there has been little research on the impact of agrochemicals, mainly pesticides, on coral reefs. An early study in Panama reported the herbicides 2,4-D and 2,4,5-T, as well as phenoxy acids, in "normal" appearing (non-bleached) corals (Glynn et al. 1984). Insecticides may have some effects on coral reef crustaceans, but this has not been studied.

20.3.3.4 Other Pollutants

Solid waste residues (mainly plastics), DDT, polychlorinated biphenols (PCBs) and polycyclic aromatic hydrocarbons

(PAHs) have been found in oysters at Los Cóbanos, the main coral habitat site in El Salvador (Michel and Henry 1997). These pollutants reach Los Cóbanos by current transport (Reyes-Bonilla and Barraza 2003). Spongberg (2004, 2006) found low concentrations of PCBs in sediments and sipunculids at Bahía Culebra, a coral habitat in Costa Rica, and minimal levels in other tropical areas (Spongberg and Witter 2008). In a study of the occurrence of pharmaceutical and personal care products (PPCPs), Spongberg et al. (2011) found low amounts in all coastal areas sampled in Costa Rica.

20.3.4 Climate Change

Indirect effects of climate change would likely cause coastal erosion, increased fresh water runoff, and elevated nutrient levels. Sea level is presently rising (Cazenave and Llovel 2010), and if the increase is more rapid than the rate of coral reef accretion, then the prospect of reef drowning could become a reality (sensu Neumann and Macintyre 1985). Vertical reef growth in the ETP has kept pace with sea level rise during Holocene time (Glynn and Macintyre 1977; Cortés et al. 1994). However, if sea level rise and precipitation increase in some areas (Allan 2011), coastal erosion would probably lead to heavy sedimentation stress and coral mortality (Newmann and Macintyre 1985; Macintyre 2007). Also, water column turbidity and increased nutrient runoff could have additional negative effects on corals and reef growth.

Finally, circulation in the Pacific is predicted to decrease as a consequence of changes in wind patterns (Collins et al. 2010; Gastineau and Soden 2011). This could further isolate the ETP from central/west Pacific coral populations by weakening current flow and propagule transport, thus diminishing larval immigration and genetic connectivity.

20.4 Conclusions and Recommendations

Aside from local-scale disturbances, such as sedimentation in Costa Rica (Cortés and Jiménez 2003; Alvarado et al. 2009) or coral extraction in El Salvador (Reyes-Bonilla and Barraza 2003), human impacts on ETP coral communities and reefs are relatively minor compared to other regions of the world (Fig. 20.7). The main disturbance affecting coral habitats in the ETP is the El Niño phase of ENSO, with its accompanying sudden occurrence of high sea temperature anomalies (see Chap. 8, Glynn et al.). This phenomenon is a natural event outside immediate human control. Nonetheless, some studies indicate that corals are more resilient and can recover faster from severe natural disturbances in areas of reduced human stressors. For example, Carilli et al. (2009, 2010) found that corals at Caribbean reef sites subject to anthropogenic stressors (sedimentation, nutrient loading, human population size, and fishing pressure) responded less favorably to heat-stress bleaching than corals at non-polluted sites. At sites with a relatively high pollution index, skeletal growth was suppressed, thermal tolerance was reduced, and recovery time was longer in the reef-building coral *Orbicella faveolata*. Also, in Pacific Costa Rica offshore reefs subject to less human influence have recovered more rapidly from El Niño disturbances than coastal reefs (Guzmán and Cortés 2001, 2007; Cortés et al. 2010). Clearly, further study is needed to separate cross-shelf (onshore-offshore) environmental differences from anthropogenic effects.

Roff and Mumby (2012) hypothesized that Caribbean reefs have a greater propensity to collapse compared with Indo-Pacific reefs. They attributed several factors responsible for this difference—(a) high abundances of macroalgae and frequent competition with corals, (b) low herbivore abundances that could limit macroalgal proliferation, and (c) low abundances of reef-building acroporid corals. A similar regional comparison would be useful for ETP reefs, which may be more sensitive to disturbances because of the low redundancy of reef-building coral diversity (some reefs are constructed by only a few species of zooxanthellate corals; Cortés 1997, 2011). However, because of the relatively high numbers of *Pocillopora* species in the ETP (12 nominal morphospecies are presently recognized, see Appendix 5.A in Chap. 5, Glynn et al.), the most frequent reef-builder in the region (see also Chap. 6, Toth et al.), some reefs composed of different species of the genus hosting unique *Symbiodinium* clades, appear to exhibit a higher resilience to bleaching disturbances (Glynn et al. 2001; Baker et al. 2008; LaJeunesse et al. 2010; see Chap. 13, Baker et al.). This possibility demands caution, however, since genetically delineated analyses (Pinzón and LaJeunesse 2011; Pinzón et al. 2013; see Chap. 14, Pinzón), and demonstrated morphological plasticity (Paz-García et al. 2015a, b) suggest that the species richness of ETP *Pocillopora* may be considerably reduced, perhaps to as few as three or four species.

Local impacts are the main cause of coral reef degradation in the ETP, which include excessive terrigenous sedimentation, pollution, human-enhanced harmful algal blooms, coral extraction, and overexploitation of associated reef organisms. Improved and integrated coastal zone management should result in a reduction of terrigenous sediments, pollutants, and nutrients. The creation and appropriate management of Marine Protected Areas should help improve fish stocks and minimize the extraction of coral reef organisms (see Chap. 21, Alvarado et al.).

Climate-change impacts to ETP coral reefs can be divided into three classes: (a) sea temperature rise, (b) changes in storm activity, and (c) ocean acidification (Fig. 20.7). One of

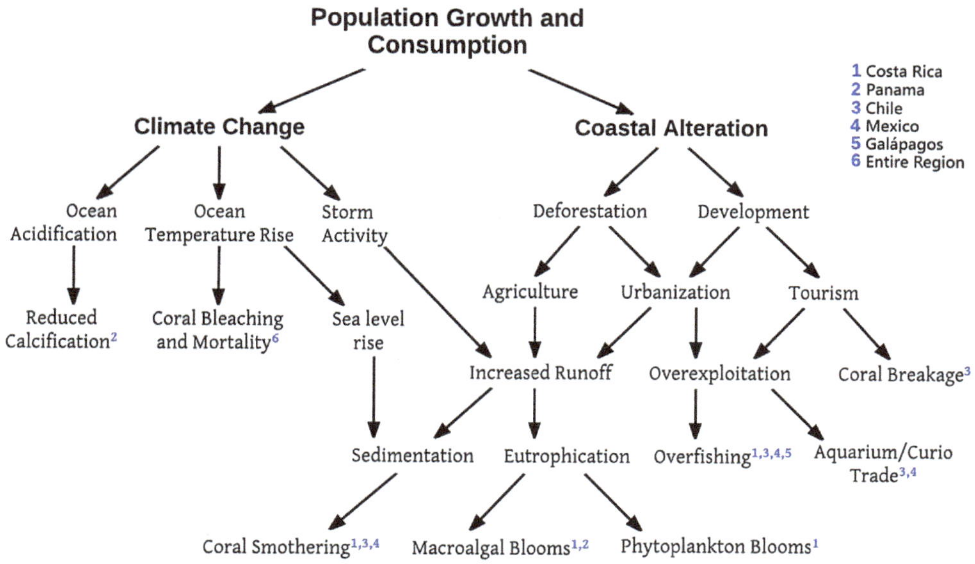

Fig. 20.7 Schema of direct (coastal alteration) and indirect (climate change) impacts affecting coral habitats throughout the eastern Pacific region. Observed site-specific effects are denoted for several countries

the consequences of higher temperatures may be an increase in the frequency and intensity of El Niño events (Collins et al. 2010), which would result in more frequent and intense coral bleaching and mortality. Another likely effect of temperature rise is more widespread coral bleaching, as observed historically across the ETP in years of very strong El Niño activity (Glynn 2000). Sea level rise, also a result of rising sea temperatures (Cazenave and Llovel 2010), can be expected to cause more coastal erosion and increasing sedimentation as observed in El Salvador and Costa Rica. Changes in storm patterns have been documented in several regions of the world (Huntington 2006; Allan 2011), and this has resulted in increased freshwater runoff, including sediment and nutrient loading, reaching reef areas. Finally, as the pH of seawater decreases, coral calcification rates are expected to decline as documented in Panama and the Galápagos Islands (Manzello 2010b; Manzello et al. 2014).

In conclusion, it is evident that ETP coral reefs, relatively small and few in number, are in urgent need of protection from all sectors of society that benefit from the environmental services and aesthetic pleasures they offer. Hopefully this review will encourage studies to better clarify human influences on coral communities, and thus promote effective strategies for their protection.

Acknowledgements We thank Peter W. Glynn, Derek Manzello and Ian C. Enochs for the opportunity to contribute this chapter on human impacts to the ETP book, and Bernhard Riegl for his critical review. JC thanks Enrique Barraza for his insights and relevant publications for El Salvador. We are also grateful for the detailed reviews of James Hendee, Peter W. Glynn and an anonymous reviewer. Finally, Ben Grassian greatly improved the quality of the figures.

References

Acuña JA, Murillo MM (1987) La contaminación por hidrocarburos de petróleo en la Isla del Caño. Ing Cienc Quím 11:95–98

Acuña-González JA, Vargas-Zamora JA, Gómez-Ramírez E, García--Céspedes J (2004) Hidrocarburos de petróleo, disueltos y dispersos, en cuatro ambientes costeros de Costa Rica. Rev Biol Trop 52 (Suppl 2):43–50

Álava JJ, Palomera C, Bendell L, Ross PS (2014) Pollution as an emerging threat for the conservation of the Galápagos Marine Reserve: environmental impacts and management perspectives. In: Denkinger J, Vinueza L (eds) The Galápagos marine reserve: a dynamic social-ecological system. Springer, New York, pp 247–283

Allan RP (2011) Human influence on rainfall. Nature 470:344–345

Alperman TJ (2001) The fisheries of ornamental fishes in Guanacaste, Costa Rica, with special emphasis on the population dynamics of the Cortez Rainbow Wrasse, *Thalassoma lucasanum* (Gill 1863). MSc thesis, University Bremen, Bremen, Germany, p 84

Alvarado JJ, Cortés J, Fernández C, Nivia J (2005) Coral communities and reefs of Ballena Marine National Park, Pacific coast of Costa Rica. Ciencias Marinas 31(4):641–651

Alvarado JJ, Fernández C, Cortés J (2009) Water quality conditions on coral reefs at the Marino Ballena National Park, Pacific Costa Rica. Bull Mar Sci 84:137–152

Alvarado JJ, Ayala A, Álvarez del Castillo-Cárdenas PA, Fernández C, Aguirre-Rubí J, Buitrago F, Reyes Bonilla H (2011) Coral communities of San Juan del Sur, Pacific Nicaragua. Bull Mar Sci 87:129–146

Álvarez-Filip L, Reyes-Bonilla H (2006) Comparison of community structure and functional diversity of fishes at Cabo Pulmo coral reef, western Mexico between 1987 and 2003. Proceedings of 10th International Coral Reef Symposium, vol 1, Okinawa, pp 216–225

Aranceta-Garza F, Balart EF, Reyes-Bonilla H, Cruz-Hernández P (2012) Effect of tropical storms on sexual and asexual reproduction in coral *Pocillopora verrucosa* subpopulations in the Gulf of California. Coral Reefs 31:1157–1167

Arnold SN, Steneck RS, Mumby PJ (2010) Running the gauntlet: inhibitory effects of algal turfs on the processes of coral recruitment. Mar Ecol Prog Ser 414:91–105

Baker AC, Starger CJ, McClanahan TR, Glynn PW (2004) Coral reefs: corals' adaptive response to climate change. Nature 430:741

Baker AC, Glynn PW, Riegl B (2008) Climate change and coral reef bleaching: an ecological assessment of long-term impacts, recovery trends and future outlook. Estuar Coast Shelf Sci. 80:435–471

Bender MA, Knutson TR, Tuleya RE, Sirutis JJ, Vecchi GA, Garner ST, Held IM (2010) Modeled impact of anthropogenic warming on the frequency of intense Atlantic hurricanes. Science 327:454–458

Bignami S, Enochs IC, Manzello DP, Sponaugle S, Cowen RK (2013) Ocean acidification alters the otoliths of a pantropical fish species with implications for sensory function. Proc Nat Acad Sci USA 110:7366–7370

Branch TA, Watson R, Fulton EA, Jennings S, McGilliard CR, Pablico GT, Ricard D, Tracey SR (2010) The trophic fingerprint of marine fisheries. Nature 468:431–435

Burke L, Spalding M (2011) Reefs at risk. World Resources Institute, Washington, DC, p 124

Cabral-Tena RA, Reyes-Bonilla H, Lluch-Cota S, Paz-García DA, Calderón-Aguilera LE, Norzagaray-López O, Balart EF (2013) Different calcification rates in males and females of the coral *Porites panamensis* in the Gulf of California. Mar Ecol Prog Ser 476:1–8

Calderón-Aguilera LE, Reyes-Bonilla H, Carriquiry JD (2007) El papel de los arrecifes coralinos en el flujo de carbono en el océano: estudios en el Pacífico mexicano. In: Gaxiola-Castro G, Hernández-de la Torre B (eds), Carbono en ecosistemas acuáticos de México. SEMARNAT-INE-CICESE, México, pp 215–226

Carilli JE, Norris RD, Black BA, Walsh SM, McField M (2009) Local stressors reduce coral resilience to bleaching. PLoS ONE 4(7): e6324. doi:10.1371/journal.pone.0006324

Carilli JE, Norris RD, Black B, Walsh SM, McField M (2010) Century-scale records of coral growth rates indicate that local stressors reduce coral thermal tolerance threshold. Global Change Biol 16:1247–1257

Cazenave A, Llovel W (2010) Contemporary sea level rise. Ann Rev Mar Sci 2:145–173

CBD/STAP (Secretariat of the Convention on Biological Diversity and the Scientific and Technical Advisory Panel) (2012). Impacts of marine debris on biodiversity: current status and potential solutions. Tech Ser No. 67, p 61

Cerón RM, Cerón JG, Aguilar C, Montalvo C, Carballo C, Cárdenas B, Ortinez A, Cocom M, Carrillo J (2012) Atmospheric dry deposition in the proximity of oil-fired power plants at Mexican Pacific coast. J Env Protect 3:1228–1237

Chabanet P, Adjeroud M, Andréfouët S, Bozec YM, Ferraris J, García-Charton JA, Schrimm M (2005) Human-induced physical disturbances and their indicators on coral reef habitats: a multi-scale approach. Aquat Liv Res 18:215–230

Cheung WW, Watson R, Morato T, Pitcher TJ, Pauly D (2007) Intrinsic vulnerability in the global fish catch. Mar Ecol Prog Ser 333:1–12

Chollett I, Mumby PJ, Cortés J (2010) Upwelling areas do not guarantee refuge for coral reefs in a warming ocean. Mar Ecol Progr Ser 416:47–56

Cisneros-Montemayor AM, Christensen V, Arreguín-Sánchez F, Sumaila UR (2012) Ecosystem models for management advice: an analysis of recreational and commercial fisheries policies in Baja California Sur, Mexico. Ecol Model 228:8–16

Collins M, An S-I, Cai W, Ganachaud A, Guilyardi E, Jin F-F, Jochum M, Lengaigne M, Power S, Timmermann A, Vecchi G, Wittenberg A (2010) The impact of global warming on the tropical Pacific Ocean and El Niño. Nat Geosci 3:391–397

Cortés J (1990) The coral reefs of Golfo Dulce, Costa Rica: distribution and community structure. Atoll Res Bull 344:1–37

Cortés J (1997) Biology and geology of coral reefs of the eastern Pacific. Coral Reefs 16(Suppl):S39–S46

Cortés J (2011) Eastern tropical Pacific coral reefs. In: Hopley D (ed) The encyclopedia of modern coral reefs: structure, form and process. Springer, Berlin, pp 351–358

Cortés J, Jiménez C (2003) Corals and coral reefs of the Pacific of Costa Rica: history, research and status. In: Cortés J (ed) Latin American coral reefs. Elsevier, Amsterdam, pp 361–385

Cortés J, Murillo MM (1985) Comunidades coralinas y arrecifes del Pacífico de Costa Rica. Rev Biol Trop 33:197–202

Cortés J, Risk MJ (1985) A reef under siltation stress: Cahuita, Costa Rica. Bull Mar Sci 36:339–356

Cortés J, Macintyre IG, Glynn PW (1994) Holocene growth history of an eastern Pacific fringing reef, Punta Islotes, Costa Rica. Coral Reefs 13:65–73

Cortés J, Jiménez CE, Fonseca AC, Alvarado JJ (2010) Status and conservation of coral reefs in Costa Rica. Rev Biol Trop 58(Suppl 1):33–50

Dapp D, Arauz R, Spotila JR, O'Connor MP (2013) Impact of Costa Rican longline fishery on its bycatch of sharks, stingrays, bony fish and olive ridley turtles (*Lepidochelys olivacea*). J Exp Mar Biol Ecol 448:228–239

D'Croz L (1988) Survey of marine pollution in the Bay of Panamá. UNEP Reg Seas Stud 97:115–123

D'Croz L, Robertson DR (1997) Coastal oceanographic conditions affecting coral reefs on both sides of the Isthmus of Panama. Proceedings of 8th International Coral Reef Symposium, vol 2, Panama, pp 2053–2058

Dixson DL, Munday PL, Jones GP (2010) Ocean acidification disrupts the innate ability of fish to detect predator olfactory cues. Ecol Lett 13:68–75

Domingues CM, Church JA, White NJ, Gleckler PJ, Wijffels SE, Barker PM, Dunn JR (2008) Improved estimates of upper-ocean warming and multi-decadal sea-level rise. Nature 453:1090–1093

Dominici-Arosemena A, Brugnoli-Olivera E, Cortés-Núñez J, Molina-Ureña H, Quesada-Alpízar M (2005) Community structure of eastern Pacific reef fishes (Gulf of Papagayo, Costa Rica). Tecnociencia 7:19–41

Doney SC, Fabry VJ, Feely RA, Kleypas JA (2009) Ocean acidification: the other CO_2 problem. Annu Rev Mar Sci 1:169–192

Dupont S, Ortega-Martínez O, Thorndyke M (2010) Impact of near-future ocean acidification on echinoderms. Ecotoxicology 19:449–462

Eakin CM, Glynn PW (1996) Low tidal exposures and reef mortalities in the eastern Pacific. Coral Reefs 15:120

Edgar GJ, Snell HL, Lougheed LW (2003) Impacts of the *Jessica* oil spill: an introduction. Mar Pollut Bull 47:273–275

Edgar GJ, Banks SA, Brandt M, Bustamante RH, Chiriboga Á, Earle SA, Garske LE, Glynn PW, Grove JS, Henderson S, Hickman CP, Miller KA, Rivera F, Wellington GM (2010) El Niño, grazers and fisheries interact to greatly elevate extinction risk for Galápagos marine species. Global Change Biol 16:2876–2890

Edmunds PJ, Davies PS (1989) An energy budget for *Porites porites* (Scleractinia) growing in a stressed environment. Coral Reefs 8:37–44

Elsner JB, Kossin JP, Jagger TH (2008) The increasing intensity of the strongest tropical cyclones. Nature 455:92–95

Erftemeijer PLA, Riegl B, Hoeksema BW, Todd PA (2012) Environmental impacts of dredging and other sediment disturbances on corals: a review. Mar Poll Bull 64:1737–1765

Fabricius KE (2005) Effects of terrestrial runoff on the ecology of corals and coral reefs: review and synthesis. Mar Poll Bull 50:125–146

Fernández C, Cortés J (2005) *Caulerpa sertularioides*, a green alga spreading aggressively over coral reef communities in Culebra Bay, North Pacific of Costa Rica. Coral Reefs 24:10

Fernández-Álamo MA, Färber-Lorda J (2006) Zooplankton and the oceanography of the eastern tropical Pacific: a review. Prog Oceanogr 69:318–359

Fernández-García C, Cortés J, Alvarado JJ, Nivia-Ruiz J (2012) Physical factors contributing to the benthic dominance of the alga *Caulerpa sertularioides* (Caulerpaceae, Chlorophyta) in the upwelling Bahía Culebra, north Pacific of Costa Rica. Rev Biol Trop 60 (Suppl 2):93–107

Friedlander AM, Ballesteros E, Beets J, Berkenpas E, Gaymer CF, Gorny M, Sala E (2013) Effects of isolation and fishing on the marine ecosystems of Easter Island and Salas y Gómez, Chile. Aquat Conserv Mar Freshw Ecosyst 23:515–531

Frieler K, Meinshausen M, Golly A, Mengel M, Lebek K, Donner SD, Hoegh-Guldberg O (2012) Limiting global warming to 2 °C is unlikely to save most coral reefs. Nat Clim Chan 3:165–170

Galil BS, Boero F, Campbell ML, Carlton JT, Cook E, Fraschetti S, Gollasch S, Hewitt CL, Jelmert A, Macpherson E, Marchini A, McKenzie C, Minchin D, Occhipinti-Ambrogi A, Ojaveer H, Olenin S, Piraino S, Ruiz GM (2014) 'Double trouble': the expansion of the Suez Canal and marine bioinvasions in the Mediterranean Sea. Ethol Ecol Evol 26:152–171

Gastineau G, Soden BJ (2011) Evidence for a weakening of tropical surface wind extremes in response to atmospheric warming. Geophys Res Lett 38(9):L09706. doi:10.1029/2011GL047138

Germain N, Hartmann H, Fernández Rivera-Melo FJ, Reyes-Bonilla H (2015) Ornamental reef fish fisheries: new indicators of sustainability and human development at a coastal community level. Ocean Coast Manag 104:136–149

Glynn PW (1976) Some physical and biological determinants of coral community structure in the eastern Pacific. Ecol Monogr 46:431–456

Glynn PW (1994) State of coral reefs in the Galápagos Islands: natural vs anthropogenic impacts. Mar Poll Bull 29:131–140

Glynn PW (1996) Coral reef bleaching: facts, hypotheses and implications. Global Change Biol 2:495–509

Glynn PW (2000) Effects of the 1997–98 El Niño-Southern Oscillation on eastern Pacific corals and coral reefs: an overview. Proceedings of 9th International Coral Reef Symposium, vol 2, Bali, pp 1169–1174

Glynn PW (2003) Coral communities and coral reefs of Ecuador. In: Cortés J (ed) Latin American Coral Reefs. Elsevier, Amsterdam, pp 449–472

Glynn PW (2004) High complexity food webs in low-diversity eastern Pacific reef–coral communities. Ecosystems 7:358–367

Glynn PW, Ault JS (2000) A biogeographic analysis and review of the far eastern Pacific coral reef region. Coral Reefs 19:1–23

Glynn PW, Macintyre IG (1977) Growth rate and age of coral reefs on the Pacific coast of Panama. Proceedings of 3rd International Coral Reef Symposium, vol 2, Miami, pp 251–259

Glynn PW, Stewart RH (1973) Distribution of coral reefs in the Pearl Islands (Gulf of Panama) in relation to thermal conditions. Limnol Oceanog 18:367–379

Glynn PW, Wellington GM, Birkeland C (1979) Coral reef growth in the Galápagos: limitation by sea urchins. Science 203:47–49

Glynn PW, von Prahl H, Guhl F (1982) Coral reefs of Gorgona Island, with special reference to corallivores and their influence on community structure and reef development. Inst Invest Mar Punta Betín 12:185–214

Glynn PW, Howard LS, Corcoran E, Freay AD (1984) The occurrence and toxicity of herbicides in reef building corals. Mar Pollut Bull 15:370–374

Glynn PW, Lirman D, Baker AC, Leyte-Morales GE (1998) First documented hurricane strikes on eastern Pacific coral reefs reveal only slight damage. Coral Reefs 17:368

Glynn PW, Maté JL, Baker AC, Calderón MO (2001) Coral bleaching and mortality in Panama and Ecuador during the 1997–1998 El Niño-Southern Oscillation event: spatial/temporal patterns and comparisons with the 1982-1983 event. Bull Mar Sci 69:79–109

Glynn PW, Wellington GM, Wieters EA, Navarrete SA (2003) Reef-building coral communities of Easter Island (Rapa Nui), Chile. In: Cortés J (ed) Latin American coral reefs. Elsevier, Amsterdam, pp 473–494

Glynn PW, Enochs IC, Afflerbach JA, Brandtneris VW, Serafy JE (2014) Eastern Pacific reef fish responses to coral recovery following El Niño disturbances. Mar Ecol Prog Ser 495:233–247

González EEC, Polo EMG (1982) Efectos de la sedimentación en el crecimiento de algunas especies de corales de Panamá. Thesis, University of Panamá, Panamá, p 74

Guzmán HM, Cortés J (1989) Growth rates of eight species of scleractinian corals in the eastern Pacific (Costa Rica). Bull Mar Sci 44:1186–1194

Guzmán HM, Cortés J (2001) Changes in reef community structure after fifteen years of natural disturbances in the eastern Pacific (Costa Rica). Bull Mar Sci 69:133–149

Guzmán HM, Cortés J (2007) Reef recovery 20 years after the 1982–83 El Niño massive mortality. Mar Biol 151:401–411

Guzmán HM, Holst I (1994) Inventario biológico y estado actual de los arrecifes coralinos a ambos lados del Canal de Panamá. Rev Biol Trop 42:493–514

Guzmán HM, Jarvis KE (1996) Vanadium century record from Caribbean reef corals; a tracer of oil pollution in Panama. Ambio 25:523–526

Guzmán HM, Cortés J, Glynn PW, Richmond RH (1990) Coral mortality associated with dinoflagellate blooms in the eastern Pacific (Costa Rica and Panama). Mar Ecol Prog Ser 60:299–303

Hansen J, Ruedy R, Sato M, Lo K (2010). Global surface temperature change. Rev Geophys 48(4). doi:10.1029/2010RG000345

Heisler J, Glibert PM, Burkholder JM, Anderson DM, Cochlan W, Dennison WC, Dortch Q, Gobler CJ, Heil CA, Humphries E, Lewitus A, Magnien R, Marshall HG, Sellner K, Stockwell DA, Stoecker DK, Suddleson M (2008) Eutrophication and harmful algal blooms: a scientific consensus. Harmful Algae 8:3–13

Hernández L, Balart EF, Reyes-Bonilla H (2008) Effect of hurricane John (2006) on the invertebrates associated with corals in Bahía de la Paz, Gulf of California. Proceedings of 11th International Coral Reef Symposium, vol 2, Ft Lauderdale, pp 301–304

Heuer RM, Grosell M (2014) Physiological impacts of elevated carbon dioxide and ocean acidification on fish. Amer J Physiol Reg Integ Compar Physiol 307:R1061–R1084

Hoegh-Guldberg O, Mumby PJ, Hooten AJ, Steneck RS, Greenfield P, Gomez E, Harvell CD, Sale PF, Edwards AJ, Caldeira K, Knowlton N, Eakin CM, Iglesias-Prieto R, Muthiga N, Bradbury RH, Duby A, Hatziolos ME (2007) Coral reefs under rapid climate change and ocean acidification. Science 318:1737–1742

Hoey AS, Bellwood DR (2011) Suppression of herbivory by macroalgal density: a critical feedback on coral reefs? Ecol Lett 14:267–273

Hughes TP, Baird AH, Bellwood DR, Card M, Connolly SR, Folke C, Grosberg R, Hoegh-Guldberg O, Jackson JBC, Kleypas J, Lough JM, Marshall P, Nyström M, Palumbi SR, Pandolfi JM, Rosen B, Roughgarden J (2003) Climate change, human impacts, and the resilience of coral reefs. Science 301:929–933

Huntington TG (2006) Evidence for intensification of the global water cycle: review and synthesis. J Hydrol 319:83–95

Iglesias-Prieto R, Reyes-Bonilla H, Riosmena-Rodríguez R (2003) Effects of 1997-1998 ENSO on coral reef communities in the Gulf of California, Mexico. Geofis Int 42:467–472

Jackson JBC, Kirby MX, Berger WH, Bjorndal KA, Botsford LW, Bourque BJ, Bradbury RH, Cooke R, Erlandson J, Estes JA, Hughes TP, Kidwell S, Lange CB, Lenihan HS, Pandolfi JM, Peterson CH, Steneck RS, Tegner MJ, Warner RR (2001) Historical overfishing and the recent collapse of coastal ecosystems. Science 293:629–639

Jiménez CE (1997) Corals and coral reefs of Culebra Bay, Pacific coast of Costa Rica: anarchy in the reef. In: Proceedings of 8th International Coral Reef Symposium, vol 2, Panama, pp 329–334

Jiménez C (2001) Arrecifes y ambientes coralinos de Bahía Culebra, Pacífico de Costa Rica: aspectos biológicos, económico-recreativos y de manejo. Rev Biol Trop 49(Suppl 2):215–231

Jiménez CE, Cortés J (2003) Growth of seven species of scleractinian corals in an upwelling environment of the eastern Pacific (Golfo de Papagayo, Costa Rica). Bull Mar Sci 72:187–198

Jompa J, McCook LJ (2003) Contrasting effects of turf algae on corals: massive *Porites* spp. are unaffected by mixed-species turfs, but killed by the red alga *Anotrichium tenue*. Mar Ecol Prog Ser 258:79–86

Kleypas JA, Buddemeier RW, Archer D, Gattuso J-P, Langdon C, Opdyke BN (1999) Geochemical consequences of increased atmospheric carbon dioxide on coral reefs. Science 284:118–120

Knutson TR, McBride JL, Chan J, Emanuel K, Holland G, Landsea C, Held I, Kossin JP, Srivastava AK, Sugi M (2010) Tropical cyclones and climate change. Nat Geosci 3:157–163

LaJeunesse TC, Reyes-Bonilla H, Warner ME (2007) Spring "bleaching" among *Pocillopora* in the Sea of Cortez, eastern Pacific. Coral Reefs 26:265–270

LaJeunesse TC, Smith R, Walther M, Pinzón J, Pettay DT, McGinley M, Asschaffenburg M, Medina-Rosas P, Cupul-Magaña AL, López-Pérez RA, Reyes-Bonilla H, Warner ME (2010) Host-symbiont recombination versus natural selection in the response of coral-dinoflagellate symbioses to environmental disturbance. Proc Roy Soc London B 277:2925–2934

Laursen L (2015) Nicaragua defies canal protests. Nature 517:7–8

Lirman D, Glynn PW, Baker AC, Leyte-Morales GE (2001) Combined effects of three sequential storms on the Huatulco coral reef tract, Mexico. Bull Mar Sci 69:267–278

Logan CA, Dunne JP, Eakin CM, Donner SD (2014) Incorporating adaptive responses into future projections of coral bleaching. Global Change Biol 20:125–139

Lougheed LW, Edgar GJ, Snell HL (2002) Impacts of the *Jessica* oil spill on the Galápagos environment. Charles Darwin Foundation, Puerto Ayora, Galápagos, Final report, p 127

Macfadyen G, Huntington T, Cappell R (2009) Abandoned, lost or otherwise discarded fishing gear. UNEP/FAO, Rome, UNEP Regional Seas Reports and Studies, p 115

Macintyre IG (2007) Demise, regeneration, and survival of some Western Atlantic reefs during the Holocene Transgression. In: Aronson RB (ed) Geological approaches to coral reef ecology. Springer, New York, pp 181–200

MacNeil MA, Graham NAJ, Cinner JE, Wilson SK, Williams ID, Maina J, Newman S, Friedlander AM, Jupiter S, Polunin NVC, McClanahan TR (2015) Recovery potential of the world's coral reef fishes. Nature 520:341–344

Manzello DP (2010a) Ocean acidification hot spots: spatiotemporal dynamics of the seawater CO_2 system of eastern Pacific coral reefs. Limnol Oceanog 55:239–248

Manzello DP (2010b) Coral growth with thermal stress and ocean acidification: lessons from the eastern tropical Pacific. Coral Reefs 29:749–758

Manzello DP, Kleypas JA, Budd DA, Eakin CM, Glynn PW, Langdon C (2008) Poorly cemented coral reefs of the eastern tropical Pacific: possible insights into reef development in a high-CO_2 world. Proc Nat Acad Sci USA 105:10,450–10,455

Manzello DP, Enochs IC, Bruckner A, Renaud PG, Kolodziej G, Budd DA, Carlton R, Glynn PW (2014) Galápagos coral reef persistence after ENSO warming across an acidification gradient. Geophys Res Lett 41. doi:10.1002/2014GL062501

Maté JL (2003) Corals and coral reefs of the Pacific coast of Panama. In: Cortés J (ed) Latin American coral reefs. Elsevier, Amsterdam, pp 387–417

McCook LJ, Jompa J, Diaz-Pulido G (2001) Competition between corals and algae on coral reefs: a review of evidence and mechanisms. Coral Reefs 19:400–417

McCullough D (1977) The path between the seas: the creation of the Panama Canal, 1870–1914. Simon & Schuster, New York, p 698

Meyer A, Huete-Peréz JA (2014) Nicaragua Canal could wreak environmental ruin. Nature 506:287–289

Michel J, Henry CB Jr (1997) Oil uptake and depuration in oysters after use of dispersants in shallow water in El Salvador. Spill Sci Tech Bull 4:57–70

Miglietta MP, Rossi M, Collin R (2008) Hydromedusa blooms and upwelling events in the Bay of Panama, tropical east Pacific. J Plankt Res 30:783–793

Molina OA (1996) Comparación de la cobertura de los arrecifes coralinos antes y después del derrame de petróleo. Los Cóbanos, Sonsonate 1993–1995. Internal report, Escuela de Biología, Univ El Salvador, San Salvador, p 36

Munday PL, Pratchett MS, Dixson DL, Donelson JM, Endo GGK, Reynolds AD, Knuckey R (2013) Elevated CO_2 affects the behavior of an ecologically and economically important coral reef fish. Mar Biol 160:2137–2144

Nakamura M, Ohki S, Suzuki A, Sakai K (2011) Coral larvae under ocean acidification: survival, metabolism, and metamorphosis. PLoS ONE 6(1):e14521

Nance HA, Klimley P, Galván-Magaña F, Martínez-Ortíz J, Marko PB (2011) Demographic processes underlying subtle patterns of population structure in the Scalloped Hammerhead Shark. *Sphyrna lewini*. PLoS ONE 6(7):e21459. doi:10.1371/journal.pone.0021459

Neumann AC, Macintyre IG (1985) Reef response to sea level rise: keep-up, catch-up, or give-up. Proceedings of 5th International Coral Reef Congress, vol 1, Tahiti, pp 105–110

Nicholls RJ, Cazenave A (2010) Sea-level rise and its impact on coastal zones. Science 328:1517–1520

Ochoa-López E, Reyes-Bonilla H, Ketchum-Mejía J (1998) Effects of sedimentation on coral communities of southern Socorro Island, Revillagigedo Archipelago, México. Cienc Mar 24:233–240

Pandolfi JM, Bradbury RH, Sala E, Hughes TP, Bjorndal KA, Cooke RG, McArdle D, McClenachan L, Newman MJH, Paredes G, Warner RR, Jackson JBC (2003) Global trajectories of the long-term decline of coral reef ecosystems. Science 301:955–958

Pauly D, Zeller D (2014) Accurate catches and the sustainability of coral reef fisheries. Curr Opin Environ Sust 7:44–51

Pauly D, Christensen V, Dalsgaard J, Froese R, Torres F (1998) Fishing down marine food webs. Science 279:860–863

Paz-García DA, Balart EF, García-de-Léon FJ (2012) Cold water bleaching of *Pocillopora* in the Gulf of California. Proceedings of 12th International Coral Reef Symposium, vol 9A–10, Cairns, pp 1–6

Paz-García DA, Aldana-Moreno A, Cabral-Tena RA, García-de-León FJ, Hellberg ME, Balart EF (2015a) Morphological variation and different branch modularity across contrasting flow conditions in dominant *Pocillopora* reef-building corals. Oecologia. doi:10.1007/s00442-014-3199-9

Paz-García DA, Hellberg ME, García-de-León FJ, Balart EF (2015b) Switch between morphospecies of *Pocillopora* corals. Am Nat 186 doi:10.1086/682363

Pelejero C, Calvo E, Hoegh-Guldberg O (2010) Paleo-perspectives on ocean acidification. Trends Ecol Evol 25:332–344

Pennington JT, Mahoney KL, Kuwahara VS, Kolber DD, Calienes R, Chavez FP (2006) Primary production in the eastern tropical Pacific: a review. Prog Oceanog 69:285–317

Pérez-España H, Abarca-Arenas LG, Jiménez-Badillo ML (2006) Is fishing-down trophic web a generalized phenomenon? The case of Mexican fisheries. Fish Res 79:349–352

Pinzón JH, LaJeunesse TC (2011) Species delimitation of common reef corals in the genus *Pocillopora* using nucleotide sequence phylogenies, population genetics and symbiosis ecology. Mol Ecol 20:311–325

Pinzón JH, Sampayo EM, Cox EF, Chauka LJ, Chen CA, Voolstra CR, LaJeunesse TC (2013) Blind to morphology: genetics identifies several widespread ecologically common species and few endemics among Indo-Pacific cauliflower corals (*Pocillopora*, Scleractinia). J Biogeogr 40:1595–1608

Purcell SW, Mercier A, Conand C, Hamel JF, Toral-Granda MV, Lovatelli A, Uthicke S (2011) Sea cucumber fisheries: global analysis of stocks, management measures and drivers of overfishing. Fish Fisher 14:34–59

Rasher DB, Hay ME (2010) Chemically rich seaweeds poison corals when not controlled by herbivores. Proc Natl Acad Sci USA 107:9683–9688

Reyes-Bonilla H (2003) Coral reefs of the Pacific coast of México. In: Cortés J (ed) Latin American coral reefs. Elsevier, Amsterdam, pp 331–349

Reyes-Bonilla H, Barraza JE (2003) Corals and associated marine communities from El Salvador. In: Cortés J (ed) Latin American coral reefs. Elsevier, Amsterdam, pp 351–360

Reyes-Bonilla H, López-Pérez RA (2009) Corals and coral reef communities in the Gulf of California. In: Johnson M, Ledezma J (eds) Atlas of marine ecosystems in the western Gulf of California. University Arizona Press, Tucson, Arizona, pp 43–55

Reyes-Bonilla H, Herrero-Pérezrul MD, Fernández-Rivera-Melo FJ (2008) Aspectos económicos de recursos pesqueros no tradicionales residentes en zonas arrecifales del Golfo de California. In: Urciaga-García J, Lluch-Cota D (eds) Economía ambiental en el noroeste de México. UABCS-CIBNOR, La Paz, México, pp 245–264

Reyes-Bonilla H, Walther-Mendoza M, Ramírez-Ortiz G (2012a) Biodiversidad y turismo ecológico. In: Ibáñez-Pérez RM, Ivanova-Boncheva A (eds) Medio ambiente y política turística en México. Instituto Nacional de Ecología, México, pp 135–148

Reyes-Bonilla H, Cupul-Magaña AL, Loreto-Viruel RM, Alvarez del Castillo-Cárdenas PA, Vázquez-Vera L, Torrejón-Arellano N, Narro-Robles C, González-Espinosa PC (2012b) Evaluación de la capacidad de carga para buceo en áreas naturales protegidas del Mar Caribe y el Golfo de California, México. Biodiversidad y turismo ecológico. In: Ibáñez-Pérez RM, Ivanova-Boncheva A (eds), Medio ambiente y política turística en México. Instituto Nacional de Ecología, México, pp 149–162

Reyes-Bonilla H, Calderón-Aguilera LE, Mozqueda-Torres MC, Carriquiry JD (2014) Presupuesto de carbono en arrecifes coralinos de México. Interciencia 39:645–650

Riegl B (2003) Climate change and coral reefs: different effects in two high-latitude areas (Arabian Gulf, South Africa). Coral Reefs 22:433–446

Riegl B, Piller WE (2003) Possible refugia for reefs in times of environmental stress. Int J Earth Sci 92:520–531

Ríos-Jara E, Galván-Villa CM, Rodríguez-Zaragoza FA, López-Uriarte E, Muñoz-Fernández VT (2013) The tourism carrying capacity of underwater trails in Isabel Island National Park, Mexico. Environ Manag 52:335–347

Rixen T, Jiménez CE, Cortés J, (2012) Impact of upwelling events on the sea water carbonate chemistry and dissolved oxygen concentration in the Gulf of Papagayo (Culebra Bay), Costa Rica: implications for coral reefs. Rev Biol Trop 60 (suppl 2) S0034-77442012000600013

Roff G, Mumby PJ (2012) Global disparity in the resilience of coral reefs. Trends Ecol Evol 27:404–413

Rogers C (1990) Responses of coral reefs and reef organisms to sedimentation. Mar Ecol Prog Ser 62:185–202

Sala E, Aburto-Oropeza O, Reza M, Paredes G, López-Lemus LG (2004) Fishing down coastal food webs in the Gulf of California. Fisheries 29:19–25

Sánchez-Noguera C (2012) Entre historias y culebras: más que una bahía (Bahía Culebra, Guanacaste, Costa Rica). Rev Biol Trop 60 (Suppl 2):1–17

Shearer TL, Rasher DB, Snell TW, Hay ME (2012) Gene expression patterns of the coral *Acropora millepora* in response to contact with macroalgae. Coral Reefs 31:1177–1192

Smith TB, Glynn PW, Maté JL, Toth LT, Gyory J (2014) A depth refugium from catastrophic coral bleaching prevents regional extinction. Ecology 95:1663–1673

Somero GN (2010) The physiology of climate change: how potentials for acclimatization and genetic adaptation will determine "winners" and "losers". J Exp Biol 213:912–920

Sorice MG, Oh C-O, Ditton RB (2007) Managing scuba divers to meet ecological goals for coral reef conservation. Ambio 36:316–322

Spongberg AL (2004) PCB contamination in surface sediments in the coastal waters of Costa Rica. Rev Biol Trop 52(Suppl 2):1–10

Spongberg AL (2006) PCB concentrations in intertidal sipunculan (Phylum Sipuncula) marine worms from the Pacific coast of Costa Rica. Rev Biol Trop 54(Suppl 1):27–33

Spongberg A, Davis P (1998) Organochlorinated pesticide contaminants in Golfo Dulce. Costa Rica. Rev Biol Trop 46(Suppl 6):111–124

Spongberg AL, Witter JD (2008) A review of PCB concentrations in tropical media, 1996–2007. Rev Biol Trop 56(Suppl 4):1–9

Spongberg AL, Witter JD, Acuña J, Vargas J, Murillo M, Umaña G, Gómez E, Perez G (2011) Reconnaissance of selected PPCP compounds in Costa Rican surface waters. Water Res 45:6709–6717

Toth LT, Aronson RB, Vollmer SV, Hobbs JW, Urrego DH, Cheng H, Enochs IC, Combosch DJ, van Woesik R, Macintyre IG (2012) ENSO drove 2500-year collapse of eastern Pacific coral reefs. Science 337:81–84

Vargas-Ángel B (1996) Distribution and community structure of the reef corals of Ensenada de Utría, Pacific coast of Colombia. Rev Biol Trop 44:643–651

Vargas-Ángel B (2001) Corals and coral reefs of the Pacific coast of Colombia with special reference to spatial and temporal patterns of environmental disturbances. PhD dissertation, University Miami, Coral Gables, Florida, p 209

Vargas-Montero M, Freer-Bustamante E, Guzmán JC, Vargas JC (2008) Florecimientos de dinoflagelados en la costa Pacífica de Costa Rica. Hidrobiológica 18(Suppl):15–23

Víquez R, Hargraves PE (1995) Annual cycle of potentially harmful dinoflagellates in the Golfo de Nicoya, Costa Rica. Bull Mar Sci 57:467–475

von Prahl H, Guhl F, Grogl M (1979) Gorgona. Futura Grupo Editorial Ltd., Bogotá, Colombia, p 279

von Prahl H, Vargas-Ángel B (1990) Tasa de crecimiento del coral *Pocillopora damicornis* en un arrecife costero del Pacífico colombiano. Rev Cienc Univ Valle 2:37–43

Wellington GM, Glynn PW (2007) Responses of coral reefs to El Niño-Southern Oscillation sea-warming events. In: Aronson RB (ed) Geological approaches to coral reef ecology, vol 192 Ecol Stud, Springer, New York, pp 342–385

Whelan P (1989) Oil spill in Galápagos. Noticias de Galápagos 47:2

Wikelski M, Wong V, Chevalier B, Rattenborg N, Snell HL (2002) Galápagos Islands: marine iguanas die from trace oil pollution. Nature 417:607–608

Wild C, Hoegh-Guldberg O, Naumann MS, Colombo-Pallotta MF, Ateweberhan M, Fitt WK, Iglesias-Prieto R, Palmer C, Bythell JC, Ortiz JC, Loya J, van Woesik R (2011) Climate change impedes scleractinian corals as primary reef ecosystem engineers. Mar Freshw Res 62:205–215

Wilson TD (2008) Economic and social impacts of tourism in Mexico. Lat Amer Persp 2008:37–52

Zapata F, Vargas-Ángel B (2003) Corals and coral reefs of the Pacific coast of Colombia. In: Cortés J (ed) Latin American coral reefs. Elsevier, Amsterdam, pp 419–447

Zapata FA, Jaramillo-González J, Navas-Camacho R (2011) Extensive bleaching of the coral *Porites lobata* at Malpelo, Colombia, during a cold water episode in 2009. Bol Inv Mar Cost 40:185–193

Coral Reef Conservation in the Eastern Tropical Pacific

21

Juan José Alvarado, Octavio Aburto-Oropeza, Rubén Abad, Enrique Barraza, Margarita Brandt, Jaime Cantera, Priscila Estrada, Carlos F. Gaymer, Ana Gloria Guzmán-Mora, James J. Herlan, and Juan L. Maté

Abstract

This chapter reviews and evaluates coral reef conservation strategies along the eastern tropical Pacific (ETP), a narrow biogeographic region on the Pacific American coast that extends from southern Baja California, Mexico to northern Peru, including several oceanic islands. The ETP is a natural laboratory, a model for understanding the development of coral biotopes in a changing environment. We evaluate conservation strategies in seven countries in the ETP region (Peru was not included for apparent lack of coral habitats). A survey of current Marine Protected Areas (MPAs) highlights great variation in the number, scale and management approaches. Generally, MPAs with no-take areas are relatively uncommon, with multi-purpose areas favored. The Cabo Pulmo MPA in Baja California, Mexico demonstrates that when a local community is involved in the creation and enforcement measures of an MPA, conservation success can be achieved. Despite such apparent

J.J. Alvarado (✉)
Centro de Investigación en Ciencias del Mar y Limnología, Escuela de Biología, Universidad de Costa Rica, 2060-1000 San José, Costa Rica
e-mail: juan.alvarado@ucr.ac.cr

O. Aburto-Oropeza · P. Estrada
Marine Biology Research Division, Scripps Institution of Oceanography, 9500Gilman Drive, La Jolla, CA 92093, USA
e-mail: maburto@ucsd.edu

P. Estrada
e-mail: priscilaep@yahoo.com.mx

R. Abad
Museo de Zoología QCAZ, Sección Invertebrados, Pontificia Universidad Católica del Ecuador, Quito, Ecuador
e-mail: rdabad90@hotmail.com

E. Barraza
Ministerio de Medio Ambiente y Recursos Naturales, Col. y Calle Las Mercedes, San Salvador, El Salvador
e-mail: ebarraza@marn.gob.sv

M. Brandt
Colegio de Ciencias Biológicas y Ambientales, Universidad San Francisco de Quito, Diego de Robles s/n y Pampite, Cumbayá, Quito, Ecuador
e-mail: mbrandt@usfq.edu.ec

J. Cantera
Departamento de Biología, Universidad del Valle, Calle 100, No 13-00Cali, Colombia
e-mail: jaime.cantera@correounivalle.edu.co

Carlos F.Gaymer
Departamento de Biología Marina, Millennium Nucleus for Ecology and Sustainable Management of Oceanic Islands (ESMOI) and Centro de Estudios Avanzados en Zonas Áridas (CEAZA), Universidad Católica del Norte, Larrondo 1281, Coquimbo, Chile
e-mail: cgaymer@ucn.cl

Ana GloriaGuzmán-Mora
Marine Program, Conservation International-Costa Rica, Los Yoses, 2365 2050 San Pedro, Costa Rica
e-mail: aguzman@conservation.org

J.J. Herlan
Millennium Nucleus for Ecology and Sustainable Management of Oceanic Islands (ESMOI), Departamento de Biología Marina, Universidad Católica del Norte, Larrondo 1281, Coquimbo, Chile
e-mail: jherlan@ucn.cl

Juan L.Maté
Smithsonian Tropical Research Institute, Balboa, Apartado Postal 0843-03092 Ancón, Republic of Panamá
e-mail: matej@si.edu

P.W. Glynn et al. (eds.), *Coral Reefs of the Eastern Tropical Pacific*,
Coral Reefs of the World 8, DOI 10.1007/978-94-017-7499-4_21

successes, inadequate and confusing legal practices have generally forestalled effective protection of coral ecosystems in the ETP. However, in several instances nongovernmental organizations (NGOs) have assisted with planning, negotiation and stakeholder engagement. Nonetheless these findings underscore how the establishment of an MPA does not guarantee that conservation goals will be achieved. This calls for a new approach that incorporates contributions from ecological studies along with a high investment in capacity development and training to ensure that the goals of MPAs better complement effective fisheries and ecosystem management within and outside their borders.

Keywords
No-take areas • Paper parks • Fisheries management • Seascape • Latin American reefs

21.1 Introduction

Coral reefs are considered to be one of the most diverse ecosystems globally, even when compared with tropical rain forests (Reaka-Kudla 1997; Stella et al. 2011). Coral reefs are valuable natural resources that support a vast biodiversity, and are of immense value to society—biologically, culturally and economically (Spurgeon 1992; Pendleton 1995; Moberg and Folke 1999; Brander et al. 2007). In addition to wetlands, mangroves and tropical forests, coral reefs have been identified as one of Planet Earth's four chief biomes with the highest values or services. While the monetary value of coral reef environments is calculated to be 352,249 US$ $ha^{-1}year^{-1}$ (de Groot et al. 2012), there is still widespread neglect of this ecosystem with an increase in disturbances globally (Wilkinson 2006; Hoegh-Guldberg 2014) that affect their diversity and services (Andersson 2007; Carpenter et al. 2008).

In order to preserve coral reefs, four main strategies have been adopted: (1) *ex situ* protection (i.e. aquariums), (2) the establishment of Marine Protected Areas (MPAs), (3) the implementation of management tools, and (4) enforcement activities. MPAs are patches in the marine environment, allocated for the protection of biodiversity, the management of fisheries, and for research, all activities that can generate recreational, economic, aesthetic and educational benefits (Hoyt 2005; Edgar et al. 2007). The establishment of MPAs is to assure the permanence of the spectrum of marine biodiversity, from the genetic variability of individuals and populations to ecosystems, thereby providing goods and services for future generations (Lubchenco et al. 2003). Globally, only 18.7 % of coral reefs are under MPA status. Of this number, most allow multi-purpose activities (12 %), 5.3 % extractive uses, and only 1.4 % includes 'no-take' areas (Mora et al. 2006). In Latin America and the Caribbean, there are approximately 760 MPAs (of which 13 % are no-take, 76 % take, and 12 % multi-purpose), covering 309,562 km^2 (5 % no-take, 17 % take, 78 % multi-purpose, in terms of total area) (Guarderas et al. 2008). In the ETP, Mora et al. (2006) reported the presence of 24 MPAs with coral habitats, covering 6899 km^2.

Legal tools for protecting coral habitats include national and international initiatives that promote the conservation and effective management of coastal ecosystems and endangered habitats. These include the United Nations Convention on the Law of the Sea (UNCLOS), Agenda 21 (Chaps. 15 and 17) together with the International Coral Reef Initiative (ICRI), the RAMSAR Convention on Wetlands, the Convention on Biological Diversity (CBD), the Convention on International Trade in Endangered Species (CITES), and the UNESCO Convention on the Protection of the World Cultural and Natural Heritage (Davidson 2002, 2006; Zedler and Kercher 2005). The ICRI seeks to implement the best management strategies to conserve the world's coral reef resources (Davidson 2006), however, across the ETP only five countries are members (Mexico, Costa Rica, Colombia, Panama and France). According to CITES (Appendix II), globally there are 231 species of corals listed in threatened categories, and this is high relative to the number of listed reef invertebrates, fishes or macroalgae (Davidson 2006). As stated by the IUCN Red List, 32.8 % of coral species were upgraded to a higher risk extinction category and between 30 and 40 % of coral reef species within the ETP are threatened or near threatened (Carpenter et al. 2008).

Worldwide, there are currently 46 UNESCO marine World Heritage Sites (WHSs). Of these, 11 contain coral reefs, and six are located within the ETP (Costa Rica: Guanacaste Conservation Area and Isla del Coco National Park; Panama: Coiba National Park and its special zone of marine protection; Ecuador: Galápagos Islands; Colombia: Malpelo Fauna and Flora Sanctuary; and Mexico: Islands and Protected areas of the Gulf of California) (Abdulla et al. 2013). The latter four WHSs represent 1.7 % of the total area of the ETP (239,031 km^2) according to the biogeographic provinces of Spalding et al. (2007).

Even if the ETP coral reef region represents 'the minimum expression of coral reef development' (Cortés 1997), it supports a unique flora and fauna that has persisted for thousands of years under marginal and stressful environmental conditions (Cortés 1997, 2003; see Chap. 7, Cortés et al.). The ETP is a unique biogeographic region—a natural laboratory—and can be used as a model to understand the fate of coral biotopes in a changing environment (Manzello 2010a, b). In addition, studies undertaken in the ETP region can provide clues for addressing the future conservation of coral reefs worldwide. At present, approximately 20 % of the region's coral reefs are included within some sort of conservation initiative (Mora et al. 2006), and 10 % of the coral species are under elevated risk of extinction according to IUCN criteria (Carpenter et al. 2008). This is driven by several factors, including the issue that many MPAs in the region allow extractive uses, and only a few are 'no-take' areas. The goal of this chapter is to review the coral reef conservation strategies in the ETP, with a focus on the implementation of MPAs and other legal instruments to protect coral reefs. In this sense, a chief objective is to assess if these tools are effective in the region, and whether there is a need to consider different types of conservation initiatives.

21.2 Country Initiatives

21.2.1 Mexico

Protected areas in Mexico are managed at different administrative levels, including federal, state or municipal entities. At present, only federally protected areas receive financial support. The National Commission for Protected Areas (Comisión Nacional de Áreas Naturales Protegidas-CONANP) is the main organization tasked with oversight, specifically by facilitating the protection of biodiversity and managing natural resources. CONANP coordinates with three other Mexican government agencies (Comisión Nacional de Acuacultura y Pesca-CONAPESCA, Procuraduría Federal de Protección al Ambiente-PROFEPA and Comisión Nacional para el Conocimiento y Uso de la Biodiversidad-CONABIO) to ensure that marine areas are appropriately managed for the benefit of both ecosystems and society.

Mexico's current administrative and organizational arrangement for managing protected areas has caused confusion and delays in effective implementation. The complexity of the governmental structure is compounded by Mexican law, which requires a four and a half year period between the declaration of an MPA and its implementation. In addition, only 68 % of Mexico's MPAs have some type of management plan, and some of these lack clarity on how different species would be protected. In addition, there are no clear strategies on how to manage potential conflicts between stakeholders (Rivera 2011). It is evident there is no appropriate design and planning of the Mexican MPAs; for example, important biological criteria have not been considered (López-Pérez and López-García 2008).

Limited financial support has hindered the goals of MPAs by curtailing the human resources needed to implement, update and improve management plans, and for carrying out monitoring and enforcement activities. For instance, CONAPESCA has fewer than ten staff members in the Gulf of California, an area with the highest number of MPAs in the Mexican Pacific, and of considerable global marine significance with three of the most profitable fisheries states in Mexico (Sonora, Sinaloa, and Baja California Sur). In a given year, there are nearly 18,000 artisanal boats (*pangas*) that work in the Gulf, totaling nearly 36,000 labor days for people involved in fisheries in this region.

The lack of enforcement within MPAs makes for a porous system: rule-breakers regularly engage in illegal fishing, with estimates of up to 50 % of total fishing activity being illegal or unreported, and pervasive fishing in no-take zones in the Upper Gulf of California and the Colorado River Delta. Furthermore, the current governance structure for MPAs has made enforcement almost impossible. This diminishes the ability of MPAs to be successful, and undermines confidence in the capacity of government institutions. Without better communication to distribute management resources among agencies, the ongoing establishment of MPAs by CONANP will continue to burden national agencies such as CONAPESCA and PROFEPA.

In recent years, several new MPAs have been established and conservation sites of significant priority have been identified for future expansion. However, without an assessment of MPA effectiveness, the value of such initiatives for ecosystem resilience or resource protection is debatable. The investments of governments in MPAs include significant human and economic resources for planning, but often do not include implementation and evaluation. Perhaps for such reasons, several scholars have described some MPAs as 'paper parks': administratively real, but practically useless (Rife et al. 2012).

Another issue is that most MPAs do not adequately include habitats within their territories to be effective buffers against overfishing and other anthropogenic stressors. The relatively limited coverage of marine ecosystems—especially deep-sea zones—within Mexico's MPAs suggests the need to incorporate pelagic and benthic environments in management plans to effectively conserve biodiversity. Also, for corals some of these areas do not represent the whole range of distribution for some species. For example, the Huatulco National Park includes approximately only 60 % of the species of corals in the State of Oaxaca (n = 17 species total), and excludes some populations that are important in maintaining regional connectivity (CONABIO et al. 2007). Other key stressors for MPAs in general include intensive trawl fishing and pollution from urban areas and

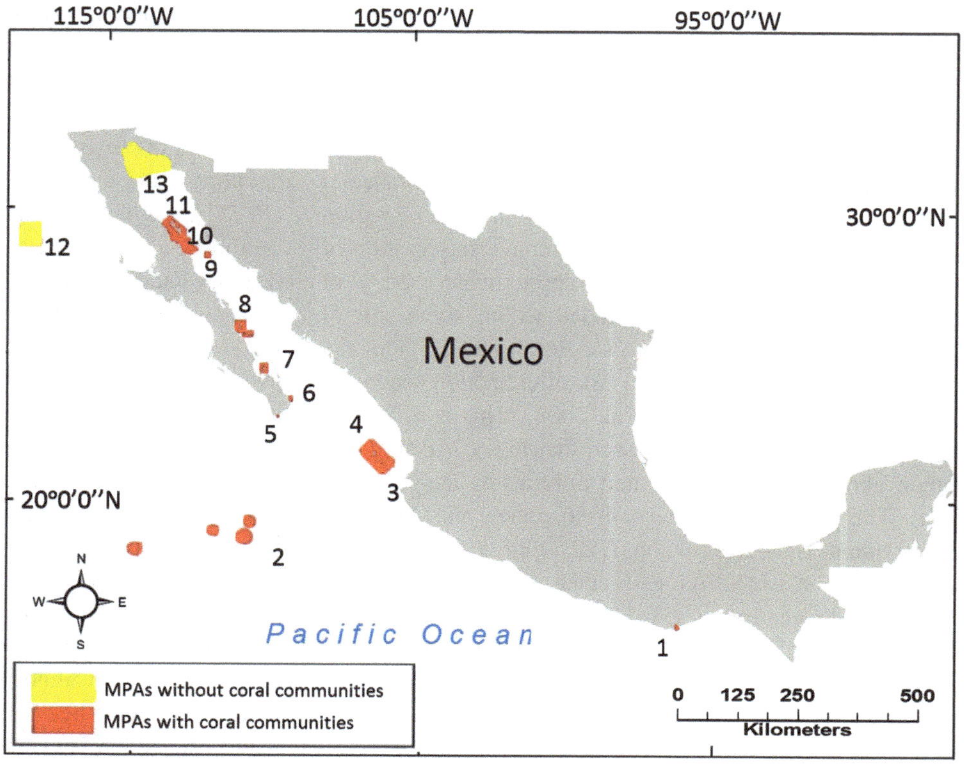

Fig. 21.1 Marine protected areas of western Mexico. *1* Huatulco National Park; *2* Revillagigedo Archipelago Biosphere Reserve; *3* Islas Marietas National Park; *4* Islas Marías Biosphere Reserve; *5* Cabo San Lucas Protected Area for Flora and Fauna; *6* Cabo Pulmo National Park; *7* Espíritu Santo Archipelago National Park; *8* Loreto Bay National Park; *9* Isla San Pedro Martir Biosphere Reserve; *10* San Lorenzo Archipiélago National Park; *11* Bahía de Los Ángeles Biosphere Reserve; *12* Isla Guadalupe Biosphere Reserve; *13* Upper Gulf and Colorado Delta Biosphere Reserve

agriculture. Yet, it will be important to balance the needs of industry and support for artisanal fishing with conservation goals. One approach would be to designate a specialized division for overseeing fishing activities inside MPAs. In the Gulf of California, there are eleven multi-purpose MPAs that are situated close to fishing grounds and human settlements (Fig. 21.1), with only a few currently designated as 'no take areas'. With the exception of Cabo Pulmo National Park, Mexico's MPAs have not met conservation or sustainability goals (Rife et al. 2012).

Cabo Pulmo National Park is often considered one of the most successful MPAs, not only in Mexico but worldwide, based on the species richness, diversity, resilience and productivity of its ecosystems (Calderón-Aguilera et al. 2007; Aburto-Oropeza et al. 2011; Rife et al. 2012; Reyes-Bonilla et al. 2014). The park is also considered a model for ecosystem-based management in rural settings, where people often rely on local natural resources for their basic livelihood. It is a place where boat captains, dive masters and local people come together to enforce the regulations of the park for both visitors and themselves. Additionally, the community participates in activities that include surveillance, fauna protection (e.g. sea turtle nesting sites), and beach programs. In 2009, a comparative study assessed changes in fish diversity and biomass at the park (between 1999 and 2009), relative to other MPAs and unprotected areas in the Gulf of California. This study concluded that Cabo Pulmo National Park has been an effective marine reserve with the recovery of reef fish biomass within its boundaries. Notably, the park exhibited the largest absolute recovery of biomass in a marine reserve (>460 %), and the most rapid relative increase in biomass of top predators, with a 30 % annual increase of this trophic category (Aburto-Oropeza et al. 2011).

The ecological reasons for such a large increase in fish biomass include several factors that can likely serve to protect other reefs within MPAs: (1) the reserve is large and thus can support permanent populations of large reef fishes with large home ranges; (2) the coral habitat is largely intact and healthy; (3) the reserve includes spawning areas for large predators; and (4) it is located in an area of upwelling and high productivity driven by the hydrodynamic forces generated by long basalt dykes, and its location in a transition zone near the entrance of the Gulf of California and the open waters of the Pacific Ocean.

The Cabo Pulmo MPA illustrates that recovery of a degraded fish community is possible when located in the right area and managed correctly, even to the level that is comparable to remote habitats that were never impacted by fishing and other local human activities. In recent years, ecotourism has become an important alternative livelihood for the residents of Cabo Pulmo (Aburto-Oropeza et al. 2011).

In 2008, a study was conducted to identify priority sites for the conservation of reef-building corals in Oaxaca State, Mexico (López-Pérez and López-García 2008). This study reported the presence of 17 species of reef-building corals across 28 sites. Because of their size and extent of development, the Oaxacan reefs are considered to be among the most important Mexican Pacific coral reefs. Additionally, the study found that Huatulco National Park (located in Oaxaca) does not include all the coral reef species existing in Oaxaca and excludes key populations that are necessary to maintain connectivity among coral reef systems of Central America and Mexico. Identifying priority conservation sites and establishing a network of MPAs are crucial for the protection of Mexican coral reef ecosystems (López-Pérez and López-García 2008).

In Mexico, coral reefs have limited legal protection; ocean law and policy are fundamental to ocean conservation. The *General Law on Ecological Equilibrium and Environmental Protection* (LGEEPA) defines the establishment, administration and evaluation of Mexican MPAs. Threatened coral species that are listed under the Official Mexican Norm NOM-059-SEMARNAT-2010 have some degree of protection. Coral species are protected under the law NOM-022-SEMARNAT-2003, which specifies the preservation, conservation and restoration of coastal wetlands.

In order to protect Mexico biodiversity and foster compatible economic development interests in coastal areas, MPAs with no-take zones are needed, especially where Pacific reefs and mangroves are located, increasing the core zone of several existing MPAs. Proper enforcement of existing MPAs is also crucial (Ezequiel et al. 2011; Rife et al. 2013). Additionally, MPAs should involve local communities in management and decision-making processes alongside government authorities (Nava and Ramírez-Herrera 2011). Numerous studies advocate for a move from 'top-down' (state or agency control) to more 'bottom-up' (control by local communities) or locally-managed approaches for coastal protection and management, mainly in situations where little or no data are available (White et al. 1994, 2000; Johannes 1998a, b; Pomeroy and Rivera-Guieb 2006; Ramírez-Herrera et al. 2010).

21.2.2 El Salvador

In the early 1970s the government of El Salvador established an authority to manage natural resources, encompassing wildlife and forests on a national scale. Subsequently, the National Parliament declared some additional natural protected areas, including a mangrove ecosystem. After the Earth Summit in Rio de Janeiro (Brazil) in 1992, authorities created an institutional framework for environmental topics as well as wildlife legislation. It was not until 2005 that a new law allowed the Ministry of the Environment and Natural Resources to declare Natural Protected Areas (NPAs). In 2007–2008, environmental authorities, following a series of technical meetings with stakeholders, established the first NPA at Complejo Los Cóbanos (Fig. 21.2, site LC). The non-governmental organization (NGO) FUNDARRECIFE promoted scientific surveys, local environmental education and awareness, and was a critical partner in the implementation of Complejo Los Cóbanos.

Los Cóbanos NPA covers 20,600 ha, and includes oceanic, benthic (coral communities), mangrove ecosystems, and a tropical dry forest. Early studies at Los Cóbanos confirmed

Fig. 21.2 Marine protected areas on the Pacific coast of El Salvador. "Complejo Los Cóbanos" (LC). Estuarine Ramsar sites: "Complejo Bahía de Jiquilisco" (BJ), "Complejo Jaltepeque" (EJ). Western rocky shores, La Libertad department (LL); La Unión (LU), Gulf of Fonseca (GF). Scleractinian zooxanthellate coral communities (*filled circle*), other non-reef building cnidarian species (*filled triangles*)

the presence of large colonies of *Porites lobata* (>2 m diameter), a highly sought after conch (*Lobatus galeatus*), macroalgal cover, important fisheries species, and low pollutant levels (Barraza 2011). This is the only locality in El Salvador where reef-building coral species have been reported: *P. lobata*, *Psammocora obtusangula* and *Psammocora stellata* occur at shallow depths (0–2.5 m), covering at least 1.0 km^2 within the area. Surveys have demonstrated that *P. lobata* is the most abundant species. The recovery of dead pocilloporid fragments in the area suggests the presence of coral patches of at least four species of *Pocillopora* (Reyes-Bonilla and Barraza 2003). Other hard and soft corals occur in this NPA, as well as many coral-associated invertebrates and fishes. The NPA also provides an important foraging area for hawksbill turtles (*Eremochelys imbricata*), nesting grounds for other species of sea turtles, and more recently humpback whales have been sighted offshore during the dry season (November–March). Current research at Complejo Los Cóbanos includes biodiversity surveys of reef fishes (Barraza 2013) as well as coral health (unpub. data).

Local environmental consciousness has evolved within the community as well as nationally with the support of environmental NGOs. One effect has been to support the protection of Los Cóbanos, where authorities exercise national regulations to protect natural resources. A few of El Salvador's key legal instruments include:

- The Wildlife Conservation Law (from 1993), which requires authorization to collect or hunt protected species, determines species' conservation status and defines the penalties of law-breaking activities. All hard corals are protected by law, and cannot be killed or harmed, and their collection for scientific purposes requires authorization.
- The 1998 Environment Law declares mangroves and coral habitats as sensitive ecosystems, therefore human disturbances are not allowed. Also environmental permits for projects close to coral habitats are required. This has led to the improvement of the management of solid and liquid wastes as well as the mitigation of other impacts in the surrounding areas.
- The 2005 Natural Protected Areas Law has legalized human settlements within all NPAs, and prohibits new infrastructures. This law also requires permission for economic activities as well as hunting and collecting.

Currently, El Salvador's Ministry of the Environment is developing a management plan for Los Cóbanos. Implementing this plan will require approval and agreement with local communities, mainly fishers who harvest invertebrates and fishes. After the legal declaration of this NPA, some arrests occurred in a combined action between environmental representatives of the Ministry, NGOs, and police. The most important cases were related to the poaching of holothurians (sea cucumbers, currently threatened by extinction within national legislation), shipwrecks, and the collection of the intertidal mollusc *Chiton stokesi*. In all cases, offenders were jailed and released after judges reviewed their cases (within 72 h). These actions and local environmental awareness reduce the poaching of marine wildlife within NPAs.

Other threats that may affect Los Cóbanos are: the unplanned use of land for tourism, the use of illegal nets and other fishing devices, pollution from human settlements, inadequate agricultural practices, and increased sedimentation resulting from livestock activities. Some positive achievements after the declaration of this NPA are: increased environmental awareness, a decrease in illegal extractions, improved local coordination for the nesting and incubation of sea turtle eggs, and the assignment of permanent environmental wardens.

21.2.3 Nicaragua

Four MPAs have been established along the Pacific coast of Nicaragua: (1) Natural Reserve (NR) Estero Padre Ramos; (2) NR Juan Venado; (3) Wildlife Refuge (WR) Río Escalante-Chacocente; and (4) WR La Flor (http://www.marena.gob.ni/) (Fig. 21.3). None of these, however, is concerned with the protection of the recently identified coral biotopes in Nicaragua (Alvarado et al. 2010, 2011a; see Chap. 5, Glynn et al.). Their conservation objectives are directed towards sea turtle nesting sites. There is now an urgent need for the establishment of MPAs around the area of Punta Gigante (11°23′07.9″N; 86°02′12.4″W), where vibrant coral reef formations have been recently discovered (Alvarado et al. 2010, 2011a).

In conservation terms, the small number of MPAs on the Nicaraguan Pacific coast could compromise the latitudinal (north-south) dispersal and connectivity of ETP coral populations. This potential problem could be critically increased with the construction of the Nicaragua transoceanic canal (Fig. 21.3). Canal construction, which began in December 2014, will involve the excavation of hundreds of km of terrain from coast to coast. No environmental or economic feasibility studies have been revealed to the public. The probable canal route cuts through a coastal Caribbean natural reserve, traverses Lake Nicaragua, and is estimated to destroy around 400,000 ha of rainforest, wetlands, and marine ecosystems (Meyer and Huete-Pérez 2014). Multiple autonomous indigenous communities also are threatened. In terms of coral biotopes, canal construction will likely threaten the Caribbean reefs of Corn Island and Cayos Perlas, and poses a major threat to coral communities adjacent to the proposed Pacific terminus, especially in the vicinity of San Juan del Sur. These coral habitats are already exposed to stressful conditions during upwelling (December–April), and to sedimentation and salinity fluctuations in the wet season (June–

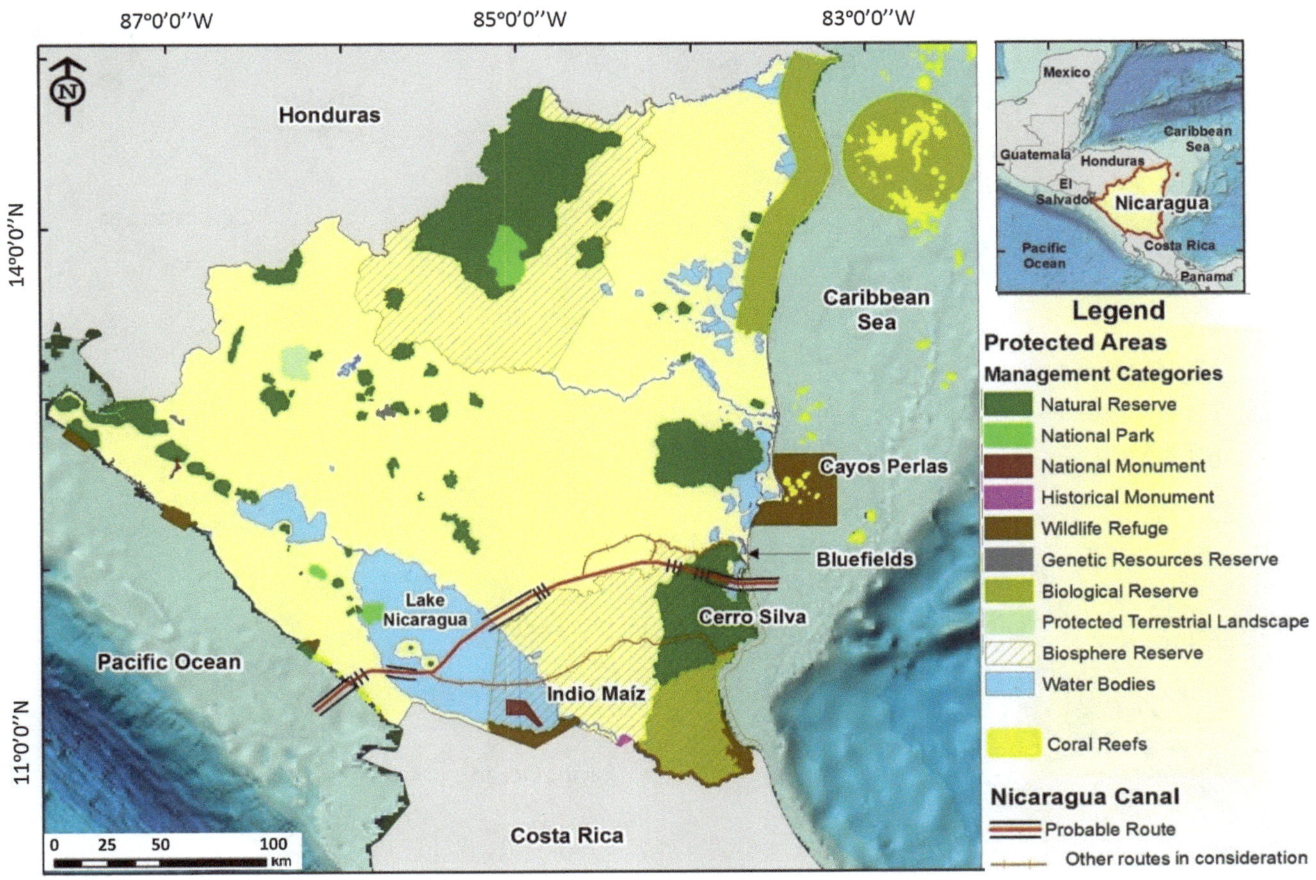

Fig. 21.3 Nicaragua's protected areas and probable route of interoceanic canal (modified from http://www.marena.gob.ni/, and Meyer and Huete-Pérez 2014)

November). The construction of the Nicaraguan canal could likely be a limiting factor for the continued existence of coral habitats on the Pacific coast of Nicaragua.

21.2.4 Costa Rica

Costa Rica has 167 protected areas (Environmental Ministry) and eight Responsible Fishing Areas (Fishery Authority) that together cover approximately 50 % of the Pacific and Caribbean coastlines. Twenty-six of these areas are MPAs classified as National Parks, National Wildlife Refuges, National Wetlands, Biological Reserves, Absolute Natural Reserves, Marine Management Areas, and Responsible Fishing Areas (RFA), which accord some level of protection for 16,048 km^2 of national waters (Alvarado et al. 2012; Salas et al. 2012; Fargier et al. 2014). Costa Rica possesses 970 km^2 of coral communities and coral reefs (0.3 % of the world's total coral reef area), and 80 % of coral habitats are located inside MPAs. It is estimated that 93 % of the coral areas are at risk (Spalding et al. 2001). On the Pacific coast, 11 MPAs contain coral reef habitats (Fig. 21.4; Table 21.1), covering an area of 4648 km^2. Ten of these are no-take MPAs, and one is a multi-purpose protected area (Golfo Dulce RFA). With the exception of the latter, the MPAs were created as extensions of their terrestrial counterparts with few studies on the merits of marine protection in their respective areas. Over time, however, studies have increased to describe and evaluate the areas in terms of supporting improved conservation management actions (Cortés et al. 2010). Nevertheless, some MPAs are considered to be 'paper parks' because control and surveillance actions are limited, mainly due to an insufficient number of park custodians. In some cases, park personnel lack the capacity to effectively manage MPAs, or their capacity to respond to threats is limited. Insufficient training and knowledge of the marine environment also affect performance and therefore protection.

The government of Costa Rica is a signatory of several international treaties (CITES, CBD, AGENDA 21, UNCLOS, RAMSAR, among others) and has several laws–such as its Wildlife Law, a Fisheries and Aquaculture Law, National Coastguard Creation Law and Environment Organic Law—for protecting and managing coral reef ecosystems. Despite this legal framework, Costa Rica does not currently have a specific legal instrument for the protection of coral reefs (AIDA 2012; Fonseca 2013). During the past ten years, however, a multidisciplinary group has

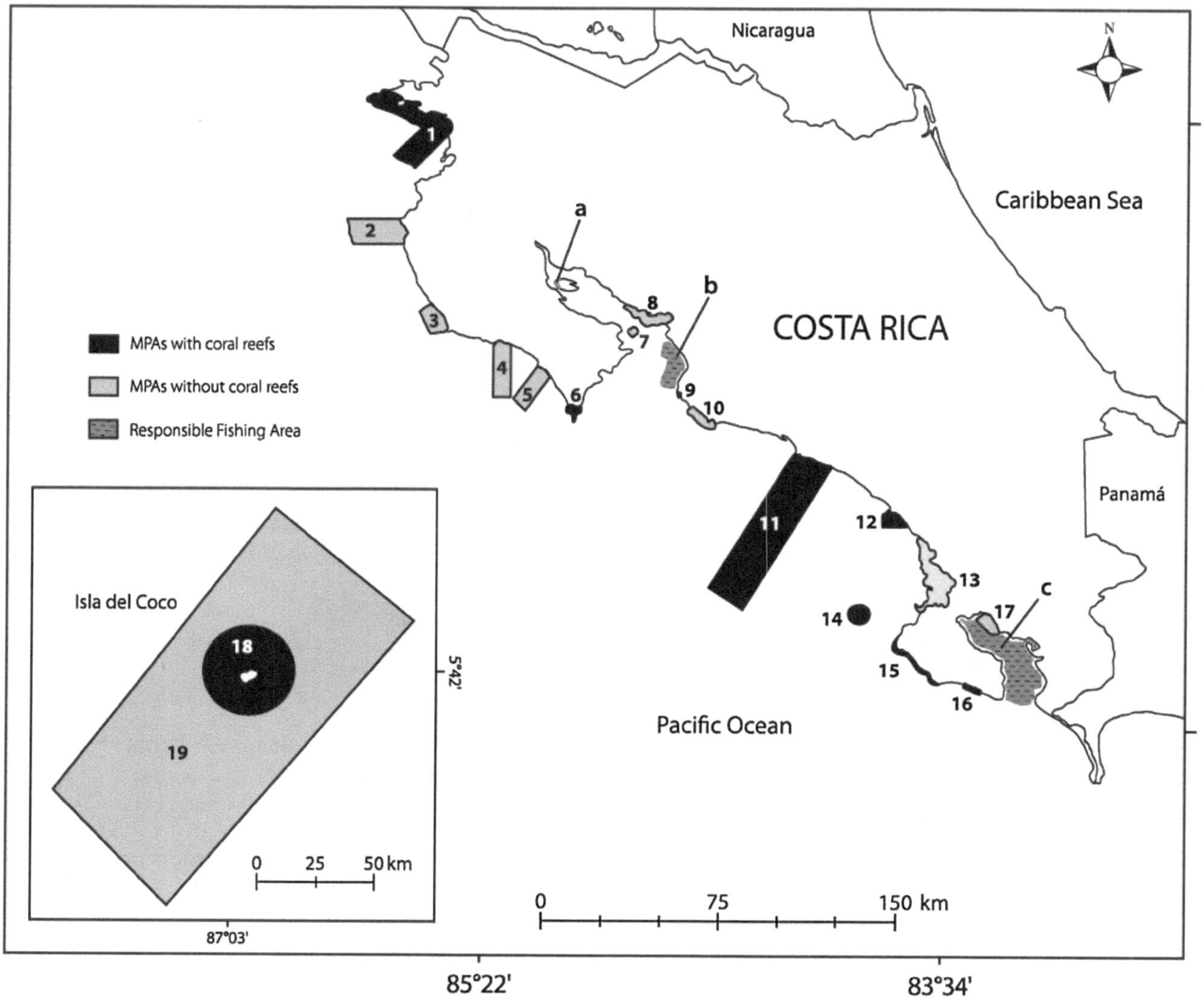

Fig. 21.4 Marine protected areas on the Pacific coast of Costa Rica. *1* Santa Rosa NP; *2* Marino Las Baulas NP; *3* Ostional NWR; *4* Caletas-Ario NWR; *5* Camaronal NWR; *6* Cabo Blanco ANR; *7* Isla San Lucas NWR; *8* Estero de Puntarenas NW; *9* Playa Blanca NW; *10* Playa Hermosa-Punta Mala NWR; *11* Manuel Antonio NP; *12* Marino Ballena NP; *13* Térraba-Sierpe NW; *14* Isla del Caño BR; *15* Corcovado NP; *16* Río Oro NWR; *17* Piedras Blancas NP; *18* Isla del Coco NP; *19* Montes Submarinos MMA; *a* Chira RFA; *b* Tarcoles RFA; *c* Golfo Dulce RFA. *NP* National Park; *NWR* National Wildlife Refuge; *BR* Biological Reserve; *ANR* Absolute Natural Reserve; *NW* National Wetland; *MMA* Marine Management Area; *RFA* Responsible Fishing Area

been working on the creation of an Executive Decree for the protection of coral reef environments (Fonseca 2013). Recent efforts supported by NGOs, such as Conservation International (CI)—Costa Rica, and AIDA, along with the University of Costa Rica (CIMAR), have initiated an evaluation of ecosystem services provided by coral reefs. This group also has leverage to improve a draft of the decree for approval by the Ministry of Environment. This decree will regulate activities that may affect the condition and development of coral reefs, such as sedimentation, extraction, anchor damage, coastline infrastructure development, and also the coordination of an improved mapping effort of the country's coral reefs, and the establishment of an ecological and health monitoring program.

Although there is great awareness in Costa Rica of the importance of coral reefs, there is still much to protect. In years past, conservation deficiencies were identified in terrestrial, freshwater and marine ecosystems. Coral reef ecosystems have become a key focus for marine conservation in recent years (Alvarado et al. 2011b). While the latest conservation gap analysis demonstrated improvements in protecting coral reefs, it also indicated that in many areas there is a lag in updating their health status, and not all regions have adequate baseline information or continuous monitoring programs. In turn, it was found that there are still many coral habitats in need of protection. To remedy this deficiency, new strategies will be required to avoid social conflicts with the presence of MPAs and also to avoid the

Table 21.1 Marine protected areas of the ETP with coral habitats and/or coral reefs

Country and sites	MPA category	IUCN category[a]	Marine area (km^2)	Year established	Other category	Take–no take	Source
Mexico							
Bahía de Los Ángeles, Canal de Ballenas y Salsipuedes	Biosphere Reserve	VI	3879.57	2007		No take 0.05 %	CONABIO et al. (2007); http://www.protectedplanet.net/
Loreto	National Park	II	2065.81	1996	Ramsar site, world heritage site	No take 0.07 %	CONABIO et al. (2007); http://www.protectedplanet.net/
Isla Espíritu Santo	National Park	II	486.55	2007		No take 1.4 %	CONABIO et al. (2007); http://www.protectedplanet.net/
Cabo Pulmo	National Park	II	71.11	2000	Ramsar site, world heritage site	No take 35.1 %	CONABIO et al. (2007); http://www.protectedplanet.net/
Islas Marietas	National Park	II	13.83	2005	Ramsar site, world heritage site, man and biosphere	No take 0.8 %	CONABIO et al. (2007); http://www.protectedplanet.net/
Islas Marías	Biosphere Reserve	VI	3172.57	2000	World heritage site	No take 2.4 %	CONABIO et al. (2007); http://www.protectedplanet.net/
Archipiélago de San Lorenzo	National Park	II	584.43	2005	World heritage site	No take 15.1 %	CONABIO et al. (2007); http://www.protectedplanet.net/
Cabo San Lucas	Fauna and Flora Protection Area	VI	39.96	1973	World heritage site	Take 100 %	CONABIO et al. (2007); http://www.protectedplanet.net/
San Pedro Martir	Biosphere Reserve	VI	298.76	2002	Ramsar site, world heritage site, man and biosphere	No take 3.7 %	CONABIO et al. (2007); http://www.protectedplanet.net/
Huatulco	National Park	II	46.77	1998	Ramsar site		CONABIO et al. (2007); http://www.protectedplanet.net/
Revillagigedo	Biosphere Reserve	I	6231.26	1994	Ramsar site, world heritage site	No take 0.07 %	CONABIO et al. (2007); http://www.protectedplanet.net/
El Salvador							
Los Cóbanos	Not defined yet	Not yet defined	206.8	2008		100 % take until management plan is approved	Diario Official, 12 February 2008, Volume 378, number 29
Costa Rica							
Santa Rosa	National Park	II	461.2	1971	UNESCO World Heritage site	No-take (100 %)	Alvarado et al. (2012), Salas et al. (2012)
Cabo Blanco	Absolute Natural Reserve	I	16.7	1963		No-take (100 %)	Alvarado et al. (2012), Salas et al. (2012)
Playa Blanca	National Wetland	IV	0.05	1994		No-take (100 %)	Alvarado et al. (2012), Salas et al. (2012)
Manuel Antonio	National Park	II	1243.7	1972		No-take (100 %)	Alvarado et al. (2012), Salas et al. (2012)
Marino Ballena	National Park	II	52.2	1992		No-take (100 %)	Alvarado et al. (2012), Salas et al. (2012)

(continued)

Table 21.1 (continued)

Country and sites	MPA category	IUCN category[a]	Marine area (km^2)	Year established	Other category	Take–no take	Source
Isla del Caño	Biological Reserve	I	55.3	1978		No-take (100 %)	Alvarado et al. (2012), Salas et al. (2012)
Corcovado	National Park	II	19.8	1975		No-take (100 %)	Alvarado et al. (2012), Salas et al. (2012)
Río Oro	Wildlife Refuge	II	16.9	1999		No-take (100 %)	Alvarado et al. (2012), Salas et al. (2012)
Piedras Blancas	National Park	II	13.2	1991		No-take (100 %)	Alvarado et al. (2012), Salas et al. (2012)
Golfo Dulce	Responsible Fishing Area	V	773.97	2010		Take (100 %)	Salas et al. (2012)
Isla del Coco	National Park	II	1994.7	1978	Ramsar site, UNESCO World Heritage site	No-take (100 %)	Alvarado et al. (2012), Salas et al. (2012)
Panama							
Golfo de Chiriquí	National Park	II	194.93	1994		No-take and take	ANAM–PMDSCH (2010)
Isla Montuosa y Banco Haníbal	Special Zone of Marine Protection	Not yet matched with IUCN	1782.33	2004[b]	UNESCO World Heritage site	Take	ANAM (2009), Cunningham et al. (2013)
Coiba	National Park	II	2,024.63	1991	UNESCO World Heritage site	No-take	Glynn and Maté (1997), Maté (2003), Guzmán et al. (2004), ANAM (2009)
Sur de Veraguas	Special Management Zone	Not yet matched with IUCN	1791.35	2008		Take	Maté (unpublished)
Cerro Hoya	National Park	II	36.33	1984		Take	Maté (2003)
Isla Iguana	Wildlife Refuge	IV	0.94	1981		No-take	Guzmán et al. (1991), Maté (2003)
Taboga	Wildlife Refuge	IV	0.45	1984		Take	Maté (2003)
Archipiélago de Las Perlas	Special Management Zone	Not yet matched with IUCN	1356.18	1984		Take	Maté (2003), Guzmán et al. (2008)
Colombia							
Gorgona	Natural National Park	II	616.87	1984			Muñoz and Zapata (2012); http://www.protectedplanet.net/
Malpelo	Flora and Fauna Sanctuary	IV	8571.5	1995	UNESCO world heritage site		http://www.protectedplanet.net/
Ensenada de Utría	Natural National Park	II	543	1986			http://www.protectedplanet.net/
Bahía Málaga	Natural National Park	II	470.94	2010			http://www.protectedplanet.net/
Ecuador							
Galápagos	Marine Reserve	VI	138,000	1998	UNESCO world heritage site	Multiple-purpose	Danulat and Edgar (2002)

(continued)

Table 21.1 (continued)

Country and sites	MPA category	IUCN category[a]	Marine area (km^2)	Year established	Other category	Take–no take	Source
Galera-San Francisco	Marine Reserve	VI	546	2008		Multiple-purpose	NAZCA (2013)
Machalilla	National Park	II	144.3	1979	Ramsar site	Multiple-purpose	MAE (2007a, b); http://www.protectedplanet.net/
El Pelado	Marine Reserve	VI	130	2012		Multiple-purpose	http://www.protectedplanet.net/
Puntilla Santa Elena	Coastal-Marine Faunal Production Reserve	VI	472.74	2008		Unknown, probably multiple-purpose	Samaniego (2010)
Isla Santa Clara	Wildlife Refuge	IV	73.43	1999	Ramsar site	Multiple-use	Hurtado et al. (2010)
Chile							
Motu Motiro Hiva Marine Park (Salas y Gómez Island)	Marine Park	I	150,000	2010		No-take (74 %)	Friedlander et al. (2013); http://www.protectedplanet.net/ http://www.mpatlas.org/region/nation/CHL/
Coral Nui Nui	Marine and Coastal Protected Area	Not yet matched with IUCN	0.0855	1999		Multiple-use	Sierralta et al. (2011); http://www.mpatlas.org/region/nation/CHL/
Motu Taura	Marine and Coastal Protected Area	Not yet matched with IUCN	0.09	1999		No-take (100 %)	Sierralta et al. (2011); http://www.mpatlas.org/region/nation/CHL/
Hanga Oteo	Marine and Coastal Protected Area	Not yet matched with IUCN	1.65	1999		Multiple-use	Sierralta et al. (2011); www.mpatlas.org/region/nation/CHL/

[a]Primary management objectives (after Marine Protected Areas 2001)
I Strict Nature Reserve-Wilderness Area: protected area managed mainly for science or wilderness protection
II National Park: managed mainly for ecosystem protection and recreation
III Natural Monument: managed mainly for conservation of specific natural features
IV Habitat/Species Management Area: managed mainly for conservation through management intervention
V Protected Landscape/Seascape: managed mainly for landscape/seascape conservation and recreation
VI Managed Resource Protected Area: managed mainly for the sustainable use of natural ecosystems
[b] Year established by government of Panama (2004), by UNESCO World Heritage Site (2005)

creation of 'paper parks' that lack effective management. Several NGOs, including CI, Costa Rica Para Siempre, Keto, MarViva, and the German Society for International Cooperation, and the United Nations Development Program (UNDP), in cooperation with national universities, have been working in coordination with the government of Costa Rica to address these conservation deficiencies.

21.2.5 Panama

Coral reefs and coral communities on the Pacific coast of Panama are well represented within the regulatory scheme of MPAs. Eight marine areas provide protection to the largest and most important coral habitats in the Gulf of Chiriquí, a non-upwelling area in western Panama with seasonal shoaling of the thermocline, and the Gulf of Panama, a seasonal upwelling environment in the eastern half of Panama. These areas include three National Parks (Marino Golfo de Chiriquí, Cerro Hoya, and Coiba), two Wildlife Refuges (Isla Iguana, and Isla Taboga), one Special Zone of Marine Protection (Isla Montuosa and Banco Haníbal), and two Special Management Zones (Sur de Veraguas, and Archipiélago de Las Perlas) (Table 21.1, Fig. 21.5). Maté (2003) provided an overview of five of these coral reef ecosystems that include the national parks and the wildlife refuges, as they were the only ones established at that time (between 1984 and 1991). New coral habitats were not included in the protection

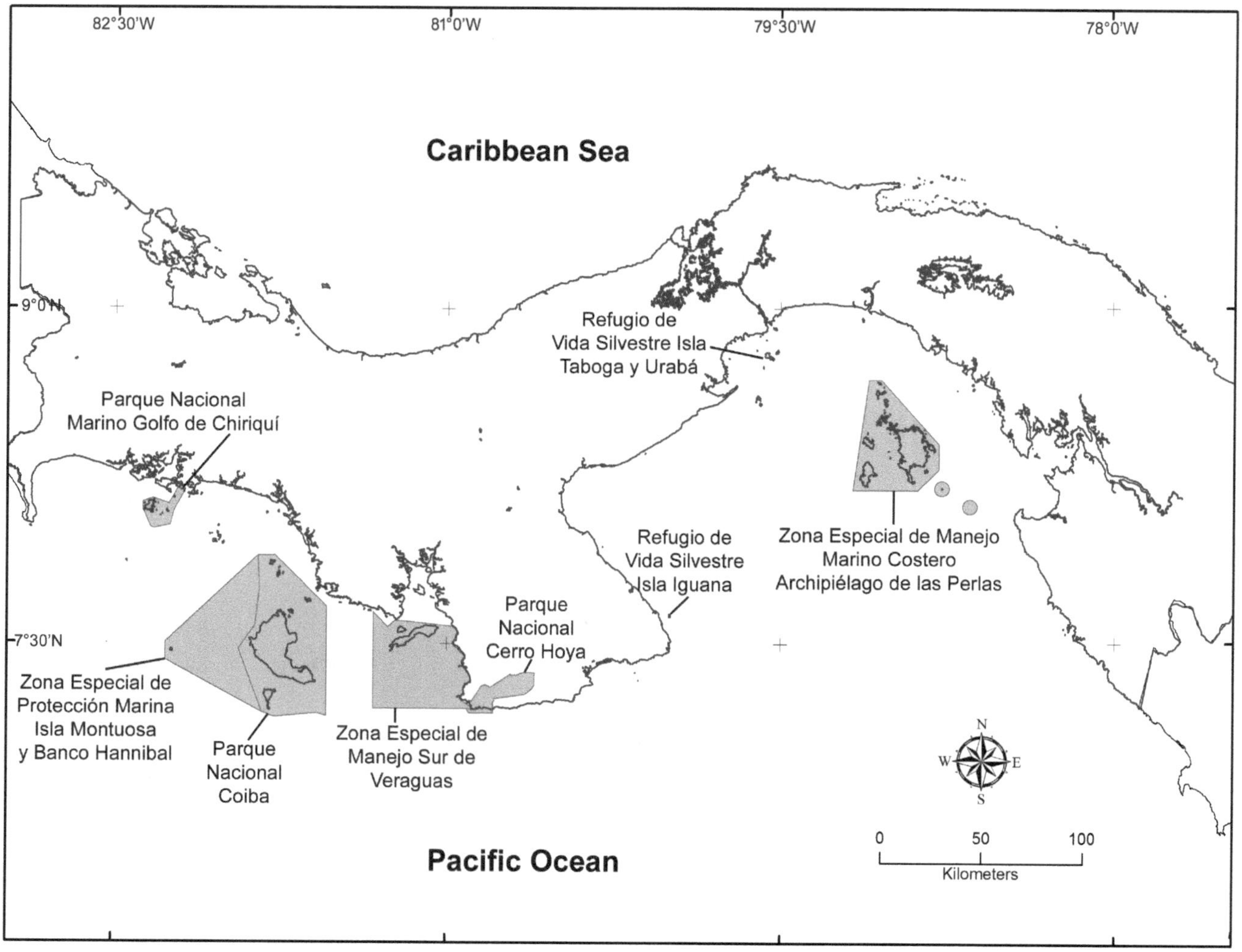

Fig. 21.5 Marine protected areas along the Pacific coast of Panama

scheme until 2004, under two new categories directed towards fisheries management, namely the Special Zone of Marine Protection, and the Special Management Zones, the latter first appearing in 2008 (Table 21.1).

The two largest MPAs in Panama with coral reef ecosystems are the Coiba National Park and the Special Zone of Marine Protection, the Archipiélago de Las Perlas. In 2004, the legal status of Coiba was upgraded to national law. In addition, a new area was established to the west of Coiba Island, and designated as the Special Zone of Marine Protection of Montuosa Island and Hannibal Bank (Zona Especial de Protección Marina Isla Montuosa y Banco Haníbal). This special zone is considered as a 'buffer zone' for the Coiba National Park. Both areas were inscribed together in 2004 in the UNESCO World Heritage List of Natural Sites as Coiba National Park and its Special Zone of Marine Protection under current criteria (ix) and (x). Coiba and its Special Zone encompass 3807 km^2 of marine area with approximately 40.5 km^2 of coral habitat (Table 21.1). Las Perlas MPA covers 1356 km^2 of area with approximately 3.1 km^2 of coral habitat.

While national parks and wildlife refuges have equivalents in the IUCN Protected Areas Categories System as National Park, and Habitat/Species Management Areas respectively, the Special Management Zones and the Special Zone of Marine Protection do not have equivalent classifications. The first three types of management areas under Panamanian regulations fall within the competency of the Ministry of Environment of Panama (MiAmbiente) and the latter under Panama's Aquatic Resources Authority (ARAP). Until March (2015), when MiAmbiente was created to replace the Panamanian National Authority for the Environment (ANAM), the Special Management Zones were under the jurisdiction of ARAP (see Maté 2003). MiAmbiente is now charged to reclassify this category to an equivalent in the IUCN Protected Areas Categories System.

Only the three national Parks (Cerro Hoya, Coiba, and Marino Golfo de Chiriquí) already have approved management plans. Yet their implementation has been relatively poor due to insufficiently-trained staff, inadequate budgets, and a lack of necessary equipment. The Special Management Zone of the Archipiélago de Las Perlas has had a management plan since 2011, which was developed through a participatory process, but authorities have been unwilling to approve it due to competing political and economic interests. MiAMBIENTE acquired jurisdiction over this area in 2015 and at the beginning of 2016 began a process to revise and validate the management plan for approval.

Nevertheless, Coiba and its Special Zone as well as Las Perlas have a relatively strong legal framework provided by national laws. These protected zones will be the main areas covered in this section. Both areas are archipelagos but only Las Perlas includes human settlements. Since information on Coiba and Las Perlas reef-building coral cover and condition is presented in Chap. 5 (Glynn et al.), we here note only some brief complementary remarks. The Coiba MPA consists of 39 main islands with no resident communities except for a small contingent of park rangers and law enforcement officers. The coral communities and reefs are considered among the most extensive in the eastern Pacific, comprising more than 17 km^2 of high biological diversity and endemism (Guzmán et al. 2004). For example, 30 species of octocorals, 20 scleractinian corals, and four hydrocorals have been reported from Coiba (Glynn and de Weerdt 1991; Glynn and Maté 1997; Glynn and Ault 2000; Breedy and Guzmán 2003, 2004, 2005; Maté 2003; Guzmán et al. 2004; Lindner et al. 2004). Two octocorals, *Pacifigorgia catedralensis* and *Leptogorgia christiae,* are endemic to the park and have been found only at a single location there (Breedy and Guzmán 2004, 2008; Guzmán and Breedy 2008). The octocorals *Pacifigorgia rubinoffi*, *Pacifigorgia smithsoniana*, and the stylasterid lace coral *Distichopora robusta* are mainly concentrated around Coiba, even though they also occur at a few other localities in the Gulf of Chiriquí (Breedy and Guzmán 2003; Lindner et al. 2004). *Pacifigorgia smithsoniana* is also found in the Special Management Zone of the Archipiélago de Las Perlas, Gulf of Panama.

The Special Zone of Marine Protection includes Hannibal Bank, an elongated, triangular guyot (Cunningham et al. 2013), which is considered one of the most important fishery areas in Panama for both commercial (snappers, groupers, and dolphin fish), and sport fishing (billfish and tuna) species (ANAM 2009; Vega et al. 2011). Approximately 40.5 km^2 of the Bank is firm substrata between 54 and 416 m in depth. The relatively shallow rugose areas probably contain corals that support a high faunal diversity (Cunningham et al. 2013).

In terms of resources and activities carried out in the park, the Coiba zoning plan for the marine environment consists of two types of marine areas: a Marine Reserve Zone (equivalent to a 'no-take zone') and a Resource Management Zone. The Marine Reserve Zone covers 23.5 % of the park, and includes most of the shallower substrates where corals are located. It is in this area, which includes the shorelines of the islands to a distance of 1.8 km offshore, where most of the directed recreation and ecotourism occur (snorkeling, diving, and beach activities). The Resource Management Zone includes the deeper areas of the park, which support marine fishery resources that can be harvested by artisanal and sport fishing. Numerous rock pinnacles with diverse coral communities are also present in the area, and are frequented for scuba diving and research. The Resource Management Zone also supports whale watching activities. A new fisheries plan has established regulations for fishing gear, catch limits, fishing effort, and the size of species that can be harvested.

Coral habitats with reefs and coral communities on rocky outcrops are of high interest for sport diving, and are frequented regularly by visitors, particularly those sites located in the northeastern area of the park. The Coiba Management Plan identifies 31 sites where these activities take place, and sets limits on the number of boats and divers allowed access at any given time. In addition, 40 mooring buoys have been installed at these sites to reduce the impact of anchor damage while also providing diving safety lines.

The management plan for the Special Zone of Marine Protection of Montuosa Island and Hannibal Bank has not yet been prepared although the World Heritage Committee (WHC) has long requested one. To wit: the World Heritage Committee Decision 37 COM 7B.31 of 2013, 'Request the State Party to urgently finalize a draft Management Plan for the Special Zone of Marine Protection, to adopt and initiate its implementation, and to start with the independent Management Effectiveness Evaluation in order to inform the effective management for both Coiba National Park and its Special Zone of Marine Protection'.

A UNESCO-IUCN monitoring mission conducted at Coiba Island in January 2014 (see Douvere and Herrera 2014) identified several issues that are negatively affecting the long-term conservation and effective management of the area. These threats reflect problems facing additional MPAs in Panama, including the Pearl Islands Archipelago, which is described in more detail below. The main threats identified by the mission include: (a) The small number of adequately trained staff, which results in a lack of effective management and enforcement; (b) A risk that fisheries, in particular illegal sport and commercial fisheries, poses a threat to the health of the Outstanding Universal Value (OUV) of the MPA. Sport fisheries is of particular concern as it seems to increase rapidly, targeting zones of spawning/nursing and where rare coral/high endemism occurs; (c) A lack of reliable information on the scope of resource extraction; (d) An increase in tourism, if not well managed, will pose a threat to the long-term conservation of the property's OUV; and (e) Coastal development within the boundaries of the MPA is inconsistent with the property's OUV and should be relocated/restricted to areas outside the park's boundaries. As a result, WHC Decision 38 COM 7B.84 of 2014, 'Expresses its concern about the negative impact of fisheries, and in particular illegal and sport fisheries, on the OUV of the property, and urges the State Party to complete and implement the management plan for the Special Zone of Marine Protection (SZMP) as a matter of priority, which should include clear regulations related to fisheries management, i.e. no-take zones and seasonal closures of critical areas, such as Hannibal Bank, Montuosa Island'.

Coiba was for many years a penal colony until its closure in 2004. Livestock maintained at the prison became feral, and is one of the primary threats to the newly established park. The livestock (cattle) have caused the deterioration of large areas of top soil, producing loose sediments that have washed onto nearby coral reefs at Playa Blanca, the largest coral reef complex in the park. Recent efforts have been made to begin removing the island's remaining livestock. The UNESCO-IUCN monitoring reactive mission has acknowledged this effort, with the caveat '...considers important that the World Heritage Committee assesses closely its implementation in view of having all livestock removed from the property by end of 2014, as suggested by the State Party'.

The second area to be considered is the Special Management Zone of the Archipiélago de Las Perlas (Table 21.1, Fig. 21.5), which contains the largest aggregation of coral reefs in the seasonal upwelling Gulf of Panama (Glynn and Maté 1997). This Management Zone also includes two satellite areas to the southeast, namely Galera Island and Bajo Trollope. There are 250 rock islands and islets in the management zone with 39 octocoral and 19 known scleractinian species (Glynn and Maté 1997; Maté 2003; Guzmán et al. 2008). Areas with the highest coral cover are located both to the north (Contadora and Saboga Islands) and south (Pedro González, San Pedro, and San Telmo Islands). In contrast to Coiba Island, with no permanent residents, there are long-established communities on some of the larger islands, both in the northern and southern parts of the archipelago. However, most of the islands are uninhabited (Guzmán et al. 2008).

In contrast to Coiba, which was designated a national park to protect biodiversity and promote conservation, the Las Perlas MPA was created to implement an integrated coastal management approach focused on fisheries, tourism, and long-term landscape conservation (Guzmán et al. 2008). Important strategies being implemented to control local sedimentation include the protection of the main watersheds in the Del Rey Hydrological Reserve, and the creation of the San Jose Natural Private Reserve (Guzmán et al. 2008).

In addition to the traditionally recognized anthropogenic stressors, recent government authorization for the private prospecting of archeological artifacts in the waters of the archipelago has introduced new challenges near coral habitats. Large blowers with air supplied from nearby ships are being used to remove (blow-off) sediments to expose artifacts. Local communities are pressuring authorities to ban such operations near coral areas.

Even though the management plan for Las Perlas has not yet been adopted, 20 mooring buoys have been installed in areas identified in the proposed plan to reduce the impact of anchor damage to highly valued benthic ecosystems. The buoys were intended for the recreational and commercial boats supporting scuba and surface divers. However, improper use of the buoys by large ships is causing significant damage to the buoys as well as to surrounding benthic communities. Other activities to support the proposed management plan could include the incorporation of high diversity and high coral cover areas into no-take zones, which would contribute significantly to conservation efforts (Guzmán et al. 2008). A Conservation and Monitoring Unit was active at Las Perlas a few years ago, but has recently been terminated.

With the exception of published information on some of the other five marine areas with some level of protection (Guzmán et al. 1991; Glynn and Maté 1997; Maté 2003; Díaz 2005), there is little new information relating to management practices or strategies being prepared for implementation. Threats and local strategies that may be implemented at some of these sites do not differ significantly from those previously noted for the Coiba and Las Perlas MPAs.

21.2.6 Colombia

The Colombian Pacific coastline is about 1300 km in length, covers an area of approximately 367,823 km^2, and contains a population of around 1,370,000 inhabitants ($\sim$4 ind km^{-2}) (Romero 2009). There is little socio-economic development in the area, with only two coastal cities with large populations, namely Buenaventura and Tumaco (Fig. 21.6). This area, known as the Chocó, is characterized by geological variability, moderate atmospheric temperatures, high rainfall and relative humidity, and high terrestrial biodiversity (Díaz and Gast 2009).

Coral habitats comprise 14.4 km^2 of area, of which 3 % are oceanic and 97 % continental. Most of the Colombian Pacific coral areas are managed for conservation (Fig. 21.6). In fact, the only unprotected coral areas occur in the northern Chocó at Punta Piñas (Gulf of Cupica), La Tebada and Cabo Marzo (Fig. 21.6, see also Chap. 5, Figs. 5.18, 5.19). The major marine protected areas with coral habitats (structural reefs or otherwise) are located at Gorgona Island Natural National Park (NNP), the Malpelo Island Flora and Fauna Sanctuary (FFS), and Ensenada Utría NNP. At the Uramba Bahía Málaga NNP, isolated coral colonies have been found on a rocky reef at the entrance to the bay (Escobar and Neira 1992). Another NNP on the Colombian Pacific coast,

Fig. 21.6 Marine protected areas and other sites with and without coral habitats on the Pacific coast of Colombia

Sanquianga Park with 800 km^2 of protected mangroves and estuaries, is devoid of corals. All of these MPAs are under the jurisdiction of the Special Administrative Unit of the Natural National Parks System of the Ministry of Environment and Sustainable Development (MADS). Despite the level of protection from anthropogenic disturbances, all of these areas are subject to natural disturbances, such as freshwater runoff, sedimentation, and sporadic ENSO activity.

The Gorgona NNP is located in the southwestern sector of the Colombian Pacific. It is 616.8 km^2 in area, of which 14.8 km^2 (2.4 %) is land, including Gorgona Island, Gorgonilla Islet and rocky promontories (El Viudo and El Horno). Gorgona Island is 9.3 km long by 2.6 km wide, and is 35 km west of Punta Reyes on the mainland. Gorgona Island served as a high security prison for 25 years (1959–1984).

The Gorgona Park was established by executive resolution 141 in July 1984. New boundaries were established by resolution 1265 in October 1995. This declaration prohibits all activities other than conservation, research, education, recreation, and cultural studies. Restoration of the natural environment since the closure of the prison, and enforcement of protective park measures, is also permitted. It therefore meets IUCN Management Category II (National Park), which mandates ecosystem protection and recreation (Table 21.1). To date, three management plans have been developed for the park. In the first (2004–2010), coral reefs were established as the most significant NNP conservation target (UAESPNN 2004, 2005), and were prioritized for research, monitoring and conservation practices. A second plan (2007–2011) was adopted by resolution 053 in January 2007. Presently, a third management plan (2012–2017) focuses on increasing the level of basic knowledge of coral reefs, strengthening monitoring, and developing more robust outreach activities (Muñoz and Zapata 2012). To accomplish these goals, the proposed research guidelines include measures to expand knowledge and information on biodiversity, exploration of poorly known areas, species' biology, oceanographic conditions, ecological processes, natural and anthropogenic perturbations as well as monitoring the health state of coral reefs.

The Malpelo Island MPA consists of a group of islands and islets in the Colombian Pacific, 506 km west of Buenaventura. The principal island is 2.9 km long by 1.3 km wide with a maximum height of nearly 380 m above sea level. The main island is surrounded by 11 islets, four in the north (Los Tres Mosqueteros and D'Artagnan), two on the east side of the island (Vagamares and La Torta), and five at the island's south end (Los Tres Reyes, La Gringa and Escuba) (see Fig. 5.20, Chap. 5).

The island's marine life is diverse and vibrant, and is located within an International Conservation Hotspot. In 1995, by Resolution 1292, Malpelo Island was designated a Flora and Fauna Sanctuary. In 1996, the marine area was extended by Resolution 1423 to six nautical miles from the shoreline. In 2002, Resolution 0761 re-defined the marine boundaries. In 2003, the International Maritime Organization declared the Sanctuary a Particularly Sensitive Sea Area (PSSA), and in 2005 Resolution 1589 extended the park's boundaries from 654.5 to 8575 km^2. In July of 2006, the island was declared a World Natural Heritage Site by UNESCO under Natural Criteria VII and X. The IUCN Management Category is IV, i.e. a Habitat-Species Management Area. At present, the island is managed by a Special Administrative Unit of the Natural National Parks System, and is a key component of the National Fisheries Management Plan for Colombia's Pacific Region. The Park Unit and the Malpelo Foundation of Colombia protect the island. Since 2005 UNESCO, together with Conservation International, designated much of the equatorial eastern Pacific a Marine Biological Corridor of Conservational and Sustainable Development. This extensive network of MPAs includes coastal and oceanic islands between Costa Rica, Panama, Colombia and Ecuador, connecting the main coral habitats of the region (e.g., Cocos, Coiba, Malpelo, Gorgona, and the Galápagos Islands). The extended, protected area around Malpelo was declared a no-take zone.

Despite its designation as a marine sanctuary, some threats, such as illegal artisanal and industrial fishing, persist within and around Malpelo's marine protected zone. Hand, seine and long-line fishing (with 1500–5000 hooks per line) contribute to overfishing of commercial species, especially sharks. Tourism, however, is carefully controlled and the concessions operating at Malpelo are mandated to undertake infrastructure maintenance and training of staff to effectively manage resources. El Niño disturbances have possibly affected Malpelo's coral habitats less than those at Gorgona Island, although the effects of the 1982–83 El Niño disturbance at Malpelo were not reported (Vargas-Ángel et al. 2001). Although there are recent documented records of *Acanthaster planci* on Malpelo Island (Narváez and Zapata 2010), sea star outbreak events are unknown.

Located on the north coast of the Colombian Pacific, the Utría NNP is part of the Chocó Biogeographic Province. It is considered to be of high conservation priority due to its biological diversity and species endemism. This MPA encompasses an elongate embayment 543 km^2 in area, and includes three villages: El Valle, Bahía Solano and Nuquí. The bay lies astride a geological fault, the Utría Fault. The coral reefs in the Utría NNP are unique in the Colombian Pacific because of their close association with extensive mangrove forests. The Utría NNP contains a 0.33 km^2 (33 ha) estuarine-mangrove ecosystem with at least five species of mangroves, which is fundamentally critical for the reproduction, protection, and development of a large number of marine species of both ecological and commercial value. In recent years, heavy exploitation has threatened some fish species in the MPA due to overfishing, the presence of semi-industrial fishing vessels, and the use of inappropriate fishing gear.

Utría National Park was created in the Department of Chocó according to Agreement 052 in 1986, and approved by Executive Resolution No. 090 in 1987. The original 543 km^2 area was expanded to include an additional 220 km^2 of protected forest. This MPA is characterized by high rainfall and species rich terrestrial and marine ecosystems. Most of the area is covered by a tropical rainforest with a high degree of endemism, and steep terrain with soils subject to erosive processes.

The objectives of the Utría NNP are stated in the mission purpose of the National Parks Unit, namely 'Ensuring the in situ conservation of biophysical and cultural diversity present in representative ecosystems of the country'. In addition, the conservation values are of strategic importance to the country's biodiversity and to local Emberá (native American) communities sharing the territory, and to ensure a sufficient supply of ecosystem services for sustainable human development that promotes conservation of culturally significant natural habitat.

Utría NNP is located in an area with limited human intervention, which in general terms creates good connectivity among the park's various landscape units. The park's primary goal is the conservation of representative ecosystems of the region. A second goal is the conservation of migratory species of high national and international ecological value, and a third goal is the conservation of environmental goods and services important to human populations that have settled within the park's boundaries. Finally, the fourth goal is the preservation of the cultural values of the indigenous Emberá nation.

The main threats currently facing the Utría MPA are: (1) Over-exploitation by an increasing fishing effort, and the impact of artisanal fishers using illegal fishing gear and ignoring catch size limits; (2) Logging in surrounding areas leading to the depletion of species of high commercial value and a general degradation of natural resources; (3) Mega projects being implemented or proposed that will threaten the long-term conservation of the target values by an expected increase in habitat fragmentation, sedimentation, human population influx from outside the region, and impaired cultural practices of indigenous and Afro-Colombian communities. The key conservation challenges for the park are: (1) Inadequate facilities for staff and research support; (2) Weak coordination among the conservation institutions involved; (3) A lack of stakeholder engagement and involvement in resource management; (4) Insufficient budgetary resources allocated by the Ministry of Environment to comply fully with the mission and management, enforcement and evaluation.

Despite the support offered by SIMAC and the importance of coral habitats in the Colombian Pacific, weaknesses in protecting the MPAs remain. Among the most critical issues are: financial constraints to carry out scientific research projects; insufficient research on some basic issues in the formulation of conservation and management programs; and a lack of information on other conservation target values (such as soft bottom and rocky shore habitats), which may be closely related to coral ecosystems. Finally, climate change, the effects of which are beyond the direct control of MPAs, could have a significant impact in the future.

The three established MPAs with coral communities and reefs (Utría, Gorgona and Malpelo) together contribute to conservation, but face the challenge of moving beyond 'paper parks' (Rife et al. 2012) in order to reach effective conservation and sustainability goals. Building capacity for effective governance and monitoring, supported by scientific knowledge from research and academic institutions and environmentally-sound national development strategies, can help solve major problems in the conservation of coral habitats.

21.2.7 Ecuador

Protected areas in Ecuador are subject to the control and supervision of the Ministry of Environment of Ecuador (MAE), which determines their rules of management.

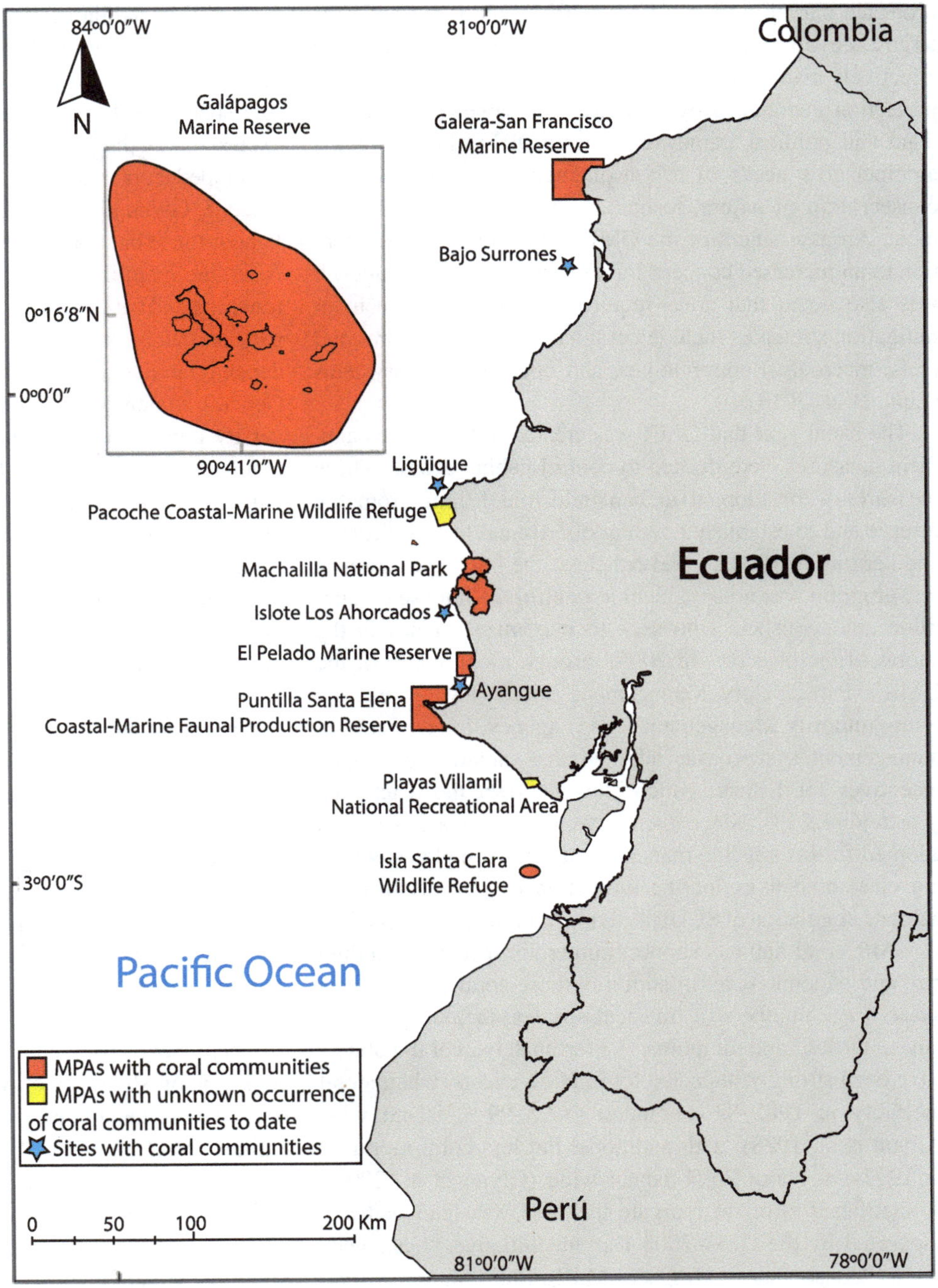

Fig. 21.7 Marine protected areas in Ecuador with or without coral communities (i.e., zooxanthellate scleractinian corals). In addition, sites outside protected areas with coral communities are shown (teal stars, not drawn to scale). MPAs with coral communities do not necessarily possess corals across all the reserve but coral formations have been identified at certain sites (see Figs. 5.24, 5.25, 5.29, and 5.30, Chap. 5). Pocilloporid corals have been reported for Isla Santa Clara Wildlife Refuge (René Espinoza, pers. comm.), and *Pocillopora damicornis* at Aconcito, located at Puntilla Santa Elena Coastal-Marine Faunal Production Reserve (Jesenia Zambrano, pers. comm.). Only the Galápagos Marine Reserve and the Machalilla National Park (including Isla de la Plata) are known to contain significant coral reefs. Map courtesy of Jael Martínez, with modifications by Juan Manuel Álava, Cecilia Sáenz, and Diana Moanga

Approximately 20 % of Ecuadorian territory, composed of 50 reserves, is within the bounds of protected areas. Of these, 14 protect coastal and marine environments (Hurtado et al. 2010) and include three marine reserves (Galápagos Marine Reserve, Galera-San Francisco Marine Reserve and El Pelado Marine Reserve, Fig. 21.7), five that contain both marine and terrestrial ecosystems (Machalilla National Park, Pacoche Coastal-Marine Wildlife Refuge, Puntilla Santa Elena Coastal-Marine Faunal Production Reserve, Isla Santa Clara Wildlife Refuge and Playas Villamil National Recreational Area, Fig. 21.7), and, six that exclusively cover mangals (Manglares Cayapas-Mataje Ecological Reserve, Manglares Churute Ecological Reserve, Manglares Estuario Río Esmeraldas Wildlife Refuge, Manglares Muisne-Cojimíes Wildlife Refuge, Manglares El Salado Faunal Production Reserve, and Manglares El Morro Wildlife Refuge) (Registro Official No. 202 2012, MAE 2014)

The Galápagos Marine Reserve (GMR) was named a UNESCO World Heritage site in 2001 (Danulat and Edgar 2002). Established in 1998 by the Organic Law on the Conservation and Sustainable Development of the Province of Galápagos (LOREG), it became the first Ecuadorian protected marine reserve (Villarta 2013), and is currently the largest, covering 138,000 km^2 in area. The GMR has been effectively managed in general terms because it benefits from protection guaranteed by LOREG. The same applies to the legal and political framework since LOREG highlights the archipelago's needs of a consolidated vision for both the conservation of natural resources and sustainable development. An assessment of the GMR habitats has called attention to an increased concern for areas near fishing villages. It was also noted that areas requiring better management or mitigation strategies include coral habitats, rocky shores and reefs, macroalgal communities, and open water ecosystems (Luna et al. 2013).

The same year that GMR was created, different management measures were drafted to control fishing access. These measures were adopted to ban industrial fishing from the reserve and to establish a system of artisanal fishing licenses and permits (PARMA). Nevertheless, the implementation of this program was not sufficient to control the intense competition among fishers who seek to maximize catches in the shortest time possible. In 2009, through a consensus of the GMR's Participatory Management Board (PMB), and the Inter-Authority Management (AIM) agency, a new fisheries management system was adopted that included economic incentives for fishers, which slowed down the catch rate (Castrejón 2013). Since then several measures have been adopted for responsible management, for example establishing clear objectives for the mitigation of impacts to coral habitats, regulation of SCUBA diving, and sustainable fishing.

GMR coral habitats support numerous species, including rare and endemic taxa. In addition, these coral communities attract large numbers of transient species, including sharks, tunas, turtles, and dolphins. Unfortunately, coral habitats have been strongly impacted by El Niño events, where coral mortality in 1982–83 amounted to 95–99 % island wide (Glynn et al. 1988), with additional but less coral mortality in 1997–98, about 26 % island wide (Glynn et al. 2001). A significant effort to mitigate this loss was launched and supported by the 2004–2008 Darwin Initiative: 'Galápagos Coral Conservation: Impact Mitigation, Mapping and Monitoring'. The main purpose of this project was to work with the Ecuadorian government in establishing a coral biodiversity baseline, and preserving coral habitats. Results from the four-year initiative are: (1) an increased understanding of the biodiversity and vulnerability of coral communities, (2) the installation of permanent mooring buoys to protect corals from anchor damage, and (3) workshops and specialized training for personnel engaged in monitoring and protecting coral habitats (Dawson et al. 2009).

The Machalilla National Park (hereafter, Machalilla), is the only protected area in continental Ecuador that includes large terrestrial and adjacent marine areas. In addition, it contains numerous cnidarian species' assemblages. The coral reefs of Machalilla cover about 0.25 km^2 (25 ha), and represent the southern-most limit of known reef development on the eastern Pacific mainland. Although the number of endemic species is low, the faunal richness of Machalilla may exceed that of the GMR (Rivera and Martínez 2011). The best developed coral reefs are present on Isla de la Plata (see Chap. 5, Glynn et al., Fig. 5.25), Isla Salango and La Playita. In addition to anomalous oceanographic conditions such as El Niño, the main threats to the reef zones of these islands are trawling, and careless anchoring. Furthermore, both inside the park and along its buffer zone, several human communities depend on the available marine and terrestrial resources. General strategies have been established for in situ biodiversity conservation, including monitoring, enforcement of regulations, environmental education, and well-managed tourism (MAE 2007a, b). Nonetheless, the most serious risk to Machalilla's marine fauna is the irresponsible and uncontrolled exploitation by commercial and artisanal fisheries.

Following is a list of anthozoans proposed for special protection at Machalilla, based on an eco-regional assessment to establish a baseline of marine biodiversity: gorgonians (octocorals, e.g. sea fans), black corals (*Antipathes galapagensis* and *Antipathes panamensis*) and zooxanthellate (reef-building) corals of the genera *Pavona* and *Pocillopora*. The criteria considered in the selection of these taxa were: (1) degree of threat; (2) endemism, i.e. whether (a) a key species, (b) a foundation species, (c) a migratory species, (d) an umbrella species, i.e. a species that indirectly protects others by association, (e) a flagship species, i.e. a high profile species that indirectly benefits others by association; (3) their vulnerability to El Niño events; and (4) their habitat specificity (MAE 2007b).

There have been attempts to design management and conservation strategies of marine biodiversity at Machalilla, but there are legal conflicts between MAE and other public entities (MAE 2007b). As long as these conflicts remain unresolved, MAE cannot achieve an analysis of the fishing activity in the area. Another problem is that uncertainty exists concerning the boundaries of Machalilla; two maps exist with different geographic configurations. Despite this, some strategies were proposed in order to preserve the marine biodiversity of the park. First, commercial and semi-commercial fishing were banned throughout the area, and efforts have been made to ensure that vessels comply with the law to remain at a distance of eight miles from the protected area. Second, measures were implemented to control and monitor the catch of species for sale from all waters of the park, as well as to establish permanent or temporary restrictions on fishing. Finally, construction projects have been restricted on the coastline to avoid adverse effects (MAE 2007a).

The Galera-San Francisco Marine Reserve is located on the north coast of Ecuador within the Chocó region. This is the largest marine reserve on continental Ecuador, comprising 546 km^2 and with a 37-km long coastline. There are a variety of marine habitats in the reserve, including: mangroves, estuaries, rocky reefs, and coral patch reefs. Within this reserve there are abundant gorgonian populations, several species of soft and stony corals, and the largest known black coral (antipatharian) population on the Ecuadorian continental shelf. The southernmost section of the reserve, "El Faro, is known to possess a significant coral formation" (Priscilla Martínez, pers. comm.). Importantly, it is noted that neighboring the reserve are seven local communities that depend mainly on fishing and agriculture for their livelihood; as a consequence, the above-named habitats are threatened from indiscriminate fishing, loss of habitat, deforestation, pollution, and human overpopulation. Some members of these communities, however, have played a key role in declaring the area a marine reserve. They have cooperated with the government and the NAZCA Institute in the development of conservation strategies (Zurita and Luna 2008; NAZCA 2013).

The NAZCA institute has been working for six years with local communities to promote conservation efforts in protected areas. The institute began analyzing different aspects of relevant communities, such as fishing activities and impacts, along with a socio-economic study of resident populations, and options for responsible management. The MAE, Muisne's township, and international institutions—The Nature Conservancy (TNC), The Lighthouse Foundation, The Jeffrey Cook Trust, Fundación Futuro Latino-Americano (FFLA), and ECOLEX—have all participated in conducting these analyses. Currently, the NAZCA institute continues working with the communities, promoting sustainable fishing, providing seafloor bathymetric data, and subtidal monitoring; these efforts are designed to help develop a new management plan to protect the Galera-San Francisco reserve.

El Pelado Marine Reserve (REMAPE), established on August 2012, covers 130 km^2 in area, and consists of rocky reefs with coral patches. This reserve harbors a variety of fishes and invertebrates, some of which are of high value to the local fishing and tourism activities. Several coral species classified by IUCN as of 'least concern' and 'near threatened', such as *Pocillopora damicornis, Pocillopora capitata*, and *Pocillopora eydouxi*, are present in the reserve. Species belonging to Category II of CITES, namely antipatharians and azooxanthellate scleractinian corals (the latter members of the family Dendrophylliidae) are also abundant (REMAPE Management Plan).

In addition to fisheries and tourism, other economic activities near REMAPE include agriculture and shrimp farming. Two of the greatest threats to the conservation of marine resources identified within REMAPE are the continuous use of prohibited fishing gear, as well as the intrusion of artisanal and commercial fishing boats (MAE 2013). Nevertheless, in order to define the uses within the protected area, and to ensure that the products and services are sustainable, a zoning scheme was proposed as a conservation tool where several uses are differentiated across six different zones (MAE 2013).

Similar in size to REMAPE, the Pacoche refuge (Coastal Marine Wildlife Refuge Pacoche) is located about 50 km north of Machalilla. It extends 6.4 km inland, and in the marine realm includes rocky reefs, sand beaches, and seamounts. It is unknown if coral habitats are present in this protected area. However, the southern-most reserve known to support scleractinian and gorgonian habitats is the Coastal-Marine Faunal Production Reserve Puntilla Santa Elena (Samaniego 2010).

A major disadvantage that both the Galera-San Francisco Marine Reserve and Machalilla face is an ambiguous and confusing legal framework. Current laws do not support the conservation and preservation of these protected areas. In addition, the existing laws are neither enforced by the government nor by local authorities (NAZCA 2013). With the exception of the GMR, none of the MPAs on continental Ecuador has specific guidelines to protect coral reefs or coral habitats. Protection at Machalilla is critical as this park contains the most extensive and diverse coral reefs on the continent.

Internationally, several nations have been considering conservation initiatives for mangroves and coral reefs in the Americas. In 2011, during the Ramsar Convention, the NAZCA institute represented Ecuadorian organizations relative to the conservation of coral reefs. One of the positive outcomes of the convention was the collective effort of nations to develop a strategic plan for the conservation, management, and sustainable use of the natural resources of mangroves and coral reefs across the equatorial eastern Pacific seascape (NAZCA 2011).

21.2.8 Chile

The coastal environment of Chile, which is strongly influenced by the Humboldt (Peru) Current, is characterized by temperate waters (Thiel et al. 2007). This restricts coral development to Rapa Nui (Easter Island) and Salas y Gómez Island, two sub-tropical oceanic islands located more than 3000 km west of the Chilean mainland (Fig. 21.8). Both islands are located on the Salas y Gómez Ridge (Gálvez-Larach 2009), are approximately 415 km apart, and belong to the same marine ecoregion. Numerous studies over the years have revealed their extraordinary biodiversity (Rehder 1980; Castilla and Rozbaczylo 1987; Disalvo et al. 1988; Parin et al. 1997; Mironov et al. 2006; Randall and Cea 2011). The global importance of these islands is now recognized, given their high levels of endemism. For example, 77 and 73 % of the biomass of the fish faunas of Rapa Nui

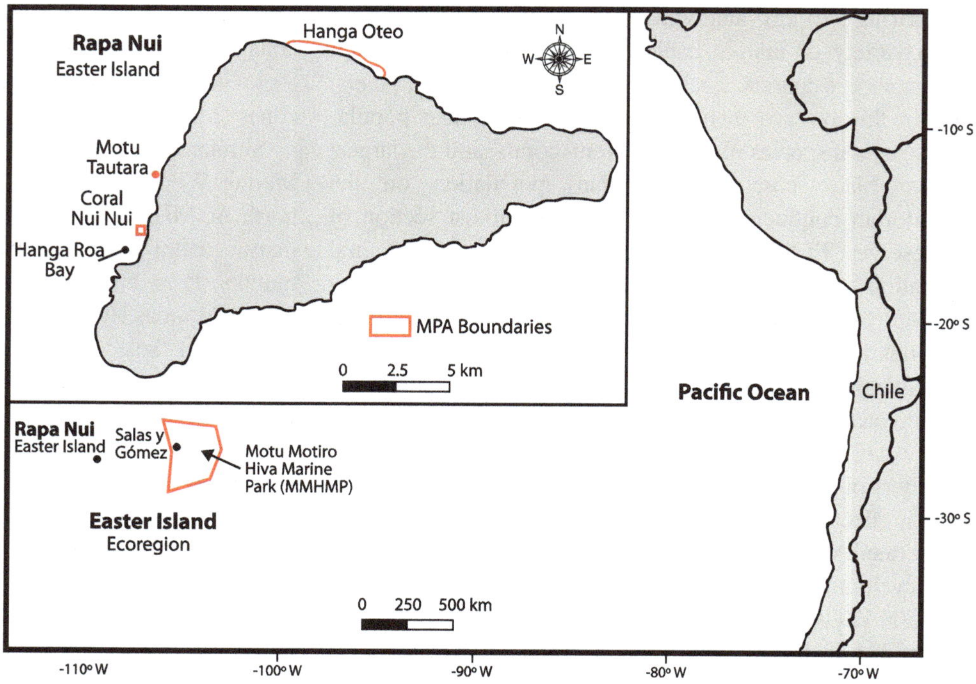

Fig. 21.8 Rapa Nui (Easter Island) and Salas y Gómez Island, Chile, MPA boundaries

and Salas y Gómez, respectively, represent endemic species (Randall and Cea 2011; Friedlander et al. 2013).

Conservation of the Rapa Nui ecoregion has not yet been fully implemented. In 1999, three small areas (Coral Nui Nui, Motu Tautara and Hanga Oteo) were designated as Marine and Coastal Protected Areas at Rapa Nui (Sierralta et al. 2011). Although these MPAs include coral cover, their main features are geological rock arches, caves, and lava platforms, which offer more scenic than biological value. Furthermore, these areas merely exist on paper and lack management plans.

Recent conservation initiatives have focused on developing the first marine reserve at Hanga Roa Bay, based on robust biological evidence, socio-economic considerations, and educational value (Figs. 21.8, 21.9; Gaymer et al. 2011, 2013). This site supports high coral cover. Development of the marine reserve has been delayed due to conflicts between Chilean authorities and the Rapa Nui community; the latter has requested more involvement in the creation of the marine reserve (Gaymer et al. 2014).

In contrast to Rapa Nui, the Chilean government created the Motu Motiro Hiva Marine Park (MMHMP) at Salas y Gómez Island and surroundings, which at 150,000 km^2 is one of the largest no-take MPAs in the world (Toonen et al. 2013; Wilhelm et al. 2014). The MMHMP was created to help protect biodiversity from the current threats of illegal fishing, which could alter the connectivity of marine life between the two islands, and impede potential replenishment of fish and other marine life through larval dispersal and recruitment to the depleted fisheries of Rapa Nui (Friedlander et al. 2013; Gaymer et al. 2014). Robust and extensive coral populations occur around Salas y Gómez Island and Scott Reef, and most likely on shallow unexplored seamounts. Unfortunately, no management plan exists for the MMHMP, resulting in only limited and sporadic enforcement by the Chilean Navy.

Although all coral species of Rapa Nui enjoy international protection by CITES, illegal trading within Chile is a common-practice. For instance, corals are still sold at Rapa Nui and on the Chilean mainland as decorative artifacts, and parts of various handcrafted products. While no coral mass-bleaching/mortality episodes have been reported at Rapa Nui or Salas y Gómez, bleaching of *Porites lobata* and *Pocillopora* spp. associated with heavy rains was reported

Fig. 21.9 Coral community dominated by *Porites lobata* and *Pocillopora* spp., Hanga Roa Bay, Rapa Nui (15 m depth). (courtesy of E. Sorensen, October 2011)

by Cea and DiSalvo (1982), and coincidentally during a period of elevated sea water temperatures (in the southern hemisphere) associated with a La Niña event (Wellington et al. 2001; Hubbard and García 2003). The latter bleaching event occurred when sea surface temperatures exceeded 26 °C. Local coral bleaching was also recently reported by Friedlander et al. (2013). The southern-most distribution of coral habitats at Rapa Nui and Salas y Gómez corresponds to subtropical waters that are several degrees cooler than those at equatorial eastern Pacific localities.

The most immediate threat to the health of Rapa Nui coral habitats is ostensibly the over-exploitation of large predatory fishes, including sharks, and fishing down the food web with the removal of herbivorous fishes (Mumby et al. 2012; Friedlander et al. 2013). It would be profitable to encourage future research to explore the trophic interactions of herbivores, macroalgae, and corals to determine if the decrease of fish herbivores may lead to a phase shift, with macroalgae outcompeting corals for space and eventually dominating benthic communities (Hughes et al. 2003; Friedlander et al. 2013).

Although conservation measures cannot prevent stresses related to climate change, such as coral bleaching, MPAs may increase community resilience by lessening the interactive effects of anthropogenic and climate-related stressors (Hughes et al. 2003). It would be advantageous for conservation research to focus on species' dispersal, coral bleaching, and overfishing, a critical triad of factors that can influence marine community structure and dynamics.

21.3 Regional Approaches

Regional approaches to management, such as networks and seascapes, have been applied recently to promote the consolidation of MPAs, improve management of marine areas by encouraging sustainable human activities, and improving the resilience of vulnerable ecosystems. In the ETP, the seascape approach seeks to achieve the protection and effective management of 2 million km^2 of key ecosystems and species within the core MPAs of Cocos Island (Costa Rica), Coiba Island (Panama), Malpelo and Gorgona Islands (Colombia), and the Galápagos Islands (Ecuador), including also nearshore islands on the continental shelf, and the promotion and application of conservation tools to improve management in the surrounding national waters of the four relevant equatorial ETP nations (Fig. 21.10).

The high species richness of these MPAs will help preserve the best developed coral habitats of the ETP, known for their high connectivity (see Chaps. 5, 16, Glynn et al., and Lessios and Baums respectively). Various coral-associated species, especially apex fish predators, sea turtles and marine mammals, move among these MPAs, using exclusive economic zones (EEZs, from mean low water mark to 200 nmi offshore) and coastal areas as feeding grounds, migratory routes and reproductive sites. In this context, coral communities are essential for providing habitat for the life cycles of diverse species vis-à-vis trophic sustenance, shelter, and aggregation sites.

The success of a regional approach for managing key ecosystems, such as coral communities and reefs, depends on active engagement with MPA managers and more effective consolidation, enforcement of fishing regulations, and the protection of species connectivity between different sites (Edgar et al. 2011). The ETP Seascape should be seen as an important step to reach broad conservation and management objectives that incorporate the initiatives of neighboring countries, and promotes better regional policies. The great ecological value of this seascape approach combined with its economic and cultural importance, is the reason why the ETP is on a short list of key oceanic areas termed "Hope Spots". This nomination should encourage managers and governments to improve and secure protection measures, and to foster appropriate management of the ETP and its critical role as a safeguard of the health of the world's oceans (http://mission-blue.org/hope-spots-new/).

21.4 General Overview

Anthropogenic threats differ from place to place. For instance, trawling and shrimp farming are destructive activities that damage not only Mexican Pacific reefs but also all marine ecosystems (Ezequiel et al. 2011). The following anthropogenic threats were, and still remain, prevalent in the ETP: illegal fishing, entanglement of lost nets, careless anchoring, diving related damage, extraction for handicrafts, agricultural runoff (fertilizers, pesticides), deforestation, pollution (sewage, industrial wastes), human overpopulation, and legal conflicts between public and governmental entities. The conflict between Chilean authorities and the Rapa Nui community regarding development of marine reserves is a clear example of how social conflicts can interfere with conservation goals (Gaymer et al. 2014). Conservation success stories inform us that the involvement of local communities, as well as local authorities, in the planning and creation of MPAs is critical to guarantee that short- and long-term conservation plans will succeed.

Two main strategies have been employed in the conservation of ETP coral habitats: (1) establishment of MPAs, and (2) implementation of legal tools and enforcement efforts. There are presently 45 MPAs extending along the Pacific Coast of tropical America (including oceanic islands), from the upper Gulf of California in the north to northern Peru just below the equator. MPAs with coral habitats cover a total area of 190,693.5 km^2 (Table 21.1). Mexico and Costa Rica

Fig. 21.10 Eastern tropical Pacific seascape connects EEZs of Costa Rica, Panama, Colombia and Ecuador (illustration courtesy of Conservation International)

rank first in terms of the number of MPAs, 11 in each country, and covering 16,891 and 4648 km^2, respectively. Panama has eight MPAs that include coral habitats with a total area of 7187 km^2. Ecuador possesses six MPAs (139,398 km^2), and Colombia and Chile four (10,202 and 150,021 km^2, respectively). El Salvador has one, covering 207 km^2.

According to IUCN MPA classification, five belong to category I (Strict Nature Reserve), 20 to category II (National Park), four to category IV (Habitat/Species Management Area), one to category V (Protected Landscape/Seascape), eight to category VI (Managed Resource Protected Area), and six are not yet classified (Table 21.1). Thus, the majority of MPAs allow some level of economic activities inside their areas, with the no-take zones being only a small fraction of the total protected area.

Established in 1963, the Cabo Blanco Absolute Natural Reserve in Costa Rica is the oldest MPA with coral habitats while the most recent, El Pelado Marine Reserve in Ecuador, was established in 2012 (Table 21.1). The greatest conservation impetus occurred in the 2000s, when 13 MPAs were established. This was followed by the establishment of 11 MPAs in the 1990s, and with fewer than 10 each in the 1970s and 1980s. During the first decade of the Twentieth Century there was mounting pressure to establish MPAs in the ETP with the goal of achieving the 2012 targets of the Convention on Biological Diversity: '*10 % of all marine and coastal ecological regions should be conserved in representative MPAs by 2012*; *by 2012, in the marine area, a global network of comprehensive, representative and effectively-managed national and regional protected area systems is established; and by 2012, all protected areas are effectively and equitably managed, using participatory and science-based site planning processes that incorporate clear biodiversity objectives, targets, management strategies and monitoring and evaluation protocols*' (http://www.cbd.int/sp/targets/rationale/target-11/).

ETP protection is vital, not only for its unique biodiversity and ecological resilience, but also because it harbors numerous coral habitats with many rare and endemic species. For continued protection, both national and international cooperation must occur.

21.5 Future Needs

Declaring MPAs does not in itself guarantee coral habitat protection and recovery. A new philosophical approach is needed to implement MPAs, taking advantage of all available scientific knowledge and the monetary investment that has been generated worldwide. In order to not only protect the biodiversity of these areas, it is also necessary to ensure that MPAs complement effective fisheries management outside their borders (Hilborn et al. 2004). Ideally, a successful MPA management plan should help increase the resilience of ecosystems and vulnerable communities to climate change and anthropogenic threats. We recommend that the managers of established and new MPAs enrich their knowledge of local communities and traditions, land-use, waste disposal, and other related activities. In order to assure the success of an MPA it is necessary to identify (as early as possible) the nature of all relevant stressors. Once identified, it is important to formulate and implement a mitigation plan. One of the reasons why MPAs of the ETP are not reaching their conservation goals is because of their complex and multilevel administration, i.e. a lack of efficiency. When decisions are made and resources are allocated by a central authority, there is better control of the budget, the staff is better informed of the day-to-day management, effective and transparent communication is easier to achieve, and coordinated interactions become simpler. Thus, the multilevel administration of some MPAs should be changed to a centralized administration. Bureaucratic sluggishness could perhaps be minimized by involving NGOs and academic entities in certain activities, such as assessment of MPA effectiveness, and updating management plans.

Research on the effects of sea warming and ocean acidification should continue on ETP coral communities, where disruptive ENSO thermal events also occur (e.g., Manzello et al. 2008, 2014). Further, national and international collaboration and support are essential to limit greenhouse gas emissions, the primary driver of Anthropocene climate change. Currently, coral habitats in the ETP are not sufficiently incorporated into MPAs. Regrettably, the majority of present-day MPAs have relatively small no-take zones. No-take zone areas should be expanded to benefit ecosystem recovery. Finally, it is apparent that ocean law and policy, which are fundamental to conservation, need to be up-graded and further developed in the ETP.

Acknowledgements We thank the editors of this book for inviting us to participate in this project. We also appreciate the support of Ian Parker who reviewed the final draft, Wilfredo Fuentes and Carolina Canales who kindly prepared the map of El Salvador, and Lucas Sandino for providing valuable information on Nicaragua. CFG and JJH were supported by the Chilean Millennium Scientific Initiative, NC120030 Grant. We appreciate the comments made by P.W. Glynn, and two anonymous reviewers who enriched this work. We thank D. Moanga, M. Chavez, R. Romero-Chavez (SIGMAR-CIMAR), and V. Jiménez for their help in improving the figures.

References

Abdulla A, Obura D, Bertzky B, Shi Y (2013) Marine Natural Heritage and the world heritage list: interpretation of World Heritage criteria in marine regions, analysis of biogeographic representation of sites, and a roadmap for addressing gaps. IUCN, Gland, Switzerland, p 68

Aburto-Oropeza O, Erisman B, Galland GR, Mascarenas-Osorio I, Sala E, Ezcurra E (2011) Large recovery of fish biomass in a no-take marine reserve. PLoS ONE 6(8):e23601. doi:10.1371/journal.pone.0023601

AIDA (Asociación Interamericana para la Defensa del Ambiente) (2012) Los arrecifes de coral en Costa Rica: valor económico, amenazas y compromisos legales internacionales que obligan a protegerlos. Asociación Interamericana para la Defensa del Ambiente y Conservación Internacional, San José, Costa Rica

Alvarado JJ, Reyes-Bonilla H, Álvarez del Castillo Cárdenas PA, Buitrago F, Aguirre-Rubí J (2010) Coral reefs of the Pacific coast of Nicaragua. Coral Reefs 29:201

Alvarado JJ, Ayala A, Álvarez del Castillo-Cárdenas PA, Fernández C, Aguirre-Rubí J, Buitrago F, Reyes-Bonilla H (2011a) Coral communities of San Juan del Sur, Pacific Nicaragua. Bull Mar Sci 87:129–146

Alvarado JJ, Herrera B, Corrales L, Asch J, Paaby P (2011b) Identificación de las prioridades de conservación de la biodiversidad marina y costera en Costa Rica. Rev Biol Trop 59:829–842

Alvarado JJ, Cortés J, Esquivel MF, Salas E (2012) Costa Rica's marine protected areas: status and perspectives. Rev Biol Trop 60:129–142

ANAM (2009) Plan de manejo del Parque Nacional Coiba, Autoridad Nacional del Ambiente. In: Maté JL, Tovar D, Arcia E, Hidalgo Y (eds) Autoridad Nacional del Ambiente: Ciudad de Panamá, República de Panamá, p 168

ANAM–PMDSCH (2010) Borrador final del plan de manejo del Parque Nacional Marino Golfo de Chiriquí, Provincia de Chiriquí. Autoridad Nacional del Ambiente – Programa Multifase de Desarrollo Sostenible de Chiriquí. Consultoría: 'Elaboración del Plan de Manejo del Parque Nacional Marino Golfo de Chiriquí y Desarrollo de Actividades Prioritarias'. Realizado por Consultores Ecológicos Panameños, S. A. (CEPSA), de octubre de 2009 a marzo de 2010. Panamá, República de Panamá

Andersson JEC (2007) The recreational cost of coral bleaching—a stated and revealed preference study of international tourists. Ecol Econ 62:704–715

Barraza JE (2011) Áreas marinas protegidas en El Salvador: un arrecife rocoso y un estuario. Revista Parques No. 1. FAO. (www.revistaparques.org)

Barraza JE (2013) Censo submarino de peces del arrecife y área natural protegida 'Complejo Los Cóbanos', El Salvador, Pacífico Oriental Tropical. Libro de Resúmenes. 2° Simposio Latinoamericano de Ictiología. Antigua Guatemala, Guatemala, 5–7 de noviembre 2013

Brander LM, van Beukering P, Cesar HSJ (2007) The recreational value of coral reefs: a meta-analysis. Ecol Econ 63:209–218

Breedy O, Guzmán HM (2003) A new species of *Pacifigorgia* (Coelenterata: Octocorallia: Gorgoniidae) from Panama. Zootaxa 128:1–10

Breedy O, Guzmán HM (2004) A new species of *Pacifigorgia* (Coelenterata: Octocorallia: Gorgoniidae) from Pacific Panamá. Zootaxa 541:1–15

Breedy O, Guzmán HM (2005) A new species of *Leptogorgia* (Coelenterata: Octocorallia: Gorgoniidae) from the shallow waters of the eastern Pacific. Zootaxa 899:1–11

Breedy O, Guzmán HM (2008) *Leptogorgia ignita*, a new shallow-water coral species (Octocorallia: Gorgoniidae) from the tropical eastern Pacific. J Mar Biol Ass UK 88:893–899

Calderón-Aguilera LE, Reyes-Bonilla H, Carriquiry JD (2007) El Papel de los arrecifes coralinos en el flujo de carbón en el océano: estudios en el Pacífico mexicano. In: Hernández de la Torre B, Gaxiola-Castro G (eds) Carbono en ecosistemas acuáticas de México. México, SEMARNAT-INE-CICESE, ISBN: 978-968-817-855-3, pp 215–226

Carpenter KE, Abrar M, Aeby G, Aronson RB, Banks S, Bruckner A, Chiriboga Á, Cortés J, Delbeek JC, DeVantier L, Edgar GJ, Edwards AJ, Fenner D, Guzmán HM, Hoeksema BW, Hodgson G, Johan O, Licuanan WY, Livingstone SR, Lovell ER, Moore JA, Obura DO, Ochavillo D, Polidoro BA, Precht WF, Quibilan MC, Reboton C, Richards ZT, Rogers AD, Sanciangco J, Sheppard A, Sheppard C, Smith J, Stuart S, Turak E, Veron JEN, Wallace C, Weil E, Wood E (2008) One-third of reef-building corals face elevated extinction risk from climate change and local impacts. Science 321:560–563

Castilla JC, Rozbaczylo N (1987) Invertebrados marinos de Isla de Pascua y Salas y Gómez. In: Castilla JC (ed) Islas Oceánicas Chilenas. Conocimiento Científico y necesidades de investigación. Ediciones Universidad Católica de Chile, Santiago, pp 191–215

Castrejón M (2013) La reforma del sistema de licencias PARMA: el primer paso para eliminar la carrera por pescar en la Reserva Marina de Galápagos. Informe Galápagos 2011-2012. DPNG, CGREG, FCD Y GC, pp 136–140

Cea A, DiSalvo LH (1982) Mass expulsion of zooxanthellae by Easter Island corals. Pac Sci 36:61–63

CONABIO, CONANP, TNC, PRONATURA (2007) Análisis de vacíos y omisiones en conservación de la biodiversidad marina de México: océanos, costas e islas. Comisión Nacional para el Conocimiento y Uso de la Biodiversidad, Comisión Nacional de Áreas Naturales Protegidas, The Nature Conservancy-Program México, Pronatura, AC México, DF

Cortés J (1997) Biology and geology of eastern Pacific coral reefs. Coral Reefs 16:S39–S46

Cortés J (2003) Coral reefs of the Americas: an introduction to Latin American coral reefs. In: Cortés J (ed) Latin American coral reefs. Elsevier Science, Amsterdam, pp 1–7

Cortés J, Jiménez CE, Fonseca AC, Alvarado JJ (2010) Status and conservation of coral reefs in Costa Rica. Rev Biol Trop 58(S1):33–49

Cunningham S, Guzmán HM, Bates R (2013) The morphology and structure of the Hannibal Bank fisheries management zone, Pacific Panama using acoustic seabed mapping. Rev Biol Trop 61:1967–1979

Danulat E, Edgar GJ (eds) (2002) Reserva Marina de Galápagos. Línea Base de la Biodiversidad. Fundación Charles Darwin/Servicio Parque Nacional Galápagos, Santa Cruz, Galápagos, Ecuador, p 484

Davidson MG (2002) Protecting coral reefs: the principal national and international legal instruments. Harvard Environ Law Rev 26:499–546

Davidson MG (2006) Legal protections for coral reefs. In: Precht WF (ed) Coral reef restoration handbook. CRC Taylor & Francis group, Boca Raton, pp 143–166

Dawson T, Henderson S, Banks S (2009) Galápagos coral conservation: impact mitigation, mapping and monitoring. In: Guest editorial, pp 3–5

de Groot R, Brander L, van der Ploeg S, Costanza R, Bernard FR, Braat L, Christie M, Crossman N, Ghermandi A, Hein L, Hussain S, Kumar P, McVittie A, Portela R, Rodriguez LC, ten Brink P, van Beukering P (2012) Global estimates of the value of ecosystems and their services in monetary units. Ecosys Ser 1:50–61

Díaz ML (2005) Refugio de vida silvestre Isla Iguana: guía de campo. 2nd edn Imprenta Post

Díaz JM, Gast F (2009) El Chocó biogeográfico de Colombia. Banco de Occidente, Bogotá, Colombia

DiSalvo LH, Randall JE, Cea A (1988) Ecological reconnaissance of the Easter Island sublittoral marine environment. Nat Geogr Res 4:451–473

Douvere F, Herrera B (2014) Mission report: Coiba National Park and its Special Zone for Marine Protection (Panama) (N1138), 12–17 Jan 2014

Edgar GJ, Russ GR, Babcock RC (2007) Marine protected areas. In: Connell SD, Gillanders BM (eds) Marine ecology. Oxford Univ Press, Melbourne, pp 533–555

Edgar GJ, Banks SA, Bessudo S, Cortés J, Guzmán HM, Henderson S, Martínez C, Rivera F, Soler G, Ruiz D, Zapata FA (2011) Variation in reef fish and invertebrate communities with level of protection from fishing across the eastern tropical Pacific seascape. Global Ecol Biogeogr 20:730–743

Escobar J, Neira R (1992) Primer registro de un coral hermatípico (*Pocillopora capitata* Verril, 1864) y su fauna asociada en el área de Bahía de Málaga, Pacífico colombiano. VII Sem Nac Cien Tec Mar y Congr Centroam Car Cien Mar, pp 370–376

Ezequiel E, Aburto-Oropeza O, Carvajal M de los A, Cudeney-Bueno R, Torre J, Nava HT, Ramirez-Herrera M (2011) Government conservation policies on Mexican coastal areas: is 'top-down' management working? Rev Biol Trop 59:1487–1501

Fargier L, Hartmann HJ, Molina-Ureña H (2014) "Marine areas of responsible fishing": a path toward small-scale fisheries co-management in Costa Rica? Perspectives from Golfo Dulce. In: Amezcua F, Bellgraph B (eds) Fisheries management of Mexican and Central American estuaries. Springer, Berlin, pp 155–179

Fonseca A (2013) Arrecifes coralinos y pastos marinos en Costa Rica. Factores de deterioro y acciones de protección. Ambientico 230–231:9–15

Friedlander AM, Ballesteros E, Beets J, Berkenpas E, Gaymer CF, Gorny M, Sala E (2013) Effects of isolation and fishing on the marine ecosystems of Easter and Salas y Gómez Islands, Chile. Aquat Conserv Mar Fresh Ecosyst 23:515–531

Gálvez-Larach M (2009) Montes submarinos de Nazca y Salas y Gómez: una revisión para el manejo y conservación. Lat Am J Aquat Res 37:479–500

Gaymer CF, Cárcamo PF, Friedlander AM, Palma AT, Bodini A, Muñoz A, García M, Sorensen E, Petit I, Zañartu L, Rapu B, Gutierrez C, Hoffens A (2011) Implementación de una reserva marina en la bahía de Hanga Roa: estudio de línea base. Informe final proyecto SUBPESCA N° 4728-11-LE11

Gaymer CF, Tapia C, Acuña E, Aburto J, Cárcamo PF, Bodini A, Stotz W (2013) Base de conocimiento y construcción de capacidades para el uso sustentable de los ecosistemas y recursos marinos de la ecoregión de Isla de Pascua. Informe Final Proyecto SUBPESCA N° 4728-33-LE12

Gaymer CF, Stadel AV, Ban NC, Cárcamo PF, Ierna J, Lieberknecht LM (2014) Merging top-down and bottom-up approaches in marine protected areas planning: experiences from around the globe. Aquat Conserv Mar Freshw Ecosyst 24(S2):128–144

Glynn PW, Ault JS (2000) A biogeographic analysis and review of the far eastern Pacific coral region. Coral Reefs 19:1–23

Glynn PW, de Weerdt WH (1991) Elimination of two reef-building hydrocorals following the 1982–83 El Niño warming event. Science 253:69–71

Glynn PW, Maté JL (1997) Field guide to the Pacific coral reefs of Panama. In: Proceedings of 8th International Coral Reef Symposium, vol 1, Panama, pp 145–166

Glynn PW, Cortés J, Guzmán HM, Richmond RH (1988) El Niño (1982–83) associated coral mortality and relationship to sea surface temperature deviations in the tropical eastern Pacific. In: Proceedings of 6th International Coral Reef Symposium, vol 3, Townsville, pp 237–243

Glynn PW, Maté JL, Baker AC, Calderón MO (2001) Coral bleaching and mortality in Panama and Ecuador during the 1997–1998 El Niño-Southern Oscillation event: spatial/temporal patterns and comparisons with the 1982–1983 event. Bull Mar Sci 69:79–109

Guarderas AP, Hacker SD, Lubchenco J (2008) Current status of marine protected areas in Latin America and the Caribbean. Conserv Biol 22:1630–1640

Guzmán HM, Breedy O (2008) *Leptogorgia christiae* (Octocorallia: Gorgoniidae) a new shallow water gorgonian from Pacific Panama. J Mar Biol Ass UK 88:719–722

Guzmán HM, Robertson DR, Díaz ML (1991) Distribuci6n y abundancia de corales en el arrecife del Refugio de Isla Iguana, Pacífico de Panamá. Rev Biol Trop 39:225–231

Guzmán HM, Guevara CA, Breedy O (2004) Distribution, diversity and conservation of coral reefs and coral communities in the largest marine protected area of Pacific Panama (Coiba Island). Environ Conser 31:111–121

Guzmán HM, Benfield S, Breedy O, Mair JM (2008) Broadening reef protection across the Marine Conservation Corridor of the eastern tropical Pacific: distribution and diversity of reefs in Las Perlas Archipelago, Panama. Environ Conser 35:46–54

Hilborn R, Stokes K, Maguire J-J, Smith T, Botsford LW, Mangel M, Orensanz J, Parma A, Rice J, Bell J, Cochrane KL, Garcia S, Hall SJ, Kirkwood GP, Sainsbury K, Stefansson G, Walters C (2004) When can marine reserves improve fisheries management? Ocean Coast Manag 47:197–205

Hoegh-Guldberg O (2014) Coral reef sustainability through adaptation: glimmer of hope or persistent mirage? Curr Opin Environ Sustain 7:127–133

Hoyt E (2005) Marine protected areas for whales, dolphins and porpoises: a world handbook for cetacean habitat conservation, 2nd edn. Earthscan, London, U.K., p 425+

Hubbard DK, Garcia M (2003) The corals and coral reefs of Easter Island—a preliminary look. In: Loret J, Tanacredi JT (eds) Easter Island: scientific exploration into the world's environmental problems in microcosm. Kluwer Academic/Plenum Pubs, New York, pp 53–77

Hughes TP, Baird AH, Bellwood DR, Card M, Connolly SR, Folke C, Grosberg R, Hoegh-Guldberg O, Jackson JBC, Kleypas J, Lough JM, Marshall P, Nyström M, Palumbi SR, Pandolfi JM, Rosen B, Roughgarden J (2003) Climate change, human impacts, and the resilience of coral reefs. Science 301:929–933

Hurtado M, Hurtado-Domínguez MA, Hurtado-Domínguez LM, Soto L, Merizalde MA (2010) Áreas costeras y marinas protegidas del Ecuador. Ministerio del Ambiente (MAE) – Fundación Natura, p 352

Johannes RE (1998a) Government-supported, village-based management of marine resources in Vanuatu. Ocean Coast Manag 40:165–186

Johannes RE (1998b) The case for data-less marine resource management: examples from tropical nearshore fisheries. Trends Ecol Evol 13:243–246

Lindner A, Cairns SD, Guzmán H (2004) *Distichopora robusta* sp. nov., the first shallow-water stylasterid (Cnidaria: Hydrozoa: Stylasteridae) from the tropical eastern Pacific. J Mar Biol Assoc UK 84:943–947

López-Pérez RA, López-García A (2008) Identificación de sitios prioritarios para la conservación de corales formadores de arrecife en el estado de Oaxaca, México. Hidrobiol 18:239–249

Lubchenco J, Palumbi SR, Gaines SD, Andelman S (2003) Plugging a hole in the ocean: the emerging science of marine reserves. Ecol Appl 13:S3–S7

Luna S, Banks S, Koch V, Ruiz D, Tirado N, Vera M, Schuhbauer A, Keith I, Acuña D, Suárez J, Parra M, Jiménez G, García C, Baque J, Delgado J (2013) Especies, comunidades y ecosistemas: el rol de la ciencia para el manejo y la conservación de la Reserva Marina de Galápagos. Informe Galápagos 2011–2012. DPNG, CGREG, FCD Y GC, pp 131–135

MAE (Ministerio del Ambiente del Ecuador) (2007a) Plan gerencial del Parque Nacional Machalilla, 2008–2010. Proyecto GEF: Ecuador Sistema Nacional de Áreas Protegidas (SNAP-GEF), Quito, Ecuador

MAE (Ministerio del Ambiente del Ecuador) (2007b) Actualización del Diagnóstico del Parque Nacional Machalilla: elementos prioritarios de la diversidad biológica y cultural. Proyecto GEF: Ecuador Sistema Nacional de Áreas Protegidas (SNAP-GEF), Quito, Ecuador

MAE (Ministerio del Ambiente del Ecuador) (2013) Plan de manejo de la Reserva Marina El Pelado. Draft document

MAE (Ministerio del Ambiente del Ecuador) (2014) Guía informativa de las áreas naturales protegidas del Ecuador, Quito, Ecuador, p 266

Manzello DP (2010a) Coral growth with thermal stress and ocean acidification: lessons from the eastern tropical Pacific. Coral Reefs 29:749–758

Manzello DP (2010b) Ocean acidification hot spots: spatiotemporal dynamics of the seawater CO_2 system of eastern Pacific coral reefs. Limnol Oceanogr 55:239–248

Manzello DP, Kleypas JA, Budd DA, Eakin CM, Glynn PW, Langdon C (2008) Poorly cemented coral reefs of the eastern tropical Pacific: possible insights into reef development in a high-CO_2 world. Proc Natl Acad Sci USA 105:10450–10455

Manzello DP, Enochs IC, Bruckner A, Renaud PG, Kolodziej G, Budd DA, Carlton R, Glynn PW (2014) Galápagos coral reef persistence after ENSO warming across an acidification gradient. Geophys Res Lett 10.1002/2014GL062501

Marine Protected Areas: tools for sustaining ocean ecosystems (2001) Committee on the evaluation, design, and monitoring of marine reserves and protected areas in the United States. National Academy Press, Wash DC, p 272

Maté TJL (2003) Corals and coral reefs of the Pacific coast of Panamá. In: Cortés J (ed) Latin American coral reefs. Elsevier Press, Amsterdam, pp 387–417

Meyer A, Huete-Pérez JA (2014) Nicaragua canal could wreak environmental ruin. Nature 506:287–289

Mironov AN, Molodtsova TN, Parin NV (2006) Soviet and Russian studies on seamount biology. http://www.isa.org.jm/en/scientific/workshops/2006/Mar06

Moberg F, Folke C (1999) Ecological goods and services of coral reef ecosystems. Ecol Econ 29:215–233

Mora C, Andréfouët S, Costello MJ, Kranenburg C, Rollo A, Veron J, Gaston KJ, Myers RA (2006) Coral reefs and the global network of marine protected areas. Science 312:1750–1751

Mumby PJ, Steneck RS, Edwards AJ, Ferrari R, Coleman R, Harborne AR, Gibson JP (2012) Fishing down a Caribbean food web relaxes trophic cascades. Mar Ecol Prog Ser 445:13–24

Muñoz CG, Zapata FA (2012) Plan de manejo de los arrecifes coralinos del Parque Nacional Natural Gorgona, Pacífico colombiano. Parques Nacionales Naturales y WWF-Colombia. Cali, Colombia

Narváez K, Zapata FA (2010) First record and impact of the crown-of-thorns starfish, *Acanthaster planci* (Spinulosida:

Acanthasteridae) on corals of Malpelo Island, Colombian Pacific. Rev Biol Trop 58:139–144

Nava H, Ramírez-Herrera MT (2011) Government conservation policies on Mexican coastal areas: is 'top-down' management working? Rev Biol Trop 59:1487–1501

NAZCA (Instituto de Investigaciones Marinas) (2011) Noticias. Reunión de la iniciativa regional para la conservación y uso racional de manglares y arrecifes de coral en las Américas. Quito, Ecuador

NAZCA (Instituto de Investigaciones Marinas) (2013) Marine protected areas. Galera-San Francisco. Quito, Ecuador

Parin NV, Mironov AN, Nesis KN (1997) Biology of the Nazca and Salas y Gómez submarine ridges, an outpost of the Indo-West Pacific fauna in the eastern Pacific Ocean: composition and distribution of the fauna, its communities and history. Adv Mar Biol 32:145–242

Pendleton LH (1995) Valuing coral reef protection. Ocean Coast Manag 26:119–131

Pomeroy RS, Rivera-Guieb R (2006) Fishery comanagement: a practical handbook. CAB International, Oxfordshire. Available at the International Development Research Centre (IDRC) (Downloaded: 14 Oct 2009, www.idrc.ca/openebooks/184-1/)

Ramírez-Herrera MT, Novella R, Bassols-Barrera N (2010) Conservación del paisaje y geositios de la costa Norte de Michoacán, México. Rev Geogr Norte Gd 46:105–121

Randall JE, Cea A (2011) Shore fishes of Easter Island. Univ Hawaii Press, Honolulu, p 164

Reaka-Kudla ML (1997) The global biodiversity of coral reefs. In: Reaka-Kudla ML, Wilson DE, Wilson EO (eds) Biodiversity II: understanding and protecting our biological resources. Joseph Henry Press, Washington, DC, pp 83–108

Registro Official No. 202 (2012) Marcela Aguiñaga Vallejo (Ministra de Ambiente). Área Nacional de Recreación Playas Villamil. Acuerdo 163

Rehder HA (1980) The marine mollusks of Easter Island (Isla de Pascua) and Salas y Gómez. Smithson Contrib Zool 289:1–176

Reyes-Bonilla H, Barraza JE (2003) Corals and associated marine communities from El Salvador. In: Cortés J (ed) Coral reefs of Latin America. Elsevier, Amsterdam, pp 351–360

Reyes-Bonilla H, Alvarez del Castillo Cárdenas PA, Calderón Aguilera LE, Erosa Ricárdez CE, Fernández Rivera Melo FJ, Frausto TC, Luna Salguero BM, Moreno Sánchez XG, Mozqueda Torres MC, Norzagaray López CO, Petatán Ramírez D (2014) Servicios ambientales de arrecifes coralinos: el caso del Parque Nacional Cabo Pulmo, B.C.S. In: Urciaga García JI (ed) Desarrollo regional en Baja California Sur. Una perspectiva de los servicios ecosistémicos. UABCS, La Paz, pp 38–63

Rife AN, Erisman B, Sanchez A, Aburto-Oropeza O (2012) When good intentions are not enough… insights on networks of 'paper park' marine protected areas. Conserv Lett 6:200–212

Rife AN, Aburto-Oropeza O, Hastings PA, Erisman B, Ballantyne F, Wielgus J, Sala E, Gerber L (2013) Long-term effectiveness of a multi-use marine protected area on reef fish assemblages and fisheries landings. J Environ Manag 117:276–283

Rivera M (2011) Evaluación de las áreas marinas protegidas de México. Tesis de Doctorado, Centro Interdisciplinario de Ciencias Marina – IPN, La Paz, México

Rivera F, Martínez P (2011) Guía fotográfica de corales y octocorales: Parque Nacional Machalilla y Reserva de Producción Faunística Marino Costera Puntilla de Santa Elena. With the contribution of Odalisca Breedy, NAZCA and Conservación Internacional, Ecuador, p 78

Romero J (2009) Geografía económica de Colombia. Documentos de trabajo sobre economía regional 116

Salas E, Ross-Salazar E, Arias A (eds) (2012) Diagnóstico de áreas marinas protegidas y áreas marinas para la pesca responsable en el Pacífico costarricense. Fundación MarViva. San José, Costa Rica

Samaniego J (2010) ¿Para qué sirven las áreas marinas protegidas? Rev Ecuador Terra Incognita 67:20–26

Sierralta L, Serrano R, Rovira J, Cortés C (2011) Las áreas protegidas de Chile. Antecedentes, Institucionalidad, Estadísticas y Desafíos. Ministerio del Medio Ambiente, Chile, Santiago

Spalding MD, Ravilious C, Green EP (2001) World atlas of coral reefs. Univ California Press, Los Angeles, p 424

Spalding MD, Fox HE, Allen GR, Davidson N, Ferdaña ZA, Finlayson M, Halpern BS, Jorge MA, Lombana A, Lourie SA, Martin KD, McManus E, Molnar J, Recchia CA, Robertson J (2007) Marine ecoregions of the world: a bioregionalization of coastal and shelf areas. BioScience 57:573–583

Spurgeon JP (1992) The economic valuation of coral reefs. Mar Pollut Bull 24:529–536

Stella JS, Pratchett MS, Hutchings PA, Jones GP (2011) Coral-associated invertebrates: diversity, ecological importance and vulnerability to disturbance. Oceanogr Mar Biol Annu Rev 49:43–104

Thiel M, Macaya E, Acuña E, Arntz E, Bastías H, Brokordt K, Camus P, Castilla JC, Castro LR, Cortéz M, Dumont CP, Escribano R, Frenández M, Lancellotti DA, Gajardo JA, Gaymer CF, Gómez I, González AE, González HE, Haye PA, Illanes JE, Iriarte JL, Luna-Jorquera G, Luxoro C, Manríquez PH, Marín V, Muñoz P, Navarrete SA, Pérez E, Poulin E, Sellanes J, Sepúlveda A, Stotz W, Tala F, Thomas A, Vargas CA, Vásquez JA, Vega A (2007) The Humboldt current system of northern-central Chile: oceanographic processes, ecological interactions and socio-economic feedback. Oceanogr Mar Biol Annu Rev 45:195–344

Toonen RJ, Wilhelm TA, Maxwell SM, Wagner D, Bowen BW, Sheppard CRC, Taei SM, Teroroko T, Moffitt R, Gaymer CF, Morgan L, Lewis N, Sheppard ALS, Parks J, Friedlander AM, The Big Ocean Think Tank (2013) One size does not fit all: the emerging frontier in large-scale marine conservation. Mar Pollut Bull 77:7–10

UAESPNN (2004) Parque Nacional Natural Gorgona, plan de manejo versión institucional. Bogotá, Colombia

UAESPNN (2005) Plan de Manejo Parque Nacional Natural Gorgona. Resumen ejecutivo. Cali (Colombia): Unidad Administrativa Especial del Sistema de Parques Nacionales Naturales de Colombia Dirección Territorial Suroccidental. Bogotá, Colombia

Vargas-Ángel B, Zapata FA, Hernández H, Jiménez JM (2001) Coral and coral reef responses to the 1997–98 El Niño event on the Pacific coast of Colombia. Bull Mar Sci 69:111–132

Vega AJ, Robles YA, Cipriani R (2011). Estudios biológico pesqueros en el Golfo de Chiriquí, Pacífico de Panamá. Informe Inédito de la Universidad de Panamá, Chiriquí, República de Panamá

Villarta M (2013) Evaluación de la efectividad del manejo de la Reserva Marina de Galápagos: principales conclusiones y recomendaciones. Informe Galápagos 2011–2012. DPNG, CGREG, FCD Y GC, pp 123–130

Wellington GM, Glynn PW, Strong AE, Navarette SA, Wieters E, Hubbard D (2001) Crisis on coral reefs linked to climate change. EOS 82:1–5

White A, Hale LZ, Renard Y, Cortesi L (1994) Collaborative and community management of coral reefs: lessons from experience. Kumarian, Bloomfield, Connecticut, USA

White AT, Vogt HP, Arin T (2000) Philippine coral reefs under threat: the economic losses caused by reef destruction. Mar Poll Bull 40:598–605

Wilhelm TA, Sheppard CR, Sheppard AL, Gaymer CF, Parks J, Wagner D, Lewis N (2014) Large marine protected areas—advantages and challenges of going big. Aquat Conserv Mar Freshw Ecosyst 24(S2):24–30

Wilkinson C (2006) Status of coral reefs of the world: summary of threats and remedial actions. In: Côté IM, Reynods JD (eds) Coral reef conservation. Univ Press, Cambridge, pp 3–39

Zedler JB, Kercher S (2005) Wetland resources: status, trends, ecosystem services, and restorability. Annu Rev Env Resour 30:39–74

Zurita P, Luna, S (2008) Case Study 2: private incentives to conserve Ecuador's coast. Instituto Nazca de Investigaciones Marinas, EcoCiencia, Ministerio del Ambiente, The Nature Conservancy, Conservación Internacional, Proyecto GEF: Ecuador Sistema Nacional de Áreas Protegidas, BirdLife Internacional, Aves & Conservación, 7 pp. http://institutonazca.org/publications/

Field Guide to Select Eastern Pacific Corals and Associated Coral Reef Biota

22

Juan L. Maté, Margarita Brandt, Benjamin Grassian and Ángel Chiriboga

Abstract

Eastern tropical Pacific (ETP) coral habitats host functionally complex biological communities. Corals and their associated biota, from charismatic reef fishes to cryptic invertebrate infauna, drive and maintain diverse ecosystem functioning in reef environments. This chapter provides a photographic guide to key and otherwise notable species present in ETP coral habitats. The objectives of this inclusion are to both facilitate field identification and to make accessible key information (i.e. habitat, relative abundance, ecological role, geographic and depth distributions, and pertinent references) available for these coral-associated species. Represented in this guide are 11 species of algae, 102 invertebrates of which 53 are cnidarians, and 28 fishes, with bias towards comprehensive inclusion of cnidarians (corals) and fishes. Effort was made to include cryptic species easily overlooked, warranting select ex situ photographs, as well as to visually describe polymorphism and behavior encountered in the field.

Keywords

Habitat • Abundance • Distribution • Ecological role • Species-specific references

J.L. Maté (✉)
Smithsonian Tropical Research Institute,
Apartado Postal 0843-03092, Balboa, Ancon, Balboa,
Republic of Panama
e-mail: matej@si.edu

M. Brandt
Colegio de Ciencias Biológicas y Ambientales,
Universidad San Francisco de Quito,
Diego de Robles s/n y Pampite,
Cumbayá, Quito, Ecuador
e-mail: mbrandt@usfq.edu.ec

B. Grassian
Graduate School of Oceanography, University of Rhode Island,
215 S Ferry Rd, Narragansett, RI 02882, USA
e-mail: bgrassian@gmail.com

Á. Chiriboga
PIXEL, De los Eucaliptos y De los Pinos, Cumbayá, Quito,
Ecuador
e-mail: chiribogaai@gmail.com

P.W. Glynn et al. (eds.), *Coral Reefs of the Eastern Tropical Pacific*,
Coral Reefs of the World 8, DOI 10.1007/978-94-017-7499-4_22

22.1 Introduction

This pictorial guide is designed to ease the field identification of some key or otherwise notable species present in eastern Pacific coral habitats with emphasis on species noted in other chapters of this book. Here included are 11 species of algae, 102 invertebrates, and 28 fishes. The photographs are accompanied with information on habitat, relative abundance, ecological role, geographic and depth distributions, as well as key references. While the majority of the images are of relatively large, readily observed reef organisms under mostly natural settings, some species are shown against a neutral (black or white) background. Many of these smaller species, less than 1.5 cm in greatest length, are members of the cryptic reef fauna, which shelters in coral rubble, fine sediments and limestone cavities, and are rarely seen. Most of these species were removed from their usual concealed habitats and photographed through a microscope. Additionally, all are associated with reef-building corals, and most (excluding obligate coral symbionts), occur in other non-coral habitats such as rock, sand, and algal substrates.

All species included here occur in at least some part of the ETP region. Endemic refers to species present exclusively in the ETP, and particular localities where they occur are noted. This selection of images is biased toward cnidarians and fishes. Other coral-associated invertebrate taxa can be found in Brusca (1980), Kerstitch and Bertsch (2007), Hickman (1998a, b, 2008), Behrens and Hermosillo (2005), and the sources listed in the reference section of this chapter.

For a small number of species two photographs are provided to better illustrate cryptic fauna, polymorphism, or behavior. Information snippets are duplicated in these instances, apart from photo details and photograph-specific descriptions. Credit is noted for those who photographed and allowed reproduction of the images. We are very grateful for their generous cooperation in offering images and information for this guide.

22.2 Field Guide

PHYLUM PHAEOPHYTA

Sargassaceae: ***Sargassum obtusifolium*** J. Agardh

H	Rocky platforms, coral communities, moderate to strong circulation
A	Common
ER	Primary producer, spatial competitor
GD	Galápagos Iss, Ecuador; Easter Is (=Rapa Nui, Chile). Central Pacific
DD	0–10 m
KR	Garske (2002), Guiry (2014a, b)
PD	Alejandro Pérez-Matus. Easter Is (=Rapa Nui, Chile). 10 m depth. 26 April 2014

Dictyotaceae: ***Zonaria stipitata*** Tanaka and K. Nozawa

H	Rocky platforms, dead coral substrates
A	Common
ER	Primary producer, spatial competitor
GD	Easter Is (=Rapa Nui, Chile). Indo-Pacific
DD	2–9 m
KR	Santelices and Abbott (1987)
PD	Alejandro Pérez-Matus. Easter Is (=Rapa Nui, Chile). 9 m depth. 26 April 2014

(continued)

Dictyotaceae: ***Padina* sp.**	
H	Rock, dead coral or rubble substrates
A	Common
ER	Primary producer, spatial competitor
GD	Mexico to Ecuador; Revillagigedo Iss, Mexico; Galápagos Iss, Ecuador
DD	Intertidal to 5 m
KR	Guiry in Guiry and Guiry (2015)
PD	Juan Maté. Contadora Is, Panama. 4 m depth. 1 May 2015

PHYLUM CHLOROPHYTA	
Caulerpaceae: ***Caulerpa sertularioides*** (S.G. Gmelin)	
H	Coastal and estuarine environments, sandy areas and seagrass beds, on mangrove prop roots and in coral habitats
A	Common, sporadic occurrence
ER	Primary producer, spatial competitor
GD	Mexico to Panama. Indo-Pacific. Caribbean Sea
DD	0–12 m
KR	Sweat (2009), Smith et al. (2010)
PD	Tyler Smith. Secas Is, Panama. 4 m depth. 14 March 2005

H Habitat; *A* Relative Abundance; *ER* Ecological Role; *GD* Geographic Distribution; *DD* Depth Distribution; *KR* Key References; *PD* Photo Details

Caulerpaceae: ***Caulerpa chemnitzia*** (Esper) J.V. Lamouroux	
H	Coral reefs
A	Uncommon
ER	Primary producer, spatial competitor
GD	Revillagigedo Iss, Mexico; Panama; Galápagos Iss, Ecuador. Indo-Pacific. Atlantic
DD	3–5 m
KR	Guiry in Guiry and Guiry (2015)
PD	Juan Maté. Contadora Is, Panama. 3.5 m depth. 2 May 2015

(continued)

Halimedaceae: ***Halimeda discoidea*** Decaisne	
H	Rocky areas, coral rubble, and loosely consolidated substrates
A	Uncommon
ER	Primary producer
GD	Gulf of California, Mexico to Panama. Indo-Pacific. Atlantic
DD	1–50 m
KR	Hillis-Colinvaux (1980), Pedroche et al. (2005), Norris (2010)
PD	Brian Wysor. Isla Canal de Afuera, Panama. 20–25 m depth. 8 February 2011

Corallinaceae: ***Porolithon castellum*** E.Y. Dawson	
H	Coral reefs and rocky outcrops
A	Common
ER	Primary producer, contributes to reef building
GD	Endemic: Costa Rica
DD	4 m
KR	Dawson (1960), Fernández-García et al. (2011)
PD	Cindy Fernández-García. Caño Is, Costa Rica. 4 m depth. 19 July 2009

Corallinaceae: ***Amphiroa misakiensis*** Yendo	
H	Coral reefs and rocky outcrops
A	Common
ER	Primary producer
GD	Baja California, Mexico to Peru. Widespread from Pacific to Atlantic
DD	1–20 m
KR	Norris and Johansen (1981), Fernández-García et al. (2011)
PD	Cindy Fernández-García. Puerto Ángel, Mexico. 1–3 m depth. 13 February 2009

H Habitat; *A* Relative Abundance; *ER* Ecological Role; *GD* Geographic Distribution; *DD* Depth Distribution; *KR* Key References; *PD* Photo Details

Hapalidiaceae: ***Lithothamnion muelleri*** Lenormand ex Rosanoff

H	Rubble beds
A	Common
ER	Primary producer, forms rhodolith beds
GD	Costa Rica. Indo-Pacific. Antarctic. Atlantic
DD	Intertidal to 85 m
KR	Guiry in Guiry and Guiry (2015)
PD	Cindy Fernández-García. Culebra Bay, Costa Rica. 5–6 m depth. 1 August 2009

Rhodomelaceae: ***Acanthophora spicifera*** (M.Vahl) Børgesen

Coated with epiphytic cyanobacteria

H	Reef flats, tidepools, rocky intertidal benches. Attaches to rocks, and dead coral skeletons
A	Rare to abundant
ER	Primary producer, competitor for space, epiphytic on other algae
GD	El Salvador to Panama. Indo-Pacific, Hawaii. Caribbean. Florida
DD	1–22 m
KR	Fernández-García et al. (2011), Guiry in Guiry and Guiry (2015)
PD	Tyler Smith. Uva Is, Panama. 3 m deph. 9 March 2003

Peyssonneliaceae: ***Peyssonnelia boergesenii*** Weber-van Bosse C.

H	Rock and coral substrates
A	Uncommon to abundant
ER	Primary producer, occasional spatial competitor that overgrows live corals
GD	Galápagos Iss, Ecuador. Indo-Pacific. Caribbean
DD	2 to at least 15 m
KR	Denizot (1968), Ballantine and Ruiz (2005), Glynn et al. (2009)
PD	Bernhard Riegl. Wellington Reef (Darwin Is), Galápagos Iss, Ecuador. 12 m depth. 6 March 2007

PHYLUM PORIFERA

Aplysinidae: ***Aplysina clathrata*** Cruz-Barraza, Carballo, Rocha-Olivares, Ehrlich and Hog

H	Rocky outcrops, crustose coralline algal areas
A	Common
ER	Filter-feeder, spatial competitor
GD	Endemic: Mexico and Panama
DD	5–20 m
KR	Cruz-Barraza et al. (2012)
PD	José Luis Carballo. Isabel Is (Nayarit), Mexico. 18 m depth. 29 September 2009

H Habitat; *A* Relative Abundance; *ER* Ecological Role; *GD* Geographic Distribution; *DD* Depth Distribution; *KR* Key References; *PD* Photo Details

Aplysinidae: ***Aplysina chiriquensis*** Díaz, van Soest, Rützler and Guzman	
H	Coral communities on rocky outcrops
A	Common
ER	Filter-feeder, spatial competitor
GD	Endemic: Costa Rica; Panama; Colombia; Ecuador (mainland and Galápagos Iss)
DD	1.5–35 m
KR	Díaz et al. (2005), van Soest (2015a)
PD	Juan Maté. Jicarón Is, Panama. 15 m depth. 21 July 2013

Aplysinidae: ***Aplysina gerardogreeni*** Gómez and Bakus	
H	Coral communities, rocky outcrops
A	Common
ER	Filter-feeder, spatial competitor
GD	Endemic: Mexico; Panama
DD	6–33 m
KR	Gómez and Bakus (1992), Caballero-George et al. (2010), van Soest (2015b)
PD	José Luis Carballo. Los Arcos, Bahía de Banderas, Mexico. 8 m depth. 10 June 2005

Aplysinidae: ***Aplysina revillagigedi*** Cruz-Barraza, Carballo, Rocha-Olivares, Ehrlich and Hog	
H	Clear waters, on vertical walls and rocky substrates, high water movement
A	Common
ER	Filter-feeder, spatial competitor
GD	Endemic: known only from Revillagigedo Iss, Mexico
DD	4–38 m
KR	Cruz-Barraza et al. (2012)
PD	José Luis Carballo. Socorro Is, Revillagigedo Iss, Mexico 9 m depth. 8 November 2009

Chalinidae: ***Haliclona (Soestella) caerulea*** (Hechtel)	
H	Associated with red alga *Jania adherens*. Overgrowing pocilloporid corals, dead coralline substrates, mollusc shells and bases of *Pocillopora*
A	Common
ER	Filter-feeder, spatial competitor
GD	Endemic: Mexico; Panama
DD	Intertidal to 6 m
KR	Wulff (1996), Ávila and Carballo (2004)
PD	José Luis Carballo. San Agustín, Mexico. 4 m depth. 14 April 2005

H Habitat; *A* Relative Abundance; *ER* Ecological Role; *GD* Geographic Distribution; *DD* Depth Distribution; *KR* Key References; *PD* Photo Details

Halichondriidae: ***Axinyssa isabela*** Carballo and Cruz-Barraza	
H	Coral communities, rocky outcrops
A	Common at single site where present
ER	Filter-feeder, spatial competitor
GD	Endemic: Mexico
DD	4–20 m
KR	Carballo and Cruz-Barraza (2008)
PD	José Luis Carballo. Isabel Is, Mexico. 15 m depth. 2 September 2009

Niphatidae: ***Amphimedon texotli*** Cruz-Barraza and Carballo	
H	Among pocilloporid corals
A	Common at single site where present
ER	Filter-feeder, spatial competitor
GD	Endemic: Mexico
DD	Intertidal to 15 m
KR	Cruz-Barraza and Carballo (2008)
PD	José Luis Carballo. La Entrega, Mexico. 5 m depth. 14 April 2005

PHYLUM CNIDARIA	
Pocilloporidae: ***Pocillopora capitata*** Verrill	
H	Coral reefs and coral communities
A	Rare to common
ER	Reef-building species
GD	Widespread from Mexico to Ecuador, including Galápagos Iss. Indo-Pacific
DD	Low mean tidal level to 20 m
KR	Cortés and Guzmán (1998), Glynn and Ault (2000)
PD	Ángel Chiriboga. Floreana Is, Galápagos Iss, Ecuador. 10 m depth. 18 April 2015

Pocilloporidae: ***Pocillopora damicornis*** (Linnaeus)	
H	Coral reefs and coral communities
A	Abundant
ER	Prominent reef-building species
GD	Widespread from Mexico to Easter Is (=Rapa Nui, Chile). Indo-Pacific
DD	Low mean tidal level to 60 m. Exposed on reef flats during spring low tides
KR	Glynn and Ault (2000)
PD	Juan Maté. Uva Is, Panama. 3 m depth. 19 September 2003

H Habitat; *A* Relative Abundance; *ER* Ecological Role; *GD* Geographic Distribution; *DD* Depth Distribution; *KR* Key References; *PD* Photo Details

Pocilloporidae: ***Pocillopora damicornis*** (Linnaeus) Close-up showing exposed polyps and absence of verrucae	
H	Coral reefs and coral communities
A	Abundant
ER	Prominent reef-building species
GD	Widespread from Mexico to Easter Is (=Rapa Nui, Chile). Indo-Pacific
DD	Low mean tidal level to 60 m, exposed on reef flats during spring tides
KR	Glynn and Ault (2000)
PD	Ángel Chiriboga. Santiago Is, Galápagos Iss, Ecuador. 9 m depth. 2 August 2003

Pocilloporidae: ***Pocillopora effusus*** Veron	
H	Coral reefs and coral communities with strong wave action
A	Rare to common
ER	Coral reef and coral community component
GD	Endemic: Mexico, Clipperton Is, Galápagos Iss, Ecuador; El Salvador; unconfirmed report from Panama
DD	Subtidal to 15 m
KR	Glynn and Ault (2000), Veron (2000, 2002), Hoeksema et al. (2008)
PD	Ángel Chiriboga. Wolf Is, Galápagos Iss, Ecuador. 15 m depth. 13 October 2004

Pocilloporidae: ***Pocillopora elegans*** Dana	
H	Coral reefs and coral communities
A	Rare to abundant
ER	Prominent reef-building species
GD	Widespread from Mexico to Ecuador. Indo-Pacific
DD	Low mean tidal level to 20 m, exposed on reef flats during spring low tides
KR	Wells (1983), Glynn and Ault (2000), Hoeksema et al. (2014a)
PD	Juan Maté. Uva Is, Panama. 4 m depth. 17 March 2005

Pocilloporidae: ***Pocillopora eydouxi*** Milne Edwards and Haime	
H	Coral reefs and coral communities
A	Rare to common
ER	A reef-building species
GD	Widespread from Mexico to Easter Is (=Rapa Nui, Chile). Indo-Pacific
DD	2–20 m
KR	Cortés and Guzmán (1998), Glynn and Ault (2000), Glynn et al. (2007), Hickman (2008), Hoeksema et al. (2014b)
PD	Jorge Cortés. Cocos Is, Costa Rica. 6 m depth. 18 March 2012

H Habitat; *A* Relative Abundance; *ER* Ecological Role; *GD* Geographic Distribution; *DD* Depth Distribution; *KR* Key References; *PD* Photo Details

Pocilloporidae: ***Pocillopora inflata*** Glynn	
H	Coral reefs and coral communities
A	Rare to uncommon
ER	Coral community component
GD	Widespread from Mexico to Ecuador. Central Pacific
DD	2–10 m
KR	Glynn (1999), Glynn and Ault (2000), Obura and Stone (2003), Chiriboga et al. (2008a, b), Hickman (2008)
PD	Ángel Chiriboga. Wolf Is, Galápagos Iss, Ecuador. 11 m depth. 13 November 2003

Pocilloporidae: ***Pocillopora meandrina*** Dana	
H	Coral reefs and coral communities
A	Rare to abundant
ER	Coral reef and coral community component
GD	Widespread from Mexico to Easter Is (=Rapa Nui, Chile); El Salvador; Clipperton Is. Indo-Pacific
DD	1–10 m
KR	Cortés and Guzmán (1998), Glynn and Ault (2000); Maté (2003)
PD	Juan Maté. Canal de Afuera Is, Panama. 4 m depth. 21 July 2003

Pocilloporidae: ***Pocillopora verrucosa*** (Ellis and Solander)	
H	Coral reefs and coral communities
A	Rare to abundant
ER	Prominent reef-building species
GD	Widespread from Mexico to Easter Is (=Rapa Nui, Chile); Galápagos Iss, Ecuador
DD	1–60 m
KR	Glynn and Ault (2000), possibly synonymous with *P. elegans*, see Reyes-Bonilla et al. (2002), Hickman (2008)
PD	Ángel Chiriboga. Wolf Is, Galápagos Iss, Ecuador. 11 m depth. 13 November 2003

Poritidae: ***Porites arnaudi*** Reyes-Bonilla and Carricart-Ganivet	
H	Coral reefs and coral communities
A	Uncommon to locally abundant
ER	Reef-building species
GD	Revillagigedo Iss, Mexico; Clipperton Is. Possibly south-central Pacific
DD	7–37 m
KR	Bonilla and Carricart-Ganivet (2000), Veron (2000)
PD	Juan Maté. Clipperton Is. 20 m depth. 28 April 2008

H Habitat; *A* Relative Abundance; *ER* Ecological Role; *GD* Geographic Distribution; *DD* Depth Distribution; *KR* Key References; *PD* Photo Details

Poritidae: ***Porites evermanni*** Vaughan	
H	Coral reefs
A	Uncommon to common
ER	Reef-building species
GD	Costa Rica; Mexico; Galápagos Iss, Ecuador. Indo-Pacific
DD	2–12 m
KR	Veron (2000), Boulay et al. (2014)
PD	Jaime Nivia Ruiz. Cocos Is, Costa Rica. 7 m depth. 11 October 2012

Poritidae: ***Porites lobata*** Dana	
Mounding colony morphology, common at shallow depths	
H	Coral reefs and coral communities
A	Abundant
ER	Prominent reef-building species
GD	Widespread from Mexico to Easter Is (=Rapa Nui, Chile); Marchena Is, Galápagos Iss, Ecuador. Indo-Pacific
DD	Low mean tidal level to 60 m. Exposed on reef flats during spring low tides
KR	Glynn and Ault (2000), Glynn et al. (2007)
PD	Joshua Feingold. Marchena, Galapagos Iss, Ecuador. 12 m depth. 12 June 2012

Poritidae: ***Porites lobata*** Dana	
Platy colony morphology, more common at greater depths, in subdued light	
H	Coral reefs and coral communities
A	Abundant
ER	Prominent reef-building species
GD	Widespread from Mexico to Easter Is (=Rapa Nui, Chile); Galápagos Iss, Ecuador. Indo-Pacific
DD	In clear water, 15–60 m
KR	Glynn and Ault (2000)
PD	Ángel Chiriboga. Darwin Is, Galápagos Iss, Ecuador. 15 m depth. 14 November 2004

Poritidae: ***Porites panamensis*** Verrill	
H	Coral reefs, rubble and rhodolith beds, coral communities, and commonly cryptic
A	Rare to abundant
ER	Coral reef and coral community component
GD	Endemic: Mexico to Ecuador; not present at Clipperton Is; Galápagos Iss, Ecuador; other EP oceanic islands
DD	Low mean tidal level to 36 m
KR	Cortés and Guzmán (1998), Glynn and Ault (2000), López- Pérez et al. (2007)
PD	Juan Maté. Pearl Is, Panama. 4 m depth. 29 August 2014

H Habitat; *A* Relative Abundance; *ER* Ecological Role; *GD* Geographic Distribution; *DD* Depth Distribution; *KR* Key References; *PD* Photo Details

Poritidae: ***Porites sverdrupi*** Durham	
H	Rhodolith beds
A	Rare
ER	Rhodolith community associate
GD	Endemic: Gulf of California, Mexico
DD	3–30 m
KR	Glynn and Ault (2000), Chiriboga et al. (2008a, b)
PD	Joshua Feingold. Catalina Is, Baja California Sur, Mexico. 9 m depth. 5 August 2014

Agariciidae: ***Pavona chiriquiensis*** Glynn, Maté and Stemann	
H	Coral reefs, coral communities, and rocky substrates
A	Common to rare, rare on coral reefs
ER	Coral community associate
GD	Costa Rica; Panama; Colombia; Ecuador (mainland and Galápagos Iss). Possibly present in Central Pacific
DD	3–30 m
KR	Glynn and Ault (2000), Glynn et al. (2001), Obura (2002), Maté (2003)
PD	Juan Maté. Uva Is, Panama. 6 m depth. 5 April 2012

Agariciidae: ***Pavona clavus*** (Dana)	
H	Coral reefs, coral communities, and rocky substrates
A	Common to abundant
ER	Coral reef-building and coral community associate
GD	Widespread from Mexico to Ecuador (mainland and Galápagos Iss). Indo-West Pacific
DD	2–20 m
KR	Cortés and Guzmán (1998), Glynn and Ault (2000)
PD	Juan Maté. Uva Is, Panama. 3 m depth. 12 September 2001

Agariciidae: ***Pavona frondifera*** (Lamarck)	
H	Coral reefs and coral communities
A	Rare
ER	Coral reef and coral community associate
GD	Costa Rica and Panama. Indo-West Pacific
DD	3–20 m
KR	Glynn and Maté (1997), Cortés and Guzmán (1998), Glynn and Ault (2000), Maté (2003)
PD	Juan Maté. Saboga Is, Panama. 3 m depth. 28 June 2000

H Habitat; *A* Relative Abundance; *ER* Ecological Role; *GD* Geographic Distribution; *DD* Depth Distribution; *KR* Key References; *PD* Photo Details

Agariciidae: ***Pavona gigantea*** Verrill	
H	Coral reefs and coral communities
A	Abundant, especially in cool-water areas
ER	Coral reef-building and coral community associate
GD	Mexico to Ecuador (mainland and Galápagos Iss); El Salvador; Clipperton Is (absent in recent surveys). South Central Pacific
DD	Low mean tidal level to 30 m
KR	Durham and Barnard (1952), Glynn and Ault (2000), Obura and Stone (2003)
PD	Ángel Chiriboga. Galápagos Iss, Ecuador. 15 m depth. 5 February 2004

Agariciidae: ***Pavona maldivensis*** (Gardiner)	
H	Coral reefs and coral communities
A	Rare to common
ER	Coral reefs and coral community associate
GD	Mexico to Ecuador (Galápagos Iss); Clipperton Is. Indo-West Pacific
DD	5–20 m
KR	Glynn and Ault (2000), Hickman (2008)
PD	Ángel Chiriboga. Galápagos Iss, Ecuador. 14 m depth. 8 September 2005

Agariciidae: ***Pavona varians*** Verrill	
H	Coral reefs, rhodolith beds and coral communities
A	Rare to abundant
ER	Coral reefs and coral community associate
GD	Mexico to Ecuador (Galápagos Iss) and Clipperton Is. Indo-West Pacific
DD	2–80 m
KR	Glynn and Ault (2000), Hickman (2008)
PD	Juan Maté. Uva Is, Panama. 7 m depth. 7 March 2012

Agariciidae: ***Leptoseris scabra*** Vaughan	
H	Coral communities and rocky substrates
A	Rare to common
ER	Coral community associate
GD	Costa Rica (Cocos Is); Ecuador (Galápagos Iss); Clipperton Is; Easter Is (=Rapa Nui, Chile)
DD	5–25 m
KR	Glynn and Ault (2000), Glynn et al. (2003), Guzmán and Cortés (2007)
PD	Jorge Cortés. Cocos Is, Costa Rica. 12 m depth. 20 March 2012

H Habitat; *A* Relative Abundance; *ER* Ecological Role; *GD* Geographic Distribution; *DD* Depth Distribution; *KR* Key References; *PD* Photo Details

Agariciidae: ***Gardineroseris planulata*** (Dana)	
H	Coral reefs, rhodolith beds, coral communities
A	Rare to common
ER	Coral reefs and coral community associate
GD	Mexico to Ecuador (mainland and Galápagos Iss). Indo-West Pacific
DD	2–30 m
KR	Glynn and Ault (2000), Fong and Glynn (2001)
PD	Irving Bethancourt. Brincanco Is, Panama. 5 m depth. 4 April 2012

Siderastreidae: ***Psammocora profundacella*** Gardiner	
H	Coral reefs, rocky substrates, usually firmly attached
A	Uncommon to common
ER	Coral reef associate
GD	Clipperton Is; Panama. Indo-West Pacific
DD	2–30 m
KR	Glynn and Ault (2000), Benzoni et al. (2007, 2010), Glynn et al. (2012)
PD	Juan Maté. Uva Is, Panama. 4 m depth. 2 April 2012

Siderastreidae: ***Psammocora stellata*** Verrill	
H	Coral reefs, rhodolith beds and rubble substrates, usually unattached
A	Common
ER	Coral community associate
GD	Mexico to Ecuador (mainland and Galápagos Iss). Indo-West Pacific
DD	1–14 m
KR	Cortés and Guzmán (1998), Glynn and Ault (2000), Glynn et al. (2012)
PD	Tyler Smith. Secas Is, Panama. 3 m depth. 12 March 2003

Siderastreidae: ***Psammocora stellata*** Verrill	
H	Coral reefs, rhodolith beds and rubble substrates, usually unattached
A	Common
ER	Coral community associate
GD	Mexico to Ecuador (mainland and Galápagos Iss). Indo-West Pacific
DD	1–14 m
KR	Cortés and Guzmán (1998), Glynn and Ault (2000), Glynn et al. (2012)
PD	Jorge Cortés. Cocos Is, Costa Rica. 10 m depth. 21 March 2012

H Habitat; *A* Relative Abundance; *ER* Ecological Role; *GD* Geographic Distribution; *DD* Depth Distribution; *KR* Key References; *PD* Photo Details

Fungiidae: ***Fungia curvata*** (Hoeksema)	
H	Coarse sediment, good circulation
A	Rare at most locations, abundant at a few localities in Mexico
ER	Free-living, deep coral community associate
GD	Mexico to Ecuador (mainland and Galápagos Iss). Indo-West Pacific
DD	3–73 m
KR	Wells (1983), Hoeksema (1989), Glynn and Ault (2000), Maté (2003)
PD	Juan Maté. Secas Is, Panama. 12 m depth. 19 March 2010

Fungiidae: ***Fungia distorta*** (Michelin)	
H	Soft bottoms, coral rubble, and seagrass beds
A	Abundant to rare
ER	Free-living, coral community member
GD	Mexico to Ecuador (mainland and Galápagos Iss). Indo-West Pacific
DD	1–85m
KR	Hoeksema (1989), Feingold (1996), Glynn and Ault (2000)
PD	Joshua Feingold. Corona del Diablo Is, Floreana Is, Galápagos Iss, Ecuador. 15 m depth. 12 June 2012

Dendrophylliidae: ***Tubastraea coccinea*** Lesson	
H	Undersides of corals, rocky substrates, subdued light
A	Abundant
ER	Azooxanthellate, coral community associate
GD	Widespread throughout ETP. Indo-Pacific. Invasive in the Atlantic
DD	1.5–54 m
KR	Cairns (1991), Fenner and Banks (2004), Glynn et al. (2008)
PD	Ángel Chiriboga. Floreana Is, Galápagos Iss, Ecuador. 12 m depth. 19 May 2006

Dendrophylliidae: ***Tubastraea faulkneri*** Wells	
H	Rocky ledges, undersides of massive corals
A	Uncommon, abundant at a few localities in Galápagos Iss
ER	Azooxanthellate, coral community associate
GD	Galápagos Iss, Ecuador. Indo-Pacific, especially western Pacific
DD	3–8 m
KR	Wells (1983), Cairns (1991), Hickman (2008)
PD	Ángel Chiriboga. Baltra Is, Galápagos Iss, Ecuador. 3 m depth. 3 August 2004

H Habitat; *A* Relative Abundance; *ER* Ecological Role; *GD* Geographic Distribution; *DD* Depth Distribution; *KR* Key References; *PD* Photo Details

Dendrophylliidae: ***Tubastraea floreana*** Wells	
H	Cryptic habitats, on ceilings of caves, ledges and rock overhangs
A	Rare, critically endangered [Hickman et al. (2007)]
ER	Azooxanthellate, coral community associate
GD	Endemic: Galápagos Iss, Ecuador
DD	2–46 m
KR	Glynn and Wellington (1983), Cairns (1991), Hickman (2008)
PD	Ángel Chiriboga. Floreana Is, Galápagos Iss, Ecuador. 29 m depth. 7 February 2004

Dendrophylliidae: ***Tubastraea tagusensis*** Wells	
H	Cryptic, on ceilings of caves, ledges and rock overhangs, upwelling environments
A	Common to locally abundant
ER	Azooxanthellate, coral community associate
GD	Galápagos Iss, Ecuador. Recently introduced to Brazil (Figueria de Paula and Creed 2004). Indo-Pacific. Atlantic
DD	1–43
KR	Wells (1983), Cairns (1991), Hickman (2008)
PD	Ángel Chiriboga. Isabela Is, Galápagos Iss, Ecuador. 3 m depth. 11 November 2004

Rhizangiidae: ***Astrangia equatorialis*** Durham and Barnard	
H	Cryptic, on basalt and calcareous substrates
A	Uncommon throughout Galápagos Iss
ER	Azooxanthellate, coral community associate
GD	Endemic: Galápagos Iss, Ecuador
DD	1–15 m
KR	Hickman (2008), Cairns (2014)
PD	Ángel Chiriboga. Floreana Is, Galápagos Iss, Ecuador. 12 m depth. 7 June 2004

Rhizangiidae***: Oulangia bradleyi*** Verrill	
Polyp expanded	
H	Coral communities, rocky outcrops
A	Common, often solitary
ER	Coral community associate
GD	Endemic: Mexico to Colombia; Cocos Is, Costa Rica; Galápagos Iss and Machalilla Is, Ecuador
DD	1–37 m
KR	Wells (1983), Cairns (1991)
PD	Ángel Chiriboga. Isabela Is, Galápagos Iss, Ecuador. 11 m depth. 3 December 2004

H Habitat; *A* Relative Abundance; *ER* Ecological Role; *GD* Geographic Distribution; *DD* Depth Distribution; *KR* Key References; *PD* Photo Details

Rhizangiidae: ***Oulangia bradleyi*** Verrill	
Polyp retracted	
H	Coral communities, rocky outcrops
A	Common, often solitary
ER	Coral community associate
GD	Endemic: Mexico to Colombia; Cocos Is, Costa Rica; Galápagos Iss, Ecuador
DD	1–37 m
KR	Wells (1983), Cairns (1991)
PD	Ángel Chiriboga. Floreana Is, Galápagos Iss, Ecuador. 15 m depth. 7 May 2004

Caryophylliidae: ***Polycyathus isabela*** Wells	
H	Coral communities, rocky outcrops
A	Rare to uncommon
ER	Coral community associate
GD	Endemic: Galápagos Iss, Ecuador
DD	10–23 m
KR	Wells (1983), Cairns (1991), Chiriboga et al. (2007),
PD	Ángel Chiriboga. Isabela Is, Galápagos Iss, Ecuador. 11 m depth. 3 December 2004

Caryophylliidae: ***Phyllangia consagensis*** (Durham and Barnard)	
Tan-brown color morph	
H	Coral communities, rocky outcrops
A	Common
ER	Coral community associate
GD	Endemic: Gulf of California, Mexico; Galápagos Iss, Ecuador
DD	1–59 m
KR	Wells (1983), Cairns (1991)
PD	Ángel Chiriboga. Isabela Is, Galápagos Iss, Ecuador. 8 m depth. 24 August 2003

Caryophylliidae: ***Phyllangia consagensis*** (Durham and Barnard)	
Pink color morph	
H	Coral communities, rocky outcrops
A	Common
ER	Coral community associate
GD	Endemic: Gulf of California, Mexico; Galápagos Iss, Ecuador
DD	1–59 m
KR	Wells (1983), Cairns (1991)
PD	Ángel Chiriboga. Isabela Is, Galápagos Iss, Ecuador. 11 m depth. 12 December 2004

H Habitat; *A* Relative Abundance; *ER* Ecological Role; *GD* Geographic Distribution; *DD* Depth Distribution; *KR* Key References; *PD* Photo Details

Caryophylliidae: ***Cladopsammia eguchii*** (Wells)	
H	Coral communities, rocky outcrops
A	Common
ER	Coral community associate
GD	Panama; Galápagos Iss, Ecuador; Cocos Is, Costa Rica. Widespread in Pacific Ocean
DD	1–27 m
KR	Hickman et al. (2005), Cairns (2015)
PD	Ángel Chiriboga. Floreana Is, Galápagos Iss, Ecuador. 15 m depth. 19 May 2006

Caryophylliidae: ***Cladopsammia gracilis*** (Milne Edwards and Haime)	
H	Cryptic, under rocks and corals, rocky outcrops
A	Common
ER	Coral community associate
GD	Galápagos Iss, Ecuador. Indo-West Pacific
DD	3–165 m
KR	Hickman et al. (2005)
PD	Ángel Chiriboga. Isabela Is, Galápagos Iss, Ecuador. 20 m depth. 29 November 2004

Zoanthidae: ***Zoanthus* cf. *vietnamensis*** Pax and Müller	
H	Coral reef substrates, rocky shores, and tide pools
A	Common
ER	Zooxanthellate zoanthid
GD	Galápagos Iss, Ecuador; Panama. Central Indo-Pacific
DD	0–4 m
KR	Reimer and Hickman (2009)
PD	Ana M. Palacio. Saboga Is, Panama. 4 m depth. 28 August 2015

Gorgoniidae: ***Antipathes galapagensis*** Deichmann	
H	Coral communities, rocky outcrops
A	Common
ER	Coral community associate
GD	Endemic: Galápagos Iss, Ecuador
DD	15–50 m
KR	Deichmann (1941), Molodtsova (2014)
PD	Ángel Chiriboga. Floreana Is, Galápagos Iss, Ecuador. 18 m depth. 25 May 2004

H Habitat; *A* Relative Abundance; *ER* Ecological Role; *GD* Geographic Distribution; *DD* Depth Distribution; *KR* Key References; *PD* Photo Details

Gorgoniidae: ***Eugorgia daniana*** Verrill	
H	Coral communities, rocky outcrops
A	Common
ER	Coral community associate
GD	Costa Rica; Colombia; El Salvador; Galápagos Iss and mainland Ecuador (Gulf of Guayaquil); Mexico; Panama; San Diego, California, USA
DD	3–70 m
KR	Breedy et al. (2009), Rivera and Martínez (2011)
PD	Ángel Chiriboga. Darwin Is, Galápagos Iss, Ecuador. 24 m depth. 26 April 2005

Gorgoniidae: ***Leptogorgia alba*** (Duchassaing and Michelotti)	
H	Coral communities, rocky outcrops
A	Abundant
ER	Coral community associate
GD	Costa Rica; Colombia; El Salvador; Galápagos Iss and mainland Ecuador; Mexico; Panama
DD	1–30 m
KR	Breedy and Guzmán (2007), Hickman (2008), Rivera and Martínez (2011)
PD	Ángel Chiriboga. Darwin Is, Galápagos Iss, Ecuador. 24 m depth. 14 November 2003

Gorgoniidae: ***Pacifigorgia dampieri*** Williams and Breedy	
H	Coral communities, rocky outcrops
A	Common
ER	Coral community associate
GD	Endemic: Galápagos Iss, Ecuador
DD	3–26 m
KR	Williams and Breedy (2004), Hickman (2008)
PD	Ángel Chiriboga. Las Cuevas Norte, Floreana Is, Galápagos Iss, Ecuador. 15 m depth. 19 May 2006

Gorgoniidae: ***Pacifigorgia rubripunctata*** Williams and Breedy	
H	Coral communities, rocky outcrops
A	Common
ER	Coral community associate
GD	Endemic: Galápagos Iss, Ecuador
DD	3–20 m
KR	Williams and Breedy (2004), Hickman (2008)
PD	Ángel Chiriboga. La Botella, Floreana Is, Galápagos Iss, Ecuador. 12 m depth. 7 May 2004

H Habitat; *A*: Relative Abundance; *ER* Ecological Role; *GD* Geographic Distributionl; *DD* Depth Distribution; *KR* Key References; *PD* Photo Details

Gorgoniidae: ***Heterogorgia hickmani*** Breedy and Guzmán	
H	Coral communities, rocky outcrops
A	Common
ER	Coral community associate
GD	Ecuador (mainland and Galápagos Iss). Western Atlantic (Brazil)
DD	7.5–26 m
KR	Breedy and Guzmán (2005, 2011), Hickman (2008), Rivera and Martínez (2011)
PD	Ángel Chiriboga. Corona del Diablo Is, Floreana Is, Galápagos Iss, Ecuador. 18 m depth. 19 April 2005

Gorgoniidae: ***Heterogorgia hickmani*** Breedy and Guzmán	
Close up view of expanded polyps	
H	Coral communities, rocky outcrops
A	Common
ER	Coral community associate
GD	Ecuador (mainland and Galápagos Iss). Western Atlantic (Brazil)
DD	7.5–26 m
KR	Breedy and Guzmán (2005, 2011), Hickman (2008), Rivera and Martínez (2011)
PD	Ángel Chiriboga. Floreana Is, Galápagos Iss, Ecuador. 12 m depth. 8 February 2004

Plexauridae: ***Muricea fruticosa*** Verrill	
H	Coral communities, rocky outcrops
A	Common
ER	Coral community associate
GD	Endemic: Panama; Galápagos Iss, Ecuador; Gulf of California, Mexico
DD	15–93 m
KR	Hickman (2008), Rivera and Martínez (2011), van Ofwegen (2014)
PD	Ángel Chiriboga. Nameless Is, Galápagos Iss, Ecuador. 21 m depth. 9 August 2003

Milleporidae: ***Millepora boschmai*** de Weerdt and Glynn	
H	Coral rubble and sand substrates, forereef slope and base
A	Ostensibly extinct in ETP
ER	Deep coral reef and coral community associate
GD	Restricted to the Gulf of Chiriquí, Panama
DD	4–10 m
KR	de Weerdt and Glynn (1991)
PD	Peter Glynn. Uva Is, Panama. 4 m depth. 22 February 1992

H Habitat; *A* Relative Abundance; *ER* Ecological Role; *GD* Geographic Distribution; *DD* Depth Distribution; *KR* Key References; *PD* Photo Details

Milleporidae: ***Millepora exaesa*** Forskål	
H	Coral reefs
A	Uncommon
ER	Reef building potential
GD	Clipperton Is. Indo-Pacific
DD	5–25 m
KR	Glynn et al. (1996), Glynn and Ault (2000)
PD	Peter Glynn. Clipperton Is. Overgrowing dead *Porites* colony. 8 m depth. 23 April 1994

Milleporidae: ***Millepora intricata*** Milne Edwards	
H	Coral communities, rocky outcrops, occasionally present on coral reefs
A	Abundant in coral communities, uncommon to common on coral reefs
ER	Coral community associate, contributes to reef building, and provides shelter refuges
GD	Panama (restricted to the Gulf of Chiriquí). Indo-West Pacific
DD	2–30 m
KR	de Weerdt and Glynn (1991), Glynn and Ault (2000), Maté (2003)
PD	Irving Bethancourt. Gulf of Chiriquí, Panama. 22 m depth. 3 April 2012

Stylasteridae: ***Distichopora robusta*** Lindner, Cairns and Guzmán	
H	Rocky outcrops, strong currents
A	Common locally at two sites
ER	Coral community associate
GD	Endemic: Gulf of Chiriquí, Panama
DD	5–25 m
KR	Maté (2003), Lindner et al. (2004)
PD	Kevan Mantell. Cathedral Rock, Jicarón Is, Panama. 22 m depth. 20 January 2011

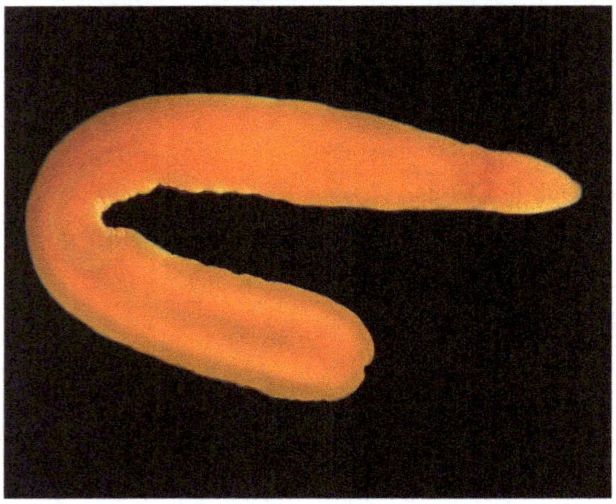

PHYLUM NEMERTEA

Lineidae: ***Micrura ignea*** Schwartz and Norenburg	
H	Coral rubble substrates
A	Rare to uncommon
ER	Predator of molluscs, crustaceans, and annelids
GD	Gulf of Chiriquí, Panama (first record in ETP). Caribbean Sea
DD	0–30 m
KR	Schwartz and Norenburg (2005)
PD	Leslie Harris. Granito de Oro Is, Panama. 4 m depth. 22 February 2003

H Habitat; *A* Relative Abundance; *ER* Ecological Role; *GD* Geographic Distribution; *DD* Depth Distribution; *KR* Key References; *PD* Photo Details

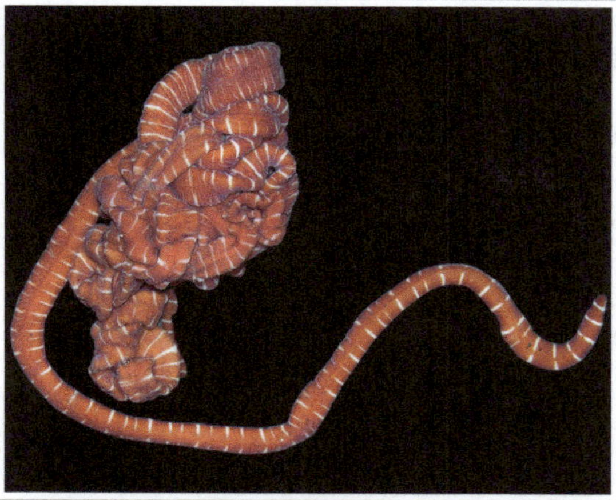

Valenciniidae: ***Baseodiscus mexicanus*** (Bürger)	
H	Tidal mud flats and coral reefs
A	Uncommon to common, secretive on coral reefs
ER	Predator
GD	Endemic: Mexico; Panama; Galápagos Iss, Ecuador
DD	Intertidal to 6 m
KR	Gibson (2015)
PD	Arthur Anker. Mud flats at Pacific terminus of Panama Canal, Panama. 3 m depth. 21 May 2007. Specimen 4 m in length

PHYLUM ANNELIDA

Amphinomidae: ***Pherecardia striata*** (Kingberg)	
H	Cryptic in coral rubble and reef frameworks
A	Common to abundant
ER	Predator and scavenger, preys on injured *Acanthaster planci*
GD	Panama. Indo-Pacific
DD	Up to 10 m
KR	Glynn (1981, 1984), WoRMS (2015)
PD	Tyler Smith. Uva Is, Panama. 4 m depth. 11 March 2003

Serpulidae: ***Spirobranchus ?incrassatus*** (Kroyer)	
H	Benthic hard substrates, tubes embeded in coral skeleton
A	Common
ER	Filter-feeder
GD	Southern California, USA to Peru
DD	3–10 m
KR	ten Hove (2014)
PD	Juan Maté. Contadora Is, Panama. 3 m depth. 29 August 2014

Polynoidae: ***Iphione ovata*** Kinberg	
H	Coral rubble, broken shells, rocky substrates
A	Rare to common, cryptic
ER	Carnivore, preys on invertebrates
GD	Gulf of California, Mexico to Panama; Clipperton Is. Indo-Pacific. Atlantic
DD	1–26 m
KR	Pettibone (1986)
PD	Leslie Harris. Uva Is, Panama. 4 m depth. 9–17 March 2005

H Habitat; *A* Relative Abundance; *ER* Ecological Role; *GD* Geographic Distribution; *DD* Depth Distribution; *KR* Key References; *PD* Photo Details

Hesionidae: ***Ophiodromus* sp.**	
H	Sandy and coral rubble substrates, also in seaweed holdfasts and on echinoderms
A	Rare to uncommon, cryptic
ER	Omnivore, predator, scavenger
GD	Gulf of California, Mexico to Panama. North Atlantic. Mediterranean Sea
DD	3–7 m
KR	Brusca (1980)
PD	Leslie Harris. Uva Is, Panama. 3–5 m depth. 9–17 March 2005

Eunicidae: ***Eunice* sp.**	
H	Sand and rock substrates, coral reefs
A	Rare to uncommon, cryptic
ER	Generalist predator and scavenger
GD	El Salvador; Panama. Indo-Pacific. Atlantic
DD	Sublittoral zone to 200 m, usually 40–70 m
KR	Brusca (1980), Fauchald (1992)
PD	Leslie Harris. Granito de Oro Is, Panama. 4 m depth. 9–17 March 2005

Sabellidae: ***Bispira* sp.** [***Bispira melanostigma*** (Schmarda) or ***Bispira monroi*** (Hartman)]	
H	Among algae and epifauna, coral reefs and coral communities
A	Common
ER	Suspension feeder, omnivore
GD	Panama. Indo-Pacific. Atlantic. Antarctica. Arctic Sea
DD	Shallow water environments in Panama, 3–7 m
KR	Brusca (1980), Capa and López (2004)
PD	Leslie Harris. Coibita Is, Panama. 4 m depth. 9–17 March 2005

Polynoidae: ***Lepidasthenia gigas*** (Johnson)	
H	Coral rubble, sand, rocky outcrops, tidal pools, inside tubes of terebellid polychaetes
A	Common
ER	Commensal of terebellid polychaetes, free-living or commensal, preys on invertebrates
GD	Southern California; Panama; Galápagos Iss, Ecuador
DD	Intertidal to 4 m
KR	Fauchald (1977)
PD	Leslie Harris. Coiba Is, Panama. 4 m depth. 9–17 March 2005

H Habitat; *A* Relative Abundance; *ER* Ecological Role; *GD* Geographic Distribution; *DD* Depth Distribution; *KR* Key References; *PD* Photo Details

Sabellidae: ***Lanicola eduardoi*** Capa and Hutchings	
H	Sand tubes on undersides of corals and rocks
A	Common
ER	Detritivore, surface deposit-feeder
GD	Endemic: Panama
DD	Intertidal to 10 m
KR	Capa and Hutchings (2006)
PD	Leslie Harris. Coibita Is, Panama. 4 m depth. 9–17 March 2005

PHYLUM MOLLUSCA

Pediculariidae: ***Jenneria pustulata*** (Lightfoot)	
H	Coral communities and coral reefs
A	Uncommon unless in feeding aggregations
ER	Corallivore, specializing on *Pocillopora* spp., occasionally feeding on *Porites* and *Pavona*
GD	Gulf of California to Ecuador; Galápagos Iss, Ecuador; Easter Is (=Rapa Nui, Chile)
DD	Low mean tidal level to 15 m
KR	Keen (1958), Glynn (1985)
PD	Peter Glynn. Uva Is, Panama. 3 m depth. February–April 1981

Muricidae: ***Coralliophila violacea*** (Kiener)	
H	Coral reefs and coral communities
A	Common, abundant in aggregations
ER	Corallivore, often on *Porites lobata*
GD	Easter Is (=Rapa Nui, Chile). Indo-Pacific
DD	Subtidal to 20 m
KR	Keen (1958)
PD	Alejandro Pérez-Matus. Easter Is (=Rapa Nui, Chile). 15 m depth. 8 May 2011

Epitoniidae: ***Asperiscala billeeana*** (DuShane and Bratcher)	
Several feeding on *Tubastraea* plus deposited eggs	
H	Parasitic on *Tubastraea coccinea*
A	Rare to common
ER	Inhabits and feeds on the solitary coral *Tubastraea coccinea*
GD	Gulf of California, Mexico to Ecuador; Galápagos Iss. Indo-Pacific
DD	Subtidal to 25 m
KR	Keen (1958), Gittenberger and Gittenberger (1995), Finet (2001)
PD	Joshua Feingold. Champion Is, Galápagos Is, Ecuador. 8 m depth. 12 June 2012

H Habitat; *A* Relative Abundance; *ER* Ecological Role; *GD* Geographic Distribution; *DD* Depth Distribution; *KR* Key References; *PD* Photo Details

Eulimidae: ***Echineulima thaanumi*** (Pilsbry)	
H	Echinoderms, occasionally observed attached to *Diadema mexicanum* and *Euapta godeffroyi*
A	Rare to Common
ER	Parasitic, castrates host by devitalization and ingestion of gonadal tissue
GD	Colombia; Panama. Hawaii. Australia
DD	2–6 m
KR	Eakin (1991), Bouchet (2015a)
PD	Tyler Smith. Uva Is, Panama. 3 m depth. 3 March 2010

Tergipedidae: ***Phestilla melanobrachia*** Bergh	
H	Coral communities and coral reefs
A	Rare to common (when present on coral hosts)
ER	Nocturnal grazer on *Porites lobata* tissues
GD	Costa Rica; Gulf of California and Jalisco, Mexico. Tropical Indo-West Pacific
DD	5 m
KR	Behrens and Hermosillo (2005), Bieler and Bouchet (2014)
PD	Terrence Gosliner. Costa Rica. 5 m depth. 15 April 1993

Tergipedidae: ***Phestilla lugubris*** (Bergh)	
H	Coral reefs
A	Rare to common (when present on coral hosts)
ER	Grazes on *Porites lobata* tissues at night
GD	Baja California, Mexico; mainland Costa Rica and Cocos Is; Galápagos Iss, Ecuador; Malpelo Is, Colombia. Tropical Indo-West Pacific
DD	2–6 m
KR	Behrens and Hermosillo (2005), Bouchet (2014)
PD	Terrence Gosliner. Gulf of Chiriquí, Panama. 5 m depth. 20 July 2014

Chromodorididae: ***Doriprismatica sedna*** Ev. Marcus and Er. Marcus	
H	Coral communities and coral reefs
A	Common
ER	Spongivore, assumed to feed only on sponges
GD	Gulf of California, Mexico to Panama; Galápagos Iss, Ecuador. Caribbean; Bahamas; Gulf of Mexico
DD	1–10 m
KR	Camacho-García et al. (2005), Johnson and Gosliner (2012)
PD	Leslie Harris. Coiba Is, Panama. 4 m depth. 17 February 2002

H Habitat; *A* Relative Abundance; *ER* Ecological Role; *GD* Geographic Distribution; *DD* Depth Distribution; *KR* Key References; *PD* Photo Details

Plakobranchidae: ***Elysia diomedea*** Bergh

H	Filamentous algae
A	Common
ER	Herbivore, feeding on *Padina* spp. and *Codium* spp.
GD	Endemic: Gulf of California, Mexico to Ecuador
DD	Intertidal to 20 m
KR	Behrens and Hermosillo (2005), Rudman (2006), Bouchet (2015b)
PD	Leslie Harris. Coiba Is, Panama. 3 m depth. 9–17 March 2005

Octopodidae: ***Octopus mimus*** Gould

H	Rocky intertidal, coral reefs, kelp forests
A	Rare to abundant
ER	Opportunistic predator on crustaceans, molluscs, fishes, and echinoderms
GD	Endemic: Northern Peru to Chile, Panama (first report for the tropical Eastern Pacific)
DD	1–200 m
KR	Norman (2000), Norman et al. (2014)
PD	Juan Maté. Contadora Is, Panama. 4 m depth. 29 August 2014

Octopodidae: ***Octopus oculifer*** Hoyle

H	Coral reefs and rocky intertidal
A	Rare to abundant
ER	Opportunistic predator on crustaceans and molluscs
GD	Endemic: Galápagos Iss, Ecuador
DD	0–50 m
KR	Norman (2000), Jereb et al. (2014), Norman et al. (2014)
PD	Joshua Feingold. Floreana Is, Galápagos Iss, Ecuador. 8 m depth. 12 June 2012

Octopodidae: ***Octopus* sp.** (Cuvier)

Pygmy octopus, probably an undescribed species

H	Cryptic in coral rubble and coral reef cavities
A	Uncommon to common
ER	Predator
GD	Gulf of Chiriquí, Panama
DD	2–5 m
KR	Brusca (1980), Norman and Hochberg (2005)
PD	Tyler Smith. Uva Is, Panama. 4 m depth. 16 September 2002

H Habitat; *A* Relative Abundance; *ER* Ecological Role; *GD* Geographic Distribution; *DD* Depth Distribution; *KR* Key References; *PD* Photo Details

PHYLUM ARTHROPODA

Hymenoceridae: ***Hymenocera picta*** Dana	
H	Coral reefs and coral communities
A	Rare to locally common
ER	Sea star predator
GD	Panama; Galápagos Iss, Ecuador; Clipperton Is. Central and Eastern Pacific. Red Sea. East Africa. Indonesia. Northern Australia
DD	1–20 m
KR	Glynn (1977), Hickman and Zimmerman (2000), Poupin et al. (2009), Prakash and Kumar (2013)
PD	Joshua Feingold. Punta Carrion, Santa Cruz Is, Galápagos Iss, Ecuador. 9 m depth. 9 June 2012

Alpheidae: ***Alpheus lottini*** Guérin-Méneville	
H	Obligate symbiont of pocilloporid corals
A	Abundant
ER	Pocilloporid commensal and grazer. Male-female pairs defend host coral from predators; forage on coral mucus and accumulated organic debris
GD	Gulf of California, Mexico to Ecuador (mainland to Galápagos Iss); Easter (=Rapa Nui, Chile); Clipperton Is; Malpelo Is, Colombia. Indo-West Pacific. Red Sea
DD	1.5–100 m
KR	Kim and Abele (1988), Poupin and By (2003)
PD	Leslie Harris. Coibita Is, Panama. 3–7 m depth. 9–17 March 2005

Hippolytidae: ***Thor algicola*** Wicksten	
H	Live coral, coral rubble, algae
A	Common, often present on live *Pocillopora*spp.
ER	Possibly an omnivore
GD	Gulf of California, Mexico to Ecuador; Revillagigedo Iss, Mexico; Clipperton Is
DD	0–20 m
KR	Wicksten (1987), Poupin (2010)
PD	Leslie Harris. Coibita Is, Panama. 4 m depth. 9–17 March 2005

Gonodactylidae: ***Neogonodactylus* sp.** [likely ***N. bahiahondensis*** (Schmitt)]	
H	Coral communities and coral reefs, coral rubble
A	Common
ER	Predator
GD	Costa Rica to Ecuador (mainland)
DD	1–20 m
KR	Schmitt (1940)
PD	Leslie Harris. Uva Is, Panama. 4 m depth. 9–17 March 2005

H Habitat; *A* Relative Abundance; *ER* Ecological Role; *GD* Geographic Distribution; *DD* Depth Distribution; *KR* Key References; *PD* Photo Details

Palaemonidae: ***Harpiliopsis depressa*** (Stimpson)	
H	Coral communities, inhabits *Pocillopora* spp.
A	Abundant
ER	Obligate commensal on pocilloporid corals
GD	Baja California to Ecuador (including Galápagos Iss); Clipperton Is. Indo-Pacific
DD	1–15 m
KR	Hickman and Zimmerman (2000), Fransen and De Grave (2015)
PD	Leslie Harris. Coibita Is, Panama. 4 m depth. 9–17 March 2005

Inachidae: ***Stenorhynchus debilis*** (Smith)	
H	Coral reefs, preference for pocilloporid corals
A	Common
ER	Associated with aggregations of the echinoid *Astropyga*
GD	Gulf of California, Mexico to Ecuador; Clipperton Is; Galápagos Iss, Ecuador
DD	1–20 m
KR	Alvarado (2008), Poupin (2010), Davie (2015a)
PD	Juan Maté. Saboga Is, Panama. 4 m depth. 2 May 2015

Mithracidae: ***Mithraculus denticulatus*** (Bell)	
H	Coral communities and coral rubble substrates, associated with algae
A	Common
ER	Omnivore: grazer, scavenger, predator
GD	Gulf of California, Mexico to Peru; Galápagos Iss, Ecuador
DD	Intertidal zone to 24 m
KR	Brusca (1980), Ng and Davie (2013)
PD	Leslie Harris. Uva Is, Panama. 4 m depth. 9–17 March, 2005

Domeciidae: ***Domecia hispida*** Eydoux and Souleyet	
H	Coral communities, induces cavity formation in coral host
A	Common
ER	Coral symbiont, filter-feeder, grazer
GD	Mexico; Panama; Galápagos Iss, Ecuador; Clipperton Is. Indo-Pacific. Florida
DD	1–40 m
KR	Poupin (2010), Davie (2015b)
PD	Leslie Harris. Uva Is, Panama. 4 m depth. 9–17 March 2005

H Habitat; *A* Relative Abundance; *ER* Ecological Role; *GD* Geographic Distribution; *DD* Depth Distribution; *KR* Key References; *PD* Photo Details

Trapeziidae: ***Trapezia bidentata*** (Forskål) =***Trapezia ferruginea*** (Latrielle)	
H	Among branches of pocilloporid corals
A	Common to abundant
ER	Obligate symbiont of *Pocillopora* spp.; feeds on coral mucus, detritus and polyps; benefits coral health and repels corallivores
GD	Endemic: Gulf of California, Mexico to Colombia; Revillagigedo Iss, Mexico; Clipperton Is; Cocos Is, Costa Rica; Malpelo Is, Colombia; Galápagos Iss, Ecuador
DD	1–54 m
KR	Castro (1996), Castro et al. (2004), Poupin et al. (2009)
PD	Joseph Poupin. Clipperton Is. 20 m depth. February–March 2005

Trapeziidae: ***Trapezia bidentata*** (Forskål) =***Trapezia ferruginea*** (Latrielle)	
H	Among branches of pocilloporid corals
A	Common to abundant
ER	Obligate symbiont of *Pocillopora* spp.; feeds on coral mucus, detritus and polyps; benefits coral health and repels corallivores
GD	Endemic: Gulf of California, Mexico to Colombia; Revillagigedo Iss, Mexico; Clipperton Is; Cocos Is, Costa Rica; Malpelo Is, Colombia; Galápagos Iss, Ecuador
DD	1–54 m
KR	Castro (1996), Castro et al. (2004), Poupin et al. (2009)
PD	Kevan Mantell. Coiba Is, Panama. 4 m depth. 23 March 2009

Trapeziidae: ***Trapezia corallina*** (Gerstaecker)	
H	Among branches of pocilloporid corals
A	Common to abundant
ER	Obligate symbiont of *Pocillopora* corals; feeds on coral mucus, detritus and polyps; benefits coral health and deters corallivores
GD	Endemic; Gulf of California, Mexico to Ecuador; Revillagigedo Iss, Clipperton Is, Galápagos Iss
DD	1–54 m
KR	Castro (1996), Castro et al. (2004), Poupin et al. (2009)
PD	Joseph Poupin. Clipperton Is. 20 m depth. February–March 2005

Porcellanidae: ***Pachycheles biocellatus*** (Lockington)	
H	Live *Pocillopora* spp. branches and coral rubble substrates
A	Common to abundant
ER	Suspension feeder
GD	Mexico to Ecuador, including Galápagos Iss, Ecuador, Clipperton Is; Revillagigedo Iss; Malpelo Is
DD	Lower intertidal zone to 54 m
KR	Poupin et al. (2009)
PD	Alexandra Hiller. Pearl Iss, Panama. Low intertidal zone. March 2015

H Habitat; *A* Relative Abundance; *ER* Ecological Role; *GD* Geographic Distribution; *DD* Depth Distribution; *KR* Key References; *PD* Photo Details

Porcellanidae: ***Petrolisthes donadio*** Hiller and Werding	
H	Under coral colonies and rubble, under rocks in intertidal zone, occasionally underneath *Diadema mexicanum*
A	Abundant
ER	Suspension feeder
GD	Endemic: Gulf of California, Mexico to Ecuador; Galápagos Iss; Malpelo Is, Colombia
DD	Lower intertidal to 35m depth
KR	Hiller and Werding (2007)
PD	Alexandra Hiller. Río Mar, Panama. Intertidal zone. 3 March 2014

Cirolanidae: ***Cirolana parva*** Hansen	
H	Sandy and rubble coral reef substrates, cryptic
A	Common to highly abundant
ER	Scavenger and predator; prey size small (settling larvae) to large (moribund invertebrates and fishes)
GD	Mexico to Ecuador. Indo-Pacific: Hawaii; India. Caribbean: Puerto Rico; Florida. Gulf of Mexico
DD	1–7 m
KR	Menzies and Glynn (1968), Brusca et al. (1995)
PD	Peter W Glynn. Uva Is, Panama. 4 m depth. 11 March 2004

Balanidae: ***Megabalanus peninsularis*** (Pilsbry)	
H	On massive corals and rock substrates
A	Common to abundant
ER	Suspension feeder, spatial competitor
GD	Endemic: Gulf of California, Mexico to Peru; Galápagos Iss, Ecuador
DD	Subtidal to 20 m
KR	Hickman and Zimmerman (2000)
PD	Joshua Feingold. Wellington Reef (Darwin Is), Galápagos Iss, Ecuador. 12 m depth. 8 June 2012

PHYLUM ECHINODERMATA	
Oreasteridae: ***Nidorellia armata*** (Gray)	
H	Coral communities, rocky shores
A	Common
ER	Consumes benthic algae, small gastropods, and sessile invertebrates (e.g., tube worms)
GD	Gulf of California, Mexico to Peru; Galápagos Iss, Ecuador. Hawaii
DD	Intertidal to 73 m
KR	Brusca (1980), Mah (2014)
PD	Ángel Chiriboga. Isabela Is, Galápagos Iss, Ecuador. 6 m depth. 22 April 2005

H Habitat; *A* Relative Abundance; *ER* Ecological Role; *GD* Geographic Distribution; *DD* Depth Distribution; *KR* Key References; *PD* Photo Details

Oreasteridae: ***Pentaceraster cumingi*** (Gray)	
H	Coral rubble substrates, sandy bottoms
A	Common, often in large aggregations
ER	Consumes coralline algae and some echinoderms, including *Eucidaris galapagensis*
GD	Gulf of California, Mexico to Peru; Galápagos Iss, Ecuador. Hawaiian Iss
DD	Extreme low intertidal zone to 180 m
KR	Brusca (1980), Dee et al. (2012)
PD	Tyler Smith. Uva Is, Panama. 4 m depth. 30 July 2007

Acanthasteridae: ***Acanthaster planci*** (Linnaeus)	
H	Coral communities and coral reefs
A	Rare, uncommon, rarely common
ER	Corallivore
GD	Gulf of California, Mexico to Panama; Malpelo Is and Gorgona Is, Colombia; rare in northern Galápagos Iss and Clipperton Is
DD	Subtidal to 15–20 m
KR	Barham et al. (1973), Glynn (1973, 1974, 1976, 1977), Vogler et al. (2008)
PD	Kevan Mantell. Gulf of Chiriquí, Panama. 16 m depth. 3 March 2014

Heliasteridae: ***Heliaster cumingii*** (Gray)	
H	Coral and rocky reefs
A	Abundant
ER	Predator. Feeds on *Salmacina* polychaetous annelids, barnacles, and tunicates
GD	Endemic: Galápagos Iss, Ecuador
DD	1–15 m
KR	Glynn and Wellington (1983), Hickman (1998a, b), Mah (2015)
PD	Joshua Feingold. Urvina Bay, Isabela Is, Galápagos Iss, Ecuador. 3m depth. 16 June 2012

Cidaridae: ***Eucidaris thouarsii*** (L. Agassiz and Desor)	
H	Intertidal and subtidal zones, coral communities, secretive
A	Common
ER	Grazes on encrusting algae, scavenges dead metazoans
GD	Upper Gulf of California, Mexico to mainland Ecuador
DD	From intertidal zone to 20 m
KR	Lessios (2005), Alvarado et al. (2010), Kroh (2015a)
PD	Juan Maté. Secas Is, Panama. 3 m depth. 20 August 2014

H Habitat; *A* Relative Abundance; *ER* Ecological Role; *GD* Geographic Distribution; *DD* Depth Distribution; *KR* Key References; *PD* Photo Details

Cidaridae: ***Eucidaris galapagensis*** Döderlein	
H	Intertidal and subtidal, coral communities, rock substrates
A	Abundant in central and southern Galápagos Iss, Ecuador
ER	Grazes on encrusting algae, reef building corals, and barnacles, prominent bioeroder
GD	Galápagos Iss, Ecuador; Cocos Is, Costa Rica; Clipperton Is
DD	From lower intertidal zone to 20 m
KR	Glynn et al. (1979), Brandt et al. (2012), Dee et al. (2012), Kroh (2015b)
PD	Joshua Feingold. Española Is, Galápagos Iss, Ecuador. 1.5 m depth. 23 May 2005

Diadematidae: ***Diadema mexicanum*** A. Agassiz	
H	Subtidal in coral communities and coral reefs, rocky low intertidal zone (tide pools)
A	Common to abundant
ER	Primarily herbivorous but also grazes on metazoans, including corals; prominent bioeroder
GD	Gulf of California, Mexico to mainland Ecuador and Galápagos Iss; Clipperton Is; Revillagigedo Iss, Mexico; Cocos Is, Costa Rica
DD	0–113 m
KR	Lessios et al. (2001), Lessios (2005), Alvarado (2010), Kroh (2014a)
PD	Juan Maté. Panama. Granito de Oro Is, 3 m depth. 29 March 2008

Diadematidae: ***Astropyga pulvinata*** (Lamarck)	
H	Coral reefs, rocky and sandy substrates
A	Rare, occasionally abundant below 10 m
ER	Primarily herbivorous but also grazes on benthic invertebrates, including corals
GD	Endemic: Panama to Ecuador
DD	0–20 m
KR	Lessios (2005), Alvarado (2008), Poupin (2010), Kroh (2014b)
PD	Juan Maté. Uva Is, Panama. 7 m depth. 18 August 2014

Toxopneustidae: ***Toxopneustes roseus*** (A. Agassiz)	
H	Secretive, sheltering under coral rubble during the day, rocky substrates
A	Uncommon, occasionally abundant, best observed at night
ER	Grazes on non-geniculate coralline algal crusts and dead coral
GD	Gulf of California, Mexico to Ecuador (mainland and Galápagos Iss)
DD	Low intertidal zone to 20 m
KR	Hickman (1998a, b), Lessios (2005), Kroh (2014c)
PD	Ángel Chiriboga. Wolf Is, Galápagos Iss, Ecuador. 5 m depth. 13 October 2004

H Habitat; *A* Relative Abundance; *ER* Ecological Role; *GD* Geographic Distribution; *DD* Depth Distribution; *KR* Key References; *PD* Photo Details

Toxopneustidae: ***Lytechinus semituberculatus*** (Valenciennes in L. Agassiz)	
H	Coral and rocky reefs, high productivity waters
A	Abundant
ER	Feeds on algae and detritus. Grazes on coralline pavements
GD	Endemic: Galápagos Iss and mainland Ecuador
DD	1–18 m
KR	Glynn and Wellington (1983), Hickman (1998a, b), Zigler and Lessios (2004), Alvarado et al. (2010)
PD	Joshua Feingold. Concha de Perla Lagoon, Isabela Iss, Galápagos Iss. 4 m depth. 13 June 2012

Toxopneustidae: ***Tripneustes depressus*** A. Agassiz	
H	Coral communities, reefs, rocky substrates, moderate wave action
A	Common in Galápagos, densities fluctuate temporally and spatially; patchy distribution across the archipelago
ER	Feeds on coralline algae (Hickman 1998a, b), and hypothesized to facilitate the persistence of this algal guild by also grazing on potentially over-topping foliose algae like *Ulva* spp. (Brandt et al. 2012)
GD	Gulf of California, Mexico to Ecuador (mainland and Galápagos Iss)
DD	Littoral to 73 m
KR	Hickman (1998a, b), Brandt et al. (2012)
PD	Margarita Brandt. León Dormido Is, Galápagos Iss, Ecuador. 14 m depth. 4 March 2015

Stichopodidae: ***Isostichopus fuscus*** (Ludwig)	
H	Coral rock and sand substrates
A	Rare to common, formerly abundant before overharvesting
ER	Detritus and deposit feeder
GD	Endemic: Mexico to South America; Cocos Is, Costa Rica; Socorro Is, Mexico; Galápagos Iss, Ecuador
DD	Subtidal to 40 m
KR	Alvarado et al. (2010), Paulay (2014)
PD	Alex Hearn. Fernandina Is, Galápagos Iss, Ecuador. 15 m depth. 29 November 2002

Synaptidae: ***Euapta godeffroyi*** (Semper)	
H	Cryptic in coral frameworks, intertidal pools and rubble substrates
A	Common to abundant
ER	Deposit feeder
GD	Gulf of California, Mexico; Costa Rica; Panama; Easter Is (=Rapa Nui, Chile). Red Sea. South Africa. Maldive Iss. Hawaii Iss. South Pacific Iss. China. New Caledonia. East Indies. Philippines. Australia
DD	1–77 m
KR	Alvarado et al. (2010), Granja-Fernández et al. (2013)
PD	Ana M. Palacio. Saboga Is, Panama. 3 m depth. 26 August 2015

H Habitat; *A* Relative Abundance; *ER* Ecological Role; *GD* Geographic Distribution; *DD* Depth Distribution; *KR* Key References; *PD* Photo Details

PHYLUM CRANIATA

Lutjanidae: ***Lutjanus viridis*** (Valenciennes)

H	Coral reefs and rocky reefs
A	Common, small groups to medium and large-sized schools
ER	Carnivorous, consuming benthic macrofauna
GD	Mexico to Ecuador, most common at offshore islands including Malpelo Is, Colombia; Galápagos Iss, Ecuador; Tres Marías and Revillagigedo Iss, Mexico
DD	0–60 m
KR	Robertson (2012), Robertson and Allen (2015)
PD	Ángel Chiriboga. Galápagos Iss, Ecuador. 15 m depth. 6 July 2004

Mullidae: ***Mulloidichthys dentatus*** (Gill)

H	Shallow sand, mud, rock and coral substrates
A	Common, often in schools (>10 ind)
ER	Omnivore, forages in sand and rubble
GD	Southern California to Peru, including the Galápagos Iss, Ecuador; rare north of southern Gulf of California, Mexico
DD	2–60 m
KR	Robertson (2012), Robertson and Allen (2015)
PD	Gerald Allen. Clipperton Is. Unrecorded depth. 15 October 2007

Balistidae: ***Sufflamen verres*** (Gilbert and Starks)

H	Coral and rock-reef rubble patches
A	Common, often 10 to 100 individuals per patch reef
ER	Benthic invertivore; excavates dead and live coral patches in search of prey
GD	Gulf of California, Mexico to Ecuador (mainland and Galápagos Iss)
DD	3–35 m
KR	Robertson (2012), Robertson and Allen (2015)
PD	Ross Robertson. 25 February 2011. Exsitu photograph

Balistidae: ***Pseudobalistes naufragium*** (Jordan and Starks)

H	Coral reefs, shallow sandy and rubble substrates
A	Occasional
ER	Feeds on sea urchins, small crustaceans and molluscs, often blowing into sand to uncover prey or overturn sea urchins
GD	Bahía de Santa Maria, Baja California, Mexico to Chile; Cocos Is, Costa Rica; Galápagos Iss, Ecuador; Malpelo Is, Colombia
DD	3–35 m
KR	Robertson (2012), Robertson and Allen (2015)
PD	Tyler Smith. Uva Is, Panama. 4 m depth. 13 March 2005

H Habitat; *A* Relative Abundance; *ER* Ecological Role; *GD* Geographic Distribution; *DD* Depth Distribution; *KR* Key References; *PD* Photo Details

Tetraodontidae: ***Arothron meleagris*** (Lacepède)	
Yellow variety	
H	Coral communities and coral reefs
A	Common
ER	Bentho-pelagic. Feeds mainly on tips of branching corals and to a lesser extent on sponges, molluscs, bryozoans, tunicates, foraminiferans, algae, and detritus
GD	SW Gulf of California, Mexico to Ecuador; all oceanic islands. Indo-Pacific
DD	1–35 m
KR	Robertson (2012), Palacios et al. (2014), Robertson and Allen (2015)
PD	Ángel Chiriboga. Wolf Is, Galápagos Iss, Ecuador. 15 m depth. 13 November 2003

Tetraodontidae: ***Arothron meleagris*** (Lacepède)	
Spotted variety	
H	Coral communities and coral reefs
A	Common
ER	Bentho-pelagic. Feeds mainly on tips of branching corals and to a lesser extent on sponges, molluscs, bryozoans, tunicates, foraminiferans, algae, and detritus
GD	SW Gulf of California, Mexico to Ecuador; all oceanic islands. Indo-Pacific
DD	1–35 m
KR	Robertson (2012), Palacios et al. (2014), Robertson and Allen (2015)
PD	Ángel Chiriboga. Wolf Is, Galápagos Iss, Ecuador. 15 m depth. 13 November 2003

Tetraodontidae: ***Arothron hispidus*** (Linnaeus)	
H	Outer coral reef slopes, rocky reefs; juveniles in estuarine weedy areas
A	Common
ER	Omnivores, fleshy and calcareous algae, detritus, diverse invertebrates
GD	Gulf of California, Mexico to Colombia; Galápagos Iss, Ecuador; all oceanic islands except Clipperton Is. Indo-Pacific: Red Sea; East Africa
DD	1–90 m
KR	Robertson (2012), Robertson and Allen (2015)
PD	Irving Bethancourt. Coiba Is, Panama. 4 m depth. 5 April 2012

Diodontidae: ***Diodon holocanthus*** Linnaeus	
H	Shelters in shaded recesses of coral and rocky reefs, boulder-strewn slopes, rock walls, and sand-rubble substrates
A	Occasional
ER	Nocturnal carnivore, consuming molluscs, sea urchins, and a variety of crustaceans
GD	Southern California to Peru and all offshore islands. Circumtropical
DD	1–100 m
KR	Robertson (2012), Robertson and Allen (2015)
PD	Ángel Chiriboga. Floreana Is, Galápagos Iss, Ecuador. 38 m depth. 25 May 2004

H Habitat; *A* Relative Abundance; *ER* Ecological Role; *GD* Geographic Distribution; *DD* Depth Distribution; *KR* Key References; *PD* Photo Details

Scaridae: ***Scarus ghobban*** Forsskål	
H	Most common on reefs with reef-building corals; rocky, boulder-strewn coral reefs and slopes
A	Abundant to common, solitary or in small schools
ER	Grazes on algae and occasionally live corals
GD	Gulf of California, Mexico to Ecuador (mainland and Galápagos Iss). Indo-Pacific: eastern Africa across Indian Ocean to western Pacific; Japan; Red Sea. Eastern Mediterranean
DD	1–150 m
KR	Robertson (2012), Robertson and Allen (2015)
PD	Tyler Smith. Uva Is, Panama. 4 m depth. 15 September 2002

Scaridae: ***Scarus rubroviolaceus*** Bleeker	
Female	
H	Coral reefs and rocky substrates (adults); shallow areas with abundant algae (juveniles)
A	Common to occasional, solitary and small schools
ER	Grazes on benthic algae
GD	Gulf of California, Mexico to Ecuador (mainland and Galápagos Iss). Indo-Pacific: East Africa to eastern Australia; Hawaiian Iss
DD	1–35 m
KR	Robertson (2012), Robertson and Allen (2015)
PD	A. Roger Steene. Galápagos Iss, Ecuador. Unrecorded depth. 15 October 2007

Scaridae: ***Scarus rubroviolaceus*** Bleeker	
Terminal male phase	
H	Coral and rocky reefs, rubble-strewn substrates
A	Common to occasional, solitary and small schools
ER	Grazes on benthic algae, occasionally on live massive corals
GD	Gulf of California, Mexico to Ecuador (mainland and Galápagos Iss). Indo-Pacific: East Africa; Australia; Hawaiian Islands
DD	1–35 m
KR	Robertson (2012), Robertson and Allen (2015)
PD	Tyler Smith. Uva Is, Panama. 4 m depth. 20 March 2008

Scaridae: ***Scarus compressus*** (Osburn and Nichols)	
Terminal male phase	
H	Coral reefs and rocky outcrops
A	Uncommon to common
ER	Herbivore
GD	Endemic: central Gulf of California, Mexico to Ecuador; Revillagigedo Iss, Mexico; Cocos Is, Costa Rica; Galápagos Iss, Ecuador
DD	3–25 m
KR	Robertson (2012), Robertson and Allen (2015)
PD	Tyler Smith. Uva Is, Panama. 3 m depth. 14 September 2004

H Habitat; *A* Relative Abundance; *ER* Ecological Role; *GD* Geographic Distribution; *DD* Depth Distribution; *KR* Key References; *PD* Photo Details

Scaridae: ***Scarus perrico*** Jordan and Gilbert	
Terminal male phase	
H	Coral reefs and rocky outcrops
A	Rare to common
ER	Herbivore
GD	Endemic: Gulf of California, Mexico to northern Peru; Galápagos Iss, Ecuador; Cocos Is, Costa Rica; Revillagigedo Iss, Mexico; Malpelo Is, Colombia
DD	1–20 m
KR	Allen et al. (2012), Robertson (2012), Robertson and Allen (2015)
PD	Tyler Smith. Uva Is, Panama. 4 m depth. 14 September 2004

Acanthuridae: ***Acanthurus xanthopterus*** Valenciennes	
H	Coral reefs, sand slopes, and lagoons; adults in deeper, protected bays and lagoons; juveniles in shallow, protected inshore waters
A	Occasional, often in small (<10 ind) schools
ER	Bentho-pelagic, feeds on diatoms, detrital sand film, filamentous algae, hydroids, and fish carrion
GD	Lower Gulf of California, Mexico to Panama; Clipperton Is; Galápagos Iss, Ecuador. Indo-Pacific
DD	1–100 m
KR	Robertson (2012), Robertson and Allen (2015)
PD	Ángel Chiriboga. Marchena Is, Galápagos Iss, Ecuador. 15 m depth. 25 April 2005

Acanthuridae: ***Acanthurus triostegus*** (Linnaeus)	
H	Coral and rocky reefs
A	Common, occasionally in large schools (>25 ind to ~100 ind)
ER	Herbivore (benthic macroalgae)
GD	Gulf of California, Mexico to Ecuador; all oceanic islands. Widespread in the tropical Indo-Pacific: East Africa to Hawaii
DD	0–90 m
KR	McIlwain et al. (2012), Robertson (2012), Robertson and Allen (2015)
PD	Tyler Smith. Uva Is, Panama. 3 m depth. 6 March 2003

Acanthuridae: ***Acanthurus nigricans*** (Linnaeus)	
H	Coral reefs and rocky outcrops
A	Common, small schools (<10 ind)
ER	Herbivore
GD	Southern Gulf of California, Mexico; southern Mexico; Costa Rica to Ecuador; Revillagigedo Iss, Mexico; Cocos Is, Costa Rica; Galápagos Iss, Ecuador
DD	0–70 m
KR	Robertson (2012), Robertson and Allen (2015)
PD	Tyler Smith. Uva Is, Panama. 3 m depth. 20 September 2004

H Habitat; *A* Relative Abundance; *ER* Ecological Role; *GD* Geographic Distribution; *DD* Depth Distribution; *KR* Key References; *PD* Photo Details

Acanthuridae: ***Prionurus laticlavius*** (Valenciennes)	
H	Coral reefs and rocky outcrops
A	Abundant, commonly in large schools (>25 ind)
ER	Herbivore
GD	Southwestern Gulf of California, Mexico to Ecuador; stragglers in Revillagigedo Iss, Mexico; Cocos Is, Costa Rica; Galápagos Iss, Ecuador; Malpelo Is, Colombia
DD	0–30 m
KR	Dominici-Arosemena et al. (2005), Robertson and Allen (2006), Abesamis et al. (2012), Robertson (2012)
PD	Peter Glynn. Uva Is, Panama. 3 m depth. 16 August 1985

Chaetodontidae: ***Johnrandallia nigrirostris*** (Gill)	
H	Coral communities, coral reefs and boulder-strewn substrates
A	Abundant
ER	Omnivore, feeds on algae, gastropods, and small crustaceans; fish cleaner, observed feeding on *Paranthias colonus* and *Mulloidichthys dentatus*
GD	Gulf of California, Mexico to Panama; Cocos Is, Costa Rica; Galápagos Iss, Ecuador; Malpelo Is, Colombia
DD	5–40 m
KR	Robertson (2012), Robertson and Allen (2015)
PD	Ángel Chiriboga. Darwin Is, Galápagos Iss, Ecuador. 15 m depth. 14 November 2003

Chaetodontidae: ***Chaetodon humeralis*** Günther	
H	Coral reefs, boulder-strewn substrates
A	Common, often in pairs
ER	Omnivore, algae and benthic invertebrates
GD	Southern California to Chile; Cocos Is, Costa Rica; Galápagos Iss, Ecuador; Malpelo Is, Colombia
DD	3–55 m; most common in 5–12 m
KR	Robertson (2012), Robertson and Allen (2015)
PD	Ángel Chiriboga. Santa Fé Is, Galápagos Iss, Ecuador. 15 m depth. 12 May 2004

Pomacanthidae: ***Holacanthus passer*** Valenciennes	
H	Coral reefs, forereef slopes, rocky boulder-strewn areas
A	Abundant, often solitary, also in large, non-polarized schools
ER	Diurnal grazer on sessile invertebrates and algae, especially sponges, occasional planktivore; cleaning behavior observed in juveniles
GD	Southern California to Peru; Galápagos Iss, Ecuador
DD	1–80 m
KR	Robertson (2012), Robertson and Allen (2015)
PD	Ángel Chiriboga. Floreana, Galápagos Iss, Ecuador. 15 m depth. 15 December 2004

H Habitat; *A* Relative Abundance; *ER* Ecological Role; *GD* Geographic Distribution; *DD* Depth Distribution; *KR* Key References; *PD* Photo Details

Pomacanthidae: ***Pomacanthus zonipectus*** (Gill)	
H	Coral reefs, high topographic relief of reefs; rocky boulder-strewn areas
A	Common, usually solitary or in pairs
ER	Diurnal grazer on sessile invertebrates and algae, especially sponges, occasional planktivore; cleaning behavior observed in juveniles
GD	Southern Gulf of California to Peru; all oceanic islands
DD	1–50 m
KR	Robertson (2012), Robertson and Allen (2015)
PD	Margarita Brandt. Isla de la Plata, Machalilla, Ecuador. 6 m depth. 14 January 2014

Zanclidae: ***Zanclus cornutus*** (Linnaeus)	
H	Seaward rocky coral substrates, inner reef lagoons, reef flats
A	Common to occasional
ER	Carnivore, small encrusting invertebrates
GD	Southern Gulf of California to Peru; all oceanic islands. Indo-Pacific: East Africa across Indian Ocean to Hawaiian Islands
DD	3–180 m
KR	Robertson (2012), Robertson and Allen (2015)
PD	Irving Bethancourt. Coiba Is, Panama. 15 m depth. 8 March 2012

Labridae: ***Novaculichthys taeniourus*** Lacepède	
H	Coral reefs and rocky reefs, sand rubble substrates and large massive corals
A	Uncommon to common
ER	Carnivore, molluscs, sea urchins, brittle stars, polychaetes, and crabs; overturns corals and rocks to expose prey
GD	Gulf of California to Panama; Galápagos Iss, Ecuador. Indo-Pacific
DD	3–30 m
KR	Robertson (2012), Robertson and Allen (2015)
PD	A. Roger Steene. Indonesia. Unrecorded depth. 15 October 2007

Labridae: ***Thalassoma grammaticum*** Gilbert	
H	Coral reefs and rocky outcrops
A	Abundant on oceanic islands
ER	Predator on hard-shelled invertebrates
GD	Endemic: widespread from Gulf of California to northern Peru; all oceanic islands
DD	0–65 m
KR	Gomon (1995), Robertson (2012), Robertson and Allen (2015)
PD	Tyler Smith. Uva Is, Panama. 3 m depth. 15 September 2002

H Habitat; *A* Relative Abundance; *ER* Ecological Role; *GD* Geographic Distribution; *DD* Depth Distribution; *KR* Key References; *PD* Photo Details

Pomacentridae: ***Stegastes acapulcoensis*** (Fowler)	
H	Coral reefs and rocky outcrops
A	Common to abundant
ER	Herbivore, cultivates and defends algal gardens
GD	Endemic: Gulf of California; Colombia to Costa Rica; Cocos Is, Costa Rica; Galápagos Iss, Ecuador; Malpelo Is, Colombia; Revillagigedo Iss, Mexico
DD	1–16 m
KR	Dominici-Arosemena et al. (2005), Robertson and Allen (2006), Robertson (2012)
PD	Peter Glynn. Uva Is, Panama. 3 m depth. 13 August 1985

Carangidae: ***Gnathanodon speciosus*** (Forsskål)	
H	Reefs, lagoons and forereef substrates, juveniles often shelter among scyphozoan tentacles
A	Common, often in schools (>10 ind)
ER	Carnivore, feeds by rooting through rubble substrates for small invertebrates and fishes
GD	Southwestern coast of Baja California Sur, Mexico and Gulf of California to Ecuador (including Galápagos Iss); Malpelo Is, Colombia; Revillagidedo Iss, Mexico. Indo-Pacific
DD	0–40 m
KR	Robertson (2012), Robertson and Allen (2015)
PD	Ross Robertson. Aquarium of the Pacific. 7 November 2014. Exsitu photo. Unrecorded depth

Muraenidae: ***Gymnothorax castaneus*** (Jordan and Gilbert)	
H	Recesses in coral reefs, rocky and boulder-strewn areas
A	Uncommon to common
ER	Carnivore, crustaceans and fishes, nocturnal forager
GD	Widespread from Gulf of California to Ecuador; Cocos Is, Costa Rica; Galápagos Iss, Ecuador; Malpelo Is, Colombia; Revillagigedo Iss, Mexico
DD	1–35 m
KR	Robertson (2012), Robertson and Allen (2015)
PD	Tyler Smith. Uva Is, Panama. 4 m depth. 31 July 2007

Muraenidae: ***Muraena lentiginosa*** Jenyns	
H	Coral reefs and rocky outcrops
A	Common
ER	Nocturnal predator, invertebrates and fishes
GD	Endemic: Gulf of California, Mexico to northern Peru; Cocos Is, Costa Rica; Galápagos Iss, Ecuador; Malpelo Is, Colombia; Revillagigedo Iss, Mexico
DD	5–25 m
KR	Robertson (2012), Robertson and Allen (2015)
PD	Kevan Mantell. Uva Is, Panama. 3 m depth. 2 April 2012

H Habitat; *A* Relative Abundance; *ER* Ecological Role; *GD* Geographic Distribution; *DD* Depth Distribution; *KR* Key References; *PD* Photo Details

Muraenidae: ***Echidna nebulosa*** (Ahl)	
H	Rocky and intertidal coral reef flats, shallow lagoons and seaward reefs
A	Common
ER	Predator, preys mainly on crustaceans
GD	Gulf of California, Mexico to Galápagos Iss, Ecuador. Indo-Pacific: East Africa across Indian Ocean
DD	1–30 m
KR	Chen et al. (1994), Robertson (2012), Robertson and Allen (2015)
PD	Tyler Smith. Uva Is, Panama. 4 m depth. 31 March 2011

Carcharhinidae: ***Triaenodon obesus*** (Rüppell)	
H	Coral communities, rocky substrates, and coral reefs
A	Uncommon to abundant
ER	Predator, preys on diverse invertebrates and fishes
GD	Lower Gulf of California, Mexico to Colombia; all oceanic islands except Clipperton Is
DD	2–300 m
KR	Robertson (2012), Robertson and Allen (2015)
PD	Kevan Mantell. Wahoo Rock, Panama. 12 m depth. 1 April 2012

H Habitat; *A* Relative Abundance; *ER* Ecological Role; *GD* Geographic Distribution; *DD* Depth Distribution; *KR* Key References; *PD* Photo Details

Acknowledgements We thank Peter Glynn for the invitation to produce this field guide and for contributing substantially to final stages of the manuscript preparation.We also thank Gerald Allen, Arthur Anker, Iliana Baums, Irving Bethancourt, José Luis Carballo, Jorge Cortés, Joshua Feingold, Cindy Fernández-García, Peter Glynn, Terrence Gosliner, Leslie Harris, Alex Hearn, Alexandra Hiller, Kevan Mantell, Alejandro Pérez-Matus, Ana M. Palacio, Joseph Poupin, Bernhard Riegl, Ross Robertson, Felix Rodríguez, Jaime Nivia Ruiz, Tyler Smith, A. Roger Steene, Janet Voight, Nancy Voss, Evie Wieters, and Brian Wysor, for kindly sharing or assisting in securing photographs herin presented. Roy Caldwell, Leslie Harris, James D. Reimer, Janet Voight, Nancy Voss, and Brian Wysor provided further assistance to species identification. Ernesto Peña helped with digitalization of photographs and Michael P.C. Fuller with final editing and formatting. We also gratefully acknowledge the logistical support offered by the Smithsonian Tropical Research Institute (Panama), the Galápagos National Park Service and the Charles Darwin Foundation (Ecuador). Permits to work and conduct field collections in Panama were provided by the Ministry of Environment, former ANAM (Autoridad Nacional del Ambiente) and ARAP (Autoridad de Recursos Acuáticos de Panama).

References

Abesamis R, Choat JH, Clements KD, McIlwain J, Myers R, Nanola C, Rocha LA, Russell B, Stockwell B (2012) *Prionurus laticlavius*. The IUCN red list of threatened species. Version 2014.2. www.iucnredlist.org

Allen G, Robertson R, Rivera R, Edgar G, Merlen G, Zapata F, Barraza E, Lea B, Rocha LA (2012) *Scarus perrico*. The IUCN red list of threatened species. Version 2015.1. www.iucnredlist.org

Alvarado JJ (2008) Seasonal occurrence and aggregation behavior of the sea urchin *Astropyga pulvinata* (Echinodermata: Echinoidea) in Bahía Culebra, Costa Rica. Pac Sci 62(4):579–592

Alvarado JJ, Solís-Marín FA, Ahern C (2010) Echinoderm (Echinodermata) diversity in the Pacific coast of Central America. Mar Biodiv 40:45–56

Ávila E, Carballo JL (2004) Growth and standing stock biomass of a mutualistic association between the sponge *Haliclona caerulea* and the red alga *Jania adherens*. Symbiosis 36:225–244

Ballantine DL, Ruiz H (2005) Two *Peyssonnelia* species (Peyssonneliaceae, Rhodophyta) from Puerto Rico including *Peyssonnelia flavescens* sp. nov. Phycologia 44(3):328–334

Barham EG, Gowdy RW, Wolfson FH (1973) *Acanthaster* (Echinodermata, Asteroidea) in the Gulf of California. Fish Bull U.S. Natl Mar Fish Surv 71:927–942

Behrens DW, Hermosillo A (2005) Eastern Pacific nudibranchs. A guide to the opisthobranchs from Alaska to Central America. Sea Challengers, Monterey, California, p 137

Benzoni F, Stefani F, Stolarski J, Pichon M, Mitta G, Galli P (2007) Debating phylogenetic relationships of the scleractinian *Psammocora*: molecular and morphological evidences. Contrib Zool 76:35–54

Benzoni F, Stefani F, Pichon M, Galli P (2010) The name game: morpho-molecular species boundaries in the genus *Psammocora* (Cnidaria, Scleractinia). Zool J Linn Soc 160:421–456

Bieler R, Bouchet P (2014) *Phestilla melanobrachia* Bergh, 1874. Accessed through: World Register of Marine Species at http://www.marinespecies.org/aphia.php?p=taxdetails&id=536543

Bonilla HR, Carricart-Ganivet JP (2000) *Porites arnaudi*, a new species of stony coral (Anthozoa: Scleractinia: Poritidae) from oceanic islands of the eastern Pacific Ocean. Proc Biol Soc Washington 113 (2):561–571

Bouchet P (2014) *Phestilla lugubris* (Bergh, 1870). Accessed through: World Register of Marine Species at http://www.marinespecies.org/aphia.php?p=taxdetails&id=536542

Bouchet P (2015a) *Echineulima thaanumi* (Pilsbry, 1921). In: MolluscaBase (2015). Accessed through: World Register of Marine Species at http://www.marinespecies.org/aphia.php?p=taxdetails&id=566019

Bouchet P (2015b) *Elysia diomedea* (Bergh, 1894). In: MolluscaBase (2015). Accessed through: World Register of Marine Species at http://www.marinespecies.org/aphia.php?p=taxdetails&id=494440

Boulay JN, Hellberg ME, Cortés J, Baums IB (2014) Unrecognized coral species diversity masks differences in functional ecology. Proc R Soc B 281:20131580

Brandt M, Witman JD, Chiriboga Á (2012) Influence of a dominant consumer species reverses at increased diversity. Ecology 93:868–878

Breedy O, Guzmán HM (2005) A new species of alcyonacean octocoral from the Galápagos Archipelago. J Mar Biol Ass UK, 85:801–807

Breedy O, Guzmán HM (2007) A revision of the genus *Leptogorgia* Milne Edwards & Haime, 1857 (Coelenterata: Octocorallia: Gorgoniidae) in the eastern Pacific. Zootaxa 1419:1–90

Breedy O, Guzmán HM (2011) A revision of the genus *Heterogorgia* Verrill, 1868 (Anthozoa: Octocorallia: Plexauridae) Zootaxa 2995:27–44

Breedy O, Guzmán HM, Vargas S (2009) A revision of the genus *Eugorgia* Verrill, 1868 (Coelenterata: Octocorallia: Gorgoniidae) Zootaxa 2151:1–46

Brusca RC (1980) Common intertidal invertebrates of the Gulf of California (2nd ed). Univ Arizona Press, Tucson, Arizona, p 427

Brusca RC, Wetzer R, France SC (1995) Cirolanidae, Crustacea: Isopoda: Flabellifera of the tropical eastern Pacific. Proc San Diego Soc Nat Hist 30:1–96

Caballero-George C, Bolaños J, Ochoa E, Carballo JL, Cruz JA, Arnold AE (2010) Protocol to isolate sponge-associated fungi from tropical waters and an examination of their cardioprotective potential. Curr Trends Biotechnol Pharm 4(4):881–889

Cairns SD (1991) A revision of the ahermatypic Scleractinia of the Galápagos and Cocos Islands. Smith Cont Zool 504:1–44

Cairns S (2014) *Astrangia equatorialis* Durham & Barnard, 1952. Accessed through: World Register of Marine Species at http://www.marinespecies.org/aphia.php?p=taxdetails&id=287120

Cairns S (2015) *Cladopsammia eguchii* (Wells, 1982). Accessed through: World Register of Marine Species at http://www.marinespecies.org/aphia.php?p=taxdetails&id=286957

Camacho-García Y, Gosliner TM, Valdés A (2005) Field guide to the sea slugs of the tropical eastern Pacific. California Acad Sci, p 129

Capa M, Hutchings PA (2006) Terebellidae (Polychaeta) from Coiba National Park, Panamanian Pacific, including description of four new species and synonymy of the genus *Paraeupolymnia* with *Lanicola*. Zootaxa 1375:1–29

Capa M, López E (2004) Sabellidae (Annelida: Polychaeta) living in blocks of dead coral in the Coiba National Park, Panama. J Mar Biol Ass UK 84:63–72

Carballo J, Cruz-Barraza J (2008) First record of *Axinyssa* Lendenfeld 1897 (Demospongiae Halichondrida) from the east Pacific Ocean with the description of *Axinyssa isabela* sp nov Zootaxa 1784:58–68

Castro P (1996) Eastern Pacific species of *Trapezia* (Crustacea, Brachyura: Trapeziidae), sibling species symbiotic with reef corals. Bull Mar Sci 58(2):531–554

Castro P, Ng PKL, Ahyong ST (2004) Phylogeny and systematics of the Trapeziidae Miers, 1886 (Crustacea: Brachyura), with the description of a new family. Zootaxa 643:1–70

Chen H-M, Shao K-T, Chen CT (1994) A review of the muraenid eels (Family Muraenidae) from Taiwan with descriptions of twelve new records. Zool Stud 33:44–64

Chiriboga Á, Hickman C, Edgar G, Banks S, Vera M (2007) *Polycyathus isabela*. The IUCN red list of threatened species. Version 2014.3. www.iucnredlist.org

Chiriboga Á, Guzmán H, Cortés J, Hickman C, Edgar G (2008a) *Pocillopora inflata*. The IUCN red list of threatened species. Version 2015.2. www.iucnredlist.org

Chiriboga Á, Edgar G, Reyes-Bonilla H (2008b) *Porites sverdrupi*. The IUCN red list of threatened species. Version 2015.1. www.iucnredlist.org

Cortés J, Guzmán HM (1998) Organismos de los arrecifes coralinos de Costa Rica: VI. Descripción, distribución geográfica e historia natural de los corales zooxantelados (Anthozoa: Scleractinia) del Pacífico. Rev Biol Trop 46:55–92

Cruz-Barraza JA, Carballo JL (2008) Taxonomy of sponges (Porifera) associated with corals from the Mexican Pacific Ocean. Zool Stud 47:741–758

Cruz-Barraza JA, Carballo JL, Rocha-Olivares A, Ehrlich H, Hog M (2012) Integrative taxonomy and molecular phylogeny of genus *Aplysina* (Demospongiae: Verongida) from Mexican Pacific. PLoS ONE 7(8): e42049. doi:10.1371/journal.pone.0042049

Davie P (2014) *Trapezia bidentata* (Forskål, 1775). Accessed through: World Register of Marine Species at http://www.marinespecies.org/aphia.php?p=taxdetails&id=443875

Davie P (2015a) *Stenorhynchus debilis* (Smith, 1871). Accessed through: World register of marine species at http://www.marinespecies.org/aphia.php?p=taxdetails&id=441967

Davie P (2015b) *Domecia hispida* Eydoux & Souleyet, 1842. Accessed through: world register of marine species at http://www.marinespecies.org/aphia.php?p=taxdetails&id=209146

Dawson EY (1960) New records of marine algae from Pacific Mexico and Central America. Pac Nat 1(19/20):31–52

De Weerdt WH, Glynn PW (1991) A new and presumably now extinct species of *Millepora* (Hydrozoa) in the eastern Pacific. Zool Meded, Leiden 65:267–276

Dee L, Witman JD and Brandt M (2012) Refugia and top-down control of the pencil urchin *Eucidaris galapagensis* in the Galápagos marine reserve. J Exper Mar Biol Ecol 416–417:135–143

Deichmann E (1941) Coelenterates collected on the presidential cruise of 1938. Smithson. Miscell Coll 99:1–17

Denizot M (1968) Les algues floridées encroutantes (à l'éxclusion des Corallinacées) Paris: Laboratoire de Cryptogamie, Muséum National d'Histoire Naturelle. SS 310

Díaz MC, van Soest RWM, Rützler K, Guzmán HM (2005) *Aplysina chiriquiensis*, a new pedunculate sponge from the Gulf of Chiriquí, Eastern Pacific (Aplysinidae, Verongida). Zootaxa 1012:1–12

Dominici-Arosemena A, Brugnoli-Olivera E, Cortés Núnez J, Molina-Urena H, Quesada-Alpizar M (2005) community structure of Eastern Pacific reef fishes (Gulf of Papagayo, Costa Rica). Tecnociencia 7:19–41

Durham JW, Barnard JL (1952) Stony corals of the eastern Pacific collected by the Velero III and Velero IV. Allan Hancock Pac Exped 16:1–110

Eakin CM (1991) The damselfish-algal lawn symbiosis and its influence on the bioerosion of an El Niño impacted coral reef, Uva Island, Pacific Panama. PhD dissert, Univ Miami, Coral Gables, p 158

Fauchald K (1977) Polychaetes from intertidal areas in Panama, with a review of previous shallow-water records. Smith Contrib Zool. 221:1–81

Fauchald K (1992) A review of the genus *Eunice* (Polychaeta: Eunicidae) based upon type material. Smith Contrib Zool. 523:1–422

Feingold J (1996) Coral survivors of the 1982–83 El Niño-Southern Oscillation, Galapagos Islands, Ecuador. Coral Reefs15:108

Fenner D, Banks K (2004) Orange cup coral *Tubastraea coccinea* invades Florida and the Flower Garden Banks, northwestern Gulf of Mexico. Coral Reefs 23:505–507

Fernández-García C, Riosmena-Rodríguez R, Wysor B, Tejada OL, Cortéz J (2011) Checklist of the Pacific marine macroalgae of Central America. Botanica Marina 54:53–73

Finet Y (2001) The marine mollusks of the Galápagos Iss.: a documented faunal list. Editions du Muséum d'Histoire naturelle, Genève, p 237

Figueria de Paula A, Creed JC (2004) Two species of the coral *Tubastraea* (Cnidaria, Scleractinia) in Brazil: a case of accidental introduction. Bull Mar Sci 74:175–183

Fransen C, De Grave S (2015) *Harpiliopsis depressa*. Accessed through: world register of marine species at http://marinespecies.org/aphia.php?p=taxdetails&id=220131

Fong P, Glynn PW (2001) Population abundance and size-structure of an eastern tropical Pacific reef coral after the 1997–1998 ENSO: a simulation model predicts field measures. Bull Mar Sci 69:187–202

Garske LE (2002) Macroalgas Marinas. In: Danulat E, Edgar GJ (eds) (2002): Reserva Marina de Galápagos. Línea Base de la Biodiversidad. Fundación Charles Darwin/Servicio Parque Nacional Galápagos, Santa Cruz, Galápagos, Ecuador, pp 419–439

Gibson R (2015) *Baseodiscus mexicanus* (Bürger, 1893). In: Norenburg J, Gibson R (eds) World Nemertea database. Accessed through: world register of marine species at http://www.marinespecies.org/aphia.php?p=taxdetails&id=156000

Gittenberger A, Gittenberger E (1995) A hitherto unnoticed adaptive radiation: epitoniid species (Gastropoda: Epitoniidae) associated with corals (Scleractinia). Contrib Zool. 74:125–203

Glynn PW (1973) *Acanthaster*: effect on coral reef growth in Panama. Science 180:504–506

Glynn PW (1974) The impact of *Acanthaster* on corals and coral reefs in the eastern Pacific. Environ Conserv 1:295–304

Glynn PW (1976) Some physical and biological determinants of coral community structure in the Eastern Pacific. Ecol Monogr 46:431–456

Glynn PW (1977) Interactions between *Acanthaster* and *Hymenocera* in the field and laboratory. In: Proceedings of 3rd International Coral Reef Symposium, vol 1. Miami, pp 209–215

Glynn PW (1981) *Acanthaster* population regulation by a shrimp and a worm. In: Proceedings of 4th International Coral Reef Symposium, vol 2. Manila, pp 607–612

Glynn PW (1984) An amphinomid worm predator of the Crown-of-Thorns sea star and general predation on asteroids in eastern and western Pacific coral reefs. Bull Mar Sci 35:54–71

Glynn PW (1985) Corallivore population sizes and feeding effects following El Niño (1982–1983) associated coral mortality in Panama. In: Proceedings of 5th International Coral Reef Congress, vol 4. Tahiti, pp 183–188

Glynn PW (1999) *Pocillopora inflata*, a new species of scleractinian coral (Cnidaria: Anthozoa) from the tropical eastern Pacific. Pac Sci 53:168–180

Glynn PW, Wellington GM (1983) Corals and coral reefs of the Galápagos Islands. Univ California Press, Berkeley p 330

Glynn PW, Maté TJL (1997) Field guide to the Pacific coral reefs of Panamá. In: Proceedings of 8th International Coral Reef Symposium, vol 1. Panama, pp 145–166

Glynn PW, Ault S (2000) A biogeographic analysis and review of the far eastern Pacific coral reef region. Coral Reefs 19:1–23

Glynn PW, Wellington GM, Birkeland C (1979) Coral growth in the Galápagos: limitation by sea urchins. Science 203:47–49

Glynn PW, Veron JEN, Wellington GM (1996) Clipperton Atoll (eastern Pacific): oceanography, geomorphology, reef-building coral ecology and biogeography. Coral Reefs 15:71–99

Glynn PW, Maté JL, Stemann T (2001) *Pavona chiriquiensis*, a new species of zooxanthellate scleractinian coral (Cnidaria: Anthozoa: Agariciidae) from the eastern tropical Pacific. Bull Biol Soc Washington 10:210–225

Glynn PW, Wellington GM, Wieters EA, Navarrete SA (2003) Reef-building coral communities of Easter Island, Chile. In: Cortés J (ed) Latin American coral reefs. Elsevier Science B.V., Amsterdam, pp 473–494

Glynn PW, Wellington GM, Riegl B, Olson DB, Borneman E, Wieters EA (2007) Diversity and biogeography of the scleractinian coral fauna of Easter Island. Pac Sci 61:67–90

Glynn PW, Colley SB, Maté JL, Cortés J, Guzmán HM, Bailey RL, Feingold JS, Enochs IC (2008) Reproductive ecology of the azooxanthellate coral *Tubastraea coccinea* in the equatorial eastern Pacific: Part V. Dendrophylliidae. Mar Biol 153:529–544

Glynn PW, Riegl B, Correa AMS, Baums IB (2009) Rapid recovery of a coral reef at Darwin Island, Galápagos Islands. Galapagos Res 66:6–13

Glynn PW, Colley SB, Maté JL, Baums IB, Feingold JS, Cortés J, Guzmán HM, Afflerbach JC, Brandtneris VW, Ault JS (2012) Reef coral reproduction in the equatorial eastern Pacific: Costa Rica, Panamá, and the Galápagos Islands (Ecuador). VII Siderastreidae, *Psammocora stellata* and *Psammocora profundacella*. Mar Biol 159(9):1917–1932. doi:10.1007/s00227-012-1979-5

Gómez P, Bakus G (1992) *Aplysina gerardogreeni* and *Aplysina aztecus* (Porifera: Demospongiae) new species from the Mexican Pacific. Anal Inst Cienc Mar Limnol, Univ Autón Mex 19:175–180

Gomon MF (1995) Labridae. Viejas, doncellasas, señoritas. In: W Fischer, F Krupp, W Schneider, C Sommer, KE Carpenter, V Niem (eds) Guia FAO para identificación de especies para los fines de la pesca. Pacifico Centro-Oriental. 3 Vols. FAO, Rome, pp 1201–1225

Granja-Fernández R, Antonio-Pérez V, López-Pérez RA (2013) *Euapta godeffroyi* (Holothuroidea: Synaptidae): Filling the distribution gap between Mexico and Costa Rica, eastern tropical Pacific. Hidrobiológica 23(1):130–132

Guiry MD (2014a) AlgaeBase. World-wide electronic publication. In: Guiry MD, Guiry GM National University of Ireland, Galway. http://www.algaebase.org

Guiry MD (2014b) *Sargassum* (*Sargassum*) *obtusifolium* J. Agardh, 1848. In: Guiry MD, Guiry GM (2014). AlgaeBase. World-wide electronic publication, National University of Ireland, Galway (taxonomic information republished from AlgaeBase with permission of MD Guiry). Accessed through: world register of marine species at http://www.marinespecies.org/aphia.php?p=taxdetails&id=376057

Guiry MD, Guiry GM (2015) AlgaeBase. World-wide electronic publication, National University of Ireland, Galway. http://www.algaebase.org

Guzmán HM, Cortés J (2007) Reef recovery 20 years after the 1982–1983 El Niño massive mortality. Mar Biol 151:401–411

Hickman CP (1998a) Guía de campo sobre estrellas de mar y otros equinodermos de Galápagos. Sugar Spring Press, Lexington, Virginia, USA, p 83

Hickman P Jr (1998b) A field guide to sea stars and other echinoderms of Galápagos. Sugar Spring Press, Lexington, Virginia, p 83

Hickman CP Jr (2008) A field guide to corals and other radiates of Galápagos. Sugar Spring Press, Lexington, VA, p 161

Hickman CP Jr, Zimmerman TL (2000) A field guide to crustaceans of Galápagos. Sugar Spring Press, Lexington, Virginia, p 156

Hickman CP Jr, Chiriboga Á, Ober WC (2005) A field guide to corals of Galápagos. Sugar Spring Press, Lexington, Virginia, p 54

Hickman C, Chiriboga A, Edgar G, Guzmán H, Banks S (2007) *Tubastraea floreana*. The IUCN red list of threatened species. Version 2015.2. www.iucnredlist.org

Hiller A, Werding B (2007) Redescription of *Petrolisthes edwardsii* (de Saussure) and description of a new, sibling species from the eastern Pacific based on different colour, morphology and genetic identity (Crustacea: Anomura: Porcellanidae). Org Divers Evol 7:181–194

Hillis-Colinvaux L (1980) Ecology and taxonomy of *Halimeda*: primary producer of coral reefs. Adv Mar Biol 17:1–327

Hoeksema BW (1989) Systematics and ecology of mushroom corals (Scleractinia: Fungiidae). Zool Verh (Liden), p 254

Hoeksema B, Rogers A, Quibilan M (2008) *Pocillopora effusus*. The IUCN red list of threatened species. Version 2015.2. www.iucnredlist.org

Hoeksema BW, Rogers A, Quibilan MC (2014a) *Pocillopora elegans*. The IUCN red list of threatened species. Version 2015.2. www.iucnredlist.org

Hoeksema BW, Rogers A, Quibilan MC (2014b) *Pocillopora eydouxi*. The IUCN red list of threatened species. Version 2015.2. www.iucnredlist.org

Jereb P, Roper CFE, Norman MD, Julian KF (eds) (2014) Cephalopods of the world. An annotated and illustrated catalogue of cephalopod species known to date. Volume 3. Octopods and Vampire Squids. FAO Species Catalogue for Fishery Purposes. No. 4, Vol. 3. Rome, FAO. 2014. 370 p. 11 colour plates

Johnson RF, Gosliner TM (2012) Traditional taxonomic groupings mask evolutionary history: a molecular phylogeny and new classification of the chromodorid nudibranchs. PLoS One. 7:e33479

Kaiser KL (1997) The recent molluscan marine fauna of the Islas Galápagos. The Festivus, supplement, 9:iii + 67

Keen AM (1958) Sea shells of tropical west America, 1st ed. Stanford Univ Press, Palo Alto, p 624

Kerstitch A, Bertsch H (2007) Sea of Cortez marine invertebrates, 2nd ed. Sea Challengers, Monterrey, CA, p 22

Kim W, Abele LG (1988) The snapping shrimp genus *Alpheus* from the eastern Pacific (Decapoda: Caridea: Alpheidae). Smith Contrib Zool, 454:1–113

Kroh A (2014a). *Diadema mexicanum* A. Agassiz, 1863. In: Kroh A, Mooi R (2014) World Echinoidea Database. Accessed through: Kroh A and Mooi R (2014) World Echinoidea database at http://www.marinespecies.org/echinoidea/aphia.php?p=taxdetails&id=513223

Kroh A (2014b) *Astropyga pulvinata* (Lamarck, 1816). In: Kroh A, Mooi R (2014) World Echinoidea database. Accessed through: Kroh A, Mooi R (2014) World Echinoidea Database at http://www.marinespecies.org/echinoidea/aphia.php?p=taxdetails&id=513138

Kroh A (2014c). *Toxopneustes roseus* (A. Agassiz, 1863). In: Kroh A, Mooi R (2014) World Echinoidea database. Accessed through: world register of marine species at http://www.marinespecies.org/aphia.php?p=taxdetails&id=513609

Kroh A (2015a) *Eucidaris thouarsii* (L. Agassiz & Desor, 1846). In: Kroh A, Mooi R (2015) World Echinoidea database. Accessed through: World register of marine species at http://www.marinespecies.org/aphia.php?p=taxdetails&id=513266

Kroh A (2015b) *Eucidaris galapagensis* Döderlein, 1887. In: Kroh A, Mooi R (2015) World Echinoidea database. Accessed through: World register of marine species at http://www.marinespecies.org/aphia.php?p=taxdetails&id=597545

Lessios HA (2005) Echinoids of the Pacific waters of Panama: status of knowledge and new records. Rev Biol Trop 53/Suppl 3:147–170

Lessios HA, Kessing BD, Pearse JS (2001) Population structure and speciation in tropical seas: global phylogeography of the sea urchin *Diadema*. Evolution 55(5):955–975

Lindner A, Cairns SD, Guzmán HM (2004) *Distichopora robusta* sp. nov., the first shallow-water stylasterid (Cnidaria: Hydrozoa: Stylasteridae) from the tropical eastern Pacific. J Mar Biol Ass UK. 84(6):943–947

López-Pérez RA, Mora-Pérez MG, Leyte-Morales GE (2007) Coral (Anthozoa: Scleractinia) recruitment at Bahías de Huatulco, western México: implications for coral community structure and dynamics. Pac Sci 61:355–369

Mah C (2014) *Nidorellia armata* (Gray, 1840). In: Mah CL (2014) World Asteroidea database. Accessed through: world register of marine species at http://www.marinespecies.org/aphia.php?p=taxdetails&id=254919

Mah C (2015) *Heliaster cumingi*. In: Mah CL (2015) World Asteroidea database. Accessed through: world register of marine species at http://marinespecies.org/aphia.php?p=taxdetails&id=255072

Maté TJL (2003) Corals and coral reefs of the Pacific coast of Panamá. In: Cortés J (ed). Latin American Coral Reefs. Elsevier Press B.V., Amsterdam, 387–417

McIlwain J, Choat JH, Abesamis R, Clements KD, Myers R, Nanola C, Rocha LA, Russell B, Stockwell B (2012). *Acanthurus triostegus*. The IUCN red list of threatened species. Version 2014.3. www.iucnredlist.org

Menzies RJ, Glynn PW (1968) The common marine isopod Crustacea of Puerto Rico. A handbook for marine biologists. Stud Fauna Curaçao Other Caribbean Islands 27:1–133

Molodtsova T (2014) *Antipathes galapagensis* Deichmann, 1941. Accessed through: World register of marine species at http://www.marinespecies.org/aphia.php?p=taxdetails&id=283898

Ng PKL, Davie P (2013) WoRMS Brachyura: World list of marine Brachyura (version 2013-03-04). In: Roskov Y, Kunze T, Paglinawan L, Orrell T, Nicolson D, Culham A, Bailly N, Kirk P, Bourgoin T, Baillargeon G, Hernandez F, De Wever A (eds) Species 2000 & ITIS catalogue of life. Species 2000: Reading, UK, 11 Mar 2013. Digital resource at www.catalogueoflife.org/col/

Norman M (2000) Cephalopods: a world guide. Hollywood import & export, Inc. p 320

Norman MD, Hochberg FG (2005) The current state of *Octopus* taxonomy. Phuket Mar Biol Cent Res Bull 66:127–154

Norman MD, Finn JK, Hochberg FG (2014) Family Octopodidae. In: Jereb P, Roper CFE, Norman MD, Finn JK (eds) Cephalopods of the world. An annotated and illustrated catalogue of cephalopod species known to date. Volume 3. Octopods and vampire squids. FAO species catalogue for fishery purposes [Rome, FAO]. 4(3) pp 36–2151–352

Norris JN (2010) Marine algae of the northern Gulf of California: Chlorophyta and Phaeophyceae. Smith Contrib Bot 94:1–276

Norris JN, Johansen HW (1981) Articulated coralline algae of the Gulf of California, Mexico, I: *Amphiroa* Lamouroux. Smithsonian Contrib Mar Sci 9:1–28

Obura D (2002) Coral species diversity. In: Obura D, Stine GS (eds) Phoenix Island: summary of marine and terrestrial assessment. New England Aquarium, Massachusetts, pp 10–12

Obura DO, Stone G (2003) Phoenix Islands. Summary of marine and terrestrial assessments conducted in the Republic of Kiribati, June 5-July 10, 2002. Primal Ocean Proj Tech Rep NEAq-03-03, New England Aquarium, Boston, p 13

Palacios MM, Muñoz CG, Zapata FA (2014) Fish corallivory on a pocilloporid reef and experimental coral responses to predation. Coral Reefs 33:625–636

Paulay G (2014) *Isostichopus fuscus* (Ludwig, 1875). Accessed through: world register of marine species at http://www.marinespecies.org/aphia.php?p=taxdetails&id=529516

Pedroche FF, Silva PC, Aguilar-Rosas LE, Dreckmann KM, Aguilar-Rosas R (2005) Catálogo de las algas marinas bentónicas del Pacífico de México.I. Chlorophycota. Ensenada, México: Universidad Autónoma de Baja California, pp 17–146

Pettibone MH (1986) Review of the Iphioninae (Polychaeta: Polynoidae) and revision of *Iphione cimex* Quatrefages, *Gattyana deludens* Fauvel, and *Harmothoe iphionelloides* Johnson (Harmothoinae). Smith Contrib Zool 428:1–43

Poupin J (2010) Biodiversité de l'Indo-Pacifique tropical français: 2514 espèces de crustacés décapodes et stomatopodes. Rapport scientifique de l'Institut de Recherche de l'Ecole Navale, Octobre 2010. p 76

Poupin J, By I (2003) Crustacea Decapoda and Stomatopoda of Easter Island and surrounding areas. A documented checklist with historical overview and biogeographic comments. Atoll Res Bull 500:1–50

Poupin J, Bouchard J-M, Albenga L, Cleva R, Hermozo-Salazar M, Solís-Weiss V (2009) Les Crustacés décapodes et stomatopodes, inventaire, écologie, et zoogéographie. In: Charpy L (coord.), Clipperton: environnement et biodiversité d'un microcosme océanique. IRD, Marseille, Muséum National d'Histoire Naturelle, Paris, Patrimoines Naturels, 68:163–216

Prakash S, Kumar TTA (2013) Feeding behavior of Harlequin Shrimp *Hymenocera picta* Dana, 1852 (Hymenoceridae) on sea star *Linckia laevigata* (Ophidiasteridae). J Threatened Taxa 5:4819–4821

Reimer JD, Hickman CP Jr (2009) Preliminary survey of zooxanthellate zoanthids (Cnidaria: Hexacorallia) of the Galapagos and associated symbiotic dinoflagellates (*Symbiodinium* spp.). Galapagos Res 66:14–19

Reyes-Bonilla H, Carriquiry JD, Leyte-Morales GE, Cupul-Magaña AL (2002) Effects of the El Niño-Southern Oscillation and the anti-El Niño event (1997–1999) on coral reefs of the western coast of Mexico. Coral Reefs 21:368–372

Rivera F, Martínez P (2011) Guía fotográfica de corales y octocorales: Parque Nacional Machalilla y Reserva de Producción Faunística Marino Costera Puntilla de Santa Elena. With the contribution of Odalisca Breedy, NAZCA and Conservación Internacional, Ecuador, p 78

Robertson DR (2012) Shorefishes of the tropical Eastern Pacific identification guide App. Ver 2.1

Robertson DR, Allen GR (2006) Shore fishes of the tropical eastern Pacific: an information system. Smith Trop Res Inst, Balboa, Panama

Robertson DR, Allen GR (2015) Shore fishes of the Tropical Eastern Pacific online information system. Version 2.0 Smith Trop Res Inst, Balboa, Panama

Rudman WB (2006) (May 29) *Elysia diomedea* (Bergh, 1894). In Sea Slug Forum. Australian Museum, Sydney. Available from http://www.seaslugforum.net/factsheet/elysdiom

Santelices B, Abbott IA (1987) Geographic and marine isolation: an assessment of the marine algae of Easter Island. Pac Sci 41:1–20

Schmitt WL (1940) The stomatopods of the west coast of America based on collections made by the Allan Hancock Expeditions, 1933–1938. Allan Hancock Found Pac Exped, 5:129–225. 217–219

Schwartz MA, Norenburg JL (2005) Three new species of *Micrura* (Nemertea: Heteronemertea) and a new type of heteronemertean larva from the Caribbean Sea. Carib J Sci 41:528–543

Smith TB, Fong P, Kennison R, Smith J (2010) Spatial refuges and associational defenses promote harmful blooms of the alga *Caulerpa sertularioides* onto coral reefs. Oecologia 164:1039–1048

Sweat LH (2009) *Caulerpa sertularioides*. Smithsonian Marine Station at Fort Pierce (On line) http://www.sms.si.edu/irlspec/Cauler_sertul.htm

ten Hove H (2014) *Spirobranchus spinosus* Moore, 1923. In: Read G, Fauchald K (eds) (2014) World Polychaeta Database. Accessed through: World Register of Marine Species at http://www.marinespecies.org/aphia.php?p=taxdetails&id=334836

van Ofwegen L (2014) *Muricea fruticosa* Verrill, 1869. Accessed through: World Register of Marine Species at http://www.marinespecies.org/aphia.php?p=taxdetails&id=287562

van Soest R (2015a) *Aplysina chiriquiensis* Díaz, van Soest, Rützler & Guzmán 2005. In: Van Soest RWM, Boury-Esnault N, Hooper JNA, Rützler K, de Voogd NJ, Alvarez de Glasby B, Hajdu E, Pisera AB, Manconi R, Schoenberg C, Janussen D, Tabachnick KR, Klautau M, Picton B, Kelly M, Vacelet J, Dohrmann M, Díaz M-C, Cárdenas P (eds) World Porifera database Accessed through: Van Soest RWM, Boury-Esnault N, Hooper JNA, Rützler K, de Voogd NJ, Alvarez de Glasby B, Hajdu E, Pisera AB, Manconi R, Schoenberg C, Janussen D, Tabachnick KR, Klautau M, Picton B, Kelly M, Vacelet J, Dohrmann M, Díaz M-C, Cárdenas P (2015) World Porifera database at http://wwwmarinespeciesorg/porifera/poriferaphp?p=taxdetails&id=234188

van Soest R (2015b) *Aplysina gerardogreeni* Gomez & Bakus 1992 In: Van Soest RWM, Boury-Esnault N, Hooper JNA, Rützler K, de Voogd NJ, Alvarez de Glasby B, Hajdu E, Pisera AB, Manconi R, Schoenberg C, Janussen D, Tabachnick KR, Klautau M, Picton B, Kelly M, Vacelet J, Dohrmann M, Díaz M-C, Cárdenas P (2015) World Porifera database at http://wwwmarinespeciesorg/porifera/poriferaphp?p=taxdetails&id=169650

Veron JEN (2000) Corals of the World. Vol. 1–3. Australian Institute of Marine Science and CRR, Queensland, Australia

Veron JEN (2002) New species described in corals of the world. Vol. 11. Aust Inst Mar Sci, Townsville, Queensland, Australia

Vogler C, Benzie J, Lessios H, Barber P, Worheide G (2008) A threat to coral reefs multiplied? Four species of crown-of-thorns starfish. Biol Lett 4:696–699

Wells JW (1983) Annotated list of the scleractinian corals of the Galápagos. In: PW Glynn, GM Wellington, Corals and coral reefs of the Galápagos Islands. Univ California Press, Berkeley-Los Angeles-London, pp 212–295

Wicksten MK (1987) A new species of hippolytid shrimp from the west coast of Mexico. Bull Southern California Acad Sci 86:27–33

Williams G, Breedy O (2004) The Panamic genus *Pacifigorgia* (Octocorallia: Gorgoniidae) in the Galápagos Archipelago. Proc California Acad Sci 55:54–87

WoRMS (2015) *Pherecardia striata* (Kinberg, 1857). In: Read G, Fauchald K (eds) (2015) World Polychaeta database. Accessed through: world register of marine species at http://marinespecies.org/aphia.php?p=taxdetails&id=209699

Wulff JL (1996) Do the same sponge species live on both the Caribbean and eastern Pacific sides of the Isthmus of Panama? Bull Royal Belgian Inst Nat Sci 66:165–173

Zigler KS, Lessios HA (2004) Speciation on the coasts of the new world: Phylogeography and the evolution of bindin in the sea urchin genus *Lytechinus*. Evolution 58:1225–1241

General Index

A
Abrasion, 50, 275, 354, 370, 380, 381, 392, 556
Abundance of infaunal bivalves, 384
Academy Bay, 254, 268
Acclimation, 505, 511, 512
Acclimatization, 417
Acclimatize, 416, 417, 502
Accretion, 3, 27, 77, 115, 129, 136, 151, 178, 179, 184, 188, 189, 197, 283, 371, 391, 394, 421, 518, 538, 553
Acroporid, 25, 26, 43, 137, 273, 326, 327, 330, 448, 527, 557
Acrothoracican barnacles, 381
Adaptations, 348, 428
ADCP (Acoustic Doppler Current Profiler), 87
Agariciids, 132, 450, 509
Agassiz A, 1, 5
Ages of 71 fossil assemblages, 41, 48
Agriculture, 554–557, 583
Agrochemicals, 454, 549, 554–586
Ahermatypic, 206, 211, 231
Albatross, 5, 6, 24, 208, 220
Algae, 16, 121, 122, 125, 129, 136, 139, 152, 156, 158, 160, 179, 181, 184, 189, 196, 205, 207, 208, 231, 234, 236, 278, 279, 292–297, 304, 305, 340, 342, 343, 345, 350, 352–358, 361, 372, 386, 392, 394, 406, 425, 453, 468, 504, 519, 520, 525, 527, 556, 557, 593, 596, 613, 616–623, 625–629
Algal bloom, 129, 343, 359, 552
Algal habitats, 340, 345
Algal lawn, 273, 274, 278, 387, 391
Algal refuges, 343, 345, 346
Algal turfs, 139, 183, 186, 297, 343, 344, 584
Alleles, 481
Allelochemical, 355
Allelopathy, 354, 556
Alpheid, 321, 329, 331
Alternate stable states, 357, 358, 359, 361
Amphinomid, 216, 217, 298, 300, 612
Amphipods, 228, 229, 230, 301
Ana Cecilia Fonseca Escalante, 369
Anakena, 155, 156
Anchor damage, 145, 152, 549, 551, 552, 575, 577, 578, 582
Anemones, 214, 330, 414
Angelfishes, 326
Annelids, 204, 215, 217, 230, 295, 302, 303, 317, 611, 621
Anomalies, 86–99, 102, 103, 151, 187, 252, 255–258, 369, 270, 272, 275, 278, 281, 332, 360, 426, 454, 456, 511, 518, 544, 553, 554, 557
Anomura, 222
Antecedent topography, 62, 71, 178, 182, 262, 346
Anthropogenic, 17, 76, 103, 204, 208, 237, 238, 254, 281, 349, 351, 357, 436, 470, 518, 520, 536, 549, 550, 557, 567, 578, 579, 585–587
Aragonite, 80, 178, 340, 371, 406, 417, 471, 512, 518, 526, 553
Arcturus, 6
ARC *Malpelo*, 22
Arizpe O, 3, 13
Asexual, 19, 21, 128, 132, 139, 198, 278, 282, 292, 306, 330, 331, 332, 371, 384, 394, 395, 396, 422, 436, 452, 462, 466, 468, 469, 470, 584
Asexual coral reproduction, 331
Asexual pocilloporid and *Gardineroseris planulata* recruits, 468
Asexual propagation, 194, 292, 306, 332, 394, 395, 422, 462, 472
Asexual recruitment, 128, 139, 462, 466, 468
Asexual reproduction, 19, 330, 331, 462, 466, 472, 584
Aspidosiphonids, 382
Assemblages, 22, 27, 29, 40, 43, 46, 48, 114, 128, 134, 136, 155, 183, 193, 204, 225, 227, 273, 278, 372, 373, 382, 383, 385, 395, 396, 470, 582
Asteroids, 20, 222, 304, 323
Atmospheric forcing, 60, 81
ATPase6, 487
Autotrophy, 235, 296, 300, 486, 509, 540
Avoidance, 324, 502, 504, 511, 512
Azooxanthellate, 4, 7, 26, 125, 133, 149, 231, 280, 439, 441, 447, 449, 461, 526, 583, 605, 606
Azuero Peninsula, 191, 494

B
Bacteria, 297, 303, 340
Bacterioplankton, 303
Bahía Cupica, 138
Bahía Honda, 11, 12, 205, 441, 466
Balistids, 293, 304, 345, 372, 395, 521
Baltra Island, 41
Barnacles, 136, 253, 276, 278, 293, 302, 305, 380, 381, 520, 621, 622
Bartolomé, 182, 194, 209
Basalt, 16, 24, 121, 124, 136, 154, 184, 186, 190, 194, 205, 282, 305, 322, 449, 468, 522, 568, 606
Baseline, 14, 20, 23, 25, 238, 528, 575, 581, 583
Beebe William, 6, 323
Benthos, 78, 230, 231, 236, 283, 292, 296, 300, 304, 3078, 343, 470
Beta stage, 374
Biodiversity, 15, 28, 29, 152, 204, 206, 207, 208, 211, 230, 232, 233, 234, 235, 236, 238, 309, 370, 501, 510, 566, 567, 569, 578–585, 588, 587
Bioeroders, 179, 222, 370–373, 378–382, 386, 387, 397, 521

P.W. Glynn et al. (eds.), *Coral Reefs of the Eastern Tropical Pacific*,
Coral Reefs of the World 8, DOI 10.1007/978-94-017-7499-4

Bioerosion, 19, 21, 23, 27, 50, 115, 119, 132, 135, 160, 178, 187, 193, 194, 197, 214, 227, 253, 254, 263, 274, 279, 280, 282, 308, 322, 360, 370–373, 375, 377–382, 386–396, 397, 406, 436, 466, 470, 518, 520, 521, 523, 525–528
Biogeography, 14, 27, 316
BioMar (CDRS), 23
Biomass, 75, 131, 135, 281, 292, 294, 295, 297, 298, 299, 300, 302, 304, 305, 306, 308–310, 323, 344, 347, 349, 352, 353, 355, 373, 381, 456, 525, 526, 568, 584
Biotic processes controlling algal biomass, 347, 352
Biotopes, 114, 115, 125, 565, 566, 570
Bioturbation, 303, 321, 421
Bite marks, 125, 209, 325, 326, 396
Bivalve densities, 383
Bivalve mollusks, 23, 304, 372, 382, 466
Bjerknes J, 86, 88, 102
Bleaching, 116
Boreholes, 183, 466
Borers, 379–383, 397, 521
Bouquet (gonad), 437, 430–441, 445
Brachyuran, 121, 159, 220, 277, 297
Branching coral, 178, 331, 346, 521, 526
Bray-Curtis index, 50
Broadcast spawner, 438, 450
Broadcast-spawning corals, 437
Brooding, 133, 436, 438–440, 444, 448, 449, 451, 452, 460, 462, 470, 471, 481
Brown algae, 207, 208, 342, 343, 348
Browsers, 323, 353
Bryozoans, 16, 110, 115, 231, 281, 293, 295, 305, 324, 497, 520, 625
Budding, 441, 462
Burrow
  Burrowing, 340, 381, 397, 466
  Burrowing shrimps, 381
  Burrows, 216, 217, 229, 380, 381, 385
Butterflyfishes, 324, 325, 330

C

Cabo Corrientes, 74, 135
Cabo Pulmo, 14, 115, 181, 182, 216, 217, 292, 304, 305, 308, 377, 379, 392, 394, 441, 465, 468, 470, 523, 553, 565–568, 572
Cabo Pulmo National Park, 567, 568, 572
$CaCO_3$ budget, 517, 521, 522, 524, 526–528
Calcification, 196, 208, 236, 253, 283, 303, 324, 331, 354, 406, 417, 452, 470, 522, 553, 558
Calcite, 340
Calices, 423
California
  California Current, 68
  California Current Water, 68
  California region, 43
Cañada Coronados reef, 49
Canales de Afuera, 135, 494, 595, 600
Canales de Tierra, 188, 191, 197, 341
Caño Island, 29, 180, 182–184, 273, 324, 328, 383, 392, 394, 447, 452, 459, 508
Carangids, 345
Carbon, 75, 293, 294, 296, 300, 321, 412, 540
Carbonate
  Bioerosion, 390
  Budget, 370, 391, 519, 520, 522, 523, 527
  Chemistry and ocean acidification, 78
  Production, 304, 306, 391, 522, 523
Caribbean, 2, 18, 27, 40, 43, 46, 50–53, 64, 108, 151, 188, 198, 205, 208, 223, 228, 236, 237, 252, 253, 259, 281, 297, 302, 343, 353, 354, 375, 377, 378, 386, 393, 408, 430, 448, 450, 452, 456, 457, 460, 478, 518, 519, 521, 554, 557, 570, 571, 594, 596, 611, 615, 620
Caridean, 159, 220, 302, 329, 330, 332
Carnivore, 281, 283, 291, 293, 297, 298, 304, 310, 612, 625, 629, 630
Carnivorous, 292, 298, 298, 301, 624
Cascade, 152, 292, 305, 308, 309, 371, 391, 454
Castilla JC, 24
CBD, 550, 566, 571
CCA, 340, 343–346, 350, 520, 527
Cea Egaña A., 24
Cellular automaton model, 361
Cement, 340, 520, 523, 525, 526
Cementation, 178, 181, 253, 372, 467, 520, 522, 525–527
Cenozoic
  Fossil assemblages, 40, 48
  Hermatypic corals, 43
  History, 39
Central American Seaway, 53
Central Pacific El Niño, 90, 93
Cetaceans, 28, 206
Chaetodontids, 281, 293
Champion, 180, 182, 192, 193, 279, 280, 387, 506, 614
Charles Darwin Research Station, 15, 25, 160, 255
Chasmoliths, 371
Chemical defenses, 347, 349, 354, 355
Chemical dissolution, 375, 380, 381, 522
Chiapas, 116, 207
Chile, 1, 3, 13, 15, 23, 25, 28, 40, 108–110, 152, 153, 160, 206, 278, 416, 507, 574, 584–588, 594, 598–601, 603, 614, 616, 617, 623, 624, 628
Chlorophyll, 75, 262, 276, 295, 348, 505, 537
Chlorophyte, 372
Chocó, 579, 580, 583
Chocó Biogeographic Province, 580
Chubs, 353
CIMAR (Costa Rica), 15, 16, 575
Circulation, 43, 59, 60, 71, 72, 74, 277, 460, 480, 535, 536, 541, 557, 594, 605
Circumglobal, 487
Cirripedes, 395
CITES, 566, 571, 583, 585
Civil unrest, 27
Clades, 53, 295, 317, 407–409, 411, 413, 414, 426, 448, 524
Cleaning, 331
Climate
  Change, 518, 549, 550, 552, 553, 557, 558, 585, 587
  Change on reef algae, 340
  Impacts, 76
  Models, 99
Clionaid, 373, 377, 396, 521, 525, 526
Clipperton Atoll, 112, 113, 121, 122, 130, 159, 178, 180, 182, 186, 194–196, 204, 206, 207, 219, 222, 231, 238, 259, 278, 292, 323, 351, 410, 423, 478, 480, 490, 507, 509, 510, 541, 543
Clonal, 237, 331, 361, 452, 462, 470
Clones, 361, 466, 467
CO_2, 75, 76, 78, 236, 371, 417, 518, 519, 522, 525–528
Coastal alterations, 554

Coastal-Marine Faunal Production Reserve Puntilla Santa Elena
Coastal-Marine Wildlife Refuge Pacoche, 583
Cocos Islands, 160, 179, 425
Coenosarc, 329
Coiba
Management Plan, 567
MPA, 134
National Park, 20, 28, 566, 567, 568, 571–582, 587
Cold-water bleaching, 268–270, 272, 406, 407, 416
Colombia, 550
Colombian Pacific coastline, 579
Colonies, 3, 9, 144, 422–424, 426, 452, 453, 504, 550
Colonization, 225, 235, 238, 274
Colony densities
of recently recruited and juvenile coral taxa, 463
of sexual and asexually-derived coral recruits, 469
Colony morphology, 132, 237, 331, 430, 449, 458, 468, 503, 555, 601
Competition, 108, 158, 160, 234, 329, 339, 341, 347, 352, 355, 358, 361, 428, 430, 470, 501, 504, 556, 557
Competition between coral and algae, 354
Competitors, 348, 430
Concepción Bay, 3
Connectivity, 27, 110, 119, 120, 123, 182, 294, 349, 416, 426, 569, 583
Conservation, 2, 13, 20, 23, 27–29, 237, 316, 370, 550, 551, 566–571, 575, 576, 579–581, 584, 586, 587
Conservation and management initiatives, 28
Constraints in detecting *Symbiodinium*, 409
Consumers, 208, 291, 292, 294, 295, 297–300, 304, 309, 310, 315, 341, 349, 352, 361
Consumption, 220, 233, 296, 297, 299, 300–302, 304, 308, 309, 315, 328, 329, 332, 352, 353, 355, 470, 525
Contadora, 135, 179, 182, 186, 188, 190, 191, 197, 229, 230, 233, 296, 346, 350, 352, 355, 356, 361, 362, 463, 464, 494, 578, 595, 613, 617
Convention
on Biological Diversity, 570
on International Trade in Endangered Species (CITES), 550
Convergence, 62, 63
Copepods, 228, 293, 301, 305
Coprophagous, 296, 302, 309
Coprophagy, 302
Coral-algal symbiosis, 406, 407, 414, 417
Coral growth, 2, 13, 17, 22, 27, 80, 120, 121, 129, 132, 134, 141, 151, 183, 184, 195, 196, 282, 321, 331, 383, 406, 412, 417, 421, 422, 503, 507, 508, 512, 520, 524–526, 528, 579
Coral Nui Nui, 559
Coral reef development, 1, 124, 144, 178, 183, 194, 197, 254, 341, 521, 550, 580
Coral species from the Colombian Pacific, 137
Coral species presence (1)/absence (0) data matrix, 160
Coral species richness, 125, 143, 147, 153, 159, 281
Coral-algal symbiosis, 294, 406, 407, 409, 414, 417
Corallimorpharians, 358
Coralliths, 27, 506
Corallivores, 222, 232, 273, 274, 278, 298, 304, 308, 309, 316, 317, 323, 330, 394, 470
Corallivory, 21, 27, 108, 296, 298, 315–317, 323, 330, 332, 436, 466, 521
Corallum, 140, 153, 394
Corals and Coral Reefs of the Galápagos Islands, 511
Corals from middle Pleistocene to Recent, 52
Corals from Pacific Costa Rica, 126
Corona del Diablo, 151, 605, 610
Cosexual, 450, 452
Cosexuality, 450
Cosmopolitan, 206, 372, 480, 490
Costa Rica
coastal current, 74, 120, 263, 441, 479
dome, 62, 70, 73–75, 341
mainland, 17, 125–128, 130, 216, 392
Couthouy J.P., 2, 4
Crab zoea, 229
Crabs, 11, 220, 232, 234, 293, 296, 302, 306, 316, 328, 329, 331, 380, 381, 629
Cretaceous, 48, 184
Cromwell Current, 149, 511
Crossland C., 6, 19
Crustaceans, 9, 16, 20, 110, 115, 122, 125, 135, 204–206, 216, 218, 220, 230, 236, 238, 273, 293, 294, 296–298, 301–303, 308, 324, 369–371, 380, 394, 521, 576, 611, 616, 625, 630, 631
Crustose coralline algae, 129, 184, 185, 193, 278, 279, 293, 294, 305, 323, 339, 342, 355, 361, 467, 519, 527
Cryptic
cryptofauna, 228
cryptos, 304
cryptic corallivores, 326
Culebra, 2, 17, 26, 126, 128, 129, 131, 182, 392, 457, 506, 596
Currents, 40, 59, 60, 62, 68, 99, 144, 204, 408, 458, 478, 480, 504, 537, 577, 580, 611
Cyanobacteria, 293, 294, 340, 342, 344–346, 349, 350, 355, 356, 372
Cycling, 349
Cyclone formation, 65
Cytochrome B (CytB), 486
Cytochrome Oxidase I (COI), 220, 483, 485, 486, 489, 490, 492–494, 496, 498

D

Damselfish lawn, 278, 391
Damselfishes, 225, 234, 273, 274, 278, 293, 297, 308, 326, 330, 353, 387
Dana JD, 4, 5, 6, 52, 194, 263, 277, 317, 321, 406, 599, 600, 617
Darwin CR, 3, 15, 22, 147, 152, 194, 326, 360, 390, 414, 472, 482, 509, 518, 545
Darwin Island, 147, 149, 151, 152, 194, 268, 282, 325, 326, 390
DD (Degree Days), 264–269, 282
Decadal, 86, 94, 96, 537, 541, 542, 544
Decapods, 205, 206, 220, 230
Decompose, 293, 296
Decomposition, 300, 349
Defend
defended, 321
defensive, 233, 295, 308, 331
Deforestation, 134, 184, 468, 470, 472, 567, 569, 580, 581
Denature Gradient Gel Electrophoresis (DGGE), 409, 410, 412, 427
Departamento de Biología Marina (Panama), 18
Depauperate, 2, 196, 208, 223, 225, 237, 238, 309, 316, 332, 408
Depositional, 48, 198, 283, 526
Deposits, 4, 5, 43, 48, 53, 121, 156, 296, 300, 303, 316, 324, 504, 538
Depth refugia
locations and characteristics, 507
oceanic islands, 509
species-specific aspects, 509
vertical connectivity, 510
Detritivores, 292, 296, 297, 302, 303, 309
Detritivory, 302
Detritus, 273, 293, 294, 296, 304, 308, 329, 619, 623, 625
Devil's Crown, 186, 192, 193, 282, 323, 354, 506, 523
DGGE, 409, 410, 412, 427
Diets, 297, 298, 302, 324, 354, 394
Dinoflagellates, 272, 425, 578
DiSalvo LH, 24, 25, 152, 154, 156, 206, 252, 569

Disconformities, 183
Discontinuities, 183
Dispersal, 19, 26, 28, 40, 52, 108, 110, 121, 153, 159, 204, 227, 263, 277, 324, 331, 384, 394, 396, 408, 426, 427, 436, 449, 454, 459, 462, 470, 471, 479, 487, 491, 497, 554, 568, 569
Dissolution, 50, 78, 236, 375, 381, 518, 521, 522, 525, 528
Distribution of *Symbiodinium* in the ETP
  host systematic distribution, 412
  environmental control, 195, 414
  biogeographic and taxonomic gaps, 414
  stability versus change, 416
Disturbances, 17, 19, 21, 26–28, 102, 108, 110, 119, 131, 132, 135, 136, 143, 157, 159, 160, 195, 198, 235, 238, 253, 254, 259, 270, 273, 278, 281, 282, 297, 331, 345, 347, 349, 352, 356, 358, 360, 361, 384, 393, 395, 409, 416, 436, 450, 452, 471, 502, 504, 509, 510, 511, 512, 521, 527, 550, 554, 563, 564, 579, 583
Divergence, 17, 63, 70, 71, 73, 75, 89, 2266, 317, 430, 478, 481, 486, 487, 491
Diversity
  diversity foci ('hot spots'), 20
DNA, 407, 408, 411, 427, 480, 481
DNA fingerprinting, 409, 412
Doldrums Expedition, 11
Durham JW, 3, 7, 9, 13, 41–46, 48, 50, 53, 117, 121, 125, 602, 603, 606, 607

E

Early Pliocene, 43, 48, 49, 52, 99
Easter Island, 2, 3, 13, 23–25, 28, 108, 110, 153, 160, 180, 182, 195, 206, 231, 238, 304, 351, 416, 477–480, 486, 490, 492, 493, 509, 567, 568, 576
Eastern Pacific
  barrier, 19, 277, 478, 502, 503
  coral fauna, 40, 50, 52–54, 139, 316
  environmental conditions, 27
  faunal gap, 110
  warm pool, 59, 62, 68, 75, 258
Eastern tropical Pacific, 1–4, 6, 27, 28, 50, 60, 68, 79, 90, 102, 108, 110, 120, 121, 130, 177, 178, 204, 205, 227, 266, 280, 292, 303, 305, 317, 326, 328, 339, 340, 369, 405, 406, 421, 422, 424, 427, 428, 441, 471, 477, 479, 512, 517, 518, 523, 525, 527, 535–538, 542–544, 570, 575
Eastern tropical Pacific warm pool, 59, 62, 68, 80
Echinoderms, 16, 20, 110, 115, 122, 204, 205, 206, 222, 223, 236, 277, 298, 303, 304, 323, 369, 386, 490, 613, 615, 616, 621
Echinoids, 19, 40, 54, 179, 222, 230, 293, 297, 369, 370, 371, 386, 387, 391–394, 521, 523
Ecological processes controlling algal
  populations and communities, 339
Ecopath, 297, 306, 308
Ecosim, 309
Ecosystems, 16, 21, 28, 78, 116, 124, 126, 197, 204, 206, 208, 218, 227, 228, 231, 235, 236, 263, 291, 292, 294, 297, 298, 301, 309, 310, 315–317, 332, 358, 360, 363, 370, 371, 373, 378, 416, 425, 472, 502, 518, 550–556, 559, 562, 564–566, 569, 571, 580
Ecotourism, 552, 561
Ecuadorian mainland, 23, 147, 192, 222, 508
Eddy, 74, 75, 94, 504
Edmondson C.H., 2, 263
Eighth International Coral Reef Symposium, 19
Ekman
  barrier, 110, 280
  pumping, 71, 73, 74, 89
El Niño, 13, 17, 19–23, 40, 65, 80, 85–88
El Niño seawater warming, 512
El Niño-Southern Oscillation, 179, 254, 406, 407, 422, 501, 511, 518, 535
El Pelado Marine Reserve (REMAPE), 559, 567, 570
El Puma, 14
El Salvador, 1, 3, 16, 110–113, 123, 124, 141, 182, 183, 204, 205, 207, 211, 214, 220, 222, 224, 343, 480, 494, 495, 508, 553, 554, 557, 570, 576, 582, 596, 599, 600, 603, 609, 613
Elevated thermal bleaching effects, 456
Embryos, 439
Endangered species, 20, 550
Endemics, 40, 51, 108, 111, 206, 480, 503
Endemism, 153, 206, 223, 227, 238, 408, 503, 561, 564, 566, 567
Endoliths, 370, 373, 383, 394
Endosymbiont, 119, 132, 151, 263, 486, 502, 503, 525, 578
Enigmatic species, 25
Ensenada de Utría, 136, 138, 139, 182, 191, 581
ENSO
  indices, 86, 87, 254
  mechanisms, 86, 88, 90, 535
Eocene, 39–41, 43, 46, 48–50, 53, 54, 478, 527
EOF analysis, 91, 96
EP El Niño, 91, 92, 94
Epibenthos, 308, 470
Epilithic bioeroders, 370, 382
Epiphytes, 208, 342, 346, 347, 355, 356
Equatorial currents, 59, 73, 99, 277, 479, 537
Equatorial Surface Water, 60, 68, 71
Equatorial Undercurrent, 68, 69, 73, 263, 511, 537, 543, 544
Erosion, 125, 152, 156, 179, 194, 231, 234, 235, 279, 281, 304, 327, 369–372, 375, 379, 381, 382, 385–387, 391–395, 521, 522, 527, 575, 578, 579, 581, 583
Estuary, 16, 116, 124, 480, 562, 567
Etching, 369, 370, 375
ETP Seascape, 569
EUC (Equatorial Under Current), 73, 74, 149, 151, 263, 544
Eunicids, 217, 380
Eutrophication, 292, 310, 416, 452, 575, 581, 582
Evolution, 10, 14, 39, 50, 87, 89, 90, 92, 93, 291, 423, 425, 427, 477
Excavate
  sponges, 370, 375, 466, 517, 526
  excavators, 323, 353
Exploitation, 305, 564, 566, 576
Extant, 53, 181, 486, 503, 510, 528
Extinction pulses, 54
Extinctions, 19, 26, 27, 51, 108, 208, 270, 281, 283, 323, 502, 511, 524
Extraction, 13, 25, 152, 190, 554, 556, 561, 569, 575, 576, 583

F

Facilitation, 347, 352, 372
Facilitative interactions, 339, 355
Faunal shifts, 51
Faviid, 153
Feces, 296, 302, 324, 332
Fecundity, 132, 435, 436, 443, 444, 459–461, 470–472
Fecundity estimates, 459, 460
Feeders, 196, 216, 218, 230, 231, 276, 291, 293, 296, 297, 298, 300–305, 308, 309, 310, 324, 371, 406
Feeding, 2, 19, 22, 121, 139, 228, 229, 232, 235, 273, 279, 292, 296–298, 300–306, 308, 309, 315–317, 321–332, 354, 380, 381, 383, 386, 393–396, 456, 459, 509, 521, 525, 569, 577, 614, 616, 628

Fernandina, 150, 151, 192, 194, 197, 390, 623
Fertilization, 277, 439, 448, 454, 456, 470, 525
Filamentous turfs, 294, 343
Fish invertivores, 298, 393, 394
Fisheries, 18, 152, 301, 527, 550, 551, 553, 555, 561–563, 566–568, 571, 576, 582
Fishers, 305, 309, 554, 564, 566, 577, 580
Fishery, 555, 561
Fishes, 6, 9, 16, 19–21, 23, 40, 59, 60, 110, 115, 121, 122, 124, 125, 135, 136, 139, 153, 168, 205, 206, 208, 222, 223, 225–227, 230–238, 274, 277–279, 292, 294–298, 300–302, 304, 308–310, 315, 317, 323, 326, 327, 329, 332, 345, 346, 349, 350, 352–354, 358, 370–373, 384, 387, 393, 395, 396, 466, 470, 477, 486, 490, 491, 495, 521, 552, 554, 567, 569, 576, 577–580, 584, 616, 620, 630, 631
Fishing, 143, 152, 238, 280, 304, 309, 323, 352, 391, 393, 551, 552, 554, 561, 563, 564, 566, 567, 569, 576, 578, 583
Flavors of ENSO, 90
Fleshy macroalgae, 208
Floreana, 9, 24, 148, 150, 151, 180, 192, 193, 275, 291, 304, 305, 308, 309, 390, 441, 446, 468, 469, 509, 523, 598, 605–610, 616, 625, 628
Foraging, 321, 370, 380, 384, 386, 392–396, 436, 466, 469, 470, 554
Foraminiferans, 281, 538, 625
Fossil, 4, 39, 40, 41, 43, 46, 48–54, 116, 178, 179, 181, 192, 195, 372, 512, 518, 520, 527, 528, 538, 540, 542, 545
Fragmentation, 136, 139, 275, 292, 331, 371, 372, 384, 394–396, 430, 436, 436, 439, 450, 452, 453, 459, 462, 466–471, 521, 564, 577, 579
Fragments, 5, 16, 27, 113, 125, 137, 179, 190, 218, 228, 234, 298, 324–326, 331, 332, 345, 379, 384, 393–395, 412, 438, 452, 462, 466–469, 554, 577
Frameworks, 2, 3, 15, 16, 23, 25, 26, 88, 110, 114, 115, 121, 129, 130, 132, 135, 136, 144, 147, 149, 151–154, 160, 177–179, 182–184, 186, 188–198, 228, 234, 235, 237, 274, 275, 279, 280, 281–283, 295, 298, 304, 305, 310, 330, 345, 357, 361, 379, 386, 387, 415, 466, 504, 506, 520, 521, 527, 528, 541, 523, 555, 561, 566, 567, 623
Frondose, 344, 346
FST statistic, 477, 481
Funding organizations, 14, 15
Functional differences in *Symbiodinium*, 411
Functional group, 155, 159, 204, 303, 304, 306, 308, 343
Functional-form groups of predominant
  macroalgae, 344
Fungi, 205, 278, 372
Fungiid corals, 114, 323, 436, 509
Future MPA needs, 627
Future of coral-algal symbiosis
  interactions between high temperature and high CO_2, 417
  responses of reef coral symbiosis to climate change, 405, 406, 417
  Ratchet hypothesis, 417
Future research agendas of fossil coral fauna, 53

G

Gálapagos Islands, 2–4, 6, 13, 15, 20, 22, 23, 25, 28, 43, 108, 109, 110, 112, 130, 141, 143, 145, 147, 160, 192–194, 196, 197, 205–210, 214, 216, 223, 230, 237, 252, 264, 265, 267, 268, 271, 274, 275, 279, 281, 304, 308, 320, 322, 324, 328, 353, 354, 380, 384, 387, 390, 418, 441, 447, 449, 452, 454, 455, 459, 469, 471, 509, 519, 521, 522, 527, 563, 569, 577, 582, 584
Galápagos Marine Reserve, 23, 558, 565
Galapagos Research (journal), 23
Galera-San Francisco Marine Reserve, 559, 565, 566
Gamete bundles, 448, 450
Gametes, 436, 437, 448–450, 454–460, 460, 470, 471, 481
Gametocyte description scheme, 437, 438
Gametogenesis, 132, 436, 437, 439, 443, 448, 450, 455, 456, 458–460, 470, 471
Gamma stage (sponges), 374
Gastrochaenid, 383–385
Gastrodermis, 437–440
Gastropods, 129, 219, 277, 293, 296–300, 302–305, 327, 328, 332, 343, 394, 520, 620, 628
Gastrovascular cavity, 409, 439, 440, 445
Gene flow
  along the coast of the eastern Pacific, 13, 23, 74, 181, 183, 211, 263, 409, 478, 480, 491, 497
  between conspecific populations, 418
  between the central and the eastern Pacific, 19, 40, 122, 123, 478, 480, 481, 486, 490, 491, 497, 503
  between the outer eastern Pacific islands, 123, 481, 490
  between the outer eastern Pacfiic, 481, 487, 490
  islands and the mainland coast, 126, 142, 178, 478
  FST for microsatellite data, 481
  ΦST for sequence data, 481
Genes, 425, 477, 478, 481, 487, 490, 491, 497, 505
Genet, 430, 462
Genetic drift
  exchange, 481
  markers, 490
  structure, 426
Genomes, 423, 481
Genotyping, 466
Geochemical tracer (Geochemical records), 277, 537, 540
Geographic pattern of coral species richness, 112
Glacial, 99, 236, 537, 538
Global
  comparison of echinoid bioerosion, 387, 393
  SST anomaly pattern, 96
  tropical moored buoy array, 87
  warming, 76–78, 86, 93, 99, 103, 283, 518, 527, 536, 544, 545
Golfo Dulce, 128, 129, 159, 183, 184, 188, 191, 205, 216, 218
Gonad diameters, 447
Gonad stages, 437, 438
Gonads
  Stage I, 437, 438
  Stage II, 437, 438
  Stage III, 437, 438
  Stage IV, 437, 438
Gonochoric, 440, 448, 449, 451, 452
Gonochorism, 448, 449, 450
Gorgona Island, 6, 20, 26, 135, 136, 138, 139, 192, 195, 223, 225, 230, 298, 306, 317, 324, 327, 331, 380–382, 385, 386, 391, 393–395, 562, 563, 569, 581, 582
Gorgona Island Natural National Park, 20
Gould A.A., 4
Grazers, 152, 179, 297, 298, 304, 323, 370, 372, 373, 380, 387, 391, 395, 521
Grazing, 119, 136, 152, 156, 194, 235, 279, 298, 322, 323, 324, 325, 326, 328, 343, 345, 353, 354, 356, 371, 373, 384, 386, 387, 391, 392, 393, 394, 395, 396, 453
Green algae, 340, 342, 343, 358
Grounding, 577
Guadalupe Island, 43
Guanacaste Conservation Area, 550
Guatemala, 16, 68, 124, 182, 183, 208, 211, 220, 317, 343, 415, 416
Guilds, 208, 292, 297, 298, 303, 304, 309, 352, 393
Guineafowl puffer, 226, 323
Gulf of California, 2, 3, 4, 9, 10, 13, 14, 26, 28, 43, 46, 48, 50, 51, 52, 53, 54, 74, 108, 110, 111, 114, 115, 118, 120, 181, 204, 218,

223, 227, 231, 263, 270, 278, 292, 303, 304, 317, 322, 324, 327, 330, 352, 377, 392, 393, 406, 408, 409, 414, 423, 424, 426, 468, 470, 480, 497, 511, 550, 551, 552, 569, 577, 578, 579, 582, 595, 602, 607, 610, 612, 613, 614, 615, 616, 617, 618, 619, 620, 621, 622, 623, 624, 625, 626, 627, 628, 629, 630, 631
Gulf of California Water, 68
Gulf of Chiriquí, 6, 9, 19, 20, 111, 132, 133, 160, 178, 184, 187, 188, 189, 191, 195, 197, 205, 211, 223, 227, 232, 235, 265, 271, 274, 280, 294, 298, 301, 304, 306, 317, 325, 328, 330, 341, 350, 352, 353, 391, 406, 444, 454, 455, 457, 462, 490, 507, 508, 509, 510, 522, 527, 541, 559, 561, 611, 615, 616, 621
Gulf of Papagayo, 16, 17, 131, 159, 182, 183, 184, 341, 480, 508, 511, 577
Gulf of Tehuantepec, 74, 110, 131, 181, 293, 341, 406, 454, 511

H

Habitats, 14, 25, 27, 28, 111, 112, 125, 128, 140, 149, 152, 157, 216, 220, 222, 225, 227, 231, 234, 328, 330, 332, 340, 345, 346, 371, 372, 381, 382, 390, 414, 416, 467, 503–505, 551, 552, 561, 562, 566, 570, 576, 577, 593–631
Hancock A, 6–7, 9, 207, 236
Hanga Oteo, 559, 568
Haplotype, 426, 427, 481, 486, 495, 496
Heat shock protein, 423
Heat tolerant, 405, 411
Herbicides, 582
Herbivores, 170, 221, 273, 294, 297–298, 350, 352–354, 393, 576, 569
Herbivory, 274, 277, 294, 309, 323, 339, 342, 343–346, 352–356, 362, 501
Hermaphrodites, 442, 448, 449, 450, 451
Hermaphroditic development, 449
Hermaphroditism, 448, 450, 452
Hermatypic, 151, 206, 211, 216, 231, 509, 512
Hermit crabs, 293, 299, 302, 305, 328–329
Heterotrophy, 235, 300, 301, 509, 525, 540
Heterozygosity, 481
Hiatuses, 283, 541
Histological preparations, 436–440, 443, 455
HMS Beagle, 3
Holobiont, 14, 52, 293, 294, 304, 409, 411, 412
Holocene, 110, 113, 177, 178, 179, 181, 184, 188, 189–190, 191–192, 197, 538, 583
Holoplankton, 296, 301
Holothurians, 222, 296, 302, 303, 305, 554
Honduras, 16, 110, 182, 480
Hopkins Stanford Galápagos Expedition, 10
Hosts, 9, 14, 233, 234, 273, 281, 302, 327, 328, 330, 347, 408, 409, 411, 412, 414, 416
Huatulco, 6, 9, 14, 16, 116, 119, 120, 159, 220, 275, 276, 373, 392, 393, 445, 454, 462, 467, 551, 582
Humboldt Current, 68, 74, 145, 567
Hurricane, 65, 94, 115, 119, 258, 276, 351, 472, 579
Hybridization, 450, 457, 481
Hydrocarbons, 454, 582
Hydroclimatic, 543
Hydrodynamic, 407, 414, 552
Hydrological, 267, 543, 562
Hysteresis, 357, 358

I

Ichthyoplankton, 228, 229
Iguana Island, 28, 180, 186, 191
Immigration, 51, 223, 227, 481, 583
Indirect sources of damage, 580–583
Infauna, 380, 382
Inhibition, 196, 349, 581
Insecticides, 582
Interactions, 17, 22, 28, 51, 97, 110, 136, 152, 291, 303, 304, 306, 308, 323, 325, 331, 332, 354–356, 370, 408, 417, 503, 569
Interdecadal, 537, 544
Intermediate disturbance hypothesis, 233
Internal bioeroder, 179
Bioerosion, 23, 27, 50, 115, 119, 178–179, 253, 254, 279–280, 370, 371, 373, 378, 381–382, 387, 391, 392, 393–394, 395, 521
Internal transcribed spacer 2, 408, 485, 493
International Coral Reef Initiative (ICRI), 550
Intertropical Convergence Zone (ITCZ), 62, 81, 142, 536, 537
Intra–colony refuges, 503–504
Intra-reef and inter-reef refuges, 504–505
Invasive species, 580
Invertebrate, 122, 132, 135, 205, 206, 280, 283, 294, 296, 298, 303, 308, 317, 352, 354, 358, 370, 394, 550, 567
Invertivores, 229, 232, 292, 293, 298, 304, 309, 394, 576
Isla del Coco, 17, 130–131, 205–210, 216–218, 228, 231–232, 237, 392, 462, 480, 490
Isla del Coco National Park, 231, 558
Isla Isabela, 149, 151, 192–194, 324, 466, 467, 537
Isla Saboga, 179, 182, 189, 190–191, 267–268, 349
Isla Wolf, 544
Islands and protected areas of the Gulf of California, 3–4, 9–10, 13, 26, 51–54, 114–116, 181, 182, 204, 216, 227, 231, 259, 263, 278, 317, 392, 426, 470, 582
Isolation, 10, 108, 223, 408, 481, 487, 497
Isolation of ETP reefs, 50, 80, 236–237, 300, 302, 342–343, 345, 349, 352–355, 361, 373, 375, 386, 406, 408, 409, 414, 417, 521
Isopoda, 229
Isothermal, 506
Isozyme, 480, 486, 487, 497
Isthmus of Panama, 18, 62, 223, 408, 486
ITS2, 424, 427, 485, 490
IUCN MPA classification, 570

K

K/T surviving genera, 50
Kelvin waves, 85, 88, 89, 90, 93, 151, 511
Keystone predators, 330–331
Keystone species, 330
Kiritimati, 482–485, 486, 487

L

La Azufrada reef, 192
La Découverte, 10
La Niña, 27, 86–90, 151, 156, 197, 252, 254–256, 258, 262–263, 267, 270, 276, 277, 279, 283, 392, 471, 538, 545
La Princesse, 10
Lagoon, 16, 116, 121, 194, 206, 282, 505, 520
Large-scale temporal patterns, 40–50
Larvae, 26, 120, 228, 263, 277, 278, 354, 355, 373, 381, 422, 436, 440, 462, 470, 471, 478, 480, 486, 487, 490, 497, 510
Larval settlement, 183, 238, 375, 525
Las Perlas MPA, 131, 560, 562
La Sociedad Mexicana de Arrecifes Coralinas (SOMAC), 15
Late Miocene, 39, 40, 48, 50, 51, 53, 54
Late Paleocene, 39, 40, 46, 50
Late Pleistocene, 40, 43, 46, 48, 50, 54
Late Pleistocene *Porites panamensis* reef, 49
Latin American Coral Reefs (2003), 16, 28, 110
Law, 550, 551, 553, 554, 555, 560, 561, 566

Leptoserid, 280
LIA (Little Ice Age), 183, 184
Light, 40, 48, 111, 191, 194, 228, 235, 283, 301, 348–349, 439, 454, 504–505, 509, 581
Limbaugh C, 11
Limiting resources: N vs. P, 349, 351
Linkages, 94, 98, 257, 295, 386, 391
Links, 304, 308, 454, 538
Lipids, 470
Lithophage, 301, 384
Lithophagid, 383, 384, 385, 386, 394, 395
Lithophaginids, 383
Little Ice Age, 129, 236
Live coral cover, 121, 125, 133–135, 140, 148, 275, 306, 371, 391, 394, 522, 527
Lobsters, 220, 298, 309, 323, 387, 576
Long-distance dispersal, 26, 53, 54, 108, 131, 486
Long-term recovery trajectories, 360
Los Cóbanos NPA, 569
Lunar cycle, 277, 296, 450, 453, 454

M

Machalilla, 23, 28, 143–146, 159, 192, 559, 565–567, 606, 629
Machalilla National Park, 23, 145, 146, 559, 565, 566
Macintyre's hypothesis
near-shore eastern Pacific areas, 511
Macroalgae, 125, 155, 156, 159, 160, 196, 207–209, 236, 253, 277, 293, 296, 304, 308, 340–355, 357, 361, 406, 504, 550, 569, 581, 583, 627
Macroborers, 371, 372, 379, 380, 382, 383, 395
Macroendolith, 373
Macrofauna, 230, 231, 303, 373, 624
Macro-herbivores, 308, 352
Macroinvertebrates, 115, 354
Madden-Julian Oscillation events, 88
Malpelo Island, 13, 20, 21, 22, 28, 110, 112, 113, 140–144, 159, 160, 180, 192, 211, 270, 272, 317, 320, 478, 509, 562–564
Malpelo Island Flora and Fauna Sanctuary, 562
Malpelo Isla Oceánica de Colombia, 20, 113
Management, 2, 4, 27–29, 110, 125, 129, 130, 152, 235, 237, 416, 527, 550–564, 566–571, 577, 580, 581, 583
Manglares, 565
Mangroves, 20, 135, 550, 553, 554, 562, 564, 567
Marginal environmental conditions in the ETP, 406
Marine Biological Corridor of Conservation and Sustainable Development, 563
Marine Conservation Corridor, 28
Marine Protected Areas
Colombia, 558, 563
Costa Rica, 130, 556, 557
Ecuador, 558, 565
El Salvador, 553, 557
Nicaragua, 16
Panama, 558, 560
Rapa Nui (Easter Island), 2, 3, 23, 25, 110, 111
Western Mexico, 552, 549, 557
Massive coral, 9, 25, 121, 139, 151, 158, 160, 178, 179, 183, 184, 187, 192, 194–196, 205, 223, 235, 278, 282, 296, 301, 324, 325, 332, 341, 360, 381, 383, 385, 392, 395, 397, 471, 486, 525, 526, 605, 620, 626, 629
Matapalo fringing reef, 128
Maturation, 397, 436, 437, 448–450, 452, 457
Maturity, 452, 453
Mayor AG, 2, 263
Mechanical bioeroder, 160, 179, 521
Mechanical bioerosion, 375
Mechanisms of PDV, 97
Medusae, 112, 441, 443, 446
Meiofauna, 17, 129, 230, 293, 298, 303, 305, 309, 340
Meiosis, 439
Meridional, 71, 73, 252, 257, 536, 538, 544
Meroplankton, 296, 30
Mesenterial filaments, 440
Mesenteries, 436, 438, 440, 441, 445
Mesoglea, 437–440, 444–447
Mesophotic, 211, 227, 231, 232, 505, 509
Metabolism, 259
Metabolites, 294
Metamorphic, 454, 467, 470
Metamorphosis, 277, 467, 525, 579
Metapopulations, 512
Meta-refugia, 510–512
Metazoans, 10, 20, 27, 115, 135, 208, 230, 233, 263, 273, 274, 277, 278, 293, 296, 299, 304, 309, 342, 345, 621, 622
Mexican Coastal Current, 74
Mexico, 1–3, 6, 10, 13–17, 27–29, 40, 43, 46, 48, 49, 52, 60, 62, 68, 108–111, 114, 116, 118–120, 124, 159, 160, 178–183, 188, 195–198, 204–211, 214, 216, 217, 219–224, 227, 232, 237, 253, 259, 263, 265, 270, 272, 278, 293, 304, 306, 317, 331, 340, 351, 373, 375, 378, 387, 388, 392–394, 397, 406, 407, 409, 410, 414–416, 422, 428, 436, 440–442, 444, 448, 454, 456, 457, 462, 466–468, 470, 472, 480, 491, 508, 522, 550–553, 557, 569, 576, 577, 580–582, 584, 594–607, 609, 610, 612–628, 630, 631
Mexico's MPAs, 551, 552
Microatolls, 129, 184–186
Microborers, 372, 521
Microborings, 525, 526
Microendolith communities, 372
Microendoliths, 370–373, 395
Micrograzers, 297
Microhabitats, 223, 231, 503, 504
Micro-herbivores, 352
Micronutrient iron, 75
Microsatellites, 408–410, 423, 428, 430, 462, 480, 481, 482, 485, 486, 490, 491
Migration, 51, 120, 121, 184, 223, 227, 228, 296, 329, 408, 428, 437, 438, 481, 487, 583
Miguel del Barco, 3
Miocene, 40, 41, 43, 46, 48–51, 53, 54, 116, 478
Miocene Imperial Valley species, 53
Mitigation, 152, 554, 566, 571
Mitochondria, 448
Mitochondrial DNA, 480, 490
Mitosis, 437, 439
Mixed sexuality, 435
Mixed surface layer, 78
Mode of development, 26, 448, 449, 452, 472
Modeling, 27, 102, 256, 297, 309, 352, 361
Models, 43, 76–78, 81, 86, 88–90, 93, 98–103, 179, 223, 233, 235–237, 256, 276, 297, 304, 308–310, 327, 343, 344, 352, 357, 361, 372, 391, 392, 412, 417, 481, 518, 522, 524, 528, 538–540, 544, 545, 551, 552, 584
Modern eastern Pacific coral fauna, 50
Molecular tools and markers, 410

Molluscs, 4, 9, 11, 16, 20, 23, 40, 110, 115, 122, 125, 152, 153, 204–207, 216, 218–220, 230, 231, 236, 238, 273, 277, 278, 281, 293, 298, 301, 302, 324, 326, 238, 354, 370–372, 377, 378, 382, 466, 486, 521, 528, 554, 576, 579, 597, 611, 616, 624, 625, 629
Monge-Nájera Julián, 17
Monitoring, 21, 22, 25, 102, 122, 131, 152, 238, 352, 457, 468, 469, 551, 556, 561–564, 566, 567, 570
Monodonta, 299, 318, 327
Monogenotypic, 466
Monomyces, 450
Monophyletic, 426, 487, 490, 491
Monospecific, 52, 114, 121, 272
Montuosa (Island), 558–562
Morphologies, 3, 27, 52, 53, 132, 140, 208, 227, 235, 237, 268, 294, 303, 315, 317, 324, 330, 331, 343, 344, 355, 394, 422, 423, 428, 439, 448, 449, 468, 487, 503, 581, 601
Morphospecies, 19, 231, 422, 423, 425, 430, 457, 481, 583
Mortality, 3, 13, 17, 19, 20, 22, 23, 26, 86, 103, 111, 115, 116, 129, 130, 132, 135, 142, 156, 157, 160, 183, 186, 187, 189, 192, 194, 196, 197, 233–236, 238, 252, 254, 257, 263, 265, 267, 270–283, 295, 298, 305, 306, 309, 310, 321, 326–329, 331, 332, 345, 347, 348, 352, 354, .355, 360, 370, 371, 373, 378, 380–387, 391–395, 406, 407, 414, 416–418, 462, 470, 471, 502–505, 508, 509, 511, 512, 518, 520, 522–524, 526, 527, 544, 566, 568, 575, 578, 579, 582, 583
Motu Motiro Hiva Marine Park (MMHMP), 568
Motu Tautara, 157, 568
MPAs, 14, 21, 25, 28, 130, 131, 133–135, 143, 145, 309, 550–571, 577
MtDNA, 480, 481, 486, 487, 491, 497
Mucus, 134, 135, 217, 233, 273, 274, 277, 293, 296, 300, 302, 303, 309, 315, 316, 325, 329–331, 347, 458, 578, 581, 617, 619
Mulegé, 3
Multicentury, 540, 541
Multidecadal, 93, 537, 542, 544
Mutation, 477, 491
Mutualism, 277, 417, 425
M/Y *Beagle III, 25*

N

Naos Island, 27, 254, 255, 344, 458
National Monitoring System of Coral Reefs in Colombia (SIMAC), 21, 564
NAZCA Institute, 567
Nektonic, 298, 310
Nematocytes, 440
Neogene, 43
Neritic, 347, 348, 506, 508, 511, 512
Net carbonate production, 387, 390
Net heat flux, 61
New records of fossil corals, 48
NGOs, 23, 152, 553, 554, 556, 559, 571
Nicaragua, 1, 16, 27, 29, 41, 43, 62, 110–113, 123–128, 131, 141, 159, 160, 178, 182, 183, 208, 209, 211, 214, 216, 220, 222, 224, 231, 317, 415, 416, 441, 554, 555, 580, 584
Nicaraguan Pacific coast, 554
Nicaraguan transoceanic canal, 554
Nitrate, 75, 76, 81, 295
Nitrogen, 75, 292, 340, 342, 347, 349, 351, 356
NO_3, 76
Nocturnal, 228, 300–303, 615, 625, 630
Nongovernmental Organizations (NGOs), 152, 554, 556, 559, 571
Nonupwelling, 19, 235, 448, 454, 522
North Equatorial Counter Current (NECC), 26, 62, 73–75, 108, 111, 112, 121, 123, 131, 132, 145, 159, 263, 278, 479, 481
North Equatorial Current (NEC), 62, 74, 262, 479
No-take (Zones), 550, 551, 553, 555, 557, 558, 559, 561–563, 568, 570, 571
Nudibranchs, 328
Nutricline, 259, 276, 346
Nutrient limitation, 349–351, 355, 356, 360
Nutrients, 2, 27, 62, 68, 75, 77, 81, 132, 183, 196, 204, 227, 232, 237, 253, 259, 262, 263, 276, 277, 291–297, 300, 303, 305, 309, 310, 324, 340–344, 346–351, 355–358, 360–362, 371–373, 382, 390, 391, 406, 425, 452–454, 468, 470, 471, 502, 505, 521, 525, 537, 540, 541, 578–584

O

Oaxaca, 6, 9, 14, 16, 116, 120, 181, 204, 237, 337, 392, 414, 427, 448, 551, 553
Ocean acidification (OA), 78, 81, 151, 196, 235, 283, 371, 372, 406, 417, 502, 518, 519, 525, 526, 528, 536, 578, 583
Oceánides (*Investigaciones Marinas*), 13
Oceanographic conditions, 19, 22, 108, 131, 135, 141, 187, 254, 277, 440, 441, 505, 512, 563, 566
Oceanographic features relevant to gene flow
  California Current, 2, 60, 68, 73–75, 116, 441, 479, 480
  Central American Gap, 480, 497
  Costa Rica Current, 480
  "Eastern Pacific Barrier" (EPB), 19, 277, 478, 502, 503
  Expanses of deep water, 478, 479
  Land bridge, 110, 478
  North and South Equatorial Currents, 479
  North Equatorial Counter Current, 26, 111, 123, 479, 481
  Peru Oceanic Current, 480
  Reversing gyres (Panama Bight), 480
  River estuaries and mangrove areas, 480
  Stepping stone, 123, 159, 478, 480, 486, 491
  Tidal flux and eddies, 480
  Upwelling, 2, 13, 16, 27, 75, 87, 93, 95, 99, 159, 182, 196, 235, 236, 256, 277, 350, 459, 522, 527, 559, 562
Octocorals, 23, 29, 133, 134, 136, 211, 215, 561, 566, 576
Oil pollution, 580, 582
Oligocene, 43, 46, 50
Oligotrophic, 75, 292, 294, 300, 508, 509
Omnivores, 216, 281, 283, 297, 300, 302, 625
Onslow Island, 22, 186, 192, 193, 274, 275, 282, 390
Oocytes, 436–439, 442–446, 448, 452
Oogenesis, 437, 452
Ophiuroids, 222, 296, 305
Opisthobranchs, 219, 220, 297
Oral pore (planula larva), 440, 444
Oscillators of ENSO, 88
Outbreaks, 119, 151, 160, 279, 303, 308, 316, 321, 322, 332, 352, 371
Ovaries, 436, 440–442, 451
Overfishing, 27, 152, 156, 208, 279, 292, 308, 309, 352, 354, 357, 359, 472, 527, 551, 563, 564, 569
Overview of diversity in *Symbiodinium*, 408
Ovoviviparous, 491
Ovum diameters, 437, 440, 444, 447, 460
Ovum volumes, 439, 444
Oxygen minimum layer, 75, 79

P

"Pacific Central American faunal gap", 16, 110, 503
Pacific decadal oscillation (PDO), 94, 537
Pacific decadal variability (PDV), 94, 96–98, 103, 537
Pacific endemic zooxanthellate Scleractinia, 113
Pacific Mexican region, 10, 75, 117, 121, 181, 220, 231, 373, 375, 378, 448, 551, 552, 569

Pacific Ocean, 10, 46, 60, 62, 64, 86, 88, 97, 103, 135, 208, 354, 255, 259, 262, 275, 283, 317, 377, 502, 509, 542, 552, 608
Pagurid crabs, 380
Paleoclimate, 94, 97–99, 103, 536–541, 545
Paleoclimatology, 27, 538, 540, 544
Palmer RH, 6, 9, 43, 50
Panama
  Bight, 74, 132, 142, 145, 270, 441, 480, 508
  Canal, 18, 19, 218, 255, 582
  Jet, 62, 75
Panmixia, 478
Papagayo Jet, 62, 74
Parasites, 308
Parasitism, 151, 281, 298, 306
Parrotfishes, 208, 235, 293, 297, 325, 345, 353, 393, 395, 527
PCBs, 581, 582
pCO_2, 80, 371, 372, 406, 417, 470, 518, 519, 522–525, 527
PCR, 408–414, 416, 418, 424
Pearl Islands, 4, 29, 133, 134, 179, 180, 189, 191, 267, 274, 277, 279, 300, 301, 304, 308, 326, 327, 330, 352, 362, 462, 561
Peru, 2, 6, 46, 60, 68, 71, 74, 75, 108, 143, 207, 211, 214, 218, 263, 278, 455, 491, 497, 508, 511, 614
Peru Current
  Coastal upwelling, 68, 74, 75, 341, 511
Pesticides, 467, 569, 582
pH, 27, 78, 80, 81, 128, 151, 196, 232, 236, 300, 372, 406, 416, 468, 470, 518, 525, 526, 537, 544, 545, 579, 580, 584
Phase shifts, 236, 292, 347, 357, 358, 361, 392
Phenotypes, 430
Phosphate, 75, 545
Phosphorus, 294, 351
Photic zone, 259, 478, 505
Photoinhibition, 349
Photosynthesis, 75, 345, 348, 452, 510
Photosynthetically active radiation (PAR), 348
Photosystems, 411
Phototrophic, 7, 196, 228, 373, 449, 486
Phylogenetic reconstruction of Clade C *Symbiodinium*, 427
Phylogenetic relations between genetically-delineated types of *Pocillopora*, 424
Phylogeographic, 322, 426
Physiochemical, 372
Phytoplankton, 62, 75–77, 132, 134, 262, 277, 293, 295, 304, 305, 308, 341, 348, 349, 502, 505, 508, 578, 582
Piscivores, 298, 299, 301, 306, 308, 309, 324, 576
Pitcairn Island, 153
Planktivores, 301–304
Planktivorous, 301, 302, 304, 308, 576
Plankton, 19, 60, 149, 227, 228, 259, 262, 292, 295–298, 300–304, 309, 348, 422, 579
Plankton blooms, 27, 119, 128, 196, 578
Planula
  diameters, 440, 444, 447, 461
  stages, 440, 444
  volumes, 444, 447
Pleistocene, 40, 48–54, 195, 491
Pliocene extinction, 54
Plio-Pleistocene Caribbean faunas, 53
PO_4, 341
Pocillopora species diversity
  genetic markers, 422, 423, 426, 481, 486
  high connectivity, 147, 427, 569
  mitochondrial genomes, 423
  morphological traits, 423
  phenotypic plasticity, 430
  phenotypic variation, 422, 428, 430
  phylogenetic patterns, 423
Pocillopora-Symbiodinium associations, 425, 426
Pocillopora type 1, 423, 424, 426, 428, 430, 490
Pollution, 263, 452, 470, 527, 551, 554, 567, 569, 576, 580–583
Polychaete worms, 20, 230, 293, 308, 324, 332, 370, 379, 383, 521
Polychlorinated biphenols, 581
Polymerase chain reaction, 87
Polyplacophoran, 382
Polypoid, 441
Pomacanthids, 281, 293
Pomacentrids, 277, 297, 298
Population structure of *Pocillopora* and *Porites* 6, 43, 155, 157, 158, 428, 487
Precambrian, 372
Precipitation
  anomalies, 98
  rate, 64
Predation, 115, 187, 208, 214, 227, 233–235, 263, 273, 292, 298, 301, 306, 308, 309, 321, 322, 324, 327, 329, 355, 373, 380, 386, 387, 394, 395, 501
Predators, 152, 156, 216, 229, 230, 233, 280, 292, 297–299, 304, 308, 309, 323, 324, 328, 330, 332, 371, 373, 383–385, 387, 391, 392, 394–396, 459, 466, 552, 569, 579, 617
Pre-Holocene, 40, 53
Press disturbance, 502
Prey, 19, 115, 135, 218, 233–235, 273, 278, 291, 292, 298, 299, 301–304, 306, 308, 310, 321, 323–326, 328, 330, 332, 380, 384, 391, 393, 394, 396, 466, 620, 624, 629
Primary productivity, 75, 81, 196, 259, 344, 345, 371, 382, 578
Production, 14, 70, 75–77, 115, 194, 196, 259, 262, 276, 277, 292–296, 299, 303–305, 306, 309, 324, 329, 331, 332, 340, 349, 370, 375, 380, 383, 386, 387, 390, 392, 394, 443, 452, 454, 459, 470, 519, 520–522, 523, 526, 527, 528, 565, 567, 581
Productivity, 75, 149, 151, 196, 228, 259, 276, 277, 279, 343
Projected changes in the amplitude of ENSO variability, 101, 102
Prokaryotic, 278, 294, 342
Propagules, 19, 136, 159, 309, 478, 481
Protandrous, 452
Proxy, 87, 97, 98, 276, 425, 538
Puerto Escondido, 10, 43
Puerto Villamil, 151
Puffer, 139, 225, 235, 273, 323, 324, 331, 490
Pufferfishes, 235, 304, 324, 392, 393, 394, 395
Pulse disturbance, 502
Punta Chivato fringing patch reef, 49
Punta Pitt, 150, 541–543, 577
Pycnoclines, 68, 75, 76, 505

Q

Quantitative PCR (qPCR), 408, 424

R

Rabbitfishes, 353
Radioisotopes
  13^C, 540
  14^C, 329, 540, 544
  18°, 540
Rafting, 110, 422, 487
Rainfall, 62, 68, 81, 94, 97, 125, 128, 136, 254, 259, 275, 281, 406, 454, 536, 541, 543, 562, 564, 581
Ramet, 462
Ranges of all known reef coral species, 51
Rapa Nui (Easter Island), 153, 238, 304, 567, 568, 576
Rapa Nui ecoregion, 568

Ratchet hypothesis, 417
rDNA, 408, 410, 424, 426
Real-time PCR, 409, 410, 412, 414
Recruitment, 19, 21, 24, 26, 27, 111, 128, 136, 139, 151, 152, 231, 274, 278–280, 282, 309, 323, 347, 361, 371–373, 380–387, 391–397, 436, 452, 456, 459, 462, 463, 468–472, 478, 510, 523, 525, 568, 581
Recycling, 292, 296, 297, 300, 309, 324, 340, 349, 371, 372
Red algae, 231, 238, 278, 340, 342, 343, 344, 347
Reef accretion, 3, 77, 115, 151, 179, 188, 191, 196, 197, 198, 236, 283, 371, 518, 538, 579, 583
Reef-building coral, 16, 26, 110, 129, 153, 237, 275, 408, 554, 561, 583
Reef framework development, 136
Reef framework structure, 179, 181
Refuge
  refugia, 134, 273, 323, 501–512, 580
  refugium, 501–512
Regional approaches to management, 569
Regional refugia, 510, 511
Regional SST time series, 95
Relative Dominance Model, 343, 344
Reproduction, 2, 19, 21, 27, 112, 129, 132, 237, 263, 327, 329, 330, 331, 332, 378, 392, 407, 416, 436, 440, 441, 452–459, 462, 466, 470, 471, 472, 503, 525, 564, 581, 584
Reproductive, 2, 129, 132, 256, 378, 430, 436, 440, 443, 450, 452–456, 460, 466, 470–472, 481, 502, 503, 569
Reproductively, 2, 133, 423, 452–455, 471
Reserve, 14, 23, 129, 144, 152, 191, 352, 353, 554–570
Resilience, 19, 27, 29, 160, 236, 237, 296, 300, 301, 373, 378, 417, 502, 503, 510, 512, 518, 551, 552, 569, 571, 583
Resistant, 115, 151, 157, 193, 194, 411, 502, 503, 505, 511, 512, 520, 524–526
Restriction Fragment Length Polymorphisms (RFLP), 409, 410
Revillagigedo Archipelago, 65, 180–182, 208, 259, 278, 478, 581
Revista de Biología Tropical, 17
Rhizomes, 347
Rhodoliths, 231, 232, 236
Rhodophyte, 372
Ribosomal DNA, 407, 408
Ribosomal internal transcribed spacers, 480
Richness, 9, 13, 24, 50, 54, 110, 111, 114, 120, 125, 132, 134, 135, 141, 143, 147, 153, 158, 159, 207, 220, 223, 225, 227, 231–237, 281, 291, 316, 467, 552, 566, 569, 583
River, 4, 27, 75, 252, 292, 480, 551
Rocky reef, 207, 223, 225, 234, 304, 309
Rossby waves, 89, 90, 93, 99
Real-time PCR (rt-PCR), 409, 410, 411, 413, 416
Rubble, 23, 121, 132, 135, 136, 138, 139, 194, 216, 218, 222, 225, 228, 230, 234, 275, 277, 279, 282, 283, 294, 298, 302, 304, 323, 332, 345, 345, 346, 347, 358, 393, 449, 468, 594–596, 601, 604, 605, 610–613, 616–626, 629, 630
Rubble habitat, 345
Runoff, 24, 74, 75, 136, 259, 276, 281, 292, 348, 454, 468, 536, 541, 563, 569, 579–583
R/V *Spencer F Baird*, 11
R/V *Tethys*, 18

S

Saboga, 135, 179, 182, 186, 189–191, 267, 268, 274, 302, 342, 346, 349, 390, 441, 445, 451, 460, 562
Salas y Gómez Island, 24, 109, 112, 152, 153, 206, 559, 568
Salinities, 252, 454
San Cristóbal, 140, 194, 360, 480, 539, 541–543, 577
Sand, 3, 23, 125, 136, 151, 179, 217, 218, 230, 231, 234, 235, 298, 328, 340, 344, 346, 491, 508, 539, 567, 577
Santa Cruz, 22, 255, 441, 445, 460, 469, 617
Santiago, 150, 194, 448, 599
Saturation state (Ω), 78, 80, 178, 181, 196, 235, 359, 405–407, 417, 471, 512, 518, 527, 579
Scarid, 208, 225, 277, 297, 299, 302, 316, 325, 326, 353, 393, 395, 521, 626, 627
Scavengers, 292, 297, 300, 304, 305, 309, 310, 324
Scavenging, 135, 300
Scleractinian, 4, 7, 9, 16, 20, 23, 26, 29, 110, 111, 112, 117, 118, 125, 130, 132, 134–138, 140, 147, 153, 160, 186, 211, 227, 235, 236, 293, 308, 323, 324, 372, 381, 392, 407, 409, 410, 413, 414, 418, 421, 422, 425, 426, 428, 430, 446, 448, 450, 461, 503, 508, 512, 524, 561, 562, 567
Scrapers, 323, 353, 393, 521
Scripps Institution of Oceanography, 11
Scuba, 5, 11, 120, 121, 207, 340, 341, 561–563, 566, 577
Sea cucumber, 292, 293, 308, 309, 554, 576
Sea of Cortez, 218, 317, 328, 480, 488, 489, 491, 495–497
Sea surface salinity, 542
Sea urchins, 23, 163, 139, 152, 156, 208, 238, 274, 279, 305, 308, 323, 345, 346, 352, 354, 357, 373, 386, 387, 390, 395, 490, 491, 517, 521, 624, 625, 629
Seagrass, 208, 219, 296, 340, 347, 363, 594, 605
Sea-level rise, 2, 198
Seasonal, 454–456
Seasons, 152, 184, 257, 267, 348, 406, 455, 458, 459, 536, 538
Seattle, 39, 41, 43, 48
SeaWiFS chlorophyll, 77
South Equatorial Current (SEC), 59, 263, 479
Sedimentation, 27, 129, 134, 135, 156, 184, 196, 259, 263, 354, 375, 396, 468, 554, 556, 562–564, 576, 580, 581, 583
Sediments, 17, 75, 78, 115, 121, 125, 129, 135, 184, 196, 215, 227, 230, 231, 233, 279, 282, 296, 303, 321, 322, 327, 328, 331, 340, 347, 371, 375, 380, 386, 391, 393, 394, 395, 467, 468, 519, 520, 522, 562, 581
Selection, 151, 256, 328, 396, 416, 418, 422, 426–428, 452, 478, 502, 525, 566
Sequential cosexual hermaphroditic cycles, 435
Sessile, 136, 217, 229, 234, 235, 295, 298, 301, 304, 331, 344, 373, 441, 512, 620, 628, 629
Sex ratios, 435, 450–452
Sexes, 450, 453, 457
Sexual recruitment, 111, 128, 139, 282, 462, 466, 468, 471, 525
Sexual reproduction, 19, 129, 132, 330, 331, 436, 440, 441, 456, 462, 466, 471, 472, 503, 584
Sexuality, 435, 448–451, 472
Shading, 60, 189, 228, 301, 348, 354, 373, 424, 503–505, 581
Sharks, 28, 222, 292, 293, 298, 308, 309, 395, 564, 566, 569, 576
Ship groundings, 575, 577
Shrimps, 220, 232, 302, 329–331, 380, 381
Simultaneous hermaphrodite, 448, 450
Sipunculans, 218, 369, 371, 380–382, 395, 521
Size and age estimates, 453
Skeletal density, 391, 417, 522, 525
Skeletons, 80, 132, 179, 192, 194, 216, 274, 279, 302, 304, 322, 330, 348, 352, 370, 371, 372, 375, 376, 380–384, 386, 393, 394, 397, 422, 466, 503, 520, 521, 525, 538, 540, 544, 596
Smithsonian Tropical Research Institute, 15, 18, 211, 255, 283, 472
Socorro (Island), 10, 120, 121, 181, 373, 581, 597, 623
Sources of damage, 576, 580
Spawn, 450, 456–458, 471
Spawners, 122, 448, 459, 462
Spawning, 120, 132, 133, 296, 436, 440, 444, 448, 450, 452, 454–462, 470–472, 481, 486, 487, 552, 561

Special Zone of Marine Protection Archipiélago de Las Perlas, 560–562
Specialist, 9, 354, 363, 372
Species
Richness, 7, 13, 39, 49, 50, 54, 108, 110–112, 114, 118, 120, 125, 132, 134, 135, 138, 141, 143, 147, 153, 159, 216, 220, 222, 223, 225, 227, 230, 232, 235, 237, 238, 281, 291, 552, 569, 583
Richness and sampling intensity, 49
Sperm, 439, 446, 448, 449, 452, 454, 456, 458, 486
Spermaries, 436–452, 458
Spermatocytes, 437–439, 443
Spermatogenesis, 439
Spermatozoa, 437, 441, 446, 448
Sponges, 159, 179, 204, 205, 208, 210, 232, 237, 238, 278, 293, 295, 298, 301, 305, 323, 324, 358, 370–375, 377, 378, 379, 396, 466, 521, 525–527, 615, 625, 629
Squires DR, 3, 10, 25, 26, 40, 42, 52, 110
Sea surface salinity (SSS), 542
Sea surface temperature (SST), 76, 86, 148, 187, 252, 257, 264, 352, 407, 450, 456, 536, 569
SST change over the twenty-first Century, 101
Standardized PC time series, 96
Stasis in species composition, 51
Stochastic forcing, 85, 90, 102
Stoddart DR, 2, 4
Stratification, 68, 77, 80, 81, 99, 102, 151
Subaerial exposure, 21, 136, 138, 139, 196, 283, 575
Subspecies, 317, 323, 487
Subsurface ocean temperature anomalies, 93
Succession, 4, 181, 278, 372, 381, 383, 384, 397, 452
Suitable substrate, 275, 331, 347, 362
Sula III Expedition, 20
Surface currents
Ocean waves, 62
Salinity, 62
Temperature, 62
Water masses, 60, 68, 262
Surgeonfishes, 225, 353
Survivorship, 151, 152, 275, 331, 418, 467
Suspension feeder, 196, 218, 231, 276, 291, 297, 298, 300–302, 308–310, 371, 613, 619, 620
Sustainability, 521, 552, 564
Symbiodinium diversity in the ETP
Molecular tools and markers, 408–410
Symbiont species diversity, 425
Symbionts, 20, 27, 75, 135, 220, 235, 263, 272, 273, 281, 294, 309, 330, 331, 406, 407, 409–414, 416–418, 423, 425, 426, 525, 526
Symbioses, 406, 409, 416, 417
Synchronization, 454, 457
SY *St. George*, 6

T
Tabuaeran (Fanning Island), 110, 112, 114, 482, 486
Taimana Rapa Nui, 154
TAO-TRITON, 87, 88
Taxonomy, 14, 20, 43, 206, 207, 228, 230, 294, 342, 372, 421, 425
Taxonomy of the "*Pocillopora* complex", 474
Tebada, 136–138, 182, 191
Tectonic, 121, 406, 414
Tehuantepec Bowl, 62, 73, 74
Tehuantepec Jet, 62
Terrigenous, 5, 129, 468, 538, 539, 580, 583
Territorial, 233, 234, 235, 273, 274, 278, 308, 309, 315, 330, 331, 315, 387
Territories, 234, 273, 274, 278, 308, 323, 330, 551
Tetraodontids, 139, 225, 226, 230, 277, 281, 298, 316, 323, 326, 332, 372, 393, 625
Thermal disturbance
refuges, 503, 504, 512
stress, 503, 512
threshold, 504
Thermally tolerant, 407, 411, 424, 525
Thermocline, 59, 62, 68–76, 80, 81, 86–90, 92–94, 99, 102, 142, 151, 231, 232, 251, 259, 268, 274, 277, 295, 339, 341, 346, 349, 350, 371, 372, 406, 407, 502, 504, 505, 506, 509, 511, 512, 518, 545, 559
Thermotolerance, 417, 418
Threatened, 78, 236, 550, 553, 554, 564, 567, 576
Tidal range, 64, 67, 196, 414, 505
Tides, 21, 59, 62, 108, 129, 136, 138, 139, 140, 187, 196, 294, 340, 457, 459, 595
Time-scale, 453
Tissues, 267, 272–274, 277, 280, 298, 304, 315, 316, 317, 322–325, 328, 329, 330, 375, 380, 383, 393, 394, 438, 443, 450, 452, 454, 458, 470, 503, 526, 615
TOGA, 86, 87
Topographic complexity, 279, 371, 392, 521
Tourism, 13, 145, 152, 292, 552, 554, 561, 562, 564, 566, 567, 577, 580, 582
Tracer, 535, 540, 541, 544, 545
Translocation, 329, 341, 412, 426
Transpacific, 122, 477, 478, 480, 486, 487, 490, 491, 497
Trans-Pacific species, 40
Trapeziid, 234, 302, 316, 321, 329, 331, 619
Triggerfishes, 139, 293, 326, 394–396
Trophic, 19, 27, 86, 115, 135, 227, 228, 231, 235, 238, 253, 259, 291–304, 305, 306, 308–310, 315, 324, 330, 332, 340, 341, 383, 456, 552, 569, 576
Trophic cascades, 291, 292, 305, 308, 309, 371
Trophodynamics, 291, 295, 300, 302, 303, 309, 310, 330
Tropical convection and rainfall, 62
Tropical cyclone tracks, 67
Tropical cyclones, 59, 65, 258–261, 352
Tropical mainland Mexico, 108, 114, 116–119, 160
Tropical surface water, 60, 62, 68, 71, 441
Tropics, 4, 60, 62, 95, 98, 99, 103, 231, 480, 502, 517, 519, 535, 536, 543
Tsuchiya Jets, 73, 74
Turbidity, 135, 136, 177, 178, 191, 194, 196, 259, 262, 283, 406, 414, 416, 426, 452, 505, 508, 512, 518, 583
Turf algae, 136, 139, 209, 278, 350, 358, 453
Twentieth century SST trends, 100

U
UNESCO
Convention on the Protection of the World Cultural and Natural Heritage, 550
Marine World Heritage Sites (WHSs), 28, 550
United Nations Convention on the Law of the Sea (UNCLOS), 550, 555
Universidad Autónoma de Baja California Sur (La Paz), 13, 14
Universidad Autónoma de Baja California Sur (Ensenada), 13, 14
Universidad de Costa Rica (San Pedro), 15, 211
Universidad del Mar (Oaxaca), 14
Universidad de Panamá (University of Panama), 43
Universidad del Valle, 15, 20, 21, 22, 160
Upper ocean hydrographic structure, 68
Upwelling, 1, 2, 13, 16, 17, 19, 27, 59, 62, 68, 70, 71, 74, 75, 78, 81, 87, 89, 90, 93, 94, 95, 99, 102, 116, 119, 125–129, 132, 134, 136, 138, 142, 149, 151, 159, 160, 177, 182–184, 187, 189, 190,

191, 192, 194, 196, 197, 205, 208, 235, 236, 256, 259, 265, 267, 277, 279, 283, 292–295, 301, 303, 305, 309, 310, 316, 317, 339, 341, 348–350, 352, 353, 369, 371, 372, 391, 405–408, 417, 441, 448, 450, 454–459, 462, 468, 470, 472, 480, 502, 505, 507–512, 518, 519, 521, 522, 527, 535–538, 540, 541, 544, 545, 552, 553, 562, 578–580, 606
Upwelling refuges, 508, 511
Urchins (sea), 23, 119, 132, 136, 139, 152, 158, 160, 179, 186, 192, 194, 208, 234, 238, 274, 279, 305, 308, 322, 323, 345, 346, 352, 354, 357, 371, 373, 380, 386, 387, 388, 390–395, 483, 486, 488, 490, 491, 494, 517, 521, 527, 624, 625, 629
Urvina Bay, 151, 180, 181, 193, 194, 541, 542–544, 621
Utría National Park, 558, 564

V
Vancouver Island, 50
Vaughan T.W., 1, 2, 6, 24, 40, 41, 43–46, 53, 117, 263
Velero III, 6–10
Verrill A.E., 1, 4, 6, 7, 25, 26, 43, 44, 46, 117, 126, 137, 138, 162, 211, 214, 254
Vesicles, 437, 438, 439, 440, 446, 459
Vicariance, 40, 52, 487
Villalobos A.F., 10
Viviparous
Von Prahl H., 3, 13, 19–21, 26, 138, 139, 220, 252, 253, 263, 280

W
Walker circulation, 76, 86, 93, 99, 102
Warming, 13, 17, 23, 25, 50, 62, 76, 86, 99, 102, 103, 129, 142, 156, 160, 198, 252, 254, 263, 265, 273, 283, 305, 348, 416, 417, 462, 503–505, 509–512, 518, 522, 523, 524–526, 536, 541–543, 578–580
Washington-Seattle, 43
Water mass characteristics, 68
Wave action, 188, 195, 196, 235, 259, 275, 332, 430, 466, 468, 505, 579, 599, 623
Webs (trophic), 291, 308
Wellington GM, 3, 12, 19, 22–25, 110, 119, 132, 148, 151, 152, 156, 179, 192–194, 197, 226, 228, 234, 238, 254, 264, 274, 279, 281, 303, 308, 306, 328, 352, 353, 357, 371, 390, 462, 480, 506, 511, 540, 541, 544, 569, 578, 579, 596, 606, 620, 621, 623
Wellington reef, 148, 149, 390, 596, 620
Wells JW, 3, 4, 12, 22, 24, 41, 43, 44, 46, 52, 108, 111, 117, 126, 130, 136, 154, 160, 236, 238, 275, 599, 605–608
Wenman Island, 148, 150, 192, 282, 390
World Heritage Site (WHS), 20, 23, 28, 550, 558, 565
Wilkes C, 4
Wind-driven currents, 292, 293
Wind vectors, 63, 65
Worms, 16, 20, 159, 215–218, 236, 292, 293, 298, 305, 308, 324, 332, 370, 379, 381, 383, 397, 521, 612, 620
Wyrtki K, 68, 88, 89, 257, 263, 275

Y
Yonge CM, 2, 263, 384

Z
Zonation, 11, 22, 24, 118, 119, 136, 139, 141, 155, 178, 185, 192, 198, 235, 308, 326, 330, 345, 383, 414
Zooplanktivores, 293, 301
Zooplanktivory, 301
Zooplankton, 129, 227–229, 293, 295, 296, 300–302, 304, 305, 308, 502, 508, 509, 578
Zooxanthellae, 119, 151, 157, 236, 293, 294–296, 300, 309, 324, 328, 330, 332, 437, 438, 440–444, 446, 448, 449, 454, 470, 524, 525, 581
Zooxanthellate, 4, 16, 23, 24, 108, 110–112, 114, 116, 117, 121–125, 130, 132, 136, 140, 144, 147, 152, 153, 159, 160, 198, 206, 231, 238, 252, 254, 263, 267, 270, 273, 280, 293, 304, 317, 324, 411, 413, 436, 439, 448, 450, 455, 459, 462, 471, 486, 503, 508, 509, 512, 526, 527, 553, 565, 578, 579, 583, 608

Taxonomic Index

A

Acanthaster, 13, 14, 19, 233, 273, 300, 306, 308, 316, 317, 321–323, 331, 352, 386
Acanthaster ellisii, 317, 486
Acanthaster planci , 19, 115, 121, 132, 160, 216–223, 232, 233, 273, 274, 278, 298, 299, 303, 306, 315, 317, 318, 330–332, 352, 480, 483, 486, 488, 490, 564, 580, 621
Acanthaster planci skeletal elements, 322
Acanthasteridae, 316, 617
Acanthophora, 343, 344, 347, 358
Acanthophora spicifera, 125, 160, 208, 294, 350–353, 355, 356, 596
Acanthuridae, 224, 297, 302, 353, 627, 628
Acanthurus, 155
Acanthurus nigricans, 483, 487, 488, 627
Acanthurus triostegus, 353, 483, 487, 488, 492, 627
Acanthurus xanthopterus, 297, 353, 627
Acropora, 1, 26, 43, 274, 325, 331, 417, 448, 456
Acropora cervicornis, 384
Acropora valida, 26, 132, 137, 139, 280, 416, 471
Adelogorgia telones, 214
Agaricia, 448
Agariciidae, 43, 126, 137, 436, 437, 439, 448–450, 460, 472, 602–604
Aiptasia, 308, 309, 324, 414
Alpheidae, 617
Alpheus, 302
Alpheus lottini, 222, 273, 299, 308, 315, 318, 321, 329, 330, 617
Alpheus simus, 381
Aluterus scriptus, 299, 319, 326
Alveopora, 43, 46, 162
Amphimedon texotli, 374, 598
Amphinomidae, 216, 217, 612
Amphipoda, 229, 296
Amphiroa beauvosii, 278
Amphiroa misakiensis, 595
Aniculus elegans, 299, 318, 328, 329
Annelida, 214, 215, 301, 315, 612
Antennariidae, 224, 305
Antipathes galapagensis, 608
Aplodactylidae, 224
Aplysina chiriquensis, 597
Aplysina clathrata, 210, 596
Aplysina gerardogreeni, 597
Aplysina revillagigedi, 597
Aplysinidae, 596, 597
Apogonidae, 224, 298, 301, 305
Appendicularia, 229
Aquaumbra klapferi, 212
Aquaumbridae, 212
Arbacia stellata, 364, 491, 494
Arothron, 299, 304
Arothron hispidus, 299, 300, 320, 324, 373, 396
Arothron meleagris, 225, 226, 273, 298–300, 306, 315, 323, 324, 331, 392–394, 396, 483, 488, 492, 521, 625
Arthropoda, 301, 315, 317, 318, 326, 617
Asperiscala billeeana, 614
Aspidosiphon, 218
Aspidosiphon elegans, 382
Aspidosiphon steenstrupii, 382
Aspidosiphonidae, 217
Asteroidea, 224
Astrangia equatorialis, 606
Astreopora, 43, 46, 162
Astrocoeniina, 126
Astropyga pulvinata, 622
Aulostomidae, 224, 225
Axinyssa isabela, 598

B

Balanidae, 620
Balistes polylepis, 319, 325, 395
Balistidae, 139, 224, 226, 230, 298, 299, 301, 302, 319, 325, 332, 394, 624
Baseodiscus mexicanus, 217, 218, 612
Bispira monroi, 613
Bispira sp. *(Bispira melanostigma)*, 613
Bivalvia, 219
Blenniidae, 224, 296, 353
Bodianus diplotaenia, 299, 304, 309, 387
Brachyura, 220, 222
Bryozoa, 206, 301
Bythitidae, 224, 305

C

Calcarea, 208, 210
Calcinus obscurus, 299, 318, 328
Calliostoma mcleani, 318
Calotomus carolinus, 319, 325, 483, 487
Cantherhines dumerilii, 319, 326
Canthigaster, 320, 324
Canthigaster amboinensis, 320, 324
Canthigaster janthinoptera, 324
Canthigaster punctatissima, 324
Canthigaster valentini, 320
Carangidae, 225, 298, 630

P.W. Glynn et al. (eds.), *Coral Reefs of the Eastern Tropical Pacific*, Coral Reefs of the World 8, DOI 10.1007/978-94-017-7499-4

Carapidae, 224, 305
Carcharhinidae, 224, 299, 332, 631
Caridea, 222, 296
Carijoa multiflora, 212
Carijoa riisei, 211, 212, 215
Caryophylliidae, 450, 607, 608
Caulerpa, 343, 347, 354, 357
Caulerpa chemnitzia, 595
Caulerpa racemosa, 129, 160, 277
Caulerpa sertularioides, 128, 129, 160, 208, 277, 346, 347, 350–353, 356, 359, 556, 594
Caulerpaceae, 594, 595
Cavernulina cf. *darwini*, 216
Centrostephanus, 387, 388
Centrostephanus coronatus, 387, 388, 391
Cephalopoda, 219, 220
Chaenopsidae, 224, 225
Chaetodon, 273, 299, 300, 319, 325, 628
Chaetodon humeralis, 299, 319, 324, 325, 628
Chaetodontidae, 224, 298, 299, 316, 319, 323, 324, 326, 628
Chaetognatha, 229, 296
Chaetophora, 347
Chalinidae, 597
Chalinula nematifera, 374
Chlorophyta, 207, 294, 342, 343, 594
Chordata, 315, 319
Chromis atrilobata, 225, 226, 301
Chromodorididae, 615
Cidaridae, 621, 622
Cirolana parva, 620
Cirolanidae, 620
Cirratulidae, 217, 379
Cirrhitidae, 224, 225, 298
Cladophora, 344
Cladopsammia eguchii, 608
Cladopsammia gracilis, 144, 608
Clavulariidae, 211, 212
Cliona, 125, 373, 377, 521, 522
Cliona albimarginata, 375, 377
Cliona amplicavata, 378
Cliona aprica, 377
Cliona californiana, 210, 374, 376, 378
Cliona caribbea, 375, 378, 526
Cliona carteri, 375
Cliona celata, 377
Cliona euryphylla, 378
Cliona flavifodina, 375, 378
Cliona medinae, 378
Cliona mucronata, 376, 378
Cliona orientalis, 526
Cliona peponaca, 377
Cliona pocillopora, 378
Cliona raromicrosclera, 374, 378
Cliona tropicalis, 376
Cliona vallartense, 374
Cliona vermifera, 370, 373, 375, 376, 378
Cliona viridis, 375
Cliothosa tylostrongylata, 378
Cnidaria, 4, 210
Cochlodinium catennatum, 276
Colpomenia, 347
Colpophyllia, 44, 46
Copepoda, 229, 230, 296, 303
Corallinaceae, 343, 591
Coralliophila monodonta, 299, 318, 327
Coralliophila violacea, 152, 614
Coscinaraea, 46
Craniata, 624
Crustacea, 135, 220, 298
Ctenochaetus, 353, 483, 487, 488
Ctenochaetus marginatus, 353, 483, 487, 488
Cycloseris, 114, 163
Cycloseris curvata, 114, 117, 126, 129, 132, 137, 147, 151, 270, 280
Cycloseris vaughani, 507
Cymodoce, 318, 329

D

Demospongiae, 208, 210
Dendrophylliidae, 436, 449, 583, 605, 606
Dendropoma maximum, 331
Diadema, 156, 279, 308, 391
Diadema antillarum, 388, 389, 527
Diadema mexicanum, 160, 186, 187, 208, 209, 223, 274, 279, 280, 300, 308, 318, 354, 386, 388, 391–393, 488, 492, 494, 521, 615, 620, 622
Diadema savignyi, 388, 389
Diadema setosum, 388
Diadematidae, 622
Diaseris distorta, 114, 117, 126, 129, 132, 137, 147, 151, 163, 193, 270, 323, 436, 438, 447, 449, 451–453, 460, 466
Dichocoenia, 45, 46, 48, 52
Dictyota, 343
Dictyotaceae, 594
Diodon holocanthus, 299, 327, 488, 492, 495, 625
Diodontidae, 299, 625
Diploastrea, 44, 46
Diploria, 44, 46, 52
Distichopora robusta, 210, 577, 611
Domecia hispida, 618
Domeciidae, 618
Doriprismatica sedna, 615
Dorvilleidae, 217, 379

E

Echidna nebulosa, 631
Echineulima thaanumi, 615
Echinodermata, 4, 222, 298, 315, 317, 318
Echinoidea, 224
Echinometra lucunter, 389
Echinometra mathaei, 389
Echinometra sp. A, 389
Echinometra vanbrunti, 222, 354, 387, 488, 494
Echinometra viridis, 389
Echinostrephus aciculatus, 388
Echinothrix diadema, 388, 483, 490, 492
Ellisella limbaughi, 213
Ellisellidae, 213
Elysia diomedea, 616
Epinephelus clippertonensis, 225
Epitoniidae, 614
Eucidaris, 23, 152, 279, 308, 309, 322, 323, 354, 384, 390, 391
Eucidaris galapagensis, 22, 152, 160, 179, 192–194, 208, 274, 275, 279, 280, 297, 303, 308, 317, 318, 322, 323, 330, 354, 387, 388, 390, 392, 393, 490, 492, 521, 621, 622
Eucidaris thouarsii, 222, 354, 387, 388, 392, 494, 621
Eugorgia ampla, 212
Eugorgia aurantiaca, 212
Eugorgia beebei, 211, 212
Eugorgia bradleyi, 212
Eugorgia daniana, 212, 609

Eugorgia excelsa, 212
Eugorgia multifida, 212
Eugorgia mutabilis, 212
Eugorgia nobilis, 212
Eugorgia panamensis, 212
Eugorgia purpurascens, 212
Eugorgia querciformis, 212
Eugorgia rubens, 212
Eugorgia siedenburgae, 212, 232
Eulimidae, 615
Eumuricea acervata, 213
Eumuricea horrida, 213
Eumuricea squarrosa, 213
Eumuricea tubigera, 213
Eunice sp., 216, 613
Eunicidae, 216, 217, 296, 379, 380, 613
Eusmilia, 45, 46, 52

F
Faviidae, 44, 46, 153
Favites, 44, 46
Flabelligeridae, 217, 379
Foraminifera, 205, 230, 281, 372, 538, 625
Forcipiger flavissimus, 319, 324, 484, 492
Fungia, 45, 46
Fungia curvata, 45, 605
Fungiidae, 45, 126, 436, 448–450, 460, 605
Fungiina, 126

G
Galaxaura, 343
Gardineroseris planulata, 43, 46, 48, 117, 124, 126, 128–130, 136–138, 141, 187, 192, 273, 274, 280, 299, 321, 325, 413, 416, 446–449, 451, 453, 460, 464, 468, 469, 471, 504, 507, 508, 512, 524, 604
Gastrochaena, 370, 383
Gastrochaena ovata, 384, 385
Gastrochaena rugulosa, 386
Gastrochaenidae, 383
Gastropoda, 135, 219, 220, 230, 298, 303
Gelidiopsis intricata, 278
Gnathanodon speciosus, 304, 630
Gobiesocidae, 224, 225
Gobiidae, 224, 225, 296, 298
Goniastrea, 44, 46
Goniopora, 45, 46
Gonodactylidae, 617
Gonyaulax monilata, 276
Gorgoniidae, 212, 608–610
Gymnodinium, 552
Gymnothorax castaneus, 299, 630

H
Haemulidae, 224, 296, 298
Halichondriidae, 598
Haliclona (Soestella) caerulea, 597
Halimeda, 340, 343, 347, 519, 520
Halimeda discoidea, 208, 595
Halimedaceae, 595
Halymenia, 343, 344, 347
Hapalidiaceae, 596
Harpiliopsis, 330
Harpiliopsis depressa, 299, 330, 618
Heliopora coerulea, 26
Hesionidae, 613
Heterogorgia hickmani, 213, 610
Heterogorgia papillosa, 214
Heterogorgia tortuosa, 214
Heterogorgia verrucosa, 213
Heterokontophyta, 342
Hippolytidae, 617
Holacanthus limbaughi, 225
Holacanthus passer, 299, 300, 319, 326, 628
Holocentridae, 224, 225, 298, 301, 305
Holothuroidea, 135, 224
Homoscleromorpha, 208, 210
Hymenocera, 13
Hymenocera picta, 299, 300, 308, 332, 617
Hymenoceridae, 617
Hypnea pannosa, 208

I
Inachidae, 618
Iphione ovata, 216, 217, 612
Isostichopus fuscus, 222, 223, 308, 623

J
Jenneria pustulata, 219, 273, 299, 303, 315, 318, 327, 470, 614
Johnrandallia nigrirostris, 299, 319, 324, 325, 628

K
Kyphosidae, 224, 302, 353

L
Labridae, 224–226, 298, 299, 301, 629
Labrisomidae, 224, 225, 305
Lanicola eduardoi, 614
Latiaxis hindsii, 318, 328
Laurencia, 343, 344, 352
Leiosolenus, 383, 384
Lepidasthenia gigas, 613
Leptastrea, 44, 46
Leptastrea purpurea, 44, 153, 164
Leptastrea transversa, 44, 153, 164
Leptogorgia aequatorialis, 212
Leptogorgia alba, 212, 215, 609
Leptogorgia californica, 212
Leptogorgia christiae, 212, 577
Leptogorgia clavata, 212
Leptogorgia cofrini, 212
Leptogorgia cortesi, 212
Leptogorgia cuspidata, 212
Leptogorgia diffusa, 212
Leptogorgia ena, 212
Leptogorgia exigua, 212
Leptogorgia filicrispata, 212
Leptogorgia flexilis, 212
Leptogorgia florae, 212
Leptogorgia fruticosa, 212
Leptogorgia ignita, 212
Leptogorgia labiata, 213
Leptogorgia laxa, 212
Leptogorgia obscura, 212
Leptogorgia parva, 212
Leptogorgia peruana, 212

Leptogorgia pumila, 212
Leptogorgia ramulus, 212
Leptogorgia regis, 212
Leptogorgia rigida, 212
Leptogorgia taboguillae, 212
Leptogorgia tricorata, 212
Leptoria, 44, 46
Leptoseris, 43, 46, 326
Leptoseris papyracea, 26, 43, 117, 126, 129, 132, 137, 164, 270, 272, 280, 416, 507
Leptoseris scabra, 43, 122, 126, 130, 147, 164, 282, 416, 464, 507, 603
Leptoseris solida, 43, 164, 507
Lineidae, 611
Lithophaga, 276, 370, 385, 386, 522
Lithophaga (Leiosolenus) laevigata, 384
Lithophaga aristata, 383, 384, 386
Lithophaga attenuata, 383, 386
Lithophaga calyculata, 384
Lithophaga corrugata, 386
Lithophaga hancocki, 384
Lithophaga hastasia, 384
Lithophaga laevigata, 383, 385
Lithophaga malaccana, 386
Lithophaga plumula, 384, 386
Lithophaga teres, 386
Lithothamnion muelleri, 596
Lutjanidae, 224, 226, 298, 299, 624
Lutjanus viridis, 226, 299, 624
Lyngbya, 349
Lytechinus, 623
Lytechinus semituberculatus, 152, 354, 623

M

Madracis, 44
Madracis mirabilis, 327
Meandrina, 45, 46
Megabalanus peninsularis, 620
Melichthys niger, 319, 326
Microdesmidae, 224
Microspathodon dorsalis, 319, 326
Micrura ignea, 611
Millepora, 1, 16, 26, 110–113, 326
Millepora boschmai, 111, 211, 271, 508, 524, 610
Millepora dichotoma, 114
Millepora exaesa, 16, 112, 611
Millepora intricata, 111, 113, 132, 280, 299, 414, 436, 446, 462, 508–510, 611
Millepora platyphylla, 114, 132, 270, 280, 508, 524
Millepora tenera, 114
Millepora tuberosa (exaesa), 114
Milleporidae, 210, 610, 611
Mithracidae, 618
Mithraculus denticulatus, 618
Mollusca, 4, 218, 301, 317, 326, 614
Monacanthidae, 224, 298, 299, 319, 326
Montastrea, 44
Montipora, 1, 43, 326
Montipora aequituberculata, 26
Montipora fragosa, 25, 26, 29
Mullidae, 225, 298, 624
Mulloidichthys dentatus, 624, 628
Muraena lentiginosa, 630
Muraenidae, 224, 225, 298, 299, 305, 630, 631
Muricea albida, 214
Muricea appressa, 214
Muricea appressa cf. *flavescens*, 214
Muricea aspera, 214
Muricea austera, 214
Muricea californica, 214
Muricea crassa, 214
Muricea equinata, 214
Muricea formosa, 214
Muricea fruticosa, 214, 610
Muricea fruticosa var. *miser*, 214
Muricea hebes, 214
Muricea purpurea, 214, 215
Muricea retusa, 214
Muricea robusta, 214
Muricea tenella, 214
Muricidae, 614
Muricopsis (Babelomurex) zeteki, 328
Muricopsis zeteki, 318
Mycetophyllia, 45, 46
Myripristis gildi, 225
Myrocongridae, 224
Mysidacea, 229, 296
Mytilidae, 383

N

Nemertea, 214, 217, 218, 230, 611
Neogonodactylus sp., 617
Nicholsina denticulata, 353
Nidorellia armata, 318, 323, 620
Niphatidae, 598
Novaculichthys taeniourus, 299, 484, 489, 629

O

Ochrophyta-Phaeophyceae, 207
Octopodidae, 616
Octopus sp., 616
Ophidiidae, 224, 305
Ophiodromus sp., 613
Ophiuroidea, 135, 224
Opisthobranchia, 219, 220
Oplegnathidae, 224
Oreasteridae, 620, 621
Ostraciidae, 224, 225
Ostreobium, 372, 525
Oulangia bradleyi, 606, 607
Oulophyllia, 45, 46

P

Pachycheles biocellatus, 619
Pacifigorgia adamsii, 213
Pacifigorgia agassizii, 213
Pacifigorgia bayeri, 213
Pacifigorgia cairnsi, 213
Pacifigorgia catedralensis, 213, 577
Pacifigorgia cribrum, 213
Pacifigorgia curta, 213
Pacifigorgia dampieri, 213, 609
Pacifigorgia darwinii, 213

Pacifigorgia douglasii, 213
Pacifigorgia englemanni, 213
Pacifigorgia exilis, 213
Pacifigorgia eximia, 213
Pacifigorgia ferruginea, 213
Pacifigorgia firma, 213
Pacifigorgia flavimaculata, 213
Pacifigorgia gracilis, 213
Pacifigorgia irene, 213, 215
Pacifigorgia labiata, 213
Pacifigorgia lacerata, 213
Pacifigorgia marviva, 213
Pacifigorgia media, 213
Pacifigorgia pulchra, 213
Pacifigorgia rubicunda, 213
Pacifigorgia rubinoffi, 213, 577
Pacifigorgia rubripunctata, 213, 609
Pacifigorgia rutila, 213
Pacifigorgia samarensis, 213
Pacifigorgia sculpta, 213
Pacifigorgia senta, 213
Pacifigorgia smithsoniana, 213, 577
Pacifigorgia stenobrochis, 213, 215
Pacifigorgia symbiotica, 213
Pacifigorgia tabogae, 213
Pacifigorgia tupperi, 213
Padina, 343, 616
Padina sp., 594
Palaemonella, 302
Palaemonidae, 618
Panulirus, 309, 387
Pavona, 4, 9, 43, 110, 461, 510, 512
Pavona chiriquiensis, 43, 126, 127, 130, 137, 147, 299, 408, 445, 447–449, 451, 453, 457–460, 471, 508, 524, 602
Pavona clavus, 43, 48, 51, 117, 124, 126, 129, 130, 132, 137, 141, 143, 144, 150, 183, 192, 193, 228, 267, 269, 280, 282, 299, 301, 410, 413, 445, 447–449, 451–455, 457–460, 466, 469, 471, 503, 507–509, 526, 541, 602
Pavona clivosa, 4
Pavona duerdeni, 43, 117, 132, 166
Pavona frondifera, 43, 126, 130, 137, 602
Pavona gigantea, 4, 16, 46, 111, 113, 119–121, 125, 128, 129, 132, 136–138, 143, 166, 183, 192, 193, 211, 301, 308, 326, 410, 413, 414, 416, 440, 442, 444, 445, 447–449, 451, 453–460, 469, 471, 507, 508, 526, 541, 603
Pavona maldivensis, 43, 117, 122, 126, 137, 147, 416, 519, 603
Pavona minuta, 43, 117, 122, 126, 130, 132, 166
Pavona varians, 121, 124, 136, 173, 299, 325, 395, 413, 438, 441, 445, 447–449, 451–460, 462, 463, 466, 469, 471, 503, 508, 555, 603
Pediculariidae, 614
Pennatulidae, 216
Pentaceraster cumingi, 299, 318, 323, 621
Petrolisthes donadio, 620
Peyssonnelia boergesenii, 596
Peyssonneliaceae, 596
Phaeophyta, 594
Pharia pyramidata, 318
Phascolosomatidae, 381
Pherecardia, 216
Pherecardia striata, 216, 299, 300, 308, 612
Phestilla, 328
Phestilla lugubris, 299, 328, 615
Phestilla melanobrachia, 299, 328, 615
Phestilla sibogae, 328
Phestilla spp., 219, 318, 328
Phyllangia americana, 327
Phyllangia consagensis, 607
Pione carpenteri, 376, 378
Pione cf. *vastifica*, 375, 377
Pione lampa, 375, 526
Pione mazatlanensis, 378
Pione vastifica, 375, 377
Placopsammia darwini, 4
Plakobranchidae, 616
Plectonema, 372
Plesiastrea, 45, 46
Plexauridae, 211, 610
Pocillopora, 3, 4, 6, 9, 11, 12, 23, 52
Pocillopora capitata, 46, 51, 52, 117, 126, 137, 280, 387, 598
Pocillopora damicornis, 46, 48, 120, 132, 136, 138, 143, 179, 184, 191, 267, 268, 299, 301, 331, 380, 385, 387, 410, 412, 414, 417, 423, 430, 440, 442, 448, 450, 454–456, 457, 462, 468, 470, 471, 481, 491, 509, 510, 525, 555, 582, 583, 598, 599
Pocillopora danae, 45
Pocillopora effusus, 113, 117, 137, 143, 599
Pocillopora elegans, 16, 132, 160, 267, 277, 280, 299, 387, 416, 447, 449, 450, 454, 457, 462, 471, 508, 599
Pocillopora eydouxi, 25, 121, 128, 155, 281, 299, 450, 457, 583, 599
Pocillopora inflata, 113, 117, 128, 131, 457, 600
Pocillopora ligulata, 45, 137, 144
Pocillopora meandrina, 46, 48, 117, 144, 430, 444, 448, 454, 457, 600
Pocillopora verrucosa, 46, 48, 115, 120, 144, 152, 153, 159, 160, 195, 238, 375, 414, 444, 448, 454, 457, 467, 600
Pocillopora woodjonesi, 45, 117, 137
Pocilloporidae, 126, 137, 327, 329, 448, 460, 525, 598–600
Polycyathus isabela, 607
Polynoidae, 216, 217, 612, 613
Polyplacophora, 219, 220
Pomacanthidae, 224, 225, 299, 319, 326, 628, 629
Pomacanthus zonipectus, 299, 326, 629
Pomacentridae, 225, 226, 301, 326, 353
Pomatogebia, 381
Pomatogebia cocosia, 381
Pomatogebia rugosa, 381
Porcellanidae, 619, 620
Porifera, 4, 208
Porites, 4, 6, 10, 26
Porites arnaudi, 15, 45, 113, 117, 121, 123
Porites australiensis, 45, 117
Porites baueri, 45, 113, 117, 118, 167
Porites californica, 4
Porites duerdeni, 167
Porites evermanni, 117, 121, 126, 129, 132, 137, 143, 147, 214, 439, 466, 469, 503, 601
Porites lichen, 45, 117, 167
Porites lobata, 9, 45, 46, 48, 53, 117, 118, 121, 124–126, 129, 130, 132, 136, 137, 141, 142, 144, 151, 154–156, 160, 167, 184–187, 191, 193–195, 206, 211, 268–270, 273, 275, 276, 281, 282, 296, 324–328, 380, 382, 384, 386, 394, 395, 410, 413, 416, 428, 437, 438, 447–460, 466, 469, 486, 487, 490, 491, 503, 508–510, 521, 555, 569, 585, 601
Porites lutea, 45, 117, 167
Porites panamensis, 4, 45, 46, 49, 51–54, 111, 113, 114, 116, 117, 120, 126, 128, 211, 214, 270, 280, 299, 324, 412, 414, 439, 440, 442–444, 448, 449, 452, 454, 455, 460, 462, 466, 469, 471, 522
Porites rus, 45, 126, 129, 132, 270, 280, 416

Porites sverdrupi, 46, 53, 54, 114, 116, 602
Porites (Synarea) rus, 26
Poritidae, 126, 137, 436, 448, 449, 460, 600–602
Porolithon castellum, 595
Priacanthidae, 225, 301
Prionurus laticlavius, 628
Prionurus punctatus, 353
Prognathodes falcifer, 319
Psammocora, 4, 46, 115, 120, 125, 129, 138, 144, 181, 282, 416, 437, 439, 459
Psammocora brighami, 46, 117
Psammocora obtusangula, 46, 125, 126, 137, 168, 569
Psammocora profundacella, 46, 117, 132, 138, 168, 449
Psammocora stellata, 4, 46, 117, 125, 126, 136, 138, 144, 151, 160, 183, 184, 191, 193, 194, 218, 299, 323, 354, 441, 446, 449, 451, 457, 460, 509, 569, 604
Psammocora superficialis, 46, 117, 126, 144, 410
Psammogorgia arbuscula, 214
Psammogorgia fucosa, 214
Psammogorgia hookeri, 214
Psammogorgia sp., 214
Psammogorgia teres, 214
Psammogorgia variabilis, 214
Pseudobalistes naufragium, 225, 299, 301, 325, 384, 395, 466, 469, 624
Pseudogramma axelrodi, 225
Ptilosarcus undulatus, 216

Q
Quoyula madreporarum, 219, 299, 327

R
Reliquiaecava robillardi, 219
Renilla ametystina, 216
Renillidae, 216
Rhizangiidae, 606, 607
Rhodomelaceae, 596
Rhodophyta, 207, 294, 342, 343, 345

S
Sabellidae, 217, 379, 613, 614
Sargassaceae, 594
Sargassum, 343, 344, 594
Sargassum obtusifolium, 594
Scaridae, 208, 225, 297, 299, 302, 316, 325, 326, 353, 393, 626, 627
Scarus compressus, 626
Scarus ghobban, 626
Scarus perrico, 299, 325, 353, 627
Scarus rubroviolaceus, 297, 299, 353, 393, 491, 626
Sciaenidae, 225
Scleractinia, 4, 6, 9, 16, 20, 23, 26, 29, 45, 46, 110, 211, 423
Scorpaenidae, 225, 298, 305
Scyllarides astori, 309
Scytalium, 216
Serpulidae, 217, 612
Serranidae, 225, 298, 299, 305
Siderastrea, 27, 46, 52, 53, 327
Siderastrea glynni, 27, 46, 52, 53, 132
Siderastrea siderea, 46, 52, 327
Siderastreidae, 126, 137, 436, 448, 604
Siphonodictyon, 375
Siphonodictyon crypticum, 375
Sipuncula, 214, 217, 218, 381
Solenastrea, 45, 46, 52
Solenastrea fairbanksi, 45, 49
Spengleria, 384
Spheciospongia incrustans, 378
Spionidae, 217, 230, 379, 380
Spirobranchus ?incrassatus, 612
Stegastes, 308
Stegastes acapulcoensis, 226, 234, 273, 274, 299, 308, 326, 353, 630
Stegastes arcifrons, 274, 326
Stegastes baldwini, 225
Stegastes beebei, 274
Stenorhynchus debilis, 221, 618
Stephanocoenia, 44, 46
Stichopodidae, 623
Stylasteridae, 210, 611
Stylatula cf. *elongata*, 216
Stylophora, 45, 329, 411
Stylophora pistillata, 168, 411
Sufflamen verres, 299, 300, 395, 624
Symbiodinium, 14, 19, 27, 119, 263, 272, 293–295, 406–418, 423–426, 502, 524, 526
Syngnathidae, 225

T
Tergipedidae, 615
Tetraodontidae, 139, 225, 226, 230, 298, 316, 323, 326, 332, 393, 625
Thalassoma grammaticum, 299, 629
Thalassoma lucasanum, 226, 300, 301, 558
Thalassoma robertsoni, 225
Thoosa calpulli, 378
Thoosa mismalolli, 373, 378
Thoosa purpurea, 376, 378
Thor algicola, 617
Thor amboinensis, 299, 330
Toxopneustes depressus, 354
Toxopneustes roseus, 222, 389, 622
Toxopneustidae, 622, 623
Trachyphyllia, 45, 46
Trapezia, 221, 273, 305, 308, 321, 331
Trapezia areolata, 329
Trapezia bidentata, 619
Trapezia cymodoce, 329
Trapezia digitalis, 221
Trapezia ferruginea, 329, 619
Trapeziidae, 316, 619
Triaenodon obesus, 298, 299, 395, 631
Tripneustes depressus, 222, 223, 623
Tripterygiidae, 225
Trizopagurus, 300

Trizopagurus magnificus, 221, 299, 300, 328, 380
Tubastraea, 436, 437, 439, 448
Tubastraea coccinea, 123, 133, 276, 411, 438, 439, 441, 447, 448, 449, 460, 550, 605
Tubastraea faulkneri, 605
Tubastraea floreana, 606
Tubastraea tagusensis, 606

U
Upogebia, 381

V
Valenciniidae, 217, 612
Veretillidae, 216
Virgularia galapagensis, 216
Virgulariidae, 216

Z
Zanclidae, 225, 298, 326, 629
Zanclus cornutus, 326, 487, 629
Zonaria stipitata, 594

Kolibri 360 GmbH
Ferdinand-Jühlke-Straße 7
99095 Erfurt, Deutschland
produktsicherheit@kolibri360.de